Handbook of Silicon Photonics

SERIES IN OPTICS AND OPTOELECTRONICS

Series Editors: **E Roy Pike**, Kings College, London, UK
Robert G W Brown, University of California, Irvine, USA

Recent titles in the series

Handbook of Silicon Photonics
Laurent Vivien and Lorenzo Pavesi (Eds.)

Microlenses: Properties, Fabrication and Liquid Lenses
Hongrui Jiang and Xuefeng Zeng

Laser-Based Measurements for Time and Frequency Domain Applications: A Handbook
Pasquale Maddaloni, Marco Bellini, and Paolo De Natale

Handbook of 3D Machine Vision: Optical Metrology and Imaging
Song Zhang (Ed.)

Handbook of Optical Dimensional Metrology
Kevin Harding (Ed.)

Biomimetics in Photonics
Olaf Karthaus (Ed.)

Optical Properties of Photonic Structures: Interplay of Order and Disorder
Mikhail F Limonov and Richard De La Rue (Eds.)

Nitride Phosphors and Solid-State Lighting
Rong-Jun Xie, Yuan Qiang Li, Naoto Hirosaki, and Hajime Yamamoto

Molded Optics: Design and Manufacture
Michael Schaub, Jim Schwiegerling, Eric Fest, R Hamilton Shepard, and Alan Symmons

An Introduction to Quantum Optics: Photon and Biphoton Physics
Yanhua Shih

Principles of Adaptive Optics, Third Edition
Robert Tyson

Optical Tweezers: Methods and Applications
Miles J Padgett, Justin Molloy, and David McGloin (Eds.)

Thin-Film Optical Filters, Fourth Edition
H Angus Macleod

Laser Induced Damage of Optical Materials
R M Wood

Principles of Nanophotonics
Motoichi Ohtsu, Kiyoshi Kobayashi, Tadashi Kawazoe, Tadashi Yatsui, and Makoto Naruse

Handbook of Silicon Photonics

Edited by
Laurent Vivien
Lorenzo Pavesi

CRC Press is an imprint of the
Taylor & Francis Group, an **informa** business
A TAYLOR & FRANCIS BOOK

Cover Image: Various silicon photonics devices realized within the HELIOS consortium of the European Commission and fabricated by CEA-Leti. LED plots are courtesy of Eveline Rigo.

CRC Press
Taylor & Francis Group
6000 Broken Sound Parkway NW, Suite 300
Boca Raton, FL 33487-2742

First issued in paperback 2020

ISBN 13: 978-0-367-57648-6 (pbk)
ISBN 13: 978-1-4398-3610-1 (hbk)

Library of Congress Cataloging-in-Publication Data

Handbook of silicon photonics / edited by Laurent Vivien and Lorenzo Pavesi.
pages cm. -- (Series in optics and optoelectronics)
Includes bibliographical references and index.
ISBN 978-1-4398-3610-1 (hardback)
1. Optoelectronic devices--Handbooks, manuals, etc. 2. Silicon--Optical properties--Handbooks, manuals, etc. 3. Photonics--Handbooks, manuals, etc. I. Vivien, Laurent.

TK8304.H35 2013
621.36'5--dc23 2012050941

Visit the Taylor & Francis Web site at
http://www.taylorandfrancis.com

and the CRC Press Web site at
http://www.crcpress.com

Contents

Preface

In 1965, Gordon Moore announced his famous law that the number of transistors per chip will double every 19 months.[1] Since then, Moore's law has driven the development of microelectronics. Microelectronic evolution has followed the motto "smaller, cheaper, faster" by using very-large-scale integration of a basic building block, the transistor. Today, there are processors that contain billions of transistors, each one with dimensions of a few tenths of nanometers. It is interesting to note that in 1969, S. T. Miller from Bell Labs already suggested that integration should drive the development of photonics as well.[2] However, the number of photonic components for optical integrated circuits has not grown as for microelectronics over the years. Current integrated photonic circuits contain only a hundred different components with the burden of a high production cost. Instead of looking for performance improvements by increasing the integration level, photonic research was concentrated on single device scale refinements. What is the reason for this? Table P.1 shows a comparison between microelectronic and photonic in term of building blocks, materials, and technology. While for microelectronics the key feature is standard (a single building block is repetitively manufactured with a single material and a single production process), for photonics, there is a diversity of materials, elementary devices, and manufacturing processes. This is the main reason why microelectronics kept pace with integration while photonics did so with isolated device optimization. The rationale of silicon photonics is to apply the paradigm of microelectronics to photonics by manufacturing various devices in a single material—silicon—and using a single manufacturing process—the CMOS process (see Table P.1). In this way, the level of integration of photonic devices can be increased, which in turn is reflected in an increase in the performance of the photonic integrated circuit. At the same time, mass manufacturing of integrated photonic circuits yields a low price of production per unit.[3]

During recent years, additional materials, mainly germanium and III–V semiconductors, have been introduced in silicon photonics to reach efficient building blocks. The complexity of each device has gotten higher and higher to take into account light polarization and performance independence in a broad wavelength range. Manufacturing technology has shifted from a pure CMOS process, still keeping in mind CMOS process compatibility. This was achieved by producing the photonic devices using typical CMOS manufacturing tools and trying to depart at least from standard CMOS processes. In fact, most of microelectronics industries are now developing programs on silicon photonics within their production line using 200- and 300-mm CMOS tools.

The development of silicon photonics is also driven by other strong pushes: on one hand, the Internet and the move of high bandwidth as close as possible to the user; on the other hand, the need to control power dissipation (i.e., heat) induced by the long-distance high-data-rate transmission of digital signals. Currently, in high-performance computers, there is already a need to manage both the exchange of data between hundreds of thousands of cores in the multiprocessors (aggregate data rate of the order of 1 Tbps) and the exchange of information between the multiprocessor and the other boards (memories, peripherals, etc.) at a rate in excess of 40 Gbps. In addition, these developments spawn other applications for silicon photonics in technologies as diverse as telecommunications, information processing, displays, sensing, metrology, medicine, consumer electronics, and energetics. The richness of applications will continuously push silicon photonics to new innovative

TABLE P.1

Comparison of Various Technologies

	Microelectronics	Photonics	Silicon Photonics
Building blocks	Transistor	Laser, photodetector, modulator, optical fiber, waveguide, DWDM, etc.	Laser, waveguides, photodetectors, modulator, microresonators, etc.
Material	Silicon[a]	Semiconductors, glasses, polymers, insulators, etc.	Silicon[a]
Manufacturing technology	CMOS	Epitaxy, deposition, glass drawing, ionic diffusion, etc.	CMOS compatible process[b]

[a] In microelectronics and silicon photonics, the substrate material is silicon. Many other materials are used as well, for example, Germanium, III–V semiconductors, rare earths, metals, etc. See, e.g., M. Heyns and W. Tsai, Ultimate scaling of CMOS logic devices with Ge and III–V materials, *MRS Bulletin*, vol. 34 (2009), pp. 485–492.

[b] Use of CMOS tools to fabricate photonic device with a maximum emphasis to adjust the technology to the CMOS process.

concepts. One can also notice that the opposite is also true: the richness of silicon photonics will push more and more applications.

It is interesting to note that these developments occurred in only a few years. Indeed, in many semiconductors textbooks, silicon is introduced as an indirect band-gap semiconductor that is important only for electronic applications.[4] For this reason, we think that this is the right time to step back and present a coherent and comprehensive overview of silicon photonics from the basics and fundamentals to integrated systems and applications. This handbook covers a broad spectrum from the material to applications, emphasizing passive and active photonic devices, fabrication, integration, and the convergence with CMOS technology. Each chapter is written by world experts in the field. Authors come from both academia as well as industries.

Specifically, the handbook starts with a set of chapters (Chapters 1 to 7) where the basics of silicon as an optical material are introduced. In Chapter 1, the basic properties and the growth mechanisms of group IV materials are reviewed. After a detailed introduction on the various growth methods, the growth of silicon and silicon/germanium alloys is discussed. Then the main properties of group IV elements are reported, with a specific emphasis on silicon and germanium. The chapter ends with a section on the strain control and engineering in Si/Ge and Ge/Sn heterostructures. Chapter 2 introduces silicon-based waveguides. First, it presents a narrow submicrometer cross-section photonic wire waveguide and discusses the problem of losses and polarization control. Then, nanostructured silicon waveguides are considered both in a slot waveguide geometry and in a subwavelength grating waveguide geometry. The chapter ends with a discussion of micrometer-sized waveguides, which are silicon- or medium index silica–based. Chapter 3 addresses the problem of light coupling from optical fiber to waveguide. After defining the problem and the proper metrics, edge-coupling solutions are discussed. Then, surface couplers based on waveguide gratings are described. Finally, free space coupling is presented. Chapter 4 is about multichannel silicon photonic devices: integrated grating technologies, waveguide-coupled microring resonator–based multiplexers and demultiplexers, and interferometric-based structures. Chapter 5 deals with nonlinear optics in silicon. Since silicon is a centrosymmetric material, most nonlinear effects are based on third-order optical nonlinearities: four-wave mixing, two-photon absorption, self- and cross-phase modulation, stimulated Raman scattering. In this chapter, there is also a review of the various approaches to generate second-order nonlinear effects. The last part of the chapter is dedicated to the applications derived from third-order nonlinear effects. Chapter 6 is about long-wavelength, i.e., medium infrared, silicon photonic circuits. The chapter covers all the associated topics, from waveguiding to light generation, from nonlinear optical properties in the long wavelength region to detectors, and from the different material platforms (silicon on insulator, silicon on sapphire, silicon on porous silicon, silicon on air) to the heterogeneous integration of III–V and silicon. Chapter 7 introduces silicon-based photonic crystals and metamaterials and motivates their use in future on-chip optical interconnects. First, a discussion of all optical switches and memories based on photonic crystal nanocavities is presented. Then, photonic crystal lasers where active III–V materials are optically coupled to silicon are reviewed. Finally, slow light generation and applications are discussed.

These chapters are followed by Chapters 8 to 10, where the different building blocks needed to drive silicon photonic integrated circuits are presented. Chapter 8 treats silicon-based light sources. An exhaustive discussion of the light emission processes in indirect gap semiconductors is followed by a review of the physics of low-dimensional silicon structures. Then, two alternative systems are introduced: Si/Ge alloys or Si/SiGe nanostructures and rare earth-doped silicon or silicon nanostructures. This is followed by a discussion of band-engineered Ge on Si lasers. The chapter ends with a review of Purcell effects in silicon nanocrystals as a way to enhance radiative recombination rates. Chapter 9 discusses the physics and device applications of silicon-based optical modulators. First, the physical mechanisms to obtain optical modulation in silicon are presented: carrier accumulation, injection, or depletion-based devices are detailed and compared. Then, Franz–Keldysh and quantum-confined Stark effects in Ge and SiGe materials are discussed. Chapter 10 reports on photodetectors suitable to be integrated in silicon photonics. It traces the development of Ge photodetectors, their performance and integration into Si photonics circuits. Si-based photodetectors developed for near-IR applications are also discussed. Through the chapter, a sensitive issue is the design, modeling, and fabrication of waveguide integrated photodetectors.

Once the building blocks are available, the next issue is their integration in complex photonic integrated circuits. This is the subject of Chapters 11 to 13. Chapter 11 deals with hybrid and heterogeneous integrations of III–V active materials in silicon photonics. First, the integration schemes and their fabrication technologies are introduced. Then, an overview of the various demonstrated devices that make a complete photonic toolbox is addressed. Chapter 12 discusses the fabrication of silicon photonic devices by providing a basic foundation on micro- and nano-fabrication technologies from photonic perspectives. Cleanrooms, lithography, depositions, etching, microfabrication tools, processes, and materials are discussed. Also, issues such as parameter variations on wafers and dimensional control in CMOS processing are presented. Chapter 13 attacks the problem of the convergence between photonics and CMOS at the single-wafer level. The various alternatives are introduced and critically discussed. Both die-to-die integration via wire bonding or flip-chip, wafer-to-wafer bonding, and single-wafer fabrication are presented.

Applications of silicon photonics in different fields from data communication or optical interconnects are reported in Chapters 14 and 15. Chapter 14 shows the different kinds of devices when silicon photonics is applied to biology or life science. Planar silicon waveguide molecular affinity sensors and the strategies of surface functionalization and bioconjugation are discussed. The manipulation and transport of biomolecules using silicon nanostructures is reviewed. The status of bioimaging using silicon nanoparticles is reported as well. Chapter 15 describes silicon-based photovoltaics. High-efficiency monocrystalline solar cells, multicrystalline silicon solar cells, modification of the solar spectrum, and the use of silicon quantum dots are reviewed.

We hope that the efforts of all chapter authors, of the publisher, and of ourselves will be useful to both specialists and newcomers. We tried to balance introductory concepts with more specialized arguments. Each chapter has been led by one expert who organized the chapter and invited other experts to contribute to specific sections. Our editorial philosophy has been to give the authors considerable freedom in addressing their topics. Thus, the chapters are a blend of different styles and approaches. They are all self-contained with little reference to the others. Still, we tried to avoid repetition. It is definitely a challenge to provide up-to-date reference books in a field as broad and as rapidly changing as silicon photonics. All authors have responded to this challenge with enthusiasm and professionalism. We would like to thank all the chapter coordinators, authors, and contributors for the great job they did.

In addition to thanking the authors, we would like to thank John Navas and Rachel Holt for their help, assistance, and patience. Last but not least, we would like to thank the worldwide silicon photonic community, which is contributing to make this research a very exciting and rapidly evolving field where a nurturing blend of basic science and technological applications that push the frontier forward each day.

Laurent Vivien
CNRS-Université Paris Sud

Lorenzo Pavesi
Unversità di Trento

Notes

1. G. E. Moore, Cramming more components onto integrated circuits, *Electronics*, vol. 38, no. 8. (19 April 1965), pp. 114–117.
2. S. T. Miller, Integrated optics: an introduction, *The Bell System Technical Journal* vol. 48, no. 7 (September 1969), pp. 2059–2069.
3. T. Baehr-Jones, T. Pinguet, P. L. Guo-Qiang, S. Danziger, D. Prather, and M. Hochberg, Myths and rumors of silicon photonics, *Nature Photonics* vol. 6 (2012), pp. 206–208.
4. "If God wanted ordinary silicon to efficiently emit light, he would not have given us gallium arsenide," said Elias Towe of Carnegie Mellon University (Pittsburgh) (quoted from *IEEE Spectrum* "Linking with Light," August 2002).

Contributors

Nikola Alic
Electrical and Computer Engineering Department
University of California, San Diego
La Jolla, California

Aleksei Anopchenko
Nanoscience Laboratory
Department of Physics
University of Trento
Trento, Italy

Toshihiko Baba
Department of Electrical and Computer Engineering
Yokohama National University
Yokohama, Japan

Ryan C. Bailey
Department of Chemistry
University of Illinois at Urbana–Champaign
Urbana, Illinois

Matthias Bauer
ASM America Inc.
Phoenix, Arizona

Simona Binetti
Milano-Bicocca Solar Energy Research Center (MIB-SOLAR)
Department of Materials Science
University of Milano-Bicocca
Milan, Italy

Przemek J. Bock
Information and Communication Technologies
National Research Council Canada
Ottawa, Ontario, Canada

Wim Bogaerts
Photonics Research Group
Department of Information Technology
Ghent University–imec
Ghent, Belgium

John E. Bowers
Department of Electrical and Computer Engineering
University of California, Santa Barbara
Santa Barbara, California

Ozdal Boyraz
EECS Department
University of California, Irvine
Irvine, California

Massimo Cazzanelli
Nanoscience Laboratory
Department of Physics
University of Trento
Trento, Italy

Yimin Chao
Energy Materials Laboratory
School of Chemistry
University of East Anglia
Norwich, United Kingdom

Pavel Cheben
Information and Communication Technologies
National Research Council Canada
Ottawa, Ontario, Canada

Gavin Conibeer
School of Photovoltaic and Renewable Energy Engineering
University of New South Wales
Sydney, Australia

André Delâge
Information and Communication Technologies
National Research Council Canada
Ottawa, Ontario, Canada

Adam Densmore
Information and Communication Technologies
National Research Council Canada
Ottawa, Ontario, Canada

Philippe M. Fauchet
Department of Electrical Engineering and Computer Science
Vanderbilt University
Nashville, Tennessee

Jean-Marc Fedeli
Commissariat à l'énergie atomique et aux énergies alternatives
MINATEC Campus
Grenoble Cedex, France

Ning-Ning Feng
Kotura Inc.
Monterey Park, California

Shaoqi Feng
Photonic Device Laboratory
Department of Electronic and Computer Engineering
Hong Kong University of Science and Technology
Hong Kong, China

Hiroshi Fukuda
Microsystem Integration Laboratories
NTT Corporation
Atsugi, Japan

Michael W. Geis
Lincoln Laboratory
Massachusetts Institute of Technology
Lexington, Massachusetts

Faezeh Gholami
Electrical and Computer Engineering Department
University of California, San Diego
La Jolla, California

Tom Gregorkiewicz
Van der Waals-Zeeman Institute
University of Amsterdam
Amsterdam, The Netherlands

Matthew E. Grein
Lincoln Laboratory
Massachusetts Institute of Technology
Lexington, Massachusetts

Martijn J. R. Heck
Department of Electrical and Computer Engineering
University of California, Santa Barbara
Santa Barbara, California

Adam T. Heiniger
Institute of Optics
University of Rochester
Rochester, New York

Shujuan Huang
School of Photovoltaic and Renewable Energy Engineering
University of New South Wales
Sydney, Australia

Yuewang Huang
EECS Department
University of California, Irvine
Irvine, California

Zoran Ikonic
Institute of Microwaves and Photonics
School of Electronic and Electrical Engineering
University of Leeds
Leeds, United Kingdom

Massimo Izzi
ENEA Research Center Casaccia
Rome, Italy

Siegfried Janz
Information and Communication Technologies
National Research Council Canada
Ottawa, Ontario, Canada

Erich Kasper
Institut für Halbleitertechnik
Universität Stuttgart
Stuttgart, Germany

Robert W. Kelsall
Institute of Microwaves and Photonics
School of Electronic and Electrical Engineering
University of Leeds
Leeds, United Kingdom

Martin Kittler
Joint Lab IHP/BTU
IHP GmbH
Frankfurt, Germany

Andrew P. Knights
Department of Engineering Physics
McMaster University
Hamilton, Ontario, Canada

Steven J. Koester
Electrical and Computer Engineering
University of Minnesota–Twin Cities
Minneapolis, Minnesota

Christian Koos
Institute of Photonics and Quantum Electronics
Karlsruhe Institute of Technology
Karlsruhe, Germany

Radovan Kopecek
International Solar Energy Research Center (ISC)
Konstanz, Germany

Alessia Le Donne
Milano-Bicocca Solar Energy Research Center (MIB-SOLAR)
Department of Materials Science
University of Milano-Bicocca
Milan, Italy

Ting Lei
Photonic Device Laboratory
Department of Electronic and Computer Engineering
Hong Kong University of Science and Technology
Hong Kong, China

Qiang Lin
Institute of Optics
University of Rochester
Rochester, New York

Jifeng Liu
Thayer School of Engineering
Dartmouth College
Hanover, New Hampshire

Guo-Qiang Lo
Institute of Microelectronics
Agency for Science, Technology and Research (A*STAR)
Singapore

David J. Lockwood
NRC Institute for Microstructural Sciences
Ottawa, Ontario, Canada

Francisco López Royo
Nanophotonics Technology Center
Universitat Politècnica de València
Valencia, Spain

Xianshu Luo
Institute of Microelectronics
Agency for Science, Technology and Research (A*STAR)
Singapore

Theodore M. Lyszczarz
Lincoln Laboratory
Massachusetts Institute of Technology
Lexington, Massachusetts

Delphine Marris-Morini
Institut d'Electronique Fondamentale
Université Paris-Sud–CNRS
Orsay, France

Goran Z. Mashanovich
Silicon Photonics Group
Optoelectronics Research Centre
University of Southampton
Southampton, United Kingdom

Shinji Matsuo
NTT Photonics Laboratories
Atsugi, Japan

Michelle McCann
International Solar Energy Research Center (ISC)
Konstanz, Germany

Al Meldrum
Physics Department
University of Alberta
Edmonton, Alberta, Canada

Jurgen Michel
Microphotonics Center and Department of Materials Science and Engineering
Massachusetts Institute of Technology
Cambridge, Massachusetts

David A. B. Miller
Ginzton Laboratory
Stanford University
Stanford, California

Milan M. Milošević
Advanced Technology Institute
Faculty of Engineering and Physical Sciences
University of Surrey
Guildford, United Kingdom

Osamu Nakatsuka
Department of Crystalline Materials Science
Graduate School of Engineering
Nagoya University
Nagoya, Japan

Masaya Notomi
NTT Basic Research Laboratories
Atsugi, Japan

Kengo Nozaki
NTT Basic Research Laboratories
Atsugi, Japan

Michael Oehme
Institut für Halbleitertechnik
Universität Stuttgart
Stuttgart, Germany

Stefano Ossicini
DiSMI, University of Modena and Reggio Emilia
Reggio Emilia, Italy

Joerg Pfeifle
Institute of Photonics and Quantum Electronics
Karlsruhe Institute of Technology
Karlsruhe, Germany

Thierry Pinguet
Luxtera, Inc.
Carlsbad, California

Andrew W. Poon
Photonic Device Laboratory
Department of Electronic and Computer Engineering
Hong Kong University of Science and Technology
Hong Kong, China

Alexei Prokofiev
Ioffe Physico-Technical Institute
Russian Academy of Sciences
Saint Petersburg, Russia

Stojan Radic
Electrical and Computer Engineering Department
University of California, San Diego
La Jolla, California

Graham T. Reed
Silicon Photonics Group
Optoelectronics Research Centre
University of Southampton
Southampton, United Kingdom

Manfred Reiche
MPI Mikrostrukturphysik
Halle, Germany

Gunther Roelkens
University of Ghent, INTEC
Ghent, Belgium

Saba Saeed
Van der Waals-Zeeman Institute
University of Amsterdam
Amsterdam, The Netherlands

Xinzhu Sang
Institute of Information Photonics and Optical Communications
Beijing University of Posts and Telecommunication
Beijing, China

Aimé Sayarath
Photonic Device Laboratory
Department of Electronic and Computer Engineering
Hong Kong University of Science and Technology
Hong Kong, China

Rebecca K. Schaevitz
Ginzton Laboratory
Stanford University
Stanford, California

Jens H. Schmid
Information and Communication Technologies
National Research Council Canada
Ottawa, Ontario, Canada

Jun-Feng Song
Institute of Microelectronics
Agency for Science, Technology and Research (A*STAR)
Singapore

and

State Key Laboratory on Integrated Opto-Electronics
College of Electronic Science and Engineering
Jilin University
Changchun, China

Richard A. Soref
University of Massachusetts
Boston, Massachusetts

Steven J. Spector
Lincoln Laboratory
Massachusetts Institute of Technology
Lexington, Massachusetts

David J. Thomson
Silicon Photonics Group
Optoelectronics Research Centre
University of Southampton
Southampton, United Kingdom

Tai Tsuchizawa
Microsystem Integration Laboratories
NTT Corporation
Atsugi, Japan

Leonid Tsybeskov
Department of Electrical and Computer Engineering
New Jersey Institute of Technology
Newark, New Jersey

Mario Tucci
ENEA Research Center Casaccia
Rome, Italy

Diedrik Vermeulen
Photonics Research Group
Department of Information Technology
Ghent University–imec
Ghent, Belgium

Qi Wang
Energy Materials Laboratory
School of Chemistry
University of East Anglia
Norwich, United Kingdom

Xiaoxin Wang
Thayer School of Engineering
Dartmouth College
Hanover, New Hampshire

Maciek Wojdak
Department of Electronic and Electrical Engineering
University College London
London, United Kingdom

Dan-Xia Xu
Information and Communication Technologies
National Research Council Canada
Ottawa, Ontario, Canada

Koji Yamada
Microsystem Integration Laboratories
NTT Corporation
Atsugi, Japan

Irina N. Yassievich
Ioffe Physico-Technical Institute
Russian Academy of Sciences
Saint Petersburg, Russia

Jung U. Yoon
Lincoln Laboratory
Massachusetts Institute of Technology
Lexington, Massachusetts

Shigeaki Zaima
Department of Crystalline Materials Science
Graduate School of Engineering
Nagoya University
Nagoya, Japan

Sanja Zlatanovic
Electrical and Computer Engineering Department
University of California, San Diego
La Jolla, California

1

Group IV Materials

Erich Kasper, Michael Oehme, Matthias Bauer, Martin Kittler, Manfred Reiche, Osamu Nakatsuka, and Shigeaki Zaima

CONTENTS

1.1 Introduction

Erich Kasper and Michael Oehme

It is interesting to note that the fantastic growth in semiconductor electronics is based on only a few concepts. The first transistor was made from germanium (Ge) because of its rather low processing temperatures in comparison to silicon (Si). What was considered a disadvantage turned out to become the unbeatable lead of Si technology since silicon dioxide and related materials allow fabrication techniques such as parallel and planar processing (Grimmeiss and Kasper, 2009). Dramatic shrinkage in feature sizes and a corresponding increase in transistor numbers led to integrated circuits (IC), which are now, with their incredibly high complexity, the undisputed backbone of microelectronics. The indirect semiconductors Si and Ge suffered in their potential for optoelectronics from low radiative emission efficiency and low absorption at the indirect band gap. The direct semiconductors, mainly from group III/V, were strong in that respect and they started to dominate optoelectronics with clear band gap emission, light emission diodes (LED) of high quantum efficiency, and laser diodes with low threshold currents even at room temperature. This had a strong impact on many areas including long-distance information transmission via optical fiber cables. With more mature fabrication technology, high-power lasers and LEDs entered manufacturing applications and lighting in vehicles and homes. Although optoelectronic integrated circuits (OEIC) are familiar on group III/V substrates, their integration complexity is much lower than that of their IC counterparts.

The historical development of microelectronics and optoelectronics differed without any doubt in simplifying brevity characterized by priority on complexity for microelectronics and priority on performance for optoelectronics. There are now several serious driving forces from applications, systems, and technology toward a merger of optoelectronics with microelectronics and to an increase in complexity of photonics. Silicon-based photonics is a strong contender in this area because of very sophisticated fabrication schemes, commercial availability of superior waveguide substrates (silicon on insulator), and a renaissance of near-infrared responsive silicon/germanium heterostructures. An example given in the following refers to information transfer on a chip scale. Many believe that the speed limit of system on chips (SOC) can only be overcome with photonic signal transmission.

Progress in Si-based photonics from Ge/Si heterostructures attracts worldwide attention (Jalali and Fathpour, 2006). Photonics and optoelectronics play an essential role in many areas of applications (Soref, 2010) as in telecommunication, information technology, and optical interconnect systems. A key challenge to obtain a convergence of classical Si-based microelectronics and optoelectronics is the manufacturing of photonic-integrated circuits integrable into classical Si-based integrated circuits. This integration would be greatly enhanced if similar facilities and technologies could be used. Therefore, one approach is the development of optoelectronic components and devices made from group IV–based materials such as silicon germanium alloys, pure germanium, or germanium tin alloys.

An example of an integrated photonic circuit (PIC) is shown in Figure 1.1. From an external source, the light is coupled (Taillaert et al., 2003, 2006) in the waveguide by a grating coupler. Grating couplers are used for the interface between glass fibers and on chip waveguides. The grating pattern depends on the wavelengths and is etched on the illuminated surface area. Taper structures provide the connectivity of the waveguide and grating area because of the large dimension contrast. For its simplicity, we propose a silicon-on-insulator (SOI) (Jalali et al., 1998) substrate. Rectangular waveguides like rib

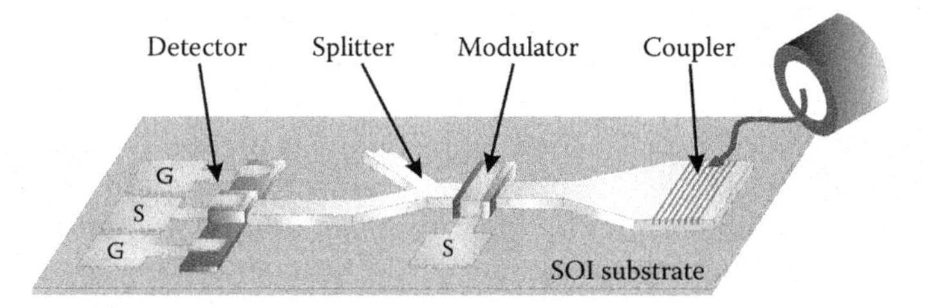

FIGURE 1.1
Scheme of a photonic circuit on SOI. Shown are the main components (i) light source, as example laser light coupled in via a grating coupler, (ii) silicon waveguide and passive waveguide devices, as example a splitter, and (iii) active waveguide devices, as examples a modulator and a detector.

or ridge waveguides provide strong optical confinement at low device fabrication requirements. One etching process is sufficient for a rib or ridge waveguide. Very good optical confinement is achieved, because of the large contrast of refractive index of silicon (n = 3.5) to silicon oxide (n = 1.45) or air (n = 1) (Barkai et al., 2007; Wang et al., 2004; Rickman and Reed, 1994). Directional couplers or beam splitters are used for on-chip routing. Directional couplers also offer various other applications such as switching and polarization beam splitters (Liang and Tsang, 2005). The spectral range covers near-infrared wavelengths between 1.2 and 1.6 μm. Due to its high refractive index and low absorbance in the near-infrared regime, silicon is considered as a very good optical waveguiding material.

Furthermore, silicon and silicon-on-insulator are well-known material systems and provide known processing techniques (Soref and Lorenzo, 1986; Soref and Bennett, 1987). For the modulation of the signal, the optical parameters of the waveguide have to be manipulated, e.g., the refraction index or the absorption coefficient by an applied field. The Mach Zehnder waveguide interferometer consists of two parallel branches. With an applied voltage, carriers are injected into the waveguide and the refractive index changes. The changing propagation velocity leads to a phase shift $\Delta\phi$ between the two branches. The phase shift depends on the applied voltage, the wavelength, and the length of the branches. Due to a voltage limitation and a fixed wavelength, only the length of the modulator increases the phase shift. For a reasonable phase shift, the length of the modulator must be in the millimeter range (10–15 mm). Compared with the waveguide and other optical devices (several micrometers), the modulator becomes too large (Liao et al., 2005; Liu et al., 2004).

In our concept, both the modulator and the detector are made from SiGe/Si, the structure of which could vary from a simple single barrier heterostructure to a multiquantum-well (MQW) or a quantum dot (QD) superlattice (SL) (Oehme et al., 2006, 2010; Yu et al., 2006; Klinger, 2009; Schmid et al., 2010; Kaschel et al., 2011). These active optical devices are grown with modern epitaxy techniques (Oehme et al., 2008a). The modulated and detected wavelength is fixed at 1.3 and 1.55 μm for standard telecommunication purposes. For on-chip processing, the near-infrared spectrum is convenient where silicon absorption is weak or negligible, and Ge may absorb well. The modulation of the absorption in heterostructures is based on the quantum-confined stark effect (QCSE). The electric field in the depletion layer shifts the absorption edge to lower energies. QCSE modulators are much shorter than Mach–Zehnder modulators, which is the main reason for our choice in integrated circuit solutions. The next sections treat the most important deposition methods, the material properties, and the management of strain. The reader will be especially referred to group IV specific solutions, which are mainly related to the strong technological position of silicon in microelectronics (Section 1.2) and the indirect character of band-gap transitions.

The latter caused early attempts to overcome the low radiation efficiency. Historically, the first suggestion was based on the "Brillouin zone folding" concept (Gnutzmann and Clausecker, 1974) from adding a short-period periodicity by a superlattice. Much later, the experimental realization of a quasi-direct transition was proven (Zachai et al., 1990), but the transition strength was orders of magnitude below the competing III/V compounds. Further reduction of the dimensionality (porous Si, quantum dots, nanocrystals) followed with remarkable results (Landolt-Boernstein, 2007). Another fascinating approach considers crystal defects as natural nanostructures—a misfit dislocation line is a quantum wire with nanometer core dimension (see Section 1.3 for detailed discussion). In recent years, the structural toolbox was enlarged by strain management methods, which, especially with tensile-strained Ge and GeSn, could deliver direct semiconductors for the infrared spectrum. The strain adjustment methodology and the role of interface structure therein are explained in Section 1.4.

1.2 Methods of Growth and Deposition

Michael Oehme and Matthias Bauer

A variety of deposition methods is available, which range from long-ago known processes such as electroplating, solution growth, and evaporation to more sophisticated processes with ion assistance and vapor phase surrounding. From the latter, we discuss epitaxy, chemical vapor deposition (CVD), and involvement in device processing.

1.2.1 Epitaxy

Epitaxy is a Greek word and describes the oriented growth of a single crystalline film on a substrate. The growing layers follow the orientation of the substrate. The driving force of the crystalline growth is minimizing of energy because the perfect crystal is the energetically favorable solid. A requirement for epitaxy is a clean substrate surface; otherwise, the information about orientation of the substrate is disordered or even destroyed, and the consequences are crystalline defects up to polycrystalline or amorphous growth.

Modern epitaxy techniques like molecular beam epitaxy (MBE) or CVD allow the growth of very thin crystalline silicon, silicon germanium, or germanium layers with well-defined dopants by very low growth temperatures. In this temperature regime, volume diffusion is negligible. Furthermore, the critical thickness of SiGe heterostructures is increasing. However, in this low temperature growth regime surface segregation of dopant or alloy atoms has turned out to be a dominant mechanism for profile smearing. Knowledge of surface segregation behavior is necessary for device design and realization based on low temperature epitaxy.

Epitaxial growth proceeds by the attachment of atoms on correct positions of a given lattice. The orientation and atomic distances of the lattice are usually predicted by a single crystalline substrate with a clean surface. The most common source of atomic steps is the unintentional misorientation of commercial wafers with angles of typically between 0.1° and 0.5°. Even with nominally oriented substrates, terrace widths of 15 to 75 nm are expected.

The build in of atoms from the vapor phase to the crystal typically follows a three-step scheme with adsorption, diffusion, and incorporation into surface steps (see Figure 1.2).

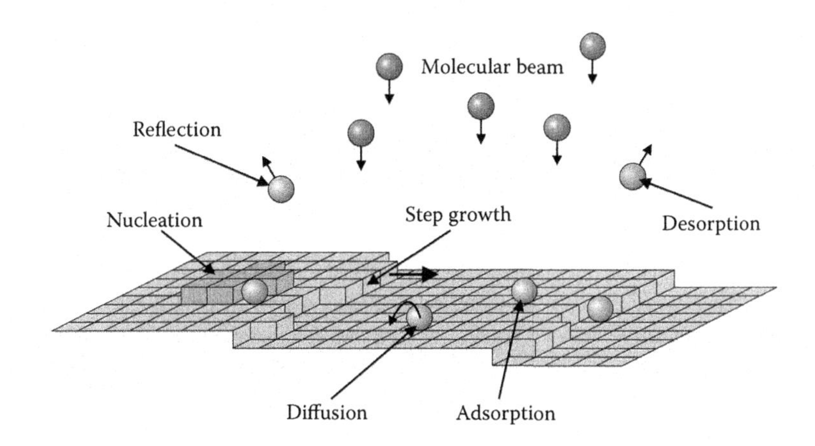

FIGURE 1.2
Overview of the basic steps in the growth of an epitaxial layer from the vapor phase. The layer growth from the vapor phase typically follows a three-step scheme with adsorption, diffusion, and incorporation into surface. Independence of growth temperature step-flow or nucleation mechanism is favored.

The adsorbed atoms called adatoms are in a precursor state for later incorporation in the lattice. However, even adsorbed, the adatom may escape by a later desorption step caused by thermal vibrations. A regular network of surface places is available for the adatoms. Indeed, the adatoms jump rather easily from one of these places to another, described as surface diffusion. This surface diffusion is described by the Brownian movement (Kasper and Herzog, 1976). The adatoms will reach a step nearby and incorporate into the crystal if not desorbed. When all adatoms reach the already existing misorientation steps, then the atomic steps move laterally forward by adatom capture. This happens at higher temperatures, as a rough estimate one can consider temperatures above a half of melting temperature (e.g., Si 1734 K). At lower temperatures, the slowly moving adatoms nucleate into two-dimensional islands. A critical nucleus is defined by a size where growth by capture of adatoms is more probable than decay of the nucleus. With high supersaturation the size of the critical nucleus is smaller, being two atoms under most MBE conditions. The step flow mechanism is based on the lateral movement of preexisting misorientation steps, the (2D) two-dimensional nucleation mechanism creates steps by a nucleation process. These nuclei annihilate after one monolayer so that 2D nucleation is a periodic process.

1.2.2 Molecular Beam Epitaxy

An MBE system involves the generation of molecular beams of matrix material such as silicon, germanium, or tin and doping species and their interaction with the substrate surface to form a single crystal deposit under ultrahigh vacuum (UHV) conditions. Atomic or molecular beams of the necessary species are directed toward the heated substrate and grow into epitaxial layers as depicted in Figure 1.2. The atomic or molecular fluxes of elemental constituents are evaporated or sublimated in special electron beam evaporators or in radiatively heated effusion cells. For precise control of the beam fluxes over the substrate, all sources exhibit rapidly acting mechanical shutters and the flux of the deposition materials can be measured directly with a flux monitor, e.g., a quadruple mass spectrometer or indirectly over the effusion cell temperature. A schematic view of a typical growth chamber with substrate holder, substrate heater, and substrate rotation, evaporation sources for matrix materials (Si, Ge, C, Sn), and dopants (Sb, B), monitoring, and analyzing facilities is shown in Figure 1.3. Gas-source MBE (GSMBE) uses gaseous compounds (mainly hydrogen-, chlorine-, or organometallic compounds) directed by nozzles toward the substrate.

The quality of the vacuum conditions and the purity of the constituent fluxes are fundamental specification of the MBE because the semiconductor properties are sensitive to impurities. Crystal quality and defect density are drastically limited by insufficient partial pressure (Kasper and Kibbel, 1984). For the growth of epitaxial layers with high purity and crystalline perfection, a UHV equipment with a basis pressure lower than 10^{-10} mbar is needed. Vacuum conditions during the layer growth are very important because very hot components in the reactor need special materials with a low vapor pressure such as tantalum or molybdenum. From the above discussion, the essential features (see Figure 1.4) of a group IV–based MBE system are

1. UHV conditions during growth
2. Sources of molecular beams
3. Substrate cleaning scheme with low thermal budget
4. In situ analysis of process conditions and real-time monitoring for close loop control

The main advantage of MBE is that all growth parameters can be controlled independently, in contrast to other epitaxy techniques such as CVD.

To improve vacuum quality and to allow cluster processing, the growth chamber is connected to a storage chamber and a load lock with additional options to cluster with

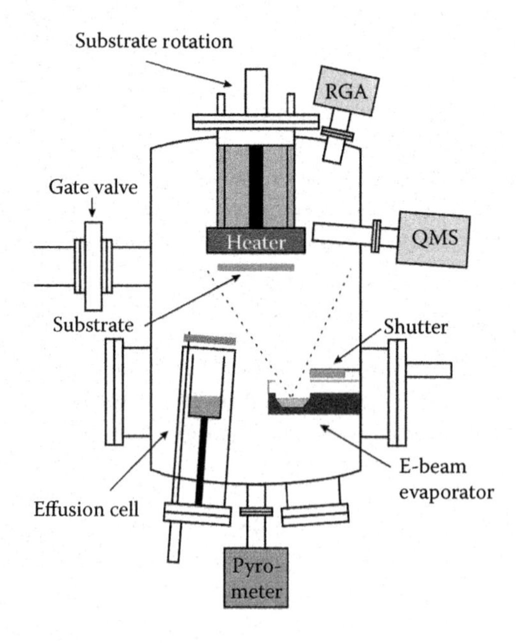

FIGURE 1.3
Schematic illustration of an MBE growth chamber. (Based on Oehme, M. et al., *Thin Solid Films*, 517, pp. 137–139, 2008.)

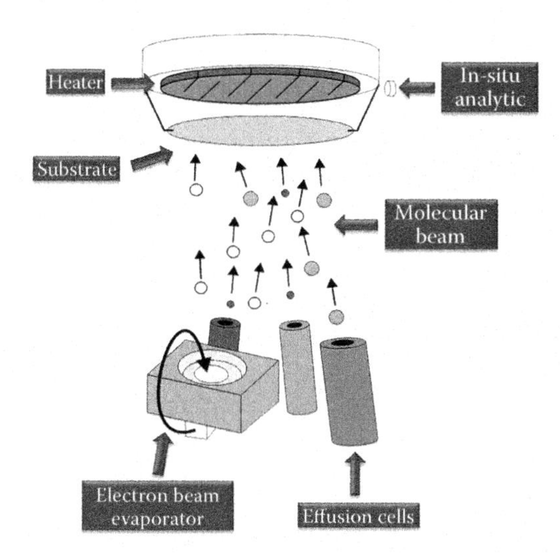

FIGURE 1.4
Basic scheme of the essential features of a group IV–based MBE system.

analytical and processing equipment by a transfer system. Commercial, 200-mm group IV MBE equipment from DCA Instruments is shown in Figure 1.5. The machine can be upgraded to wafers with a diameter of 300 mm.

Typically, the growth chamber is evacuated with a pumping cascade consisting of an oil-free fore-vacuum pump, a turbo molecular pump, a high-speed titanium sublimation pump cooled with liquid nitrogen and a cryopump. The growth chamber is always under UHV except for maintenance work or source refilling. After venting, the chamber requires a bakeout up to 250°C over more than 24 h. Under growth conditions, cooling water pipes

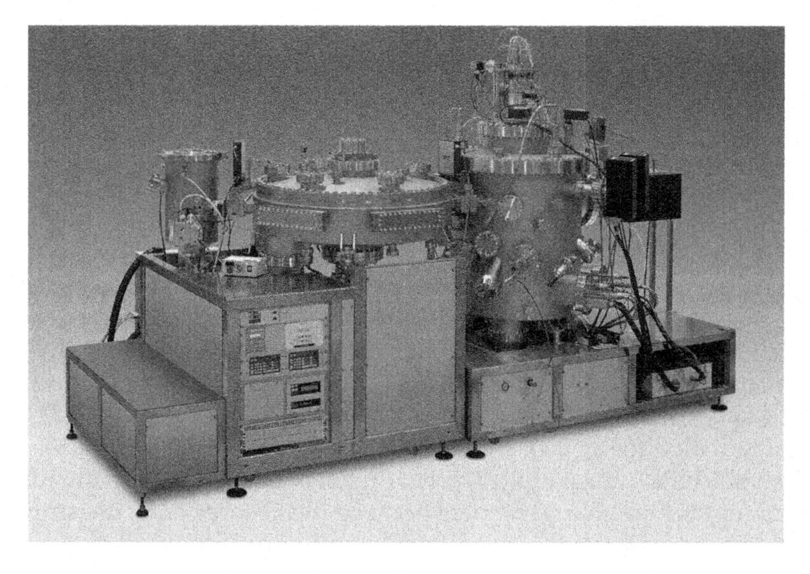

FIGURE 1.5
Commercial 200-mm group IV MBE equipment (DCA Instruments).

welded to the chamber walls effectively reduce outgassing. Also all sources of molecular beams and the substrate heater exhibit cooled walls to reduce the thermal radiation.

The evaporation source for silicon is an electron beam evaporator (EBE), to avoid a crucible attack because of the extreme reactivity of molten Si. The focused electron beam partially melts the Si surface and can be scanned over the whole crucible. The EBE is screened by cold walls with defined aperture, which limit the beam cone and reduce the thermal load of the chamber walls. All metal surfaces are covered with high-purity Si plates to reduce the risk of interaction with electrons and ions. For the evaporation of Ge, there are two possibilities, with a single effusion cell or an electron beam evaporator. The construction of Ge-EBE is similar to the Si-EBE. However, the turbulence in Ge is much more severe than in Si because Ge has a much smaller thermal conductivity than Si, resulting in complete melting of the Ge reservoir. Another possibility is the Ge evaporation from a single effusion cell with a pyrolithical BN crucible. The new material for group IV photonics is the alloy GeSn. The melting point of the element tin is very low (232°C) so it can be evaporated with a conventional effusion cell with a pyrolithical BN crucible. The flux monitoring of Si, Ge, and Sn is performed by a quadruple mass spectrometer (QMS). The QMS measures the intensity of the Si isotope with the mass number 30 because this mass is free of fragments of other residual species. The isotopes with mass 74 for Ge and mass 120 for Sn are used. The QMS signals are proportional to the fluxes, respectively, to the growth rates on the substrate surface.

p- and n-Type doping over a wide range of concentration and with abrupt doping transitions is necessary for novel device structures. High and moderate levels of dopants of both types are needed to grow devices. In a group IV MBE, several doping techniques are used. As a p dopant source, boron effusion cell are used. Boron is evaporated from a high temperature source with special graphite crucibles. Temperatures between 1500°C and 2000°C are required to achieve doping levels of 10^{15} cm^{-3} up to several 10^{20} cm^{-3}.

n-Type doping during group IV MBE growth is problematic because of low incorporation coefficients and surface segregation. Several doping methods were developed to circumvent these difficulties. Sb and P are usually chosen for n-type doping materials. Since elemental P cannot be used for P doping due to its high vapor pressure, special sources are needed. However, experiments with these sources and with very high P doping concentrations generate memory effects in the chamber, which introduce a high background doping in the complete system. From a practical viewpoint, Sb is the more attractive n-type doping material because it can be directly coevaporated from a controlled source during growth. However, the extreme temperature sensitivity of surface segregation requires special doping strategies such as prebuildup or flash-off techniques.

Substrate temperature is the most important growth parameter because it influences all adatom processes of the surface, the crystalline growth, the surface morphology, the abruptness of doping transitions, and the relaxation processes in heterostructures. In an MBE system, the substrate is heated radiatively from the backside by a current powered heater, which is made of meander-shaped pyrolytical graphite or doped SiC. At the typical growth temperatures, only a small overlap of the spectral emission of the radiative heater and the spectral absorption of silicon is given (see Figure 1.6). At 700°C, the maximum thermal radiation for a black emitter has a wavelength of 3 μm. In this wavelength range, intrinsic silicon is transparent. The heating of substrate is caused mainly by radiative absorption of the free charge carriers because the absorption is the sum of phonon-assisted band to band absorption and optical absorption by free carriers.

Standard pyrometric measurements cannot be applied during film growth because interference effects create artifacts and obscure the evaluation via Planck's law. A combination

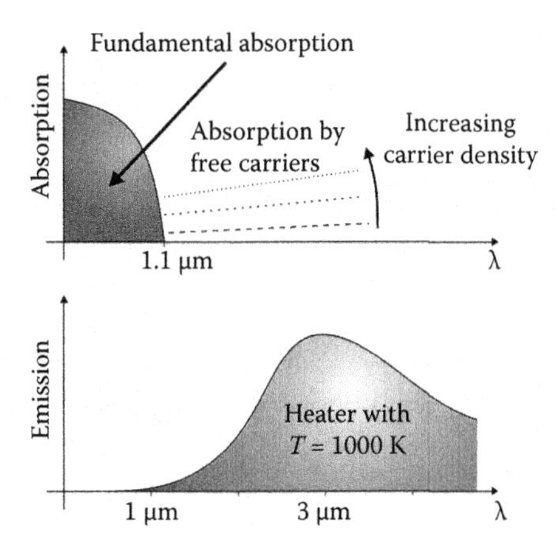

FIGURE 1.6
Comparison between the absorption spectrum of silicon and the emission spectrum of a radiative heater at 1000 K (after Planck's law), which showed only a small overlap. (From Oehme, M. et al., *Appl. Surf. Sci.* 254, pp. 6238–6241, 2008.)

of a pyrometer and a reflectometer will result in a real-time correction of emissivity ε, thus allowing the determination of temperatures even for arbitrary multilayer stacks. This measurement method is called reflection supported pyrometric interferometry (RSPI) (Bauer et al., 2000). An RSPI system for temperature measurements in the range from around 450°C to over 900°C with a resolution of about 0.1°C is used. In the very low temperature regime (150°C–500°C), the substrate temperature is measured with a Pt/PtRh thermocouple incorporated in a special Si substrate. The temperature dependency of a high (4×10^{18} cm^{-3}) and a low (1×10^{13} cm^{-3}) doped substrate as a function of heater temperature is shown in Figure 1.7. For temperatures higher than 650°C, no differences for these two types of substrates can be seen because the intrinsic carrier concentration in Si is 10^{18} cm^{-3}

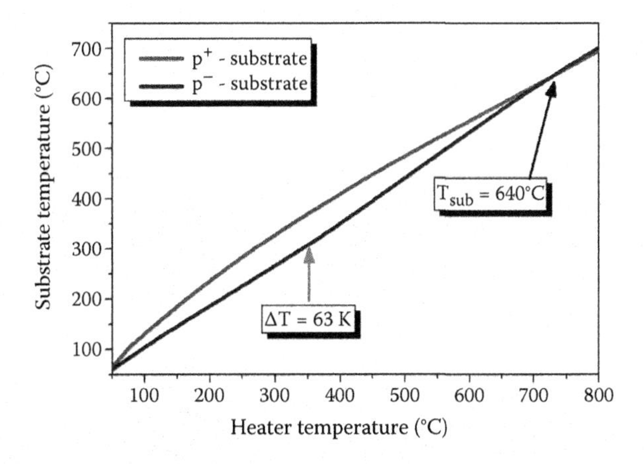

FIGURE 1.7
Temperature dependency of a high- and low-doped substrate as a function of heater temperature measured with an RSPI system for high temperatures above 450°C and with a thermocouple incorporated in the substrate surface for very low temperatures. (From Oehme, M. et al., *Appl. Surf. Sci.* 254, pp. 6238–6241, 2008.)

at 700°C (Sze, 1969). However, at lower heater temperatures, the temperature difference between high- and low-doped substrates increases because of the changing of the free carrier densities. The consequence is that different calibration curves for high- and low-doped substrates have to be used.

Process parameters are monitored in this MBE system with different in situ analyzing methods, such as vacuum monitoring, temperature monitoring and controlling, flux calibration, and controlling or measurement of layer thickness. These parameters are used for process control or for regulating sources and substrate temperatures. The RSPI measurement system has been proven to be capable of providing in situ real-time information concerning temperature and film thickness for numerous applications in semiconductor manufacturing.

1.2.3 Epitaxial Growth of Active Optical Devices

The optoelectronic part in an IC should not be higher than 1 μm above the silicon substrate surface level because of the small depth of focus of lithography systems requiring planarization steps. One solution is a selective growth of thin heterostructures in predefined areas or a uniform growth on the silicon surface covered by a patterned oxide. The latter growth is called differential epitaxy (Kasper et al., 1987) because it produces epitaxial islands in the oxide windows surrounded by polycrystalline layers on the oxide (see Figure 1.8). The poly layers can be removed by either preferential etching or chemical mechanical polishing (CMP). They can also be used for resistors or absorbers.

The challenges of the MBE growth of germanium-based optoelectronic devices integrated on Si are the lattice mismatch between Si and Ge as well as the doping structures of the active devices. Modern Ge heterojunction diodes consist of special double heterostructure contacts that enclose the intrinsic or active region of the p-i-n diode (Oehme et al., 2010). A schematic cross section of the epitaxy structure is shown in Figure 1.9 (top). This layer structure can be used in a vertical (Figure 1.9, left) and in a lateral (Figure 1.9, right) device concept. The MBE growth of the layers is divided into three stages: first, the lattice mismatch between Si and Ge has to be adjusted; second, the doping transitions between the contact regions (1×10^{20} cm^{-3} doped) and the undoped absorption region must be abrupt; third, the heterostructure contacts with good electrical properties should be realized. The Ge heterojunction devices are produced in both concepts in a quasi-planar technology with the backside contact lead to the front side with a buried layer technology.

The layer growth starts with a 500-nm-thick p^+ Si/Ge heterocontact. This contact consists of a 400-nm-thick and very high B-doped (1×10^{20} cm^{-3}) Si-buried layer and a 100-nm virtual substrate (VS) made from pure Ge, which is also B-doped with a doping concentration of 1×10^{20} cm^{-3}. The VS accomplishes the accommodation of the lattice constant of Ge

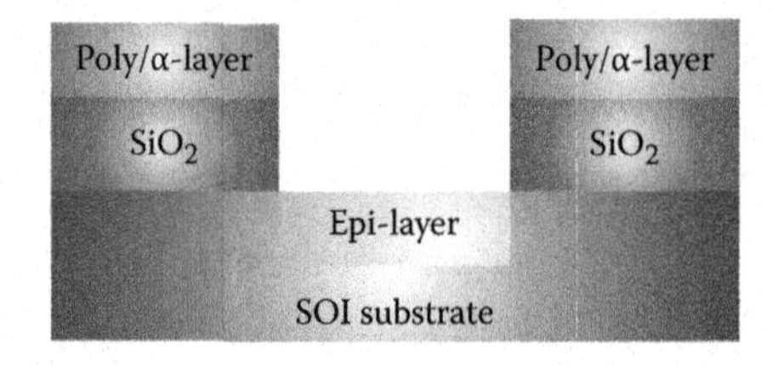

FIGURE 1.8
Differential epitaxy growth of epitaxial layers in the oxide windows surrounded by polycrystalline layers on the oxide.

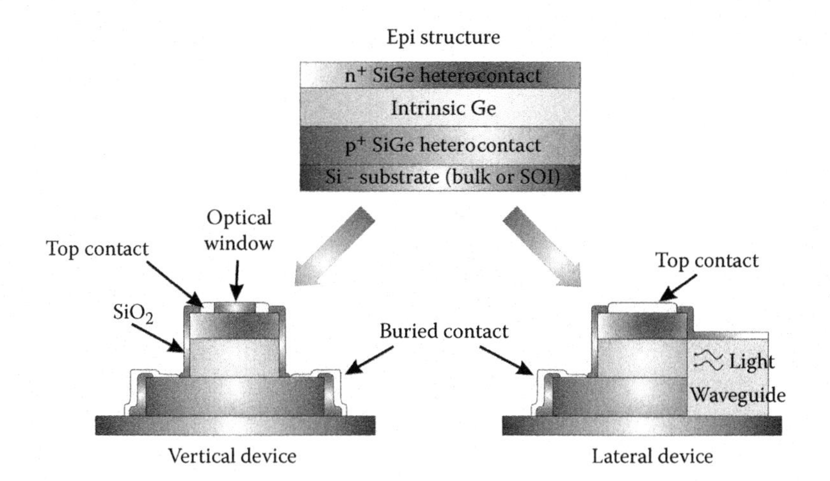

FIGURE 1.9
A vertical epitaxy structure can be used in a vertical or in a lateral device concept.

(4.2% larger than Si) to the underlying Si by a dense network of misfit dislocations at the Si interface. The technical realization of the VS is performed by a low-temperature MBE growth of Ge (growth temperature 330°C) followed by a high-temperature annealing step at 850°C for 5 min. For the sake of easier integration with Si microelectronics, this VS is chosen to be very thin (100 nm). As a tradeoff, we accept a higher density of threading dislocations (TD) (in the 10^7 cm^{-2} to 10^8 cm^{-2} range) as obtained if thicker-graded SiGe buffer layers are used. A cross-section transmission electron microscopy (TEM) of fully relaxed ultra-thin VS is shown in Figure 1.10. During the high annealing step, Si atoms from the substrate diffuse in the Ge layer. This interdiffusion generates a SiGe alloy. A sharp doping transition between the buried layer and the intrinsic region must be achieved, where the doping level should decrease abruptly over more than 4 decades. To achieve an abrupt transition of B-doping, a low growth temperature is needed, so that the segregation of boron is negligible. The intrinsic region is grown at 330°C, followed by a second annealing step at 700°C during a growth interruption. This annealing increases the crystal quality and generates a tensile strain in the Ge layer.

The n^+-doped top contact is finally realized as a highly Sb-doped Ge/Si heterojunction contact with a very high Sb doping concentration in the range greater than 10^{20} cm^{-3}. For a

FIGURE 1.10
Cross-section TEM of a fully relaxed Ge on an ultrathin VS with low TD density.

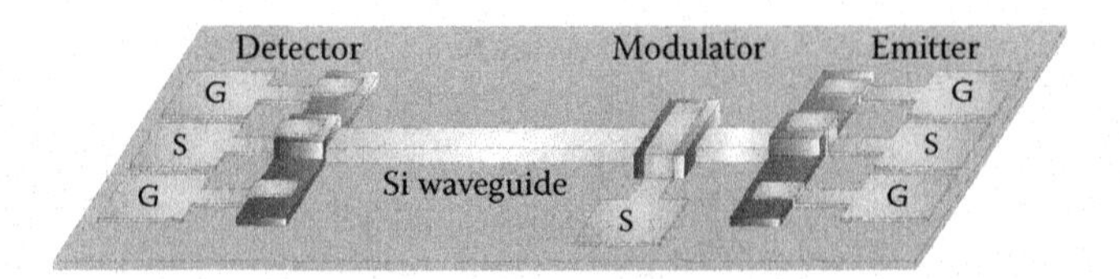

FIGURE 1.11
Scheme of an integrated waveguide circuit on SOI substrate with Si waveguide components, Ge/Si modulator/detector, and an emitter.

sharp doping transition on the pn junction, the substrate temperature is further decreased to 160°C during the growth interruption because of the extremely large segregation length of antimony in pure Ge (Jorke, 1995). For the Sb-doped Ge layer, a prebuildup growth strategy with constant flux is used. The Ge part of the top contact with thickness of 100 nm is grown at this low temperature, too. For the Si cap layer (also 100 nm thick), the growth temperature is increased to 350°C.

This Si/Ge heterojunction p-i-n diode can be used—depending on the chosen operating point and device design—serves as a broadband high-speed photodetector up to 49 GHz (Klinger et al., 2009), Franz-Keldysh effect modulator (Schmid et al., 2010) or light-emitting diode (Arguirov et al., 2011; Schulze et al., 2011; Oehme et al., 2011).

During the past decades, group IV epitaxy has been an extremely valuable tool for the investigation of novel layer structures with a thickness control on an atomic scale. The progress in special substrate technology (SOI) and in heterostructure engineering (Si/Ge heterojunctions) provides excellent new tools to push the monolithic integration of optoelectronic and microelectronic functions on Si in a very systematic manner.

A complete monolithic integration concept for the chip-to-chip connection with detector, modulator, and an infrared emitter on an SOI substrate is shown in Figure 1.11. Si waveguides connect the single devices because the infrared wavelengths are not absorbed in silicon.

1.2.4 Chemical Vapor Deposition

Deposition of layers from that vapor which contains the atoms or molecules of which the layer is composed is called physical vapor deposition (PVD). Evaporation, the earlier described MBE, and sputtering belong to this class of deposition processes. The formation of layers is called CVD if the vapor contains chemical complexes—the precursors—which have to undergo chemical reactions to form the layer materials.

CVD is a very common process in microelectronics manufacturing. The basics of silicon CVD are described in many textbooks, and the reader is referred to Chang, C.Y. and Sze, S.M., ULSI Technology, McGraw-Hill (1996).

Rapid thermal and reduced pressure CVD (RTCVD and RPCVD, respectively) have rapidly emerged as the main production technologies for group IV alloys. Group IV–based alloys, namely SiGe, SiGeC, SiC, SiSn, SiGeSn, and GeSn, are well known to be key materials for extending the capabilities of the silicon-based technologies that dominate the microelectronics industry. These alloys are fully compatible with Si technology and display advantageous electrical, optical, chemical, and mechanical properties. Even III/V epitaxy of GaAs or InP has been demonstrated on a RTCVD cross flow reactor (Thomas et al., 2010) for integration into Si via growth of relaxed Ge on Si substrates (Figure 1.12). The development of SiGeC and SiGeSn growth techniques has been very competitive in

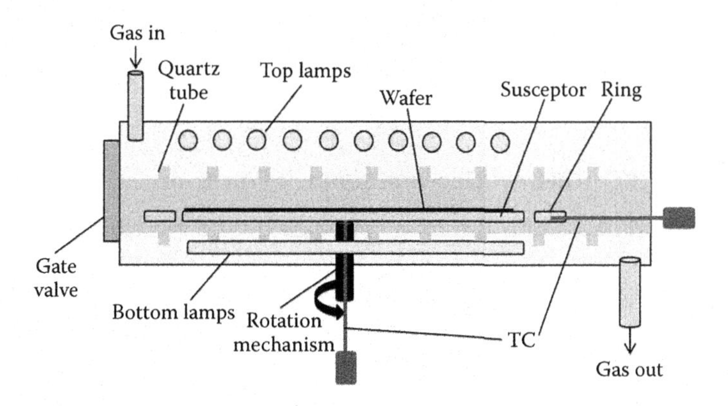

FIGURE 1.12
Schematic view of a horizontal flow reactor.

the past, and group IV material systems have been and are still studied by a variety of techniques over pressure ranges from 10^{-9} torr (via solid-source MBE) to 760 torr (via atmospheric pressure (AP) CVD techniques). Certainly, RTCVD has been a very effective way of depositing SiGe(C) and (Si)GeSn epitaxial layers. At the same time, SiGe (for HBT, eSiGe S/D, SiGe channel, optical detectors) can be effectively deposited by RTCVD.

There have been a number of comprehensive reviews on Si/SiGe epitaxy by different techniques, including CVD (Meyer, 2001; Dutartre, 2001). This section will focus on some practical aspects of CVD such as reactor design, gas panel design, temperature measurement techniques suited for low temperature operation (comparing thermocouples (TC) and pyrometers), temperature control using arrays of linear lamps and multiple sensors (multiple input multiple output; MIMO), and uniformity optimization techniques. New low-temperature precursors (Si_3H_8, Ge_2H_6, $SnCl_4/SnD_4$) and precursor selection (matching) techniques as well as operation in mass flow-dominated regimes are used to maximize precursor conversion. Additionally, deposition uniformity is optimized by gas flow profile tuning. However, challenges still remain to ensure substitutional incorporation of dilute species (such as C in Si or Sn in Ge) into the crystal lattice while minimizing interstitial incorporation and preventing precipitation.

ASM was the first company to introduce a production RPCVD epitaxial reactor in the late 1980s called the "Epsilon One." Key features included: (1) single-wafer cross flow chamber, (2) IR lamp heating, (3) cold walls via forced air and water cooling, (4) reduced pressure, (5) loadlock for moisture contamination reduction, and (6) a susceptor allowing wafer rotation for improved temperature control and deposition uniformity. Within several years, Applied Materials released the "Centura-HTF." Key differences between the two systems mainly center on the quartz chamber design and the temperature control strategy. The former uses a rectangular-shaped chamber enforced by quartz ribs and an array of linear lamps, whereas the latter uses a round dome-shaped chamber with a radially symmetric array of spot lamps. Additionally, the Epsilon employs thermocouples (TCs) for temperature control, whereas the Centura uses optical pyrometers to monitor the wafer and susceptor temperatures. At low temperatures, the signal-to-noise ratio of pyrometers suffers from the fact that the spectral emission is proportional to T^4, according to Boltzman's law. For high-temperature applications, ASM integrated a quite elegant "touch-free" transport system based on the Bernoulli effect into the wafer handler robot end effector. By design, this system must operate at relatively high pressure (e.g., ATM). A reduced pressure wafer handling system is beneficial and desirable in conjunction with an ex situ preclean module to avoid surface contamination during the transport from one chamber into another (Meyer, 2001).

1.2.4.1 Principle of CVD: Thermal Decomposition of Precursors

CVD is the deposition of a solid film on a substrate by the reaction of vapor-phase reactants (precursors). The reaction is driven by thermal energy transferred to the substrate, typically via IR heating lamps. The sequential steps of this process are as follows:

1. Transport of precursor gas into the reactor by forced convection
2. Diffusion to surface and adsorption of precursor molecules on surface
3. Surface reactions (decomposition/recombination) and incorporation into solid film
4. Desorption of by-product molecules and diffusion into the gas phase
5. Evacuation of gaseous by-products from reactor

RTCVD was originally defined as a CVD technique capable of rapid switching of process temperature also referred to as "limited reaction processing." In RTCVD, a stable gas flow is established with the wafer at low temperature and deposition is modulated by rapidly heating and cooling the wafer. However, this has severe drawbacks in terms of process control. Hence, the preferred current method involves fast gas flow switching. This approach requires a gas supply panel with fast gas flow switching capability, high-velocity carrier stream, small diameter lines, small dead volumes, fast switching valves, responsive MFCs, and balanced deposition and vent line pressures. The high thermal conductivity of modern RTCVD graphite susceptors used for wafer rotation minimizes thermal impedance, thereby allowing wafer temperature to ramp quickly. The benefits of wafer rotation and effective absorption of IR heating on film uniformity clearly exceeds the penalty of some lost temperature response relative to a susceptor free design.

The second common terminology for CVD reactor design is based on reduced pressure operation, referred to as RPCVD, with pressures ranging from 1 torr to a few hundred torr.

1.2.4.2 Key Components of Industrial Epitaxial CVD Toolset

Modern epitaxial toolsets consists of three key units: (1) mechanical wafer transfer system (including load lock, wafer load + rotation) and automation (robot + software control (MMI/GUI) + factory host communication); (2) reactor module including a quartz chamber, wafer heating, and a temperature control system; (3) gas delivery system (mixing + transport + distribution). This discussion will focus on basic reactor design including gas delivery and wafer heating, the most critical and interesting features from an epi process perspective.

The gate valve (left) isolates the quartz chamber from the transfer module. The premixed process gases enter the reactor through a gas injection flange. The upper and lower infrared lamps heat the SiC-coated graphite susceptor and the SiC-coated piece surrounding the susceptor from the outside through the cold IR—transparent quartz chamber. To allow reduced pressure processing, the chamber is reinforced by quartz ribs. Reflectors reflect the IR light to minimize the thermal load of the surrounding hardware (O-ring seals, flanges) and to minimize power consumption of the heating system. The effluent gas exits the chamber through the exhaust flange (right). The two most common techniques to monitor process temperature are (a) thermocouples placed inside the susceptor and inside the SiC-coated graphite piece surrounding the susceptor and (b) pyrometers that measure the backside of the susceptor. The metal rear exhaust flange, the front metal gate valve, and injector flange are typically water cooled to prevent overheating and damage to the O-ring seals. The quartz chamber and the quartz "bulbs" of the IR lamps are air-cooled. The key advantages/strengths of RTCVD include high productivity through high growth rates, high flexibility due to a broad temperature range (200°C–1200°C) with ability to change temperature quickly, broad pressure ranges (760–1 torr), broad range of carrier gas flows (0.5–100+ slm), and a broad range of available "gas sticks" for carrier gases, deposition precursors, electrical dopants, and etch precursors. This flexibility allows processing of various alloys in the kinetic (reaction rate) limited or the mass flow-dominated regime. The quartz chamber and SiC-coated graphite susceptor allows chamber cleaning with HCl/Cl_2 in between depositions (in between wafer unload and subsequent wafer load). This minimizes background doping and makes n- and p-type doping within same chamber possible. This is in contrast to UHVCVD techniques that face severe issues with n-type doping that consequently requires a dedicated reactor. Furthermore, the inherently simpler RTCVD tool design reduces maintenance downtime relative to more complex UHV technologies. Only intrusive maintenance into the gas panel and gas supply system requires some dry-down time, which can be accelerated by repeated pump/purge cycles with dry carrier gas.

For kinetically (reaction rate) limited processes, the layer thickness uniformity is controlled via temperature uniformity across the wafer, whereas for mass flow limited or dominated processes with strong precursor depletion, the gas flow distribution profile across the wafer determines the thickness uniformity on the rotating wafer. This gas flow profile (gas stream distribution) across the chamber and the wafer can be tailored (optimized) to obtain the desired radial thickness profile.

For high volume manufacturing, besides within-wafer uniformity and wafer-to-wafer repeatability, chamber matching from tool to tool is important. Maintaining deposition temperature and uniformity, as well as maintaining partial pressure of reactants (process gas composition and distribution), while eliminating *all* possible sources of uncontrolled variation enables the user to maintain target thickness, composition, and uniformities for a robust production process. Monitoring and logging all digital and analog tool parameters to a factory host computer allows "data mining" and close monitoring of tool operation and early identification of malfunctioning components.

Table 1.1 lists the precursors typically used in modern CVD epitaxial reactors. All bulk carrier gases as well as high-volume HCl used for chamber cleaning need point-of-use gas purification. Some other precursors such as hydrides (SiH_4, SiH_3CH_3, etc.) or Cl_2 can be purified as well. Typically, the deposition precursors with high Cl content decompose at higher temperature requiring a higher process temperature.

One of the main advantages of CVD techniques over PVD techniques is the capability of selective epitaxial growth (SEG) in predefined areas, enabling self-aligned processes. To achieve selectivity two main approaches are possible: (a) co-flow processes, i.e., simultaneous flow of deposition and etchant precursors, and (b) cyclic deposition and etch (CDE) processes. For conventional SiGe SEG with co-flow of process, the gases commonly used gases are $SiCl_2H_2/GeH_4/HCl$ and $SiH_4/GeH_4/HCl$.

Due to the convoluted relationship of temperature- and pressure-dependent chemical surface and gas reactions, the development and optimization of new processes is a little more cumbersome compared with MBE techniques.

TABLE 1.1

Commonly Used Process Gases for Group IV Epitaxy

Function of Gas	Gases Used	References
Carrier gases	H_2, N_2, inert: Ar/He	(Bauer et al., 2008a; Meunier-Beillard et al., 2004)
Etchant precursor	HCl, Cl_2, HBr	(Bauer et al., 2008a)
Deposition precursors		
Silicon	$SiCl_3H$, $SiCl_2H_2$, SiH_4, Si_2H_6, Si_3H_8, Si_4H_{10}, Si_5H_{12}, $Si_2Cl_2H_4$	(Kanoh et al., 1993; Fischer et al., 2006; Sturm et al., 2007; Bauer and Thomas, 2010a; Shinriki et al., 2012)
Germanium	$GeCl_4$, GeH_4, Ge_2H_6	(Bramblett et al., 1995; Gencarelli et al., 2012)
Tin	$SnCl_4$, (SnD_4)	(Vincent et al., 2011a)
Carbon	SiH_3CH_3, $CH_2(SiH_3)_2$, C_3H_6, $SiH(CH_3)_3$	(Bauer and Thomas, 2010a)
p-Dopants	B_2H_6, BCl_3, $GaCl_3$, TMGa	
n-Dopants	PH_3, AsH_3, SbH_3	

In the following section, the *choice of precursors* and the *choice of process conditions* (temperature, pressure, and carrier gas flow) will be discussed. Depending on the detailed target application, specifically the composition and thickness of epitaxial layer, the most suitable precursors and process conditions (temperature, pressure, carrier gas/dilution) have to be chosen. Generally the highest possible deposition rate (minimum deposition time) without compromising epi quality is chosen to maximize productivity. Increasing temperature and partial pressures of reactant precursors can sometimes lead to particle nucleation and defective films, which should be avoided, either by keeping the reactant partial pressure low enough," or by applying a CDE strategy (Shinriki et al., 2012). Some of the key challenges due to low solid solubility are dopant surface segregation, incorporation of C and Sn in interstitial (nonsubstitutional) crystal lattice sites, and the formation of precipitates (such as β-SiC in Si or Sn-clusters in Ge). To overcome these challenges, nonequilibrium growth conditions such as low temperature depositions, and high growth rates are required simultaneously. This typically requires higher order hydrides such as Si_3H_8 or Ge_2H_6 that allow high growth rates at low deposition temperature. The basic model to explain substitutional carbon incorporation in Si developed for PVD (MBE) (Osten et al., 1996, 1999) has been extended for CVD by explicitly linking growth rate to temperature (Mitchell, 2000). The termination of the (Si/Ge/Sn) surface of epitaxial films with H or Cl plays an important role as well. It has been shown that with CVD techniques steeper alloy transitions can be realized compared with MBE due to surface coverage with H (or Cl) (Gruetzmacher, 1993). Chlorine, hydrogen, and segregating dopants that terminate the surface act as surfactants and can impact growth rate, morphology (smoothness), and impurity incorporation (C, Sn, dopants). Temperature-dependent desorption of the dominating surface species (H/Cl) from the surface are the dominant growth rate limiting steps for most conventional precursors due to restriction of available surface sites for precursor adsorption. Another important aspect of surfactants is a reduction in adatom surface mobility.

The impact of dopants on growth rate can be dramatic. For instance, a steep *decrease* of growth rate is observed due to surface contamination by n-type dopants with SiH_4 in the absence of Cl or a steep *increase* in growth rate by dopant hydrides with a Cl terminated surface (e.g., $SiCl_2H_2$ at high pressure) (Agnello et al., 1993). The partial pressures of the reactants depend on precursor flow rate, system pressure, and diluent flow (carrier gas flow). Gas flow velocity (defined by pressure and carrier gas flow) also influences the boundary layer over the wafer surface.

In the majority of epi reactors, the susceptor temperature is controlled during the process and can be considered as constant and repeatable (Dutartre, 2001). The wafer rests on a SiC-coated graphite susceptor, and both pieces are lamp heated from both sides inside a cold wall chamber in a gas environment. In such an equipment, the actual wafer temperature depends explicitly on its own optical properties, leading to thermal loading effects. The equilibrium temperature of the wafer corresponds to the exact balance of radiative heating from lamps, IR radiative emission, radiative heat exchange with susceptor, conductive heat exchange with susceptor and conductive heat exchange with the quartz chamber. For susceptor (wafer) temperature control, two options mainly exist: thermocouples (TCs) and optical pyrometers. At higher temperatures and on bare wafers, pyrometers allow a direct wafer temperature measurement, but coating on the chamber walls must be carefully monitored because such deposits can interfere with the measurement. Due to the low emissivity at low temperature ($\sim T^4$) thermocouples offer an advantage at low temperature, where conductive heat exchange is the dominant heat transport term. Pyrometer-controlled systems must often operate with fixed power at low temperature to overcome the poor signal-to-noise ratio (e.g., to avoid interference of the infrared heating system with the temperature measurement). In both cases a temperature difference generally exists between the susceptor and the wafer in particular for patterned wafers and multilayer stacks (e.g., SOI). This temperature difference can be minimized by the susceptor design and further manipulated by adjusting the top-to-bottom lamp power ratio. To avoid emissivity-related temperature errors, pyrometers often measure the backside of the susceptor rather than the true wafer surface temperature. Similarly, TCs also infer wafer temperature via calibrated measurements taken in close proximity to the susceptor. TC aging (T-drift over time) is the main issue encountered with this configuration; however, this phenomenon is predictable and can be mitigated by a "burn-in" process.

Key differences in the lamp-heating system arrangement for the Epsilon and Centura reactors are the following: (a) given the square shape of the Epsilon chamber, arrays of linear lamps are located above and below the susceptor. Modern equipment allows individual control of lamps by dedicated silicon controlled rectifiers (SCR's). (b) Following the round shape of the Centura, chamber spot lamps are arranged in concentric circles above and below the susceptor.

In general, power-controlled SCRs are advantageous because they can sense fluctuations (sag) in the incoming voltage and immediately compensate by increasing the current, before the temperature sensor detects a temperature deviation caused by the incoming power fluctuation.

In the Epsilon the SCRs are grouped into multiple groups—or zones—each controlled by TCs. The total power output to the center SCRs (center zone) is controlled by the center TC (single input–multiple output: SIMO). All SCRs and TCs (zones) taken together represent a multiple input–multiple output (MIMO) control problem. To optimize temperature uniformity for new process development in which temperature, pressure, and carrier gas flow are altered, an instrumented wafer with embedded TCs can be used to map the thermal uniformity generated by the lamps. Although good temperature uniformity is desirable over the whole susceptor even if it is not rotated, the final temperature optimization for achieving uniform film thickness and composition of a reaction rate-limited process must be performed on a rotating wafer.

In the Centura, several pyrometers are used to measure the radial temperature distribution across the backside of the susceptor. The output is then correlated with the radial positions of the circular lamp zones. Again, this represents a similar MIMO control problem.

In general, different SCR ratios are generally used for very fast temperature ramps such as those used in a flash-bake comparable to conventional rapid thermal anneal toolsets. The PID values for the different zones can be optimized individually to maintain a stable temperature and to deliver a fast and crisp response to changes in temperature set points. Steep temperature ramps are routinely achieved in modern CVD equipment, perhaps only limited by thermal wafer gradients leading to thermal stress relieve in the wafer by forming undesired crystallographic slip lines. In contrast, wafer cool down (lamps off) is limited by the thermal mass of the susceptor. Due to the rotational symmetry of the Centura chamber fewer variables need to be optimized to obtain good temperature uniformity that simplifies temperature tuning. Due to the square-shaped Epsilon chamber, laminar gas flow profiles are simpler to control and optimize.

For kinetically driven processes (reaction rate limited), temperature uniformity dictates process optimization and stability because gas flow distribution is less relevant due to an oversupply of precursor. Therefore, different methods to optimize T uniformity will be discussed in the following section.

1. Instrumented wafer: a special, commercially available, test wafer is instrumented with up to 34 embedded thermocouples and attached wiring that does not allow for rotation of the wafer. The temperature, pressure, and carrier gas flow of the target process should be applied, but reactive process gases should be avoided, to prolong the lifetime of the instrumented wafer and to avoid deposition on the wafer. This technique is very fast, does not require processing of any wafers, and can be used to screen a wide parameter space (temperature, pressure, carrier gas flow, and type (e.g., H_2, N_2, He, Ar). This method can be automated to screen a wide parameter space and minimize the within wafer temperature nonuniformity by rebalancing the power distribution to the lamp system. This delivers a good starting point for (2) or (3).
2. Dopant activation: dopant activation/deactivation of an ion implanted wafer (different species, dose, energy) can be used with or without wafer rotation. This technique is commonly used to optimize the performance of rapid thermal anneal (RTA) tools (Acharya et al., 2001). This is a popular technique at IDMs and foundries, which can easily generate test wafers with ion implantation.
3. Film deposition: deposition of a poly or amorphous layer on 100-nm-thick SiO_2 in the reaction rate-limited regime, e.g., using SiH_4 or $SiCl_2H_2$ (with an α-Si seed). If the precursor (SiH_4) partial pressure is well controlled (SiH_4 mass flow, H_2 carrier flow, and system pressure), this technique allows for matching wafer temperatures between tools by comparing the growth rates and thickness patterns between tools. Also, since the activation energy and the relative thickness change are known, it can be used to determine the proper temperature offsets to obtain a uniform deposition on a nonrotated wafer (T-offset calculator). However, this should not be overemphasized because product wafers are deposited using wafer rotation. To obtain the best possible uniformity on a rotating wafer, the uniformity on a nonrotating wafer can be compromised. Further, even for nominal reaction rate limited processes, sometimes the radial temperature profile is deliberately degraded to compensate for some depletion. Depending on availability, (1), (2), (3), or a combination of any two is used.

Gas flow distribution (gas flow profile) can be used to control and optimize film uniformity and precursor utilization for mass flow limited (dominated) reactions such as high-temperature $SiCl_3H$- and $SiCl_2H_2$-based deposition processes as well as low-temperature Si_3H_8-based growth processes. Mass flow-limited processes typically achieve the highest precursor conversion rate, because a temperature in the flat region of the Arrhenius plot for a particular precursor is used.

A gas delivery system mixes the process gases, transports the mixture to the reactor, and distributes the gas mixture through a set of variable injectors forming a desired gas flow distribution profile into the quartz reactor. The carrier gas can be selected from a group of three carrier gases (e.g., H_2, N_2, Ar/He) through valving, and the flow rate is controlled through a mass flow controller (MFC). Typical gas panel configurations include multiple process gas sticks/MFC with a subset of gases chosen from Table 1.1: SiH_4/Si_2H_6, Si_3H_8, GeH_4/Ge_2H_6, SiH_3CH_3, SnD_4, dopant circuits including diluent (for p- and n-type dopants); $SiCl_3H$, $GeCl_4$, SnCl4, $SiCl_2H_2$, Cl_2, HCl. The assignment of gases to the gas sticks is flexible and can be tailored toward the specific process needs. Basic configurations include MFCs for compressed gases as well as Piezocons + MFCs for liquid sources (e.g., $Si_3H_8/SiCl_3H/GeCl_4/SnCl_4/BCl_3/GaCl_3$). For a fixed total gas flow, the partial pressure of the reactants can be accurately calculated (partial pressure A = total pressure × mass flow A/sum of all mass flows).

With careful tuning of vent and deposition line pressures, transient pressure spikes can be minimized while switching between lines. The pressure in each line depends on the total gas flow (carrier flow and process gases) and the restriction of the outlet. The restriction in the deposition lines is defined by the combined restriction of the (parallel) multiport injector (MPI), while the pressure in the vent line is defined by a similar (single) restrictor. Ideally, if the total flow through a gas line is kept constant at all times, pressure fluctuations should be minimized, if not eliminated. In general, a crisp response to flow changes and tight control of short deposition steps (pulses) is desired and can be obtained by minimizing the internal volume of the gas supply system.

Oxygen and moisture control is very essential to obtain high quality SiGe(C) and (Si) GeSn epitaxial alloys at low temperature. A lot (25 wafers) is loaded through a vacuum load-lock followed by a series of evacuation and back-fill cycles. HF-dipped wafers or pre-cleaned wafers are transferred through an ultraclean environment (<200 ppb O_2 and H_2O levels) into the reactor (Meyer, 2001). The H_2O/O_2 requirements were first discussed by Smith and Ghidini (1982) and Ghidini and Smith (1984). Purified carrier and process gases provide an ultraclean processing environment, eliminating the need for vacuum in the process chamber. Slow but unavoidable [O] diffusion through permeable O-ring seals and H_2O from wafer transfer/load result in some H_2O/O_2 in the growth ambient. Therefore, growing films fast with a small but constant [O] source actually reduces [O] incorporation into the growing film.

The capability to realize SEG is one of the greatest strengths of CVD compared with PVD methods (MBE). SEG with deposition only on exposed Si areas is very desirable because it enables self-aligned device architectures that are simpler or even novel device structures that are not possible without SEG. In that context, the integration of SEG on patterned wafers, thermal budget limitations, ex situ and in situ pre-epitaxy wafer surface treatment (wet clean, vapor clean, H_2 bake) are all closely linked and should be discussed in the following sections.

As a first rule of thumb, the thermal budget has to be reduced at each technology node according to the size reduction and the (dopant) gradients required. Based on B diffusion, often the limiting factor in many applications, the thermal budget ($\sim t^*\exp(-E_a/kt)$) reduction in between two consecutive device nodes is approximately 15°C–20°C (B-diffusion length × 0.7 or $D_{boron}/2$), assuming identical times. This rule applies for established epi applications with an identical integration scheme. For new applications, thermal budget limitations can be even more strict. The trend seems to be clear, the more advanced the technology and the later the epi integration in the process flow, the lower the allowed thermal budget for pre-epi H_2 bake and epi deposition. Another rule of thumb is the following: (1) if epitaxy is integrated very early in the process flow, almost no thermal budget limitations exist; (2) if epitaxy is integrated late in the process flow, most parts of the devices are already present and the global thermal budget of epi is strictly limited. Choosing a high-temperature (HT) nonselective base epi as an example, a thin chemical oxide or native oxide can be removed using a conventional in situ high-temperature H_2 bake and perfect epitaxial quality is generally ensured. The wafers can even be taken out of the box as is, because they generally received a good passivation chemical oxide from the wafer manufacturer.

In essence, SEG is performed on patterned/structured wafers where process results are a direct consequence of not only the process conditions, i.e., temperature, pressure, carrier gas flow, precursor flow and gas velocity, but also the nature of the patterned wafer, such as feature size, aspect ratios, density, exposed Si area, etc. Process metrics generally focus on epitaxial crystallinity, selectivity threshold/limit, thickness, and doping in different windows (loading), exact morphology of epi deposition (mainly faceting at the edge of deposition) and surface roughness.

Different SEG processes have been published, most of them based on the $SiCl_2H_2$-HCl-H_2 system (Bauer and Thomas, 2008). Apart from the DCS choice, the addition of HCl is necessary to inhibit the deposition on dielectric (e.g., CVD grown Si_3N_4, SiO_2). Conventional SEG processes utilize the fact that a nucleus has to be formed on dielectric surfaces, associated with an incubation time, whereas epitaxial growth occurs instantaneously on bare Si surfaces.

In its most simplified form, SEG from $SiCl_2H_2$-HCl-H_2 can be described as follows:

$$SiCl_2H_2(g) => SiCl_2(g) + H_2(g)$$

$$SiCl_2(g) + \_ = SiCl_2(s)$$

$$SiCl_2(s) + H_2\,(g) <=> Si(s) + 2\,HCl(g)$$

1. DCS decomposes into $SiCl_2$ and H_2.
2. $SiCl_2$(s) adsorbed on the surface can be reduced to Si and HCl vapor by H_2.
3. By adding HCl, the equilibrium can be shifted into the opposite direction, and Si(s) can be etched forming $SiCl_2$(g).

The presence of Cl on the surface allows the formation of HCl and dopant chlorides, which act to "clean" the surface of dopants, which would otherwise contaminate this region. The net effect is an enhancement in growth rate. Other innovative approaches toward SEG (e.g., for SiCP/SiP) are based on fast repetitive cycles of nonselective CVD followed by selective chemical vapor etch (Bauer et al., 2007a, 2007b, 2008a, 2008b, Bauer and Thomas, 2010b, 2012a, 2012b).

In the following, the need for pre-epi surface preparation using ex situ and in situ cleaning as well as H_2 baking is discussed. By definition, epitaxy is the growth of a monocrystalline layer that adopts the underlying atomic structure of the substrate lattice. Therefore, no amorphous film on the surface can be present, such as native SiO_2 that naturally exists on silicon surfaces. Along the same lines, to achieve high-quality epi, it is essential to remove any particles and contamination from the Si surface prior to growth initiation. Contaminants would prevent surface migration of Si atoms and form precipitates, yielding lattice defects. Perfect, defect-free epitaxy can be achieved if the following conditions are met: (a) no surface oxide; (b) a good crystal quality even after dry etching or ion implantation; (c) no precipitates of dopants or other impurities (e.g., β-SiC); and (d) no surface contamination or particles.

With a bake, a SiO_2-covered surface can be cleaned through the following two reactions:

$$SiO_2(s) + Si(s) => 2\ SiO(g)$$

$$SiO_2(s) + H_2(g) => SiO(g) + H_2O(g)$$

Integration of SEG typically requires at least a sequence of HF-last plus a moderate (900°C) to mild (750°C) in situ H_2 bake. The use of vapor phase cleaning processes was explored during the 1990s (Meyer, 2001, and references therein), along with flash bakes (Brabant, 2006), H-radical exposure, and UV illumination (Arena, 2008), and in situ addition of GeH_4/HCl (Bauer and Thomas, 2012c). A good HF last clean yields an [O] surface dose ≤5E13 cm^{-2}. Until recently, an ex situ HF clean followed by RP H_2 bake at 800°C was sufficient to produce a suitable surface for epi growth, but this requires strict Q-time control. For more sensitive processes, an in situ preclean solution is sometimes desirable. If a pre-epitaxial clean is performed in a dedicated preclean module, the wafer transfer should be under reduced pressure to minimize surface recontamination (reoxidation through traces of H_2O, O_2) during wafer transfer into the epitaxial reactor. The latest generation equipment in the industry allows integration of preclean under such ambient control. Highly sensitive tests for pre-epitaxial surface preparation are minority carrier transport devices and photoluminescence spectroscopy. To maximize productivity in a single-wafer process, it is desirable to minimize the H_2 prebake time and to grow at the highest possible temperature to achieve high growth rates. Other factors such a faceting, loading, relaxation, precipitation of carbon, or tin should also be taken into account when choosing deposition parameters. Depending on the precursor, some flexibility exists in "trading off" lower growth temperatures with higher reactor pressures and reduced carrier gas flow (dilution).

1.2.4.3 Growing SiGe/Ge/(Si)GeSn Alloys on Si Platform

For optoelectronics and photonics, Ge-based alloys are of great importance. Pure Ge can be grown on Si either by growing ~10-μm-thick graded buffer (with very severe cross hatch, requiring CMP) or by a two-step growth approach consisting of a low-temperature wetting or seed layer step followed by a higher temperature bulk growth step (Kasper et al., 1998; Luan et al., 1999). An optional (cyclic) thermal anneal can be inserted to enhance crystallinity. The wetting layer is critical in achieving high quality thick Ge films, as it forms a continuous smooth film free of voids from which the bulk layer subsequently grows (Bauer et al., 2008c). Following a 2- to 3-min seed layer deposition with high GeH_4 partial pressure (Figure 1.13), a thick relaxed Ge (r-Ge) bulk layer (Figure 1.14) is deposited by quickly ramping up the temperature while maintaining a continuous GeH_4 flow rate. Layer thickness profiles up to 350 nm have been verified using X-ray reflectivity (XRR) and spectroscopic ellipsometry (SE). Thickness profiles of Ge films exceeding 0.5 μm have been measured by FTIR. A smooth surface morphology can be maintained if growth conditions are properly optimized. AFM images reveal a RMS roughness of 0.85 Å for the seed layer (Figure 1.13) and 1.6 Å for the thick r-Ge layer (Figure 1.14).

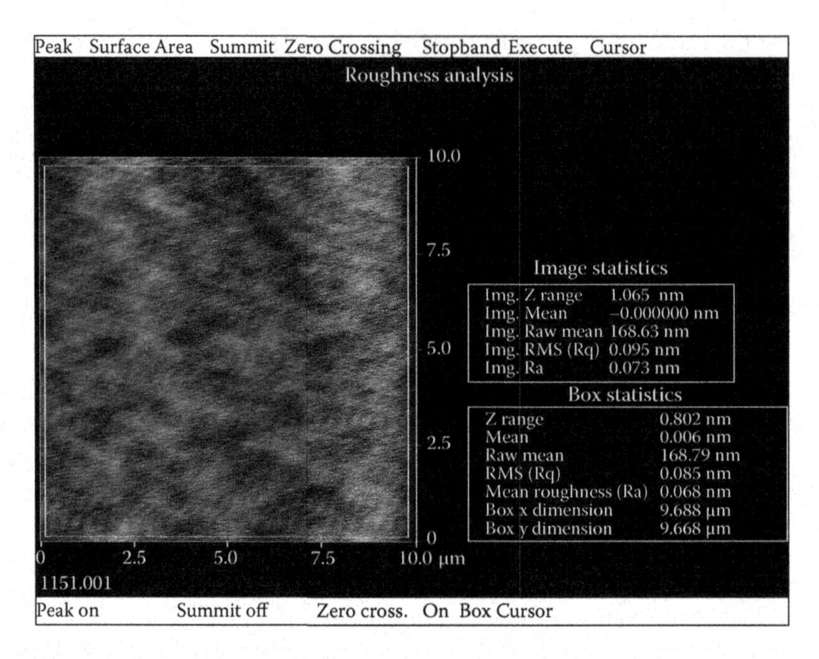

FIGURE 1.13
AFM picture showing 0.85 Å RMS roughness for a Ge seed layer.

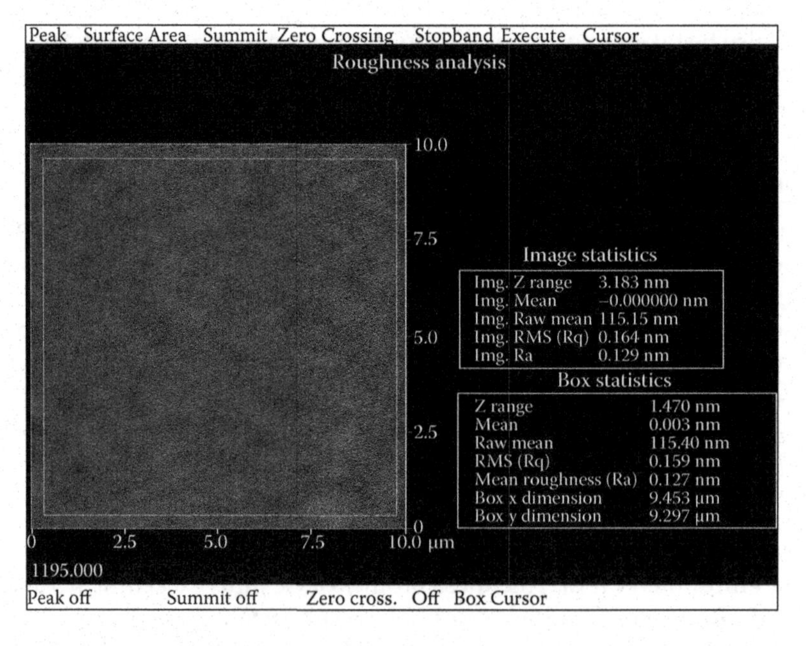

FIGURE 1.14
AFM picture showing 1.6 Å RMS roughness on 1.05-μm-thick r-Ge.

Ge doping with $AsH_3/PH_3/B_2H_6$ has also been demonstrated (Bauer et al., 2008c). AsH_3 and to a lesser extent PH_3 act as surfactants keeping the surface atomically smooth. Likewise, r-Ge can be grown on Si using GeH_4 or Ge_2H_6, r-GeSn can be grown directly on Si using Ge_2H_6 and $SnD_4/SnCl_4$. Compressively strained GeSn can be grown on r-Ge, while tensile-strained Ge can be grown on r-GeSn.

Misfit dislocation segments (in a highly mismatched hetero-epitaxial system) cannot end within the crystal, so they are connected through two threading dislocation arms with the epi film surface (wafer). A very high and regular density of misfit dislocations is required (1 misfit dislocation line every 25 Si atoms to accommodate for the 4.2% lattice mismatch between Ge and Si). Crystalline quality of Ge films can be measured by XTEM or low-temperature defect decoration with HCl, producing etch pits, shaped like inverted pyramids. To reduce the number of threading dislocations and misfit dislocation loops, the misfit lines should either extend (via glide) to the wafer edge or the threading dislocations should terminate at internal patterned surfaces. Reduction in threading dislocation density can be achieved by SEG of Ge into trenches or windows, sometimes referred to as aspect ratio trapping (ART).

SiGe alloys of various compositions are typically grown using SiH_4/GeH_4 or $SiCl_2H_2/GeH_4$ gas mixtures in H_2, sometimes with HCl added to maintain selectivity. The Ge concentration in the alloy can be controlled by adjusting the GeH_4 to $SiH_4/SiCl_2H_2$ flow rate ratios. Both precursor combinations yield increasing [Ge] incorporation with decreasing growth temperature. Higher Ge content as well as lower growth temperature both enhance selectivity, decreasing the required amount of additional HCl in the process gas mixture to maintain selectivity. Irrespective of the precursor combination chosen, the Si portion of the precursor gases ($SiH_4/SiCl_2H_2$), is typically set at a fixed value (i.e., 20, 50, or 100 sccm) and the Ge component (GeH_4 diluted to 10% in H_2) is adjusted to the targeted [Ge] concentration. For instance, higher Ge concentration is achieved by increasing the GeH_4 flow rate (at fixed $SiH_4/SiCl_2H_2$ flows), which increases the growth rate and allows for a reduction in growth temperature. Two specific applications are discussed: (a) the SEG of very high [Ge] content films and (b) the isothermal (nonselective) epitaxial growth of SiGe and Si at low temperature and high growth rate (such as for the growth of a HBT stack's or SiGe/Si superlattices).

To grow highly strained SiGe alloys, growth temperature must be low and deposition time short to maintain layer strain, which requires an aggressive tuning strategy. This can be achieved using conventional $SiCl_2H_2$ and GeH_4 chemistry. Fixing the GeH_4 flow at a relatively high value (temperature-dependent Ge growth rate is known from GeH_4 Arrhenius plot) and adding an increasing amount of $SiCl_2H_2$, the SiGe growth rate decreases. This is related to an increasing Cl coverage and decreasing Ge coverage on the SiGe surface. It has been reported for SiGe alloys in particular, that the growth rate is proportional to the [Ge] content (Tomasini et al., 2006; Bauer et al., 2004). The growth kinetics as a function of [Ge] content and temperature (for 20 torr, 200 sccm GeH_4 10% in H_2) together with the $GeH_4/SiCl_2H_2$ gas ratio is shown in Figure 1.15.

Although the growth rate data were generated by growing bi-axially tensile-strained SiGe on r-Ge on Si, the kinetics hold true for growth of compressively strained SiGe on Si (as used for SiGe channels in PMOS). Due to the much higher strain for growth of high [Ge] concentration SiGe ([Ge] $\geq$50%) on Si as opposed to Ge (or r-Ge on Si), the critical thickness limitation is far more stringent. These constraints dictate that the SiGe epi film be grown as fast as possible, not only from a throughput perspective, but also to keep the thermal budget limited (the time at T_{dep} as short as possible) to avoid relaxation (Figure 1.16). To maintain adequate control, the ideal SiGe deposition time should be on the order of a few tens of

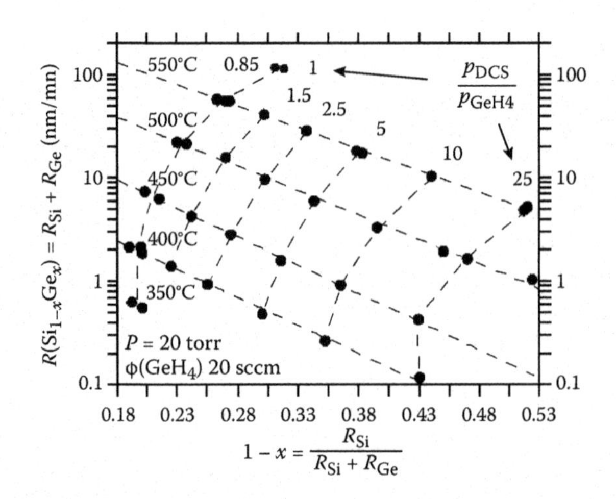

FIGURE 1.15
Kinetics of SiGe growth with $SiCl_2H_2/GeH_4$ 20 torr.

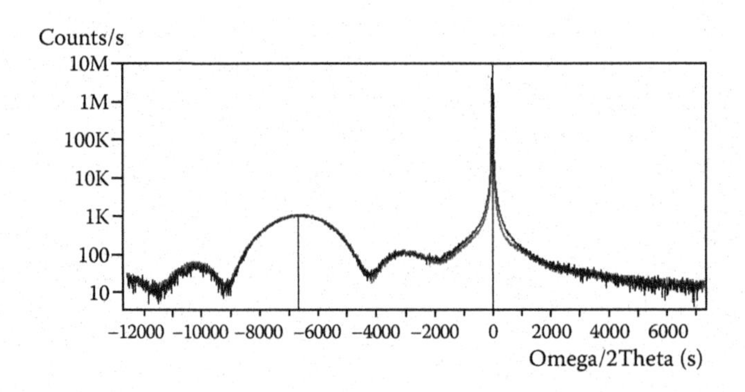

FIGURE 1.16
The 7.5-nm $Si_{0.32}Ge_{0.68}$ grown at 500°C, 10 torr, with $SiCl_2H_2/GeH_4$ chemistry within less than 1 min.

seconds to a minute. To preserve SiGe channel strain, the i-Si cap layer is deposited using a Si_3H_8/Cl_2-based chemistry in a cyclic deposition/etch process (Bauer and Thomas, 2010c).

The typical increase in [Ge] concentration with decreasing temperature for standard Si precursors ($SiH_4/SiCl_2H_2$) is due to their lower reactivity relative to that of GeH_4. For Si_3H_8, the behavior is just the opposite, [Ge] decreases with decreasing temperature and the growth rate increases with decreasing GeH_4 flow because the Si_3H_8 source is more reactive.

To achieve high GR for moderate SiGe alloy concentrations (<50%) at low temperature, as required for modern HBTs, Si_3H_8 can be employed as the Si source. Since the reactivity of Si_3H_8 is much higher than that of SiH_4 or $SiCl_2H_2$, only small amounts (20–50 mg/min) of Si_3H_8 in conjunction with GeH_4 are required to achieve high SiGe growth rates at low temperature. This philosophy equally applies to the deposition of intrinsic Si and SiGeC:B films, allowing fast isothermal processes for high quality epi in HVM settings. Growth of SiGe with Si_3H_8/GeH_4 has been studied and short period superlattices demonstrated (Bauer and Thomas, 2010c).

If the thermal reaction mismatch between a Si and a Ge source gas is small, as applies to "Si_2H_6 and GeH_4" chemistries, the temperature dependence of Ge incorporation is minimized, thus simplifying process targeting for [Ge] and layer thickness. Likewise, Ge_2H_6 is thermally well reaction matched to Si_3H_8, allowing even lower processing temperatures. The most efficient deposition (with a high conversion rate) of Si from Si_3H_8 (or Ge from Ge_2H_6; Gencarelli et al., 2012) is done in the mass flow dominated (depleted) regime. For Ge_2H_6, the transition temperature between mass flow– and reaction rate–dominated regime occurs between 350°C and 450°C for high and low partial pressures, respectively. If the Ge_2H_6 partial pressure is kept below ~1 Pa SEG can be obtained (300°C–450°C) with epi growth rates ranging 1–50 nm/min. For Ge_2H_6 partial pressures (substantially) exceeding 1 Pa direct deposition of conformal amorphous Ge layers directly on dielectric or metallic surfaces can be achieved.

Si and Ge are fully miscible over the entire composition range. If the (metastable) critical thickness is exceeded, the strained SiGe layer begins to relax through the formation of misfit dislocations. In contrast, due to the low solid solubility of C in Si (and Sn in Ge), excess nonsubstitutional [C] incorporation causes the formation of mobile interstitial C atoms that can form SiC precipitates and stacking faults (Bauer et al., 2005). Due to temperature- and growth rate-dependent carbon surface segregation, the carbon surface concentration can exceed the bulk carbon concentration by orders of magnitude (Oehme et al., 2000) and is sufficiently suppressed at temperatures ≤550°C. The growth of Sn in Ge suffers from similar issues. The incorporation of carbon into substitutional lattice sites in Si, while avoiding undesired interstitial carbon and β-SiC precipitates, requires non-equilibrium growth conditions such as low temperature (500°C–550°C) and high growth rates. Likewise, the incorporation of Sn (up to 8%) into substitutional lattice sites in Ge, while avoiding interstitial Sn and harmful precipitates, requires non-equilibrium growth conditions such as low temperature (~350°C) and high growth rates (~10 nm/min) (Vincent et al., 2011a; Gencarelli et al., 2012). SEG of GeSn can be achieved using Ge_2H_6 and $SnCl_4$ at 760 torr (Gencarelli et al., 2012). For the case of Si deposited on epitaxial Ge (grown with GeH_4 or Ge_2H_6), trisilane (Si_3H_8) Si precursor minimizes Ge up-diffusion (surface segregation) compared with SiH_4 or $SiCl_2H_2$ at identical temperatures, suggesting a unique growth mechanism (Vincent et al., 2009).

1.3 Group IV Materials and Properties

Manfred Reiche and Martin Kittler

1.3.1 Bulk Materials

The transistor function was experimentally observed by Bardeen and Brattain (1947) in n-type polycrystalline germanium. By the early 1950s, all investigations of semiconductor properties of germanium and silicon were preferred by single crystals grown by Czochralski-zone (Bojarczuk et al., 2003) or float-zone (FZ) techniques (Huff, 2002; Depuydt et al., 2006). Single crystals of SiGe compounds with different Ge concentrations were also grown by different techniques (Yildiz et al., 2005).

The electronic, optical, thermal, and mechanical properties of Si and Ge were studied over a long period and tabulated in different reference books (such as Madelung et al., 1982; EMIS, 1988; Hull, 1999; Kasper, 1995; Kasper and Lyutovich, 2000; Claeys and Simoen, 2007). Table 1.2 summarizes the important properties of Si and Ge. Like most of the elements of group IV, silicon and germanium crystallize in the diamond cubic structure

TABLE 1.2

Fundamental Properties of Si and Ge at 300 K

	Silicon	Germanium
Lattice constant (nm)	0.5431	0.5658
Number of atoms (cm^{-3})	5×10^{22}	4.4×10^{22}
Density (g/cm^3)	2.329	5.3234
Electron affinity	4.05	4.0
Dielectric constant	11.7	16.2
Melting point (°C)	1412	937
Thermal conductivity ($W\ cm^{-1}\ °C^{-1}$)	1.3	0.58
Electron mobility ($cm^2\ V^{-1}\ s^{-1}$)	1450	3900
Hole mobility ($cm^2\ V^{-1}\ s^{-1}$)	450	1900
Effective electron masses (in units of m_o)		
Longitudinal m_l	0.92	1.6
Transverse m_t	0.19	0.08
Effective hole masses (in units of m_o)	0.54	0.33
Heavy m_{hh}		
Light m_{lh}	0.15	0.043
Spin-orbit split m_{so}	0.23	0.095
Electron affinity	4.05	4.0
Energy gap	1.12	0.661
Intrinsic carrier concentration (cm^{-3})	1×10^{10}	2.0×10^{13}
Effective conduction band density of states (cm^{-3})	2.8×10^{19}	1.0×10^{19}
Effective valence band density of states (cm^{-3})	1.8×10^{19}	5.0×10^{18}
Refractive index	3.42	4.0
Radiative recombination coefficient ($cm^3\ s^{-1}$)	1.1×10^{-14}	6.4×10^{-14}

characterized by the space group $Fd3m(O_h^7)$. The outer valence configuration is $3s^23p^2$ for Si and $4s^24p^2$ for Ge. When bonds are formed, the promotion of an s-electron to a p-state to form sp^3 hybrids takes place. This explanation accounts for the nature of the bonding since sp^3 hybrid orbitals form tetrahedral bonding patterns matching the symmetry found in the diamond structure. Although both elements exhibits a s^2p^2 valence band configuration, differences exist for the band structure (Yu and Cardona, 1996). The main differences arise from the core charge configuration. Thus, the valence band levels in Ge are less tightly bound compared with Si, and this is reflected in the more metallic character of the Ge band structure. Another significant difference is the presence of 3d electrons within the core of Ge. Differences in the band structure between both elements appear mainly in the conduction band arrangement (Figure 1.17). Here, the ordering of the lowest conduction bands is different from that of Ge. Si has the Γ_{15} band lower than the Γ_2 band. Moreover, while Ge and Si are both indirect semiconductors, the conduction band minimum in Ge occurs at the L-point, while it is near the X-point in Si. In addition, the valence band configurations in both elements are almost identical. The direct band gap Γ_1 of Ge amounts to 0.8 eV corresponding to a wavelength of 1.55 μm. Accordingly, Ge becomes an interesting material for light emitters. Indeed, lasing at about 1.6 μm of the direct transition in strained Ge layers on Si was demonstrated after optical pumping (Liu et al., 2010). Electroluminescence with strong dominance of the direct transition at 1.55 μm was found for unstrained Ge layers on Si (Arguirov et al., 2011).

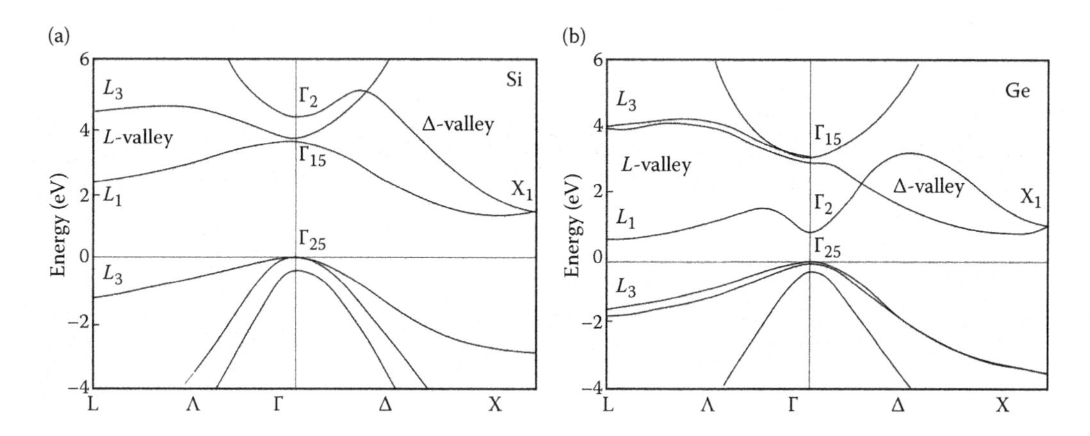

FIGURE 1.17
Band structure of (a) silicon and (b) germanium according to Schäffler (2001) and Yu and Cardona (1996).

The optical parameters and their relation to the band structure were analyzed in numerous papers. The dependence of the refractive index n and extinction coefficient k on the photon energy are shown in Figure 1.18a,b. The optical absorption coefficients for both semiconductors are presented in Figure 1.18c. The absorption coefficients α depend strongly on the wavelength of the light. For wavelengths λ shorter than λ_c, which corresponds to the band-gap energy ($\lambda_c = hc_o/E_g$), α increases fastly according to the fundamental absorption. The rise of α depends on the band-band transition. The rise is large for direct band-band transitions ($E_g^{dir} = 0.81$ eV at 300 K). For indirect band-to-band transitions for Si ($E_g^{dir} = 1.12$ eV at 300 K) and Ge ($E_g^{dir} = 3.03$ eV at 300 K) the rise is smaller.

1.3.2 Layer Systems

1.3.2.1 Amorphous, Porous, and Nanocrystalline Silicon

Amorphous silicon (a-Si) and amorphous silicon alloys have been used in different applications, mainly utilizing the possibility of fabricating large area thin films with sufficiently semiconducting properties even at low temperatures (EMIS, 1989). The structural disorder results in high electron scattering and band-tails with localized states and defects. Bound hydrogen, as it is present in a plasma-deposited material, saturates dangling bonds of Si and effectively reduces defect density. Typical properties of hydrogenated amorphous silicon (a-Si:H) are summarized in Table 1.3. Optical data are listed by Palik (1998) for $0.01 \leq \lambda \leq 148$ μm. Amorphous silicon was applied for waveguides (Cocorullo et al., 1996; Rao et al., 2010), three-dimensional tapers (Harke et al., 2008), and other passive devices. The application of a-Si:H also offers the realization of stacked waveguides by deposition of multiple layers, making the integration of passive photonic devices into the back-end of IC processes feasible. In photovoltaics a-Si is applied for thin film solar cells on glass and for HIT (heterojunction with intrinsic thin layer) solar cells, where a heterojunction between a-Si and c-Si replaces the common p–n junction (Tsunomura et al., 2009). The use of nanocrystalline and microcrystalline Si for thin film solar cells has also been studied in detail (Bergmann, 1999).

The optical properties of porous silicon (π-Si) are very different from those of bulk silicon. Porous silicon shows an increased band gap and efficient room temperature

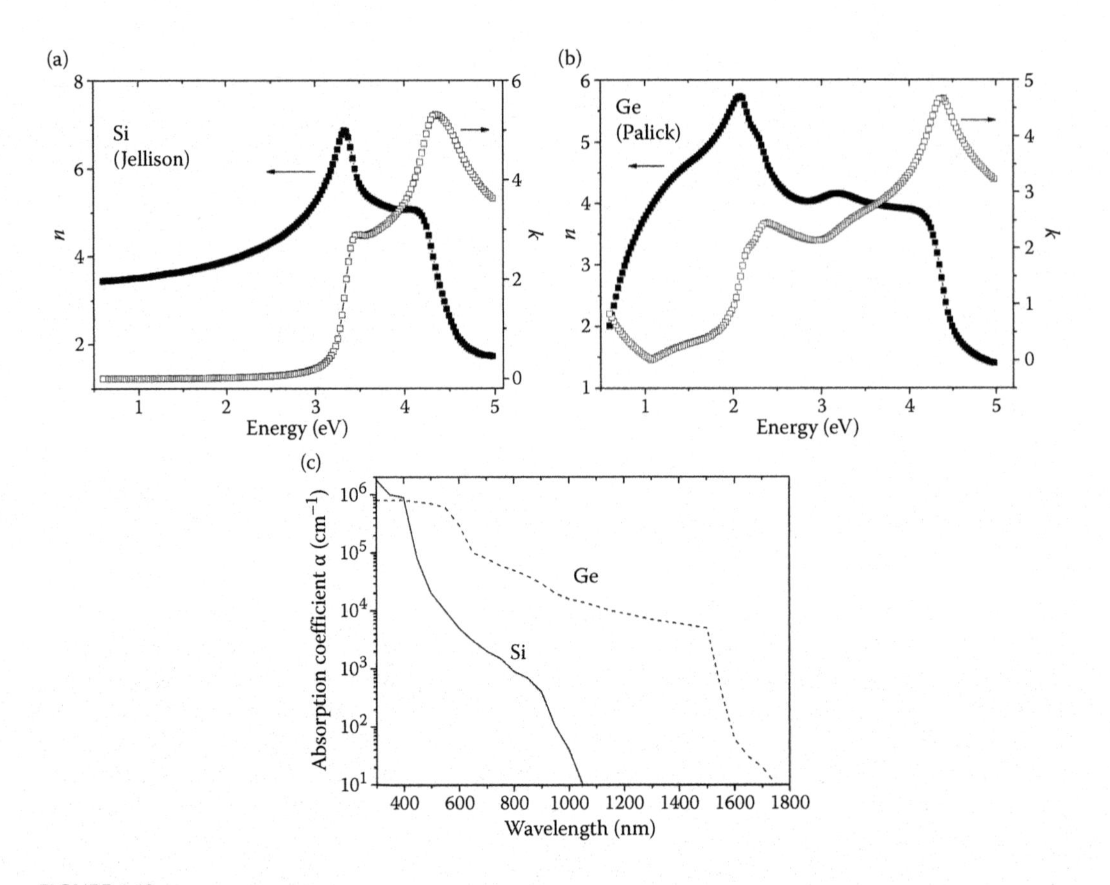

FIGURE 1.18
Refractive index n (a) and extinction coefficient k (b) for silicon and germanium. Data according to Jellison et al. (1993) and Palik (1998). The absorption coefficient vs. wavelength is shown in (c) (EMIS, 1988).

TABLE 1.3

Properties of Undoped a-Si:H at Room Temperature

Property	Typical Values
Hydrogen content	8%–15%
Optical band gap	1.7–1.8 eV
Valence band tail slope	42–50 meV
Conduction band tail slope	≅25 meV
Density of states at Fermi level	5×10^{14} eV^{-1} cm^{-3}
Electron drift mobility	≥1 cm^2 V^{-1} s^{-1}
Hole drift mobility	≥8×10^{-3} cm^2 V^{-1} s^{-1}

Sources: Street, R.A. (Ed.), 1991, *Hydrogenated Amorphous Silicon*, Cambridge University Press, Cambridge; Luft, W., Y.S. Tsuo (Eds.), 1993, *Hydrogenated Amorphous Silicon Alloy Deposition Processes*, Marcel Dekker, London; Tanaka, K., E. Maruyama, T. Shimada, H. Okamoto (Eds.), 1999, *Amorphous Silicon*, Wiley & Sons, New York.

photoluminescence in the visible range. The formation of π-Si was first described by Uhlir (1956) when investigating the electropolishing of Si in a HF containing electrolyte. The room-temperature visible luminescence was studied for the first time by Canham in 1990. Porous silicon consists essentially of an interconnected crystalline highly elastically distorted silicon skeleton, which is almost isotropic (Gösele and Lehmann, 1994). EPR measurements showed the presence of defects typically for dangling bonds in Si and at the Si/SiO_2 interface (von Bardeleben et al., 1993). The measurements indicate also that π-Si is still mainly crystalline with a clear crystallographic correlation to the bulk Si and that the observed defects act as nonradiative recombination centers. The properties of π-Si and their dependence on the fabrication conditions are presented in numerous publications. A more recent summary is given, for instance, in Lehmann (2002) and Pavesi and Turan (2010).

Silicon nanocrystals embedded in an oxide host are an alternative to π-Si. Different methods have been applied to realize such arrangements, namely ion implantation into high-quality oxides, sputtering of Si-rich oxides, or reactive evaporation of Si rich oxides. With these methods, the size of the Si nanocrystals is controlled by the Si content in the SiO_2 matrix and the crystallite density cannot be controlled independently. Another approach is the fabrication of amorphous SiO/SiO_2 superlattices by PECVD and thermal annealing for phase separation and crystallization (Park et al., 2002; Zacharias et al., 2003). The size of the nanocrystals is predetermined by the SiO layer thickness. The resulting photoluminescence depends on the size of the clusters.

1.3.2.2 SiGe Alloys

Si and Ge form a continuous substitutional solid solution with a cubic diamond structure (space group Fd3m) under normal pressure (Olesinski and Abbaschian, 1984). The lattice constant increases continuously as the Ge concentration in the $Si_{1-x}Ge_x$ alloy increases from a = 0.5431 nm (x = 0) to a = 0.5658 nm (x = 1). There is a negative deviation of the experimental data of the lattice constant from Vegard's law (Dismukes et al., 1964). The properties of SiGe alloys vary between those of both elements, making it possible to adjust specific properties (band-gap engineering, etc.). The variation of electronic and optoelectronic properties with the concentration are listed in various reference books (Kasper and Lyutovich, 2000; Schäffler, 2001; Adachi, 2009). Selected properties are summarized in Table 1.4. Besides single crystals (Yildiz et al., 2005), thin layers of a single-crystalline material grown by different deposition techniques (Kasper, 1995) are mostly applied in optoelectronics and photovoltaics. Here, Si is the predominant substrate material. The increasing lattice constant with increasing x causes a lattice mismatch $\Delta a/a$ for up to 4% resulting in a drastical decrease of the critical thickness for pseudomorphic $Si_{1-x}Ge_x$ growth (People and Jackson, 1990). Relaxed, or in other words unstrained, $Si_{1-x}Ge_x$ layers are only obtained at large layer thicknesses if they are deposited directly on a Si substrate (Perry et al., 1993). Relaxation occurs by the dislocation generation (threading dislocations). Thinner layers are biaxially strained. Relaxed layers with low dislocation densities were obtained by applying graded buffer layers. Composition and strain in $Si_{1-x}Ge_x$ layers are mostly identified by X-ray diffraction (XRD) or Raman spectroscopy. The simultaneous measurement of strain and composition by Raman spectroscopy is a critical issue (Reiche et al., 2011). The composition dependence of the fundamental absorption edge in $Si_{1-x}Ge_x$ was first determined by Braunstein et al. using optical absorption (Braunstein, 1963) and more recently by Weber and Alonso (1989) using photoluminescence. The absorption coefficient of strained SiGe layers was also studied for different Ge concentrations (Polleux

TABLE 1.4

Fundamental Properties of Unstrained $Si_{1-x}Ge_x$ Alloys

Lattice constant (nm)	$0.5431 + 0.01992x + 0.0002733x^2$
Number of atoms (cm^{-3})	$(5.00 - 0.58x) \times 10^{22}$
Density (g/cm^3)	$2.329 + 3.493x - 0.499x^2$
Dielectric constant	$11.7 + 4.5x$
Thermal conductivity (W/cm K)	$\approx 0.046 + 0.084x$ $(0.2 < x < 0.85)$
Effective electron mass (in units of m_0)	
Longitudinal m_l	≈ 0.92 $(x < 0.85)$ ≈ 1.59 $(x > 0.85)$
Transverse m_t	≈ 0.19 $(x < 0.85)$ ≈ 0.08 $(x > 0.85)$
Electron affinity	$4.05 - 0.05x$
Direct (E_0) and indirect gap energies	$E_0 = 4.06 - 3.265x$ $E_g^X = 0.520 + 0.632x - 0.300x^2$ $E_g^L = 2 - 1.34x$ $E_1 = 3.395 - 1.44x + 0.153x^2$ $E_1 + \Delta_1 = 3.428 - 1.194x + 0.062x^2$
Effective conduction band density of states (cm^{-3})	$\approx 2.8 \times 10^{19}$ $(x < 0.85)$ $\approx 1.0 \times 10^{19}$ $(x > 0.85)$
Index of refraction	$3.42 + 0.37x + 0.22x^2$

Sources: Schäffler, F., 2001, In *Properties of Advanced Semiconductor Materials: GaN, AlN, InN, BN, SiC, SiGe*, edited by Levinshtein, M.E., Rumyantsev, S.L., Shur, M.S., Wiley & Sons, New York; Adachi, S., 2009, *Properties of Semiconductor Alloys: Group-IV, III-V and II-VI Semiconductors*, Wiley & Sons, Chichester.

and Rumelhard, 2000). The optical constants in the IR–UV region have been measured by different authors (Humlicek, 1995; Jellison et al., 1993) (Figure 1.19). All measurements have shown that the E_1 and $E_1 + \Delta_1$ interband transitions are most sensitive to strain and composition. E_1 is a direct transition between Λ_1 conduction band and Λ_3 valence bands, while $E_1 + \Delta_1$ denotes the spin-orbit split. Both transitions are located in the energy range between 2 and 3.5 eV for the whole concentration range. The dependence of E_1 and $E_1 + \Delta_1$ on the composition is shown in Table 1.4. An additional shift of the critical point energies is obtained by strain. A biaxial (100) strain ε_s results in shifts (Pollak, 1990; Pickering et al., 1993)

$$\delta E_1 = E_1(\varepsilon_s) - E_1(0) = \frac{\Delta_1}{2} + E_H - \frac{1}{2}\sqrt{\Delta_1^2 + 4E_s^2} \tag{1.1}$$

and

$$\delta(E_1 + \Delta_1) = [E_1 + \Delta_1](\varepsilon_s) - [E_1 + \Delta_1](0) = \frac{-\Delta_1}{2} + E_H + \frac{1}{2}\sqrt{\Delta_1^2 + 4E_s^2}, \tag{1.2}$$

where E_H and E_s are the hydrostatic shift and uniaxial shear, respectively.

SiGe layers have been widely applied for different active and passive optoelectronic devices including light emitters and detectors as well as waveguides (Soref, 1996; Splett et al., 1994; Yin et al., 2006; Zimmermann, 2009). Recent investigations demonstrated novel

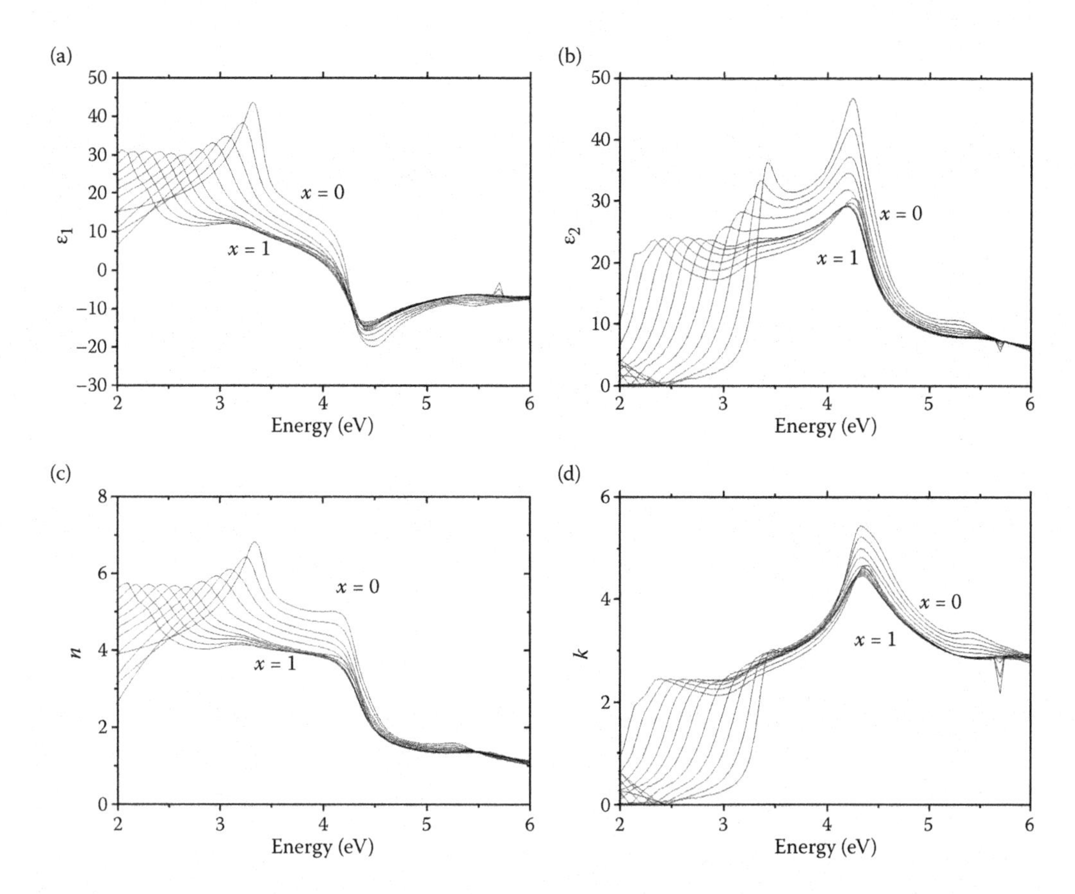

FIGURE 1.19
Optical constants in the IR–UV region of $Si_{1-x}Ge_x$ alloys. (a) Permittivity, real part; (b) permittivity, imaginary part; (c) refractive index; and (d) absorption index. Compiled from data of Jellison et al. (1993), Humlicek (1995), Palik (1998), and Landolt-Boernstein (2007).

light-emitting devices where SiGe microresonators were used. The emitter and modulator functions are described by the Purcell effect and Franz-Keldysh effect (Wada, 2008). Another approach are Ge/SiGe quantum well structures (Chaisakul et al., 2011).

1.3.2.3 Miscellaneous Layer Systems

Besides SiGe, binary alloys of Si or Ge with carbon or tin as well as ternary alloys (SiGeC, SiGeSn) are of increasing importance for optoelectronics. All binary alloys, except of Ge_xSn_{1-x} ($0 \leq x \leq 0.88$) are indirect band gap materials. In the concentration range $0 \leq x \leq 0.88$ Ge_xSn_{1-x} is a direct band gap semiconductor and becomes semimetallic for $x \leq 0.67$ (Adachi, 2009; Kouvetakis et al., 2006).

Crystalline SiC thin films have attracted much attention in recent years because of their potential application in many kinds of optoelectronic devices, such as solar cells, image sensors, and photodiodes (Harris, 1995; Edmond et al., 1998). SiC is an appropriate material for the fabrication of high-temperature, high-frequency, and high-power electronic devices, due to its wide band gap (2.3–3.2 eV), high breakdown field (>2×10^6 V cm^{-1}), high thermal conductivity (5–7 W cm^{-1} K^{-1}), and high electron saturation velocity (2×10^7 cm s^{-1}). Attempts have been made to transform single crystalline SiC into porous materials and to obtain quantum confinement of carriers in nanocrystals similarly to π-Si. Porous SiC has been reported to exhibit much more improved photoluminescence than bulk material (Chang and Sakai, 2004). An improved photoluminescence was also reported for hydrogenated amorphous and microcrystalline silicon carbide (a-SiC:H, μc-SiC:H) deposited by plasma-enhanced chemical vapor deposition (Hamakawa, 1999).

The lattice constant a of Ge_xSn_{1-x} decreases monotonically as a function of composition from 0.6493 nm (α-Sn) to 0.5658 nm (Ge). A positive deviation from Vegard's law was observed, which is in contrast to SiGe alloys (Kouvetakis et al., 2006). The fundamental absorption edge is expected to change from the Ge L_6^- indirect to the Sn Γ_7^- direct gap depending on x. The band structure near the direct–indirect transition region in relaxed or epitaxially grown Ge_xSn_{1-x} has been studied using different methods (D'Costa et al., 2006). The interband transition energies and their dependence on x were summarized by Adachi (2009).

The large lattice mismatch between Ge and Si enables strain engineering applications of SiGe alloys but causes also a number of strain management problems in alloys grown directly on Si. In applications that require thick layers, for instance photovoltaics, the incorporation of active SiGe layers was not possible due to the generation of strain-relieving threading dislocations. Furthermore, strain and band structure are coupled and are controlled by composition. In view of these limitations, the interest in ternary group IV semiconductors has grown over the last 20 years.

Ternary alloys of the type $Si_{1-x-y}Ge_xC_y$ (with $y < 0.01$) can be grown lattice-matched on Si. The smaller carbon addition compensates the compressive strain in SiGe binary alloy layers and thus relieves thickness limitations in SiGe growth. The lattice constant of $Si_{1-x-y}Ge_xC_y$ alloys is given as (Eberl et al., 1992)

$$a_{\mathrm{SiGeC}} = a_{\mathrm{Si}} + (a_{\mathrm{Ge}} - a_{\mathrm{Si}})x + (a_{\mathrm{C}} - a_{\mathrm{Si}})y \tag{1.3}$$

The band structure of relaxed and strained $Si_{1-x-y}Ge_xC_y$ was reviewed by Yang et al. (2002). The representation of the band gap versus lattice constant and composition showed that the absorption edges of the materials cover the near-infrared, visible, and near-ultraviolet ranges (Soref, 1991). The refractive index is given by the relation (Schubert and Rana, 2007)

$$n_{\mathrm{Si}_{1-x-y}\mathrm{Ge}_x\mathrm{C}_y} = n_{\mathrm{Si}} + (n_{\mathrm{Ge}} - n_{\mathrm{Si}})\left(x - \frac{2.1}{0.9}y\right) \tag{1.4}$$

$Ge_{1-x-y}Si_xSn_y$ grown on $Ge_{1-y}Sn_y$-buffered Si substrates is the first practically group IV ternary alloy, since C can only be incorporated in very small amounts (≤1 at.%) into the Ge–Si network (Soref et al., 2007). The most significant feature of $Ge_{1-x-y}Si_xSn_y$ is the possibility of independent adjustment of lattice constant and band gap. For the same value of the lattice constant one can obtain band gaps differing by more than 0.2 eV, even if the Sn concentration is in the range $y < 0.2$. The lattice constant of $Ge_{1-x-y}Si_xSn_y$ alloys is given by the relation (Sun et al., 2010)

$$a_{\mathrm{Ge}_{1-x-y}\mathrm{Si}_x\mathrm{Sn}_y} = a_{\mathrm{Ge}}(1 - x - y) + a_{\mathrm{Si}}x + a_{\mathrm{Sn}}y \tag{1.5}$$

The band structure of $Ge_{1-x-y}Si_xSn_y$ alloys was recently studied (Sun et al., 2010; Fan, 2010; D'Costa et al., 2010; D'Costa, 2009). Using a 6-band Hamiltonian, the band gap of $Ge_{1-x-y}Si_xSn_y$ alloys is given as a function of the composition as (Fan, 2010)

$$E_g^{\Gamma}(Ge_{1-x-y}Si_xSn_y) = 0.7985(1-x-y) + 4.185x - 0.413y \\ - 0.14(1-x-y)x - 2.55(1-x-y)y - 3.05xy \quad (1.6)$$

Data on spin-orbit splitting, conduction and valence band edge are presented by Sun et al. (2010).

1.3.3 Compliant Materials

1.3.3.1 Silicon-on-Insulator

Silicon-on-insulator (SOI) refers to a technology that uses a stack of a thin single crystalline silicon layer on top of an insulator layer (buried oxide, BOX), both deposited on a substrate. Silicon-on-Sapphire (SOS) was invented in 1963 at North American Aviation (now Boeing). The first published report is by Manasevit and Simpson (1964), which deposited heteroepitaxially layers of essentially single-crystalline silicon on (1–102) oriented sapphire substrates. A few years later, RCA Laboratories, Princeton, NJ, worked on making SOS as a manufacturable technology. The research was driven by requirements for circuitry immune to soft errors in communication and weather satellites. The high defect density in the silicon layers and the low minority carrier lifetime were main reasons to develop alternative SOI materials. Among other developments, separation by implantation of oxygen (SIMOX) was introduced in the late 1970s (Izumi et al., 1978). SIMOX uses a high-dose implantation of oxygen into a bulk silicon wafer and a subsequent high-temperature annealing step. A SiO_x layer ($x \approx 2$) is formed below a thin single-crystalline Si layer during the annealing. The thickness of the Si and SiO_x layer depend on the implantation conditions. The high production costs of SIMOX wafers and possible defect generation even during high-temperature annealing enforces the development of alternative technologies. Based on investigations of Lasky and others (Lasky, 1985; Shimbo et al., 1986), semiconductor wafer direct bonding methods were evolved to produce SOI wafers. For wafer bonding, the surfaces of two silicon wafers, one with a thermally grown SiO_2 layer on top, are brought into atomically contact. Adhesion forces stick both wafers together. During a subsequent annealing at elevated temperatures, strong Si–O atomic bonds are formed via the interface, making a nondetachable compound between both wafers. Finally, one of the silicon wafers is thinned down by mechanical grinding and polishing. Since the early 1990s, this bonding and etch back SOI wafers (BESOI) have been applied in numerous technologies. The advantages of BESOI wafers are variable thicknesses of the high-quality BOX layer and that Si wafers with different crystal orientation can be combined. The disadvantage even for high-performance electronic device applications is the minimum achievable thickness of the top silicon layer. This is because mechanical thinning allows only to realize thicknesses of more than about 1.5 μm and thickness tolerances (uniformity) in the order of about 0.5 μm. The invention of Bruel in 1994 resulted in substantial progress by combining hydrogen-induced layer splitting with wafer bonding methods. This technique, commercialized for the first time as the so-called Smart Cut technique (a trademark of SOITEC, Bernin, France), uses the implantation of hydrogen in silicon wafer (at doses about two orders of magnitude lower than for oxygen for SIMOX) followed by a wafer bonding to another (oxidized) wafer. The implanted hydrogen forms gas bubbles during an annealing at moderate temperatures resulting in cracks parallel to the surface and finally that a thin layer breaks from the implanted wafer and is transferred to the oxidized wafer. Most of SOI wafers used in today's device processing are fabricated by wafer bonding combined with hydrogen-induced layer splitting (Reiche, 2009). The technique combines the advantages of wafer bonding methods (high-quality oxides with variable thickness) with implantation methods to realize thin Si layers with minimum thickness tolerances. The thickness of the transferred silicon layer is adjusted by the energy of the hydrogen implantation and therefore limited to values below 1.5 μm.

A modification of SOI is strained silicon on insulator (SSOI) produced by the transfer of a thin tensily strained silicon layer to an oxidized Si substrate using an analogous process of wafer bonding in combination with hydrogen-induced layer splitting (Reiche et al., 2010). Tensile strain removes the 6-fold degeneracy in the conduction band, lowering the energy of the two Δ-valleys having electrons with a lower in-plane effective mass with respect to the four Δ-valleys having electrons with a higher in-plane effective mass. This energy splitting reduces intervalley scattering and preferential occupation of the two Δ-valleys with lower in-plane effective mass electrons results in higher electron mobility. Tensile strain also removes the valence band degeneracy at the Γ-point and shifts the spin-orbit band, thus improving the hole mobility.

SOI material is widely used in photonics, and applications are discussed in numerous publications. Reviews are given in recently published books (Reed and Knights, 2004; Kasap and Capper, 2006; Kasper and Yu, 2011; Fathpour and Jalali, 2011) and review articles (Jalali and Fathpour, 2006). Besides the original application of realization of waveguides, SOI is the material of choice for a number of emerging technologies in photonics such as the realization of photonic circuits, integration of photonics into CMOS device processes, nonlinear optics, and heterogeneous integration of III–V materials.

1.3.3.2 Germanium-on-Insulator

Germanium-on-insulator (GOI) combines the advantages of SOI with those of Ge with respect to charge carrier mobility, optical band gap, and absorption coefficient as well as lattice mismatch with GaAs, relevant for integration of III–V-based optoelectronics. GOI substrates have been realized by various approaches. Bojarczuk et al. (2003) reported epitaxially grown crystalline oxide $(La_xY_{1-x})_2O_3$ as a BOX layer followed by solid-phase epitaxial growth to form the top Ge layer. Specific wafer bonding techniques have been applied to bond bulk Ge wafers on Si or with an oxidized layer in between (Radu et al., 2005). Wafer bonding of different bulk wafers, however, result in increasing strain in the interface even during subsequent annealing, which, especially in the case of large wafer diameters, result in damage or debonding (Reiche, 2009). Therefore, the transfer of thin Ge layers by the combination of wafer bonding and hydrogen-induced layer splitting is the preferred method. Germanium condensation by thermally oxidizing a SiGe-on-insulator substrate is one technique (Tezuka et al., 2001). To achieve a high Ge concentration, aggressive oxidation is required, and the thickness of the final Ge layer is limited. Another approach is the transfer of thin layers from a Ge bulk wafer (Tracy et al., 2004; Chao et al., 2006; Akatsu et al., 2006). A smarter and more competitive way is the transfer of Ge layers grown by CVD methods on silicon substrates (Akatsu et al., 2006). The crucial problem of Ge layers grown epitaxially on Si is the high threading dislocation density. The dislocation density is reduced to about 6×10^6 cm^{-2} by applying graded buffer layers or by a two-step growth process consisting of a low-temperature seed followed by the growth of a relaxed Ge layer at higher temperatures (Depuydt et al., 2007).

The preparation of photonic crystals on GOI substrates has been reported recently (El Kurdi et al., 2008; Ngo et al., 2008).

1.3.3.3 Compound Semiconductors-on-Insulator

The combination of III–V compounds with group IV semiconductors offers new perspectives in optoelectronics and in the integration of optoelectronics and microelectronics. Caused by the lattice mismatch, the heteroepitaxial growth of III–V compounds on Si results in high densities of threading dislocations. Therefore, thick buffer layers are required. Alternatives are similar process as for the fabrication of SOI, i.e., the transfer of III–V compound layers on oxidized silicon substrates by combinations of wafer bonding with hydrogen-induced layer splitting. The successful transfer of thin GaAs, InP, InGaAs, and other compounds was already reported (Di Cioccio et al., 2004; Yokoyama et al., 2010).

1.3.4 Promising Materials

Increasing the efficiency of photodetectors and solar cells, the realization of group IV–based light emitters, and new and more efficient passive devices as well as the symbiosis of optoelectronics and microelectronics are challenges in physics, materials science, and engineering. All these issues are combined with a better understanding of existing materials and even with the integration of new materials. A number of reviews are given in recently published books (Huang, 2010; Dakin and Brown, 2006; Quillec, 1996). Table 1.5 summarizes basic properties of dielectrics such as Si_3N_4, Al_2O_3, and HfO_2, including SiO_2 as reference. More details can be found in a review by Houssa (2004). Examples for application of these dielectrics will be sketched below. Moreover, novel Si-based light emitters making use of the luminescence properties of dislocation networks will be treated.

Silicon nitride (Si_3N_4) is widely used as antireflection coating on top of c-Si solar cells (Sopori, 2003). The layer deposition is done by PECVD with hydrogen excess, leading supplementary to passivation of the electrical activity of crystal defects contained in solar Si (Sopori et al., 2001; Vyvenko et al., 2000). Furthermore, Er-doped Si_3N_4 in metal–insulator–metal structures causes enhanced light emission at 1.55 μm (Gong et al., 2009). Alumina (Al_2O_3) is applied in photovoltaics for surface passivation of Si solar cells (Schmidt et al., 2008). Porous alumina is used for making photonic crystals by self-organization of highly ordered structures during etching (Masuda and Fukuda, 1995).

High-k materials, such as HfO_2, have attracted strong interest during the last decade (for a review, see, e.g. Dabrowski et al., 2009). Microelectronics is the main driver for this developments, for example, to replace the very thin leaky silicon oxide in the FET gate stack by

TABLE 1.5

Basic Properties of Important Dielectrics according to Houssa (2004)

Material	Relative Dielectric Constant	Refractive Index	Band-Gap Energy (eV)	Conduction Band Offset with Si (eV)	Valence Band Offset with Si (eV)
SiO_2	3.9	1.46	8.9	3.1	4.7
Si_3N_4	7.4	2.1	5.1	2.2	1.8
Al_2O_3	9–11	1.79, 1.87	6.2, 8.8	2.2, 2.8, 2.4	2.9, 4.9
HfO_2	15–26	2.24, 2.45	5.6, 5.9	2.0, 1.3	2.5, 3.4

Source: Houssa, M. (Ed.), 2004, *High-k Gate Dielectrics*, Institute of Physics, Bristol.

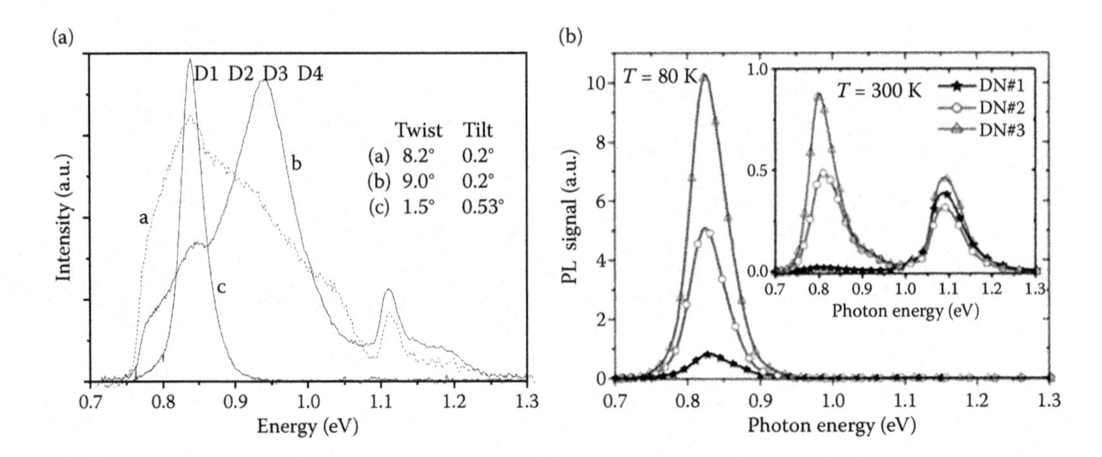

FIGURE 1.20
The impact of dislocation type and density (represented by the misorientation) on the cathodoluminescence (a) and photoluminescence spectra of three different dislocation networks (b). Measurements at $T = 80$ K (i) and $T = 80$ K and 300 K (ii).

a "thick" high-k material layer (Wong, 2004). Another application may be the use of high-k materials as tunneling layers for light emitters. It was demonstrated that a MOS tunneling diode allows injection of excess carriers toward a regular dislocation network close to the interface between the thin SiO_2 and Si to excite electroluminescence at 1.55 μm (Kittler et al., 2006). Replacement of the very thin SiO_2 by a HfO_2 layer enhances the tunneling current of the MIS diode and improves the efficiency for light emission.

The advantages of nanostructures, as the quantum confinement of carriers, result in a rapid increase of research in this field (Wang and Neogi, 2010). Si quantum structures in a dielectric matrix can be used for Si band-gap engineering, which is of interest for third-generation solar cells (Green et al., 2005). An effective band gap larger than 1.6 eV was observed for a stack of alternating c-Si (about 2 nm thick) and SiO_2 layers. The a-Si state of the layers after PECVD deposition was transformed into crystalline silicon by light-induced solid-to-solid crystallization (Mchedlidze et al., 2008).

Light emission of silicon in the visible and near-infrared regions by porous silicon and silicon nanocrystals has been extensively studied. Native nanostructures, such as dislocations, open up also new possibilities to realize silicon light emitters for the wavelength region 1.3 μm $< \lambda <$ 1.6 μm. The core of dislocations is in the order of 1 nm. Defined arrangements of dislocations were produced by specific wafer bonding techniques (Reiche and Kittler, 2011) and analyzed by photoluminescence, cathodoluminescence, and electroluminescence (Kittler et al., 2007, 2008). Modifying the tilt and twist misorientation between both wafers, well controlled during bonding, allows the generation of different types and densities of dislocations in two-dimensional arrays. Measurements of such samples provided direct evidence that the wavelength of light emitted from dislocations can be tailord to some extent by misorientation (Figure 1.20). The dislocation related luminescence lines D1 (at about 0.81 eV) and D3 (at about 0.93 eV) have the largest intensity in the spectra due to the variation of the twist angle from 8.2° to 9° (Figure 1.20a). Additional photoluminescence measurements show that screw dislocations dominantly affect the intensity of the 1.5-μm emission (D1 line). The intensity of D1 is lowest in the spectrum of the sample having the lowest density of screw dislocations (DN#1 in Figure 1.20b), whereas higher intensities appear in samples having higher densities of screw dislocations (DN#2, DN#3). Furthermore, electroluminescence measurements on MOS and pn-LEDs confirmed the previous results. Increasing or lowering the electric field in pn-LEDs causes a red or blue shift of the D1 line (Kittler et al., 2008). This observation of the Stark effect, characterized by a coefficient α = 0.0186 meV/kV/cm^2, may allow the realization of Si-based emitter and modulator combined in a single device.

1.4 Heterostructure Interfaces and Strain

Osamu Nakatsuka and Shigeaki Zaima

The strain engineering technology in Si metal-oxide-semiconductor field effect transistors (MOSFET) is very important because the strained Si channel realizes the effective enhancement of carrier mobility (Mii et al., 1991; Welser et al., 1992; Ismail et al., 1992). Meanwhile, strained Ge is also an attractive candidate from the perspective of electronic and optoelectronic devices (Ishikawa et al., 2003; Liu et al., 2007; Takeuchi et al., 2008; Shimura et al., 2009; Fang et al., 2007). One of the methods to induce strain in semiconductor thin films is using the buffer layer as stressor in global area with a wafer-scale size and another is using micro-scale stressor, which induces strain only into the local area such as a channel region in MOSFET.

1.4.1 Formation Technology of Strained Group IV Materials

Global strain is realized using a virtual substrate that consists of a buffer layer formed on a whole substrate. In general, that provides biaxial, large, and uniform in-plane strain. On the other hand, local strain is realized with local stressor such as $Si_{1-x}Ge_x$ source/drain stressor (Ghani et al., 2003), shallow trench isolation (STI) (Tiwari et al., 1997), or silicon nitride liner (Goto et al., 2004), etc. That usually provides uniaxial, small, and complicated strain structures, whereas we can induce a local strain selectively into n- and p-type MOSFETs.

Global strain for biaxially tensile strained Si layer is usually realized using a strain relaxed $Si_{1-x}Ge_x$ alloy layer. Strain relaxation in a heteroepitaxial layer is strongly influenced with solid solubility, interdiffusion, segregation, and generation of defects. It is important to understand the crystalline and chemical phenomena in epitaxial layers and their interfaces. The strain relaxed $Si_{1-x}Ge_x$ layer is usually deposited on a Si substrate with step-graded or successively changed composition of Ge (Meyerson et al., 1988; Fitzgerald, 1992, 1997). The strain relaxation in graded $Si_{1-x}Ge_x$ buffer layers is step by step, and the density of threading dislocations can be effectively reduced to 10^5 cm^{-2} (LeGoues et al., 1992). On the other hand, the ion implantation technique for strain relaxation is also proposed, in which ion implantation of H or He into a Si substrate before the $Si_{1-x}Ge_x$ growth enhances the formation of strain-relaxed $Si_{1-x}Ge_x$ epitaxial layer (Hull et al., 1990; Hollander et al., 1999; Sawano et al., 2004). Defects of vacancies or small dislocation loops contribute to the strain relaxation, and we can localize these defects in the $Si_{1-x}Ge_x$ layers or Si substrates.

The oxidation-induced Ge condensation technique is also interested to realize strain-relaxed $Si_{1-x}Ge_x$ on insulator (SGOI) or Ge on insulator (GOI) structure (Tezuka et al., 2001; Nakaharai et al., 2003). SGOI or GOI wafer can be obtained with the oxidation of a $Si_{1-x}Ge_x$

layer grown on Si on insulator (SOI) substrate because there is the deference of oxidation rate of Si and Ge. Using this technique, we can realize strained Si on insulator (SSOI) or strained Si on Si (SSOS) structures (Langdo et al., 2003; Isaacson et al., 2005).

1.4.2 Strain Relaxation and Dislocation Structure of $Si_{1-x}Ge_x$ Heteroepitaxial Layers

In general, the strain is induced using the misfit of lattice constant or the deference of thermal expansion coefficient between stressor and a strained layer. Therefore, there are usually misfit dislocations at the interface between the stressor and the strained layer, and those essentially contribute to the strain and strain relaxation in the heterostructure. Defects related to dislocations strongly influence not only crystalline structure but also electrical properties, and we have to precisely control the structure and distribution of dislocations. The non-uniformity of dislocations or strain field related to dislocations affects the variability of the carrier mobility (Ismail et al., 1994; Sugii et al., 1999).

In the case of heteroepitaxial growth of the diamond structure such as $Si_{1-x}Ge_x$, a 60° dislocation is generally introduced at the interface when the thickness of the heteroepitaxial layer increases over the critical thickness to relax the elastic strain energy in the heterostructure (Hirth and Lothe, 1992). The schematic diagram of 60° dislocation in diamond structure such as Si is shown in Figure 1.21. In the strain relaxation of 60° dislocation, the Burgers vector along to the <110> direction slips on the {111} plane. There is a concern that the surface roughening called "cross-hatch pattern" takes place after the strain relaxation of a $Si_{1-x}Ge_x$ layer on a Si(001) substrate (Fitzgerald, 1992, 1997). In addition, the mosaicity in a $Si_{1-x}Ge_x$ layer is induced due to the tilting and/or rotation of microscale domains (Mooney, 1996; Timbrell et al., 1990). The Burgers vector of 60° dislocation inclines to the (001) plane at the $Si_{1-x}Ge_x$/Si hetero-interface, and there is an extra-half plane that is parallel to the (001) plane (Mooney et al., 1994). Therefore, the $Si_{1-x}Ge_x$ (001) plane essentially inclines to the Si(001) plane, and there is asymmetry around the dislocation line of a 60° dislocation. The cross-hatch pattern roughness can be removed with planarization using chemical mechanical polishing (CMP) treatment. However, it is known that the fluctuation of strain in the $Si_{1-x}Ge_x$ layer cannot be eliminated with CMP process (Sawano et al., 2005). In addition, the strain fluctuation in a $Si_{1-x}Ge_x$ buffer layer causes the strain fluctuation in a

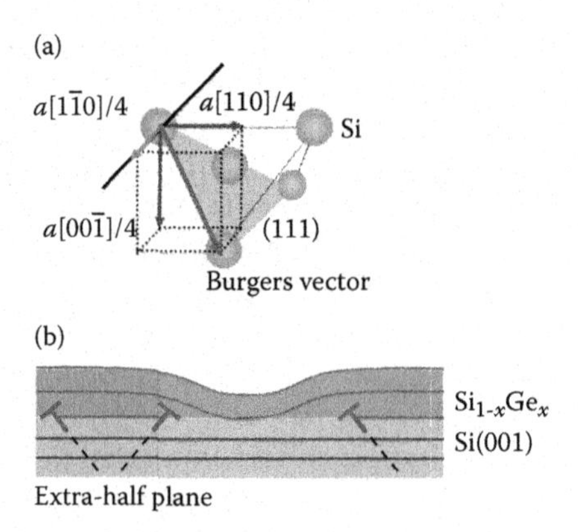

FIGURE 1.21
(a) The schematic diagram of a 60° dislocation with Burgers vector in diamond structure of Si. (b) The schematic diagram of strain relaxed $Si_{1-x}Ge_x$ layer on a Si(001) substrate with 60° dislocations.

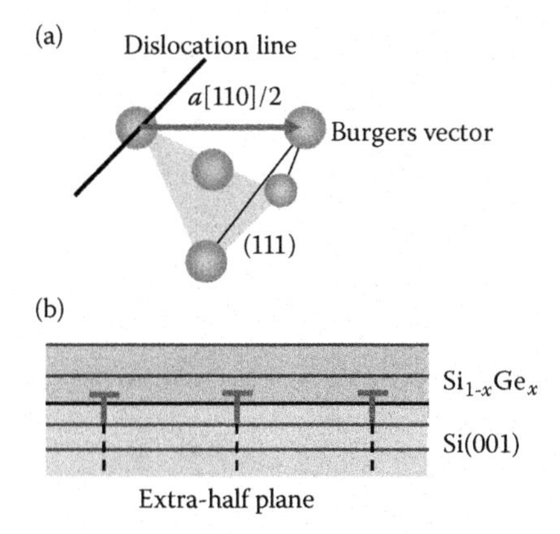

FIGURE 1.22
(a) The schematic diagram of a pure-edge dislocation with Burgers vector in diamond structure of Si. (b) The schematic diagram of strain relaxed $Si_{1-x}Ge_x$ layer on a Si(001) substrate with pure-edge dislocations.

strained Si layer (Sawano et al., 2003), and that leads to the variability of crystalline properties such as the oxidation rate of strained Si layer (Nishisaka et al., 2004).

The problem caused with 60° dislocations can be solved with using pure-edge dislocation, which has the Burgers vector parallel to the $Si_{1-x}Ge_x$/Si(001) heterointerface on the 001 plane. Figure 1.22 shows the schematic diagram of a pure-edge dislocation in diamond structure. When the pure-edge dislocation is introduced at the hetero interface, the strain in $Si_{1-x}Ge_x$ on Si(001) is uniformly relaxed without any tilting and rotation.

In general, there is a difficulty to introduce pure-edge dislocations at the $Si_{1-x}Ge_x$/Si(001) interface. There are some reports that pure-edge dislocation is locally formed with a reaction of 60° dislocations (Kvam et al., 1990; Narayan and Sharan, 1991). However, it is rarely seen that pure-edge dislocations dominantly occupy at the hetero-interface in the case of group IV semiconductor material. On the other hand, there are some reports that pure-edge dislocations are preferentially formed in Ge growth on Si(001) at a low temperature or with surfactant atoms (Eaglesham, 1991; Sakai et al., 1997). In those conditions, migration of Ge atoms on Si surface is restricted.

Using such Ge/Si(001) heterointerface structure, we can form a $Si_{1-x}Ge_x$ layer whose strain is relaxed predominantly with pure-edge dislocations. The formation process includes (1) formation of thin Ge layer on Si(001) substrate, (2) postdeposition annealing (PDA) for strain relaxation of the Ge layer, (3) high-temperature annealing (HTA) for solid-phase intermixing of Ge and Si to form a $Si_{1-x}Ge_x$ layer (Yamamoto et al., 2004; Taoka et al., 2005). Figure 1.23 shows TEM images of a thin Ge layer formed on a Si(001) substrate before and after PDA. The segments of dislocations enlarge along the [110] direction with PDA. Analysis of the contrast reveals that these dislocations are pure-edge dislocations whose Burgers vector is perpendicular to the dislocation line on (001) heterointerface (Taoka et al., 2005). The pure-edge dislocation network structure is formed after PDA as shown in Figure 1.23d. These pure-edge dislocations are regularly arrayed with the distance of each dislocation of about 10 nm. Such dislocation structure is also observed in the low temperature growth of a Ge layer on a Si substrate (Yamamoto et al., 2004; Sakai et al., 2005). Cross-sectional TEM observation reveals that defects including component of a pure-edge dislocation always exists at the Ge/Si(001) interface. These pure-edge dislocations have a Burgers vector on a (001) plane, and the component of pure-edge dislocations becomes dominant after PDA at 680°C (Taoka et al., 2005).

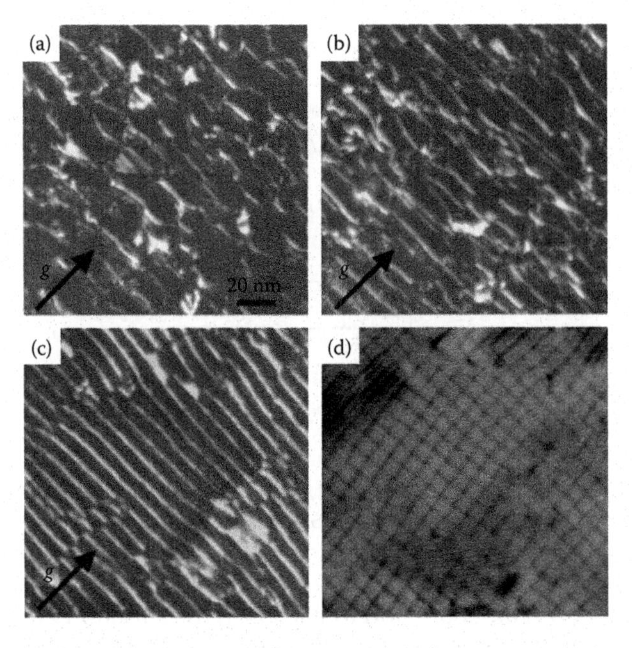

FIGURE 1.23
Plane-view dark-field TEM images of (a) as-grown, (b) 430°C annealed, and (c) 680°C annealed Ge (20 nm)/Si(001) samples [36]. These images were taken by the weak-beam method under the condition of $g/3g$ with g = [220]; (d) plane-view bright-field TEM image taken from the same area as (c), showing the network morphology of pure-edge dislocations.

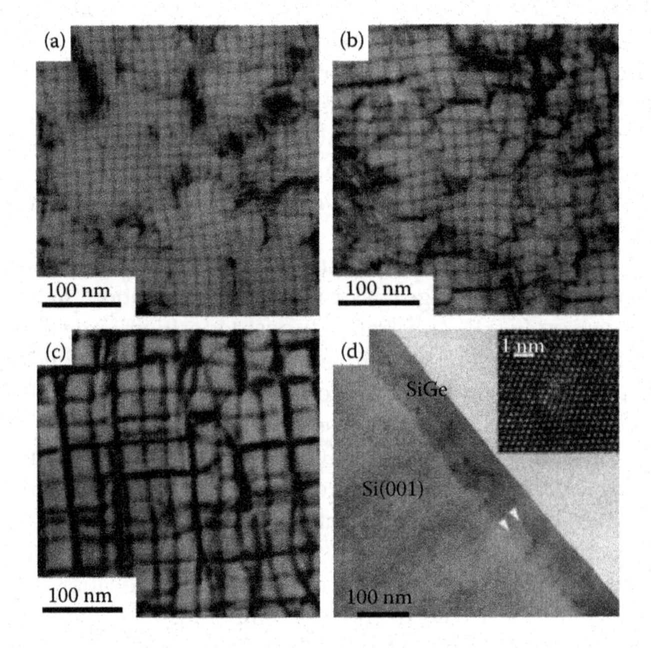

FIGURE 1.24
Plane-view TEM images of Si (17 nm)/Ge (35 nm)/Si(001) samples after HTA at (a) 950°C, (b) 1000°C, and (c) 1100°C. (d) Cross-sectional TEM image of the sample shown in (c) [36]. Inset shows a closeup high-resolution TEM image of an end-on dislocation indicated by a lower arrowhead.

When this Ge layer with pure-edge dislocation network is used as a template, we can control the dislocation structure of strain-relaxed $Si_{1-x}Ge_x$ thin layer on a Si(001) substrate (Sakai et al., 2005). After the HTA above 950°C for Si(17 nm)/Ge(35 nm)/Si(001) stacked structure, the solid-phase intermixing between Si and Ge takes place, and a $Si_{1-x}Ge_x$ alloy layer can be formed. Plane-view TEM images of samples after HTA at various temperatures are shown in Figure 1.24a–c. The regular dislocation network structure similar to Figure 1.23 can be observed even after HTA. Note that the period of pure-edge dislocations increases with the annealing temperature. Strain in the Ge layer on Si(001) substrate before PDA is almost fully relaxed with the introduction of pure-edge dislocations. The tensile strain is introduced with diffusion of Si atoms into the layer during the annealing because of the lattice constant of $Si_{1-x}Ge_x$ alloy is smaller than Ge. Then, some dislocations climb from the interface to the surface and escape out from the $Si_{1-x}Ge_x$ layer with solid-phase intermixing during HTA. As a result, the density of dislocations at the interface decreases and the period of dislocations decreases. We can also observe the stacked dislocation structure as shown with arrows in Figure 1.24d. This structure indicates the occurrence of interaction among climbing dislocations during annealing, resulting in an energetically preferred morphology.

1.4.3 Strain Structures in $Si_{1-x}Ge_x$ Hetero Layers

Engineering of strain structures with a submicron scale is a key issue to develop electronic and optoelectronic micro-devices with application of strain structure. Also, the characterization of strain structure in the microscopic scale is important to control submicron scale strain structures. There are various analysis methods for the estimation of local strain structure. Conversion beam electron diffraction (CBED) (Huang et al., 2006) or nanobeam diffraction (NBD) (Sato et al., 2010) based on TEM measurement has the highest spatial resolution. Recently, micro Raman spectroscopy has become very attractive for relatively high strain and spatial resolution (Jain et al., 1996). XRD measurement has the highest strain resolution and realizes the direct measurement of strain components without the destruction of the sample. However, the spatial resolution of conventional XRD measurement is generally poor compared with those of TEM and Raman methods. On the other hand, X-ray microdiffraction using a synchrotron radiation source provides high strain-resolution analysis ($\Delta\varepsilon$~10^{-5}) simultaneously with submicron space resolution (Matsui et al., 2002; Takeda et al., 2006). Microdiffraction is effective to evaluate the microscale local strain structure for the strain relaxed step-graded $Si_{1-x}Ge_x$ layers (Mooney et al., 1999; Eastman, 2001).

Here, we introduce the analysis of microscopic mosaic structure in a strain-relaxed $Si_{1-x}Ge_x$ layer due to 60° dislocations using microdiffraction measurement (Mochizuki et al., 2006). The microdiffraction was performed using hard X-ray source with the synchrotron radiation at the beamline BL13XU at the super photon ring-8 GeV (SPring-8) (Kimura, 2001). We can use a submicron size incident X-ray beam with enough intensity. The strain-relaxed $Si_{1-x}Ge_x$ layer was prepared with the two-step strain relaxation method (Sakai et al., 2001; Egawa et al., 2004). The strain in the $Si_{1-x}Ge_x$ layer is effectively relaxed with highly dense 60° dislocations at the first $Si_{1-x}Ge_x$ and a Si substrate. Figure 1.25a shows the XRD two-dimensional reciprocal space map (XRD-2DRSM) around Si 004 and $Si_{1-x}Ge_x$ 004 reciprocal lattice points using microdiffraction system. When there is a mosaicity in the $Si_{1-x}Ge_x$ layer, the lattice planes of individual crystalline domains are tilting to the substrate orientation as mentioned before. As a result, the Bragg reflection pattern from the $Si_{1-x}Ge_x$ layer broadens due to the mosaicity compared with that of a Si substrate. In addition, we

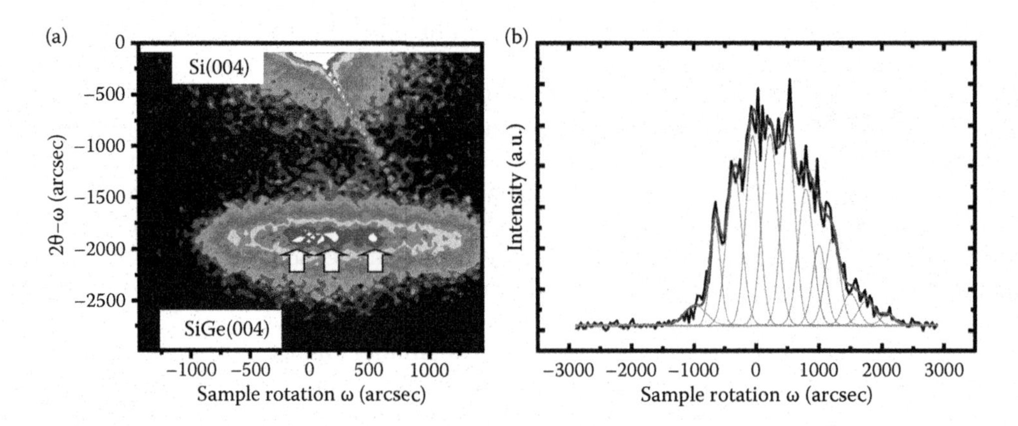

FIGURE 1.25
(See color insert.) (a) Two-dimensional reciprocal-space maps around the Si 004 and SiGe 004 Bragg reflections for the strain relaxed $Si_{1-x}Ge_x$/Si(001) sample prepared with two-step strain relaxation method [49]. (b) Cross-sectional profile at 2θ of –1890 arcsec of the SiGe 004 Bragg reflection of the sample shown in (a).

can clearly observe discrete peaks in the microdiffraction pattern of the $Si_{1-x}Ge_x$ layer. This microscopic structure cannot be observed with conventional XRD measurement with a submillimeter size X-ray beam. Figure 1.25b shows the cross-sectional profile at a 2θ of –1890 arcsec of the $Si_{1-x}Ge_x$ 004 Bragg reflection as shown in Figure 1.25a. A complex fine structure consisting of multipeaks can be clearly observed. There are several crystal domains tilting individually at different angles in the observed region, approximately $1.0 \times 9.1\ \mu m^2$, and each discrete diffraction peak comes from such a submicrometer-sized crystal domain. We can also estimate the size and tilting angles of each domain from the full-width at half maximum (FWHM) and positions, respectively, of each diffraction peak.

Figure 1.26 also shows the XRD-2DRSM using microdiffraction for the $Si_{1-x}Ge_x$/Si(001) sample whose strain is relaxed preferentially with pure-edge dislocations (Mochizuki, 2006). The strain relaxed $Si_{1-x}Ge_x$/Si(001) sample was prepared with HTA solid-phase intermixing method. Only single sharp peak can be observed like a 2D-RSM pattern with a

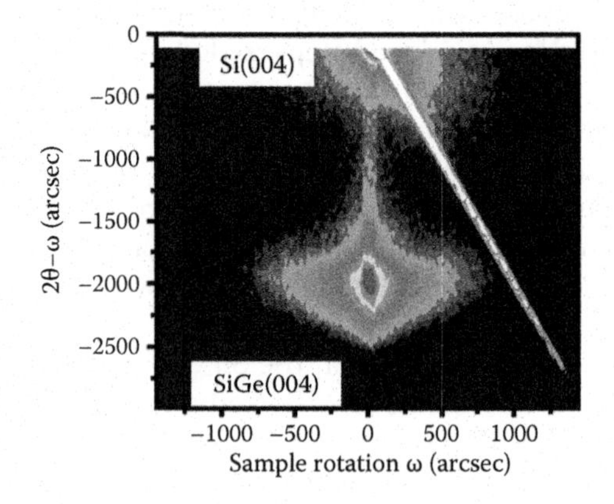

FIGURE 1.26
(See color insert.) Two-dimensional reciprocal-space maps around the Si 004 and SiGe 004 Bragg reflections for the strain relaxed $Si_{1-x}Ge_x$/Si(001) sample prepared with high temperature solid-phase mixing method [49].

conventional XRD measurement, meaning that there is no mosaicity in the $Si_{1-x}Ge_x$ layer even with a microscopic scale.

1.4.4 Strain Relaxation of $Ge_{1-x}Sn_x$ Hetero Layers

One of group IV semiconductor materials, $Ge_{1-x}Sn_x$ is an attractive material as a buffer layer for biaxially tensile-strained Ge (Takeuchi et al., 2008; Shimura et al., 2009; Fang et al., 2007) and a source/drain stressor for uniaxially compressive strained Ge channel (Vincent et al., 2011b). In addition, $Ge_{1-x}Sn_x$ alloy with a high Sn content over 10% promises to be direct-transition semiconductor from indirect-transition (Oguz et al., 1983; Jenkins and Dow, 1987). Direct transition $Ge_{1-x}Sn_x$ is very attractive for not only optoelectronic applications but also high electron mobility devices, because of relatively smaller effective electron mass of Γ-point (Menendez and Kouvetakis, 2004).

Engineering the strain and strain relaxation in $Ge_{1-x}Sn_x$ thin films on substrate is a key issue for realizing high crystalline quality of $Ge_{1-x}Sn_x$ alloy. There are some reports of the $Ge_{1-x}Sn_x$ growth with MBE (He and Atwater, 1996, 1997) and CVD method (Bauer et al., 2002; Kouvetakis et al., 2006). However, the Ge–Sn system is eutectic contrary to Si–Ge of complete solubility in the solid state, and there is a concern of Sn precipitation from $Ge_{1-x}Sn_x$ alloy because of a thermoequilibrium solid solubility as low as a few percentage of Sn in Ge. The strain relaxation of $Ge_{1-x}Sn_x$ layer on a Si or Ge substrate is mainly caused by propagation of misfit dislocations. In addition, the strain relaxation of a $Ge_{1-x}Sn_x$ layer also takes place by Sn precipitation. Therefore, it is required to control the strain relaxation of a $Ge_{1-x}Sn_x$ layer without Sn precipitation for the lattice constant engineering of $Ge_{1-x}Sn_x$ alloy.

Using virtual Ge (v-Ge) substrate is effective to achieve the strain relaxation of $Ge_{1-x}Sn_x$ with enhancing the propagation misfit dislocations at the interface between $Ge_{1-x}Sn_x$ and buffer layers (Takeuchi et al., 2007). Engineering the lattice mismatch between $Ge_{1-x}Sn_x$ and buffer layers is a key issue to properly control of the strain relaxation with avoiding Sn precipitation. There is a critical misfit strain, which gives an indication of the maximum value to suppress Sn precipitation at the interface (Takeuchi et al., 2008). Therefore, the control of misfit strain between $Ge_{1-x}Sn_x$ and buffer layers with compositionally step-graded $Ge_{1-x}Sn_x$ stack structure is effective to enhance the strain relaxation without Sn precipitation. Figure 1.27 shows cross-sectional TEM image of the $Ge_{1-x}Sn_x$ stacked layers with a compositionally step-graded method taking into account the critical misfit strain for Sn. Lateral propagation of misfit dislocations can be clearly observed at the top-$Ge_{1-z}Sn_z$/

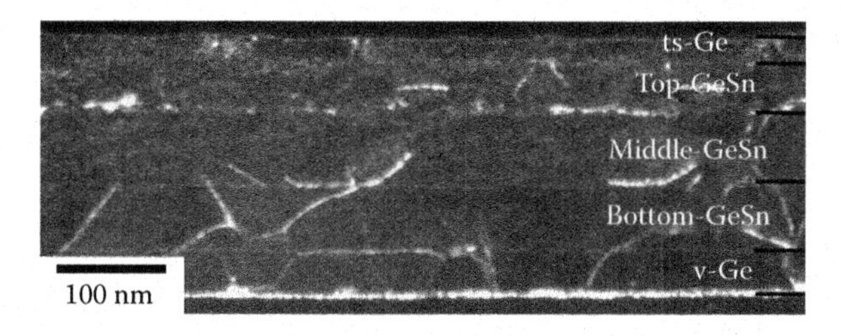

FIGURE 1.27
Cross-sectional TEM image of strained Ge/top-$Ge_{1-z}Sn_z$/middle-$Ge_{1-y}Sn_y$/bottom-$Ge_{1-x}Sn_x$/v-Ge sample prepared with compositionally graded growth method [5]. This image was taken by the weak-beam method at $g/3g$ with g_{004}.

middle-$Ge_{1-y}Sn_y$/bottom-$Ge_{1-x}Sn_x$ interfaces. It is highly likely that these misfit dislocations lying on the interfaces contribute to the strain relaxation of each layer.

In the conventional constant Sn content growth of $Ge_{1-x}Sn_x$, threading dislocations are often pinned with reactions that take place between dislocations during annealing. These pinned dislocations at the interface do not further contribute to the strain relaxation. Thus, strain relaxation is limited. On the other hand, in the step-graded method, since the upper layer takes over the pinned threading dislocations, the dislocations can propagate again at the interface due to misfit stress between the upper and lower layers. In addition, since the lower layer experiences subsequent annealing steps and takes the misfit stress caused by the upper layer, the strain relaxation of the lower layers enhances step by step.

We can also control the strain relaxation and Sn precipitations with the growth temperature of $Ge_{1-x}Sn_x$ layers. Figure 1.28 shows summaries of peak positions of $Ge_{1-x}Sn_x$ 224 reciprocal lattice points measured with XRD-2DRSM for $Ge_{1-x}Sn_x$/v-Ge substrate samples as grown and after PDA at 430~500°C (Shimura et al., 2010). In all $Ge_{1-x}Sn_x$ layers grown on v-Ge at as high as, 200°C, Sn contents are not higher than 5.5% due to the Sn precipitation after PDA. On the other hand, in a $Ge_{1-x}Sn_x$ layer grown at the lower temperature of 150°C, the Sn content and the degree of strain relaxation of $Ge_{1-x}Sn_x$ layers are achieved to the highest values. The Sn precipitation is effectively suppressed by enhancing the strain relaxation due to the propagation of misfit dislocations in the case of the lower temperature growth. The reason why the low temperature growth enhances strain relaxation of $Ge_{1-x}Sn_x$ layers can be deduced as follows: Lower temperature growth introduce many more point defects into epitaxial layers (Kasper et al., 1998). Point defects cause climb of dislocations and enhances annihilation of threading dislocation arms. This phenomenon enhances the strain relaxation of the $Ge_{1-x}Sn_x$ layer due to the fusion of dislocations and the creation of a new misfit dislocation at the interface. Furthermore, condensation of point defects in a $Ge_{1-x}Sn_x$ layer leads to the creation of prismatic dislocation loops. These dislocations also reduce the strain in $Ge_{1-x}Sn_x$ layers.

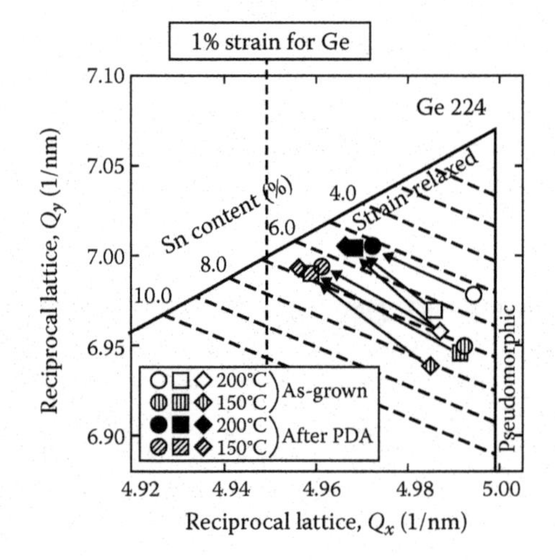

FIGURE 1.28
Summary of peak positions of the $Ge_{1-x}Sn_x$ 224 reciprocal lattice points estimated from XRD-2DRSM for $Ge_{1-x}Sn_x$ layers grown at 200°C and 150°C on virtual Ge substrates.

1.5 Summary

The strain and strain relaxations of group IV semiconductor materials, $Si_{1-x}Ge_x$, Ge, and $Ge_{1-x}Sn_x$, have been discussed. The strain relaxation behavior of heteroepitaxial layers is strongly related to the dislocation structure at the interface between the heteroepitaxial layer and the substrate. Designing and engineering dislocation structures with consideration of the Burgers vector, dislocation networks, critical misfit strain, and defect introduction, are critical issues to realize uniform and stable heteroepitaxial layers in global and local scales.

References

Acharya, N., V. Kirtikar, S. Shooshtarian, H. Doan, P.J. Timans, K.S. Balakrishnan, K.L. Knutson, 2001, Uniformity optimization techniques for rapid thermal processing systems, *Transactions on Semiconductor Manufacturing* 14, pp. 218–226.

Adachi, S., 2009, *Properties of Semiconductor Alloys: Group IV, III-V and II-VI Semiconductors*, Wiley & Sons, Chichester.

Agnello, P.D., T.O. Sedgwick, J. Cotte, 1993, Growth rate enhancement of heavy n- and p-type doped silicon deposited by atmospheric-pressure chemical vapor deposition at low temperatures, *J. Electrochem. Soc.* 140, pp. 2703–2709.

Akatsu, T., C. Deguet, L. Sanchez, F. Allibert, D. Rouchon, T. Signamarcheix, C. Richtarch, A. Boussagol, V. Loup, F. Mazen, J.-M. Hartmann, Y. Campidelli, L. Clavelier, F. Letertre, N. Kernevez, C. Mazure, 2006, Germanium-on-insulator (GeOI) substrates—a novel engineered substrate for future high performance devices, *Mater. Sci. Semicond. Proc.* 9, pp. 444–448.

Arguirov, T., M. Kittler, M. Oehme, N.V. Abrosimov, E. Kasper, J. Schulze, 2011, Room temperature direct band-gap emission from an unstrained Ge p-i-n LED on Si, *Solid State Phenom.* 178–179, pp. 25–30.

Bardeen, J., W.H. Brattain, 1947, Physical principles involved in transistor action, *Phys. Rev.* 75, pp. 1208–1225.

Barkai, A., Y. Chetrit, O. Cohen et al., 2007, Integrated silicon photonics for optical networks, *J. Opt. Netw.* 6, pp. 25–47.

Bauer, M., J. Taraci, J. Tolle et al., 2002, Ge-Sn semiconductors for band-gap and lattice engineering. *Appl. Phys. Lett.* 81, pp. 2992–2294.

Bauer, M., M. Oehme, M. Sauter, G. Eifler, E. Kasper, 2000, Time resolved reflectivity measurements of silicon solid phase epitaxial regrowth, *Thin Solid Films* 364, pp. 228–232.

Bauer, M., P. Brabant, T. Landin, 2008c, Germanium deposition, US#7,329,593; 7,479,443.

Bauer, M., P. Tomasini, C. Arena, 2004, Low temperature SiGe layer deposition with high [Ge], content using reduced pressure chemical vapor deposition from $SiCl_2H_2/GeH_4$ precursors, In *SiGe: Material, Processing and Devices, Proceedings of the First International Symposium*, edited by D. Harame et al., ECS, Pennington, NJ, pp. 1145–1152.

Bauer, M., S.G. Thomas, 2008, In-situ n-type doping during selective epitaxial growth of Si and Si:C alloys with high electrical dopant levels, *Proceedings of the 4th International SiGe Technology and Device Meeting (ISTDM)*, Hsinchu, Taiwan, May 11–14, pp. 18–19.

Bauer, M., S.G. Thomas, 2010a, Novel chemical precursors and novel CVD strategies enabling low temperature epitaxy of Si and Si:C alloys, *Thin Solid Films* 518, pp. S200–S208.

Bauer, M., S.G. Thomas, 2010b, Selective epitaxial growth (SEG) of highly doped Si:P on Source/Drain areas of NMOS devices using $Si_3H_8/PH_3/Cl_2$ chemistry, *ECS Trans.* 33, pp. 629–636.

Bauer, M., S.G. Thomas, 2010c, Selective epitaxial growth (SEG) of SiGe/Si using a Cl_2 based cyclic deposition/etch (CDE) process, *Proceedings of the 5th International SiGe Technology and Device Meeting (ISTDM)*, Stockholm, Sweden.
Bauer, M., S.G. Thomas, 2012a, Low temperature Si:C co-flow and hybrid process using Si_3H_8/Cl_2, *Thin Solid Films* 520, pp. 3133–3138.
Bauer, M., S.G. Thomas, 2012b, Low temperature selective epitaxial growth of SiCP on Si(110) oriented surfaces, *Thin Solid Films* 520, pp. 3144–3148.
Bauer, M., S.G. Thomas, 2012c, Low temperature catalyst enhanced etch process with high etch rate selectivity for amorphous silicon based alloys over single-crystalline silicon based alloys, *Thin Solid Films* 520, pp. 3139–3143.
Bauer, M., S. Zollner, N.D. Theodore, M. Canonico, P. Tomasini, B.-Y. Nguyen, C. Arena, 2005, Si_3H_8 based epitaxy of biaxially stressed silicon films doped with carbon and arsenic for CMOS applications, *MRS Symp. Proc.* 864, pp. 143–148.
Bauer, M., V. Machkaoutsan, C. Arena, 2007a, Highly tensile strained silicon–carbon alloys epitaxially grown into recessed source drain areas of NMOS devices, *Semicond. Sci. Technol.* 22, pp. S183–S187.
Bauer, M., V. Machkaoutsan, Y. Zhang, D. Weeks, J. Spear, S.G. Thomas, P. Verheyen, C. Kerner, F. Clemente, H. Bender, D. Shamiryan, R. Loo, A. Hikavyy, T. Hoffmann, P. Absil, S. Biesemans, 2008b, SiCP selective epitaxial growth in recessed source/drain regions yielding to drive current enhancement in n-channel MOSFET, *ECS Trans.*, 16, pp. 1001–1013.
Bauer, M., Y. Zhang, D. Weeks, P. Brabant, J. Italiano, V. Machkaoutsan, S.G. Thomas, 2008a, Throughput considerations for in-situ doped embedded silicon carbon stressor selectively grown into recessed source drain areas of NMOS devices, *ECS Trans.*, 13, pp. 287–298.
Bauer, M., Y. Zhang, D. Weeks, V. Machkaoutsan, S.G. Thomas, 2007b, Defect free embedded silicon carbon stressor selectively grown into recessed source drain areas of NMOS devices, *ECS Trans.* 6, pp. 419–427.
Bergmann, R.B., 1999, Crystalline Si thin-film solar cells: a review, *Appl. Phys. A* 69, pp. 187–194.
Bojarczuk, N.A., M. Copel, S. Guha, V. Narayanan, E.J. Preisler, F.M. Ross, H. Shang, 2003, Epitaxial silicon and germanium on buried insulator heterostructures and devices, *Appl. Phys. Lett.* 83, pp. 5443–5445.
Bramblett, T.R., Q. Lu, N.-E. Lee, N. Taylor, M.-A. Hasan, J.E. Greene, 1995, Ge(001) gas-source molecular beam epitaxy on Ge(001)2xl and Si(00l)2x1 from Ge2H6: Growth kinetics and surface roughening, *J. Appl. Phys.* 77, pp. 1505–1513.
Braunstein, R., 1963, Lattice vibration spectra of germanium-silicon alloys, *Phys. Rev.* 130, pp. 879–887.
Bruel, M., 1994, Process for the production of thin semiconductor material films, US Patent 5,374,564 patent application.
Canham, L.T., 1990, Silicon quantum wire array fabrication by electrochemical and chemical dissolution of wafers, *Appl. Phys. Lett.* 57, pp. 1046–1048.
Chaisakul, P., D. Marris-Morini, G. Isella, D. Chrastina, X. Le Roux, S. Edmond, J.-R.Coudevylle, E. Cassan, L. Vivien, 2011, Polarization dependence of quantum-confined Stark effect in Ge/SiGe quantum well planar waveguides, *Opt. Lett.* 36, pp. 1794–1796.
Chang, S.S., A. Sakai, 2004, Luminescence properties of spark-processed SiC, *Mater. Lett.* 58, pp. 1212–1217.
Chao, Y.-L., R. Scholz, M. Reiche, U. Gösele, J.C.S. Woo, 2006, Characteristics of germanium-on-insulators fabricated by wafer bonding and hydrogen-induced layer splitting, *Jpn. J. Appl. Phys.* 45, pp. 8565–8570.
Claeys, C., E. Simoen (Eds.), 2007, *Germanium-Based Technologies*, Elsevier, Amsterdam.
Cocorullo, G., F.G. Della Corte, I. Rendina, C. Minarini, A. Rubino, E. Terzini, 1996, Amorphous silicon waveguides and light modulators for integrated photonics by low-temperature plasma-enhanced chemical vapor deposition, *Opt. Lett.* 21, pp. 2002–2004.
D'Costa, V.R., 2009, *Electronic and Vibrational Properties of GeSn and SiGeSn Alloys*, VDM Verlag, Saarbrücken.

D'Costa, V.R., C.S. Cook, A.G. Birdwell, C.L. Littler, M. Canonico, S. Zollner, J. Kouvetakis, J. Menendez, 2006, Optical critical points of thin-film $Ge_{1-y}Sn_y$ alloys: a comparative $Ge_{1-y}Sn_y$/ $Ge_{1-x}Si_x$ study, *Phys. Rev. B*, 73, pp. 125207-1–125207-16.

D'Costa, V.R., Y.-Y. Fang, J. Tolle, J. Kouvetakis, J. Menendez, 2010, Ternary GeSiSn alloys: new opportunities for strain and band gap engineering using group-IV semiconductors, *Thin Solid Films* 518, pp. 2531–2537.

Dabrowski, J., S. Miyazaki, S. Inumiya, G. Kozlowski, G. Lippert, G. Lupina, Y. Nara, H.-J. Müssig, A. Ohta, Y. Pei, 2009, The influence of defects and impurities on electrical properties of high-k dielectrics, *Mater. Sci. Forum* 608, pp. 55–109.

Dakin, J.P., R.G.W. Brown (Eds.), 2006, *Handbook of Optoelectronics*, CRC Press, Boca Raton, FL.

Depuydt, B., A. Theuwis, I. Romandic, 2006, Germanium: from the first application of Czochralski crystal growth to large diameter dislocation-free wafers, *Mater. Sci. Semicond. Proc.* 9, pp. 437–443.

Depuydt, B., M. De Jonghe, W. De Baets, I. Romandic, Theuwis, C. Quaeyhaegens, C. Deguet, T. Akatsu, F. Letertre, 2007, Germanium materials, In *Germanium-Based Technologies: From Materials to Devices*, edited by Claeys, C., Simoen, E., Elsevier, Amsterdam.

Di Cioccio, L., E. Jalaguir, F. Letertre, 2004, Compound semiconductor heterostructures by Smart Cut: SiC on insulator, QUASIC substrates, InP and GaAs heterostructures on silicon, In *Wafer Bonding: Applications and Technology*, edited by Alexe, M., Gösele, U., Springer, Berlin.

Dismukes, J.P., L. Ekstrom, R.J. Paff, 1964, Lattice parameter and density in germanium-silicon alloys, *J. Phys. Chem.* 68, pp. 3021–3027.

Dutartre, D., 2001, Silicon epitaxy: new applications. In *Silicon Epitaxy, Semiconductors and Semimetals*, Vol. 72, edited by D. Crippa, D.L. Rode, M. Masi, Academic Press, New York, pp. 397–457.

Eaglesham, D.J., M. Cerullo, 1991, Low-temperature growth of Ge on Si(100), *Appl. Phys. Lett.* 58, pp. 2276–2278.

Eastman, D.E., C.B. Stagarescu, G. Xu et al., 2002, Observation of columnar microstructure in step-graded Si1-xGex/Si films using high-resolution X-ray microdiffraction, *Phys. Rev. Lett.* 88, pp. 156101.

Eberl, K., S.S. Iyer, S. Zollner, J.C. Tsang, F.K. Legoues, 1992, Growth and strain compensation effects in the ternary $Si_{1-x-y}Ge_xC_y$ alloy system, *Appl. Phys. Lett.* 60, pp. 3033–3035.

Edmond, J., H. Kong, G. Negley, M. Leonard, K. Doverspike, W. Weeks, A. Suvorov, D. Waltz, C. Carter, 1998, SiC-based UV photodiodes and light emitting diodes, In *SiC Materials and Devices*, edited by Park, Y.S., Academic Press, San Diego, CA.

Egawa, T., A. Sakai, T. Yamamoto et al., 2004, Strain-relaxation mechanisms of SiGe layers formed by two-step growth on Si(001) substrates, *Appl. Surf. Sci.* 224, pp. 104–107.

El Kurdi, M., T.-P. Ngo, X. Checoury, S. David, P. Boucaud, J.F. Damlencourt, O. Kermarrec, Y. Campidelli, D. Bensahel, 2008, GeOI photonic crystal cavities probed by room-temperature photoluminescence, *5th IEEE Intern. Conf. on Group IV Photonics, IEEE*, Cardiff.

EMIS (Ed.), 1988, Properties of Silicon, INSPEC, London.

EMIS (Ed.), 1989, Properties of Amorphous Silicon, INSPEC, London.

Fan, W.J., 2010, Band structures and optical gain of direct-bandgap tensile strained Ge/$Ge_{1-x-y}Si_xSn_y$ type I quantum wells, *2010 Int. Conf. on Electronic Devices, Systems and Applications, IEEE*, Kuala Lumpur.

Fang, Y.Y., J. Tolle, R. Roucka et al., 2007, Perfectly tetragonal, tensile-strained Ge on Ge1-ySny buffered Si(100), *Appl. Phys. Lett.* 90, pp. 061915.

Fathpour, S., B. Jalali (Eds.), 2011, *Silicon Photonics for Telecommunications and Biomedicine*, CRC Press, Boca Raton, FL.

Fischer, P.R., M. Bauer, S.R.A. Van Aerde, T.G.M. Oosterlaken, M. Yan, W.A. Verweij, B.W.M. Bozon, P.M. Zagwijn, 2006, Low temperature silcore deposition of undoped and doped silicon films, *ECS Trans.* 3, pp. 203–215.

Fitzgerald, E.A., S.B. Samavedam, 1997, Line, point and surface defect morphology of graded, relaxed GeSi alloys on Si substrates, *Thin Solid Films* 294, pp. 3–10.

Fitzgerald, E.A., Y.-H. Xie, D. Monroe et al., 1992, Relaxed GexSi1-x structures for III–V integration with Si and high mobility two-dimensional electron gases in Si, *J. Vac. Sci. Technol. B* 10, pp. 1807–1819.

Gencarelli, F., B. Vincent, L. Souriau, O. Richard, W. Vandervorst, R. Loo, M. Caymax, M. Heyns, 2012, Low-temperature Ge and GeSn chemical vapor deposition using Ge_2H_6, *Thin Solid Films* 520, pp. 3211–3215.

Ghani, T., M. Armstrong, C. Auth et al., 2003, *IEDM Tech. Dig.*, p. 978.

Ghidini, G., F.W. Smith, 1984, *J. Electrochem. Soc.* 131, pp. 2924–2928.

Gnutzmann U., K. Clausecker, 1974, Theory of direct optical-transitions in an optical indirect semiconductor with a superlattice structure, *Appl. Phys.* 3, pp. 9–14.

Gong, Y., S. Yerci, R. Li, L. Dal Negro, J. Vuckovic, 2009, Enhanced light emission from erbium doped silicon nitride in plasmonic metal-insulator-metal structures, *Opt. Express.* 17, pp. 20642–20650.

Gösele, U., V. Lehmann, 1994, Porous silicon quantum spronge structures: formation mechanism, preparation methods and some properties, In *Porous Silicon*, edited by Feng, Z.C., Tsu, R., World Sci. Publ., Singapore.

Goto, K., S. Satoh, H. Ohta et al., 2004, *IEDM Tech. Dig.*, p. 209.

Green, M.A., E.-C. Cho, Y. Cho, Y. Huang, E. Pink, T. Trupke, A. Lin, T. Fangsuwannarak, T. Puzzer, G. Conibeer, R. Corkish, 2005, All-silicon tandem cells based on "artificial" semiconductor synthesised using silicon quantum dots in a dielectric matrix, *20th Eur. Photovoltaic Solar Energy Conference*, Barcelona.

Grimmeiss, H., E. Kasper, 2009, Today's mainstream microelectronics—the result of technological, market and human enterprise, *Adv. Electronic Mater.* 608, pp. 1–16, TransTechPublications, Zurich.

Gruetzmacher, D.A., T.O. Sedgwick, A. Powell, M. Tejwani, S.S. Iyer, J. Cotte, F. Cardone, 1993, Ge segregation in SiGe/Si heterostructures and its dependence on deposition technique and growth atmosphere, *Appl. Phys. Lett.* 63, pp. 2531–2533.

Hamakawa, Y., 1999, Recent advances in amorphous and microcrystalline silicon basis devices for optoelectronic applications, *Appl. Surf. Sci.* 142, pp. 215–226.

Harke, A., T. Lipka, J. Amthor, O. Horn, M. Krause, J. Müller, 2008, Amorphous silicon 3-D tapers for Si photonic wires fabricated with shadow masks, *IEEE Photon. Technol. Lett.* 20, pp. 1452–1454.

Harris, G.L., 1995, *Properties of Silicon Carbide*, IEE, INSPEC, London.

He, G., H.A. Atwater, 1996, Synthesis of epitaxial SnxGe1-x alloy films by ion-assisted molecular beam epitaxy, *Appl. Phys. Lett.* 68, pp. 664–666.

He, G., H.A. Atwater, 1997, Interband transitions in SnxGe1-x alloys, *Phys. Rev. Lett.* 79, pp. 1937–1940.

Hirth, J.P., J. Lothe, 1992, *Theory of Dislocations*, Krieger, Florida.

Hollander, B., S. Mantl, R. Liedtke, 1999, Enhanced strain relaxation of epitaxial SiGe layers on Si(100) after H(+) ion implantation, *Nucl. Instrum. Methods B* 148, pp. 200–210.

Houssa, M. (Ed.), 2004, *High-k Gate Dielectrics*, Institute of Physics, Bristol.

Huang, J., M.J. Kim, P.R. Chidambaram et al., 2006, Probing nanoscale local lattice strains in advanced Si complementary metal-oxide-semiconductor devices, *Appl. Phys. Lett.* 89, pp. 063114.

Huang, Y.M.E., 2010, Optoelectronic materials, *Mater. Sci. Forum*, 663–665.

Huff, H.R., 2002, An electronics division retrospective (1952–2002) and future opportunities in the twenty-first century, *J. Electrochem. Soc.* 149, pp. S35–S58.

Hull, R. (Ed.), 1999, *Properties of Crystalline Silicon*, INSPEC, London.

Hull, R., J.C. Bean, J.M. Bonar et al., 1990, Enhanced strain relaxation in Si/GexSi1−x/Si heterostructures via point-defect concentrations introduced by ion implantation, *Appl. Phys. Lett.* 56, pp. 2445–2447.

Humlicek, J., 1995, Optical spectroscopy of SiGe, In *Properties of Strained and Relaxed Silicon Germanium*, edited by Kasper, E., INSPEC, London.

Isaacson, D.M., G. Taraschi, A.J. Pitera, N. Ariel, E.A. Fitzgerald, T.A. Langdo, 2005, *Adv. Mater. Micro- Nano-Syst.*, http://hdl.handle.net/1721.1/7373.

Ishikawa, Y., K. Wada, D.D. Cannon, J. Liu, H.-C. Luan, L.C. Kimerling, 2003, Strain-induced band gap shrinkage in Ge grown on Si substrate, *Appl. Phys. Lett.* 82, pp. 2044–2046.

Ismail, K., B.S. Meyerson, S. Rishton, J. Chu, S. Nelson, J. Nocera, 1992, High-transconductance n-type Si/SiGe modulation-doped field-effect transistors, *IEEE Elec. Dev. Lett.* 13, pp. 229–231.

Ismail, K., F.K. LeGoues, K.L. Saenger et al., 1994, Identification of a mobility-limiting scattering mechanism in modulation-doped Si/SiGe heterostructures, *Phys. Rev. Lett.* 73, pp. 3447–3450.

Izumi, K., M. Doken, H. Ariyoshi, 1978, CMOS devices fabricated on buried SiO2 layers formed by oxygen implantation into silicon, *Electron. Lett.* 14, pp. 593–594.

Jain, S.C., H.E. Maes, K. Pinardi, I. De Wolf, 1996, Stresses and strains in lattice-mismatched stripes, quantum wires, quantum dots, and substrates in Si technology, *J. Appl. Phys.* 79, pp. 8145–8165.

Jalali, B., S. Fathpour, 2006, Silicon photonics, *J. Lightwave Technol.* 24, pp. 4600–4615.

Jalali, B., S. Yegnanarayanan, T. Yoon, T. Yoshimoto, I. Rendina, F. Coppinger, 1998, Advances in silicon-on-insulator optoelectronics, *IEEE J. Select. Top. Quant. Electron.* 4, pp. 938–947.

Jellison, G.E., T.E. Haynes, H.H. Burke, 1993, Optical functions of silicon-germanium alloys determined using spectroscopic ellipsometry, *Opt. Mater.* 2, pp. 105–117.

Jenkins, D.W., J.D. Dow, 1987, Electronic properties of metastable GexSn1-x alloys, *Phys. Rev. B* 36, pp. 7994–8000.

Jorke, H., 1995, Segregation of Ge and dopant atoms during growth of SiGe layers, In *Properties of Strained and Relaxed Silicon Germanium*, EMIS Datareviews, No. 12, INSPEC (IEE), London, pp. 80–189.

Kanoh, H., O. Sugiura, M. Matsmura, 1993, Chemical vapor deposition of amorphous silicon using tetrasilane, *Jpn. J. Appl. Phys.* 32, pp. 2613–2619.

Kasap, S., P. Capper (Eds.), 2006, *Springer Handbook of Electronic and Photonic Materials*, Springer, Heidelberg.

Kaschel, M., M. Schmid, M. Oehme, J. Werner, J. Schulze, 2011, Germanium photodetectors on Silicon-on-insulator grown with differential molecular beam epitaxy in silicon wells, *Solid State Electron.* 60, pp. 105–111.

Kasper, E. (Ed.), 1995, *Properties of Strained and Relaxed Silicon Germanium*, INSPEC, London.

Kasper, E., H. Kibbel, 1984, Ultrahochvakuum-Epitaxie von Silizium, *Vakuum-Technik.* 33, pp. 13–22.

Kasper, E., H.-J. Herzog, 1976, UHV-Epitaxie von Si und SiGe auf Si-Substraten, *Wiss. Ber. AEG-Telefunken.* 49, pp. 213–228.

Kasper, E., H.-J. Herzog, K. Wörner, K., 1987, Monolithic integration using differential Si-MBE, *J. Crystal Growth* 81, pp. 458–462.

Kasper, E., J. Yu, 2011, *Silicon-Based Photonics*, Pan Stanford Publ., Singapore.

Kasper, E., K. Lyutovich (Eds.), 2000, *Silicon Germanium and SiGe:Carbon*, INSPEC, London.

Kasper, E., K. Lyutovich, M. Bauer, M. Oehme, 1998, New virtual substrate concept for vertical MOS transistors, *Thin Solid Films* 336, pp. 319–322.

Kasper, E., M. Oehme, 2008, High speed germanium detectors on Si, *Phys. Stat. Sol. (C).* 5, pp. 3144–3149.

Kimura, S., Y. Kagoshima, T. Koyama et al., 2004, Article *AIP Conf. Proc.* 705, pp. 1275.

Kittler, M., M. Reiche, T. Mchedlidze, T. Arguirov, G. Jia, W. Seifert, S. Suckow, T. Wilhelm, 2008, Stark effect at dislocations in silicon for modulation of a 1.5 μm light emitter, In *Silicon Photonics III*, edited by Kubby, J.A., Reed, G.T., SPIE, San Jose.

Kittler, M., M. Reiche, X. Yu, T. Arguirov, O.F. Vyvenko, W. Seifert, T. Mchedlidze, G. Jia, T. Wilhelm, 2006, 1.5 μm emission from a silicon MOS-LED based on a dislocation network, *IEDM Tech. Dig.* 2006, pp. 845–848.

Kittler, M., X. Yu, T. Mchedlidze, T. Arguirov, O. Vyvenko, W. Seifert, M. Reiche, T. Wilhelm, M. Seibt, O. Voß, A. Wolff, W. Fritzsche, 2007, Regular dislocation networks in silicon as a tool for nanostructure devices used in optics, biology, and electronics, *Small* 3, pp. 964–973.

Klinger, S., M. Berroth, M. Kaschel, M. Oehme, E. Kasper, 2009, Ge on Si p-i-n photodiodes with a 3-dB bandwidth of 49 GHz, *IEEE Photon. Technol. Lett.* 21, pp. 920–922.

Kouvetakis, J., J. Menendez, A.V.G. Chizmeshya, 2006, Tin-based group IV semiconductors: new platforms for opto- and microelectronics, *Ann. Rev. Mater. Res.* 36, pp. 497–554.

Kvam, E.P., D.M. Maher, C.J. Humphreys, 1990, Variation of dislocation morphology with strain in GexSi1-x epilayers on (100)Si, *J. Mater. Res.* 5, pp. 1900–1907.

Landolt-Boernstein, 2007, *New Series, Vol. III34C3, Semiconductor Quantum Structures: Optical Proprties of Group IV Semiconductors*, Springer Verlag.

Langdo, T.A., M.T. Currie, A. Lochtefeld, 2003, SiGe-free strained Si on insulator by wafer bonding and layer transfer, *Appl. Phys. Lett.* 82, pp. 4256–4258.

Lasky, J.B., 1985, Wafer bonding for silicon-on-insulator technologies, *Appl. Phys. Lett.* 48, pp. 78–80.
LeGoues, F.K., B.S. Meyerson, J.F. Morar, P.D. Kirchner, 1992, Mechanism and conditions for anomalous strain relaxation in graded thin films and superlattices, *J. Appl. Phys.* 71, pp. 4230–4243.
Lehmann, V., 2002, *The Electrochemistry of Silicon*, Wiley-VCH, Weinheim.
Liang, T.K., H.K. Tsang, 2005, Integrated polarization beam splitter in high index contrast silicon-on-insulator waveguides, *IEEE Photon. Technol. Lett.* 17, pp. 1–3.
Liao, L., D. Samara-Rubio, M. Morse et al., 2005, High speed Mach Zehnder modulator, *Opt. Express.* 13, pp. 3129–3135.
Liu, A., R. Jones, L. Liao et al., 2004, A high-speed silicon optical modulator based on a metal-oxide-semiconductor capacitor, *Nature* 427, pp. 615–618.
Liu, J., X. Sun, D. Pan, X. Wang, L.C. Kimerling, T.L. Koch, J. Michel, 2007, Tensile-strained, n-type Ge as a gain medium for monolithic laser integration on Si, *Opt. Express.* 15, pp. 11272–11277.
Liu, J., X. Sun, R. Camacho-Aguilera, L.C. Kimerling, J. Michel, 2010, Ge-on Si laser operating at room temperature, *Opt. Lett.* 35, pp. 679–681.
Luan et al., 1999, High-quality Ge epilayers on Si with low threading-dislocation densities, *Appl. Phys. Lett.* 75, pp. 2909–2911.
Luft, W., Y.S. Tsuo (Eds.), 1993, *Hydrogenated Amorphous Silicon Alloy Deposition Processes*, Marcel Dekker, London.
Madelung, O., M. Schulz, H. Weiss (Eds.), 1982, *Semiconductors*, Springer-Verlag, Berlin.
Manasevit, H.M., W.I. Simpson, 1964, Single-crystal silicon on a sapphire substrate, *J. Appl. Phys.* 35, pp. 1349–1351.
Masuda, H., K. Fukuda, 1995, Ordered metal nanohole arrays made by a two-step replication of honeycomb structures of anodic alumina, *Science* 268, pp. 1466–1468.
Matsui, J., Y. Tsusaka, K. Yokoyama et al., 2002, Microscopic strain analysis of semiconductor crystals using a synchrotron X-ray microbeam, *J. Cryst. Growth* 237–239, pp. 317–323.
Mchedlidze, T., T. Arguirov, S. Kouteva-Arguirova, M. Kittler, R. Rölver, B. Berghoff, D.L.B.S. Bätzner, 2008, Light-induced solid-to-solid phase transformation in Si nanolayers of Si-SiO_2 multiple quantum wells, *Phys. Rev. B.* 77, pp. 1613041-1–161304-4.
Menendez, J., J. Kouvetakis, 2004, Type-I Ge/Ge1-x-ySixSny strained-layer heterostructures with a direct Ge bandgap, *Appl. Phys. Lett.* 85, pp. 1175–1177.
Meunier-Beillard, P., M. Caymax, K. Van Nieuwenhuysen, G. Doumen, B. Brijs, M. Hopstaken, L. Geenen, W. Vandervorst, 2004, N_2 as carrier gas: an alternative to H_2 for enhanced epitaxy of Si, SiGe and SiGe:C, *Appl. Surf. Sci.* 224, pp. 31–35.
Meyer, D.J., 2001, Si-based alloys: SiGe and SiGe:C, In *Silicon Epitaxy, Semiconductors and Semimetals*, Vol. 72, edited by D. Crippa, D.L. Rode, M. Masi, Academic Press, New York, pp. 345–395.
Meyerson, B.S., J.F. Morar, F.K. LeGoues, 1988, Cooperative growth phenomena in silicon/germanium low-temperature epitaxy, *Appl. Phys. Lett.* 53, pp. 2555–2557.
Mii, Y.J., Y.H. Xie, E.A. Fitzgerald et al., 1991, Extremely high electron mobility in Si/GexSi1−x structures grown by molecular beam epitaxy, *Appl. Phys. Lett.* 59, pp. 1611–1613.
Mitchell, T.O., 2000, Growth and characterization of epitaxial silicon carbon random alloys on (100) silicon, PhD thesis, Publication No. AAI9961932, ISBN: 9780599657953, Stanford University, Stanford, CA.
Mochizuki, S., A. Sakai, N. Taoka et al., 2006, Local strain in SiGe/Si heterostructures analyzed by X-ray microdiffraction, *Thin Solid Films* 508, pp. 128–131.
Mooney, P.M., 1996, Strain relaxation and dislocations in SiGe/Si structures, *Mater. Sci. Eng. Rep.* 17, pp. 105–146.
Mooney, P.M., F.K. LeGoues, J. Tersoff, J.O. Chu, 1994, Nucleation of dislocations in SiGe layers grown on (001)Si, *J. Appl. Phys.* 75, pp. 3968–3977.
Mooney, P.M., J.L. Jordan-Sweet, I.C. Noyan, S.K. Kaldor, P.C. Wang, 1999, Observation of local tilted regions in strain-relaxed SiGe/Si buffer layers using x-ray microdiffraction, *Appl. Phys. Lett.* 74, pp. 726–728.

Nakaharai, S., T. Tezuka, N. Sugiyama, Y. Moriyama, S. Takagi, 2003, Characterization of 7-nm-thick strained Ge-on-insulator layer fabricated by Ge-condensation technique, *Appl. Phys. Lett.* 83, pp. 3516–3518.

Narayan, J., S. Sharan, 1991, Mechanism of formation of 60° and 90° misfit dislocations in semiconductor heterostructures, *Mater. Sci. Eng. B.* 10, pp. 261–267.

Ngo, T.P., M. El Kurdi, X. Checoury, P. Boucaud, J.F. Damiencourt, O. Kermarrec, D. Bensahel, 2008, Two-dimensional photonic crystals with germanium on insulator obtained by a condensation method, *Appl. Phys. Lett.* 93, pp. 241112-1–24112-3.

Nishisaka, M., Y. Hamasaki, O. Shirata, T. Asano, 2004, Cross-hatch related oxidation and its impact on performance of strained-Si MOSFETs, *Jpn. J. Appl. Phys.* 43, pp. 1886–1890.

Oehme, M., J. Werner, E. Kasper, M. Jutzi, M. Berroth, 2006, High bandwidth Ge p-i-n photodetector integrated on Si, *Appl. Phys. Lett.* 89, pp. 071117.

Oehme, M., J. Werner, M. Gollhofer et al., 2011, Room-temperature electroluminescence from GeSn light-emitting pin diodes on si, *IEEE Photon. Technol. Lett.* 23, pp. 1751–1753.

Oehme M., J. Werner, M. Kaschel, O. Kirfel, E. Kasper, 2008a, Germanium waveguide photodetectors integrated on silicon with MBE, *Thin Solid Films*, 517, pp. 137–139.

Oehme, M., J. Werner, O. Kirfel, E. Kasper, 2008b, MBE growth of SiGe with high Ge content for optical applications, *Appl. Surf. Sci.* 254, pp. 6238–6241.

Oehme, M., M. Bauer, C.P. Parry, G. Eifler, E. Kasper, 2000, Carbon segregation in silicon, *Thin Solid Films* 380, pp. 75–77.

Oehme, M., M. Kaschel, J. Werner, O. Kirfel, M. Schmid, B. Bahouchi, E. Kasper J. Schulze, 2010, Germanium on silicon photodetectors with broad spectral range, *J. Electrochem. Soc.* 157, pp. H144–H148.

Oguz, S., W. Paul, T.F. Deutsch, B.-Y. Tsaur, D.V. Murphy, 1983, Synthesis of metastable, semiconducting Ge–Sn alloys by pulsed UV laser crystallization, *Appl. Phys. Lett.* 43, pp. 848–850.

Olesinski, R.W., G.J. Abbaschian, 1984, The Ge–Si (germanium-silicon) system, *Bull. Alloy Phase Diagrams* 5, pp. 180–183.

Osten, H.J., M. Kim, K. Pressel, P. Zaumseil, 1996, Substitutional versus interstitial carbon incorporation during pseudomorphic growth of $Si_{1-y}C_y$ on Si(001). *J. Appl. Phys.* 80, pp. 6711–6715.

Osten, H.J., J. Griesche, S. Scalese, 1999, Substitutional carbon incorporation in epitaxial $Si_{1-y}C_y$ alloys on Si(001) grown by molecular beam epitaxy, *Appl. Phys. Lett.* 74, pp. 836–838.

Palik, E.D. (ed.), 1998, *Handbook of Optical Constants of Solids*, Elsevier, New York.

Park, N.M., S.H. Kim, G.Y. Sung, S.J. Park, 2002, Growth and size control of amorphous silicon quantum dots using SiH4/N2 plasma, *Chem. Vapor Deposit* 8, pp. 254–256.

Pavesi, L., R. Turan (Eds.), 2010, *Silicon Nanocrystals*, Wiley-VCH, Weinheim.

People, R., S.A. Jackson, 1990, Structurally induced states from strain and confinement, In *Strained-Layer Superlattices: Physics*, edited by Pearsall, T.P., Academic Press, Boston.

Perry, C.H., F. Lu, F. Namavar, 1993, Raman scattering studies of $Si_{1-x}Ge_x$ epitaxial layers grown by atmospheric pressure chemical vapor deposition, *Solid State Commun.* 88, pp. 613–617.

Pickering, C., R.T. Carline, D.J. Robbins, W.Y. Leong, S.J. Barnett, A.D. Pitt, A.G. Cullis, 1993, Spectroscopic ellipsometry characterization of strained and relaxed $Si_{1-x}Ge_x$ epitaxial layers, *J. Appl. Phys.* 73, pp. 239–250.

Pollak, F.H., 1990, Effects of homogeneous strain on the electronic and vibrational levels in semiconductors, In *Strained-Layer Superlattices: Physics*, edited by Pearsall, T.P., Academic Press, Boston.

Polleux, J.L., C. Rumelhard, 2000, Optical absorption coefficient determination and physical modelling of strained SiGe/Si photodetectors, *IEEE Int. Symp. High Performance Electron Devices for Microwave and Optoelectronic Applications, IEEE.*

Quillec, M. (Ed.), 1996, *Materials for Optoelectronics*, Springer, Berlin.

Radu, I., M. Reiche, M. Zoberbier, M. Gabriel, U. Gösele, 2005, Low-temperature wafer bonding via DBD surface activation, In *Semiconductor Wafer Bonding YIII: Science, Technology, and Applications*, edited by Hobart, K.D., Bengtsson, S., Baumgart, H., Suga, T., Hunt, C.E., Electrochemical Society, Quebec.

Rao, S., F.G. Della Corte, C. Summonte, 2010, Low-loss amorphous silicon waveguides grown by PECVD on indium tin oxide, *J. Eur. Opt. Soc.* 5, pp. 10039s-1–10039s-7.

Reed, G.T., A.P. Knights, 2004, *Silicon Photonics: An Introduction*, Wiley & Sons, Chichester.

Reiche, M., 2009, Wafer bonding in silicon electronics, *Phys. Stat. Sol. C.* 6, pp. 633–644.

Reiche, M., M. Kittler, 2011, Structure and properties of dislocations in silicon, In *Crystalline Silicon*, edited by Basu, S., InTech, Rijeka.

Reiche, M., O. Moutanabbir, J. Hoentschel, U. Gösele, S. Flachowsky, M. Horstmann, 2010, Strained silicon devices, *Solid State Phenom.* 156–158, pp. 61–68.

Reiche, M., O. Moutanabbir, P. Storck, F. Laube, R. Scholz, B. Holländer, D. Buca, S. Mantl, 2011, $Si_{1-x}Ge_x$ alloys ($0.05 < x < 0.95$): strain, composition, and optical properties, *J. Appl. Phys.*

Rickman, A.G., G.T. Reed, 1994, Silicon-on-insulator optical rib waveguides: loss, mode characteristics, bends, and y-junctions, *IEE Proc. Optoelectron.* 141, pp. 391–393.

Sakai, A., K. Sugimoto, T. Yamamoto et al., 2001, Reduction of threading dislocation density in SiGe layers on Si (001) using a two-step strain-relaxation procedure, *Appl. Phys. Lett.* 79, pp. 3398–3400.

Sakai, A., N. Taoka, O. Nakatsuka, S. Zaima, Y. Yasuda, 2005, Pure-edge dislocation network for strain-relaxed SiGe/Si(001) systems, *Appl. Phys. Lett.* 86, pp. 221916.

Sakai, A., T. Tatsumi, K. Aoyama, 1997, Growth of strain-relaxed Ge films on Si(001) surfaces, *Appl. Phys. Lett.* 71, pp. 3510–3512.

Sato, T., H. Matsumoto, K. Nakano et al., 2010, *J. Phys.* 241, p. 012014.

Sawano, K., N. Usami, K. Arimoto, S. Koh, K. Nakagawa, Y. Shiraki, 2005, Observation of strain field fluctuation in SiGe-relaxed buffer layers and its influence on overgrown structures, *Mater. Sci. Semicond. Process* 8, pp. 177–180.

Sawano, K., S. Koh, T. Shiraki, 2004, Fabrication of high-quality strain-relaxed thin SiGe layers on ion-implanted Si substrates, *Appl. Phys. Lett.* 85, pp. 2514–2516.

Sawano, K., S. Koh, Y. Shiraki, N. Usami, K. Nakagawa, 2003, In-plane strain fluctuation in strained-Si/SiGe heterostructures, *Appl. Phys. Lett.* 83, pp. 4339–4341.

Schäffler, F., 2001, Silicon-germanium ($Si_{1-x}Ge_x$), In *Properties of Advanced Semiconductor Materials: GaN, AlN, InN, BN, SiC, SiGe*, edited by Levinshtein, M.E., Rumyantsev, S.L., Shur, M.S., Wiley & Sons, New York.

Schmid, M., M. Oehme, M. Kaschel, J. Werner, E. Kasper, J. Schulze, 2010, Franz-Keldysh effect in germanium on silicon p-i-n photodetectors, *7th International Conference on Group IV Photonics, LEOS*, Beijing, China, pp. 329–331.

Schmidt, J., A. Merkle, R. Brendel, B. Hoex, M.C.M. Van De Sanden, W.M.M. Kessels, 2008, surface passivation of high-efficiency silicon solar cells by atomic-layer-deposited Al_2O_3, *Prog. Photovolt. Res. Appl.* 16, pp. 461–466.

Schubert, M.F., F. Rana, 2007, SiGeC/Si electrooptic modulators, *J. Lightwave Technol.* 25, pp. 866–874.

Schulze, J., M. Oehme, J. Werner, 2011, MBE grown Ge/Si p-i-n layer sequence for photonic devices, *Thin Solid Films* 520, pp. 3259–3261.

Shimbo, M., K. Furukawa, K. Fukuda, K. Tanzawa, 1986, Silicon-to silicon direct bonding method, *J. Appl. Phys.* 60, pp. 2987–2989.

Shimura, Y., N. Tsutsui, O. Nakatsuka, A. Sakai, S. Zaima, 2009, *Jpn. J. Appl. Phys.* 48, p. 04C130.

Shimura, Y., N. Tsutsui, O. Nakatsuka, A. Sakai, S. Zaima, 2010, Low temperature growth of Ge(1-x)Sn(x) buffer layers for tensile-strained Ge layers, *Thin Solid Films* 518, pp. S2–S5.

Shinriki, M., K. Chung, S. Hasaka, P. Brabant, H. He, T.N. Adam, D. Sadana, 2012, Gas phase particle formation and elimination on Si (100) in low temperature reduced pressure chemical vapor deposition silicon-based epitaxial layers, *Thin Solid Films* 520, pp. 3190–3194.

Smith, F.W., G. Ghidini, 1982, Reaction of oxygen with Si(111) and (100): critical conditions for the growth of SiO_2, *J. Electrochem. Soc.* 129, pp. 1300–1306.

Sopori, B., 2003, Silicon nitride processing for control of optical and electronic properties of silicon solar cells, *J. Electron. Mater.* 32, pp. 1034–1042.

Sopori, B., Y. Zhang, N.M. Ravindra, 2001, Silicon device processing in H-ambients: H-diffusion mechanisms and influence on electronic properties, *J. Electron. Mater.* 30, pp. 1616–1627.

Soref, R., 2010, Silicon photonics: a review of recent literature, *Silicon* 2, pp. 1–6.

Soref, R.A., 1991, Optical band gap of the ternary semiconductor $Si_{1-x-y}Ge_xC_y$, *J. Appl. Phys.* 70, pp. 2470–2472.
Soref, R.A., 1996, silicon-based group IV heterostructures for optoelectronic applications, *J. Vac. Sci. Technol. A* 14, pp. 913–918.
Soref, R.A., B.R. Bennett, 1987, Electrooptical effects in silicon, *IEEE J. Quantum Electron.* 23, pp. 123–129.
Soref, R.A., J. Kouvetakis, J. Tolle, J. Menendez, V.R. D'Costa, 2007, Advances in SiGeSn technology, *J. Mater. Res.* 22, pp. 3281–3291.
Soref, R.A., J.P. Lorenzo, 1986, All-silicon active and passive guidedwave components for $\lambda = 1.3$ and 1.6 μm, *IEEE J. Quantum Electron.* 22, pp. 873–879.
Splett, A., T. Zinke, K. Petermann, E. Kasper, H. Kibbel, H.-J. Herzog, H. Presting, 1994, Integration of waveguides and photodetectors in SiGe for 1.3 μm operation, *IEEE Photon. Technol. Lett.* 6, pp. 59–61.
Street, R.A. (Ed.), 1991, *Hydrogenated Amorphous Silicon*, Cambridge University Press, Cambridge.
Sturm, J.C., K.H. Chung, N. Yao, E. Sanchez, K.K. Singh, D. Carlson, S. Kuppurao, 2007, Chemical vapor deposition epitaxy of silicon and silicon-carbon alloys at high rates and low temperatures using neopentasilane, *ECS Trans.* 6, pp. 429–436.
Sugii, N., K. Nakagawa, S. Yamaguchi, M. Miyao, 1999, Role of Si1-xGex buffer layer on mobility enhancement in a strained-Si n-channel metal-oxide-semiconductor field-effect transistor, *Appl. Phys. Lett.* 75, pp. 2948–2950.
Sun, G., R.A. Soref, H.H. Cheng, 2010, Design of a Si-based lattice-matched room-temperature GeSn/GeSiSn multi-quantum-well mid-infrared laser diode, *Opt. Express* 18.
Sze, S.M., 1969, *Physics of Semiconductor Devices*, Wiley-Interscience.
Taillaert, D., W. Bogaerts W., Baets R., 2003, Efficient coupling between submicron SOI-waveguides and single-mode fibres, *Proc. Symp. IEEE/LEOS Benelux Chapter*, Enschede, pp. 289–292.
Taillaert, D., F. van Laere, M. Ayre et al., 2006, Grating coupler for coupling between optical fibres and nanophotonic waveguides, *Jpn. J. Appl. Phys.* 45, pp. 6071–6077.
Takeda, S., S. Kimura, O. Sakata, A. Sakai, 2006, Development of high-angular-resolution microdiffraction system for reciprocal space map measurements, *Jpn. J. Appl. Phys.* 45, L1054.
Takeuchi, S., A. Sakai, K. Yamamoto, O. Nakatsuka, M. Ogawa, S. Zaima, 2007, Growth and structure evaluation of strain-relaxed Ge1-xSnx buffer layers grown on various types of substrates, *Semicond. Sci. Technol.* 22, p. S231–S234.
Takeuchi, S., Y. Shimura, O. Nakatsuka, S. Zaima, M. Ogawa, A. Sakai, 2008, Growth of highly strain-relaxed Ge(1-x)Sn(x)/virtual Ge by a Sn precipitation controlled compositionally step-graded method, *Appl. Phys. Lett.* 92, p. 231916.
Tanaka, K., E. Maruyama, T. Shimada, H. Okamoto (Eds.), 1999, *Amorphous Silicon*, Wiley & Sons, New York.
Taoka, N., A. Sakai, T. Egawa, O. Nakatsuka, S. Zaima, Y. Yasuda, 2005, Growth and characterization of strain-relaxed SiGe buffer layers on Si(001) substrates with pure-edge misfit dislocations, *Mater. Sci. Semicond. Process* 8, pp. 131–135.
Tezuka, T., N. Sugiyama, T. Mizuno, M. Suzuki, S. Takagi, 2001, A novel fabrication technique of ultrathin and relaxed SiGe buffer layers with high Ge fraction for sub-100 run strained silicon-on-insulator MOSFETs, *Jpn. J. Appl. Phys.* 40, pp. 2866–2874.
Thomas, S.G., P. Tomasini, M. Bauer, B. Vyne, Y. Zhang, M. Givens, J. Devrajan, S. Koester, I. Lauer, 2010, Enabling Moore's law beyond CMOS technologies through heteroepitaxy, *Thin Solid Films* 518, pp. S53–S56.
Timbrell, P.Y., J.-M. Baribeau, D.J. Lockwood, J.P. McCaffrey, 1990, An annealing study of strain relaxation and dislocation generation in Si1-xGex/Si heteroepitaxy, *J. Appl. Phys.* 67, pp. 6292–6300.
Tiwari, S., M.V. Fischetti, P.M. Mooney, J.J. Welser, 1997, *IEDM Tech. Dig.*, p. 939.
Tomasini, P., M. Bauer, N. Cody, C. Arena, 2006, Kinetics of Si incorporation into a Ge matrix for $Si_{1-x}Ge_x$ layers grown by chemical vapor deposition, *J. Appl. Phys.* 99, 074904.
Tracy, C.J., P. Fejes, N.D. Theodore, P. Maniar, E. Johnson, A.J. Lamm, A.M. Paler, I.J. Malik, P. Ong, 2004, Germanium-on-insulator substrates by wafer bonding, *J. Electron. Mater.* 33, pp. 886–892.

Tsunomura, Y., Y. Yoshimine, M. Taguchi, T. Baba, T. Kinoshita, H. Kanno, H. Sakata, E. Maruyama, M. Tanaka, 2009, Twenty-two percent efficiency HIT solar cell, *Solar Energy Mater. Solar Cells* 93, pp. 670–673.

Uhlir, L.T., 1956, Electrolytic shaping of germanium and silicon, *Bell Syst. Tech. J.* 35, pp. 333–347.

Vincent, B., F. Gencarelli, H. Bender, C. Merckling, B. Douhard, D.H. Petersen, O. Hansen, H.H. Henrichsen, J. Meersschaut, W. Vandervorst, M. Heyns, R. Loo, M. Caymax, 2011a, Undoped and in-situ B doped GeSn epitaxial growth on Ge by atmospheric pressure-chemical vapor deposition, *Appl. Phys. Lett.* 99, 152103.

Vincent, B., W. Vandervorst, M. Caymax, R. Loo, 2009, Influence of Si precursor on Ge segregation during ultrathin Si reduced pressure chemical vapor deposition on Ge, *Appl. Phys. Lett.* 95, 262112.

Vincent, B., Y. Shimura, S. Takeuchi et al., 2011b, Characterization of GeSn materials for future Ge pMOSFETs source/drain stressors, *Microelectron. Eng.* 88, pp. 342–346.

Von Bardeleben, H.J., D. Stievenard, A. Grosman, C. Ortega, J. Siejka, 1993, Defects in porous p-type Si: an electron-paramagnetic-resonance study, *Phys. Rev. B* 47, pp. 10899–10902.

Vyvenko, O., O. Krüger, M. Kittler, 2000, Cross-sectional electron-beam-induced current analysis of the passivation of extended defects in cast multicrystalline silicon by remote hydrogen plasma treatment, *Appl. Phys. Lett.* 76, pp. 697–699.

Wada, K., 2008, A new approach of electronics and photonics convergence on Si CMOS platform: how to reduce device diversity of photonics for integration, *Adv. Opt. Technol.* pp. 1–7.

Wang Y., Z. Lin, X. Cheng, C. Zhang, F. Gao, F. Zhang, 2004, Scattering loss in silicon-on-insulator rib waveguides fabricated by inductively coupled plasma reactive ion etching, *Appl. Phys. Lett.* 85, pp. 3995–3996.

Wang, Z.M., A. Neogi (Eds.), 2010, *Nanoscale Photonics and Optoelectronics*, Springer, New York.

Weber, J., M.I. Alonso, 1989, Near-band-gap photoluminescence of Si–Ge alloys, *Phys. Rev. B.* 40, pp. 5683–5693.

Welser, J., J.L. Hoyt, J.F. Gibbons, 1992, *IEDM Tech. Dig.*, p. 1000.

Wong, H., 2004, Thermal stability and electronic structure of hafnium and zirconium oxide films for nanoscale MOS device applications, *Proc. 5th IEEE Internat. Caracas Conf. on Devices, Circuits, and Systems, IEEE*, Dominican Republic.

Yamamoto, T., A. Sakai, T. Egawa, O. Nakatsuka, S. Zaima, Y. Yasuda, 2004, Dislocation structures and strain-relaxation in SiGe buffer layers on Si(001) substrates with an ultra-thin Ge interlayer, *Appl. Surf. Sci.* 224, pp. 108–112.

Yang, M., C. Chang, J.C. Sturm, 2002, Band alignments and band gaps in SiGeC/Si structures, In *Silicon-Germanium Carbon Alloys: Growth, Properties and Applications*, edited by Pantelides, S.T., Zollner, S., Taylor & Francis, New York.

Yildiz, M., S. Dost, B. Lent, 2005, Growth of bulk SiGe single crystals by liquid phase diffusion, *J. Cryst. Growth.* 280, pp. 151–160.

Yin, T., A.M. Pappu, A.B. Apsel, 2006, Low-cost, high-efficiency, and high-speed SiGe phototransistors in commercial BiCMOS, *IEEE Photon. Technol. Lett.* 18, pp. 55–57.

Yokoyama, M., T. Yasuda, H. Takagi, H. Yamada, Y. Urabe, N. Fukuhara, M. Hata, M. Sugiyama, Y. Nakano, M. Takenaka, S.I. Takagi, 2010, High quality thin body III-V-on-insulator channel layer transfer on Si wafer using direct wafer bonding, *ECS Trans.* 33, pp. 391–401.

Yu, J., E. Kasper, M. Oehme, 2006, 1.55-μm resonant cavity enhanced photodiode based on MBE grown Ge quantum dots, *Thin Solid Films.* 508, pp. 396–398.

Yu, P.Y., M. Cardona, 1996, *Fundamentals of Semiconductors*, Springer, Berlin.

Zachai R., K. Eberl, G. Abstreiter, E. Kasper, H. Kibbel, 1990, Direct transition energies in strained 10-monolyer Ge/Si superlattices, *Phys. Rev. Lett.* 64, pp. 1055–1058.

Zacharias, M., L.X. Yi, J. Heitmann, R. Scholz, M. Reiche, U. Gösele, 2003, Size-controlled Si nanocrystals for photonic and electronic applications, *Solid State Phenom.* 94, pp. 95–104.

Zimmermann, H., 2009, *Integrated Silicon Optoelectronics*, Springer, Heidelberg.

2

Guided Light in Silicon-Based Materials

Koji Yamada, Tai Tsuchizawa, Hiroshi Fukuda, Christian Koos, Joerg Pfeifle, Jens H. Schmid, Pavel Cheben, Przemek J. Bock, and Andrew P. Knights

CONTENTS

2.1 Introduction

Koji Yamada

Silicon photonics is an emerging technology for optical telecommunications and optical interconnects in microelectronics. Based on state-of-art silicon semiconductor technology, silicon photonics would provide an inexpensive highly integrated photonic platform. Similar to existing silica-based and III–V semiconductor–based photonic systems, the silicon-based photonic system requires optical waveguides to build and connect photonic devices. For this purpose, the waveguides must have features that allow accommodating passive and dynamic photonic devices, such as wavelength filters and modulators. The waveguide materials must also be able to enable active functions, such as light emission and detection. Furthermore, for monolithic integration of photonic and electronic devices, a very important advantage of silicon photonics, the waveguides should lay on silicon substrates or be fabricated together with silicon electronic devices. This requirement is also very important to integrate both passive and active photonic devices, which possibly include electronic driving circuitry. Of course, the waveguides should guarantee a sufficiently low propagation loss and small bending radius for implementing these photonic functions on a chip, whose typical size is a few square centimeters.

These requirements are very difficult to meet with other than silicon-based waveguides. For example, the fabrication process for conventional silica-based waveguides requires high-temperatures exceeding 1000°C, which would seriously damage the electronic structure of modulators, detectors, and other electronic devices. Moreover, silica-based waveguides have a large bending radius on the order of millimeters, making it impossible to integrate photonic devices on a square centimeter chip. The III–V compound semiconductor-based waveguides and photonic devices have geometries smaller than those in the silica-based system; however, on a silicon substrate, it is very difficult to epitaxially grow the high-quality III–V materials needed for the construction of practical photonic devices. Etching and other fabrication procedures are completely different from those in silicon processes. Polymer waveguides made of organic materials cause less damage to electronic devices. However, their use is limited to the uppermost layers formed after the electronic circuits are completed or to other regions separated from the electronic devices, because they cannot withstand the temperatures used in electronic device fabrication. Moreover, they might not withstand the high running temperatures of silicon electronic devices. Thus, a simple conclusion can be drawn: silicon-based waveguides are preferred over other types of waveguides for silicon photonics.

This chapter mainly reviews silicon-based waveguides, which have advanced to a level where they can satisfy most of these severe requirements. The wavelength region of interest for propagating electromagnetic waves is that from 1.2 to 1.7 μm, which is widely used in telecommunications and optical interconnection applications. Section 2.2 reviews the silicon photonic wire waveguide as a standard waveguide platform for high-density photonic integration. Thanks to its small geometrical dimensions, such as a submicrometer core and micrometer bending, the waveguide can provide various ultracompact photonic devices and facilitate their integration in large numbers over a small area. The small geometrical dimensions would also be advantageous in implementing high-speed optoelectronic functions. However, the small geometrical dimensions of photonic wire waveguides give rise to serious problems in optical coupling with external fibers. The problem will be discussed in Chapter 3, where efficient coupling structures are presented. The small geometrical dimensions also give rise to serious polarization dependence. This problem is discussed in Section 2.3, where various polarization manipulation techniques and solutions for polarization independent photonic systems are reviewed. Section 2.4 reviews two types of nanostructured silicon waveguide. One is the slot waveguide, which consists of two strips of silicon that enclose a subwavelength low-index slot region. Discontinuities of the electric field at the high-index-contrast interfaces between silicon and low-index slot material lead to strong interactions between guided light and the material inside the slot. The other is the subwavelength grating waveguide, which consists of periodic dielectric structures with a periodicity well below the wavelength of light. In these waveguides, various useful optical effects, such as electro-optic, thermo-optic, and Kerr nonlinear effects, can be efficiently implemented. Sections 2.5 and 2.6 discuss waveguides with micrometer-size cores, in which the relaxed fabrication tolerance reduces the propagation loss, the fiber coupling loss, and the polarization dependence. Section 2.5 covers the conventional silicon rib waveguide, which, due to its relaxed fabrication tolerance, has already been applied in some commercial products. Section 2.6 covers medium index silica-based waveguides fabricated with a low-temperature process, which enables integration of low-loss, high-performance passive devices and high-speed, compact dynamic/active devices.

These waveguides can be fabricated on a silicon substrate without damaging other silicon/germanium–based opto-electronic structures, and combinations of these waveguides provide solutions to practical silicon-based photonic systems.

2.2 Silicon Photonic Wire Waveguides

Koji Yamada

2.2.1 Guided Modes

A schematic of the cross-sectional structure of a silicon photonic wire waveguide is shown in Figure 2.1a. The waveguide consist of a silicon core and silica-based cladding. The core dimension should be determined so that a single-mode condition is fulfilled. In this waveguide, the refractive index contrast between the core and the cladding is as large as 40%, which allows very tight confinement of light in the waveguide core. The core dimension is around half of the wavelength of electromagnetic waves propagating in silicon, or a few hundred nanometers for 1310- to 1550-nm telecommunications-band infrared light. Generally, the core shape is made flat along the substrate to reduce the etching depth in

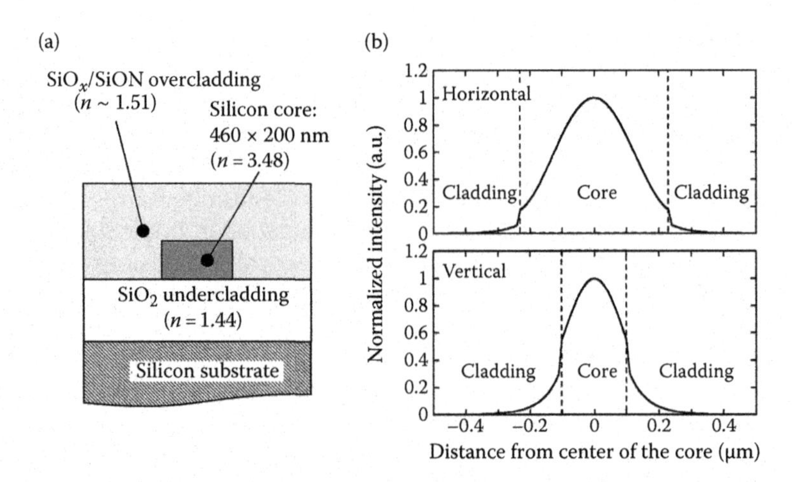

FIGURE 2.1
(a) Cross-sectional structure and (b) optical intensity distribution of a typical silicon photonic wire waveguide. (From Koji Yamada, *Silicon Photonics II, Silicon Photonic Wire Waveguides: Fundamentals and Applications*, Springer-Verlag, 2011. With permission.)

practical fabrications. In many cases, the height of the core is typically half of the width, and a typical core geometry is a 400 × 200-nm^2 rectangle.

Figure 2.2a,b show numerically calculated effective indices n_{eff} of guided modes for 1550 nm infrared light in various core geometries [1]. Calculations were performed by the film mode matching method (FMM) [2] and the indices of silicon and silica were set at 3.477 and 1.444, respectively. The mode notations are taken from [3], where the E^x and E^y modes represent the transverse electric (TE)-like and transverse magnetic (TM)-like modes, respectively. As shown in Figure 2.2a for waveguides of a silicon core thickness of 200 nm, single-mode conditions are fulfilled when the core width is narrower than 460 nm for TE-like guided modes, where the dominant electric field is parallel to the substrate. The field profile of the TE-like fundamental mode is shown in Figure 2.1b for a 460 × 200-nm^2 core [1]. For a TM-like mode, where the dominant electric field is perpendicular to the substrate, the single-mode condition is fulfilled by a core larger than that for the TE-like mode. The effective indices of TE and TM fundamental modes show a large difference. In other words, the, 200-nm-thick flat core produces large polarization dependence. For waveguides with 300-nm-thick, silicon core widths satisfying single-mode conditions are narrower than those for 200-nm-thick silicon core, as shown in Figure 2.2b. In a 300 × 300-nm^2 core, the refractive indices are identical for the TE and TM fundamental modes; that is, the polarization dependence is eliminated.

Figure 2.3 shows the calculated wavelength dependence of the effective refractive indices for the waveguides with a 400 × 200-nm^2 core [1]. Calculations were performed by the FMM and the material dispersions of refractive indices were considered. As shown in Figure 2.3, the single-mode condition is violated in the wavelength region below 1420 nm for the TE-like mode. For the 1310-nm telecommunications wavelength band, a smaller core should therefore be used to satisfy the single-mode condition.

As shown in Figure 2.2, the effective indices of silicon photonic wire waveguides are extremely sensitive to the core geometry. The group index n_g, which is an essential parameter in designing delay-based devices, such as optical filters, is also affected significantly by the core geometry. Figure 2.4 shows calculated group indices and their sensitivities to

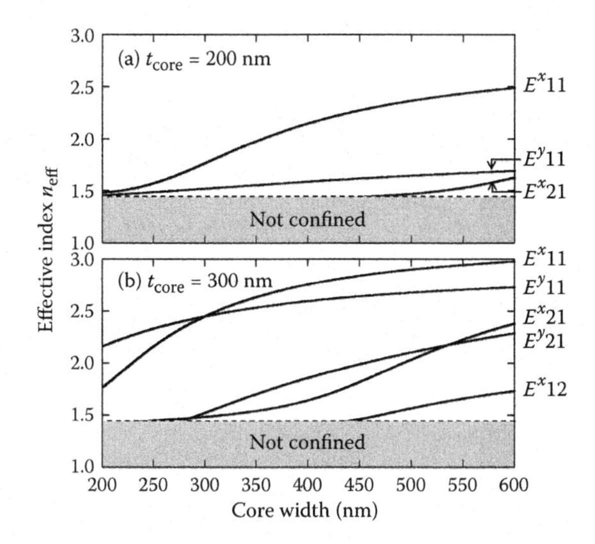

FIGURE 2.2
Core width dependence of the effective indices of silicon photonic wire waveguides for core thickness of (a) 200 nm and (b) 300 nm. (From Koji Yamada, *Silicon Photonics II, Silicon Photonic Wire Waveguides: Fundamentals and Applications*, Springer-Verlag, 2011. With permission.)

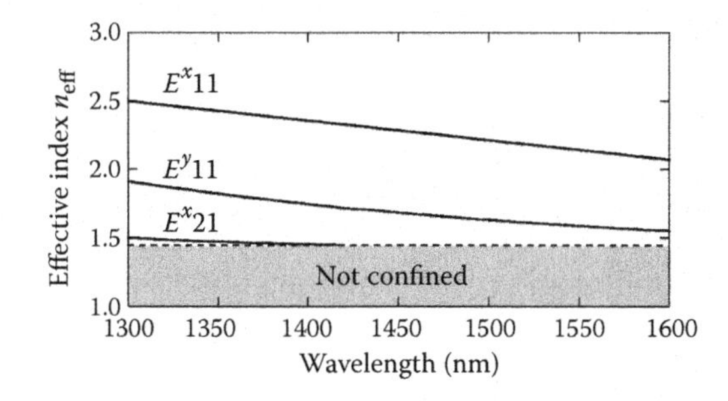

FIGURE 2.3
Wavelength dependence of the effective indices of silicon photonic wire waveguides. (From Koji Yamada, *Silicon Photonics II, Silicon Photonic Wire Waveguides: Fundamentals and Applications*, Springer-Verlag, 2011. With permission.)

the core width $dn_g/n_g dw$, where w is the waveguide width [1]. For TE-like modes, for which most of the photonic functions are designed, the sensitivity to the core width $dn_g/n_g dw$ is around 2×10^{-4} nm^{-1} for a 400 × 200 nm^2 core. For wavelength filters for dense wavelength division multiplexing (DWDM), the group index error should be of the order of 1×10^{-4} or less. The index restriction corresponds to a core width accuracy of 0.5 nm or less, which is essentially unattainable with current microfabrication technology. Fortunately, there are optimum geometries giving very low sensitivities to the core width. For example, waveguides with a 385 × 200-nm^2 core are robust against errors in core width, as are waveguides with a very wide core. In an arrayed waveguide grating (AWG) filter, waveguides with 750 × 200-nm^2 cores are used to reduce phase errors due to the variation of core width [4]. When we use such a wide-core waveguide, however, higher-order modes stimulated in bending and other asymmetric structures become a concern.

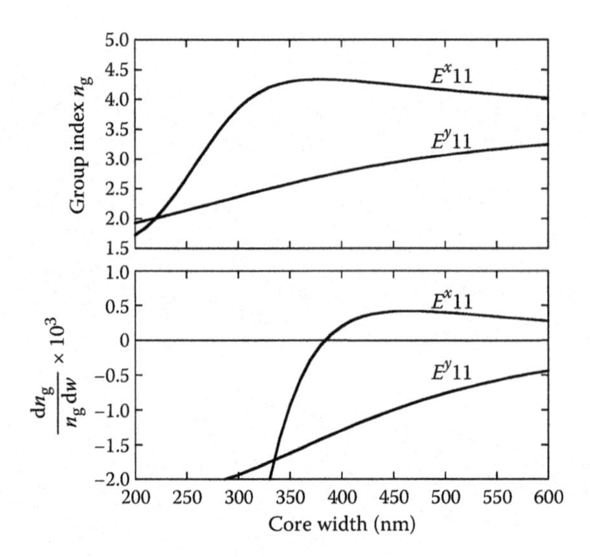

FIGURE 2.4
Core width dependence of the group indices and their derivatives for waveguides with 200-nm-thick cores. (From Koji Yamada, *Silicon Photonics II, Silicon Photonic Wire Waveguides: Fundamentals and Applications*, Springer-Verlag, 2011. With permission.)

Figure 2.4 also suggests that the structural birefringence is incredibly large and that the problem of polarization dependence in a silicon photonic wire waveguide is practically unsolvable. In a waveguide with a 400 × 200-nm^2 core, the group indices are 4.33 for the TE-like fundamental mode and 2.78 for the TM-like fundamental mode. The difference in the group indices gives a polarization mode dispersion of 51.7 ps/cm, which seriously limits the applicable bandwidth in high-speed data transmission. The polarization-dependent wavelength in delay-based filter devices, such as AWGs and ring resonators, would be incredibly large. A square core would not be a solution to this problem because the group index is very sensitive to the core geometry. In other words, polarization diversity, in which each polarization is processed independently, is necessary for eliminating the polarization dependence in photonic devices based on silicon photonic wire waveguides.

The propagation loss of photonic wire waveguides is mainly determined by scattering due to surface roughness of the core. The effect of the surface roughness on the scattering loss in dielectric waveguides has been theoretically studied and formulated by Payne and Lacey [5], and the upper bound of the scattering loss α_{max}, as given in [6], is expressed as

$$\alpha_{max} = \sigma^2\kappa/k_0 d^4 n_1 \tag{2.1}$$

where σ, k_0, d, and n_1 are the root-mean-square roughness, wave vector of the light in vacuum, half-width of the core, and effective index of a silicon slab with the same thickness as the core, respectively. κ is a parameter that depends on the waveguide geometry and the statistical distribution (Gaussian, exponential, etc.) of the roughness. Fortunately, its value is of the order of unity for most practical waveguide geometries [5]. Thus, the scattering loss is inversely proportional to the fourth power of d. In other words, it will seriously increase in photonic wire waveguides with an ultrasmall core. A roughness of only 5 nm, for instance, would cause a 60-dB/cm scattering loss in a 400-nm-wide core made of a 200-nm-thick silicon slab whose effective index is 2.7. To achieve a practical scattering loss of a few decibels per centimeter, the surface roughness should be about 1 nm or less.

2.2.2 Fabrication

A silicon photonic wire waveguide is typically fabricated as follows. First, a resist mask layer is formed on an SOI substrate. A hard mask, which is often made of SiO_2 or silicon nitride, may be added to improve the selectivity. Next, waveguide patterns are defined by electron beam (EB) lithography or excimer laser deep ultraviolet (DUV) lithography [7], which are capable of forming 100-nm patterns. Ordinarily, EB and DUV lithography technologies are used in the fabrication of electronic circuits, for which they are optimized for patterning of straight and intersecting line patterns. Therefore, no consideration has been given to curves and roughness in the pattern edges, which are important factors in fabricating low-loss optical waveguides. To reduce propagation losses of the waveguides, it is necessary to reduce the edge roughness to around 1 nm or less. This means that particular care must be taken in the data preparation for EB shots or DUV masks [8]. The writing speed of the EB lithography must also be considered in practical fabrication. For practical purposes, it is probably necessary to use EB lithography with a variable-shaped beam.

After resist development and SiO_2 etching for a hard mask, the silicon core is formed by low-pressure plasma etching with an electron-cyclotron resonance plasma or inductive-coupled plasma. Finally, an overcladding layer is formed with a SiO_2-based material or polymer resin material. To avoid damaging the silicon layer, the cladding layers must be deposited by a low-temperature process, such as the plasma-enhanced chemical vapor deposition (PE-CVD) method [9]. In particular, for waveguides associated with an electronic structure, it is essential to use a low-temperature process, so as not to damage the electronic devices. Since a silicon photonic wire waveguide has a very small mode profile, spot-size conversion is essential for connecting the waveguide to external components, such as single-mode optical fibers. Highly efficient spot size converters (SSCs) with silicon reverse adiabatic tapers [10, 11] or grating couplers [12] have been developed. These SSCs are compatible with the fabrication process described above.

Figure 2.5a shows a scanning electron microscope (SEM) image of the core of a silicon photonic wire waveguide with a cross section of 400 × 200 nm^2 [11]. The geometrical shape closely matches the design values, and the perpendicularity of the sidewalls is also very good. Figure 2.5b shows a photograph of the 80-nm-wide taper tip in the inverse-taper

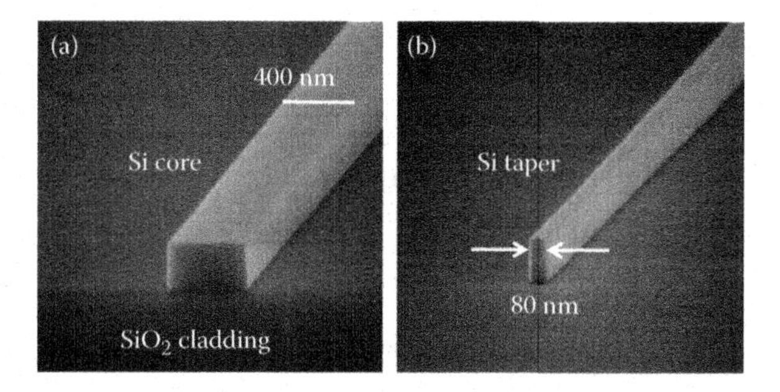

FIGURE 2.5
SEM images of a silicon photonic wire waveguide system. (a) Core of silicon photonic wire waveguide. (b) Silicon taper for SSC.

SSC. The taper may be covered with a silica waveguide core to improve coupling efficiency to fibers. More details on fabrication are provided in Chapter 3.

2.2.3 Propagation Performance

Figure 2.6 shows a typical transmission loss of silicon photonic wire waveguides with inverse-taper SSCs [13]. High-NA optical fibers with 4.3-μm mode field diameter (MFD) are used for external coupling. As shown in Figure 2.6, the propagation loss for the TE-like mode has been improved to around 1 dB/cm. In waveguides with flat cores, the propagation losses for TM-like modes are generally better than those for TE-like modes. Oxidation of the core sidewalls may further reduce the propagation losses [14]. The propagation loss of around 1 dB/cm is already at a practical level, since photonic devices based on silicon photonic wire waveguides typically require propagation length of 1 mm or less. In addition to the sidewall roughness of the core, the core width also affects propagation losses. Figure 2.7 shows the relation between measured propagation loss and core width [15]. As shown, the propagation loss is reduced by increasing the core width, because the effect of sidewall roughness is reduced in a wide core. When the core width exceeds 460 nm, the waveguide can also guide a higher order mode, which may degrade the performances of photonic devices.

The coupling loss between optical fiber and a silicon photonic wire waveguide is represented by the intercept of the vertical axis in Figure 2.6. The loss value at the intercept

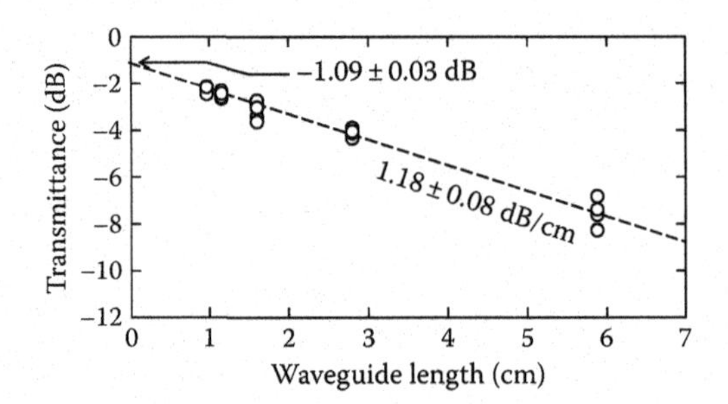

FIGURE 2.6
Measured transmission loss of a silicon photonic wire waveguide with SSCs.

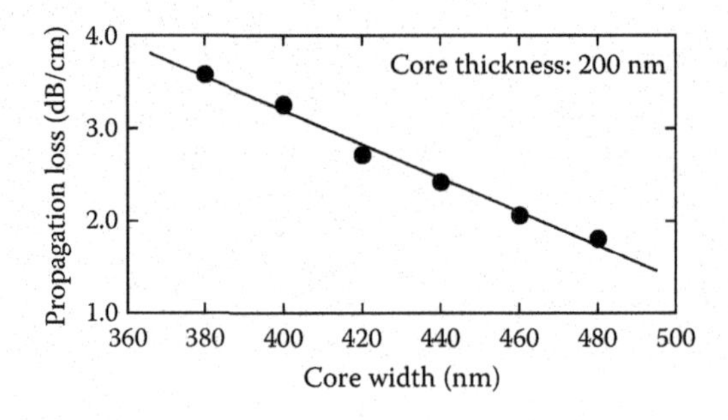

FIGURE 2.7
Relation between measured propagation loss and core width.

includes two waveguide/fiber interfaces; therefore, in this case, one interface has a 0.5-dB coupling loss at a wavelength of 1550 nm.

Figure 2.8 shows the transmission spectrum of a silicon photonic wire waveguide with SSCs [1]. The spectrum remains flat over a 200-nm wide bandwidth, and no absorption dip is observed. The flat spectrum means that the SiO_2-based material used in the SSC does not contain impurities with N–H bonds. Although absorption by residual O–H bonds exists at wavelengths of around 1400 nm, the resulting losses are not large. It is also possible to eliminate O–H bonds by thermal annealing.

Figure 2.9 shows bending losses of single-mode waveguides for TE-like modes [16]. For the bending radius of over 5 μm, bending losses are negligible. Even for an ultra-small bending radius of around 2 μm, a waveguide with a flat core maintains loss lower than 0.1 dB per 90° bend. A waveguide with a square core shows a larger bending loss for a bending radius below 5 μm. For TM-like modes, especially in waveguides with flat cores, bending losses are generally larger than those for TE-like modes. Bending losses measured in various research groups are summarized in [17].

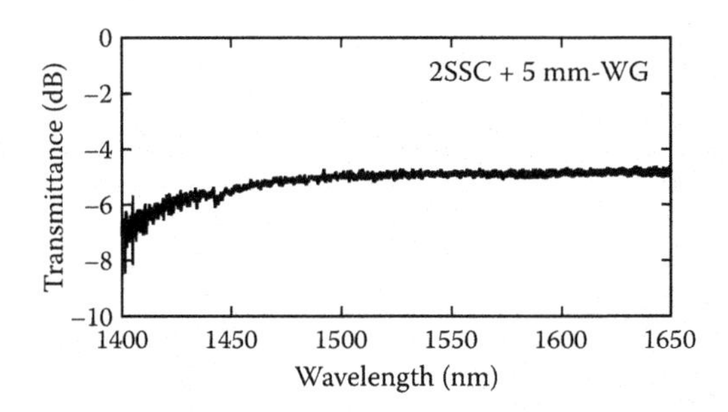

FIGURE 2.8
Measured transmission spectrum of a typical silicon photonic wire waveguide with SSCs. (From Koji Yamada, *Silicon Photonics II, Silicon Photonic Wire Waveguides: Fundamentals and Applications*, Springer-Verlag, 2011. With permission.)

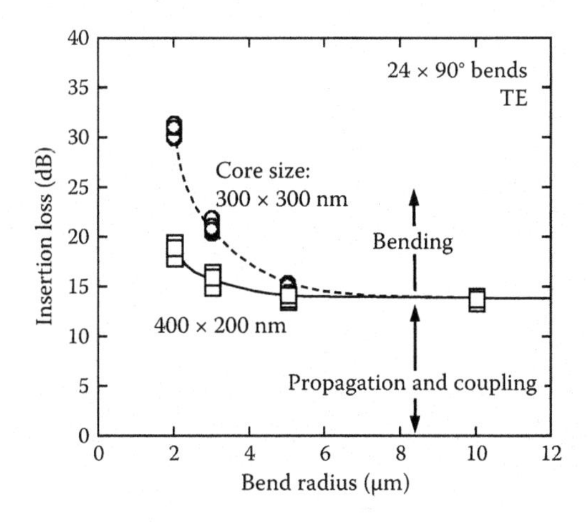

FIGURE 2.9
Relation between measured insertion losses and bending radius.

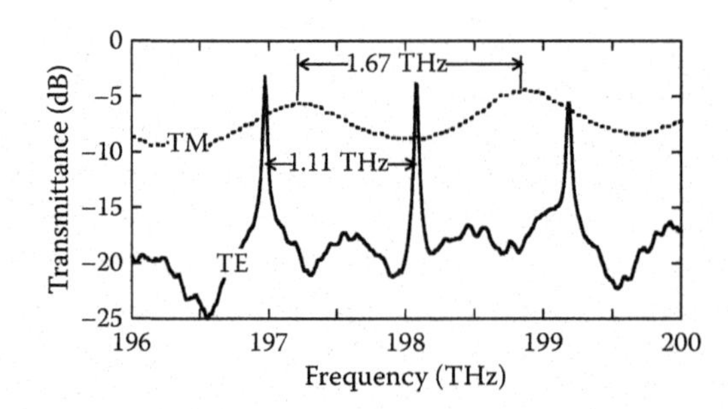

FIGURE 2.10
Measured drop port spectra of a ring resonator with 10-mm radius. (From Koji Yamada, *Silicon Photonics II, Silicon Photonic Wire Waveguides: Fundamentals and Applications*, Springer-Verlag, 2011. With permission.)

The birefringence of the waveguide can be evaluated from the free spectral ranges (FSRs) of ring resonators. Figure 2.10 shows measured transmission spectra at around, 197 THz (λ = 1.514 μm) for a ring resonator of 10-μm radius [1]. The FSR in TM-like modes (1.67 THz) is significantly larger than the one in TE-like modes (1.11 THz). Using the FSR, we can roughly express the group index of the waveguide as $n_g = c/2\pi R\Delta f$, where c, R, and Δf are the speed of light in a vacuum, the radius of the ring resonator, and the FSR in hertz, respectively. Thus, the group indices are estimated to be 4.30 and 2.86 for TE and TM modes, respectively, which agree well with the design values previously mentioned.

2.3 Polarization Diversity in Photonic Wire Waveguides

Hiroshi Fukuda

Silicon (Si) photonic wire waveguides have great potential as a platform for ultra-small photonic circuits [18–20]. Several kinds of functional devices based on Si wire have been demonstrated [21, 22]. However, polarization mode dispersion (PMD), polarization dependent loss (PDL), and polarization-dependent wavelength shift (PDλ) caused by large structural birefringence are not negligible. These drawbacks narrowly limit their application range.

There are several approaches to making a polarization-independent photonic circuit. The simplest way is to use a square core. However, for high-index-contrast waveguides, such as Si photonic wire, a slight fabrication error of few nanometers is critical and results in birefringence. For instance, for waveguide length of 5 cm, the differential group delay between cores with a 300 ± 5-nm width and 300-nm height reaches 6.6 ps, which degrades high-speed signals with the data rate of 40 Gbps. Furthermore, it is hard to remove the stress-induced birefringence, which also causes the PDL, PMD, and PDλ. In addition, the fluctuation of core width varies the group index and results in PDλ in wavelength filters. The difference in the resonant wavelength between transverse electric (TE) and transverse magnetic (TM) modes is larger than 100 GHz for a 10-μm-radius ring resonator with 300 ± 1-nm-wide core. Therefore, accuracy of under a nanometer is required for devices used in polarization-independent dense wavelength division multiplexing (DWDM) systems. This is a big obstacle to mass production.

Another solution is to use a polarization diversity system consisting of polarization splitters and rotators. When TE and TM components are separated by a splitter and the TM component is rotated 90° by a rotator, we need to fabricate functional devices (e.g., a filter) for only the TE mode, not for both polarizations. In recent years, several kinds of polarization control devices in Si photonics have been demonstrated. The three main architectures are mode-coupling–, mode-evolution–, and grating-coupler–based ones.

This section first reviews these three architectures for polarization diversity. It then describes the principle and experimental results for a mode-coupling–based polarization diversity system in detail.

2.3.1 Polarization Manipulation in Si Photonic Wire Waveguides

2.3.1.1 Grating-Coupler–Based Polarization Manipulation

An advantage of the grating-coupler–based polarization independent circuit is its small footprint. In addition, it is not necessary to fabricate a polarization splitter/rotator because the two orthogonal polarizations are divided at the entrance of the photonic circuit.

A two-dimensional (2-D) grating coupler, which couples orthogonal modes from a fiber into identical modes of two waveguides, has been developed [23–25]. Figure 2.11 shows a schematic view of a 2-D grating splitter, which decomposes the fiber polarizations into two linear polarizations that are coupled to TE mode of both waveguides. This device efficiently achieves polarization splitting with a small footprint. Using a focusing grating coupler, the size of the 2-D-grating coupler/polarization diversity structure can be made about 20 μm^2. In addition, a 2-D periodic grating can couple both fiber polarizations to the TE mode of their own photonic wire waveguide. Therefore, no additional on-chip TE/TM polarization conversion is needed. However, the coupling efficiency of the grating coupler is as small as −7 dB, which makes it impractical as a communication device. More details on grating couplers are provided in Chapter 3.

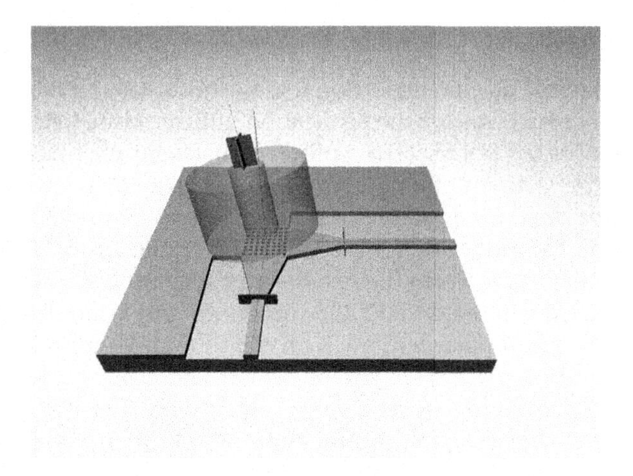

FIGURE 2.11
Schematic of the grating-based polarization splitter.

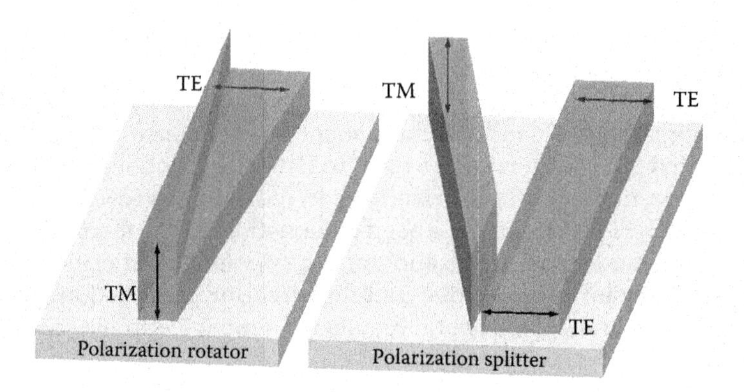

FIGURE 2.12
Schematic of the mode-evolution–based polarization splitter and rotator.

2.3.1.2 Mode-Evolution–Based Polarization Manipulation

Mode-evolution–based polarization control devices have an advantage of wide bandwidth. A polarization splitter and rotator (PSR) with an asymmetric core cross section for silicon-nitride (SiN) waveguides has been demonstrated [26–29]. A schematic diagram of mode-evolution–based polarization rotator is shown in Figure 2.12. Here, the layers are asymmetrically and oppositely tapered. The principal axis of the structure and polarization state of the fundamental mode rotates in unison along the transition. In contrast to a pure twist, the mode set changes, yet a large difference in the rates of propagation of the guided modes is maintained.

A mode-evolution–based rotator in Si photonic wire waveguide achieved polarization extinction ratio around 15 dB. The mode-evolution–based rotator has a better polarization rotation effect with a larger wavelength operation window. The wavelength dependence is not significant in the range from 1450 to 1750 nm. However, such three-dimensional structures are difficult to fabricate and to adapt to mass production.

2.3.1.3 Mode-Coupling–Based Polarization Manipulation

Mode-coupling–based PSR has low insertion losses and is easy to fabricate because there is no need for complex processes for the fabrication of three-dimensional structures.

The mode-coupling–based polarization splitter is just a simple directional coupler (DC) [30]. A Si photonic wire with an oblong core produces large PMD due to structural birefringence. In a DC fabricated using such Si photonic wire waveguides, the difference in the coupling length for the TM mode and for the TE mode is quite large. Thus, just a simple DC works as a polarization splitter. From the results of propagation simulations, a DC consisting of Si wires with a 200-nm height and 400-nm width and a 480-nm gap between them separates the two orthogonal polarizations with only 10-μm-long propagation.

An SEM image and transmission spectra of the bar and cross ports for input light with TE and TM modes are shown in Figure 2.13a–c, respectively. The transmittance is defined as the output power ratio between the splitter and a simple Si photonic wire with the same length as the splitter. The polarization extinction ratios (PERs) of the cross and bar ports in the C-band are 10 and 13 dB, respectively. The excess loss is less than 0.5 dB for each port. The PER for the cross port can be improved using a two-stage configuration. Details are discussed in [30].

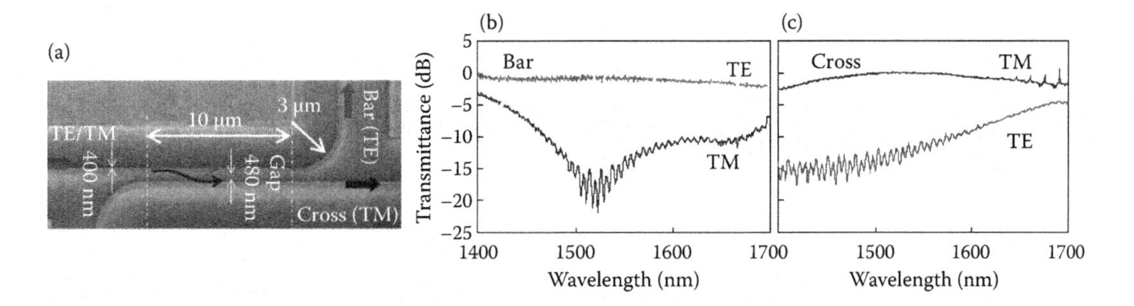

FIGURE 2.13
(a) SEM image of a mode-coupling–based polarization splitter. (b,c) Measured transmission spectra for the mode-coupling–based polarization splitter.

The mode-coupling–based polarization rotator has an off-axis double-core structure. The Si core confines light weakly, and the second core controls the polarization of the light. A schematic diagram of the cross-section of mode-coupling–based rotator is shown in Figure 2.14 [31]. The eigenmode axes of such a double-core structure are tilted toward the substrate. Thus, propagation through the waveguide produces a rotation of the polarization plane when the polarization of the incident light is parallel (TE) or orthogonal (TM) to the substrate. The material of the second core is SiO_xN_y, which has a refractive index of 1.60. The cross section of the Si core has to be just square to equally split the optical power into the two eigenmodes and ought to be a little smaller than that of a normal Si photonic wire waveguide for single-mode propagation. It was set to 200 × 200-nm^2 because the input and output waveguides are normal Si photonic wires with a 200-nm height and 400-nm width. When the size of the second core is 840 × 840 nm^2, the effective indices of the two orthogonal eigenmodes are 1.542 and 1.525. Under these conditions, an only 45-µm-long device provides a polarization rotation of 90°.

A schematic diagram of a fabricated mode-coupling–based rotator is shown in Figure 2.15a. The Si photonic wires are adiabatically connected to the 200 × 200-nm^2 Si core by 10-µm-long tapers. The polarization of the input light is TM. The transmission spectra filtered for the TE and TM components are shown in Figure 2.15b. When the length of the rotator is 35 µm, the TM component is greatly suppressed and the TE component is transmitted with an excess loss of about 1 dB. The spectral ripples are caused by the polarization rotation of the ordinary Si photonic wire waveguides used for input and output, and they complicate the estimation of the actual extinction ratio.

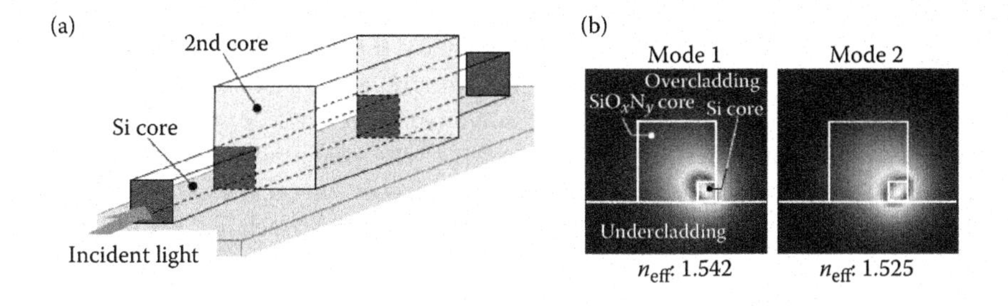

FIGURE 2.14
(a) Schematic diagram of a mode-coupling–based polarization rotator and (b) its eigenmodes.

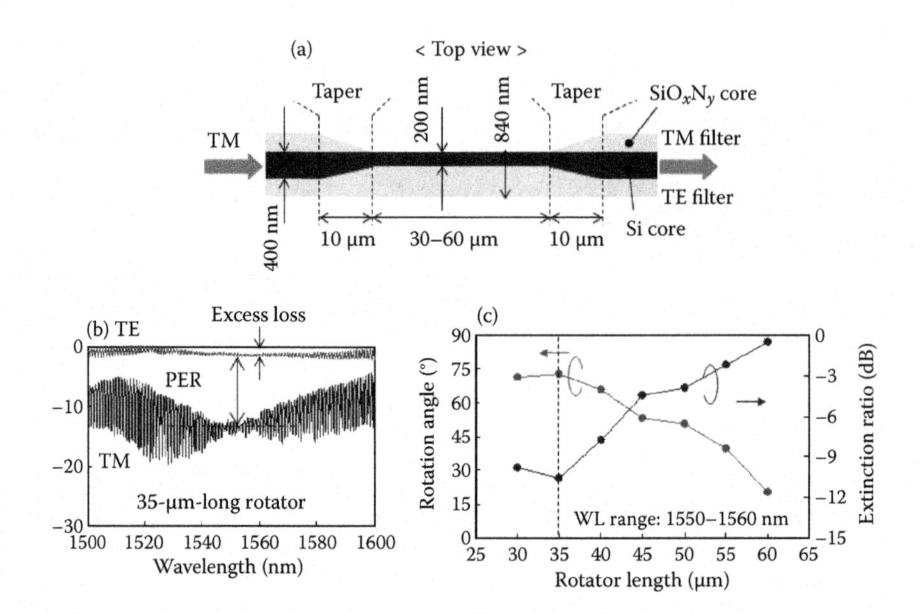

FIGURE 2.15
(a) Schematic diagram of a mode-coupling–based polarization rotator. (b) Transmission spectra of a mode-coupling–based polarization rotator with a 35-µm length. (c) Polarization rotation angle estimated from the measured Poincaré map and polarization extinction ratio calculated from the measured polarization rotation angle.

The measurement of the state of polarization with a Poincaré sphere enables to obtain the actual extinction ratio. Figure 2.15c show the rotation angle estimated from the Poincaré map along with the extinction ratio calculated from the rotation angle as a function of rotator length. The maximum rotation angle is 72°, and the maximum extinction ratio is about 11 dB when the rotator length is 35 µm.

2.3.2 Polarization Diversity Circuit

A polarization independent arrayed waveguide grating (AWG) demultiplexer in SOI nanophotonic waveguides has been presented [24]. With a 2-D fiber coupler grating, it is possible to couple from standard single-mode fiber to Si photonic wire waveguides while simultaneously decomposing the variable fiber polarization into separate circuits. Polarization independent behavior is achieved by propagating both polarizations through the same AWG in opposite directions. Insertion loss is less than 6.9 dB and crosstalk is 15 dB with PDL of 0.66 dB.

A polarization-transparent add/drop filter with the mode-evolution–based polarization diversity has already been demonstrated [28]. The circuit consists of polarization splitters and rotators and microring resonators. The waveguides are made of Si-rich SiN (n = 2.193 at 1550 nm). The polarization crosstalk of 32 dB for was achieved a two-stage PSR.

A mode-coupling–based polarization diversity circuit has also been demonstrated for Si photonic wire waveguides [32]. A photograph and schematic diagram of the circuit are shown in Figure 2.16. The circuit consists of polarization splitters, rotators, and a ring resonator. The splitters and the rotators are the same as those mentioned in Section 2.3.1.3. The filter is designed for TE polarization. Therefore, for light with TM polarization, the ring works as only a directional coupler with a very high coupling efficiency.

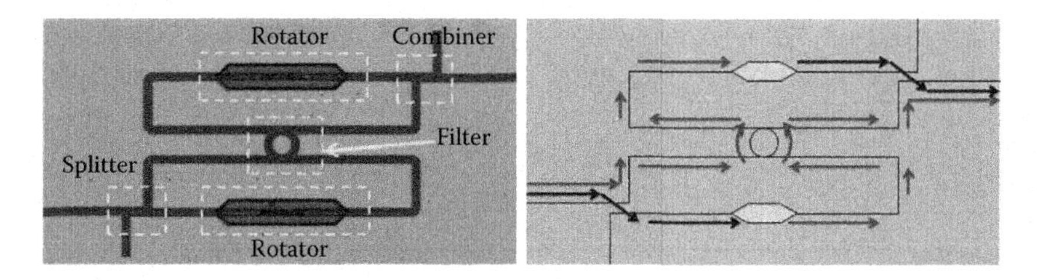

FIGURE 2.16
Photograph and schematic diagram of a mode-coupling–based polarization diversity circuit.

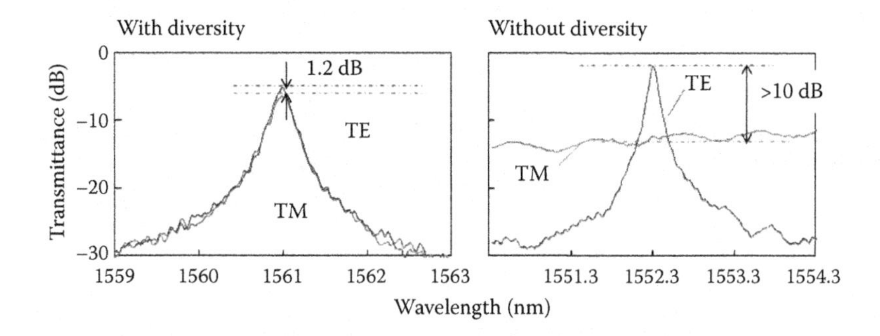

FIGURE 2.17
Transmission spectra of wavelength filters with and without polarization diversity.

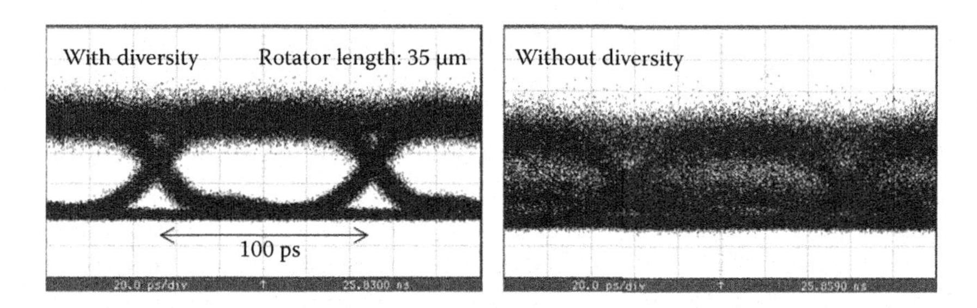

FIGURE 2.18
Measured eye diagram for the polarization-independent wavelength filter with and without polarization diversity.

Transmission spectra for wavelength filters with and without the polarization diversity configuration are shown in Figure 2.17. Without the diversity, the ring does not work as a wavelength filter. On the other hand, with polarization diversity, the ring works well for both polarizations, and the PDL is only 1.2 dB. This means that these elements do actually constitute a polarization-independent circuit. The high-speed response of the circuit has been obtained for the signal with the data rate of 10 Gbps (NRZ, PRBS 2^{31}-1, mark ratio 0.5). The light from the polarization scrambler was launched into the input waveguide of the filter and the output waveform from the filter was measured with an O/E converter and a sampling oscilloscope. Measured eye diagrams with and without the polarization diversity configuration are shown in Figure 2.18. Without the diversity configuration, the eyes are collapsed because of large PDL. On the other hand, with the diversity configuration, the eyes are well defined and opened. This means that the polarization diversity circuits are already good enough for practical use in high-speed data transmission.

2.4 Nanostructured Waveguides for Advanced Functionalities

2.4.1 Slot Waveguides

Christian Koos and Joerg Pfeifle

Slot waveguides typically consist of two strips of high-index material that enclose a subwavelength low-index slot region. Discontinuities of the electric field at the high index-contrast interfaces lead to strong interaction of the guided light with the material inside the slot [33, 34]. Slot waveguides are the basis of many silicon-organic hybrid structures, that combine the advantages of CMOS-based waveguide fabrication with the wealth of optical properties that is provided by functional organic cladding materials [35–37]. This section will first introduce fundamental principles of silicon-based slot waveguides, give design guidelines and provide an overview on the current state-of-the art in slot waveguide fabrication. On this basis, applications of slot waveguides in optical signal processing and sensing will be discussed.

2.4.1.1 Wave Propagation in Slot Waveguide Structures

A typical cross section of a silicon-on-insulator (SOI) slot waveguide is depicted in Figure 2.19a. The waveguide core consists of two high-index silicon strips with height h and width w_{st} that are embedded into a low-index top cladding. Discontinuities of the horizontal electric field component (E_x) at the high index-contrast interfaces lead to strong field enhancement in the slot region. The E-field profile shown in Figure 2.19b is simulated with the refractive indices $n_{Si} = 3.48$, $n_{cl} = 1.5$, and $n_{SiO2} = 1.44$ at $\lambda = 1550$ nm. As in standard SOI waveguides, the buried oxide of the SOI wafer acts as a buffer region that optically isolates the waveguide from the bulk silicon handle wafer. The optical signal propagating in a single waveguide mode is described in slowly varying envelope approximation [38],

$$\mathbf{E}(\mathbf{r},t) = \mathrm{Re}\left\{ A(z,t)\frac{\mathcal{E}(x,y)}{\sqrt{\mathcal{P}}}\mathrm{e}^{\mathrm{j}(\omega t-\beta z)}\right\} \tag{2.2}$$

$$\mathbf{H}(\mathbf{r},t) = \mathrm{Re}\left\{ A(z,t)\frac{\mathcal{H}(x,y)}{\sqrt{\mathcal{P}}}\mathrm{e}^{\mathrm{j}(\omega t-\beta z)}\right\} \tag{2.3}$$

where $A(z,t)$ is the complex envelope, $\mathcal{E}(x,y)$ and $\mathcal{H}(x,y)$ denote the vectorial electric and magnetic mode profiles, and $\beta(\omega)$ is the associated propagation constant. In this definition, the power of the signal averaged over some optical periods is given by $|A(z,t)|^2$. The quantity $\mathcal{P}$ is used for power normalization of the numerically computed mode fields,

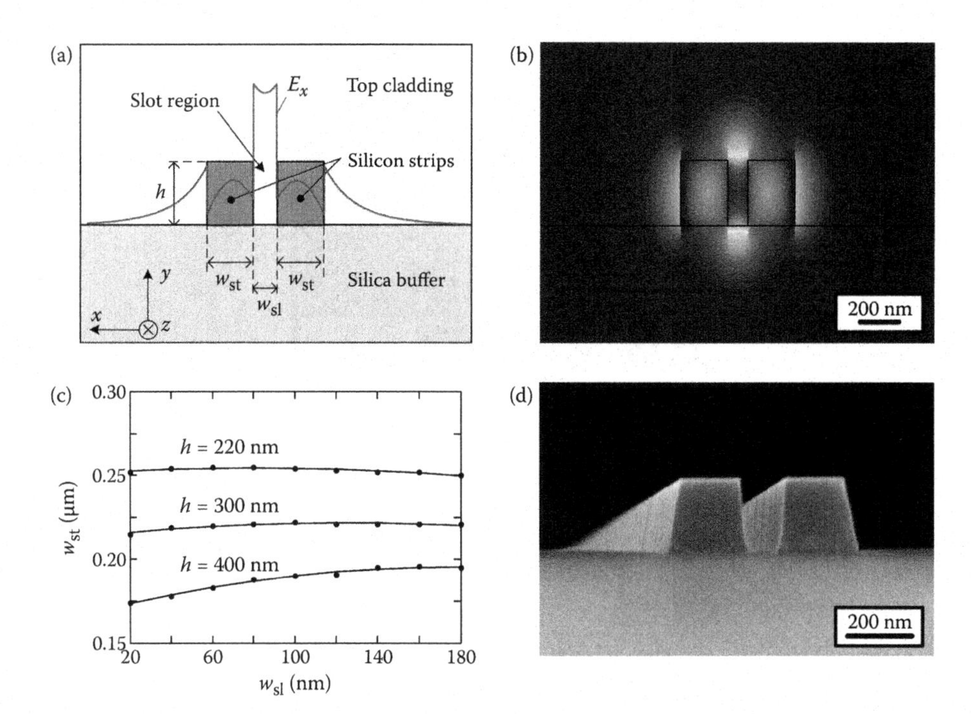

FIGURE 2.19
(See color insert.) (a) Schematic cross section of a silicon-on-insulator (SOI) slot waveguide (height h, width w_{st}, refractive index n_{Si} = 3.48 at λ = 1550 nm). (b) Dominant electric field component (E_x) of the fundamental TE mode field for n_{cl} = 1.5, w_{st} = 200 nm, h = 220 nm, and w_{sl} = 120 nm. (c) Maximum strip width w_{st} for a single-mode waveguide as a function of slot width w_{sl} and waveguide height h for n_{cl} = 1.5 and λ = 1550 nm. (d) Scanning electron microscope (SEM) view of a cleaved slot waveguide front facet. The waveguide was fabricated with 193-nm-deep UV lithography and dry etching.

$$\mathcal{P} = \frac{1}{4}\int_{-\infty}^{\infty}\int_{-\infty}^{\infty}\left[(\mathcal{E}\times\mathcal{H}^*)+(\mathcal{E}^*\times\mathcal{H})\right]\cdot\mathbf{e}_z dxdy \tag{2.4}$$

where $\mathbf{e}_z$ is unit vector in the direction of propagation.

The modal fields $E(x,y)$ and $H(x,y)$ are classified by the terms transverse electric (TE) and transverse magnetic (TM). Here, TE refers to a waveguide mode with a dominant electric field component in x-direction (parallel to the substrate plane), whereas the dominant electric field component of a TM mode is oriented parallel to the y-axis (perpendicular to the substrate plane).

For a waveguide with a vertical slot, the fundamental TE mode experiences a strong field enhancement within the slot region (see Figure 2.19a,b). Maxwell's equations state that in the absence of free charges, the normal component $D_x = \varepsilon_0 n^2 E_x$ of the dielectric displacement must be continuous at an interface. This leads to a discontinuity in the corresponding component of the electric field (E_x) with higher amplitude on the low-index side of the interface. In a slot waveguide, this effect is exploited at both sidewalls to obtain an electric mode field that is strongly confined to the slot region (see Figure 2.19b). Slot widths are much smaller than the penetration depths of the evanescent field leading to large intensities throughout the slot region. For many applications, it is desirable to design single-mode

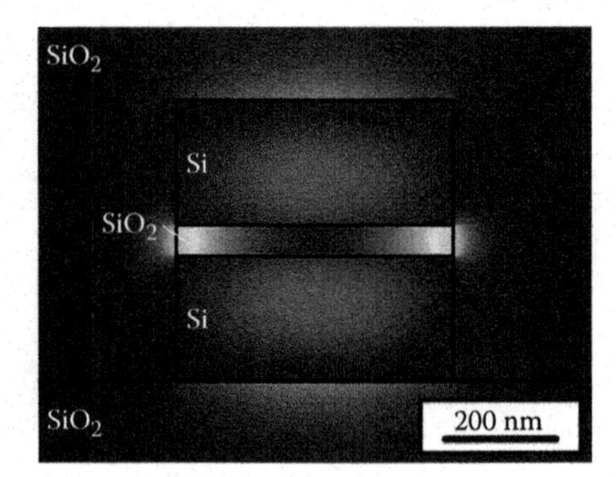

FIGURE 2.20
Mode field of a waveguide with a horizontally oriented slot waveguide: Dominant electric field component (E_y) of the fundamental TM mode field.

slot waveguides. For a given slot width w_{sl} and a given waveguide height h, there is a maximum strip width w_{st} up to which the waveguide is single mode. A numerical analysis reveals the relationship shown in Figure 2.19c for the case of a vertical-slot waveguide with a cladding refractive index of $n_{cl} = 1.5$. Field enhancement can also be exploited for waveguides with a horizontally oriented slot, where the dominant electric field component of the fundamental TM mode is perpendicular to the top and bottom interfaces of the slot as shown in Figure 2.20. The depicted mode was calculated for a wavelength of $\lambda = 1.55$ μm. The silicon strips (refractive index $n_{Si} = 3.48$) are 500 nm wide and 220 nm thick. Silica ($n_{SiO2} = 1.44$) is used as cladding, slot, and substrate material. Single-mode conditions for horizontal-slot structures can be found in [39]. Waveguide structures that comprise multiple slot regions can be used to further increase the confinement of the mode field to the low-index cladding—both for vertically and horizontally oriented slots [40, 41].

2.4.1.2 Slot Waveguide Fabrication

Slot waveguides have been fabricated in both silicon-on-insulator and silicon nitride (Si_3N_4) material systems. Waveguides with vertically oriented slots are defined lithographically by electron beam lithography [34, 42] or high-resolution deep-UV photolithography [36, 43, 44] and structured with dry etching techniques. Figure 2.19d depicts a cross-section of a fabricated silicon-on-insulator slot waveguide. Deviations from the ideal structure comprise slightly inclined sidewalls that exhibit roughness in the form of vertical grooves due to the strongly anisotropic etching process. While high-index-contrast waveguides are generally prone to roughness-induced scattering losses [45], this applies all the more to slot waveguides having two additional dry-etched vertical interfaces, at which particularly strong electric fields occur. Experimentally measured loss figures depend on the slot width and the cladding material. Slot widths as small as 50 nm and propagation losses between 7 and 11 dB/cm have been achieved with electron beam lithography [46]. Using 193-nm-deep UV lithography, slot widths down to 100 nm are possible with minimum loss figures between 4 and 6 dB/cm [43]. As a comparison, significantly lower loss figures of approximately 1.2 dB/cm can be achieved for standard strip waveguides fabricated with similar processes as shown in Section 2.2.2. Apart from smooth sidewalls, homogeneous filling of the slot region with functional material is crucial to obtain low-loss waveguides. Successful demonstrations comprises spin coating of polymer materials [42], gas phase deposition of small organic molecules [36], or inverted processes that rely on structuring the slot material and growing the silicon strips around it [44]. Waveguides with horizontally oriented slots are less affected by sidewall roughness. The slot can be defined by thin-film deposition techniques, which enables better control of slot dimensions and smoother interfaces between the low-index slot region and the high-index waveguide core. Roughness of dry etched sidewalls is of no significant consequence since field enhancement does not occur at these interfaces. However the range of usable slot materials is limited by the necessity to deposit high-quality silicon layers thereon. Loss figures below 2 dB/cm have been demonstrated for a slotted rib waveguide consisting of an 8.3-nm-thick SiO_2 layer, which was sandwiched between a crystalline and an amorphous silicon layer [47].

2.4.1.3 Linear Signal Propagation in Slot Waveguides

Slot waveguides introduce the slot width as an additional degree of freedom in the design of waveguide geometry. This allows for normal, anomalous, or zero group velocity dispersion at telecommunication wavelengths [48, 49]. In addition, linear signal propagation is strongly influenced by the properties of the cladding materials. This allows for tailoring the waveguide properties and enables optical sensing. A quantitative measurement for the interaction of the guided light with the cladding material is the field confinement factor,

$$\Gamma_{\mathrm{cl}} = \frac{n_{\mathrm{cl}} \iint_{D_{\mathrm{cl}}} |\mathcal{E}(x,y)|^2 \, \mathrm{d}x\mathrm{d}y}{c\mu_0 \iint_{D_{\mathrm{tot}}} \mathrm{Re}\{\mathcal{E}(x,y) \times \mathcal{H}^*(x,y)\} \cdot \mathbf{e}_z \, \mathrm{d}x\mathrm{d}y} \tag{2.5}$$

where the integrals in the numerator and in the denominator extend over the cladding domain D_{cl} and over the entire cross section D_{tot}, respectively. If the refractive index n_{cl} of the cladding changes by an amount Δn_{cl}, the propagation constant of the mode changes according to

$$\Delta\beta = k_0 \Gamma_{\mathrm{cl}} \Delta n_{\mathrm{cl}} \tag{2.6}$$

where k_0 denotes the free-space wave number. Numerically calculated field confinement factors for different horizontal-slot waveguide geometries can be found in [50]. Note that there are different definitions of field confinement factors in the literature. Equation 2.4 represents a field confinement factor that has been generalized to the case of high-index-contrast waveguides. As opposed to the conventional definition, Γ_{cl} does not any more correspond directly to the fraction of optical power that propagates in the cladding region of the waveguide and can even take up values slightly larger than unity for high-index-contrast waveguide [51].

Interaction of the guided light with the cladding material can, e.g., be used for athermalization and trimming of silicon photonic structures. Silicon has a large thermo-optic coefficient of $1.84 \times 10^{-4}\ \mathrm{K}^{-1}$, which, together with thermal expansion, makes resonators and filter structures prone to temperature-induced wavelength shifts. This temperature dependence can be considerably reduced using slot waveguides with a cladding material of opposite temperature coefficient [52, 53]. In addition, photosensitive cladding materials such as PMMA allow for postfabrication trimming of waveguide structures by UV exposure [53].

Slot waveguides can also be used for biochemical sensing [54, 55]. In the case of refractometric sensing, a chemical analyte changes the refractive index of a liquid, gaseous, or solid cladding material in the slot region of the waveguide. In the case of volume sensing, the refractive index changes homogeneously throughout the cladding. An index change of Δn_{cl} leads to a change $\Delta n_e = \Gamma_{cl}\, \Delta n_{cl}$ of the mode's effective refractive index and hence to a phase shift of the transmitted light,

$$\Delta\Phi = -k_0 \Gamma_{cl} \Delta n_{cl} L, \tag{2.7}$$

where L denotes the length over which the waveguide is affected by the index change.

For surface sensing, the waveguides are functionalized by an ultrathin layer of receptor molecules. If an analyte binds to the receptors, the thickness of the molecular layer changes, thereby affecting the effective index of the propagating optical mode. To calculate the sensitivity, the integration region D_{cl} in the numerator of Equation 2.5 must be replaced by a thin domain close to the waveguide surface that is affected by the binding of the analyte. For single- and multiple-slot configurations, optimum waveguide geometries for maximum sensitivity can be found in [50] and [56].

More details on sensing are given in Chapter 14.

2.4.1.4 Electro-Optic Functionalities

Efficient electro-optic modulators can be realized by combining slot waveguides with electro-optic cladding materials [37] or liquid crystals [42, 57]. If conventional slot waveguides are used, the silicon strips are electrically connected to metal transmission lines by thin conductive silicon slabs, as shown in Figure 2.21a. A voltage U applied to the electrical transmission lines induces a strong electric field and a large electro-optic index change within the slot region where the optical mode field is concentrated. If $E_{el} = U/w_{sl}$ denotes

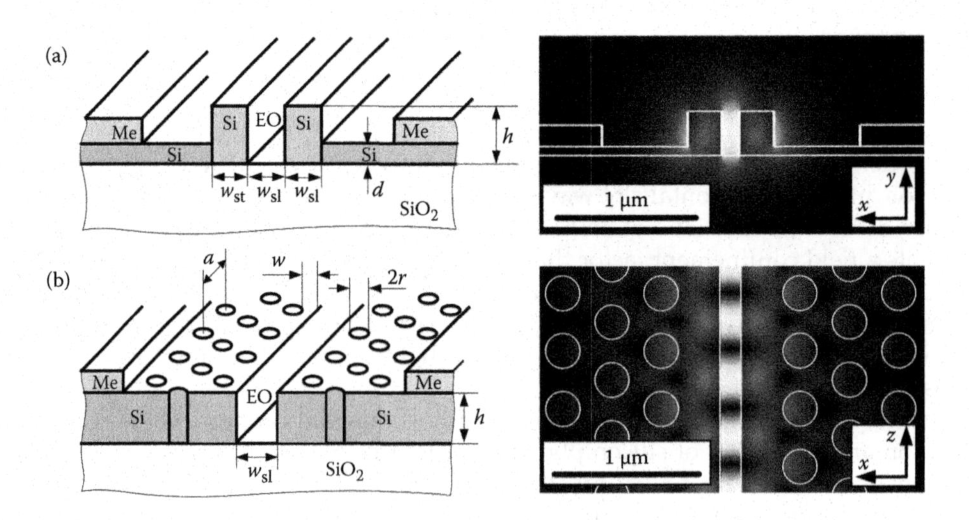

FIGURE 2.21
(See color insert.) Silicon-organic hybrid (SOH) electro-optic modulator concepts: The electric field distributions of the fundamental TE modes are depicted on the right-hand side. (a) Traveling-wave configuration. (b) Photonic crystal (PhC) slot waveguide structure.

the x-component of the externally applied electric field strength inside the slot, the refractive index change is given by

$$\Delta n_{\mathrm{cl}} = \frac{1}{2} n_{\mathrm{cl}}^3 r_{33} E_{\mathrm{el}}, \tag{2.8}$$

where r_{33} denotes the electro-optic coeficient of the deployed cladding material. For state-of-the art electro-optic polymers, typical values for r_{33} range between 50 and 170 pm/V [58]. The phase shift accumulated along a waveguide of length L is then given by Equation 2.7.

For conventional waveguide slot structures as depicted in Figure 2.21a, electrical and optical group velocities can be matched by suitable waveguide designs. Modulation bandwidths of more than 100 GHz are possible, only limited by the RC time constant which originates from the fact that the slot capacitance has to be charged via the resistance of the slab regions. More details on the design of the electrical transmission line can be found in [59]. As an alternative to the silicon slab regions, photonic band-gap structures (PBG) as shown in Figure 2.21b can be used to optically isolate the slot waveguide from the metal transmission lines while still maintaining an electrical contact. By proper design of the PBG structures, the group velocity of the optical signal can be reduced to, e.g., 4% of the vacuum velocity of light over a bandwidth of 1 THz. This allows for a drastic reduction in device length, but the bandwidth of the device is then limited to ~80 GHz by walk-off effects between the electrical signal and the slow light [60].

More details on optical modulation are given in Chapter 9.

2.4.1.5 Kerr Nonlinearities and All-Optical Signal Processing

The strong confinement of light in the slot region lends to itself nonlinear all-optical signal processing. If a Kerr nonlinear material is used as top cladding material, ultrafast effects like self-phase modulation, cross-phase modulation, or four-wave mixing can be exploited. As a quantitative measure of nonlinear interaction within the waveguide, the nonlinearity parameter is commonly used,

$$\gamma = \frac{k_0 n_2}{A_{\mathrm{eff}}} \tag{2.9}$$

In this equation, k_0 denotes the free-space wavenumber, n_2 is the Kerr coefficient of the material, and A_{eff} is the effective area of third-order nonlinear interaction. To calculate A_{eff}, the theory of nonlinear propagation in low-index contrast optical fibers has been adapted to high-index-contrast slot waveguides [38],

$$A_{\mathrm{eff}} = \frac{\mu_0}{\varepsilon_0 n_{\mathrm{cl}}^2} \frac{\left| \iint_{D_{\mathrm{cl}}} \mathrm{Re}\left\{ \mathcal{E}(x,y) \times \mathcal{H}^*(x,y) \right\} \cdot \mathbf{e}_z \, \mathrm{d}x \, \mathrm{d}y \right|^2}{\iint_{D_{\mathrm{tot}}} \left| \mathcal{E}(x,y) \right|^4 \mathrm{d}x \, \mathrm{d}y} \tag{2.10}$$

To increase waveguide nonlinearity, A_{eff} has to be minimized. For a vertical-slot waveguide of given slot width w_{sl}, optimum parameters must exist for the height h and the strip width w_{st}: for very large cross sections of the strips, the fraction of power in the cladding will become

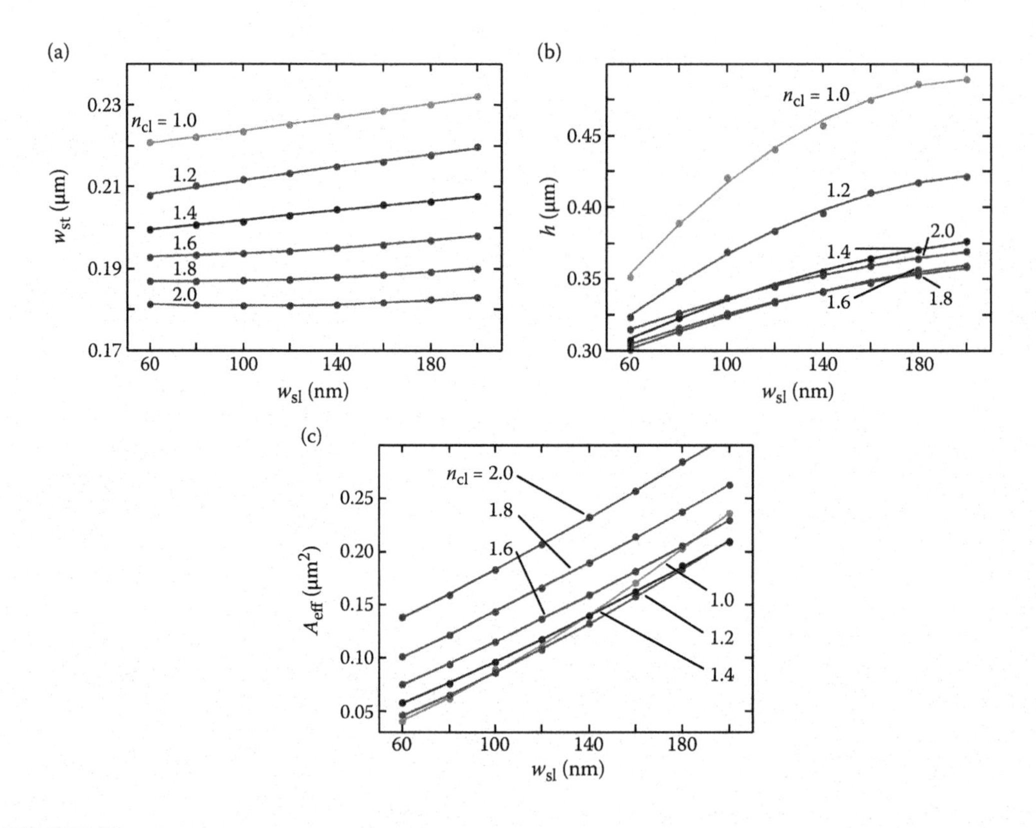

FIGURE 2.22
Parameters of slot waveguides with optimized geometry for minimum effective area A_{eff}. (a) Optimal strip width w_{st} as a function of slot width w_{sl} for different linear refractive indices n_{cl} of the top cladding. (b) Optimal strip height h. (c) Minimized effective area A_{eff} for nonlinear interaction.

small and nonlinear interaction will be weak. For very small cross sections, the mode will only be weakly guided and extend far into the nonlinear cladding such that the intensity will be low. Figure 2.22 shows optimized parameters for the strip height h and the width w_{st} as a function of the slot width w_{sl} and the cladding refractive index n_{cl}. Reducing the slot width increases the intensity inside the slot region. For horizontal slot widths $w_{sl} \geq 60$ nm, the effective nonlinear interaction area A_{eff} always decreases with w_{sl}, and effective areas below 0.1 μm² can be achieved for cladding indices $n_{cl} < 1.8$. Using highly nonlinear cladding materials, nonlinearity parameters up to 7 W^{-1} mm^{-1} are possible for silicon-organic hybrid (SOH) slot waveguides [38]. In contrast to nonlinear silicon-on-insulator waveguides, SOH slot waveguides do not suffer from two-photon absorption and free carrier accumulation, and allow for sub-ps switching intervals [61]. Such structures have been used for, e.g., demultiplexing data streams of 170 Gbit/s [36].

2.4.2 Subwavelength Grating Waveguides

Jens H. Schmid, Pavel Cheben, and Przemek J. Bock

In this section, we discuss a nanostructured silicon waveguide type that has been invented only relatively recently, the subwavelength grating (SWG) waveguide. A review of principles and applications of this waveguide type has been given in [62].

Subwavelength gratings (SWGs) are periodic dielectric structures with a periodicity well below the wavelength of light. Diffraction effects are suppressed for light propagating through SWG structures. Instead SWGs act as homogeneous effective media with spatially averaged refractive index [63]. The use of SWGs is well established in free space optics [64]. They have found applications, for example, as an alternative to antireflective (AR) optical thin film coatings on bulk dielectric surfaces [65] and mirrors [66]. More recently, it has been demonstrated that this AR effect can also be used on planar optical waveguide facets [67].

The case of a one-dimensional SWG is illustrated in Figure 2.23a. The structure is comprised of alternating slabs of dielectric materials with refractive indices n_1 and n_2 with a periodicity $\Lambda < \lambda/(2\max(n_1, n_2))$, which ensures that diffraction of the incident light of wavelength λ is suppressed independently of the angle of incidence. For the case of light incident on the grating from the top in Figure 2.23a, the effective index of the SWG depends on the polarization of the light. According to the effective medium theory (EMT), the effective refractive index can be approximated [63] by $n_\parallel = (fn_1^2 + (1-f)n_2^2)^{1/2}$ and $n_\perp = (fn_1^{-2} + (1-f)n_2^{-2})^{-1/2}$ for a wave with the electric field parallel and perpendicular to the slabs, respectively, where $f = a/\Lambda$ is the volume fraction of the material with index n_1. The effective index is thus a polarization dependent weighted average of the indices of the constituent materials. The optical properties of the same structure shown for light incident perpendicular to the dielectric slabs with wavevector k, are well known from the study of photonic crystals [68]. Light propagates through the periodic structure as a Bloch wave with a dispersion relation that looks schematically as depicted in Figure 2.23b. At the Bragg condition $k = \pi/\Lambda$, the slope of the dispersion curve flattens, corresponding to a standing wave with a group velocity of zero, and a band of forbidden frequencies exists, the "photonic band gap." A great deal of experimental work has focused on exploiting the photonic band gap effect for making optical waveguides, cavities, and related device structures, but no subwavelength periodic waveguides have experimentally been reported until recently. Unlike photonic band-gap waveguides, the SWG waveguides [62, 69, 70] operate in the long wavelength regime of the dispersion diagram, as indicated in Figure 2.23b, where dispersion is approximately linear, consistent with the concept of an effective homogeneous medium.

The spatial averaging effect of SWGs introduces a new degree of freedom in the design of silicon integrated photonic circuits, where the refractive index contrast is otherwise

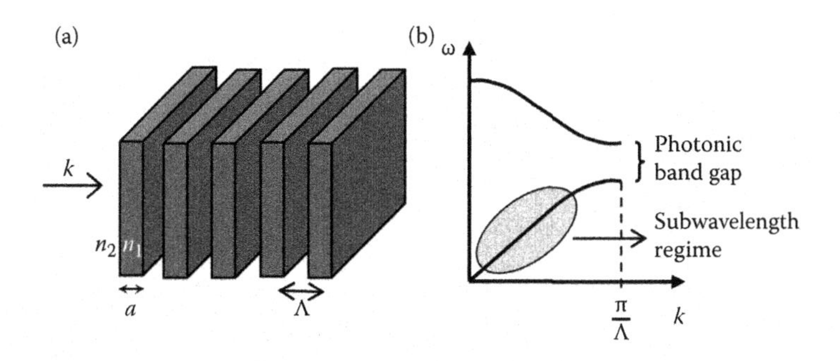

FIGURE 2.23
(a) Schematic picture of a subwavelength grating (SWG) consisting of alternating slabs of dielectric materials with indices n_1 and n_2, with pitch Λ, and duty ratio a/Λ. (b) Dispersion diagram of light propagating through the periodic structure, indicating the photonic band gap and the subwavelength region.

set by the material constants of the constituent materials, i.e., usually silicon and silicon dioxide with refractive indices of 3.5 and 1.44 (at the wavelength $\lambda = 1.55$ μm), respectively. The SWG effect can be used to modify the refractive index of the waveguide core medium locally on a chip simply by changing the grating duty cycle.

A basic building block of silicon photonic waveguide devices is the photonic wire waveguide, a single-mode silicon channel waveguide with typical cross sections of approximately 400–500 nm in width and 200–300 nm in height. The structure shown in Figure 2.24a exemplifies refractive index engineering of a silicon photonic wire waveguide using the SWG spatial averaging effect. Here, periodic gaps are etched into the silicon waveguide core, with a periodicity of less than one half of the effective wavelength of the light in the waveguide. In this way, a new type of waveguide is formed, which is called the subwavelength grating (SWG) waveguide. Because of the spatial averaging effect, the optical properties of the SWG waveguide shown in Figure 2.24a are similar to the one shown in Figure 2.24b, namely a photonic wire waveguide of identical cross section but with a reduced refractive index value of the waveguide core. This is theoretically confirmed by comparing the calculated dispersion curves of the two structures as depicted in Figure 2.24c for a waveguide cross section of 0.45 × 0.26 μm and a SWG duty ratio of 50% for a periodicity of 300 nm. The SWG waveguide dispersion, calculated using the MIT photonic band software [71], exhibits the typical behavior expected for a periodic waveguide with a flattening of the dispersion at the Bragg condition (at $\beta_{Bragg} = 10.5$ μm^{-1}). Near the operating wavelength of $\lambda = 1.55$ μm, which lies in the subwavelength regime of the diagram as indicated in the figure, there is a good match between dispersions of the SWG waveguide and the effective photonic wire with a core index of $n = 2.65$.

A scanning electron micrograph of a fabricated SWG waveguide is shown in Figure 2.25a. In this case, electron beam lithography and inductively coupled plasma reactive ion etching (ICP-RIE) was used to make the waveguide structures. Typical propagation losses of SWG waveguides were measured to be in the 2- to 3-dB/cm range with a low polarization dependent loss (PDL) of <0.5 dB/cm [69]. These loss numbers are comparable to photonic wire waveguides, while PDL is improved.

In fact, SWG waveguides and photonic wire waveguides can be integrated on a chip, and one type of waveguide can be transformed into the other using an adiabatic coupler structure as shown in Figure 2.25b. The operating principle of the coupler [72] is a gradual modification of the effective index along its length to match the photonic wire waveguide on one end and the SWG waveguide on the other. This is achieved by chirping the grating

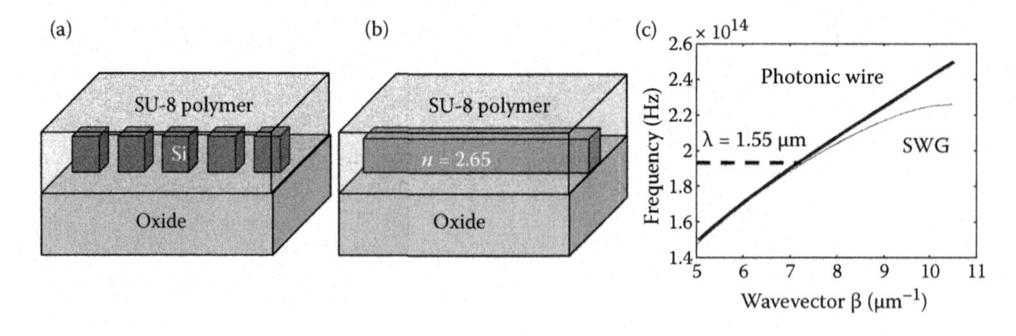

FIGURE 2.24
(a) Schematic picture of a silicon SWG waveguide. (b) Equivalent effective wire waveguide with a spatially averaged core refractive index. (c) Calculated dispersion diagrams of the structures shown in (a) and (b) showing a good match for the operating wavelength of 1.55 μm (TE polarization).

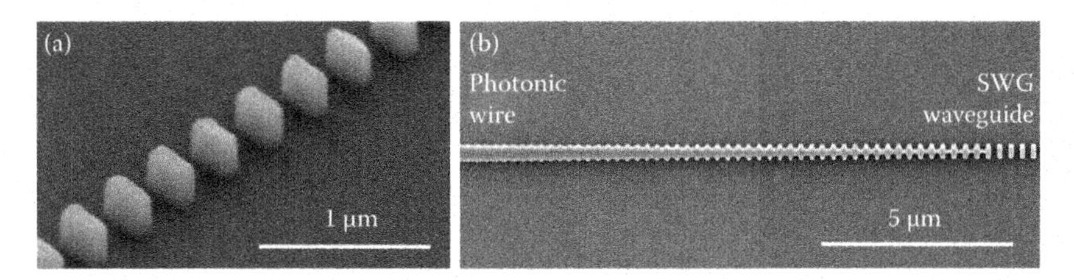

FIGURE 2.25
Scanning electron micrographs of (a) a fabricated SWG waveguide and (b) SWG to photonic wire coupling structure.

period and duty cycle over a length of 50 μm. The intrinsic loss of this coupler structure was measured to be 0.23 dB for TE-polarized light and 0.48 dB for TM [70]. In addition to enabling the seamless integration of wire and SWG waveguides on a chip, essentially the same structure can be used for making highly efficient photonic wire fiber-chip couplers [70, 72]. In this case, the SWG waveguide is extended to the chip edge where it is coupled to a lensed single-mode fiber. A fiber to waveguide coupling efficiency of −0.9 dB for TE and −1.2 dB for TM polarization was demonstrated, including robust tolerance to misalignment, which constitutes a marked improvement over photonic wire inverse taper structures [73] typically used for in-plane fiber-chip coupling to photonic wires. Closed form expressions for the scattering coefficients in SWG waveguides were recently derived [74].

An interesting application of SWG waveguides is waveguide crossings, such as the ones shown in the SEM image in Figure 2.26a. The ability to intersect waveguides with low loss and crosstalk is often considered a prerequisite for designing complex high density photonic circuits. Previous designs of photonic wire crossings have made use of a double etch structure, widening the channel to ridge waveguides in the crossing area [75], and shape optimization of the wire crossing with a genetic algorithm [76]. In both cases crossing losses in the range of 0.1–0.2 dB with crosstalk of −30 dB or better were demonstrated for TE polarized light. The reduced effective core index of SWG waveguides is advantageous for the design of crossings for two reasons. First, a waveguide crossing over another waveguide causes a lesser perturbation in a low index contrast system than in a high index contrast system, and therefore scattering and diffraction losses at the crossing point are reduced. Second, the effectively low index contrast of an SWG waveguide leads to a more delocalized mode profile compared with a photonic wire waveguide of identical cross section. The overlap of a delocalized SWG waveguide mode with the index perturbation presented by the waveguide that is being intersected is smaller than that of the more confined mode of a wire waveguide.

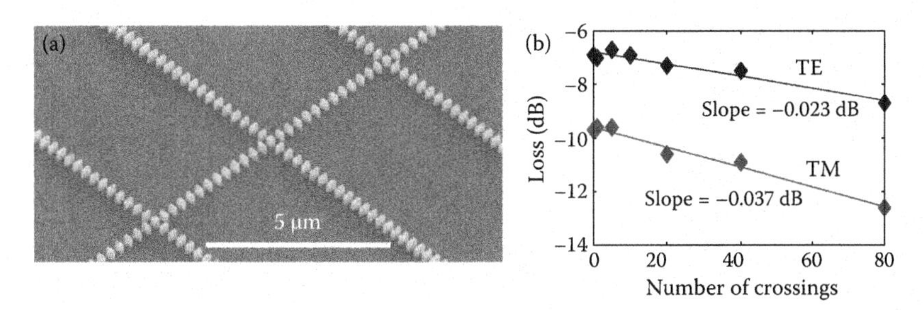

FIGURE 2.26
(a) Scanning electron micrograph of three concatenated SWG waveguide crossings. (b) Measured transmission loss of concatenated crossing structures as a function of the number of crossings, with linear fits.

The loss per SWG waveguide crossing was measured using concatenated crossing structures to be −0.023 dB and −0.037 dB for TE and TM polarized light, respectively, as shown in Figure 2.26b, with a crosstalk better than −40 dB [77]. These remarkable results confirm the suitability of SWG structures for designing highly efficient waveguide crossings.

SWG waveguides have been used to address another important issue with silicon photonic circuits, which is the temperature dependence of their optical output signals caused by the comparatively high thermo-optic (TO) material coefficient of silicon ($dn_{Si}/dT = 1.8 \times 10^{-4}$ K^{-1}). The temperature dependence of silicon wire waveguides can be reduced using a polymer overcladding with a negative TO coefficient to compensate for the silicon thermo-optic effect [78, 79]. Athermal operation of waveguides, i.e., $dn_{eff}/dT = 0$, where n_{eff} is the mode effective index and T is the ambient temperature, is achieved if waveguide dimensions are chosen such that the respective overlaps of the mode with the silicon core and the polymer cladding results in a cancellation of their combined contributions to the waveguide effective TO coefficient. This can be accomplished in narrow silicon wires with a fairly delocalized mode [80], or with slot waveguides, in which a large fraction of the modal field is confined to a narrow gap filled with the low-index polymer [52]. With SWG waveguides the spatial index averaging effect can be exploited for mitigating the silicon thermo-optic effect by filling the gaps with polymer material of negative TO coefficient for an appropriate grating duty ratio. The effect has been experimentally demonstrated in [81], where athermal waveguide behavior was observed for SWG waveguides with a composite core consisting of silicon and SU-8 polymer with a thermo-optic material coefficient of $dn_{SU8}/dT = -1.1 \times 10^{-4}\ K^{-1}$. SEM pictures of three SWG waveguides with different duty cycles are shown in Figure 2.27a. In Figure 2.27b, the measured and calculated waveguide effective thermo-optic coefficient are shown as a function of SWG duty ratio. For increasing volume ratio of silicon material in the waveguide core the plots show a sign reversal from negative to positive thermo-optic coefficients. This means that the thermal behavior of the SWG waveguide is dominated by the polymer material for low-duty ratios and by silicon for high-duty ratios. Athermal waveguide operation is achieved for an SWG duty cycle of 66%, for TE polarization.

Other applications of SWG structures in silicon photonics have recently been suggested and demonstrated. For example SWGs have been used to enhance the performance and reduce the fabrication complexity of surface grating fiber-chip couplers [82–85]. Subwavelength binary blazed grating couplers have been discussed in the literature for applications as beam splitters [86]. In [70], a novel design of a microspectrometer device on a chip is described in

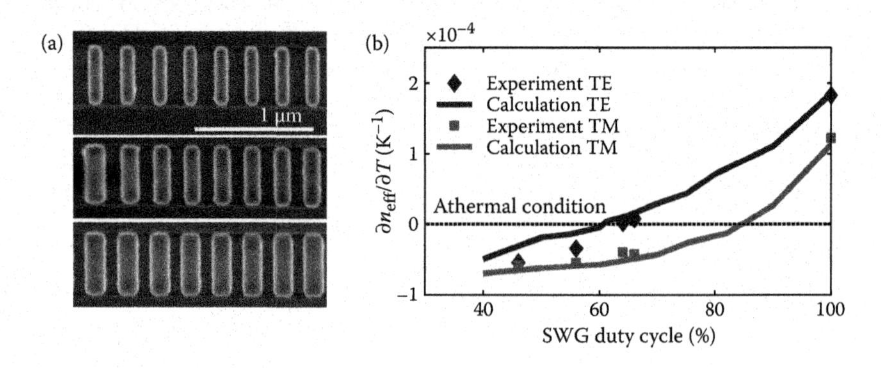

FIGURE 2.27
(a) SEM micrographs of SWG waveguides with Si duty ratios of 46%, 56%, and 66%. (b) Experimental and theoretical results for the effective thermo-optic coefficient of SU-8-clad silicon SWG waveguides as a function of grating duty cycle.

which an SWG structure fulfills a dual purpose by acting as an effective slab waveguide for diffracted light and as a lateral cladding for a channel waveguide. Other applications of SWG waveguides can be envisioned [62]. For example, mode profile engineering as employed for the athermal SWG waveguides can potentially be used to achieve enhanced nonlinear optical effects in a similar fashion to slot waveguides [36]. Engineering the refractive index of slab waveguides with SWGs may also prove to be a suitable method of adapting to integrated optics components such as lenses or transmission gratings, which are common in free space optics. An example of a waveguide lens has recently been discussed in the literature [87].

Acknowledgments

The authors would like to acknowledge the contributions of Jean Lapointe, Siegfried Janz, André Delâge, Adam Densmore, Boris Lamontagne, Rubin Ma, and Dan-Xia Xu.

2.5 Silicon Waveguide Structures with Moderate Dimensions

Andrew P. Knights

This section addresses silicon waveguides, which are deemed "moderate" in size. A working definition of moderate is here chosen such that the waveguide cross section is >1 μm^2. In fact, the majority of silicon photonic waveguides fall into one of two categories: (1) cross sections ≪1 μm^2; (2) cross sections of tens of square micrometers, and thus the exact definition of what is moderate is somewhat unimportant. The more recent work in silicon photonics has tended to be dominated by devices that fall into category 1, although a great deal of interesting and important functionality may be achieved with devices from category 2. Although these larger devices have inferior performance with regard to switching and modulation speed, for applications where bandwidth is not important, the increase in dimensional tolerance lends itself to a much more straightforward fabrication process. It is also worth noting that the early development of silicon photonics was achieved with waveguides much larger than 1 μm^2. From a historical perspective, it is interesting to appreciate the evolution from large to small, which has much in common with the shrinking of microelectronics as described by Moore's law [88]. Moderate (and larger) silicon waveguides share many properties with those from category 1. This section concentrates in the main on the unique aspects of category 2 structures.

2.5.1 The Waveguide Structure

2.5.1.1 Silicon-on-Insulator

Silicon-on-insulator (SOI) has remained the dominant substrate for the formation of silicon waveguides. There are several manufacturing technologies that can produce wafers with a buried oxide (BOX) of thickness up to a few microns, with a thin, single crystal silicon top layer, which may be varied in thickness from a few tens of nanometers to many microns [89]. SOI provides strong vertical optical confinement as a result of the large difference in refractive index (at wavelengths suitable for waveguiding such as 1300 and 1550 nm) between Si and SiO_2 (refractive indices of approximately 3.5 and 1.46, respectively). The provision for lateral confinement is discussed in Section 2.5.2.

The first reports of the use of SOI to form planar optical waveguides were made in the early 1990s [90] following the initial suggestion by Kurdi and Hall in 1988 [91]. The first waveguides utilized SIMOX material in which the BOX layer was formed through implantation of oxygen, but subsequently, waveguides were formed using bonded SOI, and ultimately SmartCut™ material provided by SOITEC, which remains prevalent to the current time.

2.5.1.2 Other Waveguide Structures

Optical waveguiding requires a variation in the refractive index profile. Although this is readily achieved in SOI material, other structures are also suitable. For example, there is a notable refractive index contrast between silicon and germanium, such that Si_xGe_{1-x} on Si (a wholly CMOS compatible structure) provides relatively strong optical confinement.

The contrast in refractive index as a result of electrical doping of silicon permits the fabrication of waveguides, albeit with much weaker confinement than either SOI or SiGe counterparts. These structures were discussed as alternatives to SOI waveguides in 1993 [92] however their usefulness decreased as waveguide dimensions shrank toward and beyond 1 μm^2—a regime where strong optical confinement is necessary.

2.5.2 The Single Mode Condition for Moderate and Large SOI Waveguides

Strong vertical confinement is achieved in planar SOI waveguides. To be useful in integrated circuits, however, lateral confinement must also be present. The simplest method to realize such confinement is geometrically, for example, through the definition of a rib structure, such as that shown in cross section in Figure 2.28. In this case, the effective index of the central rib is marginally larger than the adjacent thinner silicon film, which is often referred to as the *slab region*. The structure in Figure 2.28 thus provides strong vertical confinement and relatively weak horizontal confinement [93]. The use of a rib structure is also useful for integrating silicon waveguides with electrical components as will be discussed in Section 2.5.4.

Integrated optical devices demand single-mode operation. This is easily achieved in waveguide systems with low refractive index contrast, such as optical fiber and III–V semiconductor structures, which rely on small changes in constituent concentration. The large refractive index contrast in SOI waveguides implies that only those waveguides with silicon thin film thickness approximately <250 nm are unconditionally single mode [93].

This apparent incompatibility of large cross-section SOI waveguides with single-mode propagation was addressed by Soref et al. in a highly significant paper in 1991 [94]. One might reasonably expect large or moderate rib waveguides to be multimode in nature. However, Soref et al. were able to show that for certain geometries the rib waveguide is only able to support the fundamental mode, while higher-order modes leak to the surrounding slab. This effectively single-mode waveguide behavior is achieved when

$$\frac{W}{H} \leq 0.3 + \frac{r}{\sqrt{1-r^2}} \tag{2.11}$$

where W and H are the width and height of the rib and r is the ratio between the rib and the slab height (see Figure 2.28). Qualitatively, one may express this idea as single-mode

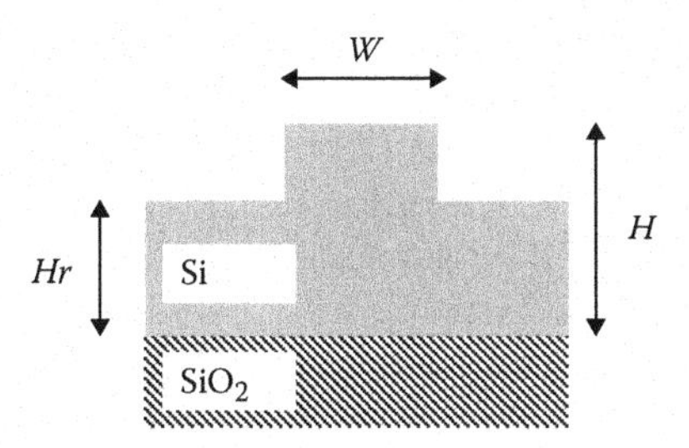

FIGURE 2.28
Cross section of a rib waveguide.

propagation being achieved for ribs where the etch depth used to form the rib is relatively shallow.

Silicon waveguides were routinely manufactured within a few years of Soref's paper by Bookham Technology with propagation losses close to 0.1 dB cm^{-1} [95]. This low propagation loss, coupled with demonstration of insignificant birefringence in SOI waveguides [93], positioned silicon as a strong contender for use in optical integrated circuits.

2.5.3 Spectral Filters

Wavelength filters have been an important component of silicon photonics since the founding of the field. These devices are used most commonly for combining many different wavelengths onto a single waveguide (multiplexing) or separating many wavelengths from a single waveguide (demultiplexing). The so-called wavelength division multiplexing (WDM) represents the most important advantage of optical communication, and through WDM, data transfer rates may be scaled by typical values of up to forty times. A recent review of wavelength filtering was provided by Bogaerts et al. [96].

There have been numerous reports on filters using waveguide elements such as ring resonators and cascaded Mach Zehnders. For category 2 type waveguides though, reported methods of filtering have concentrated on two distinctive elements: the arrayed waveguide grating (AWG) and the echelle grating. Both devices rely upon the creation of an interference pattern created through a precise difference in optical path length. A schematic representation of how an AWG works is shown in Figure 2.29. For a demultiplexing operation,

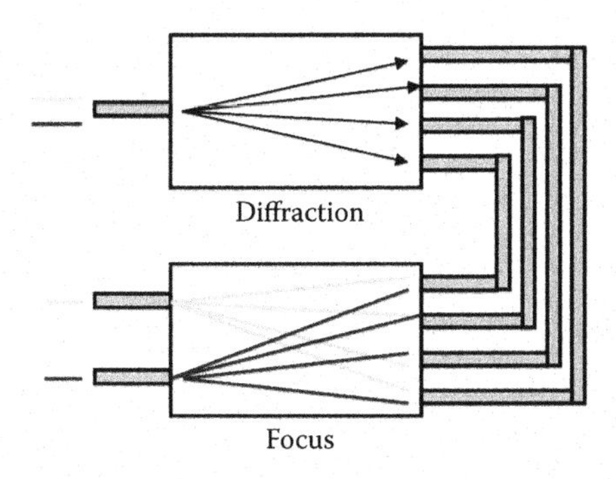

FIGURE 2.29
Schematic representation of the operation of an arrayed waveguide grating (AWG). The two signals are separated using the variation in focus position resulting from a path length variation dependent on the design of the waveguide array.

two wavelengths traveling on a single waveguide are allowed to diffract (this takes place in the slab region of the chip) and then interact with a large number of waveguides each with a slightly different path length. The light travels through this array of waveguides and into the slab region once more. The emerging interference pattern will have maxima at a focal point, whose position will be dependent on the wavelength of the light. The two wavelengths may thus be separated into two separate waveguides. This approach is scalable to many more wavelengths than two. Further, the AWG is symmetric in operation, and thus the same device may be operated as both a MUX and DEMUX. The AWG gained considerable popularity for application in telecommunications and was the most popular form of filter for operation with large and moderate dimensions [97].

2.5.4 Active Devices

The ready supply of SOI, the fundamental work on waveguiding described in Section 2.5.2, together with the emerging demand for high-bandwidth information transfer, stimulated a great deal of academic and commercial work in the 1990s aimed at using silicon devices in the fiber optic network. Fundamental functionality associated with DWDM was addressed at this time and high-performance devices were demonstrated while significant progress was made in attempts to reduce the coupling loss between single-mode optical fiber and silicon waveguides [98]. Perhaps the most important advances made at this time were related to device integration because the underlying technology has been transferred to the submicron waveguides, which today dominate silicon photonics research.

2.5.4.1 Thermo-Optic Switching

The change in refractive index of a material as a function of temperature is described as the thermo-optic effect. This is relatively large in silicon such that [99]

$$\frac{\partial n}{\partial T} = 2 \times 10^{-4} \mathrm{K}^{-1} \tag{2.12}$$

where the left hand side of Equation 2.11 is known as the thermo-optic coefficient. In an optical integrated circuit, the refractive index may be raised locally through the use of local Joule heating. It its simplest form, an integrated heater consists of a metallic strip on, or near to, a silicon rib waveguide. This effect has been exploited many times in large waveguides, with perhaps the most instructive example being that of a Mach Zehnder (MZ) interferometer such as described by Fischer et al. [99].

The utility of the MZ structure in switching is obvious. However, the thermal effect is relatively slow, on the order of kHz, while the power necessary to perform large numbers of switch operations may be crippling to a complex circuit.

2.5.4.2 Modulation via the Plasma Dispersion Effect

The presence of free carriers affects both the real and imaginary component of the refractive index of silicon. This permits the use of integrated electronic devices such as *p-n* and *p-i-n* diodes to be used in switching (e.g., in MZ structures) or for variable attenuation of a propagating signal. The degree of refractive index modification was summarized in the famous publication of Soref and Bennett [100], and for a wavelength of 1550 nm, it is summarized as

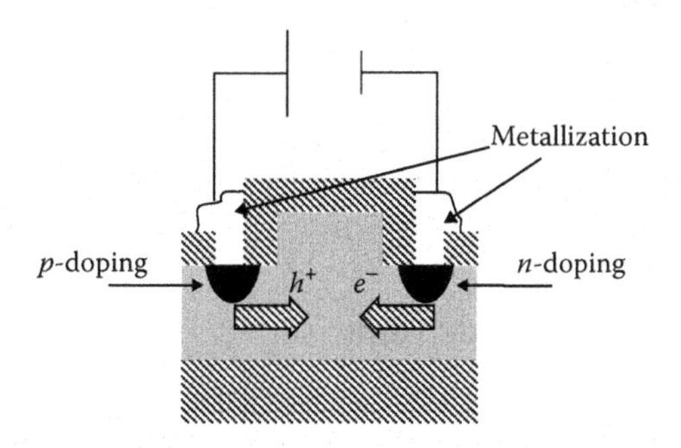

FIGURE 2.30
Cross section of a rib waveguide integrated with a *p-i-n* diode to form a variable optical attenuator.

$$\Delta n = -[8.8 \times 10^{-22}\Delta N_e + 8.5 \times 10^{-18}(\Delta N_h)^{0.8}] \tag{2.13a}$$

$$\Delta\alpha = 8.5 \times 10^{-18}\Delta N_e + 6.0 \times 10^{-18}\Delta N_h \tag{2.13b}$$

where Δn refers to the real component of refractive index and $\Delta\alpha$ refers to the optical absorption; N_e and N_h refer to the concentration of electrons and holes, respectively.

The integration of electronic and photonic functionality in moderate and large waveguides was commercially pioneered by Bookham Technology. Bookham offered a multichannel variable optical attenuator (VOA) and a MUX + VOA for telecommunication application [101].

At the time of writing, the most prominent commercial supplier of such devices is Kotura Inc. They provide both 4- and 8-channel VOAs with exceptional performance such as an attenuation range >25 dB, with off-state loss on the order of 1 dB and bandwidth on the order of MHz. Figure 2.30 is a schematic representation of a cross section of a device similar to that employed by Kotura and reported in [102]. The *p-i-n* diode is formed such that the highly doped p and n regions are either side of the silicon rib, or slab region, while the virtually intrinsic part is coincident with the volume in which light propagation takes place, thus there is little attenuation in the off-state. The length of the diode is typically on the order of several millimeters ensuring that significant on-state attenuation can be achieved.

For both previous sections, more details can be found in Chapter 9 on optical modulation.

2.6 Silica Waveguides for the Si-Based Photonic Platform

Tai Tsuchizawa

Recently, various photonic devices have been developed based on silicon photonic wire waveguides. They have excellent features, such as compactness, low power consumption, fast operation, and easy multichannel integration [103–105]; however, they are very sensitive to fabrication errors, and reducing propagation loss and the polarization dependence is not easy because of the large refractive index contrast between the core and the cladding. Consequently, it is difficult to apply them to practical passive devices, especially to telecommunications devices in which severe specifications are required. On the other hand, silica-based waveguides are very suitable for constructing high-performance passive photonic devices with low propagation loss and small polarization dependence because silica's low index contrast relaxes geometrical tolerances [106, 107]. Therefore, integrating a Si-based dynamic and active devices with a silica-based passive device would be a realistic approach to achieving practical Si photonics. In this section, silica waveguides used in the Si-based photonic platform are introduced.

2.6.1 Deposition of Silica Waveguide Films

A serious obstacle to integrating Si and silica waveguide devices is the thermal degradation of the Si devices during the fabrication of silica waveguides. Silica waveguide fabrication generally involves processes with temperatures exceeding 1000°C for silica film deposition [107]. Such high temperatures oxidize the Si core and destroy the PIN structures for Si modulation devices and Ge photodetectors. For Si and silica device integration, a low-temperature silica film deposition method is imperative. Silica films can be formed at low temperature by PE-CVD, and many types of PE-CVD equipment have been developed for Si electrical device processes. However, improvement is necessary for waveguide film deposition because waveguide films require accurate control of the refractive index. Here, we describe a low-temperature waveguide film formation method using electron-cyclotron-resonance plasma-enhanced chemical vapor deposition (ECR–PE-CVD) is introduced. The ECR plasma dissociates gas molecules efficiently and provides moderate energy to the substrate surface. This enables fast deposition of high-quality and index-controlled silica films, such as silicon oxynitride (SiO_xN_y) and silicon-rich oxide (SiO_x) films at low temperatures.

In ECR–PE-CVD, a mixture of O_2, N_2, and SiH_4 gases is used for the SiO_xN_y films, and O_2 and SiH_4 gases are used for SiO_x films. The O_2 and N_2 gases are introduced into the plasma chamber and the SiH_4 gas is introduced into the deposition chamber as shown in Figure 2.31. ECR plasma is generated in the plasma chamber using microwaves (2.45 GHz) and a magnetic field (875 G). The plasma is transported to the deposition chamber by a divergent magnetic field and irradiated to a wafer in the deposition chamber. The energy of the ions irradiated to a wafer is about 10–20 eV [108]. This moderate energy induces a reaction on the wafer surface so that high-quality films are formed at low temperatures. As the ECR–PE-CVD system does not add bias to the substrate, the wafer temperature during the film deposition can be kept below 200°C even without wafer cooling. To control the

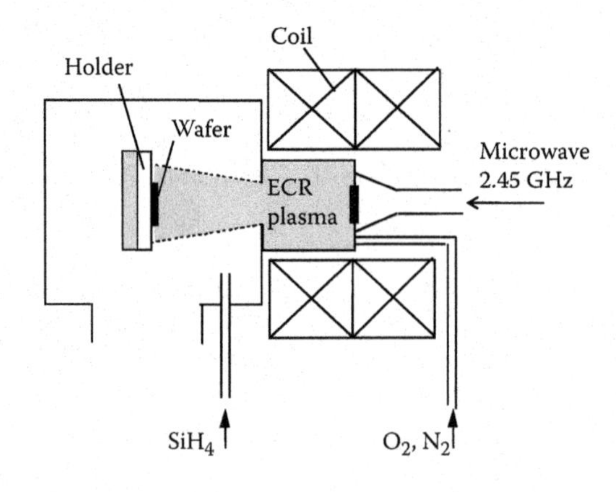

FIGURE 2.31
Schematic diagram of ECR–PE-CVD apparatus.

refractive index of the deposition films, the flow rate of O_2 and N_2 for SiO_xN_y films and the flow rate of O_2 for SiO_x films are adjusted with the flow of SiH_4 maintained at a fixed rate. The gas pressure during the deposition is kept at about 0.13 Pa.

To evaluate the optical loss of the deposited films, waveguides with a refractive index contrast of 3% have been fabricated. The refractive index of the waveguide core is 1.515, and the size is 3 × 3-μm^2. The overcladding is SiO_2, whose refractive index is 1.47 and the thickness is about 5 μm. These films were formed on 15-μm-thick thermal SiO_2 on a Si wafer.

2.6.2 Characteristics of SiO_xN_y and SiO_x Waveguides

SiO_xN_y is an attractive material for waveguide films because the refractive index can be continuously adjusted over a large range between 1.46 for SiO_2 and 2.0 for Si_3N_4. This feature provides large flexibility in optical waveguide design. Figure 2.32 shows the refractive index and the deposition rate of SiO_xN_y film deposited by ECR–PE-CVD with various flow rate ratios of O_2 and N_2. The refractive index can be changed continuously over a wide range from 1.47 to 1.95 by adjusting only the flow rate ratio of O_2 and N_2. The deposition rate is 150–200 nm/min, which is high enough for waveguide films that need a thickness of a few micrometers. In conventional PE-CVD, SiO_xN_y films are normally deposited using a mixture gas with $N_2O/NH_3/SiH_4$ gas mixture to control the film index [105, 109, 110]. N_2 is not used because it has a high dissociation energy and is not easily dissociated even in a plasma. However, in PE-CVD with ECR plasma, the refractive index can be widely controlled using only O_2 and N_2. This is because ECR plasma has a higher electron temperature, which means N_2 molecules dissociate efficiently so that N atoms are drawn into the films.

The transmission characteristics of the deposited films were evaluated by fabricating a Δ3% waveguide. Figure 2.33 shows the cross-sectional structure of the waveguide and a scanning electron microscopy image of a fabricated SiO_xN_y core. The waveguide losses not only largely depend on the transmission characteristics of the film but also on the core shape distortion and sidewall roughness that occur during fabrication. To get vertical and smooth sidewalls of SiO_xN_y core, optical lithography and reactive ion etching with CF_4 and SF_6 can be utilized.

Figure 2.34 shows the transmission spectra of a fabricated SiO_xN_y waveguide with 3-cm-length at wavelengths from 1300 to 1650 nm. Figure 2.35 shows the propagation loss of SiO_xN_y waveguides at wavelengths of 1510, 1550, and 1590 nm. The strong resonance absorptions observed near 1450 and 1515 nm are due to the overtone of the O–H and N–H

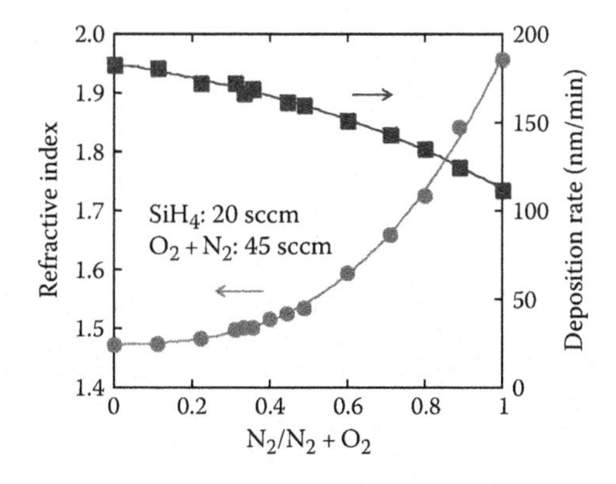

FIGURE 2.32
Refractive index and deposition rate of SiO_x with changing O_2 and N_2 flow ratio.

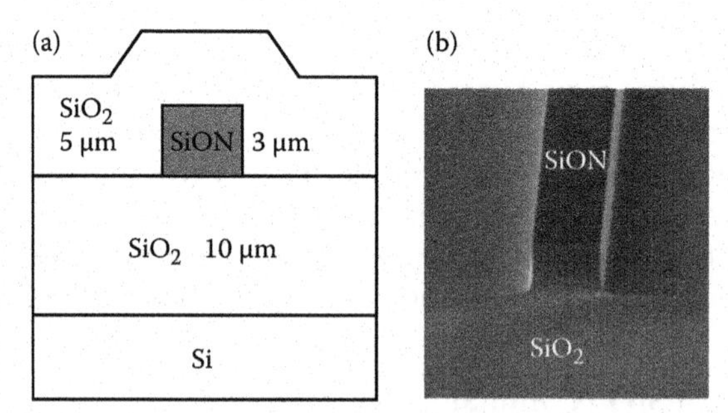

FIGURE 2.33
(a) Cross-sectional structure of a SiO_xN_y waveguide and (b) a SEM image of a fabricated SiO_xN_y core.

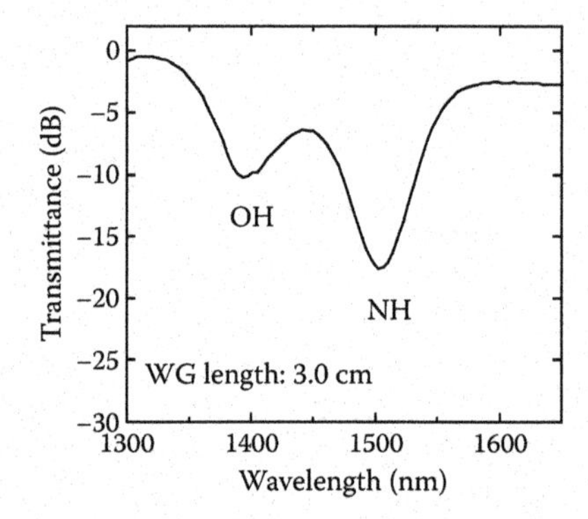

FIGURE 2.34
Transmission spectra of a fabricated SiO_xN_y waveguide with 3 cm length.

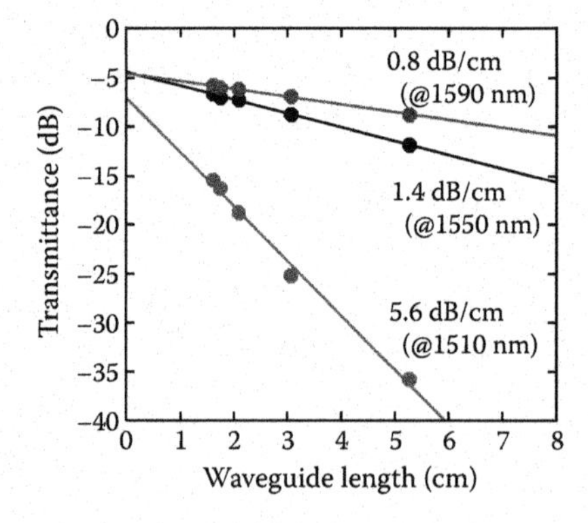

FIGURE 2.35
Propagation loss of a 3% SiON waveguide.

stretching vibration. The O–H and N–H exist because the PE-CVD silica films contain hydrogen, and they increase the propagation loss, which is over 5 dB/cm at 1510 nm. The loss due to the N–H is a serious problem for SiO_xN_y waveguides intended for use in telecommunications devices operating at wavelengths around 1550 nm. It is known that N–H bonds can be eliminated by annealing the films at temperatures above 1100°C [109, 110], but such temperatures are too high for the integration with Si devices. The N–H absorption is a large demerit of the SiO_xN_y film. However, it is sure that SiO_xN_y is still a good material

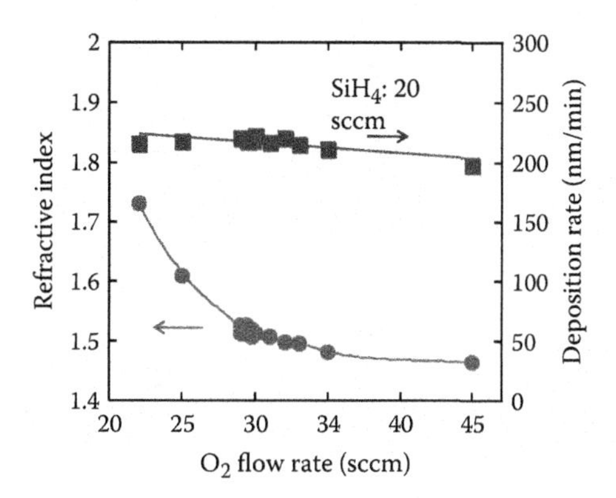

FIGURE 2.36
Refractive index and deposition rate of SiO_x with changing O_2 flow rate.

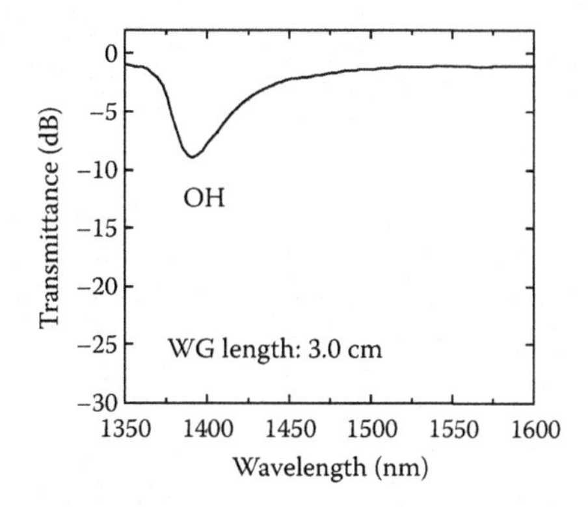

FIGURE 2.37
Transmission spectra of a fabricated SiO_x waveguide with 3 cm length.

for optical films not intended for use at the telecommunications wavelength. Actually, waveguides made with SiO_xN_y play an important role in the 850-nm wavelength region and are used in some integrated devices with Si [110, 111].

Si-rich oxide, SiO_x without N, is beneficial to avoid the absorption due to N–H. ECR–PE-CVD also produces good SiO_x waveguide film whose refractive index can be adjusted. Figure 2.36 shows the refractive index and deposition rate of SiO_x by ECR–PE-CVD with a SiH_4/O_2 gas mixture. The SiO_x film index can be controlled over a wide range from 1.47 to 1.60 by adjusting only the O_2 flow rate. The deposition rate is about 150 nm/min, which is high enough for depositing waveguide films. Figure 2.37 shows the transmission spectra of a SiO_x waveguide with 3 cm length. The absorption due to N–H around 1510 nm certainly disappears. This means that the SiO_x is a suitable material for telecommunications applications. Figure 2.38 shows a transmittance result for SiO_x waveguides for the TE mode as a function of waveguide length. The propagation loss at 1550 nm is about 0.59 dB/cm. This value for the Δ3% waveguide is low enough for making a practical device.

2.6.3 Monolithic Integration of Si- and Silica-Based Waveguide Devices

Using SiO_x films deposited at a low temperature, silica-based high-performance passive devices can be integrated with silicon-based active devices. As an example, integration

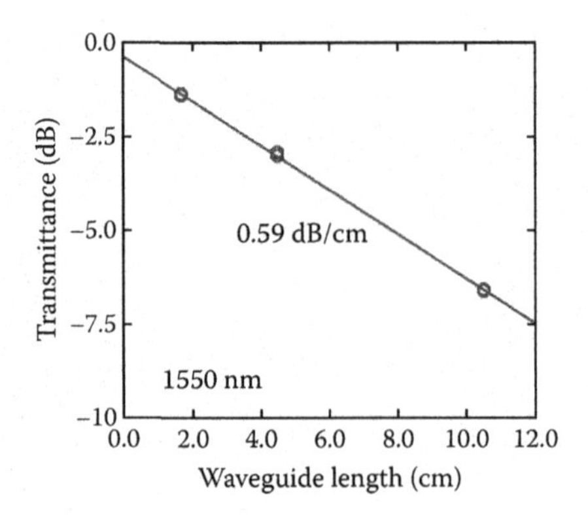

FIGURE 2.38
Propagation loss of a 3% SiO_x waveguide.

of PIN-type Si variable optical attenuators (VOAs) and a SiO_x arrayed waveguide grating (AWG) has been demonstrated [111], which is an important component of variable-attenuator multiplexers/demultiplexers (VMUXs/DEMUXs) in photonic networks. Figure 2.39 shows an optical microscope image of a fabricated VOA-AWG integrated device. The Si VOA is based on the rib-type Si waveguide with a 600 × 200-nm core and 100-nm-thick slab. The p^+ and n^+ regions are defined in the slab section and 5-mm long, and they are about 3 μm apart. The AWG, which is for wavelength demultiplexing in 16 channels with 200-GHz spacing, was made with Δ3% SiO_x waveguides. The whole device is only about 15 × 8 mm^2 in size.

The SiO_x and Si waveguide have very different dimensions. The core of the SiO_x waveguide, which composes the AWG, is about 3 × 3-μm^2. The Si wire waveguides, which compose the VOA, have a 0.2-μm core height, 0.6-μm core width, and 0.1-μm slab thickness. The large difference in core size causes a large coupling loss between the waveguides. This is another problem in the integration of Si and silica waveguides. To compensate for the mode-size mismatch and connect the Si and silica waveguides with low loss reduce, spot-size converters (SSCs) should be implemented. An SSC consists of a Si inverse taper and a low index waveguide with a large core covering the taper [112]. The low-index waveguide over the Si taper is made with the same SiO_x waveguide as for the AWG, which has about a 3-μm-square core and index contrast of about 3%. Each output port of the AWG is connected to a Si PIN-type VOA through SSCs.

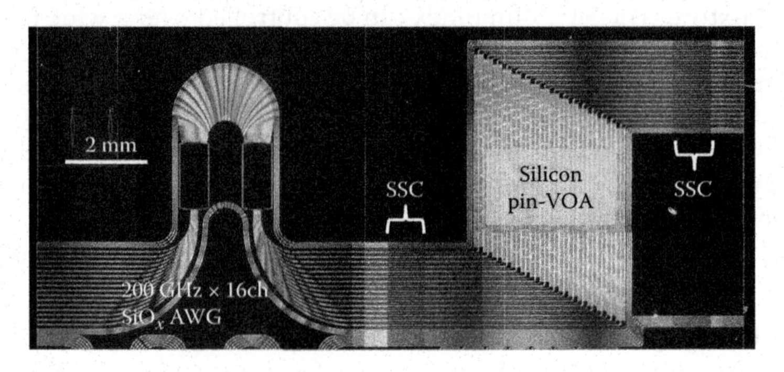

FIGURE 2.39
Optical microscope image of a fabricated VOA–AWG integrated device.

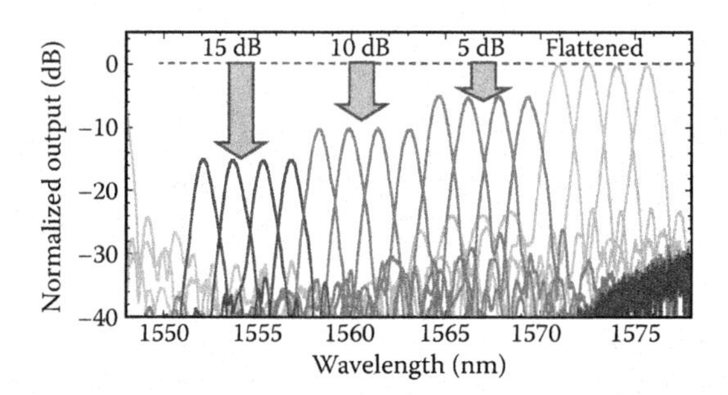

FIGURE 2.40
Demultiplexing and intensity adjustment for every channel of the VOA–AWG integrated device.

The integrated device was made as follows. First, Si-wire rib waveguide cores were fabricated on an SOI wafer. Then, to make a lateral PIN VOA, p^+ and n^+ regions were formed by ion implantation of boron and phosphorous, respectively. Next, the top Si layer in the region where an AWG is to be formed was removed by etching. Then, a SiO_x layer was deposited by low-temperature silica deposition with ECR plasma and a AWG and SSCs were fabricated. Then, a SiO_2 overcladding on the wafer was deposited using ECR-PE-CVD. Finally, contact holes were formed in the overcladding, followed by the formation of Al electrodes for the VOAs.

Owing to the Si-VOA and SiO_x-AWG, this device exhibits wavelength demultiplexing and high-speed power-level adjustment on each channel. Figure 2.40 shows the intensity adjustment results for every channel for the integrated VOA–AWG device. The response time of the VOAs is as fast as 15 ns. The complete function of these VOAs means that integration of silica-based device does not degrade performance of silicon-based active devices.

References

1. K. Yamada, in *Silicon Photonic Wire Waveguides: Fundamentals and Applications*, ed. by L. Pavesi, and D. Lockwood. Silicon Photonics II (Topics in Applied Physics 119, Springer-Verlag, 2011) p. 1.
2. A. S. Sudbo, Film mode matching: A versatile numerical method for vector mode field calculations in dielectric waveguides, *Pure Appl. Opt.* 2, 211–233 (1993).
3. K. Okamoto, *Fundamentals of Optical Waveguides* (Academic Press, 2000) p. 29.
4. W. Bogaerts, P. Dumon, D. Van Thourhout, D. Taillaert, P. Jaenen, J. Wouters, S. Beckx, V. Wiaux, and R. Baets, Compact wavelength-selective functions in silicon-on-insulator photonic wires, *IEEE J. Select. Topics Quant. Electron.* 12, 1394–1401 (2006).
5. F. P. Payne, and J. P. R. Lacey, A theoretical analysis of scattering loss from planar optical waveguide, *Opt. Quant. Electron.* 26, 977–986 (1994).
6. S. Janz, in *Silicon-Based Waveguide Technology for Wavelength Division Multiplexing*, ed. by L. Pavesi, and D. Lockwood. Silicon Photonics (Topics in Applied Physics 94, Springer-Verlag, 2004) p. 323.
7. P. Dumon, W. Bogaerts, V. Wiaux, J. Wouters, S. Beckx, J. V. Campenhout, D. Taillaert, B. Luyssaert, P. Bienstman, D. Van Thourhout, and R. Baets, Low-loss SOI photonic wires and ring resonators fabricated with deep UV lithography, *Photon. Technol. Lett.* 16, 1328–1330 (2004).

8. T. Watanabe, K. Yamada, T. Tsuchizawa, H. Fukuda, H. Shinojima, and S. Itabashi, Si wire waveguide devices, *Proc. SPIE* 6775, 67750K (2007).
9. T. Tsuchizawa, K. Yamada, T. Watanabe, S. Park, H. Nishi, R. Kou, H. Shinojima, and S. Itabashi, Monolithic integration of silicon-, germanium-, and silica-based optical devices for telecommunications applications, *IEEE J. Select. Topics Quant. Electron.* 17, 516–525 (2011).
10. T. Shoji, T. Tsuchizawa, T. Watanabe, K. Yamada, and H. Morita, Low loss mode size converter from 0.3 μm square Si wire waveguides to single mode fibres, *Electron. Lett.* 38, 1669–1670 (2002).
11. T. Tsuchizawa, K. Yamada, H. Fukuda, T. Watanabe, J. Takahashi, M. Takahashi, T. Shoji, E. Tamechika, S. Itabashi, and H. Morita, Microphotonics devices based on silicon microfabrication technology, *IEEE J. Select. Topics Quant. Electron.* 11, 232–240 (2005).
12. D. Taillaert, F. van Laere, M. Ayre, W. Bogaaerts, D. van Thourhout, P. Bienstman, and R. Baets, Grating couplers for coupling between optical fibers and nanophotonic waveguides, *Jpn. J. Appl. Phys.* 45, 6071–6077 (2006).
13. K. Yamada, T. Tsuchizawa, T. Watanabe, H. Shinojima, H. Nishi, S. Park, Y. Ishikawa, K. Wada, and S. Itabashi, Silicon photonics based on photonic wire waveguides, in *14th OptoElectronics and Communications Conference (OECC) IEEE*, ThG3. ISBN: 978-1-4244-4103-7, IEEE Catalog Number: CFP0975G-CDR.
14. K. K. Lee, D. R. Lim, L. C. Kimerling, J. Shin, and F. Cerrina, Fabrication of ultralow-loss Si/SiO_2 waveguides by roughness reduction, *Opt. Lett.* 26, 1888–1890 (2001).
15. T. Tsuchizawa, K. Yamada, H. Fukuda, T. Watanabe, S. Uchiyama, and S. Itabashi, Low-loss Si wire waveguides and their application to thermooptic switches, *Jpn. J. Appl. Phys.* 45, 6658–6662 (2006).
16. K. Yamada, T. Tsuchizawa, T. Watanabe, J. Takahashi, H. Fukuda, M. Takahashi, T. Shoji, S. Uchiyama, E. Tamechika, S. Itabashi, and H. Morita, Silicon wire waveguiding system: Fundamental characteristics and applications, *Electron. Commun. Jpn. Pt. 2*, 89, 42–55 (2006).
17. Y. A. Vlasov, and S. J. McNab, Losses in single-mode silicon-on-insulator strip waveguides and bends, *Opt. Express* 12, 1622–1631 (2004).
18. T. Tsuchizawa, K. Yamada, H. Fukuda, T. Watanabe, J. Takahashi, M. Takahashi, T. Shoji, E. Tamechika, S. Itabashi, and H. Morita, Microphotonics devices based on silicon micro-fabrication technology, *IEEE J. Select. Top. Quant. Electron.* 11, 232–240 (2005).
19. A. Sakai, G. Hara, and T. Baba, Propagation characteristics of ultrahigh-Δ optical waveguide, *Jpn. J. Appl. Phys.* 40, L384 (2001).
20. K. K. Lee, D. R. Lim, H.-C. Luan, A. Agarwal, J. Foresi, and L. C. Kimerling, Effect of size and roughness on light transmission in a Si/SiO_2 waveguide: Experiments and model, *Appl. Phys. Lett.* 77, 1617–1619 (2000).
21. K. Yamada, T. Shoji, T. Tsuchizawa, T. Watanabe, J. Takahashi, and S. Itabashi, Silicon-wire–based ultrasmall lattice filters with wide free spectral ranges, *Opt. Lett.* 28, 1663–1664 (2003).
22. H. Fukuda, K. Yamada, T. Shoji, M. Takahashi, T. Tsuchizawa, T. Watanabe, J. Takahashi, and S. Itabashi, Four-wave mixing in silicon wire waveguides, *Opt. Express* 13, 4629–4637 (2005).
23. D. Taillaert, H. Chong, P. I. Borel, L. H. Frandsen, R. M. De La Rue, and R. Baets, A compact two-dimensional grating coupler used as a polarization splitter, *IEEE Photon. Technol. Lett.* 15, 1249–1251 (2003).
24. W. Bogaerts, D. Taillaert, P. Dumon, D. V. Thourhout, and R. Baets, A polarization-diversity wavelength duplexer circuit in silicon-on-insulator photonic wires, *Opt. Express* 15, 1567–1578 (2007).
25. F. V. Laere, W. Bogaerts, P. Dumon, G. Roalkens, D. V. Thourhout, and R. Baets, Focusing polarization diversity grating couplers in silicon-on-insulator, *J. Lightwave Technol.* 27, 612–618 (2009).
26. M. R. Watts, H. A. Haus, and E. P. Ippen, Integrated mode-evolution–based polarization splitter, *Opt. Lett.* 30, 967–969 (2005).
27. M. R. Watts, H. A. Haus, and E. P. Ippen, Integrated mode-evolution–based polarization rotators, *Opt. Lett.* 30, 138–140 (2005).

28. T. Barwicz, M. R. Watts, M. A. Popović, P. T. Rakich, L. Socci, F. X. Kaertner, E. P. Ippen, and H. I. Smith, Polarization-transparent microphotonic devices in the strong confinement limit, *Nat. Photon.* 1, 57–60 (2006).
29. J. Zhang, M. Yu, G. Lo, and D. Kwong, Silicon-waveguide–based mode evolution polarization rotator, *IEEE. J. Select. Top. Quant. Electron.* 16, 53–60 (2010).
30. H. Fukuda, K. Yamada, T. Tsuchizawa, T. Watanabe, S. Shinojima, and S. Itabashi, Ultrasmall polarization splitter based on silicon wire waveguides, *Opt. Express* 14, 12401–12408 (2006).
31. H. Fukuda, K. Yamada, T. Tsuchizawa, T. Watanabe, S. Shinojima, and S. Itabashi, Polarization rotator based on silicon wire waveguides, *Opt. Express* 16, 2628–2635 (2008).
32. H. Fukuda, K. Yamada, T. Tsuchizawa, T. Watanabe, S. Shinojima, and S. Itabashi, Silicon photonic circuit with polarization diversity, *Opt. Express* 16, 4872–4880 (2008).
33. V. R. Almeida, Q. Xu, C. A. Barrios, and M. Lipson, Guiding and confining light in void nanostructure, *Opt. Lett.* 29, 1209–1211 (2004).
34. Q. Xu, V. R. Almeida, R. R. Panepucci, and M. Lipson, Experimental demonstration of guiding and confining light in nanometer-sizelow-refractive-index material, *Opt. Lett.* 29, 1626–1628 (2004).
35. T. W. Baehr-Jones, and M. J. Hochberg, Polymer silicon hybrid systems: A platform for practical nonlinear optics, *J. Phys. Chem. C* 112, 8085–8090 (2008).
36. C. Koos, P. Vorreau, T. Vallaitis, P. Dumon, W. Bogaerts, R. Baets, B. Esembeson, I. Biaggio, T. Michinobu, F. Diederich, W. Freude, and J. Leuthold, All-optical high-speed signal processing with silicon-organic hybrid slot waveguides, *Nat. Photon.* 3, 216–219 (2009).
37. J. Leuthold, W. Freude, J.-M. Brosi, R. Baets, P. Dumon, I. Biaggio, M. Scimeca, F. Diederich, B. Frank, and C. Koos, Silicon organic hybrid technology: A platform for practical nonlinear optics, *Proc. IEEE*, 97, 1304–1316 (2009).
38. C. Koos, L. Jacome, C. Poulton, J. Leuthold, and W. Freude, Nonlinear silicon-on-insulator waveguides for all-optical signal processing, *Opt. Express* 15, 5976–5990 (2007).
39. P. Muellner, and R. Hainberger, Structural optimization of silicon-on-insulator slot waveguides, *IEEE Photon. Technol. Lett.* 18, 2557–2559 (2006).
40. X. G. Tu, X. J. Xu, S. W. Chen, J. Z. Yu, and Q. M. Wang, Simulation demonstration and experimental fabrication of a multiple-slot waveguide, *IEEE Photon. Technol. Lett.* 20, 333–335 (2008).
41. N.-N. Feng, J. Michel, and L. Kimerling, Optical field concentration in low-index waveguides, *IEEE J. Quant. Electron.* 42, 885–890, Sept. (2006).
42. T. Baehr-Jones, M. Hochberg, G. Wang, R. Lawson, Y. Liao, P. Sullivan, L. Dalton, A. Jen, and A. Scherer, Optical modulation and detection in slotted Silicon waveguides, *Opt. Express* 13(14), 5216–5226 (2005).
43. R. Ding, T. Baehr-Jones, W. J. Kim, X. G. Xiong, R. Bojko, J. M. Fedeli, M. Fournier, and M. Hochberg, Low-loss strip-loaded slot waveguides in silicon-on-insulator, *Opt. Express* 18, 25061–25067 (2010).
44. E. Jordana, J. M. Fedeli, P. Lyan, J. P. Colonna, P. Gautier, N. Daldosso, L. Pavesi, Y. Lebour, P. Pellegrino, B. Garrido, J. Blasco, F. Cuesta-Soto, and P. Sanchis, Deep-UV lithography fabrication of slot waveguides and sandwiched waveguides for nonlinear applications, *4th IEEE International Conference on Group IV Photonics*, 217–219 (2007).
45. C. G. Poulton, C. Koos, M. Fujii, A. Pfrang, T. Schimmel, J. Leuthold, and W. Freude, Radiation modes and roughness loss in high index-contrast waveguides, *IEEE J. Select. Top. Quant. Electron.* 12, 1306–1321 (2006).
46. T. Baehr-Jones, M. Hochberg, G. Wang, R. Lawson, Y. Liao, P. Sullivan, L. Dalton, A. Jen, and A. Scherer, Optical modulation and detection in slotted silicon waveguides, *Opt. Express* 13, 5216–5226 (2005).
47. R. M. Pafchek, J. Li, R. S. Tummidi, and T. L. Koch, Low loss Si-SiO2-Si 8-nm slot waveguides, *IEEE Photon. Technol. Lett.* 21, 353–355 (2009).
48. S. Mas, J. Caraquitena, J. V. Galan, P. Sanchis, and J. Marti, Tailoring the dispersion behavior of silicon nanophotonic slot waveguides, *Opt. Express* 18, 20839–20844 (2010).

49. L. Zhang, Y. Yue, Y. Xiao-Li, J. Wang, R. G. Beausoleil, and A. E. Willner, Flat and low dispersion in highly nonlinear slot waveguides, *Opt. Express* 18, 13187–13193 (2010).
50. F. Dell'Olio, and V. M. N. Passaro, Optical sensing by optimized silicon slot waveguides, *Opt. Express* 15, 4977–4993 (2007).
51. G. J. Veldhuis, O. Parriaux, H. J. W. M. Hoekstra, and P. V. Lambeck, Sensitivity enhancement in evanescent optical waveguide sensors, *J. Lightwave Technol.* 18, 677 (2000).
52. J.-M. Lee, D.-J. Kim, G.-H. Kim, O.-K. Kwon, K.-J. Kim, and G. Kim, Controlling temperature dependence of silicon waveguide using slot structure, *Opt. Express* 16, 1645–1652 (2008).
53. L. J. Zhou, K. Okamoto, and S. J. B. Yoo, Athermalizing and trimming of slotted silicon microring resonators with UV-sensitive PMMA upper-cladding, *IEEE Photon. Technol. Lett.* 21, 1175–1177 (2009).
54. C. A. Barrios, K. B. Gylfason, B. Sanchez, A. Griol, H. Sohlstrom, M. Holgado, and R. Casquel, Slot-waveguide biochemical sensor, *Opt. Lett.* 32, 3080–3082 (2007).
55. C. A. Barrios, Optical slot-waveguide based biochemical sensors, *Sensors*, 9, 4751–4765 (2009).
56. A. Kargar, and C. Y. Chao, Design and optimization of waveguide sensitivity in slot microring sensors, *J. Opt. Soc. Am. A Opt. Image Sci. Vis.* 28, 596–603 (2011).
57. L. Alloatti, D. Korn, R. Palmer, D. Hillerkuss, J. Li, A. Barklund, R. Dinu, J. Wieland, M. Fournier, J. Fedeli, H. Yu, W. Bogaerts, P. Dumon, R. Baets, C. Koos, W. Freude, and J. Leuthold, 42.7 Gbit/s electro-optic modulator in silicon technology, *Opt. Express* 19(12), 11841–11851 (2011).
58. Y. Enami, C. T. Derose, D. Mathine, C. Loychik, C. Greenlee, R. A. Norwood, T. D. Kim, J. Luo, Y. Tian, A. K.-Y. Jen, and N. Peyghambarian, Hybrid polymer/sol-gel waveguide modulators with exceptionally large electro-optic coefficients, *Nat. Photon.* 1, 180–185 (2007).
59. J. Witzens, T. Baehr-Jones, and M. Hochberg, Design of transmission line driven slot waveguide Mach-Zehnder interferometers and application to analog optical links, *Opt. Express* 18, 16902–16928 (2010).
60. J.-M. Brosi, C. Koos, L. C. Andreani, M. Waldow, J. Leuthold, and W. Freude, High-speed low-voltage electro-optic modulator with a polymer-infiltrated silicon photonic crystal waveguide, *Opt. Express* 16, 4177–4191 (2008).
61. T. Vallaitis, S. Bogatscher, L. Alloatti, P. Dumon, R. Baets, M. L. Scimeca, I. Biaggio, F. Diederich, C. Koos, W. Freude, and J. Leuthold, Optical properties of highly nonlinear silicon-organic hybrid (SOH) waveguide geometries, *Opt. Express* 17, 17357–17368 (2009).
62. J. H. Schmid, P. Cheben, P. J. Bock, R. Halir, J. Lapointe, S. Janz, A. Delâge, A. Densmore, J.-M. Fédeli, T. J. Hall, B. Lamontagne, R. Ma, I. Molina-Fernández, and D.-X. Xu, Refractive index engineering with subwavelength gratings in silicon microphotonic waveguides, accepted for publication in *IEEE Photon. J.* (2011).
63. S. M. Rytov, Electromagnetic properties of a finely stratified medium, *Sov. Phys. JETP* 2, 466–475 (1956).
64. J. N. Mait, and W. W. Prather, Selected papers on subwavelength diffractive optics, *SPIE Milestone Series*, V. Ms 166 (2001).
65. H. Kikuta, H. Toyota, and W. Yu, Optical elements with subwavelength structured surfaces, *Opt. Rev.* 10, 63–73 (2003).
66. C. F. R. Mateus, M. C. Y. Huang, L. Chen, C. J. Chang-Hasnain, and Y. Suzuki, Broad-band mirror (1.12–1.62 μm) using a subwavelength grating, *IEEE Photon. Technol. Lett.* 16, 1676–1678 (2004).
67. J. H. Schmid, P. Cheben, S. Janz, J. Lapointe, E. Post, and D.-X. Xu, Gradient-index antireflective subwavelength structures for planar waveguide facets, *Opt. Lett.* 32, 1794–1796 (2007).
68. J. D. Joannopoulos, S. G. Johnson, J. N. Winn, and R. D. Maede, *Photonic Crystals: Molding the Flow of Light*, 2nd edn. (Princeton University Press: Princeton and Oxford, 2008).
69. P. J. Bock, P. Cheben, J. H. Schmid, J. Lapointe, A. Delâge, S. Janz, G. C. Aers, D.-X. Xu, A. Densmore, and T. J. Hall, Subwavelength grating periodic structures in silicon-on-insulator: A new type of microphotonic waveguide, *Opt. Express* 18, 20251–20262 (2010).

70. P. Cheben, P. J. Bock, J. H. Schmid, J. Lapointe, S. Janz, D.-X. Xu, A. Densmore, A. Delâge, B. Lamontagne, and T. J. Hall, Refractive index engineering with subwavelength gratings for efficient microphotonic couplers and planar waveguide multiplexers, *Opt. Lett.* 35, 2526–2528 (2010).
71. http://ab-initio.mit.edu/wiki/index.php/MIT_Photonic_Bands.
72. P. Cheben, D.-X. Xu, S. Janz, and A. Densmore, Subwavelength waveguide grating for mode conversion and light coupling in integrated optics, *Opt. Express* 14, 4695–4702 (2006).
73. V. R. Almeida, R. R. Panepucci, and M. Lipson, Nanotaper for compact mode conversion, *Opt. Lett.* 28, 1302–1304 (2003).
74. W. Śmigaj, P. Lalanne, J. Yang, T. Paul, C. Rockstuhl, and F. Lederer, Closed-formed expression for the scattering coefficients at the interface between periodic media, *Appl. Phys. Lett.* 98, 111107/1–11107/3 (2011).
75. W. Bogaerts, P. Dumon, D. Van Thourhout, R. Baets, Low-loss, low-cross-talk crossings for silicon-on-insulator nanophotonic waveguides, *Opt. Lett.* 32, 2801–2803 (2007).
76. P. Sanchis, P. Villalba, F. Cuesta, A. Håkansson, A. Griol, J. V. Galán, A. Brimont, and J. Martí, Highly efficient crossing structure for silicon-on-insulator waveguides, *Opt. Lett.* 34, 2760–2762 (2009).
77. P. J. Bock, P. Cheben, J. H. Schmid, J. Lapointe, A. Delâge, D.-X. Xu, S. Janz, A. Densmore, and T. J. Hall, Subwavelength grating crossings for silicon wire waveguides, *Opt. Express* 18, 16146–16155 (2010).
78. J.-M. Lee, D.-J. Kim, H. Ahn, S.-H. Park, and G. Kim, Temperature dependence of silicon nanophotonic ring resonator with a polymeric overlayer, *J. Lightwave Technol.* 25, 2236–2243 (2007).
79. W. N. Ye, J. Michel, L. C. Kimerling, Athermal high-index contrast waveguide design, *IEEE Photon. Technol. Lett.* 20, 885–887 (2008).
80. J. Teng, P. Dumon, W. Bogaerts, H. Zhang, X. Jian, X. Han, M. Zhao, G. Morthier, and R. Baets, Athermal silicon-on-insulator ring resonators by overlaying a polymer cladding on narrowed waveguides, *Opt. Express* 17, 14627–14633 (2009).
81. J. H. Schmid, M. Ibrahim, P. Cheben, J. Lapointe, S. Janz, P. J. Bock, A. Densmore, B. Lamontagne, R. Ma, W. N. Ye, D.-X. Xu, Temperature-independent silicon subwavelength grating waveguides, *Opt. Lett.* 36, 2110–2112 (2011).
82. R. Halir, P. Cheben, S. Janz, D.-X. Xu, I. Molina-Fernández, J. G. Wangüemert-Pérez, Waveguide grating coupler with subwavelength microstructures, *Opt. Lett.* 34, 1408–1410 (2008).
83. R. Halir, P. Cheben, J. H. Schmid, R. Ma, D. Bedard, S. Janz, D.-X. Xu, A. Densmore, J. Lapointe, I. Molina-Fernández, Continuously apodized fiber-to-chip-surface coupler with refractive index engineered subwavelength structure, *Opt. Lett.* 35, 3243–3245 (2010).
84. X. Chen, and H. K. Tsang, Nanoholes grating couplers for coupling between silicon-on-insulator waveguides and optical fibers, *IEEE Photon. J.* 1, 183–190 (2009).
85. L. Liu, M. Pu, K. Yvind, and J. M. Hvam, High-efficiency, large-bandwidth silicon-on-insulator grating coupler based on a fully-etched photonic crystal structure, *Appl. Phys. Lett.* 96, 051126 (2010).
86. J. Feng, and Z. Zhou, Polarization beam splitter using a binary blazed grating coupler, *Opt. Lett.* 32, 1662–1664 (2007).
87. D. H. Spadoti, L. H. Gabrielli, C. B. Poitras, and M. Lipson, Focusing light in a curved space, *Opt. Express* 18, 3181–3186 (2010).
88. G. Moore, Cramming more components onto integrated circuits, *Electronics* 38, 114–117 (1965).
89. G. Celler, and S. Cristoloveanu, Frontiers of silicon-on-insulator, *J. Appl. Phys.* 93, 4955–4978 (2003).
90. J. Schmidtchen, A. Splett, B. Schuppert, K. Petermann, and G. Burbach, Low-loss single-mode optical waveguides with large cross-section in silicon-on-insulator, *Electron. Lett.* 27, 1486–1488 (1991).
91. B. N. Kurdi, and D. G. Hall, Optical waveguides in oxygen implanted buried oxide silicon on insulator structures, *Opt. Lett.* 13, 175–177 (1988).
92. R. A. Soref, Silicon based optoelectronics, *Proc. IEEE* 81, 1687–1706 (1993).
93. G. T. Reed, and A. P. Knights, *Silicon Photonics: An Introduction* (Wiley, 2004).

94. R. Soref, J. Schmidtchen, and K. Petermann, Large single mode rib waveguides in GeSi-Si and Si-on-SiO_2, *J. Quant. Elect.* 27, 1971–1974 (1991).
95. T. Bestwick, ASOC™, A silicon based integrated optical manufacturing technology, *Proc. IEEE Electron. Comp. Technol. Conf.* 566 (1998).
96. W. Bogarerts, S. K. Selvaraja, P. Dumon, J. Brouckaert, K. De Vos, D. VanThourhout, and R. Baets, Silicon-on-insulator spectral filters fabricated with CMOS technology, *IEEE J. Select. Top. Quant. Electron.* 16, 33–44 (2010).
97. P. D. Trinh, S. Yegnanarayanan, F. Coppinger, and B. Jalali, Silicon-on-Insulator (SOI) phased-array wavelength multi/demultiplexer with extremely low-polarization sensitivity, *IEEE Photon. Technol. Lett.* 9, 940–942 (1997).
98. I. Day, I. Evans, A. Knights, F. Hopper, S. Roberts, J. Johnston, S. Day, J. Luff, H. Tsang, and M. Asghari, Tapered silicon waveguides for low insertion loss highly efficient high speed electronic variable attenuators, *IEEE OFC* 1, March 24–27, 2003, 249–251.
99. U. Fischer, T. Zinke, B. Schuppert, and K. Petermann, Single-mode optical switches based on SOI waveguides with large cross-section, *Electron. Lett.* 50, 406–408 (1994).
100. R. Soref, and B. Bennett, Electrooptical effects in silicon, *IEEE J. Quant. Elect.* 23, 123–129 (1987).
101. I. E. Day, S. W. Roberts, R. O'Carroll, A. Knights, P. Sharp, G. F. Hopper, B. J. Luff, and M. Asghari, Single-chip variable optical attenuator and multiplexer subsystem integration, *IEEE OFC Conf.*, 72–73 (2002).
102. D. Zheng, B. Smith, and M. Asghari, Improved efficiency Si-photonic attenuator, *Opt. Express* 16, 16754–16765 (2008).
103. D. M. Morini, L. Vivien, G. Rasigade, J. M. Fedeli, E. Cassan, X. L. Roux, P. Crozat, S. Maine, A. Lupu, P. Lyan, P. Rivallin, M. Halbwax, and S. Laval, Recent progress in high-speed silicon-based optical modulators, *Proc. IEEE* 97, 1199–1215 (2009).
104. R. Kasahara, K. Watanabe, M. Itoh, Y. Inoue, and A. Kaneko, Extremely low power consumption thermooptic switch (0.6mW) with suspended ridge and silicon-silica hybrid waveguide structure, P.2.02, ECOC, 21–25 Sept. 2008, Brussels, Belgium (2008).
105. N. Izhaky, M. T. Morse, S. Koehl, O. Cohen, D. Rubin, A. Barkai, G. Sarid, R. Cohen, and M. J. Paniccia, Development of CMOS-compatible integrated silicon photonics devices, *IEEE J. Select. Top. Quant. Electron.* 12, 1688–1698 (2006).
106. A. Himeno, K. Kato, and T. Miya, Silica-based planar lightwave circuits, *IEEE J. Select. Top. Quant. Electron.* 4, 913–924 (1998).
107. M. Kawachi, Silica waveguides on silicon and their application to integrated-optic components, *Opt. Quantum Electron.* 22, 391–416 (1990).
108. S. Matsuo, and M. Kiuchi, Low temperature chemical vapor deposition method utilizing an electron cyclotron resonance plasma, *Jpn. J. Appl. Phys.* 22, L210–212 (1983).
109. R. Germann, H. W. M. Salemink, R. Beyeler, G. L. Bona, F. Horst, I. Massarek, and B. J. Offrein, Silicon oxynitride layers for optical waveguide applications, *J. Electrochem. Soc.* 147, 2237–2241 (2000).
110. G. L. Bora, R. Germann, and B. J. Offrein, SiON high refravtive-index waveguide and planar lightwave circuits, *IBM J. Res. Dev.* 47, 239–249 (2003).
111. T. Tsuchizawa, T. Watanabe, K. Yamada, H. Fukuda, S. Itabashi, J. Fujikata, A. Gomyo, J. Ushida, D. Okamoto, K. Nishi, and K. Ohashi, Low-loss silicon oxynitride waveguides and branches for the 850-nm-wavelength region, *Jpn. J. Appl. Phys.* 47, 6739–6743 (2008).
112. H. Nishi, T. Tsuchizawa, T. Watanabe, H. Shinojima, S. Park, R. Kou, K. Yamada, and S. Itabashi, Monolithic integration of a silica-based arrayed waveguide grating filter and silicon variable optical attenuators based on p-i-n carrier-injection structure, *Appl. Phys. Express* 3, 102203 (2010).

3

Off-Chip Coupling

Wim Bogaerts and Diedrik Vermeulen

CONTENTS

3.1 Introduction

The advantage of silicon photonics is its potential for large-scale integration. The key enabler there is the very high index contrast: as discussed in the previous chapter, light can be confined into a submicron waveguide code and guided in bends with radii of only a few micrometers. In addition, the potential of wafer-scale processing, integration of active components, electronics, and even III–V semiconductors (topics discussed further in this book). However, the submicron waveguide core introduces a significant problem of its own: coupling light into and out of the chip to the outside world. The standard for transporting light at telecom wavelengths is single-mode fiber where the optical mode has a mode-field diameter of 10.4 μm at 1550 nm and 9.2 μm at 1310 nm. This is illustrated in Figure 3.1. When joined together at the interface, coupling efficiency from one waveguide to the other is of the order of 0.1% (−30 dB).

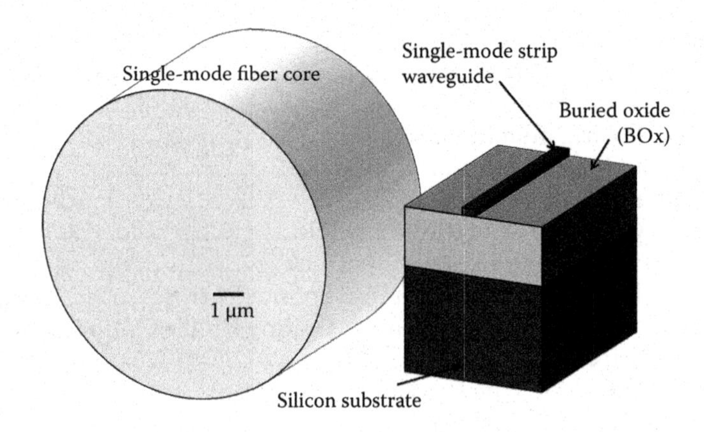

FIGURE 3.1
Optical mode size of a single-mode fiber and a silicon photonic wire waveguide, drawn to scale.

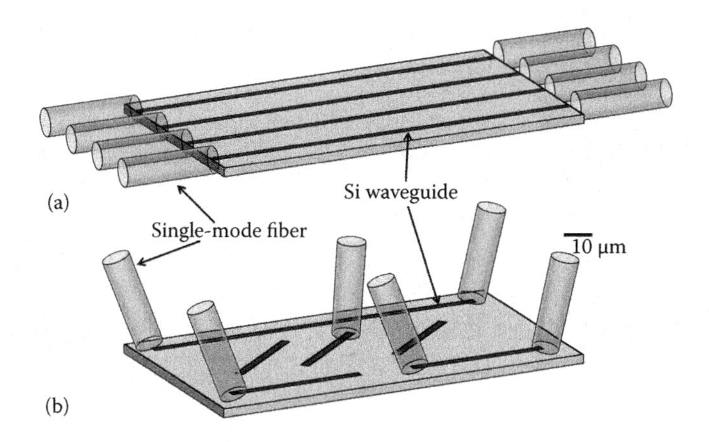

FIGURE 3.2
The two most common coupling solutions: (a) in-plane couplers and (b) out-of-plane couplers.

Bringing electrical signals off-chip is a relatively trivial challenge: a galvanic contact should be established, which can be done through wire-bonding schemes or flip-chip bumps. Electrical signals can be easily routed over various levels, materials, and cross sections. However, things become somewhat more complex when very high-speed signals need to be transferred: at that point, careful RF design is required and the effects of termination, impedance matching, and crosstalk come into play. Similar constraints apply for photonics: light cannot just change direction, and when coupling between waveguides, one needs to take care of mode matching, reflections and crosstalk.

Therefore, efficiently converting light from a fiber into a submicron on-chip silicon waveguide and vice versa due to reciprocity is a nontrivial challenge. This chapter will discuss the various approaches to address this problem.

The most common solutions for coupling are *in-plane* coupling, also often referred to as *edge* coupling, and *grating couplers,* which enable the coupling of light from the surface of the chip. These two solutions are schematically represented in Figure 3.2.

In the next section, we will discuss some generic concepts relating to fiber-to-chip waveguide coupling and introduce a number of metrics to evaluate the different solutions. In Section 3.3, we focus on in-plane mode converters for edge-coupling solutions, while in Section 3.4, we give a detailed overview of grating couplers for surface coupling. Section 3.5 steps away from fiber-based coupling solutions, looking at the options of coupling light from free space. In the conclusion in Section 3.6, we recapitulate the different solutions and put them next to one another based on the criteria we defined in Section 3.2.

3.2 Fiber-to-Waveguide Coupling

Before discussing the most common approaches of fiber-to-waveguide coupling, we will introduce some generic concepts that are common to all solutions. First of all, we will look at the generic problem of coupling between two waveguide modes and discuss the mechanisms with which this can be efficiently accomplished. Based on that, we will look at the technical metrics that determine the efficacy of a specific coupling solution, so we can compare the different implementations on a somewhat objective basis.

3.2.1 Coupling Mechanisms

If we consider the coupling between two single-mode optical waveguides, the modes in both waveguides should be matched both in real space and in k-space (impulse). Real-space matching means that the spatial distribution (i.e., the mode profile) of one waveguide should be converted into the mode profile of the other waveguide, such that the overlap integral is maximized [1]. If the modal overlap is not good, this will result in coupling to nonguided radiation modes, or if one of the waveguides is multimode, this will result in coupling to unwanted guided modes, which can cause interference and mode beating within the circuit.

Matching in k-space means that the difference in propagation vector between the guided modes of both waveguides is efficiently transformed by the coupler. This includes differences in magnitude and direction. This is the optical equivalent to impedance matching used in the microwave regime, and the effects of imperfect matching are similar: additional losses, scattering, and especially back-reflection.

It is important to note that *coupling between optical modes* is bidirectional: unless strong nonlinear or nonreciprocal effects come into play, the coupling efficiency of a particular coupler in one direction is the same as in the other direction. In the case of single-mode waveguides, mode-to-mode coupling efficiency is equivalent to the overall power coupling efficiency. In the case of multimode waveguides, this is no longer the case, and there we should make a distinction between the power coupled to the desired guided mode and between excitation of unwanted higher order modes. Furthermore, if the vertical symmetry of the waveguide is broken, coupling between the different polarizations that are guided by the waveguide can occur.

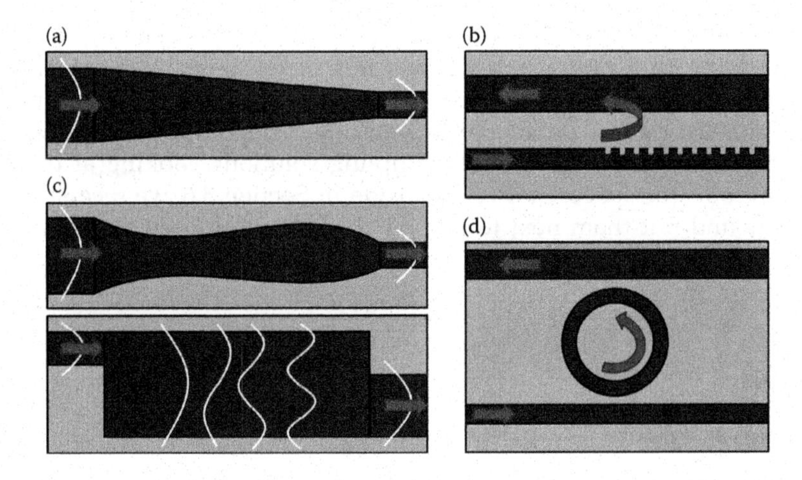

FIGURE 3.3
Coupling mechanisms: (a) adiabatic coupling, (b) diffractive coupling, (c) multimode coupling, and (d) resonant coupling.

We can discern several common mechanisms to handle both the spatial and the impulse mismatch, and they have all been demonstrated to couple light from high-contrast on-chip waveguides to optical fibers. They are illustrated in Figure 3.3.

3.2.1.1 Adiabatic Transition

The simplest method to match the modes of two different waveguide cross sections is by a long adiabatic *taper* between the two waveguides. The length of the taper is mainly determined by the mismatch that has to be overcome and by the distance in k-space of the nearest modes, which act as parasitic coupling channels. In the case of silicon photonic wires and single-mode fibers, this mismatch is quite large, and long taper sections are needed. Most edge-coupling solutions, discussed further in this chapter in Section 3.3, are based on this principle.

3.2.1.2 Diffraction

Periodic gratings scatter light at each grating tooth, and when all scattering contributions are in phase, constructive interference occurs. This effect can be used for coupling between two waveguides: by carefully tuning the scattering cross section and the period of the grating, the total scattered field can be tailored to match the target waveguide mode as close as possible. In addition, the periodicity of the grating has its own impulse, which transforms the impulse mismatch between the two waveguide modes. Because of this, the grating can also be used to change the direction of the light: for off-chip coupling, the light of the on-chip waveguide can be diverted out-of-plane. Such diffractive gratings or *grating couplers* are now the most commonly used technique for surface coupling to silicon photonic chips.

3.2.1.3 Multimode or Multipath Interference

A similar mechanism to diffractive coupling is multimode or multipath coupling. In this approach, the light from the single-mode waveguide is distributed over multiple channels, which can be either different paths, or the modes of a multimode waveguide. By tuning the phase delays and magnitude of the different contributions, the resulting field at the entrance of the target waveguide corresponds closely to the waveguide mode, enabling efficient coupling. A special case of this mechanism is *evanescent coupling*, as used in directional couplers: in that case, the light in two closely spaced waveguides is distributed over two supermodes (even and odd) and depending of the coupler length, a fraction of light is coupled from one waveguide to the other.

3.2.1.4 Resonant Coupling

A final approach is to couple both waveguides to a resonant structure. Here, the coupling should not necessarily be very high, as long as both coupling efficiencies are matched to the other loss mechanisms in the resonator. When properly designed and close to the resonance wavelength, the resonator will efficiently couple the light from the input to the output. Coupling between the resonator and the waveguides can be through various mechanisms, but the most common is *evanescent coupling*.

3.2.2 Coupling Metrics

When looking at coupling solutions from a fiber to an on-chip waveguide, there are a number of metrics, apart from the coupled power, by which to evaluate the coupling efficiency.

3.2.2.1 Coupling Efficiency

For most applications, it is imperative to lose as little light as possible when making the transition from the fiber to the chip (and back). Power coupling efficiency, and its complement *insertion loss* (IL), are usually expressed in percentage or decibels per interface. With coupling efficiency comes also the question of where the other fraction of the light is going. Couplers should have a low *back-reflection* into the fiber or on-chip waveguide, and non-coupled light should be channeled or absorbed to avoid *crosstalk*. We are generally interested in the optical power in a specific guided mode, so the coupling efficiency should always be calculated for the desired modes.

3.2.2.2 Polarization

A single-mode fiber actually supports two guided modes with orthogonal polarization, which are usually degenerate. Because of that degeneracy, the actual polarization state in the fiber is unknown in many situations. From a chip point of view, the fiber polarization can be broken down in a transverse electric (TE) and a transverse magnetic (TM) component. The silicon waveguides can also support both polarizations, but unlike in the fiber, the optical properties of both polarizations are very different. Except for a few situations, it is important to have only a single polarization in the silicon waveguide. Still, from the fiber's point of view, the silicon circuit should function irrespective of the polarization in the fiber. Therefore, coupling both fiber polarizations correctly is an important requirement for a good coupling solution. A *polarization diversity* approach, which is discussed further in Section 3.2.3, is an attractive solution to this problem. The most common metric to evaluate the polarization dependence is *polarization dependent loss* (PDL), which is expressed in decibels, either per interface or for the entire chip.

3.2.2.3 Wavelength and Bandwidth

Many potential applications require that the fiber-chip coupling solution has a broad coupling spectrum. The coupler should be able to handle the wavelength components in this spectrum with sufficient uniformity. The operational wavelength and bandwidth depends on the specifications set by the application, but in general one specifies the 1- or 3-dB bandwidth, expressed in nanometers.

3.2.2.4 Tolerances

As with many integrated systems, packaging is one of the most complex and costly steps of the entire fabrication process. The coupling to the outside world should therefore be tolerant and robust, to facilitate the pigtailing process. Important tolerance metrics for fiber coupling are positional alignment (typically expressed in micrometer for a 1-dB insertion loss penalty), angular tolerances, and temperature dependence. Position, angle, and temperature can also affect the coupling wavelength and bandwidth.

3.2.2.5 Density

The off-chip connection density or pin density is an important metric, as the data volumes that need to be transported off-chip are increasing dramatically. While a single fiber can handle a huge bandwidth, the density with which they can be integrated can be important for high-bandwidth applications, or for other applications that require many fibers (e.g., a fiber sensor readout system). Typical fiber arrays consist of a 1D row of fibers with a pitch of 127 or 250 μm. Coupling solutions with a 2D arrangement or a higher density will need custom developments and specialty fibers.

3.2.2.6 Chip Area

Given the costs induced by the fab, photomasks, and processing steps, the amount of chip area is a costly commodity. The footprint of a coupler can therefore be a significant factor in the cost of the chip, and in some cases, the reduction in scale could be severely limited by the fiber coupler solution. A good example is given in [2], where 90% of the silicon photonics chip area is consumed by the couplers to an 8-fiber array.

3.2.2.7 Processing Complexity

Like CMOS processing, the fabrication of silicon photonics is evolving toward process flows where the performance of all components is closely matched. Changes to this process flow to implement a specific coupling solution might impact the performance of other components. This is not just for the photonic building blocks, but also for integrated electronics. Therefore, coupler solutions should not violate the design and processing rules of the fabrication flow. For instance, [3] describes the issue of coupling structures penetrating the guard and seal ring and how that affects the choice for a specific coupler. Process complexity also adds to the overall cost and yield equation. Therefore, coupling solutions that require few or no additional processing steps are preferred to solutions that require substantial dedicated (post-)processing.

When discussing various coupling solutions in the next sections, we will evaluate them against these criteria.

3.2.3 Polarization Diversity

Because of their high index contrast, silicon photonic wire waveguides are usually very birefringent. The propagation constant for the guided TE mode and the guided TM mode are usually very different [4], and in some situations, only one polarization is guided. This means that silicon photonic circuits are usually designed for operation in a single polarization. As already mentioned in the previous section, in many cases, the incoming polarization in the fiber is unknown, and a good fiber-chip coupler solution needs to process the light from the fiber in such a way that the operation of the system is as polarization insensitive as possible.

One such approach is polarization diversity, illustrated in Figure 3.4. Light from the fiber is coupled to the chip and split, and the polarizations are physically separated and processed into two different circuits. If a polarization converter is included in the circuit, then the two polarizations could be made identical on the chip, such that the photonic circuits can also be identical. At the output, the outgoing polarizations are combined again using

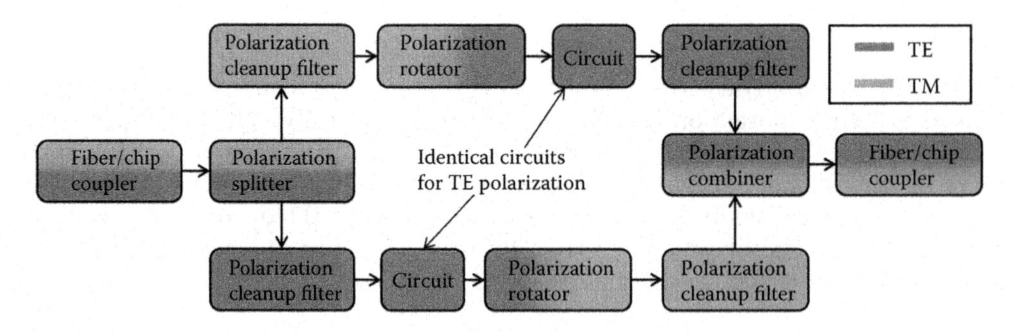

FIGURE 3.4
Principle of a polarization diversity approach using a single-polarization on-chip circuit: the polarizations of the incoming light are separated, processed independently, and combined again at the output.

the same approach. From the point of view of the incoming and outgoing fiber link, the chip appears to be polarization insensitive, while on-chip all circuits operate at a single polarization.

The polarization diversity has some drawbacks, of course: first, twice the number of circuits are needed. Due to inevitable variations in the fabrication process, the two circuits will not necessarily function in an identical way. Also, the polarization splitters and rotators might not be 100% efficient, introducing additional losses and even interferometric crosstalk within the chip.

In the next two sections, we will discuss the potential of polarization diversity with both edge couplers and grating couplers.

3.3 Edge-Coupling Solutions

The most straightforward way of coupling light off a chip is by routing it to the edge and butt-coupling it to an optical fiber that is aligned to the axis of the waveguide. This is conceptually illustrated in Figure 3.5. Although it is not necessary to change the propagation direction of the light, the propagation constants should be matched to avoid reflections and the mode in the waveguide should somehow be expanded to match the mode in the optical fiber.

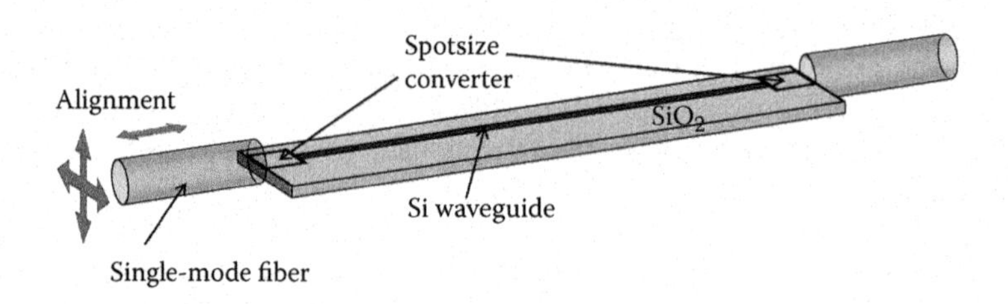

FIGURE 3.5
Edge coupling from a chip to an optical fiber.

3.3.1 On-Chip Mode Conversion

3.3.1.1 Adiabatic Tapers

The easiest approach to accomplish this is adiabatic tapering. By slowly expanding the size of the waveguide core, light will stay in the guided fundamental mode, and the mode size can be expanded until it matches that of a fiber. When the variation is sufficiently slow, there is no problem of reflections or scattering, and if the axis of the taper is straight and aligned with the fiber, there is no need to change the direction of the light: the fiber can be butt-coupled to the larger-core waveguide on the chip.

Figure 3.6 illustrates a number of possible adiabatic taper solutions for silicon photonic wires. The simplest possible taper just flares out the silicon core in the plane of the chip, which can be easily lithographically defined. This expands the mode in the horizontal direction, but in the vertical direction, there is still a significant mismatch.

Creating a vertical taper structure is less trivial: an approach as in Figure 3.6b requires that the layer thickness needs to be gradually changed along the taper's length. This is incompatible with the planar processing technologies used for the fabrication of silicon photonics. An alternative is shown in Figure 3.6c: by stacking tapered layers, an increasingly thicker waveguide core is built in a staircase geometry [5]. Still, this requires processing of thick silicon layers, which is not necessarily compatible with a CMOS-compatible process flow.

An alternative, much more attractive approach to the 3D tapering problem is shown in Figure 3.6d. Instead of widening the silicon waveguide core, it is narrowed down into a sharp tip. For narrower width, the core will no longer be able to confine the light tightly, and the mode will expand. A second waveguide core, processed in a lower-index overlay layer (silicon oxynitride [6] or polymers [7]), will take over the confinement of the light.

This inverted tapering approach can be understood more clearly from Figure 3.7, where the modal width and height are plotted as functions of the silicon wire width. We see that for narrow wires, the silicon loses confinement and the mode expands both horizontally and vertically.

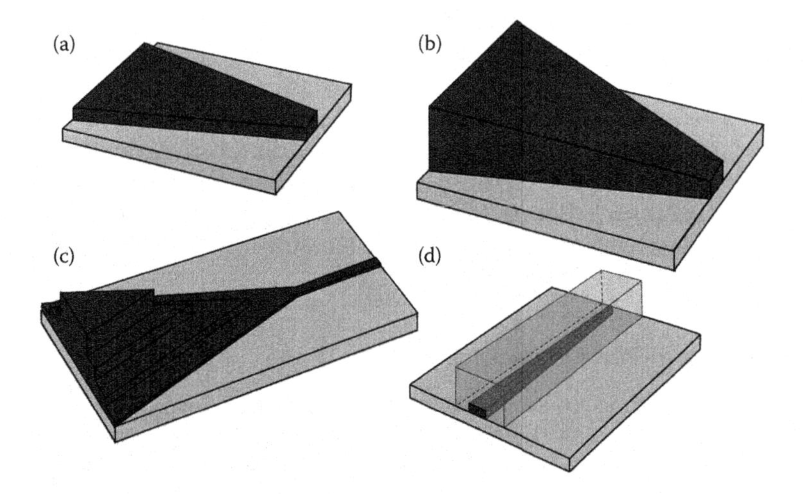

FIGURE 3.6
Adiabatic taper geometries for silicon wires: (a) in-plane (2D) taper, (b) linear 3D taper, (c) staircase taper, and (d) inverted taper.

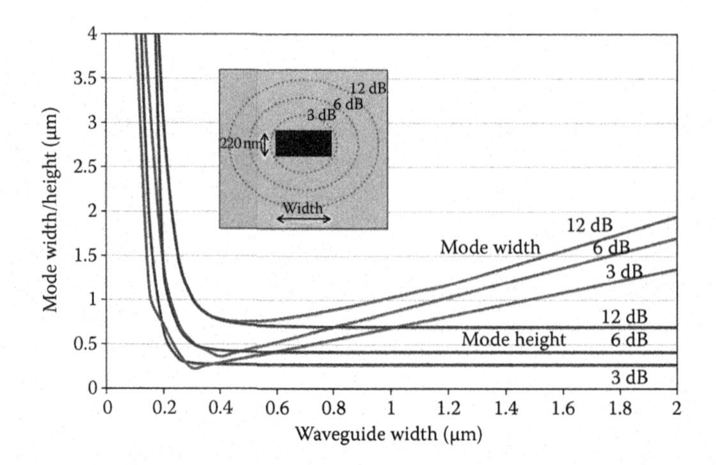

FIGURE 3.7
Width and height (3-, 6-, and 12-dB contour of the optical intensity) of the fundamental guided TE mode as function of the width of a 220-nm-thick silicon wire waveguide. (Based on *Ultra-Compact Integrated Optical Filters in Silicon-on-Insulator by Means of Wafer-Scale Technology,* P. Dumon and R. Baets, Ghent University, 2007, resimulated in Fimmwave. With permission.)

Inverted taper-based coupling solutions have been demonstrated with good performance [9]. Insertion losses can be easily lower than 1 dB per coupler [10,11] and because the mode conversion process is adiabatic, the couplers support a wide wavelength range.

However, inverted tapers are not necessarily easy to make. While the patterning can be done in the same process layer as typical silicon wire waveguides, a narrow tip is required. The best inverted tapers are therefore defined by thermally oxidizing the silicon tip [12], thereby achieving a tip width of less than 15 nm.

Also, the inverted taper is only part of the process. It also requires an overlay waveguide with a large core, which is not necessarily easy to define in a planar silicon photonics process, especially not when several metallization levels are also required. As discussed further, a typical overlay waveguide is a few micrometers thick, which will create a rather large topography on the wafer.

Inverted tapers also require a quite large footprint for the adiabatic mode expansion. Typical taper lengths are between 100 and 300 μm. Most people use linear tapers, but using optimized taper profiles, the taper length could be reduced somewhat: Close to the wire, where the mode is quite compact, the transition can be faster than in the narrower sections [13], creating a kind of inverted horn [14]. However, the footprint is not just dictated by the taper itself: the coupler needs to accommodate the mounting of the fiber, and waveguides need to be routed to the edge of the chip.

3.3.1.2 Nonadiabatic In-Plane Mode Conversion

Instead of using an adiabatic conversion between the fiber mode and the waveguide mode, one can try to focus the light of the fiber directly into the optical waveguide. A common solution to this is the use of lensed fibers, where the end facet of the fiber is curved to provide a focusing effect. However, such fibers are expensive, although the coupling efficiency to a silicon waveguide is much better than that of regular fibers. To obtain high coupling efficiency, the numerical aperture (NA) of the lens should be close to the NA of

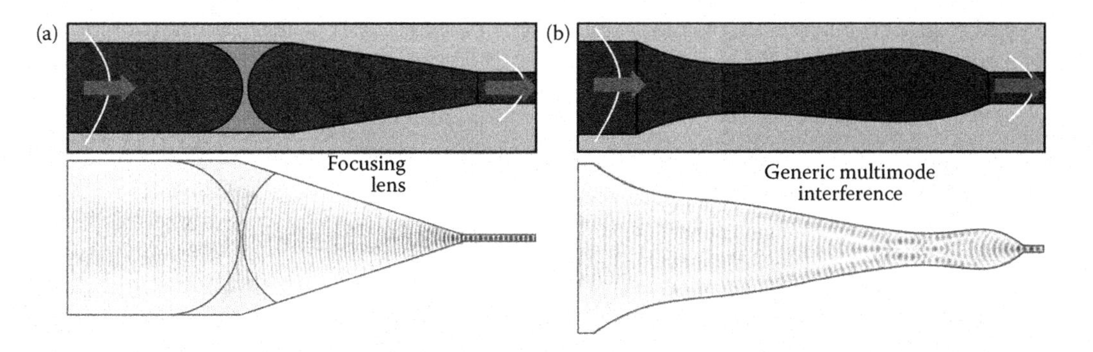

FIGURE 3.8
Nonadiabatic in-plane converters. (a) Lens-assisted mode expansion. (Field plot courtesy of K. Van Acoleyen and R. Baets, in *8th IEEE International Conference on Group IV Photonics*, 2011, pp. 157–159.) (b) Multimode interference–based mode expansion. (Field plot courtesy of B. Luyssaert, *Compact Planar Waveguide Spot-Size Converters in Silicon-on-Insulator*, Ghent University, 2005.)

the waveguide. Even then, it will be difficult to obtain a good modal overlap between the focused fiber mode and the photonic wire mode: due to the high refractive index contrast, the optical mode in the photonic wire has a sharp discontinuity at two of its four interfaces. From a practical packaging point of view, the limited depth of focus now requires an additional alignment step. Even though lensed fibers are regularly used for specialty packaging (e.g., lasers), they add a significant cost to the component.

Instead of performing the lensing in the fiber, one could also consider building the lens on the chip. This is illustrated in Figure 3.8a. By partially etching a region of a broad silicon waveguide, a planar lens can be constructed, which focuses light into a photonic wire [15]. The performance of this type of structure is quite broadband, with coupling efficiencies between −0.5 and −2 dB.

An alternative approach is the use of multimode interference. This concept was first proposed by Spühler in [17], where the coupling was optimized by sending the light through a sequence of multimode waveguide sections. The reflections and phase delays of the different modes were tuned such that the resulting field at the output was fiber-matched. An example is shown in Figure 3.8b. The concept can be extended to waveguides with nonrectangular shapes, where the distinction between the multimode approach and an optimized in-plane lens gradually disappears.

Both approaches discussed here consist of planar mode expanders. As already mentioned, this solves only part of the fiber-coupling problem. However, as we will see in Section 3.4 on grating couplers, in-plane mode expanders are still a very useful component.

3.3.2 Coupling On-Chip Tapers to Fibers

With inverted tapers, and to a lesser extent multimode and lens-based tapers, we can now couple the light from a submicron silicon waveguide to a larger waveguide mode, which is better fiber matched. The remaining problem is now to couple this mode to the actual optical fiber. For this, the core of the fiber should be aligned perfectly with the on-chip waveguide. This requires special packaging methods. First, the on-chip waveguide has to be terminated in an optically smooth facet, to avoid reflections and unwanted scattering. Second, the fiber should be mounted such that the optical axis is aligned with the waveguide and the facets are as close together as possible. As the fiber, with its cladding, has a diameter of about 125 μm this means the chip has to be cleaved or polished at the correct position, or mounting structures for the fiber have to be micromachined in silicon chip.

In Section 3.2.2, we mentioned that the commonly used fiber has a mode size of about 10 μm. This is quite large to incorporate on a silicon chip, given the amount of topography that would be required for processing. Also, such a large mode needs to be decoupled from the silicon substrate, which is located typically only a few microns below the silicon waveguide level. Therefore, the common solution for end-facet coupling of silicon nanophotonic chips is to use a small-core fiber with a higher numerical aperture. Such fibers have a mode size that is around 3 μm. Transition between high-NA fibers and standard single-mode

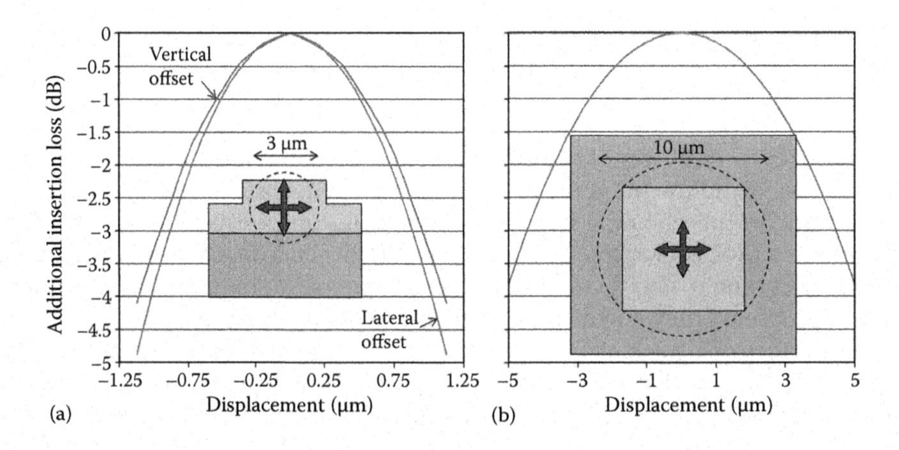

FIGURE 3.9
Misalignment penalty on the insertion loss due to fiber misalignment. (a) A 3-μm fiber aligned to a 4.5-μm silicon rib waveguide, (b) a standard single-mode fiber aligned with a 6-μm² waveguide. (Measurement data courtesy of L. Zimmermann, T. Tekin, H. Schroeder, P. Dumon, and W. Bogaerts, *IEEE LEOS Newsletter*, vol. 22, no. December, pp. 4–14, 2008.)

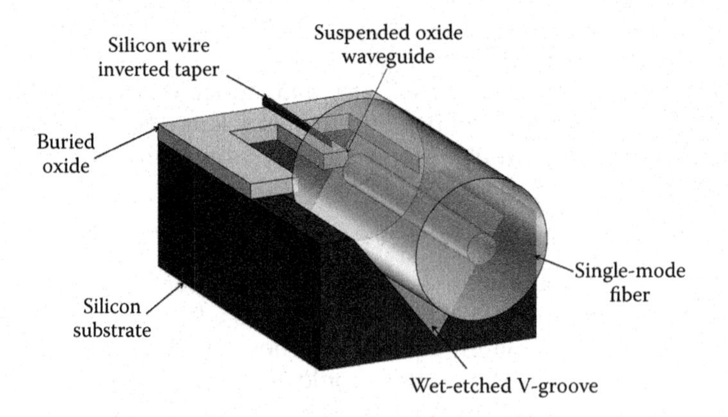

FIGURE 3.10
V-groove fiber alignment embedded with inverted taper spot-size converter. (Redrawn from J. Galan et al., in *Group IV Photonics, 2009. GFP'09. 6th IEEE International Conference on*, 2009, no. c, pp. 148–150.)

fibers can be accomplished with tapered fibers or even just by a butt-coupled splice. The latter introduces an additional coupling loss of about 0.5 dB.

One of the important metrics for fiber coupling is alignment tolerance. As plotted in Figure 3.9a, the alignment tolerance for a high-NA fiber and matched waveguide is only 0.5 μm in all directions for a 1-dB penalty [18]. A standard single-mode fiber, coupled to an on-chip waveguide of about 6 μm wide waveguide, has about double the alignment tolerance, but with an overall lower insertion loss (Figure 3.9b).

Aligning optical fibers with a horizontal waveguide is therefore far from trivial. One possible solution is depicted in Figure 3.10, using silicon V-grooves integrated with the spot-size converter [19]. Using selective wet etching, V-grooves can be micromachined in the silicon substrate, providing a lithographically aligned support for optical fibers. However, the alignment precision is determined largely by the control of the wet etch process, which can introduce variation larger than the 1-dB alignment tolerance of the fiber. More about packaging and fiber pigtailing is discussed in Chapter 11.

3.3.3 Polarization Splitters and Rotators

As discussed in Section 3.2.3, any good coupling solution should be able to handle both fiber polarizations, and make sure they are being processed on the chip in the right way. The mechanism of the inverted tapers is in essence polarization-agnostic: the adiabatic tapering converts both the TE and the TM mode of the fiber to the equivalent mode in the silicon waveguide and vice versa. However, as we already mentioned, the silicon waveguides themselves are very polarization sensitive: for most on-chip components, the TE response will be significantly different from the TM response. Therefore, the polarizations should be split, and for an efficient polarization diversity circuit, they should also be rotated.

3.3.3.1 Polarization Splitters

To split the TE and the TM mode in silicon wire waveguides, one can again use various mechanisms. As we want to split two modes with a very different propagation constant, we can use multimode interference in a directional coupler to split the two polarizations: By designing the waveguides such that the beat length in the directional coupler for TE and TM are a non-integer fraction (e.g., 3/2, 5/3, ...), the length of the coupler can be designed such that the TE mode is coupled to one waveguide and TM to the other waveguide. However, the beat length of both modes is very dependent on the geometric parameters of the waveguides and the interferometric nature of the component also makes it wavelength-sensitive, partially negating one of the benefits of the inverted taper coupler.

A more elegant approach is to make use of the difference in confinement between the TE and the TM modes using a directional coupler. By making the gap between the waveguide in the directional coupler sufficiently large, we can make sure that only the lowly confined TM mode couples to the neighboring waveguide while the highly confined TE mode stays in the same waveguide. This approach, illustrated in Figure 3.11a, is simple in its concept, and since the confinement is not very wavelength dependent, is a more wavelength insensitive approach to split the polarizations than the purely multimode interference based directional coupler [20,21].

An alternative is again an adiabatic transition. In a so-called mode-evolution polarization splitter, the shape of the waveguide is gradually changed such that the TE and the TM light preferentially reside in a different section of the waveguide. The TE mode is the

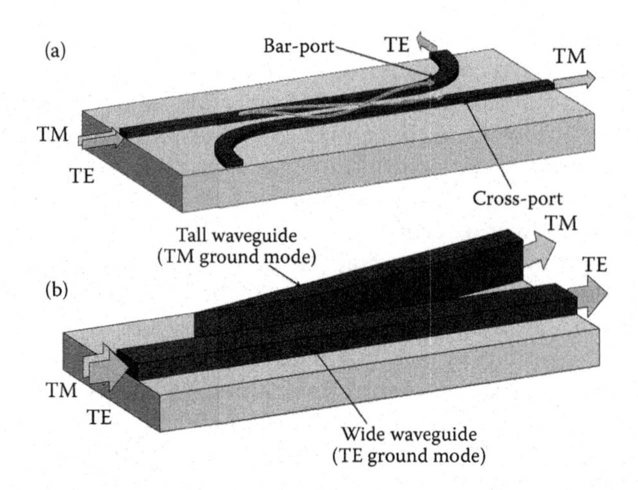

FIGURE 3.11
Polarization splitter components. (a) Directional coupler (redrawn from H. Fukuda, K. Yamada, T. Tsuchizawa, T. Watanabe, H. Shinojima, and S.-I. Itabashi, *Optics Express*, vol. 14, no. 25, pp. 12401–12408, 2006) and (b) mode evolution–based polarization splitter based on M. R. Watts, H. A. Haus, and E. P. Ippen (*Optics Letters*, vol. 30, no. 9, pp. 967–969, 2005).

ground mode of the original waveguide, where the core width is larger than the height. Likewise, the ground mode of a thicker but narrower waveguide will have a TM nature. By adiabatically introducing such a vertically oriented core in the vicinity, TM light will migrate to the second core and TE will stay in the original core.

3.3.3.2 Polarization Filter

A polarization splitter is never perfect, and a commonly used figure of merit besides the insertion loss is the ratio of transmittance (typically expressed in decibels) between the polarizations in the output ports or the *polarization extinction ratio* (PER).

To improve the PER one can use a polarization clean-up filter [23]. A straightforward solution would be to cascade several polarization splitters with a resulting PER improvement in proportion with the number of polarization splitters. An alternative that is especially interesting to filter out the TM mode is to use a waveguide section such as a rib waveguide, which only guides the TE mode and in that way radiates away the unwanted TM mode. Similarly, a waveguide bend with a sufficiently small bending radius will have a higher loss for the TM mode than for the TE mode.

3.3.3.3 Polarization Rotators

The polarization splitter is only part of the solution for a polarization diversity circuit. One of the polarizations (mostly TM) now needs to be converted to the main on-chip polarization (typically TE). For this purpose, a polarization rotator is needed. Typically, one prefers to perform the routing functionalities in the TE mode since the bending radius can be taken very small (around 5 μm) without introducing exuberant losses. Again, the rotation can be accomplished using different mechanisms, with adiabatic transitions and multimode interference being the most common solutions. As with the polarization splitters, adiabatic

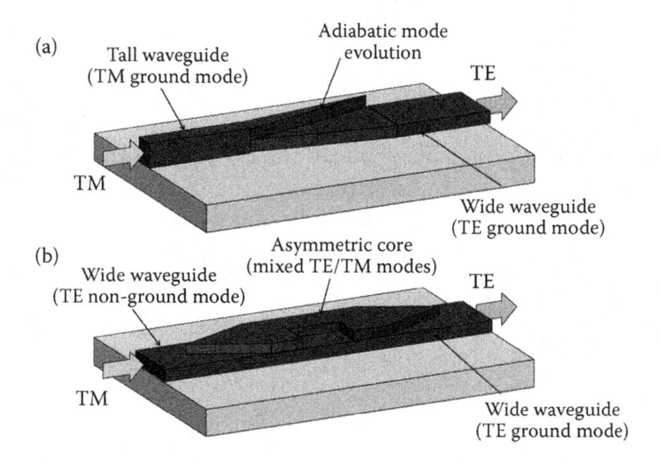

FIGURE 3.12
Polarization rotator geometries. (a) Adiabatic mode-evolution rotator (redrawn from M. R. Watts and H. A. Haus, *Optics Letters*, vol. 30, no. 2, pp. 138–140, 2005) and (b) multimode interference polarization rotator (redrawn from D. Vermeulen, S. Selvaraja, P. Verheyen, W. Bogaerts, D. V. Thourhout, and G. Roelkens, in *Group IV Photonics*, 2010, pp. 42–44).

polarization rotators are based on the principle that the TE mode is the ground mode of a "horizontal" waveguide core, whereas the TM mode is the ground mode of a "vertical" waveguide core. An adiabatic polarization rotator tries to gradually "rotate" the core. Note that simply changing the width and height of the core will not work: the TE-polarized light will remain in the TE mode, even if it is no longer the ground mode, and likewise for the TM-polarized light. A symmetry breaking is required to actually rotate the polarization. As twisting a rectangular core is difficult to accomplish in a planar fabrication process, alternative geometries are required. An example is shown in Figure 3.12a [24–26].

Polarization rotation can also be accomplished using multimode interference. The simplest form is just using two modes, as in a directional coupler. To accomplish polarization rotation, the TE and TM polarization should be coupled into a waveguide with two orthogonal modes that have a 50/50 mixed polarization state. Again, one needs to break the symmetry of the waveguide for this purpose. This was proposed in [28], using a partially etched cladding positioned with an offset to the silicon waveguide core. A more fabrication-friendly approach, which does not require a special cladding material, is shown in Figure 3.12b [27]. A partial etch of the waveguide also creates an asymmetric waveguide with 2 hybrid TE/TM modes with different propagation constants [29]. By choosing the length of this waveguide correctly, the beating between the two modes will result in a coupling from the TM to the TE mode (and the other way around). The efficiency of these components can be very high, up to −0.5 dB or higher, and very broadband. Typically one characterizes a polarization rotator using the insertion loss and the *polarization conversion efficiency* (PCE), i.e., the fraction of transmitted light that ends up in the correct polarization.

3.3.4 Edge Coupling: Conclusion

The most straightforward coupling solution for silicon photonic chips is edge coupling. This has been the standard for photonic integrated circuits in many material systems, and it can be applied to silicon as well. Inverted tapers provide a simple and efficient coupling mechanism that, when combined with on-chip polarization splitters and combiners, enables broadband, polarization-insensitive coupling between fiber and silicon waveguides.

The main difficulty with edge coupling lies in the postprocessing and packaging. Silicon wafers have to be diced, and facets need to be cleaved, polished, or micromachined before the fiber can be attached. This is a significant drawback in a wafer-scale manufacturing system, as it precludes wafer-scale testing of the photonic components.

3.4 Surface Couplers

In electronics, the standard way to contact the electrical circuits is through metal pads on the top surface of the chip. This allows for wafer-scale testing by probing the individual chips on the wafer or even the entire wafer itself without need for intermediate packaging or postprocessing steps. As discussed in the previous section, edge coupling does not provide this advantage, so for photonics, an alternative strategy is needed that allows vertical coupling of light to and from the chip. This will enable wafer-scale testing.

Also, surface couplers introduce several other advantages: The density of optical "pins" can be much higher: the couplers can be positioned anywhere on the chip (especially those for testing) and do not need to be routed to the edge. Chips can be probed with fibers oriented at an angle close to the vertical.

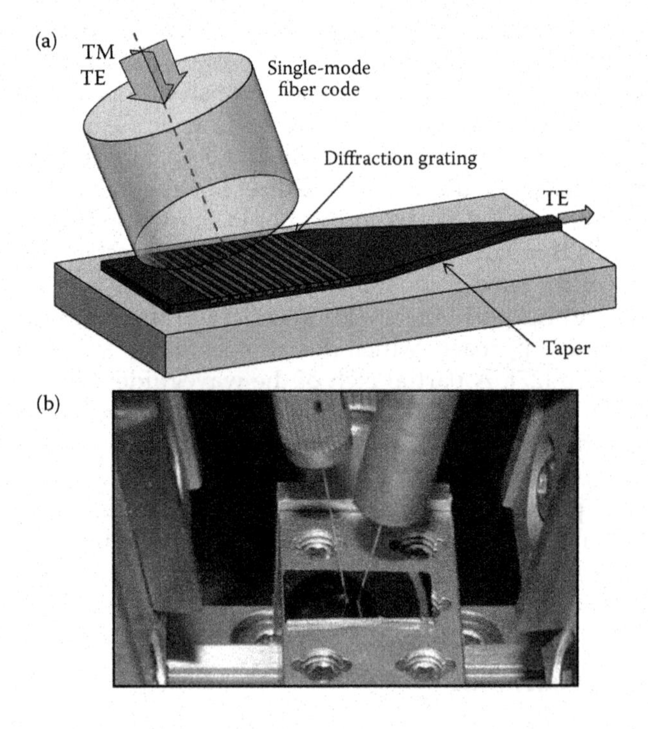

FIGURE 3.13
Grating coupler. (a) Principle: light from a (near-) vertical optical fiber is diffracted into the waveguide and (b) grating-coupler based measurement setup with two fibers.

However, there is one considerable challenge in surface coupling: light needs to make a dramatic change in direction from the on-chip waveguide to the fiber, And as with edge couplers, the mode size mismatch needs to be overcome as well.

The most common solution to this problem is the grating coupler: a diffraction grating is designed such that all the scattered contributions interfere constructively in a vertically radiated wave that is as much as possible fiber-matched. This is shown in Figure 3.13. In the past decade, the grating coupler has established itself as the standard solution, but with a large variety of flavors based on different fabrication strategies or applications. These are discussed in detail in the following section.

Grating couplers are not the only way of achieving surface coupling to a fiber. There exist some alternatives which are used for specific (research) applications, and these are briefly discussed in Section 3.4.7.

3.4.1 Grating Couplers: Operating Principle

A grating coupler consists of a periodic refractive index modulation in or close to the waveguide core. In silicon photonics, grating couplers are most often implemented as etched grooves in the silicon waveguide core [30–32]. Every such groove acts as a scatterer. Other ways to implement a refractive index grating is by adding metal lines [33] or etching subwavelength features that act as an effective medium [34–37]. When light is incident from the waveguide on the chip, some combinations of wavelength, grating period, and off-chip angle will cause all scattered contributions to be in phase, and thus a coherent phase front radiates away from the chip. This condition called the *Bragg* condition, illustrated in Figure 3.14a, occurs when the phase delay between the diffraction of two adjacent grating teeth is exactly 2π (or a multiple thereof). For the directions where this phase condition is not met, the contributions of all the grating teeth will interfere destructively.

An alternative way of calculating the operating condition for a grating coupler is in k-space, shown in Figure 3.14b. A periodic structure like a diffraction grating carries its own impulse $K = 2\pi/\Lambda$, which can be transferred to the photons in the waveguide. The conservation of momentum in the z-direction therefore dictates that $k_z = \beta + mK$ ($m = 0, \pm1, \pm2, \ldots$).

As shown in Figure 3.14b, there is a solution for $m = 1$. While in the out-of-plane direction, the nonperiodic structure of the waveguide does not pose any restrictions, except that the magnitude of the k-vector in the top and bottom cladding should match $k = n_{clad}k_0$.

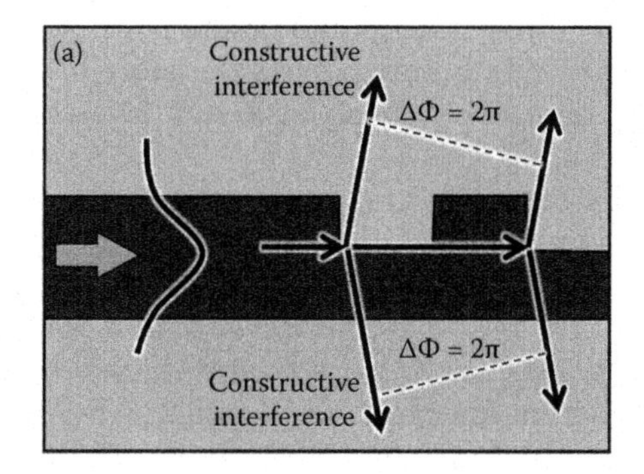

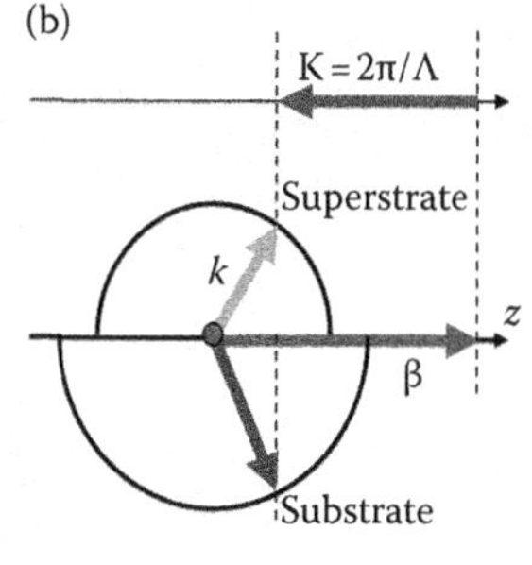

FIGURE 3.14
Operational principle of a grating coupler. (a) Phase fronts in real space and (b) k-vector diagram.

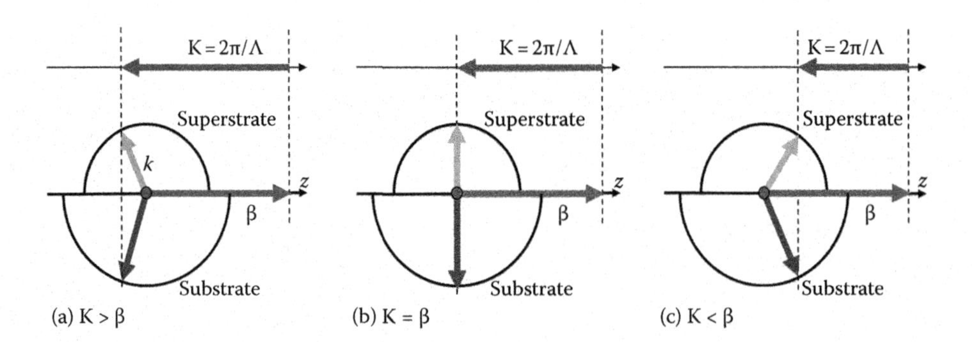

FIGURE 3.15
Grating coupler with different period (a) <2nd order, (b) exactly 2nd order, and (c) >2nd order.

We see that for light from the waveguide, there are in general two solutions: an upward and a downward radiating wave. The respective angles of these waves depend entirely on the material index of the top and bottom cladding material. The fact that there are two solutions poses a problem for the coupling efficiency: somehow, the light must be forced into the direction of the fiber, and not the substrate. Solutions for this are discussed further in Section 3.4.2.

For a given wavelength and cladding material, the period of the grating determines the outcoupling angle. For a perfectly vertical angle, we find that the grating period Λ relates to the optical wavelength λ as $\Lambda = \lambda/n_{eff}$, with n_{eff} as the average refractive index of the grating, which can be approximated by a linear relationship (a weighted average of the etched and the unetched region) according to the coupled mode theory, as the wave is propagating in the fundamental mode in the grating [38]. This situation for vertical coupling, shown in Figure 3.15b, is called a second-order grating because there is now also a solution for $m = 2$: the grating can now reflect light back into the waveguide. This makes diffraction gratings somewhat less suitable for pure vertical coupling, and most practical grating couplers use a coupling angle of 9° to 12° off the vertical. This can be either forward coupled, using a period that is larger than the pure second-order period (Figure 3.15c) or backward coupled, with a sub–second-order grating (Figure 3.15a). This problem with vertical coupling can pose a problem not only for efficient pigtailing but also for the integration of components that require true vertical coupling, such as VCSELs. Possible solutions for true vertical coupling are discussed in Section 3.4.3.

In the case of the commonly used 220-nm-thick silicon core, the effective index of the core is about 2.8 for a wavelength of 1550 nm and TE polarization. For a 70-nm etch, the effective index is about 2.5. A true second-order grating would therefore have a period of 590 nm. Forward coupling at 10° would require a period of 630 nm [39], while backward coupling would require a period of 545 nm.

The Bragg condition only describes an infinitely long grating, and does not take into account the other requirements for efficient fiber coupling. In a finite grating, every grating tooth will scatter some light, reducing the remaining light in the waveguide. This, in effect, causes an exponential decay of the power in the waveguide mode. Also, at the first teeth the grating is not yet a fully periodic structure, so some scattered light of those first teeth will not combine with the coherent flat phase front radiated by the rest of the grating. This is shown in the radiated field plot in Figure 3.16. Obviously, the exponential decay in the waveguide will also induce an asymmetry in the radiated wave. The overlap of this wave with the symmetric optical fiber mode will therefore be far from perfect. An approach to improve this overlap is discussed in Section 3.4.2.

Note that there is an equivalence in describing a grating coupler from the fiber point of view or from the waveguide point of view. As the coupling efficiencies are always expressed as mode-to-mode coupling, they are always bidirectional.

The finite size of the grating coupler has another effect: Depending on the exact parameters, there will be about 12 to 20 periods covering the size of the fiber core. Because the number of scattering point is relatively small and the scattering cross section of each tooth is relatively large, the Bragg condition will exhibit a relatively large bandwidth. This is shown in Figure 3.17b. The typical grating coupler from [31] has about 5 dB (30%) peak coupling efficiency, and a 3-dB bandwidth of about 60 nm. Similar numbers have been reported by [40–42], with coupling efficiencies going close to 50%. While the bandwidth is large compared with classical low-contrast gratings [43], it is small compared with the hundreds of nm reported for edge couplers. In many cases a larger bandwidth is required, or even several wavelength bands need to be coupled. This is discussed in some more detail in Section 3.4.6.

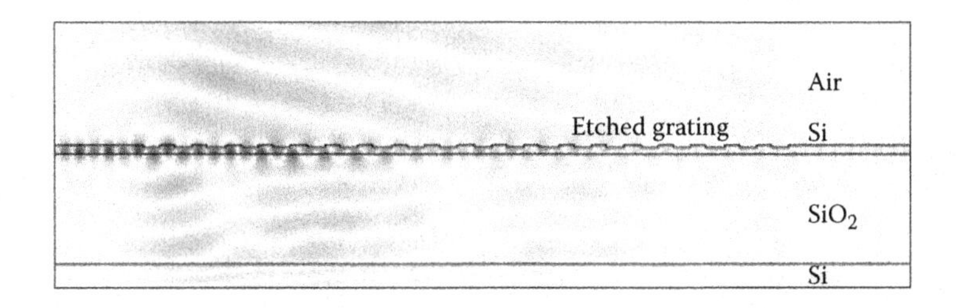

FIGURE 3.16
Field plot of an etched grating coupler, illustrating the effect of the first grating teeth and the exponential decay of light in the waveguide. The grating coupler is designed in a 220-nm silicon layer with a 70-nm etch and a period of 630 nm. (From D. Taillaert et al., *Japanese Journal of Applied Physics*, vol. 45, no. 8, pp. 6071–6077, 2006. With permission.)

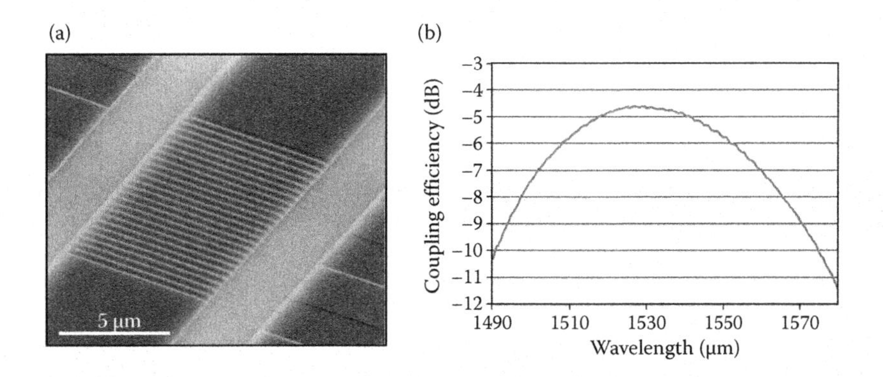

FIGURE 3.17
The grating coupler simulated in Figure 3.16. (a) SEM micrograph and (b) measured coupling spectrum for a fiber at 10°. (From D. Taillaert et al., *Japanese Journal of Applied Physics*, vol. 45, no. 8, pp. 6071–6077, 2006. With permission.)

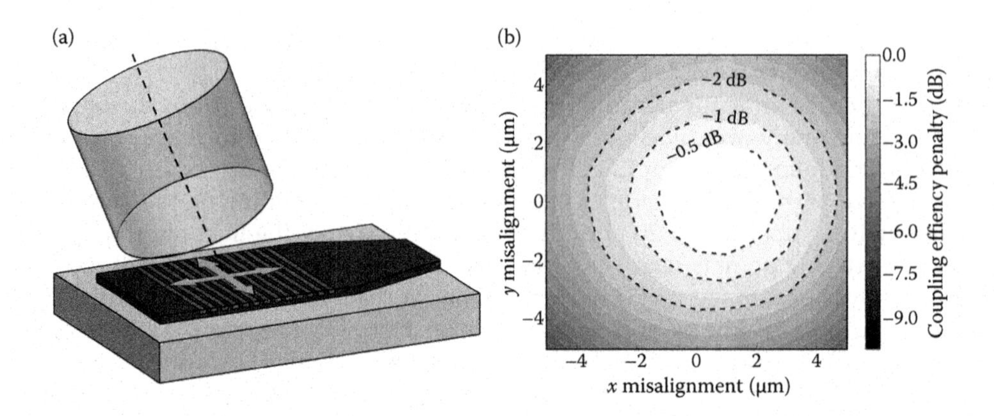

FIGURE 3.18
Misalignment tolerances for the grating coupler in Figures 3.16 and 3.17. (a) directions in which the fiber position was scanned, (b) coupling penalty as function of misalignment. (Measurement data courtesy of Michael Vanslembrouck.)

Unlike a typical edge coupler, grating couplers couple directly to a single mode fiber, with a core size of about 10 μm. Therefore, the tolerances for misalignment are higher than for the edge coupler. This is shown in Figure 3.18a. The lateral misalignment tolerance for 1-dB coupling penalty is larger than 1 μm. In the longitudinal direction, it is even better, but in Figure 3.18b, we do see a shift in peak wavelength as we move the fiber along the axis of the on-chip waveguide.

As already mentioned several times throughout this chapter, the small silicon waveguides are strongly birefringent. This means the Bragg condition for the TE and the TM waveguide mode will be very different, and a grating coupler designed for a given wavelength and coupling angle will therefore only work for a single polarization, either TE or TM. Approaches to address this issue and still enable polarization diversity on a chip are discussed in Section 3.4.4.

The 1D line gratings, as described here and shown in Figure 3.13, couples light from a fiber to a waveguide whose width is fiber-matched. This is typically 10 μm or wider. This waveguide still has to be tapered down to a single-mode wire waveguide. While the grating coupler itself can be quite compact, a linear taper should be at least 150 μm long. Solutions to this problem are discussed in Section 3.4.5.

3.4.2 Optimizing Grating Coupler Efficiency

The simple grating coupler presented in the previous section has a coupling efficiency of about 30% (−5.22 dB). While this is good enough for most research purposes, better coupling is required for most real-world applications: a high coupling loss is not only detrimental to the overall power budget, but the light that is not coupled correctly can give rise to crosstalk elsewhere on the chip.

The coupling efficiency of the grating coupler is determined mainly by *directionality* and *modal overlap*. When looking from the perspective of the waveguide, the light needs to be sent as much as possible into the right diffraction order, i.e., the first-order upward diffraction. Also, reflections have to be minimized. Once diffracted, the field profile should match as much as possible the mode of the optical fiber at the correct angle.

3.4.2.1 Improving Directionality: Bottom Mirror

In a regular (near-)second-order grating coupler, the Bragg condition has two solutions: an upward and a downward radiating wave. Somehow, that unwanted downward wave should be suppressed or redirected in the upward direction. This can be accomplished by incorporating a mirror in the substrate. In effect, a silicon-on-insulator substrate already has such a mirror embedded: the interface between the buried oxide (BOx) and the silicon wafer. This is shown in Figure 3.19a. Specular reflection at this interface is around 17% for a wave incident at 10°. So without any special tricks, there is already a fraction of downward radiated light being recycled.

However, the reflected wave will interfere with the original upward wave, so the effect of the recycling can be either positive or negative. To maximize the upward coupling efficiency, the interference should be constructive. This can be accomplished by choosing the right thickness of the buried oxide, to tune the path length difference between the upward and the reflected wave. Figure 3.19b shows the simulated coupling efficiency of the grating coupler in Figure 3.16 for a changing thickness of the BOx layer. Coupling efficiencies can vary from −2.7 dB and −7.5 dB between the best and worst oxide thickness. We also see that the 2-μm buried oxide used by many groups is not optimal for grating couplers operating at 1550 nm.

The effect of the buried oxide can be enhanced by incorporating a more efficient reflector in the substrate. Two examples are shown in Figure 3.20. Instead of a single interface, a distributed Bragg reflector can be embedded [45,46]. Of course, building such a reflector in an off-the-shelf SOI substrate is not possible. Therefore, in this case, all layers, including the waveguide layer, consist of PECVD deposited amorphous silicon and oxide. Coupling efficiency of this grating coupler is close to 70% (−1.6 dB).

While amorphous silicon can have relatively low losses [48], it is not straightforward to integrate active components in it. Therefore, an alternative way to add a bottom reflector a standard silicon substrate is to use a bonding technique: First, a top oxide and reflector

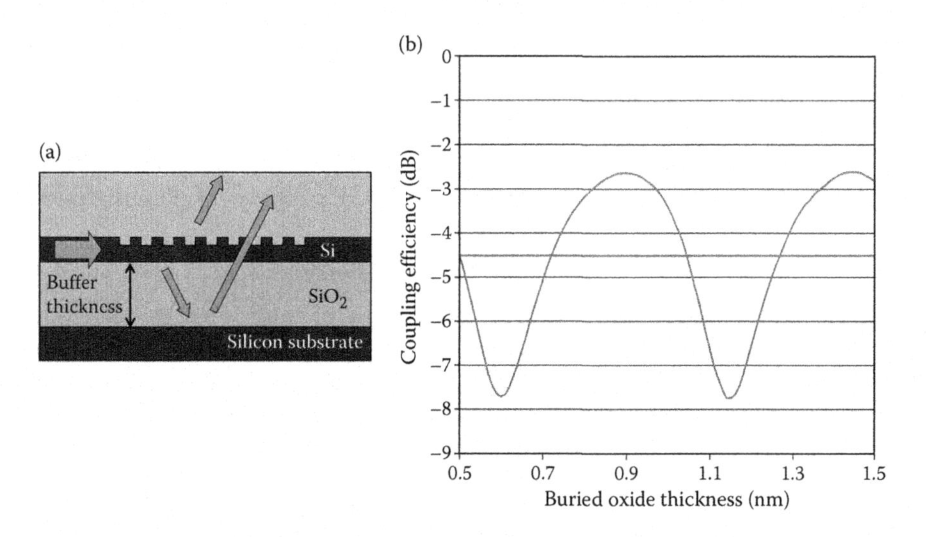

FIGURE 3.19
Reflection at the buried oxide interface: (a) principle and (b) coupling efficiency at 1550 nm and 8° angle as function of buried oxide thickness. (Simulation data courtesy of D. Taillaert, *Grating Couplers as Interface between Optical Fibres and Nanophotonic Waveguides*, Ghent University, 2005.)

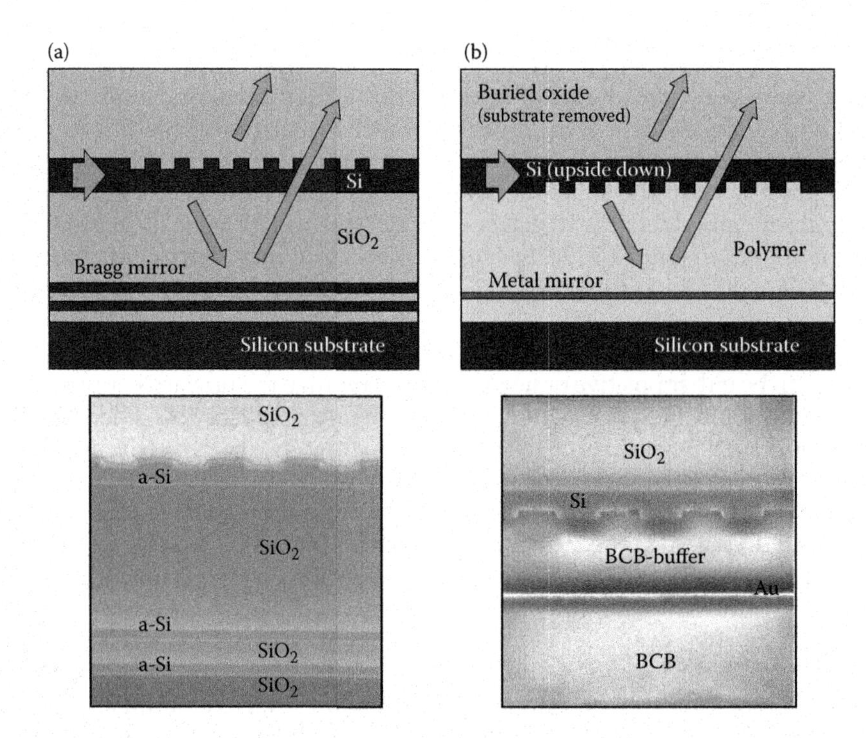

FIGURE 3.20
Grating couplers with a bottom mirror: (a) grating coupler in deposited amorphous on top of a DBR mirror stack. (SEM picture courtesy of S. K. Selvaraja et al., "Highly efficient grating coupler between optical fiber and silicon photonic circuit," in *Lasers and Electro-Optics 2009 and 2009 Conference on Quantum electronics and Laser Science Conference. CLEO/QELS, 2009. Conference on 2009*, vol. 1, pp. 1–2.) (b) Bonded grating with gold "bottom" mirror. (SEM picture courtesy of F. Van Laere et al., *Journal of Lightwave Technology*, vol. 25, no. 1, pp. 151–156, 2007.)

is deposited, and subsequently the entire structure is bonded upside down onto a carrier, after which the original substrate is removed. The top mirror then becomes a bottom reflector. An example, making use of BCB and a deposited gold mirror, is shown in Figure 3.20b [47]. The coupling efficiency, at 69%, is comparable with that of the DBR mirror in amorphous silicon. The top mirror can also be processed in silicon, as demonstrated in [49].

3.4.2.2 Improving Directionality: Grating Profile Optimization

Adding a mirror under the grating is an effective way of improving the grating coupler directionality, but it is far from straightforward from a fabrication point of view. Therefore, instead of trying to recycle the downward radiated light, we can try to suppress the downward diffraction order altogether. This can be done by optimizing the grating profile. The general principle is shown in Figure 3.21a for an etched grating: the main scattering points within one period are located at the corners of the etch profile. By now tuning the tooth width and depth, the phase difference between the upward components is 2π, while between the downward components, it is π. This way, the downward-diffracted wave will experience significant destructive interference and will be much weaker than the upward wave [50]. From a Fourier optics point of view [51,52], one can see the grating as a binary phase grating, which is different for downward-diffracted light and for upward-diffracted light. For the downward-diffracted light, the grating is a zero-phase grating, which means that the first diffraction order is suppressed. For the upward-diffracted light, the grating

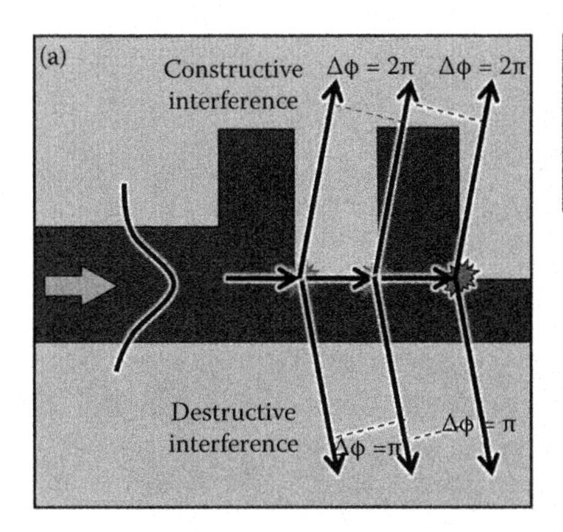

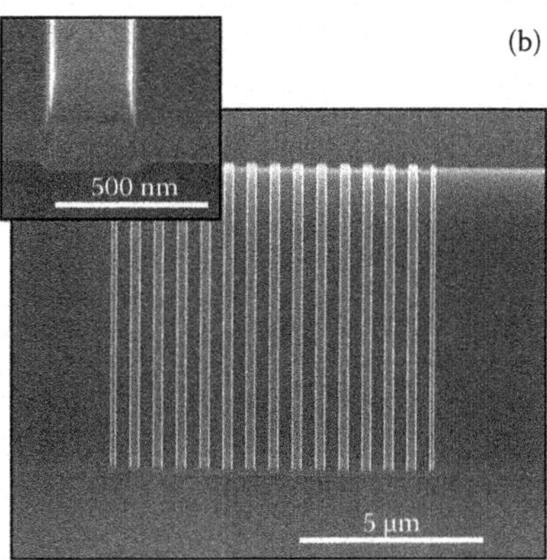

FIGURE 3.21
Improving the directionality by optimizing the grating profile: (a) principle and (b) fabricated grating. (Redrawn from D. Vermeulen et al., *Optics Express*, vol. 18, no. 17, pp. 18278–18283, 2010.)

behaves as a π-phase grating such that only uneven diffraction orders exist. This just illustrates the principle; because of the high refractive index contrast, the grating has to be numerically optimized. This results in a grating teeth that are thicker than the standard silicon waveguide core. Therefore, to fabricate such an optimized grating, additional fabrication steps are needed.

To implement a thicker grating, additional silicon needs to be added. This can be done using epitaxy, and experiments show indeed an improved grating performance from 30% to 55% [54]. However, it is not straightforward to control the tooth profile using epitaxy. Alternatively, the grating top layer can be deposited and subsequently etched [53,55]. Using an amorphous or polysilicon and embedded etch-stop layers, a better control of the grating profile is possible. Coupling efficiencies of 70% have been reported [32,56]. The grating overlay does not have to be silicon but can be any material such as SiN [57], as long as it introduces a π-phase difference. Another approach is to start with a thick silicon core and etch the grating and entrance waveguide using an inverted taper [58,59].

3.4.2.3 Improving Modal Overlap

The overlap of the diffracted wave with the optical fiber mode is determined by the lateral mode profile and the longitudinal mode profile. The lateral mode profile of the ground mode of a broad high-contrast silicon waveguide closely resembles a cosine with very small evanescent tails. The profile of the optical fiber is more Gaussian-like, with more extended tails. However, by just optimizing the silicon waveguide width, a 1D modal overlap of almost 100% can be obtained, without any further engineering.

In the longitudinal direction, we have already shown in Figure 3.16 that there is a significant mismatch between the exponentially decaying intensity profile in the waveguide and the Gaussian-like fiber mode. The maximal overlap between the exponential decaying field of a uniform grating and the Gaussian profile of the fiber is 80%. This provides much room for optimization.

The exponentially decaying field profile is induced by periodically scattering a fixed fraction of the remaining light in the waveguide. Therefore, to change the exponential profile into a more fiber-matched profile, we need to tune the amount of scattered light in each grating tooth, as shown in Figure 3.22a. The first grating teeth should only minimally scatter, most

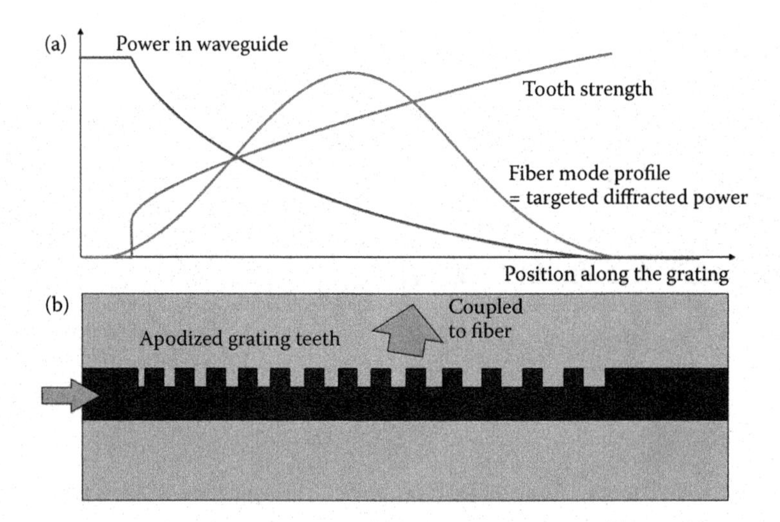

FIGURE 3.22
Optimizing modal overlap by apodizing the grating: (a) optimizing the scattering profile and (b) schematic drawing of apodization by changing the trench width. (Redrawn from D. Taillaert, P. Bienstman, and R. Baets, *Optics Letters*, vol. 29, no. 23, pp. 2749–2751, 2004.)

of the light should be scattered in the center, and the remaining light should be diffracted at the end. Also, the overall diffraction strength of such an apodized grating should be such that there is no more light remaining in the waveguide at the end of the grating. Tuning the scattering cross section of each individual tooth can be accomplished by changing the width and depth. The latter is difficult to achieve directly, unless one makes use of local etching techniques such as focused ion beam (FIB) writing or by utilizing the lag effect in the dry etching process [60]. Changing the width can be done lithographically, but as the narrowest lines are typically well below 100 nm in width, very good patterning (e-beam) is required to make such gratings. A simulated field plot of such an optimized grating is shown in Figure 3.22b, and gratings with coupling efficiency up to 75% have been demonstrated in combination with a silicon overlay or thick core waveguide grating [32,61,62].

3.4.3 Grating Couplers: True Vertical Coupling

When we considered the Bragg condition for perfect vertical coupling in Section 3.4.1, we noticed that there appeared an additional solution for the second-order diffraction: back into the silicon waveguide. While many solutions can get by with tilted fibers, vertical coupling facilitates matters considerably, and also enables coupling from vertical emitters, such as VCSELs. Reflections are not only a problem when the light comes from the waveguide. The light from the fiber will also be partially reflected, as the 0th order backward diffraction (which is specular reflection) is sent back into the fiber. Also, from the fiber point of view, a simple periodic grating will not introduce a preference for "forward" or "backward" coupling: the light of the fiber is essentially split in two. This can be alleviated by breaking the symmetry, such as in the apodized grating discussed in Section 3.4.2.3 [63] or with a slanted grating [64].

A simple solution to obtain efficient vertical coupling is to "trick" the grating into nonvertical coupling and subsequently redirecting the diffracted light in the vertical direction. This can be easily accomplished by adding a refractive interface at an angle, as shown in Figure 3.23a. Such a wedge is not straightforward to make, as it is a 3D structure. Figure

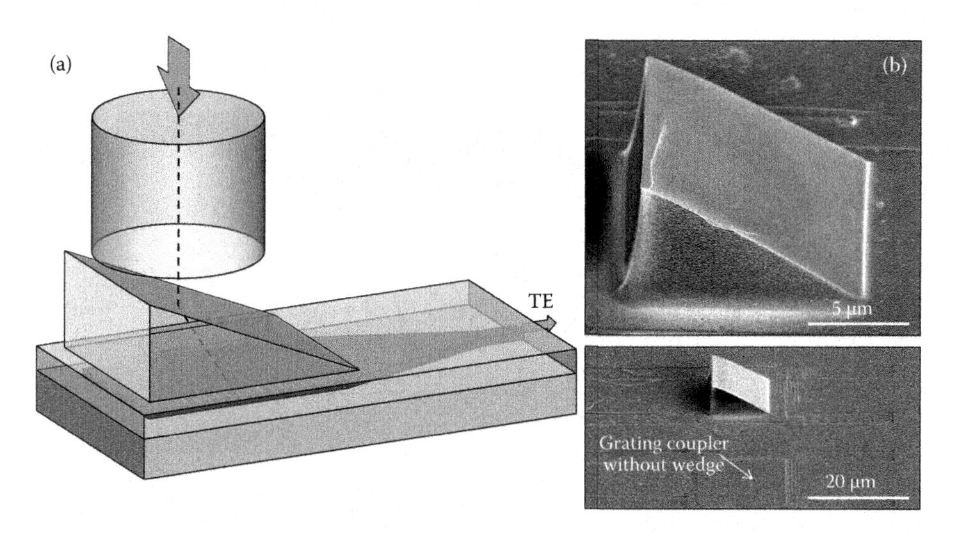

FIGURE 3.23
Vertical coupling using a wedge on the grating: (a) principle and (b) fabricated structure with a polymer wedge. SEM pictures courtesy of Jonathan Schrauwen.

3.23b shows such a wedge in polymer fabricated by imprinting a UV curable polymer with a focused-ion-beam fabricated mold [65]. Alternatively, one can use an angled fiber for vertical fiber coupling as shown in [66].

It is also possible to reduce the reflections by adding an additional reflecting structure. This might seem counterintuitive, but if the phase delays are constructed properly, both reflections can interfere destructively, resulting in a higher transmission. In Figure 3.24, this reflector is fabricated by a deep-etched slit in the waveguide [50]. To get the phase delays right, the width and the position of this slit must of course be accurately controlled. This type of reflector also introduces a symmetry breaking, which results in a highly preferential coupling from the vertical fiber into one waveguide direction.

3.4.4 Grating Couplers and Polarization

Because of the birefringence of the silicon waveguide, the Bragg conditions at a given wavelength and angle will only be satisfied for a single polarization. As the TE polarization is the most commonly used in silicon waveguides (as it is the fundamental mode when the core width is larger than the thickness), all grating couplers discussed up to this point are designed for the TE polarization. However, it is also possible to design a grating coupler for the TM polarization. For the regular 220-nm-thick SOI layer, the effective index for the TM mode is 1.85, compared with the TE index of 2.8. When using a 70-nm-deep etched grating (same value as for TE, but not optimized for TM), this will result in a grating period of 1040 nm. Coupling efficiency for such a grating is somewhat lower than for TE at about 20% [67].

However, for many applications, both fiber polarization needs to be coupled at the same time, and to enable a polarization diverse circuit, these polarizations need to be physically split into different circuits. In this section, we will discuss two solutions to this problem: using a 2D grating coupler that acts as a polarization splitter and using 1D grating couplers tuned to couple both polarizations.

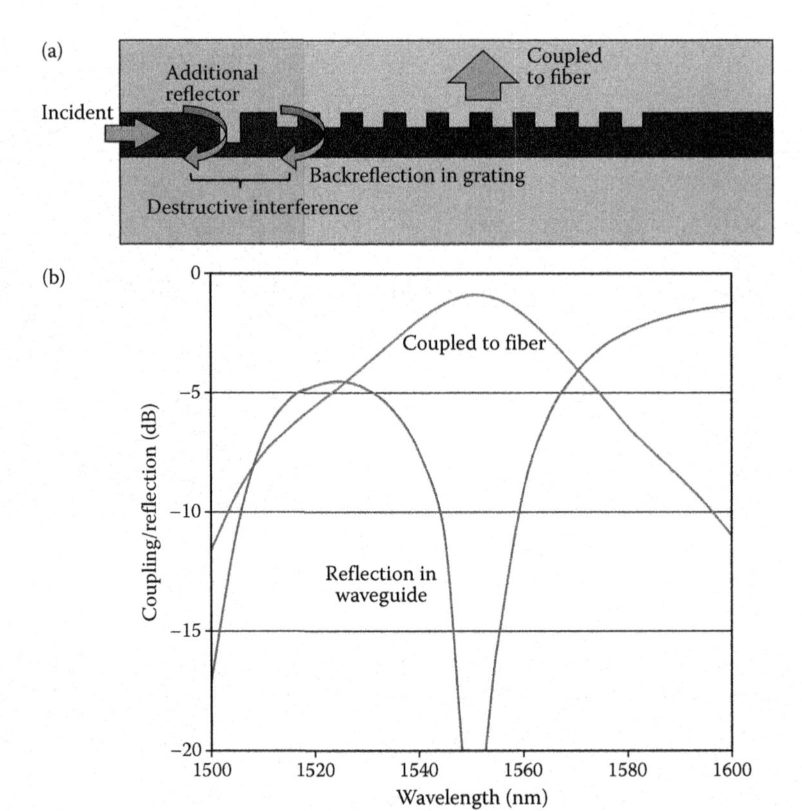

FIGURE 3.24
Vertical coupling using an additional reflector in front of the grating: (a) principle and (b) coupling efficiency and reflection as function of wavelength. (From G. Roelkens, D. V. Thourhout, and R. Baets, *Optics Letters*, vol. 32, no. 11, pp. 1495–1497, 2007. With permission.)

3.4.4.1 2D Polarization-Splitting Grating Couplers

The 1D grating shown in Figure 3.16 couples the TE polarization efficiently to the slab waveguide. However, from the point of view of a vertically oriented fiber, there is no difference between the two polarizations. Therefore, a 1D grating oriented at a right angle to the original grating coupler will couple the other fiber polarization. Overlaying the two 1D grating profiles will result in a 2D grating that can be patterned as a matrix of etched holes [68]. This principle is shown in Figure 3.25a.

However, as we already discussed, a grating designed for a perfectly vertical fiber has the drawback of introducing unwanted reflections. Therefore, we prefer to slightly tilt the fiber. Although for a 1D grating this just shifts the Bragg condition, for a 2D grating, it also breaks the symmetry between the two polarizations. This will result in a slightly reduced rejection ratio and introduce some polarization-dependent loss for the overall chip [70]. Also, the change in the Bragg condition with a tilted fiber requires that the waveguides on the chip are no longer at a right angle [71]. Alternatively, one can keep the waveguides at a right angle and use tilted grating instead. However, even then it is not possible to fully decouple the polarizations with a tilted fiber. With a vertical fiber the polarizations are perfectly decoupled, but then an additional symmetry is introduced: Each polarization can now be diffracted forward or backward. This is OK for some receiver applications [23,72], but in many situations, it is preferable to have the light of each polarization in one waveguide, not two.

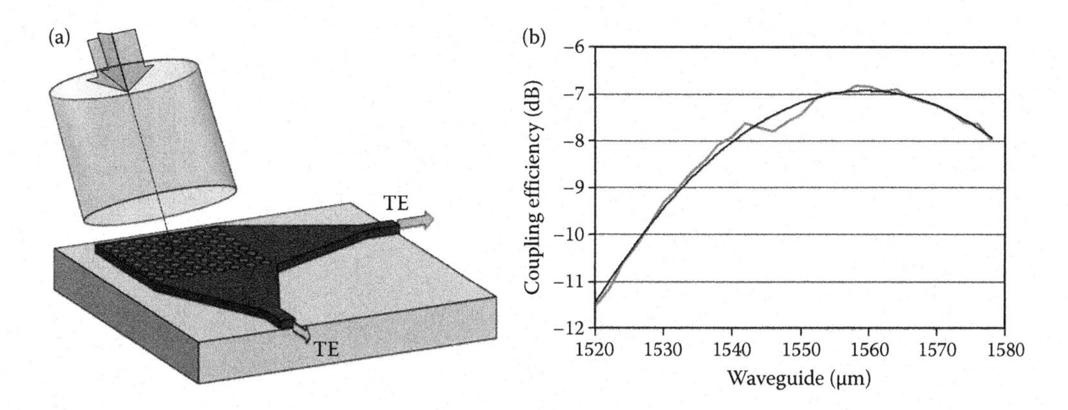

FIGURE 3.25
2D grating coupler: (a) principle and (b) coupling efficiency spectrum. (Redrawn from W. Bogaerts, D. Taillaert, P. Dumon, D. Van Thourhout, R. Baets, and E. Pluk, *Optics Express*, vol. 15, no. 4, pp. 1567–1578, 2007.)

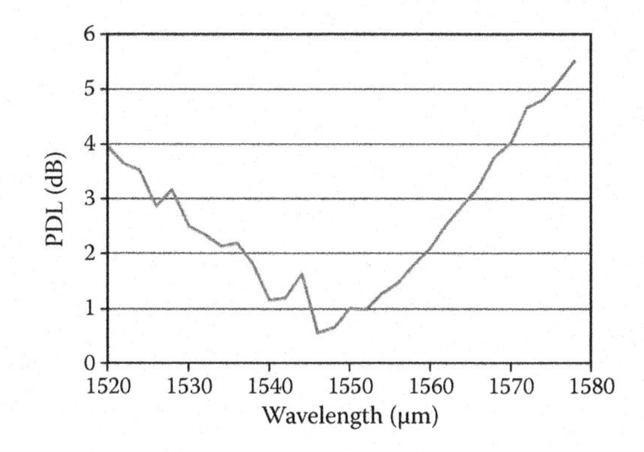

FIGURE 3.26
Polarization dependent loss for a polarization-diverse circuit with 2D grating couplers. (From R. Halir, D. Vermeulen, and G. Roelkens, *IEEE Photonics Technology Letters*, vol. 22, no. 6, pp. 389–391, 2010. With permission.)

Figure 3.25b shows the coupling efficiency for a 2D grating coupler for both the TE and the TM polarization in the fiber. We see that the coupling efficiency is slightly lower than for the 1D grating, which can be attributed to the nonoptimized etch depth.

Such grating couplers have been used to demonstrate polarization-diversity circuits with moderate polarization-dependent loss [73]. However, for tilted fibers the PDL is only low near the peak coupling wavelength. This is shown in Figure 3.26. To reduce the PDL for a broader wavelength range, one has to use more complex shape holes instead of circular holes. This will successfully reduce the PDL to a minimum over the whole bandwidth of the 2D grating coupler [32].

3.4.4.2 1D Polarization-Splitting Grating Couplers

As explained in Section 3.4.1, there is a fixed relation among the grating period, the wavelength, the effective index of the waveguide mode, and the angle of the fiber. As the effective index for the TE and TM mode of the slab waveguide is different (TM mode has a lower effective index), they will diffract at a different angle for a given grating and wavelength. One can now choose the grating design such that the angle of the TE waveguide mode is

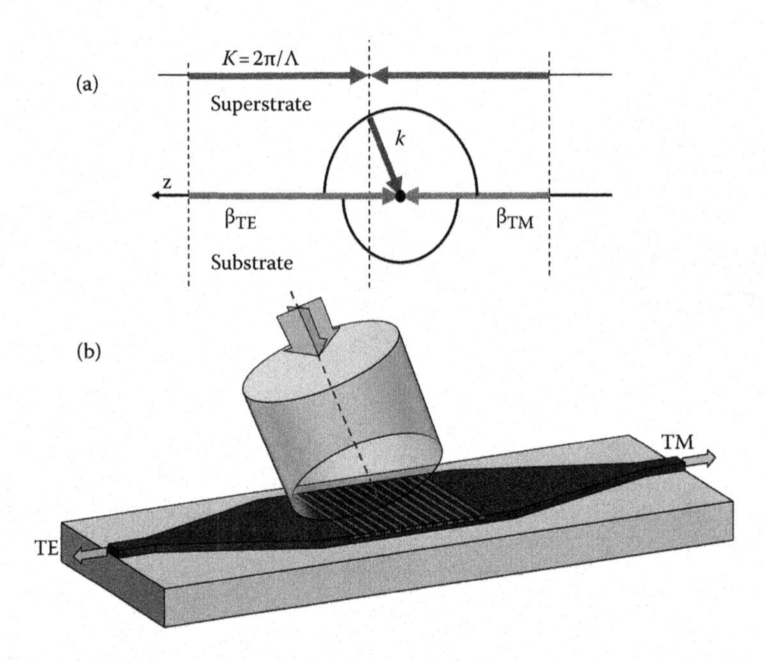

FIGURE 3.27
Principle of the 1D polarization splitting grating: (a) k-vector diagram (from the perspective of the waveguide and (b) polarization splitting function (from the perspective of the fiber).

positive (forward coupling) and that of the TM waveguide mode is negative (backward coupling), but with the same deviation from the vertical. This is illustrated in Figure 3.27a. If we now reverse the situation, and look at it from the perspective of the fiber (Figure 3.27b), the TE polarization will be coupled forward, while the TM polarization will be coupled backward [46,74–76].

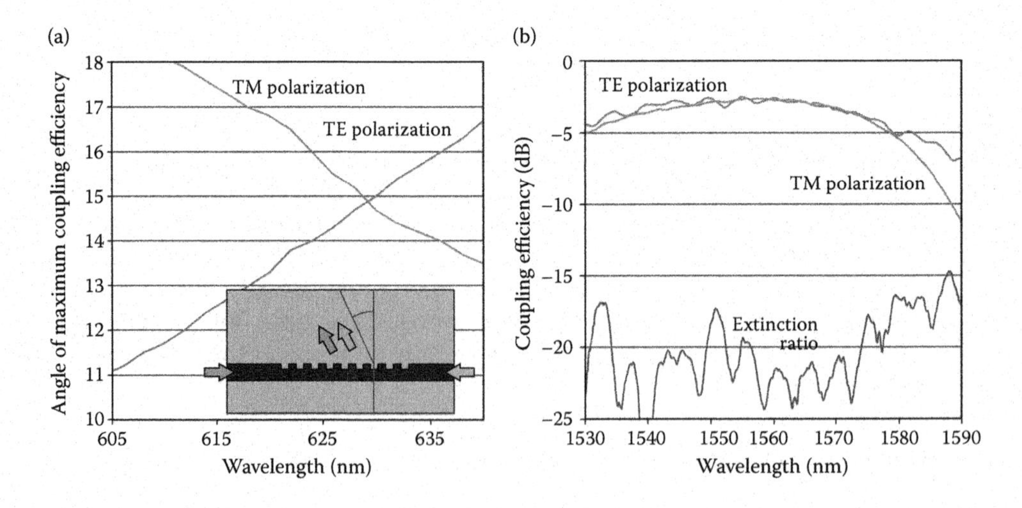

FIGURE 3.28
1D polarization splitting grating: (a) coupling angle for TE and TM mode as function of grating period, with a crossing at 630 nm and an angle of 15° and (b) measured coupling efficiency of both polarizations for this situation. (Data courtesy of Y. Tang, D. Dai, and S. He, *Photonics Technology Letters, IEEE*, vol. 21, no. 4, pp. 242–244, 2009.)

If we go through the math for a silicon slab waveguide of 260 nm thick at a wavelength of 1550 nm, with an effective index of 2.87 for TE and 2.1 for TM, we can plot the coupling angle of the fiber as a function of grating period (for 70-nm etch) in Figure 3.28a. We find that the absolute value of the coupling angle of the fiber is 15° for both polarizations when the grating period is 630 nm [75].

Figure 3.28b shows the measured coupling efficiency of such a device [75], where both polarizations show a similar insertion loss of around −3 dB, with an extinction ratio of over 10 dB.

A drawback of this approach compared with the 2D grating is that an additional polarization rotator is required to obtain identical polarizations in the two waveguides. We already discussed such components in Section 3.3.3.3.

3.4.5 Grating Couplers: Reducing the Footprint

The grating couplers we have discussed up until now couple light from a fiber into a broad waveguide. This means an additional mode size converter is required to laterally compress the light into a narrow photonic wire waveguide. An adiabatic taper for this purpose is typically 150 μm long, but we have discussed more compact taper solutions in Section 3.3.1.2.

However, it is possible to combine the function of the grating and the in-plane converter by moving away from a straight-line grating and instead use the grating diffraction itself for in-plane spot-size conversion. By curving the grating lines, the grating will act as an in-plane diffractive lens [77]. The principle is explained in Figure 3.29a. Instead of a global

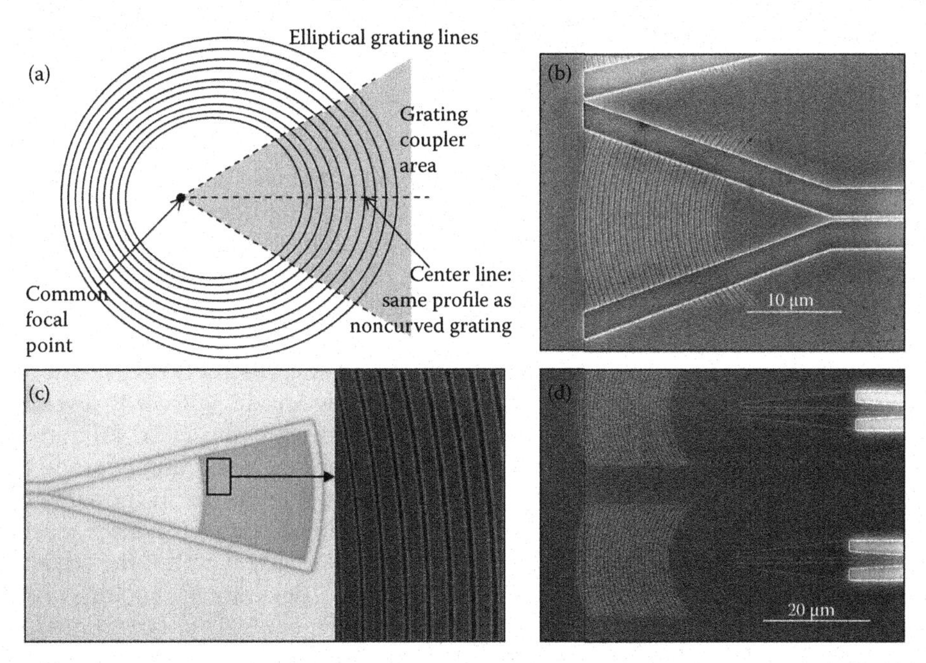

FIGURE 3.29
Curved grating coupler: (a) principle of the curves and (b) focusing grating coupler in a waveguide taper. (Image courtesy of F. Van Laere et al., "Compact focusing grating couplers for silicon-on-insulator integrated circuits," *IEEE Photonics Technology Letters*, vol. 19, no. 23, pp. 1919–1921, 2007.) (c) A similar device from A. Mekis et al. (*Selected Topics in Quantum Electronics*, no. 99, pp. 1–12, 2010. Image courtesy of Attila Mekis, Luxtera.) (d) Focusing grating coupler in a slab waveguide with a collection waveguide opening in the focal point. (Image courtesy of F. Van Laere et al., *IEEE Photonics Technology Letters*, vol. 19, no. 23, pp. 1919–1921, 2007.)

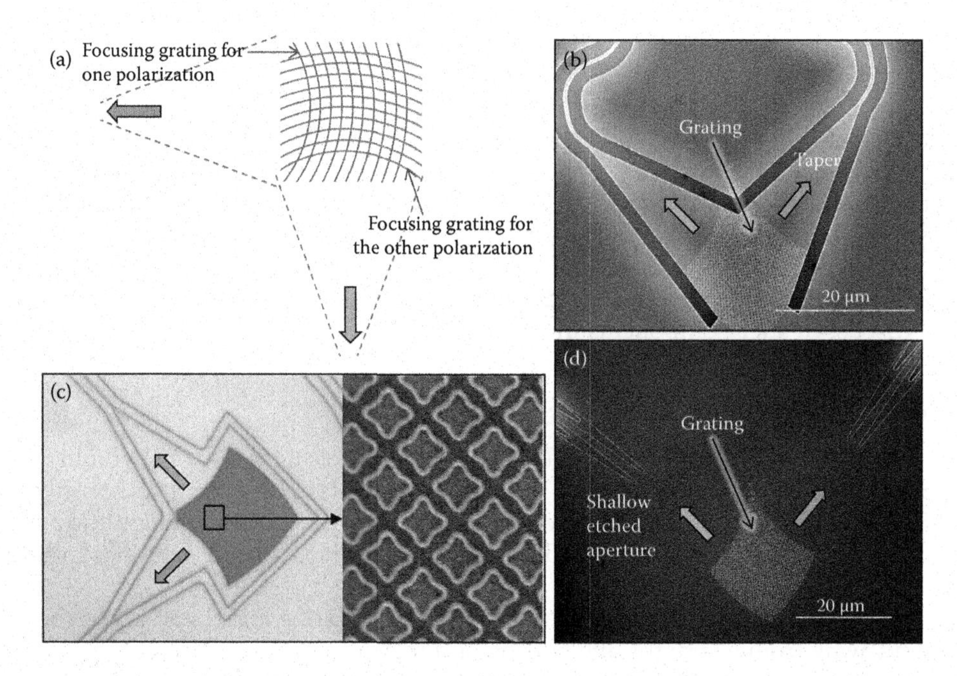

FIGURE 3.30
2D curved grating coupler: (a) principle and (b,c) SEM picture of fabricated structures. (Images courtesy of F. Van Laere, W. Bogaerts, P. Dumon, G. Roelkens, D. Van Thourhout, and R. Baets, in *2008 5th IEEE International Conference on Group IV Photonics*, 2008, vol. 1, pp. 203–205.) (d) A similar device from A. Mekis et al. ("A grating-coupler-enabled CMOS photonics platform," *Selected Topics in Quantum Electronics*, no. 99, pp. 1–12, 2010. Image courtesy of Attila Mekis, Luxtera.)

Bragg condition, the condition is calculated locally, such that the light from the fiber is diffracted toward a single point. This results in elliptical grating lines with a common focal point. The focusing distance can be quite short, but the ellipse segments should still have a significant overlap with the fiber mode. In practice, the focusing distance can be less than, 20 μm without a significant penalty on the insertion loss, and a very short taper can be used. Such a short grating is shown in Figure 3.29b.

Actually, because the grating focuses the light itself, the taper structure is not necessary: Figure 3.29d shows a curved grating etched in a slab waveguide, with only a waveguide aperture at the focal point [77]. Similar coupling efficiencies are measured for both types.

Because the Bragg condition can be tuned locally by changing the grating curves, there is no more the requirement that the fiber and the on-chip waveguide are in the same vertical plane. For this, only the orientation of the ellipses needs to be changed with respect to the coupler [78]. This will not improve the coupling efficiency in any way, but the advantage is that the grating will now no longer reflect in a specular way, back into the on-chip waveguide.

Such an off-axis grating also makes it possible to implement a 2D polarization-splitting grating coupler, by overlaying two off-axis elliptical gratings, and etching a correctly sized hole at the intersection of the grating lines [50]. Such a device is shown in Figure 3.30.

3.4.6 Grating Coupler Bandwidth

Grating couplers are diffractive structures, and the coupling conditions derived in Section 3.4.1 clearly show a wavelength-dependent behavior. Still, because the grating has only a limited number of periods (to fit in the mode size of a fiber) and the individual scatterers quite strong, the 3-dB bandwidth of the grating is still of the order of 60–80 nm. Increasing the bandwidth is possible by reducing the number of grating teeth and at the same time increasing the strength of the scatterer. In silicon, this is difficult without shrinking the grating: the bandwidth will increase, but the overall coupling efficiency will drop dramatically because of the smaller overlap with the mode of a single-mode fiber.

To have fewer teeth, but still retain a good overlap with the fiber mode, the period of the grating should increase. As the period of the grating is linked with the Bragg condition, this implies that the effective refractive index of the grating should be lowered. This is difficult in silicon, but grating couplers in silicon nitride have been demonstrated, with 1 dB bandwidths of around 100 nm [80,81].

However, there is one particular situation where a silicon grating coupler can add additional bandwidth: using both the forward and the backward diffraction. As with the 1D polarization splitter discussed earlier in Section 3.4.4.2, it is possible to design the grating in such a way that the Bragg condition for the forward and backward wave matches the same fiber angle, in this case for two different wavelengths [82]. This is shown in Figure 3.31, where the grating coupler is designed to couple a wavelength band between 1490 and 1550 nm and another band around 1310 nm.

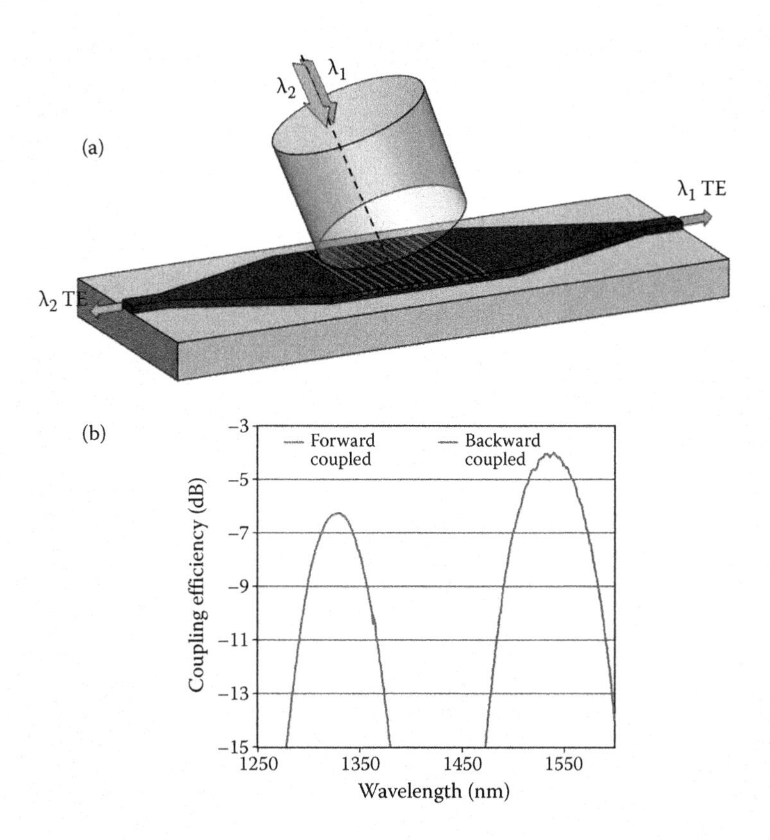

FIGURE 3.31
Grating coupler + wavelength duplexer: (a) principle and (b) measured coupling efficiency for the 1550- and 1310-nm wavelength band. (From D. Vermeulen and G. Roelkens, *2008. ECOC 2008*, vol. 2, no. September, pp. 1–2, 2008. With permission.)

3.4.7 Alternative Surface Coupling Solutions

Although grating couplers are now the most commonly used vertical coupling scheme, there are some alternatives that we will briefly discuss here.

3.4.7.1 Grating Coupler Fiber Probe

In this entire section, we considered a grating coupler that was integrated in the waveguide on the optical chip. However, there is no requirement that the grating should be physically attached to the waveguide: as long as the grating is sufficiently close to the waveguide for the waveguide mode to feel the teeth, there can be coupling.

Instead of mounting the grating coupler on the chip, it can be mounted directly on the fiber [84,85], as shown in Figure 3.32. Using a polymer curing technique and underetch, a gold grating is transferred to the fiber facet, but at an angle to keep the grating parallel with the chip surface [84]. This can even be done with polymer curing using UV light transmitted through the fiber core, which results in a very compact, self-aligned probe.

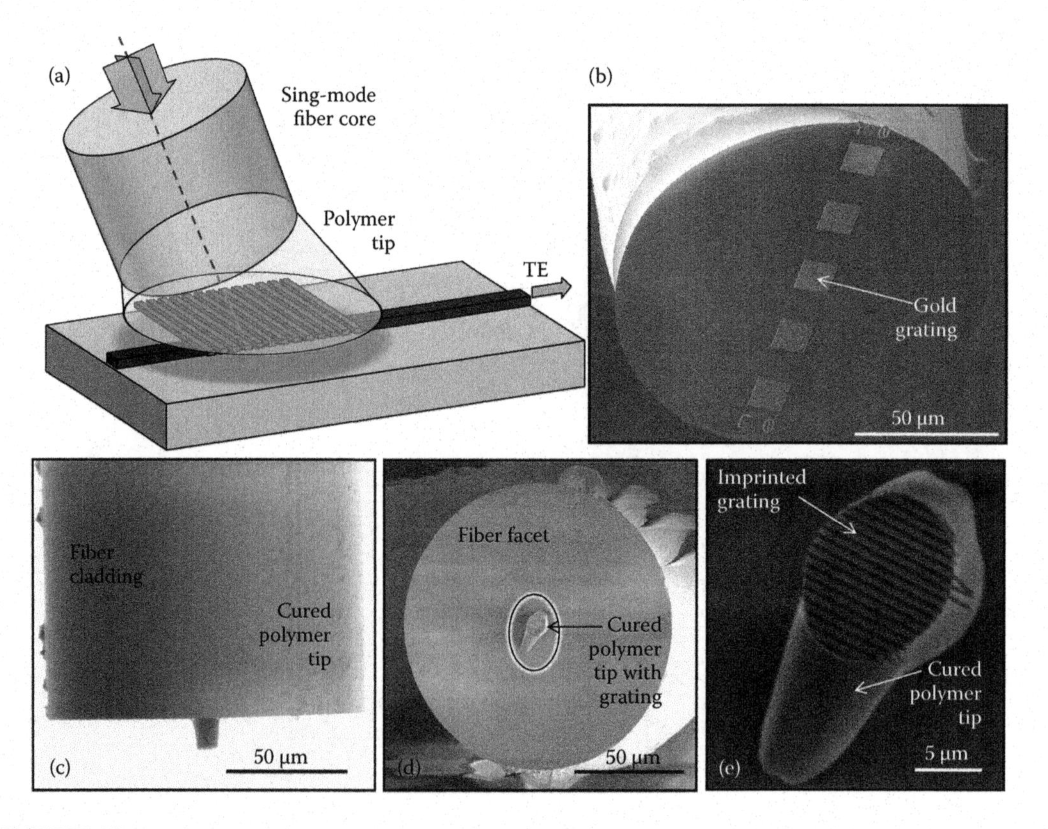

FIGURE 3.32
Grating coupler fiber probe with gold mirror on the fiber facet: (a) principle, (b) fiber facet imprinted with grating coupler, and (c–e) fiber facet with core-only cured polymer facet. (From S. Scheerlinck, P. Dubruel, P. Bienstman, E. Schacht, D. Van Thourhout, and R. Baets, *Journal of Lightwave Technology*, vol. 27, no. 10, pp. 1415–1420, 2009. With permission.)

Coupling efficiency for such a fiber is somewhat less than with a regular grating, and the fact that the distance between the grating and the waveguide cannot be controlled with the same accuracy as an etch grating, makes the coupling strength less reproducible. However, the coupling spectrum is similar to that of a regular etched grating or a gold grating fabricated on a waveguide.

The important advantage of having the grating on the fiber instead of on the waveguide is that the chip can now be probed on any location with a suitable waveguide: there is no need for dedicated fiber couplers on the chip: this means that components within a circuit can be individually probed to test their response. The requirement is of course that a sufficiently broad waveguide section is foreseen.

3.4.7.2 Evanescent Fiber Coupling

A coupling scheme with similar benefits is to use evanescent coupling between an optical fiber and an on-chip waveguide. Because the optical fiber is much larger, and protected by a cladding, it is thinned down by heating and drawing the fiber [86]. When correctly tapered down, the original cladding will now become a core with a significant fraction of light extending in the air outside the fiber. The principle is illustrated in Figure 3.33. Bringing the fiber close enough to the waveguide to enable good coupling requires very accurate positioning.

As with the grating-coupler fiber facets, this technique allows to probe everywhere on the chip. Even better, it is not required to have a broad waveguide section: a narrow wire will do. The evanescent coupling is also used a lot to probe high-quality cavities directly: this way, there is no need to design access waveguides, which generally reduce the quality factor of the cavity.

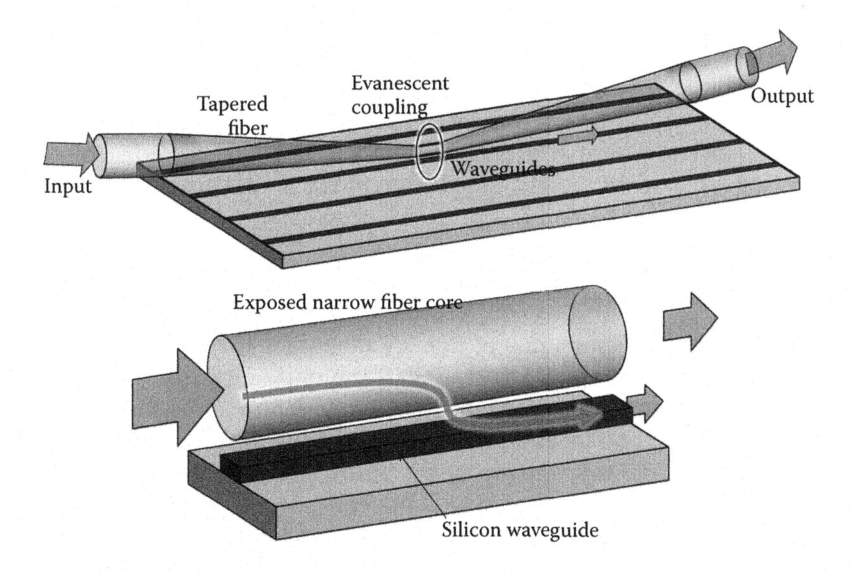

FIGURE 3.33
Evanescent coupling between fiber and on-chip waveguides. (Redrawn from C. Grillet et al., *Optics Express*, vol. 14, no. 3, pp. 1070–1078, 2006.)

3.4.7.3 Mirrors and Cantilevers

More exotic approaches to coupler light through the chip surface include the use of mirrors (either metal or by total internal reflection). However, because of the small core size of silicon waveguides, this technique is not very efficient. It has been used with larger, lower-contrast waveguides [87].

Another alternative is to bend the on-chip waveguide away from the chip surface. This can be done using techniques from MEMS fabrication, and by tuning the stresses in the chip films. This way, a waveguide can become a prestressed cantilever. Upon release from the buried oxide, the cantilever will bend upward. With proper processing, a vertical coupling can be obtained [88].

3.4.8 Surface Coupling: Conclusion

Surface coupling has significant advantages over edge coupling, not the least the potential of wafer-scale testing. Grating couplers are the most commonly used structures to enable surface coupling, and while they have limitations in bandwidth, polarization, and footprint, we have discussed different techniques to improve on these weaknesses. Note, however, that not all these tricks can be arbitrarily combined: It is difficult to combine a 2D or 1D coupler with an overlay with apodization. Also, the bidirectional operation of a 1D duplexer is difficult to combine with curved grating lines.

Still, for many applications, grating couplers are the most promising solution to enable simple and efficient coupling.

3.5 Free-Space Coupling

Up to this point, we have focused primarily on coupling light to and from optical fibers. However, in some applications fiber coupling is not necessary, and this changes the requirements for the coupling significantly. For instance, light from a chip could be coupled to a photodetector or camera, where all the light is collected and there is no requirement for a good overlap with a specific mode. Also, different trade-offs among angular spectrum, bandwidth, coupling efficiency, and footprint are possible.

3.5.1 Free-Space Edge Couplers

When coupling light to free space, one of the key metrics is the numerical aperture (NA): the angle spread in the far field over which light is emitted or captured. In general, we can consider the NA to be inversely proportional to the coupler area: the smaller the size of the emitting region (either the end facet, or the footprint of the grating coupler), the larger the NA and the wider the angular spectrum of the emitted light. Obviously, this also means that the emission into any specific direction is lower.

Both grating couplers and edge couplers can be used for free-space off-chip coupling. For edge couplers, the requirement of tapering to a fiber-matched mode disappears. It is possible to couple light from a silicon wire facet directly to free space: the submicron core will have a quite large NA, as shown in Figure 3.34a. Also, some care is needed to avoid reflections at the facet.

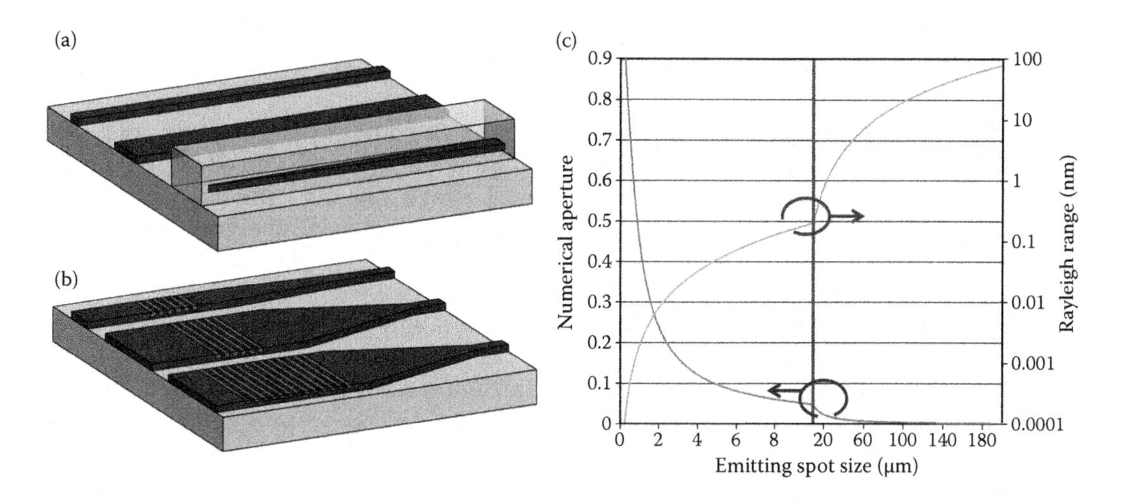

FIGURE 3.34
Tuning the numerical aperture for free-space coupling by changing the spot size. (a) edge couplers and (b) grating couplers. (c) The numerical aperture and its effect on the Rayleigh length, as function of spot size.

3.5.2 Free-Space Grating Couplers

Grating couplers can also be used for free-space coupling [89]. The coupler can be made smaller or larger depending on the requirements for numerical aperture. It is also possible to tune the size of the coupler (and thus the numerical aperture) in the two directions individually. This is shown in Figure 3.34b. In the lateral direction, the waveguide can be made narrower, and therefore, the NA will become larger in the lateral direction, but the NA in the longitudinal direction will remain the same. By making the grating teeth stronger, the light will be diffracted more at each tooth, and the exponential decay (shown in Figure 3.16) will be faster, resulting in a smaller spot radiated upward, and therefore a higher numerical aperture. As the grating coupler is a diffractive structure, it will also increase the bandwidth. Likewise, it is possible to make a narrow-beam, narrow-band coupler using a very weak grating [90].

As with grating couplers for fibers, it is also possible to make focusing grating couplers for free-space coupling. One can now even change the curved grating lines such that the light is focused above the chip, and the focal distance can be tuned for both x and y separately [91].

3.5.3 Optical Phased Arrays

Up until now, we have always considered a single coupling structure for coupling light off-chip. However, we can also employ multiple couplers in a coherent way. By splitting light over several waveguides, one can create an optical-phased array. The combined emission pattern of several couplers is then determined by the emission pattern of the individual couplers and their respective phases and amplitudes. In effect, this is a form of multi-path coupling as we have described in Section 3.2.1.3. The principle is illustrated in Figure 3.35a. The simplest configuration is a periodic array of couplers, with a fixed phase delay between the couplers. In this configuration, the phased array will behave as a much larger grating coupler. A phased array of many small, wide-angle couplers emitting in phase will behave like a much larger, narrow-angle coupler.

While integrated optical phased arrays have been implemented in AlGaAs [92], silicon photonics could improve the performance through better pattern definition and a smaller footprint.

Optical phased arrays can be implemented with edge coupling and grating couplers. Edge couplers can be organized in a line array, as shown in Figure 3.35b, while with grating couplers it is even possible to define 2D phased arrays. The performance of these 2D arrays is still limited due to the low fill factor one can achieve as all the grating couplers need to be accessed with an on-chip waveguide [93]. For a 1D configuration, one does not have this problem and a relatively high fill factor can be obtained.

By changing the phase relation between the couplers many functions are possible. A linear phase shift between subsequent couplers will keep the shape of the narrow beam, but will change the angle of emission. This way, off-chip beam steering is possible [94–98].

By playing with both amplitude and phase, the beam divergence can be tuned from a wide angle to a narrow directed beam. In principle, the emitted phase front can be tuned to a wide extent: the more couplers, the better the control.

By tuning only the phases, a lens can be programmed to focus light on a specific point in space [97]. This is especially useful for sensing applications where light is focused on a spot and the reflection is coupled back into the photonic [99].

Instead of actively tuning the phases using phase tuners, one can also think of using an optical phased array in which the phases are tuned by tuning the wavelength with the use of delay lines. In such a way, one can fabricate a free-space dispersive beam steerer [100] that can find applications in beam steering as well as in demultiplexing [101] and pulse shaping [102].

Because of reciprocity, optical phased arrays can also be used to couple light back onto the chip. This way, a chip can be made to capture light from a tilted or even distorted phase front [103]. The OPA would then act as a coherent receiver allowing phase reconstruction to improve the reception or to perform DOA (Direction Of Arrival) estimation. This can be used in sensing applications where one for example wants to sense light that is reflected from or transmitted through a scattering medium [103].

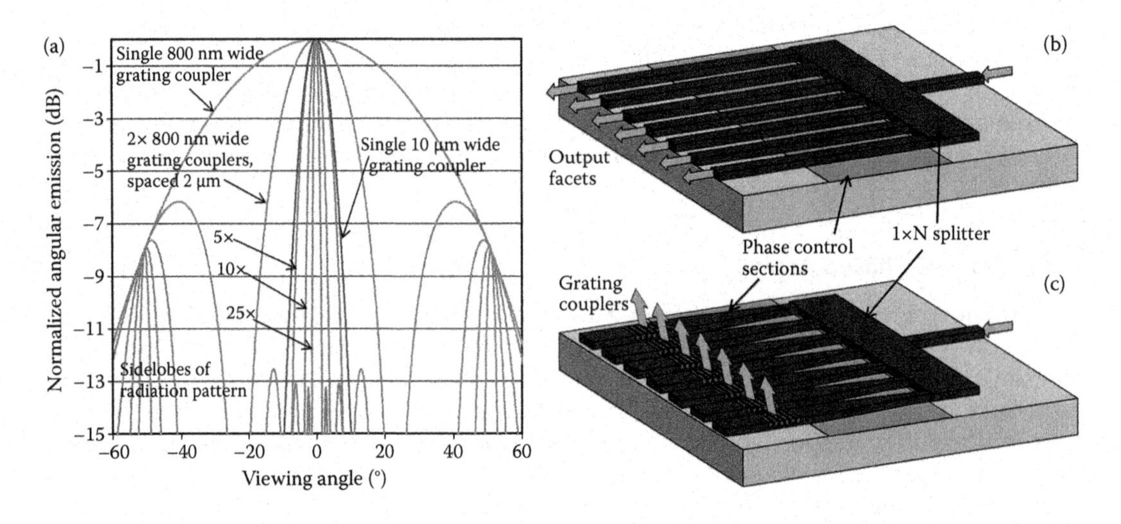

FIGURE 3.35
Optical phased arrays. (a) The total emission is a coherent superposition of the emission patterns of the individual couplers. (b) Optical phased array with edge couplers. (c) Optical phased array with grating couplers.

3.6 Conclusion

In this chapter, we have tried to give a complete overview of optical coupling solutions between silicon photonic waveguides and optical fibers (and in lesser extent free space). We have mainly focused on the on-chip structures and not gone into detail into packaging solutions. This will be discussed further in this book.

The choice of coupling solution for a particular application will be dependent on the operational specifications and trade-offs that need to be made among coupling efficiency, bandwidth, footprint, and subsequent packaging requirements.

Acknowledgments

The authors would like to thank Karel Van Acoleyen, Lars Zimmermann, Zhechao Wang, Sailing He, Yongbo Tang, Attila Mekis, Michael Vanslembrouck, Frederik Van Laere, Shankar Kumar Selvaraja, Kasia Komorowska, Liesbet Van Landschoot, Stijn Scheerlinck Jonathan Schrauwen, and Dirk Taillaert for the use of data and illustrations. Parts of the results shown here were produced in the framework of the European research project HELIOS. Diedrik Vermeulen thanks the Institute for the Promotion of Innovation by Science and Technology in Flanders (IWT) for a grant.

References

1. R. E. Wagner and W. J. Tomlinson, "Coupling efficiency of optics in single-mode fiber components," *Applied Optics*, vol. 21, no. 15, pp. 2671–2688, 1982.
2. P. Dumon et al., "Compact wavelength router based on a silicon-on-insulator arrayed waveguide grating pigtailed to a fiber array," *Optics Express*, vol. 14, no. 2, pp. 664–669, 2006.
3. A. Mekis et al., "A grating-coupler-enabled CMOS photonics platform," *IEEE Journal of Selected Topics in Quantum Electronics*, no. 99, pp. 1–12, 2010.
4. Y. Vlasov and S. McNab, "Losses in single-mode silicon-on-insulator strip waveguides and bends," *Optics Express*, vol. 12, no. 8, pp. 1622–1631, 2004.
5. I. Day et al., "Tapered silicon waveguides for low insertion-loss highly efficient high-speed electronic variable attenuators," in *Proc. IEEE Opt. Fiber Commun. Conf.*, 2003, pp. 249–251.
6. T. Tsuchizawa et al., "Microphotonics devices based on silicon microfabrication technology," *IEEE Journal of Selected Topics in Quantum Electronics*, vol. 11, no. 1, pp. 232–240, 2005.
7. S. McNab, N. Moll, and Y. Vlasov, "Ultra-low loss photonic integrated circuit with membrane-type photonic crystal waveguides," *Optics Express*, vol. 11, no. 22, pp. 2927–2939, 2003.
8. P. Dumon and R. Baets, *Ultra-Compact Integrated Optical Filters in Silicon-on-Insulator by Means of Wafer-Scale Technology*, Ghent University, 2007.
9. V. R. Almeida, R. R. Panepucci, and M. Lipson, "Nanotaper for compact mode conversion," *Optics Letters*, vol. 28, no. 15, pp. 1302–1304, 2003.
10. T. Shoji, T. Tsuchizawa, T. Watanabe, K. Yamada, and H. Morita, "Low loss mode size converter from 0.3 μm square Si waveguides to singlemode fibres," *Electronics Letters*, vol. 38, no. 25, pp. 1669, 2002.

11. L. Chen, C. R. Doerr, Y. K. Chen, and T. Y. Liow, "Low-loss and broadband cantilever couplers between standard cleaved fibers and high-index-contrast Si_3N_4 or Si waveguides," *IEEE Photonics Technology Letters*, vol. 22, no. 23, pp. 1744–1746, 2010.
12. M. Pu, L. Liu, H. Ou, K. Yvind, and J. M. Hvam, "Ultra-low-loss inverted taper coupler for silicon-on-insulator ridge waveguide," *Optics Communications*, vol. 283, no. 19, pp. 3678–3682, 2010.
13. R. Winn, "No Title," *IEEE Trans. Microwave Theory and Tech.*, vol. 23, pp. 92–123, 1975.
14. K. Kasaya, O. Mitomi, M. Naganuma, Y. Kondo, and Y. Noguchi, "A simple laterally tapered waveguide for low-loss coupling to single-mode fibers," *IEEE Photonics Technology Letters*, vol. 5, no. 3, pp. 345–347, 1993.
15. K. Van Acoleyen and R. Baets, "Compact lens-assisted focusing tapers fabricated on silicon-on-insulator," in *8th IEEE International Conference on Group IV Photonics*, 2011, pp. 157–159.
16. B. Luyssaert, *Compact Planar Waveguide Spot-Size Converters in Silicon-on-Insulator*, Ghent University, 2005.
17. M. M. Spuhler, B. J. Offrein, G.-L. Bona, R. Germann, I. Massarek, and D. Erni, "A very short planar silica spot-size converter using a nonperiodic segmented waveguide," *Lightwave Technology, Journal of*, vol. 16, no. 9, pp. 1680–1685, 1998.
18. L. Zimmermann, T. Tekin, H. Schroeder, P. Dumon, and W. Bogaerts, "How to bring nanophotonics to application-silicon photonics packaging," *IEEE LEOS Newsletter*, vol. 22, no. December, pp. 4–14, 2008.
19. J. Galan et al., "CMOS compatible silicon etched V-grooves integrated with a SOI fiber coupling technique for enhancing fiber-to-chip alignment," in *Group IV Photonics, 2009. GFP'09. 6th IEEE International Conference on*, 2009, no. c, pp. 148–150.
20. M. R. Watts, H. A. Haus, and E. P. Ippen, "Integrated mode-evolution-based polarization splitter," *Optics Letters*, vol. 30, no. 9, pp. 967–969, 2005.
21. T. Barwicz et al., "Polarization-transparent microphotonic devices in the strong confinement limit," *Nature Photonics*, vol. 1, no. 1, pp. 57–60, 2007.
22. H. Fukuda, K. Yamada, T. Tsuchizawa, T. Watanabe, H. Shinojima, and S.-I. Itabashi, "Ultrasmall polarization splitter based on silicon wire waveguides," *Optics Express*, vol. 14, no. 25, pp. 12401–12408, 2006.
23. C. R. Doerr and L. Chen, "Monolithic PDM-DQPSK receiver in silicon," in *Optical Communication (ECOC), 2010 36th European Conference and Exhibition on*, 2010, pp. 1–3.
24. M. R. Watts and H. A. Haus, "Integrated mode-evolution-based polarization rotators," *Optics Letters*, vol. 30, no. 2, pp. 138–140, 2005.
25. J. Zhang, M. Yu, G.-Q.P. Lo, D.-L. Kwong, and A. A, "Silicon-waveguide-based mode evolution polarization rotator," *IEEE Journal of Selected Topics in Quantum Electronics*, vol. 16, no. 1, pp. 53–60, 2010.
26. L. Chen, C. R. Doerr, and Y.-K. Chen, "Compact polarization rotator on silicon for polarization-diversified circuits," *Optics Letters*, vol. 36, no. 4, pp. 469–471, 2011.
27. D. Vermeulen, S. Selvaraja, P. Verheyen, W. Bogaerts, D. V. Thourhout, and G. Roelkens, "High efficiency broadband polarization rotator on silicon-on-insulator," in *Group IV Photonics*, 2010, pp. 42–44.
28. H. Fukuda, K. Yamada, T. Tsuchizawa, T. Watanabe, H. Shinojima, and S.-I. Itabashi, "Silicon photonic circuit with polarization diversity," *Quantum Electron*, vol. 16, no. 7, pp. 4872–4880, 2008.
29. Z. Wang and D. Dai, "Ultrasmall Si-nanowire-based polarization rotator," *Journal of the Optical Society of America B*, vol. 25, no. 5, pp. 747, 2008.
30. D. Taillaert, P. Bienstman, and R. Baets, "Compact efficient broadband grating coupler for silicon-on-insulator waveguides," *Optics Letters*, vol. 29, no. 23, pp. 2749–2751, 2004.
31. D. Taillaert et al., "Grating couplers for coupling between optical fibers and nanophotonic waveguides," *Japanese Journal of Applied Physics*, vol. 45, no. 8, pp. 6071–6077, 2006.
32. A. Mekis et al., "A grating-coupler-enabled CMOS photonics platform," *Selected Topics in Quantum Electronics*, no. 99, pp. 1–12, 2010.

33. S. Scheerlinck, J. Schrauwen, F. Van Laere, D. Taillaert, D. Van Thourhout, and R. Baets, "Efficient, broadband and compact metal grating couplers for silicon-on-insulator waveguides," *Optics Express*, vol. 15, no. 15, pp. 9625–9630, 2007.
34. R. Halir, P. Cheben, S. Janz, D.-X. Xu, Í. Molina-Fernández, and J. G. Wangüemert-Pérez, "Waveguide grating coupler with subwavelength microstructures," *Optics Letters*, vol. 34, no. 9, pp. 1408–1410, 2009.
35. H. K. Tsang, "Nanoholes grating couplers for coupling between silicon-on-insulator waveguides and optical fibers," *IEEE Photonics Journal*, vol. 1, no. 3, pp. 184–190, 2009.
36. L. Liu, M. Pu, K. Yvind, and J. M. Hvam, "High-efficiency, large-bandwidth silicon-on-insulator grating coupler based on a fully-etched photonic crystal structure," *Applied Physics Letters*, vol. 96, no. 5, pp. 051126, 2010.
37. R. Halir et al., "Continuously apodized fiber-to-chip surface grating coupler with refractive index engineered subwavelength structure," *Optics Letters*, vol. 35, no. 19, pp. 3243–3245, 2010.
38. X. Chen and H. K. Tsang, "Polarization-independent grating couplers for silicon-on-insulator nanophotonic waveguides," *Optics Letters*, vol. 36, no. 6, pp. 796–798, 2011.
39. W. Bogaerts et al., "Basic structures for photonic integrated circuits in silicon-on-insulator," *Optics Express*, vol. 12, no. 8, pp. 1583–1591, 2004.
40. Z. Yu et al., "High efficiency and broad bandwidth grating coupler between nanophotonic waveguide and fibre," *Chinese Physics B*, vol. 19, no. 1, pp. 014219–014215, 2010.
41. J. Bolten et al., "CMOS compatible cost-efficient fabrication of SOI grating couplers," *Microelectronic Engineering*, vol. 86, no. 4–6, pp. 1114–1116, 2009.
42. L. Vivien et al., "Light injection in SOI microwaveguides using high-efficiency grating couplers," *Journal of Lightwave Technology*, vol. 24, no. 10, pp. 3810–3815, 2006.
43. K. A. Bates, L. Li, R. L. Roncone, and J. J. Burke, "Gaussian beams from variable groove depth grating couplers in planar waveguides," *Applied Optics*, vol. 32, no. 12, pp. 2112–2116, 1993.
44. D. Taillaert, *Grating Couplers as Interface between Optical Fibres and Nanophotonic Waveguides*, Ghent University, 2005.
45. S. K. Selvaraja et al., "Highly efficient grating coupler between optical fiber and silicon photonic circuit," in *Lasers and Electro-Optics 2009 and 2009 Conference on Quantum electronics and Laser Science Conference. CLEO/QELS, 2009. Conference on 2009*, vol. 1, pp. 1–2.
46. Z. Wang, Y. Tang, L. Wosinski, and S. He, "Experimental demonstration of a high efficiency polarization splitter based on a one-dimensional grating with a Bragg reflector underneath," *Photonics Technology Letters, IEEE*, vol. 22, no. 21, pp. 1568–1570, 2010.
47. F. Van Laere et al., "Compact and highly efficient grating couplers between optical fiber and nanophotonic waveguides," *Journal of Lightwave Technology*, vol. 25, no. 1, pp. 151–156, 2007.
48. S. K. Selvaraja et al., "Low-loss amorphous silicon-on-insulator technology for photonic integrated circuitry," *Optics Communications*, vol. 282, no. 9, pp. 1767–1770, May, 2009.
49. C. Kopp, E. Augendre, R. Orobtchouk, O. Lemonnier, and J.-M. Fedeli, "Enhanced fiber grating coupler integrated by wafer-to-wafer bonding," *Journal of Lightwave Technology*, vol. 29, no. 12, pp. 1847–1851, 2011.
50. G. Roelkens, D. V. Thourhout, and R. Baets, "High efficiency grating coupler between silicon-on-insulator waveguides and perfectly vertical optical fibers," *Optics Letters*, vol. 32, no. 11, pp. 1495–1497, 2007.
51. J. W. Goodman, *Introduction to Fourier Optics*, 2nd ed. New York: McGraw-Hill, 1996.
52. A. Martínez, M. D. M. Sánchez-López, and I. Moreno, "Phasor analysis of binary diffraction gratings with different fill factors," *European Journal of Physics*, vol. 28, no. 5, pp. 805–816, 2007.
53. D. Vermeulen et al., "High-efficiency fiber-to-chip grating couplers realized using an advanced CMOS-compatible silicon-on-insulator platform," *Optics Express*, vol. 18, no. 17, pp. 18278–18283, 2010.
54. G. Roelkens et al., "High efficiency diffractive grating couplers for interfacing a single mode optical fiber with a nanophotonic silicon-on-insulator waveguide circuit," *Applied Physics Letters*, vol. 92, no. 13, pp. 131101, 2008.

55. G. Roelkens, D. Van Thourhout, and R. Baets, "High efficiency silicon-on-insulator grating coupler based on a poly-silicon overlay," *Optics Express*, vol. 14, no. 24, pp. 11622–11630, 2006.
56. D. Vermeulen, S. Selvaraja, P. Verheyen, G. Lepage, W. Bogaerts, and G. Roelkens, "High-efficiency silicon-on-insulator fiber-to-chip grating couplers using a silicon overlay," in *Group IV Photonics*, 2009.
57. T. Saha and W. Zhou, "High efficiency diffractive grating coupler based on transferred silicon nanomembrane overlay on photonic waveguide," *Journal of Physics D: Applied Physics*, vol. 42, 2009.
58. N. Na et al., "Efficient broadband silicon-on-insulator grating coupler with low backreflection," *Optics Letters*, vol. 36, no. 11, pp. 2101–2103, 2011.
59. C. Alonso-Ramos, A. Ortega-Moñux, I. Molina-Fernández, P. Cheben, L. Zavargo-Peche, and R. Halir, "Efficient fiber-to-chip grating coupler for micrometric SOI rib waveguides," *Optics Express*, vol. 18, no. 14, pp. 15189, 2010.
60. Y. Tang, Z. Wang, U. Westergren, and L. Wosinski, "High efficiency nonuniform grating coupler by utilizing the lag effect in the dry etching process," in *Optical Fiber Communication Conference*, 2010, pp. 1–3.
61. X. Chen, C. Li, C. K. Y. Fung, S. M. G. Lo, and H. K. Tsang, "Apodized waveguide grating couplers for efficient coupling to optical fibers," *Photonics Technology Letters, IEEE*, vol. 22, no. 15, pp. 1156–1158, 2010.
62. M. Antelius, K. B. Gylfason, and H. Sohlström, "An apodized SOI waveguide-to-fiber surface grating coupler for single lithography silicon photonics," *Optics Express*, vol. 19, no. 4, pp. 3592–3598, 2011.
63. X. Chen, C. Li, H. K. Tsang, and S. Member, "Fabrication-tolerant waveguide chirped grating coupler for coupling to a perfectly vertical optical fiber," *IEEE Photonics Technology Letters*, vol. 20, no. 23, pp. 1914–1916, 2008.
64. J. Schrauwen, F. Van Laere, D. Van Thourhout, and R. Baets, "Focused-ion-beam fabrication of slanted grating couplers in silicon-on-insulator waveguides," *IEEE Photonics Technology Letters*, vol. 19, no. 11, pp. 816–818, 2007.
65. J. Schrauwen, S. Scheerlinck, D. Van Thourhout, and R. Baets, "Polymer wedge for perfectly vertical light coupling to silicon," *Proceedings of SPIE*, vol. 32, no. 0, pp. 72180B–72180B8, 2009.
66. S. Scheerlinck, J. Schrauwen, G. Roelkens, D. Van Thourhout, and R. Baets, "Vertical fiber-to-waveguide coupling using adapted fibers with an angled facet fabricated by a simple molding technique," *Applied Optics*, vol. 47, no. 18, pp. 3241–3245, 2008.
67. D. Vermeulen et al., "Efficient tapering to the fundamental quasi-TM mode in asymmetrical waveguides," in *European Conference on Integrated Optics*, 2010.
68. D. Taillaert, P. I. Borel, L. H. Frandsen, R. M. De La Rue, and R. Baets, "A compact two-dimensional grating coupler used as a polarization splitter," *IEEE Photonics Technology Letters*, vol. 15, no. 9, pp. 1249–1251, 2003.
69. W. Bogaerts, D. Taillaert, P. Dumon, D. Van Thourhout, R. Baets, and E. Pluk, "A polarization-diversity wavelength duplexer circuit in silicon-on-insulator photonic wires," *Optics Express*, vol. 15, no. 4, pp. 1567–1578, 2007.
70. R. Halir, D. Vermeulen, and G. Roelkens, "Reducing polarization-dependent loss of silicon-on-insulator fiber to chip grating couplers," *IEEE Photonics Technology Letters*, vol. 22, no. 6, pp. 389–391, 2010.
71. F. Van Laere et al., "Efficient polarization diversity grating couplers in bonded InP-membrane," *IEEE Photonics Technology Letters*, 20, no. 4, pp. 318–320, 2008.
72. C. R. Doerr et al., "Monolithic polarization and phase diversity coherent receiver in silicon," *Journal of Lightwave Technology*, vol. 28, no. 4, pp. 520–525, 2010.
73. C. R. Doerr, M. S. Rasras, J. S. Weiner, and M. P. Earnshaw, "Diplexer with integrated filters and photodetector in Ge–Si using Γ–*X* and Γ–*M* directions in a grating Coupler," *IEEE Photonics Technology Letters*, vol. 21, no. 22, pp. 1698–1700, 2009.
74. Z. Wang et al., "Experimental demonstration of an ultracompact polarization beam splitter based on a bidirectional grating coupler," in *Communications and Photonics Conference and Exhibition (ACP)*, 2009, pp. 1–2.

75. Y. Tang, D. Dai, and S. He, "Proposal for a grating waveguide serving as both a polarization splitter and an efficient coupler for silicon-on-insulator nanophotonic circuits," *Photonics Technology Letters, IEEE*, vol. 21, no. 4, pp. 242–244, 2009.
76. D. Vermeulen, G. Roelkens, and D. Van Thourhout, "Grating structures for simultaneous coupling to TE and TM waveguide modes," 22 June 2009.
77. F. Van Laere et al., "Compact focusing grating couplers for silicon-on-insulator integrated circuits," *IEEE Photonics Technology Letters*, vol. 19, no. 23, pp. 1919–1921, 2007.
78. D. Vermeulen et al., "Reflectionless grating coupling for silicon-on-insulator integrated circuits," in *2011 8th IEEE International Conference on Group IV Photonics (GFP)*, 2011, vol. 1, no. 1, pp. 74–76.
79. F. Van Laere, W. Bogaerts, P. Dumon, G. Roelkens, D. Van Thourhout, and R. Baets, "Focusing polarization diversity gratings for silicon-on-insulator integrated circuits," in *2008 5th IEEE International Conference on Group IV Photonics*, 2008, vol. 1, pp. 203–205.
80. C. R. Doerr, L. Chen, Y. K. Chen, and L. L. Buhl, "Wide bandwidth silicon nitride grating coupler," *Photonics Technology Letters, IEEE*, vol. 22, no. 19, pp. 1461–1463, 2010.
81. G. Maire et al., "High efficiency silicon nitride surface grating couplers," *Optics Express*, vol. 16, no. 1, pp. 328–333, 2008.
82. G. Roelkens, D. Van Thourhout, and R. Baets, "Silicon-on-insulator ultra-compact duplexer based on a diffractive grating structure," *Optics Express*, vol. 15, no. 16, pp. 10091–10096, 2007.
83. D. Vermeulen and G. Roelkens, "Silicon-on-insulator nanophotonic waveguide circuit for fiber-to-the-home transceivers," *2008. ECOC 2008*, vol. 2, no. September, pp. 1–2, 2008.
84. S. Scheerlinck, P. Dubruel, P. Bienstman, E. Schacht, D. Van Thourhout, and R. Baets, "Metal grating patterning on fiber facets by UV-based nano imprint and transfer lithography using optical alignment," *Journal of Lightwave Technology*, vol. 27, no. 10, pp. 1415–1420, 2009.
85. S. Scheerlinck, D. Taillaert, D. Van Thourhout, and R. Baets, "Flexible metal grating based optical fiber probe for photonic integrated circuits," *Applied Physics Letters*, vol. 92, no. 3, pp. 031104, 2008.
86. C. Grillet et al., "Efficient coupling to chalcogenide glass photonic crystal waveguides via silica optical fiber nanowires," *Optics Express*, vol. 14, no. 3, pp. 1070–1078, 2006.
87. K. Watanabe, J. Schrauwen, A. Leinse, D. V. Thourhout, R. Heideman, and R. Baets, "Total reflection mirrors fabricated on silica waveguides with focused ion beam," *Electronics Letters*, vol. 45, no. 17, pp. 883, 2009.
88. P. Sun and R. M. Reano, "Vertical chip-to-chip coupling between silicon photonic integrated circuits using cantilever couplers," *Optics Express*, vol. 19, no. 5, pp. 4722–4727, 2011.
89. S. a. Masturzo, J. M. Yarrison-Rice, H. E. Jackson, and J. T. Boyd, "Grating couplers fabricated by electron-beam lithography for coupling free-space light into nanophotonic devices," *IEEE Transactions on Nanotechnology*, vol. 6, no. 6, pp. 622–626, 2007.
90. R. Waldhäusl, B. Schnabel, P. Dannberg, E. B. Kley, A. Bräuer, and W. Karthe, "Efficient coupling into polymer waveguides by gratings," *Applied Optics*, vol. 36, no. 36, pp. 9383–9390, 1997.
91. J. S. Yang et al., "Novel grating design for out-of-plane coupling with nonuniform duty cycle," *Photonics Technology Letters, IEEE*, vol. 20, no. 9, pp. 730–732, 2008.
92. F. Vasey, F. Reinhart, R. Houdre, and J. Stauffer, "Spatial optical beam steerin with an AlGaAs integrated phased array," *Applied Optics*, vol. 32, no. 18, pp. 3220–3232, 1993.
93. K. Van Acoleyen, H. Rogier, and R. Baets, "Two-dimensional optical phased array antenna on silicon-on-insulator," *Optics Express*, vol. 18, no. 13, pp. 13655–13660, 2010.
94. A. Hosseini et al., "Unequally spaced waveguide arrays for silicon nanomembrane-based efficient large angle optical beam steering," *IEEE Journal of Selected Topics in Quantum Electronics*, vol. 15, no. 5, pp. 1439–1446, 2009.
95. D. N. Kwong, Y. Zhang, A. Hosseini, and R. T. Chen, "Integrated optical phased array based large angle beam steering system fabricated on silicon-on-insulator," *Library*, vol. 7943, no. May, pp. 79430Y–79430Y6, 2011.
96. K. Van Acoleyen, W. Bogaerts, J. Jágerská, N. Le Thomas, R. Houdré, and R. Baets, "Off-chip beam steering with a one-dimensional optical phased array on silicon-on-insulator," *Optics Letters*, vol. 34, no. 9, pp. 1477–1479, 2009.

97. K. Van Acoleyen, K. Komorowska, W. Bogaerts, and R. Baets, "One-dimensional off-chip beam steering and shaping using optical phased arrays on silicon-on-insulator," *Journal of Lightwave Technology*, vol. 29, no. 99, pp. 1–1, 2011.
98. J. K. Doylend, M. J. R. Heck, J. T. Bovington, J. D. Peters, L. A. Coldren, and J. E. Bowers, "Two-dimensional free-space beam steering with an optical phased array on silicon-on-insulator," *Optics Express*, vol. 19, no. 22, pp. 21595–21604, 2011.
99. Y. Li, S. Meersman, and R. Baets, "Realization of fiber-based laser Doppler vibrometer with serrodyne frequency shifting," *Applied Optics*, vol. 50, no. 17, pp. 2809–2814, 2011.
100. K. Van Acoleyen, W. Bogaerts, and R. Baets, "Two-dimensional dispersive off-chip beam scanner fabricated on silicon-on-insulator," *IEEE Photonics Technology Letters*, vol. 23, no. 17, pp. 1270–1272, 2011.
101. J. Yang, X. Jiang, M. Wang, and Y. Wang, "Two-dimensional wavelength demultiplexing employing multilevel arrayed waveguides," *Optics Express*, vol. 12, no. 6, pp. 1084–1089, 2004.
102. V. R. Supradeepa, C.-B. Huang, D. E. Leaird, and A. M. Weiner, "Femtosecond pulse shaping in two dimensions: towards higher complexity optical waveforms," *Optics Express*, vol. 16, no. 16, pp. 11878–11887, 2008.
103. K. Van Acoleyen, E. M. P. Ryckeboer, K. Komorowska, and R. Baets, "Light collection from scattering media in a silicon photonics integrated circuit," in *IEEE Photonics 2011*, 2011, vol. 4, pp. 543–544.

4

Multichannel Silicon Photonic Devices

Ting Lei, Shaoqi Feng, Aimé Sayarath, Jun-Feng Song, Xianshu Luo, Guo-Qiang Lo, and Andrew W. Poon

CONTENTS

4.1 Introduction

Over the past decade, academic and industrial researchers worldwide have feverishly explored silicon (Si) photonics as a possible technology toward next-generation multichannel optical communications and optical interconnects for computercom [1–3]. Thanks to the mature silicon complementary metal-oxide semiconductor (CMOS) fabrication processes and foundry services [4, 5], various microscale and nanoscale photonic structures can now be fabricated with high precision and repeatability on silicon and silicon-on-insulator (SOI) wafers. As reviewed in various chapters of this handbook, silicon, as one of the most well-understood materials and the most abundant semiconductor, offers the key advantages of mature fabrication process, low cost, transparency to the telecommunication wavelength range (1300–1600 nm) due to the silicon band gap of 1.12 eV, along with a relatively high refractive index of ~3.5 for optical waveguiding and confinement. Silicon refractive index is also tunable via thermal–optic effect (10^{-4}/K) [2] and free-carrier plasma dispersion effect [6], though linear electro-optic Pockels effect is absent due to the inversion symmetry of bulk silicon crystal lattice. It is based on the above merits of silicon that various compact, passive photonic devices with reconfigurable multichannel processing capabilities have been demonstrated including devices based on Mach–Zehnder interferometers (MZIs) [7–11], arrayed waveguide gratings (AWGs) [12–17], and all sorts of microresonators [3, 18–20].

In this chapter, we discuss the fundamentals and promising applications of a few selected multichannel silicon photonic devices, with emphasis on integrated grating– and microring resonator–based multiplexers/demultiplexers (MUX/DEMUX), microring resonator–based add–drop filters and interleavers. While we believe relatively large-sized and multichannel integrated grating–based technology is more suitable for telecommunications and compact-sized microring resonator–based technology is more suitable for optical interconnects, we do not specifically focus our discussions on applications-related specifics. Instead, the focus is on the fundamentals at the graduate student level so that newcomers to silicon photonics or integrated photonics can readily make use of the principles and examples discussed here to quickly embark on their own journey in integrated photonic device design. To limit the scope, we only summarize the perspectives of the authors, and we do not attempt here to give an exhaustive review of all prior work on multichannel silicon photonic devices. Interested readers should refer to the numerous recent literature and review articles, some of which are cited in this chapter and other related chapters of this handbook. We also do not emphasize the device fabrication details as they often depend on the fabrication processes or the foundry services and also vary among fabrication facilities.

4.1.1 Multichannel Optical Communications

Optical communications systems have been widely used as information transmission systems, owing to the advantages of large data bandwidth, high security, and low power consumption. To further enhance the data transmission capacity, wavelength-division multiplexing (WDM) technology has been introduced [21]. The WDM technology is realized by multiplexing a number of optical carrier signals by different wavelengths (or frequencies) into a single communication channel, namely an optical fiber or a waveguide or even in free space for optical wireless. Each of the wavelengths is termed an optical channel.

An essential building block for WDM systems are the MUX and DEMUX, which combine and split the signals carried with different wavelengths. With the MUX and DEMUX units, different functional subsystems have been proposed and demonstrated, including optical cross-connectors (OXC) [22], optical add–drop multiplexers (OADM) [23, 24], reconfigurable optical add–drop multiplexers (ROADM) [25, 26], wavelength selective switches (WSS) [27, 28], and many others. In this chapter, we focus on MUX/DEMUX and OADM.

The functions of multiplexing and demultiplexing can be implemented using various methods, among which the concave grating (CG) [29–32] and the AWG [12–17] schemes are two widely adopted technologies. Other emerging technologies such as microring resonators [33] are also under intense investigation. CG is one of the earliest optical components demonstrated for signal multiplexing and demultiplexing, whereas AWG has improved performance from CG. On the other hand, microring resonators offer periodic sharp resonances over a compact footprint. As the WDM channel spacing becomes narrower, the challenges for MUX/DEMUX significantly increase, especially for the fabrication process in terms of uniformity and manufacturing precision.

In a dense WDM system, to reduce the channel spacing from e.g., 100–50 GHz, we need to double the number of output waveguides for an optical DEMUX. This makes the device fabrication technically much more challenging and also significantly increases the device footprint. The optical interleaver [34, 35], which splits odd and even numbers of WDM channels into two sets of complementary outputs, is one of the components that can double the channel spacing within each set of output channels and thus mitigate the aforementioned challenges. Conversely, an optical interleaver can also be used to compress the channel spacing as a de-interleaver by combining the odd and even numbers of channels.

Figure 4.1 shows the conceptual schematic of a WDM system that includes the key components, e.g., laser diode array emitting at various optical wavelength channels, MUX/DEMUX, interleavers, and photodetector array. The implementation of optical interleavers relaxes the MUX/DEMUX requirement. Thus, it is possible to use two pairs of MUX/DEMUX with doubling of the channel spacing to realize the narrower channel function. Today's WDM optical filters are required to support high data rates per channel (up to 40 Gbit/s) with high out-of-band rejection. Box-like filter profiles with flat-top response and sharp roll-off are necessary.

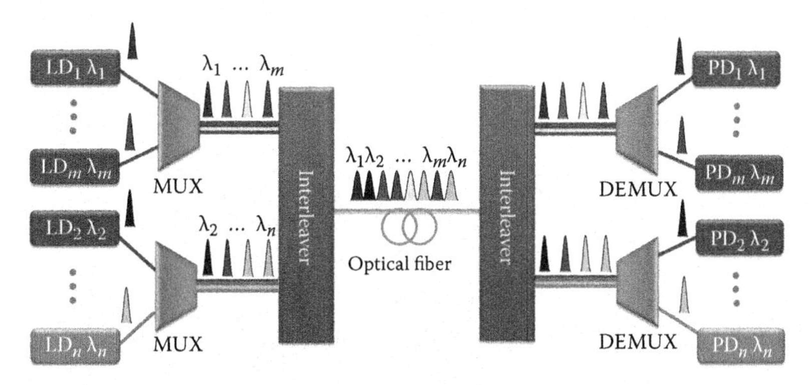

FIGURE 4.1
(See color insert.) Schematic of a WDM system including the key components. LD: laser diode, MUX: multiplexer, DEMUX: demultiplexer, PD: photodiode. $\lambda_1 \ldots \lambda_m$: odd wavelength channels, $\lambda_2 \ldots \lambda_n$: even wavelength channels.

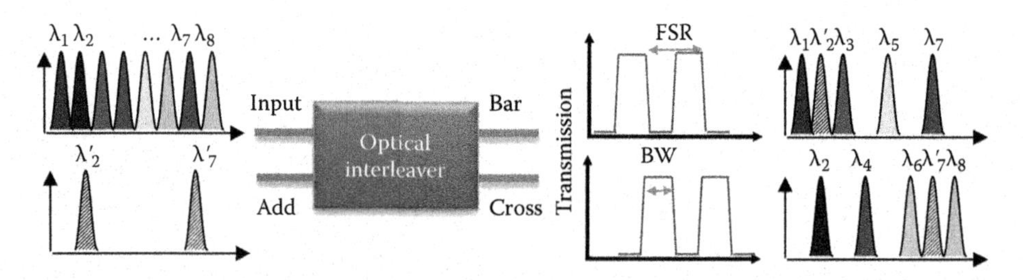

FIGURE 4.2
(See color insert.) Working principle of an optical interleaver.

Figure 4.2 schematically shows the signal multiplexing and demultiplexing using an optical interleaver. Both the bar and cross transmission spectra exhibit the same free-spectral range (FSR) value that is double the desired channel spacing. The odd (even) number of channels launched from the input-port exit in the bar- (cross-) port, while the odd (even) number of channels launched from the add-port exit in the cross- (bar-) port.

Another basic building block in a WDM system or network is the OADM. Unlike a MUX/DEMUX, which combines and splits multiple channels at the transmitting and receiving ends of an optical link, an OADM selectively drops and adds signals at certain optical channels to and from local ports (e.g., local users or premises in a metro/access network), while passing the through channels together with the added channels. The enabling component for the OADM is the optical channel add–drop filter (CADF), which drops and adds signals in a selected optical channel. Figure 4.3 schematically shows an OADM comprising multiple CADFs centered at different channels [36]. Add–drop filters have been demonstrated using various integrated photonic structures such as distributed Bragg reflectors (DBR) [37] and microring resonators [38, 39]. In this chapter, we focus on silicon microring resonator-based add–drop filters.

In essence, each WDM wavelength channel carrying a high data rate signal (10–40 Gbit/s) should be transmitted with low loss and minimal distortion. There are several key performance metrics defining a WDM optical filter that constitutes the MUX/DEMUX, interleaver, and OADM. Here in the context of channel add–drop filters we define these metrics [36], namely (a) bandwidth (1 and 3 dB), (b) channel loss or drop loss, (c) in-band rejection, (d) out-of-band rejection, and (e) roll-off ratio.

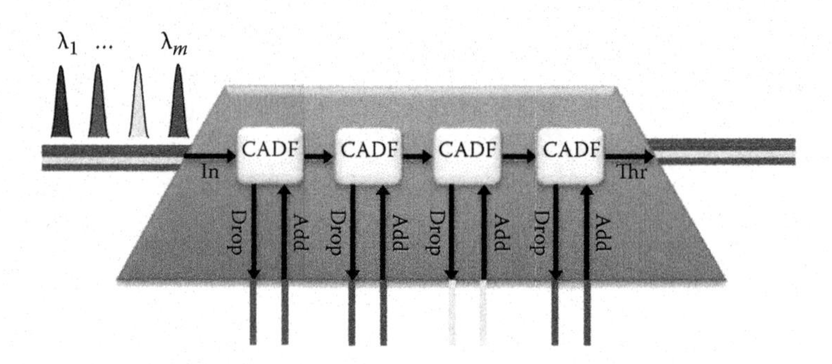

FIGURE 4.3
Schematic of an OADM comprising multiple channel add–drop filters.

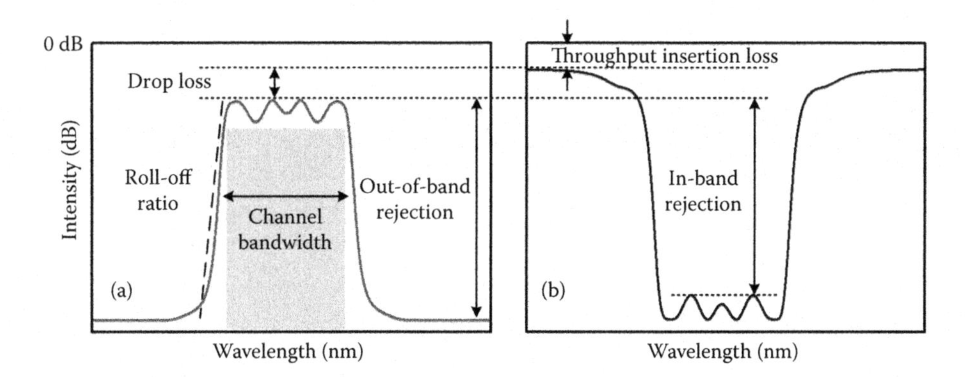

FIGURE 4.4
Schematic of the add–drop filter transmission spectra and definitions of key performance metrics. (a) Drop-port transmission spectrum. (b) Throughput-port transmission spectrum.

Figure 4.4 schematically shows the channel add–drop filter transmission spectra with labeling of these key parameters. Specifically, 1- or 3-dB bandwidth defines the optical channel selectivity. The channel loss is the in-band transmission loss or, in the case of an add–drop filter, the drop loss is the difference between the drop-port transmission and the throughput-port transmission. In-band rejection is the difference between the drop-port transmission and the in-band throughput-port transmission. Out-of-band rejection is the drop-port extinction ratio (ER). Roll-off ratio is given by the slope of the drop-port transmission band edge in the unit of dB/nm.

4.1.2 Multichannel Optical Interconnects

Besides the traditional fiber-optic communications systems, optical communications are now finding promising applications as optical interconnects in computercom [2, 40, 41]. Today's data centers are mostly if not all electrical-based due to the maturity of the microelectronics and computer industries. With the boom of the Internet and the continuous increase of the network, bandwidth demands are expected at a pace of doubling every 18 months [42], and data centers are becoming bigger everyday. It is becoming clear that the conventional electrical interconnects will soon impose hardware limits to the growth of the computercom, due to the inherent large power consumption and the resultant heat generation and thermal issues, and also the inherent bandwidth limitation of the electrical interconnects. The limitations are also associated with the dimensional scaling of the copper interconnects and the resistance-capacitance time constant associated with the shrinking dimensions.

The optical interconnects provide a potential solution, in which optical signals transmit through integrated photonic circuits [2, 43, 44]. Figure 4.5 schematically illustrates the optical interconnect system for communication between multiple central processing units (CPUs) or computers. All the key components have been demonstrated using integrated silicon, or hybrid silicon-based devices. Optical interconnects have many advantages, such as large bandwidth, potentially low power consumption, and immunity to electromagnetic interference. Multichannel WDM technology provides a way of further increasing optical interconnect data capacity. Optical interconnect technologies for computercom are in their infancy compared with the more mature telecommunications, and the possibility of a paradigm change in interconnection technology has therefore excited much research and development work [44–48].

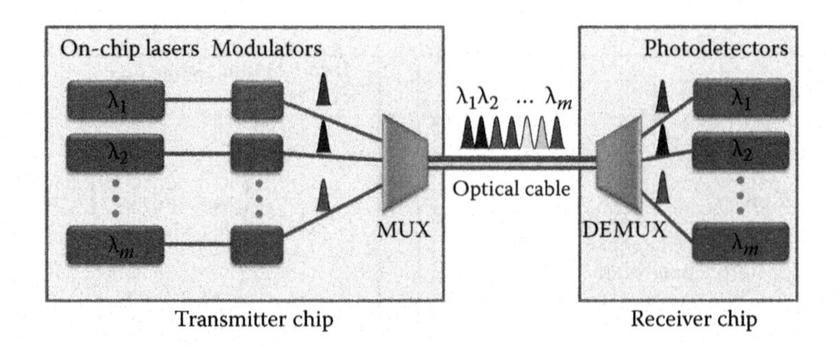

FIGURE 4.5
Schematic of a multichannel optical interconnect system using WDM technologies including on-chip transmitters/receivers and an optical cable for computercom applications.

4.1.3 Chapter Outline

The rest of the chapter is outlined as follows: Section 4.2 discusses integrated grating-based MUX/DEMUX. We discuss the principles and applications of concave gratings and arrayed-waveguide gratings. Section 4.3 illustrates waveguide-coupled microring resonators. We introduce transfer matrix modeling to model basic building blocks in the form of microring resonator-based notch and add–drop filters. Section 4.4 illustrates cascaded microresonators with tailored filtering characteristics including high-order coupled microresonators and Vernier filters. Section 4.5 illustrates two design examples of interleavers using MZI coupling to microring resonators. Section 4.6 summarizes the chapter and offers a future outlook.

4.2 Integrated Grating–Based MUX/DEMUX

4.2.1 Introduction to Integrated Grating Technologies

The most popular grating-based MUX/DEMUX components are the CG and the AWG, both of which are based on the light interfering and focusing by using Rowland circle [49, 50]. The CG comprises a Rowland circle and a grating circle. The AWG comprises two CGs with mirror symmetry. The two CGs in the AWG are usually known as star couplers. The CG has the advantage of compact footprint and thus offers high integration density. However, it is difficult to realize a large number of optical channels and narrow channel spacing within the compact footprint. The CG has been demonstrated in various material platforms, including 256-channel, 25-GHz channel spacing CGs in silica-on-silicon substrates [51], 48-channel 100-GHz channel spacing CGs in indium phosphide (InP) substrates [52], and CWDM CGs in SOI substrates [32, 53].

Compared with the CG, the AWG could have a larger number of optical channels and narrower channel spacing, with the trade-off of an enlarged device footprint. The AWG has attracted great interest and been demonstrated in various material platforms, including silica-on-silicon [54–56], InP [57, 58] and SOI [59–64]. AWGs based on silica-on-silicon have been commercialized for many years. Silica-based AWGs with exceeding 1000 channels and 25-GHz channel spacing have been demonstrated [56]. However, such silica-based AWGs are difficult for further large-scale integration. InP-based AWGs possess the advantage of monolithic integration with lasers, amplifiers, and photodetectors with the tradeoff of high-cost InP wafers. Thus, silicon-based AWGs started to gain research interest due to the low material cost and the mature electro-photonic integration capabilities.

In this section, we present examples of the CGs and the AWGs implemented on the SOI platform. We introduce the design of a low-aberration CG [65], followed with an AWG with 32 channels and 200-GHz channel spacing.

4.2.1.1 Rowland Circle and Grating Circle

Rowland circle is named after H. A. Rowland [49, 50]. In 1883, Rowland designed an optical component using a spherical grating with radius of 2*R*, which combines both diffraction and focusing functions. He observed that the focusing points of this spherical grating located at another circle with radius of *R*. Such a circle in which all the focusing points locate is named as Rowland circle, whereas the circle in which the spherical grating follows is called grating circle.

Figure 4.6 shows the schematic of the concave grating with ray optics illustration. Points *O* and *O′* represent the axis pole and the center of the Rowland circle. The radii of the grating circle and the Rowland circle are 2*R* and *R*, respectively. The incident beams and focusing points of the reflected beams are located along the Rowland circle. We assume that the incident light is launched from point *A* with coordinates (x_A, y_A) in the Rowland circle, and focused at point *B* with coordinates (x_B, y_B) after reflecting by the spherical grating. We assume point *P* with coordinates (x_p, y_p) is another point in the grating circle from which the light is reflected from point *A* to point *B*. The light paths can be calculated as follows:

$$AP = \sqrt{(x_P - x_A)^2 + (y_P - y_A)^2} \tag{4.1}$$

$$PB = \sqrt{(x_B - x_P)^2 + (y_B - y_P)^2} \tag{4.2}$$

$$OA = 2R\cos\alpha \tag{4.3}$$

$$OB = 2R\cos\beta \tag{4.4}$$

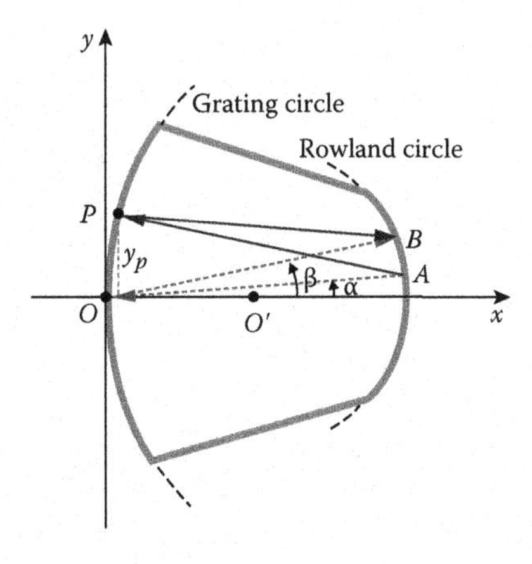

FIGURE 4.6
Principle of the concave grating with ray-optic illustration.

where α and β are the incident/reflection angles for rays AO and OB relative to the x axis. The optical interference at position B due to light paths AOB and APB depends on the light path difference Δl, which is given as

$$\Delta l = AP + PB - (AO + OB) \tag{4.5}$$

As both points A and B are located in the Rowland circle, we write their coordinates as

$$\begin{cases} x_A = 2R\cos^2\alpha \\ y_A = 2R\cos\alpha\sin\alpha \end{cases} \tag{4.6}$$

and

$$\begin{cases} x_B = 2R\cos^2\beta \\ y_B = 2R\cos\beta\sin\beta \end{cases} \tag{4.7}$$

For point P in the grating circle, we express the coordinates in terms of the distance square from the center of the grating circle as

$$(x_P - 2R)^2 + y_P^2 = 4R^2 \tag{4.8}$$

Thus, we express the path difference Δl from Equation 4.5 as

$$\Delta l = -(\sin\alpha + \sin\beta)y_P + \frac{1}{64}\frac{1}{R^3}\left(\frac{\sin^2\alpha}{\cos\beta} + \frac{\sin^2\beta}{\cos\beta}\right)y_P^4 + \text{high order} \tag{4.9}$$

For paraxial approximation with $y_p \ll R$, the high-order terms in Equation 4.9 can be ignored. Thus, we simplify Equation 4.9 as

$$\Delta l = -(\sin\alpha + \sin\beta)y_P \tag{4.10}$$

For constructive interference at a given wavelength, the light path difference is the integer number of wavelength in the grating medium. The phase-matching condition is given as

$$n(\sin\alpha + \sin\beta)y_P = im\lambda \tag{4.11}$$

or

$$n(\sin\alpha + \sin\beta)d = m\lambda \tag{4.12}$$

where integer m is the diffraction order, integer i is the grating order, λ is the light wavelength in vacuum and n is the effective refractive index of the integrated grating. $d = y_p/i$ is the distance in the y direction between adjacent constructive interference points, which is a constant. Equation 4.12 is known as the grating equation.

4.2.1.2 Low-Aberration Concave Gratings

In Equation 4.9 in Section 4.2.1.1, we ignore the high-order expansion terms using the paraxial approximation. However, in the case that the paraxial approximation is no longer satisfied, i.e., reflection point P is far from the axis, the high-order expansion terms cannot be omitted. This suggests that the light path difference for the reflection from the integrated grating is no longer an integer number of the wavelength, thus inducing aberration. To minimize such an aberration, we introduce below a method for the design of a low-aberration concave grating.

For clear illustration, the concave grating is redrawn in Figure 4.7. We assume that a light ray with multiple wavelengths is incident from point A, while points B_q ($q = 1, 2, 3, \ldots$) represent the nonaberration focus points corresponding to the incident wavelength of λ_q. The orders of the grating facets are numbered as i, j, etc. The 0th-order facet is located at the origin point O with coordinates of (0, 0). The purpose here is to design a CG that minimizes the aberration for all incident wavelengths of interest. Apparently, both the wavelength and the phase difference can be solely determined by the positions that are located along the Rowland circle. Thus, we arbitrarily choose two of the positions as the nonaberration image points, e.g., B_η and B_ξ. The optical path difference for these two points with the light ray incident from point A and reflected by the ith grating can be determined by the grating equation (Equation 4.12) as follows

$$f_{X,i}(\lambda_X, x_i, y_i) = n(\lambda_X)(r_{A,i} + r_{X,i} - r_{A,0} - r_{X,0}) + im\lambda_X \tag{4.13}$$

where the subscript $_X$ denotes either B_η or B_ξ. (x_i, y_i) are the coordinates of the ith grating center. λ_X is the free-space wavelength focused at the image point X. $n(\lambda_X)$ is the wavelength-dependent effective refractive index at λ_X. $r_{A,i}$ is the distance between the incident point A and the center of the ith grating facet. $r_{X,i}$ is the distance between the image point X and the center of the ith grating facet. $r_{A,0}$ is the distance between the incident point A and the origin point O. $r_{X,0}$ is the distance between the image point X and the origin point O.

For the nonaberration points, B_η and B_ξ here, the path difference is zero, which means

$$\begin{cases} f_{B_{\eta},i}(\lambda_X, x_i, y_i) = 0 \\ f_{B_{\xi},i}(\lambda_X, x_i, y_i) = 0 \end{cases} \tag{4.14}$$

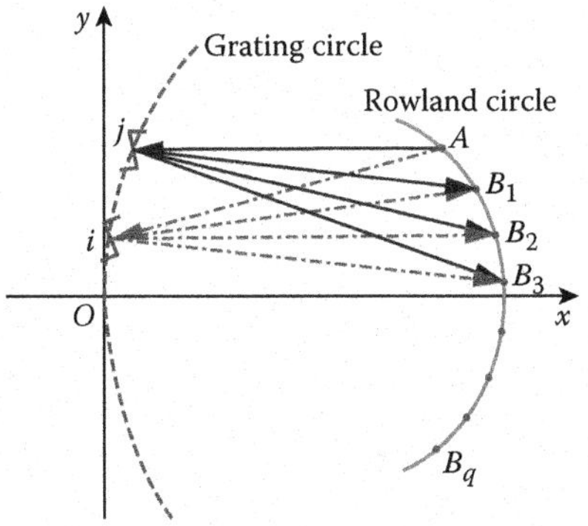

FIGURE 4.7
Schematic of a concave grating with ray-optic illustration.

By using Newtons iteration method, we can solve these equations and obtain the position coordinates of the ith grating (x_i, y_i).

To determine the other image points B_q, we use the least square method based on the fact that every image point B_q has the lowest aberration. From Equation 4.13, we obtain the total output power of the reflected light as

$$g = \sum_i f_{X,i}^2(\lambda_X, x_i, y_i) \tag{4.15}$$

For nonaberration case, we have two constraint conditions:

$$\begin{cases} \dfrac{\partial g}{\partial x_X} = 2n(\lambda_X) g^{(x)}(x_X, y_X, \lambda_X) = 0 \\ \dfrac{\partial g}{\partial y_X} = 2n(\lambda_X) g^{(y)}(x_X, y_X, \lambda_X) = 0 \end{cases} \tag{4.16}$$

Where, $g(x)$ and $g(y)$ can be described as

$$\begin{cases} g^{(x)}(x_X, y_X, \lambda_X) = \sum_i f_{X,i}(\lambda_X, x_i, y_i)(\cos\theta_{X,i} - \cos\theta_{X,0}) \\ g^{(y)}(x_X, y_X, \lambda_X) = \sum_i f_{X,i}(\lambda_X, x_i, y_i)(\sin\theta_{X,i} - \sin\theta_{X,0}) \end{cases} \tag{4.17}$$

and

$$\begin{cases} \cos\theta_{X,j} = (x_X - x_j)/r_{X,j} \\ \sin\theta_{X,j} = (y_X - y_j)/r_{X,j} \end{cases} \tag{4.18}$$

Similar to previous case, we can also use Newtons iteration method to calculate the equations and obtain the position coordinates of the image points.

4.2.2 Silicon-on-Insulator Concave Gratings MUX/DEMUX

Following such a design, we present here as an example of a four-channel CG in a commercially available 200-mm SOI substrate (with a 220-nm top silicon layer and a 2-μm buried oxide (BOX) layer). The designed grating wavelengths at imaging points B_1, B_2, B_3, and B_4 are λ_{B1} = 1531 nm, λ_{B2} = 1551 nm, λ_{B3} = 1571 nm, and λ_{B4} = 1591 nm, respectively. The radius of the Rowland circle is 94 μm, corresponding to a grating circle with radius of 188 μm. The grating period d is 3.8 μm, and the grating order is 10. The angle α between the incident light and the x axis is 41°. All the grating positions and the imaging points are determined by using the design methodology discussed above in Section 4.2.1.2.

Figure 4.8 shows the calculated deviation in the radial direction ΔR of the grating facet centers from the grating circle. It shows that for the grating facets nearby the axis pole O, the deviation is ignorable, which suggests the validity of the paraxial approximation. However, as the grating facet deviates from the axis pole, the deviation of the grating facet center from the grating circle increases. For the, 20th grating facet, the deviation is as large

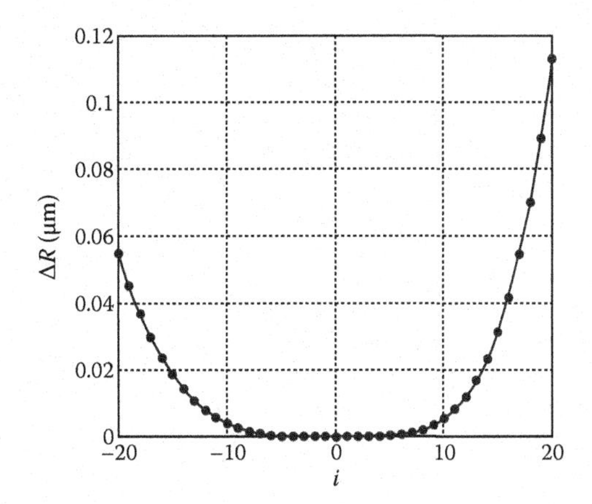

FIGURE 4.8
Calculated location deviation of the grating facet center from the grating circle. (Reprinted with permission from Song, J. et al., CWDM Planar Concave Grating Multiplexer/Demultiplexer and Application in ROADM, in *Optical Fiber Communication Conference, 2010,* Optical Society of America. Paper JThA20. © 2010 IEEE.)

as 110 nm. Thus, for low-aberration CGs, the reflection points (facet positions) are no longer in the grating circle. The asymmetric deviations for $+i$ and $-i$ positions are due to the deviated input point from the x-axis.

Figure 4.9a shows the optical micrograph of the fabricated concave grating with a footprint of ~350 × 370 μm². A 7-period Bragg grating is used as the reflection mirror in the concave grating to enhance the reflectivity. The fan shapes in Figure 4.9a vividly illustrate the input/output light distribution and interference. Figure 4.9b–e shows the zoom-in scanning electron micrographs (SEMs) of (b) the input and output waveguides, (c) the modal convertor between the slab waveguide and the channel waveguide, (d) the concave grating, and

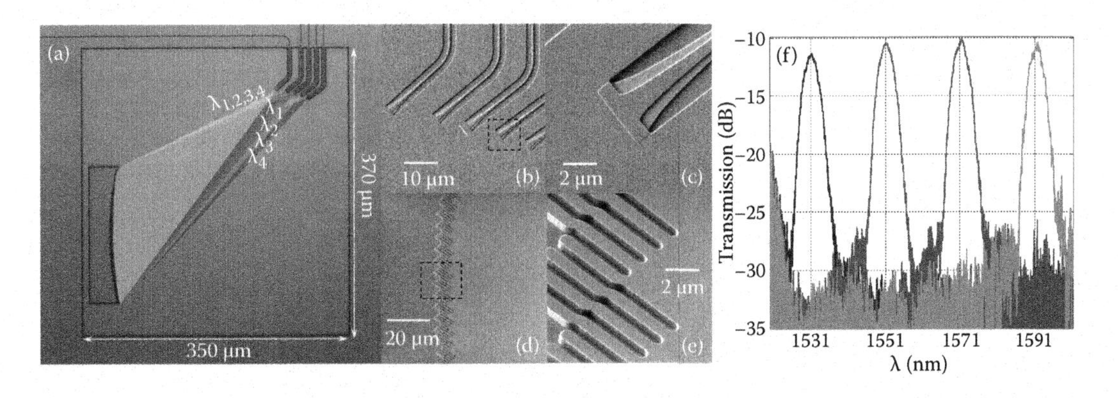

FIGURE 4.9
(a) Optical microscope picture of the fabricated four-channel concave grating. The fan shapes illustrate the input and output light distributions and interferences. Zoom-in SEMs of (b) the input and output waveguides, (c) the modal convertor between the slab waveguide and the channel waveguide, (d) the concave grating, and (e) the Bragg grating. (f) Measured transmission spectra of the four-channel concave grating. (Reprinted with permission from Song, J. et al., CWDM Planar Concave Grating Multiplexer/Demultiplexer and Application in ROADM, in *Optical Fiber Communication Conference, 2010,* Optical Society of America. Paper JThA20. © 2010 IEEE.)

(e) the Bragg grating. Figure 4.9f shows the measured quasi-TE mode transmission spectra. The peak wavelengths are 1531.16, 1551.06, 1571.6, and 1591.9 nm, respectively, which almost align to the original design targets. The total insertion loss for each channel is ~10 dB. Given the input/output coupling losses and the waveguide propagation loss, the concave grating loss is ~2.4–3.5 dB.

4.2.3 Principles of Arrayed-Waveguide Gratings

Although CG is compact, there are areas to be improved. For example, the input and output waveguides share the same interface, which could result in possible light interference thus increasing the crosstalk. Compact footprint also makes the arrangement of the waveguides difficult, thus limiting the number of possible channels. Furthermore, it is difficult to obtain high-order diffraction device in a compact CG, which is a major hurdle to attain the low crosstalk with narrow channel spacing.

Arrayed-waveguide grating has been proposed to solve these problems. Figure 4.10 shows the conceptual schematic of a 1 × N AWG, which comprises the input waveguide, the input star coupler, the arrayed waveguides, the output star coupler, and the output waveguide array. The star coupler is also called the free propagation region (FPR). In an AWG, the light that carries multiwavelength signals is incident from the input waveguide into the input star coupler. In the input star coupler, each light beam expands and splits into the arrayed waveguides. The arrayed waveguides are designed with a constant path length difference of ΔL between the neighboring waveguides. Thus, the light wave in each of the arrayed waveguides is delayed by different times but with a fixed differential time delay between the neighboring waveguides. At the output star coupler, multiple light beams exiting from the arrayed waveguides at a certain wavelength interfere with each other at the output waveguide interface, resulting in light constructive and destructive interference. Light at different wavelengths is focused on different locations at the output facet connecting to the output waveguide arrays. Thus, multiple wavelength signals are split into different output waveguides. As the phase delay determines the light interference and thus the focus point locations, the arrayed waveguides, and the star couplers require critical designs to attain the grating functions.

There are six key performance metrics describing an AWG, namely (a) central frequency, (b) channel number, (c) channel spacing, (d) free-spectral range (FSR), (e) insertion loss, and (f) crosstalk. In the following, we briefly discuss the design of the AWG by addressing the mentioned performance metrics and show a demonstration example. Interested

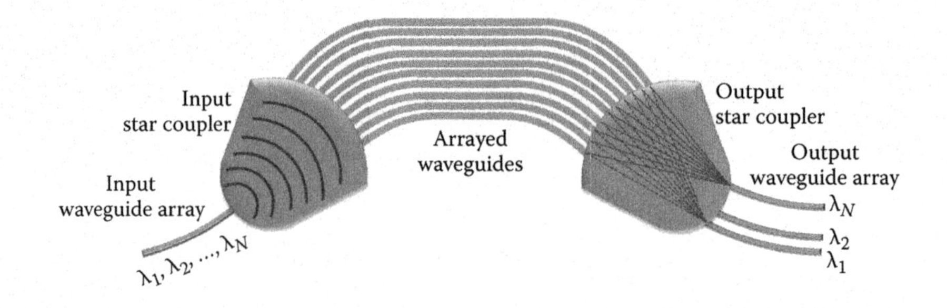

FIGURE 4.10
Conceptual schematic of a 1 × N AWG comprising the input waveguide, the input star coupler, the arrayed waveguides, the output star coupler, and the output waveguide array.

readers can refer to references such as [66–68] for more comprehensive and extensive discussions of the AWG.

In essence, an AWG is formed by putting two identical star couplers with mirror symmetry, and connecting the constructive interference points in the grating circles with the arrayed waveguides. Compared with CGs, the input and output waveguides in an AWG are separated at the ends of the two star couplers. Thus, it is suitable for large number of connecting waveguides. Furthermore, the inserted arrayed waveguides between the two star couplers increase the light paths, which result in higher-order diffraction. Such structure modifications enable performance enhancement, such as increased number of optical channels and reduced channel width.

Accounting for the arrayed waveguides, the grating equation (Equation 4.12) can be rewritten as

$$n_s \sin\theta_i d + n_c \Delta L + n_s \sin\theta_j d = m\lambda \tag{4.19}$$

where n_s is the effective refractive index of the star couplers, n_c is the effective refractive index of the phase-arrayed waveguides (including the input/output waveguides), θ_i is the angle between the x-axis and the input waveguide and θ_j is the angle between the x-axis and the output waveguide, where i and j represent the input/output waveguide orders. ΔL is the constant path length difference between the neighboring arrayed waveguides. The term $n_c\Delta L$ is the phase difference induced by the arrayed waveguides.

For the center waveguides, e.g., $\theta_i = \theta_j = 0$, we have

$$n_c \Delta L = m\lambda_0 \tag{4.20}$$

where λ_0 is the grating center wavelength in free space. This indicates that the optical path length difference between the adjacent arrayed waveguides is an integer number of the grating center wavelength in free space. Equation 4.20 thus determines the arrayed waveguide length difference.

If θ_i and θ_j are small values, we have the approximations, $\sin\theta_i \sim \theta_i$ and $\sin\theta_j \sim \theta_j$. Thus, we have the approximated grating equation

$$n_s \theta_i d + n_c \Delta L + n_s \theta_j d = m\lambda \tag{4.21}$$

The positions of the output waveguides depend on the operation wavelength and the output angle θ_j. Thus, the relationship between θ_j and the wavelength λ_j is critical. Assuming a fixed input angle θ_i, we have

$$\frac{\partial n_c}{\partial \lambda}\Delta L + n_s \frac{\partial \theta_j}{\partial \lambda} d + \frac{\partial n_s}{\partial \lambda}\theta_j d = m \tag{4.22}$$

or

$$\frac{\partial \theta_j}{\partial \lambda} = \left(\frac{n_{g,s}}{n_s} - \frac{n_{g,c}}{n_c}\right)\frac{\theta_j}{\lambda} + \left(1 - \frac{n_{g,c}}{n_c}\right)\frac{\theta_i}{\lambda} + \frac{m n_{g,c}}{n_c n_s d} \tag{4.23}$$

where the group index $n_g = n - \lambda \dfrac{\partial n}{\partial \lambda}$ is used.

Using paraxial approximation (θ_i and θ_j are small), we omit the first and second terms on the right hand side of Equation 4.20. Thus, we have

$$\frac{\partial\theta_j}{\partial\lambda} = \frac{mn_{g,c}}{n_c n_s d} d \tag{4.24}$$

or

$$\Delta\theta_j = \frac{mn_g}{n_c n_s d}\Delta\lambda \tag{4.25}$$

Equation 4.25 is called the angular dispersion equation, which suggests that the output waveguides are equally spaced subjecting to equally spaced wavelength channels.

Similarly, we can determine the positions of the input and output waveguides as follows,

$$\theta_i = i\Delta\theta_{in} \tag{4.26}$$

$$\theta_j = j\Delta\theta_{out} \tag{4.27}$$

where $\Delta\theta_{in}$ is the angle interval in the input star coupler interface and $\Delta\theta_{out}$ is the angle interval in the output star coupler interface. In the case of multiple input waveguides with multiple output waveguides, we can design the waveguides to be equally separated in angles $\Delta\theta_{in} = \Delta\theta_{out} = \Delta\theta$. The grating equation (Equation 4.19) becomes

$$n_s d(i + j)\Delta\theta + n_c\Delta L = m\lambda \tag{4.28}$$

Assuming light with wavelength λ is incident from the ith input waveguide and exits from the jth output waveguide, we can launch the same wavelength channel from the $(i + q)$th input waveguide and exits from the $(j + q)$th waveguide. This suggests that the AWG can route with various light paths.

The FSR is derived from Equation 4.28. As the operation wavelength increases by one FSR, the diffraction order m reduces by 1. For the center input/output-coupled waveguides, we have from Equation 4.20 the following expressions

$$\begin{cases} n_c\Delta L = m\lambda_0 \\ n_c'\Delta L = (m-1)(\lambda_0 + FSR) \end{cases} \tag{4.29}$$

Subtracting the expressions for the mth and $(m - 1)$th orders from each other and using the group refractive index, we have

$$(n_c - n_{g,c})FSR\Delta L = n_c\Delta LFSR - \lambda_0(\lambda_0 + FSR) \tag{4.30}$$

or

$$FSR = \frac{\lambda_0 n_c}{mn_{g,c}} \tag{4.31}$$

Although Equation 4.31 is derived using the center waveguides, it is applicable for all the waveguides, subjecting to the paraxial approximation.

The diameter of the Rowland circle 2*R*, which is the grating focal length *f*, can be determined using Equation 4.25:

$$f = 2R = \frac{d}{\Delta\theta} = \frac{n_c n_s d_2}{m n_g \Delta\lambda} \tag{4.32}$$

4.2.4 Arrayed-Waveguide Grating-Based MUX/DEMUX

In this section, we show an experimental demonstration of a DEMUX using a 1 × 32 AWG structure [64]. The schematic is the same as that shown in Figure 4.11. A group of optical signals with 32 different wavelengths ($\lambda_1 \sim \lambda_{32}$) are launched from the input waveguide. The signals were then divided into 32-output channels of the AWG. To maintain single-mode propagation in the waveguides and to reduce the crosstalk due to the fabrication imperfection-induced phase noise, ridge waveguides were adopted for the arrayed waveguides. The device was designed to operate at the grating order $m = 12$, with a path length difference $\Delta L = 8.58$ μm for the TE mode. The free propagation region focal length *f* was 183 μm. The minimum pitch between the neighboring waveguides was 1.5 μm at the fan-out section. The minimum radius of the arrayed waveguide was 10 μm with negligible bending loss. Such a small radius enables the AWG to have a compact footprint, which can effectively reduce the transmission loss. The optical channel spacing of the AWG was designed to be 200 GHz. To reduce the nonuniform optical loss, a broad FSR of 74 nm was designed, which exceeded the necessary spectral range of 51.2 nm. More than 70 arrayed waveguides were designed to further reduce the transmission loss of the AWG. We note that it is easy to extend the number of output channels to form a 1 × 46 AWG without any other modifications.

The AWG DEMUX was fabricated on an 8-inch SOI wafer (with a 300-nm silicon device layer and a 2-μm BOX layer) using standard CMOS fabrication process (248-nm-deep UV lithography and reactive-ion etching). The AWG with channel waveguides is formed after the double Si etching processes. In the input/output ends, inverse nano-tapers with a tip width of about 180 nm were used as mode-size converters. To further minimize the coupling loss between the nano-taper and the optical lensed fiber while ensuring mode confinement

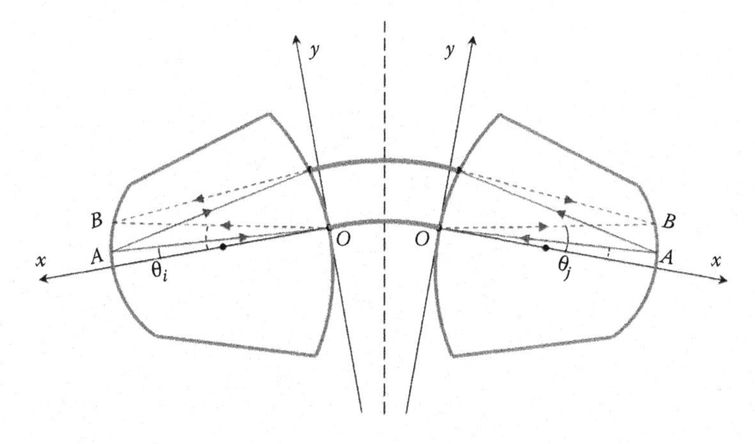

FIGURE 4.11
Schematic of an AWG comprising two star couplers with mirror symmetry.

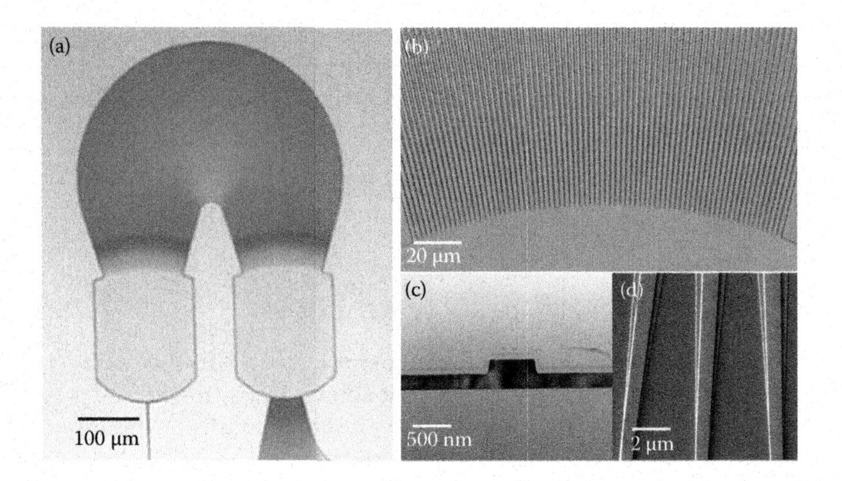

FIGURE 4.12
Optical micrographs of the fabricated (a) 1 × 32 AWG in SOI and (b) arrayed waveguides. (c) Cross-sectional view transmission electron microscopy of the arrayed waveguide. (d) Scanning electron micrograph of the transition section of the output waveguides. (Reprinted from Fang, Q. et al., *Optics Express*, 2010. **18**(5): 5106–5113. With permission from Optical Society of America.)

at the AWG output, the ridge waveguides were gradually transited into the channel waveguides at the input/output ends of the AWG. Figure 4.12a shows the optical micrograph of the entire AWG structure of ~500 × 400 μm² including the input/output waveguides and the arrayed waveguides. Figure 4.12b shows the arrayed waveguides at the fan-out section. Figure 4.12c shows the cross-sectional transmission electron microscopy (TEM) of the ridge waveguide. Figure 4.12d shows the scanning electron micrograph of the transition section between the ridge waveguide and the channel waveguide resulting from the double etching processes. Both the channel waveguide and the ridge waveguide are 500 nm in width.

Figure 4.13 shows the measured normalized TE-polarized transmission spectra from each of the 32 output waveguides. The optical power spectra are normalized to the

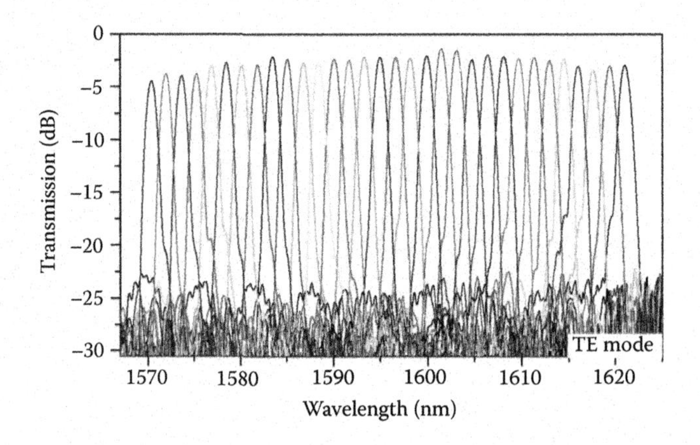

FIGURE 4.13
Measured TE-polarized optical transmission spectra of the 1 × 32 AWG (Reprinted from Fang, Q. et al., *Optics Express*, 2010. **18**(5): 5106–5113. With permission from Optical Society of America.)

TABLE 4.1

Key Measured Performance of CGs and AWGs

Devices	Material	Channel Number	Channel Spacing	Insertion Loss (dB)	Crosstalk (dB)	Footprint	Year
CG [51]	Silica	256	25 GHz	−4	−35	20 × 40 mm^2	2004
CG [52]	InP	48	100 GHz	−5.5	−30	2.5 × 8.8 mm^2	2009
CG [32]	SOI	4	20 nm	−7.5	<−30	280 × 150 μm^2	2007
CG [53]	SOI	4	20 nm	−1.9	<−25	280 × 150 μm^2	2008
CG [65]	SOI	4	20 nm	− 3.5	−20~−25	350 × 350 μm^2	2010
AWG [54]	Silica	128	25 GHz	−3.5~−5.9	−16	58 × 94 mm^2	1996
AWG [55]	Silica	256	25 GHz	−2.7~−4.7	−40	74 × 50 mm^2	2000
AWG [69]	Silica	1080	25 GHz	−4.5~−10	−19~−33	—	2002
AWG [57]	InP	15	87 GHz	−2~−4	−25	1 × 1 cm^2	1992
AWG [58]	InP	64	50 GHz	−14.4	−20	3.6 × 7 mm^2	1997
AWG [59]	SOI	64	50 GHz	−5.5	−21~−25	8.32 × 1.87 cm^2	2006
AWG [60]	SOI	16	200 GHz	−2.2	−17~−20	500 × 200 μm^2	2006
AWG [61]	SOI	50	25 GHz	−17[a]	<−10	8 × 8 mm^2	2007
AWG [62]	SOI	16	200 GHz	−3	−18	—	2008
AWG [63]	SOI	9	100 GHz	−6	<−20	10 × 9 mm^2	2009
AWG [64]	SOI	32	200 GHz	−2.5	<−18	500 × 400 μm^2	2010

[a] Including the input/output coupling losses.

transmission power spectrum of a short reference waveguide to decouple the coupling losses. The on-chip transmission loss of the AWG is ~2.5 dB and the transmission power non-uniformity among all the channels is less than 3 dB. Good crosstalk performance of more than −18 dB is obtained. In addition, the AWG also shows an additional advantage of having a wider bandwidth in its passband compared with other types of filters such as microring resonators as we discussed in Section 4.3. The 1-dB bandwidth for each channel is about 40% of the channel spacing.

Table 4.1 summarizes the key measured performance of the representative CGs and AWGs in various material platforms, with emphasis on SOI, in comparison with our examples discussed here.

4.3 Waveguide-Coupled Microring Resonators-Based MUX/DEMUX

4.3.1 Fundamentals of Microresonators

Among various silicon photonic device structures for multichannel applications, silicon microring resonators feature many advantages such as wavelength agility and compact footprint. Many silicon photonics research groups worldwide have employed silicon microring resonators as core building blocks to build optical MUX/DEMUX [70–73] and interleavers [7, 10], optical switches [47], high-speed modulators [74], and optical delay lines [75].

Here we begin with a discussion on basic concepts of optical microresonators. Figure 4.14 schematically shows a microring resonator. The light in the microresonator is partially

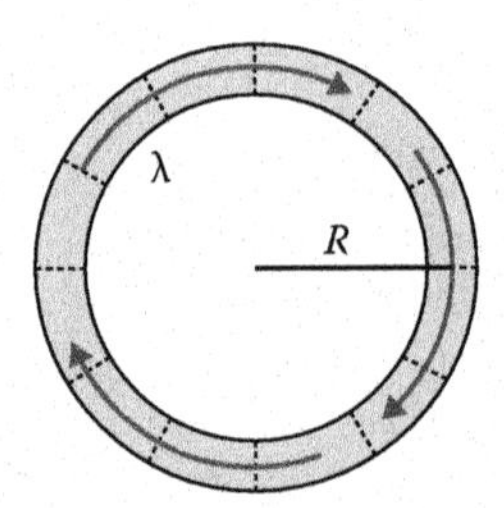

FIGURE 4.14
Schematic of a microring resonator. Dashed lines: wavefronts, arrows: traveling wave, λ: free-space wavelength, R: microring radius.

confined along the cavity sidewall by total internal reflection (TIR). The cavity field phase matches with itself upon each round trip, giving rise to an optical resonance. A microring resonator is a ring-shaped waveguide structure that typically supports singlemode resonances. It is worth mentioning that a microdisk or micropillar resonator of circular shape also supports TIR-confined optical resonances along the rim of the microdisk or micropillar. Such microdisks or micropillars typically support multiple modes known as whispering gallery modes (WGMs) after Lord Rayleigh's acoustic WGMs in circular-shaped dome of St. Paul's Cathedral [76]. Microdisks or micropillars of other shapes also support WGM-like modes [77–82]. In this chapter, we will focus on singlemode microring resonators, but it is obvious that multimode microdisks or micropillars can also be suitable for on-chip multichannel device applications.

The resonance phase-matching condition is given as follows:

$$n_{eff}L = m\lambda_m \tag{4.33}$$

where n_{eff} is the effective refractive index of the microresonator, L is the cavity round-trip length, λ_m is the mth-order resonance wavelength in free space, and m is the integer number of wavelengths along an optical round-trip length. In the case of a microring with radius R, L is approximately given as $2\pi R$.

The effective index n_{eff} is wavelength dependent in a dispersive medium. Taking into account the material dispersion, the FSR in wavelength units between adjacent orders of resonance about λ_m is given as

$$FSR = \frac{\lambda_m^2}{n_g L} \tag{4.34}$$

where n_g is the group index, which is defined as

$$n_g = n_{eff} - \lambda \frac{dn_{eff}}{d\lambda} \tag{4.35}$$

where λ is the free-space wavelength.

The cavity optical power at the resonance decays over time. Quality (Q) factor is used to characterize the microresonator loss of energy given as

$$Q = \omega_0 \frac{\text{Stored Energy}}{\text{Energy Loss Rate}} = \omega_0 \frac{U}{P_L} \tag{4.36}$$

where ω_0 is the angular resonance frequency. Following Equation 4.36, we express the cavity energy loss rate P_L as the cavity stored energy U drops over time as

$$P_L = -\frac{dU}{dt} = \frac{\omega_0}{Q} U \tag{4.37}$$

with the stored energy varying over time $U(t)$ given as

$$U(t) = U_0 e^{-\omega_0 t/Q} = U_0 e^{-t/\tau} \tag{4.38}$$

where U_0 is the initial cavity energy. We define a cavity lifetime τ as

$$\tau = \frac{Q}{\omega_0} \tag{4.39}$$

The cavity lifetime is proportional to Q and gives the time when the cavity energy drops to $1/e$ of the initial cavity energy U_0. Given the cavity energy, U is proportional to the square of the cavity electric field amplitude E, the cavity electric field at resonance frequency ω_0 is thus damped in time as follows:

$$E(t) = E_0 e^{-\omega_0 t/2Q} e^{-i\omega_0 t}. \tag{4.40}$$

where E_0 is the initial electric field amplitude. By Fourier transforming Equation 4.40 [83], we obtain

$$E(\omega) = \frac{1}{\sqrt{2\pi}} \int_0^\infty E_0 e^{-\omega_0 t/2Q} e^{i(\omega-\omega_0)t} dt \tag{4.41}$$

where $E(\omega)$ is the cavity electric field in the frequency domain. By taking the modulus square of $E(\omega)$, we obtain a frequency spectrum for the cavity field energy having a Lorentzian resonance line shape given as

$$|E(\omega)|^2 = |E_0|^2 \frac{(\omega_0/2Q)^2}{(\omega-\omega_0)^2 + (\omega_0/2Q)^2} \tag{4.42}$$

Figure 4.15 shows the calculated Lorentzian cavity resonance lineshape using Equation 4.42. At $\omega = \omega_0 \pm \omega_0/2Q$, the cavity field energy drops to one half of the maximum. The full-width at half-maximum (FWHM) of the line shape or the linewidth $\delta\omega$ is given as

$$\delta\omega = \frac{\omega_0}{Q} \tag{4.43}$$

By comparing Equation 4.39 and 4.43, we find that the cavity resonance linewidth $\delta\omega$ and the cavity lifetime τ are related as

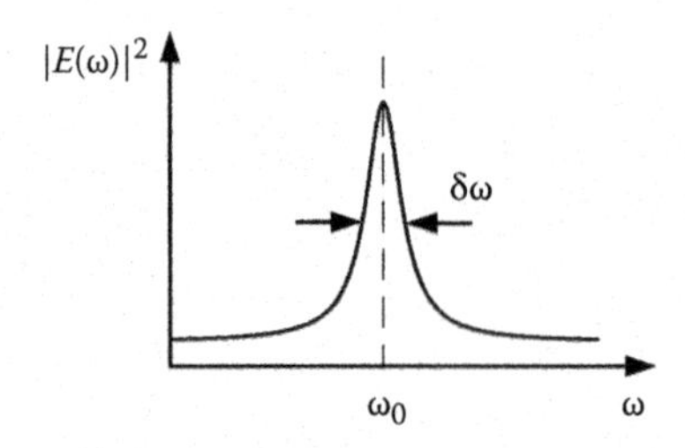

FIGURE 4.15
Modeled Lorentzian resonance line shape.

$$Q = \omega_0 \tau = \frac{\omega_0}{\delta\omega} \tag{4.44}$$

$$\delta\omega \cdot \tau = 1 \tag{4.45}$$

Thus, $\delta\omega$ and τ satisfy an "uncertainty principle" given by Equation 4.45, which says that a narrow Lorentzian cavity resonance linewidth imposes a long cavity lifetime. For example, a 25-GHz Lorentzian resonance linewidth imposes a cavity lifetime of 40 ps.

4.3.2 Waveguide-Coupling to Microring Resonators

For device applications, it is crucial that the microresonator is coupled to an optical waveguide. Figure 4.16 illustrates a microring resonator side-coupled with a bus-waveguide (a) and a microring resonator side-coupled with two parallel bus waveguides (b). Light is launched from the waveguide input port and evanescently coupled into and out of the microring resonator in the coupling region through submicrometer-sized gap spacing. To efficiently couple the light from the waveguide to the microring, phase-matching condition needs to be satisfied, namely the waveguide propagation constant should match with that of the microring.

Because of the intrinsic cavity loss, the waveguide throughput transmission exhibits quasi-periodic Lorentzian resonance dips that are spaced by a FSR given by Equation 4.34. Such filter characteristics give a notch or rejection filter. In the case that a microring resonator is both input- and output-coupled with bus waveguides, the cavity resonances can be output-coupled as Lorentzian resonance peaks while the throughput transmission exhibits the corresponding resonance dips. Such filter characteristics give a so-called

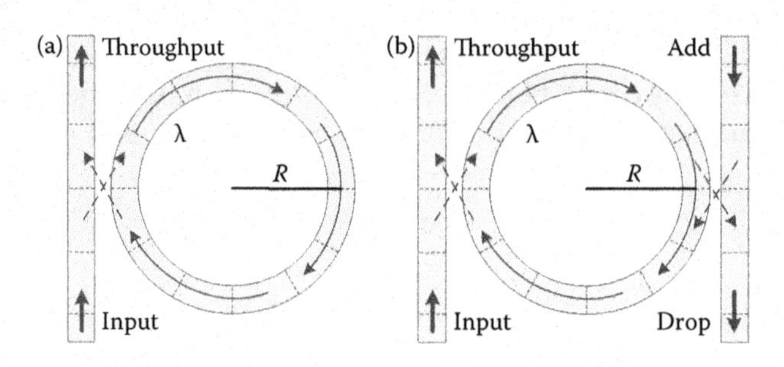

FIGURE 4.16
Schematics of waveguide coupled microring resonators. (a) Notch filter, (b) add–drop filter.

add–drop filter. The add–drop filter spectral characteristics are analogous to a Fabry–Perot or an etalon resonator with plane wave illumination, in which the Fabry–Perot transmission peaks (reflection dips) correspond to the microring resonator-based add–drop filter resonance peaks (dips) in the drop (through) port.

Figure 4.17 shows the finite-difference time-domain (FDTD) simulated mode-field distributions and transmission spectra of SOI waveguide-coupled microring resonators. The simulation assumes the transverse electric (TE) polarization. The microring radius is 2.5 μm, and the bus waveguide and microring waveguide widths are 0.5 μm. The bus waveguide and the microring waveguide are nearly phase-matched given the relatively small microring radius. The simulated electric field amplitudes at resonance wavelength of ~1560 nm show approximately threefold field enhancements, which is limited by the loaded cavity Q.

In a waveguide-coupled microring resonator, the waveguide coupling perturbs the resonator and imposes a slight shift in the resonance wavelength and additional energy loss. The energy loss per cavity round-trip thus comprises the intrinsic optical losses in the cavity, namely light scattering, radiation loss and absorption, and the extrinsic output-coupling loss by the coupled waveguide. The total Q factor of a waveguide-coupled microring resonator therefore comprises two contributions: the intrinsic cavity quality factor Q_0 and the coupling quality factor Q_c. The Q_0 factor is related to the intrinsic cavity power loss coefficient α. The Q_c factor is related to the waveguide-to-microring field coupling coefficient κ. The relations of Q, Q_0, Q_c, α, and κ are described as follows [84]

$$Q^{-1} = Q_0^{-1} + Q_c^{-1} \tag{4.46}$$

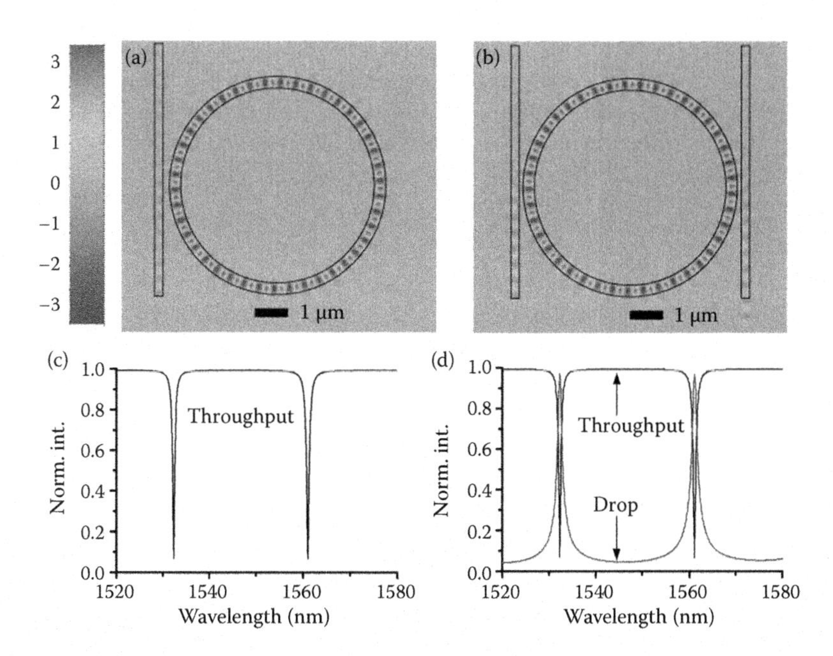

FIGURE 4.17
(a,b) FDTD-simulated mode-field distributions and (c,d) transmission spectra of a single bus waveguide-coupled microring resonator-based notch filter and a double waveguide-coupled microring resonator-based add–drop filter.

$$Q_0 = \omega_0 \tau_0 = \frac{2\pi c}{\lambda_0} \cdot \frac{n_g}{\alpha c} = \frac{2\pi n_g}{\alpha \lambda_0} \tag{4.47}$$

$$Q_c = \omega_0 \tau_c = \frac{\omega_0 \tau_{rt}}{|\kappa|^2} \tag{4.48}$$

where τ_0 is the intrinsic cavity energy $1/e$ lifetime, τ_c is the extrinsic waveguide-coupling energy $1/e$ lifetime, τ_{rt} is the round-trip light propagation time, and λ_0 is the cavity resonance wavelength in free space.

To obtain Equation 4.48, we assume the microring resonator is weakly coupled to the bus-waveguide, namely $|\kappa|^2 \ll 1$. Assuming loss-less coupling, we express the fraction of the microring light energy remaining after every round-trip as $1 - |\kappa|^2 = \exp(-\tau_{rt}/\tau_c) \approx 1 - \tau_{rt}\tau_c$.

Thus, $\tau_c = \tau_{rt}/|\kappa|^2$. Based on Equations 4.46 through 4.48, the Q factor of a waveguide-coupled microring resonator can be varied by properly designing or tuning α, κ,n_g and τ_{rt}.

4.3.3 Transfer Matrix Modeling

Here we model the transmission spectra of a waveguide-coupled microring resonator using transfer matrix approach [85]. Figure 4.18 shows the schematic of a microring resonator-based notch filter. In such case, we consider the waveguide coupler as a four-port device, as shown in the inset of Figure 4.18. Transfer matrix approach relates the fields at the bus-waveguide in the vicinity before and after the coupling (E_{in}, E_{th}) to the microring fields in the vicinity before and after the coupling (E_1, E_2).

The model assumes scalar fields and steady-state analysis. We also assume the bus-waveguide and the microring waveguide to be single-mode and phase-matched (i.e., have the same wave number β). The complex through-coupling coefficients (t, t') and the complex cross-coupling coefficients (κ, κ') are denoted in Figure 4.18. The relationships between the input and output fields at the coupler are expressed as

$$\begin{cases} E_{th} = tE_{in} + \kappa' E_1 \\ E_2 = \kappa E_{in} + t' E_1 \end{cases} \tag{4.49}$$

The complex electric fields transfer matrix (C) at the coupler is described as

$$\begin{pmatrix} E_1 \\ E_2 \end{pmatrix} = \begin{pmatrix} -\dfrac{t}{\kappa'} & \dfrac{1}{\kappa'} \\ \dfrac{\kappa\kappa' - tt'}{\kappa'} & \dfrac{t'}{\kappa'} \end{pmatrix} \begin{pmatrix} E_{in} \\ E_{th} \end{pmatrix} = C \begin{pmatrix} E_{in} \\ E_{th} \end{pmatrix} \tag{4.50}$$

where E_1 and E_2 are related as

$$E_1 = E_2 \exp\left(i\beta L - \frac{\alpha}{2} L \right) \tag{4.51}$$

where $\beta = n_{eff} 2\pi\lambda$ is the propagation constant in the microring.

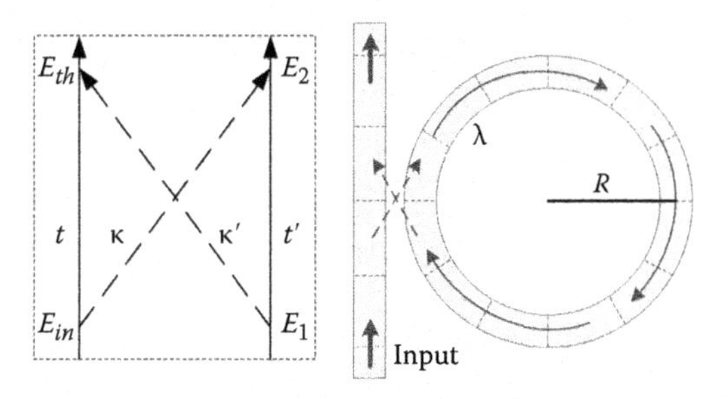

FIGURE 4.18
Schematic of a waveguide-coupled microring resonator-based notch filter. Solid arrows indicate the traveling-wave propagation directions. Dashed arrows indicate the waveguide-microring coupling. Dashed lines show the wavefronts matching between the microring and the bus-waveguide. The zoom-in shows the waveguide-microring coupling region. t, t': waveguide field transmission coefficients; κ, κ′: waveguide field coupling coefficients.

According to Stokes relation, time reversibility of the transfer matrix C at an ideal coupling between the waveguide and the microring imposes $t = t'$ and $\kappa\kappa' - tt' = -1$ [85]. The reciprocity (meaning the transmission remains the same upon interchanging the input and throughout ports) and energy conservation assuming no loss channels within the coupler give the relations $\kappa' = -\kappa^*$ and $t' = t^*$, where the "*" denotes complex conjugate. For a microring resonator-based notch filter the throughput electric field complex amplitude is given as

$$E_{th} = \frac{t - e^{(i\beta L - \alpha L/2)}}{1 - t^* e^{(i\beta L - \alpha L/2)}} E_{in} = \frac{t - ae^{i\delta}}{1 - t^* ae^{i\delta}} E_{in} \tag{4.52}$$

where $a = \exp(-\alpha L/2)$ is the microring amplitude round-trip loss factor and $\delta = \beta L$ is the microring round-trip phase shift. By taking the absolute square of Equation 4.52, we obtain the throughput transmission as

$$\left|\frac{E_{th}}{E_{in}}\right|^2 = \frac{a^2 + t^2 - 2at\cos\delta}{1 + a^2t^2 - 2at\cos\delta} \tag{4.53}$$

From Equation 4.53, we derive the microresonator-based notch filter FHWM as

$$\text{FWHM} = \frac{(1 - at)\lambda_{res}^2}{\pi n_g L\sqrt{at}} \tag{4.54}$$

The Q factor for a microresonator-based notch filter is given as

$$Q = \frac{\omega}{\delta\omega} = \frac{\lambda_{res}}{\text{FWHM}} = \frac{\pi n_g L\sqrt{at}}{(1 - at)\lambda_{res}} \tag{4.55}$$

Figure 4.19 shows the modeled throughput transmission spectrum of a microring notch filter using Equation 4.55. We assume a 20-µm-radius SOI microring coupled to a bus-waveguide with symmetric and lossless coupling coefficients of $\kappa = \kappa' = 0.2$ and

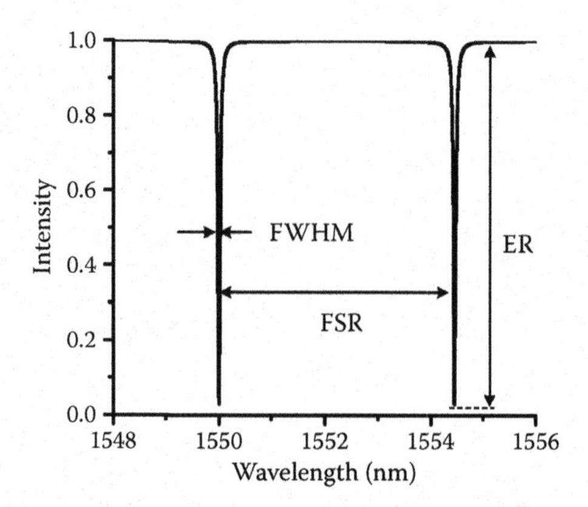

FIGURE 4.19
Calculated throughput transmission spectrum of a SOI microring resonator-based notch filter (R = 20 μm, α = 2 cm^{-1}, $\kappa_1 = \kappa_1' = 0.2$).

propagation loss coefficient of α = 2 cm^{-1}. The three filter performance metrics are, namely FSR of 4.5 nm, FWHM of 0.05 nm, and extinction ratio (ER) of 15 dB.

Another important concept regarding waveguide coupling to a microresonator is the coupling regime [86]. The coupling regime depends on the balance between the waveguide coupling and the intrinsic cavity loss. There are three coupling regimes, namely (i) undercoupling ($t > a$), (ii) critical-coupling ($t = a$), and (iii) overcoupling ($t < a$). According to Equation 4.52, the throughput optical power transmission at resonance shows minimum intensity at critical coupling regime.

The microresonator exhibits around the resonance a strong phase response, which can be leveraged for dispersion engineering in multichannel applications. We obtain the phase response from the complex throughput electric field amplitude according to Equation 4.52 as follows

$$\varphi = \tan^{-1}\left(\frac{a(1-t^2)\sin\delta}{t(1+a^2)-a(1+t^2)\cos\delta}\right) \tag{4.56}$$

Figure 4.20 shows the modeled transmission intensity and phase response according to Equations 4.52 and 4.56 in the under- and over-coupling regimes. Here we assume a 20-μm-radius microring resonator with a Q of 10^4 and an ER of 20 dB. In the undercoupling regime, there is an "anomalous" phase response at the resonance wavelength region. The anomalous phase response features an abrupt change in slope with a sign change near the center of the resonance compared with both sides of the resonance. In the overcoupling regime, there is a "normal" phase response at the resonance wavelength region. The normal phase response features a smooth change in slope with the same sign within and outside the resonance.

Figure 4.21 shows the schematic of a microring resonator coupled with two bus-waveguides in an add–drop filter configuration. E_{in}, E_{th}, E_{dr}, and E_{ad} are the input-, throughput-, drop-, and add-port electric field complex amplitudes, respectively. E_1, E_2, E_3, and E_4 are the microring electric field complex amplitudes just before and after the input and output couplers. The transmission (t_i, t_i') and cross-coupling coefficients (κ_i, κ_i'), where

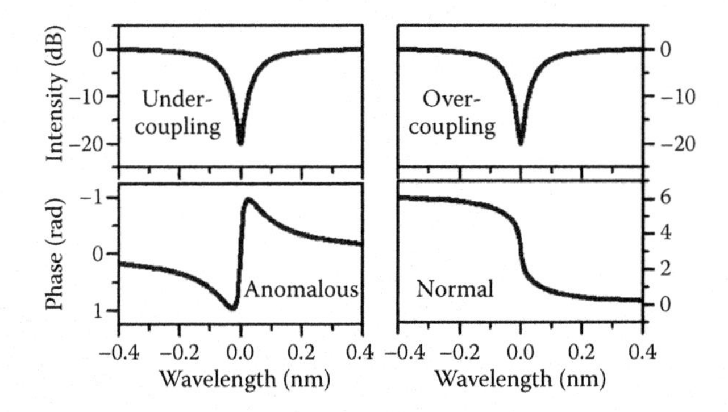

FIGURE 4.20
Modeled resonance throughput transmission intensity and phase response of a waveguide-coupled microring resonator in the undercoupling regime (left) and the overcoupling regime (right).

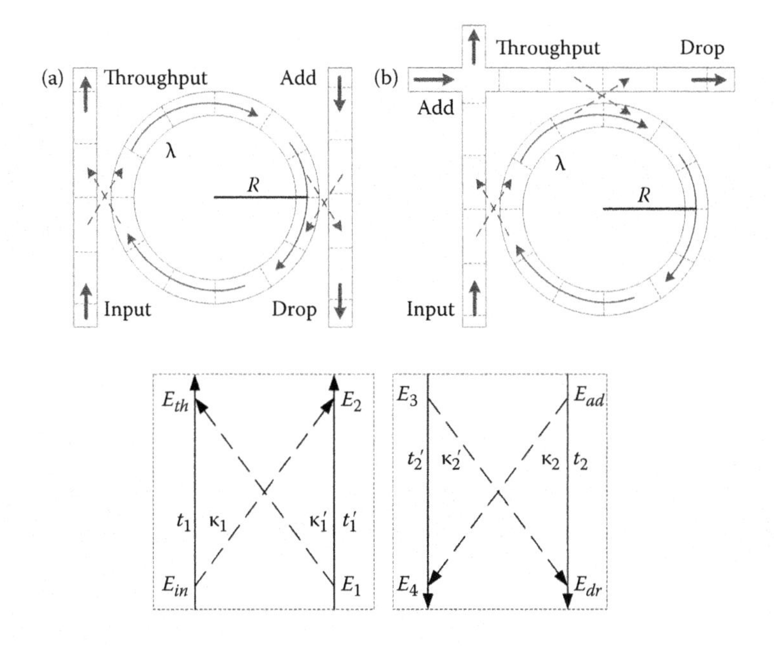

FIGURE 4.21
Schematic of a waveguide-coupled microring resonator-based add–drop filters with (a) parallel and (b) crossing bus waveguides. Solid arrows indicate the traveling-wave propagation directions. Dashed arrows indicate the waveguide-microring coupling. Dashed lines show the wavefronts matching between the microring and the bus waveguide. The zoom-in pictures show the coupling regions. t_i, t_i': waveguide field transmission coefficients, κ_i, κ': waveguide field coupling coefficients, where i = 1,2.

i = 1,2, are denoted in Figure 4.21. Thus, the relationships between the input and the output fields at the two couplers are expressed as

$$\begin{cases} E_{th} = t_1 E_{in} + \kappa_1' E_1 \\ E_2 = \kappa_1 E_{in} + t_1' E_1 \end{cases} \tag{4.57}$$

and

$$\begin{cases} E_{dr} = t_2 E_{ad} + \kappa_2' E_3 \\ E_4 = \kappa_2 E_{ad} + t_2' E_3 \end{cases} \tag{4.58}$$

The complex electric fields transfer matrices (C and C') at the input and output waveguide couplers are described as

$$\begin{pmatrix} E_1 \\ E_2 \end{pmatrix} = \begin{pmatrix} -\dfrac{t_1}{\kappa_1'} & \dfrac{1}{\kappa_1'} \\ \dfrac{\kappa_1\kappa_1' - t_1 t_1'}{\kappa_1'} & \dfrac{t_1'}{\kappa_1'} \end{pmatrix} \begin{pmatrix} E_{in} \\ E_{th} \end{pmatrix} = C \begin{pmatrix} E_{in} \\ E_{th} \end{pmatrix} \tag{4.59}$$

$$\begin{pmatrix} E_{ad} \\ E_{dr} \end{pmatrix} = \begin{pmatrix} -\dfrac{t_2'}{\kappa_2} & \dfrac{1}{\kappa_2} \\ \dfrac{\kappa_2\kappa_2' - t_2 t_2'}{\kappa_2} & \dfrac{t_2}{\kappa_2} \end{pmatrix} \begin{pmatrix} E_3 \\ E_4 \end{pmatrix} = C' \begin{pmatrix} E_3 \\ E_4 \end{pmatrix} \tag{4.60}$$

The internal fields (E_1, E_2) and (E_3, E_4) are related as

$$\begin{pmatrix} E_3 \\ E_4 \end{pmatrix} = \begin{pmatrix} 0 & e^{\left(i\beta - \frac{\alpha}{2}\right)L_1} \\ e^{\left(i\beta - \frac{\alpha}{2}\right)L_2} & 0 \end{pmatrix} \begin{pmatrix} E_1 \\ E_2 \end{pmatrix} = P \begin{pmatrix} E_1 \\ E_2 \end{pmatrix} \tag{4.61}$$

where P is the phase transfer matrix in the cavity. L_1 is the propagation path length between E_2 and E_3, and L_2 is the propagation path length between E_4 and E_1. In the case that the add–drop filter has a symmetric structure as shown in Figure 4.21a, we have $L_1 = L_2 = R\pi$. In the case that the add–drop filter has a crossing structure as shown in Figure 4.21b, we have $L_1 = R\pi 2$ and $L_2 = 3R\pi 2$.

By cascading the transfer matrices C, P, and C', we obtain the transfer matrix relation between the two fields in the input-coupled waveguide (E_{in}, E_{th}) and the two fields in the output-coupled waveguide (E_{ad}, E_{dr}) as follows

$$\begin{pmatrix} E_{ad} \\ E_{dr} \end{pmatrix} = C'PC \begin{pmatrix} E_{in} \\ E_{th} \end{pmatrix} = M_1 \begin{pmatrix} E_{in} \\ E_{th} \end{pmatrix} = \begin{pmatrix} M_{11} & M_{12} \\ M_{13} & M_{14} \end{pmatrix} \begin{pmatrix} E_{in} \\ E_{th} \end{pmatrix} \tag{4.62}$$

The overall transfer matrix M_1 comprises the complex matrix elements M_{11}, M_{12}, M_{13}, and M_{14}. For a single input, we set the boundary condition as $E_{ad} = 0$. Therefore, the field transfer functions at the throughput and drop ports are given as follows:

$$E_{th} = -\frac{M_{11}}{M_{12}} E_{in} \tag{4.63}$$

$$E_{dr} = \left(M_{13} - \frac{M_{11}M_{14}}{M_{12}} \right) E_{in} \tag{4.64}$$

From Equations 4.63 and 4.64, the throughput (*T*) and drop (*D*) transmission intensities are expressed as

$$T = \left| \frac{E_{th}}{E_{in}} \right|^2 \tag{4.65}$$

$$D = \left| \frac{E_{dr}}{E_{in}} \right|^2 \tag{4.66}$$

From Equations 4.65 and 4.66, we find that the resonance linewidth or FHWM of the microring resonator-based add–drop filter is given as

$$\text{FWHM} = \frac{(1 - at_1t_2)\lambda_{res}^2}{\pi n_g L\sqrt{at_1t_2}} \tag{4.67}$$

The *Q* factor for the microring resonator-based add–drop filter is given as

$$Q = \frac{\omega}{\delta\omega} = \frac{\lambda_{res}}{FWHM} = \frac{\pi n_g L\sqrt{at_1t_2}}{(1 - at_1t_2)\lambda_{res}} \tag{4.68}$$

Figure 4.22 shows the calculated throughput- and drop-port transmission spectra of a microring resonator-based add–drop filter using Equations 4.65 and 4.66. We choose R = 20 μm, α = 2 cm^{-1}, $\kappa_1 = \kappa_1' = \kappa_2 = \kappa_2' = 0.4$. The throughput port exhibits an inverted Lorentzian line shape while the drop port exhibits the Lorentzian line shape. The ER of the throughput resonance dip ER_t is defined as the ratio of the maximum off-resonance transmission to the resonance dip transmission. The ER of the drop-port resonance peak ER_d is

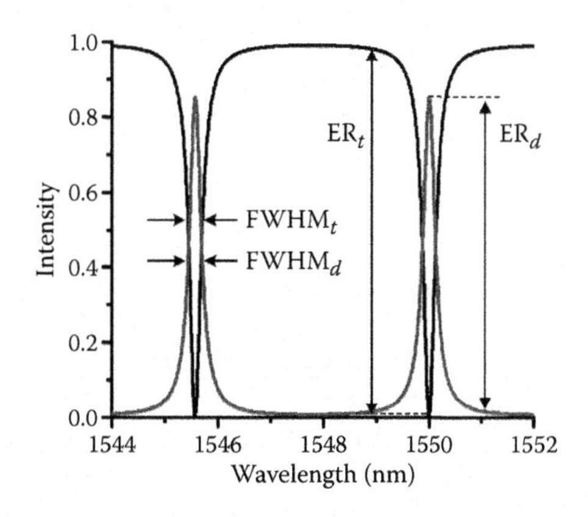

FIGURE 4.22
Calculated throughput- and drop-port transmission spectra of a microring resonator-based add–drop filter.

defined as the ratio of the maximum on-resonance transmission peak to the minimum off-resonance transmission. We note that ER_t and $FWHM_t$ can be different from ER_d and $FWHM_d$. In Sections 4.3.4 and 4.3.5, we use the transfer matrices *C*, *C′*, and *P* for microring resonator-based notch and add–drop filters as building blocks to model more sophisticated microring resonator-based devices for advanced multichannel processing functionalities.

4.3.4 1 × N Series-Cascaded Microring Resonators-Based MUX/DEMUX

Here we consider as an example a series-cascaded microring resonator-based add–drop filters as a 1 × N channel MUX/DEMUX. Figure 4.23 schematically illustrates a four-channel MUX/DEMUX. All microring resonators are side-coupled to a common bus-waveguide that serves as a common input and a common throughput. Each microring resonator is also side-coupled to an output bus-waveguide that crosses the common bus-waveguide enabling adding and dropping the resonant channel. Each microring resonator thus serves as an add–drop filter. The radius of each microring resonator is designed slightly differently to have different resonance wavelength channels ($\lambda_1\lambda_2\lambda_3\lambda_4$). The WDM input signals from the common bus-waveguide can therefore be separated into individual channels through this structure. Conversely, individual channels adding from the crossed waveguides can be multiplexed into the common bus waveguide. In practice, to compensate fabrication imperfection-induced phase error, each of the microrings is typically integrated with a thermal heat pad (see below) or an embedded diode for active fine tuning. Besides, the waveguide crossings can also lead to undesirable crosstalk. Some special designs such as multimode interference (MMI) crossings can reduce the crosstalk [87, 88].

The transfer matrix modeling of each of the cascaded waveguide crossing-coupled microring add–drop filters follows Equations 4.63 and 4.64. The 1 × N DEMUX throughput and drop-port transmission field amplitudes are given as follows:

$$E_{th} = \prod_{i}^{N}\left(-\frac{M_{i1}}{M_{i2}}\right)E_{in} \tag{4.69}$$

$$E_{j,dr} = \left(M_{j3} - \frac{M_{j1}M_{j4}}{M_{j2}}\right)\prod_{i}^{j-1}\left(-\frac{M_{i1}}{M_{i2}}\right)E_{in} \tag{4.70}$$

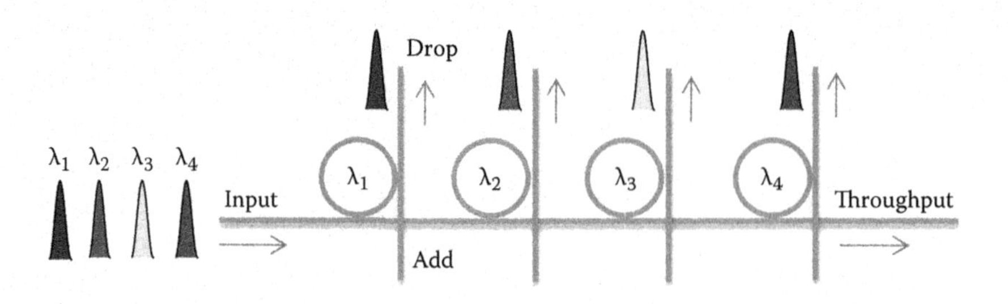

FIGURE 4.23
Schematic of a 1 × 4 DEMUX comprising four waveguide-crossing–coupled microring resonators resonant at four different wavelength channels λ_1, λ_2, λ_3, and λ_4.

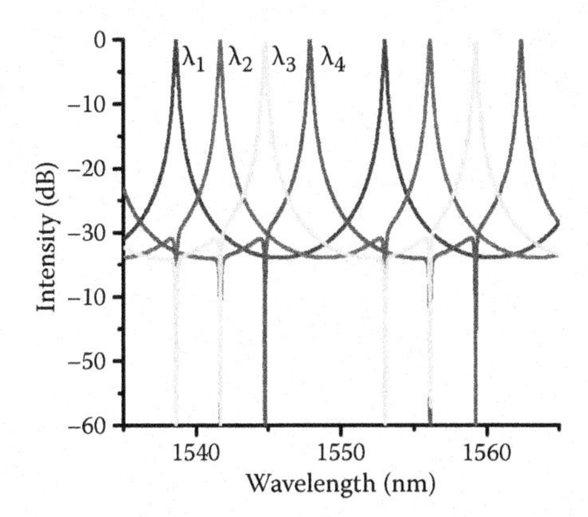

FIGURE 4.24
Calculated transmission spectra of the 1 × 4 DEMUX at the four drop ports.

where E_{th} is the overall throughput transmission field, $E_{j,dr}$ is the drop-port transmission field of the jth microring resonator. Figure 4.24 shows the modeled transmission spectra of the four drop ports from a 1 × 4 DEMUX according to Equations 4.69 and 4.70. We adopt a coupling coefficient of 0.2 for the common bus waveguide-to-microring coupling, a coupling coefficient of 0.2 for the drop waveguide-to-microring coupling, and the four microring radii around 10 μm with slight difference. We assume an effective refractive index of 2.68 for SOI. We neglect the crossing waveguide scattering loss or crosstalk in this modeling. The modeling results show that the channels are demultiplexed one by one and each channel leaves a transmission dip for the following channels. For example channel 4 drop-port transmission exhibits three transmission dips induced by the first three channels.

Figure 4.25 shows as an example the optical micrograph and the measured drop-port transmission spectra of a 4-channel microring resonator-based MUX/DEMUX on a SOI

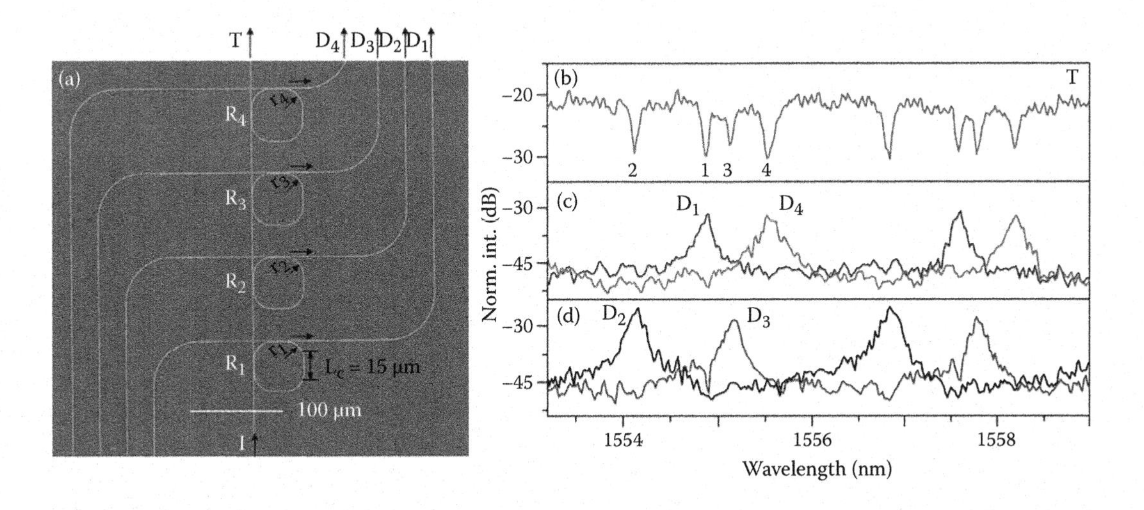

FIGURE 4.25
(a) SEM of the fabricated four-channel microring resonators-based MUX/DEMUX fabricated in a SOI substrate. (b) Measured TE-polarized throughput-port and (c, d) drop-port transmission spectra. (Reprinted from Xu, F., and A.W. Poon, *Optics Express*, 2008. **16**: 8649–8657. With permission of Optical Society of America.)

substrate [38]. The four microring resonators differ slightly in the corner radii by a fixed increment of 0.2 μm. The measured drop-port transmission spectra show the channel dips as we discussed in the modeling.

4.4 Channel Bandwidth and Spacing Tailoring in Microring Resonator–Based MUX/DEMUX

4.4.1 Channel Bandwidth Broadening Using High-Order Coupled-Microresonators

The Lorentzian line shape of a single microring resonator-based add–drop filter is, however, too sharp to be applicable for transmitting today's high-data-rate (broadband) WDM signals as it does not provide a sufficiently steep roll-off or a flat-top passband. To tailor the microring resonator-based filters for a more desirable line-shape, one possible way is to cascade multiple microring resonators to attain high-order box-like filter responses, namely flat-top, broadband, and steep roll-off. The fundamental principle is based on intercavity coupling-induced mode splitting [89], meaning each constituent microresonator mode is slightly shifted due to the coupling of another identical microresonator.

Figure 4.26 schematically shows the *N*th-order (N = 2, 3, and 4) coupled microring resonator-based add–drop filters. Following Equation 4.69, we define coupling matrices C_i, where i = 1 to N − 1, for the intercavity coupling between the ith and (i + 1)th microring.

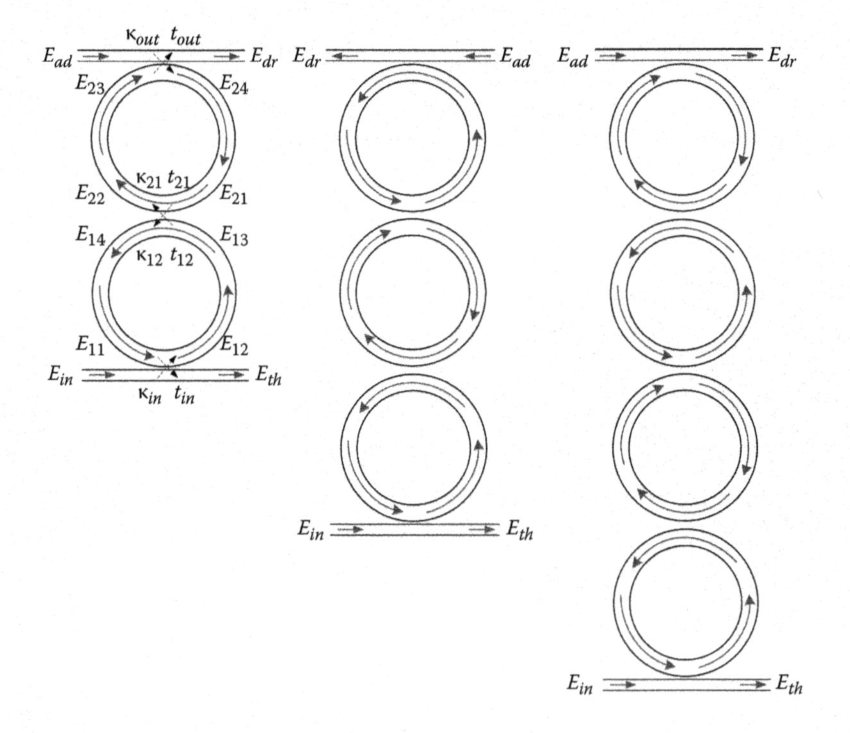

FIGURE 4.26
Schematics of the second-, third-, and fourth-order coupled-microring resonator-based add–drop filters.

The electric fields between the two adjacent resonators in the vicinity of the coupler (E_{13}, E_{14}) and (E_{21}, E_{22}) are given as

$$\begin{bmatrix} E_{21} \\ E_{22} \end{bmatrix} = C_1 \begin{bmatrix} E_{13} \\ E_{14} \end{bmatrix} \tag{4.71}$$

The input- and output-coupling transfer matrices, C_{in} and C_{out}, are given as follows:

$$\begin{bmatrix} E_{11} \\ E_{12} \end{bmatrix} = C_{in} \begin{bmatrix} E_{in} \\ E_{th} \end{bmatrix} \tag{4.72}$$

$$\begin{bmatrix} E_{ad} \\ E_{dr} \end{bmatrix} = C_{out} \begin{bmatrix} E_{23} \\ E_{24} \end{bmatrix} \tag{4.73}$$

For the electric fields inside the cavity, according to Equation 4.61, we define propagation matrices P_i, where $i = 1$ to N, for the intracavity propagation within the ith microring as

$$\begin{bmatrix} E_{13} \\ E_{14} \end{bmatrix} = P_1 \begin{bmatrix} E_{11} \\ E_{12} \end{bmatrix} \tag{4.74}$$

$$\begin{bmatrix} E_{23} \\ E_{24} \end{bmatrix} = P_2 \begin{bmatrix} E_{21} \\ E_{22} \end{bmatrix} \tag{4.75}$$

Therefore, using Equations 4.71 through 4.75, the overall transfer matrix for the second-order microring-coupled filter is given as follows:

$$\begin{bmatrix} E_{ad} \\ E_{dr} \end{bmatrix} = C_{out} P_2 C_1 P_1 C_{in} \begin{bmatrix} E_{in} \\ E_{th} \end{bmatrix} = M_2 \begin{pmatrix} E_{in} \\ E_{th} \end{pmatrix} = \begin{bmatrix} M_{21} & M_{22} \\ M_{23} & M_{24} \end{bmatrix} \begin{pmatrix} E_{in} \\ E_{th} \end{pmatrix} \tag{4.76}$$

where M_N is the transfer matrix of the Nth-order coupled-microring resonator-based filter. Considering the boundary condition of a single input, $E_{ad} = 0$, we express the electric fields at the throughput and drop port as follows:

$$E_{th} = -\frac{M_{21}}{M_{22}} E_{in} \tag{4.77}$$

$$E_{dr} = \left(M_{23} - \frac{M_{21} M_{24}}{M_{22}} \right) E_{in} \tag{4.78}$$

Likewise, we derive the overall transfer matrix of higher-order coupled-microring resonator-based filters by extending Equation 4.76. We illustrate as examples the transfer matrix relations for third-order coupled-microring resonator-based filters:

$$\begin{bmatrix} E_{ad} \\ E_{dr} \end{bmatrix} = C_{out}P_3C_2P_2C_1P_1C_{in}\begin{bmatrix} E_{in} \\ E_{th} \end{bmatrix} = M_3\begin{bmatrix} E_{in} \\ E_{th} \end{bmatrix} \tag{4.79}$$

and for fourth-order coupled-microring resonator-based filters:

$$\begin{bmatrix} E_{ad} \\ E_{dr} \end{bmatrix} = C_{out}P_4C_3P_3C_2P_2C_1P_1C_{in}\begin{bmatrix} E_{in} \\ E_{th} \end{bmatrix} = M_4\begin{bmatrix} E_{in} \\ E_{th} \end{bmatrix} \tag{4.80}$$

Here we assume the microrings are identical but the intercavity couplings can be non-uniform by varying the coupling gap spacing. The resulting coupled-microring resonators exhibit a broadband transmission spectrum due to the intercavity coupling-induced mode-splitting [89]. The resonance splitting, however, can result in ripples in the passband, which is undesirable for WDM high-bit-rate signal transmission. Apodization of the coupling coefficients is a generic way, to tailor the cascaded microring resonator filter profile. Over the years, there have been various coupling coefficients design rules proposed to tailor the filter profile for different specifications, and interested readers can refer to the many literature for details [33, 90, 91]. The gist of these design methods is to precisely tailor each of the coupling coefficient, typically assuming the precision control of the submicrometer coupling gap spacing via advanced nanolithography. Here we follow one of those design rules [33] to demonstrate as an example the flat-top filter response of a third-order coupled-microring resonator-based filter.

Figure 4.27a shows the modeled single-element microring resonator-based filter transmission spectra according to Equations 4.65 and 4.66, while Figure 4.27b shows the modeled third-order coupled-microring resonator-based filter transmission spectra according to Equation 4.79. The modeling assumes a microring radius of 5 μm, a power loss coefficient of $\alpha = 2\ \text{cm}^{-1}$, and an effective refractive index of 2.65 for modeling a SOI substrate. The waveguide-to-microring coupling coefficients are $\kappa_{in} = \kappa_{out} = 0.43$, which are the same between those in the single microring resonators and the coupled-microring resonators. In the case of the coupled-microring resonators, the intercavity coupling coefficients are set to 0.08 according to the design rule [33].

Compared with the modeled single microring resonator-based filter, which exhibits a sharp Lorentzian line shape, the modeled high-order filter exhibits a flat-top line-shape with a wide 3-dB bandwidth of 1.05 nm and a steep roll-off ratio of 26.8 dB/nm due to the

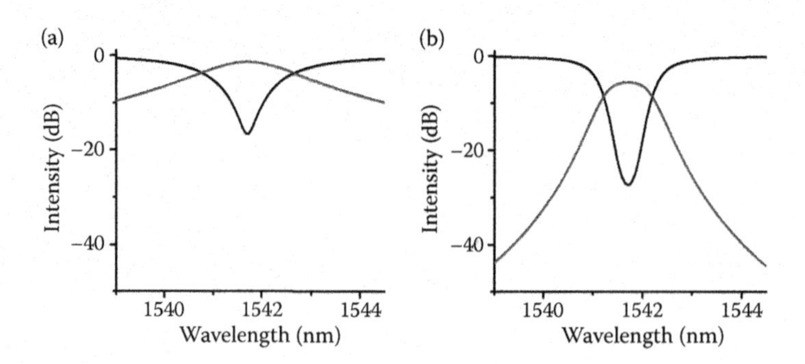

FIGURE 4.27
(a) Modeled transmission spectra of a single microring resonator-based add–drop filter. (b) Modeled transmission spectra of a third-order coupled-microring resonator-based add–drop filter.

coupling coefficients apodization. Various research groups worldwide have demonstrated high-order coupled-microring resonator-based filters in SOI and SiN-on-silica platforms. For example, S. Xiao et al. [92] demonstrated a third-order coupled-microring resonator-based filter in SOI, as shown in Figure 4.28. The microring resonators are ultracompact with 2.5-μm radius. The bus-waveguide is narrower than the microring waveguide to phase-match the straight waveguide mode to the bend waveguide mode.

Barwicz et al. [93] demonstrated a third-order coupled-microring resonator-based filter on a silicon-rich nitride platform, as shown in Figure 4.29. The high-order filter demonstrates a wide and flat pass-band of 88 GHz, a fast roll-off ratio of ~0.2 dB/GHz and a high

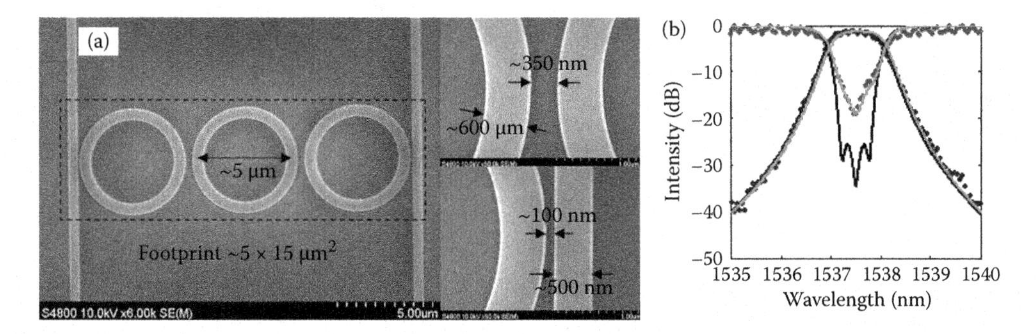

FIGURE 4.28
(a) SEM picture of a third-order coupled-microring resonator-based filter. The zoom-in pictures show the waveguide and gap widths of the microring-to-microring coupling region and the microring-to-waveguide coupling region. (b) Measured and simulated transmission spectra of a third-order coupled-microring resonator-based filter. (Reprinted from Xiao, S. et al., *Optics Express*, 2007. **15**(22): 14765–14771. With permission from Optical Society of America.)

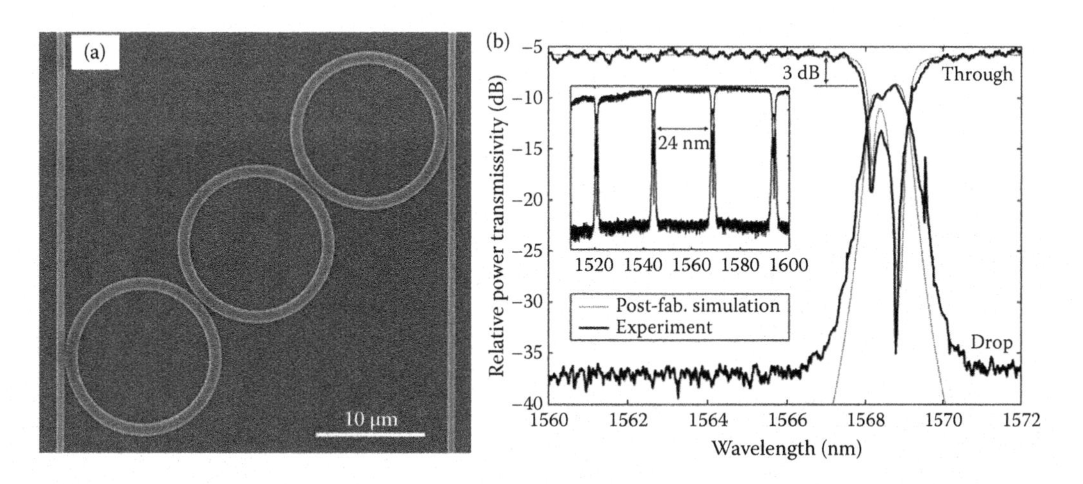

FIGURE 4.29
(a) SEM picture of a third-order coupled-microring resonator-based filter in a silicon nitride on silica substrate. (b) Measured and simulated transmission spectra. (Reprinted from Barwicz, T. et al., *Optics Express*, 2004. **12**(7): 1437–1442. With permission from Optical Society of America.)

TABLE 4.2

Key Design Parameters and Measured Performance Metrics in Some Representative Demonstrated Coupled-Microring Resonator-Based High-Order Filters

	Reference			
	[92]	[93]	[94]	[95]
Platform	SOI	SiN-on-silica	SOI	SOI
Number of coupled microring resonators	3	3	3–16	3
Footprint (μm^2)	75	~1000	70	40
FSR (nm)	32	24	18	18
1-dB bandwidth (nm)	0.85	0.7	2.48	–
3-dB bandwidth (nm)	1.15	–	–	3.3
Drop loss (dB)	–	3	<3	<0.3
In-band rejection (dB)	20	7	NA	12
Out-of-band rejection (dB)	40	30	30	18
Roll-off (dB/nm)	25	25	25	100

out-of-band rejection ratio of 7.5 dB. Due to the relatively low index contrast of SiN-on-silica waveguides, the microring bending radius is relatively large, and the overall footprint is in the order of 1 mm^2. Table 4.2 summarizes key design parameters and measured performance metrics in some representative demonstrated coupled-microring resonator-based high-order filters.

High-order coupled-microring resonators-based filters have been employed as a building block for large-channel MUX/DEMUX. Here, we show an example of 32-channel, thermal-tunable MUX/DEMUX comprising 32 third-order coupled-microring resonators coupled to the same bus-waveguide on a SOI substrate [96]. Figure 4.30a shows the optical micrograph of the fabricated high-order filter-based MUX/DEMUX. The microring radius

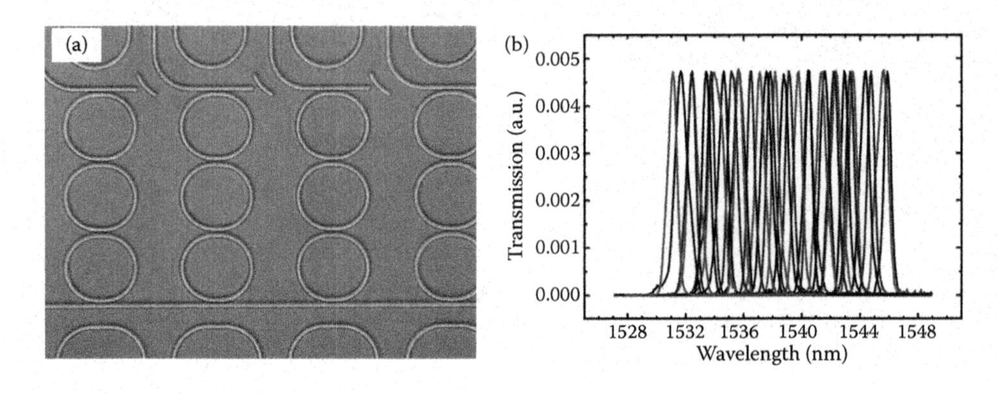

FIGURE 4.30
(a) Optical micrograph of a 32-channel MUX/DEMUX comprising 32 third-order coupled-microring resonator-based filters on a SOI substrate. (b) Measured transmission spectra of the 32-channel MUX/DEMUX. (Reprinted from Park, S. et al., *Optics Express*, 2011. **19**(14): 13531–13539.)

is typically 9 μm, which enables a FSR of 12.8 nm with 50-GHz channel spacing and a 20-dB adjacent channel crosstalk. Figure 4.30b shows the transmission spectra of the 32-channel MUX/DEMUX. The spectra exhibit a relatively broad 3-dB bandwidth of 0.56 nm.

4.4.2 FSR Expansion Using Vernier Effect

Another essential criterion for WDM communications applications concerns the FSR. For single channel filtering within the telecommunications C-band, for example, a filter FSR as wide as the entire C-band (1530–1565 nm), namely about 35 nm, is needed to select one particular wavelength channel. Furthermore, wide FSR can accommodate more optical channels. As previously discussed in Section 4.3.1, the FSR is inversely proportional to the microring optical round-trip length. A large FSR can be attained by designing small optical microresonators at the expense of significantly increasing the microring radiation loss and reducing the tolerance to fabrication imperfections.

To design a wide FSR while avoiding the use of ultrasmall optical microring resonators, one alternative design is to use Vernier effect. The idea of Vernier effect is to expand the FSR by using two coupled microring resonators with slightly different radii. The two set of resonances are slightly mismatched with each other and thus the aligned resonance peaks are preserved while the misaligned ones are suppressed, resulting in a significant expansion of FSR compared with that of the constituent microrings. Figure 4.31 schematically shows such a Vernier effect–based series-coupled double-microring resonator in the form of an add–drop filter and the concept of the FSR expansion.

The expanded FSR under Vernier operation considering two slightly mismatched sets of microring resonances can be given as follows:

$$m_1\text{FSR}_1 = m_2\text{FSR}_2 = \text{FSR}_{\text{Vernier}} \tag{4.81}$$

where m_1 and m_2 are the integer FSR expansion factors that are co-primes. FSR_1 and FSR_2 are the FSR values of the microring with radius of R_1 and the microring with radius of R_2, respectively, and $\text{FSR}_{\text{Vernier}}$ is the resultant FSR under the Vernier effect. Assuming $m_1 > m_2$ and $\text{FSR}_2 > \text{FSR}_1$, we obtain

$$\text{FSR}_{\text{Vernier}} = \frac{(m_1 - m_2)\text{FSR}_1\text{FSR}_2}{\text{FSR}_2 - \text{FSR}_1} \tag{4.82}$$

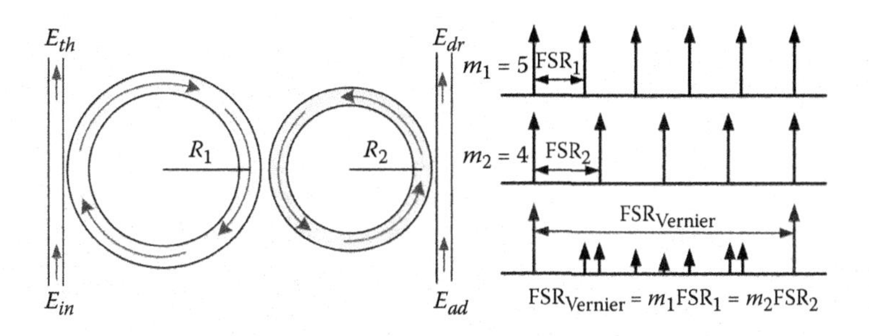

FIGURE 4.31
Schematic of a series-coupled double-microring resonator-based add–drop filter using Vernier effect. The two microrings are slightly different in size.

Following Equation 4.34, FSR_1 and FSR_2 are given as

$$\text{FSR}_1 = \frac{\lambda_{m1}^2}{n_g L_1} \tag{4.83}$$

$$\text{FSR}_2 = \frac{\lambda_{m2}^2}{n_g L_2} \tag{4.84}$$

where $L_1 = 2\pi R_1$ and $L_2 = 2\pi R_2$ are the microring round-trip lengths. In the case that $m_1 = m_2 + 1$, and assuming resonance wavelength λ_{m1} in the microring with radius R_1 is aligned with resonance wavelength λ_{m2} in the microring with radius R_2, we obtain from Equations 4.82 through 4.84, the FSR_{Vernier} as follows:

$$\text{FSR}_{\text{Vernier}} = \frac{\lambda_m^2}{n_g (L_1 - L_2)} \tag{4.85}$$

where $\lambda_m = \lambda_{m1} = \lambda_{m2}$. From Equation 4.85, we find that a large FSR can be obtained by designing the microring round-trip lengths L_1 and L_2 to be close to each other. Using Equations 4.81, 4.83, and 4.84, we can express the relationship between L_1 and L_2 for a given FSR expansion factor of $m_1 = m_2 + 1$ as follows:

$$L_2 = \frac{m_1 - 1}{m_1} L_1 \tag{4.86}$$

The desired FSR_{Vernier} can therefore be obtained by properly designing the two different microring round-trip lengths.

Figure 4.32 shows the modeled drop-port transmission spectra of a Vernier filter and the constituent microrings. The modeling is essentially the same as that of the second-order coupled microring resonators given by Equations 4.77 and 4.78, except that the two microrings have different radii of R_1 = 25 μm and R_2 = 20 μm, which gives an expanded FSR of 18 nm. Here we assume a waveguide-microring coupling coefficient $\kappa = 0.4$, an intercavity coupling coefficient $\kappa = 0.08$, a loss coefficient $\alpha = 2\ \text{cm}^{-1}$. The inset shows the zoom-in view of the series-coupled double microring transmission spectrum around 1557 nm, with 38-dB channel isolation between the aligned resonance channel and the adjacent misaligned channel.

Boeck et al. [97] demonstrated series-coupled double microring resonators on SOI platform using Vernier effect. Figure 4.33a shows the SEM of the Vernier filter. Figure 4.33b shows the measured and modeled transmission spectra with an expanded FSR. The two microring round-trip lengths are 71.1 and 56.5 μm. The resulting expanded FSR is ~36 nm with expansion factors of 5 and 4 for the two microrings as defined in Equation 4.81. The interstitial peak suppression is between 9 and 17 dB.

However, we note that Vernier filters using double microrings provide only limited interstitial mode suppression and thus relatively poor channel isolation. Besides, the resultant channel bandwidth is still relatively narrow. To design filters that encompass the advantages of wide FSR and flat-top broadband filter profiles, we remark that the Vernier effect can be applied to the higher-order microring filters. For example, in the case of third-order coupled-microring resonators, one of the three coupled microrings can be slightly smaller in size. These structures can be useful for WDM systems as they enable potentially wide FSR due to Vernier effect and a flat-top broadband filter profile with large channel isolation due to the high-order microring coupling.

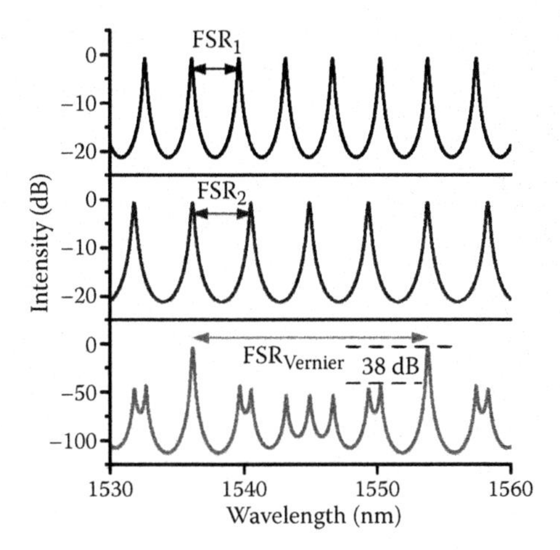

FIGURE 4.32
Modeled transmission spectra of a Vernier effect–based filter. The top set of modeled drop-port transmission resonance peaks are given by microring R_1, whereas the middle set of modeled drop-port transmission resonance peaks are given by microring R_2. The bottom set of modeled resonance peaks are given by the transmission spectrum of the series-coupled double microrings.

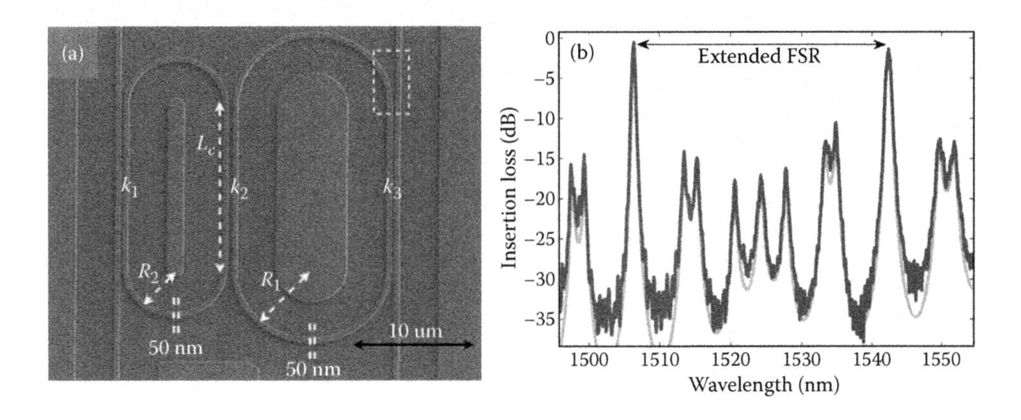

FIGURE 4.33
(a) SEM of the fabricated Vernier filter on SOI platform. (b) Measured and modeled drop-port transmission spectra of the Vernier filter with an extended FSR. (Reprinted from Boeck, R., et al., *Optics Express*, 2010. **18**(24): 25151–25157.)

4.5 Interferometer-Coupled Microring Resonators

Besides coupled microring resonators, another way we could tailor the microring resonator-based filter transmission is by coupling the microring resonator to an interferometer such as a Mach-Zehnder interferometer (MZI). The microring–interferometer coupling enables interference between the microring resonance field and the coherent nonresonant background field that is separated and recombined via the interferometer.

Such an interference mechanism results in an asymmetric resonance line shape known as Fano resonance line shape [98]. The microring resonator coupling to the interferometer also enables a resonance-assisted asymmetric interferometer periodic transmission that exhibits a relatively box-like periodic transmission. Here we discuss the basic principle of such a microring–interferometer structure spanning the scope of Fano resonance line shape tuning, and the application of periodic box-like filter profiles as an interleaver.

4.5.1 Principles of Interferometer-Coupled Microring Resonators: Fano Resonances

Figure 4.34 schematically shows the microring-assisted MZI. One arm of the MZI is coupled with a microring resonator, and another arm of the MZI is uncoupled. The microring-coupled arm experiences the amplitude and phase response of the microring resonator. The uncoupled arm experiences only the waveguide attenuation and phase delay. The light at the output ports are interference of the two arms. Thus, by varying the microring response and the uncoupled arm phase delay, we can vary the output-port transmissions.

Using the transfer matrix formalism, we express the transfer matrix relations of the two directional couplers of the MZI as follows:

$$\begin{pmatrix} E_3 \\ E_4 \end{pmatrix} = \begin{pmatrix} \cos\theta_1 & -i\sin\theta_1 \\ -i\sin\theta_1 & \cos\theta_1 \end{pmatrix} \begin{pmatrix} E_1 \\ E_2 \end{pmatrix} = F_1 \begin{pmatrix} E_1 \\ E_2 \end{pmatrix} \tag{4.87}$$

$$\begin{pmatrix} E_7 \\ E_8 \end{pmatrix} = \begin{pmatrix} \cos\theta_2 & -i\sin\theta_2 \\ -i\sin\theta_2 & \cos\theta_2 \end{pmatrix} \begin{pmatrix} E_5 \\ E_6 \end{pmatrix} = F_3 \begin{pmatrix} E_5 \\ E_6 \end{pmatrix} \tag{4.88}$$

where θ_1 and θ_2 determine the power-coupling ratios of the input and output directional couplers, F_1 and F_3 are the transfer matrices of the input and output directional couplers, E_1 and E_2 (E_3 and E_4) are the electric field amplitudes at the two input (output) ports of the input directional coupler, E_5 and E_6 (E_7 and E_8) are the electric field amplitudes at the two input (output) ports of the output directional coupler. We express the transfer matrix relation between the electric field amplitudes (E_3, E_4) in the MZI arms just before the microring and the electric field amplitudes (E_5, E_6) in the MZI arms just after the microring according to Equation 4.52 as follows,

$$\begin{pmatrix} E_5 \\ E_6 \end{pmatrix} = \begin{pmatrix} \left(\dfrac{t - e^{-(i\beta+\alpha/2)L}}{1 - te^{-(i\beta+\alpha/2)L}}\right) e^{-(i\beta+\alpha/2)L_d} & 0 \\ 0 & e^{-(i\beta+\alpha/2)(L_d+\Delta L_d)} e^{i\Delta\Phi} \end{pmatrix} \begin{pmatrix} E_3 \\ E_4 \end{pmatrix} = F_2 \begin{pmatrix} E_3 \\ E_4 \end{pmatrix} d \tag{4.89}$$

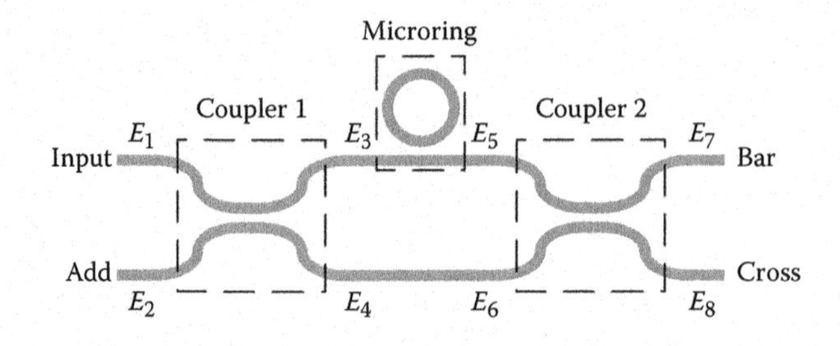

FIGURE 4.34
Schematic of the microring-coupled MZI.

where F_2 is the transfer matrix for the MZI fields before and after the coupled microring, L_d is the path length of the MZI arm with the coupled microring, ΔL is the path length difference between the two arms of the MZI, and $\Delta\Phi$ is an external tunable phase shift. Thus, the electric fields at the two output ports (E_7, E_8) are related to the electric fields at the two input ports (E_1, E_2) as follows:

$$\begin{pmatrix} E_7 \\ E_8 \end{pmatrix} = F_3 F_2 F_1 \begin{pmatrix} E_1 \\ E_2 \end{pmatrix} \tag{4.90}$$

Here we assume ideal 3-dB couplers ($\theta_1 = \theta_2 = \pi/4$), and the light is only launched from the E_1 port ($E_2 = 0$).

Figure 4.35 shows the calculated transmission spectra at the microring-coupled MZI output ports for various $\Delta\Phi$. When $\Delta\Phi = 0$ and π, the spectra exhibit asymmetric Fano line shapes, which experience sharp change around the resonance center. When $\Delta\Phi = 0.5\pi$ and 1.5π, the spectra also exhibit asymmetric Fano line shapes, however, the line shapes become close to symmetric Lorentzian line shape.

The Fano resonance line shape is due to the interference between a resonant field and a coherent background field [98]. The Fano resonance line shape in the transmission spectrum can be expressed as follows:

$$|E(\omega)|^2 = |E_0|^2 \frac{\left((\omega - \omega_0) + q(\omega_0/2Q)\right)^2}{(\omega - \omega_0)^2 + (\omega_0/2Q)^2} \tag{4.91}$$

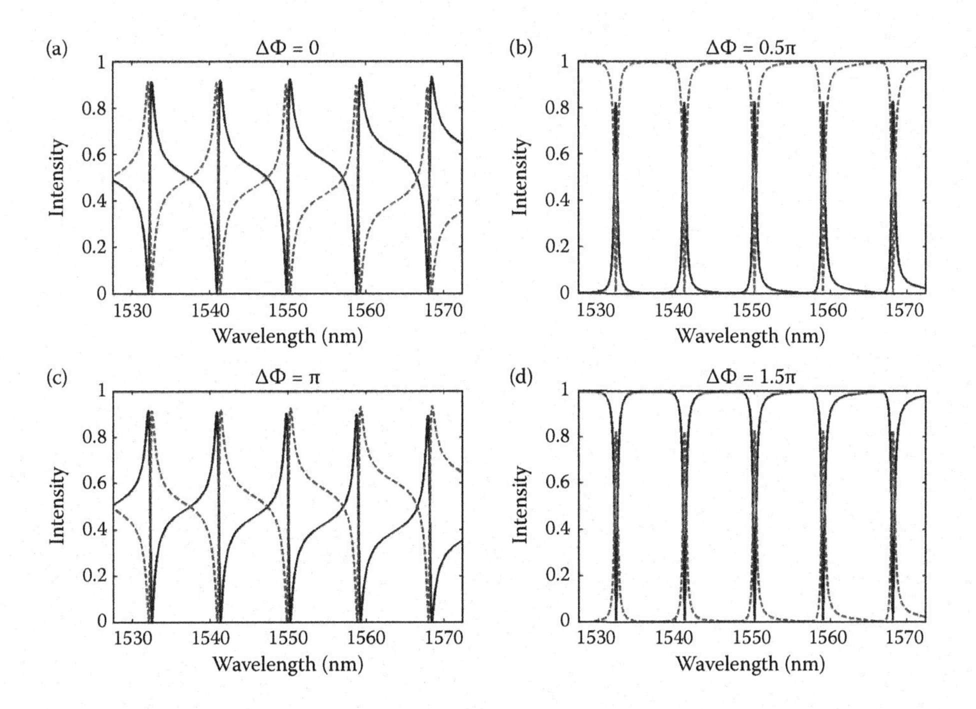

FIGURE 4.35
Modeled microring-MZI transmission spectra at the bar-port (solid) and the cross-port (dashed) for $\Delta\Phi$ = (a) 0, (b) 0.5π, (c) π, and (d) 1.5π. ($R = 10$ μm, $\Delta L = 0.5$ μm, $\alpha = 2$ cm^{-1}, $\kappa = 0.5$.)

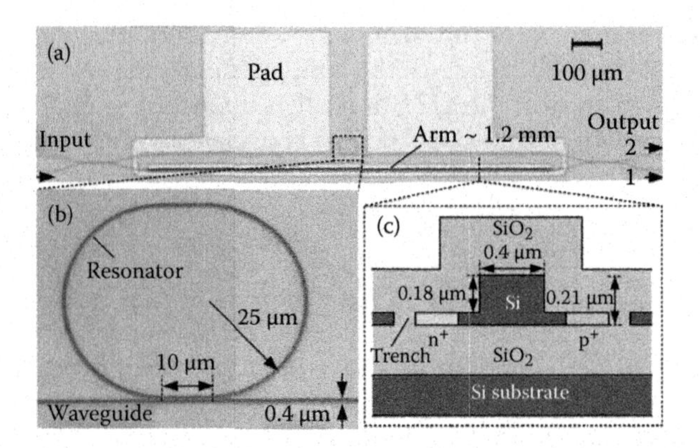

FIGURE 4.36
(a) Optical micrograph of the fabricated device. (b) Optical micrograph zoom-in view of the waveguide-coupled microring resonator. (c) Schematic cross-sectional view of the p-i-n diode embedded in the MZI nonresonance arm. (Reprinted from Zhou, L.J., and A.W. Poon, *Optics Letters*, 2007. **32**(7): 781–783. With permission from Optical Society of America.)

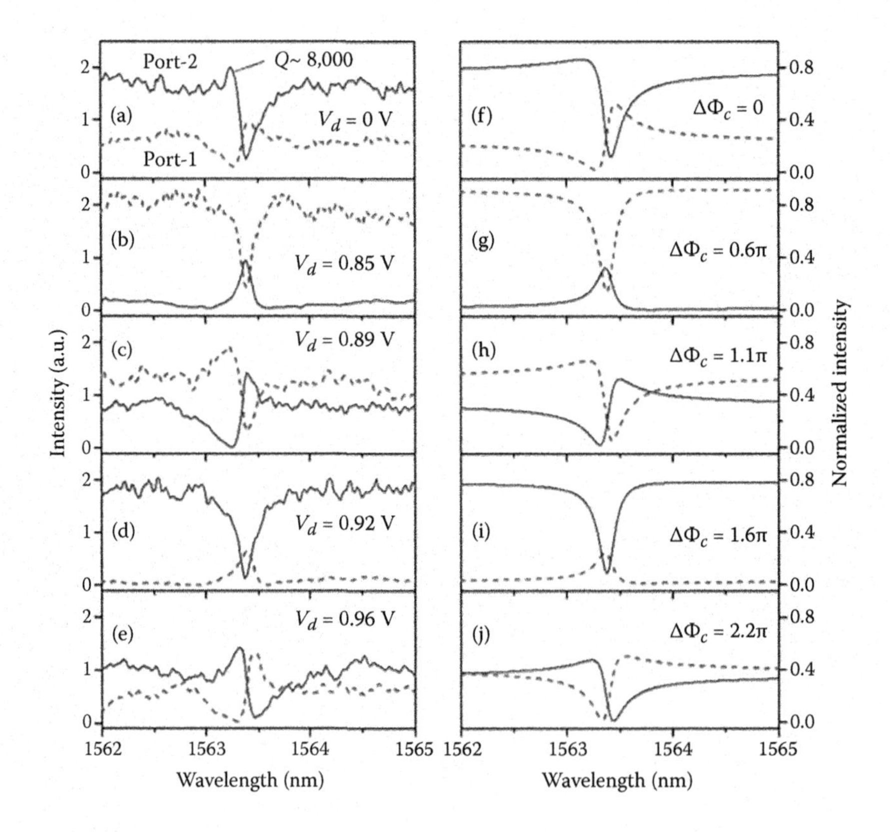

FIGURE 4.37
(a)–(e) Measured TE-polarized output-port transmission spectra for the microring resonator-coupled MZI with various bias voltage values V_d. (f)–(j) Modeled corresponding output-port transmission spectra with various phase shift values ΔΦ. (Reprinted from Zhou, L.J., and A.W. Poon, *Optics Letters*, 2007. **32**(7): 781–783. With permission from Optical Society of America.)

where q is the Fano parameter. In the case that the Fano term dominates the frequency detuning from the resonance center $q(\omega_0/2Q) \gg (\omega - \omega_0)$, the Fano line shape approaches the Lorentzian line shape.

We show an example of a Fano resonance microring-coupled MZI on a SOI substrate. Figure 4.36 shows the optical micrograph of the fabricated device. The uncoupled arm of the MZI is laterally integrated with a p-i-n diode as a phase shifter upon applying a forward bias via the free-carrier plasma dispersion effect. Figure 4.37 shows the measured and modeled tranmission spectra at the two output ports. The asymmetric resonance line shapes vary according to the applied phase difference.

4.5.2 Optical Interleavers Using Microring Resonator–Based Interferometers

Besides resonance line shape tuning, microring resonator-coupled interferometer structures have been investigated as optical interleavers by various research groups [7–11]. While an asymmetric MZI can function as an interleaver, the roll-off of the transmission passband for such a simple interleaver is relatively slow and thus the crosstalk level is high. Typically an interleaver requires a flat-top and box-like spectral response likes a MUX/DEMUX does. To solve these problems, microring resonators have been employed to flatten the top of the passband spectrum [100, 101]. The key distinction between the designs for an interleaver and for Fano resonance line shape tuning is the FSR requirements for the microring and the MZI. While the Fano resonance does not impose a specific FSR requirement, the interleaver requires the asymmetric MZI path difference to give a FSR that is twice the WDM channel spacing and the microring FSR needs to be half the MZI FSR. Figure 4.38 schematically shows the principle of the interferometer-coupled microring resonator-based interleaver. The microring resonator resonances interfering with the MZI envelopes tailor the output from the bar and cross ports into complementary channels, which could serve as interleavers with flattop and broadband transmissions.

Here we introduce two different silicon microring resonator–coupled MZI-based interleaver structures, namely the microring-assisted MZI structure, and what we term as the microring resonator MZI structures.

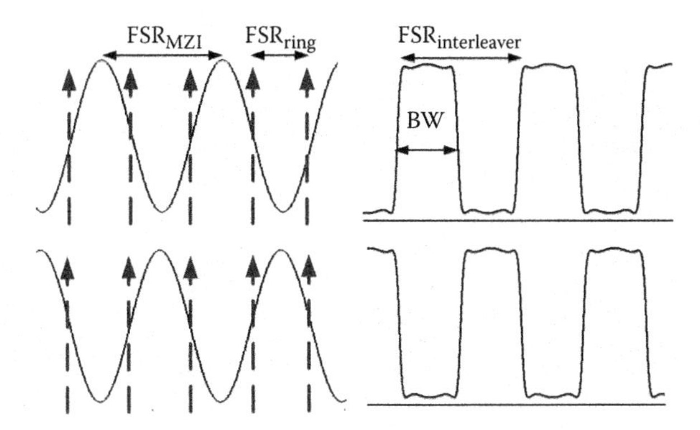

FIGURE 4.38
Principle of the microring resonator–coupled MZI-based interleaver.

4.5.2.1 Microring-Assisted MZI Interleavers

Figure 4.39a shows a conventional interleaver structure comprising a 2 × 2 asymmetric MZI and a microring resonator coupled to one of the MZI arms, referred to as auto regressive moving average (ARMA) filter [102]. The asymmetry of the MZI is introduced by a delay line in the other MZI arm. The length of the delay line determines the FSR of the interleaver. Here the optical round-trip path length of the microring resonator is double that of the delay line length. To obtain the interleaver function, an external π phase shift is needed to add to the microring resonator [100]. Such a phase difference can also be applied to the delay line by a $\pi/2$ phase change, as shown in Figure 4.39b. Alternatively, using a Y-branch to replace the 3-dB coupler in the input port, such an externally induced phase shift can be eliminated thanks to the balanced phase between the light after the Y-branch, compared with the $\pi/2$ phase difference in the 3-dB directional coupler. We name such an interleaver structure as ring-assisted MZI (RA-MZI) interleaver, as shown in Figure 4.39c. Here we only focus on the RA-MZI interleaver.

With the microring round-trip loss factor $a = 1$ (assuming lossless microring) and the waveguide transmission coefficient $t = 1/3$, we calculate the transmission responses for both ARMA and RA-MZI interleavers, as shown in Figure 4.40a,b. Other than an additional π phase shift, both structures show the same transmission properties.

Figure 4.41a shows the calculated transmission spectra upon various transmission coefficient values. As t approaches 1/3, the transmission spectrum shows nearly box-like transmission band with flat top. With increased t, the passband fluctuations increase along with sharpen roll-off ratios.

Another key structure parameter that affects the transmission properties is the optical delay line length in the other MZI arm. As mentioned, in an ideally designed interleaver, the length of the optical delay line should be half of the microring circumference. If we define this ratio as v, the deviation of v from 1/2 results in degradation of the transmission properties. Figure 4.41b shows the transmission spectra at different v values, with a fixed $t = 1/3$. As v deviates from 1/2, the extinction ratio is significantly reduced.

We show an example of the RA-MZI interleaver fabricated in a commercially available, 200-mm SOI wafer [8]. The device was defined by 248-nm-deep UV lithography and ICP

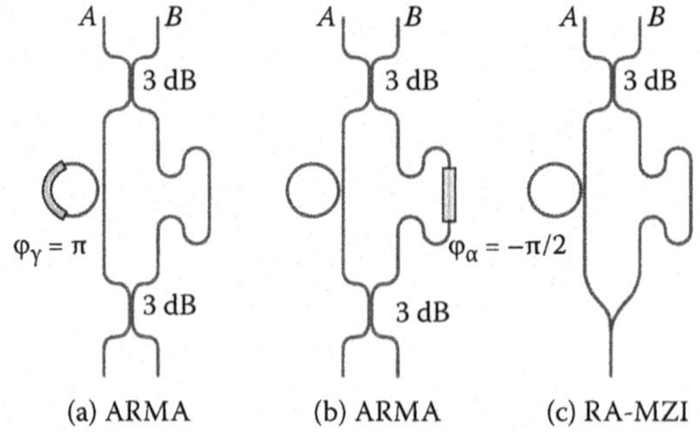

FIGURE 4.39
Schematic of the microring-assisted MZI interleavers comprises an asymmetric MZI structure and a microring resonator. (a) An external phase shift π introduced in the ring resonator. (b) An external phase shift $\pi/2$ introduced in an arm of the MZI. (c) Schematic of the RA-MZI with the Y-junction replacing the input 3-dB coupler. The gray regions in (a) and (b) denote the external phase shifts. (Reprinted from Zhang, L. et al., *Optics Letters*, 2008. **33**(13): 1428–1430. With permission from Optical Society of America.)

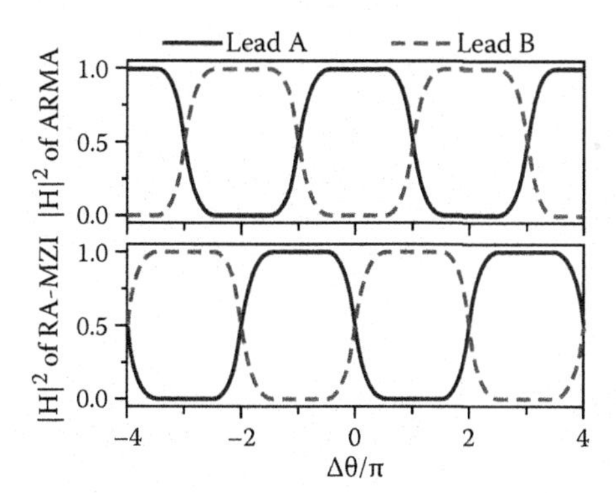

FIGURE 4.40
Calculated transmission spectra of the ARMA and RA-MZI interleavers with $t = 1/3$. (Reprinted from Zhang, L. et al., *Optics Letters*, 2008. **33**(13): 1428–1430. With permission from Optical Society of America.)

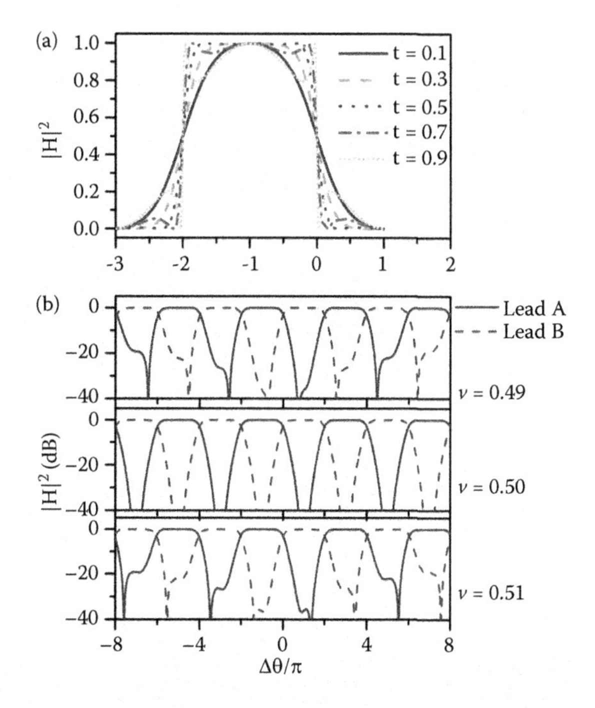

FIGURE 4.41
(a) Characteristics of the pass band with different coupling coefficient t. (b) Characteristics of the response spectra with different v at $t = 1/3$. (Reprinted from Zhang, L. et al., *Optics Letters*, 2008. **33**(13): 1428–1430. With permission from Optical Society of America.)

etching. Figure 4.42a shows the SEM of the fabricated devices. Figure 4.42b–e show the zoom-in SEMs of the 3-dB coupler, the optical delay line, the waveguide-to-microring coupling region and the input inverse taper. The silicon waveguide was ~300 × 300 nm. The square-shaped waveguide was designed to minimize the polarization dependence. The radius of the microring resonator was 40 μm, while the optical delay line in the other

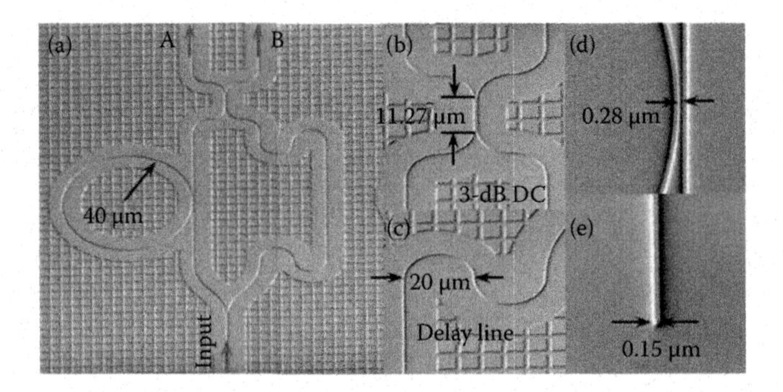

FIGURE 4.42
(a) The SEM picture of the fabricated RA-MZI. (b) 3-dB directional coupler. The coupling length is 2.79 μm and the coupling gap is 300 nm. (c) The optical delay line with hemicircles. (d) The side-coupled waveguide-to-microring coupling region with gap separation of 280 nm. (e) The inverse taper with 150 nm in width and 200 μm in length. (Reprinted from Zhang, L. et al., *Optics Letters*, 2008. **33**(13): 1428–1430. With permission from Optical Society of America.)

arm comprised four hemicircles with radius of 10 μm. Such a design ensured the ratio between the optical delay line length and the microring circumference to be 1/2. To minimize the optical fiber to waveguide coupling loss, an inversely tapered nanowaveguide was adopted.

Figure 4.43 shows the measured transmission spectra. The results show a good interleaver function. However, the extinction ratio is only ~10 dB. Furthermore, the polarization independence is not obvious in the measurements. We attribute this failure to the following reasons: (1) The fabrication imperfection could significantly affect the device properties, especially the 3-dB coupler. (2) Although the waveguide was designed with a square shape, the optical propagation loss is still polarization dependent as the roughness of the waveguide upper surface and of the sidewalls are not the same. (3) The 3-dB coupler and the waveguide-to-microring coupler are intrinsically polarization dependent.

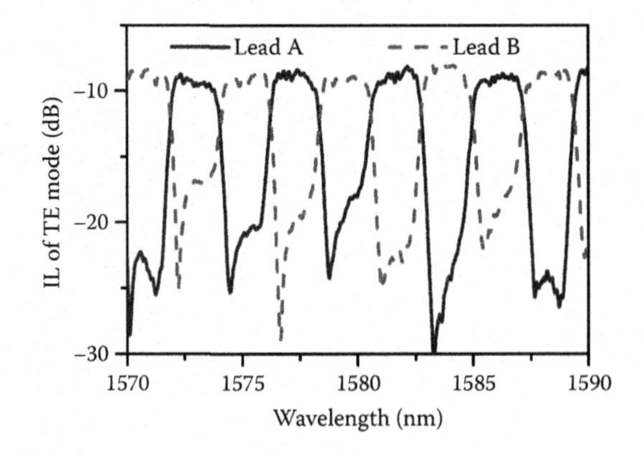

FIGURE 4.43
Measured TE-polarized transmission spectra from output ports A (solid lines) and B (dashed lines). (Reprinted from Zhang, L. et al., *Optics Letters*, 2008. **33**(13): 1428–1430. With permission from Optical Society of America.)

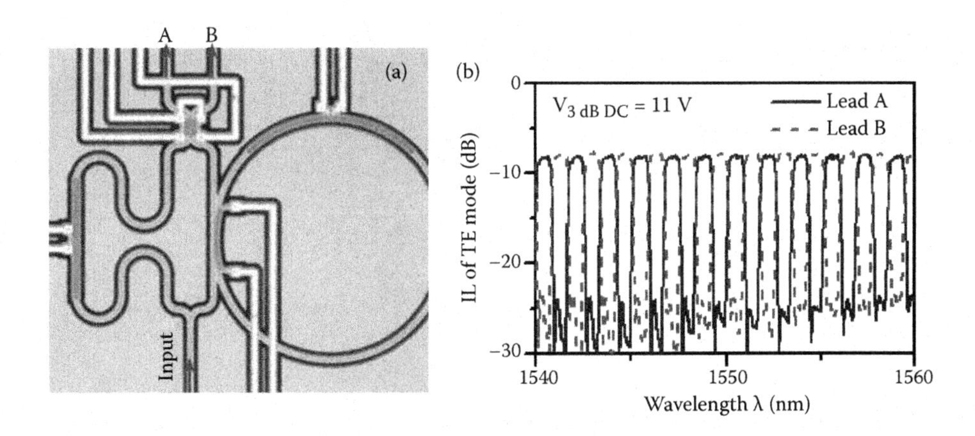

FIGURE 4.44
(a) Microscope of the full RA-MZI interleaver structure. (b) Spectra of the bar and the cross waveguides of the RR-MZI with 11 V applied voltage. (Reprinted with permission from Song, J. et al., *IEEE Photonics Technology Letters*. **20**(24): 2165–2167, © 2008 IEEE.)

The interleaver properties can be enhanced by dynamic tuning. Here we show another RA-MZI optical interleaver integrated with Ti heater for thermo-optical (TO) tuning. Figure 4.44a shows the optical micrograph of the fabricated TO-tunable optical interleaver. The optical microring, the optical delay line, the MZI 3-dB coupler, and the waveguide-to-microring coupler are all integrated with Ti heaters to enable fine phase adjustment. Figure 4.44b shows the measured transmission spectra with an applied voltage of 11 V to the 3-dB coupler. The transmission spectra are uniform across the spectra ranging from 1520 to 1600 nm. However, only the spectra range from 1540 to 1560 nm are shown here for better visualization. Furthermore, the extinction ratio increases to ~20 dB by fine-adjusting the 3-dB coupler. This also confirms the significance of the 3-dB couplers for demonstrating high-performance optical interleavers.

4.5.2.2 Microring Resonator MZI Interleavers

In an RA-MZI interleaver, the microring is side-coupled to the MZI. Here we introduce another simple interleaver structure in which the input 3-dB directional coupler is replaced by a microring resonator in add–drop configuration, while removing the coupled microring from the MZI arm. We term this structure as ring resonator MZI (RR-MZI) interleaver, which is shown in Figure 4.45a. The light that couples into the ring resonator splits into throughput and drop to the two MZI arms. The light from the output 3-dB directional coupler interferes with each other, giving rise to the interleaver function. The shortcoming of this structure is that one of the input ports lefts floating and is difficult to connect with an optical fiber. Thus, the structure lacks an add port. However, if we replace the conventional microring by using an 8-shaped crossing microring, the floating terminal can serve as an add port. We name such a structure as cross-ring MZI (CR-MZI) interleaver, as shown in Figure 4.45b. The RR-MZI and CR-MZI interleavers function identically, albeit the additional insertion loss and possible crosstalk induced by the waveguide crossing for the CR-MZI interleavers.

The transmission of the two interleavers can be modeled using transfer matrix method as discussed in Section 4.3.3. The microring response can be expressed as

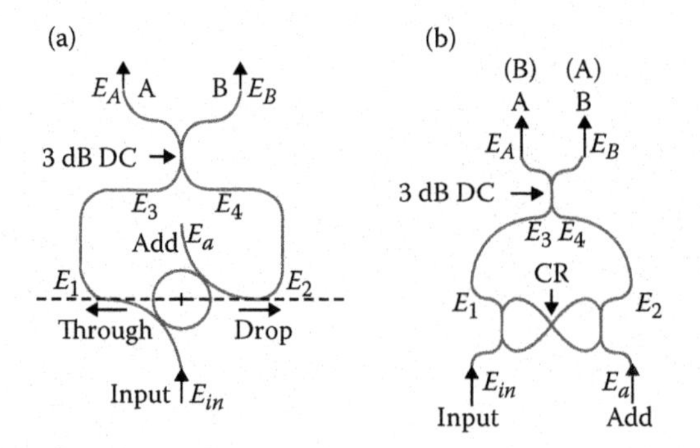

FIGURE 4.45
Schematics of the ring-resonator (RR) MZI interleaver structures. (a) An RR-MZI interleaver with an optical ring resonator replacing the input 3-dB coupler and (b) a cross-ring (CR) MZI interleaver. (Reprinted from Song, J. et al., *Optics Express*, 2008. **16**(11): 7849–7859; Song, J. et al., *Optics Express*, 2008. **16**(26): 21476–21482. With permission from Optical Society of America.)

$$\begin{pmatrix} E_1 \\ E_2 \end{pmatrix} = M \begin{pmatrix} E_{in} \\ E_a \end{pmatrix} \tag{4.92}$$

where M is the transfer matrix of the microring add–drop filter (see Equation 4.62), E_1 and E_2 are the electric field amplitudes at the input and add ports right before the microring coupling, E_3 and E_4 are the electric field amplitudes in the throughput and drop port of the MZI right after the microring coupling. Following the modified propagation matrix, we express the electric field amplitudes in the two arms of the MZI as follows:

$$\begin{pmatrix} E_5 \\ E_6 \end{pmatrix} = \begin{pmatrix} e^{-(i\beta+\alpha)L_d} & 0 \\ 0 & e^{-(i\beta+\alpha)(L_d+\Delta Ld)+\Delta\Phi} \end{pmatrix} \begin{pmatrix} E_3 \\ E_4 \end{pmatrix} = P \begin{pmatrix} E_3 \\ E_4 \end{pmatrix} \tag{4.93}$$

where P is the transfer matrix of the MZI, L_d, ΔL_d, and $\Delta\Phi$ follow the definitions in Equation 4.90. We express the 3-dB coupler transfer matrix as

$$\begin{pmatrix} E_A \\ E_B \end{pmatrix} = \begin{pmatrix} \cos\theta & -i\sin\theta \\ -i\sin\theta & \cos\theta \end{pmatrix} \begin{pmatrix} E_3 \\ E_4 \end{pmatrix} = F \begin{pmatrix} E_3 \\ E_4 \end{pmatrix} \tag{4.94}$$

here F is the transfer matrix of the 3-dB directional coupler (see Equation 4.88). Thus, we express the transfer matrix relationship between the electric field amplitudes at the two input ports and the two output ports as follows:

$$\begin{pmatrix} E_A \\ E_B \end{pmatrix} = FPM \begin{pmatrix} E_{in} \\ E_a \end{pmatrix} \tag{4.95}$$

Figure 4.46a shows the modeled transmission phase responses of the throughput and drop transmissions of the interleaver after a microring resonator, according to Equation 4.95. The phase difference between the throughput and drop port transmissions is $\pm\pi/2$.

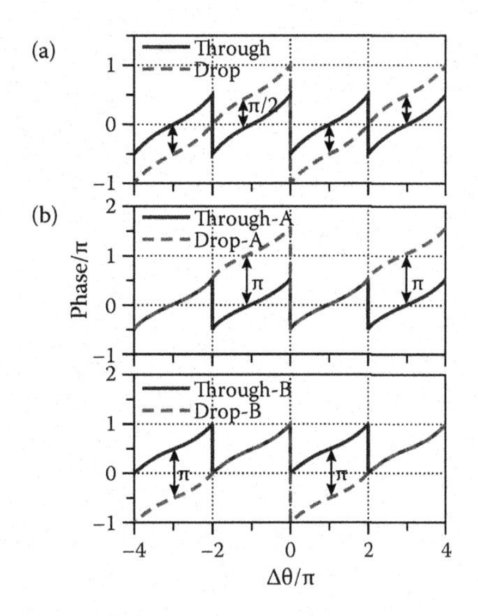

FIGURE 4.46
(a) The phase responses of the throughput (solid line) and drop (dashed line) ports. The phase difference is $\pm\pi/2$. (b) The phase responses of leads A and B. The phase difference increases to π. (Reprinted from Song, J. et al., *Optics Express*, 2008. **16**(11): 7849–7859. With permission from Optical Society of America.)

As the 3-dB coupler contributes an additional phase change of $\pi/2$, the phase difference increases to π after the light passes through it to leads A and B. Figure 4.46b shows the modeled phase responses of leads A and B with π phase difference. As light from both the through and drop ports contributes to the light output from leads A and B, we separately plot the phase responses of leads A and B, which show clearly the π phase shifts. Therefore, constructive and destructive interference occurs in leads A and B in alternate spectral bands. To obtain high extinction ratios, the light amplitude in the two MZI branches for interference should be identical, which means $2t = \kappa^2$ or $t = \sqrt{2} - 1$. In this condition, we attain a good interleaver function, as shown in Figure 4.47.

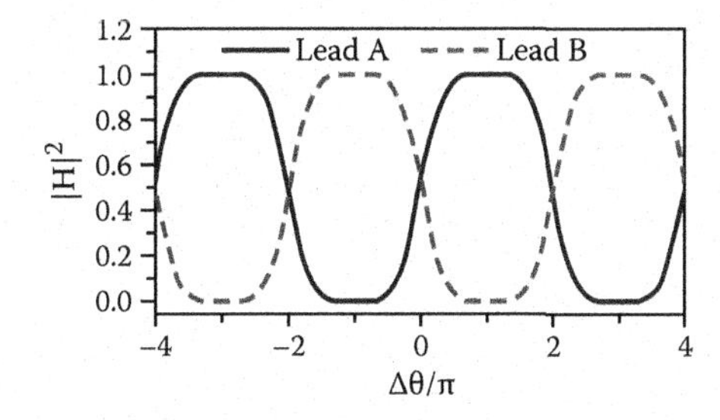

FIGURE 4.47
Relative light intensity of leads A and B as a function of frequency. (Reprinted from Song, J. et al., *Optics Express*, 2008. **16**(11): 7849–7859. With permission from Optical Society of America.)

Here we show as an example a RR-MZI interleaver in a SOI substrate. Figure 4.48 shows the SEM of the fabricated RR-MZI interleaver. We adopted a racetrack microring resonator here for design convenience. The cross section of the silicon wire was chosen to be ~300 × 300 nm. Figure 4.48b shows a fabricated 3-dB directional coupler. We chose the coupling length of ~2.79 μm to flatten the transmission spectrum. The racetrack structure is shown in Figure 4.48c. The radius of the microring is 10 μm.

Figure 4.49a,b show the measured polarization dependent loss (PDL) values and the transmission spectra from leads A and B. PDL is commonly defined as the peak-to-peak difference in transmission for light with various states of polarization. The passband exhibits flat-top transmission. The insertion loss and the PDL within the passband are approximately −10 dB and <5 dB, whereas the insertion loss is approximately −20 dB within the rejection band. This suggests that the crosstalk is only approximately −10 dB.

To illustrate dynamic control of the interleaver performances, we show as an example demonstrated TO-tunable RR-MZI and CR-MZI structures. Figure 4.50a shows the design layout of the RR-MZI interleaver. The red lines represent the optical waveguide. The blue

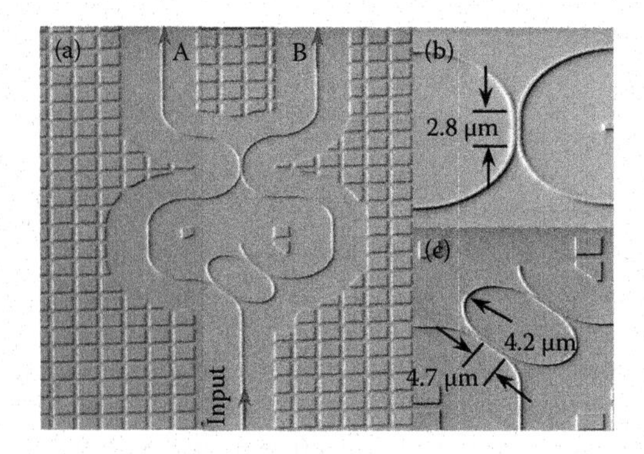

FIGURE 4.48
(a) SEM picture of the RR-MZI interleaver. (b) 3-dB directional coupler with coupling length of 2.79 μm and the coupling gap of 300 nm. (c) Add–drop type microring resonator with coupling length is 4.7 μm. (Reprinted from Song, J. et al., *Optics Express*, 2008. **16**(11): 7849–7859. With permission from Optical Society of America.)

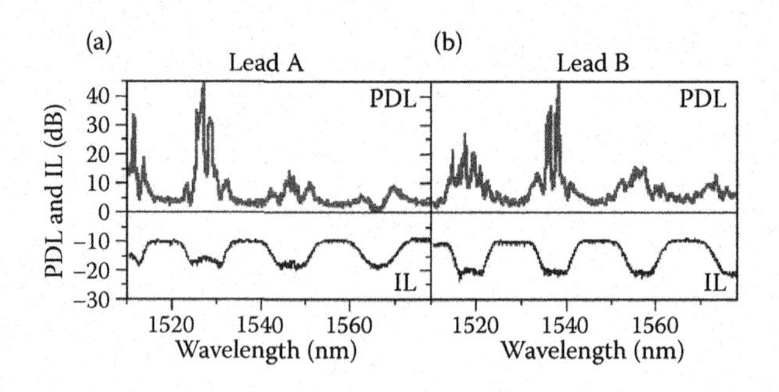

FIGURE 4.49
Measured optical polarization dependent loss (PDL) and transmission spectra from (a) lead A and (b) lead B. (Reprinted from Song, J. et al., *Optics Express*, 2008. **16**(11): 7849–7859. With permission from Optical Society of America.)

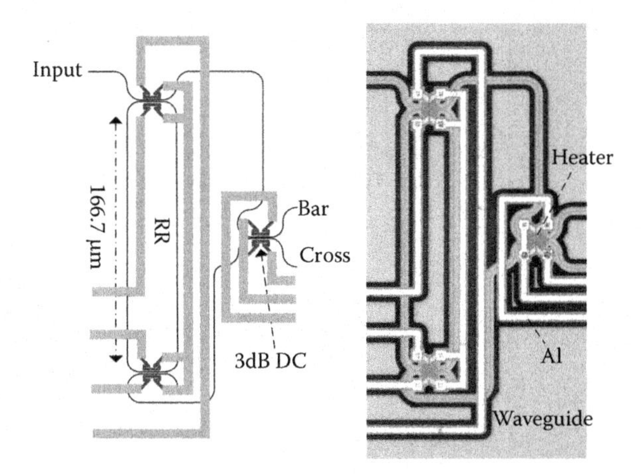

FIGURE 4.50
(a) Design layout of the full RR-MZI interleaver structure. The black lines represent the silicon waveguide, the dark gray lines represent the Ti heater, and the gray lines represent the Al. (b) Optical micrograph of the RR-MZI interleaver, where the dark regions are deep trenches. (Reprinted with permission from Song, J. et al., *IEEE Photonics Technology Letters*. **20**(24): 2165–2167, © 2008 IEEE.)

lines represent the Ti heater. The green lines represent the upper layer aluminum (Al). Figure 4.50b shows the optical micrograph of the fabricated device. To minimize the power consumption, the underneath silicon substrate has been removed for heat concentration, as shown in the dark regions in Figure 4.50b. The total length of the device was 2.9 mm.

Figure 4.51a,b shows the measured transmission spectra upon 0- and 7.5-V DC voltage supplies across the 3-dB coupler. Without dynamic tuning the 3-dB coupler, the measured crosstalk was only ~10 dB. The output intensities from the two ports are not identical as expected due to the fabrication imperfection. In the current device, the intensity difference increased from ~1 dB at 1550 nm to ~5 dB at 1590 nm. This suggests the necessity of dynamic tuning of the 3-dB coupler. Figure 4.51b shows the measured transmission spectra of the RR-MZI interleaver with 7.5-V (~25.5 mW) voltage applied to the 3-dB DC. At 1570 nm, the crosstalk is improved from −12 to −22 dB.

Figure 4.52a shows the SEM of the CR-MZI interleaver with TO-tunable 3-dB DCs. Figure 4.52b shows the optical micrograph of the final device integrated with thermal heaters. All three DCs are integrated with Ti thermal heater for dynamic tuning to compensate the fabrication imperfection induced deviation. The length of the 3-dB DC is 11 μm. The coupling length between the waveguide and the cross-ring is 15 μm. The width of the titanium wire is 500 nm and the gap separation between the wires is also 500 nm. The total length of the device is 2.9 mm.

Figure 4.53a,b show the measured transmission spectra of the CR-MZI interleavers without and with 9-V voltage supply to the 3-dB DC. Without voltage supply, the crosstalk is only approximately −5 dB. While with 9-V voltage supply to the 3-dB DC, the interleaver transmission performance is significantly enhanced. The transmission characteristics are uniform for bar and cross transmissions. The crosstalk is improved to approximately −20 dB. However, the transmission spectra clearly show the crosstalk for both short- and long-wavelength ranges were relatively small, which we attribute to the wavelength dependence of both the directional coupler and the waveguide-to-microring coupler.

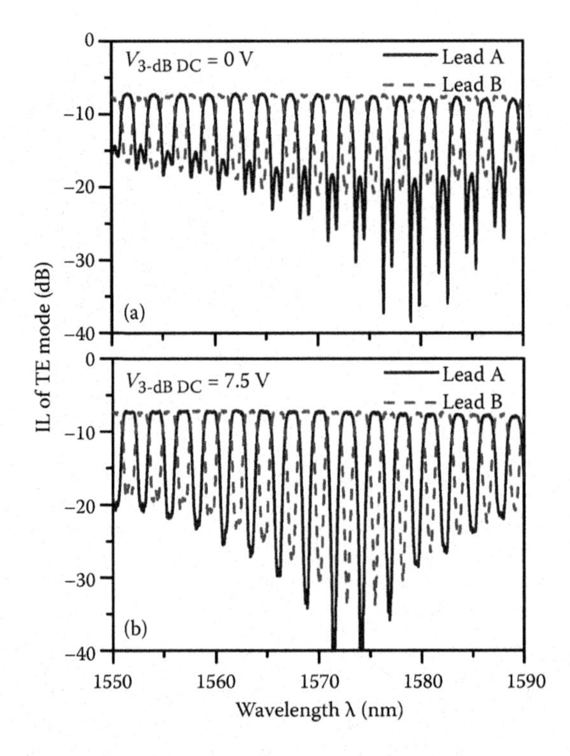

FIGURE 4.51
Measured transmission spectra of the 1 × 2 RR-MZI interleaver with voltage supplies of (a) 0 V and (b) 7.5 V to the 3-dB DC. (Reprinted with permission from Song, J. et al., *IEEE Photonics Technology Letters*. **20**(24): 2165–2167, © 2008 IEEE).

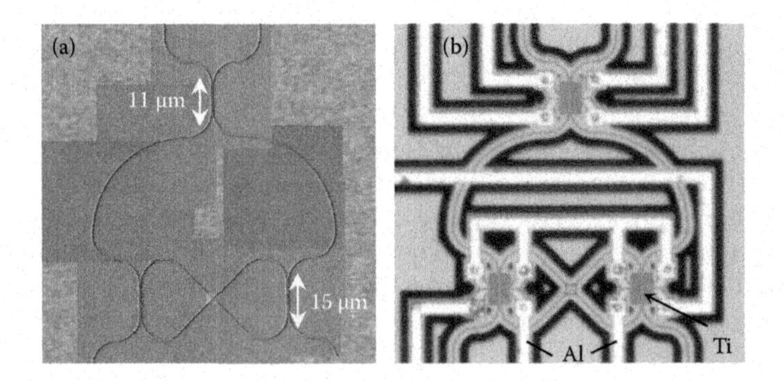

FIGURE 4.52
(a) The SEM picture of the passive cross-ring interleaver. (b) The optical micrograph of the fabricated cross-ring interleaver. (Reprinted from Song, J. et al., *Optics Express*, 2008. **16**(26): 21476–21482. With permission from Optical Society of America.)

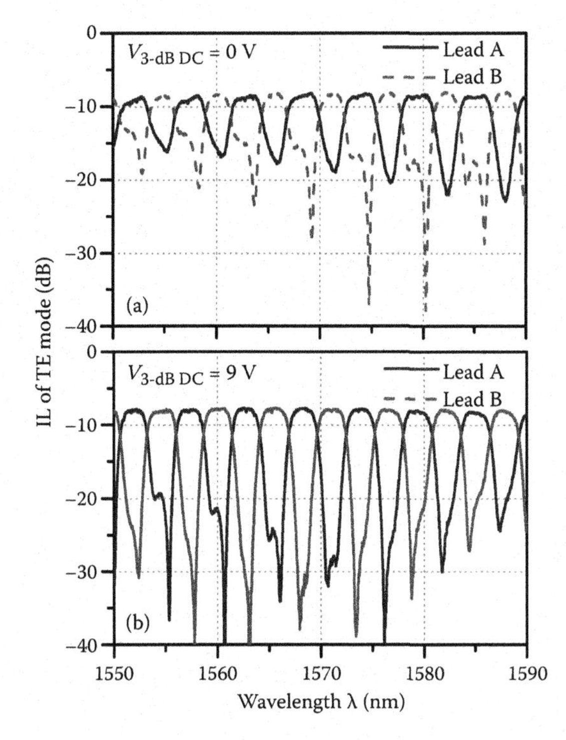

FIGURE 4.53
The measured transmission spectra from a CR-MZI interleaver with (a) 0 V and (b) 9 V applied voltages to the 3-dB coupler. (Reprinted from Song, J. et al., *Optics Express*, 2008. **16**(26): 21476–21482. With permission from Optical Society of America.).

TABLE 4.3

Key Measured Performance of Optical Interleavers in SOI

Structure	FSR (GHz)	Insertion Loss (dB)	Crosstalk (dB)	3-dB Bandwidth
Three-microring coupled asymmetric MZI [11]	250	~−8[a]	−20	0.48FSR
Ring-assisted MZI [8, 9]	200	~−8[a]	−17	0.46FSR
Ring-resonator MZI [7, 9]	340	~−10[a]	−22	0.52FSR
Cross-ring MZI [10]	680	~−8[a]	−20	0.45FSR

[a] Including the input/output coupling losses.

Table 4.3 summarizes the state-of-the-art demonstrations of the interleavers in SOI, in comparison with the examples discussed above.

4.6 Summary and Future Outlook

In this chapter, we have summarized the basic principles and some analysis and experimental results of two main types of multichannel silicon photonic devices, namely waveguide grating– and microring resonator–based devices. We have demonstrated their promising applications by examples as multiplexers, demultiplexers, add–drop filters, and interleavers. We have shown that microring resonators and MZIs have their own advantages and disadvantages. Microring resonators are more compact (approximately tens of micrometers) than MZI-based devices that are typically in the millimeter size range. While MZIs are more robust to thermal effects and fabrication errors compared with microring resonators. Needless to say their potential applications for multichannel communications systems, optical interconnections, and other technological areas such as sensing should go beyond what we have covered here.

However, despite the promise of low cost and large-scale photonic integration offered by silicon photonics, one salient feature among all the experimentally demonstrated multichannel silicon photonic devices we discussed here is that they currently show relatively inferior performance compared with their more mature counterparts, such as the silica-on-silicon platform. This leaves much room for researchers and newcomers to continue to perfect the existing technologies and explore new ground. Here we suggest some future research directions that, if properly addressed, can make multichannel silicon photonic devices more practical:

1. Fabrication tolerance. The submicron-sized silicon waveguides and the microring resonators have tight fabrication constraints. The optical performances rely strongly on fabrication accuracy. Optimization or additional compensation schemes (such as EO [6] or TO effects [1]) are required to attain the designed performances, which render more complicated designs. The need for active tuning also imposes additional power consumption, which is undesirable especially for low-power applications in WDM systems and optical interconnects.
2. Polarization dependence. Silicon optical devices with submicron waveguides are intrinsically polarization dependent, either due to the waveguide polarization-dependent loss or the polarization-dependent coupling between waveguides. Silicon microring resonators are particularly polarization dependent due to the two sets of TE and TM modes. Thus, polarization conversion structures such as optical rotators [103] or polarization combiners [104] are required, which will again complicate the device design and occupy extra footprints.
3. Optical fiber-to-waveguide coupling. The intrinsic mode mismatch between the optical fiber and the submicron silicon waveguide results in low coupling efficiency. Although various mode-size converters such as inverted tapers have been adopted, the coupling loss is with the best value of ~2-dB/facet [105], which is still a significant power loss in an optical link.

With the solutions to the above practical challenges and many other related fundamental/engineering challenges of silicon photonics unfolded, it is hopeful that, in the near future, we will find live data streaming through multichannel silicon photonic devices in the next-generation Internet and energy-efficient computers.

Acknowledgments

A.W.P. acknowledges the many contributions from his present and former students in the Photonic Device Laboratory, HKUST, over the past decade. Particularly, the thesis work by Drs. Chao Li, Linjie Zhou, Hui Chen, and Xianshu Luo has laid the basis of many silicon microring resonator–based devices reviewed in this chapter. A.W.P. acknowledges the research grants from the Research Grants Council of the Hong Kong Special Administrative Region. A.W.P. also acknowledges the fabrication support from the Nanoelectronics Fabrication Facility of HKUST.

G.Q.L. acknowledges the contributions from the research staff in Nano-Photonics Group and Semiconductor Process Technology Laboratory in Institute of Microelectronics (IME)/ASTAR in Singapore and the continuing support from the management of IME.

References

1. Reed, G.T., and A.P. Knights, *Silicon photonics: an introduction*. 2004, Chichester: Wiley.
2. Pavesi, L., and G. Guillot, *Optical interconnects: the silicon approach*. 2006, Berlin Springer-Verlag.
3. Feng, S., Lei, T., Chen, H., Cai, H., Luo, X., and Poon, A., Silicon photonics: from a microresonator perspective. *Laser & Photonics Reviews*. 2011: doi: 10.1002/lpor.201100020.
4. ePIXfab Silicon Photonics Platform (http://www.epixfab.eu/).
5. IME's Silicon Photonics MPW Prototyping (http://www.ime.a-star.edu.sg/PPSSite/index.asp).
6. Soref, R., and B. Bennett, Electrooptical effects in silicon. *IEEE Journal of Quantum Electronics*, 1987. **23**(1): p. 123–129.
7. Song, J. et al., Proposed silicon wire interleaver structure. *Optics Express*, 2008. **16**(11): p. 7849–7859.
8. Zhang, L. et al., Monolithic modulator and demodulator of differential quadrature phase-shift keying signals based on silicon microrings. *Optics Letters*, 2008. **33**(13): p. 1428–1430.
9. Song, J. et al., Thermo-optical enhanced silicon wire interleavers. *IEEE Photonics Technology Letters*, 2008. **20**(24): p. 2165–2167.
10. Song, J. et al., Effective thermo-optical enhanced cross-ring resonator MZI interleavers on SOI. *Optics Express*, 2008. **16**(26): p. 21476–21482.
11. Luo, L.W. et al., High bandwidth on-chip silicon photonic interleaver. *Optics Express*, 2010. **18**(22): p. 23079–23087.
12. Smit, M.K. New focusing and dispersive planar component based on an optical phased array. *Electronics Letters*, 1988. **24**(7): p. 385–386.
13. Takahashi, H. et al., Arrayed-waveguide grating for wavelength division multi/demultiplexer with nanometre resolution. *Electronics Letters*, 1990. **26**(2): p. 87–88.
14. Dragone, C., An N* N optical multiplexer using a planar arrangement of two star couplers. *IEEE Photonics Technology Letters*, 1991. **3**(9): p. 812–815.
15. Pearson, M.R. et al. Arrayed waveguide grating demultiplexers in silicon-on-insulator. *Proceeding of the SPIE 3953, Silicon-based Optoelectronics II*, 2000.
16. Oguma, M. et al., Passband-width broadening design for WDM filter with lattice-form interleave filter and arrayed-waveguide gratings. *IEEE Photonics Technology Letters*, 2002. **14**(3): p. 328–330.
17. Fang, Q., F. Li, and Y. Liu, Compact SOI arrayed waveguide grating demultiplexer with broad spectral response. *Optics Communications*, 2006. **258**(2): p. 155–158.

18. Xia, F. et al., Coupled resonator optical waveguides based on silicon-on-insulator photonic wires. *Applied Physics Letters*, 2006. **89**(4): p. 041122.
19. Xia, F.N. et al., Ultra-compact high order ring resonator filters using submicron silicon photonic wires for on-chip optical interconnects. *Optics Express*, 2007. **15**(19): p. 11934–11941.
20. Li, Q. et al., Design and demonstration of compact, wide bandwidth coupled-resonator filters on a silicon-on-insulator platform. *Optics Express*, 2009. **17**(4): p. 2247–2254.
21. Banerjee, A. et al., Wavelength-division-multiplexed passive optical network (WDM-PON) technologies for broadband access: a review [invited]. *Journal of Optical Networking*, 2005. **4**(11): p. 737–758.
22. Nolting, H.P., and M. Gravert, Electro-optically controlled multiwavelength switch for WDM cross connector application. *IEEE Photonics Technology Letters*, 1995. **7**(3): p. 315–317.
23. Okamoto, K., K. Takiguchi, and Y. Ohmori, 16-Channel optical add/drop multiplexer using silica-based arrayed-waveguide gratings. *Electronics Letters*, 1995. **31**(9): p. 723–724.
24. Klein, E.J. et al., Reconfigurable optical add–drop multiplexer using microring resonators. *IEEE Photonics Technology Letters*, 2005. **17**(11): p. 2358–2360.
25. Feuer, M.D., D.C. Kilper, and S.L. Woodward, ROADMs and their system applications, in *Optical Fiber Telecommunications 5*, T.L. Ivan P. Kaminow, Alan E. Willner, Editor. 2008, Academic Press. p. 293.
26. Chen, W. et al., Monolithically integrated 32x four-channel client reconfigurable optical add/drop multiplexer on planar lightwave circuit. *IEEE Photonics Technology Letters*, 2003. **15**(10): p. 1413–1415.
27. Marom, D. et al. *Wavelength-selective 1×4 switch for 128 WDM channels at 50 GHz spacing*. 2002: Optical Society of America.
28. Geuzebroek, D. et al., Compact wavelength-selective switch for gigabit filtering in access networks. *IEEE Photonics Technology Letters*, 2005. **17**(2): p. 336–338.
29. Beutler, H., The theory of the concave grating. *JOSA*, 1945. **35**(5): p. 311–350.
30. Namioka, T., Theory of the concave grating. *JOSA*, 1959. **49**(5): p. 446–460.
31. Fujii, Y., and J. Minowa, Optical demultiplexer using a silicon concave diffraction grating. *Applied Optics*, 1983. **22**(7): p. 974–978.
32. Brouckaert, J. et al., Planar concave grating demultiplexer fabricated on a nanophotonic silicon-on-insulator platform. *Journal of Lightwave Technology*, 2007. **25**(5): p. 1269–1275.
33. Little, B.E. et al., Microring resonator channel dropping filters. *Journal of Lightwave Technology*, 1997. **15**(6): p. 998–1005.
34. Cao, S. et al., Interleaver technology: Comparisons and applications requirements. *Journal of Lightwave Technology*, 2004. **22**(1): p. 281.
35. de Ridder, R.M., and C.G.H. Roeloffzen, Interleavers, in *Wavelength Filters in Fibre Optics*, H. Venghaus, Editor. 2010, Springer.
36. Popovic, M., *Theory and design of high-index-constrast microphotonic circuits*. 2008, Massachusetts Institute of Technology.
37. Murphy, T.E., J.T. Hastings, and H.I. Smith, Fabrication and characterization of narrow-band Bragg-reflection filters in silicon-on-insulator ridge waveguides. *Journal of Lightwave Technology*, 2001. **19**(12): p. 1938–1942.
38. Xu, F., and A.W. Poon, Silicon cross-connect filters using microring resonator coupled multimode-interference-based waveguide crossings. *Optics Express*, 2008. **16**: p. 8649–8657.
39. Zheng, S., H. Chen, and A.W. Poon, Microring-resonator cross-connect filters in silicon nitride: Rib waveguide dimensions dependence. *IEEE Journal on Selected Topics in Quantum Electronics*, 2006. **12**(6): p. 1380–1387.
40. Jalali, B., and S. Fathpour, *Silicon photonics for telecommunications & biomedical applications*. 2010, Boca Raton, FL: CRC Press.
41. Reed, G., *Silicon photonics: the state of the art*. 2008, Chichester: Wiley.
42. Sean Koehl, A.L.a.M.P., Integrated silicon photonics: harnessing the data explosion, in *OPN (Optics & Photonics News)*. 2010. p. 6.

43. Miller, D., Device requirements for optical interconnects to silicon chips. *Proceedings of the IEEE*, 2009. **97**(7): p. 1166–1185.
44. Ohashi, K. et al., On-chip optical interconnect. *Proceedings of the IEEE*, 2009. **97**(7): p. 1186–1198.
45. Krishnamoorthy, A.V. et al., Computer Systems Based on Silicon Photonic Interconnects. *Proceedings of the IEEE*, 2009. **97**(7): p. 1337–1361.
46. Miller, D.A.B., Device requirements for optical interconnects to silicon chips. *Proceedings of the IEEE*, 2009. **97**(7): p. 1166–1185.
47. Poon, A.W. et al., Cascaded microresonator-based matrix switch for silicon on-chip optical interconnection. *Proceedings of the IEEE*, 2009. **97**(7): p. 1216–1238.
48. Van Thourhout, D. et al., Nanophotonic devices for optical interconnect. *IEEE Journal of Selected Topics in Quantum Electronics*, 2010. **16**(5): p. 1363–1375.
49. Rowland, H.A., Preliminary notice of the results accomplished in the manufacture and theory of gratings for optical purposes. *Philosophical Magazine Series* 5, 1882. **13**(84): p. 469–474.
50. Rowland, H.A., On concave gratings for optical purposes. *London, Edinburgh, and Dublin Philosophical Magazine and Journal of Science*, 1883. **16**(99): p. 197–210.
51. Janz, S. et al., Planar waveguide echelle gratings in silica-on-silicon. *IEEE Photonics Technology Letters*, 2004. **16**(2): p. 503–505.
52. Kwon, O.K. et al., InP-based polarization-insensitive planar waveguide concave grating demultiplexer with flattened spectral response. *ETRI Journal*, 2009. **31**(2): p. 228–230.
53. Brouckaert, J. et al., Planar concave grating demultiplexer with high reflective Bragg reflector facets. *IEEE Photonics Technology Letters*, 2008. 20(4): p. 309–311.
54. Okamoto, K. et al., Fabrication of 128-channel arrayed-waveguide grating multiplexer with 25 GHz channel spacing. *Electronics Letters*, 1996. **32**(16): p. 1474–1476.
55. Hida, Y. et al., Fabrication of low-loss and polarisation-insensitive 256 channel arrayed-waveguide grating with 25 GHz spacing using 1.5% waveguides. *Electronics Letters*, 2000. **36**(9): p. 820–821.
56. Takada, K. et al., A. 25 GHz-spaced 1080-channel tandem multi/demultiplexer covering the S-, C-, and L-bands using an arrayed-waveguide grating with Gaussian passbands as a primary filter. *IEEE Photonics Technology Letters*, 2002. **14**(5): p. 648–650.
57. Zirngibl, M., C. Dragone, and C. Joyner, Demonstration of a 15 × 15 arrayed waveguide multiplexer on InP. *IEEE Photonics Technology Letters*, 1992. **4**(11): p. 1250–1253.
58. Kohtoku, M. et al., InP-based 64-channel arrayed waveguide grating with 50 GHz channel spacing and up to–20 dB crosstalk. *Electronics Letters*, 1997. **33**(21): p. 1786–1787.
59. Lin, Y.H., and S.L. Tsao, Improved design of a 64 × 64 arrayed waveguide grating based on silicon-on-insulator substrate. *IEEE Proceedings—Optoelectronics*, 2006. **153**: p. 57–62.
60. Bogaerts, W. et al., Compact wavelength-selective functions in silicon-on-insulator photonic wires. *IEEE Journal of Selected Topics in Quantum Electronics*, 2006. **12**(6): p. 1394–1401.
61. Cheben, P. et al., A high-resolution silicon-on-insulator arrayed waveguide grating microspectrometer with sub-micrometer aperture waveguides. *Optics Express*, 2007. **15**(5): p. 2299–2306.
62. Kim, D.J. et al., Crosstalk reduction in a shallow-etched silicon nanowire AWG. *IEEE Photonics Technology Letters*, 2008. **20**(19): p. 1615–1617.
63. Fang, Q. et al., Monolithic integration of a multiplexer/demultiplexer with a thermo-optic VOA array on an SOI platform. *IEEE Photonics Technology Letters*, 2009. **21**(5): p. 319–321.
64. Fang, Q. et al., WDM multi-channel silicon photonic receiver with 320 Gbps data transmission capability. *Optics Express*, 2010. **18**(5): p. 5106–5113.
65. Song, J. et al., CWDM Planar Concave Grating Multiplexer/Demultiplexer and Application in ROADM, in *Optical Fiber Communication Conference, 2010*, Optical Society of America. Paper JThA20.
66. R. Doerr, C., and K. Okamoto, Planar lightwave circuits in fiber-optic communications in *Optical Fiber Telecommunications V A: Components and Subsystems*, I.P. Kaminow, T. Li, and A.E. Willner, Editors. 2008, Academic Press.
67. Miller, F.P., A.F. Vandome, and J. McBrewster, *Arrayed waveguide grating*. 2010: VDM Publishing House Ltd.

68. Leijtens, X.J.M., B. Kuhlow, and M.K. Smit, Arrayed waveguide grating, in *Wavelength Filters in Fibre Optics*, H. Venghaus, Editor. 2010, Springer.
69. Takada, K. et al., A. 2.5 GHz-spaced 1080-channel tandem multi/demultiplexer covering the. S-, C-, and L-bands using an arrayed-waveguide grating with Gaussian passbands as a primary filter. *IEEE Photonics Technology Letters*, 2002. **14**(5): p. 648–650.
70. Popovic, M.A. et al. Tunable, fourth-order silicon microring-resonator add–drop filters, in *33rd European Conference and Exhibition on Optical Communication ECOC 2007* (2007).
71. Dong, P. et al., Low power and compact reconfigurable multiplexing devices based on silicon microring resonators. *Optics Express*, 2010. **18**(10): p. 9852–9858.
72. Klein, E.J. et al., Densely integrated microring resonator based photonic devices for use in access networks. *Optics Express*, 2007. **15**(16): p. 10346–10355.
73. Klein, E.J. et al., Reconfigurable optical add–drop multiplexer using microring resonators. *IEEE Photonics Technology Letters*, 2005. **17**(11): p. 2358–2360.
74. Xu, Q. et al., Micrometre-scale silicon electro-optic modulator. *Nature*, 2005. **435**(7040): p. 325–327.
75. Xia, F., L. Sekaric, and Y. Vlasov, Ultracompact optical buffers on a silicon chip. *Nature Photonics*, 2007. **1**(1): p. 65–71.
76. Rayleigh, L., The problem of the whispering gallery. *Philosophical Magazine* 1910. 20: p. 1001–1004.
77. Vahala, K.J., Optical microcavities. *Nature*, 2003. **424**(6950): p. 839–846.
78. Ma, N., C. Li, and A.W. Poon, Laterally coupled hexagonal micropillar resonator add–drop filters in silicon nitride. *IEEE Photonics Technology Letters*, 2004. **16**(11): p. 2487–2489.
79. Luo, X.S., and A.W. Poon, Coupled spiral-shaped microdisk resonators with non-evanescent asymmetric inter-cavity coupling. *Optics Express*, 2007. **15**(25): p. 17313–17322.
80. Li, C. et al., Silicon polygonal microdisk resonators. *IEEE Journal of Selected Topics in Quantum Electronics*, 2006. **12**(6): p. 1438–1449.
81. Che, K., and Y. Huang, Multimode resonances in metallically confined square-resonator microlasers. *Applied Physics Letters*, 2010. **96**: p. 051104.
82. Huang, Y. et al., Optical bistability in InP/GaInAsP equilateral-triangle-resonator microlasers. *Optics Letters*, 2009. **34**(12): p. 1852–1854.
83. Jackson, J.D., *Classical electrodynamics*: John Wiley & Sons.
84. Haus, H.A., *Waves and fields in optoelectronics*. 1984, New Jersey: Prentice-Hall.
85. Yariv, A., and P. Yeh, *Photonics: optical electronics in modern communications*. 6th ed. 2006: Oxford University Press, Inc. New York, USA.
86. Yariv, A., Universal relations for coupling of optical power between microresonators and dielectric waveguides. *Electronics Letters*, 2000. **36**(4): p. 321–322.
87. Xu, D.X. et al., High bandwidth SOI photonic wire ring resonators using MMI couplers. *Optics Express*, 2007. **15**(6): p. 3149–3155.
88. Hui, C., and A.W. Poon, Design of silicon MMI crossing laterally coupled microring resonators, in *Conference on Lasers and Electro-Optics and 2006 Quantum Electronics and Laser Science Conference, CLEO/QELS, 2006* (2006).
89. Smith, D.D., H.R. Chang, and K.A. Fuller, Whispering-gallery mode splitting in coupled microresonators. *Journal of the Optical Society of America B—Optical Physics*, 2003. **20**(9): p. 1967–1974.
90. Capmany, J. et al., Apodized coupled resonator waveguides. *Optics Express*, 2007. **15**(16): p. 10196–10206.
91. Chak, P., and J. Sipe, Minimizing finite-size effects in artificial resonance tunneling structures. *Optics Letters*, 2006. **31**(17): p. 2568–2570.
92. Xiao, S. et al., A highly compact third-order silicon microring add–drop filter with a very large free spectral range, a flat passband and a low delay dispersion. *Optics Express*, 2007. **15**(22): p. 14765–14771.
93. Barwicz, T. et al., Microring-resonator-based add–drop filters in SiN: fabrication and analysis. *Optics Express*, 2004. **12**(7): p. 1437–1442.
94. Xia, F. et al., Ultra-compact high order ring resonator filters using submicron silicon photonic wires for on-chip optical interconnects. *Optics Express*, 2007. **15**(19): p. 11934–11941.

95. Li, Q. et al., Design and demonstration of compact, wide bandwidth coupled-resonator filters on a silicon-on-insulator platform. *Optics Express*, 2009. **17**(4): p. 2247–2254.
96. Park, S. et al., Si micro-ring MUX/DeMUX WDM filters. *Optics Express*, 2011. 19(14): p. 13531–13539.
97. Boeck, R. et al., Series-coupled silicon racetrack resonators and the Vernier effect: theory and measurement. *Optics Express*, 2010. **18**(24): p. 25151–25157.
98. Fano, U., Effects of configuration interaction on intensities and phase shifts. *Physical Review*, 1961. **124**(6): p. 1866–1878.
99. Zhou, L.J., and A.W. Poon, Fano resonance-based electrically reconfigurable add–drop filters in silicon microring resonator-coupled Mach-Zehnder interferometers. *Optics Letters*, 2007. **32**(7): p. 781–783.
100. Wörhoff, K. et al., Design and application of compact and highly tolerant polarization-independent waveguides. *Journal of Lightwave Technology*, 2007. **25**(5): p. 1276–1283.
101. Wang, Z. et al., A high-performance ultracompact optical interleaver based on double-ring assisted Mach–Zehnder interferometer. *IEEE Photonics Technology Letters*, 2007. **19**(14): p. 1072–1074.
102. Madsen, C., and J. Zhao, *Optical filter design and analysis: A signal processing approach*. 1999: John Wiley & Sons, Inc. New York, USA.
103. Morichetti, F., A. Melloni, and M. Martinelli, Effects of polarization rotation in optical ring-resonator-based devices. *Journal of Lightwave Technology*, 2006. **24**(1): p. 573–585.
104. Bogaerts, W. et al., A polarization-diversity wavelength duplexer circuit in silicon-on-insulator photonic wires. *Optics Express*, 2007. **15**(4): p. 1567–1578.
105. Tsuchizawa, T. et al., Microphotonics devices based on silicon microfabrication technology. *IEEE Journal of Selected Topics in Quantum Electronics*, 2005. **11**(1): p. 232–240.

5

Nonlinear Optics in Silicon

Ozdal Boyraz, Xinzhu Sang, Massimo Cazzanelli, and Yuewang Huang

CONTENTS

5.1 Introduction

The intrinsic high optical nonlinearity originating from tight optical mode confinement due to inherent high index contrast between the core and cladding, along with the potential ability of dense on-chip integration with microelectronic circuits, made silicon photonics one of the rapidly growing research areas. The Kerr coefficient and Raman gain coefficient of silicon exhibits more than 200 and 1000 times larger than those of silica glass, which enable efficient nonlinear interactions of optical waves at relatively low power levels inside a short SOI waveguide. In the last several years, considerable efforts have been made to investigate the nonlinear phenomena such as self-phase modulation (SPM), cross-phase modulation (XPM), stimulated Raman scattering (SRS), four-wave mixing (FWM), nonlinear optical absorption, and so on. Almost all nonlinear properties in silicon are currently being explored to realize a variety of optical functional devices on the chip scale. Compared with the silica fibers, the silicon is a semiconductor crystal exhibiting unique features such as two photon absorption (TPA), FCD, and anisotropic and dispersive third-order nonlinearity. For example, stimulated Raman scattering (SRS) is employed to make optical amplifiers and Raman lasers. Kerr effects are successfully used for nonlinear signal processing, such as SPM and XPM. FWM as one of the Kerr effects has been used to make wavelength converters, optical parametric amplifiers and signal generators. TPA and TPA-induced free carrier generation are suitable for all-optical switching, modulation, pulse compression, and pulse characterization. In addition, nonlinear optical effects can be significantly enhanced in engineered microstructured silicon waveguides, which opens a new subfield in nonlinear optics for exploration.

Recently a new window of research on second-order nonlinear effects in silicon opened. In conventional knowledge, silicon crystal possesses a crystalline symmetry that inhibits the existence of nonzero elements in the second-order nonlinear susceptibility $\chi^{(2)}$ tensor. Breaking this symmetry can make silicon one of the rare crystals where both $\chi^{(2)}$ and $\chi^{(3)}$ effects can be cultivated simultaneously. In this novel window of research, scientists discover stress-based techniques that can break the centrosymmetry and illustrate applications such as second harmonic generation.

In this chapter, the nonlinear optical mechanisms in silicon are presented, and the related applications are discussed. The chapter starts with the source of optical nonlinear effects and description of individual effects in a practical sense. Then it discusses the second-order nonlinear effects and approaches to generate second-order nonlinear effects. The last part of the chapter is dedicated to the practical applications derived from third-order nonlinear effects.

5.2 Nonlinear Optical Effects in Silicon

5.2.1 Source of Optical Nonlinearities

Electromagnetic waves interacting with atoms involves nonlinearity. When the interaction of the light field is described by Maxwell's equations, the interactions of the light with the silicon waveguide contain nonlinear components. Silicon-on-insulator (SOI) has become a promising material for nonlinear photonic devices on chip. High intensity optical waves in the silicon waveguide core can be nonlinear interactions [1–3]. In addition, the silicon core can be embedded in nonlinear cladding material, and interaction with the evanescent part of the guided mode can be enhanced [4,5], where the fraction of optical power guided in the low-index region can be maximized by appropriate waveguide dimensions [6]. Because of its crystal inversion symmetry, the lowest-order nonlinearity in crystalline silicon originates from third-order susceptibility. In the presence of a single and two input signals, the third-order nonlinear polarization P related TPA in the silicon waveguide is given as [7]

$$P^{(3)}(\omega) = \varepsilon_0 \frac{3}{4}\chi^{(3)}(-\omega;\omega,-\omega,\omega)\left|E_\omega\right|^2 E_\omega \tag{5.1}$$

$$P^{(3)}(\omega_2) = \varepsilon_0 \frac{3}{2}\chi^{(3)}(-\omega_2;\omega_1,-\omega_1,\omega_2)\left|E_{\omega_1}\right|^2 E_{\omega_2} \tag{5.2}$$

where ε_0 is the permittivity of free space and $\chi^{(3)}$ is third-order susceptibility tensor. The former equation describes Kerr effect and degenerate TPA, and the later corresponds to nondegenerate Kerr and TPA effects. TPA and Kerr effects are related to the imaginary part, and the real part of $\chi^{(3)}$ that correspond to the optical intensity dependence of the optical absorption and the refractive index, respectively.

The strength of third-order nonlinear interaction in a waveguide is normally described by the nonlinear coefficient γ, and the real part depends on the waveguide geometry as well as on the nonlinear index coefficient n_2 of the material. To optimize the waveguide dimensions for maximal nonlinear interaction. The effective mode area for nonlinear interaction A_{eff} is an elevation based on a scalar approximation of the modal field. The actual cross-sectional power P related to the effective core area A_{eff} accounts then for the nonlinear deviation n_2P/A_{eff} from the linear effective refractive index of the waveguide mode.

Figure 5.1 shows examples for cross sections of the waveguides. The core domain D core consists of silicon ($n_{core} \approx 3.45$ for $\lambda = 1.55$ μm), the substrate domain D_{sub} is made out of silica ($n_{sub} \approx 1.44$), and the cover domain D_{cover} comprises a cladding material with refractive index $n_{cover} < n_{core}$. For the strip waveguide in Figure 5.1a, nonlinear interaction can either occur within the waveguide core ("core nonlinearity"), or the evanescent part of the guided light interacts with a nonlinear cover material ("cover nonlinearity"). The slot waveguide depicted in Figure 5.1b enables particularly strong nonlinear interaction of the guided wave with the cover material inside the slot.

To maximize the nonlinear interaction in strip or slot waveguides, a set of optimal geometry parameters w and h must be properly chosen. An increase of the waveguide cross section will decrease the intensity inside the core and weaken the nonlinear interaction, and vice versa.

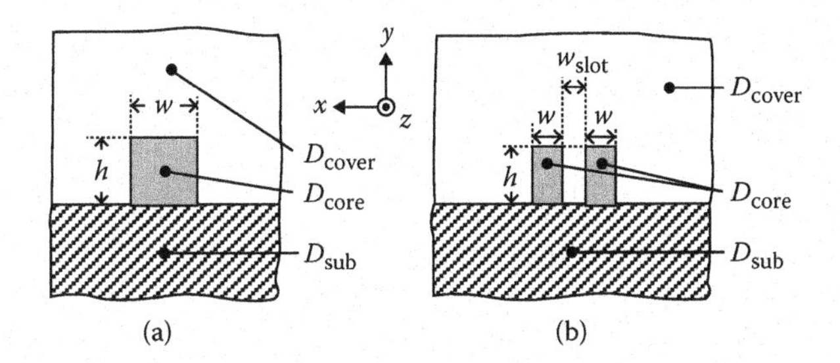

FIGURE 5.1
Waveguide cross sections: (a) Strip waveguide and (b) slot waveguide. (From C. Koos, L. Jacome, C. Poulton, J. Leuthold, and W. Freude, *Opt. Express* 15, 5976–5990, 2007. With permission.)

Third-order nonlinearities are especially important in silicon as they exhibit a wide variety of phenomena, which can be easily demonstrated for an electric field E comprising three frequency components (ω_k):

$$E(r,t) = \sum_{k=1}^{3} E_k = \frac{1}{2}\sum_{k=1}^{3}\left(E_{\omega_k k}(r,\omega_k)e^{i\omega_k t} + c.c.\right) \tag{5.3}$$

where c.c. denotes the complex conjugate. By expanding the frequency components of the third-order nonlinear term $\chi^{(3)}$, several terms at new frequencies for the third-order polarization $P^{(3)}$ [9] are generated:

$$\begin{aligned}
P^{(3)} = &\frac{3}{4}\varepsilon_0\chi^{(3)}\left[\left|E_{\omega_1}\right|^2 E_1 + \therefore\right]\text{SPM} \\
&+\frac{6}{4}\varepsilon_0\chi^{(3)}\left[\left(\left|E_{\omega 2}\right|^2 + \left|E_{\omega 3}\right|^2\right)E_1 + \therefore\right]\text{XPM} \\
&+\frac{1}{4}\varepsilon_0\chi^{(3)}\left[\left(E_{\omega_1}^2 e^{i3\omega_1 t} + c.c.\right) + \therefore\right]\text{THG} \\
&+\frac{3}{4}\varepsilon_0\chi^{(3)}\left[\frac{1}{2}\left(E_{\omega_1}^2 E_{\omega_2} e^{i(2\omega_1+\omega_2)t} + c.c.\right) + \therefore\right]\text{FWM} \\
&+\frac{3}{4}\varepsilon_0\chi^{(3)}\left[\frac{1}{2}\left(E_{\omega_1}^2 E_{\omega_2}^* e^{i(2\omega_1-\omega_2)t} + c.c.\right) + \therefore\right]\text{FWM} \\
&+\frac{6}{4}\varepsilon_0\chi^{(3)}\left[\frac{1}{2}\left(E_{\omega_1} E_{\omega_2} E_{\omega_3}^* e^{i(\omega_1+\omega_2-\omega_3)t} + c.c.\right) + \therefore\right]\text{FWM} \\
&+\frac{6}{4}\varepsilon_0\chi^{(3)}\left[\frac{1}{2}\left(E_{\omega_1} E_{\omega_2} E_{\omega_3} e^{i(\omega_1+\omega_2+\omega_3)t} + c.c.\right) + \therefore\right]\text{FWM}
\end{aligned} \tag{5.4}$$

where the symbol $\therefore$ accounts for possible permutations of frequencies. The terms in Equation 5.4 correspond to phenomenon called SPM, XPM, third-harmonic generation (THG), and FWM. Each of the terms on the right-hand side of Equation 5.4 corresponds to a nonlinear optical excitation, and only those maintaining both energy and momentum conservation result in efficient excitation. Some nonlinear processes available can be selected to some degree by choosing the energy levels and phase-matching appropriately.

5.2.2 Nonlinearity and Mode Confinement

The simulated electric field distribution in a cross section of a SOI nanowaveguide can be well confined due to the large numerical aperture. The small mode effective area increases the nonlinear coefficient of the waveguide, and the mode confinement allows to flexibly tailor the waveguide dispersion.

As an example, a glass-rod-in-air confining 800-nm light and a rectangular 1:1.5 aspect ratio embedded SOI waveguide confining 1550-nm light in the TE polarization mode are shown in Figure 5.2.

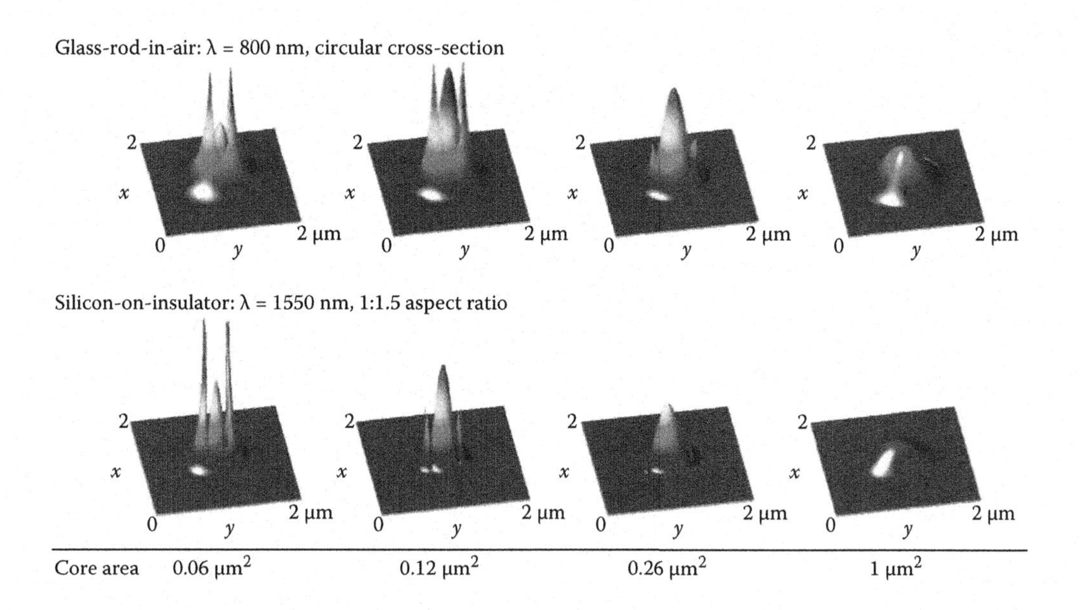

FIGURE 5.2
Mode intensity profile inside glass-rod-in-air and SOI for various submicron sizes and wavelength of 800 nm (glass-rod-in-air) and 1550 nm (SOI). (From M. A. Foster, A. C. Turner, M. Lipson, and A. L. Gaeta, *Opt. Express* 16, 1300–1320, 2008. With permission.)

For core areas of 1 μm^2, the mode is well confined to the core for both materials. As the dimensions are decreased below 1 μm, the mode area undergoes the same decrease. Eventually, as the waveguide dimensions are further reduced, the evanescent field appears to dominate, and the waveguide cannot tightly confine the light. The above condition is sooner in the glass-rod-in-air due to the lower index contrast. In the evanescent regime, the effective nonlinearity is diminished if the nonlinear material is assumed to be solely in the core. The mode confinement behavior determines the optimal size for the effective nonlinearity of the waveguide as illustrated in Figure 5.3 [8,11–16].

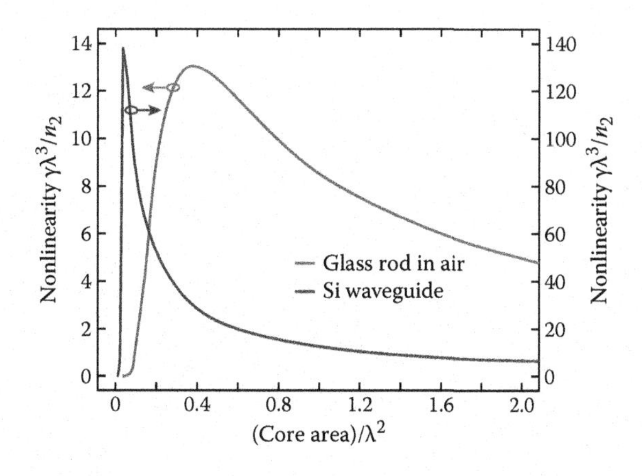

FIGURE 5.3
Nonlinearity as a function of core area of silica glass rod in air and a SOI platform. (From M. A. Foster, A. C. Turner, M. Lipson, and A. L. Gaeta, *Opt. Express* 16, 1300–1320, 2008. With permission.)

5.2.3 Nonlinear Optical Absorption

At telecommunication wavelength, two-photon absorption (TPA) in silicon is an instantaneous nonlinear loss mechanism dominating at high optical intensities, which are often required to induce Kerr and Raman nonlinearities. Free carrier absorption (FCA) induced by TPA will also introduce additional loss, so it has often been considered as a fundamental limitation for nonlinear silicon photonic devices in optical communication and information processing systems [17–21]. Here we present the main theoretical background of TPA in silicon photonics [22].

Figure 5.4 shows the schematic TPA process in silicon. When the total energy of input two photons is greater than the band-gap energy of silicon ($E_g = E_c - E_v \approx 1.12$ eV), they will be absorbed and excite an electron from the valence band to the conduction band, resulting in generation of a free carrier (electron-hole pair). Depending on the parity difference in the initial and final sates, three types of two-photon transitions can occur, which are referred to as allowed–allowed (a-a), allowed–forbidden (a-f), and forbidden–forbidden (f-f). The TPA coefficient $\beta(\omega) = [3\pi/(c\varepsilon_0\lambda n_0^2)]\text{Im}\{\chi^{(3)}\}$ is a fundamental parameter to evaluate TPA. Considering three types of two-photon processes, the total TPA coefficient is given by $\beta(\omega) = \sum_{n=0}^{2}\beta^{(n)}(\omega)$, with $n = 0$, 1, and 2 for a-a, a-f, and f-f transitions [23–25], respectively. Because $\beta^{(n)}(\omega) = 2CF_2^{(n)}(h\omega/E_g)$ incorporates a material-dependent constant C, the contribution of each type process is determined by the fundamental photon energy $h\omega$. In the expression of the above equation, $F_2^{(n)}(x) = \left[\pi(2n+1)!!/2^{n+2}(n+2)!\right](2x)^{-5}(2x-1)^{n+2}$, $x = h\omega E_g$ for parabolic electron and hole bands. Assuming that phonon energies are $\ll E_g$, the f-f process is weaker than the two other processes, but it gets its maximum at $h\omega \approx 5E_g/2$ [25]. The other two processes become stronger when photon energies are near E_g. In the degenerate TPA, two photons with the same wavelength are absorbed, and the power of the input light will be attenuated along the silicon waveguide. A lot of experimental and theoretical work related to degenerate TPA in semiconductors has been carried out. It should be noted that due to the off-diagonal elements in the third-order susceptibility tensor, the TPA exhibits optical polarization dependence, and the presence of TPA anisotropy in silicon was observed [26]. Up to now, z-scan technique and pulse transmission method are used to measure or elevate the TPA coefficient of silicon. The measured TPA coefficients at different wavelengths are summarized in Figure 5.5.

In the nondegenerate TPA process, the absorption of a signal is change by a second pump signal instead of by the signal itself, which means that a signal photon is absorbed as a

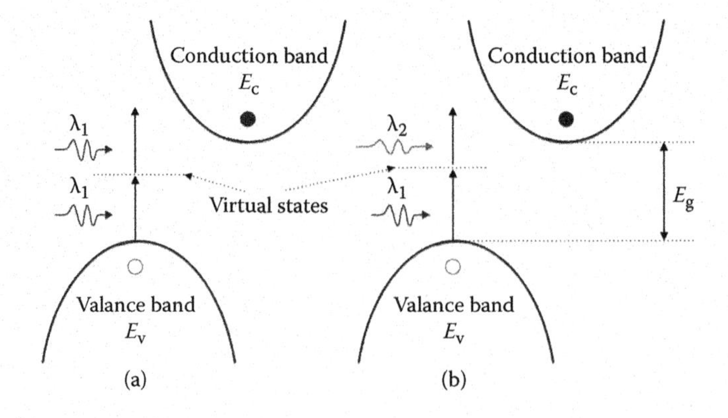

FIGURE 5.4
Schematic illustration of TPA in silicon: (a) degenerate TPA and (b) nondegenerate TPA. (From X. Sang, E. K. Tien, and O. Boyraz, *J. Optoelectron. Adv. Mater.* 11, 15–25, 2009. With permission.)

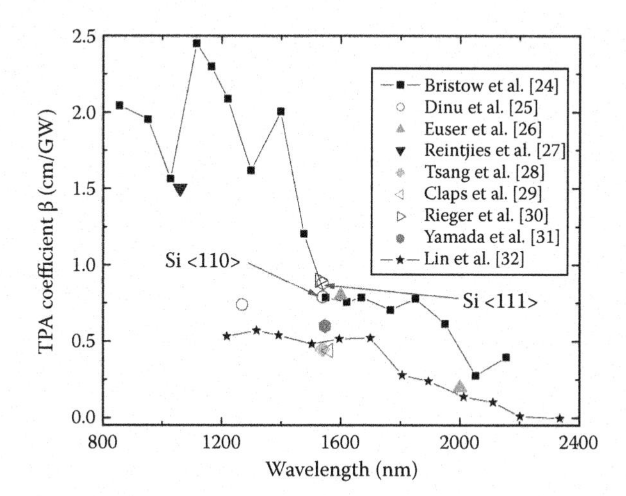

FIGURE 5.5
Measured TPA coefficients at different wavelengths from different literatures compiled in [22]. (From X. Sang, E. K. Tien, and O. Boyraz, *J. Optoelectron. Adv. Mater.* 11, 15–25, 2009. With permission.)

pump photon is present and leads to cross-absorption. From Equation 5.4, the nondegenerate TPA process is two times stronger than the degenerate TPA process. It should be pointed out that the nondegenerate TPA is polarization dependent and the linearly polarized beams show stronger polarization dependence than the circularly polarized ones [27,28].

TPA process in silicon leads to the generation of free carriers, whose density depends on the incident optical intensity, resulting in FCA loss and FCD. However, diffusion will reduce the carrier density at the center of the optical mode. The dynamics of free carriers induced by TPA can be described by the Drude model. Since the TPA-induced free carriers in the silicon waveguide vary slowly along the z propagation direction, lateral diffusion (along x axis) will dominantly impact the free carrier density. For CW signals or pulse signals whose pulse widths are wider than τ_0, free carrier density will be stabilized at a local position with value of $N(z) = \tau_0\beta I^2(z)/(2h\nu)$ after the τ_0 period. For pulse signals whose pulse widths are much narrower than τ_0, we must consider the repetition rate R of the pulse. When the pulse signals operate at low repetition rate ($R\tau_0 \ll 1$), the free carrier density is time dependent, and the peak density depends on the pulse width, just like a single pulse, as shown in Figure 5.6a.

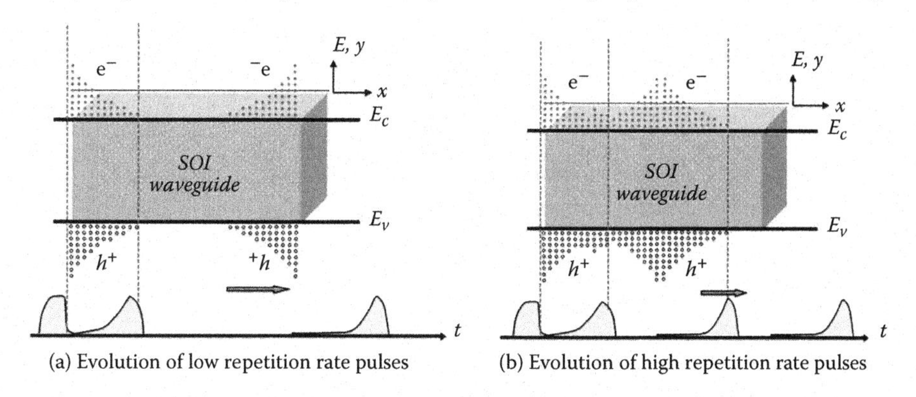

FIGURE 5.6
Evolutions of pulses inside the silicon waveguide: (a) when the repetition rate is low, (b) when the repetition rate is high. (From X. Sang, E. K. Tien, and O. Boyraz, *J. Optoelectron. Adv. Mater.* 11, 15–25, 2009. With permission.)

When the pulse repetition rate increases, those free carriers generated by the former pulses may not have enough time to recombine, which will impact the later pulses, as shown in Figure 5.6b.

5.2.4 Dispersion and Free Carrier Plasma Dispersion

5.2.4.1 Dispersion

The SOI platform has inherent advantages for all-optical devices due to the high-index contrast between the silicon core and silica cladding, allowing for strong optical confinement and large effective nonlinearities. Recently, such waveguides have been analyzed for different applications [29–31]. For these and other applications, the group-velocity dispersion (GVD) of the waveguide is important. The GVD influences the broadening of ultrafast pulses, the walk off between pump and probe pulses, the phase-matching of parametric processes, and the generation of temporal optical solitons.

As we know, the GVD in a waveguide is determined by both the intrinsic material dispersion and by a contribution from the confinement of the waveguide. For typical low-index contrast waveguides with core sizes on the order of a few microns, the GVD is characterized mainly by the material GVD with a small correction due to the waveguide contribution [32,33]. However, for large index contrast waveguides, the waveguide dispersion plays an increasing role, which allows for a high degree of GVD engineering. As the core is reduced to submicron dimensions in these high index contrast waveguides, the waveguide dispersion becomes dominant.

Making a rectangular 1:1.5 aspect ratio embedded SOI waveguide as a researching example, for various dimensions, the changes of waveguide GVD and net GVD are depicted by Figures 5.7 and 5.8, respectively [10]. Figure 5.7 demonstrates that as the core is reduced in size, the point at which the waveguide GVD changes from anomalous to normal shifts to shorter wavelengths. This behavior resulting from the waveguide dispersion is shared by all nanowires although for larger index contrasts the magnitude of the GVD increases

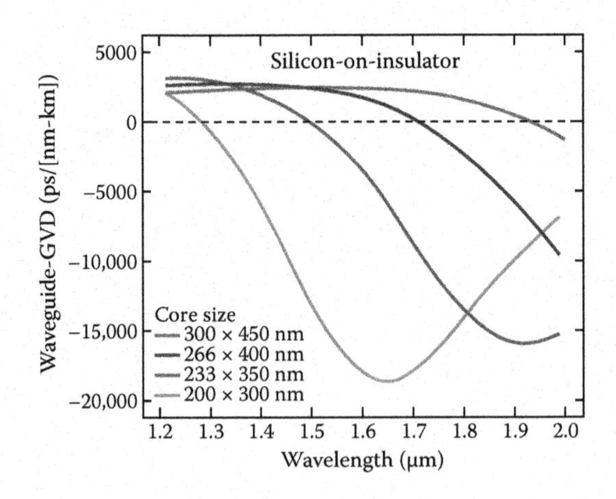

FIGURE 5.7
(See color insert.) Waveguide contribution to the group-velocity dispersion (GVD) in SOI platform of various dimensions. (From M. A. Foster, A. C. Turner, M. Lipson, A. L. Gaeta, *Opt. Express* 16, 2, 1300, 2008. With permission.)

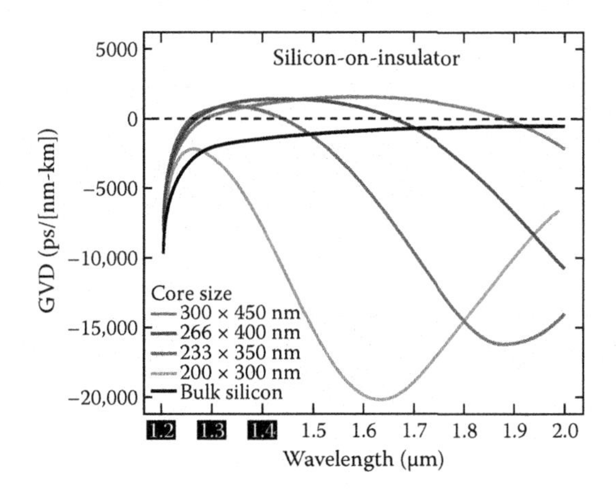

FIGURE 5.8
(See color insert.) Net group-velocity dispersion in SOI platform of various dimensions. (From M. A. Foster, A. C. Turner, M. Lipson, A. L. Gaeta, *Opt. Express* 16, 2, 1300, 2008. With permission.)

and the shifts between anomalous GVD and normal GVD occur at smaller characteristic dimensions. Figure 5.8 shows the net GVD of the photonic nanowires when the material GVD of the core is included. At the respective wavelengths of interest (800 and 1550 nm), a wide variety of net GVD values are accessible. A large range of values from 1500 to –18,200 ps/(nm·km) is achievable as a result of the large index contrast, in despite of the larger material GVD of silicon of –900 ps/(nm·km).

Therefore, GVD engineering becomes possible in waveguides by tailoring waveguide dimensions. It was shown that dispersion parameters can be tailored in channel and strip silicon nitride waveguides by computer simulations and measurements at telecom wavelengths.

5.2.4.2 Free Carrier Plasma Dispersion

High-density optical silicon integration on a monolithic substrate is an ongoing concern for the photonics community [34]. This is partly due to the fact that silicon is transparent in the infrared region of the spectrum and also because it offers a well-developed fabrication technology. To achieve high levels of functionality in a silicon waveguide, one needs to actively or dynamically control the confinement and propagation of light in it.

One method used to perform active light control in silicon is the plasma dispersion effect [35–37], which involves changing the refractive index of silicon by the creation, injection, or depletion of charge carriers. However, many existing implementations generate charge carriers by electronic means such as through a p-i-n junction, which may be undesirable in certain applications and potentially slower compared with direct optical generation. Recently, free carrier plasma dispersion effect is basically adopted to realize silicon based optical switches with fast response [38,39]. Free carriers induced by TPA are inevitable in silicon at the high optical intensity, which give rise to optical absorption and are responsible for plasma dispersion effect. The absorption loss and the refractive index change induced by free carriers are given by

$$\alpha_{FC} = \frac{e^3\lambda^2}{4\pi^2 c^2 \varepsilon_0 n}\left(\frac{\Delta N_e}{m_{ce}^2\mu_e} + \frac{\Delta N_h}{m_{ch}^2\mu_h}\right)$$
$$\approx 1.45\times 10^{-17}\left(\frac{\lambda}{1550}\right)^2 N(z,t) \tag{5.5}$$

$$\Delta n_{FC} = -\frac{e^2\lambda^2}{8\pi^2 c^2 \varepsilon_0 n}\left(\frac{\Delta N_e}{m_{ce}} + \frac{\Delta N_h}{m_{ch}}\right)$$
$$\approx -8.2\times 10^{-22}\lambda^2 N(z,t) \tag{5.6}$$

So the dispersion induced by free carriers is given by

$$\Delta D_{FC} = \frac{1}{c}\frac{\partial \Delta n_{FC}}{\partial \lambda} = -5.46\times 10^{-30}\lambda N \tag{5.7}$$

Because the TPA in silicon is intensity dependent, both FCA loss and FCD depend on the optical intensity. During propagation inside the waveguide, the optical intensity $I(z,t)$ decreases due to linear absorption and scattering, TPA, and FCA. The evolution of the optical signal intensity along the waveguide may be described as

$$\frac{dI(z,t)}{dz} = -\alpha I(z,t) - \alpha_{FC} I(z,t) - \beta I^2(z,t) \tag{5.8}$$

where α is the linear loss coefficient. Based on the above analysis, when 5-ps optical pulses at 1550-nm input a 1-cm silicon waveguide with free carrier lifetime of 300 ps and effective mode area of 0.10 μm², the evolution of FCD and FCA loss along the waveguide at two groups pulse values are shown in Figure 5.9 [22].

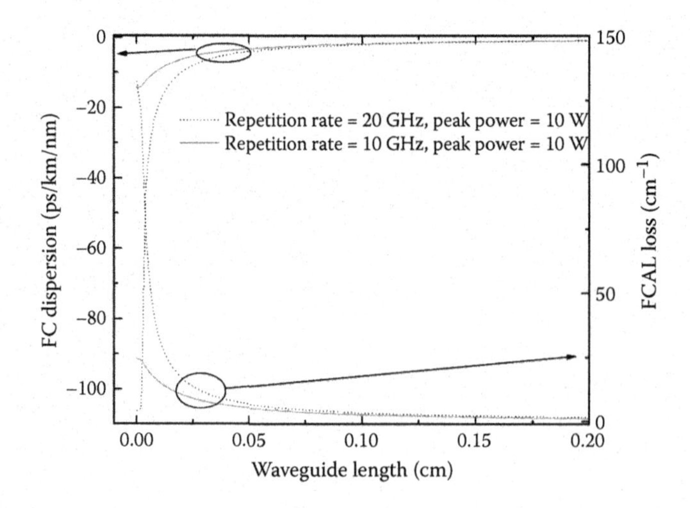

FIGURE 5.9
FCD and FCA loss evolution along the waveguide. (From X. Sang, E. K. Tien, and O. Boyraz, *J. Optoelectron. Adv. Mater.* 11, 15–25, 2009. With permission.)

5.2.5 Self-Phase Modulation

The parametric nonlinear processes often depend on the GVD, which has a profound impact on nonlinear pulse propagation, for efficient operation. Perhaps the most straight forward parametric process to observe is that of SPM in which the Kerr-induced phase shift of a propagating pulse causes a red (blue) frequency shift on the front (back) of the pulse [40]. This effect generally characterizes the initial stages of supercontinuum generation discussed previously [41]. In the silicon photonic waveguide, the efficiency of SPM not only depends on the optical intensities but also on the presence of free carriers generated by two-photon absorption (TPA) [42] and TPA introducing free carrier absorption (FCA) [43]. TPA and FCA cause optical losses, which in turn lower the peak power inside the waveguide and therefore reduce the SPM-induced spectral broadening. In particular, the lifetime of generated carriers is a critical parameter. Asymmetric spectral broadening is observed due to an additional rapid phase shift from the TPA-generated free carriers, which aids the blue shift at the back of the pulse but opposes the red shift at the front.

In Figure 5.10, there are some numerical simulations to explain that how the TPA, FCA, and FCD affect the SPM. The top row of Figure 5.8 shows the output pulse spectra at two different input intensities when all effects are included (solid curves). An important feature to note in Figure 5.8 is that, while TPA leaves the pulse spectrum symmetric, the free carriers produced by TPA make it considerably asymmetric. In particular, free carriers

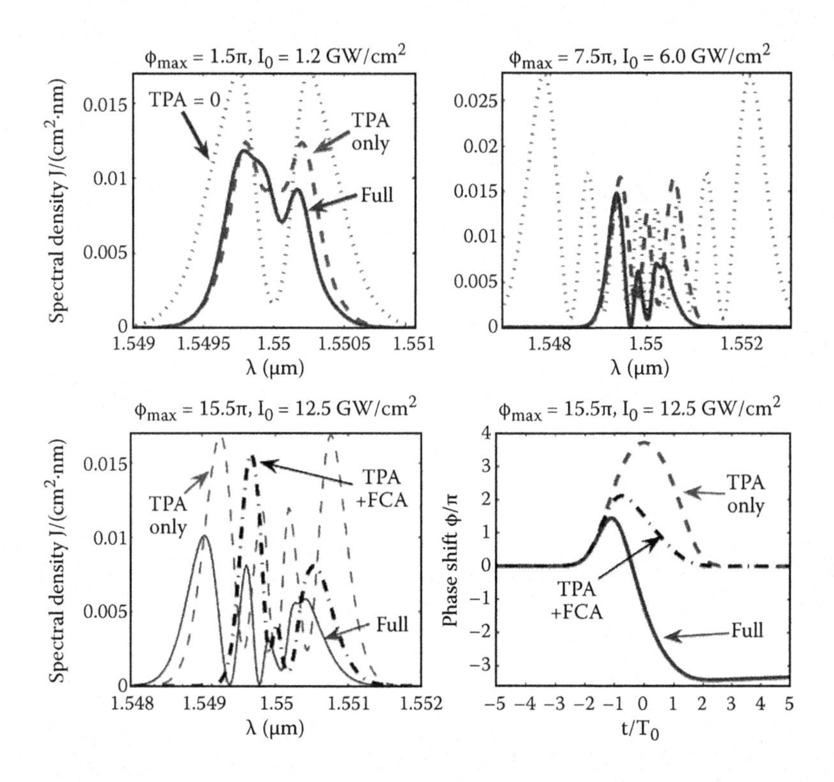

FIGURE 5.10
SPM-broadened pulse spectra (solid curves) at the end of a 2-cm-long SOI waveguide at three input intensities such that $\varphi_{max} = 1.5\pi$, 7.5π, and 15.5π. Dashed curves include TPA but neglect FCA and FCD effects; dotted curves neglect TPA as well. The nonlinear phase profiles in the three situations are shown in the last plot for $\varphi_{max} = 15.5\pi$. (From L. Yin and G. P. Agrawal, *Opt. Lett.* 32, 2031–2033, 2007. With permission.)

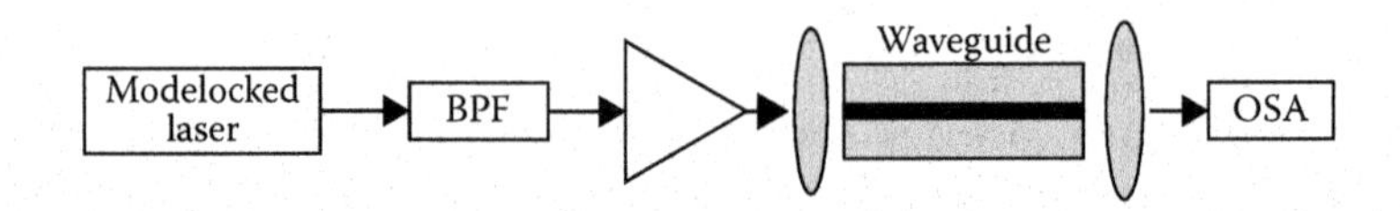

FIGURE 5.11
An experimental setup used for measuring SPM in silicon waveguides. (From O. Boyraz, T. Indukuri, and B. Jalali, *Opt. Express* 12, 829–834, 2004. With permission.)

affect mostly the "red" part of the spectrum; the "blue" part remains almost unchanged at low intensity levels. The spectrum becomes narrower and asymmetric when FCA is included without FCD. The effect of FCD is to broaden the spectrum and shift it toward the "blue" side.

Figure 5.11 shows an experimental setup used for measuring SPM in silicon waveguides, and the spectral broadening can be measured at the output. The spectra measured in the experiment setup as shown in Figure 5.11 also agree qualitatively with the numerical result [44].

Moreover, SPM has been observed for both picosecond and femtosecond pulses [3,44, 45]. For the picosecond configuration, there is a numerical and experimental measurement for the transmission of picosecond pulses through SOI submicron waveguides with excitation wavelengths between 1400 and 1650 nm and peak powers covering four orders of magnitude. We can learn that SPM induced spectral broadening is found to be significant at coupled peak powers of even a few tens of mW. The nonlinear index coefficient, extracted from the experimental data, is estimated as $n_2 \sim 5 \times 10^{-18}$ m^2/W at 1500 nm.

Additional spectral characteristics have been identified to result from dispersive-wave generation, from nonnegligible third-order dispersion and from soliton formation [46–48]. And with the incorporation of an integrated spectral filter, an all-optical regeneration device has been demonstrated [49] using silicon nanowaveguides based on the technique originally proposed by Mamyshev [50].

5.2.6 Cross-Phase Modulation

Cross-phase modulation (XPM) is particularly important since it allows control of one pulse via a nonlinear phase shift using a second "pump" pulse at a different wavelength. Several groups have reported XPM in silicon waveguides. These include one observation of XPM in 2-μm^2 scale waveguides [51], for which free carrier effects are important, while a second study examined XPM in Si wires using very high-intensity 300-fs pulses [52]. In the case of high intensity femtosecond pulses several nonlinear phenomena and pulse dispersion effects can enter into the interpretation of the results, and thus, it is essential to match experiments with a full theoretical treatment.

Figure 5.12 gives a detailed description of the XPM-induced spectral changes, in which, 200-fs pump and probe pulses (at 1527 and 1590 nm) launched into a 4.7-mm-long SOI waveguide (w = 445 nm, h = 220 nm). From these figures, it can be clearly seen that the pump and probe pulses travel at different speeds, which is called the walk-off effect, and the XPM-induced phase shift occurs as long as pulses overlap. Moreover, the asymmetric XPM-induced spectral broadening depends on pump power (blue curve), and the red curve is the probe spectral without pump.

Cross-phase modulation is comparable to SPM except that the impact of the Kerr-induced phase shift occurs on a second generally weaker propagating pulse. Devices based on this effect have been demonstrated to be efficient in silicon photonic nanowires, and the

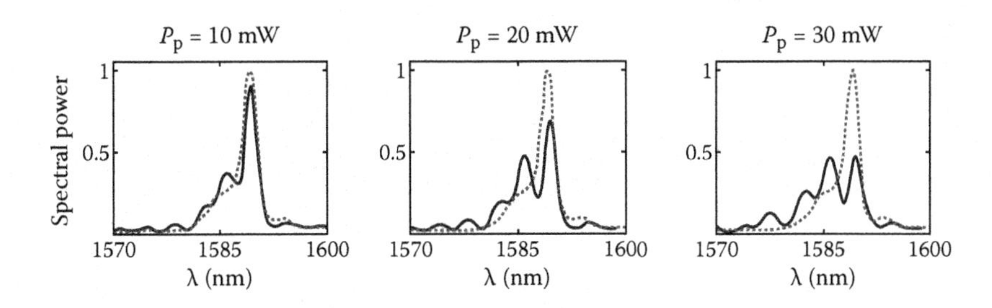

FIGURE 5.12
XPM-induced spectral changes. (From I. Hsieh, X. Chen, J. Dadap, N. Panoiu, R. Osgood, Jr., S. McNab, and Y. Vlasov, *Opt. Express* 15, 1135, 2007. With permission.)

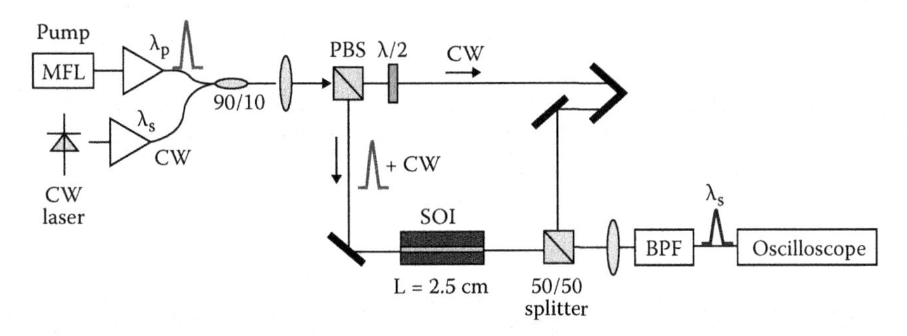

FIGURE 5.13
An experiment of XPM-induced switching. (From O. Boyraz, P. Koonath, V. Raghanathan, and B. Jalali, *Opt. Express* 12, 4094, 2004. With permission.)

effects of the GVD have been observed to greatly impact the interaction when the pulses are of significantly different wavelengths [52,53]. Figure 5.13 shows a experimental setup about XPM-induced switching, in which a Mach–Zehnder interferometer is used for optical switching, and short pump pulses (<1 ps) at 1560nm pass through the arm containing a 2.5-cm-long SOI waveguide. CW probe experiences XPM-induced phase shift in that arm. Then temporal slice of the probe overlapping with the pump is optically switched.

5.2.7 Four-Wave Mixing

When one or two pumps are incident into the silicon waveguide with a signal light, parametric amplification happens. Meanwhile, a light wave called "idler light" is also generated in the process. We call this FWM.

FWM in silicon waveguide generates idler light at the symmetric side of the pump or central pump (in the case of two pumps, which is the arithmetic average of them) with the signal light at the frequency domain. As the pump intensity is strong, the nonlinear effect originates from the nonlinear response of bound electrons of silicon semiconductor to an electromagnetic field. Except from parametric amplification, other effects like SPM, XPM, FCA, and FCD happens as well. All this whole mechanism leads to FWM, which amplifies signal light and generates idler light. The coupled mode equations are as follows [43]:

$$\frac{\partial A_p}{\partial z}+\beta_{1p}\frac{\partial A_p}{\partial t}+\frac{i\beta_{2p}}{2}\frac{\partial^2 A_p}{\partial t^2}=-\frac{1}{2}[\alpha_p+\alpha_{fp}(z,t)]A_p+i\beta_{0p}A_p+i(\gamma_e+\gamma_R)I_pA_p \tag{5.9}$$

$$\frac{\partial A_s}{\partial z}+\beta_{1s}\frac{\partial A_s}{\partial t}+\frac{i\beta_{2s}}{2}\frac{\partial^2 A_s}{\partial t^2}=-\frac{1}{2}[\alpha_s+\alpha_{fs}(z,t)]A_s+i\beta_{0s}A_s+i(2\gamma_e+\gamma_R)I_pA_s$$
$$+i\gamma_eA_p^2A_i^*+i\gamma_RA_p\int_{-\infty}^{t}h_R(t-\tau)e^{i\Omega_{sp}(t-\tau)}\left[A_p^*(z,\tau)A_s(z,\tau)+A_p(z,\tau)A_i^*(z,\tau)\right]d\tau \tag{5.10}$$

where α_j accounts for the linear scattering loss and $\alpha_{fj}(z, t) = 1.45 \times 10^{-17}(\lambda_j/1550nm)^2N_{eh}(z, t)$, $N_{eh}(z, t)$ is the density of electron-hole pairs created though pump-induced TPA. β_{0j} is the propagation constant at the frequency $\omega_j(j = p, s, i)$, β_{1j} is inverse of the group velocity, β_{2j} is the GVD parameter, and $\Omega_{sp} = \omega_s - \omega_p$ is the signal pump frequency detuning. The values of β_{2p} and β_{2s} can be inferred from the GVD curves. The idler equation can be obtained by exchanging the subscripts s and i in Equation 5.10. In the above coupled equations, it is assumed that the intensities of both the signal and idler are much smaller than that of the pump so that the pump is not depleted and higher-order nonlinear effects are negligible. The nonlinear effects are included through the parameters γ_e and γ_R representing the contributions of bound electrons and optical phonons (Raman process), respectively. The Raman response of the stimulated Raman scattering (SRS) is to be mentioned as

$$\gamma_e=\xi_e(\gamma_0+i\beta_T/2),\gamma_R=\frac{\xi_Rg_R\Gamma_R}{\Omega_R}$$

where $\gamma_0 = n_2\omega_p/c$ with $n_2 = 6 \times 10^{-5}$ cm²/GW and $\beta_T = 0.45$ cm/GW is the coefficient of TPA. The constants ξ_e and ξ_R are polarization factors related to the intrinsic symmetry of silicon. For the waveguide configuration, $\xi_e = 1$ and $\xi_R = 0$ for the TM mode, but $\xi_e = 5/4$ and $\xi_R = 1$ for TE one. Other unmentioned parameters are related to Raman process, which will be introduced later.

At the end of the silicon waveguides, the intensities of signal and idler lights are obtained. The well known signal amplification gain $G(L)$ and idler wavelength conversion efficiency $C(L)$ are

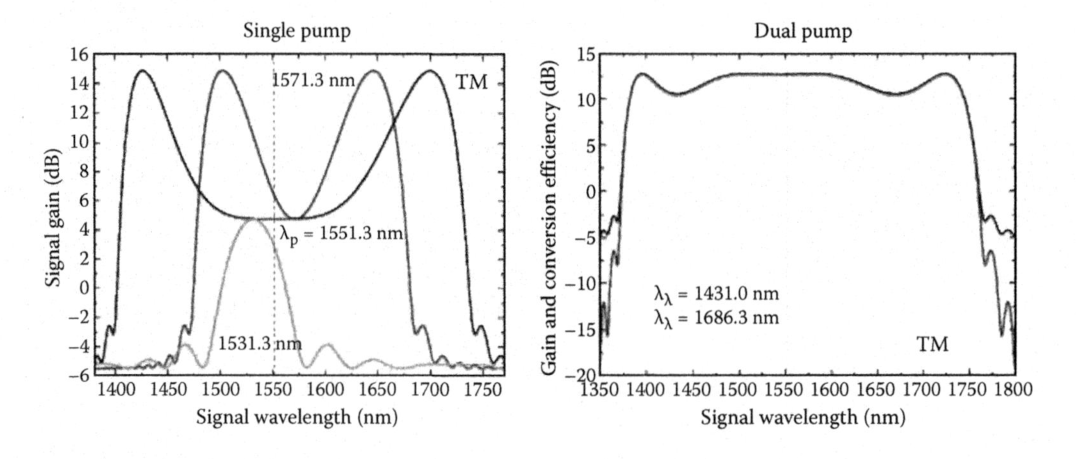

FIGURE 5.14
Single and dual configuration of FWM-based parametric amplification in silicon. (From Q. Lin, J. Zhang, P. M. Fauchet, and G. P. Agrawal, *Opt. Express* 14, 4786, 2006. With permission.)

$$G(L) = 10\lg(I_s(L)/I_s(0)),$$
$$C(L) = 10\lg(I_i(L)/I_s(0)). \tag{5.11}$$

where I is the intensity and L is propagation length of the lights.

Figure 5.14 gives the gain spectrum of single and dual configuration of FWM-based parametric amplification in silicon. It can be clearly seen that a large bandwidth can be realized by pumping an SOI waveguide with two pumps, which is possible because of a relatively short device length.

5.2.8 Simulated Raman Scattering

In the processes of Raman scattering, the incoming photons at frequency, ω_p are scattered by optical photons with frequency, ω_v, and then new photons are created at Stokes frequency, $\omega_s = \omega_p - \omega_v$. Moreover, there is a certain threshold value for the incoming optical pulse. If this value is broken, this scattering process becomes stimulated and the power of the newly generated Stokes pulse increase exponentially. For building an all-silicon laser [30,55,56] in the telecommunication wavelength among C-band, Stimulated Raman Scattering (SRS) undoubtedly become a successful approach. Also, SRS can also be used to amplify optical signal in silicon waveguides [57,58].

The dynamics of Raman-mediated pulse propagation in silicon photonic can be described by a set of coupled partial differential equations as follow:

$$\begin{aligned} &i\frac{\partial u_p}{\partial z} + i\beta_{1p}\frac{\partial u_p}{\partial t} - \frac{\beta_{2p}}{2}\frac{\partial^2 u_p}{\partial t^2} - i\frac{\beta_{3p}}{6}\frac{\partial^3 u_p}{\partial t^3} \\ &= \frac{ic\beta_{1p}k_p}{2n}\left(\alpha_{in} + \alpha_{FC}^{p}\right)u_p - \frac{\omega_p\beta_{1p}k_p}{n}\delta n_{FC}^{p}u_p \\ &- \frac{3\omega_p\beta_{1p}}{\varepsilon_0 A_0}\left(P_p\tau_p\beta_{1p}\left|u_p\right|^2 + 2P_s\tau_{sp}\beta_{1s}\left|u_s\right|^2\right)u_p - \frac{6\omega_p\beta_{1p}^2 P_s}{\varepsilon_0 A_0}\tau_R^{p}(-\omega_R)\left|u_s\right|^2 u_p \end{aligned} \tag{5.12}$$

$$\begin{aligned} &i\frac{\partial u_s}{\partial z} + i\beta_{1s}\frac{\partial u_s}{\partial t} - \frac{\beta_{2s}}{2}\frac{\partial^2 u_s}{\partial t^2} - i\frac{\beta_{3s}}{6}\frac{\partial^3 u_s}{\partial t^3} \\ &= \frac{ic\beta_{1s}k_s}{2n}\left(\alpha_{in} + \alpha_{FC}^{s}\right)u_s - \frac{\omega_s\beta_{1s}k_s}{n}\delta n_{FC}^{s}u_s \\ &- \frac{3\omega_s\beta_{1s}}{\varepsilon_0 A_0}\left(2P_p\tau_{ps}\beta_{1p}\left|u_p\right|^2 + P_s\tau_s\beta_{1s}\left|u_s\right|^2\right)u_p - \frac{6\omega_s\beta_{1s}^2 P_s}{\varepsilon_0 A_0}\tau_R^{p}(-\omega_R)\left|u_s\right|^2 u_p \end{aligned} \tag{5.13}$$

where $P_{p,s}$ are the pulse peak powers, α_{in} is the intrinsic loss, $\alpha_{FC}^{p,s}$ are the FCA coefficients at pump and Stokes frequency, corresponding, and $\delta n_{FC}^{p,s}$ are the FC-induced change of the refractive index. $T_j = A_0\int e_j^* X^{(3)} \vdots e_j e_j^* e_j \, dA/J_j^2$, and $\tau_{jl} = A_0\int e_l^* X^{(3)} \vdots e_j e_j^* e_l \, dA/J_j J_l$, $J_{s,p} = \int n^2(r_\perp)\left|e_{s,p}\right|^2 dA$ and $e_{s,p} \equiv e_{s,p}(r;\, \omega_{s,p})$ are the waveguide modes at the corresponding frequency. Besides the effective Raman susceptibility $\tau_R^{p,s}(\Omega) = A_0\int e^*(\omega_{s,p})X^R \vdots e(\omega_{s,p})e^*(\omega_{s,p})e(\omega_{s,p})\, dA/J_j^2$.

From the analysis above, SRS-mediate pulse dynamics numerically investigated including light amplification and lasing based on SBS. Raman gain and amplification beyond 11 dB has been reported for both the telecommunications window, and the mid-infrared wavelength region.

5.2.9 Enhancement and Suppression Nonlinear Interaction in Silicon

As we know, the high refractive index contrast between the SiO_2 cladding layer and the silicon waveguide core, which are defined by completely etching through a $\lambda/2n$ thick silicon waveguide layer, is finally exploited by using nanophotonic strip waveguides. Optical nonlinearities are expected to occur at appropriate optical input powers on this waveguide platform. And due to the high confinement in the silicon waveguide layer and due to the high quality factor, ultracompact cavities this expectation can be realized. Nonlinear operation in SOI waveguide circuits typically relies on the carriers created by the relatively weak two photon absorption process or on the weak Kerr effect. The nonlinear effect can be enhanced and suppressed by deceasing or increasing the effective mode area in the silicon waveguide. In addition, engineering the dispersion can be also used to influence the nonlinear process. In the last recent years, the nonlinear process in nanostructured or microstructured silicon waveguides based on photonic crystal and microring are effectively enhanced.

Recently, integrating some specific materials on the SOI waveguide platform gave the SOI waveguide an enhanced nonlinear behavior. Figure 5.15 shows an III–V/SOI nanophotonic waveguide structure, which is reported in [59,60]. The SOI nanophotonic strip waveguide structure used consists of a 220-nm-thick silicon waveguide layer on top of a 2-μm-thick SiO_2 buffer layer to prevent the leakage of light to the silicon substrate. The integration of a high refractive index III–V layer on top of the SOI nanophotonic waveguide can drastically change the modal properties of the waveguide structure.

These III–V/SOI cavities can show an enhanced nonlinear behavior because of the absorption that takes place in the III–V active layers when injecting light with a wavelength shorter than the band gap wavelength of the III–V active layers. This absorption process creates free carriers in the III–V active layer, which reduce the refractive index of the III–V active layer material due to the plasma dispersion effect band filling, and band gap shrinkage. The reduction in the guided mode effective index results in a blue shift in the resonance spectrum, and hence results in a change in transmission as a function of the density of the free carriers or as a function of the injected optical power.

The spontaneous Raman scattering in a W1 photonic crystal waveguide on SOI with the lower silica cladding is enhanced [61]. In particular, we can see a reshaping of the Raman spectrum and a more than tenfold enhancement of the Raman scattering efficiency in a W1 photonic crystal waveguide as compared with a single-mode ridge waveguide.

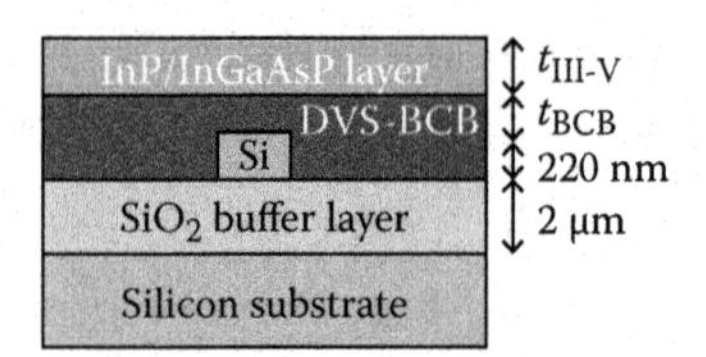

FIGURE 5.15
Cross section of the considered III–V/SOI nanophotonic waveguide structure. (Reprinted with permission from [59]. Copyright 2008, American Institute of Physics.)

By integrating some specific materials on the SOI waveguide platform, the SOI waveguide can exhibit an enhanced nonlinear behavior. However, based on the same principle, some other specific materials maybe can lead to the suppression of the nonlinear interaction in silicon, which remaining to be demonstrated in the future.

5.3 Second-Order Nonlinear Silicon Photonics

Silicon crystal possesses a crystalline symmetry [62] that inhibites the existence of nonzero elements in the second-order nonlinear susceptibility $\chi^{(2)}$ tensor [63].

When a crystal having a second-order nonlinearity is illuminated with strong electromagnetic fields, it can generate new optical fields at multiples, submultiples, sum, or difference of the pump frequencies [63]. This is an essential characteristics in the development of optoelectronic devices having access to spectral regions where traditional light sources do not have access (mid-infrared, far-infrared).

Silicon is then limited in its optical capabilities and in its employment in efficient electro-optic modulators, one-pump-beam frequency converters, optical rectifiers for THz generation as well as in nonlinear mid/far-infrared photon detection schemes [64].

Despite this fact, second-order nonlinear optics in silicon has a long history dating back since the late 1960s [65–67]. The way to circumvent the material intrinsic nonlinear limitations was to exploit the natural centro-symmetry break, that silicon experiences at its interfaces [68]. In that region the periodic cubic displacement of silicon atoms abruptly stops and in a relatively thin (few hundreds of nanometers) surface region a dipolar $\chi^{(2)} \neq 0$ is observed, arising from the electronic potential distortion [69].

In 2006, a Danish team of researchers reported on the observation of electro-optic effects related to the presence of a bulk $\chi^{(2)} \neq 0$ in a slow-wave silicon Mach–Zehnder interferometer [70]. This finding has been debated (see for example [71]) but had undoubtedly triggered new theoretical proposals [72] on how to extend silicon capabilities to the realm of bulk second-order nonlinear optics.

Recently the Nanoscience Laboratory of the University of Trento has experimentally demonstrated that micrometer-sized silicon waveguides can be mechanically modified within a CMOS-compatible growth technology, by reporting on the first bulk-dipolar second harmonic generation from crystalline silicon [73] achieving a $\chi^{(2)}$ = 240 pm/V [74].

A further motivation for the quest of a second-order nonlinear silicon photonics resides in the technological fact that second-order nonlinear optics is inherently driven at much lower power than third-order, as can be simply understood by the expansion of the nonlinear polarization field $\vec{P}^{NL}$ in terms of pump electric field $\vec{E}$, where the higher-order expansion terms rapidly decrease in their absolute value:

$$\vec{P}^{NL}(t) = \chi^{(1)}\vec{E}(t) + \chi^{(2)}\vec{E}^2(t) + \chi^{(3)}\vec{E}^3(t) + \cdots \tag{5.14}$$

Much simpler pumping schemes are involved in the $\chi^{(2)}$ processes. Only one pump beam is typically used to generate new frequencies at higher or lower energy with respect with that of the pump itself.

5.3.1 Second-Order Nonlinear Optical Effects in Silicon

5.3.1.1 Surface Dipolar Effects

When the surface of silicon is illuminated with a strong electromagnetic (EM) field, the resulting motion of atomic electrons cannot be described in the simple and usual scheme of linear oscillators, as happens in the case of weak light illumination, but the contribution of the neighbor atoms has to be taken into account, in particular for their nonquadratic distortion of the electronic potential. This situation has been experimentally studied and theoretically interpreted since 1983 by Shen et al. [75] in terms of the response of the surface electrical dipole sheet of polarization corrected with bulk quadrupolar contributions. Figure 5.16 reports the typical second-harmonic (SHG) signals obtained in reflection configuration from Si(111) and Si(100) surfaces when a 45° incident laser beam (wavelength λ = 532 nm) is made reflecting on it. The rotation angle is about normal for the sample surface.

As expected, the crystal symmetry reflects into the periodic pattern observed for the SHG signal vs. the azimuthal angle rotation. These early results have been confirmed, reproduced, and are so well understood that nowadays the surface second harmonic generation has became a technological tool for testing surface film quality during the epitaxial growth of silicon or of other materials on silicon [76,77]. Other early SHG reflection experiments claimed to have observed significant bulk dipolar $\chi^{(2)}$ contributions [66] but were later corrected and interpreted inside the traditional *surface-dipolar-bulk-quadrupolar* interpretation [78].

The amount of $\chi^{(2)}$ coming from dipolar surface effects has been determined to be of the order of 10^{-19} m^2 V^{-1} and the corresponding bulk-anisotropic contribution of the order of $<10^{-2}$ pm V^{-1} in the usual reflection configuration (see Figure 5.17) [69].

Such amounts of $\chi^{(2)}$ explain well why silicon has never been taken into account as a second-order nonlinear material. The absolute value of the nonlinearity is low if compared for example with Li_3NbO_4 or GaAs [79]. For this reason people has worked on the attempt to find some new degrees of freedom in the material itself in order to enhance such nonlinearities and exploit them in nonlinear optoelectronic devices.

Govorkov et al. [67] have thoroughly studied the effect of application of an external inhomogeneous deformation to the surface of Si in SHG experiments. They accomplished such a task by means of ion implantation of the silicon surface with different ion species and with different implantation doses. This way, they were able to enhance the SHG signal of orders of magnitude as reported in Figure 5.18.

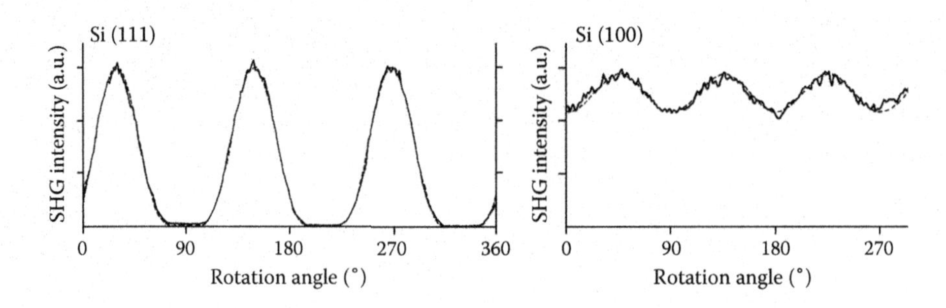

FIGURE 5.16
p-Polarized SHG under p-polarized pump: (left) Si (111) and (right) Si (100). (Reprinted with permission from H. W. K. Tom, T. F. Heinz, and Y. R. Shen, *Phys. Rev. Lett.* 51, 1983, 1983. Copyright 1983 by the American Physical Society.)

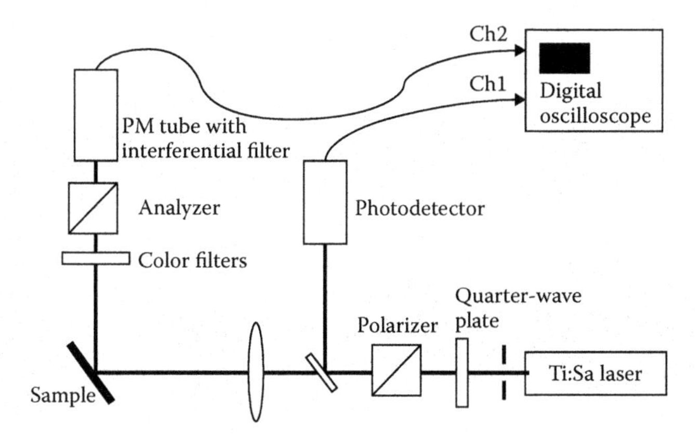

FIGURE 5.17
Typical experimental arrangement for surface SHG probing on silicon surfaces. (From M. Falasconi, L. C. Andreani, A. M. Malvezzi, M. Patrini, V. Mulloni, and L. Pavesi, *Surf. Sci.* 481, 105–112, 2001. With permission.)

In the same work, they also studied the effect of the deposition of a silicon oxide and other silicide films on the silicon surface, on the SHG signal and found again signal enhancements of the order of 50–200. They extracted an estimate of the stress field applied of the order of few gigapascals as well. Huang [80] later on confirmed the above reported results and demonstrated that reflection SHG is also a powerful tool to extract information about stress/strain in a lattice deformed layer of silicon.

Later, the possibility to exploit such effects in suitable photonic devices was studied by Schriever et al. [81–83].

Recently, an *on-chip* all-silicon harmonic generator has been demonstrated by Galli et al. [84]. In their work, they employed planar photonic crystal high-Q factor silicon nanocavities

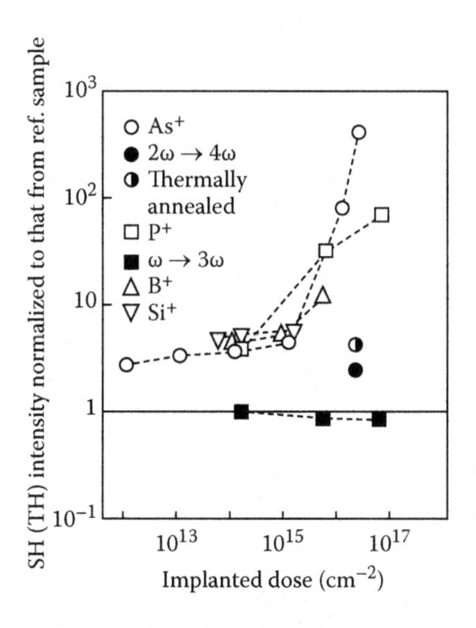

FIGURE 5.18
Intensity of the SHG reflected from differently implanted Si (111) samples. The reported values are the ratio between the measurement and the SHG from a reference Si (111) sample. The effect of implanted dose is evident. The SHG of the fundamental YAG/Nd^{3+} frequency was measured. (From S. V. Govorkov, V. I. Emel'yanov, N. I. Koroteev, G. I. Petrov, I. L. Shumay, and V. V. Yakovlev, *J. Opt. Soc. Am. B* 6, 1117, 1989. With permission.)

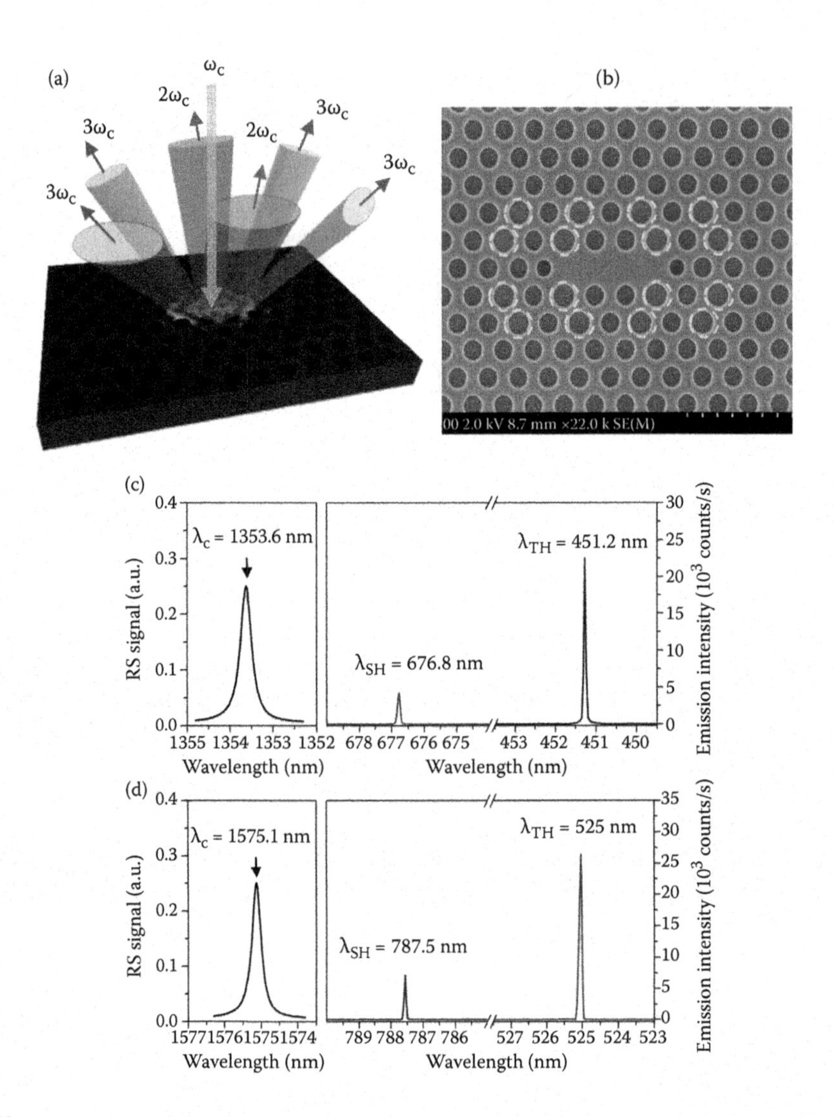

FIGURE 5.19
(a) Artist's view of the SHG and THG emission from the suspended silicon membrane containing the photonics crystal nanocavities. (b) SEM image of the sample. Experimental demonstration of (c) SHG and (d) THG by showing the resonant scattering spectra of the pump and of the collected signals. (From M. Galli, D. Gerace, K. Welna, T. F. Krauss, L. O'Faolain, G. Guizzetti, and L. C. Andreani, *Opt. Express* 18, 26613, 2010. With permission.)

excavated in a 220-nm-thick silicon suspended membrane. They pumped their samples (see Figure 5.19) with a CW laser source at 1.35 and 1.57 μm and observed simultaneous second- and third-harmonic generation in the so-called resonant light spectroscopy geometry [85].

Within this generation scheme, the third harmonic signal comes from bulk effects in crystalline silicon while the second harmonic signal is attributed to surface effects at the sidewall holes Si/air boundaries of the photonic crystal. The amount of $\chi^{(2)}$ is not estimated, but a pump to SHG conversion efficiency (η) is reported $\eta \sim 1 \times 10^{-9}$. This is to our knowledge the first experimental demonstration of an all-silicon *on-chip* second-order nonlinear device, operating on the exploitation of surface optical nonlinearities. Still this system is based on technologically complex growth techniques making it difficult to implement in real nonlinear devices.

Another surface SHG silicon-based device has been proposed by Levy et al. in [86]. Again the demonstration of simultaneous second- and third-harmonic generation has been presented, but in this work, the device is composed by Si_3N_4 silicon microrings. The advantage of such a scheme resides mainly in the CMOS-compatible growth technique that makes it a good candidate for a real implementation on a future *on-chip* all silicon harmonic converter (Figure 5.20).

The authors propose in their paper a possible "bulk origin" of the generated signal and extract a $\chi^{(2)}$ value of the of 0.04 pmV^{-1}. On the other side, the physical mechanism drawn in the paper for the origin of the observed dipolar second-order susceptibility is said to be originated from the asymmetric dipole potential formed at the $0.75 \times 1.50\text{-}\mu m^2$ waveguide surfaces, thus indicating a more probable surface-related conversion mechanism, which is supported also by their modal cross-section-simulated (and measured) profiles for SHG that are reported in Figure 5.21, where it is evident the spatial proximity (tens of nanometers) of the SHG mode profile to the sidewall interfaces.

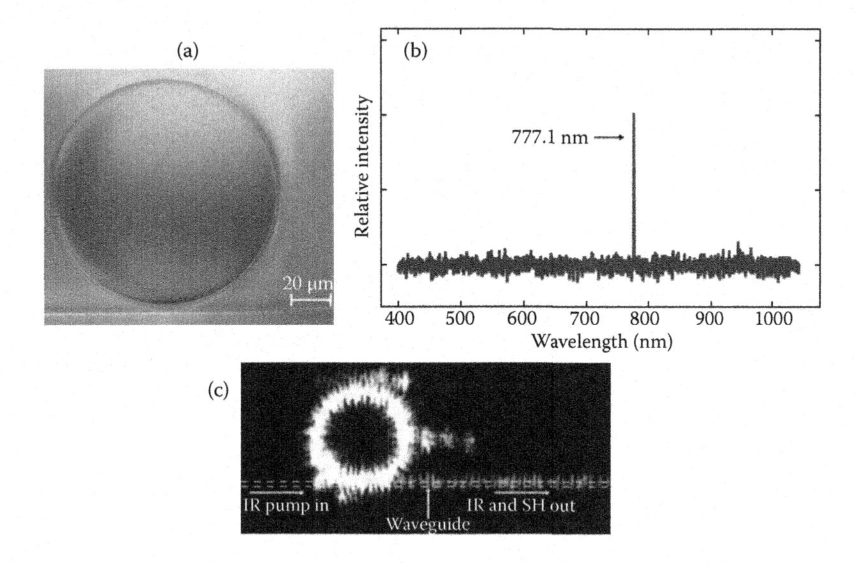

FIGURE 5.20
(a) SEM micrograph of a typical silicon nitride ring. (b) SHG signal when pumping at around 1.54 μm. (c) Visible CCD image of the red generated SHG signal. (From J. S. Levy, M. A. Foster, A. L. Gaeta, and M. Lipson, *Opt. Express* 19, 11415, 2011. With permission.)

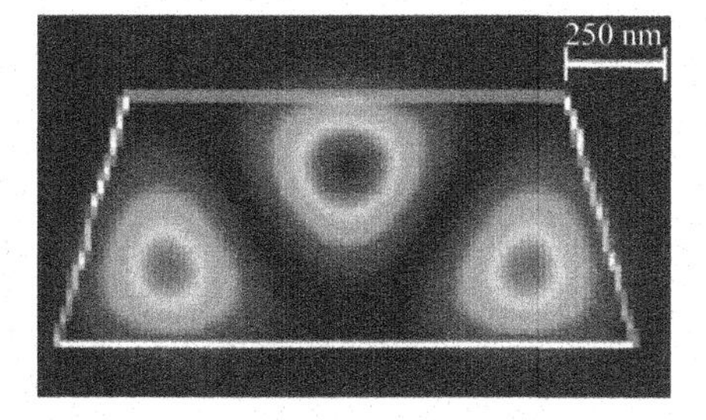

FIGURE 5.21
Modal cross-section profiles for phase-matched SHG in the silicon micro-rings. (From J. S. Levy, M. A. Foster, A. L. Gaeta, and M. Lipson, *Opt. Express* 19, 11415, 2011. With permission.)

5.3.1.2 Bulk Dipolar Effects

Bulk $\chi^{(2)}$ in silicon has always been related to residual quadrupolar effects [65,69] due to the crystalline symmetry of silicon, which in principle inhibits the existence of a bulk dipolar second-order susceptibility in the material itself [63]. Still this fact is obviously true, but some attempts to circumvent or at least to control it, have recently appeared in literature.

In 2006, for the first time, Jacobsen et al. [70] claimed to have observed bulk dipolar $\chi^{(2)}$-related effects inside a micrometric-sized silicon waveguide (Figure 5.22). In particular, the working principle of a slow-wave electro-optic modulator based on a Mach–Zehnder

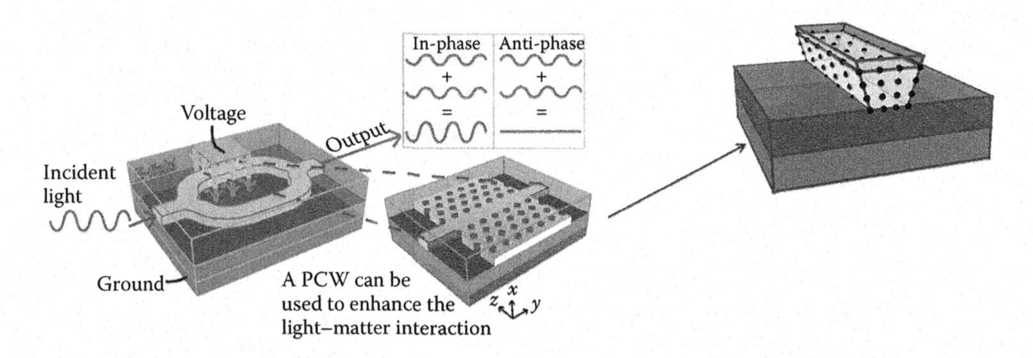

FIGURE 5.22
The incident ~1.56 μm pump light is sent to the *on-chip* Mach-Zehnder interferometer, where a strained silicon waveguide has been embedded inside a pump-resonant slow-wave structure acting as an enhancer of the non-linear interaction between the pump and the mechanically induced $\chi^{(2)}$. The waveguide is then "zoomed out" in the right-panel schematics where the crystalline break, as well as the silicon nitride overlayer thickness, are exaggerated, to be better appreciated by the reader. (Reprinted by permission from Macmillan Publishers Ltd. *Nature* [70] copyright 2006.)

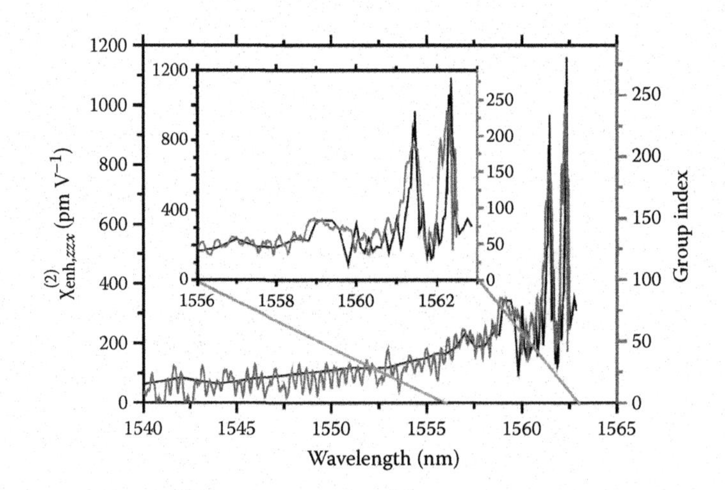

FIGURE 5.23
The straight line reports the slow-wave enhanced $\chi^{(2)}$ which clearly scales with the group index n_g (curly line) that is extracted from the parameters of the photonic crystal composing the Mach-Zehnder waveguiding structure. (Reprinted by permission from Macmillan Publishers Ltd. *Nature* [70], copyright 2006.)

on-chip device was attributed to the raise of a nonzero $\chi^{(2)}$ caused by the internal crystalline break of silicon, due to the deposition of a Si_3N_4 overlayer.

Electro-optical transmission measurements indicated the presence of a bulk $\chi^{(2)}$, which has been resonantly enhanced (via slow-wave effects) at ~1.562 μm up to ~800 pmV^{-1}, which is a very high $\chi^{(2)}$ effective value. Figure 5.23 show the main experimental results of [70].

This result has remained not yet fully reproduced and it is not yet clear how much of the electro-optic modulation there observed is attributable to the optical nonlinearity with respect to free carrier effects in silicon [71]. However, this is to our knowledge the first report of a bulk dipolar $\chi^{(2)}$ in crystalline silicon and has addressed the scientific community to study new ways to induce efficient second-order nonlinearities in silicon.

On this basis Hon et al. [72] have proposed a semiquantitative theoretical proposal on how to exploit such a strain-induced nonlinearity in a difference-frequency-generation experiment aimed at mid-infrared generation within a so-called periodically poled silicon waveguide [72]. To date, an experimental demonstration of such a proposed experiment has not yet appeared and also the scheme on how to invert the sign of the strain-induced $\chi^{(2)}$ has been strongly modified/simplified by a recent paper of Driscoll et al. [87], where the width modulation of a linear silicon waveguide has been employed to phase match the 1543-nm pump and the 1544- to 1690-nm signal in a $\chi^{(3)}$-based FWM conversion scheme generating a 1419- to 1541-nm idler. Figure 5.24 shows a draw of such structure and experiment.

Very recently, M. Cazzanelli et al. [73] have demonstrated for the first time in all optical experiment, that the silicon nonlinear optical properties can be mechanically modified by the controlled deposition of a silicon nitride overlayer. The strain field induced in a micrometer-sized (~2 × 10-μm² section) linear waveguide has been experimentally observed. It penetrates deeply and inhomogeneously below the interfacial region between silicon nitride overlayer and the silicon waveguide, in the region where the fundamental mode of the pump is efficiently guided. This result is reported in Figure 5.25.

Nonlinear transmission measurements have been performed on these strained silicon waveguides with high-power pulsed (both femtosecond- and nanosecond-long) laser sources and an efficient second-harmonic generation has been observed coming out from the guides when pumping with a tunable pump wavelength λ_{pump} = 2050–2300 nm. The experimental results are reported in Figure 5.26.

By measuring the absolute SHG power, the pump to SHG conversion efficiency has been experimentally determined and by comparison with generalized nonlinear envelope equation (GNEE) numerical solution of the nonlinear propagation of short pulses in such not-phase-matched waveguide, the exact amount of the bulk second-order nonlinearity $\chi^{(2)}$ = 240 pmV^{-1} has been obtained [74].

Such experimental results have already found a substantial confirmation in the work of Chmielak et al. [88], where an electro-optic nonlinear modulator has demonstrated in a

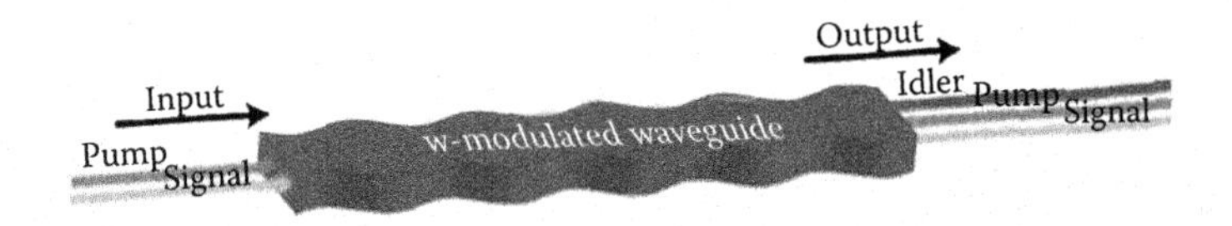

FIGURE 5.24
Width-modulated silicon nanowires demonstrated by Driscoll et al., where an idler tunable between 1419 and 1541 nm is nonlinearly (via $\chi^{(3)}$) generated. (From J. B. Driscoll, N. Ophir, R. R. Grote, J. I. Dadap, N. C. Panoiu, K. Bergman, and R. M. Osgood, Jr., *Opt. Express* 20, 9227, 2012. With permission.)

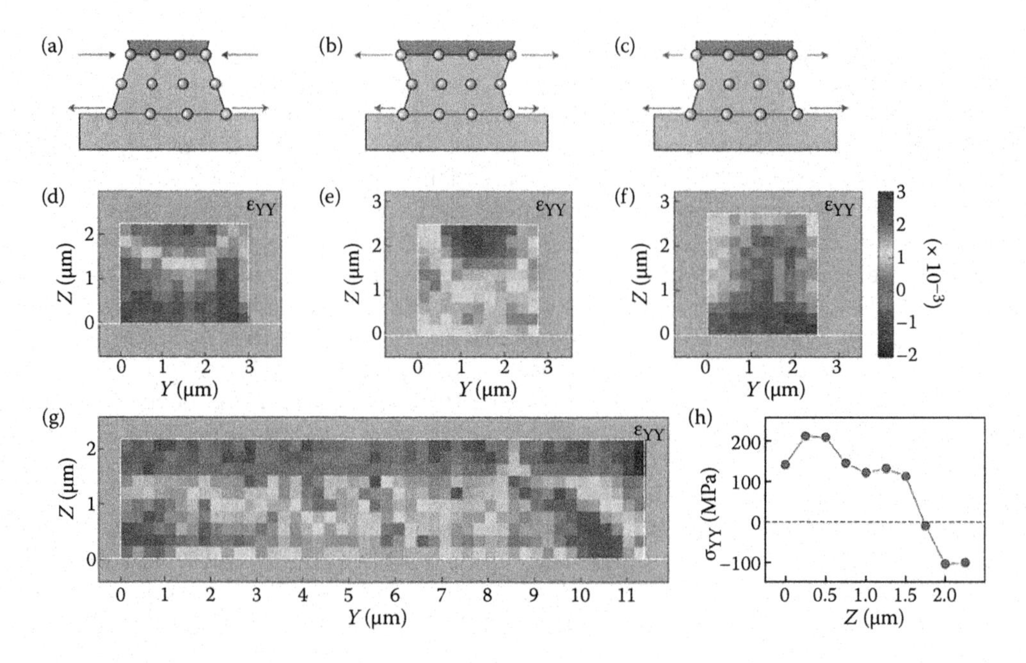

FIGURE 5.25
Summary of the experimental measurement of waveguide internal strain field via micro-Raman measurements on the waveguide facet. (a–c) Schematic representations of the mechanically strained waveguides. The lattice deformation is exaggerated to show the effect of the applied stressing silicon nitride overlayer (top layer). (d–f) Two-dimensional map of the measured strain-tensor element for a tensile/compressive-strained waveguide. (g) Two-dimensional map of the strain-tensor element for the ~10-μm-wide waveguide. (h) Line scan of the measured stress at the center of the SOI1 10.7-μm-wide waveguide. (Reprinted by permission from Macmillan Publishers Ltd. *Nature Materials* 11,148, copyright 2011.)

silicon rib waveguide and a $\chi^{(2)} = 122$ pmV^{-1} has been observed (see Figure 5.27) in a different waveguide-straining geometry.

5.4 Applications of Nonlinear Optical Effects in Silicon

5.4.1 Amplifiers and Lasers

5.4.1.1 Raman Amplification

Silicon exhibits a strong and relatively narrow (below 1 nm) Raman response. Using this response, several Raman-based all-optical devices have been developed for the SOI platform. The Raman amplification has attracted great attention. A major hurdle for a Raman amplification in silicon was the nonlinear loss resulting from TPA and TPA-induced FCA. To remove FCA and lower the power requirements for such devices, the use of silicon photonic wire can provide a benefit. Here, the CW and short-pulse Raman amplification in silicon photonic wire will be presented.

5.4.1.1.1 CW-Pumped Raman Amplification

For both the pump and the signal in the form of CW waves, the interaction can be described by solving the coupled intensity equations [89]. The influence of FCA on CW

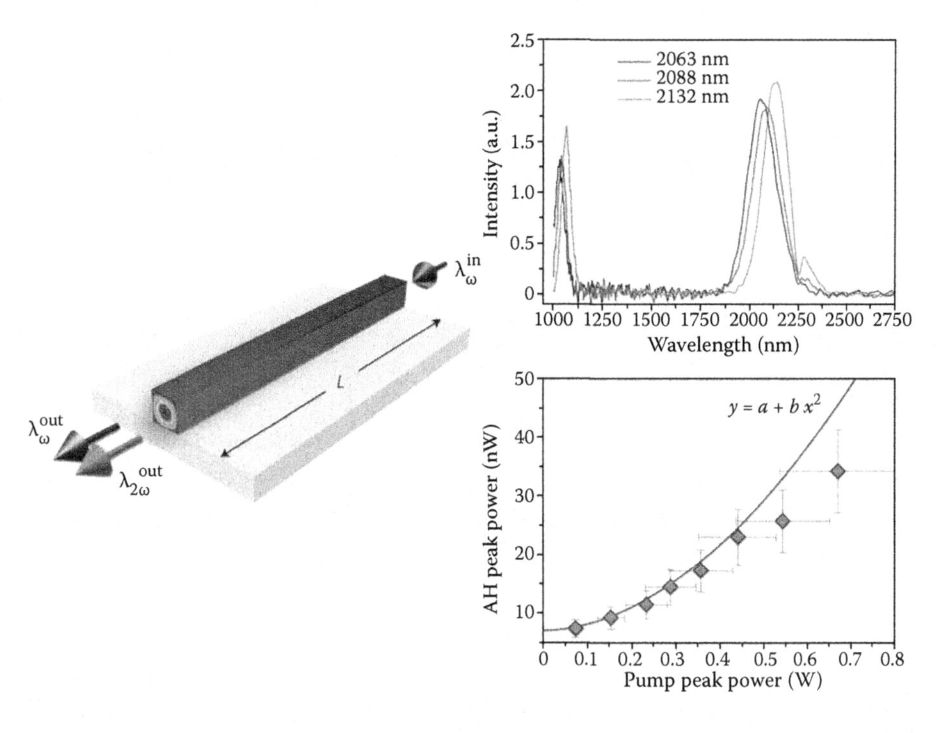

FIGURE 5.26
(top) Typical transmission spectra recorded under a femtosecond-pulsed pump where can be clearly seen both the pump at λ_{pump} around 2100 nm and the SHG at half its wavelength. (bottom) Typical quadratic behavior of the SHG signal when varying the pump power (such data refer to ns-pump conditions). (left) Schematic of nonlinear transmission experiment where the pump is seen to be coupled in the waveguide and both pump and SHG beam are collected at the waveguide output facet. (Reprinted by permission from Mcmillan Publishers Ltd. *Nature Materials* [73], copyright 2011).

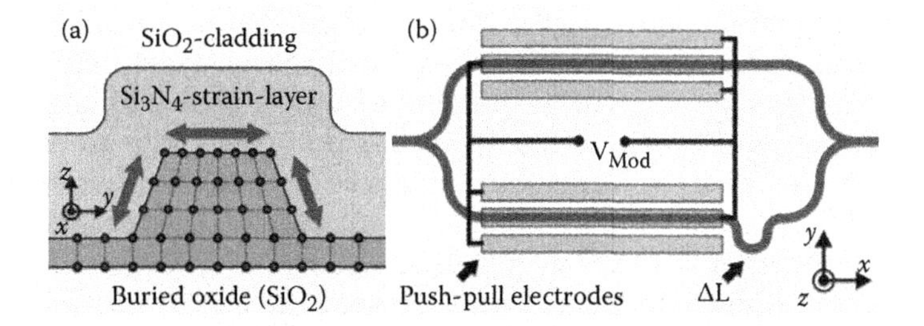

FIGURE 5.27
(a) Schematics of the straining geometry accomplished in the silicon rib waveguides embedded in the Mach-Zehnder silicon modulator. Both the waveguide top and sidewalls are strained. (b) Top-view schematics of the electro-optically driven *on-chip* interferometer. (From B. Chmielak, M. Waldow, C. Matheisen, C. Ripperda, J. Bolten, T. Wahlbrink, M. Nagel, F. Merget, and H. Kurz, *Opt. Express* 19, 17212, 2011. With permission.)

Raman amplification manifests in two ways. First, FCA leads to an overall attenuation of the signal. Secondly, compared with the linear-loss situation, it leads to a decrease in the effective length. The influence of total input intensity on the generalized effective length is illustrated by solid curves in the upper panel of Figure 5.28. Clearly, one can increase the effective length $L_{eff}(z)$ substantially by decreasing I0. A similar increase in the effective length may be realized by reducing τc because κ scales linearly with τc.

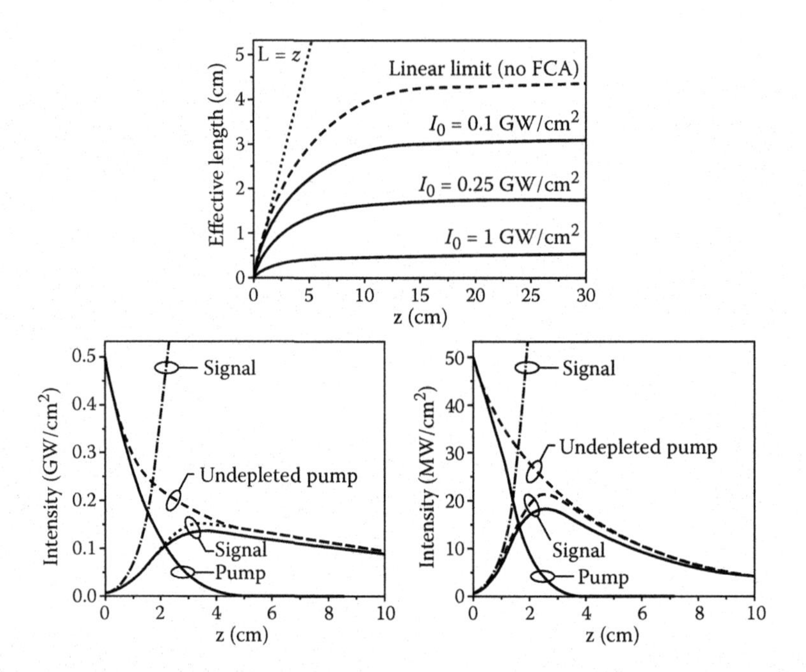

FIGURE 5.28
(top) Generalized effective length L_{eff} (z) versus propagation distance for different input intensities I_0 (solid curves). The dashed curve shows the linear loss limit. The dotted line corresponds to the lossless case. (bottom) Evolution of the pump and the signal intensities obtained by calculating numerically (solid curves) and predicted analytically by the corrected (dashed curves) and uncorrected (dotted curves) solutions. The solution corresponding to the undepleted pump approximation is shown by dashed dotted curves. (left) Ip_0 = 0.5 GW/cm² and (right) 0.05 GW/cm², respectively, with $Is_0 = 0.01Ip_0$. The other parameter values are α = 1 dB/cm, β = 0.5 cm/GW, τc = 1 ns, g_R = 76 cm/GW, λp = 1550 nm, and λs = 1686 nm. (From I. Rukhlenko, M. Premaratne, C. Dissanayake, and G. Agrawal, *Opt. Lett.* 34, 536–538, 2009. With permission.)

Because of the exponential nature in the Raman amplification, any intensity noise associated with the pump can be severely enhanced during early stages before the gain saturation. Below the transfer of noise from the pump to the signal will be studied. We use the concept of RIN to quantify this effect. The coupled intensity equations can be used to study the RIN transfer in silicon Raman amplifiers by introducing two time derivatives, as in [89].

The RIN transfer together with the average signal intensity is presented in Figure 5.29. The solid and dashed dotted-dotted curves represent the RIN transfer under the operating conditions used in the lower panel of Figure 5.14 (in both cases, τc = 1 ns). When the pump intensity is sufficiently high, it can be seen that the RIN transfer grows with the propagation distance, and a maximum value is reached just before the signal peaks, starting to decay with the pump intensity becoming low. The decaying occurs due to the signal smoothing caused by the intensity-dependent FCA. The extent of RIN transfer can obtain a constant value as the signal intensity becomes low enough to make FCA negligible.

The situation becomes more intricate in the waveguides when considering a longer free- arrier lifetime. For example, the dashed, dotted, and dotted-dashed curves in Figure 5.29 show how the situation changes for τc = 2, 3, and 4 ns, respectively. All other parameters are equal to those used for the solid curve. It shows that the magnitude of RIN transfer from the pump to the signal takes large negative values on the decibel scale,

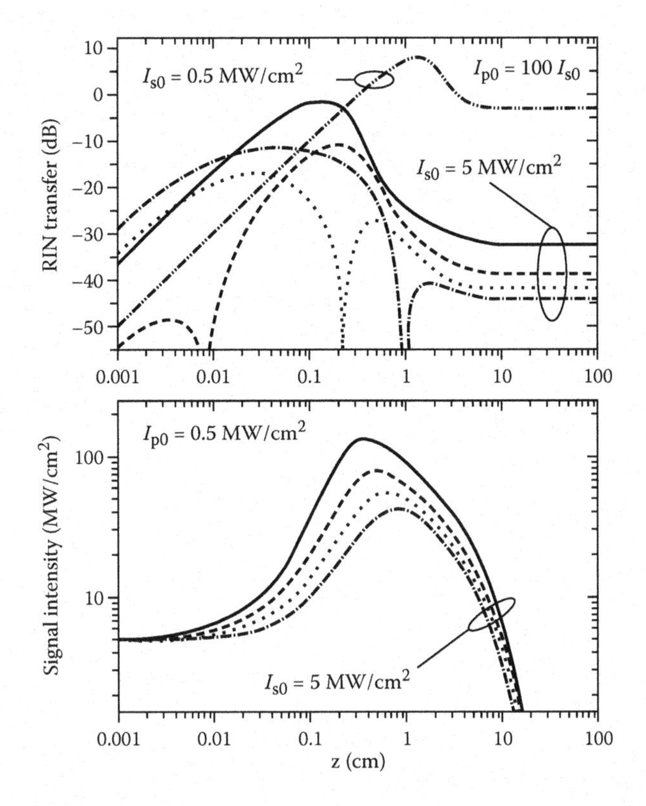

FIGURE 5.29
(top) RIN transfer and (bottom) average signal intensity as a function of propagation distance for τc = 1 ns (solid curves), 2 ns (dashed curves), 3 ns (dotted curves), and 4 ns (dashed-dotted curves). Solid and dashed dotted-dotted curves represent the RIN transfer from the pump to the signal shown in the lower panel of Figure 5.28. (I. Rukhlenko, M. Premaratne, C. Dissanayake, and G. Agrawal, *IEEE J. Sel. Top Quantum Electron.* 16, 200, 2010. © 2010 IEEE.)

indicating that the signal being much less noisy than the pump at some distance within the waveguide.

5.4.1.1.2 Pulse-Pumped Raman Amplification

The waveguide geometry is presented in Figure 5.30. When the pump pulse and the seeded Raman signal are coupled into the waveguide, the interaction of them as well as the dynamics of the generated FCs are illustrated in Figure 5.31. When the pump and the signal powers are comparable, the pump pulse is strongly depleted and its temporal profile is reshaped. Thus, after propagating a distance of 5 L_R, the central part of the pump pulse is almost completely depleted so that two temporally separated pulses emerges at the output facet of the waveguide. For the signal pulse, it is first amplified over a certain distance, and then it starts to decay because the losses compensate the Raman amplification. Therefore, there is an optimum distance at which the signal amplification reaches a maximum value.

When the temporal duration of the interacting pulses is comparable to the Raman response time of silicon (a few picoseconds), it is necessary to consider the effect of the delayed response of the nonlinear medium. Besides, the pulse walk-off effect plays an important role. Figure 5.32 shows that the group velocity mismatching between two pulses reduces their effective interaction length, which limit the efficiency of Raman

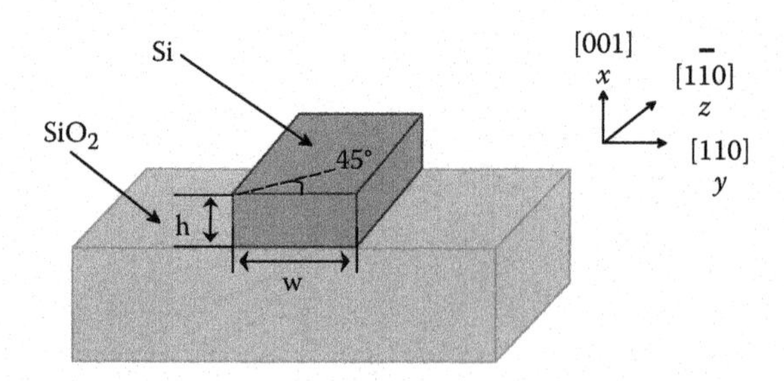

FIGURE 5.30
Waveguide geometry of the silicon photonic wire, with the input facet in the [110] plane of the silicon crystal lattice. The dashed line denotes the [100] direction. The width and height of waveguide are w and h, respectively. The length is 2 cm. (From X. Chen, "Nonlinear optics in silicon photonic wires: theory and applications," PhD thesis, Columbia University, 2007.)

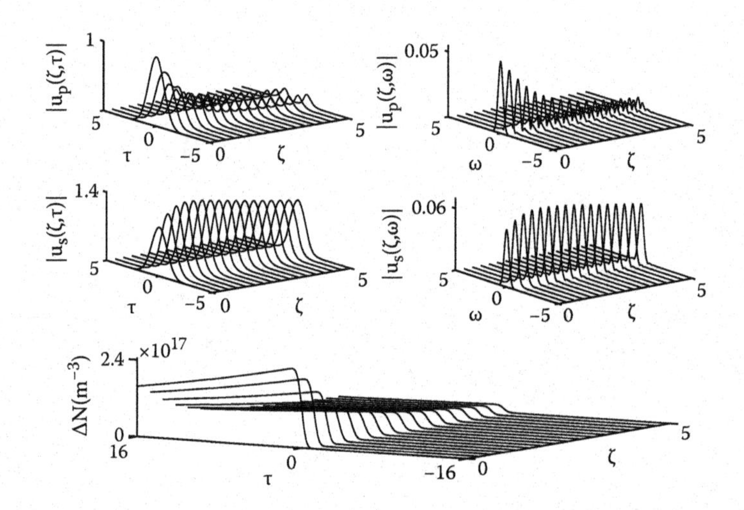

FIGURE 5.31
Raman interaction of a pump and a signal pulse in silicon wire waveguide. The pulse parameters are as follows: temporal widths, 30 ps; peak power of pump, 0.1 mW; Raman length, 3.83 mm; seed signal power, 0.05 mW. (From X. Chen, "Nonlinear optics in silicon photonic wires: theory and applications," PhD thesis, Columbia University, 2007.)

amplification. Due to the walk-off effect, the temporal profile of the two pulses is strongly asymmetric at the output facet of the waveguide.

5.4.1.2 Parametric Amplification

Optical parametric amplifiers (OPAs) through four-wave-mixing (FWM) can deliver large and flexible gain bandwidth. With a suitable small core waveguide design to shift zero dispersion wavelength toward the 1550-nm band, theoretical investigation shows that it is difficult to achieve net signal gain with a continuous-wave (CW) pump due to presence of two photon absorption (TPA) and FCA induced by TPA. By tailoring the cross-sectional

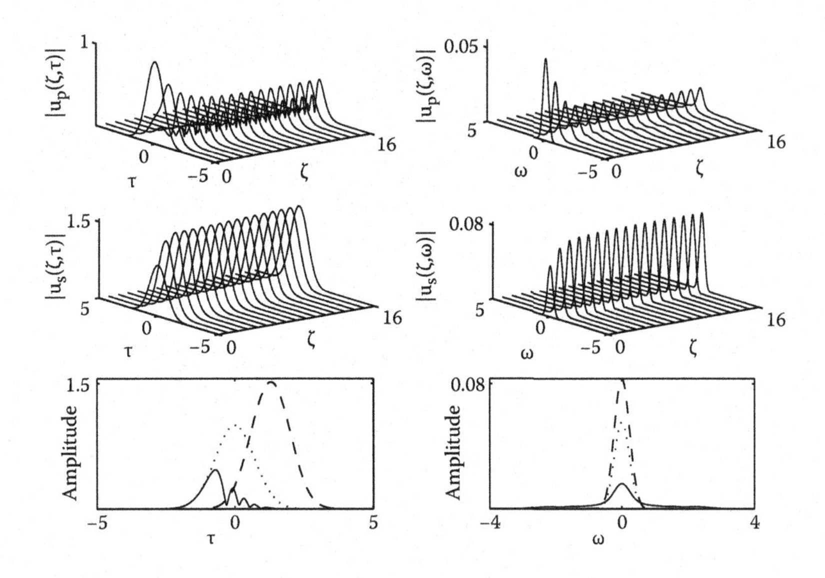

FIGURE 5.32
Raman interaction of a pump and a signal pulse in silicon wire waveguide. The pulse parameters are as follows: pulse temporal width, 2 ps; walk-off length, 4.76 cm; peak power of pump, 0.1 mW; Raman length, 3.83 mm, and the seed signal power is 0.05 mW. (From X. Chen, "Nonlinear optics in silicon photonic wires: theory and applications," PhD thesis, Columbia University, 2007.)

size and shape, the waveguide geometries, the silicon waveguide shows the anomalous group velocity dispersion (GVD), and the net gain parametric amplification using a pulsed pump in suitably designed silicon waveguides can be efficiently achieved for the high-speed optical communication and optical signal processing systems. Below the parametric gain and noise figure (NF) evolutions inside the short silicon waveguides for high repetition rates are studied.

To achieve large pump power, optical pump sources around 1550 nm are often amplified with erbium-doped fiber amplifier (EDFA), as illustrated in Figure 5.33. The amplified spontaneous emission (ASE) of the EDFA together with the pump laser's relative intensity noise (RIN) influences OPA's noise performance.

To achieve net gain through parametric amplification, the pump wavelength should be operating in anomalous GVD regimes. To demonstrate this, 1-ps wide pump pulses centered at 1550 nm with 10-GHz repetition rate and 5-W peak power are used to investigate parametric gain at different dispersion values, and the results are summarized in Figure 5.34. We clearly see that net gain can only be achieved within limited bandwidth in anomalous GVD regimes. The maximum gain achieved at the phase matching point shifts

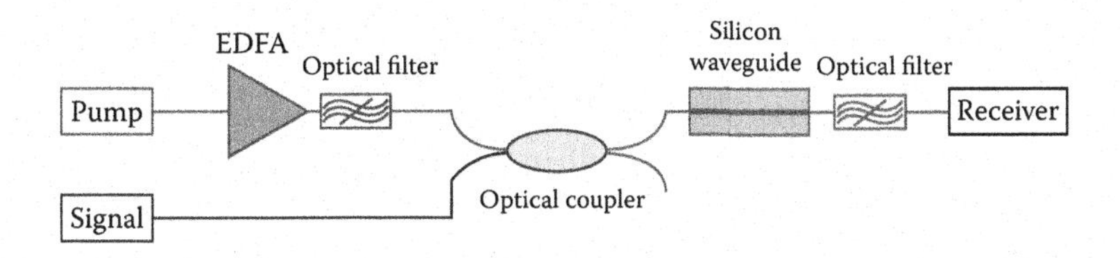

FIGURE 5.33
Schematic configuration of the OPA in silicon waveguide. (From S. Xinzhu and O. Boyraz, *Opt. Express* 16, 13122–13132, 2008. With permission.)

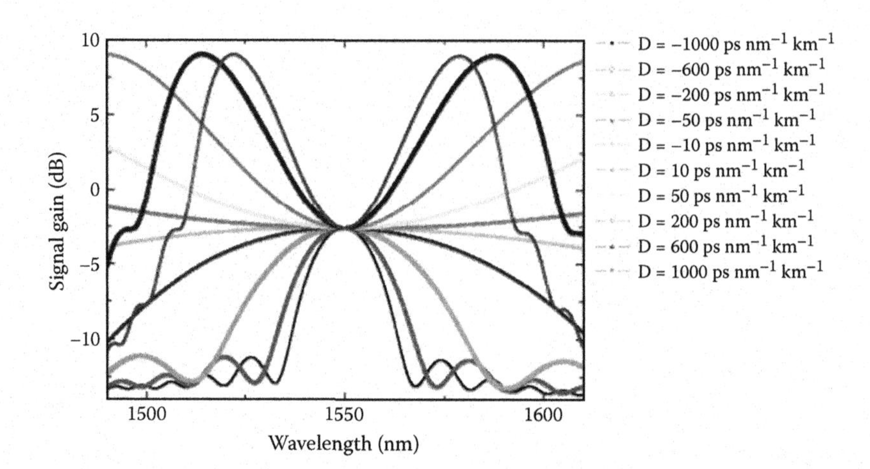

FIGURE 5.34
Parametric gain profiles at different dispersion values. (From S. Xinzhu and O. Boyraz, *Opt. Express* 16, 13122–13132, 2008. With permission.)

and the gain bandwidth decreases with increasing the dispersion value in the anomalous dispersion regimes.

In the silicon waveguide with dispersion of 600 ps/(nm km), 1-ps pulses at repetition rate of 10 GHz are used as pump. Signal gain and NF at eight wavelengths are investigated. With increasing the peak pump power, gain, and NF evolutions at eight wavelengths are shown in Figure 5.33a,b. In Figure 5.33a, the gain decreases at relatively low peak pump power for some wavelengths, which can be explained by nonlinear losses and phase mismatch. With increasing the peak pump power, the gain enters into saturation and then begins to decrease. The gain at the wavelengths near the pump wavelength enters into saturation earlier than that at the wavelengths far from the pump wavelength is due to shift of the phase matching point. We note that, unlike fiber OPAs, high peak pump power does not mean large gain due to presence of intensity-dependent TPA and FCA losses in the silicon waveguide, which is shown clearly in Figure 5.35a.

In Figure 5.35b, the NF values are illustrated with respect to the peak pump power for eight different wavelengths. NFs increase with increasing the peak pump power, which is different from the gain evolutions. These results are expected because of the fact that the nonlinear losses by TPA and FCA have significant contributions to the NF, and hence the NFs do not enter into saturation, as illustrated in Figure 5.35b.

Figure 5.36 shows the maximum gain and corresponding NF evolutions versus the repetition rate at four different pulse widths. We can know that the gain and NF have no significant changes when the repetition rate is below 500 MHz, and the FCA loss is determined by the pulse width. With further increasing the repetition rate, the gain begins to decrease and the NF begins to increase, as displayed in Figure 5.36a. The net gain can even be achieved at 80 GHz when the pulse width is 0.5 ps, but there is no net gain for 10 ps at 5 GHz. The NF performances at different pulse width are illustrated in Figure 5.36b, which shows that lower NF can be achieved at shorter pulse and the NF increases with higher repetition rate when the repetition rate is above 1 GHz.

In Figure 5.37, we know that net gain and NF are strongly influenced by the free carrier loss and hence by the free carrier lifetime. The free carrier lifetime of 300 ps has been assumed to match the value, which can be realized in passive silicon devices. To date, there

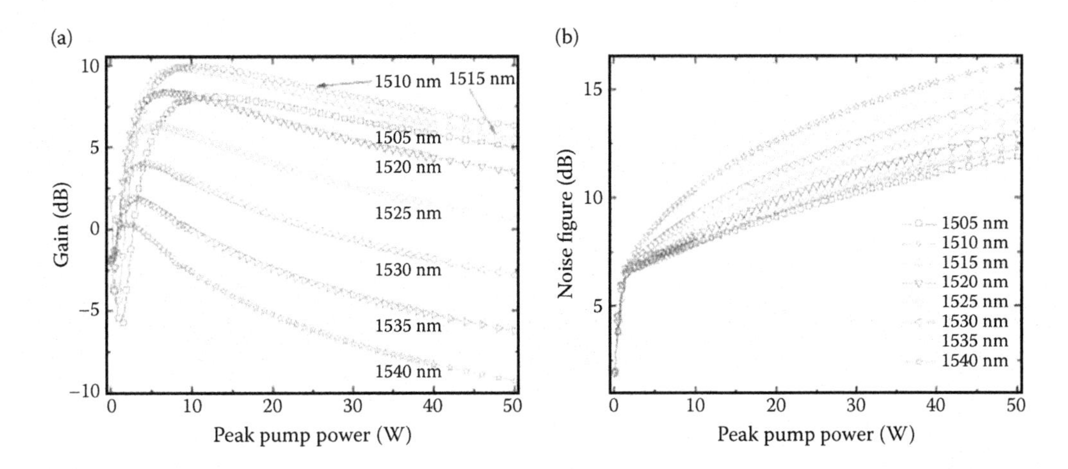

FIGURE 5.35
(a) Gain and (b) NF evolution at different wavelengths. (From S. Xinzhu and O. Boyraz, *Opt. Express* 16, 13122–13132, 2008. With permission.)

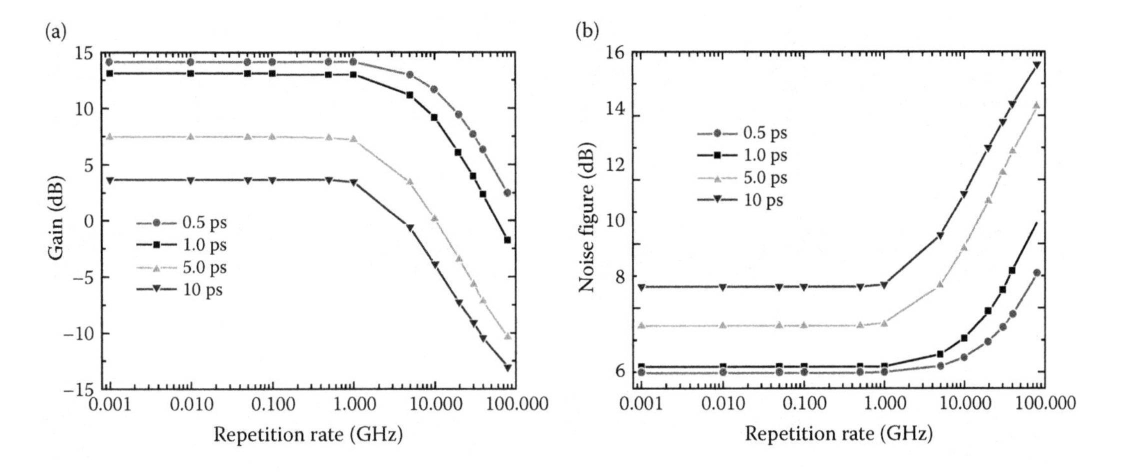

FIGURE 5.36
(a) Parametric gain and (b) NF versus the repetition rate for different pulse widths. (From S. Xinzhu and O. Boyraz, *Opt. Express* 16, 13122–13132, 2008. With permission.)

are a few approaches presented to reduce the free carrier lifetime in silicon waveguides, which can be applied to these calculations. Active carrier removal by means of the electric field of a reverse-biased p-n junction is expected to reduce the free carrier lifetime to values close to 10 ps at low intensity excitation. Additionally, ion implantation can be used to reduce the free carrier lifetime to 50 ps.

Figure 5.38 illustrates the gain profile and the corresponding noise figure, which can be achieved in a 1-cm-long silicon wire with free carrier lifetimes of 20 and 10 ps. Here, the optimum gains are achieved at peak pump powers of 4.6 W (4.6 GW/cm^2) and 2.5 W (2.5 GW/cm^2) for two free carrier lifetimes, respectively. As shown in these calculations the net parametric gain can be achieved at peak optical powers above 1 W (1 GW/cm^2).

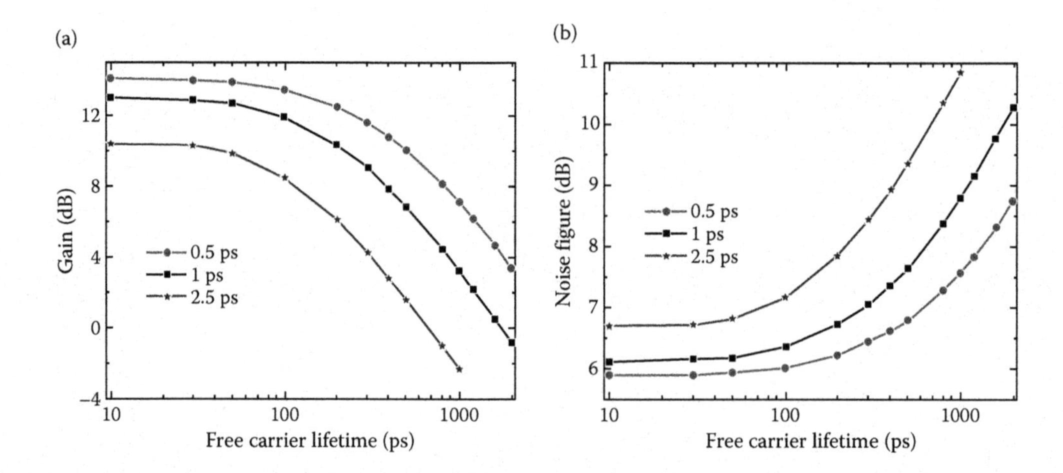

FIGURE 5.37
(a) Parametric gain and (b) NF versus the free carrier lifetime for different pulse widths. (From S. Xinzhu and O. Boyraz, *Opt. Express* 16, 13122–13132, 2008. With permission.)

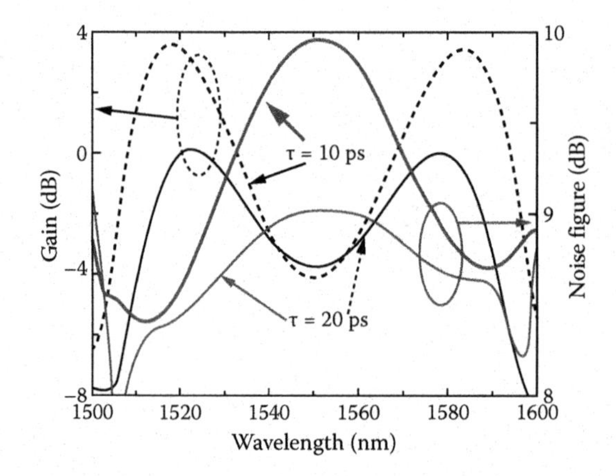

FIGURE 5.38
Gain and NF spectra of OPA in the silicon waveguide with a CW pump. (From S. Xinzhu and O. Boyraz, *Opt. Express* 16, 13122–13132, 2008. With permission.)

However, at these intensities, the generated free carriers will screen the applied electric field and null the lifetime reduction. The lifetime reduction by the ion implementation is achieved at the expense of the propagation loss, and hence, the induced linear loss might be prohibitive for these applications.

5.4.1.3 Raman Laser

Although the silicon has the potential to provide cheap integrated optical components or generate light on-chip, the indirect band-gap of silicon means that it is inefficient in spontaneously emitting photons. One solution toward the realization of a silicon laser is to exploit stimulated Raman scattering (SRS).

Figure 5.39 shows an instructive example of a silicon Raman laser on a single chip with a ring laser, where the silicon Raman laser chip comprises a 3-cm ring resonator

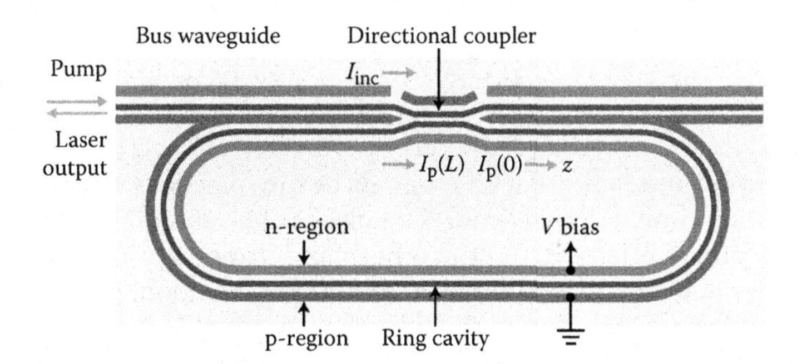

FIGURE 5.39
A silicon Raman ring laser within a ring laser cavity. A reverse bias p-i-n structure is placed across the silicon rib waveguide to "sweep out" TPA-induced carriers. (Reprinted by permission from Macmillan Publishers Ltd. *Nature Photonics* [9], copyright 2010.)

with waveguide widths of w = 1.5 μm, heights of h = 1.55 μm, and etch depths of 0.76 μm. The Raman pump and Raman-generated lasing signal are coupled in and out through a directional coupler. One of the main issues in Raman-pumped silicon lasers are the free carrier–related losses resulting from TPA. In this case, a p-i-n structure was placed along the waveguide rib to sweep out these free carriers and thus reduce losses. A reverse bias of up to 25 V was applied along the junction, which had a width of 6 μm. When a pump laser of wavelength 1550 nm was launched into the device, laser emission at 1686 nm was observed. The threshold was found to be at a pump power of 20 mW, and the slope efficiency was close to 28%. Up to 50 mW of output power was measured with a reverse bias of 25 V across the junction, whereas a maximum of only 10 mW was achieved without a reverse bias. SRS has also been useful for tuning the slowdown factor of light in silicon.

5.4.2 Nonlinear Optical Pulse Shaping

Using a standard procedure (as in [93]), the pulse envelope evolution in the presence of optical nonlinearities is conventionally estimated by solving the nonlinear Schrödinger equation [94,95]:

$$\frac{\partial E(t,z)}{\partial z} = -\frac{1}{2}\{\alpha + \alpha_{FCA}(t,z) + \alpha_{TPA}(t,z)\}E(t,z) - i\gamma\left|E(t,z)\right|^2 E(t,z) + i\frac{2\pi}{\lambda} n\Delta(t,z)E(t,z) \tag{5.15}$$

where parameters $E(t,z)$ represent the electric-field, α is the attenuation constants, γ is the effective nonlinearity, and Δn is the free carrier induced index change, respectively. The pulse shaping is determined by the nonlinear losses on the right hand side of Equation 5.16. Among two nonlinear loss terms, the TPA,

$$\alpha_{TPA} = \frac{1}{z}\ln(1 + \beta I_0 z) \tag{5.16}$$

is an intensity dependent and memory less attenuation following the pulse shape. Here α_{TPA} is the TPA coefficient (0.45 cm/GW at 1550 nm), A_{eff} is the effective area of the waveguide and I_o is the intensity. However, the FCA,

$$\alpha_{FCA}(z,t) = 1.45\times10^{-17}\left(\frac{\lambda}{1.55}\right)^2 N(z,t)(\text{cm}^{-1})d \tag{5.17}$$

has memory and is accumulative. For CW signals or time-varying signals wider than free carrier recombination time, τ_o, free carrier densities will be stabilized at a local $N(z)$ value of: $N(z) = \tau_o \cdot \beta \cdot I^2(z)/2h\upsilon$ after $\sim\tau_o$. On the other hand, time varying optical signals with pulse width smaller than τ_o will result in different time-dependent local free carrier densities and FCA along the waveguide governed by the equation:

$$\frac{dN(t,z)}{dt} = -\frac{N(t,z)}{\tau_0} + \beta\frac{I^2(t,z)}{2h\upsilon} \tag{5.18}$$

For pulse shaping, time- and space-dependent free carrier density $N(z,t)$ and the nonlinear losses are the utmost important parameters to be determined. Based on the peak intensity and the pulse energy, the nonlinear response of silicon can be divided into two separate regimes in time and in space: TPA-dominated regime and FCA-dominated regime [95]. Figure 5.40a shows the calculated results of two regimes for 20- and 500-ps optical pulses with 1-kW peak power. The waveguide is 1 cm long with 5-μm^2 effective area and 16 ns free carrier recombination time [95]. The 20-ps pulses are broadened due to TPA dominance, and the 500-ps pulses are compressed to 350 ps by FCA dominance. The FC dispersion causes a linear phase change across the pulse. Figure 5.40b illustrates the pulse compression for 100-ps optical pulses at different optical intensity with different waveguide length [95]. At low intensity, stronger TPA induced pulse broadening is over FCA induced pulse self-compression. FCA dominance can be achieved at higher intensities and facilitate pulse compression.

The schematic of experimental setup is shown in Figure 5.41a. The silicon waveguide has a p-i-n diode structure to inject carriers and hence a modulation capability. The output of the waveguide is connected to a 3/97 tap coupler, where 3% used as an output and 97%

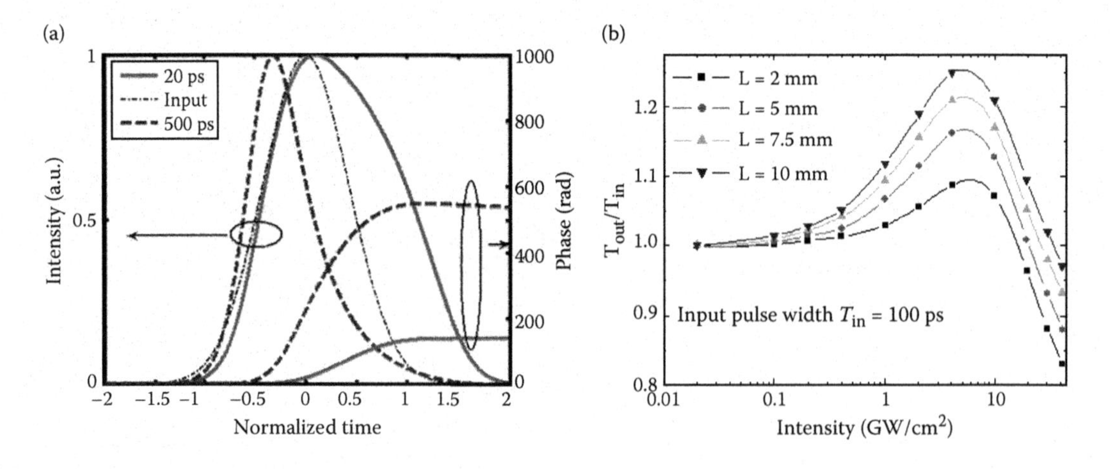

FIGURE 5.40
(a) Output pulses and phase induced by TPA and free carriers. (b) Predicted pulse compression in silicon waveguides. (Reprinted with permission from [95] Copyright 2007, American Institute of Physics.)

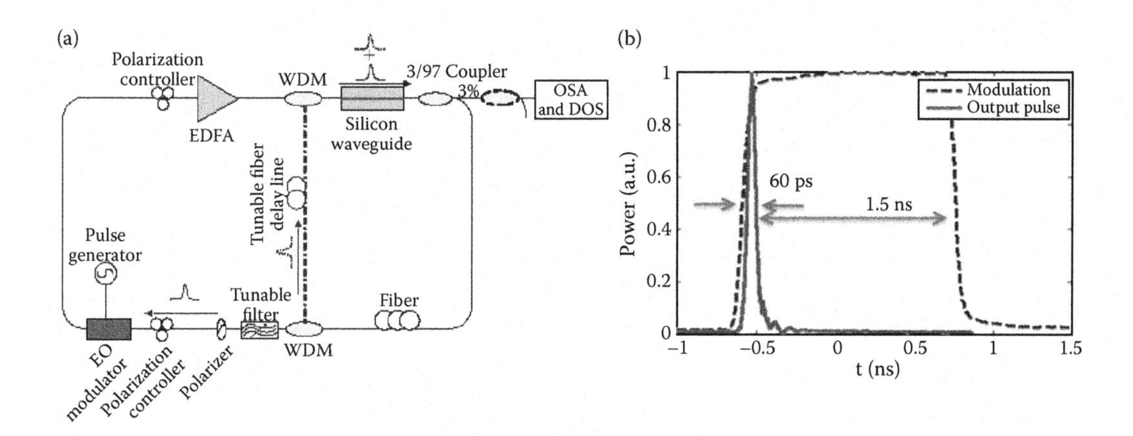

FIGURE 5.41
(a) Schematic of the experimental setup. (b) Compressed pulse for 1.5 ns input pulse. (X. Sang, E. K. Tien, N. S. Yuksek, F. Qian, Q. Song, and O. Boyraz, *IEEE Photon. Technol. Lett.* 20(13), 1184–1186, 2008. © 2008 IEEE.)

is fed into the gain medium, a high power erbium doped fiber amplifier (EDFA). A resonator is formed by launching the EDFA output back into the silicon waveguide input end. An optical bandpass filter is inserted to insure lasing at a desired wavelength and suppress undesired ASE accumulation. Since the pulse shaping by free carriers requires time varying optical signal circulating inside the laser cavity, an electro-optic (EO) modulator is connected to the waveguide to start initial pulsation. The output pulse width is expected to be minimum when the frequency of the function generator matches the fundamental cavity frequency. Here, frequency locking is achieved manually by monitoring the pulse shape and the modulation frequency, simultaneously. The output of the resonator is connected to an optical spectrum analyzer (OSA) and a photodetector followed by a 25-GHz digital sampling oscilloscope (DSO) to measure the output pulse characteristics. Figure 5.41b shows that the 1.5-ns pulse can be compressed to 60 ps [96].

Once the short pulses are generated, we are able to utilize them as pump to achieve stimulated Raman scattering and modelocking in the same silicon waveguide. Because of the wavelength selectivity of the filter and other optics inside the EDFA, another cavity for Raman Stokes pulse is needed. The original laser cavity is modified by using two WDM couplers, which separate the Stokes from the wavelength sensitive components in the pulse compression cavity and recombine before the silicon waveguide, and a tunable fiber delay line as shown in Figure 5.41a, to facilitate resonance at the pump wavelength and the Stokes wavelengths. After pulses of both waves are aligned temporally, we observe dual wavelength lasing [96]. The spectra of the dual wavelength laser are shown in Figure 5.8 [96]. The pump wavelength is selected by the bandpass filter inside the laser cavity at 1540 nm and the expected Raman Stokes signal is at wavelength of 1675 nm. Since free carriers are created and accumulate in the silicon waveguide, the Raman Stokes pulse also suffers from free carrier attenuation. Therefore, the final outcome of the Raman Stokes pulses is the combination of Raman amplification, TPA, and FCA (Figure 5.42).

In the some application, the SPM can generate the ultra broadband spectra in SOI with only a few centimeters, which have extensive applications such as the frequency metrology, all-optical signal regeneration, and so on. A recent experiment demonstrated that the pulses were compressed and the continuum was generated. The spectral slicing and the

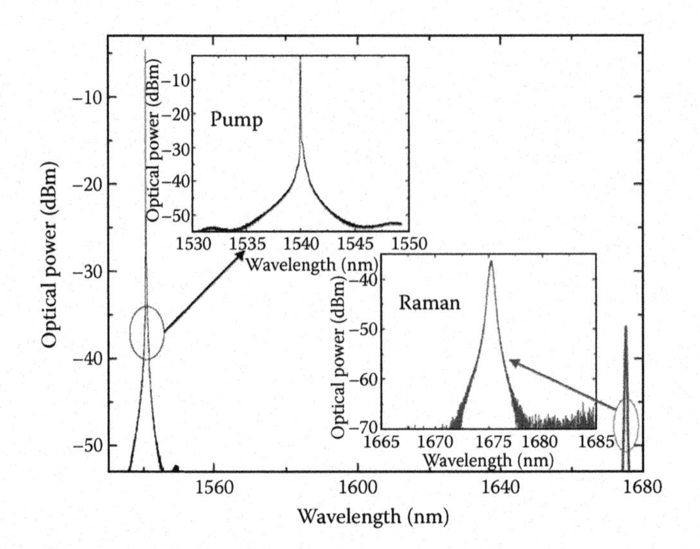

FIGURE 5.42
Spectra of pulse-compressed pump and Raman laser. (X. Sang, E. K. Tien, N. S. Yuksek, F. Qian, Q. Song, and O. Boyraz, *IEEE Photon. Technol. Lett.* 20(13), 1184–1186, 2008. © 2008 IEEE.)

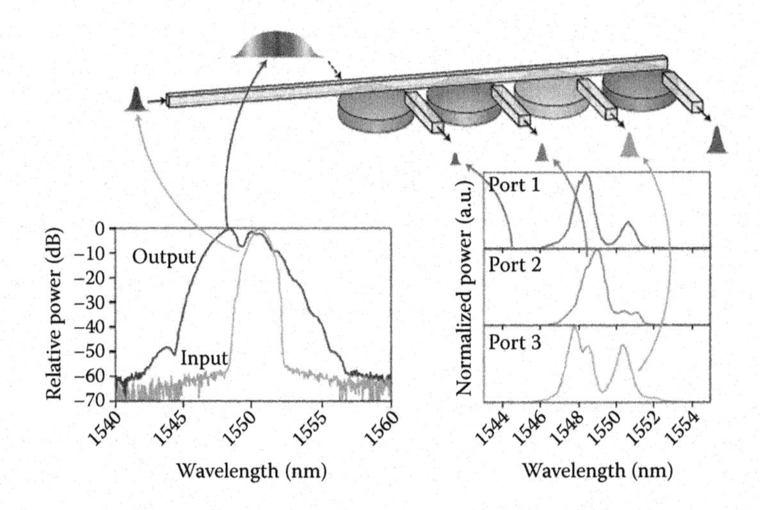

FIGURE 5.43
A short input pulse is spectrally broadened in a nonlinear silicon wire waveguide using SPM and subsequently distributed to several outputs, each at different wavelengths. (Reprinted by permission from Macmillan Publishers Ltd. *Nature Photonics* [9], copyright 2010.)

dropping of several wavelengths were on a single chip, as shown in Figure 5.43. It shows that a short laser pulse is spectrally broadened in a silicon wire waveguide of 2 cm, and it is subsequently dropped onto three outputs at different wavelengths using the ring filters [9].

5.4.3 Controlling of the Flow of Light

The control of the group velocity of pulses of light in optical materials has attracted great attention. Many of the techniques for achieving "slow light" rely on a laser-induced transmission window within a spectral absorption line of a material to produce a reduction in the group velocity. They can be used in data synchronization, tunable data buffers, and pattern correlation. Development of silicon-based nanophotonic devices for photonics-on-chip technology has recently been the subject of substantial research activity. The large Raman gain coefficient (g_R = 4.2 cm/GW) and the relatively narrow line width of silicon represent significant advantages for producing large slow-light effects. Some on-chip delay schemes in silicon have recently been explored, involving resonators [9], making use of the band-edge of silicon [97], or utilizing photonic crystal structures [98].

A slow-light structure of band model is shown in Figure 5.44a,b, a silicon strip waveguide with grated sidewalls resting on a silicon dioxide substrate. Silicon has low absorption near 1.55 μm, high refractive index, and large change in index with temperature; the index varies by >1% from 100°C to 300°C. Sensitivity, length, and dispersion figures of merit were calculated from the dispersion relation of the structure, which was obtained from fully vectorial solutions of Maxwell's equations in a plane-wave basis. Figure 5.44c shows the lowest band near the band edge. The details for simulation are presented in [97]. The results of the exact numerical calculation (symbols) overlaid on the expressions for s, L, and d obtained from the quadratic band model is shown in Figure 5.45 [97].

The large slow-light effects are generated on chip due to the narrower Raman linewidth and highly confining geometries of silicon waveguides. An optical parametric oscillator (OPO) with the central wavelength tuned to 1686.14 nm and bandwidth of 1.47 nm is used as the signal source. To generate the synchronous pump pulses, the OPO signal is detected using a fast photodiode and the photocurrent pulse is used to amplitude modulate a CW laser at 1549.97 nm to produce a pump pulse 50 ps in duration. This pump pulse is amplified using an EDFA, and the signal and pump pulses are combined using a wavelength-division multiplexer (WDM) and coupled into the Si waveguide, as shown in Figure 5.46. To measure the delay, a reference pulse is used. A Fourier transform spectral interferometry (FTSI) allows for measurement of delays as small as 10 fs. A Michelson interferometer is used to generate the signal and reference pulses. A Mach–Zehnder interferometer is used to recombine the pump and signal pulses. The continuously tunable delays are observed by varying the pump power as shown in Figure 5.47. The slope of the delay versus G plot and the theoretically predicted relation between delay and gain implies a

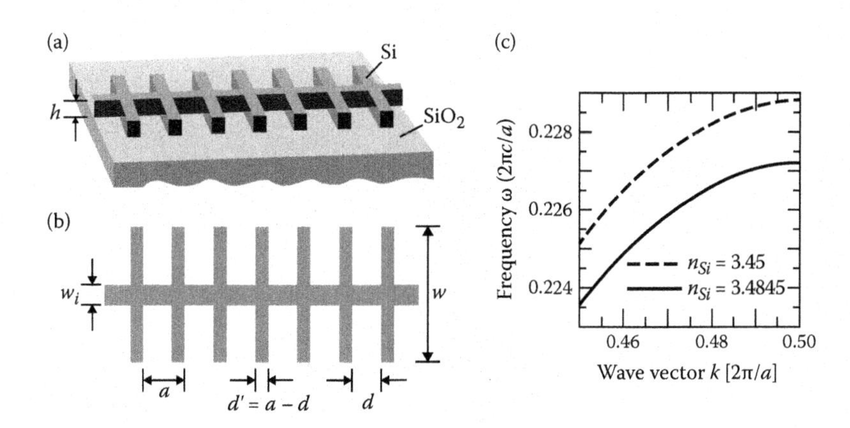

FIGURE 5.44
(a) 3D perspective view of a slow-light grating structure. (b) Top view. (c) Band structure. (From M. L. Povinelli, S. G. Johnson, and J. D. Joannopoulos, *Opt. Express* 13, 7145–7159, 2005. With permission.)

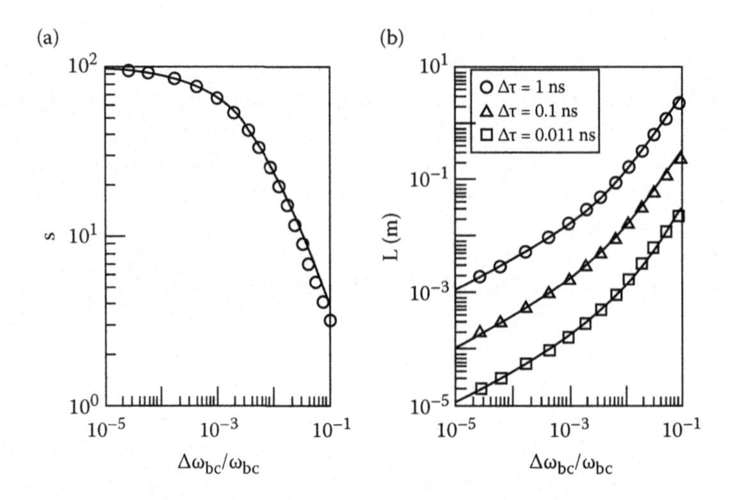

FIGURE 5.45
(a) Sensitivity figure of merit as a function of fractional frequency from the band edge. (b) Required length for different amounts of tunable time delay. Symbols/solid lines are exact/quadratic approximation calculations for a fractional index shift $\Delta n/n = -0.01$. (From M. L. Povinelli, S. G. Johnson, and J. D. Joannopoulos, *Opt. Express* 13, 7145–7159, 2005. With permission.)

Raman line width of 300 GHz. This value is larger than the Raman line width measured experimentally, and It can be concluded that this discrepancy results from the assumption of narrow pulse bandwidths compared with the Raman gain line width in the theory.

As shown in Figure 5.48a, the combination with the use of ultrashort pulses allows the direct visualization of dynamic effects inside the structure. The PhCWs were fabricated on a SOI wafer, with a 220-nm-thick silicon (Si) waveguide layer on top of a 1-μm-thick silicon dioxide (SiO_2) cladding. Holes are arranged in a hexagonal array (period a = 400 and 460 nm, hole radius r = 130 nm, and three rows of missing holes define the so-called W3 PhCW; as

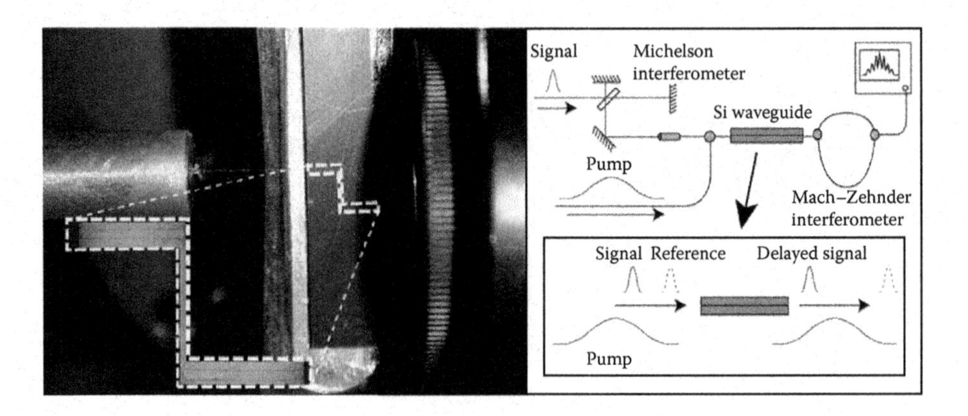

FIGURE 5.46
(left) Experimental setup for coupling into the SOI planar waveguide. (right) Schematic configuration for generating and measuring delay. The signal and pump pulses are overlapped in time going into the waveguide. (From Y. Okawachi, M. A. Foster, J. E. Sharping, and A. L. Gaeta, *Opt. Express* 14, 2317–2322, 2006. With permission.)

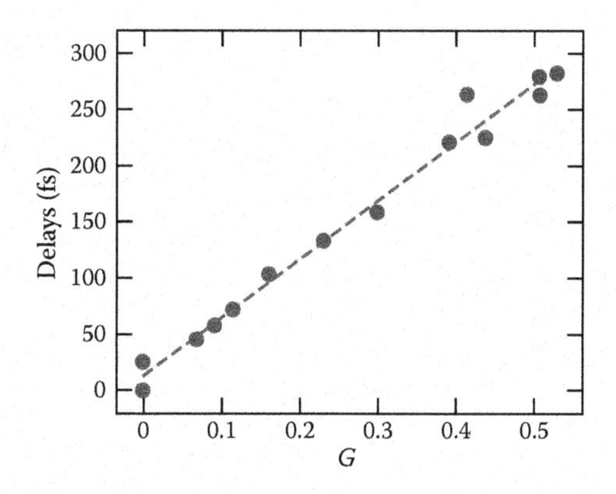

FIGURE 5.47
Total measured delay as a function of the measured Raman gain parameter (dashed line represents the best fit to the data). Continuous tuning of the delay is generated between 0 and 282 fs. We observe the theoretically predicted linear relationship between the delay and the gain parameter. (From Y. Okawachi, M. A. Foster, J. E. Sharping, and A. L. Gaeta, *Opt. Express* 14, 2317–2322, 2006. With permission.)

shown in Figure 5.49b). Linearly TE-polarized light (E // to crystal plane) is coupled to the PhCW.

Figure 5.49 presents pulse tracking measurements for this flat band. Figure 5.49a is the measured topography. Figure 5.49b–k show relationship between the optical amplitude depicted and the reference time. Several dynamic phenomena are observed. First, a first order mode pulse propagating through the structure in Figure 5.49b–e. Second, in the wake of this pulse, striking effects are visible. Fascinatingly, a localized and stationary optical field is observed, which persists for very long times. Despite the fact that the PhCW is an open structure and no resonant cavity is present, the optical field persists for more

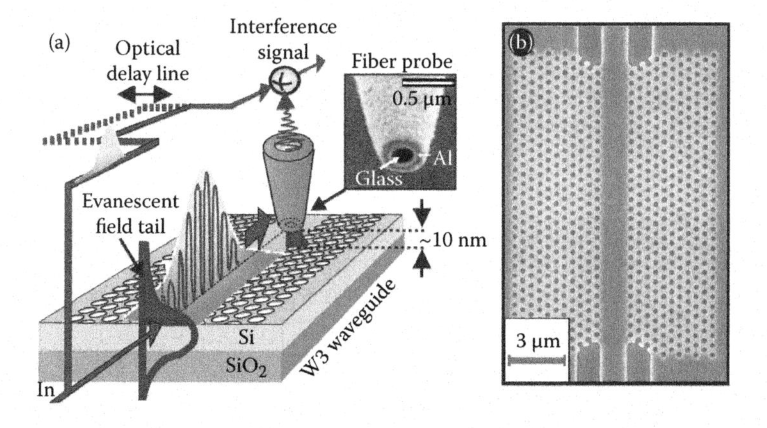

FIGURE 5.48
(a) Schematic representation of a pulse tracking experiment on a W3 PhCW. The evanescent field of a propagating pulse is picked up by a metal coated fiber probe with a subwavelength-sized aperture and interferometrically mixed with light from a reference branch. The inset shows the probe used in the experiment. The aperture size ($\Phi = 240$ nm) in the first approximation determines the optical resolution. Measurements are done by raster scanning the optical probe across the structure at a constant height (<10 nm). (b) Top view of the PhCW under study, although with a shorter device length. (Reprinted with permission from H. Gersen, T. J. Karle, R. J. P. Engelen, W. Bogaerts, J. P. Korterik, N. F. van Hulst, T. F. Krauss, and L. Kuipers, *Phys. Rev. Lett.* 94, 073903, 2005. Copyright 2005 by the American Physical Society.)

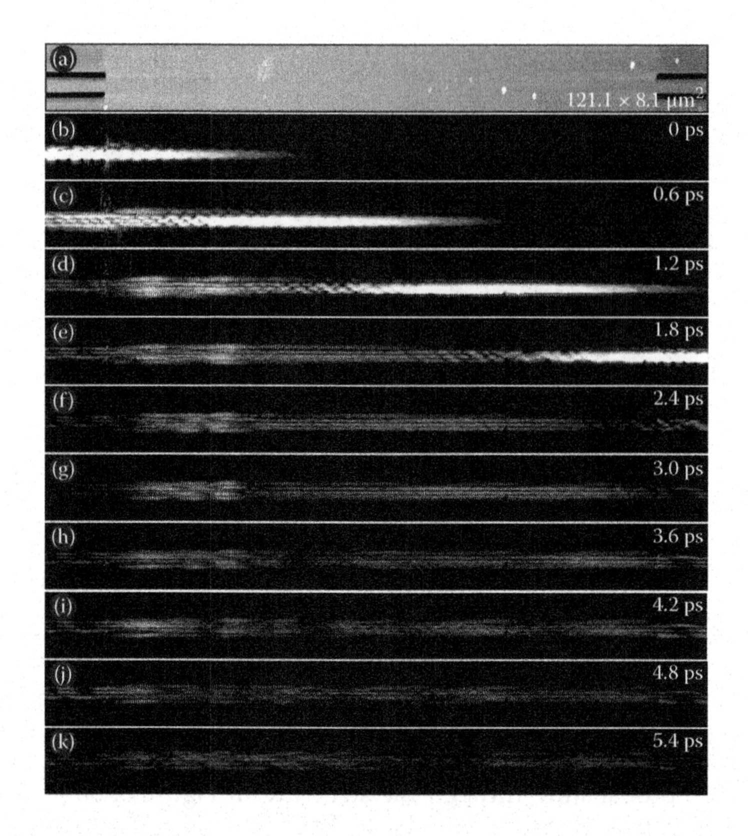

FIGURE 5.49
(See color insert.) Pulse tracking experiment at a flat band at $\omega = 0.305$ ($\lambda = 1310$ nm, $a = 400$ nm). (a) Topography of the PhCW. (b–k) The optical amplitude as a function of reference time (all frames have the same color scale). A movie displaying the pulse propagation is available. We observe a complex and stationary localized modal pattern in the first 25 μm of the PhCW, which exists for more than 3.6 ps after the excitation pulse has moved away. The field pattern moves by less than 0.9 μm in 3 ps, suggesting a group velocity of less than $c = 1000$. Note that all light observed in (f–k) belongs to the same mode located between $k = 0.4$ and $k = 0.6$. (Reprinted with permission from H. Gersen, T. J. Karle, R. J. P. Engelen, W. Bogaerts, J. P. Korterik, N. F. van Hulst, T. F. Krauss, and L. Kuipers, *Phys. Rev. Lett.* 94, 073903, 2005. Copyright 2005 by the American Physical Society.)

than 3.6 ps. This trapping time can be compared with the quality factor Q of a resonator. After roughly 3.6 ps, the spatial shape of the field alters, but even after 5 ps light with a comparable symmetry is still present in the structure. Between 1.2 and 4.2 ps, movement of this localized light field is hardly discernible: if a group velocity should be assigned, it would be at most $c = 1000$ [99].

5.4.4 Nonlinear Optical Signal Processing

5.4.4.1 Wavelength Conversion

All-optical wavelength conversion based on FWM, XPM in silicon waveguide has been studied intensively in recent years. The wavelength conversion has great application in the optical communications to overcome the wavelength-blocking issues at network nodes. However, TPA-generated free carriers accumulate, and the conversion efficiency suffers from FCA at higher data rates. Although a reverse-biased p-i-n junction can be used to actively move free carriers out of the waveguide core, two-photon-induced nonlinear loss still represents a problem in many all-optical signal processing applications. The problem can be avoided by using the silicon organic hybrid (SOH) platform [9].

A more recent experiment demonstrated FWM-based wavelength conversion of a 56-Gbit s^{-1} DQPSK (differential quadrature phase shift keying) phase-encoded signal—the first demonstration of phase-preserving wavelength conversion on a silicon chip. The experiment was performed with a SOH wire waveguide. The experimental setup and eye diagram of the quadrature channel are shown in Figure 5.50 [9].

The optical time-to-frequency conversion and time-lenses have been demonstrated to exploit the FWM in silicon waveguides and take advantage of second-order dispersive fibers. A potential application is the realization of oscilloscopes that can magnify an ultrafast pulse sequence into a lower speed signal or that can magnify a lower bit-rate signal into an ultrafast code.

The conceptual experiment discussed here gives the conceptual framework [9]. The time-to-frequency setup includes the dispersive fibers with lengths larger than the respective dispersion lengths, so that the chirp induced by the dispersion decreases with the fiber length and dispersion D. The set-up still has a coupler and a nonlinear FWM element, as shown in Figure 5.51a. Initially, a pulse sequence P_{in} and a single short-pulse P_{p} are launched into dispersive fibers with negative accumulated total dispersion $-D_1$ and $-D_2$, respectively. The frequency–time spectrograms at the various positions within the setup are plotted beneath the scheme. It can be seen from the first spectrogram that P_{in} and P_{p} are initially unchirped. After passing through the fibers, they both become chirped to different degrees. On performing FWM in the nonlinear element, an idler field is generated from the pump and the signal fields E_{p} and E_{sig}, respectively. Because the amplitude of the pump field is squared, the pump chirp increases to the same amount as the signal

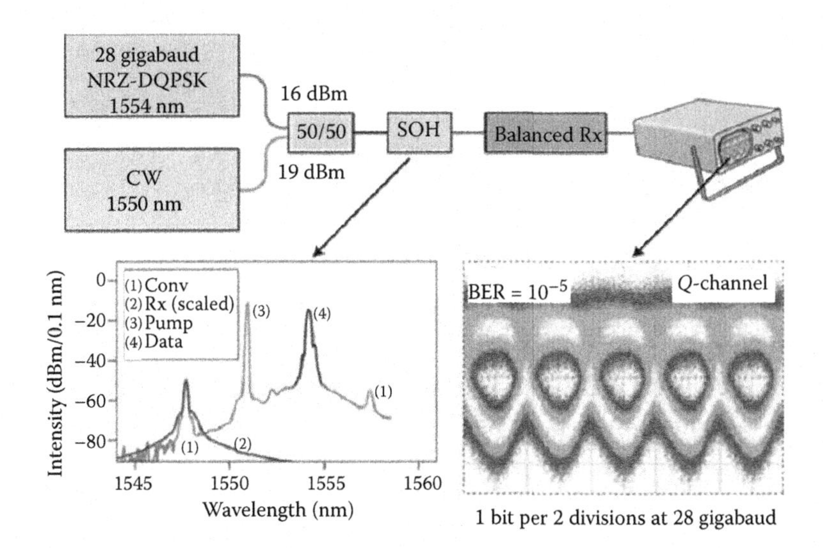

FIGURE 5.50
All-optical wavelength conversion of a 56-Gbit s^{-1} DQPSK signal from 1554 to 1550 nm using FWM in a 4-mm-long silicon organic hybrid (SOH) waveguide. The spectra shows the NRZ DQPSK signal (4), the continuous-wave signal (3), and the new wavelength converted signal (2), which was detected in a balanced received (Rx). The eye diagram shows open eyes for both the in- and quadrature-phase components. BER, bit error rate. (Reprinted by permission from Macmillan Publishers Ltd. *Nature Photonics* [9], copyright 2010.)

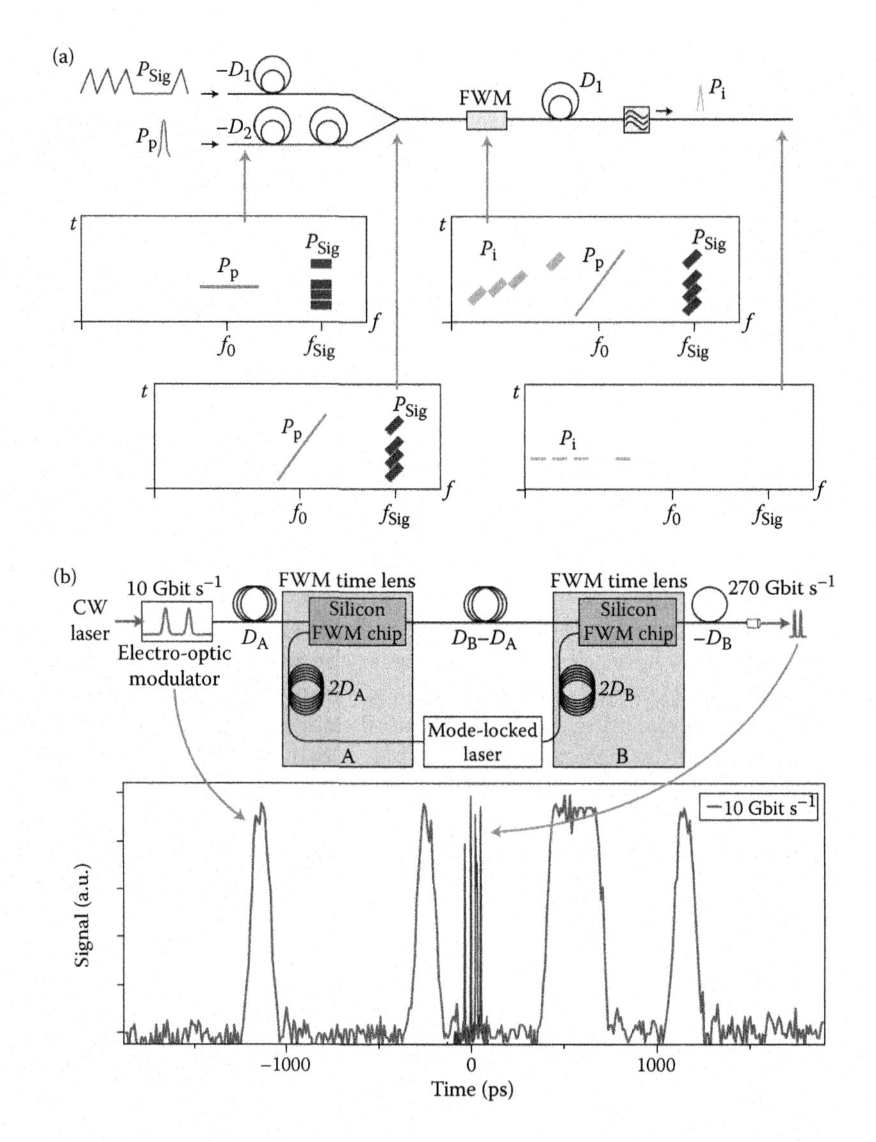

FIGURE 5.51
Time-to-frequency conversion scheme and demonstration of a time lens. (a) Operating principle of time-to-frequency conversion. The spectrograms show an ultrashort-pulse Pp and the 10111 sequence Psig. (b) Setup of a time lens comprising time-to-frequency (A) and frequency-to-time (B) schemes, each with different scaling factors, thus allowing both magnification and demagnification. The pattern located by the left arrow at the bottom was demagnified by a factor of 27 into the pattern located by the right arrow. (Reprinted by permission from Macmillan Publishers Ltd. *Nature Photonics* [9], copyright 2010.)

chirp. As a result of FWM, at any point in time, the spectral components of the idler all keep the same distance from the pump spectrum, making the idler spectrally spread as per the chirp of the involved pulses P_{sig} and P_p. Upon transmitting the idler through a fiber with a totally accumulated positive dispersion D_1, the dispersion is undone, and the pulses arrive all at the same time. Consequently, a perfect time-to-frequency conversion is achieved. Cascading time-to-frequency and frequency-to-time conversion schemes allows the generation of a temporal image of a pulse sequence that is either stretched or compressed in time. This is a useful feature that allows an optical signal either to be sampled with limited-speed electronics or its bit rate to be increased within a certain time window. Figure 5.51b shows a temporal imaging system comprising a time-to-frequency conversion scheme with accumulated dispersion D_A and a frequency-to-time converter with accumulated dispersion $-D_B$. Cascading the two schemes results in a temporal image that is demagnified by a factor D_B/D_A.

5.4.4.2 Signal Regeneration

The limitations on the rate of data transmission in optical communication systems induced by the dispersion, amplifier noise, and fiber nonlinearities can be overcome using optical regeneration techniques. As data rates increase, it is important to perform optical regeneration without relying on high-speed electronic components. Several nonlinear optical techniques have been demonstrated for all-optical signal regeneration, including SPM in fibers or integrated waveguides, XPM, and FWM. Below, as an example, the regeneration is achieved by the SPM method in an integrated silicon device.

The diagram of the SOI (SOI) waveguide device with 8-mm long and the cross-sectional dimension of 250 × 450 nm is shown in Figure 5.52. A spectral component at an offset wavelength is isolated in the drop port of a resonator with 20-μm-radius ring at the end of the waveguide. The linear propagation loss in the waveguide is 3 dB/cm. The transmission of the drop and pass ports of the filter are measured using amplified spontaneous emission as a broadband source, and the measured response is plotted in Figure 5.52b. The free spectral range of the ring resonator is 9 nm, and the bandwidth is 0.75 nm. The center wavelength and the bandwidth of the ring resonator bandpass filter are different for TE and TM polarizations.

Figure 5.53a shows the experimental setup. The femtosecond optical parametric oscillator generates the optical pulses centered at 1550 nm with 75-MHz repetition rate. The 3.5-ps pulses are produced by a 1-nm bandpass filter centered at 1552 nm. TM-polarized light is coupled into the waveguide by a tapered lens fiber and an inverse taper mode converter. Figure 5.53b shows the broadened spectra for different peak pump powers. Due to the low duty cycle of the pulses, SPM in the EDFA is unavoidable, resulting in the sidebands, even without broadening in the silicon waveguide. As the input power is increased, the pulse spectrum undergoes up to 3 nm of broadening due to SPM. At high peak powers, the phase shift induced by the generated free carriers generates a blue shift, which results in

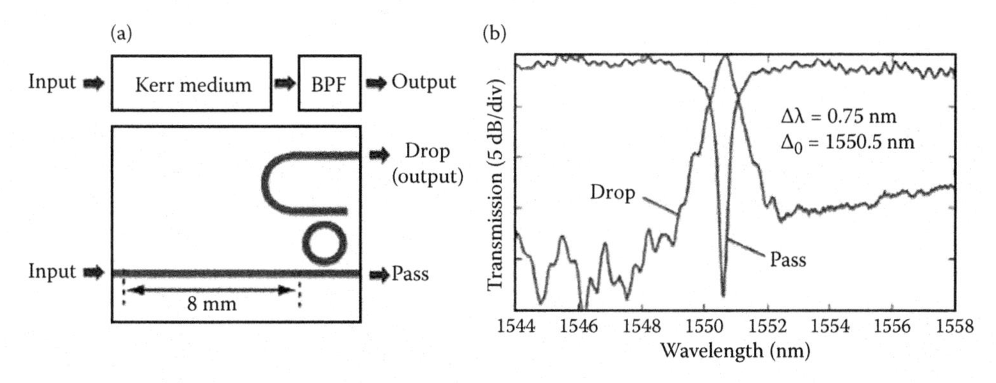

FIGURE 5.52
(a) Diagram of the regenerator device. BPF: band-pass filter. (b) Measured transmission spectra of pass and drop ports of the ring resonator for TM polarization. (From R. Salem, M. A. Foster, A. C. Turner, D. F. Geraghty, M. Lipson, and A. L. Gaeta, *Opt. Express* 15, 7802–7809, 2007. With permission.)

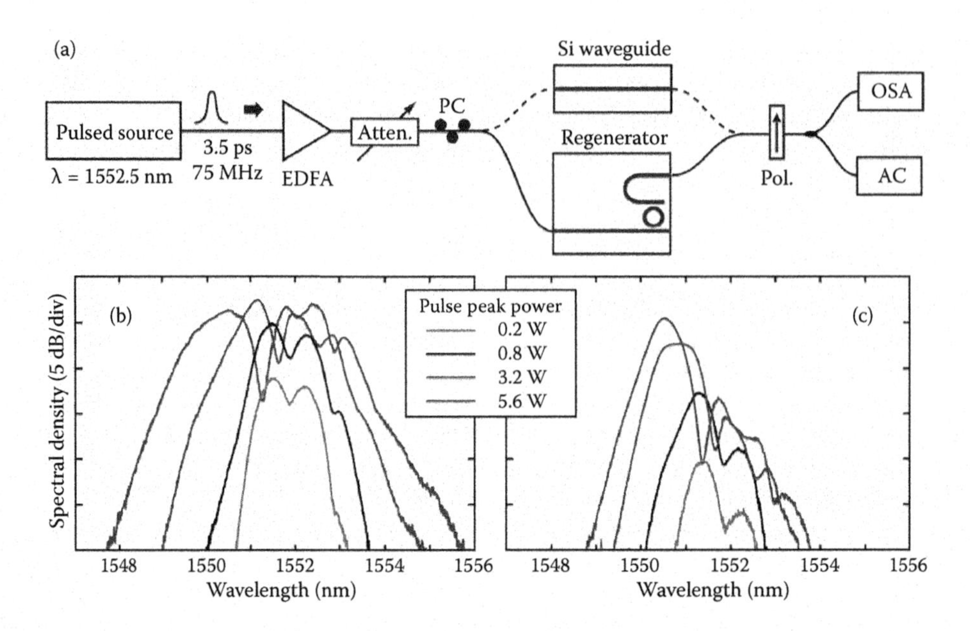

FIGURE 5.53
(a) Experimental setup used for measuring the power transfer function of the regenerator. The waveguide without ring resonator is used to measure the broadened spectrum. (b) Broadened spectra for different pulse peak powers after passing through the silicon waveguide without the ring resonator. (c) Output spectra from the drop port of the ring resonator (regenerator output). (From R. Salem, M. A. Foster, A. C. Turner, D. F. Geraghty, M. Lipson, and A. L. Gaeta, *Opt. Express* 15, 7802–7809, 2007. With permission.)

an asymmetry in the broadened spectrum. As presented in Figure 5.53b, the center wavelength of the ring bandpass filter is 1550.5 nm. Even without broadening, a small fraction of the power passes through the filter, since it has an extinction ratio of only 15 dB at the center of the input spectrum. As the peak power of the input signal increases, the spectrum broadens and more power is coupled into the drop port of the ring resonator. The spectra measured at the drop port of the ring for different pulse peak powers are shown in Figure 5.53c.

To evaluate the performance of the device shown in Figure 5.53a, a simple way to model amplitude jitter and ghost pulses is to vary the power at the logical 1's and 0's randomly by a certain percentage. The mean peak power of the 1's is adjusted to achieve optimal regeneration, and the maximum amplitude variations of the 1's and 0's are set to 15% and 10% of this optimal value, respectively, as shown in Figure 5.43a. Another way is to consider a data signal mixed with random noise. This type of numerical analysis has been done for SPM-based regenerators that use fiber as the nonlinear medium. The details are presented in [54]. A fifth-order super-Gaussian response with 2-nm bandwidth is considered, which provides the maximum improvement of the quality factor. Figure 5.54b shows plots of the eye diagram for the input and output data signal with OSNR = 14 dB. It has been shown previously that the SPM-based regenerator cannot completely eliminate the random noise due to the phase amplitude coupling in the nonlinear process.

5.4.5 Pulse Characterization

The accurate amplitude and phase information of laser pulse is necessary to reveal the dynamic process in ultrafast applications and signal processing systems. Although

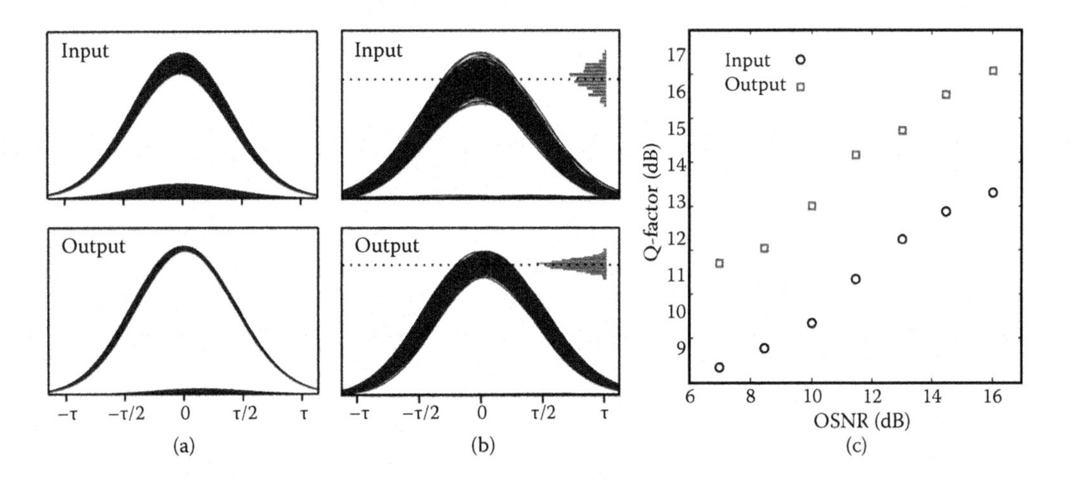

FIGURE 5.54
(a) Simulated eye diagrams of the data at the input and output of the regenerator when amplitude and ghost pulses are present. (b) Simulated eye diagrams of the noisy data at the input and output of the regenerator (histogram for the peak power of the logical 1's is shown). (c) Estimated Q-factor of the input and output for the case of noisy input. (From R. Salem, M. A. Foster, A. C. Turner, D. F. Geraghty, M. Lipson, and A. L. Gaeta, *Opt. Express* 15, 7802–7809, 2007. With permission.)

amplitude measurement is usually straightforward, the phase/chirp information of the pulse on time domain is not easy to be obtained from intensity measurements and full characterization of ultra short pulses is traditionally difficult. Silicon has the inherent advantages of strong Kerr nonlinearity, which facilitates large nonlinear interaction in short lengths. In the application of ultrafast pulse characterization, the autocorrelator has been demonstrated using two photon generated current in the silicon [100]. FWM has been utilized to demonstrate frequency-resolved optical gating (FROG) measurement [101] and an optical oscilloscope for pulsed lasers measurements [102]. Since enough nonlinear efficiency in silicon waveguides can be achieved in short length, and dispersion can be engineered for short "walk off," XPM-based FROG in silicon can facilitate precise pulse characterization at wavelengths from 1.2 to 5 μm without phase matching requirement [103].

Since the dispersive effects are negligible, the propagated probe pulses will carry the information of nonlinear phase modulation, which can be defined as [103]:

$$E_p^{SPM+XPM}(t,\tau) = E_{P0}(t)\exp\left(i\gamma L\left[\frac{2}{3}\left|E_P(t)\right|^2 + \frac{4}{3}\left|E_G(t-\tau)\right|^2\right]\right) \tag{5.19}$$

where $E_p(t)$, $E_G(t)$ are probe and pump fields with orthogonal polarization states, γ is the Kerr nonlinear coefficient, L is the length of the waveguide. The exponential term on the right hand side includes the nonlinear phase shift due to SPM and XPM.

However, by increasing the power in favor of the gate signal at the input we can simplify the Equation 5.19 by neglecting the self phase modulation component as:

$$E_{sig}^{XPM}(t,\tau) = E_P(t)e^{\frac{4}{3}i\gamma\left|E_G(t-\tau)\right|^2} \tag{5.20}$$

The spectrum of the signal is

$$I_{sig}^{XPM}(w,\tau)=\left|\int_{-\infty}^{\infty} E_P(t)e^{\frac{4}{3}i\gamma|E_G(t-\tau)|^2} e^{i\omega t}\right|^2 \quad (5.21)$$

The spectrogram is generated by measuring the spectrum with respect of delay τ between two signals. The pulse amplitude and phase information then can be retrieved from the spectrogram using the principal component generalized projections (PCGP) algorithm [104]. In the PCGP algorithm some criteria of the probe and gate is required to avoid ambiguous solutions. For example, in SHG FROG the criteria are that the probe and gate fields are the same. In the XPM FROG, the probe field is the input probe field $E_P(t)$ and the gate function becomes the XPM exponential term $\exp(4/3i\gamma|E_G(t-\tau)|^2)$ in Equation 5.20. Since the gate function is only a phase modulation on the probe pulse, the amplitude of the gate function is reset to one in each itineration while keeping the phase information from previous step. In addition, the spectrogram is filtered and normalized to compensate the energy fluctuations caused by TPA and the free carrier effect in silicon. The sensitivity of the system can be estimated by assume a π phase shift generated from XPM within the waveguide. The 500-mW peak power is required for generate π phase shift in a 1.7-cm-long waveguide with 5-μm^2 mode area. The estimated sensitivity which is defined as the product of the average power and the peak power is 2.5×10^{-6} W^2. This estimated sensitivity shows that XPM FROG in silicon can achieve similar sensitivity with SHG autocorrelation method. The sensitivity of the system can be improved by using a silicon nanowire waveguide, in which the mode area is about ~30 times smaller.

Figure 5.55 illustrates the experimental setup used for silicon XPM FROG measurements. A fiber modelocked laser at 1550 nm, which generates 540-fs pulses at 20-MHz repetition rate, is used as the source laser. The generated pulses then split into two paths by a polarization beam splitter (PBS). A polarization controller before splitting is inserted to control the splitting ratio and adjust the relative power levels of pump arm (also called the gate signal), and the probe arm. An optical time delay line constructed by two fiber collimators and a moving stage is used to produce tunable time delay for scanning. After passing the delay line, the two polarization branches are combined by a polarization beam coupler (PBC) and launched into a 1.7-cm SOI waveguide. The waveguide has 5-μm^2 modal area. A polarizer aligned with the probe pulse is used to filter the strong pump pulse and to facilitate measurement of changes on the probe arm. Then, we use an optical spectrum analyzer (OSA) to acquire the spectrum at each time delay and generate the spectrogram.

The spectrogram for measurement of a femtosecond pulse generated from the modelocked fiber laser is shown in Figure 5.56a. The spectrogram is taken within 3-ps time delay. The input pulse is then retrieved by the PCGP algorithm. Figure 5.56b shows that the error between the measured and retrieved spectrograms is 0.03%, which can be attributed to imperfections in the polarization combining and irregularities in manual scanning. After >5000 iterations, the actual pulse amplitude and phase information is generated. The retrieved pulse and the phase are shown in Figure 5.57a. We can see that the input pulse has a full width half maximum width of 540 fs. The structure of the pedestal of the pulse is also reconstructed and resolved by the FROG system. To confirm the measurement, an autocorrelation measurement is performed by measuring the two photon absorption generated current across the p-n junction over the waveguide with respect of the time delay between two arms of the probe and gate pulses. Figure 5.57 shows the pulse measured by XPM FROG with pulse measured by autocorrelation. As expected, we achieve good

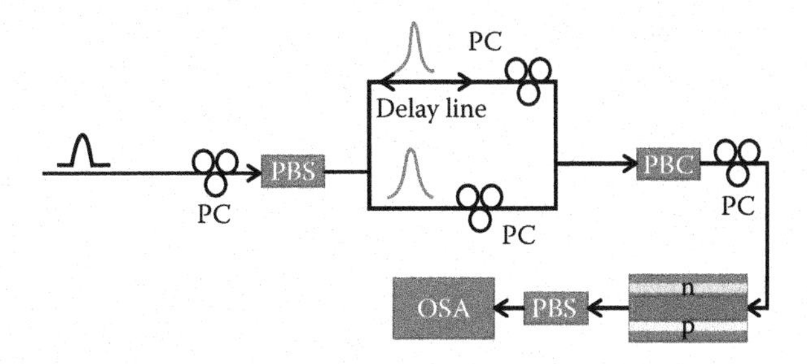

FIGURE 5.55
Schematic setup of XPM FROG in silicon. PC, polarization controller; PBS/PBC, polarization beam splitter/coupler; OSA: optical spectrum analyzer. (Reprinted with permission from [103]. Copyright 2009, American Institute of Physics.)

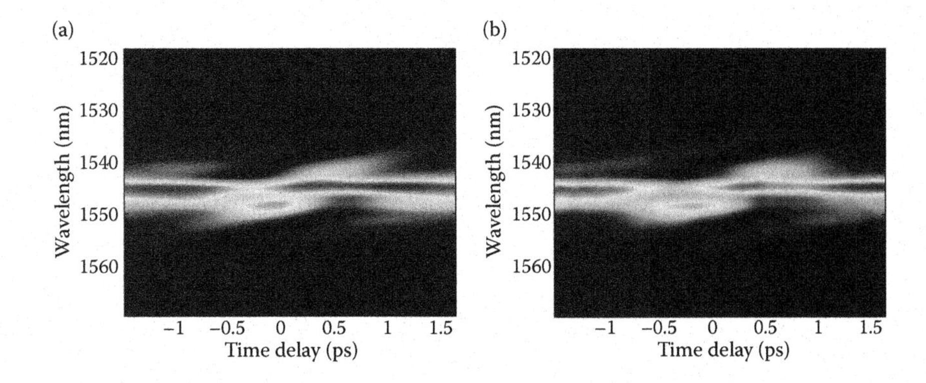

FIGURE 5.56
(a) Retrieved spectrogram and (b) measured spectrogram. (Reprinted with permission from [103]. Copyright 2009, American Institute of Physics.)

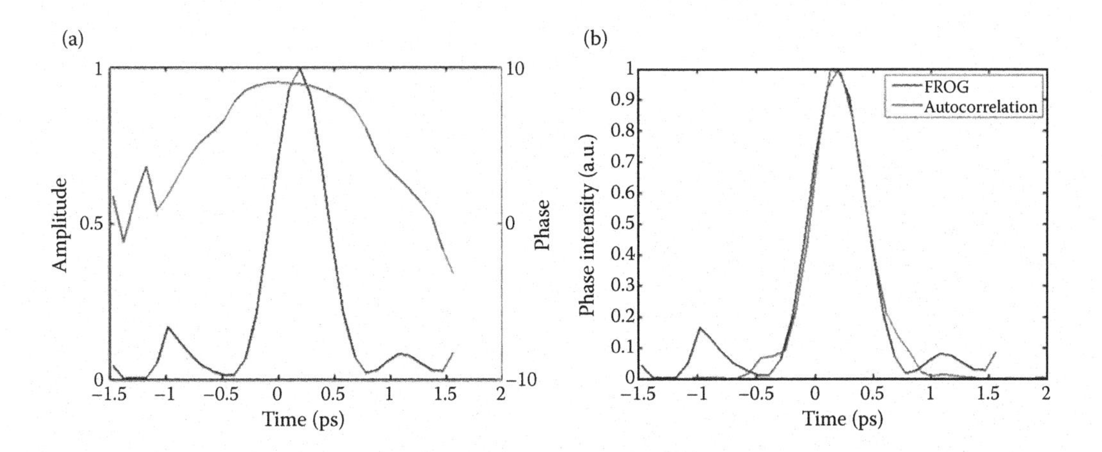

FIGURE 5.57
(a) Retrieved pulse and phase. (b) FROG measured pulse and the autocorrelator measured pulse. (Reprinted with permission from [103]. Copyright 2009, American Institute of Physics.)

agreement on the pulse envelope with the exception of lost sharp features in the autocorrelation measurements as illustrated by the red curve in Figure 5.57b.

5.5 Conclusion

Recent progress in nonlinear effects in silicon and their applications has been reviewed. Silicon waveguide has convincingly demonstrated its ability to amplify, generate, and process signals, which has proved to be a powerful candidate to realize a number of practical nonlinear optical devices. At present, many silicon-based optoelectronic devices have been successfully demonstrated. They are stable and reliable, and can be applied for constructing future all-optical networks. Moreover, compared with other material, its low cost, compactness, mass-manufacturability, and compatibility with CMOS architecture are important to bringing this technology to market.

The field of second-order nonlinear silicon photonics has demonstrated its partly unexpected potentialities. Its inherent access (on the integrated scale) to high bulk $\chi^{(2)}$ values is an important step forward for the silicon photonics field. Nonlinear conversion schemes where a near infrared pump source generates entangled mid-infrared photons via the time-reversed process of SHG (SFG) the spontaneous parametric down conversion [89] or where a near infrared pump can be converted in mid- and far-infrared spectral regions where laser sources are absent, are now available.

References

1. H. K. Tsang, C. S. Wong, T. K. Liang, I. E. Day, S. W. Roberts, A. Harpin, J. Drake, and M. Asghari, "Optical dispersion, two-photon absorption and self-phase modulation in silicon waveguides at 1.5 μm wavelength," *Appl. Phys. Lett.* 80, 416–418, 2002.
2. H. Yamada, M. Shirane, T. Chu, H. Yokoyama, S. Ishida, and Y. Arakawa, "Nonlinear-optic silicon-nanowire waveguides," *Jpn. J. Appl. Phys.* 44, 6541–6545, 2005.
3. E. Dulkeith, Y. A. Vlasov, X. Chen, N. C. Panoiu, and R. M. Osgood, "Self-phase-modulation in submicron silicon-on-insulator photonic wires," *Opt. Express* 14:5524–5534, 2006.
4. V. R. Almeida, Q. Xu, C. A. Barrios, and M. Lipson, "Guiding and confining light in void nanostructure," *Opt. Lett.* 29:1209, 2004.
5. Q. Xu, V. R. Almeida, R. R. Panepucci, and M. Lipson, "Experimental demonstration of guiding and confining light in nanometer-size low-refractive-index material," *Opt. Lett.* 29, 1626–1628, 2004.
6. P. Müllner and R. Hainberger, "Structural optimization of silicon-on-insulator slot waveguides," *IEEE Photon. Technol. Lett.* 18, 2557–2559, 2006.
7. P. Dumon, G. Priem, L. R. Numers, W. Bogaerts, D. V. Thourhout, P. Bienstman, T. K. Liang, M. Tsuchiya, P. Jaenen, S. Beckx, J. Wouters, and R. Beats, "Linear and nonlinear nanophotonic devices based on silicon-on-insulator wire waveguides," *Jpn. J. Appl. Phys.* 45, 6589–6602, 2006.
8. C. Koos, L. Jacome, C. Poulton, J. Leuthold and W. Freude, "Nonlinear silicon-on-insulator. waveguides for all-optical signal processing," *Opt. Express* 15, 5976–5990, 2007.
9. J. Leuthold, C. Koos and W. Freude, "Nonlinear silicon photonics," *Nat. Photon.* 4, 535–544, 2010.
10. M. A. Foster, A. C. Turner, M. Lipson, and A. L. Gaeta, "Nonlinear optics in photonic nanowires," *Opt. Express* 16, 1300–1320, 2008.

11. D. Akimov, M. Schmitt, R. Maksimenka, K. Dukel'skii, Y. Kondrat'ev, A. Khokhlov, V. Shevandin, W. Kiefer, and A. M. Zheltikov, "Supercontinuum generation in a multiple-submicron-core microstructure fiber: toward limiting waveguide enhancement of nonlinear-optical processes," *Appl. Phys. B* 77, 299–305, 2003.
12. A. M. Zheltikov, "The physical limit for the waveguide enhancement of nonlinear-optical processes," *Opt. Spectrosc.* 95, 410–415, 2003.
13. V. Finazzi, T. M. Monro, and D. J. Richardson, "The role of confinement loss in highly nonlinear silica holey fibers," *IEEE Photon. Technol. Lett.* 15, 1246–1248, 2003.
14. M. A. Foster, K. D. Moll, A. L. Gaeta, "Optimal waveguide dimensions for nonlinear interactions," *Opt. Express* 12, 2880–2887, 2004.
15. Y. Lize, E. Magi, V. Ta'eed, J. Bolger, P. Steinvurzel, and B. Eggleton, "Microstructured optical fiber photonic wires with subwavelength core diameter," *Opt. Express* 12, 3209–3217, 2004.
16. L. M. Tong, R. R. Gattass, J. B. Ashcom, S. L. He, J. Y. Lou, M. Y. Shen, I. Maxwell, and E. Mazur, "Subwavelength-diameter silica wires for low-loss optical wave guiding," *Nature* 426, 816–819, 2003.
17. H. Rong, R. Jones, A. Liu, O. Cohen, D. Hak, A. Fang, and M. Paniccia, "A continuous Raman silicon laser," *Nature* 433, 725–728, 2005.
18. M. A. Foster, A. C. Turner, J. E. Sharping, B. S. Schmidt, M. Lipson, and A. L. Gaeta, "Broadband optical parametric gain on a silicon photonic chip," *Nature* 441, 960–963, 2006.
19. K. K. Tsia, S. Fathpour, and B. Jalali, "Energy harvesting in silicon wavelength converters," *Opt. Express* 14, 12327–12333, 2006.
20. T. K. Liang and H. K. Tsang, "Role of free carriers from two-photon absorption in Raman amplification in silicon-on-insulator waveguides," *Appl. Phys. Lett.* 84, 2745–2747, 2004.
21. S. Fathpour, K. K. Tsia and B. Jalali, "Two-photon photovoltaic effect in silicon," *IEEE J. Quantum Electron.* 43, 1211–1217, 2007.
22. X. Sang, E. K. Tien, and O. Boyraz, "Applications of two-photon absorption in silicon," *J. Optoelectron. Adv. Mater.* 11, 15–25, 2009.
23. M. Sheik-Bahae, D. J. Hagan, and E. W. Van Stryland, "Dispersion and band-gap scaling of the electronic Kerr effect in solids associated with two-photon absorption," *Phys. Rev. Lett.* 65, 96–99, 1990.
24. H. Carcia and R. Kalyanaraman, "Phonon-assisted two-photon absorption in the presence of a dc-field: the nonlinear Franz-Keldysh effect in indirect gap semiconductors," *J. Phys. B: Atom. Mol. Opt. Phys.* 39, 2737–2746, 2006.
25. A. D. Bristow, N. Rotenberg and H. M. van Driel, "Two-photon absorption and Kerr coefficients of silicon for 850–2200 nm," *Appl. Phys. Lett.* 90, 191104-1-3, 2007.
26. M. Dinu, F. Quochi and H. Garcia, "Third-order nonlinearities in silicon at telecom wavelength," *Appl. Phys. Lett.* 82, 2954–2956, 2003.
27. M. Sheik-Bahae, J. Wang, and E. W. Van Stryland, "Nondegenerate optical Kerr effect in semiconductors," *IEEE J. Quantum Electron.* 30, 249–255, 1994.
28. T. Kagawa, and S. Ooami, "Polarization dependence of two-photon absorption in Si avalanche photodiodes," *Jpn. J. Appl. Phys.* 46, 664–668, 2007.
29. V. R. Almeida, C. A. Barrios, R. R. Panepucci, and M. Lipson, "All-optical control of light on a silicon chip," *Nature* 481, 1081–1084, 2004.
30. R. Claps, D. Dimitropoulos, V. Raghunathan, Y. Han, and B. Jalali, "Observation of stimulated Raman amplification in silicon waveguides," *Opt. Express* 11, 1731–1739, 2003.
31. Y. A. Vlasov, M. O'Boyle, H. F. Hamann, and S. J. McNab, "Active control of slow light on a chip with photonic crystal waveguides," *Nature* 438, 65–69, 2005.
32. A. W. Snyder and J. D. Love, *Optical Waveguide Theory*. New York: Kluwer Academic Publishers, 1983.
33. C. R. Pollock and M. Lipson, *Integrated Photonics*. Boston: Kluwer Academic Publishers, 2003.
34. R. A. Soref and B. R. Bennett, "Electro-optical effects in silicon," *IEEE J. Quantum Electron.* QE-23, 123, 1987.
35. C. A. Barrios, V. R. Almeida, R. Panepucci, and M. Lipson, "Electrooptic modulation of silicon-on-insulator submicrometer-size waveguide devices," *J. Lightwave Technol.* 21, 2332, 2003.

36. S. R. Giguere, L. Friedman, R. A. Soref, and J. P. Lorenzo, "Simulation studies of silicon electro-optic waveguide devices," *J. Appl. Phys.* 68, 4964, 1990.
37. P. Dainesi, A. Kung, M. Chabloz, A. Lagos, P. Fluckiger, A. Ionescu, P. Fazan, M. Declerq, Ph. Renaud, and Ph. Robert, "CMOS compatible fully integrated Mach-Zehnder interferometer in SOI technology," *IEEE Photon. Technol. Lett.* 12, 660, 2000.
38. P. Dainesi, L. Thevenaz, and P. Robert, "5 MHz 2 × 2 optical switch in silicon on insulator technology using plasma dispersion effect," *Proc. 27th ECOC* 2, 132, 2001.
39. B. Li and S. J. Chua, "Two mode interference photonic waveguide switch," *IEEE J. Lightwave Technol.* 21, 1685, 2003.
40. G. P. Agrawal, *Nonlinear Fiber Optics*. Boston: Academic Press, 1989.
41. J. M. Dudley, G. Genty, and S. Coen, "Supercontinuum generation in photonic crystal fiber," *Rev. Mod. Phys.* 78, 1135–1184, 2006.
42. L. Yin and G. P. Agrawal, "Impact of two-photon absorption on self-phase modulation in silicon waveguides," *Opt. Lett.* 32, 2031–2033, July 15, 2007.
43. Q. Lin, Oskar J. Painter, and Govind P. Agrawal, "Nonlinear optical phenomena in silicon waveguides: Modeling and applications," *Opt. Express* 15, 16621–16628, 2007.
44. I.-W. Hsieh, X. Chen, J. I. Dadap, N. C. Panoiu, and R. M. Osgood, S. J. McNab, and Y. A. Vlasov, "Ultrafast-pulse self-phase modulation and third-order dispersion in Si photonic wire-waveguides," *Opt. Express* 14, 12380–12387, 2006.
45. O. Boyraz, T. Indukuri, and B. Jalali, "Self-phase-modulation induced spectra broadening in silicon waveguides," *Opt. Express* 12, 5, 829–834, 8 March 2004.
46. L. Yin, Q. Lin, and G. P. Agrawal, "Soliton fission and suppercontinuum generation in silicon waveguides," *Opt. Lett.* 32, 391–393, 2007.
47. J. Zhang, Q. Lin, G. Piredda, R. W. Boyd, G. P. Agrawal, and P. M. Fauchet, "Optical solitons in a silicon waveguide," *Opt. Express* 15, 7682–7688, 2007.
48. I. W. Hsieh, X. G. Chen, X. P. Liu, J. I. Dadap, N. C. Panoiu, C. Y. Chou, F. N. Xia, W. M. Green, Y. A. Vlasov, and R. M. Osgood, "Supercontinuum generation in silicon photonic wires," *Opt. Express* 15, 15242–15249, 2007.
49. R. Salem, M. A. Foster, A. C. Turner, D. F. Geraghty, M. Lipson, A. L. Gaeta, "All-optical regeneration on a silicon chip," *Opt. Express* 15, 7802–7809, 2007.
50. P. V. Mamyshev, "All-optical data regeneration based on self-phase modulation effect," in *Proc. European Conference on Optical Communications* (ECOC98), 475, 1998.
51. O. Boyraz, P. Koonath, V. Raghunathan, and B. Jalali, "All optical switching and continuum generation in silicon waveguides," *Opt. Express* 12, 4094–4102, 2004.
52. R. Dekker, A. Driessen, T. Wahlbrink, C. Moormann, J. Niehusmann, and M. Först, "Ultrafast Kerr-induced all-optical wavelength conversion in silicon waveguides using 1.55 μm femtosecond pulses," *Opt. Express* 14, 8336–8346, 2006.
53. I. Hsieh, X. Chen, J. Dadap, N. Panoiu, R. Osgood, Jr., S. McNab, and Y. Vlasov, "Cross-phase modulation-induced spectral and temporal effects on co-propagating femtosecond pulses in silicon photonic wires," *Opt. Express* 15, 1135–1146, 2007.
54. Q. Lin, J. Zhang, P. M. Fauchet, and G. P. Agrawal, "Ultrabroadband parametric generation and wavelength conversion in silicon waveguides," *Opt. Express* 14, 4786–4799, 2006.
55. O. Boyraz and B. Jalali, "Demonstration of a silicon Raman laser," *Opt. Express* 12, 5269–5273, 2004.
56. H. Rong, A. Liu, R. Jones, O. Cohen, D. Hak, R. Nicolaescu, A. Fang and M. Paniccia, "An all-silicon Raman laser," *Nature* 433, 294–296, 2005.
57. D. Dimitropoulos, B. Houshmand, R. Claps, and B. Jalali, "Coupled mode theory of Raman effect in silicon-on-insulator waveguides," *Opt. Lett.* 28, 1954–1956, 2003.
58. R. L. Espinola, J. I. Dadap, R. M. Osgood, S. J. McNab, and Y. A. Vlasov, "Raman amplification in ultrasmall silicon-on-insulator wire waveguides," *Opt. Express* 16, 3716–3718, 2004.
59. G. Roelkens, L. Liu, D. VanThourhout, R. Baets, R. Nötzel, "Light emission and enhanced nonlinearity in nanophotonic waveguide circuits by III–V/silicon-on-insulator heterogeneous integration," *J. Appl. Phys.* 104, 033117-1-7, 2008.

60. G. Roelkens, L. Liu, D. VanThourhout, R. Baets, R. Nötzel, "Enhanced nonlinearity in SOI micro cavities by III–V/SOI heterogeneous integration," *ECOC 2008*, 2, 167–168, September 21–25 2008.
61. X. Checoury, M. El Kurdi, Z. Han, and P. Boucaud, "Enhanced spontaneous Raman scattering in silicon photonic crystal waveguides on insulator," *Opt. Express* 17, 3500–3507, 2009.
62. M. N. Wybourne, *Properties of Silicon*, EMIS Data Review Series 4, London: INSPEC, 1988.
63. R. W. Boyd. *Nonlinear Optics*, 2nd edition. San Diego: Academic Press, 2003.
64. J. Midwinter and J. Warner, "Up-conversion of near infrared to visible radiation in lithium-meta-niobate," *J. Appl. Phys.* 38, 519, 1967.
65. N. Bloembergen, R. K. Chang, S. S. Jha, and C. H. Lee, "Optical second-harmonic generation in reflection from media with inversion symmetry," *Phys. Rev.* 174, 813–822, 1968.
66. D. Guidotti, T. A. Driscoll, and H. G. Geritsen, "Second harmonic generation in centro-symmetric semiconductors," *Solid State Commun.* 46, 337–340, 1983.
67. S. V. Govorkov, V. I. Emel'yanov, N. I. Koroteev, G. I. Petrov, I. L. Shumay, V. V. Yakovlev, and R. V. Khokhlov, "Inhomogeneous deformation of silicon surface layers probed by second-harmonic generation in reflection," *J. Opt. Soc. Am. B* 6, 1117–1124, 1989.
68. Y. R. Shen, *The Principles of Nonlinear Optics*. New York: Ed. John Wiley and Sons, 1984.
69. M. Falasconi, L. C. Andreani, A. M. Malvezzi, M. Patrini, V. Mulloni, and L. Pavesi, "Bulk and surface contributions to second-order susceptibility in crystalline and porous silicon by second-harmonic generation", *Surf. Sci.* 481, 105–112, 2001.
70. R. S. Jacobsen, K. N. Andersen, P. I. Borel, J. Fage-Pedersen, L. H. Frandsen, O. Hansen, M. Kristensen, A. V. Lavrinenko, G. Moulin, H. Ou, C. Peucheret B. Zsigri, and A. Bjarklev, "Strained silicon as a new electro-optic material," *Nature* 441, 199–202, 2006.
71. T. Baehr-Jones, J. Witzens, and M. Hochberg, "Theoretical study of optical rectification at radio frequencies in a slot waveguide," *IEEE J. Quantum Electron.* 46, 1634–1641, 2010.
72. N. K. Hon, K. K. Tsia, D. R. Solli, and B. Jalali, "Periodically-poled silicon," *Appl. Phys. Lett.* 94, 091116, 2009.
73. M. Cazzanelli, F. Bianco, E. Borga, G. Pucker, M. Ghulinyan, E. Degoli, E. Luppi, V. Véniard, S. Ossicini, D. Modotto, S. Wabnitz, R. Pierobon, and L. Pavesi, "Second-harmonic generation in silicon waveguides strained by silicon nitride," *Nat. Mater.* 11, 148–154, 2011.
74. D. Modotto, V. V. Kozlov, S. Wabnitz, M. Cazzanelli, F. Bianco, L. Pavesi, M. Ghulinyan, and G. Pucker, presented at *Advanced Photonics, 2012 Congress*, June 17–21, 2012, Cheyenne Mountain Resort, Colorado Springs, CO, USA.
75. H. W. K. Tom, T. F. Heinz, and Y. R. Shen, "Second harmonic reflection from silicon surfaces and its relation to structural symmetry," *Phys. Rev. Lett.* 51, 1983, 1983.
76. J. I. Dadap, Z. Xu, X. F. Hu, M. C. Downer, N. M. Russell, J. G. Ekerdt, and O. A. Aktsipetrov, "Second harmonic spectroscopy of a Si(001) surface during calibrated variations in temperature and hydrogen coverage," *Phys. Rev. B* 56, 13367, 1997.
77. V. I. Gavrilenko, R. Q. Wu, M. C. Downer, J. G. Ekerdt, D. Lim, and P. Parkinson, "Optical second-harmonic spectra of silicon ad-atom surfaces: theory and experiment," *Thin Solid Films* 364, 1–5, 1999.
78. J. A. Litwin, J. E. Sipe, and H. M. van Driel, "Picosecond and nanosecond laser induced second-harmonic generation from centrosymmetric semiconductors," *Phys. Rev. B* 31, 5543, 1985.
79. R. L. Sutherland, *Handbook of Nonlinear Optics*, 2nd edition. Boca Raton, FL: CRC Press, 2003.
80. J. Y. Huang, "Probing inhomogeneous lattice deformation at interface of Si(111)/SiO2 by optical second-harmonic reflection and raman spectroscopy," *Jpn. J. Appl. Phys.* 33, 3878, 1994.
81. C. Schriever, C. Bohley, and R. B. Wehrspohn, "Strain dependence of second-harmonic generation in silicon," *Opt. Lett.* 35, 273–275, 2010.
82. C. Schriever, C. Bohley, J. Schilling, and R. B. Wehrspohn, "Strain dependence of second-harmonic generation in silicon," *Proc. SPIE* 7719, 77190Z, 2010.
83. C. Schriever, C. Bohley, and R. B. Wehrspohn, Strain-induced nonlinear optics in silicon. In *Mechanical Stress on the Nanoscale—Simulation, Material Systems and Characterization Techniques*, Germany: Wiley-VCH, 2011.

84. M. Galli, D. Gerace, K. Welna, T. F. Krauss, L. O'Faolain, G. Guizzetti, and L. C. Andreani, "Low-power continuous-wave generation of visible harmonics in silicon photonic crystal nanocavities," *Opt. Express* 18, 26613–26624, 2010.
85. M. McCutcheon, G. W. Rieger, I. W. Cheung, J. F. Young, D. Dalacu, S. Frederic, P. J. Poole, G. C. Aers, and R. Williams, "Resonant scattering and second-harmonic spectroscopy of planar photonic crystal microcavities," *Appl. Phys. Lett.* 87, 221110, 2009.
86. J. S. Levy, M. A. Foster, A. L. Gaeta, and M. Lipson, "Harmonic generation in silicon nitride ring resonators," *Opt. Express* 19, 11415–11421, 2011.
87. J. B. Driscoll, N. Ophir, R. R. Grote, J. I. Dadap, N. C. Panoiu, K. Bergman, and R. M. Osgood, Jr., "Width-modulation of Si photonic wires for quasi-phase-matching of four-wave-mixing: experimental and theoretical demonstration," *Opt. Express* 20, 9227–92242, 2012.
88. B. Chmielak, M. Waldow, C. Matheisen, C. Ripperda, J. Bolten, T. Wahlbrink, M. Nagel, F. Merget, and H. Kurz, "Pockels effect based fully integrated, strained silicon electro-optic modulator," *Opt. Express* 19, 17212, 2011.
89. I. D. Rukhlenko, M. Premaratne, and G. P. Agrawal, "Nonlinear silicon photonics: analytical tools," *IEEE J. Sel. Top. Quantum Electron.* 16, 200, 2010.
90. I. Rukhlenko, M. Premaratne, C. Dissanayake, and G. Agrawal, "Continuous-wave Raman amplification in silicon waveguides: beyond the undepleted pump approximation," *Opt. Lett.* 34, 536–538, 2009.
91. X. Chen, "Nonlinear optics in silicon photonic wires: theory and applications," PhD thesis, Columbia University, 2007.
92. S. Xinzhu and O. Boyraz, "Gain and noise characteristics of high-bit-rate silicon parametric amplifiers,"*Opt. Express* 16, 13122–13132, 2008.
93. Q. Lin, J. Zhang, G. Piredda, R. W. Boyd, P. M. Fauchet, and G. P. Agrawal, "Dispersion of silicon nonlinearities in the near infrared region," *Appl. Phys. Lett.* 91, 021111–021113, 2007.
94. E. Tien, N. S. Yuksek, F. Qian, and O. Boyraz, "Pulse compression and modelocking by using TPA in silicon waveguides," *Opt. Express* 15, 6500–6506, 2007.
95. E. Tien, F. Qian, N. S. Yuksek, and O. Boyraz, "Influence of nonlinear loss competition on pulse compression and nonlinear optics in silicon," *Appl. Phys. Lett.* 91, 201115-1–201115-3, 2007.
96. X. Sang, E. K. Tien, N. S. Yuksek, F. Qian, Q. Song, and O. Boyraz, "Dual-wavelength mode-locked fiber laser with an intracavity silicon waveguide," *IEEE Photon. Technol. Lett.* 20(13), 1184–1186, 2008.
97. M. L. Povinelli, S. G. Johnson, and J. D. Joannopoulos, "Slow-light, band-edge waveguides for tunable time delays," *Opt. Express* 13, 7145–7159, 2005.
98. Y. Okawachi, M. A. Foster, J. E. Sharping, and A. L. Gaeta, "All-optical slow-light on a photonic chip," *Opt. Express* 14, 2317–2322, 2006.
99. H. Gersen, T. J. Karle, R. J. P. Engelen, W. Bogaerts, J. P. Korterik, N. F. van Hulst, T. F. Krauss, and L. Kuipers, "Real-space observation of ultraslow light in photonic crystal waveguides," *Phys. Rev. Lett.* 94, 073903, 2005.
100. T. K. Liang, H. K. Tsang, I. E. Day, J. Drake, A. P. Knights, and M. Asghari, "Silicon waveguide two-photon absorption detector at 1.5 mu m wavelength for autocorrelation measurements," *Appl. Phys. Lett.* 81, 1323–1325, 2002.
101. A. F. Mark, S. Reza, F. G. David, C. T. Amy, L. Michal, and L. G. Alexander, "Frequency-resolved optical gating on a silicon photonic chip," in *Coherent Optical Technologies and Applications*, Optical Society of America, Boston, 2008, CMC4.
102. M. A. Foster, R. Salem, D. F. Geraghty, A. C. Turner-Foster, M. Lipson, and A. L. Gaeta, "Silicon-chip-based ultrafast optical oscilloscope," *Nature* 456, 81–84, 2008.
103. E.-K. Tien, X. Z. Sang, F. Qian, Q. Song, and O. Boyraz, "Ultrafast pulse characterization using cross phase modulation in silicon," *Appl. Phys. Lett.* 95, 051101, 2009.
104. D. J. Kane, "Real-time measurement of ultrashort laser pulses using principal component generalized projections," *IEEE J. Select. Top. Quantum Electron.* 4, 278–284, 1998.

6

Long-Wavelength Photonic Circuits

Goran Z. Mashanovich, Milan M. Milošević, Sanja Zlatanovic, Faezeh Gholami, Nikola Alic, Stojan Radic, Zoran Ikonic, Robert W. Kelsall, and Gunther Roelkens

CONTENTS

6.1 Introduction

Goran Z. Mashanovich

From the previous chapters, it could be seen that the majority of research effort in silicon photonics has been focused on the near-IR wavelength region. The long wavelength IR region, on the other hand, offers a plethora of possible applications ranging from sensing, medical diagnostics, and free space communications, to thermal imaging and IR countermeasures.

Silicon and germanium are transparent in a broad range of the long-wave IR and therefore can be used as photonics platforms in this wavelength region. Furthermore, the free carrier plasma dispersion effect should be stronger, two photon absorption is reduced, and more robust optical fibers are now available at longer wavelengths. In addition, the dimensional tolerances are more relaxed than in the near-IR. Moreover, III–V sources and detectors could be bonded on the silicon wafer, similar to approaches already demonstrated in the near-IR. Finally, silicon photonics, or group IV photonics, long wavelength platform is potentially low-cost and offers a possibility of photonic-electronic integration.

Recently, there has been an increased interest in long wavelength group IV photonics. This chapter presents recent results in the field. Possible waveguide structures have been suggested, and a review of silicon photonics passive structures designed for wavelengths ranging from 3.4 to 5.5 μm is given in Section 6.2. Different material platforms, such as silicon-on-insulator (SOI), silicon-on-sapphire, silicon on porous silicon and suspended silicon have been used.

A basic overview of optical nonlinearities and nonlinear material properties of silicon in the mid-IR region, as well as the associated anisotropy and dispersion is given in Section 6.3. The main properties of silicon waveguides contributing to efficient nonlinear interactions, supported with recent experimental results are also presented, together with experimental demonstrations of the main nonlinear phenomena, including Raman amplification, parametric amplification, and four-wave mixing.

Mid-IR and terahertz sources and detectors, including quantum cascade lasers, intracenter-based terahertz lasers, Si/SiGe quantum well, and quantum dot detectors, and micro-bolometers are discussed in Section 6.4. Finally, in Section 6.5, preliminary results on mid-IR heterogeneous integration of III–V and silicon are shown.

The long wavelength group IV photonics field is still in its infancy compared with the near-IR research, and there are a number of challenges that need to be overcome in the future before group IV photonic integrated circuits are employed in a host of applications offered by this spectral range. Nonetheless, significant progress has been made recently as described in this chapter. Therefore, we should expect to see some rather remarkable and exciting advances in this field in the near future.

6.2 Passive Devices for the Mid-Infrared Silicon Photonics

Milan M. Milošević and Goran Z. Mashanovich

6.2.1 Introduction

The fundamental challenge for mid-IR group IV photonics is the fact that the most popular material platform in the NIR range, that of SOI, cannot be used in the majority of the MIR, due to high material losses of SiO_2. Therefore, other waveguide structures need to be developed. Soref et al. proposed several waveguide structures suitable for mid- and long-IR spectral regions [1, 2]: Si rib-membrane waveguides, Si-on-Si_3N_4 (SON), Si-on-sapphire (SOS), Ge-on-Si, Ge-on-SOI, or GeSn-on-Si strip, and slot waveguides, or hollow waveguides with Bragg or antiresonant cladding [1,2]. As waveguide dimensions scale with the wavelength, to reduce the dimensions and facilitate CMOS compatibility of MIR devices, plasmonic waveguides, and plasmonic waveguided components may prove to be a suitable solution. It is expected that the plasmonic propagation loss will decrease significantly as the wavelength of operation is increased into the MIR, and far-IR [2]. In this section, we will present recently reported experimental data on passive MIR photonics devices.

6.2.2 Silicon-on-Insulator

It is well known from literature that silicon is relatively low loss (<2 dB/cm) for wavelengths up to 8 μm, while there are several multiphonon absorption peaks at longer wavelengths [1,2]. On the other hand, SiO_2 optical loss rapidly increases beyond 3.6 μm, and therefore, SOI is not a suitable candidate for longer wavelengths [3].

The SOI platform can be, however, used at shorter MIR wavelengths around 2 μm, as shown in Section 6.5. Mashanovich et al. were the first to report losses in the 3- to 4-μm wavelength range. They designed, fabricated, and characterized SOI waveguides at 3.39 μm [3] following the guidelines published in [4]. Rib waveguides with a height of 2 μm, width of 2 μm and etch depth of 1.2 μm were chosen, as their overall dimensions ensured efficient coupling from MIR optical fibers since a low-power MIR laser was used. In addition, almost polarization insensitive light propagation is achievable in these relatively large waveguides, which is important as polarization maintaining mid-IR fibers are not yet commercially available. Standard lithography and reactive ion etching (RIE) were used in fabrication [3].

The measurements were undertaken by using a 3.39-μm, CW linearly polarized He-Ne laser (Figure 6.1). Additional polarization control was provided by two ZnSe MIR polarizers. Light was coupled into a 9/125-μm single-mode MIR optical fiber [5] via a ZnSe objective lens and then it was butt coupled into the sample. Light exiting the sample was butt coupled to another single mode MIR fiber and then it was directed to a MIR detector. To confirm MIR light was propagating through the waveguides, a broadband NIR source and detector were introduced into the setup. A NIR camera was used to align the sample such that light was visible in the waveguide. After the output power was maximized on the detector, the fiber connectors were used to switch the input and output to the MIR laser and detector. This setup represent a low cost alternative to a setup with a much more expensive MIR camera [6]. To increase the signal-to-noise ratio, a chopper, and a lock-in amplifier were used. To accurately measure the output signal from the sample, a sample stage with a Peltier element was added to change the temperature of the sample and avoid a Fabry–Perót (FP) dependence of the output signal.

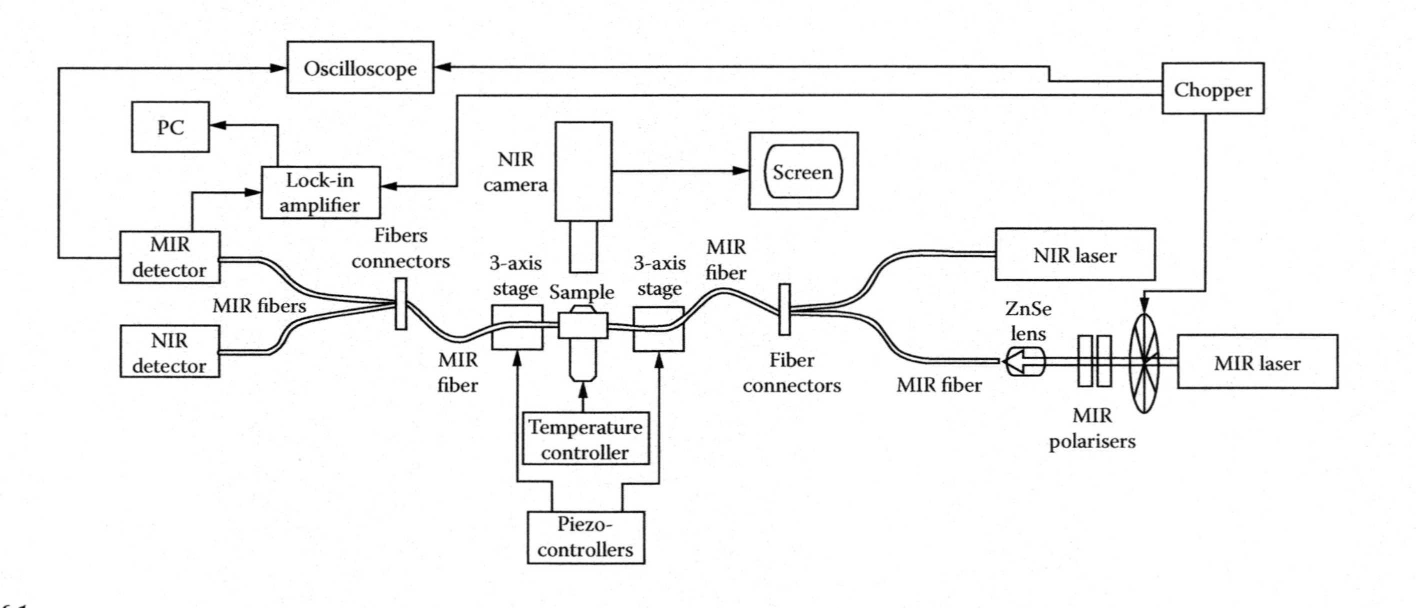

FIGURE 6.1
Experimental setup used for measurements of the SOI rib waveguides. (From G. Z. Mashanovich, M. M. Milošević, M. Nedeljkovic, N. Owens, B. Xiong, E. J. Teo, and Y. Hu, *Opt. Express* vol. 19, 7112–7119, 2011. With permission.)

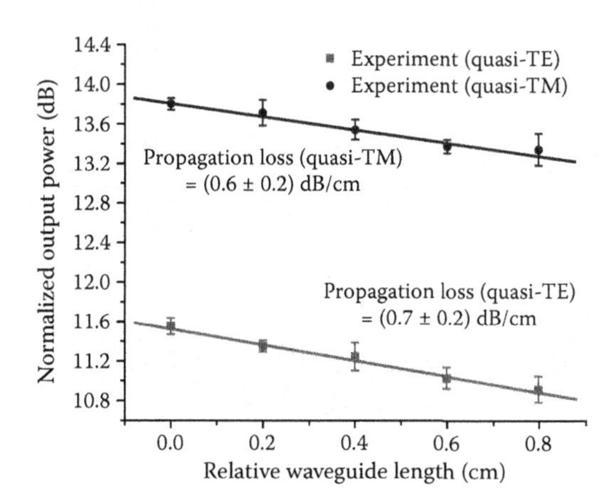

FIGURE 6.2
Propagation losses for SOI rib waveguides were similar for both TE and TM input polarizations ($H = W = 2$ μm, $D = 1.2$ μm, $\lambda = 3.39$ μm). (From G. Z. Mashanovich, M. M. Milošević, M. Nedeljkovic, N. Owens, B. Xiong, E. J. Teo, and Y. Hu, *Opt. Express* vol. 19, 7112–7119, 2011. With permission.)

The noise floor was more than 20 dB lower than the typical measured peak power and the total on chip loss was around 23 dB. The experimentally measured bend loss was 0.006 ± 0.002 dB for 90° bends with 200-μm radius.

The waveguides with a 2-μm-thick BOX had losses of 1.9–2.4 dB/cm. As the propagation loss was still relatively high, the waveguides were oxidized with a 20- to 30-nm-thick thermal oxide to reduce the surface roughness. This resulted in propagation losses of 0.6–0.7 dB/cm at 3.39 μm (Figure 6.2). Measurements were performed for different input polarizations and the results for TE/TM input polarizations were very similar (Figure 6.2).

Milošević et al. used tunable laser with central wavelength at 3.8 μm to characterize the same chips at longer wavelengths [7]. Preliminary propagation losses of 2.9 dB/cm at $\lambda = 3.73$ μm and 3.4 dB/cm at $\lambda = 3.8$ μm were obtained for unoxidized samples, while 1.5 dB/cm at $\lambda = 3.73$ μm and 1.8 dB/cm at $\lambda = 3.8$ μm were obtained for oxidized samples [7]. These results reveal very promising prospects for SOI platform at wavelengths longer than 3.6 μm. It can be seen that these losses are about 1 dB/cm higher than propagation losses for the same structure at 3.39 μm. This is not surprising taking into account that SiO_2 material loss increases beyond 3.6 μm. It is believed that the influence of SiO_2 material loss [8] on waveguide propagation losses is still tolerable at these wavelengths and for relatively large waveguides.

Shankar et al. characterized silicon photonic crystal cavities realized on SOI platform at even longer wavelength of 4.4 μm. The cavities were undercut and the authors used the resonant scattering method for the characterization (Figure 6.3) [9]. A tunable quantum cascade laser (QCL) with emission from 4.315 to 4.615 μm was used to send the light into a ZnSe objective lens with numerical aperture (NA) of 0.22 and focus it onto the sample. The sample was placed so that the cavity mode polarization was oriented at 45° with respect to the E-field of the laser spot. A visible HeNe laser beam was aligned to the path of the QCL and then used to align the rest of the optics, while Ge neutral density filter was used to reduce the light intensity. The sample was mounted on an automatic micropositioner stage, which can be scanned using computer control. The light that was coupled and re-emitted by the photonic crystal cavities was backscattered into the ZnSe objective, and then traveled through a second polarizer (the analyzer), which was cross-polarized with respect to the input polarizer, before being focused onto a thermoelectrically cooled

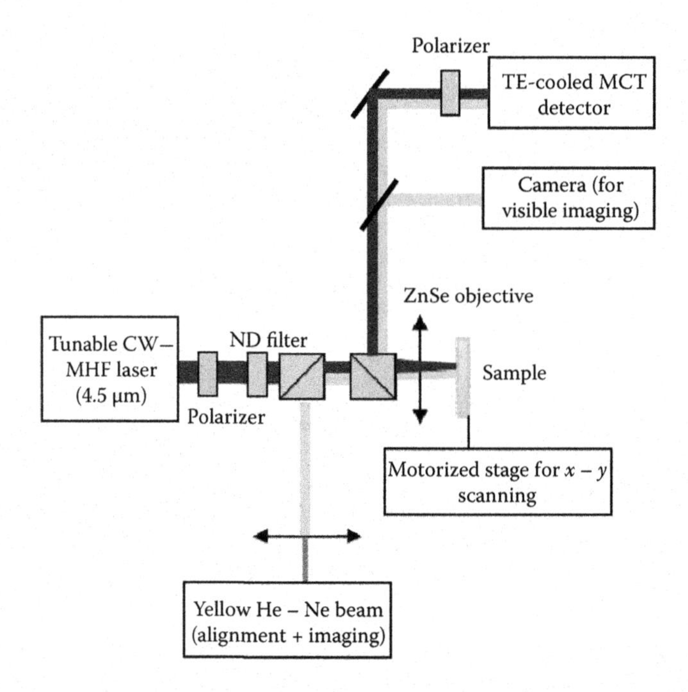

FIGURE 6.3
Schematic of experimental setup used for resonant scattering-based characterization of photonic crystal cavities. (From R. Shankar, R. Leijssen, I. Bulu, and M. Lončar, *Opt. Express* vol. 19, 5579–5586, 2011. With permission.)

mercury cadmium telluride (MCT) detector. This cross-polarization method enhanced the signal-to-background ratio of the resonantly scattered light (the signal) to the nonresonantly scattered light (the background).

Cavity modes were imaged using the technique of scanning resonant scattering microscopy (Figure 6.4) [9]. The silicon height was 500 nm, BOX thickness that was undercut

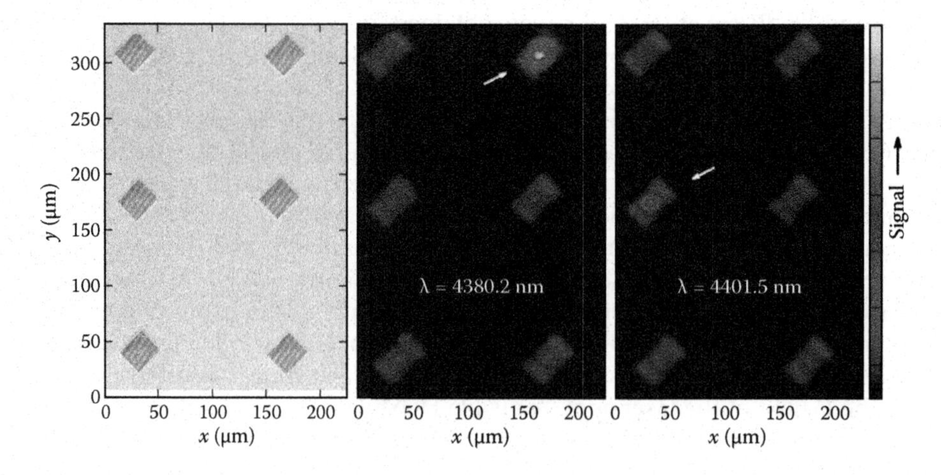

FIGURE 6.4
Mid-IR scanning resonant scattering image of an array of FIVE cavities (upper left hand structure is photonic crystal with no cavity). Scanning electron micrograph is provided for comparison. When our laser is tuned to one of the cavity resonances, and scanned over the cavity array, only the cavity in resonance with the laser lights up. (From R. Shankar, R. Leijssen, I. Bulu, and M. Lončar, *Opt. Express* vol. 19, 5579–5586, 2011. With permission.)

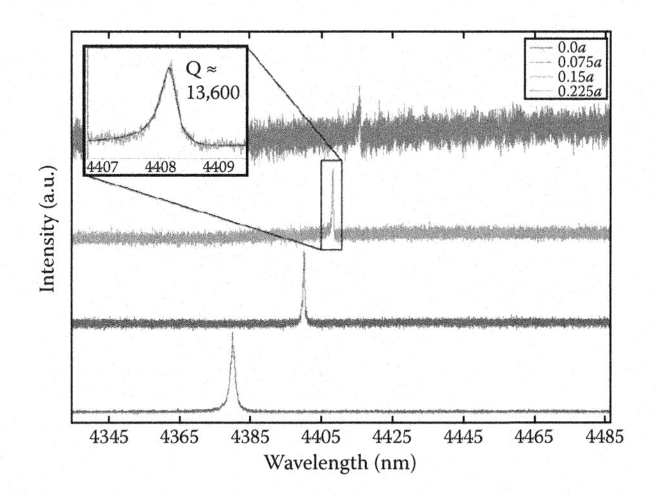

FIGURE 6.5
(See color insert.) Resonant scattering spectra of photonic crystal cavities with the air hole shift of: $s = 0$, $s = 0.075a$, $s = 0.15a$, $s = 0.225a$. Scale is linear. (From R. Shankar, R. Leijssen, I. Bulu, and M. Lončar, *Opt. Express* vol. 19, 5579–5586, 2011. With permission.)

during fabrication was 3 μm, periodicity of the photonic crystal lattice was $a = 1.34$ μm, and the air hole radius $r = 353$ nm ($r/a = 0.263$). A number of different L3 photonic crystal cavities were fabricated by varying the air hole shift, resulting in cavities with different quality factors and resonance wavelengths. Sets of cavity with both four mirror hole pairs and ten mirror hole pairs were fabricated. The devices have demonstrated a peak quality factor of 13.6 k at the wavelength of 4.4 μm (Figure 6.5).

These results are important as high quality factor and low mode volume optical resonators in the mid-IR are of interest for many applications. It is expected that wavelength-scale optical resonators, in the form of L3 photonic crystal cavities, could enable realization of chip-scale systems for trace gas sensing, optical-wireless, on-chip optical interconnects, or phased-arrays for LIDAR applications [9].

6.2.3 Silicon-on-Sapphire

The SOS platform is an alternative to SOI for the mid-IR applications as waveguides realized in this material have high confinement and low losses from $\lambda = 1.1$ to 6.2 μm.

The researchers from the University of Washington, USA, investigated submicron SOS strip waveguides at two different operating wavelengths, 4.5 and 5.5 μm [10]. Waveguide dimensions of 1.8 × 0.6 μm were chosen since they allow single mode propagation at those wavelengths (Figure 6.6). Jones et al. successfully reduced propagation losses from 9.6 to 4.3 dB/cm at 4.5 μm for TE input polarization. Devices that were tested ranged in total length from 1 mm to 1.4 cm, not including the input tapered regions, which were identical on all devices. A bend radius of 40 μm was used to design the propagation loss section, which was well in excess from 10 μm radius that was obtained from modeling.

The resistivity of the silicon for the wafers used for waveguide fabrication was specified by the manufacturer to be 100 Ω cm, suggesting that optical loss due to free-carriers will be minimal. Devices were fabricated using electron beam lithography and reactive ion etching (RIE). All waveguides tested terminated on both ends with a wider 8.0 × 0.6 μm waveguide to improve edge coupling efficiency. The chips were cleaved manually and

FIGURE 6.6
(See color insert.) A false-color scanning electron micrograph of the cleaved endfacet of a waveguide. Silicon is shown in green, and sapphire in blue. (From T. Baehr-Jones, A. Spott, R. Ilic, A. Spott, B. Penkov, W. Asher, and M. Hochberg, *Opt. Express* vol. 18, 12127–12135, 2010. With permission.)

measured. The experimental setup used to measure these devices consisted of an Nd:YAG laser that drove an optical parameteric generator/difference frequency generator (OPG/DFG) (Figure 6.7). The OPG/DFG provided 30 ps pulses of IR light at a repetition rate of 50 Hz with pulse energies around 150 μJ. It was capable of producing linearly polarized

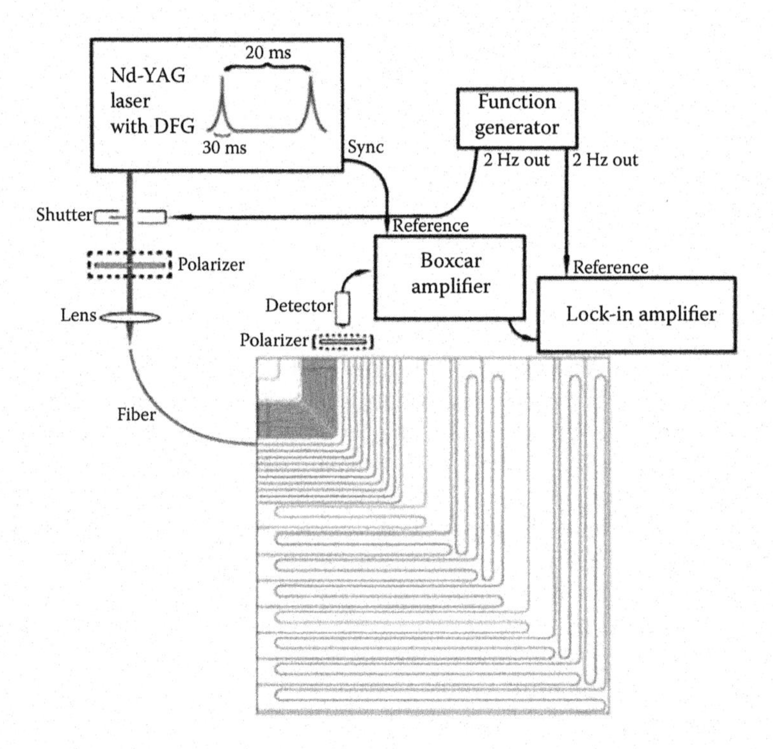

FIGURE 6.7
Experimental setup used to measure SOS waveguides. (From T. Baehr-Jones, A. Spott, R. Ilic, A. Spott, B. Penkov, W. Asher, and M. Hochberg, *Opt. Express* vol. 18, 12127–12135, 2010. With permission.)

light from 2 to 9 μm, with 4.5 μm radiation used for testing. The wavelength of 4.5 μm was chosen as the fiber cut off at longer wavelengths, and it was impossible to achieve as great a dynamic range at other wavelengths due to decreased emission power and beam stability from the OPG/DFG. The laser was coupled through a polarizer into ZnSe lens and into a 9/125-μm single-mode MIR optical fiber. The output of the chip was coupled directly to free space and then into a detector. The boxcar amplifier was used to reject the signal during the times when the laser was not providing output. The signal-to-noise ratio was enhanced by mechanically chopping the laser at 2 Hz, and using a lock-in amplifier to detect the 2-Hz-modulated signal. This resulted in an overall signal-to-noise ratio of 85 dB and the insertion losses of 12 dB.

Spott et al. demonstrated low loss low loss ridge waveguides and the first ring resonators for the mid-IR, for wavelengths ranging from 5.4 to 5.6 μm [11]. Structures were fabricated using electron-beam lithography on the SOS material system. The loss measurements were taken with the QCL CW MHP laser operating around 100 mW and 5.5 μm. A total dynamic range near 85 dB and an insertion loss of around 25 dB coupling into the waveguides were achieved. Measured waveguide loss was 4.0 ± 0.7 dB/cm. The ring resonators were also investigated. Low loss SOS ring resonators of 40 μm radius and 250 nm edge-to-edge spacing revealed Q-values up to 3.0 k and FSR of 29.7 nm in the wavelength range of 5.4–5.6 μm (Figure 6.8). Atmospheric absorption is an increasing problem at longer wavelengths in the MIR; therefore a nitrogen-pure environment was used to remove the absorption peaks, giving a much clearer indication of resonator peaks as seen in Figure 6.8.

The researchers at the University of Surrey, UK, also investigated SOS platform as a potential candidate for the MIR [7]. Submicron 2.0 × 0.6 μm SOS strip waveguides were

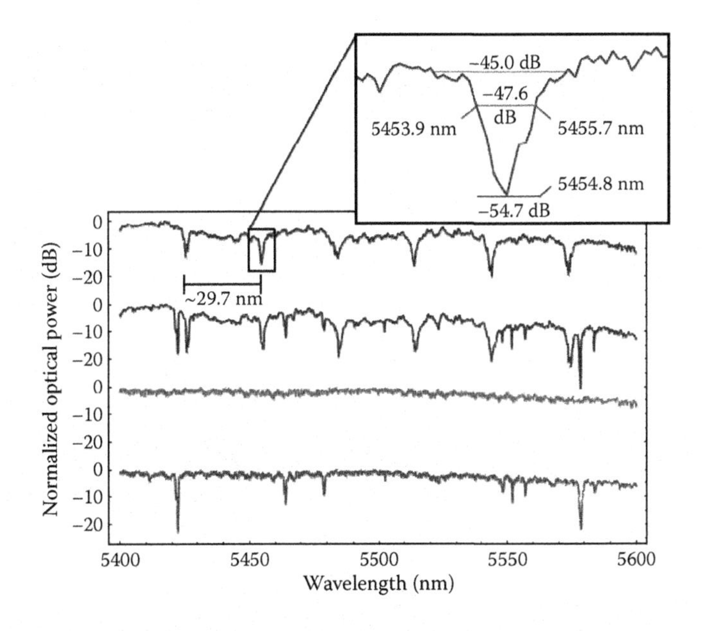

FIGURE 6.8
Transmission spectra for the primary ring: tested in a nitrogen-purged environment (top) and tested under normal atmospheric conditions (second). Transmission spectra for a regular waveguide: tested under a nitrogen-purged environment (third) and tested under normal atmospheric conditions (bottom). A Q-factor of 3.0 k and an FSR of 29.7 nm can be seen. (From Spott, Y. Liu, T. Baehr-Jones, R. Ilic, and M. Hochberg, *Appl. Phys. Lett.* vol. 97, 213501, 2010. With permission.)

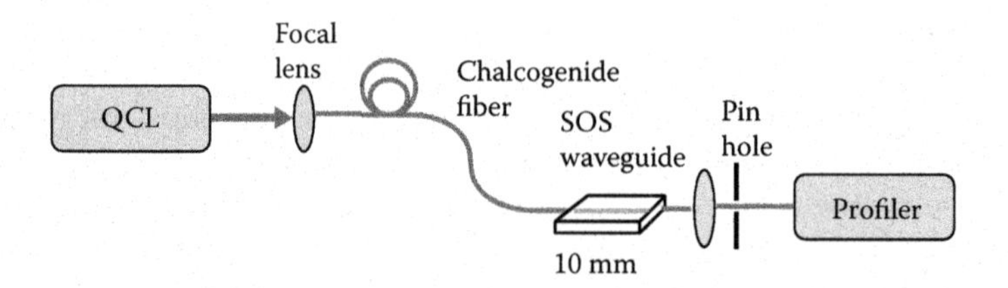

FIGURE 6.9
Experimental setup for measurements at λ = 5.18 μm. QCL = quantum cascade laser. The chalcogenide fiber was single mode As_2Se_3 fiber. (From F. Li, S. D. Jackson, C. Grillet, E. Magi, D. Hudson, S. J. Madden, Y. Moghe, C. O'Brien, A. Read, S. G. Duvall, P. Atanackovic, B. J. Eggleton, and D. J. Moss, *Opt. Express* vol. 19, 15212–15220, 2011. With permission.)

fabricated using standard lithography and RIE etching. The focused ion beam (FIB) with the beam current of 0.5 nA was used to trim waveguide facets. The resulting profile was improved dramatically thus facilitating efficient coupling to the SOS waveguides. The propagation losses of 3.6 dB/cm at 3.39 μm were achieved.

Li et al. investigated the propagation losses of submicron 1.0 × 0.3 μm SOS strip waveguides at 5.08 μm [12]. A tunable CW quantum cascade laser capable of producing linearly polarized light from 5.07 μm to 5.37 μm was used to perform measurements at 5.08 μm (Figure 6.9). The light was coupled into a very short length (50 cm) of single mode AsSe chalcogenide fiber having a mode diameter of ~8 μm at this wavelength. The laser emitted TE polarized light, which was not altered substantially by the short fiber length. The chalcogenide fiber was butt coupled to the nanowire ensuring the light was TE polarized, and the output was then imaged on a mid-IR camera using a ZnSe mid-IR objective lens with a focal length of 6 mm. The mode profiles were imaged using a Spiricon Beam Profiling Cameras (OPHIR) and averaged 250 times. Relative loss measurements at λ = 5.18 μm were obtained by measuring the intensity of the central guided mode peak on the imaging camera as a function of waveguide length. This method was highly effective at discriminating against scattered light or substrate guided light. The drawback was the ability to obtain the relative loss measurements, which yielded only propagation loss, not coupling loss.

Low propagation losses of 1.9 dB/cm were achieved at 5.08 μm using the custom made SOS wafers. Laser scribing and cleaving was used prior to testing to achieve better profile of the waveguide facets. The same chips were characterized at the operating wavelength of 1.55 and 2.08 μm achieving the propagation losses of 0.8 and 1.1–1.4 dB/cm, respectively. The low loss at these operating wavelengths can be attributed to the use of an I-line stepper mask aligner, along with the epitaxial growth process used to produce the SOS wafers, which resulted in extremely low defect density. The resolution of stepper mask aligner was ~400 nm, which is much lower than that for typical electron beam lithography machines contributing to substantially lower sidewall roughness [12].

6.2.4 Silicon-on-Porous Silicon

By replacing the oxide with a different cladding, structures that guide at longer MIR wavelengths could be obtained. Such a cladding could be air or porous silicon for example. Mashanovich et al. have characterized silicon-on-porous silicon (SiPSi) waveguides at the operating wavelength of 3.39 μm [3]. Propagation losses for SiPSi waveguides fabricated by a direct write process, were around 6 dB/cm. As large area irradiation is a preferred

FIGURE 6.10
Cross section of a Si on porous Si waveguide fabricated by proton beam irradiation over a large area. (From G. Z. Mashanovich, M. M. Milošević, M. Nedeljkovic, N. Owens, B. Xiong, E. J. Teo, and Y. Hu, *Opt. Express* vol. 19, 7112–7119, 2011. With permission.)

method for the fabrication of SiPSi waveguides (Figure 6.10) the measurements for those waveguides were also performed. These were oxidized after etching to reduce the surface roughness. Losses as low as 2.1 ± 0.2 dB/cm at 1.55 μm, and 3.9 ± 0.2 dB/cm at 3.39 μm were measured for 4 × 2 μm waveguides.

It is suspected that the observed losses are mainly due to surface roughness. For longer wavelengths it can be expected that losses decrease as the ratio of the roughness amplitude to the wavelength also decreases. However, for longer wavelengths a larger proportion of the optical mode is interacting with the sidewalls thus increasing the propagation loss. For the SiPSi waveguides, the latter contribution seems to be larger than the former, and hence, the propagation loss at 3.39 μm is also higher. The absorption loss of porous silicon at the MIR needs also to be determined to confirm this assumption.

In addition, the SiPSi propagation loss at the wavelength of 1.55 μm is slightly larger than previously reported values of 1.4–1.6 dB/cm [13], suggesting that slight variations in the fabrication process may result in a reduction of MIR propagation loss from the current value of 3.9 dB/cm to values below 3 dB/cm. Although relatively low resistivity (0.7-Ω cm) crystalline silicon is preferable for the fabrication of SiPSi waveguides, the silicon with higher resistivity is necessary to reduce the propagation loss without compromising the fabrication process.

6.2.5 Chalcogenide Waveguides

Chalcogenide glasses such as As_2S_3 possess desirable optical properties in the near- and mid-IR such as a wide transmission window (long wavelength cutoff at 9.4 μm), high refractive index (n = 2.4 at λ = 4.8 μm), high nonlinearity, and photorefractive behavior [14–17]. These attributes make chalcogenide glasses a good match for integrated near- and mid-IR applications.

Tsay et al. proposed As_2S_3 waveguides that could be used for chemical sensing applications [15]. In their paper, the absorbance of the substrate material was investigated as an important criterion for low loss. Waveguides were fabricated on three different substrate materials: SiO_2, $LiNbO_3$, and NaCl. The fabricated strip waveguides were 40 μm wide and 10 μm high. The device testing was performed using a setup that comprised a QCL MIR laser at 4.8 μm,

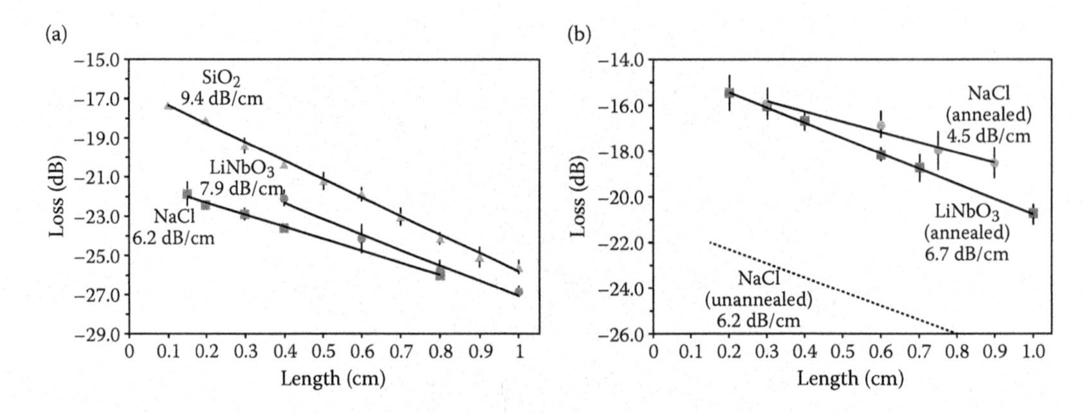

FIGURE 6.11
Propagation losses of As_2S_3 strip waveguides: (a) unannealed and (b) annealed samples. (From C. Tsay, E. Mujagić, C. K. Madsen, C. F. Gmachl, and C. B. Arnold, *Opt. Express* vol. 18, 15523–15530, 2010.)

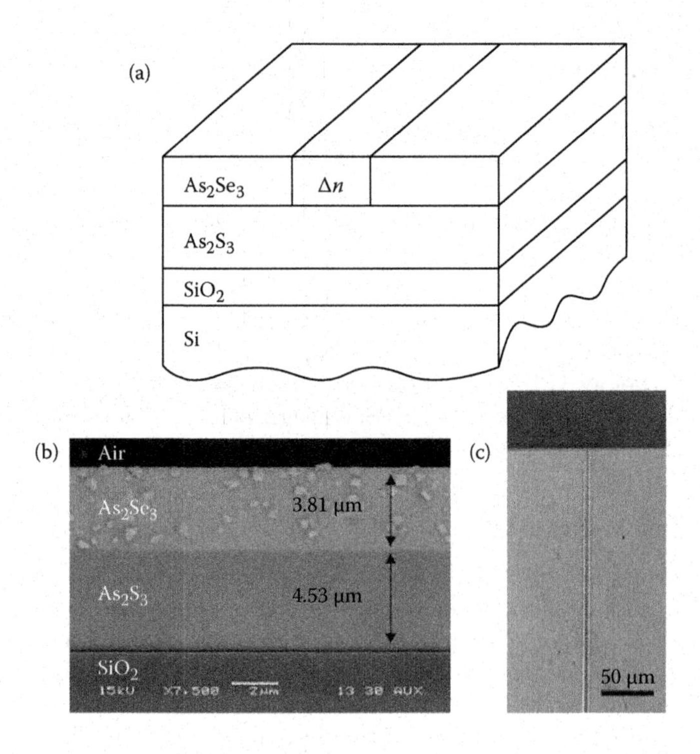

FIGURE 6.12
Multilayer film and channel waveguide. (a) Schematic diagram of multilayer film channel waveguide after deposition and laser writing. (b) SEM cross-section image of the film, with thickness of each layer. (c) Optical microscope image of channel waveguide. (From N. Hô, M. C. Phillips, H. Qiao, P. J. Allen, K. Krishnaswami, B. J. Riley, T. L. Myers, and N. C. Anheier, Jr., *Opt. Lett.* vol. 31, 1860–1862, 2006.)

liquid nitrogen cooled HgCdTe (MCT) photodetector and optics [15]. The cut-back measurement results showed that the propagation loss of the As_2S_3 waveguides depended heavily on the transmittance of the underlying substrate. For unannealed samples the propagation losses were in the 6- to 10-dB/cm range (Figure 6.11a). There was a 3.2-dB/cm reduction in waveguide attenuation when SiO_2 substrate was replaced with NaCl substrate (Figure 6.11).

From Figure 6.11b, it can be seen that the annealing step improved the losses by ~2 dB/cm. The lowest propagation loss of 4.5 dB/cm was measured for annealed waveguides on NaCl substrate.

Hô et al. investigated chalcogenide strip waveguides at the operating wavelength of 8.4 μm [16]. The 3.81-μm thick As_2Se_3 was used as the waveguide core, while the 4.53-μm As_2S_3 layer served as the lower cladding (Figure 6.12). The top 2 μm of the wafer was a layer of SiO_2, preventing possible coupling from the waveguide into the Si substrate (Figure 6.12). The photodarkening induced the local change in the refractive index of As_2Se_3 resulting in the waveguide width of 5.4 μm. Propagation losses were measured using the liquid nitrogen cooled QCL (Maxion Technologies, Inc.) emitting at 8.4 μm, and a MCT detector. The input polarization was controlled with a half-wave rhomb and a linear polarizer. The propagation losses, estimated using the cut back method, were 0.5 ± 0.1 dB/cm for TE and 1.1 ± 0.1 dB/cm for TM polarization.

Transmission measurements of $Te_2As_3Se_5$ rib waveguides on As_2S_3 have been demonstrated at 10.6 μm in [17]. The rib waveguide height of 4 μm, etch depth of 1.9 μm and the waveguide width of 15 μm were fabricated. Total insertion losses of a 1-cm-long sample was 20 dB. The authors estimated that coupling losses contributed by 10 dB and hence concluded that the propagation loss was 10 dB/cm.

6.3 Nonlinear Effects in Silicon in the Mid-Infrared Region

Sanja Zlatanovic, Faezeh Gholami, Nikola Alic, and Stojan Radic

6.3.1 Introduction

Numerous applications specific to the mid-IR region of high practical and scientific interests (e.g., as remote sensing [18], pollution monitoring, combustion dynamics measurements [19], IR countermeasures [20], and biomedical applications such as breath analyzers [21]) critically depend on the availability of high quality light sources in this region. Short of momentous funding necessary for research and development of commercial-grade sources comparable to those in the mature telecom region, nonlinear effects offer instant access to the mid-IR by means of porting existing superior quality sources from other electromagnetic spectral regions to this spectral band. Indeed, nonlinear processes such as parametric wavelength conversion allow generation of light, and access to virtually any part of the spectrum, pending on the availability of an appropriate nonlinear medium, availing an efficient conversion. In particular, the nonlinear silicon photonic devices due to the material properties show great potential for parametric light generation in the mid-IR. This, joined with maturity and the CMOS fabrication process, presents a fascinating opportunity of bringing together nonlinear optics and a mixing platform with a superior fabrication process, opening the doors to a large scope of applications, that not only have been closed for decades, but would remain out of the reach for many years to come if left to the standard development means.

Two material properties that make silicon suitable for nonlinear applications are its strong optical nonlinearity and large linear refractive index. Crystalline silicon possesses strong nonlinearity, nearly 200 times larger than that of silica, potentially allowing efficient nonlinear interactions in subcentimeter length devices. Moreover, due to the large linear refractive index of silicon (n = 3.47), optical modes can further be confined to the submicron scale cross sections in the silicon waveguides, providing yet another aspect for miniaturization as well as a nonlinear interaction enhancement. Silicon waveguides have previously been used in optical processing experiments in the near-IR. However, the applications on which this section focuses have been extended to the mid-IR spectral region. The main benefit from the operation in the mid-IR stems from the absence of parasitic nonlinear absorption that renders silicon performance in the near-IR part of the spectrum rather limited. Indeed, the performance of nonlinear silicon devices in the near-IR is greatly hindered by two-photon absorption (TPA) and the associated free-carrier absorption (FCA). However, the deleterious TPA vanishes at the wavelengths beyond 2.25 μm, corresponding to photon energies lower than half of the material band-gap [22, 23]. Consequently, the operation of nonlinear silicon devices in the mid-IR does not suffer from this strong parasitic nonlinear absorption effect. While other multi-photon absorption effects exist at the mid-IR wavelengths, their magnitude is much smaller and does not present a significant obstacle to the performance of silicon devices. It is primarily for this reason that the mid-IR potentially presents a superior region for operating silicon nonlinear devices, allowing them to unleash their full potential, unimpeded by the shorter wavelength operational obstacles.

The remainder of this section is organized as follows: we shall commence with a basic overview of optical nonlinearities and nonlinear material properties of silicon in the mid-IR region, as well as the associated anisotropy and dispersion. The main properties of the silicon waveguides contributing to efficient nonlinear interactions, supported with recent experimental results will be presented next. Finally we will conclude with experimental demonstrations of the main nonlinear phenomena, including Raman amplification [24], parametric amplification [25] and the four-wave mixing [26].

6.3.2 Nonlinear Optics in Silicon

Nonlinear optical phenomena fundamentally rely on interaction of high-intensity light beams with matter. In fact, the presence of a strong light beam can have a profound effect on the optical properties of select materials. Mathematically, this effect is described by the electric field-dependent polarizability, $\boldsymbol{P}$, as [27]

$$\boldsymbol{P}(\boldsymbol{r},t)=\varepsilon_0(\chi^{(1)}\cdot \boldsymbol{E}(\boldsymbol{r},t)+\chi^{(2)}:\boldsymbol{E}(\boldsymbol{r},t)\boldsymbol{E}(\boldsymbol{r},t)+\chi^{(3)}\vdots \boldsymbol{E}(\boldsymbol{r},t)\boldsymbol{E}(\boldsymbol{r},t)\boldsymbol{E}(\boldsymbol{r},t)+\cdots), \tag{6.1}$$

assuming instantaneous nonlinear response, where $\chi^{(n)}$ is optical susceptibility tensor of the nth order. Silicon has a cubic crystalline structure of $m3m$ class [27]. Being a centrosymmetric crystal, the second-order nonlinearity cannot exist in silicon [27], while the third-order nonlinearity is present. The third-order nonlinearity governs phenomena such as self-focusing in the spatial domain, and self-phase modulation, spontaneous, and stimulated Raman scattering, Brillouin scattering, Kerr effect, two-photon absorption (TPA), and four-wave mixing (FWM), as spatial domain temporal counterparts. The third-order susceptibility tensor of silicon contains twenty one nonzero elements, of which only four are independent due to the crystallographic structure [27]. These tensor components are namely $\chi^{(3)}_{1111}$, $\chi^{(3)}_{1122}$, $\chi^{(3)}_{1212}$, $\chi^{(3)}_{1221}$, where numerical subscripts 1, 2, and 3 are for x, y, and z

directions, respectively. Furthermore, assuming small dispersion of nonlinear susceptibility and applying Kleinman symmetry [27] leads to only two independent tensor components $\chi^{(3)}_{1111}$, $\chi^{(3)}_{1122}$ [28].

The values of χ^3 are typically derived from measured values of optical Kerr nonlinearity n_2 and TPA coefficient β_{TPA}. The relation between effective susceptibility, and these parameters are given by

$$3\omega/(4\varepsilon_0 c^2 n^2)\chi^{(3)}_{eff} = \omega/cn_2 + i/2\beta_{TPA} \tag{6.2}$$

The third-order nonlinearity in silicon is anisotropic and is responsible for nonlinear processes, based on polarization of the input light. Both Kerr coefficient and TPA coefficient exhibit anisotropy [29]. Furthermore, it has also been demonstrated that anisotropy in the Raman tensor plays a crucial role for Raman amplification [23, 30]. The magnitude of n_2 and β_{TPA} anisotropy has been measured by Zhang et al. [29]. In particular, this study has shown that both of these parameters decrease by about 12% from maximum when the direction of the polarization is changed. These results have been used to deduce anisotropy of $\chi^{(3)}_{eff}$ itself in the region between 1.2 and 2.4 μm. It has been found that two independent components of the third-order susceptibility have the ratio of $\chi^{(3)}_{1111}/\chi^{(3)}_{1122} \sim 2.36$, constant over the entire measured wavelength range.

6.3.3 Two-Photon Absorption and Kerr Coefficient

The measurements of absolute values of the Kerr and TPA coefficients for silicon were reported in the mid-IR region up to 2.35 μm [31, 32]. In the near and mid-IR region, the silicon third-order nonlinearity exhibits significant dispersion due to the photon energies close to the half of the band-gap. The data presented in Figure 6.13a,b demonstrate that the order of magnitude (1×10^{-18} m²/W) of the Kerr coefficient is consistent across the literature. The functional forms of the Kerr coefficient dispersion also agree very well. The Kerr coefficient exhibits a peak between 1800 and 1900 nm, and subsequently decreases at longer wavelengths. An estimate of the Kerr coefficient values beyond 2.35 μm is obtained by Kramers–Krönig transform of TPA data (Figure 6.13c). Kramers–Krönig transform accurately predicts the functional form of the Kerr coefficient dispersion; however, it does underestimate the magnitude of the Kerr coefficient. This discrepancy is likely due to the neglect of Raman and quadratic Stark effect contributions. The calculated Kerr coefficient dispersion curve (Figure 6.13c) shows that a slower decrease in magnitude of the Kerr coefficient is expected at longer wavelengths, wheareas the trend is maintained even deeper into the mid-IR. It is this finding that so favorably promotes silicon as a mid-IR mixing platform. Furthermore, these implications are further supported by the measured silicon nonlinearity of n_2 of 1.9×10^{-18} m²/W at 10 μm [33].

In contrast, and as shown in Figure 6.14a,b, the two-photon absorption is strong at near-IR frequencies and limits the efficiency of nonlinear effects. The functional form of TPA dispersion obtained using theory developed by Garcia and Kalayanaraman [34], taking into account three types of indirect transitions that contribute to TPA (forbidden–forbidden, allowed–forbidden, and allowed–allowed) is in a very good agreement with the measured data. The TPA coefficient has a broad peak around 1.1 μm (corresponding to the energy of the indirect band-gap) and decreases steadily for longer wavelengths. It completely vanishes at wavelengths beyond 2.25 μm. This is of crucial importance for efficient operation of nonlinear silicon devices at longer mid-IR wavelengths. While the TPA itself does not

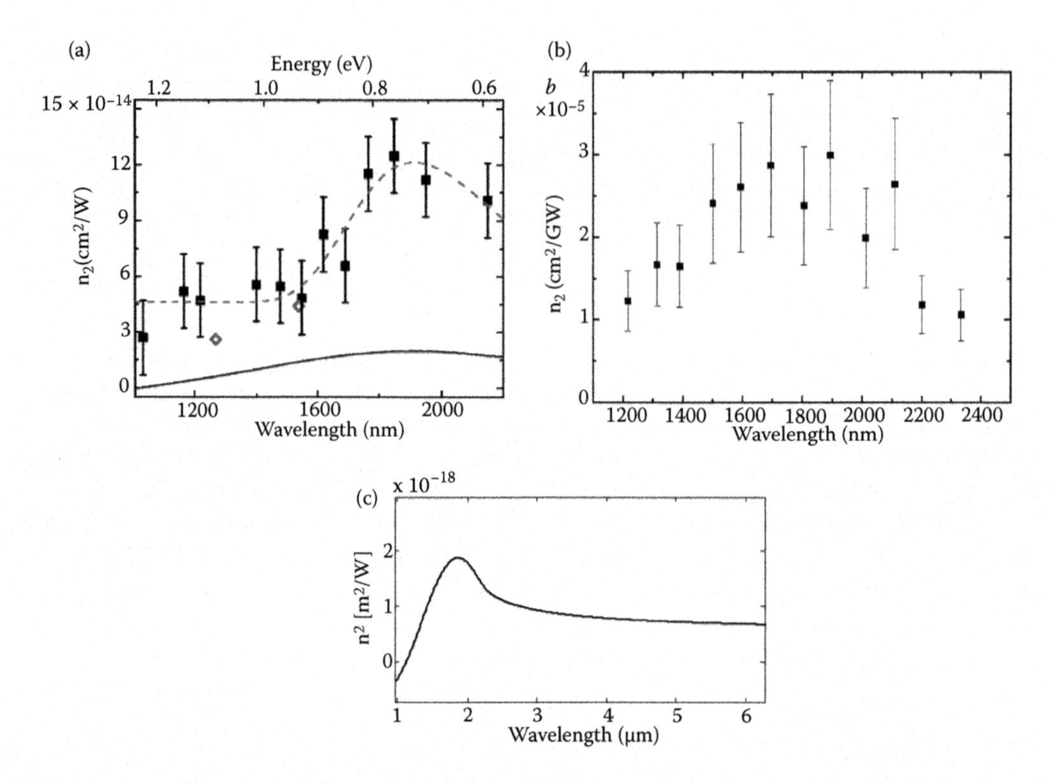

FIGURE 6.13
Wavelength dependence of the Kerr coefficient n_2 of silicon measured by (a) Bristow et al. (Reproduced from A. D. Bristow et al., *Appl. Phys. Lett.*, 90, 191104, 2007. With permission.); (b) Lin et al. (Reproduced from Q. Lin et al., *Appl. Phys. Lett.*, 91, 021111, 2007. With permission.); (c) functional form of Kerr coefficient extracted from Kramers–Krönig transformation of TPA coefficient (Reproduced from S. Zlatanovic et al., *IEEE J. Select. Topics Quantum Electron.*, 10.1109/JSTQE.2011.2119295, 2011. With permission.)

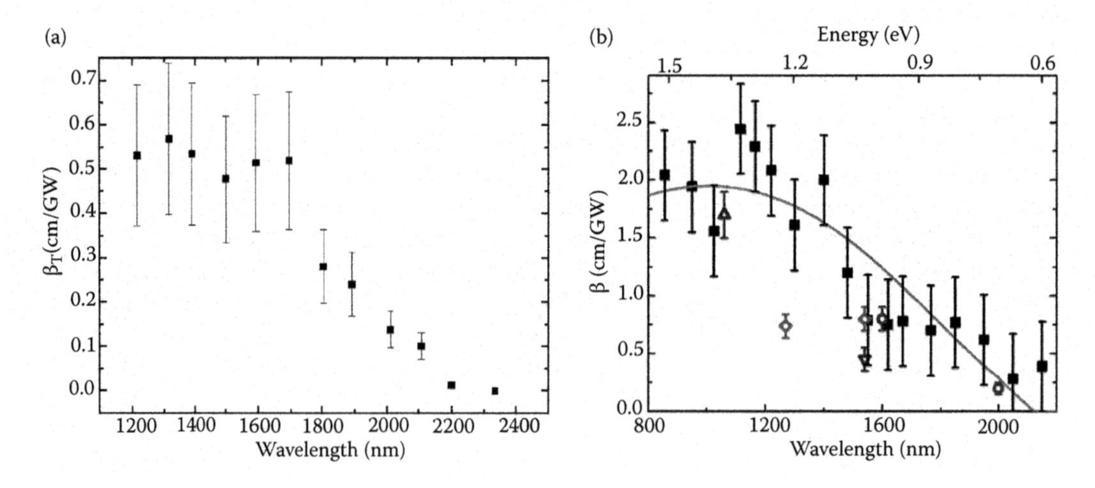

FIGURE 6.14
Measured wavelength dependence of TPA coefficient β_{TPA} of silicon according to (a) Lin et al. (Reproduced from Q. Lin et al., *Appl. Phys. Lett.*, 91, 021111, 2007. With permission.) and (b) Bristow et al. (Reproduced from A. D. Bristow et al., *Appl. Phys. Lett.*, 90, 191104, 2007. With permission.). The solid curve represents the best fit based on calculations of Garcia and Kalayanaraman [16].

put significant restriction on efficiency of nonlinear processes, it is inevitably accompanied by generation of free-carriers. Free-carrier absorption (FCA) in silicon, however, places a stringent restriction on the efficiency by limiting and preventing optical gain, as will be explained in the following subsections.

6.3.4 Three-Photon Absorption

Three-photon absorption (3PA) of silicon is the next in line parasitic absorption effect present at mid-IR frequencies. The experimental results for 3PA coefficient were obtained by Pearl et al. [35] and are shown in Figure 6.15a. The three-photon absorption exhibits a peak around 2700 nm, slowly decreases at longer wavelengths and vanishes as photon energies become lower than one-third of the band-gap energy. Pearl et al. have also measured anisotropy of 3PA coefficient and demonstrated that it varies by ~30% from a maximum value for light polarized along [011] direction.

To compare nonlinear performance of the silicon in the near-IR and mid-IR, a nonlinear figure of merit (FOM) is defined as $n_2/(\lambda\beta_{TPA})$ for region where two-photon absorption dominates. For the region beyond 2.25 μm, a corresponding 3PA nonlinear parameter is defined as $n_2/\lambda\beta_{3PA}I$. Unlike FOM that is purely material property for a given wavelength, the 3PA nonlinear parameter depends on intensity, since the three-photon absorption is higher order $\chi^{(5)}$ effect [36]. The values for both FOM and 3PA nonlinear parameter are given in Figure 6.15b and show that nonlinear performance is better in mid-IR region due to, in comparison to two-photon absorption, the smaller effect of three-photon absorption at the intensities typically used in silicon waveguides.

6.3.4.1 Free-Carrier Effects

Two- and three-photon absorption mechanisms are inevitably followed by the generation of free carriers. The free carriers interact with optical beam in two manners: (a) they increase absorption and (b) induce wavelength-dependent index change that in turn

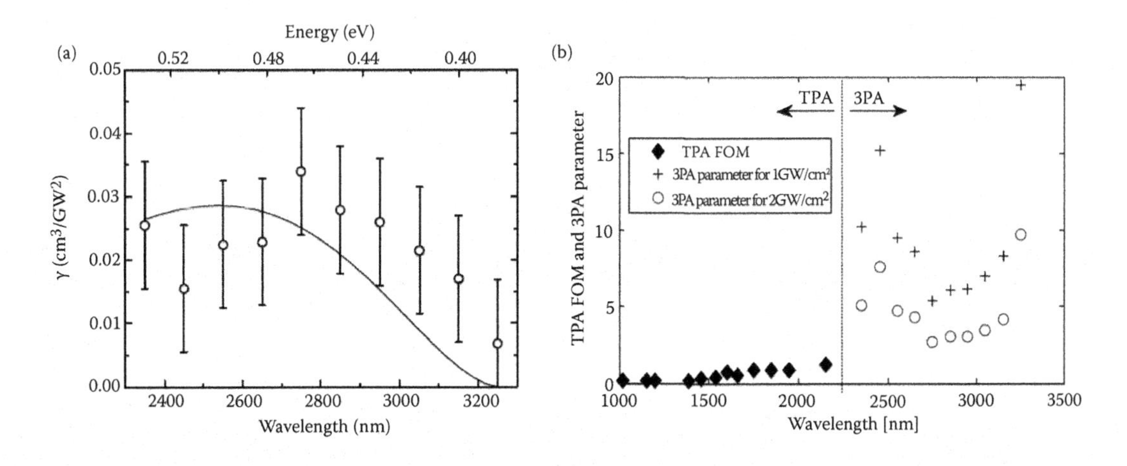

FIGURE 6.15
(a) Three-photon absorption coefficient β_{3PA} (Reproduced from S. Pearl et al., *Appl. Phys. Lett.*, 93, 131102, 2008. With permission.) and (b) nonlinear figure of merit (diamonds), 3PA nonlinear parameter for intensity of 1 and 2 GW/cm^2 (squares and circles, respectively). (Reproduced from S. Zlatanovic et al., *IEEE J. Select. Topics Quantum Electron.*, 10.1109/JSTQE.2011.2119295 , 2011. With permission.)

changes the phase. The free-carrier absorption (FCA) depends on the concentration of free carriers as

$$\alpha_{FC} = e^3 N(N/m_{ce}^* + N^{0.8}/m_{ch}^*)/(\varepsilon_0 c n \omega^2) \tag{6.3}$$

and the free carrier induced change in refractive index is

$$\delta n_{FC} = -e^2(1/(\mu_e m_{ce}^*) + 1/(\mu_h m_{ch}^*))/(2\varepsilon_0 n \omega^2), \tag{6.4}$$

where N is the carrier density, m_{ce}^*, and m_{ch}^* are effective mass of electrons and holes, respectively, and μ_e and μ_h are the mobility of electrons and holes, respectively.

The density of generated carriers (N) is directly proportional to the strength of multiphoton absorption. As a consequence, it is highly preferential to operate silicon nonlinear devices in the spectral region where the two-photon absorption vanishes, merely since two-photon absorption is much stronger effect than three-photon absorption. Furthermore, the higher order nonlinear process such as three-photon absorption and the associated free-carrier absorption are expected to be of far less significance because of the relatively low probability of simultaneous absorption of as many as three photons. Nevertheless, Equation 6.3 also implies quadratic dependence of FCA on the wavelength. Consequently, it is critical to keep the free-carrier density low to prevent excess loss at longer wavelengths.

6.3.5 Nonlinear Optical Processes in Silicon at Mid-Infrared Wavelengths

As previously mentioned, the third-order nonlinearity in silicon exhibits a wide variety of nonlinear phenomena. Optical processes such as Raman amplification self-phase modulation (SPM), cross-phase modulation (XPM), third-harmonic generation (THG) [37], and four-wave mixing (FWM) have all been experimentally demonstrated in silicon at near-IR wavelengths. In this section experimental demonstrations of nonlinear processes in silicon at mid-IR wavelengths will be summarized.

6.3.5.1 Mid-IR Raman Amplification in Silicon

Raman scattering is a third-order nonlinear process that can be described, in macroscopic terms, as an interaction of the lattice thermal vibrations at frequency ω_v with the incident pump field (ω_p). In this simplified description, thermal vibrations of the lattice induce a sinusoidal modulation of the susceptibility. The incident pump field beating with the susceptibility oscillation excites two distinct contributions: one at the sum frequency, called anti-Stokes and one at the difference frequency denoted as the Stokes wave. Particularly, in the case of stimulated Raman scattering the pump and the Stokes field are both present in the medium, and their frequency difference is equal to the atomic vibrational frequency. In this case, the atomic oscillations are enhanced by the driving force generated by two optical fields, which, in turn, increases the amplitude of the Stokes field. The described joined action forms a positive feedback that results in the amplification of the Stokes wave. In reality, the physical phenomena leading to Raman scattering in silicon are much more complex. The direct coupling of light with atomic vibrations is very weak at mid-IR wavelengths due to the large atomic mass of silicon and the lack of dipole moment [23]. In fact, the Raman scattering process in silicon is mediated by electrons. The pump and Stokes

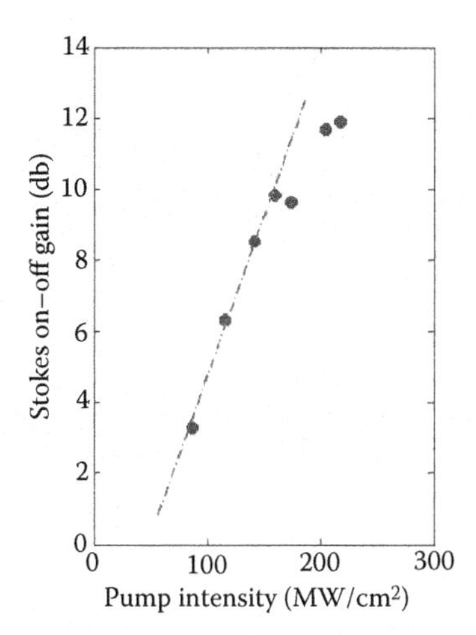

FIGURE 6.16
Raman on–off gain as a function of effective pump intensity interacting with Stokes input. (Reproduced from V. Raghunathan et al., *Opt. Express*, 15, 14355–14362, 2007. With permission.)

wave separation in silicon is ~15.6 THz and the stimulated Raman scattering gain has a relatively narrow bandwidth of 105 GHz. The stimulated Raman process can be described by

$$dI_s/dz = g_R I_p I_s, \tag{6.5}$$

where g_R is the Raman gain coefficient that is obtained from the so called Raman susceptibility using $g_R = 6\pi\,\mu_0 \chi^{(3)}_{ijkl}/(\lambda_s n_s n_p)$. The value of the Raman gain coefficient at 1550 nm is in the range of 10–20 cm/GW [38, 39] and scales inversely with wavelength as $1/\lambda_s$, leading to diminished levels at mid-IR wavelengths.

The experimental demonstration of Raman amplification in mid-IR has been reported by Raghunathan et al. [24]. In this experiment, a signal at 3.39 μm was amplified using a pump at 2.88 μm. The active medium was a 2.5-cm-long [100] silicon crystal. The Raman on–off gain obtained in this experiment was ~12 dB as shown in Figure 6.16. The saturation of Raman gain at higher pump intensities was due to damage of the sample surface.

For the measured gain in this experiment, the value of Raman gain coefficient falls in the range between 4.5 and 9 cm/GW that is predicted from the measurements at 1550 nm and the $1/\lambda_s$ dependence.

This experiment was the first successful demonstration of Raman gain in silicon at mid-IR wavelengths. In more recent attempts, the stimulated Raman scattering at mid-IR wavelengths was applied for beam cleanup [40]. In this experiment an amplified near-diffraction-limited Stokes beam was obtained using a severely aberrated pump.

6.3.5.2 Self-Phase Modulation at Mid-IR Wavelengths

Self-phase modulation (SPM) is a process in which refractive index change is induced by the mere presence of the pump. Owing to the refractive index dependence on the intensity, a strong pulsed pump induces a temporal phase shift, thus causing a frequency shift (or, equivalently photon generation at the immediately adjacent frequencies—in a continuous manner). As a consequence, the pulse spectrum broadens and develops a characteristic multiple-peak structure. The SPM was investigated in silicon waveguides at the mid-IR [41].

Figure 6.17 shows SPM spectra for pump wavelengths ranging from 1775 to 2250 nm. In these experiments, the authors used 4-mm-long silicon waveguides with cross-sectional dimensions of 220 × 600 nm that express normal dispersion in the spectral range investigated.

The figures show the characteristic spectral broadening irrespective of the pump spectral location and an increased number of spectral fringes as the pump power is increased. There are additional features that can be observed. The most noticeable spectral broadening is achieved at 1775 nm for the highest peak power of 33. 5 W, while for other wavelengths the spectral broadening is not as significant. In this particular experiment, the observed behavior is the consequence of the fact that the input pump pulse is not transform-limited, but negatively chirped. At low pump powers, the amount of effective positive chirp introduced by SPM is insufficient to overcome the initial negative chirp of the pump. Therefore the effect is less pronounced at longer wavelengths where the pump power is lower. Another feature that can be observed is that in the region of two-photon absorption (1775 and 1988 nm) the SPM signature exhibits asymmetry with fringes on the red side of the spectrum more closely spaced than on the blue side. In addition, at these wavelengths spectra have a significant spectral blue-shift at high input powers. These phenomena are well known and originate from the presence of free-carrier-induced change in refractive index [42, 43]. Therefore, they are characteristic only for the spectral region where the two-photon absorption induces free-carrier generation and are absent in the mid-IR region (2200 and 2250 nm). These SPM experiments were also used to obtain values for both nonlinear refractive index n_2 and the two-photon absorption coefficient β_{TPA} and show good agreement with values measured in bulk silicon [31, 32].

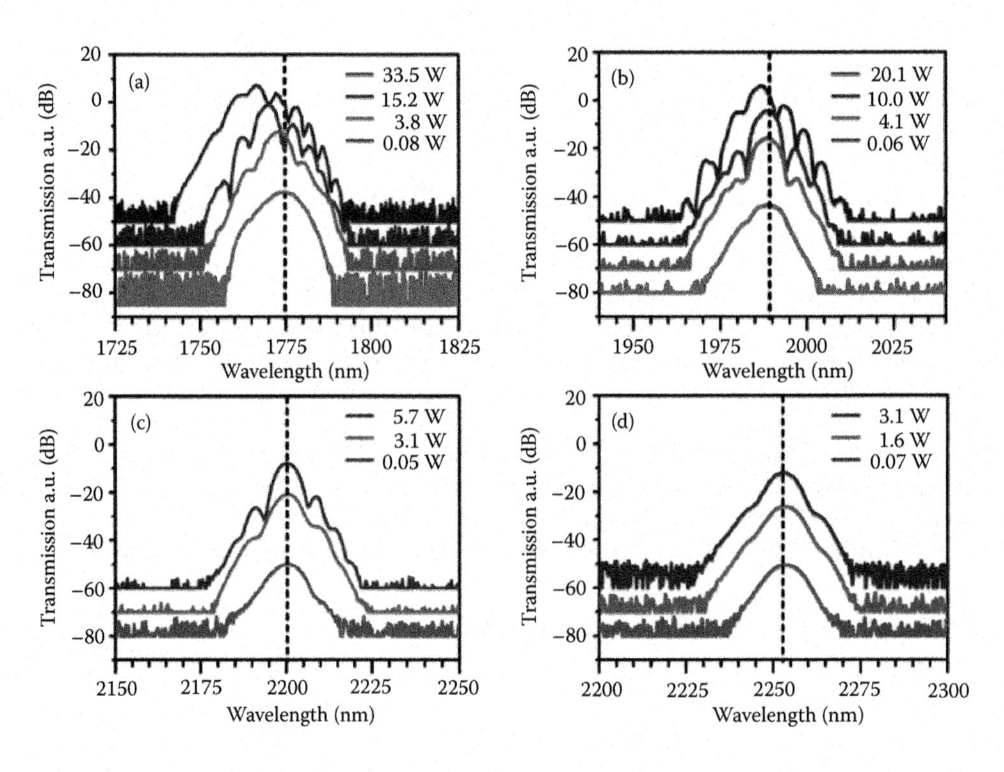

FIGURE 6.17
Experimental power-dependent transmission spectra as a function of pump center wavelength. Spectral broadening due to SPM is observed. (Reproduced from X. Liu et al., *Opt. Express*, 19, 7778–7789, 2011. With permission.)

6.3.5.3 Four-Wave Mixing and Parametric Amplification in the Mid-IR

Four-wave mixing (FWM) is a nonlinear process in which three photons (not necessarily all distinct) interact, whereby a fourth photon satisfying the overall energy conservation is generated, as a consequence. In particular in what is generally denoted to as a single pump (degenerate) four-wave mixing configuration, two pump photons are annihilated to generate two photons one at signal and one at idler frequency. The physics of the process can simply be described in terms of energy and momentum conservation (phase-matching condition) in the nondegenerate pump case as:

$$\omega_i = \omega_{p1} + \omega_{p2} - \omega_s \tag{6.6}$$

$$\Delta k = 2\gamma P_0 - \Delta k_L = 2\gamma P_0 - (k_{p1} + k_{p2} - k_s - k_i), \tag{6.7}$$

where ω_i, ω_s, ω_{p1}, ω_{p2} are idler, signal, and pumps frequencies and k_s, k_i, k_{p1}, k_{p2} are wave vectors defined as $k = n_{eff}\,\omega/c$, where n_{eff} is the effective refractive index, γ is effective nonlinearity and P_0 is pump power. In a simplified form, the effective nonlinearity can be expressed as $\gamma = n_2\omega/(c\,A_{eff})$, where A_{eff} is effective modal area. Efficient FWM requires phase matching, which can be achieved through waveguide dispersion engineering. Due to the strong optical confinement of the optical mode in silicon waveguides, the dispersion is dominated by waveguide geometry. By changing the dimensions of the waveguide, the dispersion properties can be modified to satisfy the requirements of a particular conversion process. To fulfill the phase-matching condition (the momentum conservation relationship), Δk must be equal to zero. The conversion efficiency defined as a ratio of the idler output to signal input power can be expressed in a simplified manner as [44]

$$\eta(P_0) = (\gamma P_0/g)^2 \times sinh^2(gL), \tag{6.8}$$

where

$$g = (\gamma P_0 \Delta k_L - (\Delta k_L/2)^2)^{1/2}. \tag{6.9}$$

The conversion bandwidth strongly depends on the waveguide dispersion. In addition, due to very large values of n_2 in silicon waveguides, conversion is possible in the centimeter-scale lengths, which translates into higher conversion bandwidth. As pointed out previously, the effective nonlinearity of silicon waveguides depends inversely on the effective modal area and wavelength. This property, thus, implies nonlinearity decreases at longer wavelengths due to the increased mode size.

The potential of four-wave mixing in silicon in the mid-IR was recognized by Painter et al. [45]. They theoretically analyzed parametric mixing in silicon waveguides and microresonators. In addition, an extensive theoretical analysis of the dispersion in the mid-IR for the silicon waveguides on platforms such as silicon-on-sapphire, silicon on SiO_2, and silicon with air cladding was published by Tien et al. [46]. Recently, several experiments reported four-wave mixing in silicon waveguides in the mid-IR spectral range where nonlinear absorption is not detrimental to the waveguide performance.

Wavelength conversion to mid-IR was demonstrated in 3.8-mm-long silicon waveguides on SOI platform using an ultra-compact telecom-derived pump source at 2025 nm [26, 47]. Operation at this wavelength is characterized by low TPA. The silicon waveguides had cross-sectional dimensions of 1060 × 250 nm. FWM was performed with long pump pulses of 1ns-duration which was comparable with the free-carrier lifetime in silicon and can therefore be considered a quasi-CW operation. With the pump power of 176 mW, the conversion efficiency of –22.5 dB was achieved across 240 nm bandwidth (Figure 6.18, left).

Furthermore, light was generated at the mid-IR wavelengths up to 2388 nm (Figure 6.18, right). The reported measured nonlinearity in the waveguides was $97.3(\mathrm{Wm})^{-1}$ and agrees well with measurements of n_2 reported by Bristow et al. [31].

A broadband parametric conversion to the mid-IR has also been demonstrated using the continuous wave (CW) pumps [48, 49]. In the demonstration, using pump at 1950 nm a continuously tunable wavelength conversion was observed across 324 nm and the idler was generated at the wavelength of 2384 nm.

The first demonstration of a parametric net off-chip gain was reported in silicon at mid-IR wavelengths [25] proving that nonlinear absorption is not detrimental as in the near-IR. In the demonstration, a 4-mm-long waveguide with 700 × 425 nm cross section was used. The dimensions provided zero-dispersion wavelength at 2260 nm, having anomalous dispersion at pump wavelength of 2170 nm. Using pump power of 27.9 W and pulse duration of 2 ps, the total gain of 25.4 dB has been achieved (Figure 6.19), with net off-chip gain of

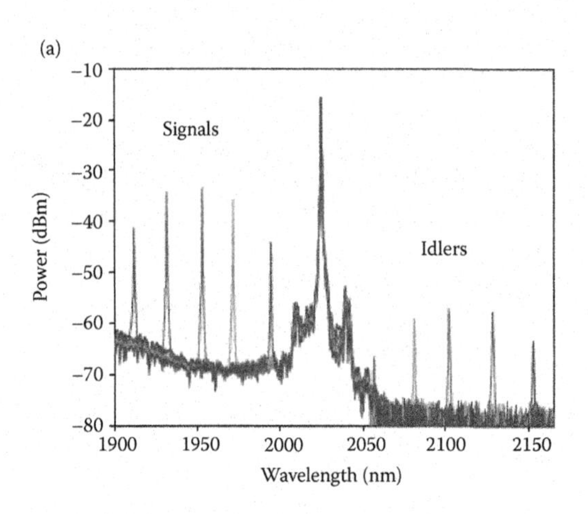

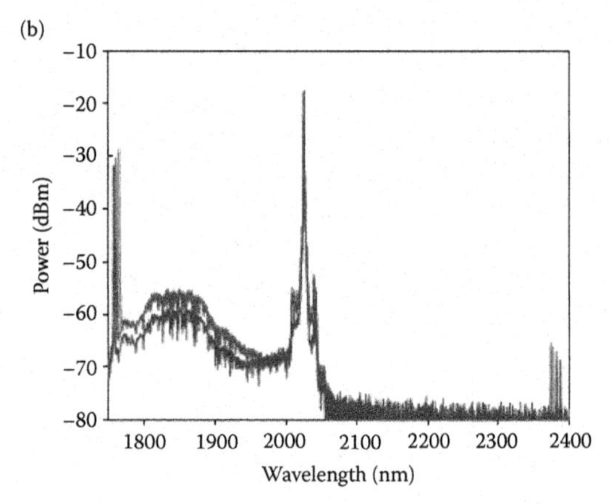

FIGURE 6.18
Measured four-wave mixing spectrum in the mid-IR across the range of 241 nm (a). Generation of mid-IR wavelength up to 2388 nm (b). (Reproduced from S. Zlatanovic et al., *Nat. Photonics*, 4, 561–564, 2010. With permission.)

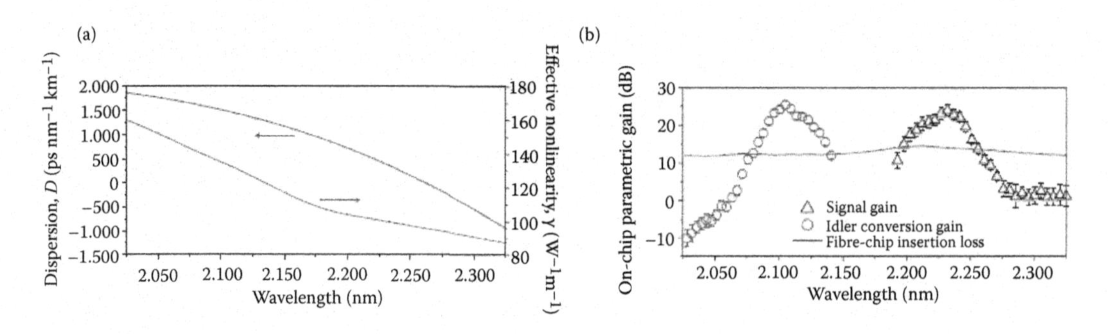

FIGURE 6.19
Engineered silicon nanophotonic waveguide characteristics (a) and mid-IR broadband on-chip parametric amplification (b). (Reproduced from X. Liu et al., *Nat. Photonics*, 4, 557–560, 2010. With permission.)

10 dB for the signal and 13 dB for the idler. The overall on-chip gain bandwidth spanned from 2060 to 2280 nm. This experiment was the first demonstration of a broadband net off-chip gain in silicon.

6.3.6 Conclusion

Silicon waveguides present a promising nonlinear platform for the mid-IR region in particular due to the inherent high nonlinearity and the negligible effect of nonlinear absorption in this part of the spectrum. Consequently, detrimental effects typical for the near-IR region, such as two-photon absorption and associated free-carrier absorption are altogether avoided by pumping silicon at mid-IR wavelengths. As discussed, operation at the mid-IR enabled the first net off-chip gain in silicon, which, in itself, represented a milestone of significant importance for the technology.

In summary, this, by no means all-encompassing, review of nonlinear properties of silicon as a nonlinear platform shows its extraordinary potential to provide access to the entire mid-IR region with sufficient conversion efficiency. More importantly, the results obtained so far single out the silicon waveguide platform as a strong contender for a vast range of applications, pertinent to this region and overviewed in this chapter.

6.4 Long Wavelength Silicon-Based Sources and Detectors

Zoran Ikonic and Robert W. Kelsall

6.4.1 Introduction

Sources and detectors are essential elements of any photonic circuit, and considerable effort has been devoted to the development of lasers and photodetectors, which either use silicon as an active material or are compatible with silicon processing. In the long wavelength range, there has been a 20-year history of detector research, resulting in devices that, although outperformed by their III/V-based counterparts as stand-alone components, can be readily implemented in a Si-based environment. Development of Si-based long wavelength lasers, which has been ongoing for at least 10 years, has proved to be a far more difficult task, particularly for operation at practical operating temperatures. This section describes different approaches to source and detector design, and the results achieved so far in these areas.

6.4.2 Quantum Cascade Lasers

Early mid-IR (MIR) lasers were based on narrow band-gap semiconductors, such as the IV–VI lead salt materials, and generally required cooling to suppress Auger recombination. The advent of the quantum cascade laser (QCL) in 1994 [50] led to a great improvement in mid-infrared source provision, providing coverage across much of the 3- to 5-μm and 8- to 14-μm atmospheric windows. In contrast to conventional semiconductor lasers, QCLs are unipolar devices, emitting photons via transitions between quantum confined subbands in a multiple quantum well heterostructure, rather than via interband recombination (see Figure 6.20). Consequently, a wide range of laser wavelengths can be achieved using

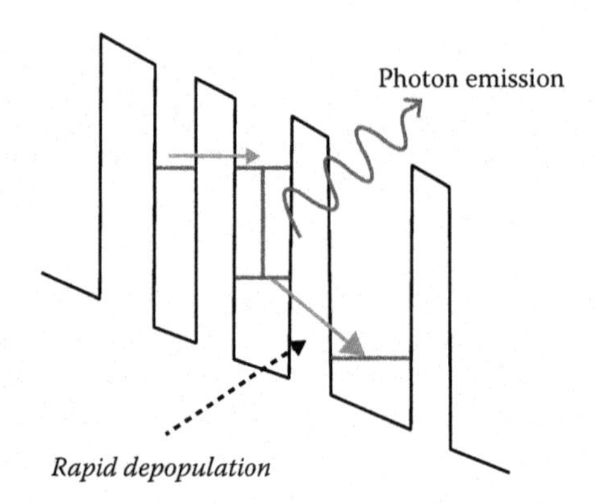

FIGURE 6.20
Simplified energy level diagram of one period of a quantum cascade laser. Population inversion between the two laser states (actually two-dimensional quantum-confined subbands) is achieved by resonant injection into the upper state from the preceding quantum well, and fast depopulation of the lower state either by phonon scattering (in the case of so-called resonant phonon QCLs) or Coulombic scattering into a pseudo-continuum of subbands (in so-called bound-to-continuum QCLs).

different combinations of materials and quantum well/barrier widths. Mid-IR QCLs are generally based on the InGaAs/InAlAs/InP heterostructure system, and devices operating at wavelengths ranging from 3.5 to 24μm have been demonstrated. Another consequence of the use of intersubband transitions to achieve photon emission is that charge carriers are not annihilated, but can be "reused" to generate additional photons by stacking identical heterostructure designs together to form a long, multiple period device. By this means, output powers of over 1W have been demonstrated for CW operation at room temperature [51]. The QCL design can be readily adapted to achieve terahertz lasing, and the first terahertz device was reported in 2002 [52]. The much smaller photon energies involved mean that thermal issues are much more important, and the maximum operating temperature to date is 186K [53]. Virtually all reported THz QCLs have been fabricated from GaAs/AlGaAs heterostructures, using a low Al mole fraction (typically 15%), although InP-based THz devices have also been demonstrated [54].

The unipolar nature of QCL operation offers great potential for silicon-based devices, since the indirect band gap of silicon (and germanium) presents no disadvantage. Early attempts to develop Si/SiGe QCLs led to reports of intersubband electroluminesence at both MIR [55] and terahertz (THz) frequencies [56]. Interestingly, both devices were p-type heterostructures—chosen primarily because of the larger valence band offset available in the Si/SiGe system: indeed these were the first reports of electroluminesence from intersubband hole transitions in any materials system at MIR & THz frequencies, respectively. However, intersubband lasing in Si/SiGe heterostructures has remained elusive. The p-type MIR structures suffer from very fast nonradiative (phonon) scattering, which appears impossible to avoid given the photon energies involved. For terahertz operation, phonon scattering can be suppressed if the photon energy is less than the Ge optical phonon energy (~37 meV—which corresponds to ~9 THz). The main difficulty in this case appears to be the band-mixing between heavy and light hole states, which leads to distorted subband dispersion curves and makes the required subband energies critically dependent on achieving the exact layer thicknesses and SiGe alloy compositions during epitaxy. However, detailed simulation work has eliminated n-type Si-rich heterostructures as a likely QCL design (at least for (001) substrate orientations), primarily due to the large effective mass of the silicon Δ valleys—which leads to very low intersubband material gains—and strong nonradiative scattering from the lowest Δ_2 subbands into strain-split Δ_4 valleys, which are closely located in energy [57]. n-type Si/SiGe structures grown on (111) substrates offer better prospects [58], but the most promising Si-based system appears to comprise Ge quantum wells and Ge-rich SiGe barriers with selective n-type doping. Recent developments in Ge-on-Si epitaxy (e.g., [59]) mean that such structures can be grown on relaxed Ge buffers on Si substrates.

For terahertz operation, the large size of the optical mode raises issues over laser cavity design. Conventional guided mode cavities would be so thick as to require impractically long epitaxial growth times. The approach developed for III–V terahertz QCLs has been, instead, to support surface plasmon modes pinned at the top and bottom of the cavity either by two metal layers (which requires wafer bonding) (so-called double metal waveguides) [60], or by a top metal layer and a doped semiconductor layer under the active region of the device (so-called single plasmon waveguides) [52]. Even so, terahertz QCL waveguide cavities are typically 10 μm thick, which requires epitaxial growth of approximately 2000 layers. This cannot be achieved by solid source Si/Ge MBE, because of the very slow growth times involved, but appears feasible via chemical vapor deposition (CVD), and possibly by gas-source molecular beam epitaxy (MBE). To date, growth of 4200 layers has been demonstrated by CVD (using a plasma enhanced approach) [61] and 1200 layers by gas source MBE [62]. These very thick quantum well stacks are well beyond the critical thickness limit for strain relaxation in a lattice-mismatched heterostructure. Therefore, strain-balanced designs must be used, in which well and barrier layers experience opposing lattice mismatch strains (compressive and tensile, respectively), and the whole structure is grown on a relaxed buffer whose alloy composition achieves net strain balance within each QCL period and therefore throughout the stack.

Unfortunately, doped silicon (and germanium) layers are much less effective than doped GaAs in pinning surface plasmon modes (at least for practically achievable doping densities during epitaxy), so single plasmon waveguides of the form shown in reference [52] do not appear to be realizable in Ge/SiGe structures [63], although the use of surface gratings to support so-called "spoof" surface plasmon modes offers potential for achieving reasonable mode confinement (up to 40%) with low waveguide losses [64]. Double metal waveguides are more lossy than the single plasmon structures, due to strong free carrier absorption in the metal layers, but provide almost complete mode confinement. Calculations show that a 10-μm Ge/SiGe QCL double metal cavity structure would have acceptable losses (~37 cm^{-1})—considerably less than the gain in appropriately designed cascade structures at low temperatures, and even offering potential for room temperature lasing [57]. The anticipated improvement in thermal performance, compared with III–V terahertz QCLs, stems from the lack of fast polar optical phonon scattering in Ge and Si (which competes with the lasing transition), and the higher thermal conductivity of the silicon substrate.

6.4.3 Intracenter-Based Terahertz Lasers

An alternative route to terahertz lasing in silicon—but also based on a unipolar device—is stimulated emission from transitions between impurity states in a bulk-doped structure. Within the simple, isotropic single-valley effective mass picture, the electronic structure of group V impurities (donors) in Si is hydrogen-like, with various transition energies falling in the THz range, and is sixfold degenerate due to the presence of six equivalent X-valleys in the conduction band. The deviation of the microscopic atomic potential in the lattice from spherical symmetry (valley-orbit interaction) modifies the electronic structure, so that the 1s state splits into the lowest, ground state 1s(A), and higher 1s(E), and 1s(T_2) states, above which come higher excited states 2p_0, 2s, 2$p_\pm$, etc., as shown in Figure 6.21. Stimulated emission in the THz range, first reported in [65], has been obtained on different transitions between donor states in silicon ("intracenter" transitions), using Si:P, Si:As, Si:Sb, and Si:Bi materials. A comprehensive review of these lasers is given in [66].

They operate only at very low temperatures, at which the impurities are not ionized. Optical pumping is performed by irradiating the doped Si by another (CO_2) laser or by frequency tunable free-electron laser. In the former case, known as photoexcitation, an electron is excited from the ground, impurity-bound 1s(A) state high into the conduction band, from where it relaxes, by a cascade of optical and acoustic phonon emission events, toward the bottom of the conduction band, and eventually gets captured by the impurity, into its excited bound states, some of which acts as the upper laser state. In the latter case, known as intracenter optical pumping, an electron is excited from the ground directly into a designated excited bound impurity state. The detailed mechanism of laser action varies from one impurity type to another. The mechanism of population inversion in Si:P (and also Si:Sb) is the carrier accumulation in a long-lived bound 2p_0 excited state, acting as the upper laser state, while the lower laser states 1s(E,T_2) are almost empty, because the lifetime of 2p_0 is ~10^{-9} s, larger by up to two orders of magnitude than that of 1s(E,T_2), which makes Si:P and Si:Sb typical four-level laser systems. Lasing in Si:As and Si:Bi has also been achieved, the mechanism therein being somewhat different, due to resonant electron-phonon interaction which makes the 2s and 2p_0 states also short-lived. The lasing transitions in these cases are 2$p_\pm$–1s(T_2,E), or from higher excited states into 2s or 2p_0.

The emission frequency of these lasers depends on the type of donors and the operating transition, and is in the 3.6- to 7-THz range, with a couple of longer-wavelength lines around 1.3–1.8 THz. There is a possibility of frequency tuning by applying stress. Uniaxial stress lowers the crystal symmetry and shifts and/or splits various levels of donor atoms, depending on the level character and stress orientation. In Si:P and Si:Sb lasers no tuning is possible because both the upper and lower laser state shift in the same way (however, the lasing threshold may be reduced by an order of magnitude [67]), while Si:Bi 2$p_\pm$–1s(T2) laser transition can be stress-tuned by 3%/kbar [66]. Values of gain amount to ~0.5 cm^{-1}

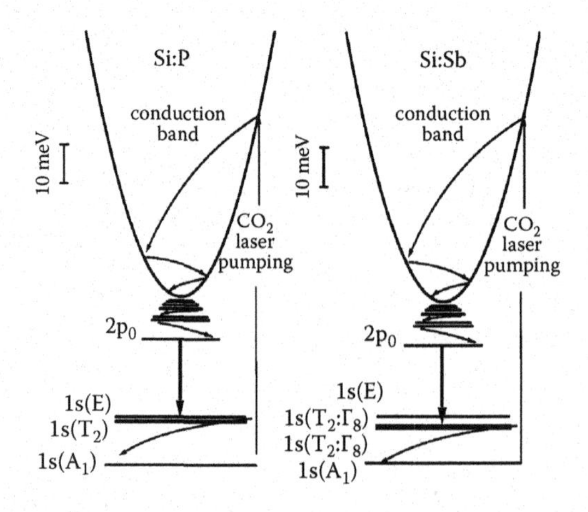

FIGURE 6.21
Energy levels and lasing scheme in Si:P and Si:Sb laser [17]. Copyright IOP Ltd. Reproduced with permission.

for photoexcitation, and 5–10 cm^{-1} for the intracenter pumping scheme [68]. For a number of reasons (to prevent thermal ionization of long-lived states, keep the electron capture probability to acceptable values [69], preserve favorable ratio of relaxation rates of relevant states, etc.) the impurity based lasers operate only at low temperatures, e.g., up to 17 K for Si:P, the record high being 30K for Si:Bi laser. The photoexcitation pumping threshold amounts to 40 kW/cm^2 for Si:P and Si:Sb, or 100 kW/cm^2 for Si:Bi, and is ~3 orders of magnitude lower for intracenter pumping. On the practical side, the Si:impurity lasers are made by cutting doped silicon ingots into parallelopipeds, with ~5 mm sides, and polishing the end facets to high accuracy, to make a high-Q resonator for internal reflection optical modes.

Another type of optically pumped impurity lasers use stimulated Raman scattering, either of pure electronic type or using intervalley phonons. They have been demonstrated in Si:P,As,Sb, emitting around 5 or 6 THz [70, 71], with the Stokes shift equal to the 1s(A)–1s(E) transition. The pumping threshold was similar as in optically pumped lasers.

The possibility of electrical excitation of impurity lasers has also been considered. Electroluminescence has been observed in P-, Ga-, and B-doped Si, by pulsed current excitation of bulk samples or multiple QW structures, even at relatively high temperatures of up to 118K for Si:B [72–74]. Emission spectra show that emission comes from intracenter transitions, the excitation mechanism being the impact ionization of neutral impurities, followed by electron relaxation, as in optically pumped schemes. It is not clear whether this can lead to selective population of upper laser state, which would be robust enough to outweigh the impact ionization. Carefully tailored excitation schemes based on tunneling in modulation δ-doped quantum cascade structures have also been proposed [75], where the donor ground state serves as the lower laser state, (out)coupled to a subband of a neighbouring quantum well, and the upper laser state is the impurity-related 2D-like continuum. The emission frequency would be tunable by precise positioning of donor δ-doped layer in the quantum well, because it influences the transition energy. So far, no electrically pumped intracenter lasers, either in bulk or QW structures, have been demonstrated.

6.4.4 Si/SiGe Quantum Well Infrared Photodetectors

Intersubband devices can be used for long wavelength detection as well as for lasing. Quantum well IR photodetectors (QWIPs) have been widely demonstrated in III–V materials for 3–5 μm and 8–14 μm detection [76], and thermal imaging cameras based on QWIP arrays are commercially available. The operating principle for QWIPs is much simpler than for QCLs: a QWIP comprises a stack of identical quantum wells, separated by barriers that are assumed to be sufficiently thick to prevent tunneling. Photon absorption excites carriers from the ground state of each well to the continuum, where they constitute a photocurrent when the device is biased, as shown in Figure 6.22. The main drawback of QWIPs is the high dark current, which arises from thermally induced carrier excitation out of the quantum wells and which generally necessitates device cooling below room temperature.

Si/SiGe QWIPs operating at MIR wavelengths have been demonstrated by several groups [77–79]. These devices were p-type, comprising strained SiGe quantum wells and Si barriers grown directly on a Si substrate. The absence of any buffer layer, and hence the lack of strain symmetrization, restricted the number of quantum wells, which could be grown to typically ~20, depending on the alloy composition of the quantum wells. p-type devices were chosen because when growing Si/SiGe directly on silicon, there is virtually no band offset in the conduction band. However, p-type QWIPs have the additional advantage of normal-incidence operation, whereas only edge-absorption is possible in n-type devices

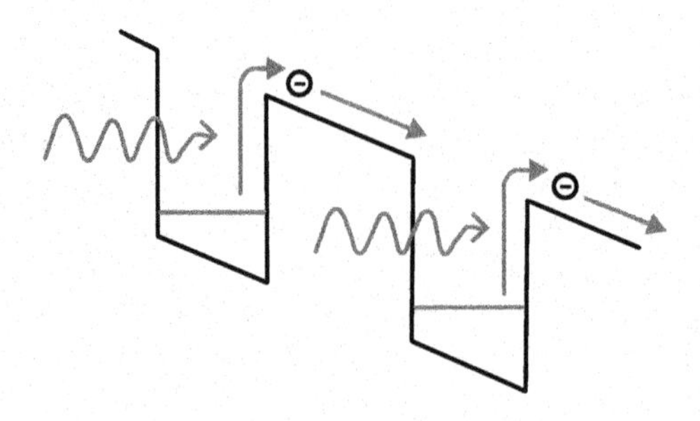

FIGURE 6.22
Energy level diagram for a 2 quantum well section of a QWIP. Incident photons excite electrons from the ground state of each quantum well into continuum states, where they are swept out of the device by an external bias.

due to the symmetry rules for the intersubband optical matrix element. Increased responsivity was obtained by use of both buried oxide and buried silicide reflector layers [80, 81].

Terahertz Si/SiGe QWIPs have also been proposed and designed [82], but there appears to have been no experimental realization of such devices. Although, in principle, further improvements in Si-based QWIPs may be obtained using strain-symmetrized designs grown on relaxed SiGe buffer layers, little attention has been paid to this area of research.

6.4.5 SiGe Quantum Dot Detectors

The operation principle of quantum dot (QD) photodetectors is the same as for QW devices. Carriers are photoexcited, via intersubband absorption, into quasi-bound or free states, generating photocurrent under an applied electric field. A particular feature of Ge or SiGe dots in Si matrix is a much smaller band offset in the conduction band than in the valence band. The band alignment becomes type II (binding potential only for holes) for Ge content >25% [83]. However, with n-doping in the Ge regions, a comparatively shallow self-consistent potential well is induced in the conduction band as well. With typical QD sizes (a couple of nanometers in height and 10–20 nm across) hole inter-sublevel transitions generally fall into the mid-IR, and electron transitions into far-IR range, and either can be used for photodetection. One advantage of using QDs instead of QWs relies on an expected larger photoconductive gain, due to a reduced capture probability. Another advantage is related to the nonzero normal incidence absorption on inter-sublevel transitions, enabling the practically convenient normal-incidence photodetectors, and this exists for various reasons (all-sided wavefunction confinement, HH-LH mixing in hole states, finite off-diagonal conduction band L-valley mass tensor components for dots grown on (100) Si substrate). For various configurations and structural parameters a photoresponse based on hole inter-sublevel transitions has been demonstrated, e.g., in the 2.8- to 4.8-μm range [84], the 3.7- to 6-μm range [85], and in n-doped structure in the range of 16–20 μm [84]. Except for the active transition linewidth, the broader response spectrum also comes from dot size dispersion, and from the fact that more quantum levels may be involved in the absorption.

6.4.6 Si/SiGe Micro-Bolometers

Another type of detector, with extremely broad spectral response, are micro-bolometers (thermistors). In fact, the response width there depends only on the ability of the detector structure to absorb incident radiation power and transfer it to the lattice. The operation of the SiGe/Si thermistor relies on thermal excitation of charge carriers from quantum-confined low-energy states, or carriers trapped at various defects, where they have low mobility, to free states where they are highly mobile. The temperature coefficient of resistance (TCR) increases with an increasing activation energy—the difference between the free and bound state energy—hence in SiGe QW or QD structures p-type doping is clearly much better than n-type. A high TCR favors larger Ge content and well width/dot size, which is limited by strain relaxation effects, and QDs are here advantageous over QWs. A TCR of approximately 3.0%/K has been obtained at room temperature in QWs with 32%Ge, while Ge QDs in a Si matrix offered a somewhat larger value of 3.4%/K, but with the noise factor increased by three orders of magnitude [86]. Amorphous SiGe:H-based bolometers, also compatible with the CMOS technology, have been successfully tested as detectors of terahertz radiation (at ~1 THz), although they have originally been designed for IR wavelengths [87].

6.4.7 Conclusions

Research activity in Si-based long-wavelength detectors has resulted in several different devices that have sufficiently good performance to be considered ready for use in silicon photonic circuits. Of these, the Si microbolometers are perhaps the most interesting devices for practical applications, because of their room temperature operation and compatibility with CMOS processing. The development of long wavelength Si-based lasers is still in relative infancy, with the only successful devices requiring optical pumping by large, high power sources, which effectively precludes their employment in integrated Si-photonic circuits. However, simulation work has shown that Si-based quantum cascade lasers that rely on intersubband transitions in n-type germanium quantum wells are more promising than any of the structures previously investigated experimentally.

6.5 Mid-Infrared Heterogeneous Integration

Gunther Roelkens

6.5.1 Introduction

Although silicon-based photonic integrated circuits are promising for passive optical functions in the short-wave and mid-IR, substantially more complex photonic integrated circuits could be realized if these waveguide structures could be co-integrated with photodetectors and light sources for SWIR/MIR. This could, for example, enable the realization of integrated spectroscopic chips, in which a silicon-based wavelength demultiplexer is combined with an integrated photodetector array. Or, it could enable the realization of integrated tunable laser sources for the SWIR/MIR, also a very useful component in spectroscopic systems. Mid-IR spectroscopy has gained significant importance in recent years as a detection technique for substances that absorb in this spectral region. Traditionally, a spectroscopic system consists of bulky equipment that is difficult to handle and incurs high cost. An integrated spectroscopic system would eliminate these disadvantages.

GaSb-based active opto-electronic devices allow realization of mid-IR light sources and detectors in the 2- to 3-μm wavelength range for such integrated systems. Hence, the integration of GaSb-based active devices onto silicon (SOI) passive waveguide circuits potentially realizes such integrated functions.

6.5.2 Heterogeneous Integration—General

In this work, a heterogeneous integration approach is adopted to intimately integrate the III–V semiconductors with the silicon waveguide circuits. The heterogeneous integration is based on an epitaxial layer transfer process using the polymer divinylsiloxane-benzocyclobutene (DVS-BCB) as a bonding agent. The process is performed by transferring the epitaxial layer to an SOI waveguide circuit wafer through a die-to-wafer bonding process, as shown in Figure 6.23.

The process starts off with the finalized SOI photonic wafer, carrying the passive optical functions. On top of the finished wafer, the DVS-BCB adhesive bonding agent is spin-coated. GaSb-based epitaxial wafers are cut into small dies (ranging from 1 mm^2 to 1 cm^2) and these dies are attached, epitaxial layers down, to the SOI wafer. Since the III–V dies do not yet contain any structure, there are no stringent alignment tolerances and integration using a fast pick-and-place machine can be envisioned. After attachment and curing of the DVS-BCB, the GaSb growth substrate is removed by a combination of mechanical grinding and wet chemical etching, until an etch stop layer is reached.

This leaves the thin epitaxial films attached to the SOI wafer, after which the opto-electronic components can be fabricated, lithographically aligned to the underlying SOI waveguide circuit. A heterogeneous integration approach has several advantages compared with a classical hybrid flip-chip integration process: the device density is larger and the integration process is lower cost, since no accurate alignment is required during the positioning of the die. The actual alignment of the opto-electronic component is achieved using lithographic techniques. Since DVS-BCB consists of an organic moiety, absorption peaks can be expected in the SWIR/MIR. Figure 6.24 illustrates the absorption spectrum of DVS-BCB in the 1- to 3-μm wavelength range. There are strong absorption peaks around 1.7 μm and 2.3 μm due to its C–H bonds. Nevertheless, this absorption is manageable with careful device design, by reducing the overlap of the optical field with this absorbing bonding layer.

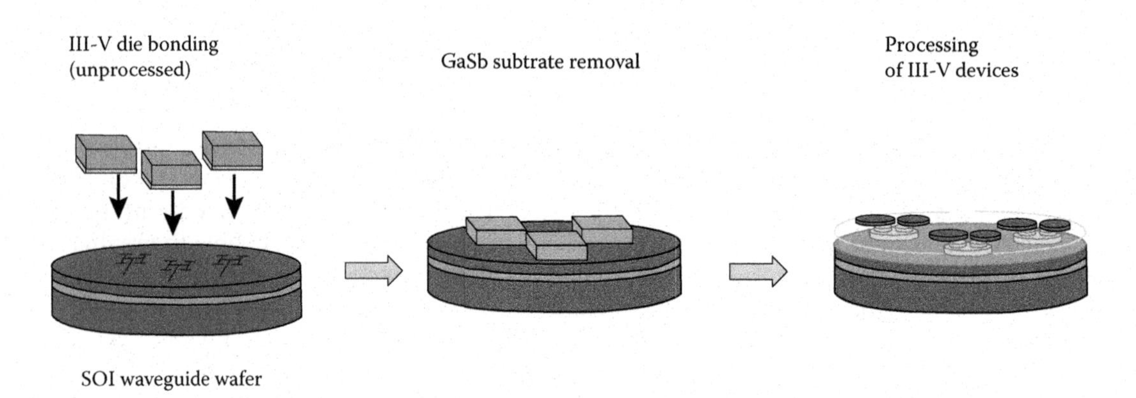

FIGURE 6.23
Schematic outline of the wafer-scale heterogeneous integration process.

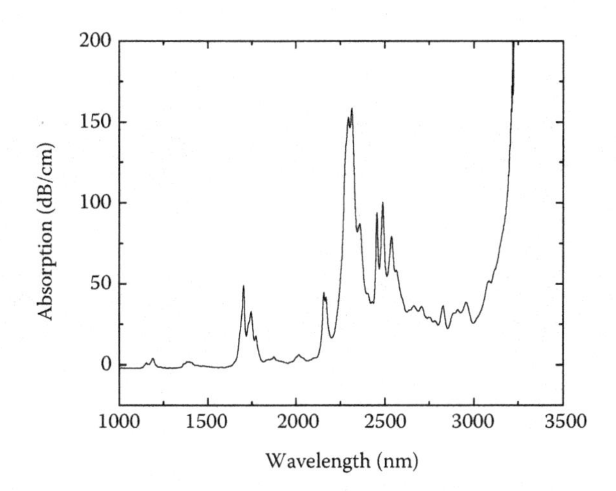

FIGURE 6.24
Absorption spectrum of DVS-BCB, measured using a photospectrometer setup.

6.5.3 Heterogeneous Integration—Detailed

Before the integration process, the SOI and GaSb epitaxy are first prepared. The SOI waveguide wafer is fabricated in a CMOS pilot line. Epitaxial growth is done at Université Montpellier 2, France, using molecular beam epitaxy. The bonding process starts by cleaning both the SOI and GaSb epitaxy die. The GaSb epitaxy is cleaned with acetone and IPA. The SOI die is cleaned using a standard clean-1 solution ($NH_4OH:H_2O_2:H_2O$) in 1:1:5 v/v at 70°C for 15 minutes. The DVS-BCB polymer (Cyclotene 3022-35) is diluted using mesitylene in 2:3 v/v to achieve a thin bonding layer. The DVS-BCB is then spin-coated at 3000 rpm for 40 seconds on the SOI waveguide die to approximately achieve a 200-nm-thick bonding layer. The sample is baked on a hot plate at 150°C for 3 minutes to remove mesitylene from the thin DVS-BCB layer. The GaSb die is then transferred onto the SOI substrate at the same temperature. The curing process is done at 250°C for 1 hour to achieve more than 95% of polymerization. N_2 is used during the curing process to prevent oxidation of the DVS-BCB when the temperature increases. After the bonding process, the GaSb substrate is removed using a combination of mechanical grinding and wet etching. The GaSb substrate is grinded mechanically, leaving approximately 50 μm. The rest of the substrate is then removed by selective wet etching using a mixture of CrO_3 and HF and water in 1:1:3 v/v at 25°C. The etching rate is ~7 μm/minute.

An InAsSb layer is used as an etch stop layer, which can be selectively removed by wet etching using a 2:1 v/v citric acid and hydrogen peroxide solution. The etch rate is ~100 nm/minute. Figure 6.25a shows an SEM cross section image of the GaSb layer bonded on an SOI waveguide (after photodetector mesa definition). A III–V/SOI bonding layer of 150–200 nm is easily achievable, which is sufficient for the optical coupling between the SOI waveguide circuit and GaSb-based opto-electronic components. Figure 6.25b represents the surface of the transferred GaSb epitaxy after the substrate removal process.

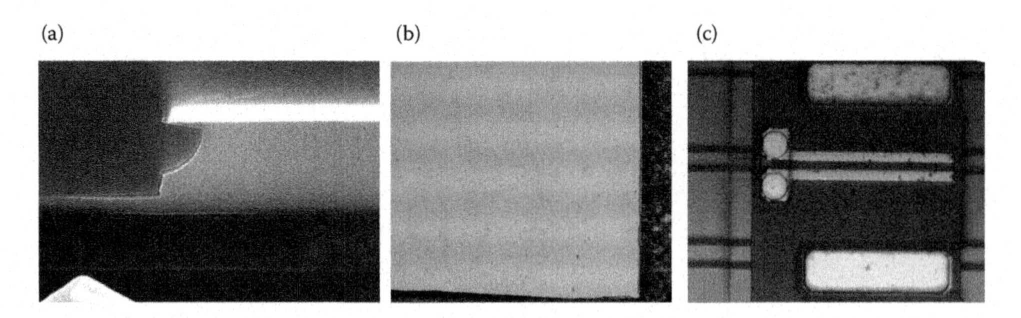

FIGURE 6.25
Cross section of the bonded layer stack, after GaSb-mesa definition (a), top view of the transferred epitaxial layer (b), and microscope image of the realized photodetectors (c).

6.5.4 Heterogeneously Integrated SWIR Photodetectors

A schematic of the realized integrated photodetectors is shown in Figure 6.26a. It consists of a 3-μm × 220 nm SOI waveguide onto which a GaInAsSb epitaxial layer stack is bonded using DVS-BCB adhesive bonding. Light is coupled from the SOI waveguide into the photodetector using evanescent coupling. Optimal coupling can be achieved by controlling the phase matching between the SOI waveguide and the photodiode waveguide. The simulation of the coupling efficiency as a function of the photodetector thickness is carried out using a full-vectorial 2D eigenmode expansion method. The simulation result is shown in Figure 6.26b for a 50-μm-long and 9-μm-wide detector. It shows that the maximum absorption is achieved when the thickness of photodiode phase matches with the Si waveguide mode.

The epitaxial stack consists of a 500-nm $Ga_{0.79}In_{0.21}As_{0.19}Sb_{0.81}$ intrinsic region, indicated as the optimum point from simulation, 50 nm of $Ga_{0.79}In_{0.21}As_{0.19}Sb_{0.81}$ and 50 nm $InAs_{0.91}Sb_{0.09}$ doped 1.0×10^{18} cm^{-3} as the p-contact layer. An $InAs_{0.91}Sb_{0.09}$ layer is chosen as n-contact

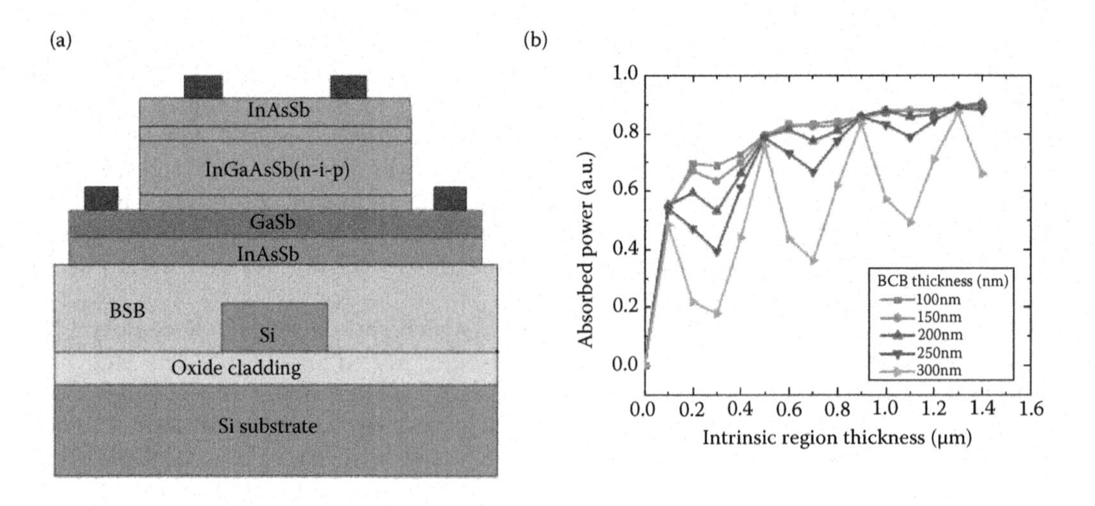

FIGURE 6.26
Schematic of the photodetector cross section (a) and simulation results indicating the influence of phase matching on the device responsivity (b).

because of its lower band-gap (0.35eV). Photoluminescence of this epitaxial stack at room temperature peaks at 2.5 μm, roughly corresponding to the cut-off wavelength of the envisioned photodetectors.

After the bonding and substrate removal, the mesa is formed by a combination of both dry (CH_4:H_2) and wet etching (citric acid:H_2O_2:H_3PO_4:H_2O 55:5:3:220 v/v) to reduce the photodetector dark current. Ti(2 nm)/Pt(35 nm) and Au(100 nm) is deposited for both contacts using e-beam evaporation. BCB (1.3 μm thick) is then spin coated on the sample and cured at 250°C for 1 hour to passivate the device. Figure 6.23a is an SEM cross-section image of the fabricated photodetector, showing undercut due to the wet etching process. The bonding thickness for the reported devices is 258 nm. The original process scheme included the removal of the InAsSb cap layer to obtain a clean surface before bonding. However, this created a hydrophilic surface, which prevented good bonding with BCB. Therefore, the InAsSb cap layer was kept in this experiment.

The experimental setup consists of a continuous wave mid-IR tunable laser, which is coupled to a single mode fiber. After polarization control, TE-polarized light from a single mode fiber is injected into the SOI waveguide through a grating coupler structure defined in the silicon waveguide layer. The gratings have −8 dB peak coupling efficiency at 2.17 μm with 200 nm 3-dB-bandwidth. The distance between the grating coupler and the device is 400 μm. Therefore, the loss due to the propagation in the waveguide can be neglected.

The photoresponse measured at 2.25 μm wavelength at different fiber input power levels is shown in Figure 6.27a. Good linearity is obtained over 12 dB input power range. Dark current at −0.1 V is 1.13 μA. This can be improved by changing the mesa etching condition, for example, by wet etching and using a passivation process such as applying ammonia sulfide before BCB passivation. Taking into account the loss of the grating coupler, we obtain a peak responsivity of 0.44 A/W corresponding to 24% quantum efficiency at 2.29 μm wavelength. The dark current at −1 V as a function of temperature is shown in Figure 6.27. It decreases significantly as the temperature decreases.

This is the first demonstration of an integrated photodetector for the short-wave IR.

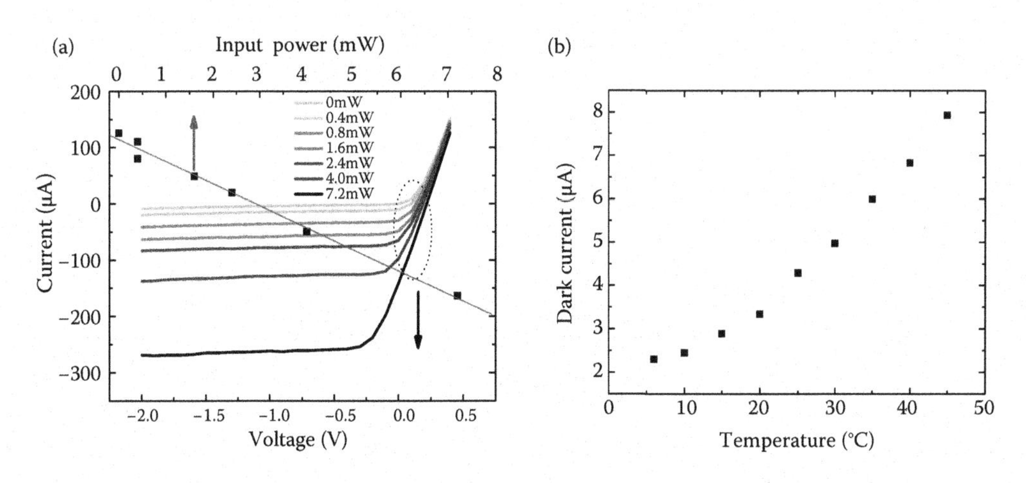

FIGURE 6.27
Experimentally obtained IV curves under illumination (a) and the dark current of the integrated photodetectors as a function of temperature (at -1 V) (b).

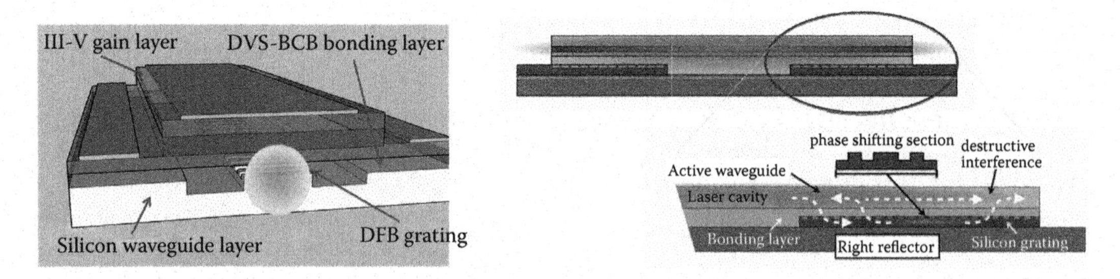

FIGURE 6.28
Proposed laser architectures for short-wave infrared single wavelength laser diodes.

6.5.5 Heterogeneously Integrated SWIR Laser Diodes

Integrated laser diodes in the short-wave IR could be also considered. The device architectures are very similar to those demonstrated on a heterogeneous InP/SOI platform, for telecommunication applications [88]. Here we will briefly describe two potential laser geometries, for integrated short-wave IR laser sources. These are illustrated in Figure 6.28.

The first approach is a hybrid implementation of a GaSb/silicon distributed feedback laser diode. In this case, the first order grating, with a quarter wavelength shift, is implemented in the silicon waveguide layer. The optical mode is predominantly confined in the silicon waveguide layer, while the evanescent tail overlaps with the GaSb-based active layers. While this approach results in a straightforward coupling to a passive waveguide circuit, the confinement in the active region is low. Also, this confinement, and hence the modal gain, is critically dependent on the BCB bonding layer thickness. This approach therefore requires stringent bonding layer thickness control to be viable. On the other end of the design space, one could consider to fully confine the laser mode to the GaSb layer stack and use a resonant grating structure to provide the wavelength selective feedback and at the same time provide outcoupling to the silicon waveguide circuit. This approach is more relaxed in terms of wafer bonding requirements, while the requirements on phase matching between the GaSb waveguide structure and the silicon waveguide layer is more stringent. Many other configurations can be envisioned to realize a heterogeneously integrated laser structure, very similar to the type of structures that are studied in InP/silicon lasers.

References

1. R. A. Soref, S. J. Emelett, and W. R. Buchwald, "Silicon waveguided components for the long-wave infrared region," *J. Opt. A* vol. 8, 840–848, 2006.
2. R. Soref, "Mid-infrared photonics in silicon and germanium," *Nat. Photonics* vol. 4, 495–497, 2010.
3. G. Z. Mashanovich, M. M. Milošević, M. Nedeljkovic, N. Owens, B. Xiong, E. J. Teo, and Y. Hu, "Low loss silicon waveguides for the mid-infrared," *Opt. Express* vol. 19, 7112–7119, 2011.
4. M. M. Milošević, P. S. Matavulj, P. Y. Yang, A. Bagolini, and G. Z. Mashanovich, "Rib waveguides for mid-infrared silicon photonics," *J. Opt. Soc. Am. B* vol. 26, 1760–1766, 2009.
5. www.irphotonics.com.

6. G. Z. Mashanovich, W. R. Headley, M. M. Milošević, N. Owens, E. J. Teo, B. Q. Xiong, P. Y. Yang, M. Nedeljkovic, J. Anguita, I. Marko, and Y. Hu, "Waveguides for mid-infrared group IV photonics," in *Proc. of IEEE Conference on Group IV Photonics*, (Institute of Electrical and Electronics Engineers, 2010), 374–376.
7. M. M. Milošević, D. J. Thomson, X. Chen, D. Cox, and G. Z. Mashanovich, "Silicon waveguides for the 3–4 μm wavelength range," in *Proc. of IEEE Conference on Group IV Photonics*, (Institute of Electrical and Electronics Engineers, 2011).
8. E. D. Palik, *Handbook of Optical Constants of Solids, Vol. 1*, London: Academic, London, 1985.
9. R. Shankar, R. Leijssen, I. Bulu, and M. Lončar, "Mid-infrared photonic crystal cavities in silicon," *Opt. Express* vol. 19, 5579–5586, 2011.
10. T. Baehr-Jones, A. Spott, R. Ilic, A. Spott, B. Penkov, W. Asher, and M. Hochberg, "Silicon-on-sapphire integrated waveguides for the mid-infrared," *Opt. Express* vol. 18,12127–12135, 2010.
11. A. Spott, Y. Liu, T. Baehr-Jones, R. Ilic, and M. Hochberg, "Silicon waveguides and ring resonators at 5.5 μm," *Appl. Phys. Lett.* vol. 97, 213501, 2010.
12. F. Li, S. D. Jackson, C. Grillet, E. Magi, D. Hudson, S. J. Madden, Y. Moghe, C. O'Brien, A. Read, S. G. Duvall, P. Atanackovic, B. J. Eggleton, and D. J. Moss, "Low propagation loss silicon-on-sapphire waveguides for the mid-infrared," *Opt. Express* vol. 19, 15212–15220, 2011.
13. E. J. Teo, A. A. Bettiol, P. Yang, M. B. H. Breese, B. Q. Xiong, G. Z. Mashanovich, W. R. Headley, and G. T. Reed, "Fabrication of low-loss silicon-on-oxidized-porous-silicon strip waveguide using focused proton-beam irradiation," *Opt. Lett.* vol. 34, 659–661, 2009.
14. M. A. Hughes, W. Yang, and D. W. Hewak, "Spectral broadening in femtosecond laser written waveguides in chalcogenide glass," *J. Opt. Soc. Am. B* vol. 26, 1370–1378, 2009.
15. C. Tsay, E. Mujagić, C. K. Madsen, C. F. Gmachl, and C. B. Arnold, "Mid-infrared characterization of solutionprocessed As_2S_3 chalcogenide glass waveguides," *Opt. Express* vol. 18, 15523–15530, 2010.
16. N. Hô, M. C. Phillips, H. Qiao, P. J. Allen, K. Krishnaswami, B. J. Riley, T. L. Myers, and N. C. Anheier, Jr., "Single-mode low-loss chalcogenide glass waveguides for the mid-infrared," *Optics Letters* vol. 31, 1860–1862, 2006.
17. C. Vigreux-Bercovici, E. Bonhomme, A. Pradel, J.-E. Broquin, L. Labadie, and P. Kern, "Transmission measurements at 10.6 μm of $Te_2As_3Se_5$ rib waveguides on As_2S_3 substrate," *Appl. Phys. Lett.* vol. 90, 011110, 2007.
18. S. Kameyama, M. Imaki, Y. Hirano, S. Ueno, S. Kawakami, D. Sakaizawa, and M. Nakajima. "Development of 1.6 μm continuous-wave modulation hard target differential absorption LIDAR system for CO2 sensing," *Opt. Lett.* vol. 34, 1513–1515, 2009.
19. U. Willer, M. Saraji, A. Khorsandi, P. Geiser, and W. Schade, "Near- and mid-infrared laser monitoring of industrial processes, environment and security applications," *Optics and Lasers in Engineering* vol. 44, 699–710. in *Optical Diagnostics and Monitoring: Advanced Monitoring Techniques and Coherent Sources*, 2006.
20. M. Bashkansky, H. R. Burris, E. E. Funk, R. Mahon, and C. I. Moore, "RF phase coded random-modulation LIDAR," *Opt. Commun.* vol. 231, 93–98, 2004.
21. Y. A. Bakhirkin, A. A. Kosterev, C. Roller, R. F. Curl, and F. K. Tittel, "Mid-infrared quantum cascade laser based off-axis integrated cavity output spectroscopy for biogenic nitric oxide detection," *Appl. Opt.* vol. 43, 2257–2266, 2004.
22. V. Raghunathan, R. Shori, O. Stafsudd, and B. Jalali, "Nonlinear absorption in silicon and the prospects of mid-infrared silicon Raman lasers," *J. Phys. Status Solidi A* vol. 203, R38–R40, 2006.
23. B. Jalali, V. Raghtmathan, R. Shori, S. Fathpour, D. Dimitropoulos, and O. Stafsudd, "Prospects for silicon mid-IR Raman lasers," *IEEE J. Select. Top. Quantum Electron.* vol. 12, 1618–1627, 2006.
24. V. Raghunathan, D. Borlaug, R. Rice, and B. Jalali, "Demonstration of a mid-infrared silicon Raman amplifier," *Opt. Express* vol. 15, 14355–14362, 2007.
25. X. Liu, R. M. Osgood, Y. A. Vlasov, and W. M. J. Green, "Mid-infrared optical parametric amplifier using silicon nanophotonic waveguides," *Nat. Photon.* vol. 4, 557–560, 2010.

26. S. Zlatanovic, J. S. Park, S. Moro, J. M. C. Boggio, I. B. Divliansky, N. Alic, S. Mookherjea, and S. Radic, "Mid-infrared wavelength conversion in silicon waveguides using ultracompact telecom-band–derived pump source," *Nat. Photon.* vol. 4, 561–564, 2010.
27. R. W. Boyd, *Nonlinear Optics*, Academic Press, 2003.
28. P. D. Maker, and R. W. Terhune, "Study of optical effects due to an induced polarization third order in the electric field strength," *Phys. Rev.* 137, A801–A818, 1965.
29. J. Zhang, Q. Lin, G. Piredda, R. W. Boyd, G. P. Agrawal, and P. M. Fauchet, "Anisotropic nonlinear response of silicon in the near-infrared region," *Appl. Phys. Lett.* vol. 91, 071113, 2007.
30. A. Liu, H. Rong, R. Jones, O. Cohen, D. Hak, and M. Paniccia, "Optical amplification and lasing by stimulated Raman scattering in silicon waveguides," *J. Lightwave Technol.* vol. 24, 1440–1455, 2006.
31. D. Bristow, N. Rotenberg, and H. M. van Driel, "Two-photon absorption and Kerr coefficients of silicon for 850–2200 nm," *Appl. Phys. Lett.* vol. 90, 191104, 2007.
32. Q. Lin, J. Zhang, G. Piredda, R. W. Boyd, P. M. Fauchet, and G. P. Agrawal, "Dispersion of silicon nonlinearities in the near infrared region," *Appl. Phys. Lett.* vol. 91, 021111, 2007.
33. J. J. Wynne, and G. D. Boyd, "Study of optical difference mixing in Ge and Si using a CO_2 gas laser," *Appl. Phys. Lett.* vol. 12, 191–192, 1968.
34 H. Garcia and R. Kalyanaraman, "Phonon-assisted two-photon absorption in the presence of a dc-field: the nonlinear Franz–Keldysh effect in indirect gap semiconductors," *J. Phys. At. Mol. Opt. Phys.* vol. 39, 2737–2746, 2006.
35. S. Pearl, N. Rotenberg, and H. M. van Driel, "Three photon absorption in silicon for 2300–3300 nm," *Appl. Phys. Lett.* vol. 93, 131102, 2008.
36. S. Zlatanovic, J. S. Park, F. Gholami, J. Chavez Boggio, S. Moro, N. Alic, S. Mookherjea, and S. Radic," Mid-infrared wavelength conversion in silicon waveguides pumped by silica-fiber-based source," *IEEE J. Select. Top. Quantum Electron.* DOI10.1109/JSTQE.2011.2119295, 2011.
37. B. Corcoran, *et al.* Green light emission in silicon through slow-light enhanced third-harmonic generation in photonic crystal waveguides. *Nat. Photon.* vol. 3, 206–210, 2009.
38. A. Liu, H. Rong, M. Paniccia, O. Cohen, and D. Hak," Net optical gain in a low loss silicon-on-insulator waveguide by stimulated Raman scattering," *Opt. Express* vol. 12, 4261–4268, 2004.
39. R. Claps, D. Dimitropoulos, V. Raghunathan, Y. Han, and B. Jalali, " Observation of stimulated Raman scattering in silicon waveguides," *Opt. Express* vol. 11, 1731–1730, 2003.
40. D. Borlaug, R. R. Rice, and B. Jalali, "Raman beam cleanup in silicon in the mid-infrared," *Opt. Express* vol. 18, 12411–12414, 2010.
41. X. Liu, J. B. Driscoll, J. I. Dadap, R. M. Osgood, S. Assefa, Y. A. Vlasov, and W. M. J. Green, "Self-phase modulation and nonlinear loss in silicon nanophotonic wires near the mid-infrared two-photon absorption edge," *Opt. Express* vol. 19, 7778–7789, 2011.
42. E. Dulkeith, Y. A. Vlasov, X. G. Chen, N. C. Panoiu, and R. M. Osgood, Jr., "Self-phase-modulation in submicron silicon-on-insulator photonic wires," *Opt. Express* vol. 14, 5524–5534, 2006.
43. J. I. Dadap, N. C. Panoiu, X. G. Chen, I. W. Hsieh, X. P. Liu, C. Y. Chou, E. Dulkeith, S. J. McNab, F. N. Xia, W. M. J. Green, L. Sekaric, Y. A. Vlasov, and R. M. Osgood., "Nonlinear-optical phase modification in dispersion-engineered Si photonic wires," *Opt. Express* vol. 16, 1280–1299, 2008.
44. G. Agrawal, *Nonlinear Fiber Optics*, Academic Press, 2006.
45. Q. Lin, T. J. Johnson, R. Perahia, C. P. Michael, and O. J. Painter, "A proposal for highly tunable optical parametric oscillation in silicon micro-resonators," *Opt. Express* vol. 16, 10596–10610, 2008.
46. E.-K. Tien, Y. Huang, S. Gao, Q. Song, F. Qian, S. K. Kalyoncu, and O. Boyraz, "Discrete parametric band conversion in silicon for mid-infrared applications," *Opt. Express* vol. 18, 21981–21989, 2010.
47. J. M. Chavez Boggio, S. Zlatanovic, F. Gholami, J. M. Aparicio, S. Moro, K. Balch, N. Alic, and S. Radic, "Short wavelength infrared frequency conversion in ultra-compact fiber device," *Opt. Express* vol. 18, 439–445, 2010.
48. R. K. W. Lau, M. Ménard, Y. Okawachi, M. A. Foster, A. C. Turner-Foster, R. Salem, M. Lipson, and A. L. Gaeta, "Continuous-wave mid-infrared frequency conversion in silicon nanowaveguides," *Opt. Lett.* vol. 36, 1263–1265, 2011.

49. C. Turner-Foster, M. A. Foster, R. Salem, A. L. Gaeta, and M. Lipson, "Frequency conversion in silicon waveguides over two-thirds of an octave," in *Conference on Lasers and Electrooptics OSA/IEEE*, Baltimore, 2009.
50. J. Faist, F. Capasso, D. L. Sivco, C. Sirtori, A. L. Hutchinson, and A. Y. Cho, "Quantum cascade laser," *Science* vol. 264, 553–556, 1994.
51. Y. Bai, S. R. Davish, S. Slivken, W. Zhang, A. Evans, J. Nguyen, and M. Razeghi, "Room temperature continuous wave operation of quantum cascade lasers with watt-level optical power," *Appl. Phys. Lett.* vol. 92, 101105-3, 2008.
52. R. Kohler, A. Tredicucci, F. Beltram, H. E. Beere, E. H. Linfield, A. G. Davies, D. A. Ritchie, R. C. Iotti, and F. Rossi, "Terahertz semiconductor-heterostructure laser," *Nature* vol. 417, 156–159, 2002.
53. S. Kumar, Q. Hu, and J. Reno, "186 K operation of terahertz quantum-cascade lasers based on a diagonal design," *Appl. Phys. Lett.* vol. 94, 131105-3, 2009.
54. L. Ajili, G. Scalari, N. Hoyler, M. Giovannini, and J. Faist, "InGaAs-AlInAs/InP terahertz quantum cascade laser," *Appl. Phys. Lett.* vol. 87, 141107-3, 2005.
55. G. Dehlinger, L. Diehl, U. Gennser, H. Sigg, J. Faist, K. Ensslin, D. Grutzmacher, and E. Muller, "Intersubband electroluminescence from silicon-based quantum cascade structures," *Science* vol. 290, 2277–2280, 2000.
56. S. A. Lynch, R. Bates, D. J. Paul, D. J. Norris, A. G. Cullis, Z. Ikonic, R. W. Kelsall, P. Harrison, P. Murzyn, D. D. Arnone, and C. R. Pidgeon, "Intersubband electroluminescence from Si/SiGe cascade emitters at terahertz frequencies," *Appl. Phys. Lett.* vol. 81, 1543–1545, 2002.
57. A. Valavanis, T. V. Dinh, L. J. M. Lever, Z. Ikonic, and R. W. Kelsall, "Material configurations for *n*-type silicon-based terahertz quantum cascade lasers," *Phys. Rev. B* vol. 83, 195321-8, 2011.
58. A. Valavanis, L. Lever, C. A. Evans, Z. Ikonic, and R. W. Kelsall, "Theory and design of quantum cascade lasers in (111) *n*-type Si/SiGe," *Phys. Rev. B* vol. 78, 035420-7, 2008.
59. V. A. Shah, A. Dobbie, M. Myronov, D. J. F. Fulgoni, L. J. Nash, and D. R. Leadley, "Reverse graded relaxed buffers for high Ge content SiGe virtual substrates," *Appl. Phys. Lett.* vol. 93, 192103-3, 2008.
60. B. S. Williams, S. Kumar, H. Callebaut, and Q. Hu, "Terahertz quantum-cascade laser at λ ~100 μm using metal waveguide for mode confinement," *Appl. Phys. Lett.* vol. 83, 2124–2126, 2003.
61. G. Isella, G. Matmon, A. Neels, E. Muller, M. Califano, D. Chrastina, H. von Kanel, L. Lever, Z. Ikonic, R. W. Kelsall, and D. J. Paul, "SiGe/Si quantum cascade structures deposited by low-energy plasma-enhanced CVD," in *Proc. 5th IEEE Int. Conf. on Group IV Photonics*, 29–31, 2008, ISBN 978-1-4244-1768-1.
62. R. W. Kelsall, Z. Ikonic, P. Harrison, S. A. Lynch, P. Townsend, D. J. Paul, D. J. Norris, S. L. Liew, A. G. Cullis, X. Li, J. Zhang, M. Bain, and H. S. Gamble, "Optical cavities for Si/SiGe tetrahertz quantum cascade emitters," *Opt. Mater.* vol. 27, 851–854, 2005.
63. Z. Ikonic, P. Harrison, and R. W. Kelsall, "Waveguide design for mid- and far-infrared p-Si/SiGe quantum cascade lasers," *Semicond. Sci. Technol.* vol. 19, 76–81, 2004.
64. A. De Rossi, M. Carrase, and D. J. Paul, "Low-loss surface-mode waveguides for terahertz Si–SiGe quantum cascade lasers," *IEEE J. Quantum Electron.* vol. 42, 1233–1238, 2006.
65. S. G. Pavlov, R. Kh. Zhukavin, E. E. Orlova, V. N. Shastin, A. V. Kirsanov, H.-W. Hübers, K. Auen, and H. Riemann, "Stimulated emission from donor transitions in silicon," *Phys. Rev. Lett.* vol. 84, 2550–2553, 2000.
66. H.-W. Hübers, S. G. Pavlov, and V. N. Shastin, "Terahertz lasers based on germanium and silicon," *Semicond. Sci. Technol.* vol. 20, S211–S221, 2005.
67. R. K. Zhukavin, V. V. Tsyplenkov, K. A. Kovalevsky, V. N. Shastin, S. G. Pavlov, U. Böttger, H.-W. Hübers, H. Riemann, N. V. Abrosimov, and N. Nötzel, "Influence of uniaxial stress on stimulated terahertz emission from phosphor and antimony donors in silicon," *Appl. Phys. Lett.* vol. 90, 051101-3, 2007.
68. R. Kh. Zhukavin, V. N. Shastin, S. G. Pavlov, H.-W. Hübers, J. N. Hovenier, T. O. Klaassen, and A. F. G. van der Meer, "Terahertz gain on shallow donor transitions in silicon," *J. Appl. Phys.* vol. 102, 093104-5, 2007.

69. E. E. Orlova, "Temperature dependence of inverse population on intracenter transitions of shallow impurity centers in semiconductors," *Semiconductors* vol. 44, 1457–1463, 2010.
70. S. G. Pavlov, H.-W. Hübers, J. N. Hovenier, T. O. Klaassen, D. A. Carder, P. J. Phillips, B. Redlich, H. Riemann, R. Kh. Zhukavin, and V. N. Shastin, "Stimulated terahertz Stokes emission of silicon crystals doped with antimony donors," *Phys. Rev. Lett.* vol. 96, 037404-4, 2006.
71. S. G. Pavlov, U. Böttger, J. N. Hovenier, N. V. Abrosimov, H. Riemann, R. Kh. Zhukavin, V. N. Shastin, B. Redlich, A. F. G. van der Meer, and H.-W. Hübers, "Stimulated terahertz emission due to electronic Raman scattering in silicon," *Appl. Phys. Lett.* vol. 94, 171112-3, 2009.
72. P.-C. Lv, R. T. Troeger, S. Kim, S. K. Ray, K. W. Goossen, J. Kolodzey, I. N. Yassievich, M. A. Odnoblyudov, and M. S. Kagan, "Terahertz emission from electrically pumped gallium doped silicon devices," *Appl. Phys. Lett.* vol. 85, 3660–3662, 2004.
73. S. A. Lynch, P. Townsend, G. Matmon, D. J. Paul, M. Bain, H. S. Gamble, J. Zhang, Z. Ikonic, R. W. Kelsall, and P. Harrison, "Temperature dependence of terahertz optical transitions from boron and phosphorus dopant impurities in silicon," *Appl. Phys. Lett.* vol. 87, 101114-3, 2007.
74. G. Xuan, S. Kim, M. Coppinger, N. Sustersic, J. Kolodzey, and P.-C. Lv, "Increasing the operating temperature of boron doped silicon terahertz electroluminescence devices," *Appl. Phys. Lett.* vol. 91, 061109-3, 2007.
75. N. A. Bekin and S. G. Pavlov, "Quantum cascade laser design based on impurity–band transitions of donors in Si/GeSi(111) heterostructures," *Physica B* vol. 404, 4716–4718, 2009.
76. B. F. Levine, "Quantum-well infrared photodetectors," *J. Appl. Phys.* vol. 74, R1–R81, 1993.
77. R. P. G. Karunasiri, J. S. Park, and K. L. Wang, "Normal incidence infrared detector using intervalence-subband transitions in $Si_{1-x}Ge_x$/Si quantum wells," *Appl. Phys. Lett.* vol. 61, 2434–2436, 1992.
78. D. J. Robbins, M. B. Stanaway, W. Y. Leong, J. L. Glasper, and C. Pickering, "$Si_{1-x}Ge_x$/Si quantum well infrared photodetectors," *J. Mater. Sci. Mater. Electron.* vol. 6, 363–367, 1995.
79. D. Krapf, B. Adoram, J. Shappir, A. Sa'ar, S. G. Thomas, J. L. Liu, and K. L. Wang, "Infrared multispectral detection using Si/Si_xGe_{1-x} quantum well infrared photodetectors," *Appl. Phys. Lett.* vol. 78, 495–497, 2001.
80. R. T. Carline, D. J. Robbins, M. B. Stanaway, and W. Y. Leong, "Long-wavelength SiGe/Si resonant cavity infrared detector using a bonded silicon-on-oxide reflector," *Appl. Phys. Lett.* vol. 68, 544–546, 1996.
81 R. T. Carline, D. A. Hope, V. Nayar, D. J. Robbins, and M. B. Stanaway, "Resonant cavity long-wave SiGe-Si photodetector using a buried silicide mirror," *Photon. Technol. Lett.* vol. 10, 1775–1777, 1998.
82. M. A. Gadir, P. Harrison, and R. A. Soref, "Arguments for p-type quantum-well $Si_{1-x}Ge_x$/Si photodetectors for the far- and very-far (terahertz)-infrared," *Superlatt. Microstruct.* vol. 30, 135–143, 2001.
83. Y.-Y. Lin, and J. Singh, "Theory of polarization dependent intersubband transitions in p-type SiGe/Si self-assembled quantum dots," *J. Appl. Phys.* vol. 96, 1059–1063, 2004.
84. K. L. Wang, D. Cha, J. Liu, and C. Chen, "Ge/Si self-assembled quantum dots and their optoelectronic device applications," *Proc. IEEE* vol. 95, 1866–1883, 2007.
85. C.-H. Lin, C.-Y. Yu, C.-C. Chang, C.-H. Lee, Y.-J. Yang, W. S. Ho, Y.-Y. Chen, M. H. Liao, C.-T. Cho, C.-Y. Peng, and C. W. Liu, "SiGe/Si quantum-dot infrared photodetectors with δ doping," *IEEE Trans. Nanotechnol.* vol. 7, 558–564, 2008.
86. M. Kolahdouz, A. Afshar Farniya, L. Di Benedetto, and H. H. Radamson, "Improvement of infrared detection using Ge quantum dots multilayer structure," *Appl. Phys. Lett.* vol. 96, 213516-3, 2010.
87. A. Kosarev, S. Rumyantsev, M. Moreno, A. Torres, S. Boubanga, and W. Knap, "Si_xGe_y:H-based micro-bolometers studied in the terahertz frequency range," *Solid State Electron.* vol. 54, 417–419, 2010.
88. G. Roelkens, L. Liu, D. Liang, R. Jones, A. Fang, B. Koch, and J. Bowers, "III–V/silicon photonics for on-chip and inter-chip optical interconnects," *Laser Photon. Rev.* (invited) vol. 4, 751–779, 2010.

7

Photonic Crystals

Masaya Notomi, Kengo Nozaki, Shinji Matsuo, and Toshihiko Baba

CONTENTS

7.1 Introduction: Why Do We Need Photonic Crystals?

Masaya Notomi

Photonic crystals are artificial dielectric structures having periodic refractive index variation in a wavelength scale, which can realize strong light confinement (John, 1987; Yablonovitch, 1987). The most extensively studied structure is a so-called photonic crystal slab (Notomi, 2010), which is typically formed by a periodic arrangement of air holes with the diameter of around 400 nm in a semiconductor (Si, GaAs, or InP) membrane of around 200 nm thickness. This structure can be fabricated by a process very similar to making Si-wire photonics since the typical dimension is the same order (note that the typical waveguide cross section for Si-wire photonics is 400×200 nm^2). In fact, the cross-sectional mode area is almost similar between a conventional Si-wire waveguide and a photonic crystal line-defect waveguide. The waveguide loss is also comparable (Notomi et al., 2007). Then, why do we call for photonic crystals?

The present chapter with four subsections is planned to present several reasons why photonic crystals may have important positions in future Si photonics (more generally, in future on-chip photonics). In principle, photonic crystals can give us the following three merits (Notomi, 2010) in comparison with other technologies, especially with Si-wire photonics. (1) Stronger light confinement in a smaller space. A photonic crystal enables an optical cavity with the mode volume of $(\lambda/n)^3$, and the cavity quality factor of 10^5–10^6, which is hardly achievable in microcavities in Si-wire photonics. In addition, it is relatively easy to control the amount of the coupling strength for intercavity or cavity-waveguide configuration in photonic crystals than in microdisc/ring cavities. (2) Great controllability of the frequency dispersion, especially to achieve slow light. Although slow light modes are achievable in Si-wire photonics by employing gratings or ring resonators, photonic crystals can give us extraordinary slow light modes with small footprint and great flexibility in design. (3) Great controllability of the spatial dispersion, such as negative refraction (Notomi, 2000) and super-prism phenomena (Kosaka et al., 1999). This is a direct manifestation of the periodicity, and thus Si-wire photonics does not exhibit this property by itself. The first merit is directly concerned with the device footprint and the energy consumption. As will be discussed in Sections 7.2 and 7.3, photonic crystals have already demonstrated several types of micron-scale devices with the energy consumption of an order of fJ/bit (Matsuo et al., 2010; Nozaki et al., 2010). The second merit is related to dispersion management and optical buffering (Baba, 2008), which will be discussed in Section 7.4. This means that if future chips require large-scale photonic integration, photonic components with ultrasmall energy consumption, or chip-scale dispersion controller/optical buffers, photonic crystals may play a very important role.

When will we need photonic crystals for on-chip photonics? It all depends on how Si photonics and on-chip photonics will progress in future. Since the introduction of photonics into a chip has just started to develop recently, so far it is unpredictable how far this trend will go forward. One possible (and optimistic) future vision for on-chip photonics technologies is shown in Figure 7.1, which represents our point of view (Notomi et al., 2011). A chip like MPU (microprocessor unit) was first evolved in size, and then divided into multiple units or cores with appropriate interconnecting techniques. Currently, Si photonics is expected to play a key role in introducing optical interconnect (Miller, 2000). The reason behind such trends could be reduced into a simple fact that as the bit rate becomes higher, the energy cost of the signal transport via photons in a chip becomes cheaper than that via electrons. Si photonics is very suited for this purpose, which is extensively described in other chapters. It is, however, predicted that in 2020 the energy cost should be an order of fJ/bit for intrachip interconnection (Miller, 2009), which is not easy to realize even with Si photonics. To meet this requirement, we have to significantly reduce the driving energy of photonic components, such as lasers, detector, modulators, etc. We foresee that photonic crystal technologies may play an important role at this stage since they can realize extremely energy-efficient devices.

In addition to that, recently, it is becoming a problematic issue how to manage intrachip network when the number of cores increases. For example, an advanced 80-core chip adopts a network-on-chip architecture and has on-chip routing processors in each core (Vangal, 2007). Driven by this trend, a photonic network on chip where the intrachip network will be managed by photonics is being discussed (Shacham et al., 2008). This is a natural extension of the optical interconnect because photonics is supposed to be more suited for networking than electronics when the bit rate becomes higher. Currently, a relatively simple photonic network on chip is discussed, which is mostly dominated by passive optical components, but more complicated devices may be required. In this scenario, we will need various tiny optical devices that are integrable in a single chip, for which photonic crystal technologies are ideally suited.

Then, what is next? Although it is still an open question, we can expect much more photonics penetration in a chip to manage the intrachip network in future. The same scenario has already happened in backbone/metro network management (in our macroscopic world) where various photonics technologies have replaced electronic counterparts. If this

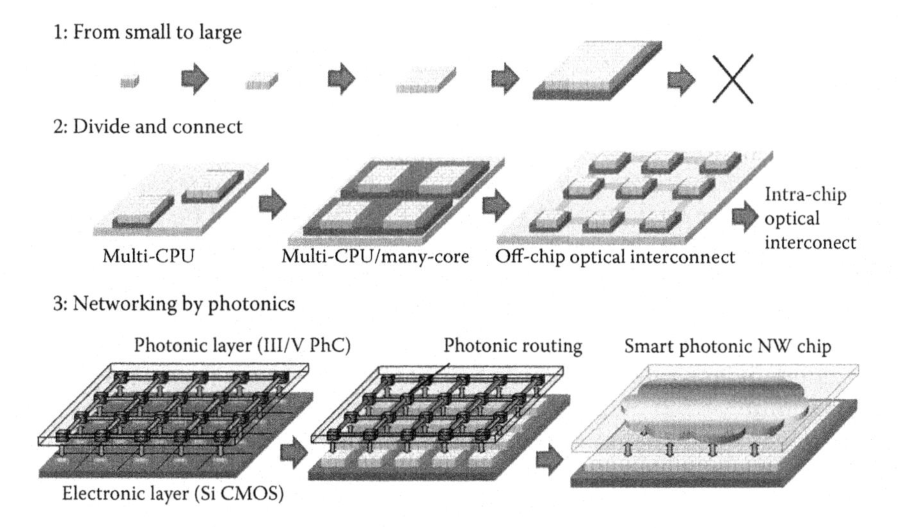

FIGURE 7.1
History and trend of MPUs.

happens in a chip, we need a fairly large-scale optical integration with various complicated functionalities, far beyond a simple optical interconnect. We expect that this transition will occur as shown in Figure 7.1. First, we may put a photonic network layer attached to a multi-core Si CMOS layer. The photonic layer may consist of lasers, waveguides, and receivers in the first place, and then we may add WDM (wavelength division multiplexing) functions and also route switching functions. If we become able to introduce dynamic routing functions with appropriate optical buffers, it will enable an on-chip photonic routing processor. If we take a step to this stage, photonic crystals may have a crucial role since we need a large number of integrable photonic devices with ultrasmall consumption energy and complex functions. Note that in this scenario, the photonic network layer does not have to be based on Si. We assume that III/V semiconductors may be more suited for constructing this photonic layer. These two layers may be bonded and communicate with each other electrically or optically in the vertical direction. Finally, we may end up with a very large Si CMOS (single core) attached to a large photonic network layer if the photonic network layer becomes efficient enough.

We are not sure how far we can go in this direction, but it is certain that ultimate information processing is not a matter of simple computation, but it may consist of a combination of memory and network. (Look at what our brain is doing. Our brain consists of huge associative memory networks.) For such an ultimate goal, electronics is suited for the former and photonics is suited for the latter. Probably, we need a dense large-scale photonic integration technology to meet such requirement. This might be a final and ultimate reason why we need photonic crystals.

7.2 All-Optical Switches and Memories Based on Photonic Crystal Nanocavities

Kengo Nozaki

7.2.1 Introduction

Photonic technologies have played a key role in telecommunication networks based on optical fiber links, but the demand for larger network capacity is still increasing. It is now recognized that data processing at nodes will limit capacity, and thus novel photonic technologies have been investigated to replace their electronic counterparts and thus offer transparency for high-bit-rate optical signals. Key functions will be provided by optical routers including optical switches, optical random access memories (RAMs), wavelength converters, and so on. These functionalities can eliminate the need for optical–electrical–optical (O-E-O) conversion at the network nodes, and might therefore achieve transparency for optical signals and reduce power consumption. In addition to telecommunication network applications, it is worth introducing these optical devices into much shorter communication links (chip-to-chip and intrachip) for information processing during them. This is a promising way of solving existing problems related to energy and speed in versatile information processors based on electronic integrated circuits. This approach is known as a photonic network on chip (PhNoC) (Notomi et al., 2011). For this purpose, photonic elements should have several characteristics including a low-power consumption, compactness, and large-scale integrability. Although Si-based photonic devices have proved attractive for achieving optical interconnections between CMOS electronic circuits, their relatively large power consumption makes them difficult for achieving large-scale integration and PhNoC. On the other hand, III–V materials make it possible to greatly reduce the operating power of various functional devices, as discussed in this section. In addition, recent works effectively demonstrate a heterogeneous integration of III–V-based devices on Si platform by means of a bonding process (Fang et al., 2006; Roelkens et al., 2006). Therefore, these devices are worth developing for constructing a III–V-based PhNoC in combination with CMOS circuits.

Optical switches and memories are fundamental components for controlling optical data signals. Optical switches in network systems are classified in terms of their switching speed and have various applications including optical line switches (ms), optical burst and packet switches (μs–ns), optical time-domain multiplexing and logical processing (ps). As the required switching time decreases, electrical control is insufficient in terms of speed and energy. All-optical switches will therefore become key components, because an optical signal can be controlled with another light, enabling the high-speed control of optical data. Optical memory is also an important element for future photonic information processing, although the realization of a practical optical memory is one of the most difficult challenges as regards replacing various electrical processors with photonic counterparts. Optical memory devices are expected to be employed for optical packet routers, since they provide a controllable optical delay while retaining the high-speed characteristic of optical signals. Optical buffering can resolve the contention caused when multiple optical packets are destined for the same output port at the same time. Optical bistability in a nonlinear cavity or a laser has therefore been studied for a long time with a view to achieving optical memory functions. However, these researches on optical switches and memories have faced problems since the required size and power consumption make the devices too large for integration. A significant problem is that light–matter interactions are intrinsically weak in all materials, so a large optical power or long interaction length is necessary if we are to realize these functions.

As described later, small optical cavities with an appropriate design might be a good way to realize all-optical switches and memories. Research in these fields is still in its early stages, although some interesting techniques have already been demonstrated. This section reviews state of the art all-optical switches and memories; the best result will be highlighted in semiconductor photonic-crystal technologies.

7.2.2 All-Optical Switches

7.2.2.1 Review of Previous All-Optical Switches

A key component in photonic processing is the all-optical switch, in which the intensity or direction of an optical signal is controlled with another light. High-speed all-optical switches employing fast optical nonlinearity (NL) have already been reported in various forms. For example, an inter-subband transition (ISBT) in semiconductor quantum wells (Gopal et al., 2002; Cong et al., 2007) or optical parametric processes (Yamamoto et al., 1998; Andrekson et al., 2008) in highly nonlinear waveguides can be applied for all-optical switches operating at a few picoseconds or less. However, they typically require a switching energy of several pJ/bit or more. This means that the power consumption would be several tens of milliwatts or more at 10 Gb/s. On the other hand, there are also all-optical switches based on semiconductor optical amplifiers (SOAs) (Nakamura et al., 2001; Nielsen et al., 2006) with a very small switching energy of less than 1 fJ/bit, but the total power is even higher because they require additional mW level power to drive them. When considering PhNoC applications, this high power consumption, along with the large size, are undesirable. It has been discussed that future interconnect technology will demand micron-scale optical components in a chip consuming energy of less than 1 fJ/bit (Miller, 2009).

Optical resonators (cavities), which can strongly confine light in very small spaces, have been seen as an attractive way to enhance optical NLs and hence have been extensively studied for use as all-optical switches (Almeida et al., 2004; Soljacic et al., 2004; Tanabe et al., 2005a; Notomi et al., 2007). Optical NLs in a cavity can be enhanced by a high cavity Q factor and a small modal volume V, because the field intensity $|E|^2$ is scaled as Q/V. Very low switching energy/power can be therefore expected if we apply cavities to all-optical nonlinear switching. Over the last few decades, different types of optical micro- and nano-cavities characterized by an ultrahigh Q/V value have been successfully demonstrated as a result of the development of nano-scale fabrication techniques, and they include photonic-crystal (PhC) cavities, micro-ring cavities, micro-toroid and micro-sphere resonators (Vahala, 2003). In these cavities, free-carrier-based NLs in semiconductors provide an effective way of realizing all-optical switching because they offer a good balance between nonlinear efficiency and switching speed. However, we still need to optimize both the cavity design and the material if we are to obtain a low switching energy and fast response for PhNoC applications.

7.2.2.2 All-Optical Switches Based on Photonic Crystal Nanocavities

Of the various types of optical cavities, PhC nanocavities with an ultrasmall size have proven to be promising candidates for developing an all-optical switch with integrability on a chip. Recent advances in nanofabrication processes have made it possible to realize an ultrahigh Q of more than 10^6 while keeping a small mode volume comparable to $(\lambda/n)^3$ (λ is the wavelength of light, n is refractive index) (Kuramochi et al., 2006; Takahashi et al., 2007; Tanabe et al., 2007). Therefore, many optical NLs that were previously studied in relation to conventional Fabry–Perot etalons are worth investigating in relation to PhC nanocavities. Although the physical mechanisms of NLs are similar to those observed in their conventional counterparts, the power consumption of NL-based optical devices in PhC nanocavities should be several orders of magnitude smaller owing to their small size.

Figure 7.2a is a schematic of our device and Figure 7.2b shows the operating principle behind all-optical switching. (Tanabe et al., 2005a; Nozaki et al., 2010). The PhC nanocavity is monolithically coupled with input/output PhC line-defect waveguides. For switching, the pump and signal light are injected into the waveguide simultaneously. The pump light generates carriers in the cavity and induces a wavelength shift in the resonant transmission spectrum, which makes it possible to control the signal light output. Switch-on (corresponding to switching from the normally off condition) or switch-off (switching from the normally on condition) operations are selected via the initial setting of the signal wavelength.

A significant limiting factor as regards the operating speed in carrier-induced all-optical switches is the carrier relaxation time (τ_c). Generally, τ_c is of nanosecond order, but PhC nanocavities offer an efficient way to reduce it. When the cavity is ultrasmall, the photo-generated carriers rapidly diffuse out from the cavity, and τ_c becomes small (Tanabe et al., 2008; Nozaki et al., 2010). Figure 7.3 shows simulated results for the carrier diffusion dynamics in PhC nanocavities. First, we simulated the carrier dynamics without a

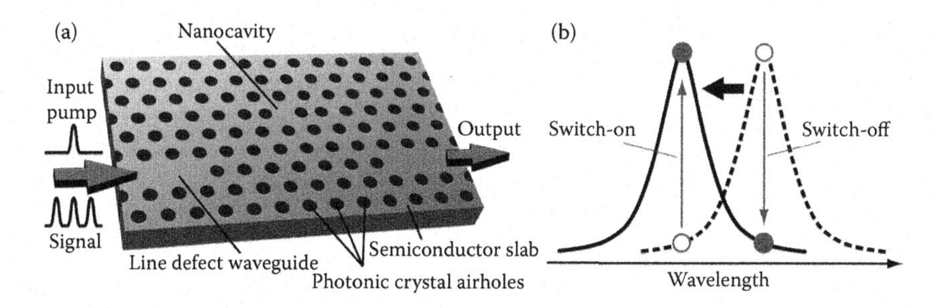

FIGURE 7.2
All-optical switch based on photonic crystal nanocavity. (a) Structural schematic of photonic crystal nanocavity coupled to input and output line defect waveguides. (b) Operating principle of all-optical switching. A pumped carrier induces a wavelength shift in the resonant transmission spectrum, resulting in the signal light being switched on or off.

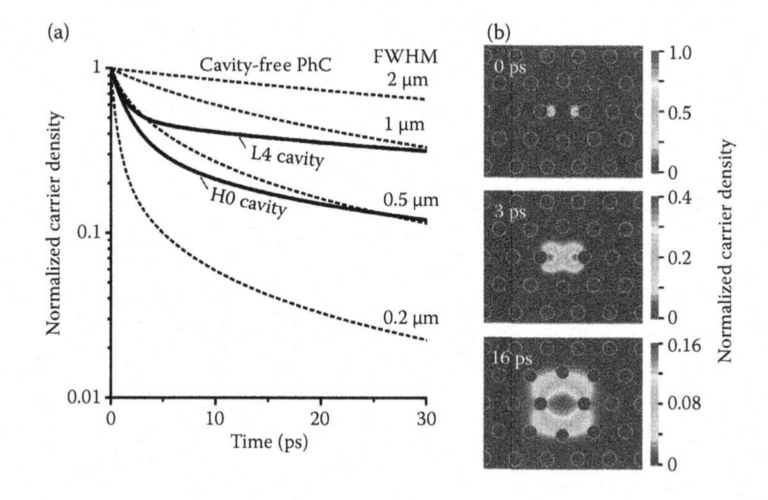

FIGURE 7.3
Carrier decay for photonic crystal cavities. (a) Simulated carrier decay for an H0 cavity, an L4 cavity, and cavity-free PhCs (dashed curve). A Gaussian carrier distribution with a different spatial width is adopted for the cavity-free PhCs. (b) Time evolution of the carrier density distribution for the H0 cavity.

cavity but assuming an initial Gaussian carrier distribution in a defect-free PhC lattice. In Figure 7.3a, the four dashed lines show the decay for different sizes of initial distribution. This clearly shows that a small amount of excitation leads to surprisingly fast diffusion. To consider more realistic cases, a four-airhole missing nanocavity (L4 cavity) and a lattice-shifted nanocavity (H0 cavity) were calculated as shown by the black lines. The fitted τ_c values are as short as 3.5 and 18 ps. Figure 7.3b shows the time evolution of the carrier distribution for the H0 cavity. Initially, the carriers are localized in the cavity mode, and then they start to spread out. Note that the nonradiative recombination process that occurs at the airhole sidewalls is incorporated in the simulation, but the carrier decay is still dominated by the ultrafast diffusion. These results imply that we can expect a switching bandwidth of nearly 100 GHz. Consequently, a PhC-nanocavity switch will offer a significant advantage in terms of high-speed response.

7.2.2.3 Switching Energy in Carrier-Induced PhC Switches

For an all-optical switch based on a combination of resonant cavity and carrier-based nonlinearity, the switching energy U_{sw} is defined as the pulse energy required for a nonlinear wavelength shift equal to the cavity spectral width. Hence, $1/Q = \Gamma\sigma_c N/n$ where Q is the cavity Q factor, Γ is the field confinement factor in the cavity, N is the carrier density, and σ_c is a coefficient of the index change with carrier density. N is given by $N = U_{sw}\eta_{abs}\eta_c/(v\hbar\omega V_m)$, where η_{abs} denotes the absorption efficiency in the cavity, η_c is the accumulation efficiency of pumped carriers for the input pulse duration and is given by $\eta_c = \exp(-t_{pulse}/\tau_{carrier})$. ν is the number of photons required for a single transition process (1 for linear absorption (LA) and 2 for two-photon absorption (TPA)). This leads to the following formula for the switching energy,

$$U_{sw} = \frac{vn\hbar\omega V_m}{\Gamma\eta_{abs}\eta_c\sigma_c Q} \tag{7.1}$$

This formula suggests that a large Q/V_m ratio is advantageous in terms of realizing a low operating power. A PhC nanocavity is therefore suitable in this respect. In addition, the platform material should be carefully chosen to enhance η_{abs} and σ_c and achieve a moderate absorption and large index change.

7.2.2.4 Si-Based PhC-Nanocavity Switches

Against this background, all-optical switching at telecommunication wavelengths (1.55 μm) in silicon PhC nanocavities was first reported in 2005 (Tanabe et al., 2005a). Figure 7.4a,b show scanning electron microscope (SEM) images and transmission spectra for a cold cavity. The device consists of a four-airhole-missing end-hole-shifted nanocavity (L4 cavity) connected to input and output waveguides. This cavity has two resonant modes (mode C with Q = 11,500 and mode S with Q = 23,000), and the modal volumes are around 0.07 μm^3. Although Si is transparent at a wavelength of 1.55 μm, carriers can be generated via the

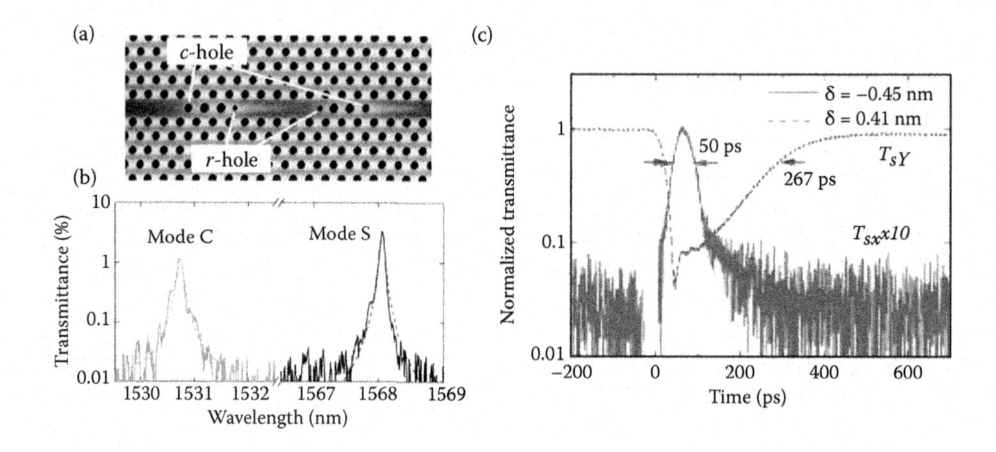

FIGURE 7.4
All-optical switching in a silicon PhC nanocavity by carrier-induced nonlinearity. (a) SEM image. (b) Transmission spectrum scanned by CW light in linear condition. (c) Switching dynamics for different signal wavelength.

two-photon absorption (TPA) process. On the other hand, the refractive index change relies on the free-carrier dispersion (FCD), which is a typical carrier-induced nonlinearity in silicon. A combination of TPA and FCD can be used to achieve optical NL. The control pulse light and signal CW light are set at mode C and mode S, respectively. Figure 7.4c shows the time-resolved output intensity for the signal mode when a 6-ps control pulse is input. Clear all-optical switching from off to on (on to off) is observed for a signal detuning of 0.45 nm (0.01 nm) from the cold cavity wavelength. The required switching energy is as small as 200 fJ/bit, which is much smaller than that of other Si switches based on the same mechanism. If we take the coupling efficiency between the cavity and waveguide into account, the actual pulse energy used for switching is less than 10 fJ/bit. The switching time ranges approximately from 50 to 300 ps. Considering the fact that the carrier relaxation time in silicon is typically long (~microseconds), this switching time is surprisingly fast, which is evidence for fast carrier diffusion as discussed in Section 7.2.2.2.

7.2.2.5 InGaAsP-Based PhC-Nanocavity Switches

The all-optical switches based on Si-PhC nanocavities described above show that a size reduction has a significant effect in reducing both switching energy and response time. However, the potential of the PhC nanocavity has not been fully realized because the nonlinearity and absorption efficiency have not been maximized. As mentioned above, a high Q/V ratio is preferable for a lower switching energy, but this is an oversimplification. In practice, the cavity Q needs to be moderately low to maintain a fast operating speed. In contrast, the modal volume V should always be as small as possible. In this sense, the H0 cavity, which was first reported by Zhang et al., should constitute a good candidate (Zhang et al., 2004). The V of an H0 cavity is calculated to be 0.025 $\mu m^3 \sim 0.26(\lambda/n)^3$, which is the smallest of the dielectric-core high-Q PhC cavities (Notomi, 2010). In fact, an all-optical switch based on a GaAs-based H0 cavity has been reported that has a low switching energy of 120 fJ and a very fast response time of 15 ps (Husko et al., 2009).

Second, the choice of NL should be considered. In principle, any type of NL phenomenon can be employed for cavity-based optical switches, but we should choose the most efficient NL in the target time scale. If the requirement is a switching repetition of 10–40 Gb/s, NLs based on real transitions (e.g., carrier-induced NLs) are more efficient than those based on virtual transitions (e.g., Kerr-based NLs), as long as the carrier relaxation is sufficiently fast. As shown by a Si-based PhC cavity, all-optical switches employing a combination of FCD and TPA are typical forms. However, a combination of band-filling dispersion (BFD) and linear absorption (LA) near the electronic band edge is more efficient (Bennett et al., 1990).

Third, the choice of materials should be considered. Since Si and GaAs have electronic band edge wavelengths of around 1100 and 850 nm, respectively, the BFD and LA would be very weak if we consider an operating wavelength of 1.55 μm. One solution is to employ a quaternary III/V semiconductor InGaAsP. Table 7.1 compares the material parameters in Si, GaAs, and InGaAsP for an operating wavelength of 1.55 μm. InGaAsP has a much stronger carrier-induced NL and absorption, which should produce a significant reduction in switching energy. The switching speed of these materials is likely to be similar, as long as it is limited by ambipolar carrier diffusion. The large surface recombination velocity in GaAs offers fast carrier depletion, and other techniques such as ion implantation or low-temperature growth for carrier recombination enhancement are also promising ways to achieve faster response operation.

Figure 7.5a shows the calculated index change and the linear absorption lifetime as a function of the incident wavelength detuning from the band-gap wavelength $\lambda_{in} - \lambda_g$. The

TABLE 7.1

Comparison of Typical Material Parameters in Si, GaAs, and InGaAsP(1.47Q)

Material	α_{LA} (cm^{-1})	β_{TPA} (cm/GW)	σ_c (×10^{20} cm^3)	D_{am} (cm^2/s)	v_{sr} (cm/s)
Si (Dinu et al., 2003; Barclay et al., 2005)	–	0.4 – 0.9	0.1 – 0.2	19	10^4
GaAs (Tsang et al., 1991; Van et al., 2002)	–	10	0.6	19	10^5
InGaAsP (1.47Q) (Tsang et al., 1991; Mork et al., 1994)	3	40 – 80	8.2	17	10^4

Note: The operating wavelength is assumed to be 1.55 μm. α_{LA} is an LA coefficient, β_{TPA} is a TPA coefficient, σ_c is the index change with carrier density, D_{am} is an ambipolar carrier diffusion coefficient, and v_{sr} is the surface recombination velocity.

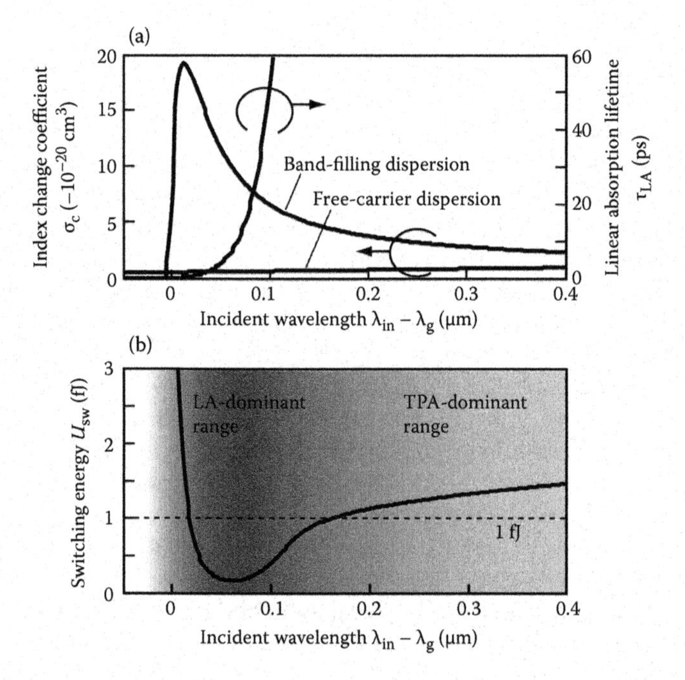

FIGURE 7.5
Material optimization for minimizing switching energy. (a) Index change with carrier density (left axis) and linear absorption lifetime (right axis) as a function of incident wavelength detuning from the band-gap wavelength. (b) Calculated switching energy for H0 cavity with parameters of V_m = 0.025 μm^3, Γ = 0.8, n = 3.44, η_c = 0.5.

dispersion curve for BFD is estimated through the Kramers-Kronig relations on the parabolic band edge shift, while the FCD is estimated by using the Drude model. The calculation procedure is detailed in (Bennett et al., 1990). Since the BFD depends strongly on the position of the electronic band edge wavelength, an appropriate InGaAsP composition should be found in relation to the resonant wavelength of the cavities. On the other hand, optical absorption relies on LA and TPA. LA also becomes stronger near the band edge wavelength, and allows efficient absorption. An important point is that excess absorption induces cavity Q degradation and an increase in switching power, so an appropriate adjustment of the composition is needed to minimize the switching energy. The calculated switching energy U_{sw} for an H0 cavity is shown in Figure 7.5b. Near the band edge, LA boosts the absorption efficiency and BFD enhances the nonlinear resonance shift, thereby effectively reducing U_{sw}. This indicates the minimum value with less than 1 fJ around $\lambda_{in} - \lambda_g = 0.05 - 0.1$ μm.

Figure 7.6a show a top-view image of an InGaAsP-H0 cavity fabricated using standard top-down processes including electron-beam lithography and Cl_2-based dry etching. Figure 7.6b shows the transmission spectrum acquired by scanning the wavelength of CW light. The periodic peaks in the spectrum are not a nanocavity mode, but appear because of interference with the Fabry–Perot resonance caused by the facet end of the waveguide. The fitting curve (black) clarifies the nanocavity mode transmission, indicating that the cavity Q factor is 6,500 and the corresponding photon lifetime is $\tau_{ph} = 5.4$ ps, which is not likely to restrict the switching recovery time.

To measure the switching dynamics, a degenerate pump-probe technique with an optical pulse width of 14 ps was employed (Szymanski et al., 2009). The center wavelength of the pump pulse was always set at the resonance, while the wavelength of the probe pulse λ_{probe} was set with detuning $\Delta\lambda_{det} = \lambda_{probe} - \lambda_{cav}$. Figure 7.6c shows the switching dynamics for different $\Delta\lambda_{det}$ values. For $\Delta\lambda_{det} = 0.0$ nm, the transmission of the probe pulse is abruptly switched *off* when the pump pulse temporally overlaps the probe pulse. On the other hand, the probe transmission is switched *on* for $\Delta\lambda_{det} = 0.3$ and -0.6 nm. This is because a pump-induced carrier NL induces a resonant blueshift. The achieved switching energies were 420 and 660 aJ for contrasts of 3 and 10 dB, respectively. These energies are over two magnitudes lower than that reported for Si or GaAs based PhC cavities. It should be noted that the switching time window is only 20–35 ps. This value is much shorter than the carrier recombination lifetime (several hundred picoseconds), which is attributable to the rapid carrier diffusion.

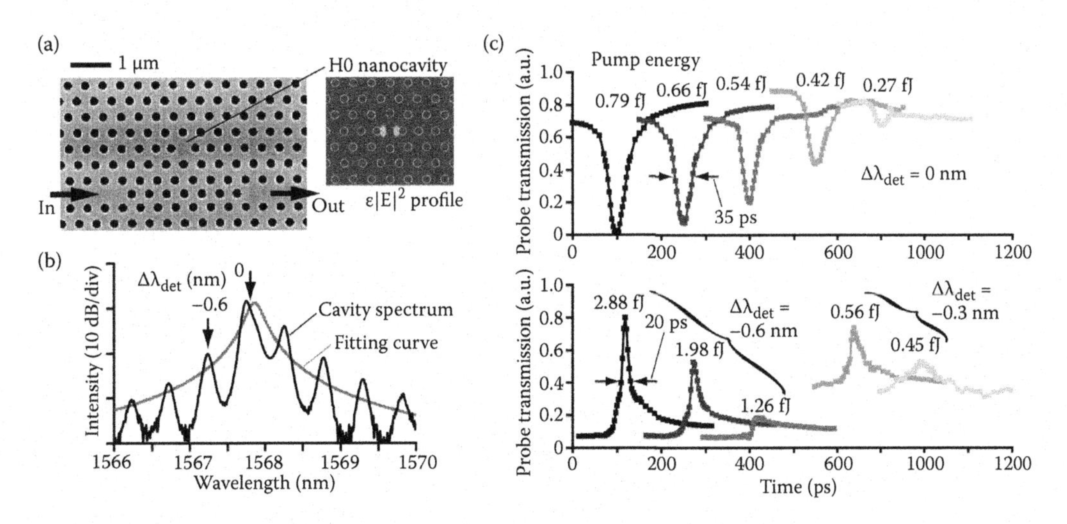

FIGURE 7.6
Switching dynamics of InGaAsP-H0 nanocavity acquired by pump-probe measurement. (a) Top-view image and simulated modal distribution. (b) Transmission spectrum scanned with a wavelength-tunable CW laser. (c) Switching dynamics measured with the pump-probe method. The upper and lower plots correspond to the switch-off and switch-on operation results, respectively. Different colors denote the results for different pump energies.

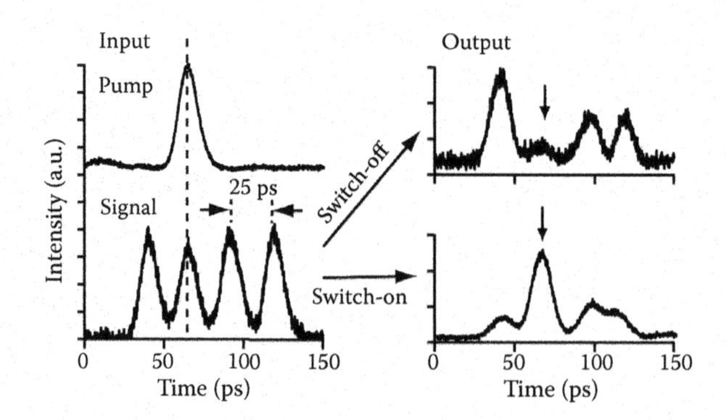

FIGURE 7.7
Gate switching for a signal pulse train at 40 Gbps. The pump pulse temporally matches the second signal pulse, and the output signals indicate the operations in the switch-off and switch-on regimes.

The all-optical switch was used to demonstrate a pulse extraction from a repetitive signal train. As shown in Figure 7.7, a signal train with four pulses with a 25-ps period (40-Gbps repetition) was generated, and injected with a pump pulse so that it overlapped the second signal pulse to switch it selectively. The output signal pulses show the result of the pulse extraction experiment, indicating that the second pulse (indicated by an arrow) was selectively switched off or on. These results are promising as regards the suitability of PhC nanocavity switches for 40-Gbps operation.

7.2.2.6 Comparison of All-Optical Switches

Figure 7.8a compares various all-optical switches in terms of their switching energy per bit and switching time. It is clearly seen that an InGaAsP-PhC–based all-optical switch can operate with approximately two or more orders of magnitude less energy than other switches and has entered the aJ energy range for the first time. In addition, the previously developed switches suffer from a trade-off between switching time and energy, that is,

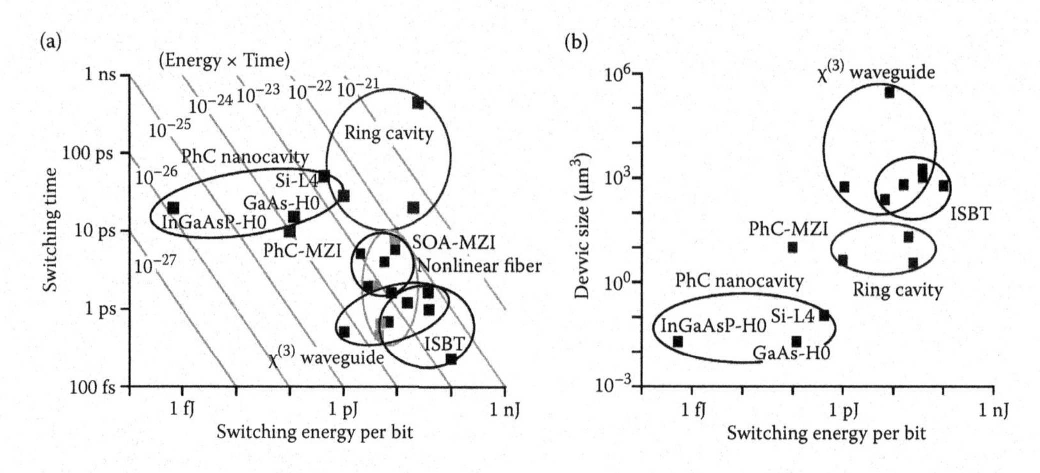

FIGURE 7.8
Comparison of all-optical switches. (a) Switching time vs. switching energy. (b) Device size vs. switching energy.

the energy-time product is limited to around 10^{24}–10^{22}. By contrast, the InGaAsP-PhC cavity clearly overcomes this limitation, denoting an energy-time product of 10^{-26}. Figure 7.8b compares on-chip all-optical switches in terms of switching energy and device size, which indicates that the PhC cavities realize the smallest size. Although all-optical switches involving third-order nonlinearity (Koos et al., 2009) and inter-subband transition (ISBT) (Gopal et al., 2002; Cong et al., 2007) can operate at much higher bit rates, a large energy consumption and size might be not acceptable for an on-chip integrated circuit. As for a PhC cavity, if we assume that there are 1,000 devices on a single chip, all operating at a bit rate of 10 Gbps, the power consumption is only a few mW. In addition, the ultrasmall size enables us to integrate 1,000 devices with a small footprint of less than 0.01 mm^2 (assuming that the footprint for a single device is less than 10 μm^2). Consequently, low-power, ultrasmall, and fast PhC-nanocavity switches are unique when compared with other switches.

7.2.3 All-Optical Memories

7.2.3.1 Various Schemes for Buffering Light

At present, the optical delay for photonic processing employs a long optical fiber to hold the optical data stream. However, this system size is very large, and this is an obstacle to future photonic integration. There has been considerable research on all-optical buffers in various forms and on a micro-scale size. Slow-light techniques, employing a dispersion-engineered waveguide, can control the group velocity of light and thereby buffer the optical signal as with a fiber delay line but with a much smaller size (Baba, 2008). There have already been some reports on PhC waveguides and coupled-resonator optical waveguides that provide an optical delay of several hundred picoseconds (Xia et al., 2007; Baba, 2008; Notomi et al., 2008) However, they cannot allow random access to signal bits with an arbitrary timing. On the other hand, there have recently been breakthroughs as regards achieving stopped light by using electromagnetically induced transparency (EIT) in a cold atomic gas, in which the material dispersion is dynamically tuned (Harris, 1997). It is difficult to implement an EIT system on a chip at room temperature, but analogous techniques using an optical cavity system have been reported (Notomi et al., 2007; Tanaka et al., 2007; Xu et al., 2007). In the system, a pump light changes the out-coupling rate of light with waveguides, resulting in a change in the cavity Q of several orders of magnitude. Such dynamic Q tuning can open/close a cavity gate with arbitrary timing, and also enable optical buffering. A significant issue with respect to these buffers is that they inevitably suffer from optical loss during buffering, and therefore the storage time is limited to less than a nanosecond.

On the other hand, all-optical RAM utilizing an optical bistability (Gibbs, 1985), in which an optical bit can be memorized for a long time and the state of the memory is accessible with arbitrary timing, has been researched for more practical use. In addition to a simple memory function, the integration of optical RAM enables us to achieve various logic functions such as flip-flops. Semiconductor microlasers employing the injection-locking method can switch the several lasing modes by launching an optical bit signal, and operate as an optical memory (Mori et al., 2006; Takeda et al., 2008; Liu et al., 2010). All-optical dispersive bistabilities related to optical nonlinear processes, such as the thermal effect, carrier effect, and optical Kerr effect, in optical resonators are also the attractive solutions. In these devices, the bistability appears at the transmission/reflection level, and their two states can be switched by the nonlinear shift of a resonant wavelength. Some groups have reported a memory operation employing a Fabry–Perot etalon (Olbright et al., 1984), microring cavity (Almeida and Lipson, 2004; Xu et al., 2006) and PhC nanocavity (Notomi et al., 2005; Tanabe et al., 2005b; Shinya et al., 2008). The main problems have been that the operating power is too high and the size too large, which make it difficult to integrate a large number of elements.

7.2.3.2 Optical Bistabilities in PhC Nanocavity

Since all-optical bistabilities in PhC nanocavities have a similar reliance on similar NL phenomena to all-optical switches, the high Q and small size of a cavity are still effective in reducing the energy/power consumption of optical memory devices. In addition, a PhC nanocavity can be effectively coupled with a single-mode PhC waveguide, which suggests an easy way to achieve a densely integrated all-optical RAM chip.

The first demonstration of optical bistability in PhC nanocavities was carried out by employing the thermo-optic (TO) effect induced by TPA in silicon (Notomi et al., 2005). As shown in Figure 7.9a, an L4 cavity with two resonant modes is employed, in which one cavity is used for a control (mode A) and the other for a signal (mode B). The CW control light injection into mode A causes the TO-induced redshift of mode B. Figure 7.9b shows the mode B output power as a function of the mode A input power, which exhibits clear bistable switching behavior. The bistable switching polarity of mode B is inverted for different wavelength detunings δB of 20 and 260 pm. The threshold power for observing bistability is as small as 40 μW. This value is much smaller than that of bulk-type TO nonlinear etalons (a few to several tens of milliwatts) (Olbright et al., 1984) and also smaller than that of TO silicon microring resonators (~0.8 mW) (Almeida and Lipson, 2004). Although the bistable phenomenon itself is similar to that of nonlinear etalons, these nonlinear PhC nanocavities are clearly distinguished from them in terms of the operating power and integrability. In addition, the ultrasmall size is also beneficial for increasing the switching speed, because the generated heat is diffused out from the nanocavity. The relaxation time of our switch is approximately 100 ns, which is much shorter than that of conventional TO

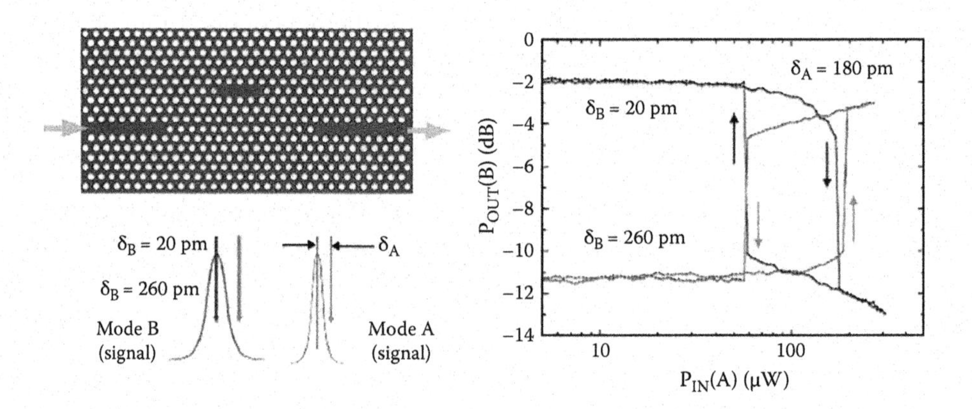

FIGURE 7.9
All-optical bistable switching in a Si PhC nanocavity by TPA-induced TO nonlinearity.

switches (~msec). The observed switching time suggests the applicability of the PhC-based bistable switch to certain network systems such as photonic burst or packet switches.

7.2.3.3 Carrier-Induced Optical Bistable Memories

Carrier effects such as BFD and FCD used for all-optical switches can also be used for all-optical bistable memories. TO-based bistability has the intrinsic issue of a slow response time, which does not allow high-rate switching and limits the application field. On the other hand, the carrier effect should be able to achieve a faster switching response of picoseconds to nanoseconds with a low operating power, as seen with carrier-induced all-optical switches.

The power consumption of all-optical bistable memories is dominated by the CW bias power, as long as the memory switching rate is not as high as GHz. The minimum bias power for a bistable memory is deduced from Equation 7.1 and is given by

$$P_{\text{bias}} = \frac{vn\hbar\omega V_{\text{m}}}{\Gamma\eta_{\text{abs}}\sigma_{\text{c}}Q\tau_{\text{c}}} \tag{7.2}$$

where τ_c is the carrier relaxation time. For the all-optical switch in Section 7.2.2, a small cavity volume and a short carrier relaxation time are desirable in terms of balancing low energy and fast response, and therefore an ultrahigh Q factor is unnecessary and a small V_m is rather important. By contrast, since CW bias power reduction is the key to realizing an all-optical memory, we should focus on the high cavity Q and large τ_c as well as the small V_m.

Carrier-based bistability and memory operation were demonstrated in a Si-PhC nanocavity consisting of an L4 cavity and with the design described in Section 7.2.2.4 (Tanabe et al., 2005b). Figure 7.10 shows memory operations performed by employing a bias light and a pair of set and reset pulses. The bias light wavelength is slightly detuned to the shorter side (0.15 nm) from the cold cavity resonance. When a set pulse is launched into the cavity, the output bias light is switched from off to on and remains on even after the

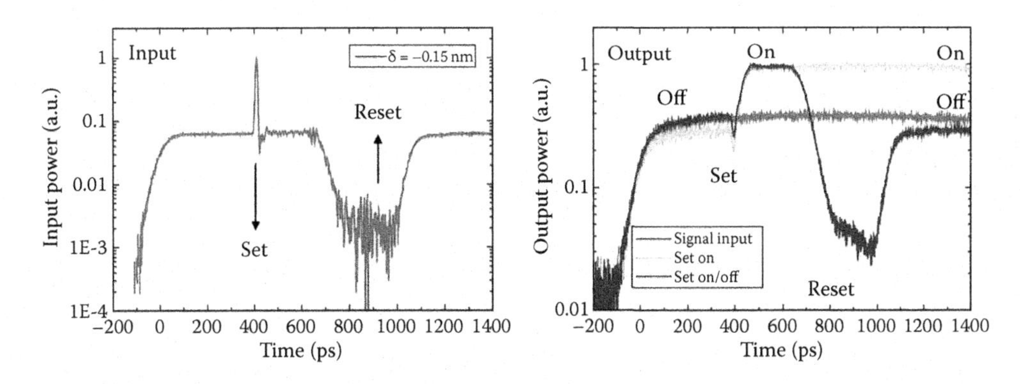

FIGURE 7.10
All-optical bistable memory operation in a silicon PhC nanocavity.

set pulse exits. When a pair of set and reset pulses is applied, the output is switched from off to on by the set pulse and then on to off by the reset pulse. This is a memory operation using optical bistability. The energy of the set pulse is less than 100 fJ, and the bias input for sustaining the on/off states is only 400 μW.

7.2.3.4 Optical Bistable Memories Realized with InGaAsP-PhC Nanocavity

As discussed in Section 7.2.2.5, a compound semiconductor InGaAsP exhibits moderate absorption and a strong carrier-induced NL compared with Si. These features are still effective for realizing a bistable memory. Namely, an InGaAsP-PhC cavity can function as an optical memory with much lower power than a Si-PhC cavity (Shinya et al., 2008).

Figure 7.11a shows an SEM image of a fabricated InGaAsP-PhC cavity consisting of a width-tuned line defect, (Kuramochi et al., 2006) in which the air holes surrounding the cavity are shifted away from the center of the line defect. The cavity mode volume is only 0.16 μm^3. The memory operation in InGaAsP PhC nanocavities is shown in Figure 7.11b. When the bias input wavelength was detuned by −0.2 nm from the cavity resonance, the output light intensity was clearly switched between the on and off states by inputting a set pulse and a negative reset pulse. The required bias power and set pulse energy were 10 μW and 24 fJ, respectively. The power consumption is an order of magnitude lower than that of Si-PhC cavities due to the strong BFD-based nonlinearity. However, the memory holding time is limited to 300 ns, as seen in Figure 7.11b, showing that the memory holding is automatically broken even without launching a reset pulse. This is because the accumulated heat and subsequent TO effect cause a red shift in the resonant wavelength and cancel the carrier effect. To solve these problems, we need to enhance the carrier effect and suppress heat accumulation.

Figure 7.12 shows the read-out operation of the bit memory. Here, two resonant modes are used; one for write-in and the other for read-out. The input powers of the write/read pulse are sufficiently above/below the bistable threshold. As shown in Figure 7.12, read-out pulses do not pass though the cavity while the cavity is off. After the write-in pulse is launched to turn the cavity on, the read-out pulses pass through the cavity. Therefore, this clearly shows that the information stored in the memory can be read by launching optical pulses.

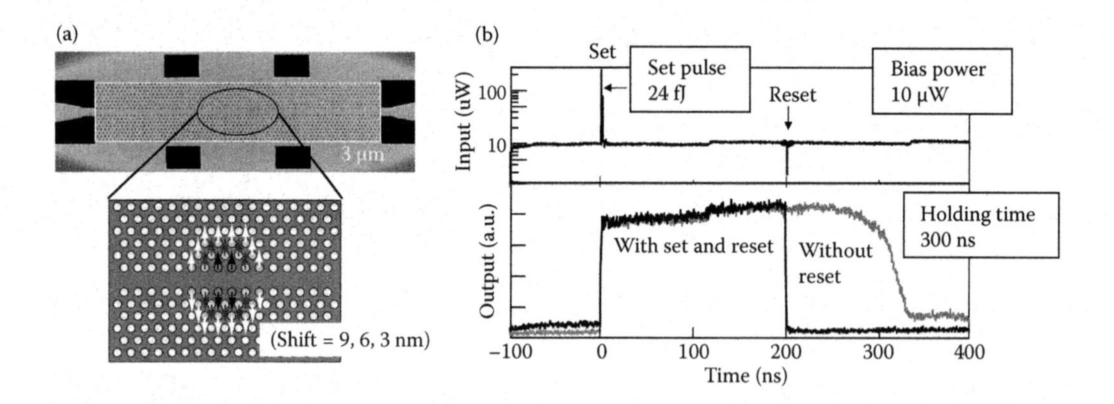

FIGURE 7.11
All-optical bit memory operation of an InGaAsP PhC nanocavity. (a) SEM image of the width-modulated line-defect cavity. (b) Bit memory operation dynamics.

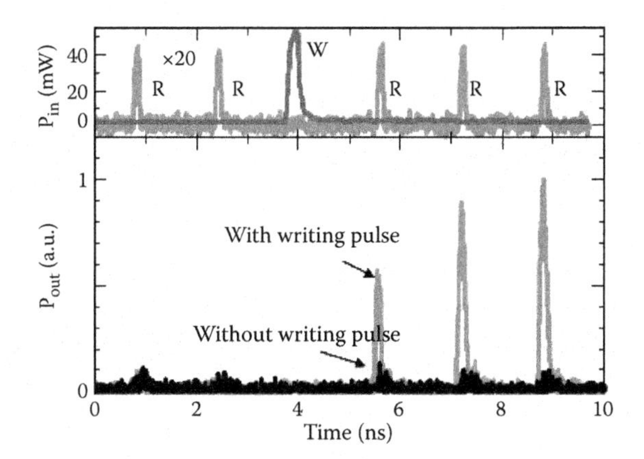

FIGURE 7.12
Write-in/read-out operation using two resonant modes.

7.2.3.5 Comparison of All-Optical Memories

Figure 7.13 compares the performance of the existing on-chip all-optical memories. The operating power and device area are the most important factors when multiple memories are integrated and simultaneously operated on a chip. The figure shows that these two are monotonically reduced together; therefore, the size reduction is a straightforward approach to power reduction. Conventional laser-based memories utilizing an injection locking technique can operate with a very low set pulse energy of a femtojoule or less, but a large total power consumption including a milliwatt-level bias power might be unacceptable for large-scale integration. PhC-cavity-based all-optical memories have a much smaller device area and operating power. With respect to realizing an ultrasmall size and power, a megabit level of integration will be possible within a footprint of less than 10^1 mm^2 and a total power consumption of several watts. There is still room to reduce the power of PhC-based memories, because they have no carrier confinement structure. The recent development of the PhC nanolaser has enabled us to confine carriers strongly in a cavity by employing a heterostructure barrier (Matsuo et al., 2010). A similar structure might help to enhance the carrier relaxation time τ_c in Equation 7.2

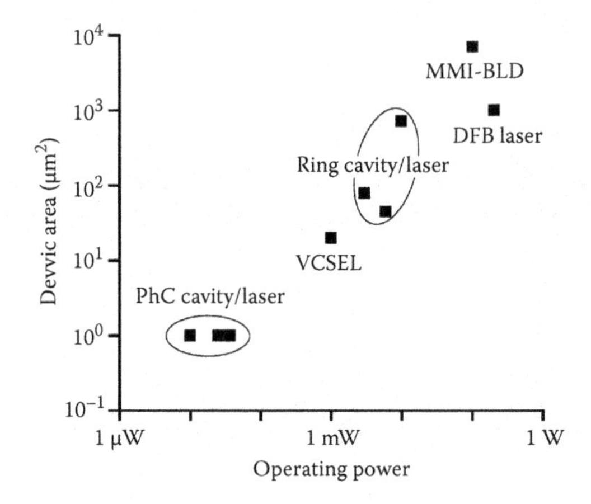

FIGURE 7.13
Device area vs. operating power for all-optical memories.

and directly save operating power. A PhC-based memory can also be incorporated in all-optical logic functions, such as flip-flop and signal regeneration circuits, by using a cavity/waveguide composite layout (Shinya et al., 2006). Consequently, such a low-power, ultrasmall, and integratable optical memory is unique when compared with other optical memories.

7.2.4 Summary and Outlook

This section confirmed that PhC nanocavities can strongly enhance optical nonlinearity and thus achieve ultralow-power all-optical functionalities. All-optical switches and bistable memories are fundamental operations of nonlinear cavities, and an ultrasmall cavity with an appropriate material selection results in a large reduction in power consumption. Since silicon-based PhC devices have the benefit of a well-developed fabrication process and allow us to achieve a high Q and small size cavity, the operating power is much lower than those of other silicon-based optical devices. However, it is still high for our target of photonic processing on chip. This is mainly due to the low absorption and nonlinearity for the light wavelength of interest. Recent progress on implementing p-i-n junctions with silicon-PhC platforms reveals the potential for developing efficient carrier injection/extraction devices, such as optical modulators and photodetectors. (Tanabe et al., 2009, 2010). The ultrasmall size of the PhC cavity is also effective for these devices, because a small junction can allow us to realize a modulator with a low charging energy into the capacitance and a photodetector with a very low dark current. On the other hand, InGaAsP-based PhC cavities exhibit more efficient all-optical nonlinearity, and make it possible to reduce the operating power to much less than that of a silicon platform. This is because the optical nonlinearity and absorption efficiency can be maximized by appropriately choosing an electronic band-edge wavelength and operating wavelength. These features contributed to the development of an unprecedented all-optical switch with an atto-joule switching energy and applicability for 40-Gb/s data signals, and an all-optical bit memory with a μW operating power. This is reason why PhC nanodevices based on III–V materials are more attractive for constructing a PhNoC than those based on Si. The achievement of these ultralow-power and ultrasmall PhC-nanocavity devices is important, because of they offer low power consumption in the mW range even in an integrated chip including thousands of devices with a footprint of less than 0.01 mm^2. A similar scenario can be applied to modulators by implementing a p-i-n junction, and also to lasers and photo-detectors. The next step for these PhC-based optical elements should be the integration and collaborative operation of many devices with the goal of achieving an optical interconnect, optical routing, and logic processing. Progress on this research has the potential to lead to chip-based all-optical data processors, which can partly replace some of the functions of integrated electronic circuits with a much higher signal bandwidth.

7.3 Photonic Crystal Lasers

Shinji Matsuo

7.3.1 Introduction

Introduction of photonic technologies greatly increases the transmission capacity and reduces the power consumption in telecommunication networks from backbone to access.

As described in Section 7.1, the demand for introducing the photonic technologies into chip-level communications (off- and on-chip interconnects) is increasing. In this application area, the construction of the high-density integrated photonic integrated circuits (PICs) is a critical issue as several thousand lasers, photodetectors, and switches must be integrated on a single chip (Matsuo et al., 2011; Notomi et al., 2011). Thus, reducing the power consumption for each element is important. For example, the energy cost for the single-bit data transport is expected to be less than 100 fJ/bit for off-chip interconnect and 10 fJ/bit for on -chip interconnect (Miller, 2009).

However, conventional lasers do not meet such requirements because the energy cost is proportional to the active volume. Figure 7.14 shows energy cost for various laser structures. Conventional edge-emitting lasers consume about 100 mW corresponding to 4 pJ/bit when device operates at 25 Gbit/s (Tadokoro et al., 2009). VCSELs (vertical-cavity surface emitting lasers) show a considerably low threshold power, but it is still around 286 fJ/bit at 35 Gbit/s (Chang, 2009). In addition, VCSELs are intrinsically unsuited for in-plane integration. There is one experimental result using photonic crystal laser (Bagheri et al., 2006). In this device, the energy cost is almost same as VCSEL because the active volume is the same range. To achieve target energy, we have to reduce the active volume in sub-μm^3 range.

This trend is easy to understand when we consider the laser relaxation oscillation frequency, which limits the modulation bandwidth. The laser relaxation oscillation frequency is given by

$$\omega_R^2 = \frac{v_g a N_p}{\tau_p} = \frac{v_g a}{q}\eta_i \frac{(I - I_{th})}{V}, \tag{7.3}$$

where v_g is the group velocity, a is the differential gain, N_p is the photon density, τ_p is the photon lifetime, η_i is the internal efficiency, I is the injection current, I_{th} is the threshold current, and V is the mode volume. This equation indicates that the modulation speed of the laser is increased with input current density. Thus, to decrease operation energy by injection current reduction, small active volume is essential.

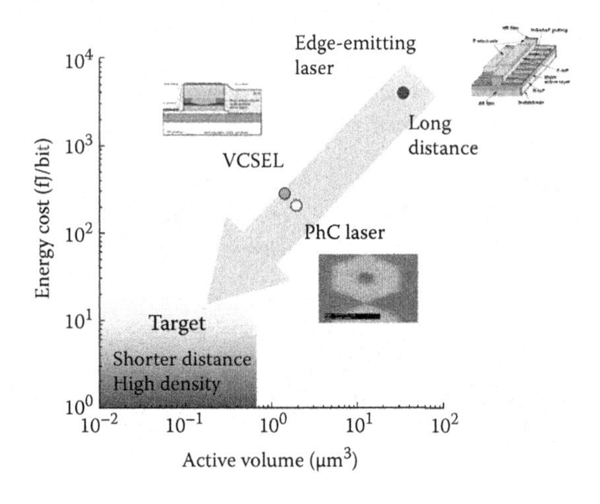

FIGURE 7.14
Energy cost of semiconductor lasers as a function of active volume for various lasers.

To realize the laser oscillation in sub-μm^3 cavity size, photonic crystal (PhC) nanocavities are ideally suited. As a result of strong light confinement achieved by the existence of photonic band gaps, one can realize ultrasmall (comparable to the wavelength) and high-Q ($Q > 10^6$) cavities in PhCs (Tanabe et al., 2007; Takahashi et al., 2007). If we introduce optical gain in them, low-energy cost lasers will be realized. In addition, a number of studies have showed that PhC cavities are intrinsically suited for in-plane integration (Notomi et al., 2008).

In this section, researches of PhC lasers are described. Especially, the importance of introducing the buried heterostructure in nanocavity is described for obtaining high-speed modulation with ultra-low power consumption.

7.3.2 Review of Previous Photonic Crystal Lasers

Photonic crystal lasers have already been developed many research groups (Park et al., 2004; Shih et al., 2007; Nomura et al., 2006; Nozaki et al., 2007). A typical PhC laser is formed in a thin membrane of a certain gain material (conventionally InGaAsP) suspended in air to achieve strong light confinement as shown in Figure 7.15. Due to strong light confinement in the cavity, low threshold power is achieved.

However, it is difficult to achieve the lasing in room temperature continuous wave (CW) operation. The poor performance is primarily due to this geometrical configuration of the PhC laser, which creates two fundamental problems.

First, the heat conduction is extremely poor because of the existence of air claddings on both top and bottom, which prevents room temperature CW operation in most cases. In addition, when InGaAsP is used for gain material, its poor thermal conductivity results in further increase of active region temperature. To achieve room temperature CW operation, the reduction of threshold input power is necessary by using quantum dot (Nomura et al., 2006) or reducing the cavity size (Nozaki et al., 2007). Even when CW operation is achieved, the output power saturates as the pumping power is increased to only a few times larger than the threshold because of the heating problem. One solution for solve this problem is introducing heat-conductive claddings such as sapphire. However, the bonding to sapphire reduces the cavity-Q value, which requires the increase of total gain, i.e., increase of the active region. Thus, threshold input power was increased 1.5 mW (Bagheri et al., 2006).

Second, the generated carriers easily escape out of the cavity because there is no carrier confinement structure. This means reduction of the efficiency of the laser, which increases the input power to obtain same carrier density. Therefore, it also increases active region temperature.

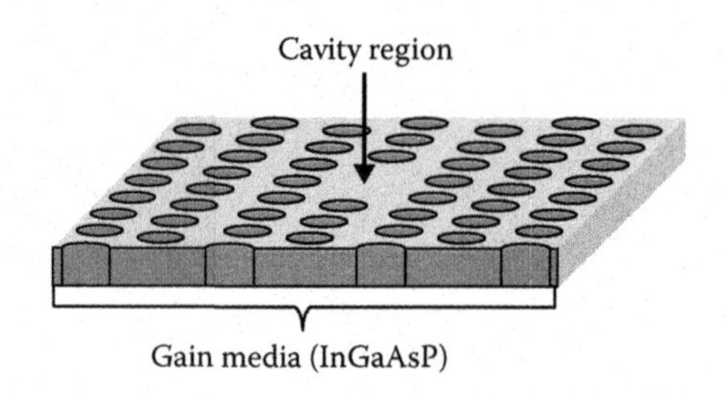

FIGURE 7.15
Typical photonic crystal laser structure using thin membrane of gain media.

As described in Section 7.5.1, it is important to reduce the active region volume for reducing the energy cost. However, typical PhC structure has a problem in the increase of active region temperature. The increase of active region temperature deplete the laser performances such as laser threshold, internal efficiency, and differential gain, which decreases the modulation speed as describe in Equation 7.3. Thus, reduction of the active region temperature is important when the laser operates at a high input power density.

7.3.3 Buried Heterostructure Photonic Crystal Laser

To solve the inherent problems of previous PhC lasers, an ultra-compact buried heterostructure (BH) PhC laser has been developed (Matsuo et al., 2009, 2010). Figure 7.16 shows the structure of BH-PhC laser, in which InGaAsP-based active region is buried within InP line-defect waveguide.

In BH-PhC lasers, the thermal resistance is greatly reduced in comparison with the previous PhC lasers. Conventionally, InGaAsP is employed for the active media of PhC lasers, which has a poor thermal conductivity (4.2 W/m/K). In BH-PhC lasers, the small active region is surrounded by InP, which has more than an order of magnitude higher thermal conductivity (68 W/m/K).

Figure 7.17a,b shows the calculated results of active region temperature for BH-PhC laser and conventional PhC laser, respectively. The calculation was done by using three-dimensional finite volume method. A 100-μW heat source in a $3 \times 0.5 \times 0.2$ μm^3 active region and the substrate temperature of 298 K were assumed. When the active region was buried in an InP layer, the active region temperature increased by 6.7 K, as shown in Figure 7.17a. In all-InGaAsP (active layer) structure, the active region temperature increased by 52.2 K, as shown in Figure 7.17b. These results indicate that the BH effectively reduces the thermal resistance of the device.

In addition, the heat is generated only in the gain region for the BH PhC laser, whereas the heat is generated in an area a few micrometers in diameter (corresponding to the pump beam spot) in the previous design as shown in Figure 7.15. This eliminates the unwanted, extra heat generated outside the cavity mode. In addition, the reduction of the heat generating volume generally enhances thermal relaxation, which is especially evident for wavelength-sized PhC cavities (Notomi et al., 2005).

Another important feature of BH is the effective confinement for both carrier and optical field. The proposed BH-PhC laser has an extremely small buried active region, which is buried within straight line defect waveguide in InP slab. In this structure, we can expect to construct a high-Q cavity because an extraordinarily high-Q (Q > 1 million) nanocavity mode was formed with a mode volume (V_{eff}) of ~$(\lambda/n)^3$ by implementing a slight local structural modulation of a line defect waveguide in Si-PhC (Tanabe et al., 2007). In the case

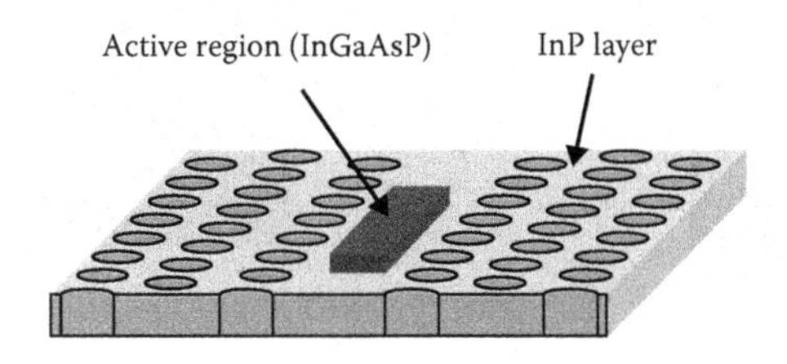

FIGURE 7.16
Structure of buried heterostructure photonic crystal laser. InGaAsP-based active region is buried with InP laser.

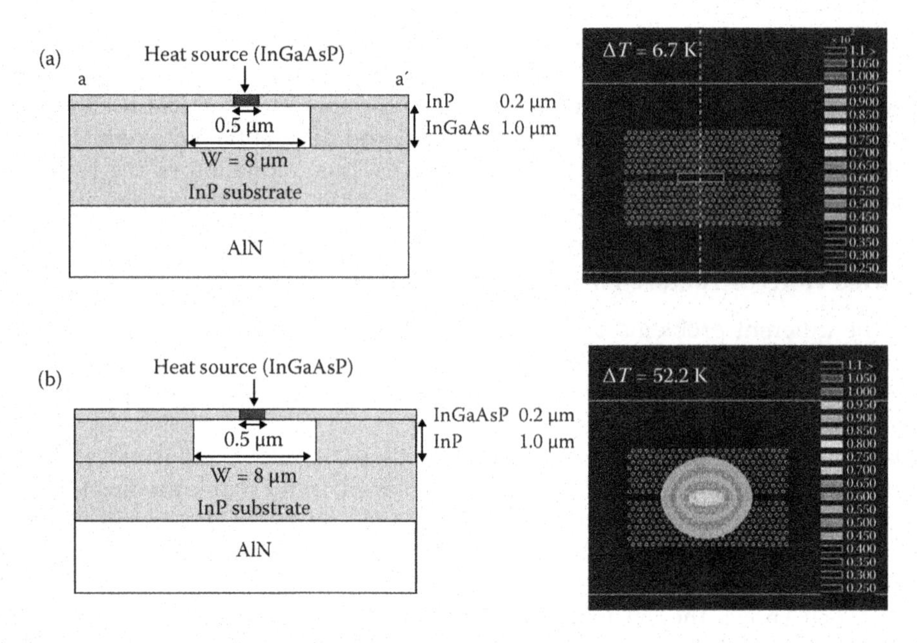

FIGURE 7.17
(**See color insert.**) Numerical simulations of active region temperature for (a) BH-PhC structure and (b) all-active PhC laser structure.

of BH, similar cavity modes can be created by refractive index difference between InGaAsP active region and InP-buried region. The field profile of the cavity mode in BH-PhC laser is shown in Figure 7.18. FDTD (finite-difference time-domain) calculation exhibits a high-Q (Q ~ 1.7 × 10^6) cavity mode with V_{eff} ~ 0.2 μm^3, where we assumed that the effective refractive index were 2.77 and 2.59 for InGaAsP-based SQW layer and InP slab, respectively. The size of the active region was 4 × 0.3 × 0.15 μm^3 (0.18 μm^3). As is clear in this figure, the area of the active region is largely overlapped with the cavity mode profile. It enables us to obtain efficient laser characteristics, such as a low threshold, high quantum efficiency, because BH effectively confines the carrier in buried active region.

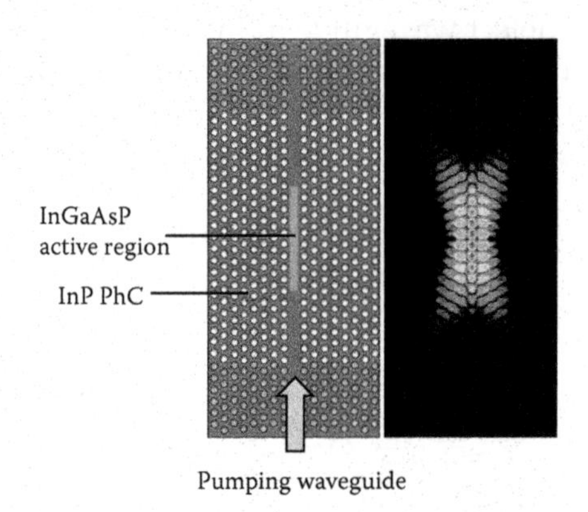

FIGURE 7.18
(**See color insert.**) Field profile of the cavity mode in BH-PhC laser using FDTD method.

7.3.4 Fabrication and Characteristics of BH-PhC Laser

As described above, BH laser structures would be ideal for PhC lasers, it had not been realized because of the difficulty in fabricating an extremely small BH with a precise position to prevent wet chemical etching during etching of the InGaAs sacrificial layer and in obtaining a smooth top surface after creating the BH for a high-Q cavity. Figure 7.19 shows SEM images for fabricated laser. The epitaxial wafer was grown by metal organic chemical vapor deposition (MOCVD) on an n-doped InP substrate. The epitaxial structure consisted of the active region, an InP layer, and a 1.0-μm-thick InGaAs sacrificial layer. The active region consists of a single quantum well (SQW), with a 1.55-μm photoluminescence (PL) peak, sandwiched by 70-nm-thick InGaAsP layers with a 1.35-μm PL peak. The active region was defined with an SiN_x mask using electron beam lithography and then etched with inductively coupled plasma reactive ion etching (ICP-RIE) and selective wet chemical etching techniques using H_2SO_4:H_2O_2:H_2O solutions. The etching was stopped at the InP layer underneath the active region. Then, an InP layer was selectively regrown by using MOCVD. After removing the SiN_x mask, we deposited a 35-nm InP layer to obtain a flat surface. This is important in terms of obtaining a high-Q cavity. As shown in Figure 7.19, the active region ($4 \times 0.3 \times 0.15$ μm^3) was fabricated within a wavelength-scale photonic crystal cavity. We obtained a flat top surface and a smooth dry-etched air-hole surface.

The light-in light-out (LL) characteristic of the fabricated device at room temperature CW operation is shown in Figure 7.20. A 1310-nm distributed feedback (DFB) laser operated in CW was used for optical pumping. As shown in the inset of Figure 7.20, we used input waveguide consisting of 3-μm-wide waveguide, taper waveguide and line defect of PhC so that the light could be input efficiently into line-defect waveguide. The estimated coupling loss between an optical fiber and a line-defect waveguide was 10 dB. In the fabricated device, the output power is emitted in the direction perpendicular to the wafer, and we measured the output power from the top surface with an objective lens corrected into single mode fiber. The device exhibits a clear kink at a threshold of 1.5 μW, which is the lowest value for room-temperature CW lasing operation of any type of laser. The 3-dB bandwidth of the output spectrum near the threshold was 0.02 nm, corresponding to a fairly high Q-factor of 77,400. Thus, the reduction of the threshold is mainly attributed to the small gain volume, high-Q cavity, and effective confinement of carrier in the gain region. The output power is increased by more than 20 dB above the threshold value. This result is in stark contrast with previous PhC lasers, where the output power saturates just a few times the value obtained at threshold. This clearly shows that BH-PhC laser does not

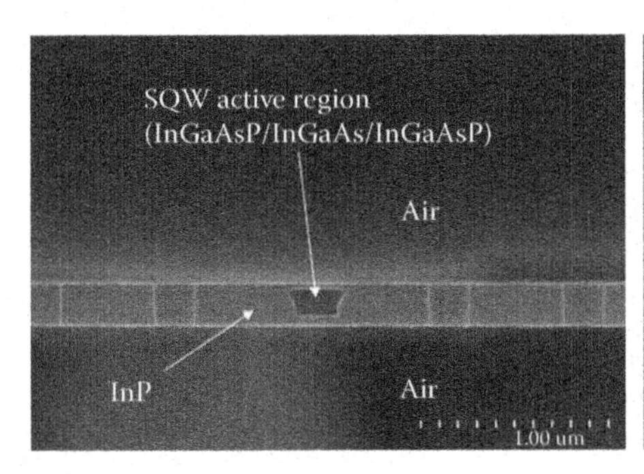

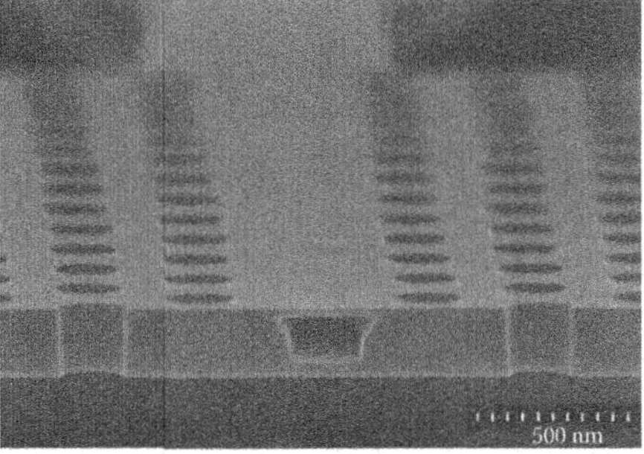

FIGURE 7.19
SEM images of fabricated BH-PhC laser.

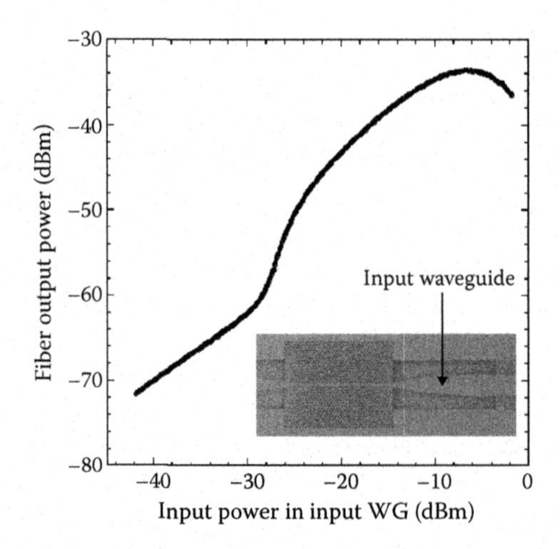

FIGURE 7.20
Light-in light-out characteristic of the fabricated device at room temperature CW operation.

suffer from the heating problem due to the large thermal conductivity of InP. The maximum single-mode fiber output power of −33.5 dBm was obtained when the input power was −6.73 dBm.

Figure 7.21 shows the temperature dependence of the lasing wavelength for both pulsed and CW operation. The input powers in input waveguide were 100 and 200 μW, respectively. The lasing wavelength dependence on temperature change was 0.095 nm/K. Therefore, from the measured lasing wavelength, increase of the active region temperature caused by CW operation was estimated to be 5.3 and 9.6 K for input powers of 100 and 200 μW, respectively. These results agree well with the calculated results of Figure 7.17a and the BH effectively reduces the thermal resistance of the device.

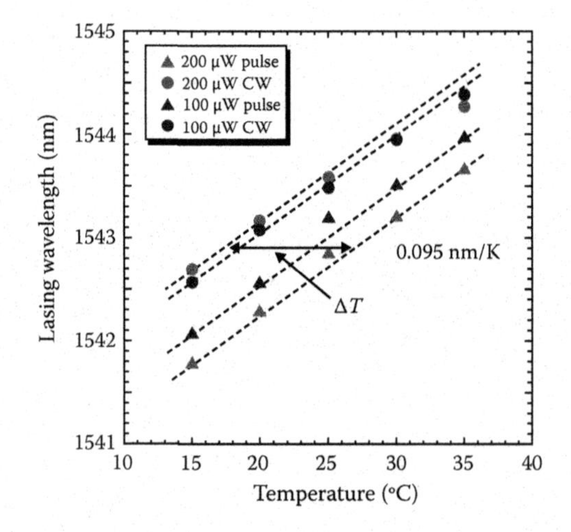

FIGURE 7.21
(See color insert.) Temperature dependence of the lasing wavelength for both pulsed and CW operation. The input powers in input waveguide were 100 and 200 μW, respectively.

7.3.5 PhC Laser Integrated with Output Waveguide

It is important to integrate the output waveguide for constructing large-scale PIC, in which lasers and detectors are connected by waveguides in PhC slab. As shown in Figure 7.18, the mode field is extended to the Γ-M direction when the cavity has no output waveguide. Thus, effective coupling can be realized with the output waveguide when the output waveguide is placed in the Γ-M direction of the PhC cavity. Figure 7.22 shows the calculated mode profiles of the devices with output waveguide for offset 3*a*, where *a* is lattice constant. As shown in this figure, the optical field of the cavity is effectively coupled to output waveguide. On the other hand, the cavity Q-factor was decreased to ~2,900, which depends on the offset of the output waveguide, W_{offset}. Furthermore, there is only one output port in this device, and this is an important characteristic when constructing a large-scale photonic integrated circuit designed to reduce crosstalk. This is in sharp contrast to other types of laser structure such as the conventional edge emitting lasers and VCSELs, in which there are two output ports.

Figure 7.23 shows an SEM image of our fabricated BH-PhC laser with output waveguide (Matsuo et al., 2011). The input waveguide consists of a 3-μm-wide waveguide, a taper

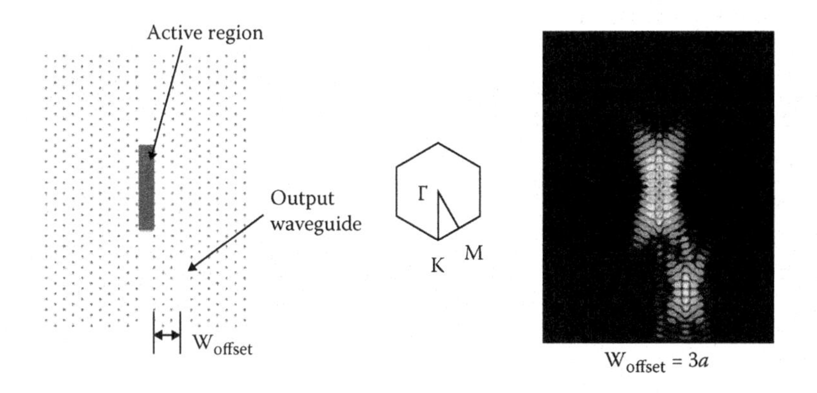

FIGURE 7.22
(**See color insert.**) FDTD calculated mode profile of PhC laser with output waveguide. Output waveguide is placed in Γ-M direction.

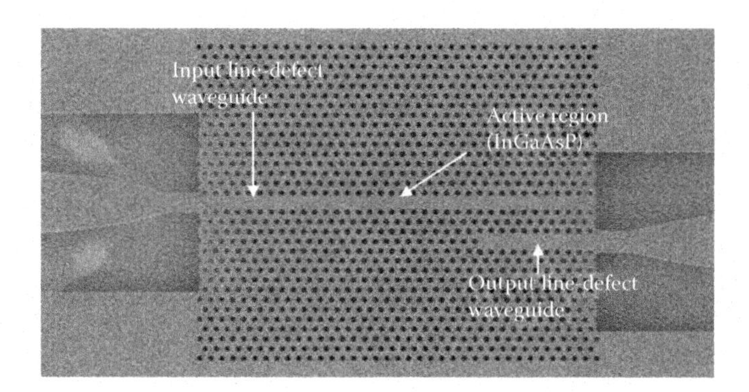

FIGURE 7.23
SEM image of BH-PhC laser with output waveguide.

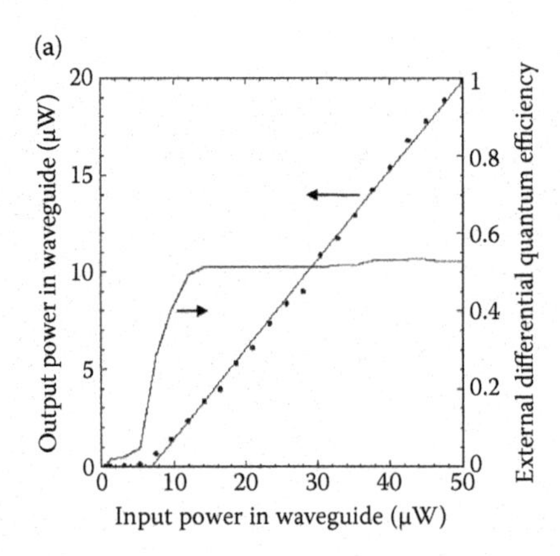

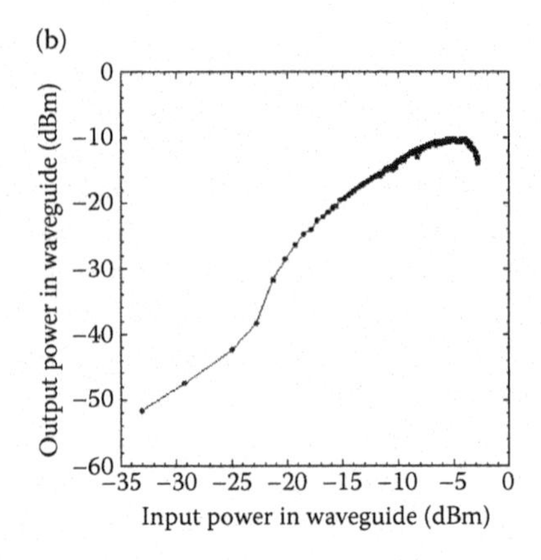

FIGURE 7.24
(a) Light-in versus light-out (LL) characteristics and external differential quantum efficiency of near threshold region on a linear scale. (b) LL characteristic for a wide input range on a logarithmic scale.

waveguide and the line defect of the PhC. Therefore, the pump light is input into the active region through a line defect waveguide and is effectively absorbed in the active region. On the other hand, the output waveguide is placed in an offset position (W_{offset} = 3a) with respect to the line defect including the active region. To increase the output power, the number of quantum wells was increased to 3, whereas the device in previous section has only a single quantum well.

Figure 7.24a shows the LL characteristic and the external differential quantum efficiency near the threshold input power range. The external differential quantum efficiency is defined as a ratio of photon number between the pumping light in the input line defect waveguide and the output light collected in the output line defect waveguide. The active region was 5.0 × 0.3 × 0.16 μm^3. A 1310-nm DFB laser was employed for optical pumping. The pump light was input into the input line defect waveguide from the fiber using a collimator lens, a 3-μm-wide waveguide and a tapered waveguide. The facet of the output waveguide was coated with anti-reflection film whereas that of the input waveguide was as cleaved. The device exhibited a clear kink at a threshold of 6.8 μW at room temperature (298 K) with CW operation. The Q-factor at the threshold was decreased to 14,300, comparing with BH-PhC laser without output waveguide. Thus, the threshold input power is increased due to the increase of the cavity Q factor and quantum well number. A high external differential quantum efficiency of 53% is obtained. This indicates that the generated carrier was effectively converted to photon because BH was confined both the carrier and photon in the wavelength-sized cavity. Furthermore, the output waveguide effectively collects the laser output.

Figure 7.24b also shows the LL characteristics with a wide input range for device. The maximum output power was –10.3 dBm for the devices with offset 3*a*. The maximum output power and external differential quantum efficiency are the highest reported values for a PhC nanocavity laser for CW operation at room temperature.

7.3.6 Lasing Wavelength Control and Wide Temperature Range Operation

In the photonic network-on-chip application, it is important to utilize the wavelength division multiplexing (WDM) technology because it dramatically decreases the power consumption and increases the total bandwidth. In the case of BH-PhC laser, to employ the WDM technology, the active volume must be controlled precisely because the equivalent

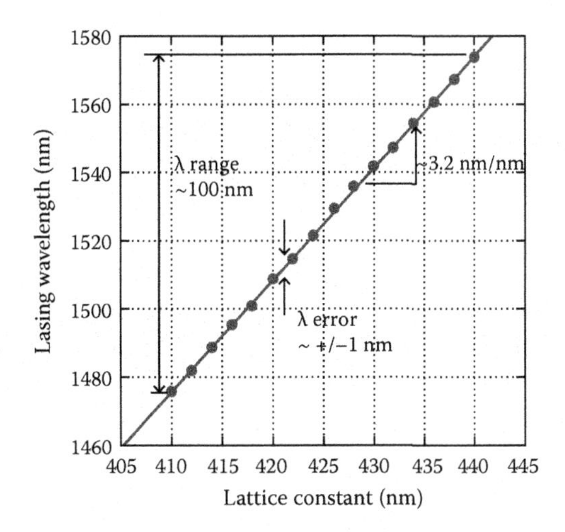

FIGURE 7.25
Lasing wavelength as a function of lattice constants. The PhC laser had a wavelength range of 100 nm by changing the lattice constant from 410 to 440 nm.

refractive index of the cavity was changed by the active volume. To investigate this, the lasing wavelengths of the PhC lasers were measured with various lattice constants as shown in Figure 7.25 (Takeda et al., 2011). The lasing wavelengths could be controlled from 1476 to 1574 nm corresponding to the lattice constants of 410 to 440 nm. The wavelength variations induced by the fabrication errors were ±1 nm even when ultra-small butt-joint structures were employed in the PhC waveguides, which might cause size fluctuations.

When the photonic network on Si-CMOS chip is constructed, the photonic devices have to operate high temperature up to 80°C because the maximum surface temperature of the CMOS chip is 80°C. The light-in/light-out characteristics of the PhC laser were measured at various stage temperatures, which were changed from room temperature to 80°C. A CW lasing of the PhC laser was confirmed up to 80°C stage temperature (Figure 7.26).

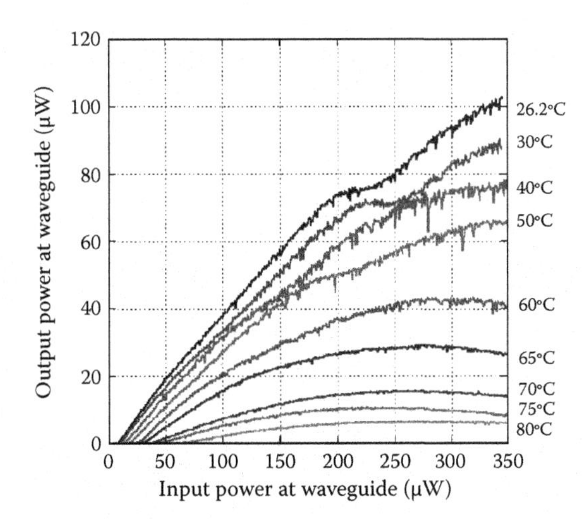

FIGURE 7.26
LL characteristic of the PhC laser under various stage temperatures. The device could be operated at 80°C with the CW pumping.

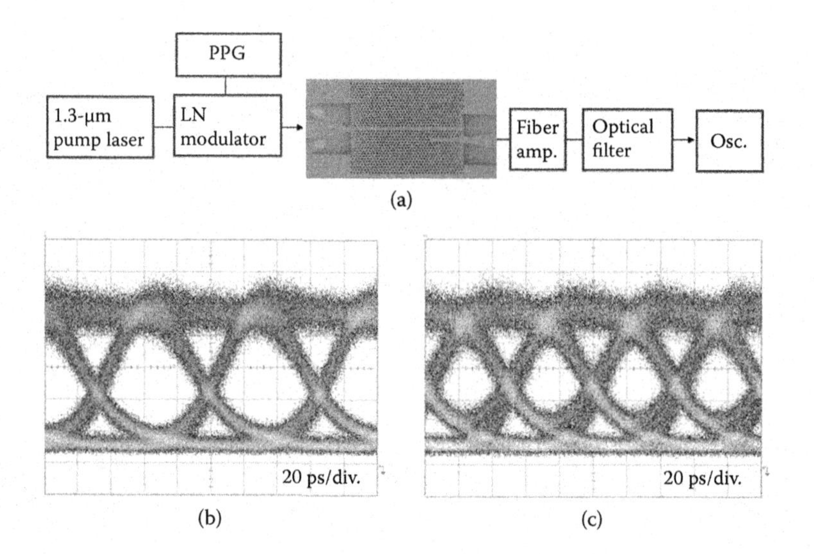

FIGURE 7.27
(a) Experimental setup for direct modulation. Eye diagrams for (b) 15 Gbit/s and (c) 20 Gbit/s NRZ signals.

7.3.7 Direct Modulation of BH-PhC Laser

This section describes experimental results for direct modulation with optical pumping. This is an important characteristic when we use the device to construct a photonic network chip on a CMOS chip. Figure 7.27a shows the experimental setup. To measure the direct modulation response of the fabricated laser, the 1.31-μm pump light was modulated by using a $LiNbO_3$ modulator from 35 to 208 μW with an NRZ signal with a pseudo-random bit sequence (PRBS) of 2^{31}–1. Figure 7.27b,c shows the eye diagrams of 1.55-μm output light modulated by 15- and 20-Gbit/s NRZ signals, respectively. In these experiments, the output signal was observed by using an optical fiber amplifier and an optical filter. As shown in these figures, clear eye opening was observed. The data transfer energy for a single bit was estimated to be 8.76 fJ by taking into consideration a confinement factor of 0.616 and an absorption coefficient of 6000 cm^{-1} for a 1.31-μm pump light when the device was operated with a 20-Gbit/s NRZ signal. This is the smallest reported energy cost for any type of semiconductor laser.

7.4 Photonic Crystal Waveguides and Slow Light

Toshihiko Baba

Photonic crystal waveguides have been studied and developed toward a platform of large-scale photonic integration. Actually, they are used in many studies as access waveguides for photonic-crystal based functional devices such as high-*Q* cavities. In recent years, however, they have been rather focused as slow light devices. In the first part of this section, various waveguide structures are introduced and the most common structure: photonic crystal slab line defect waveguide is explained in detail. Slow light in this waveguide is discussed in the second part. After summarizing applications of slow light, the mechanism of how slow light is generated and how it behaves are explained. Indicating its crucial issues on a narrow bandwidth and higher order dispersion, dispersion-compensated and low-dispersion slow light are discussed as solutions. The tunable delay, nonlinear enhancement, and compact optical modulators are presented as their functions and applications.

7.4.1 Various Structures

The multidimensional periodicity comparable to the optical wavelength in photonic crystals gives rise to the so-called photonic band-gap (PBG). In uniform photonic crystals, the light propagation is inhibited at frequencies overlapping with the PBG. Light propagates along the line defect introduced into the photonic crystal (Meade et al., 1994; Joannopoulos et al., 2008). When the equivalent modal index of the line defect is higher than that of the photonic crystal, both the total internal reflection (TIR) due to the index difference and Bragg reflection due to the PBG contribute to the optical confinement and light propagation. Which one is dominant depends on the structure and guiding condition. When the line defect has a lower equivalent index, the light propagation is only supported by the Bragg reflection.

Figure 7.28 shows some structures of photonic crystal waveguides: (a) and (b) are based on 2D photonic crystals consisting of high-index periodic microcolumns in a low-index medium. In these structures, the PBG occurs for the polarization along the columns (it is sometimes called TM in this field) (Tokushima et al., 2004). (c) and (d) are based on another type of crystals consisting of low-index holes in a high-index slab (photonic crystal slab). Here, the PBG occurs for a polarization inside the 2D plane (TE). (a) and (d) have lower index line defects, so light is laterally confined only by the PBG. (b) and (c) have higher index line defects, so light is confined by the mixture of the TIR and PBG. At the early stage of the research, the light propagation characteristics were discussed theoretically, assuming their infinite height in the vertical direction. In particular, sharp bends, ultrasmall splitters, and directional couplers in (a) attracted much attention (Mekis et al.,

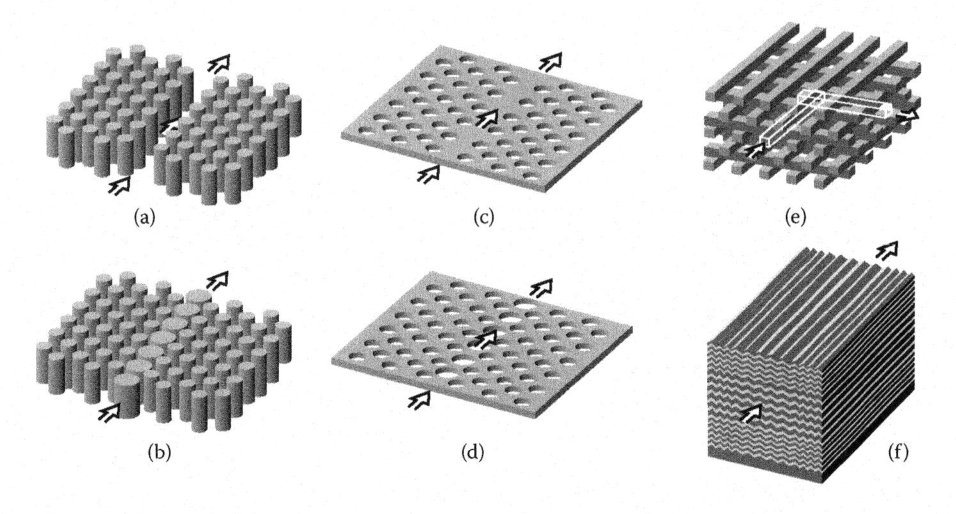

FIGURE 7.28
Structures of photonic crystal waveguides. (a) Microcolumn photonic crystal with air line defect. (b) Microcolumn photonic crystal with dielectric line defect. (c) Slab photonic crystal with missing air hole line defect. (d) Slab photonic crystal with larger hole line defect. (e) Woodpile photonic crystal with missing wood line defect. (f) Corrugated multilayer photonic crystal with modified period line defect.

1996; Yonekura et al., 1999). However, they cannot confine light vertically in actual structures. (b)–(d) maintain loss-less guided modes with the vertical confinement by the TIR. Since a simple structure is preferable for easy fabrication and low loss, (c) has become the most common (Baba et al., 1999; Chutinan and Noda, 2000a; Notomi et al., 2001; Baba et al., 2002; Yamada et al., 2002; McNab et al., 2003; Notomi et al., 2004; Sugimoto et al., 2004; Sakai et al., 2004; Kuramochi et al., 2005). Since photonic crystals operate with the Bragg reflection, the polarization dependence cannot be avoided; (c) is usually used for TE because of the PBG for this polarization.

Waveguides based on 3D crystals have also been studied: (e) called the woodpile structure exhibits the complete PBG in all directions and polarizations. The line defect is formed by removing and/or modifying "woods" (Chutinan and Noda, 2000b). Here, the relation between the equivalent index and optical confinement is the same as for (a)–(d). (f) called autocloning structure is fabricated by depositing multilayers on a 2D corrugated substrate. The waveguide is formed by modulating the periodicity in the corrugation and multilayers (Ohtera et al., 2002). In this waveguide, the TIR is dominant for the optical confinement.

7.4.2 Photonic Crystal Slab Line Defect Waveguide

The standard photonic crystal slab line defect waveguide consists of triangular lattice circular holes (pitch a, diameter $2r$) in a high-index slab (index n, thickness t). Figure 7.29 shows the band diagram assuming Si as the slab and air inside the holes and as the upper and lower claddings (Baba et al., 2002). Here, a/λ ($=\omega a/2\pi c$) is the normalized frequency, λ the wavelength, π the angular frequency, c the light velocity in vacuum, k the wave number along the line defect, and $k = 0.5(2\pi/a)$ the band-edge corresponding to the Bragg condition. Dark and medium gray regions depict the leakage conditions called light cone and slab mode, respectively. Both polarizations exhibit band curves outside of these conditions. The light gray region depicts the transmission band of the main guided mode. The optical

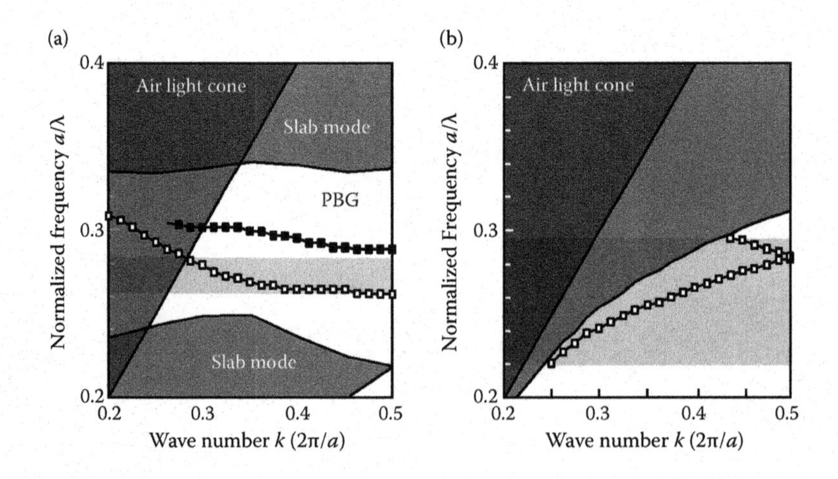

FIGURE 7.29
Photonic band diagram of air-clad photonic crystal slab line defect waveguide. (a) TE polarization. (b) TM polarization. $n = 3.5$. $2r/a = d/a = 0.571$. Slab mode depicts the condition that light leaks to the photonic crystal slab. Air light cone depicts the condition that light leaks out-of-plane. White and black square plots show symmetric and antisymmetric modes, respectively. Light gray region indicates transmission band of the symmetric mode.

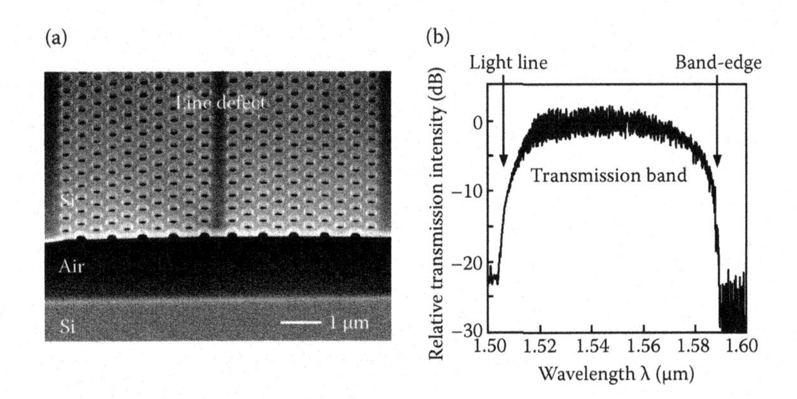

FIGURE 7.30
Photonic crystal slab line defect waveguide fabricated on SOI substrate. (a) SEM view around cleaved facet. (b) Measured transmission spectrum of symmetric guided mode.

confinement of the TE is more strengthened than TM due to the PBG. When $2r$ increases and/or n and t decrease, the PGB expands and shifts to higher frequencies while the light cone being unchanged, resulting in a narrower bandwidth. At $2r/a = 0.5$, the bandwidth is approximately 10% of the center frequency. If another medium such as silica is assumed inside the holes and as claddings, the light cone shifts to lower frequencies by a factor of its index, and severely shrinks the bandwidth.

The photonic crystal slab can be fabricated by making air holes into a high-index thin film and removing the lower cladding medium (sacrificial layer) to form an air-bridge structure. Many studies employ commercially available silicon-on-insulator (SOI) substrate with a silica box layer and top Si layer, which become the sacrificial layer and slab, respectively. Typical thicknesses of these layers are >2 and 0.22 μm, respectively. For this Si thickness, $a = 0.44$ μm, $2r = 0.25$ μm, and $a/\lambda \sim 0.28$ for $d/a = 0.5$, $2r/a = 0.571$ and $\lambda \sim 1.55$ μm. Such hole size can be easily patterned by *e*-beam lithography and also by advanced photolithography in the Si CMOS process. The holes are formed by reactive ion etching with fluorine-based gases. The air-bridge structure is formed by HF selective wet etching of the box layer. Figure 7.30a shows the scanning electron micrograph (SEM) of fabricated device. Here, waveguide facets are formed by cleavage. The transmission spectrum is measured by coupling tunable laser light to one facet and detecting output light from the other facet using power meter, as shown in Figure 7.30b. Almost flat transmission band is lying at $\lambda = 1.51$–1.58 μm. The fast oscillation in the band is caused by the Fabry–Perot resonance between the cleaved facets and can be suppressed by fiber couplers discussed later. As understood from Figure 7.29a, the transmission band in Figure 7.30b is sandwiched by the light line and band-edge. The propagation loss is mainly caused by the scattering, and depends strongly on the uniformity of the process. The disordering of a few nanometers yields a propagation loss of less than 5 dB/cm (Kuramochi et al., 2005). This value is acceptable for compact devices and small-area high-density integration.

7.4.3 Scheme for Integration

After the theoretical demonstration of low-loss sharp bends for the waveguide of Figure 7.28a, the photonic crystal waveguide was first anticipated to be the basic element of the optical wiring in large-scale photonic integrations. As aforementioned, it is difficult to guide light in this waveguide actually. Afterward, similar bends were studied for the structure in Figure 7.30. However, the local mode mismatch and excitation of unwanted modes at bends give rise to large reflection and scattering losses, resulting in a low transmission efficiency and/or narrow bandwidth. Meanwhile, the Si wire waveguide has been developed as a basic element in Si photonics. It easily achieves similar sharp bends; a high transmission efficiency of over 99% is obtainable with a wide bandwidth, even though the bend radius is several microns (Sakai et al., 2001). This radius is slightly larger than that expected for photonic crystal waveguides, but sufficiently small for most wiring patterns. Because of these situations, many studies became to employ the scheme where photonic crystal waveguides as functional devices are connected with wire waveguides as the optical wiring. For this scheme, a simple low-loss junction has been developed. Here, the width of the Si wire and the cut edge of holes of the photonic crystal waveguide are optimized, and these waveguides are simply buttjoined. The junction loss can be less than 0.5 dB when the wavelength is apart from the band-edge and the group index is moderately small (Gomyo et al., 2004).

In the measurement of Figure 7.30, the coupling loss from the external setup to the photonic crystal waveguide is typically larger than 15 dB. This large loss is caused by the small modal cross section of this waveguide, e.g., $0.5 \times 0.2\ \mu m^2$, which is smaller than the diffraction limit in air and leads a large mismatch with a focal spot formed by a lensed fiber and objective lenses. This situation is improved when this waveguide is connected with wire waveguides, because fiber couplers developed for wire waveguides can be used (Yamada et al., 2004). The spotsize converter is a low-loss coupler consisting of inversely tapered wire waveguide covered with a large-core low-index waveguide. Light from the fiber is first coupled to the low-index waveguide and then transferred to the wire waveguide. The lowest loss ever reported is 0.4 dB/coupling over a wide bandwidth. The grating coupler is a flexible coupler consisting of surface grating connected with the wire waveguide through a taper. Light from the upper fiber is incident on the grating, converted to the inplane wave, and finally coupled to the wire waveguide. The loss is typically 3 dB for the wavelength range of several tens of nanometers (Bogaerts et al., 2005). It allows the free access to any gratings formed on the same chip without making waveguide facets. This is particularly effective for testing devices before dicing the wafer.

7.4.4 Expectations for Slow Light

From this section, slow light in photonic crystal waveguides is discussed. Before going into details, let us summarize the expectations for slow light.

In photonics, extreme phenomena and their applications have been studied in terms of different physical parameters of light, such as intensity, profile, wavelength (frequency), phase, and duration of light. However, the velocity has never been focused until recently; there were no methods that widely control the velocity and delay of light. If the velocity is flexibly controlled, various applications in Figure 7.31 will be available. Optical buffers that change the delay of packet signals on demand will be the significant breakthrough in the next era network technology. Digitally tunable delay lines switching fibers or waveguides with different lengths may be applicable for this purpose. However, continuously tunable delays are only available with a mechanically movable mirror at present. Slow light has first been expected to realize this function on a chip. The main challenges for this are the long buffering time of over 100 ns, reasonably low loss, and switching time mush shorter than the packet length, e.g., nanoseconds. Similarly to this, optical memories each storing one optical bit will be desired in more advanced systems. On

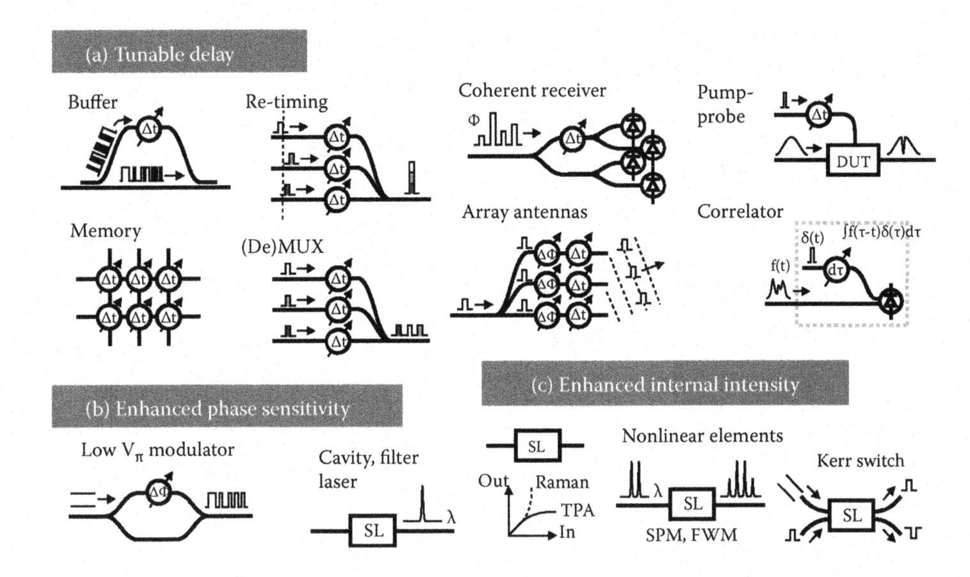

FIGURE 7.31
Potential applications of slow light using (a) tunable delay, (b) enhanced phase sensitivity, and (c) enhanced internal intensity. Here, t and τ denote time, ϕ the phase, and SL slow light device.

the other hand, re-timing, synchronization, and multi/demultiplexing of optical pulses only require picosecond order delays, although the switching time must be shorter than this. The tunable 1-bit delay makes differential-phase coherent receivers compact and symbol-rate variable. Controlling the delay simultaneously with phase achieves ideal array antennas and beam scanners, in which the phase front completely matches with the intensity profile. Pump-probe and correlation measurements are currently utilizing a mechanical delay scanner. Since they do not need the wide-range tuning nor fast switching, they can be close targets, which make the system compact and enhances the data acquisition speed.

In addition to the tunable delay, slow light enhances the phase sensitivity against the index change. As discussed later, slow light arises from the large first-order dispersion. Therefore, the phase shift in the device against the wavelength shift is enhanced by slow light. The enhancement occurs with the index change at a fixed wavelength. This improves the efficiency and/or reduces the size of optical modulators. If slow light propagates as a short optical pulse without suffering the higher order dispersion, the pulse is compressed in space in the device. This intensifies the pulse peak and enhances various optical nonlinearities. Possible applications are the wavelength converters based on self-phase modulation, four-wave mixing, and harmonic generations, ultrafast optical Kerr switch, and intensity controller utilizing stimulated Raman scattering and two-photon absorption.

7.4.5 Principle of Slow Light

Slow light has a group velocity υ_g much smaller than the velocity c in vacuum. As seen in textbooks on optical theory, υ_g is given by

$$\upsilon_g = (dk/d\pi)^{1} \tag{7.4}$$

This indicates that slow light occurs with the large first-order dispersion $dk/d\pi$. The slow down factor of υ_g against c is usually evaluated by the group index n_g given by

$$n_g \equiv c/\upsilon_g = c(dk/d\pi) \tag{7.5}$$

Large first-order dispersion can be seen in various material phenomena and structures (Thévenaz, 2008; Baba, 2008), which are on-resonance with light. The electromagnetic-induced transparency (EIT) in vapor atoms at cryogenic temperature is a representative phenomenon showing a huge n_g up to 10^8. The stimulated Raman and Brillouin scattering in optical fibers and the absorption/amplification in semiconductor optical amplifiers are more common phenomena giving small n_g of less than 10 due to their material dispersion. Photonic crystal waveguides and high-Q cavities are the structures exhibiting $n_g = 10^1 - 10^8$. They are more suitable for on-chip integration of the device. The slow light characteristics of two different types of photonic crystal waveguides on the SOI substrate are shown in Figure 7.32 (Mori et al., 2007). Since these waveguides have opposite second-order dispersion, their band-edges appear on opposite sides of the transmission band. Near the band-edges, they exhibit $n_g > 50$.

As observed from Figure 7.32, the bandwidth exhibiting slow light is usually very narrow. In any slow light arising from the resonance, n_g and the delay Δt are constrained by the frequency bandwidth Δf. Assuming $\upsilon_g \ll c$, the following relation is derived from Equation 7.3 (Baba, 2008):

$$\frac{\Delta f}{f} \cong \frac{\Delta n}{n_g} = \Delta n \frac{L}{c\Delta t} \tag{7.6}$$

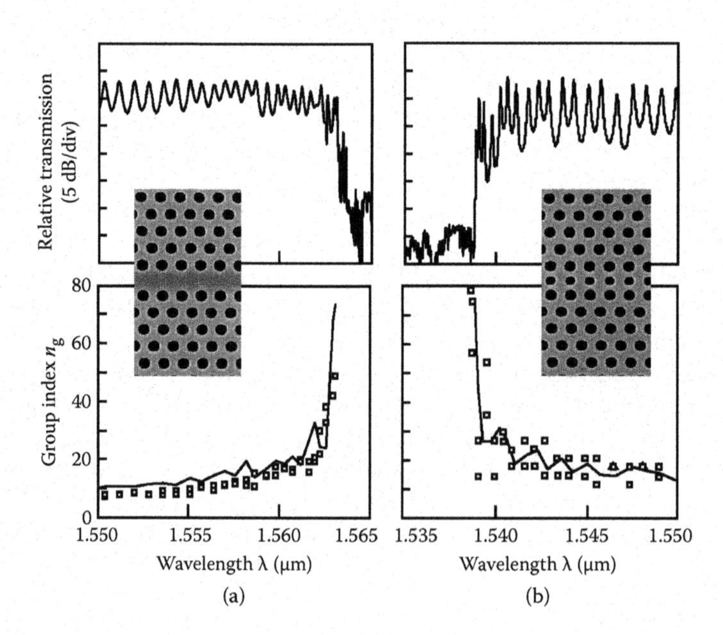

FIGURE 7.32
Observation of band-edge slow light in two types of photonic crystal waveguides. Upper and lower panels show transmission and group index spectra, respectively. Square plots and solid lines in the lower panel are evaluated from Fabry–Perot resonance and phase shift of modulated signals, respectively. (a) Single line defect waveguide. (b) Shifted small holes added to waveguide (a). Insets show SEM views of waveguides.

where f is the center frequency, L is the device length, and Δn is the change of material index or the modal equivalent index. In general, Δf is inversely proportional to n_g and Δt. This means that remarkable slow light can be observed only with a narrow bandwidth. Because of this relation, the delay-bandwidth product $\Delta t\Delta f$ and normalized delay-bandwidth product $n_g(\Delta f/f)$ are used as the criteria showing the slow light performance. Provided that optical signals are Gaussian pulses and their full-width at half maximum (FWHM) is equivalent to the pitch between signals, the buffering capacity of the device is given by 2.3 $\Delta t\Delta f$. From Equation 7.8, the following equations are derived:

$$\Delta t\Delta f \leqq \Delta n(L/\lambda),\ n_g(\Delta f/f) \leqq \Delta n \tag{7.7}$$

They indicate that Δn is a crucial parameter. When using material phenomena, Δn is determined by the material property and usually no higher than 0.01. On the other hand, in high index contrast structures, e.g., those composed of semiconductor and air, Δn can be of 0.1 order. It is seen in Figure 7.29a that the guided mode band of the photonic crystal waveguide is located at $a/\lambda\sim$ 0.28, and k changes by ~0.2($2\pi/a$) from the band-edge to the light line. They lead to a maximum Δn of ~0.7.

However, the net buffering capacity is usually degraded from that estimated from $\Delta t\Delta f$ due to the higher order dispersion. When the resonance is a simple Lorentzian, as is in EIT and high-Q cavities, the second-order dispersion (group velocity dispersion: GVD), which broadens the optical pulse, vanishes at the center frequency of the slow light band but rapidly increases in the vicinity. The third-order dispersion disrupting the pulse is large over the slow light band. In standard photonic crystal waveguides, the dispersion is asymmetric, as shown in Figure 7.32, so the GVD also becomes very large. The fractional delay $\Delta t/\Delta\tau$ is used to evaluate the net buffering capacity affected by such higher order dispersion, where $\Delta\tau$ is the output pulse width. Of course, the delay-bandwidth product and fractional delay are the criteria for fixed delays. Many important applications including optical buffers finally require a tunable delay. In photonic crystal waveguides, the tunable delay is obtained by changing the material index uniformly and shifting the photonic band. If a certain wavelength approaches to the band-edge, n_g of this wavelength is increased (Vlasov et al., 2005). However, this method also changes $\Delta\tau$ due to the large GVD and only gives a fractional delay much less than 1.0. The effective tunability must be gained with suppressing the GVD.

7.4.6 Wideband Dispersion-Free Slow Light

Two different approaches are considered, as shown in Figure 7.33, which suppress the GVD and allow the transmission of the slow light pulse in a moderate bandwidth.

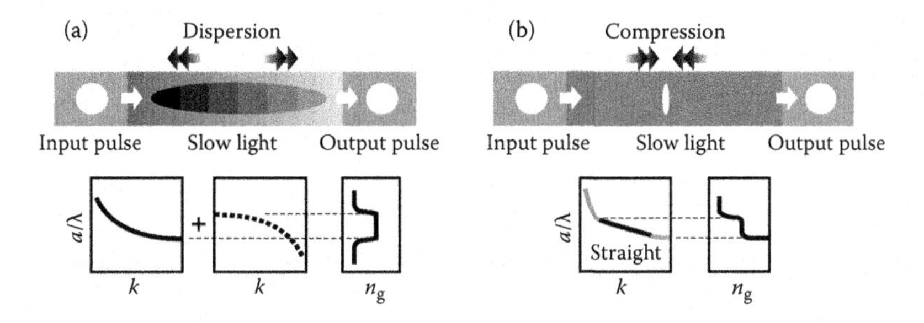

FIGURE 7.33
Two different approaches for wideband dispersion-free slow light. (a) Dispersion-compensated slow light in a chirped structure. (b) Low-dispersion slow light generated by straight photonic band.

Dispersion-compensated slow light is generated by a flat band sandwiched by the opposite GVD characteristics in modified waveguide structures (Mori and Baba, 2004; Mori and Baba, 2005). This band is gradually shifted along the waveguide by changing some structural parameters (chirped structure) (Baba et al., 2004). Then, a short optical pulse incident on the device is once dispersed in the device due to the first GVD, and each wavelength component is slowed at different position. Finally, the initial pulse shape is recovered by the second GVD. The bandwidth Δf of slow light is determined by the chirp range, but a wide chirping shortens the overlap length of each wavelength component with the slow light condition and reduces Δt. Thus the constraint of the delay-bandwidth product remains even in chirped structures. Since the internal intensity of this type of slow light is suppressed by the dispersion, it is effective for suppressing nonlinear effects. On the other hand, low-dispersion slow light is generated by the straight photonic band with a small slope (Sakai et al., 2004; Yu Petrov et al., 2004; Settle et al., 2007). In this case, a pulse incident on the device is simply compressed in space and propagates without deformation. Since the internal intensity is enhanced, it is effective for nonlinear enhancement (Hamachi et al., 2009; Corcoran et al., 2009).

7.4.7 Dispersion-Compensated Slow Light and Tunable Delay

An example structure that generates dispersion-compensated slow light is the chirped photonic crystal coupled waveguide (PCCW), as shown in Figure 7.34 (Mori and Baba,

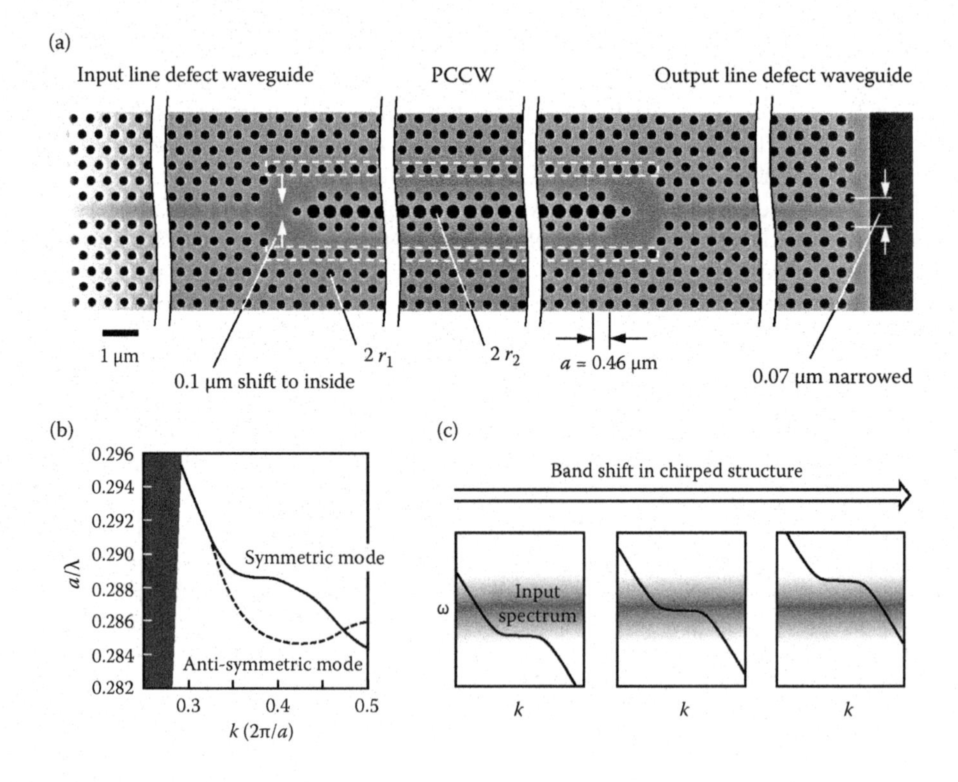

FIGURE 7.34
Chirped PCCW for dispersion-compensated slow light. (a) SEM view of PCCW on SOI substrate. Left and right are simple line defect waveguides for optical input and output. (b) Photonic band for structure (a). (c) Schematic band shift of the symmetric mode in a chirped structure.

2005). It consists of two line defect waveguides separated by three rows of holes whose size and position are partially changed. In the photonic band diagram, two bands appear, which correspond to the symmetric and anti-symmetric modes arising from the directional coupling of two line defects. Here, the symmetric mode is selectively excited by connecting the PCCW with simple line defect waveguides at symmetrical positions for optical input and output. The symmetric mode band exhibits a flat part at the inflection point sandwiched by curves with opposite bending directions, i.e., opposite GVD. The chirped structure is formed by changing the hole diameter and slab index (or equivalent modal index).

As aforementioned, Δf is determined by the chirp range and Δt is constrained by Δf. Therefore, if the chirp range is controlled externally, Δt becomes tunable. The chirp range can be controlled, for example, by nonuniform heating and carrier injection (Baba et al., 2008). Figure 7.35a shows the change of the group delay spectrum for the sloped laser heating to the PCCW without pre-chirping (Adachi et al., 2010). Without heating, Δt shows a peak of ~110 ps (n_g = 118) at λ = 1.556 μm for a device length of 280 μm. In a longer sample, larger values of Δt = 330 ps and n_g = 132 were also observed. This peak arises from the flat band of the symmetric mode in Figure 7.34b. Ideally, the peak becomes infinite at a single wavelength but broadens due to the disordering of fabricated structure. With the sloped heating, the temperature is sloped but overall temperature increases. Therefore, the peak broadens and redshifts with increasing the heating power. At P = 25 mW, Δt is maintained around 37 ps (n_g ~ 40) within a wavelength bandwidth of 12 nm (Δf = 1.5 THz). From these values, $\Delta t \Delta f$ is evaluated to be 54 and Δn ~ 0.3. In another sample of 800 μm length, Δt = 110 ps,

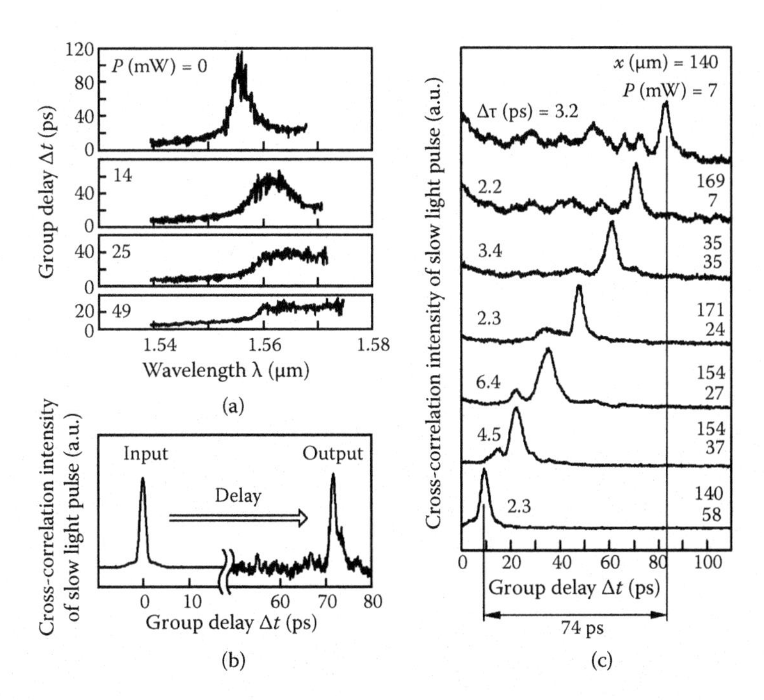

FIGURE 7.35
Dispersion-compensated slow light and its tunable delay. (a) Group delay spectrum measured using modulation phase shift method. P is the heating power. (b) Cross-correlation trace of input and output pulse, both of which are 0.9 ps in width. (c) Pulse delay tuning with changing power P and central position x of the heating.

$\Delta f = 1.0$ THz, and $\Delta t \Delta f = 110$ were recorded. The delay-bandwidth product of other slow light of EIT, high-Q cavities, coupled resonator optical waveguides, all pass filters and so on are no higher than 10 (Xia et al., 2006; Notomi et al., 2008). The much larger value in the PCCW is attributed to the large Δn in the high index contrast structure.

Figure 7.35b demonstrates the transmission of a short optical pulse (Adachi et al., 2010). Here, the incident pulse width $\Delta\tau$ is 0.9 ps, and the output pulse is observed through the cross-correlation. The device is 400 μm long and is locally heated to form an optimum chirping. The output pulse width does not change, indicating the effective dispersion compensation. The delay Δt is 72 ps and the corresponding fixed fractional delay $\Delta t/\Delta\tau$ is 80. Figure 7.35c demonstrates the tunable delay of the slow light pulse. By changing the power P and central position x of the heating, Δt is tuned from 83 to 9 ps, so the tunable range is 74 ps. In this tuning, the average output pulse is 3.3 ps wide, so the tunable fractional delay is 22. When the bandwidth of incident pulse is limited, the tunable range is expanded to maximally 110 ps.

7.4.8 Low-Dispersion Slow Light and Nonlinear Enhancement

Low-dispersion slow light is obtained by modifying the standard line defect waveguide such as narrowing the line defect, enlarging holes of the photonic crystal slab, and shrinking or shifting holes adjacent to the line defect. Fundamentally, they change the mutual position of the guided mode and slab mode bands and deform the guided mode band. The band shape becomes straight against the moderate modification. In particular, the longitudinal shift in the third rows of holes is advantageous because it forms the straight band near the band-edge, which is almost fixed for a wide range of shift s, as shown in Figure 7.36a,b (Hamachi et al., 2009). This simplifies the observation of such slow light at a target frequency. Figure 7.36c shows the transmission and group index spectra. The low-dispersion band is observed on the long wavelength side of the transmission dip at the flat band seen in the band diagram of (b). The group index of the low-dispersion band is nearly constant around 30 except at the edges of the low-dispersion band. The short pulse transmits through the device without severe dispersion, as shown above the spectra.

The nonlinear enhancement is evaluated by amplifying the incident pulse. Figure 7.37a compares the intensity response for two different samples of 470 μm length exhibiting $n_g \sim 8$ and $n_g \sim 30$. An almost linear response is observed for $n_g \sim 8$, while clear saturation appears for $n_g \sim 30$. This saturation is caused by the two-photon absorption (TPA) and accompanied free-carrier absorption (FCA) in the Si layer. Similar saturation is also observed in Si wire waveguides of centimeters length at ~10 W peak power (Yamada et al., 2005). In comparison with this, the saturation is observed with shorter lengths and at a lower power in the photonic crystal waveguide. In this figure, solid lines are calculated for typical material parameters of Si, carrier lifetime of 70 ps, and so on. The effective TPA coefficient β enhanced by the slow light effect is estimated to be 25 cm/GW for $n_g \sim 30$. This value is 42-fold higher than a reported value for the wire waveguide. This enhancement can be explained by 6.7-fold larger n_g of the photonic crystal waveguide; the ratio of n_g^2 nearly equals to the enhancement factor of β. The output pulse spectrum from the sample of $n_g \sim 30$ is also plotted in Figure 7.37b. At a pulse peak power P of 0.1 W, the spectrum exhibited a particular profile caused by the transmission spectrum in the linear regime. When P is higher than 1 W, the spectrum broadens to the short wavelength side. This is the mixed effect of the self-phase modulation (SPM) and the dynamic wavelength shift arising from the carrier plasma effect of TPA carriers. The dip at $\lambda \sim 1554$ nm is caused by the 1.5π phase shift of the SPM.

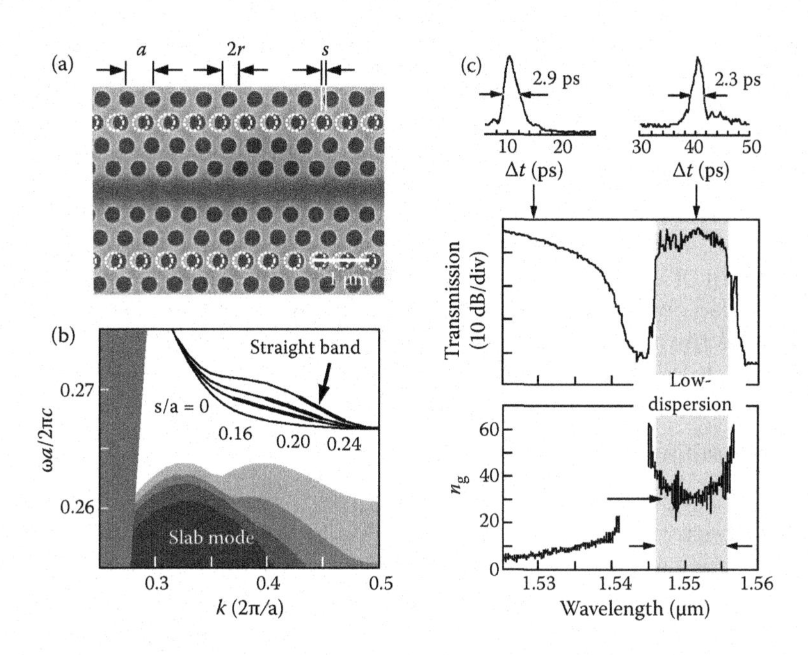

FIGURE 7.36
Hole-shift-type photonic crystal waveguide for low-dispersion slow light. (a) Top view of fabricated device on SOI substrate. (b) Calculated photonic band. (c) Transmission and group index spectra and auto-correlation waveform of short optical pulses with center wavelengths indicated by arrows.

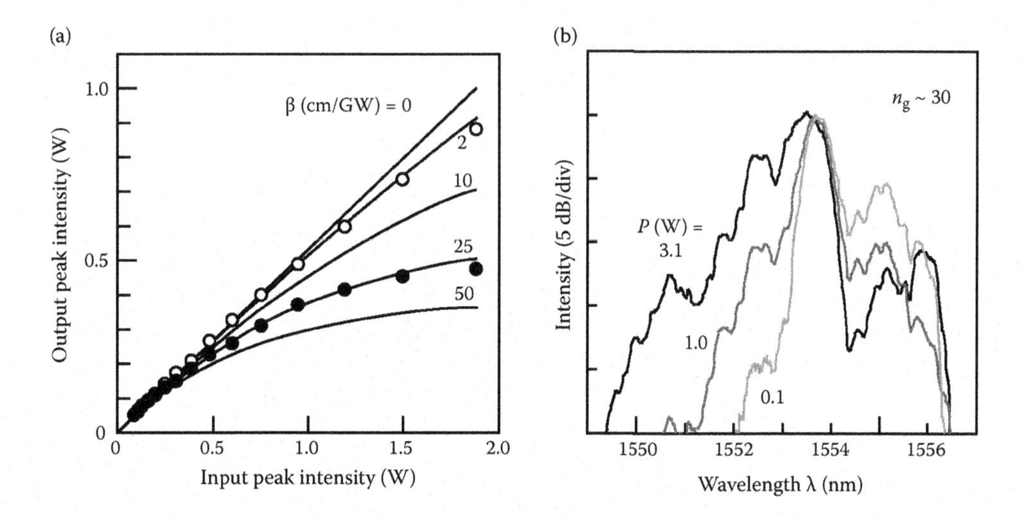

FIGURE 7.37
Observed optical nonlinearity in hole-shift-type photonic crystal waveguide. (a) Nonlinear intensity response. Circles and lines show experimental and theoretical results, respectively. White and black circles depict samples of n_g ~ 8 and 30, respectively. (b) Normalized spectrum of output pulse for different pulse peak power P.

The nonlinearity can be further enhanced by employing highly nonlinear materials. Chalcogenide glasses are promising for this because they have high Kerr nonlinearity and low TPA coefficients at $\lambda \sim 1.55$ μm than semiconductors such as AlGaAs. In addition, they can be formed by simple thermal evaporation at a temperature lower than 400°C, which is compatible with Si CMOS electronics and photonics. The photonic crystal waveguide has been fabricated into Ag-doped As_2Se_3 chalcogenide glass (Suzuki et al., 2009). This material is known to have 20-fold higher nonlinear index n_2. For 400 μm devices with $n_g \sim 5$ and 22, the 1.5π phase shift of the SPM is observed at $P = 0.78$ and 0.42 W, respectively. The four-wave mixing is also observed with a conversion efficiency from signal to idler of −14 dB. From these results, the nonlinear waveguide parameter γ_{eff} reaches 6.3×10^4 W^{-1} m^{-1} (Suzuki et al., 2010), which is 200-fold higher than that of the Si wire waveguide.

7.4.9 Slow Light Modulator

Among the several types of silicon optical modulators, based on Mach–Zehnder interferometers (MZIs), microrings, and electroabsorption type, MZI modulators are considered versatile as they are capable of both amplitude and phase modulation, as well as having a large working spectrum. Although advances in Si MZI modulators have reduced their device size to millimeter-order, further reduction in their size to submillimeter lengths is required for large-scale integration with other optical components. One way to achieve this is to incorporate slow light, as their large group indices result in a larger sensitivity of phase-shift against the material index control. The increased phase-shifter efficiency can be exploited to reduce the device length, lower the operating voltage, or both. In terms of fabrication, it is also desirable to avoid rib-waveguide structures contained in most MZI modulators, as they require vertical partial etching of the Si slab, which can be difficult to control and maintain uniformity across an entire wafer.

The photonic crystal waveguide is such a structure that satisfies both requirements. So far, 10 Gb/s modulation has been demonstrated in a photonic crystal waveguide MZI Si modulator, as shown in Figure 7.38 (Nguyen et al., 2011). It incorporates a photonic crystal waveguide phase shifter of 200 μm length, and is fabricated through the CMOS process. While most modulators operate under either forward-bias, in which speed is limited by carrier diffusion, or reverse-bias, which requires long device lengths, here the preliminary modulation experiments are

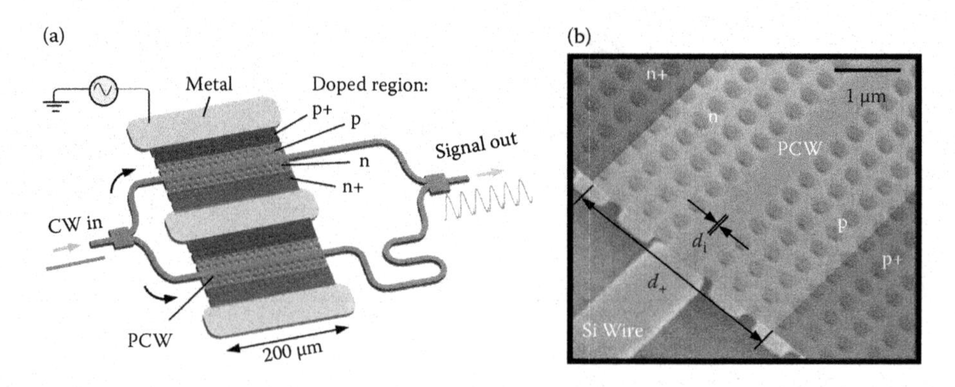

FIGURE 7.38
Photonic crystal waveguide MZI modulator. (a) Schematic of the device. (b) Magnified image of a typical photonic crystal waveguide with SiO_2 cladding removed, overlaid with schematic of p/n doping regions, with $d_i = 0$ in this case.

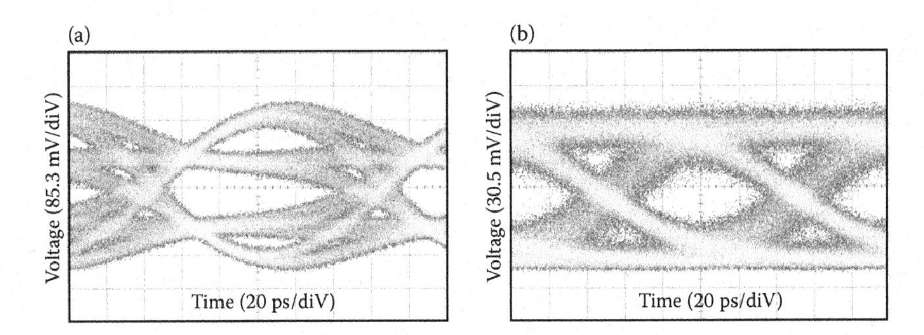

FIGURE 7.39
Eye diagrams for 10-Gb/s modulation with pre-emphasis. (a) Drive signal. (b) Modulated optical signal.

done under 0 V bias. Therefore modulation occurs predominantly by carrier injection. To characterize the modulator performance, the device is electrically driven with non-return-to-zero pseudo-random-bit-sequence signals that is 2^{31}–1 bits in length. Furthermore, the drive signal is pre-emphasized to minimize the effect of slow carrier diffusion. Even though the device is operated under semi-slow light regime ($n_g = 18 - 28$), clearly open eye patterns were observed at bitrates of 10 Gb/s using drive signals with pre-emphasis, as shown in Figure 7.39.

7.4.10 Summary

Photonic crystal waveguides, particularly on a SOI photonic crystal slab, have been well studied theoretically and experimentally. However, their effective targets in photonic integration have changed from the dense optical wiring to more valuable functions, which cannot be obtained in Si wire waveguides, such as slow light. Now it is easy to observe the light propagation including band-edge slow light in this waveguide. However, its detailed structure must be designed carefully for useful slow light, taking account of the delay-bandwidth product and suppression of the higher order dispersion. The photonic crystal coupled waveguide balances these requirements and achieves a tunable fractional delay of 22 in dispersion-compensated slow light. The hole-shift-type waveguide exhibits low-dispersion slow light and its nonlinear enhancement. Slow light is also effective for downsizing Mach-Zehnder interferometer modulators, and 10 Gb/s operation is obtainable in a device of as short as 200 μm. Further investigating photonic crystal waveguides and slow light will offer more applications and advanced performance in photonic devices and integrated circuits.

References

Adachi, J., Ishikura, N., Sasaki, H., and Baba, T., 2010. Wide range tuning of slow light pulse in SOI photonic crystal coupled waveguide via folded chirping. *IEEE J. Sel. Top. Quantum Electron.* 16, 192–199.

Almeida, V. R., Barrios, C. A., Panepucci, R. R., and Lipson, M., 2004. All-optical control of light on a silicon chip. *Nature*, 431, 1081–1084.

Almeida, V. R., and Lipson, M., 2004. Optical bistability on a silicon chip. *Optics Letters* 29, 2387–2389.

Andrekson, P. A., Sunnerud, H., Oda, S., Nishitani, T., and Yang, J., 2008. Ultrafast, atto-Joule switch using fiber-optic parametric amplifier operated in saturation. *Optics Express*, 16, 10956–10961.

Baba, T., 2008. Slow light in photonic crystals. *Nat. Photon.* 2, 465–473.
Baba, T., Fukaya, N., and Yonekura, J., 1999. Observation of light propagation in photonic crystal optical waveguides with bends. *Electron. Lett.* 35, 654–655.
Baba, T., Kawasaki, T., Sasaki, H., Adachi, J., and Mori, D., 2008. Large delay-bandwidth product and tuning of slow light pulse in photonic crystal coupled waveguide. *Opt. Express*, 16, 9245–9253.
Baba, T., Mori, D., Inoshita, K., and Kuroki, Y., 2004. Light localizations in photonic crystal line defect waveguides. *IEEE J. Sel. Top. Quantum Electron.* 10, 484–491.
Baba, T., Motegi, A., Iwai, T., Fukaya, N., Watanabe, Y., and Sakai, A., 2002. Light propagation characteristics of straight single-line-defect waveguides in photonic crystal slabs fabricated into a silicon-on-insulator substrate. *IEEE J Quantum Electron.* 38, 743–752.
Bagheri, M., Shih, M. H., Wei, Z.-J. et al., 2006. Linewidth and modulation response of two-dimensional microcavity photonic crystal lattice defect lasers. *IEEE Photon. Technol. Lett.* 18, 1161–1163.
Barclay, P. E., Srinivasan, K., and Painter, O., 2005. Nonlinear response of silicon photonic crystal microresonators excited via an integrated waveguide and fiber taper. *Optics Express*, 13, 801–820.
Bennett, B. R., Soref, R. A., and Delalamo, J. A., 1990. Carrier-induced change in refractive-index of InP, GaAs, and InGaAsP. *IEEE Journal of Quantum Electronics*, 26, 113–122.
Bogaerts, W., Baets, R., Dumon, P. et al., 2005. Nanophotonic waveguides in silicon-on-insulator fabricated with CMOS technology. *J. Lightwave Technol.* 23, 401–412.
Chang, Y.-C., and Coldren, L. A., 2009. Efficient, high-data-rate, tapered oxide-aperture vertical-cavity surface-emitting lasers. *IEEE J. Select. Top. Quant. Electron.* 15, 1–12.
Chutinan, A., and Noda, S., 2000a. Waveguides and waveguide bends in two-dimensional photonic crystal slabs. *Phys. Rev. B.* 62, 4488–4492.
Chutinan, A., and Noda, S., 2000b. Design for waveguides in three-dimensional photonic crystals. *Jpn. J. Appl. Phys.* 39, 2353–2356.
Cong, G. W., Akimoto, R., Akita, K., Hasama, T., and Ishikawa, H., 2007. Low-saturation-energy-driven ultrafast all-optical switching operation in (CdS/ZnSe)/BeTe intersubband transition. *Optics Express*, 15, 12123–12130.
Corcoran, B., Monat, C., Grillet, C., Moss, D. J., and Eggleton, B. J., 2009. Green light emission in silicon through slow-light enhanced third-harmonic generation in photonic-crystal waveguides. *Nat. Photon.* 3, 206–210.
Dinu, M., Quochi, F., and Garcia, H., 2003. Third-order nonlinearities in silicon at telecom wavelengths. *Applied Physics Letters*, 82, 2954–2956.
Fang, A. W., Park, H., Cohen, O. et al., 2006. Electrically pumped hybrid AlGaInAs-silicon evanescent laser. *Optics Express*, 14, 9203–9210.
Gibbs, H. M., 1985. *Optical Bistability: Controlling Light With Light*. Orlando, FL: Academic Press.
Gomyo, A., Ushida, J., Shirane, M., Tokushima, M., and Yamada, H., 2004. Low Optical Loss Connection for photonic crystal slab waveguides. *IEICE Trans. Electron.* E87, 328–335.
Gopal, A. V., Yoshida, H., Neogi, A. et al., 2002. Intersubband absorption saturation in InGaAs-AlAsSb quantum wells. *IEEE Journal of Quantum Electronics*, 38, 1515–1520.
Hamachi, Y., Kubo, S., and Baba, T., 2009. Slow light with low dispersion and nonlinear enhancement in a lattice-shifted photonic crystal waveguide. *Opt. Lett.* 34, 1072–1074.
Harris, S. E., 1997. Electromagnetically induced transparency. *Physics Today*, 50, 36–42.
Husko, C., De Rossi, A., Combrie, S. et al., 2009. Ultrafast all-optical modulation in GaAs photonic crystal cavities. *Applied Physics Letters*, 94, 021111.
Joannopoulos, J. D., Johnson, S. G., Winn, J. N., and Meade, R. D., 2008. *Photonic Crystal: Molding the Flow of Light 2nd Ed.*, Princeton University Press.
John, S., 1987. Strong Localization of photons in certain disordered dielectric superlattices. *Physical Review Letters*, 58, 2486–2489.
Koos, C., Vorreau, P., Vallaitis, T. et al., 2009. All-optical high-speed signal processing with silicon-organic hybrid slot waveguidesx. *Nature Photonics*, 3, 216–219.
Kosaka, H., Kawashima, T., Tomita, A., Notomi, M., Tamamura, T., Sato, T., and Kawakami, S., 1999. Superprism phenomena in photonic crystals: Toward microscale lightwave circuits. *Journal of Lightwave Technology*, 17, 2032–2038.

Kuramochi, E., Notomi, M., Hughes, S., Shinya, A., Watanabe, T., and Ramunno, L., 2005. Disorder-induced scattering loss of line-defect waveguides in photonic crystal slabs. *Phys. Rev. B*, 72, 161318(1–4).
Kuramochi, E., Notomi, M., Mitsugi, S. et al., 2006. Ultrahigh-Q photonic crystal nanocavities realized by the local width modulation of a line defect. *Applied Physics Letters*, 88, 041112.
Liu, L., Kumar, R., Huybrechts, K. et al., 2010. An ultra-small, low-power, all-optical flip-flop memory on a silicon chip. *Nature Photonics*, 4, 182–187.
Loncâr, M., Yoshie, T., Scherer, A., Gogna, P., and Qiu, Y., 2002. Low-threshold photonic crystal laser. *Appl. Phys. Lett.* 81, 2680–2682.
Matsuo, S., Shinya, A., Chen, C.-H. et al., 2011. 20-Gbit/s directly modulated photonic crystal nanocavity laser with ultra-low power consumption. *Opt. Express*, 19, 2242–2250.
Matsuo, S., Shinya, A., Kakitsuka, T. et al., 2009. Ultra-small InGaAsP/InP buried heterostructure photonic crystal laser. *LEOS, 2009 Annual Meeting*, WH3, 453–454.
Matsuo, S., Shinya, A., Kakitsuka, T., Nozaki, K., Segawa, T., Sato, T., Kawaguchi, Y., and Notomi, M., 2010. High-speed ultracompact buried heterostructure photonic-crystal laser with 13 fJ of energy consumed per bit transmitted. *Nature Photonics*, 4, 648–654.
McNab, S. J., Moll, N., and Vlasov, Y. A., 2003. Ultra-low loss photonic integrated circuit with membrane-type photonic crystal waveguides. *Opt. Express*, 11, 2927–2939.
Meade, R. D., Devenyi, A., Joannopoulos, J. D., Alerhand, O. L., Smith, D. A., and Kash, K., 1994. Novel applications of photonic band gap materials: low-loss bends and high Q cavities. *J. Appl. Phys.* 75, 4753–4755.
Mekis, A., Chen, J. C., Kurland, I., Fan, S., Villeneuve, P. R., and Joannopoulos, J. D., 1996. High transmission through sharp bends in photonic crystal waveguides. *Phys. Rev. Lett.* 77, 3787–3790.
Miller, D. A. B., 2000. Rationale and challenges for optical interconnects to electronic chips. *Proceedings of the IEEE*, 88, 728–749.
Miller, D. A. B., 2009. Device Requirements for optical interconnects to silicon chips. *Proceedings of the IEEE*, 97, 1166–1185.
Mori, D., and Baba, T., 2004. Dispersion-controlled optical group delay device by chirped photonic crystal waveguides. *Appl. Phys. Lett.* 85, 1101–1103.
Mori, D., and Baba, T., 2005. Wideband and low dispersion slow light by chirped photonic crystal coupled waveguide. *Opt. Express*, 13, 9398–9408.
Mori, D., Kubo, S., Sasaki, H., and Baba, T., 2007. Experimental demonstration of wideband dispersion-compensated slow light by a chirped photonic crystal directional coupler. *Opt. Express*, 15, 5264–5270.
Mori, T., Yamayoshi, Y., and Kawaguchi, H., 2006. Low-switching-energy and high-repetition-frequency all-optical flip-flop operations of a polarization bistable vertical-cavity surface-emitting laser. *Applied Physics Letters*, 88, 101102.
Mork, J., Mark, J., and Seltzer, C. P., 1994. Carrier heating in InGaAsP laser-amplifiers due to 2-photon absorption. *Applied Physics Letters*, 64, 2206–2208.
Nakamura, S., Ueno, Y., and Tajima, K., 2001. Femtosecond switching with semiconductor-optical-amplifier-based symmetric Mach–Zehnder-type all-optical switch. *Applied Physics Letters*, 78, 3929–3931.
Nguyen, H. C., Sakai, Y., Shinkawa, M., Ishikura, N., and Baba, T., 2011. 10 Gb/s operation of photonic crystal silicon optical modulators. *Opt. Express*, 19, 13000–13007.
Nielsen, M. L., Mork, J., Suzuki, R., Sakaguchi, J., and Ueno, Y., 2006. Experimental and theoretical investigation of the impact of ultra-fast carrier dynamics on high-speed SOA-based all-optical switches. *Optics Express*, 14, 331–347.
Nomura, M., Iwamoto, S., Watanabe, K. et al., 2006. Room temperature continuous-wave lasing in photonic crystal nanocavity. *Opt. Express*, 14, 6308–6315.
Notomi, M., 2000. Theory of light propagation in strongly modulated photonic crystals: Refraction-like behavior in the vicinity of the photonic band gap. *Physical Review B*, 62, 10696–10705.
Notomi, M., 2010. Manipulating light with strongly modulated photonic crystals. *Reports on Progress in Physics*, 73, 096501.

Notomi, M., Kuramochi, E., and Tanabe, T., 2008. Large-scale arrays of ultrahigh-Q coupled nanocavities. *Nature Photonics*, 2, 741–747.
Notomi, M., Shinya, A, Mitsugi, S., Kira, G., Kuramochi, E., and Tanabe, T., 2005. Optical bistable switching action of Si high-Q photonic-crystal nanocavities. *Opt. Express*, 13, 2678–2687.
Notomi, M., Shinya, A., Mitsugi, S., Kuramochi, E., and Ryu, H.-Y., 2004. Waveguides, resonators and their coupled elements in photonic crystal slabs. *Opt. Express*, 12, 1551–1561.
Notomi, M., Shinya, A., Nozaki, K., et al., 2011. Low-power nanophotonic devices based on photonic crystals towards dense photonic network on chip. *IET Circuits Devices & Systems*, 5, 84–93.
Notomi, M., Tanabe, T., Shinya, A., Kuramochi, E., Taniyama, H., Mitsugi, S., and Morita, M., 2007. Nonlinear and adiabatic control of high-Q photonic crystal nanocavities. *Optics Express*, 15, 17458–17481.
Notomi, M., Yamada, K., Shinya, A., Takahashi, J., Takahashi, C., and Yokohama, I., 2001. Extremely large group-velocity dispersion of line-defect waveguides in photonic crystal slabs. *Phys. Rev. Lett.* 87, 253902(1–4).
Nozaki, K., Kita, S., and Baba, T., 2007. Room temperature continuous wave operation and controlled spontaneous emission in ultrasmall photonic crystal nanolaser. *Opt. Express*, 15, 7506–7514.
Nozaki, K., Tanabe, T., Shinya, A., Matsuo, S., Sato, T., Taniyama, H., and Notomi, M., 2010. Sub-femtojoule all-optical switching using a photonic-crystal nanocavity. *Nature Photonics*, 4, 477–483.
Ohtera, Y., Kawashima, T., Sakai, Y. et al., 2002. Photonic crystal waveguides utilizing a modulated lattice structure. *Opt. Lett.* 27, 2158–2160.
Olbright, G. R., Peyghambarian, N., Gibbs, H. M., Macleod, H. A., and Vanmilligen, F., 1984. Microsecond room-temperature optical bistability and crosstalk studies in ZnS and ZnSe interference filters with visible-light and milliwatt powers. *Applied Physics Letters*, 45, 1031–1033.
Painter, O., Lee, R.-K., Scherer, A. et al., 1999. Two dimensional photonic band-gap defect mode laser. *Science*, 284, 1819–1821.
Park, H.-G., Kim, S.-H., Kwon, S.-H. et al., 2004. Electrically driven single-cell photonic crystal laser. *Science*, 305, 1444–1447.
Roelkens, G., Van Thourhout, D., Baets, R., Notzel, R., and Smit, M., 2006. Laser emission and photodetection in an InP/InGaAsP layer integrated on and coupled to a Silicon-on-Insulator waveguide circuit. *Optics Express*, 14, 8154–8159.
Ryu, H. Y., Notomi, M., Kuramochi, E., and Segawa, T., 2004. Large spontaneous emission factor (>0.1) in the photonic crystal monopole-mode laser. *Appl. Phys. Lett.*, 84, 1067–1069.
Sakai, A., Hara, G., and Baba, T., 2001. Propagation characteristics of ultrahigh-Δ optical waveguide on silicon-on-insulator substrate. *Jpn. J. Appl. Phys.* 40, L383–385.
Sakai, A., Katoh, I., Mori, D., Baba, T., and Takiguchi, Y., 2004. Anomalous low group velocity and low dispersion in simple line defect photonic crystal waveguides. *Tech. Dig. IEEE/LEOS Annual Meet.* 2, 884–885.
Settle, M. D., Engelen, R. J. P., Salib, M., Michaeli, A., Kuipers, L., and Krauss, T. F., 2007. Flatband slow light in photonic crystals featuring spatial pulse compression and terahertz bandwidth. *Opt. Express*, 15, 219–226.
Shacham, A., Bergman, K., and Carloni, L. P., 2008. Photonic networks-on-chip for future generations of chip multiprocessors. *IEEE Transactions on Computers*, 57, 1246–1260.
Shih, M. H., Bagheri, M., Mock, A. et al., 2007. Identification of modes and single mode operation of sapphire bonded photonic crystal lasers under continuous-wave room temperature operation. *Appl. Phys. Lett.*, 90, 121116.
Shinya, A., Matsuo, S., Yosia et al., 2008. All-optical on-chip bit memory based on ultra high Q InGaAsP photonic crystal. *Optics Express*, 16, 19382–19387.
Shinya, A., Mitsugi, S., Tanabe, T. et al., 2006. All-optical flip-flop circuit composed of coupled two-port resonant tunneling filter in two-dimensional photonic crystal slab. *Optics Express*, 14, 1230–1235.
Soljacic, M., and Joannopoulos, J. D., 2004. Enhancement of nonlinear effects using photonic crystals. *Nature Materials*, 3, 211–219.

Sugimoto, Y., Tanaka, Y., Ikeda, N., Nakamura, Y., Asakawa, K., and Inoue, K., 2004. Low propagation loss of 0.76 dB/mm in GaAs-based single-line-defect two-dimensional photonic crystal slab waveguides up to 1 cm in length. *Opt. Express*, 12, 1090–1096.
Suzuki, K., and Baba, T., 2010. Nonlinear light propagation in chalcogenide photonic crystal slow light waveguides. *Opt. Express*, 18, 26675–26685.
Suzuki, K., Hamachi, Y., and Baba, T., 2009. Fabrication and characterization of chalcogenide glass photonic crystal waveguides. *Opt. Express*, 17, 22393–22400.
Szymanski, D. M., Jones, B. D., Skolnick, M. S. et al., 2009. Ultrafast all-optical switching in AlGaAs photonic crystal waveguide interferometers. *Applied Physics Letters*, 95, 141108.
Tadokoro, T., Yamanaka, T, Kano, F., Oohashi, H., Kondo, Y., and Kishi, K., 2009. Operation of a 25-Gbps direct modulation ridge waveguide MQW-DFB Laser up to 85°C. *IEEE Photon. Technol. Lett.*, 21, 1154–1156.
Takahashi, Y., Hagino, H., Tanaka, Y., Song, B.-S., Asano, T., and Noda, S., 2007. High-Q nanocavity with a 2-ns photon lifetime. *Optics Express*, 15, 17206–17213.
Takeda, K, Sato, T., Shinya, A. et al., 2011. 80°C Continuous wave operation of photonic-crystal nanocavity lasers. *23rd International Conference on Indium Phosphide and Related Materials (IPRM2011)*, Tu-3.2.2.
Takeda, K., Kanema, Y., Takenaka, M., Tanemura, T., and Nakano, Y., 2008. Polarization-insensitive all-optical flip-flop using tensile-strained multiple quantum wells. *IEEE Photonics Technology Letters*, 20, 1851–1853.
Tanabe, T., Nishiguchi, K., Kuramochi, E., and Notomi, M., 2009. Low power and fast electro-optic silicon modulator with lateral p-i-n embedded photonic crystal nanocavity. *Optics Express*, 17, 22505–22513.
Tanabe, T., Notomi, M., Kuramochi, E., Shinya, A., and Taniyama, H., 2007. Trapping and delaying photons for one nanosecond in an ultrasmall high-Q photonic-crystal nanocavity. *Nat. Photon.* 1, 49–52.
Tanabe, T., Notomi, M., Mitsugi, S., Shinya, A., and Kuramochi, E., 2005a. All-optical switches on a silicon chip realized using photonic crystal nanocavities. *Applied Physics Letters*, 87, 151112.
Tanabe, T., Notomi, M., Mitsugi, S., Shinya, A., and Kuramochi, E., 2005b. Fast bistable all-optical switch and memory on a silicon photonic crystal on-chip. *Optics Letters*, 30, 2575–2577.
Tanabe, T., Sumikura, H., Taniyama, H., Shinya, A., and Notomi, M., 2010. All-silicon sub-Gb/s telecom detector with low dark current and high quantum efficiency on chip. *Applied Physics Letters*, 96, 101103.
Tanabe, T., Taniyama, H., and Notomi, M., 2008. Carrier diffusion and recombination in photonic crystal nanocavity optical switches. *Journal of Lightwave Technology*, 26, 1396–1403.
Tanaka, Y., Upham, J., Nagashima, T. et al., 2007. Dynamic control of the Q factor in a photonic crystal nanocavity. *Nature Materials*, 6, 862–865.
Thévenaz, L., 2008. Slow and fast light in optical fibers. *Nat. Photon.* 2, 474–481.
Tokushima, M., Yamada, H., and Arakawa, Y., 2004. 1.5-μm-wavelength light guiding in waveguides in square-lattice-of-rod photonic crystal slab. *Appl. Phys. Lett.* 84, 4298–4300.
Tsang, H. K., Penty, R. V., White, I. H. et al., 1991. 2-photon absorption and self-phase modulation in InGaAsP/InP multi-quantum-well wave-guides. *Journal of Applied Physics*, 70, 3992–3994.
Vahala, K. J., 2003. Optical microcavities. *Nature*, 424, 839–846.
Van, V., Ibrahim, T. A., Ritter, K. et al., 2002. All-optical nonlinear switching in GaAs-AlGaAs microring resonators. *IEEE Photonics Technology Letters*, 14, 74–76.
Vangal, S., 2007. An 80-Tile 1.28 TFLOPS Network-on-chip in 65 nm CMOS. *International Solid State Circuits Conference.*
Vlasov, Y. A., O'Boyle, M., Hamann, H. F., and McNab, S. J., 2005. Active control of slow light on a chip with photonic crystal waveguides. *Nature* 438, 65–69.
Xia, F., Sekaric, L., and Vlasov, Y., 2006. Ultracompact optical buffers on a silicon chip. *Nat. Photon.* 1, 65–71.
Xia, F. N., Sekaric, L., and Vlasov, Y., 2007. Ultracompact optical buffers on a silicon chip. *Nature Photonics*, 1, 65–71.

Xu, Q. F., and Lipson, M., 2006. Carrier-induced optical bistability in silicon ring resonators. *Optics Letters*, 31, 341–343.

Xu, Q. F., Dong, P., and Lipson, M., 2007. Breaking the delay-bandwidth limit in a photonic structure. *Nature Physics*, 3, 406–410.

Yablonovitch, E., 1987. Inhibited spontaneous emission in solid-state physics and electronics. *Physical Review Letters*, 58, 2059–2062.

Yamada, H., Shirane, M., Chu, T., Yokoyama, H., Ishida, S., and Arakawa, Y., 2005. Nonlinear-optic silicon-nanowire waveguides. *Jpn. J. Appl. Phys.* 44, 6541–6545.

Yamada, K., Notomi, M., Shinya, A., Takahashi, C., Takahashi, J., and Morita, H., 2002. Singlemode lightwave transmission in SOI-type photonic-crystal line-defect waveguides with phase-shifted holes. *Electron. Lett.* 38, 74–75.

Yamada, K., Tsuchizawa, T., Watanabe, T. et al., 2004. Microphotonics devices based on silicon wire waveguiding system. *IEICE Trans. Electron.* E87, 351–358.

Yamamoto, T., Yoshida, E., and Nakazawa, M., 1998. Ultrafast nonlinear optical loop mirror for demultiplexing 640Gbit/s TDM signals. *Electronics Letters*, 34, 1013–1014.

Yonekura, J., Ikeda, M., and Baba, T., 1999. Analysis of finite 2-D photonic crystals of columns and lightwave devices using the scattering matrix method. *J. Lightwave Technol.* 17, 1500–1508.

Yu Petrov, A., and Eich, M., 2004, Zero dispersion at small group velocities in photonic crystal waveguides. *Appl. Phys. Lett.* 85, 4866–4868.

Zhang, Z. Y., and Qiu, M., 2004. Small-volume waveguide-section high Q microcavities in 2D photonic crystal slabs. *Optics Express*, 12, 3988–3995.

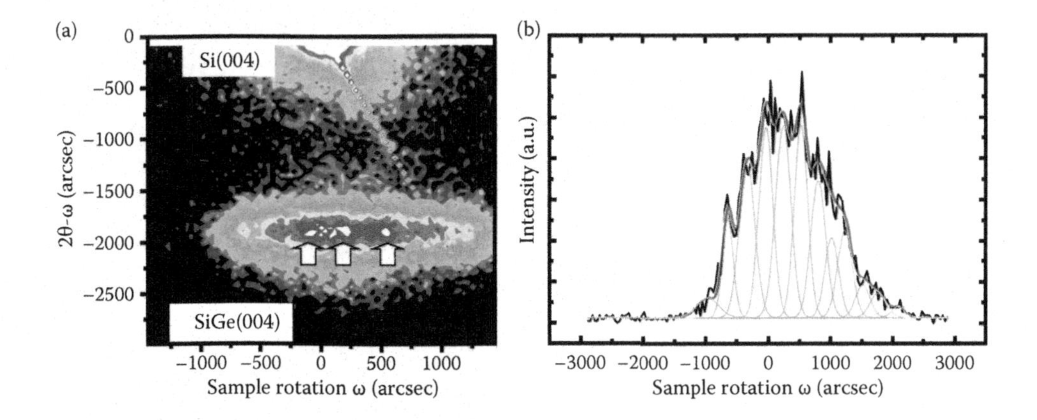

FIGURE 1.25
(a) Two-dimensional reciprocal-space maps around the Si 004 and SiGe 004 Bragg reflections for the strain relaxed $Si_{1-x}Ge_x$/Si(001) sample prepared with two-step strain relaxation method [49]. (b) Cross-sectional profile at 2θ of −1890 arcsec of the SiGe 004 Bragg reflection of the sample shown in (a).

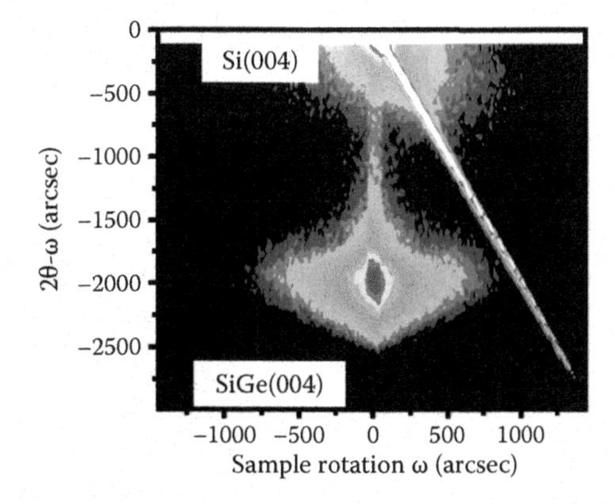

FIGURE 1.26
Two-dimensional reciprocal-space maps around the Si 004 and SiGe 004 Bragg reflections for the strain relaxed $Si_{1-x}Ge_x$/Si(001) sample prepared with high temperature solid-phase mixing method [49].

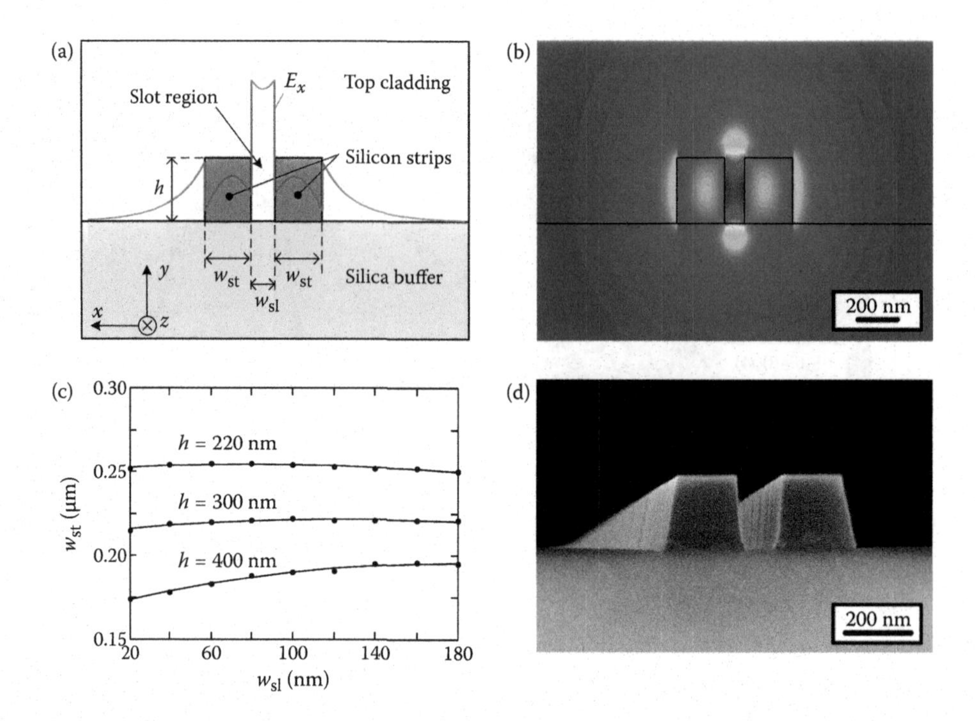

FIGURE 2.19
(a) Schematic cross section of a silicon-on-insulator (SOI) slot waveguide (height h, width w_{st}, refractive index $n_{Si} = 3.48$ at $\lambda = 1550$ nm). (b) Dominant electric field component (E_x) of the fundamental TE mode field for $n_{cl} = 1.5$, $w_{st} = 200$ nm, $h = 220$ nm, and $w_{sl} = 120$ nm. (c) Maximum strip width w_{st} for a single-mode waveguide as a function of slot width w_{sl} and waveguide height h for $n_{cl} = 1.5$ and $\lambda = 1550$ nm. (d) Scanning electron microscope (SEM) view of a cleaved slot waveguide front facet. The waveguide was fabricated with 193-nm-deep UV lithography and dry etching.

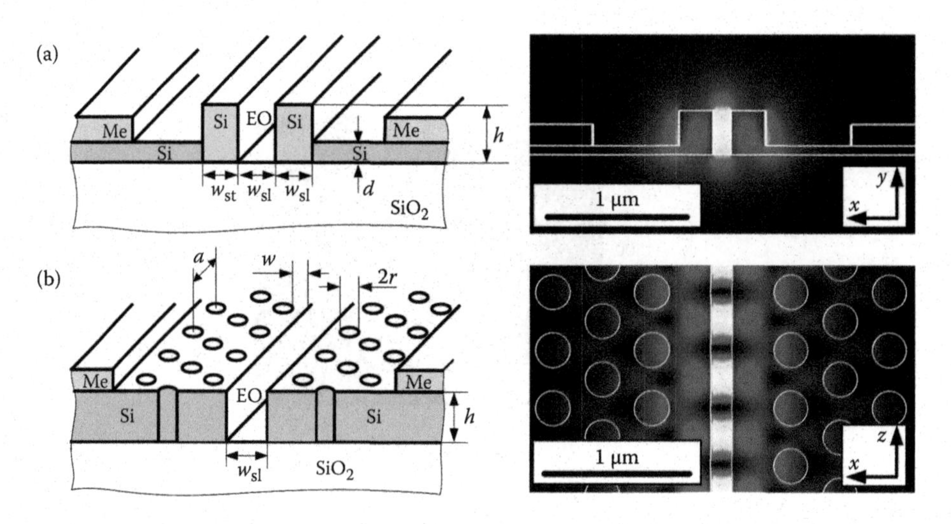

FIGURE 2.21
Silicon-organic hybrid (SOH) electro-optic modulator concepts: The electric field distributions of the fundamental TE modes are depicted on the right-hand side. (a) Traveling-wave configuration. (b) Photonic crystal (PhC) slot waveguide structure.

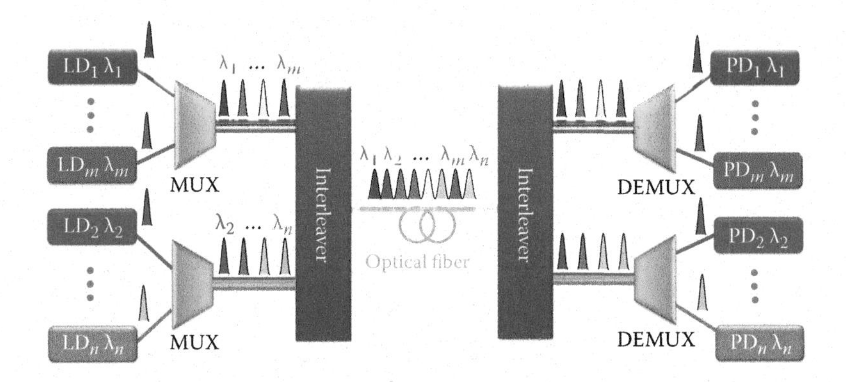

FIGURE 4.1
Schematic of a WDM system including the key components. LD: laser diode, MUX: multiplexer, DEMUX: demultiplexer, PD: photodiode. $\lambda_1 \ldots \lambda_m$: odd wavelength channels, $\lambda_2 \ldots \lambda_n$: even wavelength channels.

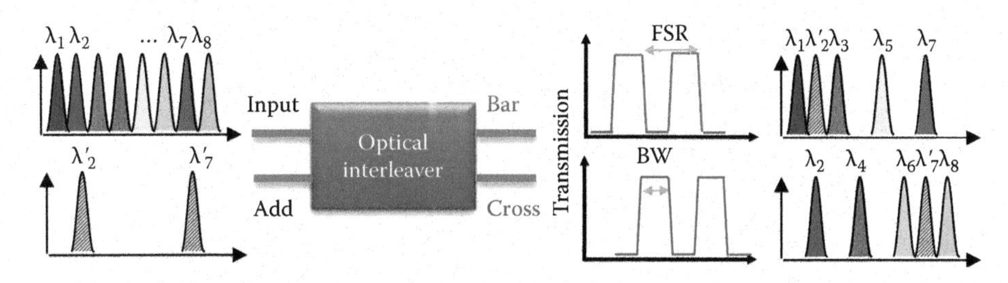

FIGURE 4.2
Working principle of an optical interleaver.

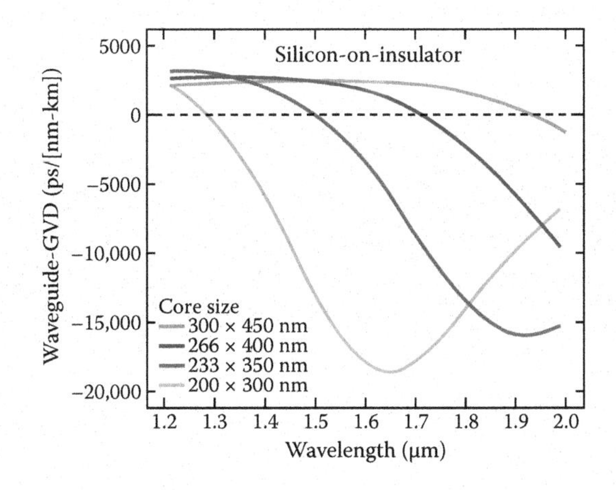

FIGURE 5.7
Waveguide contribution to the group-velocity dispersion (GVD) in SOI platform of various dimensions. (From M. A. Foster, A. C. Turner, M. Lipson, A. L. Gaeta, *Opt. Express* 16, 2, 1300, 2008. With permission.)

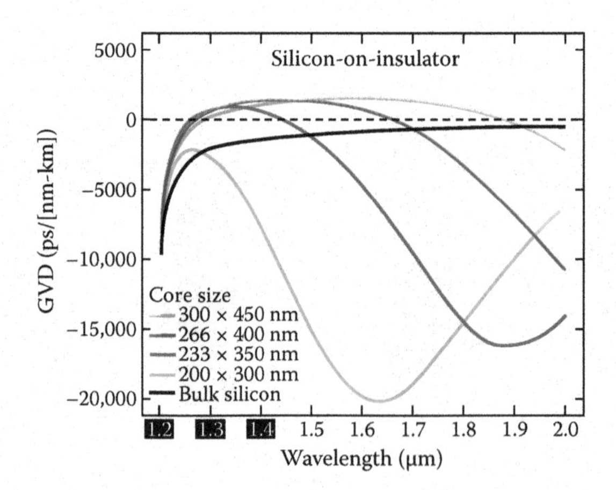

FIGURE 5.8
Net group-velocity dispersion in SOI platform of various dimensions. (From M. A. Foster, A. C. Turner, M. Lipson, A. L. Gaeta, *Opt. Express* 16, 2, 1300, 2008. With permission.)

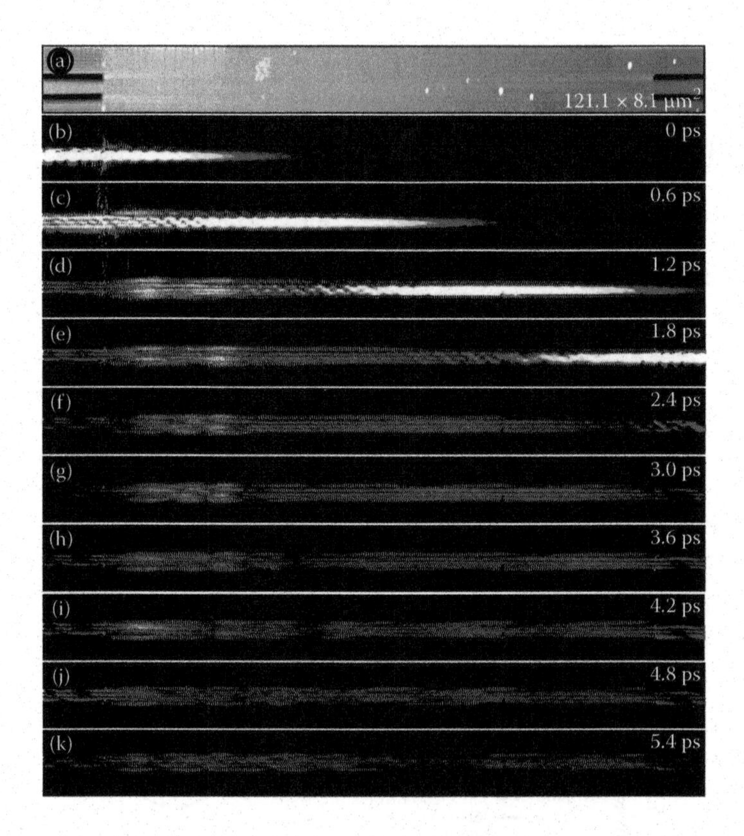

FIGURE 5.49
Pulse tracking experiment at a flat band at $\omega = 0.305$ ($\lambda = 1310$ nm, $a = 400$ nm). (a) Topography of the PhCW. (b–k) The optical amplitude as a function of reference time (all frames have the same color scale). A movie displaying the pulse propagation is available. We observe a complex and stationary localized modal pattern in the first 25 μm of the PhCW, which exists for more than 3.6 ps after the excitation pulse has moved away. The field pattern moves by less than 0.9 μm in 3 ps, suggesting a group velocity of less than $c = 1000$. Note that all light observed in (f–k) belongs to the same mode located between $k = 0.4$ and $k = 0.6$. (Reprinted with permission from H. Gersen, T. J. Karle, R. J. P. Engelen, W. Bogaerts, J. P. Korterik, N. F. van Hulst, T. F. Krauss, and L. Kuipers, *Phys. Rev. Lett.* 94, 073903, 2005. Copyright 2005 by the American Physical Society.)

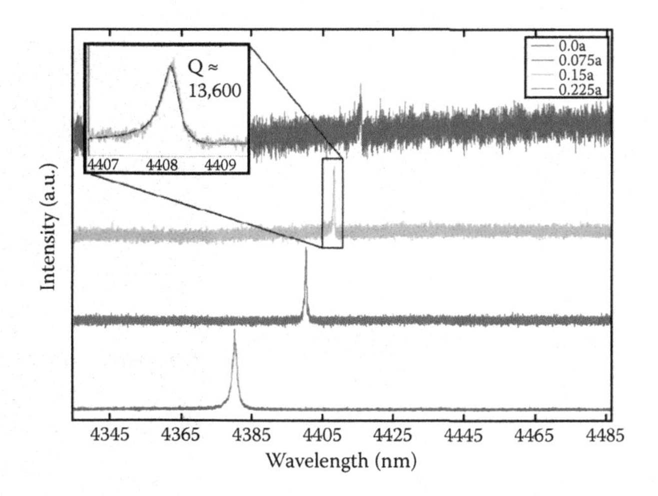

FIGURE 6.5
Resonant scattering spectra of photonic crystal cavities with the air hole shift of: $s = 0$, $s = 0.075a$, $s = 0.15a$, $s = 0.225a$. Scale is linear. (From R. Shankar, R. Leijssen, I. Bulu, and M. Lončar, *Opt. Express* vol. 19, 5579–5586, 2011. With permission.)

FIGURE 6.6
A false-color scanning electron micrograph of the cleaved endfacet of a waveguide. Silicon is shown in green, and sapphire in blue. (From T. Baehr-Jones, A. Spott, R. Ilic, A. Spott, B. Penkov, W. Asher, and M. Hochberg, *Opt. Express* vol. 18, 12127–12135, 2010. With permission.)

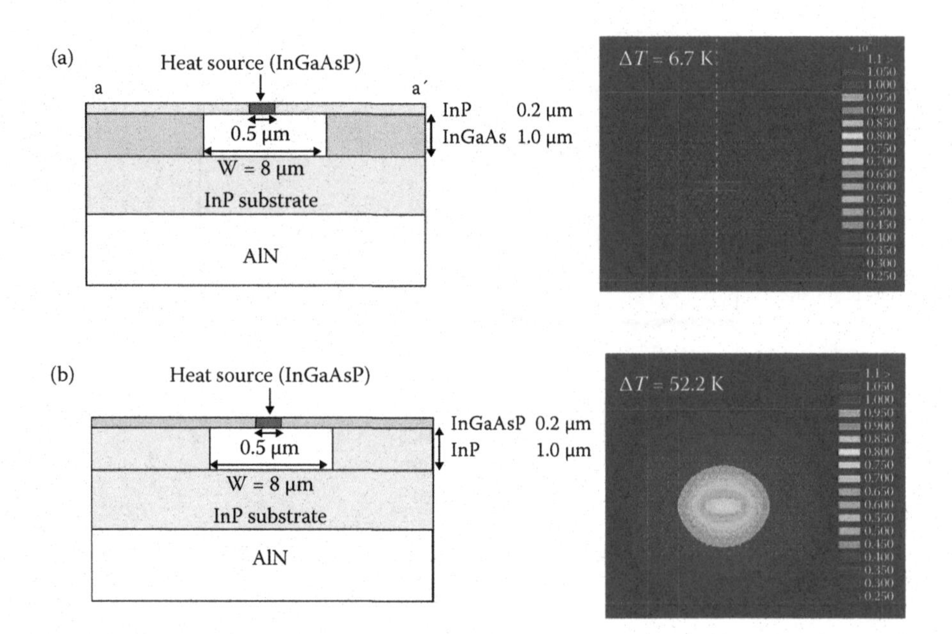

FIGURE 7.17
Numerical simulations of active region temperature for (a) BH-PhC structure and (b) all-active PhC laser structure.

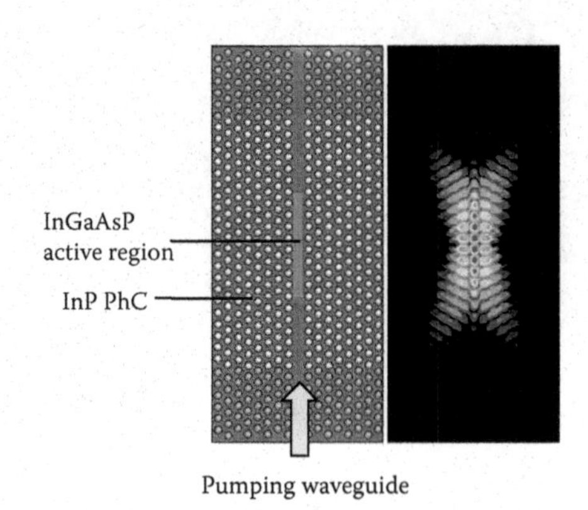

FIGURE 7.18
Field profile of the cavity mode in BH-PhC laser using FDTD method.

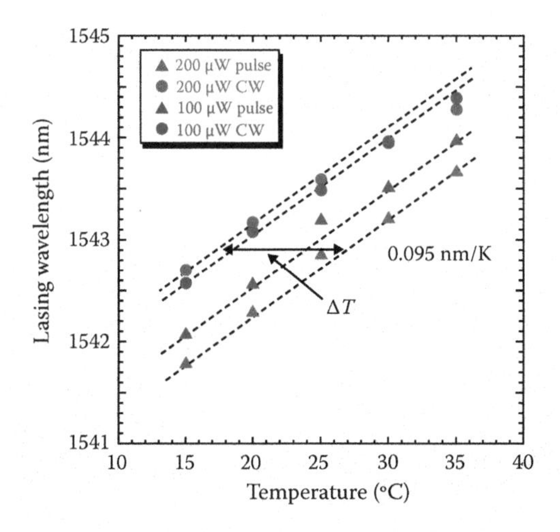

FIGURE 7.21
Temperature dependence of the lasing wavelength for both pulsed and CW operation. The input powers in input waveguide were 100 and 200 μW, respectively.

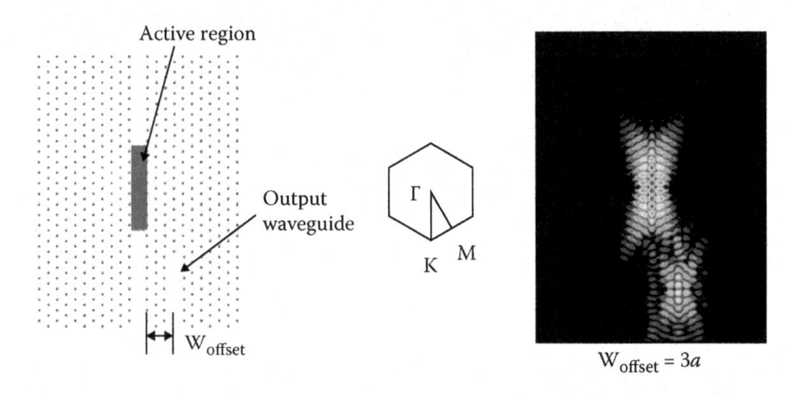

FIGURE 7.22
FDTD calculated mode profile of PhC laser with output waveguide. Output waveguide is placed in Γ-M direction.

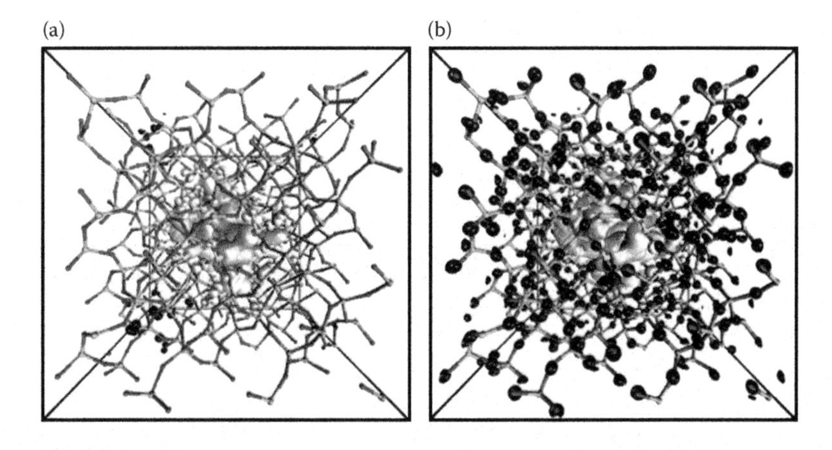

FIGURE 8.9
Stick and ball pictures of the final optimized structure of Si_{32} NC in a crystalline SiO_2 matrix. Red spheres represent the O atoms, cyan the Si atoms of the matrix, and yellow the Si atoms of the NC. On the structures, Kohn-Sham orbitals at 10% of their maximum amplitude are reported. The ones in white are the sum of all band-edges orbitals localized on the NC, while the ones in black are the states delocalized on the matrix: (a) valence states and (b) conduction states. The results show that the HOMO and the LUMO are localized on the Si NC. (After Guerra, R., E. Degoli, M. Marsili, O. Pulci, and S. Ossicini. 2010. *Physica Status Solidi B Basic Solid State Physics* 247 (8):2113–2117. With permission.)

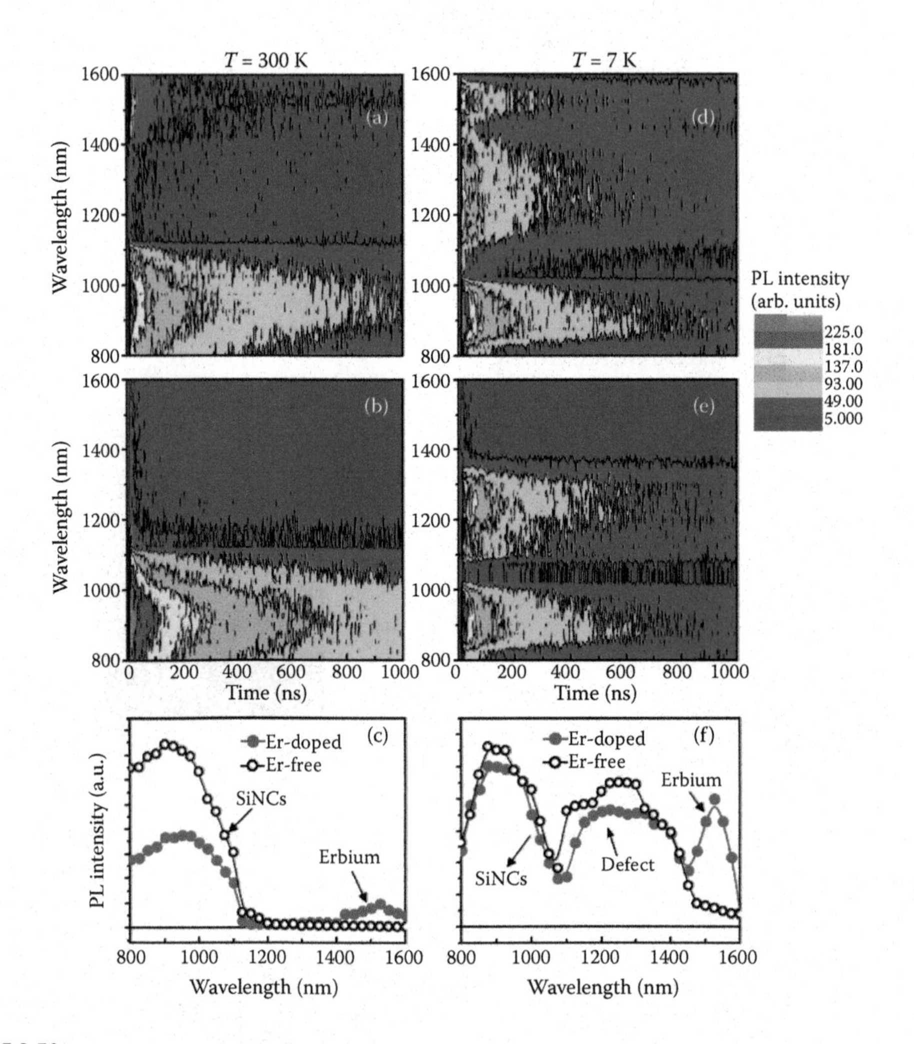

FIGURE 8.50
Time-resolved PL spectra at room temperature (a)–(c) and $T = 7$ K (d)–(f) for Er-doped (a),(d) and Er-free (b),(e) samples. The PL intensity (colored contour plots) is represented as a function of time (x axis) and detection wavelength (y axis). The results show the PL decay in the first 1000 ns after laser excitation at $\lambda_{exc} = 450$ nm. Si NC- and Er-related bands can be seen at both temperatures, while a broad band around $\lambda = 1.3$ μm can only be observed at $T = 7$ K. The spectral dependence of the integrated PL (0–200 ns) is compared for samples with and without Er^{3+} ions at room temperature (c) and at $T = 7$ K (f). (http://link.aps.org/doi/10.1103/PhysRevB.83.155323 © 2011 American Physical Society.)

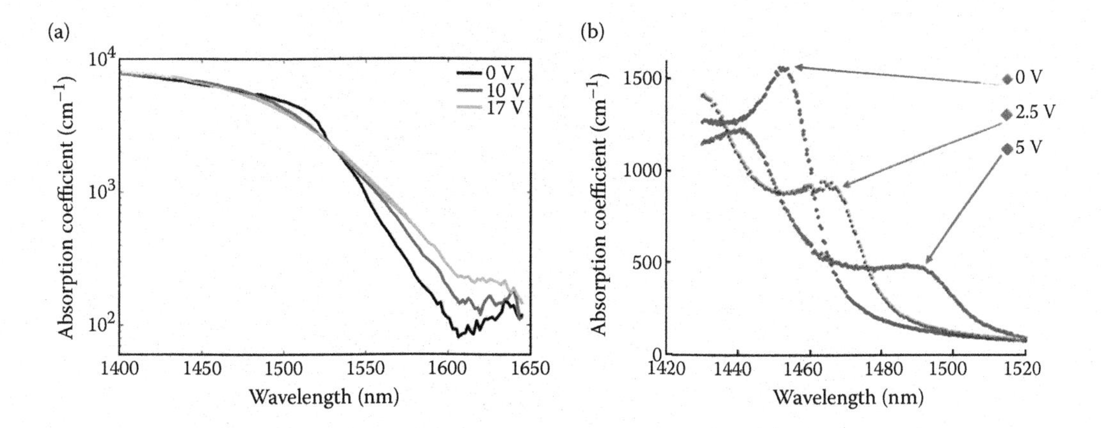

FIGURE 9.22
(a) FKE spectra from 3-μm-thick $Si_{0.994}Ge_{0.006}$ grown on Si(100) wafers. (After Y. Luo, et al, "Experimental studies of the Franz–Keldysh effect in CVD grown on GeSi epi on SOI," *Proc. SPIE*, vol. 7944, no. 79440P, 2011. With permission.) (b) QCSE spectra for Ge/$Si_{0.16}Ge_{0.84}$ quantum wells on $Si_{0.1}Ge_{0.9}$ virtual substrate showing strong exciton enhancement and step-like absorption spectra over the underlying indirect absorption. (From R. K. Schaevitz, J. E. Roth, S. Ren, O. Fidaner, and D. A. B. Miller, *IEEE J. Select. Top. Quant. Electron.*, vol. 14, no. 4, pp. 1082–1089, 2008. With permission.)

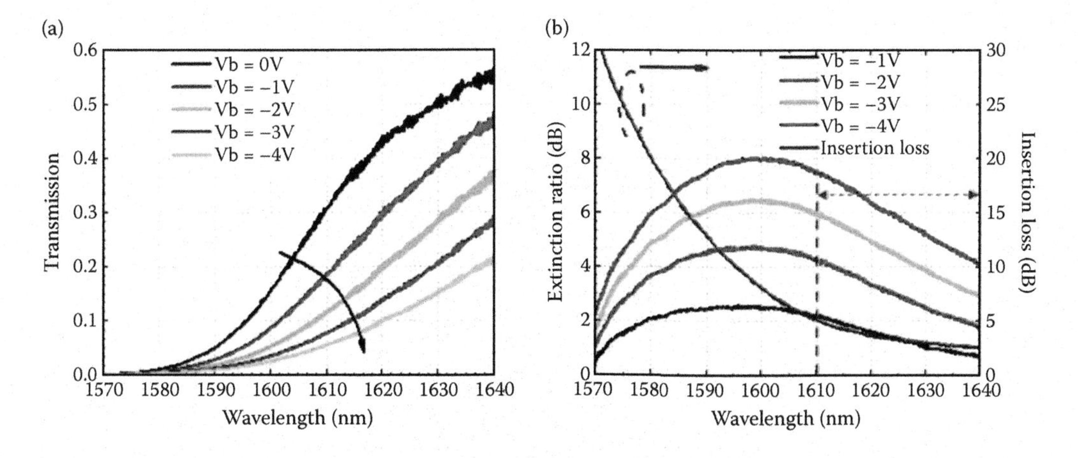

FIGURE 9.23
(a) Transmission spectrum of the Ge FKE modulator versus voltage and (b) the associated extinction ratios and insertion loss for a 45-μm-long device. (From N.-N. Feng, D. Feng, S. Liao, X. Wang, P. Dong, H. Liang, C.-C. Kung, W. Qian, J. Fong, R. Shafiiha, Y. Luo, J. Cunningham, A. V. Krishnamoorthy, and M. Asghari, *Opt. Express*, vol. 19, no. 8, pp. 7062–7067, 2011. With permission.)

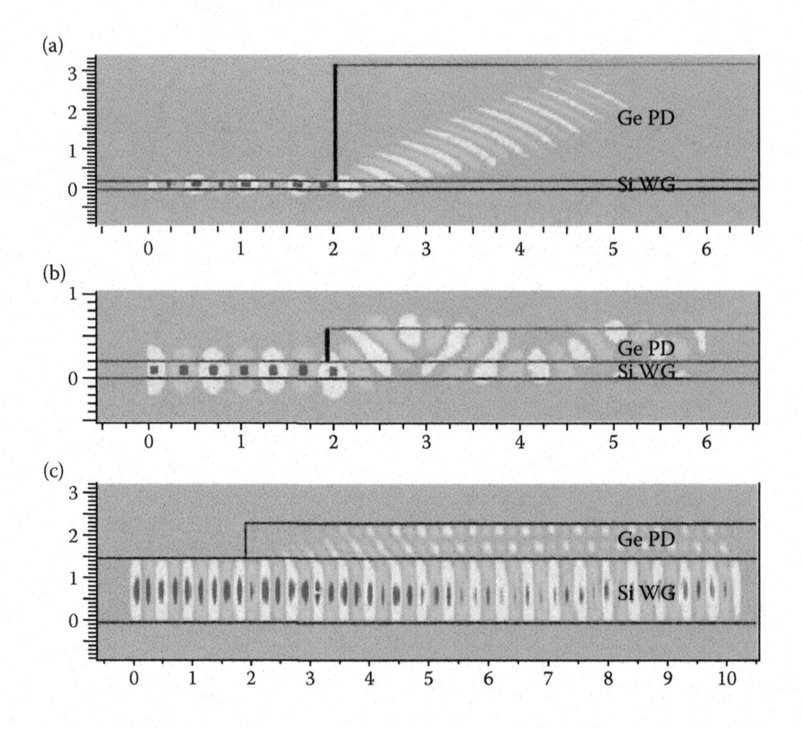

FIGURE 10.11
3D FDTD side-section view of photon propagation in bottom-waveguide–coupling structures, where (a) 3.0-μm- vs. (b) 0.4-μm-thick Ge photodetector is placed on a 0.2-μm-thick Si channel waveguide and (c) 0.8-μm-thick Ge photodetector is placed on a 1.5-μm-thick Si rib waveguide structure (Yin et al., 2007). Ge absorption coefficient of $\alpha_{Ge} = 4000\ cm^{-1}$ at $\lambda = 1550$ nm was assumed for simulation, i.e., considering 0.22% tensile strain in Ge (Liu et al., 2005b). As shown in the figures, when the thickness of Ge photodetector decreases relative to the Si waveguide, the coupling efficiency deceases due to a shift in coupling mechanism from direct coupling and scattering dominated regime in (a) to slow evanescent coupling-dominated regime in (c).

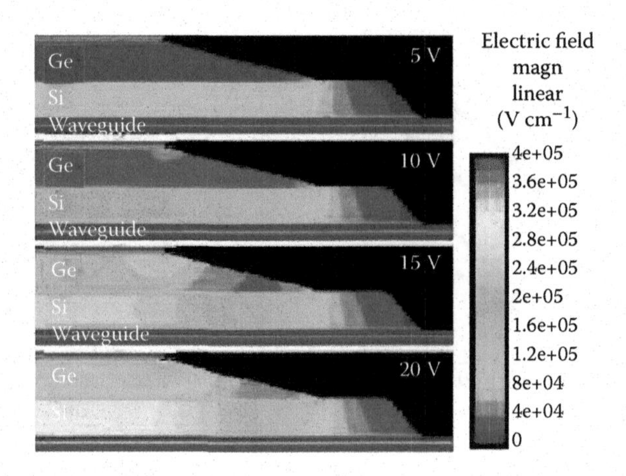

FIGURE 10.26
Electric field distribution of a waveguide-integrated beveled Ge/Si APD. (From Ang, K. W., Ng, J. W., Lim, A. E. et al., 2010. Waveguide-integrated Ge/Si avalanche photodetector with 105 GHz gain-bandwidth product. *Opt. Fiber Commun. Conf.*, OSA, Technical (CD) JWA36. Reproduced with permission.)

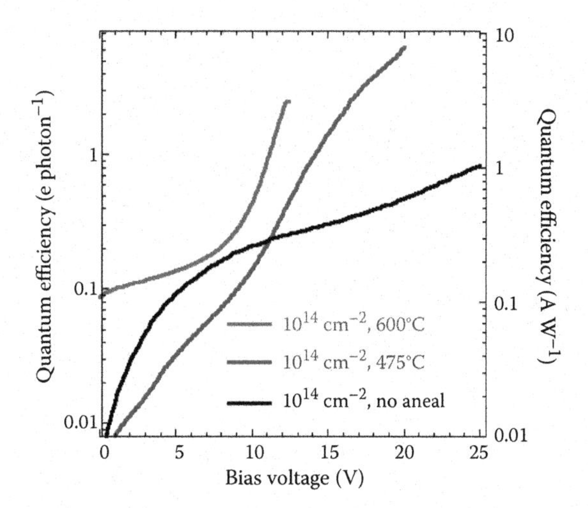

FIGURE 10.36
Photodiode quantum efficiency, the ratio of photocurrent to light absorbed, for devices implanted with Si^+ to 10^{14} cm^{-2} as a function of bias voltage, for several annealing temperatures (Geis et al., 2007a,b, 2009). The annealing temperatures are in the legend.

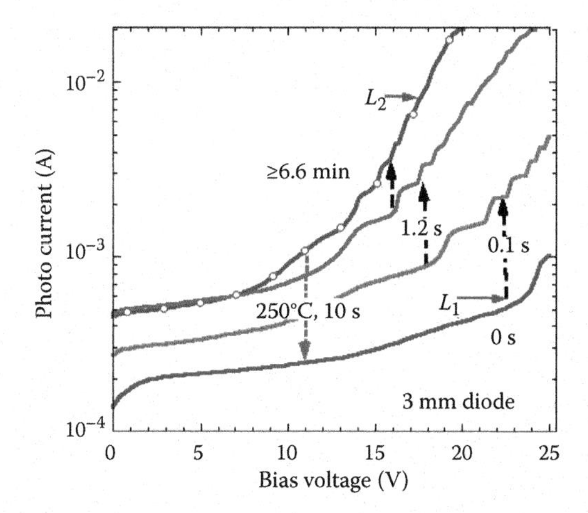

FIGURE 10.37
Sequence of measurements of photocurrent vs. bias voltage for a single photodiode at fixed optical power. At time 0 s, the diode is in state L_1. Operation in forward bias of 100 mA/cm continuously transforms the diode to the L_2 state, saturating in 6 minutes. Heating the diode to 250°C for 10 s returns it to the L_1 state (Geis et al., 2007a).

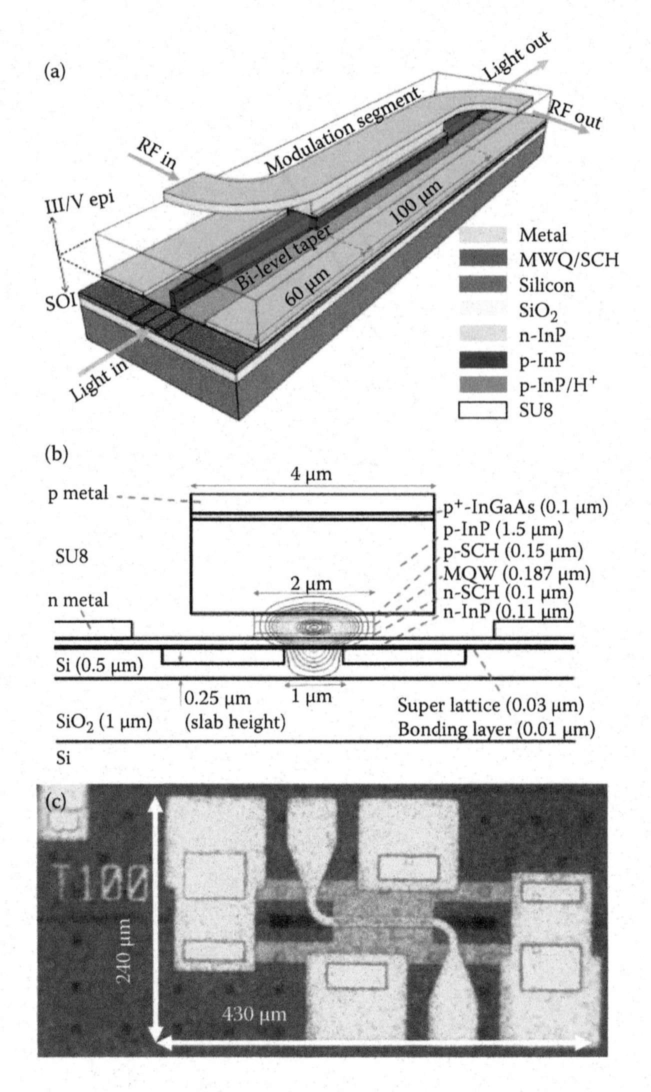

FIGURE 11.23
Schematic structure of a hybrid silicon TW-EAM (a). Cross-section of the modulation segment with a superimposed fundamental optical mode (b). Top-view photograph of a fabricated TW-EAM (c). (Tang, Y. et al., *Opt. Express* 19(7):5811, 2011.)

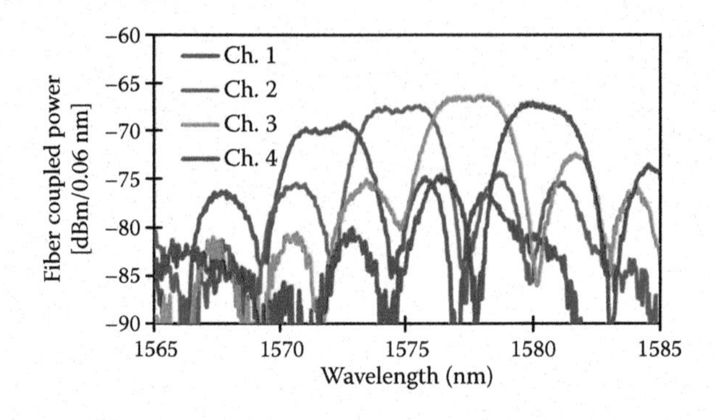

FIGURE 11.48
ASE output of the common output waveguide for an SOA bias of 75 mA. (Picture taken from Kurczveil, G. et al., *IEEE J. Sel. Top. Quant. Electron.* 17(6):1521–1527, 2011.)

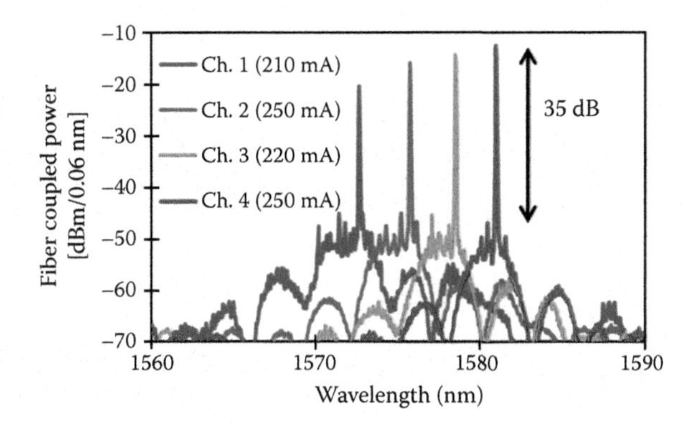

FIGURE 11.49
Optical spectrum of the four channels optimized for maximum side mode suppression ratio. (Picture taken from Kurczveil, G. et al., *IEEE J. Sel. Top. Quant. Electron.* 17(6): 1521–1527, 2011.)

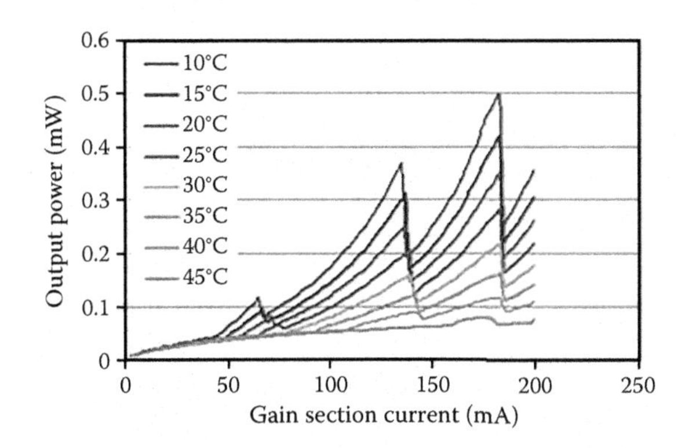

FIGURE 11.58
Output power-gain current measurement results for SG-DBR at stage temperatures from 10°C to 45°C.

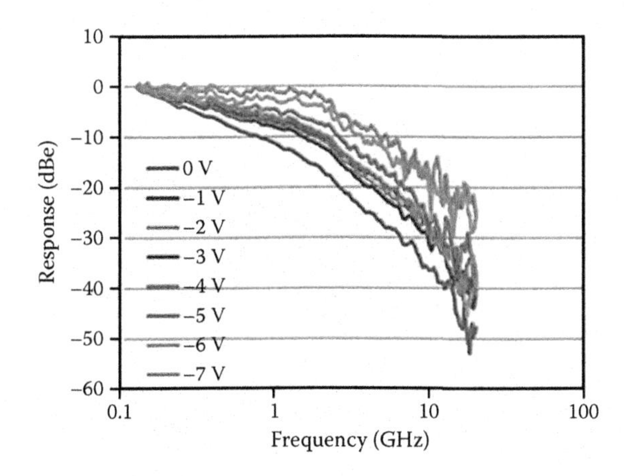

FIGURE 11.59
Electrical to optical small signal response of integrated EAMs in the SG-DBR-EAM with a 2.5-μm-wide silicon waveguide.

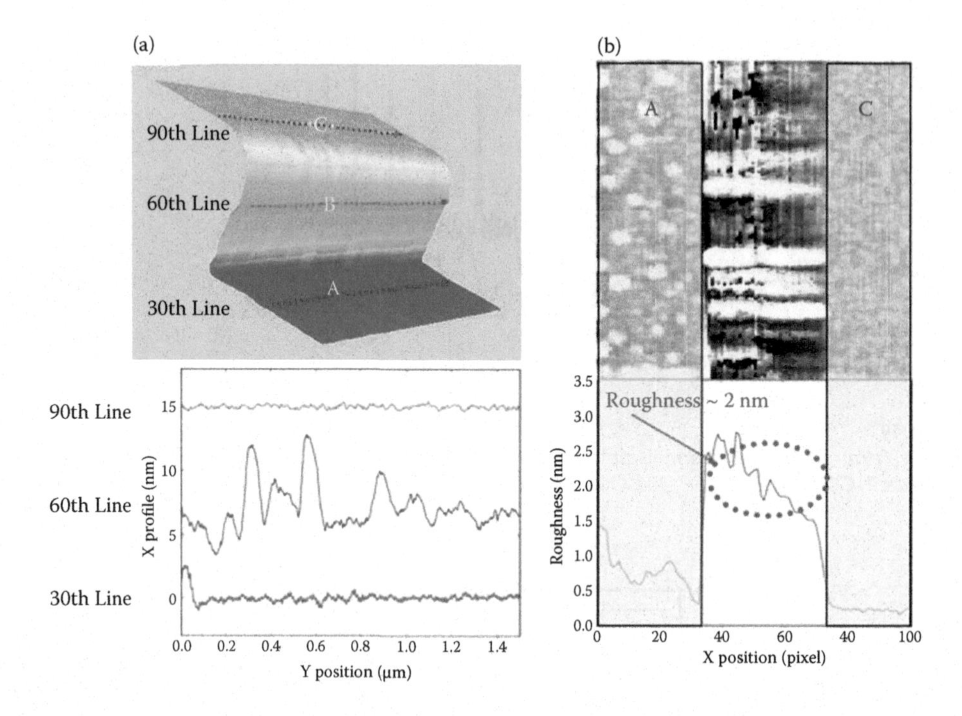

FIGURE 12.19
AFM images of low sidewall roughness of approximately 2 nm for waveguides fabricated at NTC-UPVLC [108] obtained by an optimized process. Region A (a) is the top surface of the BOX oxide, region B (b) is the sidewall, region C is the top surface of the silicon waveguide. (Images courtesy of Park Systems, KANC 4F, Iui-Dong 906-10 Suwon, South Korea 443-270 (www.parkafm.com).)

FIGURE 13.6
Photonic dies coupled by grating couplers using the flip-chip technique.

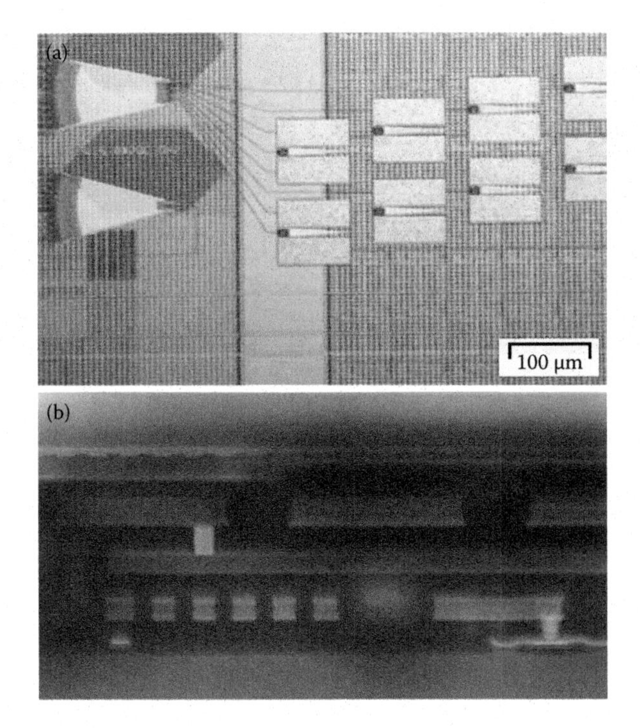

FIGURE 13.17
Photonic–electronic wafer with AWG and Ge photodiodes. Cross-cut with flip fiber coupler.

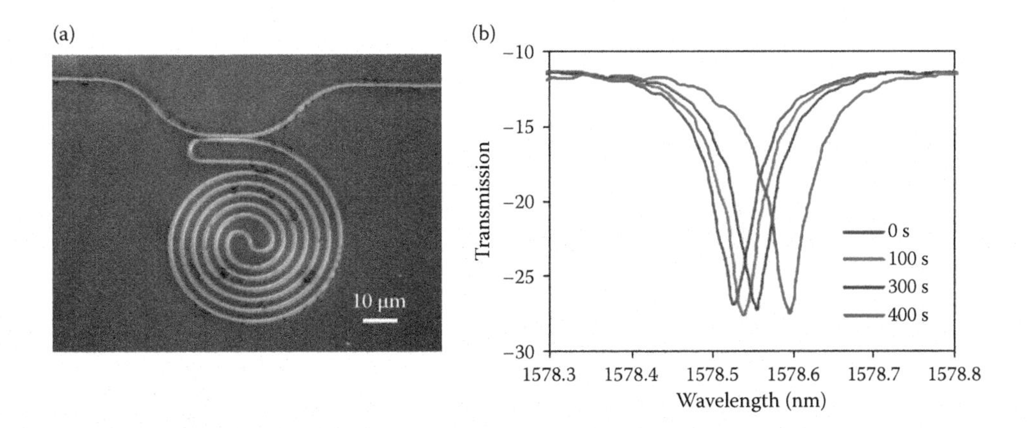

FIGURE 14.6
(a) A spiral ring resonator sensor element formed by silicon photonic wire waveguides. (b) The wavelength shift of a resonance as streptavidin protein molecules bind to the surface. The maximum shift at 400 s corresponds to approximately 10% of a protein monolayer. (From Xu, D.-X. et al. *Opt. Express* 16:15137–15148, 2008. With permission.)

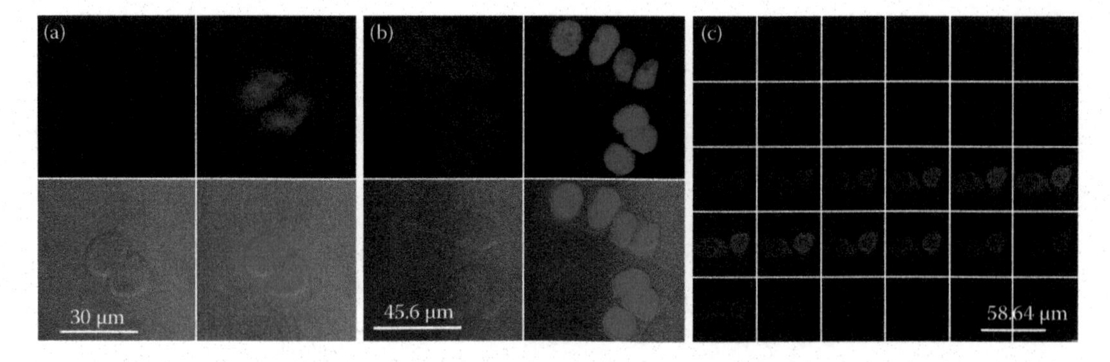

FIGURE 14.20
Confocal images of MCF-7 cells incubated for 2 h with (a) medium only, (b) 0.5% v/v PAAc-SiNPs in distilled water, (c) z-stack image. Images were collected at 0.37-μm intervals with the 488-nm laser to create a stack in the Z axis. In (a) and (b), the top left corner shows images obtained from the red channel, which is the fluorescence from SiNPs; the top right corner presents images from the blue channel, indicating nuclei staining with DAPI; at the bottom left is the bright field; and at the bottom right is a combination of the three. (From Wang, Q. et al. *J. Nanoparticle Res.* 13:405–413, 2011. With permission.)

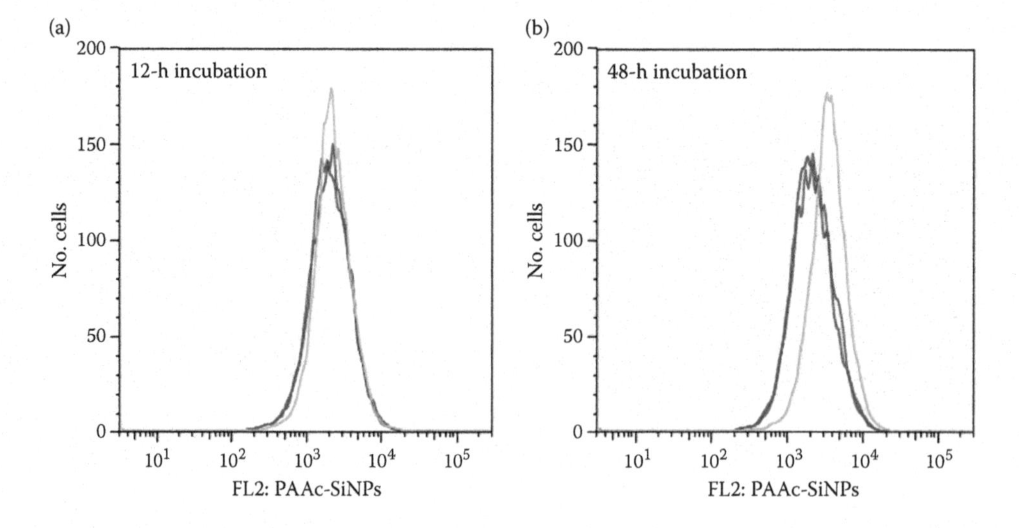

FIGURE 14.21
Uptake efficiency of SiNPs in $HepG_2$ cells with various incubation times. $HepG_2$ cells were exposed to PAAc-SiNPs for 12, 24, 36, 48, 60, and 72 h at two concentrations of 10 or 100 μg/mL. Red, control; blue, 10 μg/mL PAAc-SiNP; green, 100 μg/mL PAAc-SiNP: (a) 12 h incubation; (b) 48 h incubation.

8

Silicon-Based Light Sources

Aleksei Anopchenko, Alexei Prokofiev, Irina N. Yassievich, Stefano Ossicini, Leonid Tsybeskov, David J. Lockwood, Saba Saeed, Tom Gregorkiewicz, Maciek Wojdak, Jifeng Liu, and Al Meldrum

CONTENTS

8.1 Introduction

Aleksei Anopchenko

The search for silicon-based light sources is an active research area. Bulk silicon is an indirect band semiconductor material with a strong disparity in radiative and nonradiative recombination lifetimes (O'Mara et al., 1990; Hull, 1999)—hence, a poor light emitter. There are, however, several approaches to enhance radiative recombination in Si, which we will describe in this chapter. These approaches are diverse and include band-gap engineering, defect and strain engineering, quantum confinement, carrier confinement, and a synthesis of the above-mentioned approaches. Here, we will focus on Si photonics approaches that use nanostructured and complementary metal semiconductor oxide (CMOS) materials. We define "silicon-based" as a material fully and inherently integrated into a modern silicon technology. Therefore, we do not limit our discussion to Si only, but also include an overview of Ge and SiGe light sources. No electrically pumped interband Si laser has been demonstrated until now, but good progress in Si-based light sources has been made and will be reviewed in this chapter.

Development of Si-based light sources has a long history of discoveries and innovations that are closely related to a number of breakthroughs in silicon technology and boosted by the expansion of the World Wide Web in recent years. Table 8.1 shows in a simplified timeline frame the complex network of historical events that led to a present state of the development of Si-based sources (see also "Nature Milestones: Photons," 2010; Loebner, 1976; Schubert, 2006 for references to the original works in the table and the references in this chapter as well). There are several very good review papers and books that cover many aspects of this historical path of the Si-based sources (see, for example, Soref, 1993; Lockwood, 1998; Fiory and Ravindra, 2003; Pavesi et al., 2003; Ossicini et al., 2003; Fauchet, 2004; Reed and Knights, 2004; Shainline and Xu, 2007; Pavesi, 2008).

This chapter is written by recognized experts in the field of Si-based light sources and gives an overview of the most recent developments in the field as well as some basic and essential knowledge about optical properties of indirect band semiconductors, Si and Ge nanostructures, and optical microresonators. For a comprehensive account of Si-based light sources, the interested reader is advised to see the works cited above and in a relevant section. The chapter is organized as follows: Section 8.2 gives a brief summary of some fundamental optical properties of indirect band semiconductors and bulk Si light-emitting diodes (LEDs); the next three Sections, 8.3–8.5, give an overview of light-emitting properties of Si and SiGe nanostructures and Si nanostructures doped with rare-earth elements; Section 8.6 describes a strained Ge on Si laser, and Section 8.7 discusses an enhancement of the emission rate of Si nanocrystals (Si NCs) in optical microcavities, called Purcell effect.

TABLE 8.1

A Timeline Showing the Development of Si-Based Light Sources

Period	Year	Milestone	IR Devices	Visible Devices
Radio	1907	SiC EL (Round)		
	1916	Stimulated emission, SE (Einstein)		
	1928	SiC LED (Losev)		
Radar & TV	1936	Excitons (Frenkel)		
	1939	p-n junction (Davydov, Ohl)		
	1947	Ge transistor (Bardeen, Brattain)		
Computer	1951		Si & Ge LEDs, forward bias (Haynes, Briggs)	
	1955	Si Oxidation metallization (Moll);		Si LED, reverse bias (Newman)
	1958	IC (Kilby, Noyce)		
	1962	Infrared (Hall) & visible (Holonyak) diode lasers		
	1966	Silica optical fiber (Kao)		
	1970	Heterostructure laser (Alferov, Hayashi, Panish)		
Internet & mobile phone	1984			Si QD EL (DiMaria et al.)
	1985		Er-doped Si El (Ennen et al.)	
	1986	Er-doped fiber amplifier		

	1990			Porous Si PL (Canham)
World Wide Web	1994		RT Er-doped SiOx LEDs (Franzo et al.)	
	1995			Si-SiO_2 superlattice PL (Lu et al.)
	1996			Integrated porous Si LED (Hirschman et al.)
	1997		Iron-disilicide LEDs (Leong et al.)	
	1998	Silicon-on-Insulator		
	2000		Si/SiGe cascade laser (Dehlinger et al.)	Optical gain in Si QDs (Pavesi et al.)
	2001		Optical gain in Er-doped Si-QDs (Han et al.)	
	2002		RT bulk Si LEDs (Green et al., Ng et al.)	Si-QD LED (Franzo et al.)
	2003	Multigate FETs	Er-doped Si QD LEDs (Castagna et al.) Si Raman laser (Rong et al.)	
	2005		SE from point defects in Si (Cloutier et al.)	Si-QD field effect LED (Walters et al.)
	2006	Hybrid III-V on Si laser (Fang et al.)	Er doped microdisk laser (Kippenberg et al.)	
	2008		Cascaded Si Raman laser (Rong et al.)	
	2010		Room temperature Ge laser (Liu et al.)	
	2011			SE in Si multi-fin LEDs (Saito et al.)
	2012		Electrically pumped Ge laser (Camacho-Aguilera et al.)	

8.2 Fundamentals of Light Emission from Indirect Band Semiconductors

Alexei Prokofiev and Irina N. Yassievich

8.2.1 Introduction

In this section, the basic problems of light emission from indirect band semiconductors are considered in application to silicon. The fundamental difference between interband optical processes in indirect and direct band semiconductors is discussed.

8.2.2 Direct and Indirect Optical Processes

Consider an intrinsic semiconductor at low temperature, so that there are no electrons in the conduction band. When the light falls on the semiconductor, it can be absorbed due to the transition of an electron from the valence band to the conduction band. The light absorption obeys energy and momentum conservation laws.

Figure 8.1 schematically shows the top of the valence band and bottom of the conduction band of a direct band semiconductor (e.g., GaAs).

The energy conservation law limits the photon energy $\hbar\omega$ that should be equal or larger than the energy band-gap E_g. The momentum conservation law requires that the quasi momentum of a final electron state to be $\hbar\vec{k}_2 = \hbar\vec{k}_1 + \hbar\vec{q}_{\text{phot}}$, where $\hbar\vec{k}_1$ is the quasi momentum of an initial electron state in the valence band, and $\hbar\vec{q}_{\text{phot}}$ is the momentum of incoming photon. One can estimate the value of photon wave vector (for visible light) as $\vec{q}_{\text{phot}} = \dfrac{2\pi}{\lambda} \approx 10^7\ \text{m}^{-1}$, which is only 0.001 of a value of the wave vector corresponding to a Brillouin zone edge. So optical transitions are usually considered as direct (or vertical): $\vec{k}_2 = \vec{k}_1$. In Figure 8.1, the optical absorption process controlled by the momentum conservation is shown on example of the transition in GaAs.

In case of indirect semiconductor, e.g., Si (see Figure 8.2), the minimum of the conduction band is shifted from the maximum of the valence band in the k-space, and direct optical transitions are not possible. In the indirect optical transition the third quasiparticle should be involved, e.g., phonon, to accommodate such a difference in momentum between the initial and the final electron states.

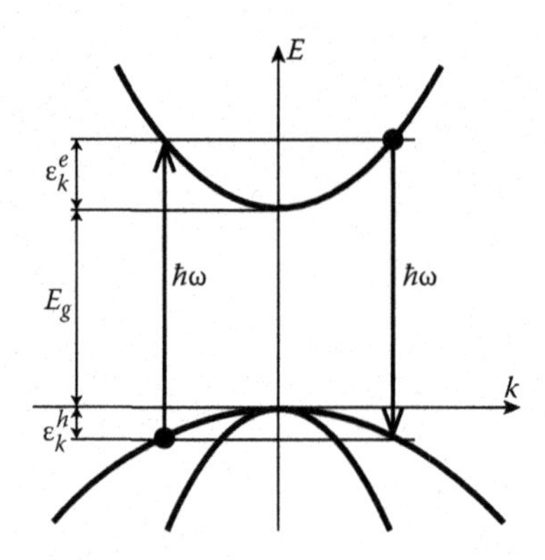

FIGURE 8.1
Optical processes in direct band-gap semiconductor. The momentum of the electron does not change.

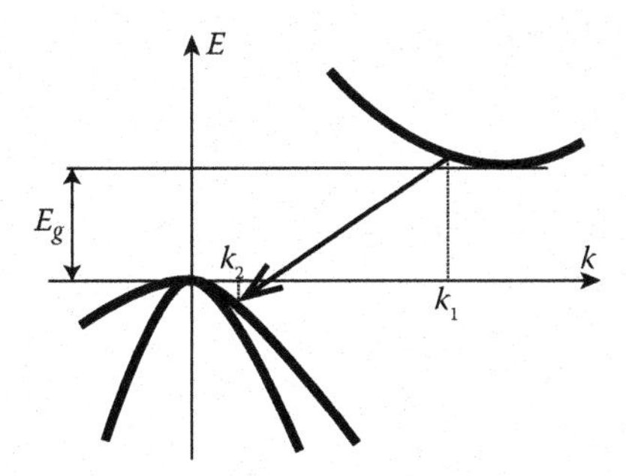

FIGURE 8.2
Scheme of interband transition of electron in indirect band-gap semiconductor. The transition is assisted by phonon emission or absorption.

If the conduction band states are populated with electrons, and there are free electron states (holes) in the valence band, electrons may emit photons by going down to the valence band, i.e., the radiative recombination of electron–hole (e-h) pair takes place. The radiative recombination rate per unit volume is described by the radiative recombination coefficient B as $R = Bnp$, where n and p are electron and hole concentrations. In indirect semiconductors, R is defined by electrons from the indirect minimum and the radiative transition is assisted by the interaction with the third body (usually with a phonon) (Figure 8.2). Tables 8.2 and 8.3 show the recombination coefficients B and the radiative lifetimes τ_R of carriers for the case $n = p = 10^{18}$ cm^{-3} using the data presented in Landsberg (1991). The direct radiative transitions are more probable by the tremendous factor about 10^6 than the indirect ones.

TABLE 8.2
Direct Semiconductors

	GaAs	GaSb	InSb
Energy gap (eV)	1.435	0.72	0.18
B (cm^3/s)	7.2×10^{-10}	2.4×10^{-10}	4.6×10^{-11}
τ_R (s)	1.3×10^{-9}	4.2×10^{-9}	2.2×10^{-8}

Source: Landsberg, Peter Theodore. 1991. *Recombination in semiconductors*. Cambridge, New York: Cambridge University Press. With permission.

TABLE 8.3
Indirect Semiconductors

	Si	Ge
Energy gap (eV)	1.12	0.66
B (cm^3/s)	1.8×10^{-15}	5.3×10^{-14}
τ_R (s)	5.6×10^{-4}	1.9×10^{-5}

Source: Landsberg, Peter Theodore. 1991. *Recombination in semiconductors*. Cambridge, New York: Cambridge University Press. With permission.

8.2.3 Probability of Spontaneous Radiative Transitions

8.2.3.1 Direct Band Semiconductors

In direct band semiconductors, the probability of spontaneous radiative transition of an electron from the initial state i of the conduction band to the final state f of the valence band is given by the Fermi's golden rule:

$$W_{i\to f} = \frac{2\pi}{\hbar^2}\left|M_{f,i}\right|^2 g(\omega), \tag{8.1}$$

where $M_{i\to f}$ is the matrix element of the electric dipole interaction and $g(\omega)$ is the density of photon states. Radiative transition with k-selection is shown in Figure 8.1. The energy of emitted photon $\hbar\omega = E_g + \varepsilon_{i\vec{k}} + \varepsilon_{f\vec{k}}$ ($\varepsilon_{i\vec{k}}$, $\varepsilon_{f\vec{k}}$ are the kinetic energies of recombining electron and hole). We use for matrix element $M_{i\to f}$ the expression:

$$M_{i\to f} = \left\langle f\left|\vec{d}\,\vec{E}\right|i\right\rangle, \tag{8.2}$$

where $\vec{d} = e\hat{\vec{r}}$ is operator of dipole momentum and $\vec{E}$ is the electric field of the light wave:

$$\vec{E} = \vec{e}_{\mathrm{L}} E_0. \tag{8.3}$$

In Equation 8.3, $\vec{e}_{\mathrm{L}}$ is the polarization vector of light, and the amplitude of electric field for the one-photon electromagnetic field in medium with refraction index n is given by

$$E_0 = \frac{1}{n}\sqrt{\frac{\hbar\omega}{2\kappa_0 V_0}}. \tag{8.4}$$

Here κ_0 is vacuum permittivity and V_0 is normalization volume. Correspondingly, we use for density of photon states the following expression:

$$g(\omega) = \frac{\omega^2 V_0 n^3}{\pi^2 c^3}, \tag{8.5}$$

where c/n is the speed of light in the semiconductor. Averaging over all possible direction of polarization, we get the final result:

$$W_{i\to f} = \frac{e^2\left|\left\langle f\left|\hat{\vec{r}}\right|i\right\rangle\right|^2 n\omega^3}{3\pi\kappa_0\hbar c^3}. \tag{8.6}$$

One can use the relation

$$\hat{\vec{p}} = \frac{im_0}{\hbar}\left[\hat{H}\hat{\vec{r}} - \hat{\vec{r}}\hat{H}\right], \tag{8.7}$$

to rewrite Equation 8.6 in an alternative form:

$$W_{i\to f} = \frac{e^2\left|\left\langle f\left|\hat{\vec{p}}\right|i\right\rangle\right|^2 n\omega}{3\pi\kappa_0\hbar m_0^2 c^3}. \tag{8.8}$$

For example, we have produced the calculation of the radiative lifetime τ_R of an electron with wave vector $\vec{k}$ when there is a corresponding hole in the valence band (see Figure 8.1). We have used Equation 8.8 taking into account only the heavy hole subband, averaging on two possible orientation of electron spin and neglecting the kinetic energies of both electron and hole when compared with E_g. We have got for III–V semiconductors of GaAs type

$$\frac{1}{\tau_R} = \frac{e^2 P^2 n E_g}{3\pi\kappa_0 \hbar^4 c^3}. \tag{8.9}$$

Here P is the constant introduced by Kane (1956), which is related to the matrix elements of the momentum operator between the Bloch amplitudes corresponding to the edges of the conduction and the valence bands by the following expression:

$$\langle S|p_x|X\rangle = \langle S|p_e|Y\rangle = \langle S|p_z|Z\rangle = iP\frac{m_0}{\hbar}. \tag{8.10}$$

All other matrix elements equal to zero due to tetragonal symmetry. In the Kane model, parameter P determines the effective masses of electrons m_c and light holes m_l near the band edges. For example, the light hole mass

$$m_l = \frac{3E_g\hbar^2}{4P^2}. \tag{8.11}$$

Using Equations 8.9 and 8.11, one gets the value of radiative lifetime $\tau_R \approx 0.4$ ns for carriers in GaAs, which is in a good agreement with the experimental value presented in Table 8.2.

8.2.3.2 Indirect Band Semiconductors

We can use again Equation 8.1 for the probability of spontaneous radiative transition of an electron from the initial state i of the conduction band to the final state f of the valence in the case of the indirect band semiconductors. However, the matrix element $M_{f,i}$ should be calculated in the second-order perturbation theory approximation, because we should take into account the interaction with phonons. The radiative transition from initial state i to final state f occurs now as a two-step process via intermediate virtual state j. Figure 8.3 demonstrates two possible ways of the radiative recombination in case of silicon. Correspondingly, $|M_{f,i}|^2$ in Equation 8.1 is given by the following expression:

$$|M_{f,i}|^2 = \left\{ \left| \sum_{j1} \frac{M_{f,j1}^{phon} M_{j1,i}^{ligt}}{E_{j1} - E_i} \right|^2 + \left| \sum_{j2} \frac{M_{f,j2}^{light} M_{j2,i}^{phonon}}{E_{j2} - E_i} \right|^2 \right\}, \tag{8.12}$$

where E_i, E_f, $E_{j\alpha}$ ($\alpha = 1, 2$) are the energies of initial, final, and intermediate (virtual) states, respectively.

Now the energy conservation leads to the following relation

$$E_f - E_i = E_g + \varepsilon_{i\vec{k}} + \varepsilon_{f\vec{k}'} = \hbar\omega \pm \hbar\Omega_{\vec{q}}, \tag{8.13}$$

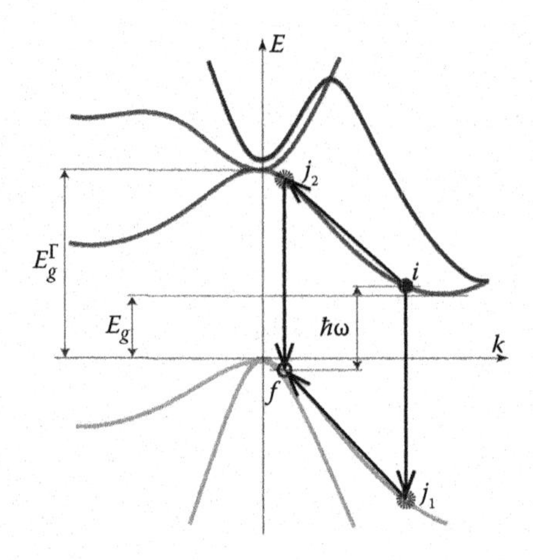

FIGURE 8.3
The scheme of radiative recombination in silicon. There are two possible ways: (1) via virtual state j_1 and (2) via virtual state j_2. The transitions from i to j_2 and from j_1 to f involve phonon emission or absorption.

where $\hbar\Omega_{\vec{q}}$ is the energy of phonon (sign "+" corresponds to phonon emission, while "−" corresponds to phonon absorption). The momentum conservation low connects momentum $\vec{k}$ of electrons from the valley ν, hole momentum $\vec{k}'$, and momentum of phonon $\vec{q}$:

$$\vec{k}+\vec{k}_{0\nu}=\vec{k}'+\vec{q} \tag{8.14}$$

where $\vec{k}_{0\nu}$ is the position of the minimum of the conduction band valley ν in k-space. There are six equivalent valleys in Si that occur along axes [100], [010], and [001] with $k_{0\nu} = 0.85 \times 2\pi/a$ (a is the lattice constant). Equation 8.14 shows that q is close to the limit value. It means the phonon energy $\hbar\Omega_{\vec{q}}$ in Equation 8.13 should be close to its limit values equal to 18 meV for TA branches, 45 meV for LA, and about 50 meV for optical phonons. The dominant optical transition in bulk silicon involves TO phonon with energy 57 meV (Shaklee and Nahory, 1970). The results are (i) the radiative transition should be accompanied by phonon emission at $T \leq 300$ K and (ii) the probability of radiative transition in indirect band semiconductors is considerably less than in direct ones, for example, in bulk Si $\tau_R \approx 5 \times 10^{-4}$ s. The details of calculation of radiative recombination rates in Si and Ge can be found in Landsberg (1991) and Yu and Cardona (2001).

At this point, we discuss the role of Coulomb interaction between electrons and holes in the recombination process. First, the Coulomb interaction leads to formation of the bound e-h states (the Wannier–Mott excitons in semiconductors) with energy E less than the energy gap: $E = E_g - E_{ex}$. Excitons are similar to hydrogen atoms. In the first approximation, we can use for E_{ex} the modified Bohr formula:

$$E_{ex}=\frac{\mu e^4}{2\hbar^2(4\pi\kappa_0\kappa)^2}, \tag{8.15}$$

where κ is relative permittivity and μ is given by the relation:

$$\frac{1}{\mu}=\frac{1}{m_e^*}+\frac{1}{m_h^*}. \tag{8.16}$$

Here m_e^* and m_h^* are effective masses of density states for electrons and holes, respectively. In silicon $m_e^*=0.33m_0$, $m_h^*=0.5m_0$, and we get E_{ex} = 18.8 meV from Equation 8.15. Shaklee and Nahory (1970) determined the value E_{ex} = 15 meV from the absorption spectrum at 2 K. Nowadays, the widely accepted value for exciton binding energy at low temperature is E_{ex} = 14.7 meV. A small increase with temperature is expected due to change of effective masses.

On the other hand, the Coulomb interaction should enhance strongly the radiative recombination of the free e-h pair due to increasing the probability of finding an electron and a hole at the same point, which is given by Sommerfield factor S_+:

$$S_+=2\pi\sqrt{\frac{E_{ex}}{E_k}}\frac{1}{1-\exp\left(-2\pi\sqrt{\frac{E_{ex}}{E_k}}\right)}, \tag{8.17}$$

where $E_k = \hbar^2k^2/(2\mu)$ is relative kinetic energy. The averaged kinetic energy of free carriers increases with temperature for nondegenerate carriers. In result, the radiative recombination coefficient B gets the dependence on temperature in this case. The temperature dependence of the coefficient B in silicon for the case $n = p = 10^{15}$ cm^{-3}, measured from 100 to 400 K in Schlangenotto et al. (1974), is presented in Figure 8.4. In Schlangenotto et al. (1974), the detailed theoretical analysis of decreasing radiative recombination in silicon with temperature is presented. It is shown also that the exciton state introduces a considerable contribution into the recombination rate, even at room temperature.

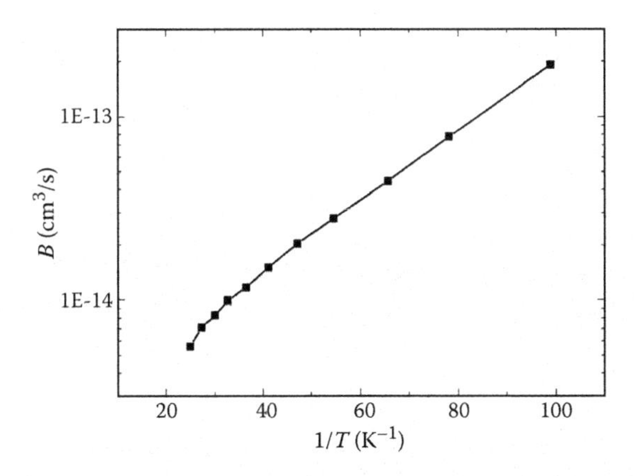

FIGURE 8.4
Experimental results on temperature dependence of B. (Data are taken from Schlangenotto, H., H. Maeder, and W. Gerlach. 1974. *Physica Status Solidi A* 21 (1):357–367. With permission.)

8.2.4 Spontaneous and Stimulated Emissions

There are three Einstein coefficients, denoted A_{21}, B_{12}, and B_{21}, which determine the light absorption and emission processes for two-level system (see Figure 8.5). For example, that may be an atom with two energy levels: ground (1) and excited (2), as Einstein (1916) considered in his paper. Spontaneous emission coefficient A_{21}, which determinates the rate of spontaneous radiative transition from the upper state 2 to the lower state 1 (i.e., without any outside influence), can be calculated from the first principles based on quantum mechanics. In Section 8.2.3.1, we have presented the example of such calculation for direct band-gap semiconductors. We have

$$\frac{dN_2}{dt} = -A_{21}N_2 = -\frac{N_2}{\tau_R}, \tag{8.18}$$

where N_2 is the number of atoms in state 2.

The process of absorption is illustrated in Figure 8.5b. It is not a spontaneous process. Electron cannot jump to the excited state unless it receives the energy required from incoming photon. Following the Einstein's treatment, the rate of absorption process is given by

$$\frac{dN_1}{dt} = -B_{12}N_1u(\omega), \tag{8.19}$$

where $u(\omega)$ is the spectral energy density of the electromagnetic field at frequency $\omega(\hbar\omega = E_2 - E_1)$:

$$u(\omega) = \hbar\omega\frac{g(\omega)}{V_0}N_\omega = \frac{\hbar\omega^3 n^3}{\pi^2 c^3}N_\omega, \tag{8.20}$$

where N_ω is number of photons with energy $\hbar\omega$ per one state, n is the refractive index of the medium. In the conditions of thermal equilibrium

$$N_\omega = \frac{1}{\exp\left(\frac{\hbar\omega}{kT}\right) - 1}. \tag{8.21}$$

Einstein realized that there is a third type of transition called stimulated emission (SE) (also known as induced emission). In this process, the incoming photons can stimulate

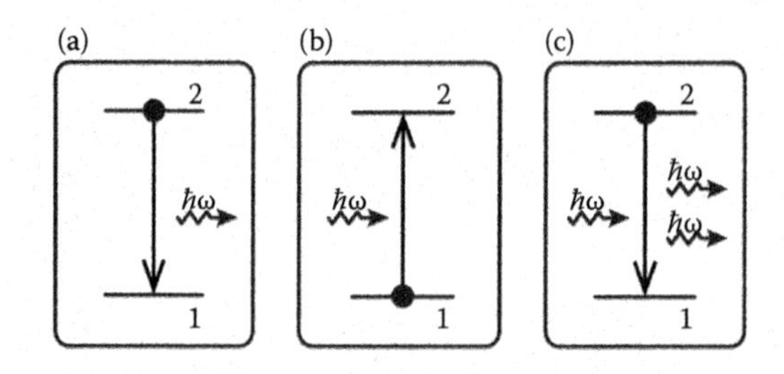

FIGURE 8.5
Three possible optical transitions in two-level system.

downward emission process as well as upward absorption transitions. The SE rate is governed by Einstein coefficient B_{21}:

$$\frac{dN_2}{dt} = -B_{21}N_2u(\omega). \tag{8.22}$$

Three Einstein coefficients are not independent:

$$A_{21} = \frac{\hbar\omega^3 n^3}{\pi^2 c^3} B_{21}, \quad g_1 B_{12} = g_2 B_{21}, \tag{8.23}$$

where g_1 and g_2 are the degeneracy of levels 1 and 2, respectively.

SE is a coherent quantum mechanical effect. The photons emitted are in phase with photons that induce the transition. This was the effect that opened the way to create lasers.

We should remark that relations 8.23 cannot be applied to the optical interband transitions in indirect band semiconductors as any radiative process is usually assisted by a phonon emission or absorption. For some considerations and discussion on the spontaneous emission rate, absorption and SE in indirect band semiconductors, the reader may see Würfel et al. (1995), Trupke et al. (2003a), Würfel (2009), and Imhof and Thraenhardt (2010).

8.2.5 Bulk Si LEDs

We will discuss the silicon LEDs working on interband optical transitions. In semiconductors, in the conditions of thermal equilibrium, the light with the quantum energy $\hbar\omega > E_g$ is absorbed as the electron population of the conduction band f_c is always less than that of the valence band f_v. Their relation is determined by the Boltzmann statistics:

$$\frac{f_c}{f_v} \propto \exp\left(-\frac{E_g}{k_B T}\right). \tag{8.24}$$

For light emission from semiconductors, it is necessary to strongly destroy this relation. This problem can be solved by using carrier injection through a p-n junction. In Figure 8.6a, the energy diagram for a semiconductor with the p-n junction in the condition of thermal

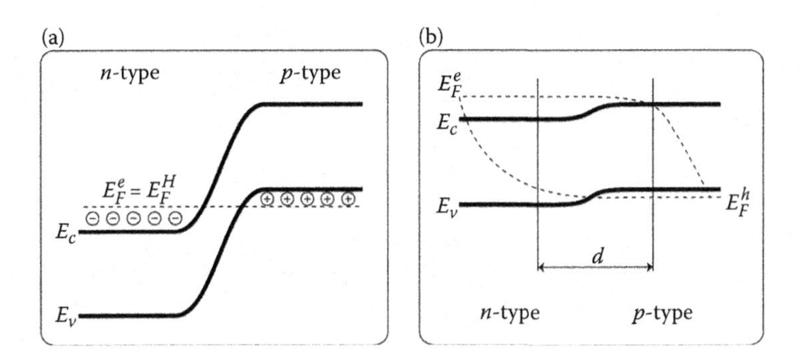

FIGURE 8.6
A *p-n* junction in thermal equilibrium (a) and under forward bias (b).

equilibrium is presented for the case when the Fermi quasi-levels E_F^e and E_F^h (controlling the concentration of carriers in the conduction and valence bands, respectively) occur in the bands. In the equilibrium $E_F^e = E_F^h$. Under forward bias, the equilibrium is destroyed and electrons are injected into the *p*-region while holes are injected into the *n*-region (see Figure 8.6b). The size of active region d can be found from the condition: $d \approx \sqrt{D\tau}$ where D is diffusion coefficient and τ is the average lifetime of nonequilibrium carriers (usually d is determined by injection of electrons into the *p*-region).

The total lifetime τ is controlled both radiative and nonradiative recombination processes:

$$\frac{1}{\tau} = \frac{1}{\tau_R} + \frac{1}{\tau_{NR}}. \tag{8.25}$$

The nonradiative pathways include recombination at bulk and surface defects and Auger recombination. The internal quantum efficiency (QE) of LED is given by

$$\eta_{in} = \frac{\tau_{NR}}{\tau_{NR} + \tau_R}. \tag{8.26}$$

In direct band semiconductors, internal QE η_{in} may approach 100%, because the radiative lifetime is shorter than the nonradiative lifetime. However, in indirect band semiconductors, such as Si, where the radiative lifetime is long, η_{in} drops by several orders of magnitude and typically lower than 10^{-5} (Michaelis and Pilkuha, 1969). To improve the internal QE of silicon nonradiative pathways should be suppressed, e.g., using Si nanostructures as it is shown in the next sections. The device performance is generally gauged by a measurement of the external QE η_{ext} of light emission, which is less than the internal QE η_{in}.

The first LED was realized by O. V. Losev on injection via a p-n boundary in SiC crystal around 1928 (see Table 8.1). The phenomenon of light emission was explained by Lehovec et al. in 1951. A few months later after Lehovec et al.'s publication, Si and Ge LEDs were demonstrated by Haynes and Briggs of Bell Laboratories. An efficient silicon LED with external QE of $\eta_{ext} = 2 \times 10^{-4}$ at room temperature was created by the group of K. P. Homewood in 2001 (Ng et al., 2001a). Similar and even larger efficiency close to 1% were observed in commercial p-i-n structures (Dittrich et al., 2001) and a textured high-quality bulk Si p-n diode (Green et al., 2001), respectively. The internal QE higher than 20% at room temperature is calculated at optimal carrier injection levels and a low surface recombination velocity (Trupke et al., 2003b; Kittler et al., 2006; Tu, 2007). Silicon-based light-emitting devices reported in the literature have exploited a variety of fabrication methods. Progress in the field of efficient silicon-based light sources is presented in this chapter.

8.2.6 Light Emission by Hot Carriers in Si

Light emission from silicon reverse-biased p-n junctions has been observed since the 1950s (Newman, 1955). In a modern version of this classical experiment, several investigators have worked by using micron-sized metal-oxide-semiconductor field-effect transistor (MOSFET) devices (see, for example, Herzog and Koch, 1988, and references therein). In addition to a recombination band with a maximum in the range of photon energies close to the width of the band-gap of bulk silicon $E_g = 1.12$ eV, the electroluminescence (EL) spectrum also exhibits wideband emission: (i) in infrared range below 0.8 eV and (ii) in visible and ultraviolet regions.

The infrared radiation is connected with direct radiative transitions of hot holes between different subbands of the valence band (direct intersubband transitions) or indirect intra-subband radiative transitions of hot carriers assisted by phonon emission or collisions with impurities. To clarify the role of various transitions, the spectra of far infrared EL from silicon bipolar transistor structures have been studied at accelerating voltage low enough that only one kind of carriers has been involved in the excitation (Kosyachenko and Mazur, 1999). The study has shown that indirect intrasubband transitions give the main input in far-infrared emission.

Although the visible and ultraviolet emission from a reverse biased silicon junction has been known since the 1950s (Newman, 1955), the exact origin of the emission has remained the subject of debate. Recently, the tip of scanning tunneling microscope (STM) has been used to locally inject electrons or holes into Si (100) and (111) wafers and to excite the emission of light (Schmidt et al., 2007). Measured spectra have revealed a contribution of direct optical transition in Si bulk. The threshold of direct optical transition in bulk silicon is about 3.2 eV. The efficiency of the no-phonon light emission in bulk silicon is low due to a fast energy relaxation of hot carriers. However, in Si NCs, the efficiency of this no-phonon emission increases by a factor of 10^3 and a considerable red shift of the luminescence band takes place due to quantum confinement (de Boer et al., 2010). For a more detailed account of Si NC luminescent properties, see the next section.

8.3 Low-Dimensional Si

Stefano Ossicini and Aleksei Anopchenko

A silicon-based light-emitting device with high efficiency is a missing part in the design of complete optoelectronic circuits based on Si technology. Bulk Si does not exhibit good optoelectronic properties because of the indirect gap, in the infrared region, and of the low probability of radiative e-h recombination (see Section 8.2). In addition, bulk Si shows significant free carrier absorption (FCA) and Auger recombination rates that impede population inversion and, hence, optical gain under strong pumping. One possibility to overcome the inability of bulk Si to be efficient light emitter is the scaling down of its structure to low-dimensional nanometer sizes. It was two decades ago when bright photoluminescence (PL) from porous silicon (PS) was discovered, and it was connected with its nanocrystalline structure and the quantum confinement effect (QCE) (Canham, 1990). This discovery seemed to remove the main obstacle to silicon photonics, and the number of research papers in this area increased enormously (Ossicini et al., 2003; Shainline and Xu, 2007; Pavesi and Turan, 2010; Khriachtchev et al., 2012).

Low dimensionality causes the zone folding of the conduction band minimum of bulk Si that is located near the X-point, thus originating in several low-dimensional structures a direct band-gap. Moreover, the QCE associated to the reduced size enlarges the energy band-gap enabling light emission in the visible range. Indeed the nanometer size of low-dimensional Si structures can also enhance the spatial localization of the e-h wave functions and their overlap, thus significantly augmenting the probability of e-h recombination.

Now low-dimensional Si can exist in three flavors: zero-dimensional (0D) systems like Si quantum dots or Si NCs, one-dimensional (1D) systems like Si quantum wires or nanowires (Si NWs), and two-dimensional (2D) systems like Si quantum wells (Si QWs) or quantum slabs. In this section, the light emission properties of these low-dimensional Si-based light sources will be discussed (for a recent review, see Degoli and Ossicini, 2009); moreover, results regarding the optical gain and the electroluminescent devices will be presented.

8.3.1 Light Emission Properties of Low-Dimensional Si

8.3.1.1 Si Nanocrystals

Embedding Si NCs in wide band-gap insulators is one way to obtain a strong QCE. Si NCs embedded in a silica matrix (see Figure 8.7) have been obtained by several techniques as ion implantation (Kanemitsu et al., 1996; Bonafos et al., 2001; Shimizu-Iwayama et al., 2001; Brongersma et al., 2000; Daldosso et al., 2003); chemical vapor deposition (Godefroo et al., 2008; Rolver et al., 2006; Hernandez et al., 2005; Gardelis et al., 2009; Iacona et al., 2000); laser pyrolysis (Kanemitsu, 1996; Delerue et al., 2006); electron beam lithography (Sychugov et al., 2005); sputtering (Charvet et al., 1999; Takeoka et al., 2000; Averboukh et al., 2002; Antonova et al., 2008); molecular beam deposition (Lu et al., 1995), and some others (Ossicini et al., 2003; Pavesi and Turan, 2010). Moreover also silicon-rich nitride matrices have been used (Warga et al., 2008; Dal Negro et al., 2006).

The obtained Si NCs exhibit bright PL in the visible region (sample dependent, with quantum yield, ranging from 1% up to 60%, in this last case using careful surface passivation and avoiding oxygen during all stages of the Si NC preparation (Jurbergs et al., 2006)). The PL decay time ranges from a few to tens of microseconds and is described by a stretched exponential function (Pavesi and Turan, 2010). Experimentally, several factors contribute to make the interpretation of measurements on these systems a difficult task. First, independent of the fabrication technique, in experimental samples, no two nanocrystals (NCs) are the same. For instance, samples show a strong dispersion in the NC size, that is difficult to determine. In this case, it is possible that the observed quantity does not correspond exactly to the mean size but instead to the most responsive NCs (Credo et al. 1999). Moreover, NCs synthesized by different techniques often show different properties in size, shape, and in interface structure. Finally, in solid nanocrystal arrays, some collective effects caused by electron, photon, and phonon transfer between the NCs can strongly

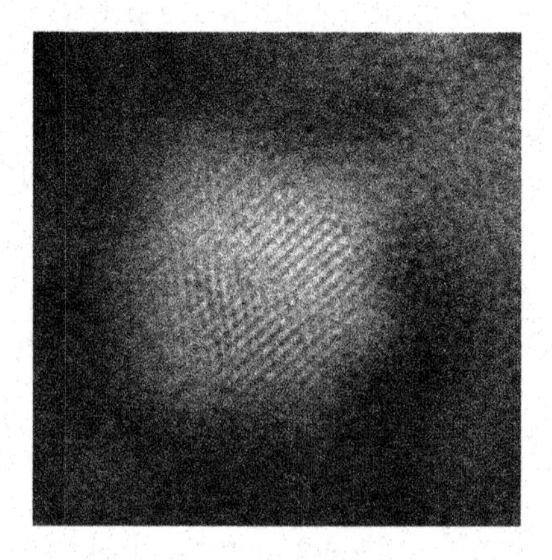

FIGURE 8.7
High resolution energy filtered TEM image of a Si NC after annealing at 1250°C. The presence of the Si(111) reticular planes is clearly visible in the nanocrystal core. The distance between adjacent (111) planes is 0.32 nm. (After Daldosso, N., M. Luppi, S. Ossicini et al. 2003. *Physical Review B* 68 (8):085327–8. With permission.)

influence the electron dynamics in comparison with the case of isolated NCs (Daldosso et al., 2003). In practice, all the conditions remarked above lead to measurements of collective quantities, making the identification of the most active configurations at the experimental level a nontrivial task.

As shown in Figure 8.8, the PL band of Si NCs is very broad. PL from a single Si NC at low temperatures has been measured (Sychugov et al., 2005), demonstrating that each Si NC exhibits atomic-like very narrow PL peaks, that the broadening observed for NC ensembles is a consequence of size and shape distributions and that Si NCs retain the indirect band-gap character of bulk Si.

The optical emission has been attributed to transitions between states localized inside the NC (Figure 8.9), as a consequence of the QCE (Sykora et al., 2008; Moskalenko et al., 2007; Hill and Whaley, 1995; Derr et al., 2009; Dovrat et al., 2009) or between defect states (Shimizu-Iwayama et al., 2001; Kanemitsu, 1996; Averboukh et al., 2002; Koponen et al., 2009; Martin et al., 2008; Wolkin et al., 1999). Although there is still some debate on which of the above mechanisms primarily determines the emission energy, some recent works have proposed that a concomitance of both mechanisms is always present, favoring one or the other depending on the structural conditions (Godefroo et al., 2008; Rolver et al., 2006; Allan et al., 1996; Hao et al., 2009; Gourbilleau et al., 2009; Zhou et al. 2003; Lin and Chen, 2009; Luppi and Ossicini, 2003; Filonov et al., 2002; Guerra and Ossicini, 2010; Guerra et al., 2009a,b, 2010, 2011).

Within this picture, it was suggested that for NC diameters above a certain threshold (of about 3 nm) the emission peak simply follows the QCE model, while interface states would assume a crucial role only for small-sized NCs. In particular these theoretical works already highlighted the dramatic sensitivity of the optoelectronic properties to the Si/SiO_2 interface configuration, especially for very small NCs ($d < 1$ nm), where a large proportion of the atoms is localized at the interface. For these sizes, NCs conditions such as passivation, symmetry, and strain considerably concur for the determination of the final optoelectronic response, producing sensible deviations from the QCE model (Guerra et al., 2009a,b, 2010, 2011; Guerra and Ossicini, 2010). Moreover, many PL experiments demonstrated that only

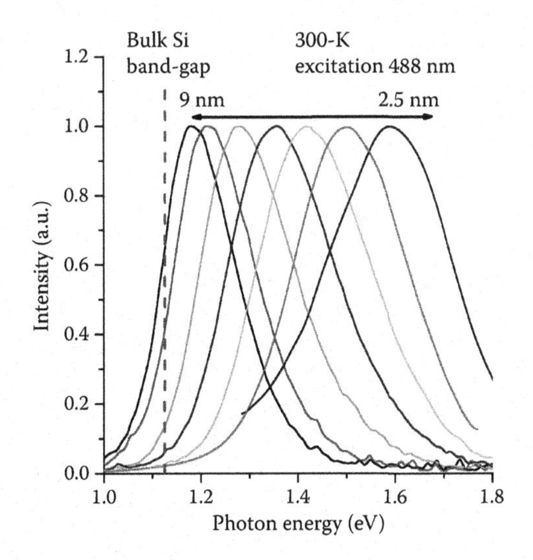

FIGURE 8.8
PL spectra of Si NCs embedded in SiO_2 thin films at room temperature. The average diameters are changed from about 9 to 2.5 nm. (After Takeoka, Shinji, Minoru Fujii, and Shinji Hayashi. 2000. *Physical Review B* 62 (24):16820–16825. With permission.)

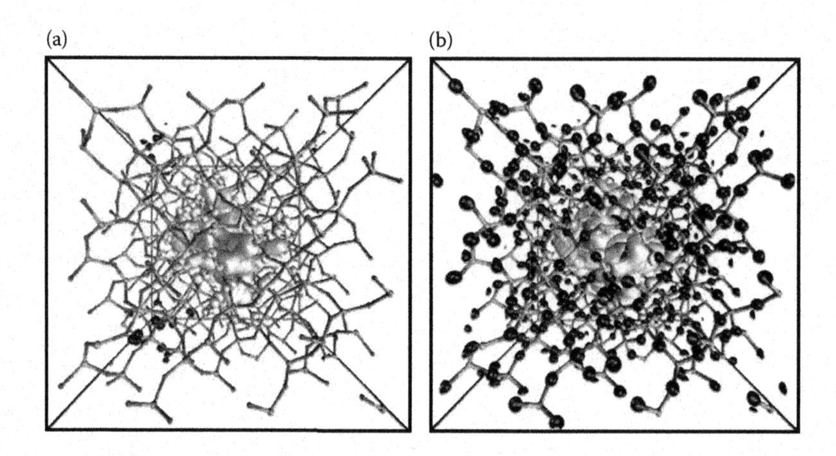

FIGURE 8.9
(See color insert.) Stick and ball pictures of the final optimized structure of Si_{32} NC in a crystalline SiO_2 matrix. Red spheres represent the O atoms, cyan the Si atoms of the matrix, and yellow the Si atoms of the NC. On the structures, Kohn-Sham orbitals at 10% of their maximum amplitude are reported. The ones in white are the sum of all band-edges orbitals localized on the NC, while the ones in black are the states delocalized on the matrix: (a) valence states and (b) conduction states. The results show that the HOMO and the LUMO are localized on the Si NC. (After Guerra, R., E. Degoli, M. Marsili, O. Pulci, and S. Ossicini. 2010. *Physica Status Solidi B Basic Solid State Physics* 247 (8):2113–2117. With permission.)

a very small fraction of the NCs in the samples contributes to the observed PL, enforcing the idea that precise structural conditions are required to achieve high absorption and emission rates. Finally, recent calculations reported especially high optical yields for smaller NCs (Guerra et al., 2011), enhancing the weight held by their contribution in real samples. It is thus clear the importance of understanding all the factors that, at these sizes, contribute to enhance (or reduce) the global optical response.

As pointed out before Si NCs remain indirect band-gap materials where structures related to momentum-conserving phonons were clearly observed. This drawback can be circumvented by introducing into the Si NCs an isoelectronic impurity (Pavesi and Turan, 2010; Bisi et al., 2000) or by simultaneous n- and p-type impurity doping (Fujii et al., 2005). In a series of intriguing papers, Fujii et al. (2003, 2004, 2005) and Fujio et al. (2008) have demonstrated that it is possible to control the PL properties of Si NCs by simultaneous doping with B and P impurities. They have shown not only that the PL intensity of codoped Si NCs is always higher than that of either P- or B-doped Si NCs, but that it is even higher than that of the undoped Si NCs. In addition, under resonant excitation conditions the codoped samples did not exhibit structures related to momentum-conserving phonons, suggesting that in this case the quasidirect optical transitions are predominant. As shown in Figure 8.10, the PL energy of the codoped sample can be far below the band-gap energy of the bulk Si crystals. Although PL peak energy shifts slightly to lower energy by doping with either P or B, below bulk band-gap PL can be realized only when both kinds of impurities are present. This is consistent with ab initio (including many body effects) theoretical calculations, which showed that the electronic properties of B- and P-codoped Si NCs are qualitatively and quantitatively different from those of either B- or P-single-doped Si NCs (Ossicini et al., 2005, 2006; Iori et al., 2007; Ramos et al., 2007, 2008).

The presence of both an n and a p impurity leads to a HOMO level that contains two electrons and to a HOMO–LUMO gap that is strongly lowered with respect to that of the corresponding undoped NC (Figure 8.11). Moreover these calculations demonstrated preferential formation of NCs with equal number of n- and p-type impurities.

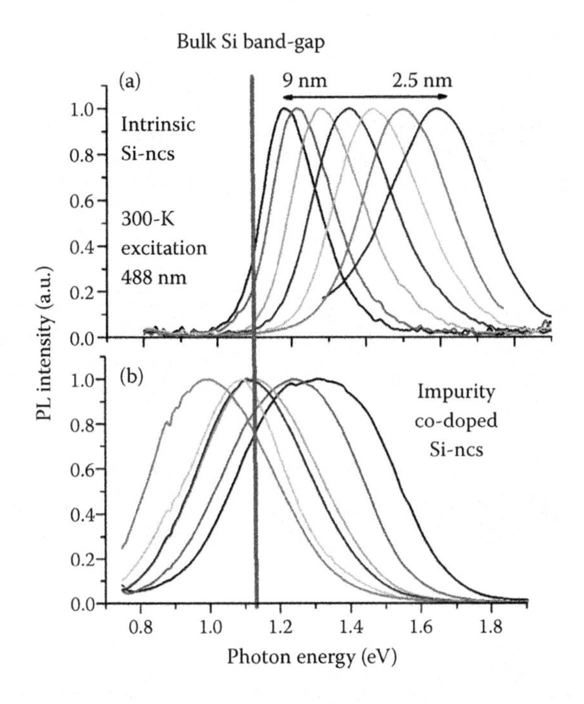

FIGURE 8.10
Normalized PL spectra of (a) intrinsic and (b) p- and n-type impurities codoped Si NCs at room temperature. The lowest possible PL energy of intrinsic Si NCs is the bulk Si band-gap, while that of codoped Si NCs is extended to 0.9 eV. (After Pavesi, L., and R. Turan: *Silicon Nanocrystals: Fundamentals, Synthesis and Applications*. 2010. Copyright Wiley-VCH Verlag GmbH & Co. KGaA. Reproduced with permission.)

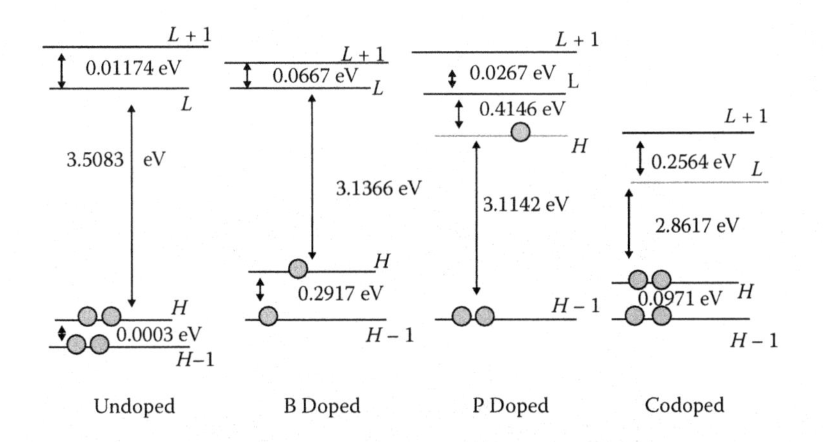

FIGURE 8.11
Calculated energy levels at the Γ point for the $Si_{35}H_{36}$-NC family. From left to right: Undoped $Si_{35}H_{36}$-NC, single-doped $Si_{34}BH_{36}$-NC, single-doped $Si_{34}PH_{36}$-NC, codoped $Si_{33}BPH_{36}$-NC. Alignment has been performed by locating, at the same energy, the fully occupied levels with the same type of localization. (After Degoli, E., and S. Ossicini. 2009. *Advanced in Quantum Chemistry Vol. 5: Theory of Confined Quantum Systems: Pt 2*, 58:203–279.)

8.3.1.2 Si Nanowires

Among the different Si nanostructures, Si NWs have recently attracted a lot of interest. Being 1D structures, they seem potentially as useful as carbon nanotubes and probably more so, due to the possibility of tailoring their chemistry, moreover it is much easier to control their electrical properties and, as long as the surface is properly passivated they are invariably semiconducting. Si NWs of different controllable sizes and growth directions (see Figure 8.12) have been synthesized through several routes (Morales and Lieber, 1998; Holmes et al., 2000; Ma et al., 2003; Kawamura et al., 2003; She et al., 2006).

Si NWs have been extensively studied and several experiments have already characterized some of their structural and electronic properties (She et al., 2006; Cui et al., 2001; Cui and Lieber, 2001; Lauhon et al., 2002; Wu et al., 2004; Menon et al., 2004). Recently, it has been possible to fabricate, for example, single-crystal Si NWs with diameters as small as 1 nm and lengths of a few tens of micrometers (Morales and Lieber, 1998; Holmes et al., 2000; Ma et al., 2003). Several applications have been demonstrated, ranging from electron devices, logic gate, nonvolatile memories, photovoltaics, photonics, biological sensors, giant piezoresistance effect, to enhanced thermoelectric performances. There exist many experimental reviews (Xia et al., 2003; Patolsky and Lieber, 2005; Li et al., 2006; Lu and Lieber, 2006; Thelander et al., 2006; Wu et al., 2008) and a very recent theoretical one (Rurali, 2010).

PL data (Ossicini et al., 2003; Bisi et al., 2000; Walavalkar et al., 2010) revealed a substantial blue shift with decreasing size of NWs. Further scanning-tunneling spectroscopy data (Ma et al., 2003) also showed a significant increase in the electronic energy gap for very thin semiconductor NWs, explicitly demonstrating quantum-size effects (Figure 8.13).

From a theoretical point of view, it has been demonstrated that only a fully microscopic, ab initio theory that correctly includes self-energy corrections to obtain the quasiparticle (QP) energies and excitonic effects, well beyond the DFT-LDA, correctly explains the size-dependent experimental gaps in Si NWs (Bruno et al., 2007). In Figure 8.14 (left), the QP electronic gaps are compared with the experimental data obtained by Ma et al. (2003) through scanning tunneling spectroscopy, for different Si NWs (grown along the [112] and [110] directions) with diameters ranging from 1.3 to 7 nm. All the experimental gaps, although 50% larger than the LDA values, fall within the [100] and [110] theoretical QP curves. Otherwise, comparing in Figure 8.14 (right) the Si NWs QP with the experimental

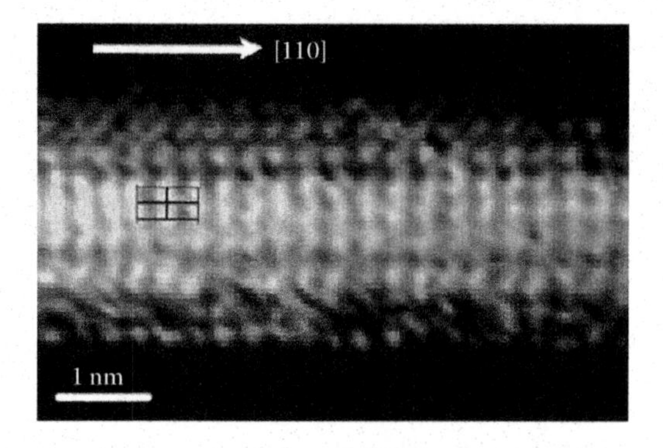

FIGURE 8.12
Constant-current STM image of a Si NW on a HOPG substrate. The wire's axis is along the [110] direction. (After Ma, D. D. D., C. S. Lee, F. C. K. Au, S. Y. Tong, and S. T. Lee. 2003. *Science* 299 (5614):1874–1877. With permission.)

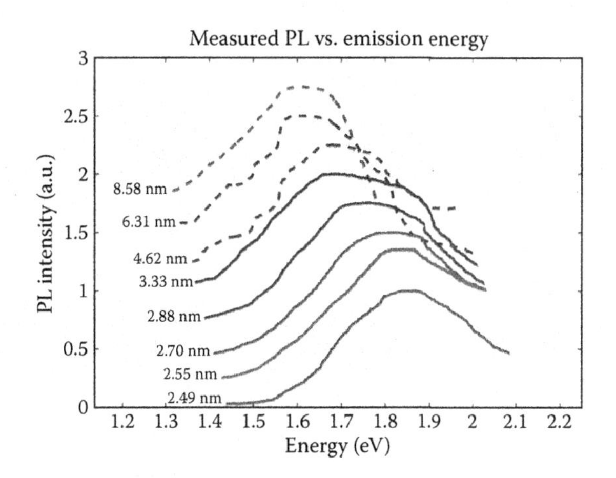

FIGURE 8.13
Normalized PL intensity for sub-10-nm Si nanopillars. The variation in diameter was obtained by changing the oxidation temperature and the diameters reported are the average pillar size measured on a sample by reflection mode TEM. Dotted (continuous) lines represent pillars with 50 nm (35 nm) initial diameters. (After Walavalkar, S. S., C. E. Hofmann, A. P. Homyk, M. D. Henry, H. A. Atwater, and A. Scherer. 2010. *Nano Letters* 10 (11):4423–4428. With permission.)

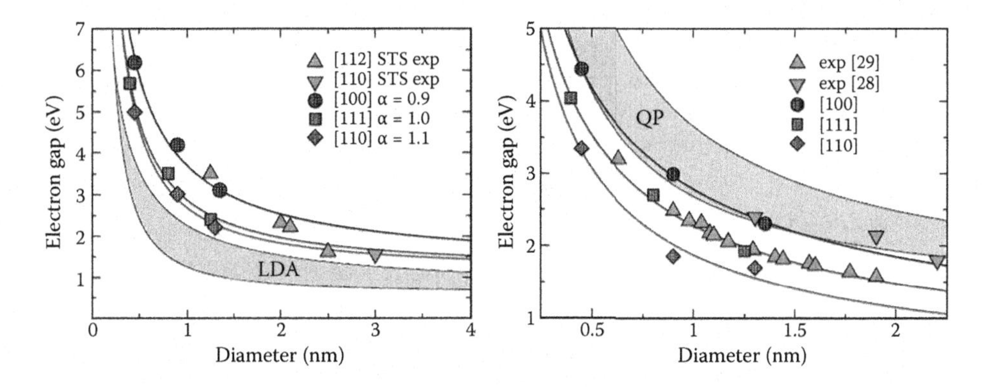

FIGURE 8.14
Left: QP gaps for [100] (circles), [111] (squares), and [110] (diamonds) Si NWs as a function of wire size compared with experimental results (triangles) from scanning tunneling spectroscopy (STS) (Ma et al., 2003). The gray region represents the LDA electronic gaps from [110] (bottom) to [100] (top) NWs. Right: Excitonic gaps for [100] (circles), [111] (squares), [110] (diamonds) Si NWs, and experiments (triangles). Down- and up-pointing triangles correspond, respectively, to the PL data from (Zhang and Bayliss, 1996) and (Wolkin et al., 1999) in PS samples. The gray region represents the QP electronic gaps, from [110] (bottom) to [100] (top) NWs. (After Bruno, M., M. Palummo, A. Marini, R. Del Sole, and S. Ossicini. 2007. *Physical Review Letters* 98 (3):036807–4. With permission.)

PL results for PS, one see that the QP gaps largely overestimate the experiment. This large discrepancy can be interpreted only in terms of the short range interactions occurring between the electrons confined in the 1D structure. Such interactions are described taking into account the effects of e-h interaction, thus the calculated Si NW excitonic gaps, corresponding to the lowest exciton with nonvanishing optical strength, are in excellent agreement with the PL experimental data from (Wolkin et al., 1999; Zhang and Bayliss, 1996).

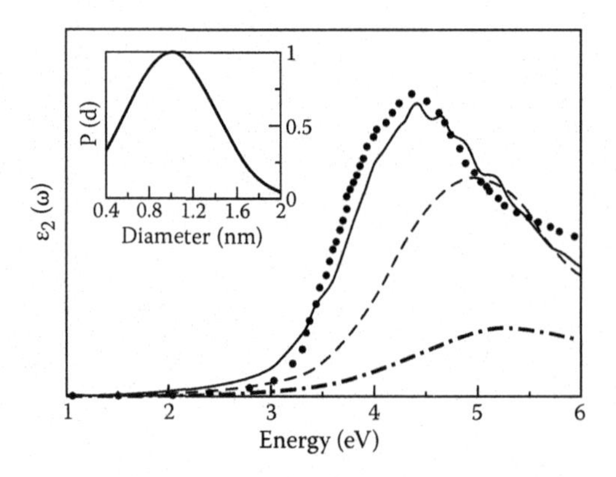

FIGURE 8.15
Excitonic (solid line) and QP (dashed line) $\varepsilon_2(\omega)$ of PS compared with the experiments (dots) (Koshida et al., 1993). The average contribution due to NW–NW interaction is also shown (dot–dashed line). In the inset is reported the Gaussian distribution of NWs as a function of their diameters, as used in the described modeling of PS. (After Bruno, M., M. Palummo, A. Marini, R. Del Sole, and S. Ossicini. 2007. *Physical Review Letters* 98 (3):036807–4. With permission.)

Indeed, the interpretation of PL spectra of PS has long been a subject of debate, now these ab initio many-body results predicted excitonic gaps and optical absorption functions that are in good agreement with experimental data; here, the PL was modelized as resulting from an array of interacting Si-NWs of varying diameters (Figure 8.15).

8.3.1.3 Si Quantum Wells

Regarding the 2D systems, the optoelectronic properties of Si/CaF_2 (d'Avitaya et al., 1995; Bassani et al., 1996; Bassani et al., 1997; Ioannou-Sougleridis et al., 2001), Si/SiO_2 (Lu et al., 1995; Lockwood et al., 1996; Takahashi et al., 1995; Novikov et al., 1997; Min et al., 1996; Kanemitsu and Okamoto, 1997; Zacharias et al., 1999; Yoo and Fauchet, 2008; Portier et al., 2003), Si/SiN_x (Warga et al., 2008; Martin et al., 2008; Baribeau et al., 1998) and Si/$Si_xO_yN_z$ superlattices (Modreanu et al., 2002) have been experimentally studied, confirming the role of quantum confinement, even if the discussion is still alive. In fact, to observe the QC properties of these systems, after the thermal crystallization process, it seems necessary to have the presence of a dielectric (high band-gap) material that separate the NCs within the layer (for reviews, see Pavesi and Turan, 2010; Heitmann et al., 2005). Thus, owing the importance of the control of the size in the QC process, these approaches have been mainly used to realize an improved size control of Si NCs. The most interesting results have been obtained within the multilayer approach in which the Si-rich silicon oxide (SiO_x, $x < 2$) sublayers are sandwiched between SiO_2 sublayers (Figure 8.16), allowing to control the size of the Si NCs formed during the growth and/or after specific annealing treatment (Zacharias et al., 1999; Portier et al., 2003).

Regarding the theoretical efforts with respect to the 2D systems, it is worth noting that also for these systems the results elucidate the role of both quantum confined and interface states (Degoli and Ossicini, 2000). For a recent review devoted to the theoretical aspects the reader is left to (Degoli and Ossicini, 2009).

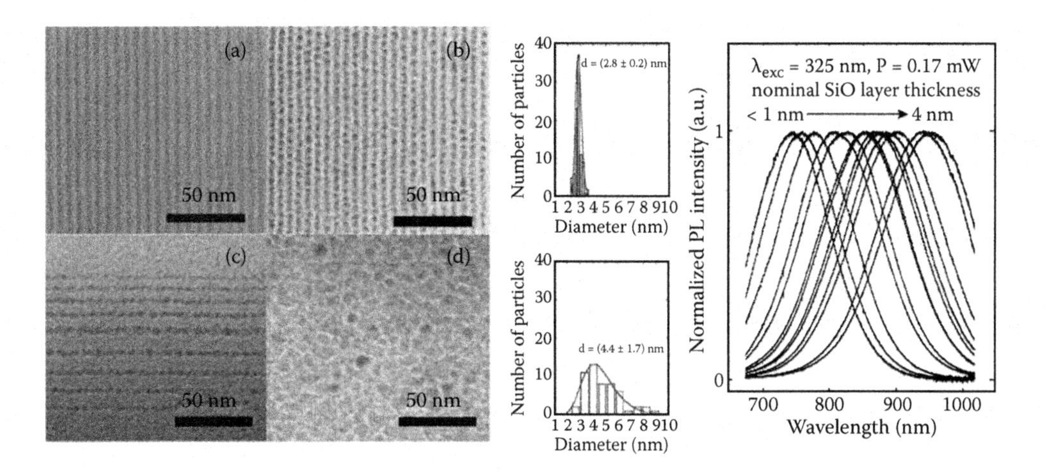

FIGURE 8.16
Left: Cross-sectional bright field TEM images of (a) an as-prepared amorphous SiO/SiO_2 superlattice, (b) the same film after 1100°C annealing under nitrogen, (c) the effect of changing the thickness of the SiO layer and the correspondent change in Si NCs size, and (d) a bulk SiO film after annealing for comparison. Right: PL spectra after 1100°C annealing normalized to peak intensity. (After Pavesi, L., and R. Turan: *Silicon Nanocrystals: Fundamentals, Synthesis and Applications*. 2010. Copyright Wiley-VCH Verlag GmbH & Co. KGaA. Reproduced with permission; Heitmann, J., F. Muller, M. Zacharias, and U. Gosele. 2005. *Advanced Materials* 17 (7):795–803. With permission.)

8.3.2 Electroluminescent Low-Dimensional Si Devices

Bulk Si LEDs were described in Section 8.2.5. Here, we will give a brief overview of low-dimensional Si LEDs. There are several good books and review papers that cover in detail many aspects of light-emitting properties of low-dimensional Si devices described in this section: Ossicini et al. (2003), Koshida (2008), Lockwood (2008), Khriachtchev (2009), and Pavesi and Turan (2010), and Tsu (2011).

The most common Si NC LED structure is a MOS-like capacitor or a MOSFET device grown in the planar CMOS technology, where Si NCs are embedded in the gate oxide layer (Figure 8.17); see, for example, Franzo et al. (2002), Walters et al. (2005), and Marconi et al. (2011). Light emission is normally collected through the device gate, which is usually formed by a degenerate semitransparent polycrystalline silicon layer. Majority of the Si NC LEDs are unipolar (one type of charge carriers; as a rule, electrons), which operate in a high electric filed/applied voltages, where charge injection obeys the Fowler–Nordheim tunneling law (for a review, see Koshida, 2008). Direct charge tunneling and bipolar injection were demonstrated in multilayered Si NCs devices with very thin silicon oxide layers (Marconi et al., 2009). Other device layouts include p-i-n hetero- and homo-junctions of undoped or boron and phosphorous doped Si NCs (Perez-Wurfl et al., 2009; Koshida, 2008).

The best Si NC LEDs have optical power of about 1 mW (Gong-Ru et al., 2008) and a long-term stability (Marconi et al., 2010). Power and QE of Si NC LEDs has increased significantly over the years, and reaches values, which are somewhat less than 1% at low injection levels at room temperature (Marconi et al., 2009; Koshida, 2008). These characteristics are yet to be improved for practical applications of Si NC LEDs. The main reason for the low LED efficiency is low and unbalanced carrier injection currents due to the insulating silicon oxide matrix. Other matrices with improved carrier injection and transport properties,

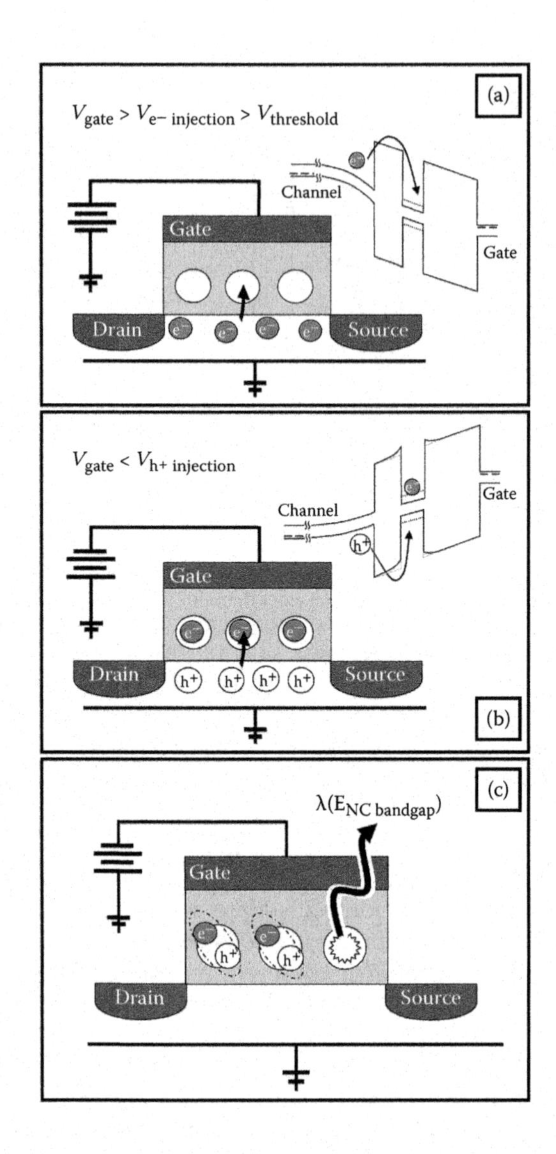

FIGURE 8.17
A schematic diagram of a silicon nanocrystal floating gate transistor structure. The array of silicon nanocrystals embedded in the gate oxide of the transistor can be sequentially charged with electrons (a) by Fowler–Nordheim tunneling and holes (b) via Coulomb field-enhanced Fowler–Nordheim tunneling to prepare excitons that radiatively recombine (c). Inset band diagrams depict the relevant injection processes. (With kind permission from Walters, R. J. 2007. *Silicon Nanocrystals for Silicon Photonics, Engineering and Applied Science*. California Institute of Technology, Pasadena, California.)

like silicon nitride (Yi, 2008) or silicon carbide (Song et al., 2008), are under intensive investigation too. However, presence of larger amount of defects in those matrices and absence of a good passivation of Si NCs result in lower device efficiencies.

Si NC device modeling, which describes electrical and light emission properties of Si NC LEDs is limited and sparse. Existing models differ in complexity and a level of generalization, which spans from ab initio (Degoli and Ossicini, 2009), quantum mechanical, and semiclassical (Flynn et al., 2010; Li et al., 2011) to electric circuit (Carreras et al., 2009) and phenomenological levels (Balberg, 2011).

Novel device architectures and carrier injection schemes were suggested that enhance charge transport and injection into the Si NCs and/or light emission properties. The hetero-junction (HJ) LED with a graded-size ensemble of Si NCs (Figure 8.18a) presents a good compromise between an ensemble of large Si NCs with high electrical conductivity, but low luminescence, and an ensemble of small Si NCs with high luminescence, but poor electrical conductivity (Anopchenko et al., 2011). A stepwise decrease in the Si NC size toward a recombination region of this LED reduces the energy difference between adjacent tunneling states when an external bias is applied and enhances EL (Figure 8.18b).

Another example of the novel device architectures is a field effect LED (Walters et al., 2005; Walters, 2007). This LED is excited by a pulsed gate voltage rather than a constant current, and excitons are created in Si NCs by the sequential injection of complementary charge carriers from a semiconductor channel (Figure 8.17). The field effect LED operates in the Fowler–Nordheim regime and has low EL efficiency.

Direct modulation rate of the field effect LED and other Si NC LEDs is limited by the radiative lifetime of Si NCs and normally does not exceed a few hundreds of kHz (Marconi et al., 2011). Nevertheless, by controlling the injection of additional carriers into the gate oxide with Si NCs, it is possible to drive the Si NC LED beyond the spontaneous emission lifetime where the light emission modulation rate is determined by the gate bias (Walters, 2007). This was shown by Carreras et al. (2008) in the field effect LED where excitons were

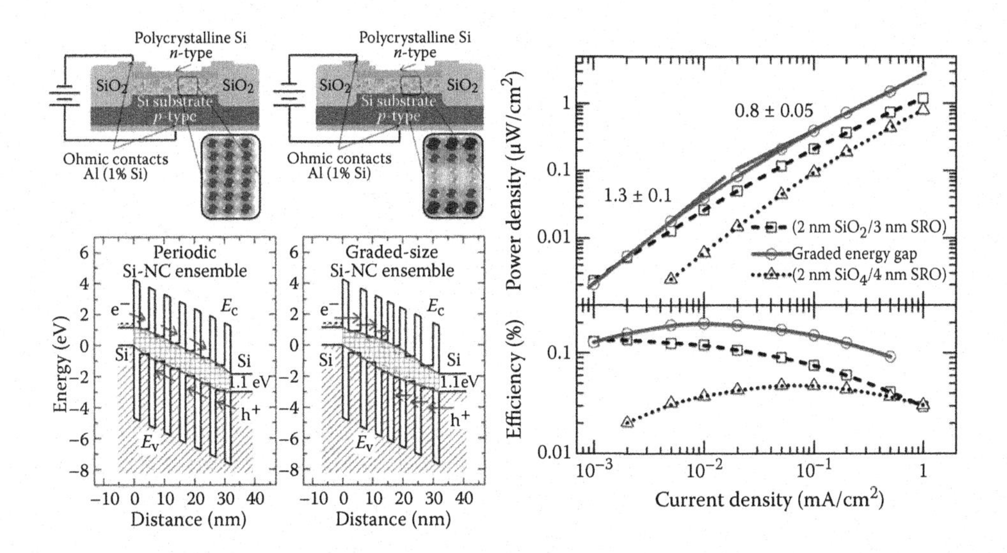

FIGURE 8.18
On the left: Schematic of multilayered Si NC LEDs with periodic (left) and graded size (right) Si NC ensembles and corresponding energy band diagrams. E_C and E_V stand for the bottom of the conduction and the top of the valence energy bands, respectively. The checkered areas highlight the difference in the energy band-gap of the active recombination region of the two LEDs. Notice an even alignment of the (electron and hole) ground states of Si NC quantum dots and wider band-gap in the middle of the active region of the graded-size LED. The arrows show direct tunneling (followed by a rapid thermalization) of electrons (e^-) and holes (h^+) from the cathode and anode, respectively. On the right: Optical power density as a function of injected electrical current density for the periodic and graded-size LEDs. Numbers indicate slope values of linear regressions. The bottom panel shows the corresponding power efficiency. (Reprinted with permission from Anopchenko, A., A. Marconi, M. Wang, G. Pucker, P. Bellutti, and L. Pavesi, *Applied Physics Letters*, 99, 181108–3, 2011. Copyright 2011 by the American Physical Society.)

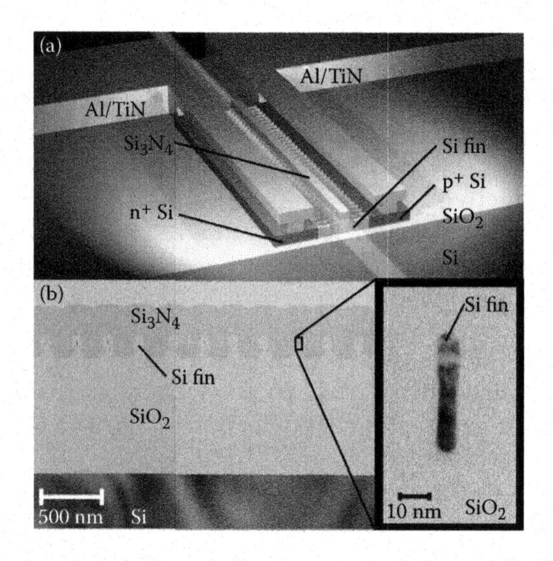

FIGURE 8.19
Si FinLED. (a) Schematic device structures. Active channel width (length) is 500 (1) μm, and cavity length (width) is 500 (0.6) μm. (b) Transmission electron microscope (TEM) images after processing. Inset shows single Si (100) fin. (Reprinted with permission from Saito, S., T. Takahama, K. Tani et al., *Applied Physics Letters*, 98, 261104–3, 2011. Copyright 2011 by the American Physical Society.)

intentionally quenched shortly after creating them in Si NCs by injecting additional carriers to induce Auger recombination.

Nonplanar CMOS and silicon-on-insulator technologies offer new opportunities for Si LEDs. Ding et al. (2011) studied LEDs with a lateral current injection into Si multiple quantum wells (Si QWs) and observed 20 times EL intensity enhancement compared to a p-i-n LED with electric field applied across the QWs. A four orders of magnitude increased conductivity in lateral direction compared with QWs with vertical contacts is shown in a similar fabricated Si NC devices (Rolver et al., 2008).

SE in a red/near-infrared region by current injections at room temperatures has been reported recently from a single and multiple Si QWs, called fin, LEDs (FinLEDs) (Saito et al., 2009, 2011). The Si QWs were formed, and the size of fins was reduced down to 5 nm, by silicon oxidation processes using silicon-on-insulator substrates. The fin-LED structure (Figure 8.19) is similar to the fin field-effect transistor, except for the impurity profiles and gate processes. By lateral current injections into Si fins, SE of near-infrared radiation have been observed at the edge of the Si_3N_4 waveguide.

8.3.3 Optical Gain in Si Nanocrystals

Si NCs in silicon oxide matrix is the most investigated candidate among other forms of low-dimensional Si for a gain medium of the Si laser. Optically pumped gain in Si NCs was first reported by Pavesi et al. (2000) and later by a number of research groups (for reviews and references, see Pelant, 2011; Belyakov et al., 2008; Daldosso and Pavesi, 2009; Pavesi et al., 2003)). The gain coefficient of Si NCs in oxide matrix is ranging from 10 to 100 cm^{-1}, which depends on sample preparation (Daldosso and Pavesi, 2009). Some key specimen requirements relating to observation of optical gain and SE in dense ensembles of Si NCs as well as gain limiting factors, like the gain profile and energy transfer among Si NCs, are reviewed in Belyakov et al. (2008).

Nitride-passivated Si NCs have attracted much attention because of a large refractive index, fast PL, electrical injection, and lower annealing temperatures, see, for example, Yi (2008). An optical gain of $<52\ cm^{-1}$ has been measured recently in nitride-passivated Si NCs under femtosecond optical pumping (Monroy et al., 2011).

Commonly used techniques for optical gain measurements in Si NCs are the variable stripe length technique (VSL) (Pavesi et al., 2000) in combination with the shifting excitation spot (SES) technique (Pelant, 2011) and pump and probe technique (Dal Negro et al., 2004). An extension of the VSL method to ultrafast gain measurements is reported in Žídek et al. (2011).

There is no common opinion about the nature of radiative processes that led to the observed gain in Si NCs, but there is a belief that both the QCE and recombination states localized on Si NC surface are equally important and contribute to both optical emission and gain in Si NCs. The surface states are associated to either silicon dimers (Filonov et al., 2002) or Si=O bonds (Wolkin et al., 1999; Luppi and Ossicini, 2003) at the interface between the Si NC and the oxide matrix (see Section 8.3.1).

Optical gain and stimulate emission were modeled qualitatively using a four-level system of rate equations shown in Figure 8.20, which describes competition between recombination processes, absorption, and SE (Dal Negro et al., 2003). Absorption of a photon occurs as a vertical electronic transition between the ground state (level 1) and the excited state (level 2). The excited electron then relaxes to a new minimum energy configuration (level 3). Emission (either stimulated or spontaneous) is represented in this diagram by a downward electronic vertical transition to level 4. Once the Si NC is in its ground state, it relaxes again to the minimum energy configuration, which corresponds to level 1. This scheme implies that absorption (transition between level 1 and level 2) occurs at shorter wavelengths than emission (transition between level 3 and level 4).

Physical interpretation of such a four-level system depends on which the emission band of luminescent Si NCs is considered, or in other words, a time scale of radiative transitions. Ultrafast quasi-direct transitions of hot carriers in Si NCs are attracting much attention recently because of high efficiency (see a comment in Section 8.2.6). Žídek et al. (2011) observed a transient optical gain up to $600\ cm^{-1}$ on a subpicosecond time scale in oxidized

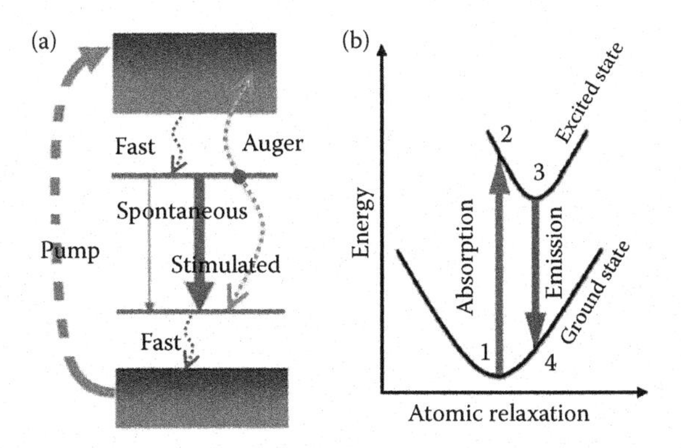

FIGURE 8.20
(a) Energy diagram for a four-level system. The various transitions are indicated by lines, those with wavy lines are nonradiative. (b) Configuration coordinate diagram associated to atomic relaxation. (From Daldosso, N., and L. Pavesi: Nanosilicon photonics. *Laser & Photonics Reviews*, 2009, 3, 508–534. Copyright Wiley-VCH Verlag GmbH & Co. KGaA. Reproduced with permission.)

Si NCs and attributed it to the QCE and efficient quasi-direct radiative transitions in a core of the Si NC (de Boer et al., 2010). The rapid cutoff of optical gain was ascribed to a very fast carrier trapping at the Si NC surface states.

An optical cavity/resonator with a high quality factor is required to realize a laser with the Si NC gain medium. Some examples of such optical microcavities to which Si NCs could be coupled will be presented in Section 8.7.

8.4 Light Emission in Si/SiGe Nanostructures

David J. Lockwood and Leonid Tsybeskov

8.4.1 Light-Emitting Properties of Bulk SiGe Alloys

Since both crystalline silicon (c-Si) and crystalline germanium (c-Ge) are indirect band-gap semiconductors, it is a common assumption that SiGe crystalline alloys should also exhibit properties of an indirect band-gap semiconductor. Actually, the entire concept of

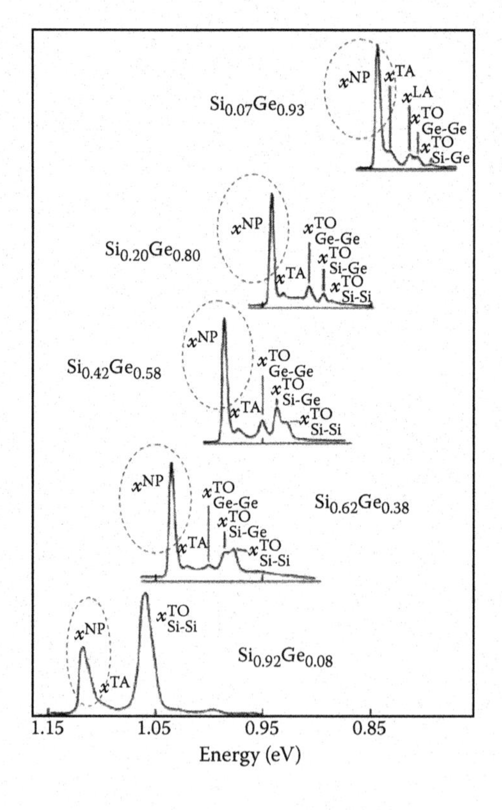

FIGURE 8.21
Low-temperature PL spectra in bulk $Si_{1-x}Ge_x$ alloys for a wide range of compositions x. (Reprinted with permission from Weber, J., and M. I. Alonso, *Physical Review B*, 40, 5683–5693, 1989. Copyright 1989 by the American Physical Society.)

wave vector (k) space is applicable to a crystal lattice with an atomically perfect long-range order. SiGe alloys are crystalline material but with strong compositional disorder (i.e., the absence of long range order in the position of Si and Ge atoms within a SiGe crystal lattice). As a result of this compositional disorder, carrier recombination selection rules in SiGe alloys are partially relaxed. This conclusion is supported by Figure 8.21, where instead of a totally suppressed zero-phonon transition, which is normally observed in low-temperature PL spectra of high-quality single-crystal Si and Ge, SiGe bulk alloys exhibit PL spectra showing both phonon-assisted and zero-phonon radiative transitions in the spectral region of 1.3–1.6 μm (see also Weber and Alonso, 1989). The relaxation of carrier recombination selection rules in SiGe alloys produces PL that is faster compared with that in Si but still orders of magnitude slower compared with the recombination time in direct band-gap semiconductors. As the temperature increases, the PL intensity quickly decreases with a thermal-quenching activation energy of ~10–15 meV, which has been attributed to the binding energy of an exciton localized on SiGe stoichiomety fluctuations (Weber and Alonso, 1989; Sturm et al., 1991; Lenchyshyn et al., 1992). Therefore, this type of luminescence in SiGe bulk crystalline alloys has little practical importance as far as room-temperature light-emitting device applications are concerned.

8.4.2 Light Emission in Si/SiGe Quantum Wells and Si/Ge Nanowires

8.4.2.1 Quantum Wells

Epitaxial (2D) growth of pure Ge on Si is complicated by the 4.2% lattice mismatch (see, for example, Paul, 2004, a review article). To avoid formation of structural defects (mainly dislocations) with a large number of nonradiative recombination centers, controlled-composition $Si_{1-x}Ge_x$ alloys with $0.1 < x < 0.2$ can be grown using techniques such as molecular beam epitaxy (MBE) or ultra-high vacuum (UHV) chemical vapor deposition (CVD). These heterostructures exhibit a low defect density at Si/SiGe interfaces, and $Si_{1-x}Ge_x$ multilayers with $0.1 < x < 0.2$ were intensively studied as a possible method to confine e-h pairs in double hetero-junction (DHJ) or QW configurations and to reduce the luminescence thermal quenching. An intense luminescence, but still restricted to low temperature, has been demonstrated in $Si_{1-x}Ge_x$ QWs and superlattices with $x \leq 0.2$ (see Figure 8.22 and also Robbins et al., 1992).

Several studies have been specifically focused on the determination of the energy band alignment (i.e., type I or type II, see Figure 8.23) in these nanostructures (NSs), but so far no final conclusion has been reached (Houghton et al., 1995; Thewalt et al., 1997; Shiraki and Sakai, 2005). It has also been shown that SiGe interdiffusion during growth of $Si/Si_{1-x}Ge_x$ QWs can be minimized using relatively low growth temperatures (< 550°C). On the other hand, SiGe epitaxial growth at temperatures below 500°C usually produces a significant density of dislocations (Shiraki and Sakai, 2005; Hull and Bean, 1999; Baribeau et al., 1990). Thus, high quality $Si/Si_{1-x}Ge_x$ QWs with $x < 0.2$ can be grown using only a relatively narrow processing growth "window," but again, the PL QE thermal quenching remains the unresolved problem for practical light emitters at 1.3–1.5 μm.

8.4.2.2 Nanowires and Nanowire HJs

Another interesting possibility is to use 1D growth in the form of nanowires (NWs) produced by vapor-liquid-solid (VLS) growth or similar techniques (Levitt, 1970; Kamins et al., 2004; Hannon et al., 2006; Zakharov et al., 2006). Compared with 2D and 3D NSs, NWs

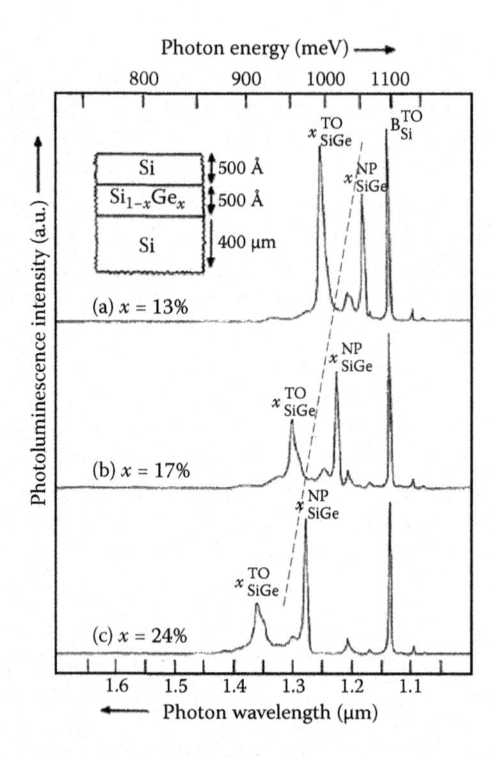

FIGURE 8.22
Low-temperature PL spectra in $Si/Si_{1-x}Ge_x/Si$ single QW (see inset) with composition $0.1 < x < 0.25$. (Reprinted with permission from Robbins, D. J., L. T. Canham, S. J. Barnett, A. D. Pitt, and P. Calcott. 1992. Near-band-gap photoluminescence from pseudomorphic $Si_{1-x}Ge_x$ single layers on silicon. *Journal of Applied Physics* 71 (3):1407–1414. Copyright (1992), American Institute of Physics.)

may have Si/Ge HJs with different positions (e.g., Ge NW-Si substrate HJs) and geometries (e.g., axial Si/Ge NW HJs, where the Si/Ge heterointerfaces are perpendicular to the NW axes and radial "core-shell" NW HJs, where the Si/Ge heterointerfaces are parallel to the NW axes). (Zakharov et al., 2006; Lauhon et al., 2002; Xiang et al., 2006).

Figure 8.24 shows scanning electron micrographs of ~40-nm-diameter Ge NWs grown on (100) and (111) single-crystal Si substrates. In both cases, the Ge NW (111) crystallographic

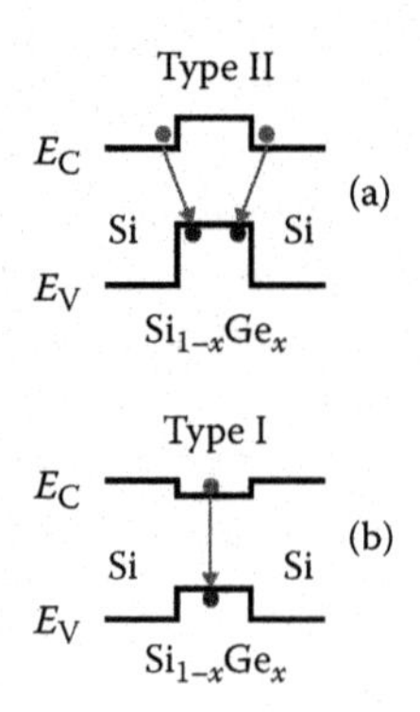

FIGURE 8.23
Schematic representation of (a) type I and (b) type II energy band alignments thought to exist in Si/SiGe NSs. E_C and E_V are the energies of the conduction and valence bands, respectively, in Si and $Si_{1-x}Ge_x$.

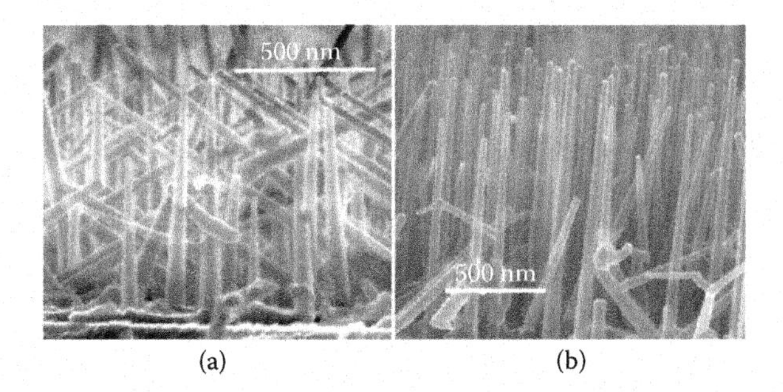

FIGURE 8.24
Scanning electron micrographs of ~40 nm diameter Ge NWs grown on (a) (100) and (b) (111)-4″ single-crystal Si substrates.

growth orientation determines the NW spatial growth direction: Ge NWs on a (100) Si substrate form an angle of ~±55° to the Si substrate normal. Such NWs require a cautious use of the laser excitation for Raman and PL measurements and an independent check of the sample temperature during photoexcitation (Kamenev et al., 2005a).

The Raman spectrum of Ge NW's grown on a (100) Si substrate exhibits two peaks (Kamenev et al., 2005a). A peak at ~520 cm^{-1} is related to Si–Si vibrations, and it originates from the c-Si substrate. The second Raman peak is at ~300 cm^{-1} and is related to Ge–Ge vibrations; its intensity strongly depends on the excitation wavelength with a clear maximum at an excitation wavelength of ~500 nm (Kamenev et al., 2005a). For wavelength excitation >500 nm, the Raman spectrum at 300 cm^{-1} is fully symmetric with a full width at half maximum (FWHM) of ~6 cm^{-1}. Such a narrow Raman peak and its intensity dependence on the excitation wavelength in Ge NWs are similar to the Raman data obtained in single crystal Ge structures (Yu and Cardona, 2001; Sui et al., 1993); they confirm the high crystallinity of the Ge NW core.

Figure 8.25 compares low-temperature PL spectra in single-crystal Si and in Ge NWs grown on (100) and (111) substrates. The main PL peak in Ge NWs (presumably the

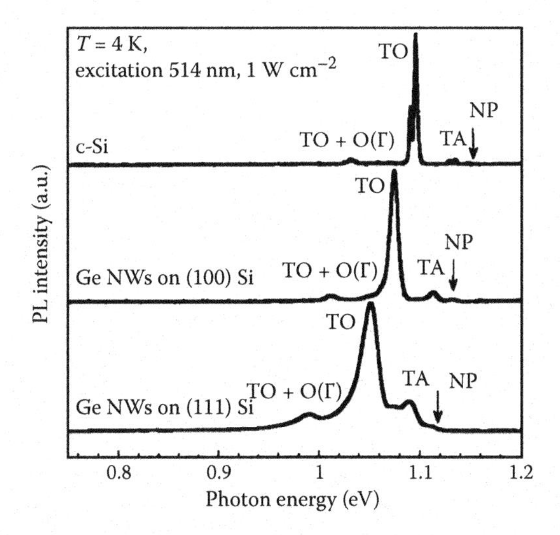

FIGURE 8.25
Low-temperature PL spectra of c-Si and Ge NWs grown on (100)- and (111)-oriented Si substrates. The NP PL line and PL bands associated with characteristic phonons are shown.

transverse optic (TO)–phonon PL band) is slightly red-shifted compared with the PL spectrum of c-Si (by ~20 meV) in the Ge NW sample grown on Si(100), and by ~45 meV for Ge NWs grown on a Si(111) substrate. The positions of no-phonon (NP) PL lines (indicated by arrows), as well as transverse acoustic (TA)Si and TOSi + O(Γ)Si phonons, are also shown (for Si phonon energies, see Davies, 1989). Note that the ~20-meV FWHM of the PL spectrum from the Ge NW sample grown on the (111) Si substrate is almost twice as broad as that for the sample of Ge NWs grown on the (100) substrate. In addition, the PL spectrum of Ge NWs grown on the (111) substrate shows an additional PL feature at 1.074 eV located between the TASi and TOSi phonon replicas. It is possible that the PL band centered at 1.074 eV is associated with one of the Si–Ge (Ge) phonons (Weber and Alonso, 1989). The PL intensities are found to be comparable for Ge NWs grown on (111) and on (100) Si substrates.

The PL intensity does not depend on the NW length for either substrate orientation. The PL signal is temperature independent for $T < 16$ K; it decreases exponentially with increasing temperature, exhibiting a PL thermal-quenching activation energy of $E_{PL} \sim 13$ meV (Kamenev et al., 2005a). This activation energy is close to the exciton binding energy in Si-rich SiGe alloys (Weber and Alonso, 1989).

The data clearly suggest that the observed PL originates at the Ge NW-Si substrate heterointerfaces and not within the Ge NW volume or the Si substrate. The PL main peak is strongly blue shifted compared with the c-Ge band-gap and red-shifted compared with c-Si band-gap. The Ge NWs average diameter of ~40 nm is too large for any significant influence of the QCE. We note that similar PL spectra are also found in SiGe nanostructures with Ge content <10% (Kamenev et al., 2004). The PL intensity shows no correlation with the Ge NW length; it is similar for all three NW lengths and both substrate orientations used in this study. Why does the Ge NW volume not show a significant PL signal near the crystalline Ge band-gap? The main reason, most likely, is the known poor properties of germanium–germanium oxide interfaces with high concentrations of nonradiative recombination centers combined with the high NW surface-to-volume ratio.

Raman measurements do not detect any significant SiGe vibrations at ~400 cm^{-1} (Kamenev et al., 2005a), pointing out that the NW-substrate interface transition region is very thin. At the same time, our PL measurements under the chosen excitation (Figure 8.25) do not show any PL from the Si substrate, which at $T < 10$ K may have QE approaching 1%. Therefore, we conclude that the internal QE of the Ge NW-Si substrate heterointerface is high.

In axial Si/Ge NW HJs, strain created by the lattice mismatch between Si and Ge can partially be relieved at the heterointerfaces by the lateral displacement of atoms in small-diameter NWs (Li et al., 2006; Ertekin et al., 2005). Thus, the fabrication and study of Si/Ge NW HJ properties could open interesting opportunities in building defect-free nanodevices.

A TEM micrograph of a typical Si/Ge NW HJ is shown in the inset of Figure 8.26. The clearly visible NW diameter expansion near the transition from Si to Ge is, most likely, related to the effect of the 4.2% lattice constant mismatch and the different thermal contractions of Si and Ge between their growth temperatures and room temperature. Raman scattering measurements (Figure 8.26) confirm the presence of Ge in the NWs (Raman peak at 300 cm^{-1}) as well as strong SiGe intermixing (a peak near 400 cm^{-1}). Also, a broad Raman peak near 500 cm^{-1} associated with Si–Si vibration in SiGe indicates a nonuniform strain (Nakashima et al., 2006), whereas the Raman signal associated with the Si substrate is found at 520 cm^{-1}.

Figure 8.27 compares the normalized Si/Ge NW PL spectra with the c-Si PL spectrum measured at the same low temperature. The Si/Ge NW PL spectra clearly exhibit at least

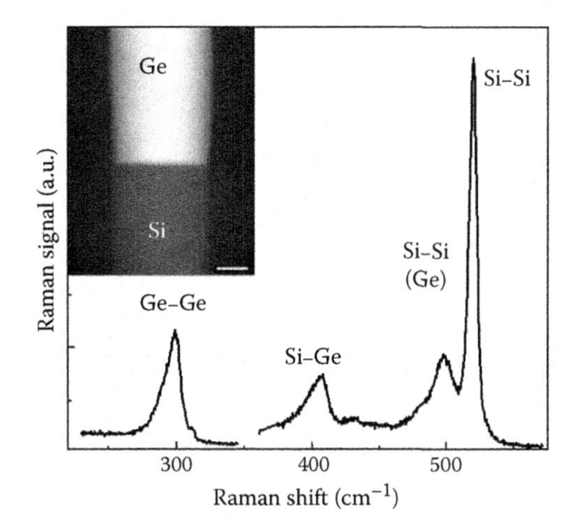

FIGURE 8.26
Raman spectrum of Si/Ge NW HJs collected under 458 nm excitation of an Ar$^+$ laser. The inset shows a TEM micrograph of a wire sample with a 20-nm-length scale bar.

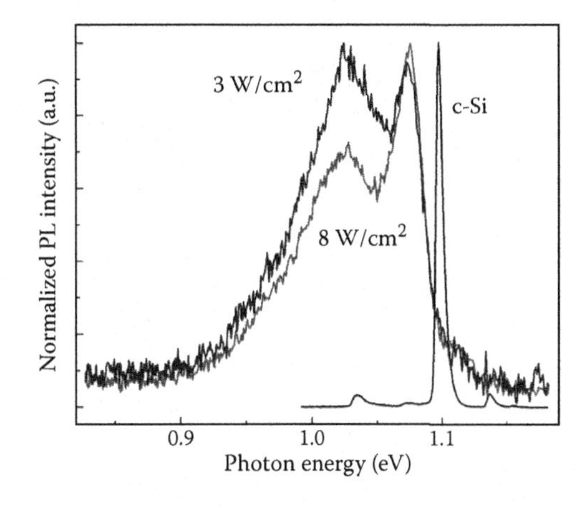

FIGURE 8.27
Low-temperature (20 K) normalized PL spectra of Si/Ge NW HJs recorded under two different excitation intensities, as indicated. The normalized PL spectrum of c-Si is shown for comparison.

two peaks (a narrower PL peak at ~1.08 eV, and a broader PL peak at ~1.025 eV) and no significant PL with photon energies close to the bulk c-Si transverse optical (TO) PL peak at ~1.1 eV. The relative intensities of the two Si/Ge NW PL peaks are found to be excitation and temperature dependent (see Figures 8.27 and 8.28). Measuring the Si/Ge NW PL temperature dependence, we find that the broader PL peak at ~1.025 eV is actually comprised of two peaks, which is most clearly seen at 60 K, with the peak positions at ~0.96 and ~1.0 eV (Figure 8.28). The ~1.08-eV PL peak exhibits an asymmetric broadening at higher temperatures (60 K), which can be well fitted using Boltzmann thermal broadening on the high photon energy side of the PL spectrum (Tsybeskov et al., 1996). The PL peaks at ~1.08 and 0.96 eV do not change their positions significantly with temperature, while the PL peak at ~1.0 eV shifts nonmonotonically and exhibits the lowest photon energy between 60 and 80 K. The PL shown in Figures 8.27 and 8.28 is associated with radiative transitions

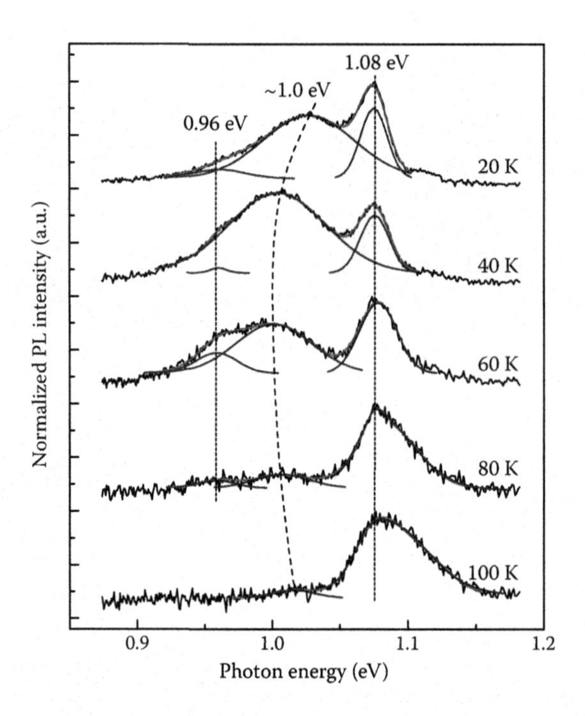

FIGURE 8.28
Normalized PL spectra of Si/Ge NW HJs measured under 8 W/cm^2 excitation intensity at different indicated temperatures with the fitted PL peaks and marked peak positions. The PL spectra are shifted vertically for clarity.

in Si/Ge NW HJs. The PL peak at ~1.08 eV is ~20 meV redshifted and strongly broadened when compared with the c-Si PL spectrum (Figure 8.27), indicating, in agreement with the Raman data and theoretical analysis in Hannon et al. (2006), significant strain in the Si/Ge NWs. The PL at ~ 1.0 eV indicates spontaneous formation of $Si_{1-x}Ge_x$ alloys between the Si and Ge parts of the NWs. Based on the strain introduced into the Si NW part, which can be estimated from the strained Si PL peak spectral position the PL peak at 1.08 eV, we find $x \leq 20\%$. Indeed, the PL peak at 1.0 eV corresponds to band-gap PL in a strained $Si_{1-x}Ge_x$ alloy on (111)Si with an alloy composition close to $x \approx 15\%–20\%$ (Kamenev et al., 2004; Paul, 2004; Brunner, 2002). Since the PL peaks at 0.96 and 1.0 eV show different temperature dependences, they are not associated with TO-phonon replica and no-phonon PL, which is frequently observed in SiGe alloys (Paul, 2004; Brunner, 2002). Most likely, the PL peaked at 0.96 eV corresponds to the band-gap of a $Si_{1-x}Ge_x$ alloy with $x \approx 0.5$ (Brunner, 2002; Paul, 2004).

The spontaneous SiGe intermixing found in VLS grown Si/Ge NW HJs is, most likely, due to residual Si atoms in the eutectic Si–Au alloy after switching to Ge deposition and to strain-induced Si and Ge interdiffusion caused by the 4.2% lattice mismatch between Si and Ge. Our PL data show that the concentration gradient from Si to Ge within the VLS grown Si/Ge NW HJ is not constant and that the $Si_{1-x}Ge_x$ alloy composition x does not change uniformly from 0 to 1. Instead, the PL data suggests the formation of $Si_{1-x}Ge_x$ alloys with preferential compositions of $x \sim 0.15–0.2$ close to the Si part of the Si/Ge NW and $x \approx 0.5$ close to the Ge part of the Si/Ge NW. This conclusion, supported by our preliminary TEM and EDX analyses Figures 8.26 and 8.29 (insets), is also in agreement with

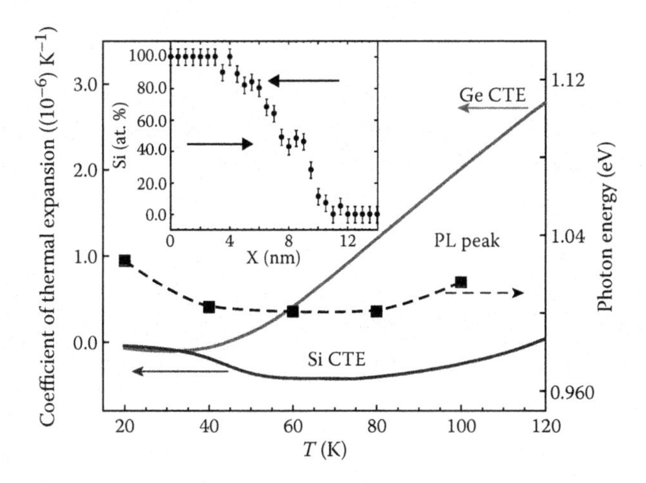

FIGURE 8.29
Temperature dependence of the PL peak near 1.0 eV in comparison with the temperature dependences of Si and Ge coefficients of thermal expansion. The inset shows representative EDX data along the NW axis with arrows pointing to the expected preferential compositions of 80 and 50 at.% of Si within the SiGe NW HJs.

the PL data and SiGe composition measurements in $Si/Si_{1-x}Ge_x$ 3D, i.e., cluster morphology nanostructures where $x \approx 0.2$ is a typical composition of the SiGe wetting layer and $x \leq 0.5$ is a stable composition close to the SiGe cluster core (Kamenev et al., 2003, 2004; Brunner, 2002; Lockwood et al., 2008).

Figure 8.29 supports this conclusion by showing that the PL peak at 1.0 eV exhibits an unusual nonmonotonic temperature dependence with a minimum at around 60–80 K. We suggest that this shift is produced by a change in Si/Ge NW strain due to the large difference in Si and Ge coefficients of thermal expansion (CsTE). Specifically, as the temperature increases from 20 to 120 K, the Si CTE first decreases and then increases, with a minimum around 60–80 K, while the Ge CTE monotonically increases (Reeber and Wang, 1996). The PL peak at ~1.0 eV, which we relate to a strained $Si_{1-x}Ge_x$ alloy $x \approx 0.15$–0.2 near the Si part of the Si/Ge NW, follows the direction of the c-Si CTE temperature dependence Figure 8.29, while the PL peak at ~0.96 eV associated with $Si_{1-x}Ge_x$ alloy near the Ge part of the NW ($x \approx 0.5$) does not show this behavior. Thus, the mismatch in Si and Ge CsTE creates an additional temperature-dependent strain in the Si/Ge NW HJs.

To conclude, light emission in Si/Ge NW HJs can be used to analyze structural properties of these nanoscale objects including strain, chemical composition, defects, etc. At the same time, light-emitting devices using NW HJs could be designed and built using a traditional junction geometry, where the Si and Ge parts of the NW could be doped by using standard p- and n-type dopants and radiative recombination under a forward bias would bring electrons and holes to the region of the Si/SiGe/Ge HJ.

8.4.3 PL and EL in Si/SiGe 3D Nanostructures

By the 1990s, a different form of SiGe NS, namely the 3D self-assembled system produced by the well-known Stranski–Krastanov or cluster–layer, growth mode in lattice mismatched materials, had been demonstrated (Eaglesham and Cerullo, 1990; Mo et al., 1990; Jesson et al., 1993; Kamins et al., 1997; Baribeau et al., 2006a,b). It has been shown that dislocation-free

SiGe growth can be achieved using a higher temperature (≥600°C), and that the nonplanar geometry is mainly responsible for the significant increase of the SiGe critical layer thickness. It has also been found that, compared with 2D Si/SiGe NSs, the PL and EL QE in 3D Si/SiGe NSs is higher (up to 1%), especially for $T > 50$ K (Schittenhelm et al., 1995; Apetz et al., 1995; Schmidt et al. Eberl, 1999; Kamenev et al., 2004). Despite many successful demonstrations of PL and EL in the spectral range of 1.3–1.6 μm, which is important for optical fiber communications, the proposed further development of 3D Si/SiGe based light emitters was discouraged by several studies indicating a type II energy band alignment at Si/SiGe heterointerfaces (Thewalt et al., 1997; Van de Walle and Martin, 1986; Schittenhelm et al., 1998; El Kurdi et al., 2006), where the spatial separation of electrons (located in Si) and holes (localized in SiGe) (see Figure 8.28) was thought to make carrier radiative recombination very inefficient. Later, it was also shown that 3D Si/SiGe NSs exhibit an extremely long (of the order of 10^{-2} s) luminescence lifetime (Kamenev et al., 2005b), which is of the order of a million times longer than in III–V semiconductors and their NSs. Thus, according to this analysis, 3D Si/SiGe NSs cannot be used to achieve efficient and commercially valuable light-emitting devices. Recently, however, it has been shown that despite the fact that the Si/SiGe heterointerface most likely exhibits type II energy band alignment, it is still possible to obtain conditions favorable for an efficient carrier radiative recombination.

8.4.3.1 Light-Emitting Properties of Si/SiGe 3D Nanostructures

In a system with strong selection rule relaxation (e.g., semiconductor bulk alloys), carrier recombination provides a higher PL QE compared with that in an indirect band-gap semiconductor, e.g., single crystal Si and Ge. PL measurements in Si/SiGe NSs reveal a significantly enhanced intensity ratio between no-phonon (NP) luminescence and phonon-assisted luminescence compared with c-Si. Thus, systematic studies of the PL spectra in 3D $Si/Si_{1-x}Ge_x$ NSs with control over the average Ge atomic concentration x provide very important information regarding changes in the carrier recombination mechanism (e.g., selection rule relaxation, conduction, and valence band alignment) as x increases from 0 (c-Si) to ~55%, which is the highest Ge composition in structural defect-free SiGe clusters (Tsybeskov and Lockwood, 2009).

In MBE-grown 3D $Si/Si_{1-x}Ge_x$ NSs with $x = 0.53$, the PL spectrum contains two bands peaked at 0.85 and 0.75 eV. There are no characteristic phonons in the Si/SiGe system with an energy of ~100 meV, and it is thus reasonable to assume that the observed PL bands are associated with carrier recombination within two different regions of the 3D SiGe NSs. The PL band peaked at 0.85 eV has a PL quenching activation energy of ~20 meV while the PL intensity as function of excitation intensity is linear over a wide range of excitation intensities. The second PL band peaked at 0.75 eV has an activation energy of ~60 meV and is nearly temperature independent up to 100 K. Because of its high QE, it can be monitored almost up to room temperature. These data suggest that 3D $Si/Si_{1-x}Ge_x$ NSs with $x = 0.53$ contain coupled subsystems with different (lower and higher) Ge concentrations. It is quite possible that a spatial localization of e-h pairs within 3D regions of higher Ge concentration, and thus having a lower band-gap, could be responsible for the observed sublinear excitation dependence of the PL band at 0.75 eV. The remaining 3D regions with a lower Ge concentration (e.g., having a higher band-gap) have a lower carrier concentration and the PL band (peaked at 0.85 eV) exhibits a linear excitation dependence. The PL thermal quenching could be associated with different mechanisms. The greater activation energy could be attributed to carrier diffusion from a 3D potential well. This assumption is justified by the expected type II band alignment in SiGe NSs with a deep (>100 meV) potential well for holes and a relatively small potential barrier for electrons (Kamenev et al., 2005b). Thus, phonon-assisted carrier tunneling can produce the observed ~60-meV activation energy for thermal quenching of the PL intensity. In these 3D $Si/Si_{1-x}Ge_x$ samples with a Ge concentration higher than 50%, a simultaneous threshold-like appearance of two clearly resolved PL peaks at 0.85 and 0.75 eV is observed. This suggests that Ge segregation might take place as x increases up to ~0.5. Since the PL peak at 0.75 eV is close to the value of the band-gap in pure c-Ge, we propose that such a segregation results in a Ge-rich core within a SiGe shell forming the 3D $Si/Si_{1-x}Ge_x$ NSs embedded within a pure Si matrix. This compositional variation has been confirmed experimentally (Baribeau et al., 2005, 2006a,b; Lockwood et al., 2007).

In contrast to MBE-grown samples, CVD growth of 3D Si/SiGe NSs does not provide precise control over the atomic composition, and, most likely, it produces more Si/SiGe interdiffusion at heterointerfaces (Baribeau et al., 2006a,b). This is well reflected in the PL spectra, where no fine structure has been found (see Figure 8.30). However, the broad and asymmetric PL peak is well fitted by two Gaussian bands, often identified as the NP and TO phonon lines (Kamenev et al., 2004) and separated by an energy of ~48 meV, which is close to the energy of characteristic SiGe phonons. Thus, there is at least a qualitative similarity between PL spectra in MBE and CVD grown 3D Si/SiGe NSs. Also, the PL efficiencies in MBE- and CVD-grown samples are comparable. Studies of the PL spectra and intensity temperature dependence show that at low temperatures the PL intensity is nearly temperature independent (Figure 8.30). At higher temperatures, the PL intensity drops exponentially, and the activation energies of PL thermal quenching are shown in Figure 8.31.

It has been known for some time that the PL spectra in 3D Si/SiGe NSs, which is similar to that in III–V QWs with type II energy band alignment (Hu et al., 1998), exhibit a blue shift as the excitation intensity increases (Wagner and Viña, 1984; Auston et al., 1975).

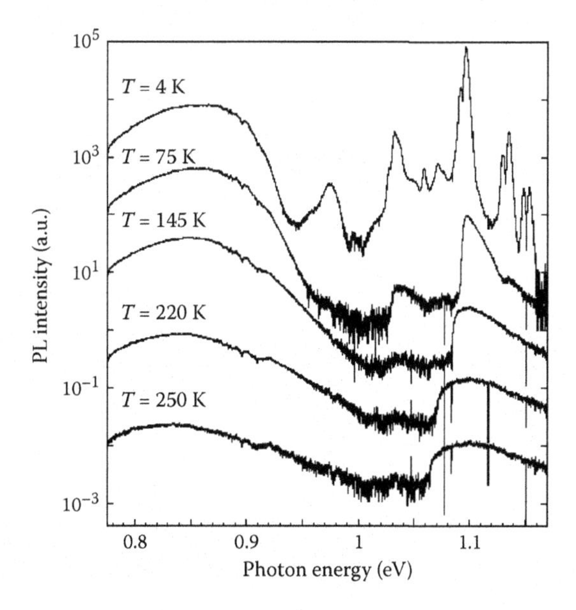

FIGURE 8.30
PL spectra in CVD grown Si/SiGe 3D NSs measured with an excitation intensity of 5 W/cm² at the indicated temperatures (the PL spectra have been shifted vertically for clarity).

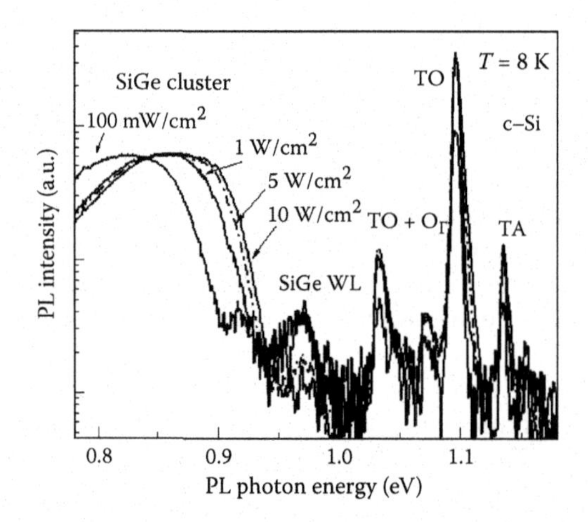

FIGURE 8.31
PL spectra in CVD grown Si/SiGe 3D multilayer samples measured at 8 K under different (indicated) excitation intensities.

In our studies, this effect is found in both MBE- and CVD-grown samples. Figure 8.31 shows PL spectra in a CVD-grown sample measured under excitation intensities varied from 0.1 to 10 W/cm^2. At the lowest excitation intensity used (0.1 W/cm^2), the PL peaks at ~0.8 eV. With increasing excitation intensity, a continuous almost-parallel PL blue shift of 30–40 meV per decade of excitation intensity increase is observed. At an excitation intensity of 10 W/cm^2, the PL peak reaches ~0.92 eV.

Figure 8.32 shows a modified Arrhenius plot of the normalized integrated PL intensity of a 3D Si/SiGe multilayer sample grown by CVD and measured at different excitation intensities. The normalized PL intensity temperature dependencies are readily fitted with two thermal quenching activation energies E_1 and E_2. In all measurements of all samples,

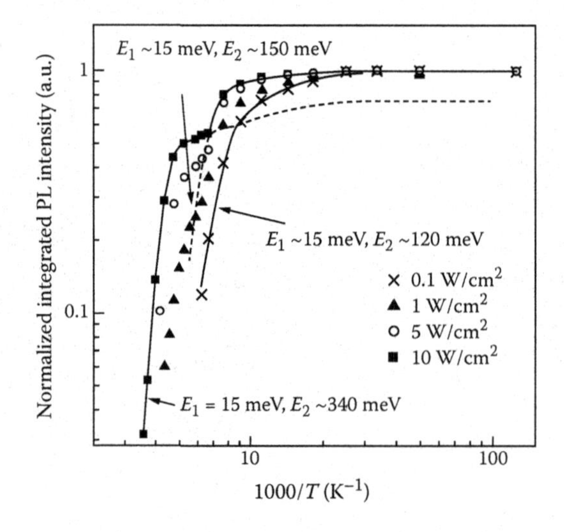

FIGURE 8.32
Typical integrated PL intensity behavior for CVD grown Si/SiGe 3D multilayer samples as a function of the reciprocal temperature measured with different excitation intensities, as indicated. The activation energies obtained from fits (shown by the different lines) to the data (shown by the different points) are also given.

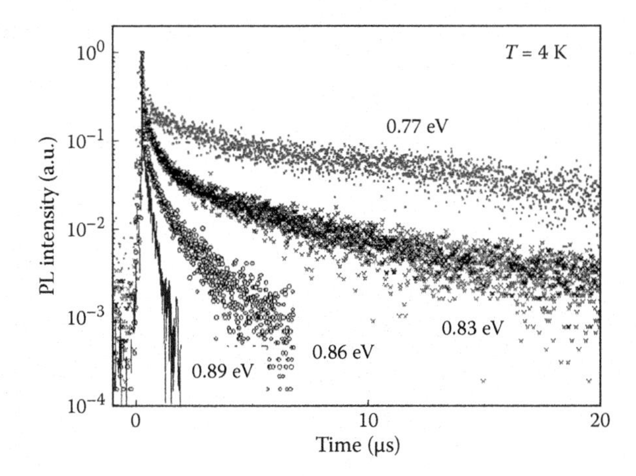

FIGURE 8.33
Typical low temperature (T = 4 K) PL dynamics for CVD grown Si/SiGe 3D multilayer samples measured at the indicated photon energies using a short (~6-ns) excitation pulse.

the PL thermal quenching activation energy $E_1 \approx 15$ meV and E_1 is independent of excitation intensity. In contrast, the activation energy E_2 depends significantly on the excitation intensity: the PL temperature dependence shows a step-like behavior, and E_2 increases dramatically from ~120 to 340 meV as the excitation intensity increases from 0.1 to 10 W/cm^2.

The PL dynamics, i.e., the PL intensity decay under pulsed laser excitation, is an important technique for studying the carrier recombination mechanism. Figure 8.33 shows the normalized low-temperature (4 K) PL decays collected from a CVD-grown Si/SiGe 3D sample. The initial PL decay is fast—close to the resolution of our detection system (<20 ns). The longer-lived PL shows a strong dependence on the detection photon energy: the PL lifetime at photon energies below 0.8 eV is found to be ~20 μs, and it drastically decreases to ~200 ns for the PL component measured at 0.89 eV.

8.4.3.2 Carrier Recombination in Si/SiGe Nanostructures

In our discussion of carrier recombination in Si/SiGe NSs, we focus on MBE and CVD grown 3D (or cluster morphology) samples with an average Ge atomic composition close to 50%. These samples show the highest observed PL QE (better than ~1% at low excitation intensity) with a PL peak wavelength close to 1.5–1.6 μm. The PL spectral distribution extending well below the band-gap of pure Ge (Figure 8.30) and the extremely long carrier radiative lifetime of ~10 ms (Kamenev et al., 2005b), as well as the ~30-meV per decade PL spectral shift toward higher photon energies as the excitation intensity increases (Kamenev et al., 2006), point out strong similarities between the PL in 3D Si/SiGe NSs and the PL in III–V QWs with type II energy band alignment (Hu et al., 1998). Generally, a type II energy band alignment at the heterointerface is a strong disadvantage for light-emitting devices due to a weak overlap between spatially separated electron and hole wavefunctions. In reality, however, the critical limitation in the efficiency of light-emitting structures is rather the presence of competing nonradiative recombination channels for excess electrons and holes. The most important nonradiative mechanism is carrier recombination via defects, especially heterointerface structural defects such as propagating dislocations and dislocation complexes. The 3D Si/SiGe NSs grown by both MBE and CVD processes show an almost undetectable density of dislocations. Thus, in these properly grown 3D Si/SiGe NSs at low excitation intensities, there is a high QE of PL with photon energy <0.9 eV, which is associated with carrier recombination in the vicinity of SiGe clusters.

It has been suggested that SiGe Stranski–Krastanov (S-K) clusters with a small (3–5 nm) height and ~10:1 base-height aspect ratio can be modeled as NSs with a type II energy band alignment and possible SiGe cluster valence-band energy quantization in the direction of growth (Brunner, 2002). Strained Si and Si-rich SiGe alloy regions near the base of the clusters (also called SiGe wetting layers) also need to be considered. Including the effect of strain, the observed PL bands at 0.92 and 0.97 eV (see Figure 8.31) indicate a composition of the $Si_{1-x}Ge_x$ transition region, which is located near the bottom of the Ge/Si pyramid-like clusters, to be close to $x \approx 0.2$, and this conclusion is supported by recent direct analytical TEM measurements (Baribeau et al., 2006a; Lockwood et al., 2007).

It has also been proposed that the broad PL band with a peak energy of ~0.8–0.9 eV is due to the recombination of carriers localized in the Ge-richest areas of the clusters, which is close to the center of "pancake"-shaped SiGe clusters (Baribeau et al., 2005). We suggest that at the lowest excitation intensity, the PL arises from e-h recombination between holes localized in the Ge-richest regions of the cluster and electrons localized in the strained SiGe alloy region near the cluster base. This immediately explains the extension of the observed PL spectrum below the pure Ge band-gap energy.

The observed excitation-independent PL thermal quenching activation energy of ~15 meV is close to the exciton binding energy in SiGe alloys and Si/SiGe superlattices. Thus, we conclude that one of the mechanisms of PL thermal quenching is the thermal dissociation of excitons. The activation energy of ~15 meV can therefore be associated with exciton localization on specific regions of the clusters associated with variations of the SiGe composition. Hence, the nonuniform SiGe cluster composition and, perhaps, variations in SiGe cluster size and shape could be responsible for the observed relatively broad PL spectra.

Using the model discussed above of type II energy band alignment at the Si/SiGe heterointerface, we focus next on nonradiative carrier recombination and the different mechanisms of e-h separation. Electron transport in 3D Si/SiGe NSs is limited by a small (≤10–15 meV) conduction band energy barrier and SiGe compositional disorder (Tilly et al., 1995). Thus, the PL thermal quenching activation energy of ~15 meV could also be associated with a small conduction band energy barrier for electrons in Si/SiGe 3D NSs.

In contrast to a low energy barrier for electrons, hole diffusion in 3D Si/SiGe multilayer NSs with a high Ge content is controlled by large (>100 meV) valence band energy barriers at Si/SiGe heterointerfaces. In this system, we consider two major mechanisms of hole transport: hole tunneling and hole thermionic emission. Hole tunneling in 3D Si/SiGe NSs with thin (5–7 nm) Si-separating layers and nearly perfect SiGe cluster vertical self-alignment could be very efficient. These NSs are usually grown by MBE and exhibit a PL thermal quenching activation energy of ~60 meV. The same PL thermal quenching activation energy is found in our CVD-grown samples with 7.5-nm-thick Si separating layers for the lowest excitation intensity. This suggests that in 3D Si/SiGe multilayer NSs with thin Si layers at low excitation intensity, the e-h separation, and nonradiative carrier recombination are mainly controlled by hole tunneling between SiGe clusters. Due to significant variations in SiGe cluster size, shape, and chemical composition, the process of hole tunneling could be assisted by phonon emission and/or absorption (Kamenev et al., 2006; Qin et al., 2001). Therefore, the observed PL thermal quenching activation energy is close to the Si TO phonon energy. In 3D Si/SiGe multilayer samples with 20-nm-thick Si layers, where SiGe cluster vertical self-alignment is practically absent (Baribeau et al., 2005), the probability of hole tunneling is reduced, and hole thermionic emission over the Si/SiGe heterointerface barrier is playing a bigger role. Thus, in these samples the PL thermal quenching activation energy is expected to be greater, as has been found experimentally (see Figure 8.32).

In this simple model, efficient hole tunneling between adjacent SiGe nanoclusters requires not only reasonably low and thin Si barriers but also a low carrier concentration (i.e., a large enough number of empty adjacent SiGe clusters). By increasing the photoexcitation intensity (i.e., the number of photogenerated carriers), hole tunneling can effectively be suppressed since fewer empty adjacent SiGe clusters can be found. At high excitation intensity, assuming (i) a negligible value of the conduction band offset compared with that in the valence band and (ii) a nearly pure Ge composition in the SiGe cluster core, the maximum anticipated PL thermal quenching activation energy should be ≤400 meV. This value sets the upper limit of the activation energy of PL intensity thermal quenching in 3D SiGe multilayer NSs, and it is close to the activation energy of $E_2 \approx 340$ meV that has been found under the highest excitation intensity (see Figure 8.32).

The same model also explains the experimentally found strong decrease in carrier radiative lifetime (~100 times) as the detection photon energy increases from 0.77 to 0.89 eV (Figure 8.33). Under a low level of excitation intensity, holes are localized within the Ge-rich core of SiGe clusters, and the hole wavefunction does not penetrate into the Si barriers. Thus, in this quasi-type II energy band alignment, electron and hole wavefunctions do not overlap, causing a long carrier radiative lifetime. Under high-level excitation, holes occupying the excited energy states in small size SiGe clusters, as well as holes leaking into the SiGe wetting layer with a lower Ge concentration, have their wavefunction significantly extended into the Si barriers. Therefore, a stronger overlap between electron and hole wavefunctions is responsible for a shorter radiative lifetime, i.e., a higher PL QE at greater photon energy. Thus, this explanation is consistent with both, the decrease of the PL decay time at greater photon energy and the PL spectral blue shift as excitation intensity increases.

Figure 8.34 depicts the quasi-type II energy band alignment at Si/SiGe heterointerfaces, where a gradual increase of Ge concentration toward the SiGe cluster core has recently been verified experimentally (Baribeau et al., 2006a; Lockwood et al., 2007). It also shows five major processes controlling e-h recombination in 3D Si/SiGe NSs with a Ge-rich core:

1. Radiative recombination between electrons localized at the Si/SiGe heterointerface and holes localized within the Ge-rich core of a SiGe cluster—this slow recombination is associated with the lower photon energy part of the PL band;
2. Radiative recombination between electrons localized at the Si/SiGe heterointerface and holes localized in excited states in small-size SiGe clusters as well as holes leaking into the Si-rich outer part of the cluster or/and SiGe wetting layer—this faster carrier recombination channel is responsible for the higher photon energy part of the PL spectrum;
3. Nonradiative carrier recombination due to e-h separation via hole tunneling in Si/SiGe NS samples with thin (<10 nm) Si barriers;
4. Nonradiative carrier recombination due to e-h separation via hole thermionic emission in samples with thicker (>15–20 nm) Si barriers;
5. Nonradiative carrier recombination due to Auger processes.

It has been pointed out that possible transformations of hole energy spectra due to quantization and/or strain might dramatically increase the rate of nonradiative Auger recombination in SiGe QWs by more than 100 times (Williams et al., 1998). Thus, the "Auger limit"

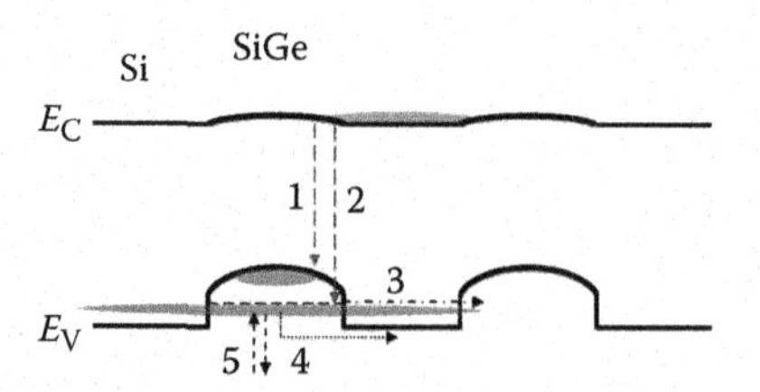

FIGURE 8.34
Schematic of the Si/SiGe cluster multilayer energy band diagram with different electronic transitions: 1, slow e-h recombination due to spatial separation and a weak wavefunction overlap; 2, a faster recombination, presumably involving holes at excited energy states in SiGe clusters; 3, hole diffusion due to cluster-to-cluster tunneling; 4, hole diffusion due to thermionic processes; and 5, Auger hole excitation with possible charge transfer.

in Si/SiGe NSs could be considerably lower compared with that in c-Si and c-Ge. On the other hand, the Auger process competes with radiative processes, and their characteristic lifetimes are ~10^{-3} s in bulk Si and ~10^{-5} s in bulk Ge (Table 8.3) and just ~10^{-7} s in the high-photon-energy part of the PL spectra in 3D Si/SiGe NSs.

Figure 8.34 also reflects a continuous change in the Ge atomic concentration by a gradually increasing energy band-gap from the cluster center toward the cluster edge and SiGe wetting layer where the Ge atomic concentration is estimated to be ~20%. The proposed energy band diagram and carrier recombination/diffusion mechanisms explain the experimental observations of (i) the PL spectral blue shift under increasing excitation intensity; (ii) the dramatic (~100 times) decrease in carrier radiative lifetime measured at photon energies from 0.77 to 0.89 eV; and (iii) the unusual PL intensity temperature dependence, which shows a different PL thermal quenching activation energy at different excitation intensities. This schematic representation also points out that moderate excitation intensity changes the overlap of e-h wavefunctions and allows a faster carrier radiative recombination. In other words, the type II energy band alignment at the Si/SiGe heterointerface can effectively be replaced by the "dynamic type I" alignment, where the electron and hole spatial separation no longer controls the recombination rate and the QE of PL and EL.

8.4.3.3 EL in Si/SiGe Nanostructures

EL in Si/SiGe NSs was demonstrated almost simultaneously with the first investigations of the PL (Brunner, 2002; Apetz et al., 1995). The structures used for the EL studies usually are Si/SiGe p-i-n diodes with SiGe clusters, SiGe alloy layers or SiGe QWs in the i-region. Figure 8.35a shows EL spectra (adapted from Apetz et al., 1995) with quite narrow EL peaks attributed to NP and TO phonon emission. The proposed EL mechanism is quite similar to the PL mechanism; i.e., it is due to radiative e-h recombination in Si/SiGe NSs with electrically injected electrons and holes. Similarly to that in PL, the EL spectrum depends on the SiGe layer (or cluster) composition, and a higher Ge concentration results in the EL spectrum red shift (i.e., a shift toward lower photon energy). Figure 8.35b (adapted from Apetz et al., 1995) shows the dependence of the integrated EL intensity on current. The EL intensity versus current is a power function $I_{EL} \sim J^n$ where I is the intensity, J is the current density, and $0.5 < n < 2$. A good quality Si/SiGe p-i-n structure usually provides current flow associated with carrier diffusion in a forward biased p-n junction (i.e., an exponential current as a function of forward bias), and the diode ideal factor is less than 2 (Apetz et al., 1995; Stoffel et al., 2003).

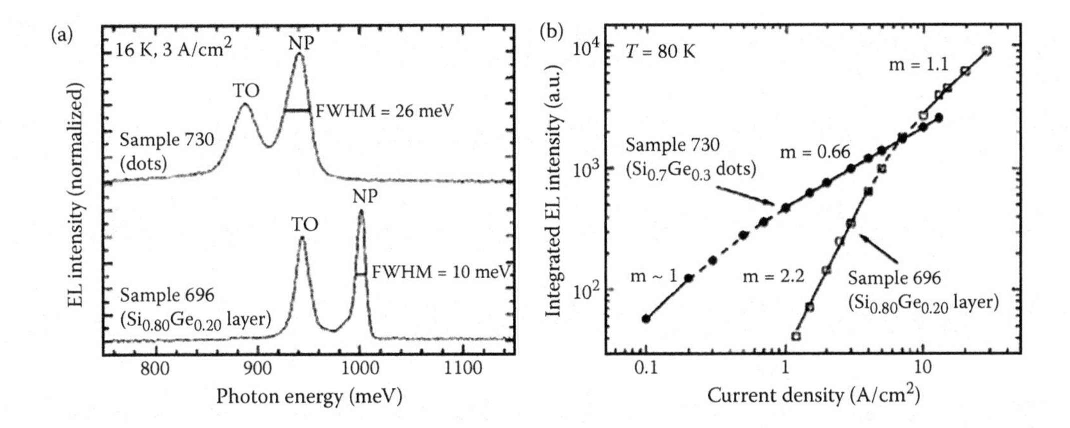

FIGURE 8.35
(a) EL spectra of samples containing 3D SiGe NSs and (b) the EL intensity as a function of current under reverse bias. (Reprinted with permission from Apetz, R., L. Vescan, A. Hartmann, C. Dieker, and H. Luth. 1995. Photoluminescence and electroluminescence of SiGe dots fabricated by island growth. *Applied Physics Letters* 66 (4):445–447. Copyright 1995 by the American Institute of Physics.)

In Si/SiGe QW–based devices, the EL intensity temperature dependence, similar to that in PL, exhibits significant thermal quenching (Apetz et al., 1995). The EL thermal quenching activation energy is greater in Si/SiGe 3D nanostructures (especially with small size SiGe clusters) compared with that in planar QWs, and the EL can be extended up to room temperature (Figure 8.36). In Stoffel et al. (2003), the EL thermal quenching activation energy is found to be as large as 200–250 meV, and this is possibly the largest reported value. The SiGe cluster EL spectra full width at half maximum is relatively narrow (~90 meV), possibly reflecting the cluster narrow size distribution. When the temperature increases from 40 K up to room temperature, the EL is reduced by a factor of ~40, but the EL peak is still observable at approximately the same photon energy. Considering the EL mechanism, the luminescence efficiency is believed to be limited (similarly to that in PL) by type II radiative recombination between holes confined in the SiGe clusters and electrons localized in the silicon spacer layers above and below the islands. Also, in the p-i-n structures with 3D Si/SiGe layers within the active region, the series resistance is found to be high (Figure 8.36). Thus, the EL power efficiency is low, in part due to the high series resistance of the Si/SiGe p-i-n diode.

Typically, as we already mentioned, for a current density of 1–100 A/cm², the EL intensity is a power function of current $I_{EL} \sim J^n$ with $0.5 < n < 2$. Thus, in p-i-n structures, the EL intensity should have an exponential dependence on the applied forward bias. In a simpler structure with two nearly ohmic contacts to Si/SiGe cluster multilayers, the integrated EL intensity is found to be nearly linear as a function of the applied voltage (Figure 8.37). There are no energy barriers due to the p-i-n junction in these samples, and the EL intensity as a function of temperature shows thermal quenching with a smaller activation energy of ~130 meV. Interestingly, in the same sample, the device current as a function of temperature shows nearly an exact anticorrelation with the EL intensity and exhibits an activation energy of ~140 meV. Since both carrier transport and EL intensity exhibit a strong dependence on temperature, an accurate estimation of the EL QE over a wide temperature region requires normalization of the EL intensity on current density, i.e., measurements of how many photons are emitted per one injected e-h pair at different temperatures.

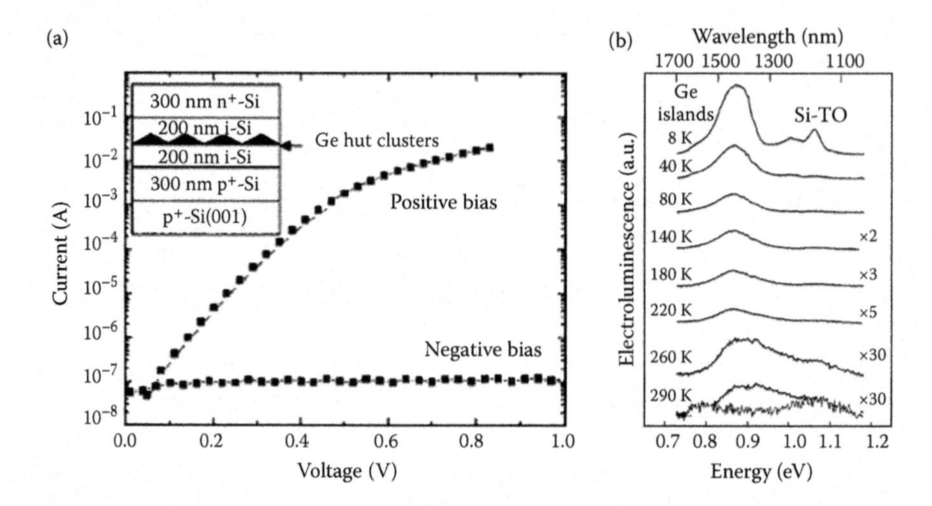

FIGURE 8.36
(a) Room-temperature current–voltage characteristic of the p-i-n diode including one Ge hut cluster layer in the i-region. The device structure is given in the inset. (b) Temperature dependence of the EL spectra at constant current of 400 mA. (Reprinted with permission from Stoffel, M., U. Denker, and O. G. Schmidt. 2003. Electroluminescence of self-assembled Ge hut clusters. *Applied Physics Letters* 82 (19):3236–3238. Copyright 2003 by the American Institute of Physics.)

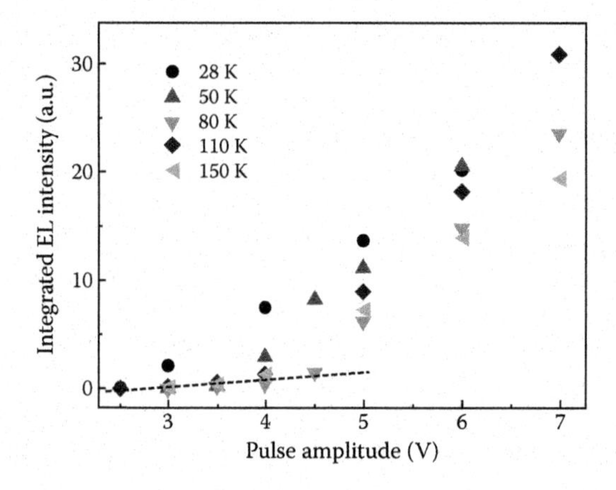

FIGURE 8.37
Integrated EL intensity over the 0.75–0.95-eV spectral region for a CVD grown Si/SiGe 3D multilayer sample as a function of the pulsed voltage amplitude measured at the indicated temperatures. The dashed straight line is a guide to the eye.

The importance of the last statement is pointed out in Peng et al. (2004), where the EL of LEDs with 5, 10, and 30 layers of Si/Ge clusters in the active region is studied. The enhanced integral EL intensity at temperature $T \geq 200$ K in the samples with 30 layers of Si/Ge clusters presents behavior opposite to the usually observed EL intensity thermal quenching, which is also found in the similar samples with 5 and 10 Si/SiGe layers. In Peng et al. (2004), this EL intensity temperature dependence is attributed mainly to the EL intensity dependence on injection current, which is temperature activated.

Acknowledgments

We would like to acknowledge the invaluable contributions over a number of years from our many collaborators on this work and whose names are given in the references to our work in this paper. We thank especially B. Kamenev of NJIT, J.-M. Baribeau, and X. Wu of NRC Canada and T. Kamins of HP Laboratories. We acknowledge partial financial support for this research provided by U.S. National Science Foundation, Intel Corporation, Semiconductor Research Corporation, and Foundation at NJIT.

8.5 Rare-Earth Doped Silicon and Silicon Nanostructures

Tom Gregorkiewicz, Saba Saeed, and Maciek Wojdak

8.5.1 Introduction

8.5.1.1 Rare Earth Ions as Optical Dopants

Doping with rare earth (RE) ions offers a possibility of creating optical systems whose emissions are characterized by sharp, atomic-like spectra with predictable and temperature-independent wavelengths. For that reason, RE-doped matrices are frequently used as laser materials (large band-gap hosts, e.g., Nd:YAG) and for optoelectronic applications (semiconducting hosts). All RE elements have a similar atomic configuration [Xe] $4f^{n+1}6s^2$ with $n = 1$–13. Upon incorporation into a solid, RE dopants generally tend to modify their electronic structure in such a way that the 4f-electron shell takes the [Xe] $4f^n$ electronic configuration, characteristic of trivalent RE ions. Very attractive optical features of RE ions follow from the fact that their emissions are due to internal transitions in the partially filled 4f-electron shell. This core shell is effectively screened by the more extended 5s and 5p orbitals. Consequently, optical, and also magnetic, properties of RE ion are relatively independent of a particular host.

8.5.1.2 Rare Earth Doping of Semiconductors

In addition to the predictable optical properties and, in particular, the fixed wavelength of emission, RE-doped semiconductor hosts offer yet one more important advantage, that is, RE dopants can be excited not only by a direct absorption of energy into the 4f-electron core but also indirectly, by energy transfer from the host. This can be triggered by optical band-to-band excitation, giving rise to PL, or by electrical carrier injection, resulting in EL. Among many possible RE-doped semiconductor systems, research interest has been mostly concentrated on Yb in InP and Er in Si. The erbium ion, Er^{3+}, has the electronic configuration [Xe] $4f^{11}$. The electronic structure is dominated by electron-electron and spin-orbit interactions within the 4f-electron shell resulting in J-multiplets with $^4I_{15/2}$, $^4I_{13/2}$, $^4I_{11/2}$ as the ground, first, and second excited states, respectively. When Er^{3+} ions are embedded in an insulator like SiO_2, the only way to excite them is optically. This is done by pumping resonantly with a laser beam directly to one of the excited states. Consequently, the commercially available Er-doped fiber amplifiers or optical generators based on SiO_2:Er are expensive, because they need to be pumped with a tunable laser of high power, due to the small cross section of Er^{3+} for direct absorption. For Si:Er, a host-mediated excitation is possible, and the research interest has been fueled by prospective applications in Si photonics, in view of the full compatibility of Er doping with CMOS technology. Upon its identification, Er-doped crystalline Si (c-Si:Er) emerged as a perfect system where the most advanced and successful Si technology could be used to manufacture optical elements whose emission coincides with the 1.5 μm minimum absorption band of silica fibers currently used in telecommunications.

8.5.2 Er-Doped Crystalline Silicon

8.5.2.1 Preparation

Erbium is not a good dopant for c-Si as its valence characteristics and ionic radius are very different from those of Si. For this reason Er^{3+} ions are randomly positioned in the Si host, with a large number of possible local environments and a variety of local crystal fields. Consequently, while photons emitted from individual Er^{3+} ions are very well defined, the ensemble spectrum from a c-Si:Er sample is inhomogeneously broadened. This leads to the situation where a photon emitted by one Er-center is not in resonance with transitions of another one and, as such, cannot be absorbed. Combined with the small absorption cross section of Er^{3+} ions, this makes realization of optical gain in c-Si:Er very challenging and to date reports on realization of optical gain (Lourenco et al., 2007) remain controversial (Ha et al., 2010).

In view of the long radiative lifetime (milliseconds) of the first $^4I_{13/2}$ excited state of Er^{3+}, a large concentration of Er^{3+} is desirable to maximize the emission intensity. This is, however, precluded by the low solid-state solubility of Er^{3+} in c-Si. Therefore, nonequilibrium methods are commonly used for preparation of Er-doped Si. The best results have been obtained with ion implantation (Benton et al., 1991), MBE (Serna et al., 1995), and sublimation MBE (SMBE) (Andreev et al., 1999; Karim et al., 2008). Sputtering and diffusion are also occasionally used for preparation of Er-doped structures (Cavallini et al., 1999). With such nonequilibrium doping techniques, Er^{3+} concentrations as high as [Er] ≈ 10^{19} cm^{-3} can be realized. Nevertheless, it had been observed that only a small part of the high Er^{3+} concentration, typically ~1%, contributes to photon emission. Possible reasons for that are segregation of Er^{3+} to the surface, clustering into metallic inclusions, and concentration quenching. In addition to these, it had been postulated that to attain optical activity, i.e., the ability to emit 1.5 μm radiation, the Er^{3+} ion must form an optical center of a particular microscopic structure. Co-doping of c-Si:Er with oxygen can enhance optical activity. This effect is optimal for an oxygen-to-erbium doping ratio of approximately 10:1 and an Er^{3+} concentration of 10^{19} cm^{-3} (Eaglesham et al., 1991; Scalese et al., 2000). Oxygen atoms play at least two roles in the c-Si:Er system. First, O can greatly lower the crystal binding energy of the interaction between O and Si and also O and Er atoms, thus enabling the incorporation of Er^{3+} into Si. Secondly, the presence of O modifies the c-Si:Er electrical properties by introducing a donor level that can facilitate efficient excitation route (Benton et al., 1991; Priolo et al., 1995).

8.5.2.2 Excitation Mechanism

In RE-doped ionic hosts and molecular systems, the excitation transfer usually proceeds by energy exchange between an RE ion, acting as an energy acceptor, and a radiative recombination center, an energy donor. In that case, the first step is the excitation of the energy donor center. Subsequently, the energy is nonradiatively transferred via the multipolar, or exchange mechanism to the 4f-shell of an RE ion, with possible energy mismatch being compensated by phonons. In a semiconducting host, the first excitation stage involves host band states (exciton generation) and is usually very efficient. The subsequent energy transfer to (and similarly from) a RE ion depends crucially on the availability of traps allowing, e.g., creation of a bound exciton state in the direct vicinity of the RE ion shown in Figure 8.38 (Vinh et al., 2009). Therefore, the excitation process changes dramatically if the RE ion itself introduces a level within the band-gap of the host material.

In general, the Er-related luminescence in Si can be induced electrically, by carrier injection, or optically with the photon energy exceeding the energy gap. In EL, Er excitation is accomplished either by collision with a hot carrier under reverse bias, or by recombination of e-h pairs in a forward biased p-n junction. The electronic collision under reverse bias was recognized as the most efficient excitation procedure for c-Si:Er. In PL, energy transfer to the 4f electron core is accomplished by nonradiative recombination of an exciton bound in the proximity of an Er^{3+} ion.

Interesting insights into the excitation process were obtained by investigating emission from an Er-implanted sample under different configurations of optical excitation (Pawlak et al., 2001). Comparison of PL recorded with a laser beam incident on the implanted side and on the substrate side of the sample gave the evidence that energy is being transported to Er^{3+} ions by excitons, and that the efficiency of this step strongly depends on the distance between the photon absorption region, where excitons are generated, and Er^{3+} ions. The effective cross section for the above mentioned indirect excitation mode is of the order of $\sigma \approx 10^{-14}$ cm^2, i.e., factor ~10^6 higher than direct resonant photon absorption by Er^{3+} ions (Vinh et al., 2004).

Excitation of Er^{3+} ions can be influenced by the presence of shallow-level doping or damages introduced during growth. In particular, alternative relaxation paths may appear at every stage of the excitation process, strongly affecting its final efficiency. For example, in Er-doped Si:P, it had been demonstrated (Gregorkiewicz et al., 1996) that application of an electric field can block energy relaxation through shallow donor phosphorus, thus channeling it to Er^{3+} ions and enhancing the excitation efficiency. This result shows that phosphorus donors and Er-related centers compete in exciton localization. In addition, it also provides direct evidence that the exciton binding energy is bigger for Er-related traps than for P, suggesting larger ionization energy of the relevant donor center.

Next to the above-mentioned excitation of Er^{3+} into its first $^4I_{13/2}$ excited state, Er^{3+} ions can be directly excited to higher lying $^4I_{11/2}$ state. This state can be reached via the second conduction sub-band of c-Si (Yassievich et al., 1993), due to energy matching.

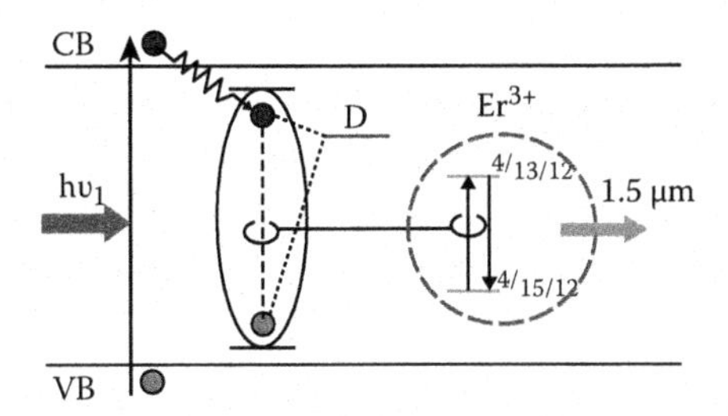

FIGURE 8.38
Model for photo-excitation of Er^{3+} ions in Si, where CB, VB, and D stand for conduction band, valence band, and donor level, respectively. (From Vinh, N. Q., Ha, N. N., and Gregorkiewicz T., *Proceedings of the IEEE* 97 (7):1269–1283 © 2009 IEEE).

8.5.2.3 De-Excitation Mechanism

Theoretically, for an RE ion in vacuum, transitions between different multiplets originating from the 4f-electron shell are forbidden for parity reasons. Upon incorporation in a matrix, the local crystal field leads to a small perturbation of these states and nonzero transition matrix elements appear. However, the effect of crystal field is strongly reduced due to screening; therefore, recombination times remain in the millisecond range. Upon temperature increase, nonradiative recombinations appear. For RE ions in insulating hosts, the nonradiative recombination is usually dominated either by multiphonon relaxation or by a variety of energy transfer phenomena to other RE ions. The presence of delocalized carriers (either free or weakly bound) in semiconductors opens new channels specific for these materials, such as back-transfer process in which the excitation process is reversed and energy dissipation to free carriers, which are then promoted to higher band states.

The back-transfer process originally proposed for InP:Yb (Taguchi et al., 1993) is generally held responsible for the high-temperature quenching of the RE PL intensity and lifetime. The low probability of radiative recombination makes the back-transfer process possible with the necessary energy being provided by simultaneous absorption of several lattice phonons. During the back-transfer, the last step of the excitation process is reversed: upon nonradiative relaxation of an RE ion, the intermediate excitation stage (the bound-exciton state) is recreated. The activation energy of such a process is equal to the energy mismatch that has to be overcome and therefore depends on the gap position of the aforementioned RE-related donor state. In addition to back-transfer process, availability of shallow centers in the host exerts a profound influence on nonradiative relaxation of RE ions. A very effective mechanism of such a nonradiative recombination is the impurity Auger process involving energy transfer to conduction-band electrons (Palm et al., 1996). Direct evidence of the importance of energy transfer to conduction-band electrons was given by an investigation of temperature quenching of PL intensity for samples with different background doping (Priolo et al., 1998).

For Er^{3+} ions in c-Si, the longest reported radiative lifetime is $\tau \approx 2$ ms at $T = 15$ K (Priolo et al., 1998). Upon temperature increase, nonradiative recombination dominates de-excitation processes. Thermally induced quenching of both the PL intensity and the effective lifetime had been experimentally observed for Er^{3+} (Thao et al., 2000). For c-Si:Er, the energy necessary to activate the back-transfer process is $\Delta E \approx 150$ meV, and therefore the participation of at least three optical phonons is required. Investigations of thermal quenching of the PL intensity and lifetime in c-Si:Er reported two activation energies: $\Delta E_1 \approx 15$–20 meV and $\Delta E_2 \approx 150$ meV (Gregorkiewicz et al., 1996). The former is usually related to exciton ionization or dissociation, and the latter is commonly taken as a fingerprint of the back-transfer process. The multiphonon-assisted back-transfer process for c-Si:Er was modeled theoretically (Prokofiev et al., 2005), in full agreement with the experimental results. The detrimental role of free carriers on the emission of c-Si:Er can also be inferred from the fact that free carriers govern the effective lifetime of the excited state of the Er^{3+} ion. That had been elegantly demonstrated in an experiment where He-Ne laser, operating in a continuous mode in parallel to a chopped Ar laser, was used to provide an equilibrium background concentration of free carriers. As a result, shortening of the Er^{3+} lifetime had been observed. The magnitude of this effect was proportional to the square root of the background illumination power (Palm et al., 1996). Since the exciton recombination dominated the relaxation, this indicates that the efficiency of the lifetime quenching is related to the free-carrier concentration.

In next section, Si/Si:Er multinanolayer structures comprised of interchanged layers of Er-doped and un-doped c-Si will be discussed. For these structures it may be speculated that the excitation mode is considerably enhanced since excitons efficiently generated in the spacer layer of undoped Si have a long lifetime and can diffuse toward Er-doped regions.

8.5.3 Erbium-Doped Silicon Nanolayer Structures

8.5.3.1 Structure

Si/Si:Er nanolayered structures can be successfully grown by sublimation molecular beam epitaxy (SMBE) (Kuznetsov et al., 1991). Figure 8.39 (inset) shows a typical Si:Er structure grown by SMBE (Andreev et al., 1999). It comprised 400 periods of 2.3 nm-thick layers of Si doped with Er, separated by 1.7 nm-thick undoped Si layers. Total thickness of the epitaxial structure was around ~1.6 μm. The average concentration of Er^{+3} ions in the epitaxial layer was determined by the secondary ion mass spectroscopy as [Er] ≈ 3.5×10^{18} cm^{-3}. For the optical activation of Er^{3+} ions, the as-grown samples were annealed at 800°C for 30 min in a continuous flow of nitrogen. Figure 8.39a shows the PL spectrum of an SMBE-grown sample featuring several sharp lines. These lines were assigned to a specific Er-defect center called Er-1, marked by arrows (Vinh et al., 2004, 2007). The width of the Er-1-related emission lines (panel b) is among the smallest ever measured for any emission band in a semiconductor matrix; it was below the experimental resolution of $\Delta E = 8$ μeV.

8.5.3.2 Optical Activity

The level of optical activity is an essential parameter of Er-doped structures determining their application potential. To quantify the intensity of the Er-related emission from multi-nanolayer and establish the optical activity level, a SiO_2:Er implanted sample (called STD) was compared with two SMBE samples labeled as SMBE01 and SMBE02 (Vinh et al., 2007, 2009), shown in Figure 8.40. The estimation of the number of excitable centers was made

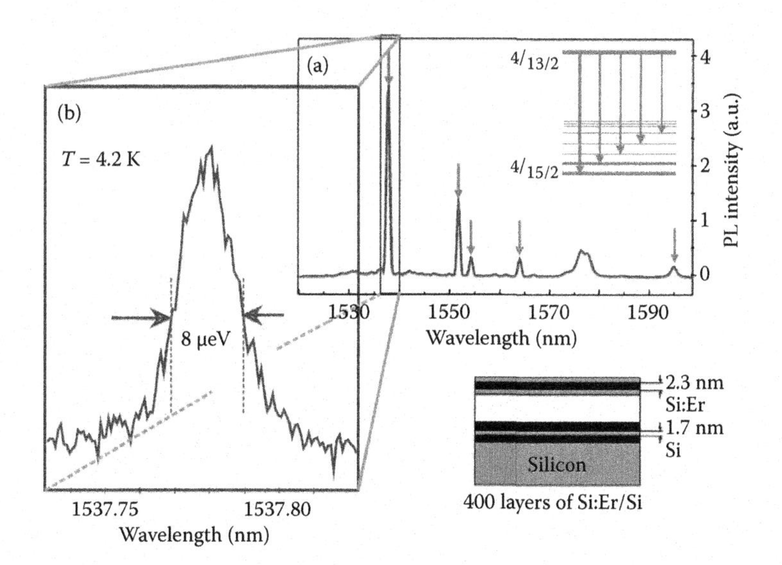

FIGURE 8.39
(a) PL spectrum of the Si/Si:Er multinanolayer structure, measured at $T = 4.2$ K under continuous excitation at 532 nm. (b) Detail of the studied Er-1 center related ultra narrow emission line at 1.54 μm (also in the figure, a sketch of the multilayer structure). (http://link.aps.org/doi/10.1103/PhysRevB.81.195206 © 2010 American Physical Society.)

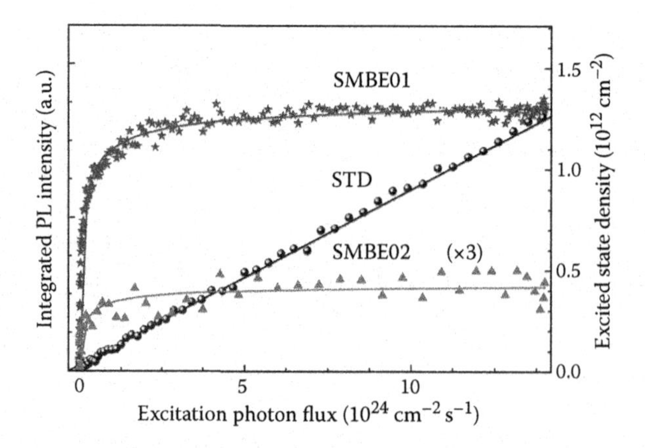

FIGURE 8.40
Integrated PL intensity dependence of multinanolayer structures and the SiO_2:Er "standard" on excitation photon flux at $T = 4.2$ K. (From Vinh, N. Q., Ha, N. N., and Gregorkiewicz, T., *Proceedings of the IEEE* 97 (7):1269–1283 © 2009 IEEE.)

by comparing the saturation level of PL intensities of STD (direct excitation at 520 nm) and SMBE samples. Taking into account the differences in the PL spectra, decay kinetics (nonradiative and radiative), extraction efficiencies, and excitation cross sections, optical activities of 2% ± 0.5% for SMBE01 and 15% ± 5% for SMBE02 were determined. Therefore, the percentage of photon-emitting Er-dopants obtained for the Si/Si:Er multinanolayer is similar to that achieved in the best Si:Er materials prepared by ion implantation. In view of the relatively long radiative lifetime of Er^{3+} ions used for concentration evaluation, the estimated percentages represented just the lower limits (Vinh et al., 2007).

8.5.3.3 Role of Oxygen

The role of oxygen in case of the Er-1 center was monitored using a tunable mid-IR free electron laser (FEL) (Tsimperidis et al., 1998; Forcales et al., 2002, 2003; Minissale et al.,

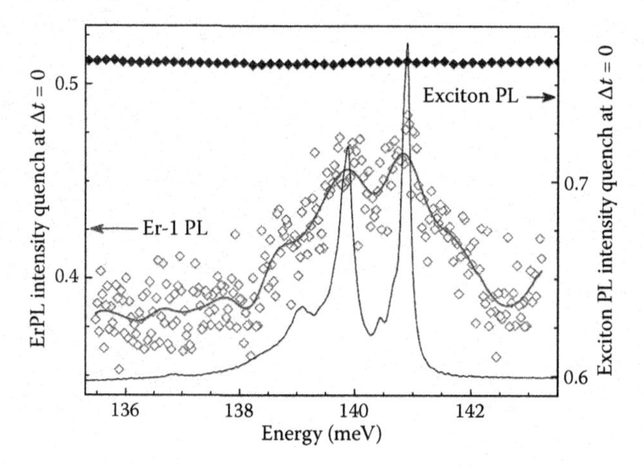

FIGURE 8.41
Results of two-color experiments at $T = 4.2$ K for Er-1 (◇) and exciton PL (◆). For comparison, the infrared absorption spectrum at $T = 55$ K (black trace) of the sample is also given. (From Vinh, N. Q., Ha, N. N., and Gregorkiewicz T., *Proceedings of the IEEE* 97 (7):1269–1283 © 2009 IEEE.)

2008). The results of two-color experiment are depicted in Figure 8.41, where FEL was used to activate the antisymmetric vibration mode of O in Si (ν_3 mode) at 8.80 μm (141 meV). Its effect was then monitored on the Er-1 emission, as induced by band-to-band pumping with a Nd:YAG laser. Figure 8.41 shows the quenching ratio of Er^{3+} PL as function of the photon energy of the FEL. A clear resonant feature was observed for FEL photon energy around 141 meV, which coincides with the oxygen-related vibrational absorption band (black trace). The explanation of this effect is that the local temperature is increased due to oxygen vibration induced by the FEL. This results in quenching of the Er^{3+} PL but negligibly increases the temperature of the whole layer, thus leaving the exciton-related PL practically unaltered. The magnitude of the quenching depends on the quantum energy $h\nu_{FEL}$, photon flux Φ_{FEL}, and timing Δt of the FEL pulse with respect to the band-to-band excitation. The resonant quenching of the Er^{3+} PL upon activation of oxygen vibrational modes evidenced the spatial correlation of both dopants (Minissale et al., 2008). On the other hand, the presence of Er^{3+} influences also the vibrational properties of oxygen. This manifests itself as a shift (or, more likely, a broadening) of the 141-meV absorption band and a change of the vibrational lifetime (Minissale et al., 2008).

8.5.3.4 Resonant Excitation

Figure 8.42 shows the PL excitation (PLE) spectrum of the Er-related emission measured for excitation energy close to the band-gap of c-Si. In addition to the usually observed contribution produced by (the onset of) the band-to-band excitation, a resonant feature, peaking at the energy around 1.18 eV, is also clearly visible. This observation is directly supported by power dependence of PL intensity, shown in the inset of Figure 8.42, for the two photon energies indicated by arrows. For the higher energy value of 1.54 eV, it exhibits the saturating behavior characteristic of the band-to-band excitation mode (Gusev et al., 2001). The dependence for excitation energy at 1.18 eV had a strong linear component superimposed on this saturating background. Such a linear dependence is expected for direct pumping. The results obtained in two-color and PLE spectroscopies explicitly demonstrate that the optical activity of Er in c-Si is related to a gap state. The ionization energy of this state was determined as $E_D \approx 218$ meV. This level provides indeed the gateway for Er^{3+} excitation as, demonstrated by 1.5 μm emission upon resonant pumping into

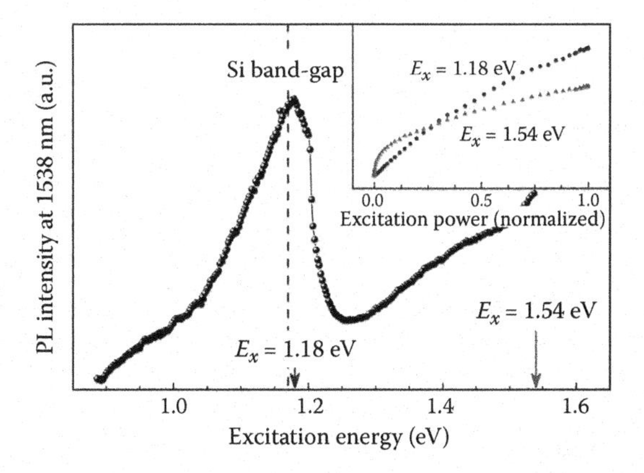

FIGURE 8.42
PLE spectrum of the 1.5 μm Er-related emission at $T = 4.2$ K. The normalized power dependence of the PL intensity is given in the inset for two excitation wavelengths indicated by arrows. (From Vinh, N. Q., Ha, N. N., and Gregorkiewicz, T., *Proceedings of the IEEE* 97 (7):1269–1283 © 2009 IEEE.)

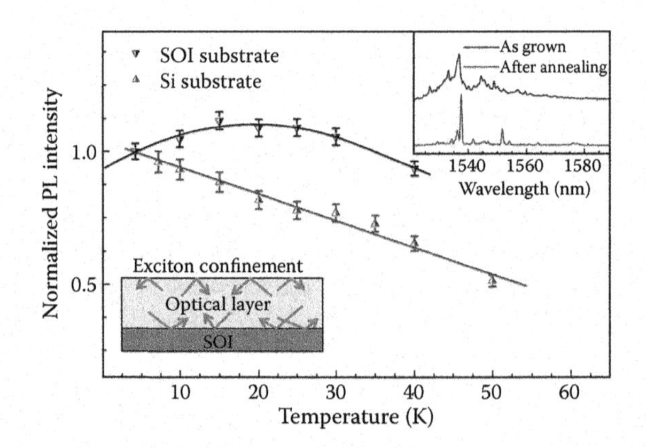

FIGURE 8.43
Temperature dependence of PL intensity from the multinanolayer structures grown on Si and SOI. The inset shows PL spectra of the sample on SOI at $T = 4.2$ K before and after annealing at 900°C for 30 min in N_2. (From Vinh, N. Q., Ha, N. N., and Gregorkiewicz, T., *Proceedings of the IEEE* 97 (7):1269–1283 © 2009 IEEE.)

the bound exciton state of the identified donor, allowing excitation of Er^{3+} while avoiding Auger quenching by free carriers.

For practical application in photonic devices, the spatial confinement of photons emitted by Er^{3+} ions is necessary. For that purpose, Si/Si:Er multinanolayer structures were grown by SMBE on a semi-insulating (SOI) substrate (Vrielinck et al., 2005). Figure 8.43 compares the PL from structure grown on Si substrate and on SOI. The PL intensity decreased monotonously when the temperature was increased for structure on Si substrate, whereas, for the structure grown on SOI, the PL intensity initially increased to attain a maximum at about 15–25 K and then decreased. The initial increase of PL intensity with temperature suggests confinement of excitons and/or free carriers within the multinanolayer structure on SOI shown schematically in the inset of Figure 8.43. The thermal energy detraps carriers/excitons bound to impurity centers or defects, thus generating free excitons, which enable energy transfer to unexcited Er-1 centers.

8.5.4 Erbium in SiO_2, Erbium-Doped SIPOS, and Idea of Silicon Nanocrystals

In Er-doped semiconductor hosts, excitation cross section of Er^{3+} PL is high due to effective band-to-band absorption, but nonradiative de-excitation processes lead to thermal quenching. Wider band-gap of the host reduces thermal quenching of Er^{3+} PL. In case of Er-doped Si, with its narrow band-gap of 1.12eV thermal quenching is severe and only at very low temperatures strong Er^{3+} PL can be observed (Favennec et al., 1989). On the other hand, in an insulating matrix-like SiO_2:Er, intense Er^{3+} PL can be observed at room temperature. However, in SiO_2:Er, only direct excitation of erbium ions is allowed: excitation wavelength need to be resonant with transitions between the levels of Er^{3+} ions and excitation cross section is very low, of the order 10^{-21} cm^2 (Miniscalco, 1991). Heavy doping with oxygen can turn silicon into semi-insulating polycrystalline silicon (SIPOS). It was found that Er-implanted SIPOS provides good thermal stability of Er^{3+} luminescence and allows nonresonant excitation of Er^{3+} ions (Lombardo et al., 1995; Polman, 1997).

Subsequent investigations showed that good thermal stability of PL from SiO_2:Er and high excitation cross section of Si:Er can be combined in one material. Such materials were realized by dispersing high concentration of Si NCs in Er-doped SiO_2 matrix (Kenyon et al., 1994; Fujii et al., 1997, 1998; Franzo et al., 1999; Priolo et al., 2001; Chryssou et al., 1999). Upon illumination, incoming photons are predominantly absorbed by band-to-band transitions in Si NCs. As the indirect band structure of Si is preserved in the nanocrystalline form (Kovalev et al., 1998), the generated e-h pairs have relatively long lifetime. This allows energy transfer to Er^{3+} ions located in the vicinity of Si NCs. In that way, an indirect channel of Er^{3+} ions excitation dispersed in SiO_2 is created and the Er-related emission at 1.5 μm appears. This nonresonant excitation process has an (effective) excitation cross section of $\sigma = 10^{-17}$–10^{-16} cm^2, which represents an increase by a factor 10^3 in comparison to resonant pumping of Er^{3+} ions in SiO_2 (Kenyon et al., 2002).

8.5.5 Sensitization of Erbium in SiO_2 by Si NCs

8.5.5.1 Preparation

Er-doped SiO_2 sensitized with Si NCs samples can be prepared by growing SiO_x:Er thin films (substoichiometric SiO_2 with silicon excess, where $x < 2$) on Si or quartz substrate followed by thermal annealing to form Si NCs. There are several methods to prepare SiO_x:Er thin films: co-sputtering, plasma-enhanced chemical vapor deposition (PECVD), and implantation. In co-sputtering, the SiO_2 target is sputtered together with silicon pieces and Er_2O_3 pellets placed on SiO_2. The fraction of the area taken by Si pieces and Er_2O_3 allows controlling Si excess and Er^{3+} concentration in the film (Fujii et al., 1997, 1998). In the PECVD deposition, film composition is controlled by flow rates of precursor gases: SiH_4, N_2O, and erbium precursor $Er(thd)_3$, evaporated from the bubbler and carried by Ar gas. The SiO_x:Er films can be also prepared by implantation of Er^{3+} and Si ions into thermal oxide on the Si wafer or quartz plates. After deposition films are annealed, which causes phase segregation: excess of Si forms Si NCs. Annealing at temperatures above 1100°C leads to growth of crystalline Si NCs, whereas at lower temperatures, amorphous Si-nanoclusters develop.

8.5.5.2 Limitations of Sensitization

Despite its apparent attraction, the sensitization of SiO_2:Er by Si NCs has a severe drawback: the low fraction of Er^{3+} ions that can couple to Si NCs, while majority of Er^{3+} ions remains optically inactive in the presence of Si NCs. This became clear after spectroscopic study of samples with the same concentration of Er^{3+}; of which one was stoichiometric SiO_2:Er, and the other contained Si NCs (Wojdak et al., 2004). For that study, the Si NCs were fabricated by annealing (1250°C, 1 h) of a 0.1-mm-thick film of SiO_x with 42 at. % Si, grown on a quartz substrate by PECVD (Priolo et al., 2001). The concentration of Si NCs was determined by transmission electron microscope coupled with an electron energy loss spectrometer tuned to the energy of the plasmon in Si and was found to be 5×10^{17}–1×10^{18} cm^{-3} with a mean Si NC diameter of $d < 3.5$ nm. The uniform Er^{3+} concentration of 2.2×10^{20} cm^{-3} over the whole films thickness was obtained by multiple Er^{3+} implantations at different doses and energies, followed by 1-h annealing at 900°C.

Figure 8.44 shows room-temperature PLE spectra of Er-related PL at $\lambda = 1.53$ μm for sample with and without Si NCs. PLE measurements were performed by scanning wavelength of tunable optical parametric oscillator (OPO), producing pulses of 5 ns duration at 20 Hz repetition rate. For Er^{3+} ions in SiO_2, only resonant excitation is allowed with emission peaks corresponding to internal transitions within the 4f electron core. The introduction of Si NCs allows indirect excitation over a broad wavelength range. The inset of Figure 8.44

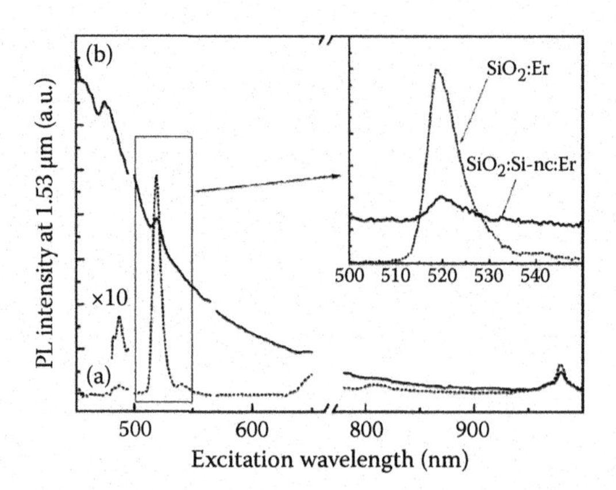

FIGURE 8.44
Room-temperature PL excitation spectra for SiO_2:Er (a) and SiO_2:Si NCs,Er (b). Inset presents details of PLE spectra at $\lambda_{exc} \approx 520$ nm, measured with maximum OPO power. (http://link.aps.org/doi/10.1103/PhysRevB.69.233315 © 2010 American Physical Society.)

shows details of PLE spectra around $\lambda_{exc} \approx 520$ nm measured with the maximum available OPO power. The corresponding photon flux during OPO pulse was $\Phi \approx 3 \times 10^{25}$ cm^{-2} s^{-1}. As can be seen, the PL in the sample of SiO_2:Er is stronger than that in the sample with Si NCs (SiO_2:Si NCs,Er). In the early stage of research, excitation path via Si NCs was believed to be much more efficient than direct excitation in SiO_2:Er, because low excitation power was used. This became clear when the power dependence of PL intensity was measured.

Figure 8.45 presents power dependence of Er^{3+} PL from samples without and with Si NCs resonantly excited at wavelength λ_{exc}= 520 nm traces (a) and (b) respectively, also the sample with Si NCs nonresonantly excited is shown in (c). The difference of traces (b) and (c) is shown in (d). The trace (c) represents PL of only those Er^{3+} ions that are coupled to Si NCs. As Er^{3+} in the sample with Si NCs can be excited both in direct and

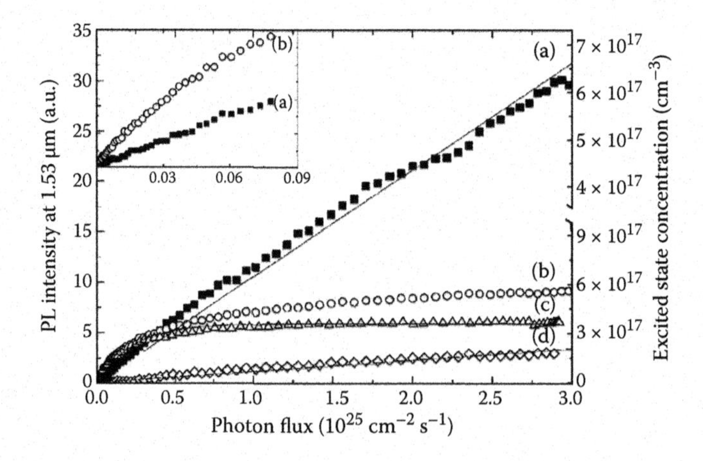

FIGURE 8.45
Time-integrated PL intensity at λ = 1.53 μm as function of the pulsed excitation density: (a) SiO_2:Er excited at λ_{exc} = 520 nm, (b) SiO_2:Si NC,Er excited at λ_{exc} = 520 nm, (c) SiO_2:Si NC,Er excited at λ_{exc} = 510 nm; (d) the difference between (b) and (c). The right-hand scale shows the concentration of Er^{3+} ions in the excited state in SiO_2:Er (upper part) and SiO_2:Si NC,Er (lower part). The inset presents a detail for the low flux regime. (http://link.aps.org/doi/10.1103/PhysRevB.69.233315 © 2010 American Physical Society.)

indirect way, an additional luminescence intensity can be seen in trace (b). The difference—trace (d)—shows the contribution from direct excitation in sample with Si NCs. It appears to be linear, however with smaller slope than directly excited PL in sample without Si NCs.

On the right-hand side of the figure, the excited state population is shown. It was estimated in the following way: for SiO_2:Er the excited state population after pumping time Δt was calculated using formula derived from single rate equation:

$$N_{Er}^{*}(t = \Delta t) = N_{Er}[1 - \exp(-\sigma\Phi\Delta t)] \quad (8.27)$$

The pumping time Δt corresponds to OPO pulse duration of 5 ns. The other parameters are N_{Er}, which is the total concentration of optically active Er^{3+} ions; σ, which is the excitation cross section; and Φ, which is the pumping photon flux. When $\sigma\Phi\Delta t \ll 1$, $N_{Er}^{*}(\Delta t) \approx N_{Er}\sigma\Phi\Delta t$. PL intensity was integrated after pumping pulse, and the result is proportional to $N_{Er}^{*}(\tau/\tau_{rad})$, where τ is observed luminescence lifetime and τ_{rad} is radiative lifetime. For SiO_2:Er, all parameters were known: $\sigma = 2 \times 10^{-20}\,cm^2$ (Miniscalco, 1991), $N_{Er} = 2.2 \times 10^{20}\,cm^{-3}$ (all implanted ions were assumed to be optically active), and $\tau/\tau_{rad} = 1$. The observed lifetime of 11 ms is the radiative lifetime, so the nonradiative recombination does not take place in sample without Si NCs.

For SiO_2:Si NCs,Er, the observed luminescence lifetime is 3 ms. In this material, radiative lifetime of Er^{3+} ions can be shorter due to the higher effective refractive index and radiative lifetime can be estimated as 9 ms. Further shortening of observed lifetime is due to nonradiative processes. As a result, in this sample three times higher excited state population needs to be obtained to produce the same PL intensity as in SiO_2:Er. In the above-mentioned experiment, the duration of pumping pulse was shorter than the time of energy transfer from Si NC to Er^{3+} ion and only one exciton per Si NC could be created. Therefore, one Si NC could excite only one Er^{3+} ion per one pumping pulse, even if there were several Er^{3+} ions coupled to one Si NC. Under continuous pumping all Er^{3+} ions coupled to one Si NC can be excited. Comparison of PL power dependence measured with pulsed and continuous pumping showed that about 20 ions can be coupled to one Si NC. Nevertheless, this means that only about 3% of Er^{3+} ions in the sample could be excited via Si NCs (Wojdak et al., 2004; Jambois et al., 2009).

The contribution of direct excitation in sample with Si NCs is represented by data set (d). It is linear like data set (a) but with a smaller slope. If we assume that direct excitation cross section is the same in both samples, it means that optically active Er^{3+} ions concentration in the sample with Si NCs is about 30% of that in sample without Si NCs. This means that presence of Si NCs turns majority of erbium ions optically inactive.

Other investigations found that 3.5% (Garrido et al., 2006) or 2% (Garrido et al., 2007) of Er^{3+} ions present in the material was coupled to Si NCs. Recently, it was reported that the fraction of Er^{3+} ions coupled to Si NCs can be improved up to 11% by material optimization (Hijazi et al., 2009), but that is still insufficient for the development of lasers based on this material.

In another experiment (Izeddin et al., 2006), room temperature PL spectrum of the SiO_2:Si NCs,Er was measured with the excitation wavelength $\lambda_{exc} = 450$ nm, i.e., not in resonance with internal transitions of Er^{3+} ions. Both Er- and Si NC-related bands centered around 1.53 and 0.9 μm were observed, shown in Figure 8.46. The inset shows the intensity of 1.5 μm Er^{3+} PL band as function of photon flux for two excitation wavelengths of 512 and 522 nm, corresponding to nonresonant and resonant ($^4I_{15/2} \rightarrow {}^4S_{3/2}$) excitation modes, respectively. The estimated average number of excitons generated per Si NC is

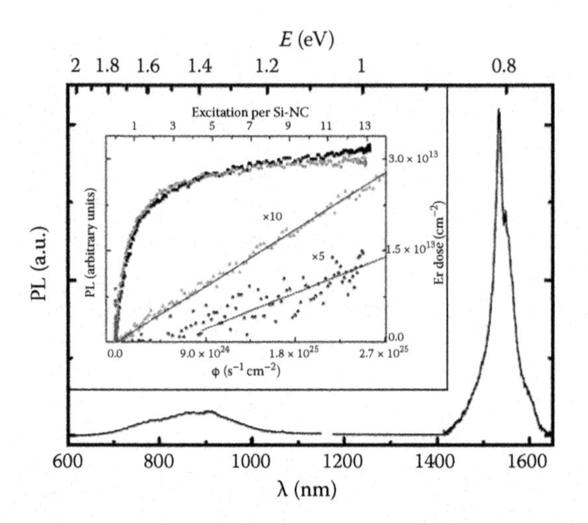

FIGURE 8.46
Room temperature PL spectrum of the investigated sample, λ_{exc} = 450 nm. In the inset: flux dependence of the PL intensity at λ = 1.5 μm for nonresonant (λ_{exc} = 512 nm, gray) and resonant (λ_{exc} = 522 nm, black) excitation. Lower scattered dots represent the difference between the two curves. Right-hand scale has been calibrated using the data obtained for the SiO_2:Er reference sample (middle one). (http://link.aps.org/doi/10.1103/PhysRevLett.97.207401 © 2010 American Physical Society.)

indicated in upper horizontal scale of the inset. Also intensity of the 1.5 μm PL band observed under resonant excitation (522 nm) in the SiO_2:Er standard sample is depicted (middle trace). Using a calibration procedure (Wojdak et al., 2004), the implanted SiO_2:Er sample serves to ascribe a particular concentration of Er^{3+} emitters to the PL intensity measured in the investigated sputtered layer. The appropriately corrected Er^{3+} density is given by the right-hand scale of the inset. It was then concluded that the Si NC-sensitized 1.5 μm PL saturates for Er^{3+} areal density of $[Er^{3+}] \approx 3.1 \times 10^{13}$ cm^{-2}, i.e., at ~0.55% of the total Er^{3+} content. There is also a possibility that the saturation level can slightly increase by excitation diffusion under continuous pumping, this represents a marginal part of the total Er^{3+} doping (Wojdak et al., 2004). At the same time, the difference between the two experimental curves (for resonant and nonresonant excitation, also shown) indicated that only ~25% of Er^{3+} dopants contributed to photon emission upon resonant pumping.

8.5.5.3 Excitation Model

To understand the excitation mechanism of Er^{3+} ions in SiO_2 sensitized with Si NCs, Izeddin et al. (2006) studied the decay dynamics of Er-related PL at 1.5 μm. The obtained detail decay dynamics is shown in Figure 8.47. Three different temporal regimes can be distinguished from decay dynamics. (i) In a period <1 μs, an intense emission appeared shortly after the laser pulse and rapidly decayed toward a (temporary) minimum. This indicates an almost instantaneous (less than a nanosecond) excitation directly into the first $^4I_{13/2}$ excited state of Er^{3+}, followed by fast relaxation. (ii) In a time window from 1 to 10 μs, PL intensity increases again, at a slower rate, and attains a maximum. This corresponds to the Si NC-to-Er energy transfer. As can be seen in the inset of Figure 8.47, the decay of Si NCs, characterized by a (stretched) exponential behavior with a time constant of $\tau_{Si\ NC} \approx 1.2$ μs, is paralleled by an increase of the Er-related emission on a similar time

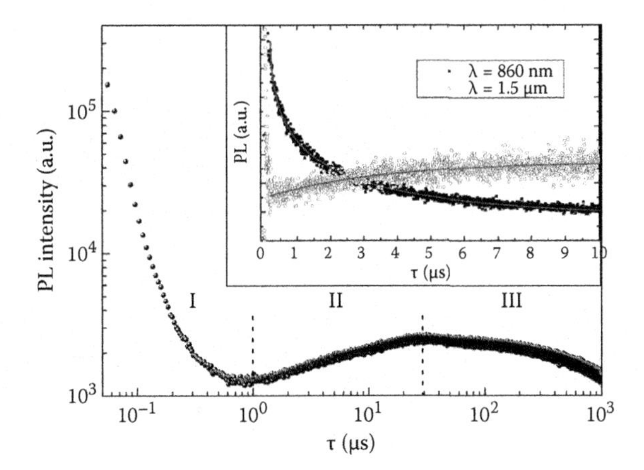

FIGURE 8.47
Double-logarithmic representation of the Er-related 1.5 μm PL decay, showing three different temporal regimes (room temperature, λ_{exc} = 450 nm). In the inset: room temperature PL temporal evolution for the first 10 μs after the excitation pulse, measured at λ = 1.5 μm for Er^{3+} and at λ = 860 nm for Si NCs. Stretched exponential decay formula has been used to fit the Si NCs decay (τ = 1.2 μs, β = 0.5) and single exponential rise for Er^{3+} (τ = 3.5 μs). (http://link.aps.org/doi/10.1103/PhysRevLett.97.207401 © 2010 American Physical Society.)

scale, $\tau_{Er}^{rise} \approx 3.5$ μs. (iii) The final decay for time greater than 10 μs. The time constant of this process, $\tau_{Er}^{D} \approx 2.4$ ms, is characteristic for the (predominantly) radiative recombination of $^4I_{13/2}$ excited state of Er^{3+} ion in SiO_2. The time integral of the Er-related PL signal was determined by the "slow" emission in regimes II and III, which, as discussed before, represents a contribution of ~0.5% of the total Er^{3+} content (Izeddin et al., 2006).

Based on the above-mentioned decay dynamics, an excitation model was proposed as shown in Figure 8.48 (Izeddin et al., 2006). Under optical pumping, hot e-h pairs are generated. In view of the large-level separation in Si NCs, these hot carriers can relax to the ground exciton state only by multiphonon processes, which is considerably slower than those assisted by a single phonon. A faster cooling of hot excitons can be facilitated by an

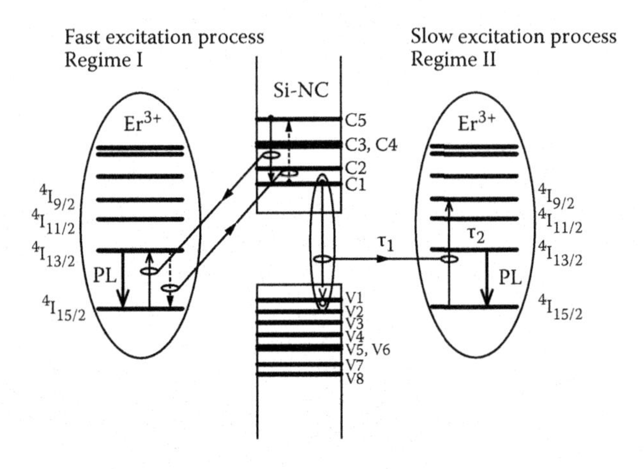

FIGURE 8.48
Schematic illustration of different excitation mechanisms proposed to be responsible for the "fast" regime I and the "slow" regimes II and III Er^{3+} PL. (http://link.aps.org/doi/10.1103/PhysRevLett.97.207401 © 2010 American Physical Society.)

Auger process of energy transfer to Er^{3+} ions, which are then excited directly to the first $^4I_{13/2}$ excited state, as shown in Figure 8.48. Such a process is similar to impact excitation by hot carriers in bulk Si, but should also be multiphonon assisted. The estimations showed that this process is effective for Er^{3+} dopants located within a distance comparable with the radius of a Si NC, ~1.3 nm. Depending on the actual distance, the effective transfer time was found to change within the 1–100-ns range. This Auger process was then considered to be responsible for appearance of the ultrafast Er^{3+} PL within nanoseconds after the laser pulse. However, parallel to the excitation process, fast Er^{3+} luminescence quenching should take place by a reverse process, in which a carrier confined in a Si NC acquires energy bringing the excited Er^{3+} ion back to its ground state. This process induces fast decay of the initial Er^{3+} PL. However, a small part of the rapidly excited ions will escape Auger quenching giving rise to the long-living component overlapping with the slow PL band and responsible for its initial amplitude– see the inset to Figure 8.47. Therefore, it is the Auger process of energy exchange between Er^{3+} ions and carriers confined in Si NCs, which gives rise to the ultrafast Er-related PL band depicted in Figure 8.47. In Si NCs, the Auger recombination rate controlling the lifetime of confined e-h pairs at high pumping levels is influenced by the alteration of both the energy structure and the density of states induced by quantum confinement. The estimations showed that about 50% of all Er^{3+} ions can be involved in this fast energy exchange process. It is therefore clear that investigations of optical gain in SiO_2 sensitized with Si NCs should concentrate on the time domain of the first microsecond after the laser pulse, when population inversion can possibly be realized (Izeddin et al., 2006).

Both fast processes of Auger excitation and quenching will continue as long as hot carriers are available in Si NCs. At a final stage, when only one e-h pair in a Si NC is left, another possible Er^{3+} excitation process will appear due to a (multi)phonon assisted Auger recombination of the last e-h pair see Figure 8.48. Recombination of excitons can excite Er^{3+} ions exclusively into one of the higher excited states, the second $^4I_{11/2}$, transition energy 1.24 eV, or the third $^4I_{9/2}$, transition energy 1.55 eV. Calculations showed that the excitation of Er^{3+} to the second or third excited states is very effective for Er^{3+} inside Si NCs or very close to their surface (at a distance of 0.2–0.4 nm). Evidence for coexistence of both slow and fast Er^{3+} excitations was further confirmed by observation of the so-called space-separated quantum cutting in Er-doped SiO_2 sensitized with Si NCs. (Timmerman et al., 2008, 2009). In this experiment, Er^{3+} ions outside the Si NCs were used as receptors for the down-converted energy. Er-related PL was monitored as a function of excitation wavelength and compared with the absorption of the sample at that wavelength. Under the chosen conditions, no direct absorption by Er^{3+} ions was possible; therefore, the Er-related PL was correlated to photon absorbed by Si NCs. Figure 8.49 shows the experimentally measured quantum yield of Er-related PL for two different samples having the same NC size and Er^{3+} concentration, but different concentration of Si NCs. The inset shows the distance distribution from the center of the Si NC to the nearest Er^{3+} ion for the two samples used. Quantum yield is defined as the ratio of the number of photons emitted by the sample to the number of absorbed photons $\eta = N_{PL}/N_{abs}$. The measured relative yield enabled the direct comparison of the different mechanisms responsible for the emission. For a single-photon generation process, the correlation between the number of absorbed and emitted photons is linear and the ratio constant: with a certain efficiency a long-wavelength photon is emitted by an Er^{3+} ion for every short-wavelength photon absorbed by a Si NC. From Figure 8.49, it is evident that this scenario is indeed followed for the lower range of excitation energies. However, a clear enhancement was seen for energies above a certain threshold, ~2.6 eV (480 nm) for both samples. At this energy, the QE of the energy transfer to Er^{3+} ions increased, due to the onset of double-photon generation process. The excess energy

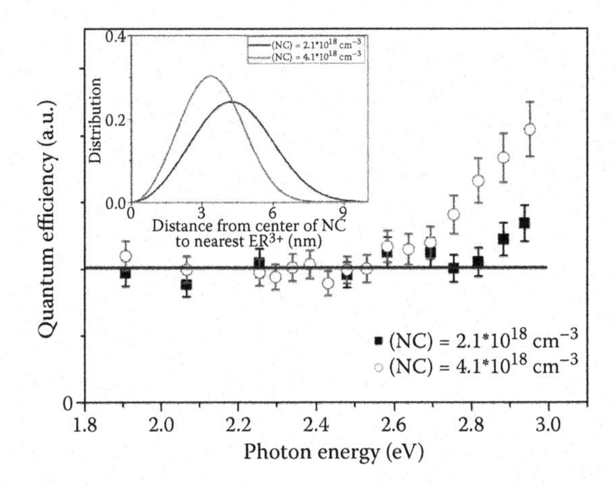

FIGURE 8.49
The QE of Er-related PL for different excitation energies for two samples. The inset shows the distance distribution from the center of a Si NC to the nearest Er^{3+} ion for the both samples. (From Timmerman, D., and Gregorkiewicz, T. 2009. *Materials Science & Engineering:* B: *Solid-State Materials for Advanced Technology* 159–160: 87–89.)

ΔE of the "hot" carrier ($\Delta E = h\nu - E_{NC}$, where $h\nu$ and E_{NC} are photon and exciton energies, respectively) was large enough to allow for an Auger process of intraband relaxation with simultaneous Er^{3+} ion excitation. As a result, two Er^{3+} ions were excited per single photon absorbed by a Si NC, with the second resulting from a conventional band-to-band Auger process, as indicated in the schematic. The onset for increase was similar for both samples, while the increase in efficiency was much more prominent in the sample where the average distance between Si NCs and Er^{3+} ions is smaller. Analyzing the energy diagram in Figure 8.49, such a process is expected to take place for photon energies exceeding the sum of the Si NC band-gap (~1.5 eV) and the Er^{3+} excitation (0.8 eV), thus above ~2.3 eV.

8.5.5.4 New Evidence for Fast Er^{3+} Ion Excitation

As described earlier, investigation of the dynamics of 1.5 μm PL emission shows three different regimes: (i) in times shorter than 1 μs, an intense emission appearing instantaneously after the laser pulse and rapidly decaying toward a temporary minimum; (ii) a slow temporal rise of PL on a microsecond time scale; (iii) final slow decay on a millisecond time scale (Izeddin et al., 2006, 2008; Al Choueiry et al., 2009). The slow rise and decay regimes (ii) and (iii) were conclusively assigned to the emission from $^4I_{13/2}$ state of Er^{3+} ions, which is populated through higher excited states ($^4I_{11/2}$ or $^4I_{9/2}$). Recently, different explanations were put forward for the fast 1.5 μm emission in regime (i), as similar fast PL at 1.5 μm was also observed in Er-free SiO_2:Si NC samples (Al Choueiry et al., 2009; Navarro-Urrios et al., 2009). Therefore it was postulated that the fast PL at 1.5 μm arise due to deep trap centers that emit in the visible and infrared regions (Al Choueiry et al., 2009). In another interpretation, the fast infrared emission was related to recombination at defect centers either in the SiO_2 matrix or at the interface with the Si NCs (Navarro-Urrios et al., 2009; Pitanti et al., 2010a).

Recently, Saeed et al. (2011) addressed the above mentioned issue and investigated PL of Er-doped and Er-free samples at room temperature and low temperature (7K). The results are shown in Figure 8.50. Figure 8.50 compares data for the Er-doped and Er-free sample at room temperature (Figure 8.13a–c) and at T = 7 K (Figure 8.13d–f). Strong emission at

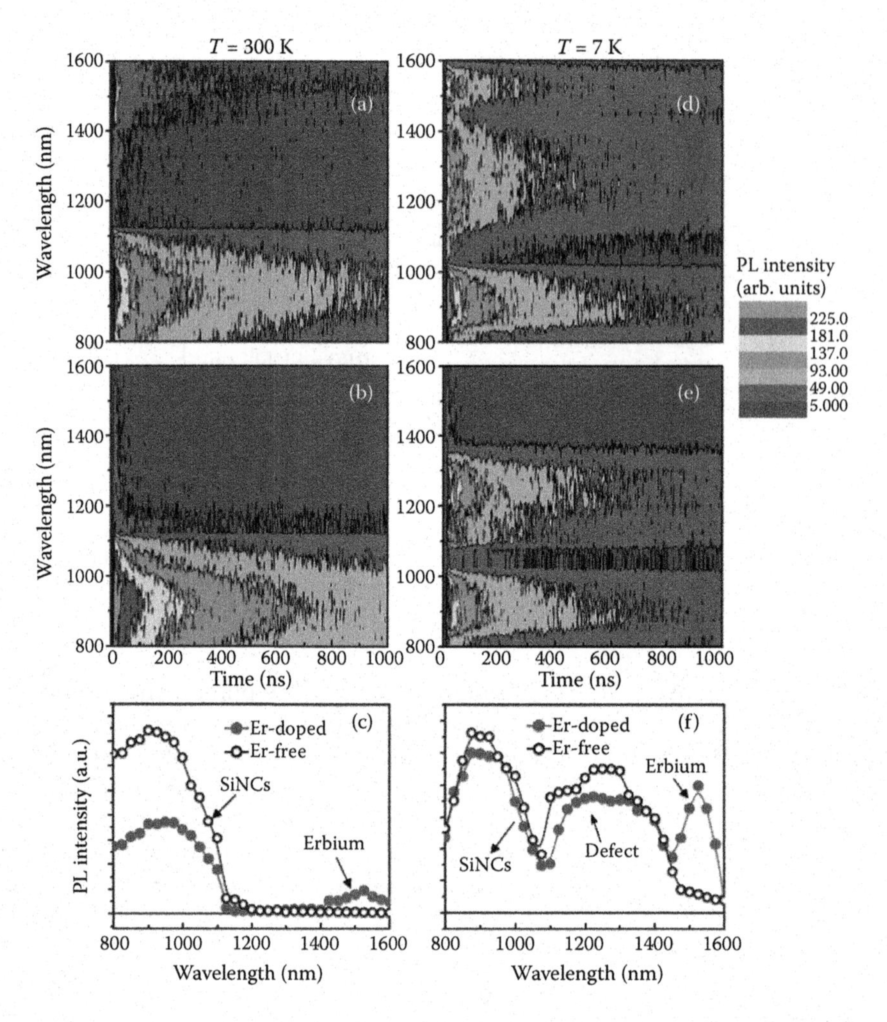

FIGURE 8.50
(See color insert.) Time-resolved PL spectra at room temperature (a)–(c) and $T = 7$ K (d)–(f) for Er-doped (a),(d) and Er-free (b),(e) samples. The PL intensity (contour plots) is represented as a function of time (x axis) and detection wavelength (y axis). The results show the PL decay in the first 1000 ns after laser excitation at $\lambda_{exc} = 450$ nm. Si NC- and Er-related bands can be seen at both temperatures, while a broad band around $\lambda = 1.3$ μm can only be observed at $T = 7$ K. The spectral dependence of the integrated PL (0–200 ns) is compared for samples with and without Er^{3+} ions at room temperature (c) and at $T = 7$ K (f). (http://link.aps.org/doi/10.1103/PhysRevB.83.155323 © 2011 American Physical Society.)

1.5 μm was clearly observed in the Er-doped sample at both temperatures, while there was no such signal in the Er-free sample. A defect-related signal was observed in both Er-doped and Er-free samples but only at low temperature. The wavelength dependence of the PL signals from Er-doped and Er-free samples integrated over a 0–200-ns time window is shown in Figure 8.50c and f. Here, the Si NC-related emission band can be seen in both samples, while the 1.5 μm emission is present exclusively in the Er-doped sample. This proved that the fast emission at 1.5 μm is indeed related to presence of Er^{3+} ions in the sample. Further, to see the contribution of defect-related band around 1.3 μm from Figure 8.50, defect band was studied in detail for which two possibilities were considered: (i) defect-related band has the same characteristics in Er-doped and Er-free sample; (ii) dynamics of

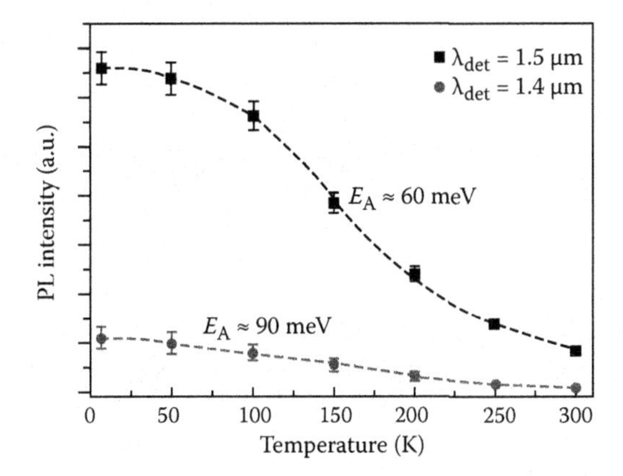

FIGURE 8.51
PL intensity of 1.5 and 1.4 μm emission as a function of sample temperature. Dotted line shows the data fitting by the Arrhenius law. Activation energies of $E_A \approx 60$ meV for 1.5 μm and $E_A \approx 90$ meV for 1.4-μm emissions are found. (http://link.aps.org/doi/10.1103/PhysRevB.83.155323 © 2011 American Physical Society.)

defect-related emission are modified due to the presence of Er^{3+} ions (Saeed et al., 2011). For the first case estimated defect-related contribution in 1.5 μm Er-related emission was only 24%, whereas for the second case, it was 40%. In any case, it was concluded that main contribution to the fast PL at 1.5 μm from Er-doped SiO_2 sensitized with Si NCs was Er-related. To investigate further the origin of this fast signal, Arrhenius plot was fitted for PL at 1.5 μm where Er^{3+} ions emit and the results were compared with the fit at 1.4 μm where Er^{3+} ions do not emit. Different activation energies of 60 meV for 1.5 μm and 90 meV for 1.4-μm band, shown in Figure 8.51, suggest that different centers are responsible for these two emissions. It was then concluded that Er-related trap centers might be responsible for the fast appearance and quenching of 1.5 μm emission.

8.5.6 Electroluminescent Devices

PL is an important characterization technique, but the technical objective is usually to develop electrically pumped devices. EL can be obtained from RE-doped materials in various ways. Most often the mechanism is impact excitation of RE ions by hot electrons accelerated by applied electric field, as was demonstrated in ZnS:Tb (Kahng, 1968; Krupka and Mahoney, 1972). The p-n junctions formed by growing an Er-doped p-type epitaxial silicon layer on an n-type silicon substrate showed an intense sharply structured EL spectrum at 1.54 μm (Ennen et al., 1985; Zheng et al., 1994; Neufeld et al., 1999). In case of SiO_2 containing Si NCs and Er^{3+} ions, it was proposed that carriers can be injected to Si NCs, which act as traps and subsequently excite Er^{3+} ions through e-h recombination (Priolo et al., 2006).

Obtaining impact excitation in RE-doped Si-based material is difficult, whereas the best luminescence of RE ions can be obtained in insulating SiO_2 matrix. The solution is then to use very thin layer of SiO_2:RE, which allows Fowler–Nordheim tunneling of hot electrons through the layer and impact excitation of RE ions. Metal-oxide-semiconductor (MOS) tunnel diode with Er-doped active SiO_2 layer was presented by Wang et al. (1997). MOS devices containing ions of gadolinium (Sun et al., 2004, 2006), terbium (Sun et al., 2005), and europium (Prucnal et al., 2007) were also demonstrated.

EL from devices based on silica doped with Er^{3+} ions and Si NCs was also investigated (Iacona et al., 2002; Nazarov et al., 2005; Sun et al., 2008a; Jambois et al., 2009). However, the role of the Si NCs in the EL of Er^{3+} and conduction mechanisms is controversial. Presence of Si NCs introduces more efficient conduction mechanisms, including variable range hopping (Fujii et al., 1996), direct tunneling, trap-assisted tunneling, Poole–Frenkel conduction (Priolo et al., 2006), or space charged limited current (Sun et al., 2008a). Si NCs allow injection of higher currents and improve device lifetime (Prezioso et al., 2008) but also reduce the population of hot electrons (DiMaria et al., 1985), which reduces impact excitation of Er^{3+} ions (Nazarov et al., 2005). On the other hand, it has been argued that presence of Si NCs can allow energy transfer from electrically excited Si NCs to Er^{3+} ions (Priolo et al., 2006), which is alternative EL mechanism for impact excitation. Generally, different processes can be dominant depending on film thickness and composition, or voltage regimes (Jambois et al., 2009).

8.5.7 Outlook

Despite immense research efforts, on Er-doped Si materials efficient amplifiers and lasers based on c-Si:Er have not yet been reported. The major problems are related to the low solubility of Er in c-Si, thermal quenching, and inhomogeneous broadening of emission spectra due to multiplicity of Er-related optical centers formed in the Si host. Also SiO_2:Er, and more generally, SiO_2:RE sensitized with Si NC have until now found only limited applications for practical devices. In that case, the major challenge has been the relatively low level of optical activity of the RE dopants.

8.6 Band-Engineered Ge-on-Si Lasers

Jifeng Liu

8.6.1 Introduction and Historical Overview

Epitaxial Ge-on-Si is a promising candidate for efficient monolithic light emitters and lasers due to its pseudo-direct gap behavior and compatibility with Si CMOS technology. Ge has already been incorporated into advanced electronic devices such as strain-enhanced high mobility CMOS transistors (Lee and Fitzgerald, 2005). Ge-on-Si–integrated photonic devices based on the direct gap transition of Ge, such as waveguide-coupled photodetectors (Ahn et al., 2007) and electro-absorption modulators (Liu et al., 2008a), have also been achieved in recent years. If a high-performance Ge-on-Si light source can be implemented, all active photonic devices on Si can be fulfilled using Ge, which greatly simplifies the monolithic electronic-photonic integration process.

The band structure of bulk Ge is schematically shown in Figure 8.52a. Although conventionally Ge is considered an indirect gap materials, the energy difference between its direct (Γ) and indirect band-gap (*L*) is only 136 meV (Madelung, 1982), or ~5 k_BT at room temperature. Realizing this small energy difference between the direct and the indirect gaps, it is natural to ask if one could enhance the direct gap emission simply by increasing the injection level such that some of the injected electrons spill into the direct Γ valley. Dr. Herbert Kroemer, a Nobel Prize laureate for his seminal contributions to HJ semiconductor devices, proposed in the 1960s that lasing from the direct gap transition of Ge could

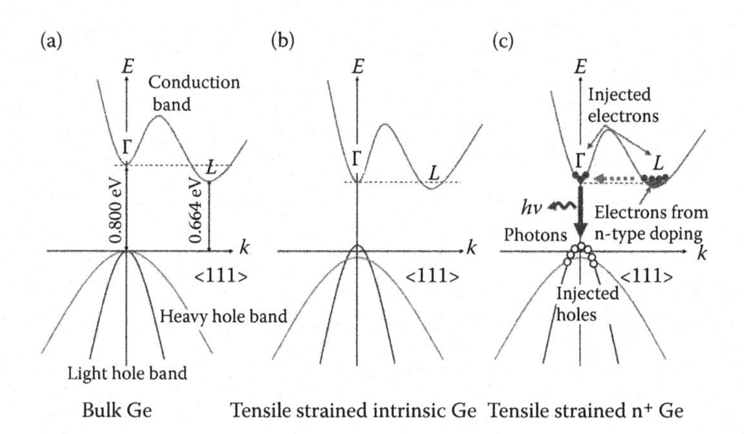

FIGURE 8.52
(a) Schematic band structure of bulk Ge, showing a 136 meV difference between the direct gap and the indirect gap (Liu et al., 2007), (b) the difference between the direct and the indirect gaps can be decreased by tensile strain (Liu et al., 2007), and (c) the rest of the difference between direct and indirect gaps in tensile strained Ge can be compensated by filling electrons into the *L* valleys via n-type doping. Because the energy states below the direct Γ valley in the conduction band are fully occupied by extrinsic electrons from n-type doping, injected electrons are forced into the direct Γ valley and recombine with holes, resulting in efficient direct gap light emission. (From Liu, J. F., Sun, X., Pan, D. et al. 2007. *Optics Express* 15:11272–11277. With permission.)

be made possible using HJ structures to inject carriers into their direct conduction valleys (Kroemer, 1963). Around the same time, Haynes and Nilsson (1964) studied PL from the direct gap transition of Ge at room temperature and revealed that the radiative recombination rate of the direct transition in Ge is 4–5 orders of magnitude higher than that of the indirect transition. In the 1970s, van Driel et al. studied direct gap PL of Ge at high injection levels. They argued that population inversion has been observed at ~100 kW cm^{-2} optical pumping intensity since fitting the PL data to theoretical model yielded a quasi-Fermi level separation 50 meV greater than the direct gap of Ge (van Driel et al., 1976). Klingenstein and Schweizer further observed a significant optical bleaching of ~70% upon high-energy pulsed laser pumping of intrinsic bulk Ge crystal at low temperatures (Klingenstein and Schweizer, 1978). More recently, room temperature continuous wave (CW) pump-probe studies of intrinsic Ge-on-insulator (GeOI) showed a similar result with up to 60% bleaching at photon energies greater than the direct band-gap (Sun et al., 2008b). However, no net optical gain has ever been reported from intrinsic Ge in the past 50 years.

The significant optical bleaching of direct gap absorption indicates that population inversion of direct gap transition indeed happens at high injection levels; yet the lack of net optical gain in intrinsic bulk Ge shows that the gain from direct gap transition cannot overcome loss mechanisms such as FCA. In the next section, we will present a theoretical modeling showing why intrinsic bulk Ge cannot achieve net optical gain, and how to overcome this issue by introducing tensile strain and n-type doping for band engineering.

8.6.2 Theoretical Modeling of Band-Engineered Ge Optical Gain Medium

The net optical gain of Ge upon carrier injection equals to direct gap optical gain subtracted by FCA losses (Liu et al., 2007; Sun et al., 2010). The gain coefficient of the direct band transition at a given photon energy $\gamma_\Gamma(h\nu)$ is related to the absorption coefficient of the direct band transition $\alpha_\Gamma(h\nu)$ by

$$\gamma_\Gamma(h\nu) = |\alpha_\Gamma(h\nu)|(f_c - f_v) \quad (8.28)$$

where $(f_c - f_v)$ is the well-known population inversion factor for direct band transitions. For bulk Ge, the direct band-to-band absorption can be expressed as

$$|\alpha_\Gamma(h\nu)| = A\left(\sqrt{h\nu - E_g^\Gamma}\right)/h\nu, \quad (8.29)$$

where $h\nu$ is the photon energy, $E_g^\Gamma = 0.8\,\text{eV}$ is the direct gap of bulk Ge, and A is a constant related to the transition matrix element and the effective mass of the material. Fitting to experimentally measured absorption spectrum yields $A = 1.9 \times 10^4\ \text{eV}^{1/2}\ \text{cm}^{-1}$ for bulk Ge.

The energy difference between the direct and the indirect gaps of Ge can be further reduced by tensile strain, as schematically shown in Figure 8.52b. With biaxial tensile stress, both direct and indirect gaps shrink, but the direct gap shrinks faster. Therefore, Ge transforms from an indirect gap material toward a direct gap material with the increase of tensile strain. Furthermore, the top of the valence band is determined by light hole band under biaxial tensile stress. The small effective mass of light hole band reduces density of states in the valence band, which in turn decreases the threshold for optical transparency and lasing.

Since the valence band becomes nondegenerate at $k = 0$ for tensile-strained Ge, Equation 8.29 is modified to fit the absorption spectra of tensile strained Ge:

$$|\alpha_\Gamma(h\nu)| = A\left(0.318\sqrt{h\nu - E_g^\Gamma(lh)} + 0.682\sqrt{h\nu - E_g^\Gamma(hh)}\right)/h\nu, \quad (8.30)$$

where $E_g^\Gamma(lh)$ and $E_g^\Gamma(hh)$ are the direct band-gaps associated with the light- and heavy-hole bands, respectively, and the factors of 0.318 and 0.682 are the relative contribution of light and heavy hole bands to the total absorption based on the ratio of their reduced effective mass.

Theoretically, Ge can be transformed into a direct gap material with ~1.8 % tensile strain (Van de Walle, 1989; Sun et al., 2010; El Kurdi et al., 2010a). However, the band-gap decreases to 0.5 eV and the corresponding emission wavelength shifts to 2500 nm in that case. A disadvantage of operating at such a long emission wavelength is that the corresponding photodetector of the optical interconnect system will suffer from a significantly larger dark current due to reduced band-gap. To obtain optical gain from the direct gap transition of Ge while maintaining the emission wavelength around 1550 nm, n-type doping has been combined with 0.2%–0.3% tensile strained to compensate the energy difference between direct and indirect gaps (Figure 8.52c) (Liu et al., 2007). In this case the required n-type doping level is in the order of 10^{19}–10^{20} cm^{-3}. For larger tensile strain the required n-doping level can be reduced as the energy difference between direct and indirect gaps becomes smaller. Since the lower energy states in indirect L valleys are already occupied by electrons from n-type doping, upon carrier injection, some injected electrons are forced to occupy the direct Γ valley and recombine with holes radiatively via efficient direct transitions. Furthermore, because the radiative recombination rate of the direct transition is 4–5 orders of magnitude higher than that of the indirect transitions (Haynes and Nilsson, 1964), the injected electrons in the Γ valley are depleted much faster than those in the L

valleys. To maintain the quasi-equilibrium of electrons in the conduction band, the electrons initially injected into in the L valleys will populate the Γ valley following intervalley scattering, as shown by the horizontal dashed arrow in Figure 8.52c. This process results in further radiative recombination via efficient direct transitions. Now the critical question is, can optical gain from the direct gap transition overcome the FCA losses?

Empirically, FCA coefficient in Ge can be expressed as (Spitzer et al., 1961; Newman and Tyler, 1957; Liu et al., 2007):

$$\alpha_f(\lambda) = 3.4 \times 10^{-25} N \lambda^{2.25} + 3.2 \times 10^{-25} P \lambda^{2.43}, \tag{8.31}$$

where $\alpha_f(\lambda)$ is the FCA coefficient (in cm^{-1}), N and P are electron and hole concentrations, respectively (in cm^{-3}), and λ is the wavelength (in nm). Note that at $\lambda = 1500 - 1600$ nm, the free hole absorption is about 4× larger than free electron absorption. As we will show later, due to the relatively small absorption cross section of free electrons optical gain in n^+ Ge can still overcome FCA losses, and lead to net optical gain.

Using Equations 8.28 through 8.31, the calculated net gain, which is the difference between the optical gain and the FCA, is shown in Figure 8.53a for unstrained and tensile strained Ge with and without n-type doping (Sun et al., 2010). Because tensile strained and unstrained Ge have different band-gaps, to compare the optical gain for each case, we choose the photon energy corresponding to the peak of each optical gain spectra at 1×10^{19} cm^{-3} injected carrier density. The theoretical modeling shows that unstrained intrinsic Ge initially exhibits a significant optical bleaching upon carrier injection, yet at an injection level of $\Delta n > 4 \times 10^{18}$ cm^{-3}, the absorption starts to increase again since the increase in optical gain can no longer catch up with that of FCA. As a result, we can only observe optical bleaching in intrinsic Ge but not optical gain. This prediction is in good agreement with the experimental results on unstrained intrinsic Ge (Klingenstein and Schweizer, 1978; Sun et al., 2008b). In n^+ Ge with an n-type doping level of 7×10^{19} cm^{-3}, on the other

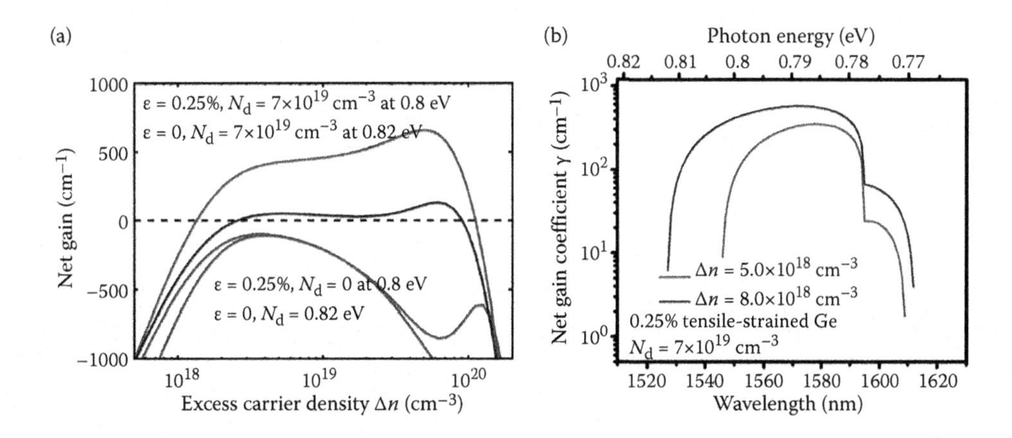

FIGURE 8.53
(a) Net optical gain versus injected carrier density for or unstrained and tensile-strained Ge with and without n-type doping from theoretical modeling. While unstrained intrinsic Ge cannot achieve net gain, band engineering by tensile strain and n-type can overcome this issue and achieve a large gain of 500 cm^{-1} (Sun et al., 2010). (b) Gain spectra for 0.25% tensile strained Ge with an n-type doping level of $N_d = 7 \times 10^{19}$ cm^{-3} at different injection levels. A broad gain spectrum is predicted for an injection level of $>5 \times 10^{18}$ cm^{-3}. (From Sun, X., Liu, J. F., Kimerling, L. C., and Michel, J. 2010. *IEEE Journal of Selected Topics in Quantum Electronics* 16:124–131. With permission.)

hand, net gain can be obtained at injected carrier densities above 10^{18} cm^{-3} even without tensile strain. A major difference from intrinsic Ge case is that the Fermi level is already raised close to Γ valley by n-type doping (Figure 8.52c), so that a much smaller amount of hole injection is needed to achieve population inversion. As a result, free hole absorption, which dominates FCA in Ge, is significantly reduced so that net gain is enabled. The modeling also shows that the net gain of n$^+$ Ge can be improved by >5× from less than 100 cm^{-1} to over 500 cm^{-1} by introducing 0.25% tensile strain due to smaller energy difference between Γ and L valleys. Figure 8.53b further shows that tensile strained n$^+$ Ge can achieve a broad net gain spectrum from 1600 to 1550 nm at an injection level >5 × 10^{18} cm^{-3}. Such a broad gain spectrum is applicable to on-chip wavelength division multiplexing (WDM). These results indicate that a combination of tensile strain and n-type doping is promising to transform Ge into an optical gain medium on Si.

Optical gain from Ge with >0.3% tensile strain have also been theoretically studied in recent years. Chang et al. (2009) modeled optical gain from 0.51% tensile-strained n-type Ge QWs pseudomorphically grown on SiGeSn buffer layers with an n-type doping of 2 × 19^{19} cm^{-3}. They concluded that a large optical gain of ~7500 cm^{-1} can be obtained at λ = 1550 nm for transverse magnetic (TM) modes at an injected surface carrier density of 10^{13} cm^{-2}. The threshold is comparable to III–V QW lasers. El Kurdi et al. (2010a) performed an elaborate 30-band k-p modeling on highly tensile-strained Ge with in-plane tensile strain >1.9%. The net gain can reach 3000 cm^{-1} at λ = 3060 nm with 3% tensile strain at an injected carrier density of $\Delta n = 1 \times 10^{18}$ cm^{-3}, exceeding the gain coefficient of GaAs at the same injection level. All these theoretically studies indicate that band-engineered Ge is a promising candidate for monolithic lasers on Si.

8.6.3 Epitaxial Growth of Band-Engineered Ge-on-Si

Now that theoretical modeling demonstrates the feasibility of optical gain and lasing from band-engineered Ge-on-Si, high-quality epitaxial material growth is the first critical step to implement the device experimentally. The greatest challenge for high quality Ge epitaxy on Si is the 4.2% lattice mismatch between these two materials. To achieve device quality epitaxial Ge on Si without using very thick SiGe buffer layers, a two-step direct Ge growth technique is applied to prevent islanding, and subsequent annealing was developed to decrease the threading dislocation density significantly (Baribeau et al., 1988; Colace et al., 1997; Luan et al., 1999). In the initial growth step, a thin epitaxial Ge buffer layer of 30–60 nm is directly grown on Si at 320°C–360°C. At such low growth temperatures the islanding of Ge is kinetically suppressed due to the low surface diffusivity of Ge. In the main growth step, the growth temperature is increased to >600°C for higher growth rates and better crystal quality. A postgrowth annealing at >750°C can reduce the threading dislocation density from 10^9 to 10^7 cm^{-2}. In selectively grown Ge mesas with lateral dimensions ~10 μm, the threading dislocation density can be further reduced to 10^7 cm^{-2} since these dislocations can glide to the edge of the mesas and annihilate (Luan et al., 1999). More details on high-quality epitaxial Ge on Si can be found in an earlier chapter of this book. Here we will focus on two critical aspects related to band-engineered Ge-on-Si light emitters: (1) introduction of tensile strain and (2) N-type doping.

8.6.3.1 *Tensile-Strained Ge*

When we consider Ge epitaxy on Si, a common impression is that *compressive* strain would be introduced to Ge during pseudomorphic growth due to its larger lattice constant compared with Si. However, this is not the case for relatively thick epitaxial Ge layers on Si for photonic devices. The thickness required in optical devices is in the order of λ/n_r, where λ is the wavelength and n_r is the real part of the refractive index. For Ge photonic devices operating at λ~1550 nm, the thickness is thus in the order of several hundred nanometers. This thickness far exceeds the critical thickness for pseudomorphic growth of Ge on Si, so the Ge film is relaxed at growth temperatures >600°C. Upon cooling to room temperature, *tensile strain*, instead of compressive strain, can be accumulated in the Ge layers due to the larger thermal expansion coefficient of Ge compared with Si (Ishikawa et al., 2003; Cannon et al., 2004; Hartmann et al., 2004). Typically 0.2%–0.3% tensile strain can be achieved using this process, which reduces the difference between the direct and indirect band-gaps of Ge from 136 to ~100 meV and red-shifts direct band-gap from 0.8 to 0.77–0.76 eV (Liu et al., 2004a; Sun et al., 2009). To further enhance the tensile strain in epitaxial Ge, in recent years, relaxed GeSn buffer layers on Si have been developed as a lattice template for tensile strained Ge (Fang et al., 2007a,b). This approach has achieved up to 0.68% tensile strain so far (Takeuchi et al., 2008). Stressor layers such as silicides (Liu et al., 2004b, 2005) or silicon nitride (Kuroyanagi et al., 2011; de Kersauson et al., 2011) have also been proven effective to enhance tensile strain in Ge.

An alternative approach to tensile strained Ge is directly applying external mechanical stress. Lim et al. (2009) proposed a micromechanical structure for introducing tensile stress to Ge light emitters. El Kurdi et al. (2010b) applied external stress to bulk Ge wafer and demonstrated 0.6% tensile strain, leading to a redshift of direct gap emission peak from 1535 to 1660 nm. Cheng et al. (2010) reported a 1.8× enhancement of direct gap PL for 0.37% tensile strain mechanically applied on bulk Ge. For device applications, microelectromechanical systems (MEMS) may be coupled with suspended thin Ge layers to achieve enhanced tensile strain in Ge light emitters.

8.6.3.2 N-Type Doping

As shown earlier, N-type doping is another critical factor to achieve efficient light emission from the direct gap transition of Ge. Table 8.4 summarizes solubility of n-type dopants in Ge (Madelung, 1982). According to these data, phosphorous is the best n-type dopant for Ge due to the largest solubility in the widest temperature range. One should note, however, that nonequilibrium growth methods such as MBE could achieve doping levels exceeding the solubility limit. A particular challenge, though, is that n^+ doping has to be implemented without introducing nonradiative recombination centers. Although

TABLE 8.4

Solubility of Different n-Type Dopants in Ge

Dopant	Maximum Solubility	Temperature Dependence
P	2×10^{20} cm^{-3} at 580°C	>1×10^{20} cm^{-3}, 500–800°C
As	8.7×10^{19} cm^{-3} at 800°C	>8.0×10^{19} cm^{-3}, 750–880°C
Sb	1.1×10^{19} cm^{-3} at 800°C	>1.0×10^{19} cm^{-3}, 750–870°C

Source: Data collected from Springer Science+Business Media: *Physics of group IV elements and III–V compounds, Landolt–Börnstein: Numerical data and functional relationships in science and technology*, vol. 17a, 1982, O. Madelung.

ion implantation is widely applied for doping processes in CMOS circuits, the damage and defects induced by high-energy ion bombardment are not suitable for light-emitting devices (Liu et al., 2008b). To minimize defect formation, two approaches have been developed for heavy n-type doping in Ge: (1) in situ doping (Liu et al., 2008b; Sun et al., 2009b; Cheng et al., 2009; Kasper et al., 2010; Hartmann et al., 2011) and (2) diffusion doping from spin-on dopant (SOD) sources (Liu et al., 2011).

For in situ doping, PH_3 is used as the precursor during CVD growth (Liu et al., 2008b; Sun et al., 2009b; Cheng et al., 2009; Hartmann et al., 2011) while Sb has been adopted for n-type doping in MBE growths (Kasper et al., 2010; Arguirov et al., 2011). In CVD, the optimal growth temperature is typically 600°C–700°C to achieve $>10^{19}$ cm^{-3} *activated* P concentration without sacrificing crystal quality. Although higher P atomic concentration can be incorporated at lower growth temperatures of ~400°C, the material quality and dopant activation are adversely affected (Hartmann et al., 2011) due to lack of atomic diffusion. On the other hand, if the growth temperature is above 750°C, P atoms in Ge tend to out-diffuse and evaporate into the gas phase (P_2 gas molecules) and reduce n-type doping concentration. So far, an active n-type doping concentration up to 3.5×10^{19} cm^{-3} has been achieved by in situ CVD growth (Camacho-Aguilera et al., 2011) with significantly enhanced direct gap light emission. In MBE growths, the Sb doping was performed at a low temperature of ~350°C and a doping level up to 10^{20} cm^{-3} was demonstrated. A notable difference from n^+ Ge grown by CVD is that the Ge film is relaxed rather than tensile-strained due to the low MBE growth temperature, i.e., thermally induced tensile strain almost exactly cancels the residual compressive lattice strain in this case.

For a simple fabrication process of photonic device integration, it is desirable to locally dope designated Ge regions for light-emitting devices so that other regions can remain intrinsic for Ge photodetector and electroabsorption modulators. Based on this motivation, SODs has been developed as an alternative approach to achieve n^+ doping in Ge (Liu et al., 2011). After coating SOD on epitaxial Ge films, the samples are baked and ashed to remove solvents and any organic components so that only P-doped SiO_2 is left in the Ge surface. A SiN layer is deposited on top to help prevent out-diffusion of phosphorous upon drive-in annealing. After rapid thermal annealing at >700°C, an average doping concentration up to 2×10^{19} cm^{-3} can be achieved in initially intrinsic Ge films. The trade-off in this case is drive-in diffusion into Ge versus out-diffusion of phosphorous dopants. Both in situ doping and diffusion doping from SOD have achieved strong room temperature PL from the direct gap transition of Ge, which is ~10 times higher than implanted samples with the same phosphorus doping concentration.

8.6.4 Band-Engineered Ge-on-Si Light Emitters

With high-quality band-engineered Ge-on-Si materials, light emission properties of Ge are expected to improve significantly. In this section, we will present spontaneous emission, optical gain and lasing from these materials and discuss on-going work toward electrically pumped Ge-on-Si lasers.

8.6.4.1 Spontaneous Emission

Although direct gap PL of intrinsic and heavily doped bulk Ge has been studied in several papers between 1960s and, 1980s (Haynes and Nilsson, 1964; van Driel et al., 1976; Klingenstein and Schweizer, 1978; Wagner et al., 1983, 1984; Wagner and Viña, 1984), a systematic study on the direct radiative transition of epitaxial Ge-on-Si band-engineered by tensile strain and/or n-type doping did not start until recent years. Direct gap spontaneous emission in Ge, regardless of optical or electrical pumping, exhibits three distinct features that are exactly *opposite* to the behavior of conventional direct gap III–V semiconductors: (1) The integrated emission intensity *increases* with *n-type doping level* (Liu et al., 2008b; Sun et al., 2009b; Cheng et al., 2009; Bessette et al., 2011). The integrated PL intensity at room temperature increases by >100× with 3×10^{19} cm^{-3} active n-type doping compared with intrinsic Ge-on-Si, consistent with our theoretical model on filling L valleys by n-type doping (see Figure 8.54a,b) (Sun et al., 2009b; Bessette et al., 2011). (2) At high injection levels, the spontaneous emission intensity increases *superlinearly* with optical pumping power (Klingenstein and Schweizer, 1978) or injection current (see Figure 8.55) (Sun et al., 2009a; Cheng et al., 2009; Kasper et al., 2010; Arguirov et al., 2011). (3) The emission intensity *increases* with *temperature* (Sun et al., 2009b; Cheng et al., 2009) following Arrhenius relation, as shown in Figure 8.56a,b. Recent studies revealed that the PL intensity keeps

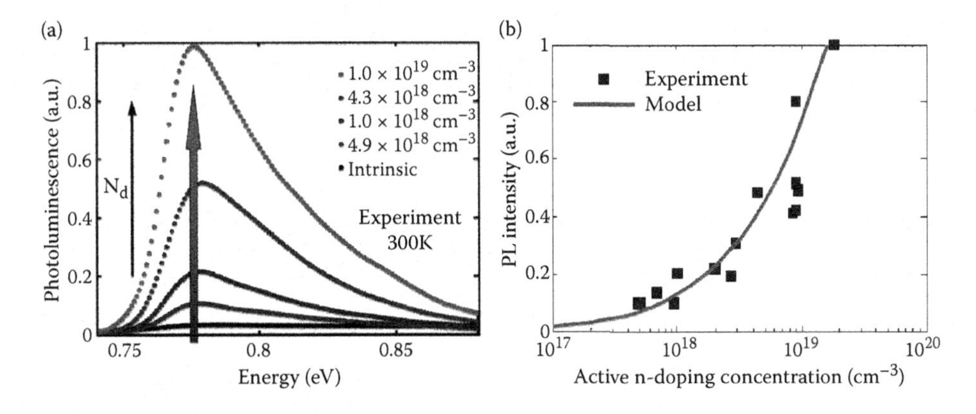

FIGURE 8.54
(a) PL spectra of 0.2% tensile strained Ge-on-Si with different n-type doping levels at room temperature (b) PL intensity versus n-type doping level. The line shows theoretical modeling and the dots shows experimental data. (From Sun, X., Liu, J. F., Kimerling, L. C., and Michel, J. 2009b. *Applied Physics Letter* 95:011911. With permission.)

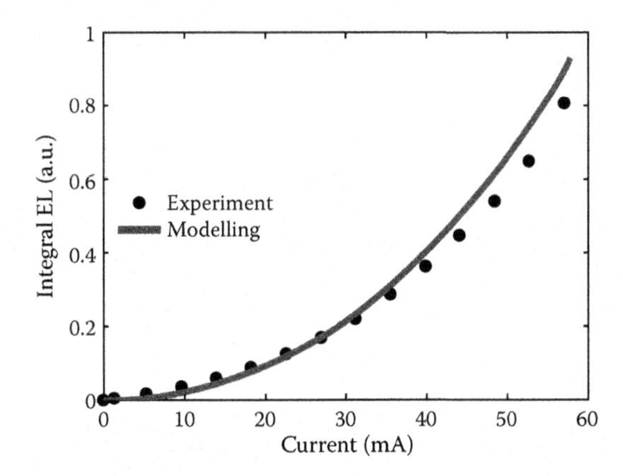

FIGURE 8.55
Integrated EL intensity versus injection current of a 0.2% tensile strain Ge-on-Si LED (Sun et al., 2009a). The EL intensity increases superlinearly with injection current. (From Sun, X, Liu, J. F., Kimerling, L. C., and Michel, J. 2009a. *Optics Letter* 34:1198–1200. With permission.)

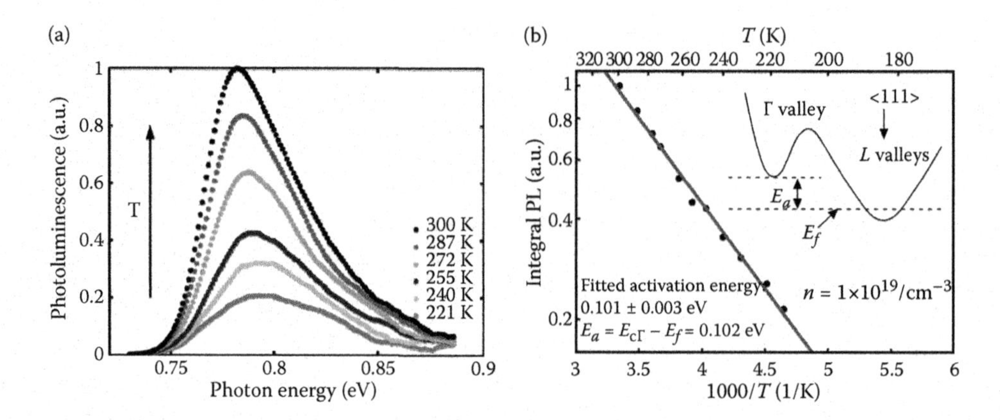

FIGURE 8.56
(a) PL spectra of 0.2% tensile strained n⁺ Ge-on-Si at different temperatures. The n-type doping level in the Ge film is 1×10^{19} cm^{-3}. The PL intensity clearly increases with temperature. (b) An Arhenius plot of integrated PL intensity versus temperature showing that the activation energy is equal the energy difference between the Γ valley and the Fermi level. (From Sun, X., Liu, J. F., Kimerling, L. C., and Michel, J. 2009b. *Applied Physics Letter* 95: 011911. With permission.)

increasing up to ~100°C (Bessette et al., 2011). These unusual phenomena are due to the fact that the energy states at direct Γ valley are higher than those in indirect *L* valley. The direct gap luminescence intensity is proportional to the number of electrons in the Γ valley, $n_e(\Gamma)$, which is the product of the total injected electron concentration, $n_e(total)$, and the faction of the electrons in the Γ valley, $f(\Gamma)$:

$$\text{Emission}_{\text{dir}} \propto n_e(\Gamma) = n_e(total)f(\Gamma) \tag{8.32}$$

At a constant injection level, i.e., constant $n_e(total)$, increasing n-type doping concentration raises the Fermi level so that the fraction of electrons occupying the higher energy Γ valley, $f(\Gamma)$, will increase, leading to enhanced direct gap emission. Similarly, thermal excitation promotes electrons from *L* valleys to higher energy states in Γ valley so that increase in temperature also enhances light emission via direct transition. When the excitation intensity increases, $n_e(total)$ scales linearly with the injection level, while $f(\Gamma)$ also increases with the injection level owing to the increase of the quasi-Fermi level. The multiplication of these two terms in Equation 8.32 results in a superlinear increases of direct gap emission with excitation level. The increase in emission efficiency with injection level and temperature are especially attractive for high power on-chip laser sources where high injection current and high thermal stability up to 80°C are required in electronic-photonic integrated circuits.

Band-engineering by tensile strain and n-type doping also modifies the peak emission wavelength of Ge. It is well understood that direct gap emission peak red shifts with the increase of tensile strain as a result of direct band-gap shrinkage (Liu et al., 2004a; El Kurdi et al., 2010b). The effect of n-type doping appears more complicated, though. Up to an n-type doping level of 1×10^{19} cm^{-3} there is no significant shift in emission peak position (see Figure 8.54a) (Sun et al., 2009b). This result is consistent with the electroreflectance data reported by Lukeš and Humlíček (1972) showing no significant change in direct band-gap for an n-type doping level up to 2×10^{19} cm^{-3}. On the other hand, at >2×10^{19} cm^{-3} n-type doping level, the emission peak starts to show a red shift (Bessette et al., 2011; Kasper et al., 2010), as shown in Figure 8.57. The PL peak shifts to 1700 nm with 10^{20} cm^{-3}

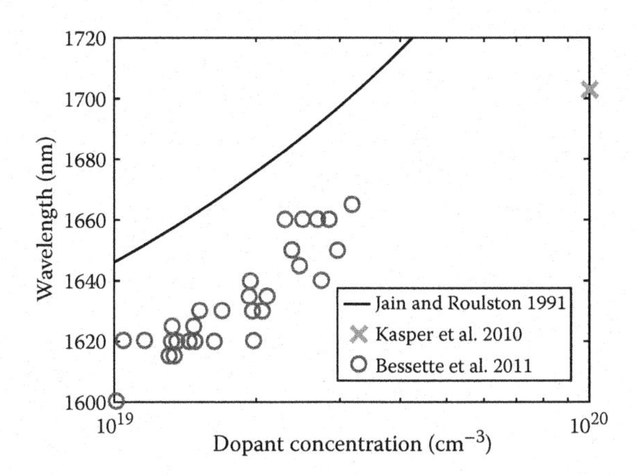

FIGURE 8.57
PL peak position of 0.2% tensile strained n^+ Ge-on-Si versus n-type doping level. The symbols show experimental data from Bessette et al. (2011) and Kasper et al. (2010), while the line shows theoretical model from Jain and Roulston (1991).

Sb doping by MBE growths (Kasper et al., 2010). However, the amount of red shift at $n > 2 \times 10^{19}$ cm^{-3} seems much smaller than previous experimental reports based on optical absorption spectra (Haas, 1962). The red shift is also smaller than the theoretical prediction proposed by Jain and Roulston (1991). It is possible that in Haas (1962) defect-induced band tails lead to an overestimate of the red shift from the absorption spectra. At the time this chapter is written, the detailed mechanism of direct band-gap shrinkage with n-type doping is still under investigation.

In recent years, there has been an increasing interest in studying light emission from microcavities due to their small footprint, potentials for on-chip WDM laser bank, and spontaneous emission enhancement by Purcell effect. Lim et al. (2008) reported enhanced emission at resonant wavelengths from intrinsic Ge-on-Si microring resonators. A quality factor of Q=620 was achieved, an order of magnitude higher than that previously reports on crystalline germanium microcavities due to strong optical confinement. Shambat et al. (2010) fabricated and optically characterized Ge-on-Si microdisks. The emission was coupled to a tapered fiber and multiple whispering gallery modes were observed. Cheng et al. (2011) further demonstrated cavity modes in EL spectra from similar Ge-on-Si microdisks. However, in all these cases when pumping power or injection current is increased the Q factor decreases, indicating that there is no optical bleaching in these devices. This result is surprising considering that optical bleaching has been observed even in intrinsic unstrained bulk Ge (Klingenstein and Schweizer, 1978) or GeOI (Sun et al., 2008b). It is likely that the lack of bleaching in these Ge-on-Si microresonators is due to defects introduced by fabrication. Reactive ion etching has been applied to fabricate microresonators in these cases, which could introduce defects on the sidewall that stronger interacts with the whispering gallery modes. Future investigations may elucidate the key to high performance Ge microresonator emitters.

8.6.4.2 Optical Gain

As we have discussed earlier, to achieve net optical gain Ge has to be band engineered by tensile strain and/or n-type doping to compensate the energy difference between its direct and indirect band-gaps. Experimentally, optical gain was first observed from selectively grown, 0.25% tensile-strained n⁺ Ge-on-Si mesas with n-type doping level of 10^{19} cm^{-3} and lateral dimensions <25 μm in pump-probe spectroscopy (Liu et al., 2009). There are two major advantages in using selectively grown Ge mesas to investigate optical gain compared with blanket Ge films on Si: (1) the threading dislocation density is lower than blanket Ge films since these dislocations can glide to the sidewalls of the mesas and annihilate (Luan et al., 1999). Consequently, nonradiative recombination is effectively reduced. (2) The SiO_2 mask layer for the selective growth naturally provides carrier confinement in the lateral directions to increase injected carrier concentration in the Ge mesa at a given pump power. A CW 1480-nm-pump laser was used in this study; therefore, all the optical bleaching and gain results are directly related to steady state pumping for device applications. The pumping photon energy is only 70 meV above the direct band-gap so that the excitation condition is similar to electrical injection using HJ structures. The transmittance spectra of the probe laser through the thickness of the Ge mesa with and without optical pumping are shown in Figure 8.58a. A significant increase in transmittance upon optical pumping was observed at wavelengths above the direct gap transition (0.765 eV). To derive the absorption spectra of the sample with and without optical pumping from the transmittance data, the transfer matrix method and Kramers-Kronig relation were applied to solve both the real refractive indices (n_r) and the absorption coefficients coefficient (α) by iterative sel-consistency regression. The derived absorptions spectra with and without optical pumping are shown in Figure 8.58b. The absorption coefficients at photon energies >0.765 eV (λ < 1620 nm) decreases significantly upon optical pumping, corresponding to the transmittance increase in Figure 8.58a. Especially, negative absorption coefficients corresponding to the onset of optical gain are observed in the wavelength range of 1600–1608 nm. The shape of the gain spectrum near the direct band edge of Ge resembles those of III–V semiconductor materials. A gain coefficient of ~50 cm^{-1} was observed at 1605 nm. As we will show later, this optical gain supports optically pumped Ge-on-Si lasers.

Very recently, optical gain has also been reported in tensile-strained n⁺ Ge photonic wires grown on GaAs (de Kersauson et al., 2011). The motivation of using GaAs substrate is that it has almost no lattice mismatch with Ge, thereby significant improving the material

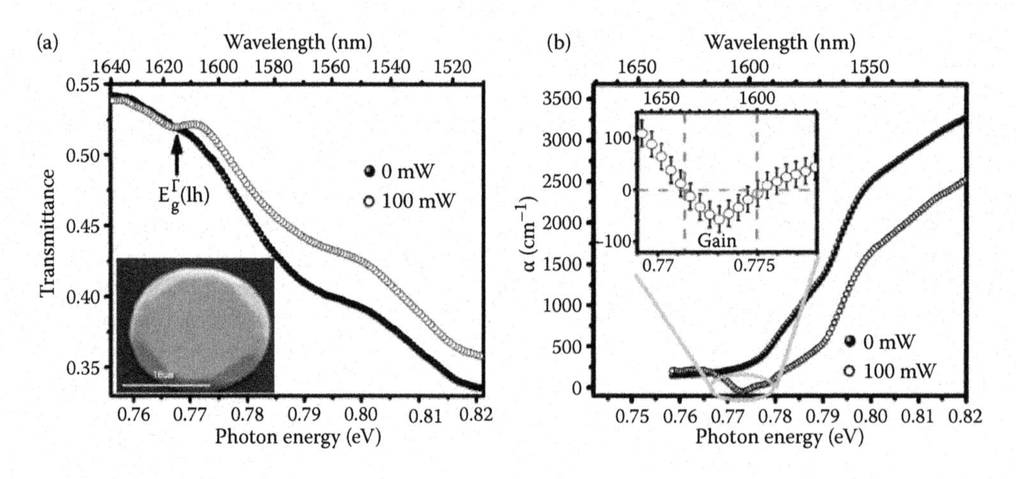

FIGURE 8.58
(a) Transmittance spectra of a 500 μm² Ge-on-Si mesa sample with n = 1.0 × 10^{19} cm^{-3} under 0- and 100-mW CW optical pumping at λ = 1480 nm. The inset shows an SEM picture of the mesa. (b) Absorption spectra of the n⁺ Ge mesa sample under 0 and 100 mW optical pumping. Negative absorption coefficients corresponding to optical gain are observed in the wavelength range of 1600–1608 nm, as shown in the inset.

quality of epitaxial Ge layers compared with Si substrate. The n-type doping level in Ge was 3×10^{19} cm^{-3}. A Si_3N_4 stressor with an initial compressive stress of 1.3 GPa was deposited and patterned to introduce tensile strain into the Ge layer. A maximum tensile strain of 0.6 % was transferred to Ge at the SiN_x/Ge interface, while the average tensile strain in the Ge region is ~0.4%. An optical gain of 80 cm^{-1} at ~1685 nm was measured by variable strip length (VSL) method. Emission line narrowing with increased optical pumping power was also observed.

Lange et al. (2009) performed ultrafast pump-probe spectroscopy on intrinsic Ge/SiGe QW structures at low temperatures and reported transient gain in the order of several hundreds per centimeter. However, due to intervalley scattering from Γ to L valleys the gain lifetime is <100 fs. Note that in this case the Ge QWs are compressively strained, which increases the energy difference between Γ and L valleys and enhances the undesirable electron scattering from direct Γ to indirect L valleys. It is possible that the scattering rate can be improved with n-type doping and tensile-strain stressors to provide a longer gain lifetime.

8.6.4.3 Optically Pumped Ge-on-Si Lasers

Based on optical gain in band-engineered Ge-on-Si, an optically pumped Ge laser has been demonstrated at room temperature (Liu et al., 2010a). The device consists of multimode Ge waveguides with mirror polished facets selectively grown on a lightly doped p-type Si wafer. The Ge material incorporates 0.24% thermally induced tensile strain and 1×10^{19} cm^{-3} in situ phosphorous doping. A cross-sectional scanning electron microscopy (SEM) picture of the Ge waveguide is shown in the inset of Figure 8.59a. The length of the waveguides was 4.8 mm to guarantee a mirror loss of $\ll$10 cm^{-1}, which is much smaller than the optical gain of Ge (Liu et al., 2009). The entire waveguide was excited by a 1064-nm Q-switched laser with a pulse duration of 1.5 ns and a maximum output of 50 μJ/pulse. Figure 8.59a shows the light emission spectra of a Ge waveguide under different pumping levels. With the increase of pumping pulse energy, the spectrum evolved from a broad spontaneous emission band to sharp emission lines featuring SE. Correspondingly, the polarization evolved from a mixed TE/TM to predominantly TE with a contrast ratio of 10:1 due to the increase of optical gain, as expected for typical lasing behavior. The multiple emission peaks at high pump power are due to multiple guided modes in the high index contrast Ge waveguide. A clear threshold behavior is demonstrated in Figure 8.59b. Figure 8.59c further shows a high resolution scan of the emission line at 1593.6 nm using a spectral resolution of 0.1 nm. Periodic peaks corresponding to longitudinal Fabry–Perot modes are clearly observed in the spectrum. The spacing of 0.060±0.003 nm is in good agreement with the calculated Fabry–Perot mode spacing of 0.063 nm for a 4.8-mm-long Ge waveguide cavity.

8.6.4.4 Toward Electrically Pumped Ge-on-Si Lasers

Developing electrically pumped Ge-on-Si lasers is the ultimate goals for practical applications. Ge-on-Si LEDs based on direct gap transition have been demonstrated at room temperature (Sun et al., 2009a; Cheng et al., 2009; Kasper et al., 2010; Arguirov et al., 2011). However, electrical-pumping Ge-on-Si lasers pose two major challenges compared with optically pumped lasers: (1) higher material gain is required to overcome optical losses in heavily doped electrodes. Optically pumped lasers do not require any electrodes, so Ge waveguides can be grown on lightly doped Si substrate with negligible FCA loss in Si. Electrically pumped lasers, on the other hand, inevitably require heavily doped electrodes

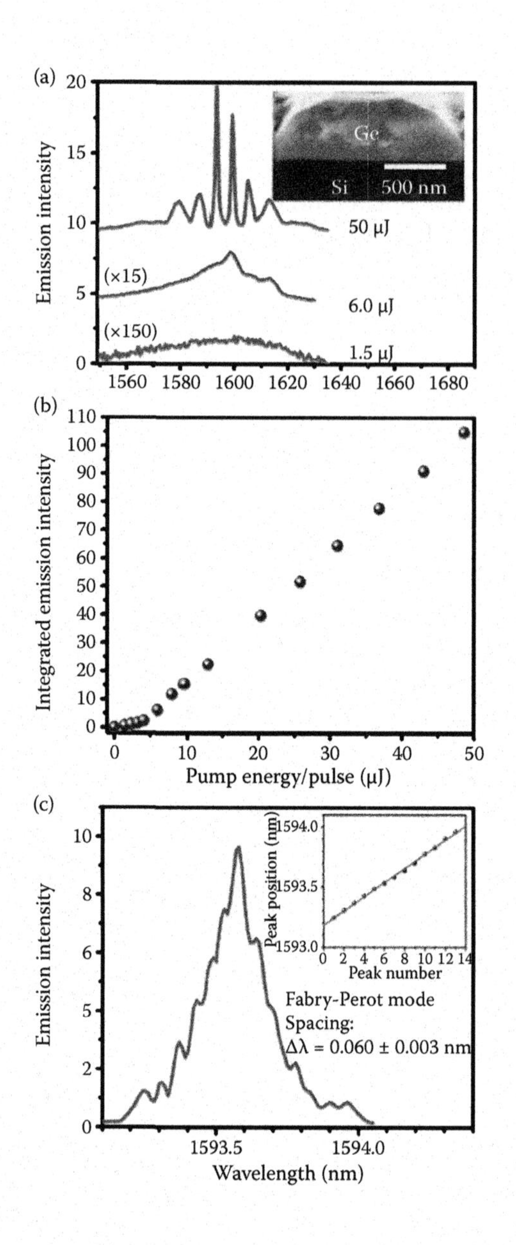

FIGURE 8.59
(a) Edge-emission spectra of a Ge-on-Si waveguide with mirror polished facets at different optical pumping levels. With the increase of pumping pulse energy, the spectrum evolves from broad spontaneous emission to sharp lines of SE. The inset shows a cross-sectional SEM picture of the Ge waveguide. (b) Integral emission intensity from the waveguide facet versus optical pump power showing the lasing threshold. (c) High-resolution scan of the emission line at 1593.6 nm. Longitudinal Fabry–Perot modes are clearly observed, and the period is consistent with the Ge waveguide cavity length of 4.8 mm.

for current injection. For Si/Ge/Si DHJ structures, the FCA loss in heavily doped Si is ~100 cm^{-1}, exceeding the optical gain of band-engineered Ge we have demonstrated so far. Further considering the mirror losses, practically a material gain coefficient of ~200 cm^{-1} in the Ge active region is needed to achieve electrically pumped lasers. Therefore, n-type doping and/or tensile strain has to be increased to achieve this goal. (2) High-efficiency injection of holes into n^+ Ge needs to be achieved. Optical pumping naturally generate pairs of electrons and holes simultaneously when an excitation photon is absorbed by Ge, so optical injections of electrons and holes are equally efficient regardless of the doping level. However, electrical injection of holes into a heavily doped n$^+$ Ge region is more challenging since the concentration of these minority carriers tends to decay rapidly with the distance from the junction due to recombination with electrons (majority carriers) in n$^+$ Ge. Adequate HJ structures have to be developed to overcome this issue.

Sun et al. (2010) theoretically investigated electrical injection in 0.2% tensile-strained intrinsic and n$^+$ Ge with 1×10^{19} cm^{-3} dopant concentration using Si/Ge/Si DHJ structures. The modeling shows that the EL efficiency of n$^+$Si/n$^+$ Ge/p$^+$ Si DHJ LEDs is ~10%, two orders of magnitude higher than tensile-strained intrinsic Ge and comparable to their III–V counterparts. The estimate of efficiency enhancement by n-type doping is consistent with previous PL studies (Sun et al., 2009b). The doping level in p$^+$ Si has a more significant effect on light emission efficiency since hole injection is the limiting factor in n$^+$ Ge. However, the optical gain at this n-type doping level is not large enough to overcome FCA in Si electrodes for electrically pumped lasers.

Recently, another electrical injection modeling was performed on n$^+$ Ge with a doping level of 5×10^{19} cm^{-3}, which can potentially achieve a net material gain of >200 cm^{-1} (Liu et al., 2011). Figure 8.60 shows a simulation result of band diagrams, quasi-Fermi levels, and carrier concentration distribution in an n$^+$ Si/n$^+$ Ge/p$^+$ Si DHJ structure upon electrical injection. The doping concentrations of n$^+$ Si, n$^+$ Ge, and p$^+$ Si are 1×10^{20}, 5×10^{19}, and 5×10^{19} cm^{-3}, respectively. The simulation takes into account Auger recombination but not the lasing dynamics. It is therefore suitable for estimating whether lasing threshold can be reached but will not be accurate in predicting the behavior after lasing starts. Interestingly, although Si/Ge interface shows typical type II band alignment, the n$^+$ doping and applied forward bias creates a band bending that overwhelms the band alignment effect, resulting in a pseudo type I alignment shown in Figure 8.60a. With 1.2 V (net) forward bias applied to the junction the separation between quasi-Fermi levels of electrons and holes is larger than the direct band-gap, which indicate that population inversion and optical gain can be achieved. Figure 8.60b further reveals that ~6×10^{18} cm^{-3} holes can be injected into the n$^+$Ge material at 1.2 V forward bias. These simulation results show that it is promising to achieve efficient hole injection for electrically pumped optical gain using an n$^+$ Si/n $^+$Ge/ p$^+$ Si DHJ structures.

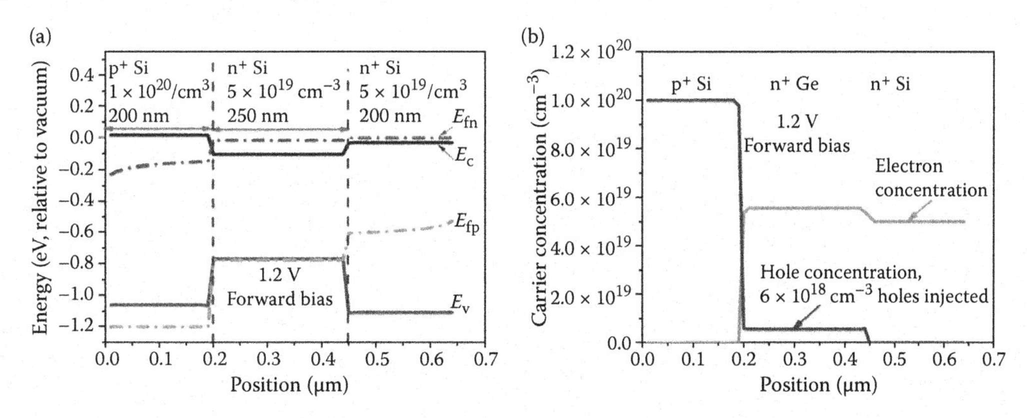

FIGURE 8.60
Electrical injection simulation of an n$^+$ Si/n$^+$Ge/p$^+$ Si DHJ structure. The doping levels and layer thicknesses are indicated in the figures. (a) Band diagram and quasi-Fermi levels across the junction. Here, E_{fn} and E_{fp} stand for quasi-Fermi levels of electrons and holes, respectively. (b) carrier density distribution.

As an important step toward electrically pumped Ge-on-Si lasers, edge-emitting n^+ Si/ n^+ Ge/p^+ Si DHJ waveguide LEDs are being investigated (Liu et al., 2010b, 2011; Camacho-Aguilera et al., 2011). Liu et al. reported room-temperature EL spectrum from an edge-emitting n^+ Ge LED grown on p^+ Si substrate. The Ge waveguide incorporated 0.25% tensile strain and 1×10^{19} cm^{-3} n-type doping, similar to the one used in optically pumped lasers. The room temperature edge-emission EL spectrum is shown in Figure 8.61a. Compared with previous PL and EL results, the spectrum is somewhat different since it shows a peak at 1530 nm and a shoulder around 1600 nm. A plausible reason is that the p/n junction is right at the Ge/Si interface near the low temperature Ge buffer layer, which tends to intermix with Si as well as accumulate less thermally induced tensile strain. Both mechanisms can result in a slightly larger band-gap that blue shifts the emission spectrum. The shoulder at 1600 nm is contributed by holes that managed to diffuse into the upper Ge layer and recombine with electrons. The upper Ge layer is grown at a higher temperature with a larger tensile strain, and it is not affected by Si/Ge intermixing at the interface. Therefore, the emission wavelength is longer. Another difference from surface emission is that this edge emission spectrum shows a "cutoff" at shorter wavelengths <1450 nm due to significant reabsorption along the waveguide.

Recently, Camacho-Aguilera et al. (2011) further improved the in situ phosphorus doping level in n^+ Ge waveguides to 3.5×10^{19} cm^{-3} and fabricated edge-emitting Ge LEDs on n^+ instead of p^+ Si substrate to avoid placing the junction near the defective buffer layer. The peak position of the room-temperature edge-emission EL is in good agreement with PL (Figure 8.61b), confirming that the blue-shifted peak in Figure 8.61a is indeed due to the low-temperature buffer layer. At an injection current of 130 mA (2.6 kA/cm^2 current density), there is still a spectrum "cutoff" at shorter wavelengths <1500 nm. When the injection current increased to 200 mA (4-k A/cm^2 current density), on the other hand, the emission at shorter wavelengths <1450 nm is significantly enhanced, indicating less reabsorption due to optical bleaching under electrical pumping. Further increasing n-type doping to $>4 \times 10^{19}$ cm^{-3} led to the demonstration of the first electrically-pumped Ge-on-Si

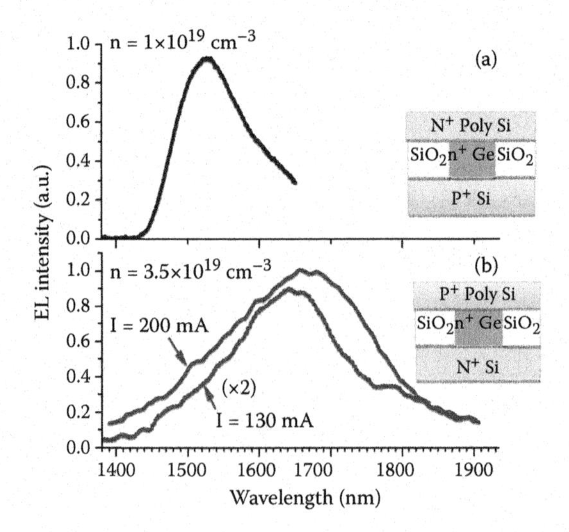

FIGURE 8.61
Edge emission spectra of Ge waveguide LEDs fabricated on (a) p^+ Si substrate and (b) n^+ Si substrate (Liu et al., 2010a,b). The latter demonstrates optical bleaching under electrical injection, an important step toward electrically pumped lasers.

laser (Camacho-Aguilera, 2012). The demonstration of lasing in a broad wavelength range of 1520–1700 nm confirmed the broad gain spectrum modeled in Figure 8.53b.

8.6.5 Conclusions

We have reviewed the history of research on direct gap light emission from Ge, and extensively discussed recent developments of band-engineered Ge-on-Si gain media and lasers since 2006. Band engineering by tensile strain and n-type doping proved an effective approach to transform Ge into an optical gain medium for monolithic lasers on Si. The experimental demonstration of room-temperature optical gain and lasing under optical and electrical pumping point to a promising future for the applications of monolithic Ge-on-Si lasers in large-scale, high-volume electronic-photonic integrated circuits utilizing WDM.

8.7 Purcell Effect in Silicon Nanocrystals

Al Meldrum

8.7.1 Free-Space Luminescence from Silicon Nanocrystals

Ensembles of silicon nanocrystals typically feature a broad, featureless luminescence spectrum peaking at wavelengths ranging from ~700 to 950 nm (see Figure 8.8) (Meldrum et al., 2006). This property is nonideal from the perspective of any light-emitting device in which spectral purity or fine control is desirable. There are several issues that limit the ability to achieve such control over the luminescence of silicon NCs. First, a narrower distribution of particle sizes and shapes, as demonstrated for many years in direct-gap nanocrystals (Murray et al., 1993; Peng et al., 1998, 2000), has not been achieved for the case of silicon. Next, the short-wavelength emission (i.e., blue and green) tends to be unstable over time and eventually shifts toward longer wavelengths, as shown most clearly in the case of porous silicon (Wolkin et al., 1999). Third, the linewidths of individual Si NCs are apparently quite broad, so even with a narrow size distribution one may not achieve "monochromatic" emission similar to that achieved with direct gap NCs (Valenta et al., 2002; Sychugov et al., 2005). Thus, the wide size dispersions in most samples, the presence of oxide-related sub-gap states on the NC surfaces, and the inherently broad homogeneous linewidth of individual NCs all combine to render the luminescence spectrum broad, featureless, and difficult to tune.

8.7.2 Introduction to Optical Microcavities

Optical microcavities offer a potential solution that could permit control over the emission spectrum and dynamics of silicon nanocrystals. Essentially, microcavities have electromagnetic resonances that can enhance the luminescence of Si NCs at certain frequencies. Thus, microcavities offer a means to externally control the emission spectrum of an ensemble of NCs that otherwise suffers both from homogeneous and inhomogeneous broadening. Here, we will briefly review the main cavity–NC coupling regimes and discuss how the Purcell effect can enhance or suppress the luminescence in ensembles of Si NCs. Next, we will briefly outline the key types and properties of microcavities to which

Si NCs have been coupled, and, finally, we will give an overview how these effects could be utilized in devices.

Cavity effects have been studied at least since the time of Lord Rayleigh, who, in 1910, published a study on the propagation of sound waves that reflected along the inner circumference of St. Paul's Cathedral in London (Rayleigh, 1910). The resonances at certain sound frequencies were henceforth called "whispering gallery modes" (WGMs); at these resonant frequencies the sound waves carried around the 100-m circumference of one of the upper galleries are amplified. Almost contemporaneously with Lord Rayaleigh's investigations of acoustic resonances in St. Paul's Cathedral, Gustav Mie developed solutions of Maxwell's equations as applied to light scattering from spherical particles (Mie, 1908). Mie's equations would lead to mathematical expressions for microsphere resonant wavelengths whose numerical solutions are still being refined today (Okada and Cole, 2010).

Essentially, spherical droplets or glass microspheres confine electromagnetic radiation within the cavity volume by total internal reflection. Constructive interference occurs at specific wavelengths as light propagates along the microsphere circumference, leading to the development of WGM optical resonances. Currently, there are many other types of optical microcavities in addition to microspheres; the main requirement is simply that electromagnetic radiation be spatially confined by reflection or interference, leading to the development of resonant electromagnetic modes.

All microcavities are characterized by two critical parameters: the mode quality factor (or finesse, which encompasses both the quality factor and the free spectral range) and the mode volume. Briefly, the quality factor is a measure of the degree to which the resonant mode of the microcavity can trap radiation. The mode quality factor is defined as $Q = 2\pi \frac{\text{energy stored initially}}{\text{energy lost per cycle}}$. Treating the cavity as an underdamped oscillator, if the energy stored initially is denoted E_0 and is normalized, then the energy lost per oscillation period, T, can be written as $E_T = E_0 e^{-2\gamma T}$ (Kao and Santosa, 2008), where γ represents the damping (i.e., $\gamma = \zeta\omega_0$, with ζ as the damping ratio) in the cavity. Thus, we have

$$Q = 2\pi \frac{1}{1 - \exp\left[-2\gamma \frac{2\pi}{\omega_l}\right]} \tag{8.33}$$

in which ω_l is the damped natural frequency whose real part is given by $\omega_l = \sqrt{\omega_0^2 - \gamma^2}$. Equation 8.33 shows that the quality factor is approximately equal to $\omega_l/2\gamma$ if γ is small (underdamping). Since the complex optical resonance frequency of the cavity can be written as $\omega_{cav} = \pm\omega_l + i\gamma$, an identical definition of the quality factor is $Q = \text{Re}(\omega_{cav})/2\text{Im}(\omega_{cav})$. Alternatively, the material-limited Q-factor of the cavity is $n/2\kappa$, where n and κ are the refractive index and extinction coefficient of the cavity medium, respectively. In other words, a cavity composed of a given medium cannot have a quality factor higher than $n/2\kappa$.

Finally, through a rather lengthy algebraic procedure derived for RLC circuit resonances specifically under conditions of low damping, it can be shown that $\omega/\Delta\omega$ $(=\lambda/\Delta\lambda = f/\Delta f) = \omega_l/2\gamma = Q$, where $\Delta\omega$ is the resonance linewidth at its half-maximum. This result relates the Q factor to the resonance frequencies and linewidth. All of the previous definitions are interchangeable as long as Q is not too small. Since the cavity losses in Equation

8.33 are exponential in time, the resonance linewidths are defined by Lorentzians in the frequency domain. Microcavity Q-factors typically range in magnitude from values of a few tens in metal-mirrored Fabry–Perot cavities to several thousand in microdisks, photonic crystals, and micropillars, and up to 10^9 or even higher in silica microspheres.

The second key microcavity parameter is the mode volume. Mathematically, the cavity mode volume is expressed as the normalized volume integral of the modal electric field, which for the case of a microsphere is

$$V_m = \int_V \varepsilon(r)\left|\phi(r)\right|^2 d^3r \tag{8.34}$$

where $\varepsilon(r)$ is the relative permittivity as a function of radius r, which will be different inside and outside the cavity, and $\phi(r)$ is the normalized electric field amplitude equal to $E(r)/E_{max}$. The normalization ensures that the mode volume is independent of the field amplitude in the cavity. Since the field propagates dynamically outside the cavity, the integral should be taken over all space; unfortunately, however, in open resonators the integral diverges as $r \to \infty$ so in practice some arbitrary integration limit is usually taken (Agha et al., 2006). With this caveat in mind, the approximate open-resonator mode volumes of small cavities such as photonic crystals and micropillars are on the order of a few cubic wavelengths, whereas for larger structures like microspheres, the mode volume can be as great as $1000(\lambda/n)^3$ or more (Gérard and Gayral, 1999). In the first two cases, the mode volume is similar to the physical dimensions of the structure; for microspheres and disks, simple numerical formulas and graphs exist for quick calculation of an approximate mode volume (Weinstein, 1969; Srinivasan et al., 2006).

8.7.3 Strong and Weak Coupling and the Purcell Effect

Next, we examine the effect of a microcavity on the light emission properties of a silicon nanocrystal. Essentially, an NC coupled to an optical microcavity can be viewed as a lightly damped coupled oscillator system. In this picture, under weak excitation conditions (i.e., no more than one photon in the cavity field and a two-level emitter), and for the cavity and NC on resonance, the luminescence spectrum can be obtained by solving the appropriate Hamiltonian for this system, yielding the luminescence spectrum $S(\omega)$ given by (Carmichael et al., 1989):

$$S(\omega) = \left|\frac{\Omega_+ - \omega_0 + i\Delta\omega/2}{\omega - \Omega_+} - \frac{\Omega_- - \omega_0 + i\Delta\omega/2}{\omega - \Omega_-}\right|^2 \tag{8.35}$$

In Equation 8.35, the frequency eigenvalues $\Omega_\pm$ are (Andreani et al., 1999):

$$\Omega_\pm = \omega_0 - i(\Delta\omega_{cav} + \Delta\omega_{NC})/4 \pm \sqrt{g^2 - (\Delta\omega_{NC} - \Delta\omega_{cav})^2/16} \tag{8.36}$$

Here, $\Delta\omega_{NC}$ and $\Delta\omega_{cav}$ are the NC and cavity linewidths, respectively, ω_0 is the NC central transition frequency (common to both the cavity and the NC), and g is the cavity–NC coupling constant. The coupling constant is the product of the NC transition dipole matrix element and the cavity field at the NC location:

$$g = \sqrt{[1/(4\pi\varepsilon_{\omega})] \cdot [\pi e^2 f/(m_0 V_m)]} \tag{8.37}$$

In Equation 8.37, f is the dipole oscillator strength, ε_{ω} is the permittivity of the medium, and m_0 is the free electron mass. Of course, a cavity with an ensemble of Si NCs will have many photons even under modest excitation. These higher-order effects would represent fairly minor corrections (Carmichael et al., 1989), so we can reasonably analyze Equation 8.35 to study the important single NC–cavity physics.

Equations 8.35 through 8.37 have several important characteristics. First, if $g > (\Delta\omega_{NC} - \Delta\omega_{cav})/4$, the eigenvalues $\Omega_{\pm}$ admit two real solutions, leading to the luminescence doublet characteristic of the "strong coupling" regime. The two peaks of the strong-coupling doublet are separated by energy $\Delta E = 2\hbar\sqrt{g^2 - \left(\frac{\Delta\omega_{NC} - \Delta\omega_{cav}}{4}\right)^2}$ under the condition of NC-cavity resonance. Thus, for strong coupling one would like the coupling constant g to be large. This implies a small cavity mode volume and a large oscillator strength, observing from Equation 8.37 that $g \propto \sqrt{f/V_m}$. The latter condition is met by Si NCs, which have a large oscillator strength (Dovrat et al., 2004). Using the definition $Q = ½\mathrm{Re}(\Omega)/\mathrm{Im}(\Omega)$, the linewidths of each of the coupled eigenstates in Equation 8.36 is given by $(\Delta\omega_{cav} + \Delta\omega_{NC})/2$. This implies that the observation of strong coupling as a resolvable doublet in the luminescence spectrum also requires narrow cavity and NC linewidths. Therefore, the key *cavity* feature for observing strong coupling is a high Q and a high Q/V ratio, while the necessary *dot* parameters include a high oscillator strength and a narrow emission linewidth. The requirement for a narrow linewidth suggests that strong coupling is not likely to be observable with silicon nanocrystals, whose emission linewidths are apparently much too wide, even at cryogenic temperatures (Sychugov et al., 2005), to permit the characteristic spectral features to be resolved.

In the weak coupling regime, the coupling coefficient is smaller than the combined cavity and NC decay rates, i.e., $(\Delta\omega_{cav} + \Delta\omega_{NC})/4 > g$. In this case, the real parts of the frequency eigenvalues of $\Omega_{\pm}$ are identical, so there is no doublet in the emission spectrum. If $(\Delta\omega_{cav} + \Delta\omega_{NC})/4 \gg g$ and $\Delta\omega_{cav} \gg \Delta\omega_{NC}$ (i.e., the coupling constant is relatively small and the NC emission is much narrower than the cavity linewidth) the two complex solutions to Equation 8.36 are (Andreani et al., 1999):

$$\Omega_+ = \omega_0 - i\left[\frac{\Delta\omega_{cav}}{2}\right], \text{and } \Omega_- = \omega_0 - i\left[\frac{\Delta\omega_{NC}}{2} + \frac{2g^2}{\Delta\omega_{cav}}\right] \tag{8.38}$$

These two solutions describe the cold cavity mode and the lineshape of the NC emission into the cavity, respectively (recalling that the observable luminescence spectrum is given in Equation 8.35 and depends on both Ω_+ and Ω_-). Equation 8.38 shows that the emission rate (or linewidth) into the cavity is enhanced by a factor of $2(2g^2/\Delta\omega_{cav})$ compared with that in a free medium. This solution essentially describes the "traditional" weak coupling situation in which we have a cavity-nanocrystal system in which the loss rates are greater than the coupling rate, and the NC linewidth is narrow compared with that of the cavity. The validity of this assumption for Si NCs will be discussed further below.

In weak coupling, one can treat the effect of the cavity purely as a modification of the local optical density of states experienced by the emitter. Then, as established more than 60 years ago (Purcell, 1946), using Fermi's golden rule, we can write the spontaneous emission rate into the cavity as

$$w_{cav} = \frac{2\pi}{\hbar^2} \int_0^\infty \left| \langle f | H | i \rangle \right|^2 \rho(\omega) \Lambda(\omega) d\omega \tag{8.39}$$

Here, the variables take their usual forms: $\rho(\omega)$ and $\Lambda(\omega)$ are the Lorentzian lineshapes for the final density of states of the cavity and the emitter, respectively, and $\langle f | H | i \rangle$ is the dipole matrix element for the emitter transition. As discussed above, the emitter linewidth is traditionally assumed to be a δ-function peaked at ω_0, in which case the Dirac delta in Equation 8.39 leads to

$$w_{cav} = \frac{2\pi}{\hbar^2} \left| \langle f | H | i \rangle \right|^2 \rho(\omega_0) \tag{8.40}$$

The dipole matrix element for the NC is

$$\left| \langle f | H | i \rangle \right| = \frac{\xi^2 \omega_0 \hbar \mu^2}{2 \varepsilon_\omega V_m} \tag{8.41}$$

where ξ^2 incorporates the polarization averaging, μ^2 is the transition dipole moment, ω_0 is the central transition frequency (recalling that the transition is, in this case, a delta function), ε_ω is the permittivity of the local medium (it could be written here as ε_{ω_0}), and V_m is the mode volume. For a Lorentzian cavity profile and a delta-function emitter profile, Equations 8.40 and 8.41 lead to

$$w_{cav} = \frac{2}{\hbar} \cdot \frac{\xi^2 \omega_0 \mu^2}{\varepsilon_\omega V} \cdot \frac{\Delta\omega_{cav}}{4(\omega_0 - \omega_{cav})^2 + \Delta\omega_{cav}^2} \tag{8.42}$$

In contrast, the spontaneous emission rate in the free-medium is given by

$$w_0 = \frac{\mu^2 \omega_0^3}{3\pi \varepsilon_\omega \hbar c^3} \tag{8.43}$$

Dividing Equation 8.42 by Equation 8.43 and setting the NC and cavity transition frequencies, ω_0 and ω_{cav}, equal leads to the standard Purcell rate enhancement:

$$F_p = \frac{w_{cav}}{w_0} = \xi^2 \frac{3Q(\lambda/n)^3}{4\pi^2 V} \tag{8.44}$$

in which we have used the relationship $Q = \omega/\Delta\omega$. Comparing Equations 8.42 through 8.44 with Equation 8.37, the enhanced spontaneous emission rate can also be written in terms of the coupling parameter: $w_{cav} = 4g^2/\Delta\omega_{cav}$ in agreement with Equation 8.38. This is the "textbook" derivation of the spontaneous emission rate enhancement, as originally shown by Purcell in 1946.

Unfortunately, Si NCs cannot be treated as δ-functions since the linewidth may be tens of meV (Valenta et al., 2002; Sychugov et al., 2005). Furthermore, a given NC in an ensemble distribution may or may not be on resonance with a cavity mode. Thus, it becomes necessary return to Equation 8.39 and do the integration without assuming a δ-function or resonance matching:

$$w_{cav}(\omega_{NC},\omega_{cav}) = \frac{4\xi^2\mu^2}{\pi^2\varepsilon_\omega V\hbar}\cdot \int_0^\infty \frac{\omega_{cav}}{\Delta\omega_{cav}^2 + 4(\omega-\omega_{cav})^2}\cdot\frac{\Delta\omega_{NC}}{\Delta\omega_{NC}^2+4(\omega-\omega_{NC})^2}\omega\cdot d\omega \tag{8.45}$$

In recent work (Meldrum et al., 2010), it was demonstrated that this integral yields

$$w_{cav}(\omega_{NC},\omega_{cav}) = \frac{2\mu_0^2}{\hbar\varepsilon_\omega V}\cdot\frac{\omega_{cav}\Delta\omega_{NC}+\omega_{NC}\Delta\omega_{cav}}{4(\omega_{NC}-\omega_{cav})^2+(\Delta\omega_v+\Delta\omega_{NC})^2}+C \tag{8.46}$$

where the C term is given by

$$C=\frac{1}{2\pi^2}\left[\begin{array}{l} +\dfrac{\omega_0(\omega_0-\omega_c)+\delta_0(\delta_0-\delta_c)}{(\omega_0-\omega_c)^2+(\delta_0-\delta_c)^2}\log\sqrt{\dfrac{\omega_0^2+\delta_0^2}{\omega_c^2+\delta_c^2}} \\ -\dfrac{\omega_0(\omega_0-\omega_c)+\delta_0(\delta_0+\delta_c)}{(\omega_0-\omega_c)^2+(\delta_0+\delta_c)^2}\log\sqrt{\dfrac{\omega_0^2+\delta_0^2}{\omega_c^2+\delta_c^2}} \\ +\dfrac{\omega_c\delta_0-\omega_0\delta_c}{(\omega_c-\omega_0)^2+(\delta_0-\delta_c)^2}\left(\tan^{-1}\dfrac{\delta_0}{\omega_0}+\tan^{-1}\dfrac{\delta_c}{\omega_c}\right) \\ -\dfrac{\omega_c\delta_0+\omega_0\delta_c}{(\omega_c-\omega_0)^2+(\delta_0+\delta_c)^2}\left(\tan^{-1}\dfrac{\delta_0}{\omega_0}-\tan^{-1}\dfrac{\delta_c}{\omega_c}\right) \end{array}\right] \tag{8.47}$$

In Equation 8.47, δ_0 and δ_c are the half linewidths of the emitter and cavity, respectively. This represents a small correction since at optical frequencies, $\omega_x \gg \delta_x$ leading to $C \approx 0$, so we will neglect this term in the discussion below. Equation 8.46 can then be rewritten as

$$w_{cav}(\omega_{NC},\omega_{cav}) = A\cdot B\cdot\frac{\Delta\omega_{cav}+\Delta\omega_{NC}}{4(\omega_{NC}-\omega_{cav})^2+(\Delta\omega_{cav}+\Delta\omega_{NC})^2} \tag{8.48}$$

where $A = 2\xi^2\mu^2/(\hbar\varepsilon_\omega V)$ and $B = (\omega_{NC}\Delta\omega_{cav} + \omega_{cav}\Delta\omega_{NC})/(\Delta\omega_{cav} + \Delta\omega_{NC})$. This shows an interesting effect for an ensemble of NCs coupled to a cavity. As we "scan" the NC emission frequency through the ensemble size distribution, the rate function w_{SE} has the form of a modified Lorentzian centered at ω_{cav}, with a linewidth of $\Delta\omega_{cav} + \Delta\omega_{NC}$. Thus, for weak excitation the ensemble emission spectrum will demonstrate cavity modes with a Q-factor of $Q_{LUM} \approx \omega_{cav}/(\Delta\omega_{cav} + \Delta\omega_{NC})$, compared with the cold cavity $Q_{cav} = \omega_{cav}/\Delta\omega_{cav}$.

The lifetime given by Equation 8.48 will vary as the ensemble size distribution (emission rate) scans over the mode resonance, with the lifetimes for any NC needing to be summed over all the modes. This situation, although it seems complicated, can be solved with a simple computational algorithm (Meldrum et al., 2010). Overall, the cavity effect is to broaden the ensemble lifetime distribution, leading to smaller values of the exponent β in the stretched exponential decay typical of Si NCs (the stretched exponential was discussed in Section 8.3.1).

Finally, the generalized Purcell enhancement factor for an arbitrary NC coupled to an arbitrary cavity mode is

$$F_p(general) = \frac{3\xi^2(\lambda_0/n)^3}{4\pi^2 V} \cdot \frac{\omega_{NC}\Delta\omega_{cav} + \omega_c\Delta\omega_{NC}}{4(\omega_{NC} - \omega_{cav})^2 + (\Delta\omega_{cav} + \Delta\omega_{NC})^2} \tag{8.49}$$

The "strength" of the Purcell effect in Equation 8.49 depends in the parameter ξ^2, which we have not yet discussed. This factor can be more practically written to incorporate a set of effects that, overall, tend to weaken the NC-cavity coupling. First, the NC dipole should ideally be polarized the same as the cavity field. If the NC orientations are random in an ensemble, the Purcell factor is reduced by a factor of 3 (i.e., $\xi^2 = 1/3$). Maximum coupling efficiency also requires that the NC be physically located at the antinode of the cavity field; if the dots are instead randomly distributed in the cavity or are not at the field antinode, the Purcell enhancement is reduced by an additional factor (Gérard and Gayral, 1999). Finally, there are two possible polarizations (TE or TM) into which the dot can emit; if these polarizations are degenerate in frequency the overall emission rate at that frequency will be increased by a factor of 2 (still, the net polarization effect would be to multiply w_{SE} by a factor of 2/3 for randomly oriented dots located at the field antinode) (Gérard et al., 1998).

8.7.4 Experimental Observations with Silicon Nanocrystals

The first experimental method for coupling Si NCs to optical microcavities utilized thin film growth and annealing to form a $\lambda/2$-thick layer of nanocrystals between a set of distributed Bragg reflectors (DBRs), as diagrammed in Figure 8.62a, or metal mirrors (Iacona et al., 2001; Spooner et al., 2003; Amans et al., 2003, 2004; Belarouci et al., 2007; Grun et al., 2011). The luminescence spectrum consists mainly of a single peak centered on the cavity resonance and located within the broad emission band of the NC ensemble. One useful feature of thin planar cavities is their large free spectral range (FSR), so that typically only one cavity resonance overlaps with the ensemble emission spectrum. In this way, it was possible to achieve good spectral purity and fine control over emission spectrum from amorphous Si NCs, with one single sharp emission peak located at any desired wavelength throughout the visible spectrum (Figure 8.63) (Hryciw et al., 2005). Er-doped Si NCs in planar cavities have also been reported, and feature narrowed and intensified emission near a wavelength of 1.54 μm (Iacona et al., 2001; Hryciw et al., 2006). A spontaneous emission rate enhancement at the planar cavity resonance has been reported, suggesting that the Purcell effect (i.e., weak coupling) may be experimentally measurable in the Si NC luminescence dynamics (Tochikiyo et al., 2003; Belarouci and Gourbilleau, 2006).

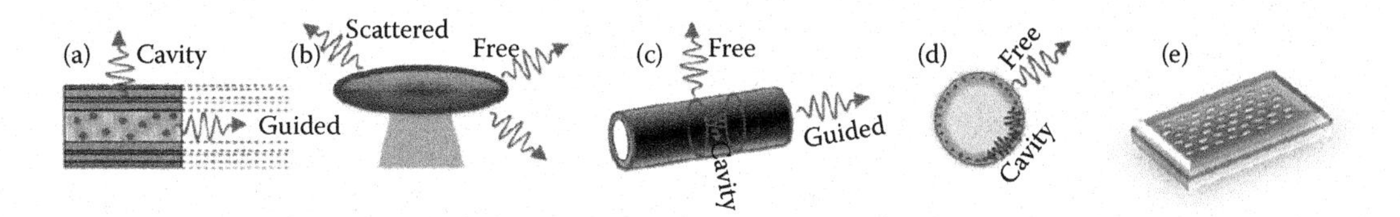

FIGURE 8.62
Illustration of several types of microcavities with silicon nanocrystals. (a) Si NCs randomly positioned in a Fabry–Perot cavity. (b) Silicon-rich oxide or nitride microdisk containing Si NCs. (c) Cylindrical cavity (like a fiber or capillary) coated with Si NCs. (d) Microsphere coated with Si NCs. (e) Photonic crystal cavity made from a set of holes e-beam patterned into a nanocrystal film.

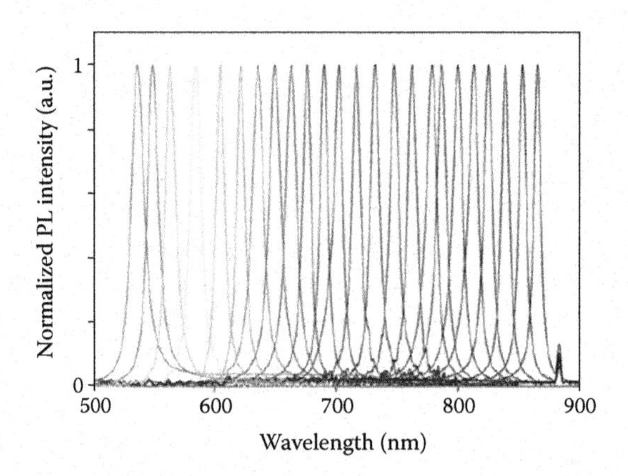

FIGURE 8.63
Tunable cavity emission using planar cavities with metal mirrors. The emitters were related to amorphous Si nanoclusters in silicon-rich oxide. (Hryciw, A. et al.: Tunable luminescence from a silicon-rich oxide microresonator. *Adv. Mater.* 2005. 17. 845–849. http://onlinelibrary.wiley.com/doi/10.1002/adma.200401230/abstract. Copyright Wiley-VCH Verlag GmbH & Co. KGaA. Reproduced with permission.)

In general, luminescence Q-factors in planar cavities have been on the order of a few hundred at best. One experimental difficulty for planar cavities is the requirement to anneal the precursor SiO_x films at temperatures near 1000°C to grow the silicon nanocrystals. This has the undesirable side effect of changing the deposited DBR layer thicknesses and reducing Q. Microcavities fabricated by etching porous silicon into a layered structure avoid annealing, although the DBR uniformity and cavity Q factors remain low (Pavesi et al., 1995; Chan et al., 2000). Metal mirrors, although much simpler to work with (Hryciw et al., 2005), are inherently lossier than DBRs and cannot survive the annealing treatments required to produce crystalline Si particles. Improvements can likely be made, for example, higher Q-factors might be achievable using nanocrystal deposition methods that do not require exposure of the structure to 1000-degree temperatures. It should additionally be possible to lithographically etch planar cavities to form micropillar structures, as has been frequently done for other cavity–NC systems (Gérard et al., 1998; Reitzenstein et al., 2006, 2010), thereby decreasing the mode volume and leading to more pronounced effects in the emission spectrum and dynamics.

More recently, the luminescence of Si NCs has been coupled to the whispering gallery modes of lithographically defined microdisks (Figure 8.62b) (Zhang et al., 2006). The Q-factors can in principle also be higher with disks than for planar cavities, with values of a few thousand reported in the PL spectrum (Ghulinyan et al., 2008; Pitanti et al., 2010b). In one study, the Purcell factor for a microdisk with silicon NCs was found to be as high as 171 (Kippenberg et al., 2009), although this value was arrived at using scattering and not luminescence. Further studies should confirm whether such large enhancements can indeed occur experimentally. Other work found luminescence rate enhancements slightly larger than unity, with Purcell factors as high as 25 being at least theoretically possible in ring resonators (Pitanti et al., 2010b).

The various Q-limiting mechanisms for microdisks and other microcavities containing Si NCs include radiation (diffraction) loss, scattering from the NCs, band-to-band absorption, and roughness-induced scattering (Kekatpure et al., 2008a). Evidently, sidewall roughness in microdisks, which have to be made using lithographic procedures, is the limiting loss mechanism in structures larger than about 4 μm in diameter (Kekatpure et al., 2008a). For smaller disks, radiation losses limit the Q-factors. Additional losses arise from confined carrier absorption (CCA) when the NCs are strongly excited. The increasing luminescence cavity linewidth that occurs as a function of pump power for Si NCs coupled to a microdisk is, in fact, a direct measure of the CCA cross section. By measuring the Q-factors as a function of pump power, the effective nanocrystal CCA cross section was found to be almost an order of magnitude higher than that for free-carrier absorption in bulk Si (Kekatpure et al., 2008b).

Cylindrical microcavities (Figure 8.62c) have not been as widely investigated due to the lack of a method for coating nonplanar structures with luminescent Si NCs. However, the recent development of a new Si NC coating process (Hessel et al., 2007) opened the door to new types of microcavity-coupled Si NC luminescent structures including optical fibers (Bianucci et al., 2009a, b) bottle resonators (Bianucci et al., 2010), and capillaries (Figure 8.64) (Rodriguez et al., 2008). This coating method produces smoother surfaces than lithography, so the main Q-factor-limiting mechanism associated with microdisks (i.e., surface roughness, as discussed above) (Kekatpure et al., 2008a) is reduced. Cavity Q-factors as high as ~10^6 were found in bottle resonators coated with a layer of Si NCs (Bianucci et al., 2010) although in PL the Q-factors for both bottle resonators and optical fibers were only a few thousand (e.g., Figure 8.65). The reasons for this apparent discrepancy can be either that the spectrometer sensitivity cannot adequately resolve the mode, as we initially thought (Bianucci et al., 2009c), or alternatively, it could be that the NC linewidth limits the luminescence Q-factor according to Equation 8.48. Given that the spectrometer resolution is *not* the limiting issue, then a comparison of the cavity and PL Q-factors places a limit on the linewidth of the Si NCs. From Equation 8.48, for PL Q-factors of a few thousand the effective NC linewidths cannot exceed a few millielectron volts. For a linewidth larger than about 20 meV, the modes observed in the spectrum would be almost entirely washed out (Meldrum et al., 2010).

Although the mode volume is large, nanocrystal-coated microspheres (Figure 8.62d) offer potentially the highest Q-factor of any type of microcavity. Thus, there have been

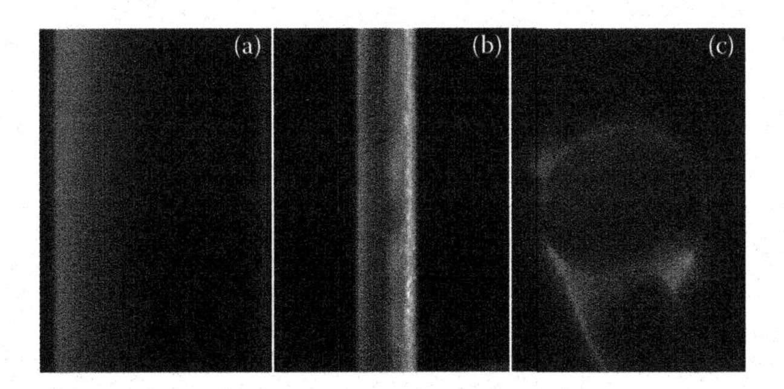

FIGURE 8.64
Fluorescence images showing the practical realization of several of the cavity types illustrated in Figure 8.62. Here, (a) shows an Si-NC-coated optical fiber (diameter = 125 μm). (Reprinted with permission from Bianucci et al. 2009a. http://jap.aip.org/resource/1/japiau/v105/i2/p023108_s1. Copyright 2009 by the American Physical Society. (b) Is a capillary whose channel (diameter nominally 50 μm) is coated with Si-NCs. (Reprinted from Physica E, 41, Bianucci, P. et al. Mode structure in the luminescence of silicon nanocrystals in cylindrical microcavities, 1107-1110. http://www.sciencedirect.com/science/article/pii/S1386947708003147. Copyright 2009, with permission from Elsevier). (c) Is a 51-μm-diameter microsphere. Some scattered blue light from the excitation laser is visible in the images. (Bianucci, P. et al.: Whispering gallery modes in silicon nanocrystal coated microcavities. *Phys. Status Solidi A*. 2009. 206. 973–975. http://onlinelibrary.wiley.com/doi/10.1002/pssa.200881274/abstract. Copyright Wiley-VCH Verlag GmbH & Co. KGaA. Reproduced with permission.)

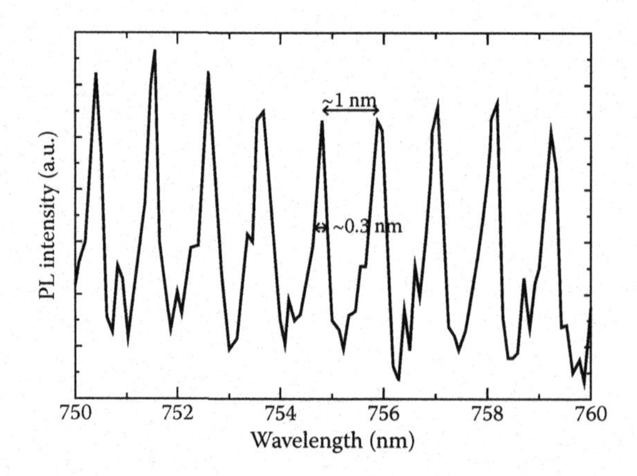

FIGURE 8.65
PL spectrum for a Si-NC-coated optical fiber (from Bianucci, P. et al., 2010. *Opt. Express* 18:8466–8481.). The Q-factors are approximately 2500 (linewidth ~0.3 nm), and the mode spacing is ~1 nm, consistent with a 125-μm-diameter fiber. (Reprinted from *Physica E*, 41, Bianucci, P. et al. Mode structure in the luminescence of silicon nanocrystals in cylindrical microcavities, 1107–1110. http://www.sciencedirect.com/science/article/pii/S1386947708003147. Copyright 2009, with permission from Elsevier.)

several efforts to coat Si NCs onto glass microspheres. One of the first attempts used thermal deposition of SiO_x thin films onto self-assembled lattices of microspheres ranging in size from 2 to 20 μm (Beltaos and Meldrum, 2007). Although these spheres could not be coupled evanescently to a tunable light source for characterization of the cold cavity, they did exhibit luminescence Q factors of ~1500. With this method, it was not possible to coat the spheres uniformly, and a significant surface roughness was observed on the films. The former issue was mostly solved by building a rotating specimen holder into the thin film deposition apparatus, and then coating a single sphere that was initially formed by melting a fiber tip (Chen et al., 2005). The latter work, intriguingly, reported features "consistent with SE" although it seems that the CCA process may prevent the unequivocal confirmation of this phenomenon (Kekatpure et al., 2008b). More recently, PL Q-factors approaching 4,000 were reported (Bianucci et al., 2011) on single microspheres using the coating method developed in reference (Hessel et al., 2007). This coated sphere showed cold cavity modes higher than the luminescence modes, attributed at least in part to the contribution of the NC linewidth to the PL Q-factor.

At the other end of the spectrum (i.e., very small mode volumes but more modest Q-factor), one can couple the Si NC luminescence into photonic crystal cavities. The basic structure is illustrated in Figure 8.62e. One study utilized thin film deposition and electron-beam lithography to pattern an array of holes in a silicon-rich SiN film. PL Q-factors, presumably from Si NCs embedded in the SiN layer, ranged from 200 to 300 in the ensemble PL (Makarova et al., 2006). Although the PL lifetimes were not measured, simulations of the structure suggested that the Purcell factor could be as high as 35 in an idealized sample. In a follow-up work, the same group studied the luminescence of Er-doped Si NCs in SiN that comprised a photonic crystal cavity made of holes in the SiN layer (Makarova et al., 2008). As discussed in Section 8.5.5, Si NCs transfer their excitation to nearby Er^{3+} ions via a sensitizer-activator process. Here, Q-factors higher than 6,000 were observed in the Er-related luminescence. Considering the small mode volume, these structures probably have the highest Q/V ratio of any Si NC-coupled microcavity to date. The estimated

Purcell factors were on the order of 2–3, although much higher values could be possible with optimization of sample structure and collection geometry (Makarova et al., 2008).

Equations 8.35 through 8.38, as well as the various experimental results attained so far, imply that strong coupling applications such as (potentially) quantum computation are not likely to be achieved with Si NCs; instead, weak coupling and the Purcell enhancement offer the best possibilities for future devices. Essentially, coupling the luminescence of Si NCs to optical microcavities suggests several applications for these materials that would not be possible without the narrow luminescence mode structure. The main objectives are related primarily to light emission for microphotonic devices and sensing applications for microfluidics.

A small cavity can considerably lower the lasing threshold of a gain medium (e.g., nanocrystals), mainly because of the small number of optical modes into which the emission can couple. Additionally, the small material volume lowers the total energy required to achieve a population inversion (Kavokin et al., 2007). Thus, a low modal lasing threshold requires, as usual, small V and large Q. Obviously, a major goal would be to achieve SE in weakly coupled Si NCs; this would be a significant breakthrough that could pave the way toward a silicon-compatible microphotonic light source (Pavesi, 2008). Coupling to microcavities may be one of the best options, but for lasing to be achieved unequivocally, it will be necessary to minimize losses associated with CCA. This might be possible by careful cavity design, for example, by controlling the location of the NCs within the cavity mode profile. A second potential application as a light source may arise from color-controlled silicon-based LEDs. Indeed, the extreme color control coupled with the use of metal mirrors (Hryciw et al., 2005) suggests that this should be possible, although achieving good efficiency is a serious problem SiO_x films owing in part to nonideal charge transport mechanisms.

In sensing applications too, practical devices are inching ever closer. Porous silicon-based Fabry–Perot cavities are being extensively explored as luminescent gas sensors (Chan et al., 2000; King et al., 2007). In these structures, adsorption in the pores modifies the refractive index of the central $\lambda/2$ layer, measurably shifting the luminescence cavity mode. Others have shown that the thermo-optic effect—usually a problem requiring temperature stabilization in sensing applications—could be employed in the development of an optical fiber switch (Tewary et al., 2007). Here, the silicon nanocrystals absorb energy from a pump laser, heating the sample (a microsphere in this case) and causing a shift of the mode resonance position via a combination of the thermo-optic and thermal expansion coefficients. The switch developed in reference (Tewary et al., 2007) had one of the best figures of merit for such devices (measured as the product of the switching power and device size).

Silicon NCs coated on the inner surface of microcapillaries or on the outer surface of optical fibers (Bianucci et al., 2009a) may also be used as refractometric sensors (Manchee et al., 2011). In this case, the WGMs supported by the high-effective-index NC film can sample the adjacent fluid medium. In the case of microcapillaries for example, the Si NC luminescence modes shift to a longer resonant wavelength as the refractive index of fluids pumped through the capillary channel is increased. Such devices are similar to LCORRs (liquid-core optical ring resonators) (White et al., 2009) but operate in fluorescence and do not require specially thinned capillaries. Furthermore, these coated fluorescent capillaries are amenable to silica-based surface functionalization methods, implying that such Si NC-microcavity coupled structures could be used as microfluidic fluorescence biosensors. An alternative sensing method could make use of the fluorescence Q-factors: if the fluid to be sampled is absorbing at the resonant wavelength, the Q-factor will decrease monotonically with concentration, providing an additional transduction method.

8.7.5 Conclusion

There are many potential applications for microcavity-coupled silicon nanocrystals in various stages of development. The spontaneous emission rate enhancement and luminescence mode spectrum, both signatures of the weakly coupled "Purcell" regime, make possible a variety of potential devices ranging from light sources to sensors. In some cases, significant technical challenges remain such as reducing the losses that prevent clear observation of SE or lasing in Si NCs coupled to a microcavity. In other cases, cavity-coupled Si NCs are challenging existing sensing technologies and may find a place in the market before long (Jane et al., 2009). Refractometic and biosensing applications are fairly new; the main technical achievements still needed include demonstrating sufficient sensitivity for desired analytes, as well as solving packaging issues (i.e., how to economically incorporate a pump LED, a luminescent microcavity, a sample container or fluid pump, and a detection system in a single unit). Innovative solutions to these issues are likely to be advanced in the coming years.

References

Agha, I. H., J. E. Sharping, M. A. Foster, and A. L. Gaeta. 2006. Optimal sizes of silica microspheres for linear and nonlinear optical interactions. *Applied Physics B* 83:303–309.

Ahn, D. H., C. Y. Hong , J. F. Liu et al. 2007. High performance, waveguide integrated Ge photodetectors. *Optics Express* 15:3916–3921.

Al Choueiry, A., A. M. Jurdyc, B. Jacquier, F. Gourbilleau, and R. Rizk. 2009. Submicrosecond fluorescence dynamics in erbium-doped silicon-rich silicon oxide multilayers. *Journal of Applied Physics* 106:053107.

Allan, G., C. Delerue, and M. Lannoo. 1996. Nature of luminescent surface states of semiconductor nanocrystallites. *Physical Review Letters* 76 (16):2961–2964.

Amans, D., S. Callard, A. Gagnaire, J. Joseph, G. Ledoux, and F. Huisken. 2003. Optical properties of a microcavity containing silicon nanocrystals. *Materials Engineering B* 101:305–308.

Amans, D., S. Callard, A. Gagnaire, J. Joseph, F. Huisken, and G. Ledoux. 2004. Spectral and spatial narrowing of the emission of silicon nanocrystals in a microcavity. *Journal of Applied Physics* 95:5010–5013.

Andreani, L. C., G. Panzarini, and J.-M. Gérard. 1999. Strong-coupling regime for quantum boxes in pillar microcavities: Theory. *Physical Review B* 60:13276–13279.

Andreev, B. A., A. Y. Andreev, H. Ellmer et al. 1999. Optical Er-doping of Si during sublimational molecular beam epitaxy. *Journal of Crystal Growth*, 201–202:534–537.

Anopchenko, A., A. Marconi, M. Wang, G. Pucker, P. Bellutti, and L. Pavesi. 2011. Graded-size Si quantum dot ensembles for efficient-emitting diodes. *Applied Physics Letters* 99 (18):181108–3.

Antonova, I. V., M. Gulyaev, E. Savir, J. Jedrzejewski, and I. Balberg. 2008. Charge storage, photoluminescence, and cluster statistics in ensembles of Si quantum dots. *Physical Review B* 77 (12):125318–5.

Apetz, R., L. Vescan, A. Hartmann, C. Dieker, and H. Luth. 1995. Photoluminescence and electroluminescence of SiGe dots fabricated by island growth. *Applied Physics Letters* 66 (4):445–447.

Arguirov, T., M. Kittler, M. Oehme, N. V. Abrosimov, E. Kasper, and J. Schulze. 2011. Room temperature direct band-gap emission from an unstrained Ge p-i-n LED on Si. *Solid State Phenomenon* 178–179:25–30.

Auston, D. H., C. V. Shank, and P. LeFur. 1975. Picosecond optical measurements of band-to-band Auger recombination of high-density plasmas in germanium. *Physical Review Letters* 35 (15):1022–1025.

Averboukh, B., R. Huber, K. W. Cheah et al. 2002. Luminescence studies of a Si/SiO_2 superlattice. *Journal of Applied Physics* 92 (7):3564–3568.
Balberg, I. 2011. Electrical transport mechanisms in three dimensional ensembles of silicon quantum dots. *Journal of Applied Physics* 110 (6):061301–26.
Baribeau, J. M., T. E. Jackman, D. C. Houghton, P. Maigné, and M. W. Denhoff. 1988. Growth and Characterization of $Si_{1-x}Ge_x$ and Ge Epilayers on (100) Si. *Journal of Applied Physics* 63: 5738–5746.
Baribeau, J. M., D. J. Lockwood, Z. H. Lu, H. J. Labbé, S. J. Rolfe, and G. I. Sproule. 1998. Amorphous Si/insulator multilayers grown by vacuum deposition and electron cyclotron resonance plasma treatment. *Journal of Luminescence* 80 (1–4):417–421.
Baribeau, J. M., R. Pascual, and S. Saimoto. 1990. Interdiffusion and strain relaxation in $(Si_mGe_n)_p$ superlattices. *Applied Physics Letters* 57 (15):1502–1504.
Baribeau, J. M., N. L. Rowell, and D. J. Lockwood. 2005. Advances in the growth and characterization of Ge quantum dots and islands. *Journal of Materials Research* 20 (12):3278–3293.
Baribeau, J. M., X. Wu, and D. J. Lockwood. 2006a. Probing the composition of Ge dots and Si/ $Si_{1-x}Ge_x$ island superlattices. *Journal of Vacuum Science and Technology A* 24 (3):663–667.
Baribeau, J. M., X. Wu, N. L. Rowell, and D. J. Lockwood. 2006b. Ge dots and nanostructures grown epitaxially on Si. *Journal of Physics: Condensed Matter* 18 (8):R139–R174.
Bassani, F., I. Mihalcescu, J. C. Vial, and F. Arnaud d'Avitaya. 1997. Optical absorption evidence of quantum confinement in Si/CaF_2 multilayers grown by molecular beam epitaxy. *Applied Surface Science* 117–118 (0):670–676.
Bassani, F., L. Vervoort, I. Mihalcescu, J. C. Vial, and F. Arnaud d'Avitaya. 1996. Fabrication and optical properties of Si/CaF_2(111) multi-quantum wells. *Journal of Applied Physics* 79 (8):4066–4071.
Belarouci, A., and F. Gourbilleau. 2006. Optical properties from SRSO/SiO_2 multilayers in planar microcavities. *Journal of Luminescence* 121:282–285.
Belarouci, A., F. Gourbilleau, and R. Rizk. 2007. Microcavity enhanced spontaneous emission from silicon nanocrystals. *Journal of Applied Physics* 101:073108.
Beltaos, A., and A. Meldrum. 2007. Whispering gallery modes in silicon-nanocrystal-coated silica microspheres. *Journal of Luminescence* 126:607–613.
Belyakov, V. A., V. A. Burdov, R. Lockwood, and A. Meldrum. 2008. Silicon nanocrystals: Fundamental theory and implications for stimulated emission. *Advances in Optical Technologies* 2008:279502–32.
Benton, J. L., J. Michel, L. C. Kimerling et al. 1991. The electrical and defect properties of erbium-implanted silicon. *Journal of Applied Physics* 70:2667–2671.
Bessette, J., R. Camacho-Aguilera, Y. Cai, L. C. Kimerling, and J. Michel. 2011. Optical Characterization of Ge-on-Si Laser Gain Media. *8th IEEE International Conference on Group IV Photonics,* paper 1.25 (September 2011, London, UK).
Bianucci, P., J. R. Rodriguez, F. Lenz, C. M. Clement, J. G. C. Veinot, and A. Meldrum. 2009a. Silicon nanocrystal luminescence coupled to whispering gallery modes in optical fibers. *Journal of Applied Physics* 105:023108.
Bianucci, P., J. R. Rodriguez, F. Lenz, J. C. G. Veinot, and A. Meldrum. 2009b. Mode structure in the luminescence of silicon nanocrystals in cylindrical microcavities. *Physica E* 41:1107–1110.
Bianucci, P., J. R. Rodriguez, C. Clements, C. M. Hessel, J. G. C. Veinot, and A. Meldrum. 2009c. Whispering gallery modes in silicon nanocrystal coated microcavities. *Physica Status Solidi A,* 206:973–975.
Bianucci, P., X. Y. Wang, J. G. C. Veinot, and A. Meldrum. 2010. Silicon nanocrystals on bottle resonators: Mode structure, loss mechanisms and emission dynamics. *Optics Express* 18:8466–8481.
Bianucci, P., Y. Y. Zhi, F. Marsiglio, J. Silverstone, and A. Meldrum. 2011. Microcavity effects in ensembles of silicon quantum dots coupled to high-*Q* resonators. *Physica Status Solidi A* 208:639–645.
Bisi, O., S. Ossicini, and L. Pavesi. 2000. Porous silicon: A quantum sponge structure for silicon based optoelectronics. *Surface Science Reports* 38 (1–3):1–126.
Bonafos, C., B. Colombeau, A. Altibelli et al. 2001. Kinetic study of group IV nanoparticles ion beam synthesized in SiO_2. *Nuclear Instruments & Methods in Physics Research B—Beam Interactions with Materials and Atoms* 178:17–24.

Brongersma, M. L., P. G. Kik, A. Polman, K. S. Min, and H. A. Atwater. 2000. Size-dependent electron-hole exchange interaction in Si nanocrystals. *Applied Physics Letters* 76 (3):351–353.
Brunner, K. 2002. Si/Ge nanostructures. *Reports on Progress in Physics* 65 (1):27–72.
Bruno, M., M. Palummo, A. Marini, R. Del Sole, and S. Ossicini. 2007. From Si nanowires to porous silicon: The role of excitonic effects. *Physical Review Letters* 98 (3):036807–4.
Camacho-Aguilera, R., J. Bessette, Y. Cai, L. C. Kimerling, and J. Michel. 2011. Electroluminescence of highly doped Ge pnn diodes for Si integrated lasers. *8th IEEE International Conference on Group IV Photonics*, paper ThC1 (September 2011, London, UK).
Camacho-Aguilera, R., Y. Cai, N. Patel, J. T. Bessette, M. Romagnoli, L. C. Kimerling, and J. Michel. 2012. An electrically pumped germanium laser. *Optics Express* 20:11316–11320.
Canham, L. T. 1990. Silicon quantum wire array fabrication by electrochemical and chemical dissolution of wafers. *Applied Physics Letters* 57 (10):1046–1048.
Cannon, D. D., J. F. Liu, Y. Ishikawa et al. 2004. Tensile strained epitaxial Ge films on Si(100) substrates with potential applications in L-band telecommunications. *Applied Physics Letters* 84:906–908.
Carmichael, H. J., R. J. Brecha, M. G. Raizen, H. J. Kimble, and P. R. Rice. 1989. Subnatural linewidth averaging for coupled atomic and cavity-mode oscillators. *Physical Review A* 40:5516–5519.
Carreras, J., J. Arbiol, B. Garrido, C. Bonafos, and J. Montserrat. 2008. Direct modulation of electroluminescence from silicon nanocrystals beyond radiative recombination rates. *Applied Physics Letters* 92 (9):091103–3.
Carreras, J., O. Jambois, S. Lombardo, and B. Garrido. 2009. Quantum dot networks in dielectric media: From compact modeling of transport to the origin of field effect luminescence. *Nanotechnology* 20 (15):155201–12.
Cavallini, A., B. Fraboni, S. Pizzini et al. 1999. Electrical and optical characterization of Er-doped silicon grown by liquid phase epitaxy. *Journal of Applied Physics* 85:1582–1586.
Chan, S., P. M. Fauchet, Y. Li, L. J. Rothberg, and B. L. Miller. 2000. Porous silicon microcavities for biosensing applications. *Physica Status Solidi A* 182:541–546.
Chang, G.-E., S.-W. Chang, and S. L. Chuang. 2009. Theory for n-type doped, tensile-strained Ge-$Si_xGe_ySn_{1-x-y}$ quantum well lasers. *Optics Express* 17:11246–11258.
Charvet, S., R. Madelon, F. Gourbilleau, and R. Rizk. 1999. Spectroscopic ellipsometry analyses of sputtered Si/SiO_2 nanostructures. *Journal of Applied Physics* 85 (8):4032–4039.
Chen, H., J.-Y. Sung, A. Tewary, M. Brongersma, J. H. Shin, and P. M. Fauchet. 2005. Evidence for stimulated emission in silicon nanocrystal microspheres. *2nd IEEE International Conference on Group IV Photonics*, pp. 99–101.
Cheng, S.-L., J. Lu, G. Shambat et al. 2009. Room temperature 1.6 μm electroluminescence from Ge light emitting diode on Si substrate. *Optics Express* 17:10019–10024.
Cheng, S.-L., G. Shambat, J. Lu et al. 2011. Cavity-enhanced direct band electroluminescence near 1550 nm from germanium microdisk resonator diode on silicon. *Applied Physics Letters* 98:211101.
Cheng, T.-H., K.-L. Peng, C.-Y. Ko et al. 2010. Strain-enhanced photoluminescence from Ge direct transition. *Applied Physics Letters* 96:211108.
Chryssou, C. E., A. J. Kenyon, T. S. Iwayama, C. W. Pitt, and D. E. Hole. 1999. Evidence of energy coupling between Si nanocrystals and Er^{3+} in ion-implanted silica thin films. *Applied Physics Letters* 75:2011–2013.
Colace, L., G. Masini, F. Galluzzi. et al. 1997. Ge/Si (001) photodetector for infrared light. *Solid State Phenomena* 54:55–58.
Credo, G. M., M. D. Mason, and S. K. Buratto. 1999. External quantum efficiency of single porous silicon nanoparticles. *Applied Physics Letters* 74 (14):1978–1980.
Cui, Y., L. J. Lauhon, M. S. Gudiksen, J. F. Wang, and C. M. Lieber. 2001. Diameter-controlled synthesis of single-crystal silicon nanowires. *Applied Physics Letters* 78 (15):2214–2216.
Cui, Y., and C. M. Lieber. 2001. Functional nanoscale electronic devices assembled using silicon nanowire building blocks. *Science* 291 (5505):851–853.
d'Avitaya, F. A., L. Vervoort, F. Bassani, S. Ossicini, A. Fasolino, and F. Bernardini. 1995. Light Emission at Room Temperature from Si/CaF_2 Multilayers. *EPL (Europhysics Letters)* 31 (1):25–30.

Dal Negro, L., M. Cazzanelli, N. Daldosso et al. 2003. Stimulated emission in plasma-enhanced chemical vapour deposited silicon nanocrystals. *Physica E: Low-dimensional Systems and Nanostructures* 16 (3–4):297–308.

Dal Negro, L., M. Cazzanelli, B. Danese et al. 2004. Light amplification in silicon nanocrystals by pump and probe transmission measurements. *Journal of Applied Physics* 96 (10):5747–5755.

Dal Negro, L., J. H. Yi, J. Michel et al. 2006. Light emission efficiency and dynamics in silicon-rich silicon nitride films. *Applied Physics Letters* 88 (23):233109–3.

Daldosso, N., M. Luppi, S. Ossicini et al. 2003. Role of the interface region on the optoelectronic properties of silicon nanocrystals embedded in SiO_2. *Physical Review B* 68 (8):085327–8.

Daldosso, N., and L. Pavesi. 2009. Nanosilicon photonics. *Laser & Photonics Reviews* 3 (6):508–534.

Davies, G. 1989. The optical properties of luminescence centres in silicon. *Physics Reports—Review Section of Physics Letters* 176 (3–4):83–188.

de Boer, W. D. A. M., D. Timmerman, K. Dohnalova et al. 2010. Red spectral shift and enhanced quantum efficiency in phonon-free photoluminescence from silicon nanocrystals. *Nature Nanotechnology* 5 (12):878–884.

Degoli, E., and S. Ossicini. 2000. The electronic and optical properties of Si/SiO_2 superlattices: Role of confined and defect states. *Surface Science* 470 (1–2):32–42.

Degoli, E., and S. Ossicini. 2009. Engineering quantum confined silicon nanostructures: Ab-initio study of the structural, electronic and optical properties. *Advances in Quantum Chemistry, Vol 58: Theory of Confined Quantum Systems: Pt 2* 58:203–279.

de Kersauson, M., M. El Kurdi, S. David et al. 2011. Optical gain in single tensile-strained germanium photonic wire. *Optics Express* 19:17925–17934.

Delerue, C., G. Allan, C. Reynaud, O. Guillois, G. Ledoux, and F. Huisken. 2006. Multiexponential photoluminescence decay in indirect-gap semiconductor nanocrystals. *Physical Review B* 73 (23):235318–4.

Derr, J., K. Dunn, D. Riabinina, F. Martin, M. Chaker, and F. Rosei. 2009. Quantum confinement regime in silicon nanocrystals. *Physica E—Low-Dimensional Systems & Nanostructures* 41 (4):668–670.

DiMaria, D. J., T. N. Theis, J. R. Kirtley, F. L. Pesavento, Dong, D. W., and S. D. Brorson. 1985. Electron heating in silicon dioxide and off-stoichiometric silicon dioxide films. *Journal of Applied Physics* 57:1214–1238.

Ding, L., M. B. Yu, X. G. Tu, G. Q. Lo, S. Tripathy, and T. P. Chen. 2011. Laterally-current-injected light-emitting diodes based on nanocrystalline-Si/SiO_2 superlattice. *Optics Express* 19 (3):2729–2738.

Dittrich, T., V. Y. Timoshenko, J. Rappich, and L. Tsybeskov. 2001. Room temperature electroluminescence from a c-Si p-i-n structure. *Journal of Applied Physics* 90 (5):2310–2313.

Dovrat, M., Y. Goshen, J. Jedrzejewski, I. Balberg, and A. Sa'ar. 2004. Radiative versus nonradiative decay processes in silicon nanocrystals probed by time-resolved photoluminescence spectroscopy. *Physical Review B* 69:155311.

Dovrat, M., Y. Shalibo, N. Arad, I. Popov, S. T. Lee, and A. Sa'ar. 2009. Fine structure and selection rules for excitonic transitions in silicon nanostructures. *Physical Review B* 79 (12):125306–5.

Eaglesham, D. J., and M. Cerullo. 1990. Dislocation-free Stranski-Krastanow growth of Ge on Si(100). *Physical Review Letters* 64 (16):1943–1946.

Eaglesham, D. J., J. Michel, E. A. Fitzgerald et al. 1991. Microstructure of erbium-implanted Si. *Applied Physics Letter* 58:2797–2799.

Einstein, A. 1916. Strahlungs-Emission und Absorption nach der Quantentheorie. 18:318–323.

El Kurdi, M., H. Bertin, E. Martincic et al. 2010b. Control of direct band-gap emission of bulk germanium by mechanical tensile strain. *Applied Physics Letters* 96: 041909.

El Kurdi, M., G. Fishman, S. Sauvage, and P. Boucaud. 2010a. Band structure and optical gain of tensile-strained germanium based on a 30 band k-p formalism. *Journal of Applied Physics* 107:013710.

El Kurdi, M., S. Sauvage, G. Fishman, and P. Boucaud. 2006. Band-edge alignment of SiGe/Si quantum wells and SiGe/Si self-assembled islands. *Physical Review B* 73 (19):195327–1953279.

Ennen, H., G. Pomrenke, A. Axmann, K. Eisele, W. Haydl, and J. Schneider. 1985. 1.54-µm electroluminescence of erbium-doped silicon grown by molecular beam epitaxy. *Applied Physics Letters* 46:381–383.

Ertekin, E., P. A. Greaney, D. C. Chrzan, and T. D. Sands. 2005. Equilibrium limits of coherency in strained nanowire heterostructures. *Journal of Applied Physics* 97 (11):114325–10.

Fang, Y.-Y., J. Tolle, R. Roucka et al. 2007a. Perfectly tetragonal, tensile-strained Ge on $Ge_{1-y}Sn_y$ buffered Si(100). *Applied Physics Letters* 90:061915.

Fang, Y.-Y., J. Tolle, J. Tice et al. 2007b. Epitaxy-driven synthesis of elemental Ge/Si strain-engineered materials and device structures via designer molecular chemistry. *Chemistry Materials 19*:5910–5925.

Fauchet, P. M. 2004. Monolithic silicon light sources. In *Silicon Photonics*, edited by L. Pavesi, and D. J. Lockwood: Springer, Berlin/Heidelberg.

Favennec, P. N., H. L'Haridon, M. Salvi, D. Moutonnet, and Y. Le Guillou. 1989. Luminescence of erbium implanted in various semiconductors: IV, III–V and II–VI materials. *Electronics Letter* 25:718–719.

Filonov, A. B., S. Ossicini, F. Bassani, and F. A. d'Avitaya. 2002. Effect of oxygen on the optical properties of small silicon pyramidal clusters. *Physical Review B* 65 (19):195317–9.

Fiory, A. T., and N. M. Ravindra. 2003. Light emission from silicon: Some perspectives and applications. *Journal of Electronic Materials* 32 (10):1043–1051.

Flynn, C., D. König, M. A. Green, and G. Conibeer. 2010. Modelling of metal–insulator–semiconductor devices featuring a silicon quantum well. *Physica E: Low-dimensional Systems and Nanostructures* 42 (9):2211–2217.

Forcales, M., T. Gregorkiewicz, I. V. Bradley, and J.-P. R. Wells. 2002. Afterglow effect in photoluminescence of Si:Er. *Physical Review B* 65:195208.

Forcales, M., T. Gregorkiewicz, M. S. Bresler, O. B. Gusev, I. V. Bradley, and J.-P. R. Wells. 2003a. Microscopic model for non-excitonic mechanism of 1.5 μm photoluminescence of the Er^{3+} ion in crystalline Si. *Physical Review B* 67:085303.

Forcales, M., M. A. J. Klik, N. Q. Vinh, J. Phillips, J.-P. R. Wells, and T. Gregorkiewicz. 2003b. Two-color mid-infrared spectroscopy of optically doped semiconductors. *Journal of Luminescence* 102–103:85–90.

Franzo, G., A. Irrera, E. C. Moreira et al. 2002. Electroluminescence of silicon nanocrystals in MOS structures. *Applied Physics a-Materials Science & Processing* 74 (1):1–5.

Franzò, G., F. Priolo, S. Coffa, A. Polman, and A. Carnera. 1994. Room-temperature electroluminescence fro Er-doped crystalline Si. *Applied Physics Letters* 64:2235–2237.

Franzò, G., V. Vinciguerra, and F. Priolo. 1999. The excitation mechanism of rare-earth ions in silicon nanocrystals. *Applied Physics A: Materials Science and Processing* 69:3–12.

Fujii, M., Y. Inoue, S. Hayashi, and K. Yamamoto. 1996. Hopping conduction in SiO_2 films containing C, Si, and Ge clusters. *Applied Physics Letters* 68:3749–3751.

Fujii, M., K. Toshikiyo, Y. Takase, Y. Yamaguchi, and S. Hayashi. 2003. Below bulk-band-gap photoluminescence at room temperature from heavily P- and B-doped Si nanocrystals. *Journal of Applied Physics* 94 (3):1990–1995.

Fujii, M., Y. Yamaguchi, Y. Takase, K. Ninomiya, and S. Hayashi. 2004. Control of photoluminescence properties of Si nanocrystals by simultaneously doping n- and p-type impurities. *Applied Physics Letters* 85 (7):1158–1160.

Fujii, M., Y. Yamaguchi, Y. Takase, K. Ninomiya, and S. Hayashi. 2005. Photoluminescence from impurity codoped and compensated Si nanocrystals. *Applied Physics Letters* 87 (21):211919–3.

Fujii, M., M. Yoshida, S. Hayashi, and K. Yamamoto. 1998. Photoluminescence from SiO_2 films containing Si nanocrystals and Er: Effects of nanocrystalline size on the photoluminescence efficiency of Er^{3+}. *Journal of Applied Physics* 84:4525–4531.

Fujii, M., M. Yoshida, Y. Kanzawa, S. Hayashi, and K. Yamamoto. 1997. 1.54 μm photoluminescence of Er^{3+} doped into SiO_2 films containing Si nanocrystals: Evidence for energy transfer from Si nanocrystals to Er^{3+}. *Applied Physics Letters* 71:1198–1200.

Fujio, K., M. Fujii, K. Sumida, S. Hayashi, M. Fujisawa, and H. Ohta. 2008. Electron spin resonance studies of P and B codoped Si nanocrystals. *Applied Physics Letters* 93 (2):021920–3.

Gardelis, S., A. G. Nassiopoulou, N. Vouroutzis, and N. Frangis. 2009. Effect of exciton migration on the light emission properties in silicon nanocrystal ensembles. *Journal of Applied Physics* 105 (11):113509–7.

Garrido, B., C. García, P. Pellegrino et al. 2006. Distance dependent interaction as the limiting factor for Si nanocluster to Er energy transfer in silica. *Applied Physics Letters* 89:163103.
Garrido, B., C. García, S.-Y. Seo et al. 2007. Excitable Er fraction and quenching phenomena in Er-doped SiO_2 layers containing Si nanoclusters. *Physical Review B* 76:245308.
Gérard, J. M., and B. Gayral. 1999. Semiconductor microcavities, quantum boxes, and the Purcell effect. In: H. Benisty, C. Weisbuch, J. M. Gérard, R. Houdré, J. Rarity, eds. 1999. *Lecture Notes in Physics Vol. 531, Confined Photon Systems.* New York: Springer. pp. 331–351.
Gérard, J. M., B. Sermage, B. Gayral, B. Legrand, E. Costard, and V. Thierry-Mieg. 1998. Enhanced Spontaneous Emission by Quantum Boxes in a Monolithic Optical Microcavity. *Physical Review Letter* 81:1110–1113.
Ghulinyan, M., D. Navarro-Urrios, A. Pitanti, A. Lui, G. Pucker, and L. Pavesi. 2008. Whispering-gallery modes and light emission from a Si-nanocrystal-based single microdisk resonator. *Optics Express* 16:13218–13224.
Godefroo, S., M. Hayne, M. Jivanescu et al. 2008. Classification and control of the origin of photoluminescence from Si nanocrystals. *Nature Nanotechnology* 3 (3):174–178.
Gong-Ru, L., P. Yi-Hao, and L. Cheng-Tao. 2008. Microwatt MOSLED Using SiO_x With Buried Si Nanocrystals on Si Nano-Pillar Array. *Journal of Lightwave Technology* 26 (11):1486–1491.
Gourbilleau, F., C. Ternon, D. Maestre, O. Palais, and C. Dufour. 2009. Silicon-rich SiO_2/SiO_2 multilayers: A promising material for the third generation of solar cell. *Journal of Applied Physics* 106 (1):013501–7.
Green, M. A., J. H. Zhao, A. H. Wang, P. J. Reece, and M. Gal. 2001. Efficient silicon light-emitting diodes. *Nature* 412 (6849):805–808.
Gregorkiewicz, T., I. Tsimperidis, C. A. J. Ammerlaan, F. P. Widdershoven, and N. A. Sobolev. 1996. Excitation and de-excitation of Yb^{3+} in InP and Er^{3+} in Si: Photoluminescence and impact ionization studies. *MRS Proceedings* 422:207.
Grun, M., P. Miska, H. Rinnert, and M. Vergnat. 2011. Optical properties of a silicon-nanocrystal-based-microcavity prepared by evaporation. *Optical Materials* 33:1248–1251.
Guerra, R., E. Degoli, M. Marsili, O. Pulci, and S. Ossicini. 2010. Local-fields and disorder effects in free-standing and embedded Si nanocrystallites. *Physica Status Solidi B Basic Solid State Physics* 247 (8):2113–2117.
Guerra, R., E. Degoli, and S. Ossicini. 2009a. Size, oxidation, and strain in small Si/SiO_2 nanocrystals. *Physical Review B* 80 (15):155332–5.
Guerra, R., I. Marri, R. Magri et al. 2009b. Silicon nanocrystallites in a SiO_2 matrix: Role of disorder and size. *Physical Review B* 79 (15):155320–9.
Guerra, R., M. Marsili, O. Pulci, and S. Ossicini. 2011. Local-field effects in silicon nanoclusters. *Physical Review B* 84 (7):075342–7.
Guerra, R., and S. Ossicini. 2010. High luminescence in small Si/SiO_2 nanocrystals: A theoretical study. *Physical Review B* 81 (24):245307–6.
Gusev, O., M. S. Bresler, P. E. Pak et al. 2001. Excitation cross section of erbium in semiconductor matrices under optical pumping. *Physical Review B* 64:075302.
Ha, N. N., K. Dohnalová., T. Gregorkiewicz, and J. Valenta. 2010. Optical gain of the 1.54 μm emission in MBE-grown Si:Er nanolayers. *Physical Review B* 81:195206.
Haas, C. 1962. Infrared absorption in heavily doped n-type germanium. *Physical Review* 125: 1965–1971.
Hannon, J. B., S. Kodambaka, F. M. Ross, and R. M. Tromp. 2006. The influence of the surface migration of gold on the growth of silicon nanowires. *Nature* 440 (7080):69–71.
Hao, X. J., A. Podhorodecki, Y. S. Shen, G. Zatryb, J. Misiewicz, and M. A. Green. 2009. Effects of Si-rich oxide layer stoichiometry on the structural and optical properties of Si QD/SiO_2 multilayer films. *Nanotechnology* 20 (48):485703–10.
Hartmann, J. M., A. Abbadie, A. M. Papon et al. 2004. Reduced pressure-chemical vapor deposition of Ge thick layers on Si (001) for 1.3–1.55 μm photodetection. *Journal of Applied Physics* 95:5905–5913.

Hartmann, J. M., J. P. Barnes, M. Veillerot, and J. M. Fedeli. 2011. Selective Epitaxial Growth of intrinsic and in-situ phosphorous-doped Ge for optoelectronics. 2011 European Materials Research Society (E-MRS), 2011 Spring Meeting. Nice, France, 9–13 May 2011. Paper I4–3.

Haynes, J. R., and N. G. Nilsson. 1964. The direct radiative transitions in germanium and their use in the analysis of lifetime. *Proceedings of VIIth International Conference on Physics of Semiconductors*. Paris: Dunod, p. 21–31.

Heitmann, J., F. Muller, M. Zacharias, and U. Gosele. 2005. Silicon nanocrystals: Size matters. *Advanced Materials* 17 (7):795–803.

Hernandez, A. V., T. V. Torchynska, Y. Matsumoto et al. 2005. Optical investigation of Si nano-crystals in amorphous silicon matrix. *Microelectronics Journal* 36 (3–6):510–513.

Herzog, M., and F. Koch. 1988. Hot-carrier light-emission from silicon metal-oxide-semiconductor devices. *Applied Physics Letters* 53 (26):2620–2622.

Hessel, C. M., M. A. Summers, A. Meldrum, M. Malac, and J. G. C. Veinot. 2007. Direct patterning,conformalcoating, anderbium doping of luminescent nc-Si/SiO_2 thin films from solution processable hydrogen silsesquioxane. *Advanced Materials* 19:3513–3516.

Hijazi, K., R. Rizk, J. Cardin, L. Khomenkova, and F. Gourbilleau. 2009. Towards an optimum coupling between Er ions and Si-based sensitizers for integrated active photonics. *Journal of Applied Physics* 106: 024311.

Hill, Nicola A., and K. Birgitta Whaley. 1995. Size dependence of excitons in silicon nanocrystals. *Physical Review Letters* 75 (6):1130–1133.

Holmes, J. D., K. P. Johnston, R. C. Doty, and B. A. Korgel. 2000. Control of thickness and orientation of solution-grown silicon nanowires. *Science* 287 (5457):1471–1473.

Houghton, D. C., G. C. Aers, S. R. Eric Yang, E. Wang, and N. L. Rowell. 1995. Type I band alignment in $Si_{1-x}Ge_x$/Si(001) quantum wells: Photoluminescence under applied [110] and [100] uniaxial stress. *Physical Review Letters* 75 (5):866–869.

Hryciw, A., C. Blois, A. Meldrum, T. Clement, R. DeCorby, and Q. Li. 2006. Photoluminescence from Er-doped silicon oxide microcavities. *Optical Materials* 28:873–878.

Hryciw, A., J. Laforge, C. Blois, M. Glover, and A. Meldrum. 2005. Tunable luminescence from a silicon-rich oxide microresonator. *Advanced Materials* 17:845–849.

Hu, J., X. G. Xu, J. A. H. Stotz et al. 1998. Type II photoluminescence and conduction band offsets of GaAsSb/InGaAs and GaAsSb/InP heterostructures grown by metalorganic vapor phase epitaxy. *Applied Physics Letters* 73 (19):2799–2801.

Hull, R. 1999. *Properties of crystalline silicon*: INSPEC, the Institution of Electrical Engineers, London, United Kingdom.

Hull, R., and J. C. Bean. 1999. *Germanium silicon: Physics and materials*: Academic Press, San Diego, CA, USA.

Iacona, F., G. Franzo, E. C. Moreira, and F. Priolo. 2001. Silicon nanocrystals and Er^{3+} ions in an optical microcavity *Journal of Applied Physics* 89:8354–8356.

Iacona, F., G. Franzo, and C. Spinella. 2000. Correlation between luminescence and structural properties of Si nanocrystals. *Journal of Applied Physics* 87 (3):1295–1303.

Iacona, F., D. Pacifici, A. Irrera et al. 2002. Electroluminescence at 1.54 μm in Er-doped Si nanocluster-based devices. *Applied Physics Letters* 81:3242–3244.

Imhof, S., and A. Thraenhardt. 2010. Phonon-assisted transitions and optical gain in indirect semiconductors. *Physical Review B* 82 (8):085303–6.

Ioannou-Sougleridis, V., A. G. Nassiopoulou, T. Ouisse, F. Bassani, and F. Arnaud d'Avitaya. 2001. Electroluminescence from silicon nanocrystals in Si/CaF_2 superlattices. *Applied Physics Letters* 79 (13):2076–2078.

Iori, F., E. Degoli, R. Magri et al. 2007. Engineering silicon nanocrystals: Theoretical study of the effect of codoping with boron and phosphorus. *Physical Review B* 76 (8):085302–14.

Ishikawa, Y., K. Wada, D. D. Cannon, J. F. Liu, H. C. Luan, and L. C. Kimerling. 2003. Strain-induced direct bandgap shrinkage in Ge grown on Si substrate. *Applied Physics Letters* 82:2044–2046.

Izeddin, I., A. S. Moskalenko, I. N. Yassievich, M. Fujii, and T. Gregorkiewicz. 2006. Nanosecond Dynamics of the Near-Infrared Photoluminescence of Er-Doped SiO_2 sensitized with Si Nanocrystals. *Physical Review Letters* 97:207401.

Izeddin, I., D. Timmerman, T. Gregorkiewicz et al. 2008. Energy transfer in Er-doped SiO_2 sensitized with Si nanocrystals. *Physical Review B* 78:035327.

Jain, S. C., and D. J. Roulston. 1991. A simple expression for band-gap narrowing in heavily doped Si, Ge, GaAs and Ge_xSi_{1-x} strained layers. *Solid-State Electronics* 34:453–465.

Jambois, O., Y. Berencen, K. Hijazi et al. 2009. Current transport and electroluminescence mechanisms in thin SiO_2 films containing Si nanocluster-sensitized erbium ions. *Journal of Applied Physics* 106:063526.

Jane, A., R. Drono, A. Hodges and N. H. Voelcke. 2009. Porous silicon biosensors on the advance. *Trends in Biotechnology* 27:230–239.

Jesson, D. E., S. J. Pennycook, J. Z. Tischler, J. D. Budai, J. M. Baribeau, and D. C. Houghton. 1993. Interplay between evolving surface morphology, atomic-scale growth modes, and ordering during Si_xGe_{1-x} epitaxy. *Physical Review Letters* 70 (15):2293–2296.

Jurbergs, D., E. Rogojina, L. Mangolini, and U. Kortshagen. 2006. Silicon nanocrystals with ensemble quantum yields exceeding 60%. *Applied Physics Letters* 88 (23):233116–3.

Kahng, D. 1968. Electroluminescence of rare-earth and transition metal molecules in II–VI compounds via impact excitation. *Applied Physics Letters* 13:210–212.

Kamenev, B. V., H. Grebel, L. Tsybeskov et al. 2003. Polarized Raman scattering and localized embedded strain in self-organized Si/Ge nanostructures. *Applied Physics Letters* 83 (24):5035–5037.

Kamenev, B. V., E. K. Lee, H. Y. Chang et al. 2006. Excitation-dependent photoluminescence in Ge/Si Stranski-Krastanov nanostructures. *Applied Physics Letters* 89 (15):153106–3.

Kamenev, B. V., V. Sharma, L. Tsybeskov, and T. I. Kamins. 2005a. Optical properties of Ge nanowires grown on Si(100) and (111) substrates: Nanowire-substrate heterointerfaces. *Physica Status Solidi a-Applications and Materials Science* 202 (14):2753–2758.

Kamenev, B. V., L. Tsybeskov, J. M. Baribeau, and D. J. Lockwood. 2004. Photoluminescence and Raman scattering in three-dimensional $Si/Si_{1-x}Ge_x$ nanostructures. *Applied Physics Letters* 84 (8): 1293–1295.

Kamenev, B. V., L. Tsybeskov, J. M. Baribeau, and D. J. Lockwood. 2005b. Coexistence of fast and slow luminescence in three-dimensional $Si/Si_{1-x}Ge_x$ nanostructures. *Physical Review B* 72 (19): 193306–4.

Kamins, T. I., E. C. Carr, R. S. Williams, and S. J. Rosner. 1997. Deposition of three-dimensional Ge islands on Si(001) by chemical vapor deposition at atmospheric and reduced pressures. *Journal of Applied Physics* 81 (1):211–219.

Kamins, T. I., X. Li, and R. S. Williams. 2004. Growth and structure of chemically vapor deposited Ge nanowires on Si substrates. *Nano Letters* 4 (3):503–506.

Kane, E. O. 1956. Energy band structure in p-type germanium and silicon. *Journal of Physics and Chemistry of Solids* 1 (1–2):82–99.

Kanemitsu, Y. 1996. Mechanism of visible photoluminescence from oxidized silicon and germanium nanocrystallites. *Thin Solid Films* 276 (1–2):44–46.

Kanemitsu, Y., and S. Okamoto. 1997. Photoluminescence from Si/SiO_2 single quantum wells by selective excitation. *Physical Review B* 56 (24):R15561–R15564.

Kanemitsu, Y., N. Shimizu, T. Komoda, P. L. F. Hemment, and B. J. Sealy. 1996. Photoluminescent spectrum and dynamics of Si^+-ion-implanted and thermally annealed SiO_2 glasses. *Physical Review B* 54 (20):R14329–R14332.

Kao, C.-K., and F. Santosa. 2008. Maximization of the quality factor of an optical resonator. *Wave Motion* 45:412–427.

Karim, A., G. V. Hansson, and M. K. Linnarson. 2008. Influence of Er and O concentrations on the microstructure and luminescence of Si:Er/O LEDs. *Journal of Physics Conference Series* 100: 042010.

Kasper, E., M. Oehme, T. Aguirov, J. Werner, M. Kittler, and J. Schulze. 2010. Room temperature direct band-gap emission from Ge p-i-n heterojunction photodiodes. Post deadline paper on the 7th IEEE International Conference on Group IV Photonics (September 2010, Beijing, China).

Kavokin, A. V., J. J. Baumberg, G. Malpuech, F. P. Laussy. 2007. *Microcavities*. Oxford: Oxford Science Publications.

Kawamura, M., N. Paul, V. Cherepanov, and B. Voigtlander. 2003. Nanowires and nanorings at the atomic level. *Physical Review Letters* 91 (9):096102–4.

Kekatpure, R. D., and M. L. Brongersma. 2008a. Fundamental photophysics and optical loss processes in Si-nanocrystal-doped microdisk resonators. *Physical Review A* 78: 023829.

Kekatpure, R. D., and M. L. Brongersma. 2008b. Quantification of free-carrier absorption in silicon nanocrystals with an optical microcavity. *Nano Letter* 8:3787–3793.

Kenyon, A. J., C. E. Chryssou, C. W. Pitt et al. 2002. Luminescence from erbium-doped silicon nanocrystals in silica: Excitation mechanisms. *Journal of Applied Physics* 91:367–374.

Kenyon, A. J., P. F. Trwoga, M. Federighi, and C. W. Pitt. 1994. Optical properties of PECVD erbium-doped silicon-rich silica: Evidence for energy transfer between silicon microclusters and erbium ions. *Journal of Physics: Condensed Matter* 6:L319–324.

Khriachtchev, L. 2009. *Silicon nanophotonics: Basic principles, present status and perspectives*: World Scientific Publishing, Hackensack, USA - London, UK.

Khriachtchev, L., S. Ossicini, F. Iacona, and F. Gourbilleau. 2012. Silicon nanoscale materials: From theoretical simulations to photonic applications. In *International Journal of Photoenergy*.

King, B. H., A. M. Ruminski, J. L. Snyder, and M. J. Sailor, 2007. Optical-fiber-mounted porous silicon photonic crystals for sensing organic vapor breakthrough in activated carbon. *Advanced Materials* 19:4530–4534.

Kippenberg, T. J., A. L. Tchebotareva, J. Kalkman, A. Polman, and K. J. Vahala. 2009. Purcell-factor-enhanced scattering from Si nanocrystals in an optical microcavity. *Physical Review Letter* 103: 027406.

Kittler, M., M. Reiche, T. Arguirov, W. Seifert, and X. Yu. 2006. Silicon-based light emitters. *Physica Status Solidi a-Applications and Materials Science* 203 (4):802–809.

Klingenstein, W., and H. Schweizer. 1978. Direct gap recombination in germanium at high excitation level and low temperature. *Solid-State Electronics* 21:1371–1374.

Koponen, L., L. O. Tunturivuori, M. J. Puska, and R. M. Nieminen. 2009. Effect of the surrounding oxide on the photoabsorption spectra of Si nanocrystals. *Physical Review B* 79 (23):235332–6.

Koshida, N. 2008. *Device applications of silicon nanocrystals and nanostructures*: Springer, New York, USA.

Koshida, N., H. Koyama, Y. Suda et al. 1993. Optical characterization of porous silicon by synchrotron radiation reflectance spectra analyses. *Applied Physics Letters* 63 (20):2774–2776.

Kosyachenko, L. A., and M. P. Mazur. 1999. Hot-carrier far infrared emission in silicon. *Semiconductors* 33 (2):143–146.

Kovalev, D., H. Heckler, M. Ben-Chorin, G. Polisski, M. Schwartzkopff, and F. Koch. 1998. Breakdown of the k-conservation rule in Si nanocrystals. *Physical Review Letters* 81:2803–2806.

Kroemer, H. 1963. A proposed class of heteojunction injection lasers. *Proceedings of the IEEE* 51:1782–1783.

Krupka D. C., and D. M. Mahoney. 1972. Electroluminescence and photoluminescence of thin films of ZnS doped with rare-earth metals. *Journal of Applied Physics* 43:2314–2319.

Kuroyanagi, R., Y. Ishikawa, T. Tsuchizawa, and K. Wada. 2011. Controlling strain in Ge on Si for EA modulators. *8th IEEE International Conference on Group IV Photonics*, paper ThC8 (September 2011, London, UK).

Landsberg, P. T. 1991. *Recombination in semiconductors*. Cambridge, New York: Cambridge University Press.

Lange, C., N. S. Köster, S. Chatterjee et al. 2009. Ultrafast nonlinear optical response of photoexcited Ge/SiGe quantum wells: Evidence for a femtosecond transient population inversion. *Physical Review B* 79:201306R.

Lauhon, L. J., M. S. Gudiksen, D. Wang, and C. M. Lieber. 2002. Epitaxial core-shell and core-multishell nanowire heterostructures. *Nature* 420 (6911):57–61.

Lee, M. J., and E. A. Fitzgerald. 2005. Strained Si, SiGe, and Ge channels for high-mobility metal-oxide-semiconductor field-effect transistors. *Journal of Applied Physics* 97:011101.

Lehovec, K., C. A. Accardo, and E. Jamgochian. 1951. Injected light emission of silicon carbide crystals. *Physical Review* 83 (3):603–607.

Lenchyshyn, L. C., M. L. W. Thewalt, J. C. Sturm et al. 1992. High quantum efficiency photoluminescence from localized excitons in $Si_{1-x}Ge_x$. *Applied Physics Letters* 60 (25):3174–3176.

Levitt, A. P. 1970. *Whisker technology*. New York: Wiley-Interscience.

Li, T. S., F. Gygi, and G. Galli. 2011. Tailored nanoheterojunctions for optimized light emission. *Physical Review Letters* 107 (20):206805–5.

Li, Y., F. Qian, J. Xiang, and C. M. Lieber. 2006. Nanowire electronic and optoelectronic devices. *Materials Today* 9 (10):18–27.

Lim, P. H., Y. Kobayashi, S. Takita, Y. Ishikawa, and K. Wada. 2008. Enhanced photoluminescence from germanium-based ring resonators. *Applied Physics Letters* 93: 041103.

Lim, P. H., S. Park, Y. Ishikawa, and K. Wada. 2009. Enhanced direct bandgap emission in germanium by micromechanical strain engineering. *Optics Express* 17:16358–16365.

Lin, S. W., and D. H. Chen. 2009. Synthesis of water-soluble blue photoluminescent silicon nanocrystals with oxide surface passivation. *Small* 5 (1):72–76.

Liu, J. F., M. Beals, A. Pomerene et al. 2008a. Waveguide integrated, ultra-low energy GeSi electroabsorption modulators. *Nature Photonics* 2:433–437.

Liu, J. F., R. Camacho-Aguilera, X. Sun et al. 2011. Ge-on-Si Optoelectronics. *Thin Solid Films* 520: 3354–3360.

Liu, J. F., D. D. Cannon, K. Wada et al. 2004a. Deformation potential constants of biaxially tensile stressed Ge epitaxial films on Si(100). *Physical Review B* 70:155309.

Liu, J. F., D. D. Cannon, K. Wada et al. 2004b. Silicidation-induced band-gap shrinkage in Ge epitaxial films on Si. *Applied Physics Letter* 84:660–662.

Liu, J. F., D. D. Cannon, K. Wada et al. 2005. Tensile strained Ge p-i-n photodetectors on Si platform for C and L band optical communications. *Applied Physics Letters* 87:011110.

Liu, J. F., X. Sun, P. Becla, L. C. Kimerling, and J. Michel. 2008b. Towards a Ge Laser for CMOS Applications. 5th IEEE International Conference on Group IV Photonics p16–18 (Sept. 2008, Sorrento, Italy).

Liu, J. F., X. Sun, R. Camacho-Aguilera, Y. Cai, J. Michel, and L. C. Kimerling. 2010b. Band-engineered Ge-on-Si lasers. 2010 International Electronic Device Meeting (IEDM), paper 6.6 (Dec. 2010, San Francisco, USA).

Liu, J. F., X. Sun, R. Camacho-Aguilera, L. C. Kimerling, and J. Michel. 2010a. Ge-on-Si laser operating at room temperature. *Optics Letters* 35:679–681.

Liu, J. F., X. Sun, L. C. Kimerling, and J. Michel. 2009. Direct-gap optical gain of Ge on Si at room temperature. *Optics Letters* 34:1738–1740.

Liu, J. F., X. Sun, D. Pan et al. 2007. Tensile-strained, n-type Ge as a gain medium for monolithic laser integration on Si. *Optics Express* 15:11272–11277.

Lockwood, D. 2008. Band gap of nanometer thick Si/SiO2 quantum wells: Theory versus experiment. Paper presented at the Proceedings of the SPIE—The International Society for Optical Engineering.

Lockwood, D. J. 1998. Light emission in silicon: From physics to devices. Academic Press.

Lockwood, D. J., J. M. Baribeau, B. V. Kamenev, E. K. Lee, and L. Tsybeskov. 2008. Structural and optical properties of three-dimensional Si1−xGex/Si nanostructures. *Semiconductor Science and Technology* 23 (6):064003–10.

Lockwood, D. J., Z. H. Lu, and J. M. Baribeau. 1996. Quantum confined luminescence in Si/SiO2 superlattices. *Physical Review Letters* 76 (3):539–541.

Lockwood, D. J., X. H. Wu, and J. M. Baribeau. 2007. Compositional redistribution in coherent $Si_{1-x}Ge_x$ islands on Si(100). *IEEE Transactions on Nanotechnology* 6 (2):245–249.

Loebner, E. 1976. Subhistories of the light emitting diode. *IEEE Transactions on Electron Devices* ED-23 (7):675–699.

Lombardo, S., S. U. Campisano, G. N. Hoven, and A. Polman. 1995. Room-tempertaure luminescence in semi-insulating polycrystalline silicon implanted with Er. *Nuclear Instruments and Methods in Physics Research, Section B: Beam Interactions with Materials and Atoms* 96:378–381.

Lourenço, M. A., R. M. Gwilliam, and K. P. Homewood. 2007. Extraordinary optical gain from silicon implanted with erbium. *Applied Physics Letters* 91:141122.

Lu, W., and C. M. Lieber. 2006. Semiconductor nanowires. *Journal of Physics D-Applied Physics* 39 (21):R387–R406.
Lu, Z. H., D. J. Lockwood, and J. M. Baribeau. 1995. Quantum confinement and light emission in SiO_2/Si superlattices. *Nature* 378 (6554):258–260.
Luan, H.-C., D. R. Lim, K. K. Lee et al. 1999. High-quality Ge epilayers on Si with low threading-dislocation densities. *Applied Physics Letters* 75:2909–2911.
Lukeš, F., and J. Humlíček. 1972. Electroreflectance of heavily doped n-type and p-type germanium near the direct gap. *Physical Review B* 6:521–533.
Luppi, M., and S. Ossicini. 2003. Multiple Si = O bonds at the silicon cluster surface. *Journal of Applied Physics* 94 (3):2130–2132.
Ma, D. D. D., C. S. Lee, F. C. K. Au, S. Y. Tong, and S. T. Lee. 2003. Small-diameter silicon nanowire surfaces. *Science* 299 (5614):1874–1877.
Madelung, O. 1982. *Physics of group IV elements and III–V compounds, Landolt–Börnstein: Numerical data and functional relationships in science and technology*, vol. 17a. Springer, Berlin-Heidelberg-New York.
Makarova, M., V. Sih, J. Warga, R. Li, L. Dal Negro, and J. Vuckovic. 2008. Enhanced light emission in photonic crystal nanocavities with Erbium-doped silicon nanocrystals. *Applied Physics Letters* 92:161107.
Makarova, M., J. Vuckovic, H. Sanda, and Y. Nishi. 2006. Silicon-based photonic crystal nanocavity light emitters. *Applied Physics Letters* 89:221101.
Manchee, C. P. K., V. Zamora, J. W. Silverstone, J. G. C. Veinot, and A. Meldrum. 2011. Refractometric Sensing with Fluorescent-Core Microcapillaries. *Optics Express* 19:21540–21551.
Marconi, A., A. Anopchenko, G. Puker, and L. Pavesi. 2010. High efficiency and long-term stability of nanocrystalline silicon based devices. Paper presented at the Group IV Photonics (GFP), 2010 7th IEEE International Conference on, 1–3 Sept. 2010.
Marconi, A., A. Anopchenko, G. Pucker, and L. Pavesi. 2011. Silicon nanocrystal light emitting device as a bidirectional optical transceiver. *Semiconductor Science and Technology* 26 (9):095019–3.
Marconi, A., A. Anopchenko, M. Wang, G. Pucker, P. Bellutti, and L. Pavesi. 2009. High power efficiency in Si-nc/SiO_2 multilayer light emitting devices by bipolar direct tunneling. *Applied Physics Letters* 94 (22):221110–3.
Martin, J., F. Cichos, F. Huisken, and C. von Borczyskowski. 2008. Electron-phonon coupling and localization of excitons in single silicon nanocrystals. *Nano Letters* 8 (2):656–660.
Meldrum, A., P. Bianucci, and F. Marsiglio. 2010. Modification of ensemble emission rates and luminescence spectra for inhomogeneously broadened distributions of quantum dots coupled to optical microcavities. *Optics Express* 18:10230–10246.
Meldrum, A., A. Hryciw, A. N. MacDonald et al. 2006. Photoluminescence in the silicon-oxygen system. *Journal of Vacuum Science & Technology, A: Vacuum, Surfaces, and Films* 24:713–717.
Menon, M., D. Srivastava, I. Ponomareva, and L. A. Chernozatonskii. 2004. Nanomechanics of silicon nanowires. *Physical Review B* 70 (12):125313–6.
Michaelis, W., and M. H. Pilkuhn. 1969. Radiative recombination in silicon p-n junctions. *Physica Status Solidi B* 36 (1):311–319.
Mie, G. 1908. Beiträge zur optik trüber medien, speziell kolloidaler metallösungen. *Annals of Physics* 330:377–445.
Min, K. S., K. V. Shcheglov, C. M. Yang, H. A. Atwater, M. L. Brongersma, and A. Polman. 1996. Defect-related versus excitonic visible light emission from ion beam synthesized Si nanocrystals in SiO_2. *Applied Physics Letters* 69 (14):2033–2035.
Miniscalco, W. J. 1991. Erbium-Doped Glasses for Fiber Amplifiers at 1500 nm. *Journal of Lightwave Technology* 9:234–250.
Minissale, S., N. Q. Vinh, W. de Boer, M. S. Bresler, and T. Gregorkiewicz. 2008. Microscopic evidence for role of oxygen in luminescence of Er^{3+} ions in Si: Two-color and pump-probe spectroscopy. *Physical Review B* 78:035313.
Mo, Y. W., D. E. Savage, B. S. Swartzentruber, and M. G. Lagally. 1990. Kinetic pathway in Stranski-Krastanov growth of Ge on Si(001). *Physical Review Letters* 65 (8):1020–1023.

Modreanu, M., M. Gartner, and D. Cristea. 2002. Investigation on preparation and physical properties of LPCVD $Si_xO_yN_z$ thin films and nanocrystalline Si/$Si_xO_yN_z$ superlattices for Si-based light emitting devices. *Materials Science and Engineering: C* 19 (1–2):225–228.
Monroy, B. M., O. Cregut, M. Gallart, B. Honerlage, and P. Gilliot. 2011. Optical gain observation on silicon nanocrystals embedded in silicon nitride under femtosecond pumping. *Applied Physics Letters* 98 (26):261108–3.
Morales, A. M., and C. M. Lieber. 1998. A laser ablation method for the synthesis of crystalline semiconductor nanowires. *Science* 279 (5348):208–211.
Moskalenko, A. S., J. Berakdar, A. A. Prokofiev, and I. N. Yassievich. 2007. Single-particle states in spherical Si/SiO_2 quantum dots. *Physical Review B* 76 (8):085427–9.
Murray, C. B., D. J. Norris, and M. G. Bawendi. 1993. Synthesis and characterization of nearly monodisperse CdE (E = S, Se, Te) semiconductor nanocrystallites. *Journal of the American Chemical Society* 115:8706–8715.
Nakashima, S., T. Mitani, M. Ninomiya, and K. Matsumoto. 2006. Raman investigation of strain in Si/SiGe heterostructures: Precise determination of the strain-shift coefficient of Si bands. *Journal of Applied Physics* 99 (5):053512–6.
Nature milestones: Photons. 2010. *Nature Materials* 9:S5–S20.
Navarro-Urrios, D., A. Pitanti, N. Daldosso et al. 2009. Energy transfer between amorphous Si nanoclusters and Er^{3+} ions in a SiO_2 matrix. *Physical Review B* 79:193312.
Nazarov, A., J. M. Sun, W. Skorupa et al. 2005. Light emission and charge trapping in Er-doped silicon dioxide films containing silicon nanocrystals. *Applied Physics Letters* 86:151914.
Neufeld, E., M. Markmann, A. Vörckel, K. Brunner, and G. Abstreiter. 1999. Optimization of erbium-doped light-emitting diodes by *p*-type counterdoping. *Applied Physics Letters* 75:647–649.
Newman, R. 1955. Visible Light from a Silicon p-n Junction. *Physical Review* 100 (2):700–703.
Newman, R., and W. W. Tyler. 1957. Effect of impurities on free-hole infrared absorption in p-type germanium. *Physical Review* 105:885–886.
Ng, W. L., M. A. Lourenco, R. M. Gwilliam, S. Ledain, G. Shao, and K. P. Homewood. 2001a. An efficient room-temperature silicon-based light-emitting diode. *Nature* 410 (6825):192–194.
Ng, W. L., M. A. Lourenco, R. M. Gwilliam, S. Ledain, G. Shao, and K. P. Homewood. 2001b. An efficient room-temperature silicon-based light-emitting diode (vol 410, pg, 192, 2001). *Nature* 414 (6862):470–470.
Novikov, S. V., J. Sinkkonen, O. Kilpela, and S. V. Gastev. 1997. Visible luminescence from Si/SiO_2 superlattices. *Journal of Vacuum Science & Technology B* 15 (4):1471–1473.
Okada, N., and J. B. Cole. 2010. Simulation of whispering gallery modes in the Mie regime using the nonstandard finite-difference time domain algorithm. *Journal of the Optical Society of America B: Optical Physics* 27:631–639.
O'Mara, W. C., R. B. Herring, and L. P. Hunt. 1990. *Handbook of semiconductor silicon technology*. Noyes Publications, Parkdridge, New Jersey, USA.
Ossicini, S., E. Degoli, F. Iori et al. 2005. Simultaneously B- and P-doped silicon nanoclusters: Formation energies and electronic properties. *Applied Physics Letters* 87 (17):173120–3.
Ossicini, S., F. Iori, E. Degoli et al. 2006. Understanding doping in silicon nanostructures. *Ieee Journal of Selected Topics in Quantum Electronics* 12 (6):1585–1591.
Ossicini, S., L. Pavesi, and E. Priolo. 2003. Light emitting silicon for microphotonics. Springer, New York.
Palm, J., F. Gan, B. Zheng, J. Michel, and L. C. Kimerling. 1996. Electroluminescence of erbium-doped silicon. *Physical Review B* 54:17603.
Patolsky, F., and C. M. Lieber. 2005. Nanowire nanosensors. *Materials Today* 8 (4):20–28.
Paul, D. J. 2004. Si/SiGe heterostructures: from material and physics to devices and circuits. *Semiconductor Science and Technology* 19 (10):R75-R108.
Pavesi, L. 2008. Silicon-based light sources for silicon integrated circuits. *Advances in Optical Technologies* 2008:416926.
Pavesi, L., L. Dal Negro, C. Mazzoleni, G. Franzo, and F. Priolo. 2000. Optical gain in silicon nanocrystals. *Nature* 408 (6811):440–444.

Pavesi, L., S. V. Gaponenko, and L. D. Negro. 2003. *Towards the first silicon laser*: Kluwer Academic Publishers.

Pavesi, L., C. Mazzoleni, A. Tredicucci, and V. Pellegrini. 1995. Controlled photon emission in porous silicon microcavities. *Applied Physics Letters* 67:3280–3282.

Pavesi, L., and R. Turan. 2010. *Silicon nanocrystals: fundamentals, synthesis and applications*: Wiley-VCH.

Pawlak, B. P., N. Q. Vinh, I. N. Yassievich, and T. Gregorkiewicz. 2001. Influence of p-n junction formation at a Si/Si:Er interface on low-temperature excitation of Er^{3+} ions in crystalline silicon. *Physical Review B* 64:132202.

Pelant, I. 2011. Optical gain in silicon nanocrystals: Current status and perspectives. *Physica Status Solidi a-Applications and Materials Science* 208 (3):625–630.

Peng, X., L. Manna, W. Yang et al. 2000. Shape control of CdSe nanocrystals. *Nature* 404:59–61.

Peng, X. G., J. Wickham, and A. P. Alivisatos. 1998. Kinetics of II–VI and III–V colloidal semiconductor nanocrystal growth: "Focusing" of size distributions. *Journal of the American Chemical Society* 120:5343–5344.

Peng, Y. H., C. H. Hsu, C. H. Kuan et al. 2004. The evolution of electroluminescence in Ge quantum-dot diodes with the fold number. *Applied Physics Letters* 85 (25):6107–6109.

Perez-Wurfl, I., X. J. Hao, A. Gentle, D. H. Kim, G. Conibeer, and M. A. Green. 2009. Si nanocrystal p-i-n diodes fabricated on quartz substrates for third generation solar cell applications. *Applied Physics Letters* 95 (15):153506–3.

Pitanti, A., M. Ghulinyan, D. Navarro-Urrios, G. Pucker, and L. Pavesi. 2010b. Probing the spontaneous emission dynamics in Si-nanocrystals-based microdisk resonators. *Physical Review Letters* 104:103901.

Pitanti, A., D. Navarro-Urrios, N. Prtljaga et al. 2010a. Energy transfer mechanism and Auger effect in Er^{3+} coupled silicon nanoparticle samples. *Journal of Applied Physics* 108:053518.

Polman, A. 1997. Erbium implanted thin film photonic materials. *Journal of Applied Physics* 82:1–39.

Pomrenke, G. S., P. B. Klein, and D. W. Langer. 1993. Editors, Proc. Symp. Rare Earth Doped Semiconductors, San Francisco, CA, MRS. Proc., 301.

Portier, X., C. Ternon, F. Gourbilleau, C. Dufour, and R. Rizk. 2003. Anneal temperature dependence of Si/SiO_2 superlattices photoluminescence. *Physica E-Low-Dimensional Systems & Nanostructures* 16 (3–4):439–444.

Prezioso, S., A. Anopchenko, Z. Gaburo et al. 2008. Electrical conduction and electroluminescence in nanocrystalline silicon-based light emitting devices. *Journal of Applied Physics* 104: 063103.

Priolo. F., G. Franzo, S. Coffa et al. 1995. The erbium-impurity interaction and its effects on the 1.54 μm luminescence of Er^{3+} in crystalline silicon. *Journal of Applied Physics* 78:3874–3882.

Priolo, F., G. Franzò, S. Coffa, and A. Carnera. 1998. Excitation and nonradiative deexcitation processes of Er^{3+} in crystalline Si. *Physical Review B* 57:4443–4455.

Priolo, F., G. Franzò, F. Iacona, D. Pacifici, and V. Vinciguerra. 2001a. Excitation and non-radiative de-excitation processes in Er-doped Si nanocrystals. *Materials Science & Engineering, B: Solid-State Materials for Advanced Technology* 81:9–15.

Priolo, F., G. Franzò, D. Pacifici, V. Vinciguerra, F. Iacona, and A. Irrera. 2001b. Role of the energy transfer in the optical properties of undoped and Er-doped interacting Si nanocrystals. *Journal of Applied Physics* 89:264–272.

Priolo, F., C. D. Presti, and G. Franzò. 2006. Carrier-induced quenching processes on the erbium luminescence in silicon nanocluster devices. *Physical Review B* 73:113302.

Prokofiev, A. A., I. N. Yassievich, H. Vrielinck, and T. Gregorkiewicz. 2005. Theoretical modeling of thermally activated luminescence quenching processes in Si:Er. *Physical Review B* 72 (4): 045214-9.

Prucnal, S., J. M. Sun, W. Skorupa, and M. Helm. 2007. Switchable two-color electroluminescence based on a Si metal-oxide-semiconductor structure doped with Eu. *Applied Physics Letters* 90:181121.

Purcell, E. M. 1946. Spontaneous emission probabilities at radio frequencies. *Physical Review* 69:681.

Qin, H., A. W. Holleitner, K. Eberl, and R. H. Blick. 2001. Coherent superposition of photon- and phonon-assisted tunneling in coupled quantum dots. *Physical Review B* 64 (24):241302–4.

Ramos, L. E., E. Degoli, G. Cantele et al. 2007. Structural features and electronic properties of group-III-, group-IV-, and group-V-doped Si nanocrystallites. *Journal of Physics: Condensed Matter* 19 (46):466211–12.
Ramos, L. E., E. Degoli, G. Cantele et al. 2008. Optical absorption spectra of doped and codoped Si nanocrystallites. *Physical Review B* 78 (23):235310–11.
Rayleigh, L. 1910. The problem of the whispering gallery. *Philosophical Magazine* 20:1001–1004.
Reeber, R. R., and K. Wang. 1996. Thermal expansion and lattice parameters of group IV semiconductors. *Materials Chemistry and Physics* 46 (2–3):259–264.
Reed, G. T., and A. P. Knights. 2004. *Silicon photonics: an introduction*: John Wiley & Sons.
Reitzenstein, S., A. Bazhenov, A. Gorbunov et al. 2006. Lasing in high-*Q* quantum-dot micropillar cavities. *Applied Physics Letters* 89:051107.
Reitzenstein, S., and A. Forchel. 2010. Quantum dot micropillars. *Journal of Physics D: Applied Physics* 43:033001.
Robbins, D. J., L. T. Canham, S. J. Barnett, A. D. Pitt, and P. Calcott. 1992. Near-band-gap photoluminescence from pseudomorphic $Si_{1-x}Ge_x$ single layers on silicon. *Journal of Applied Physics* 71 (3):1407–1414.
Rodriguez, J., P. Bianucci, A. Meldrum, and J. G. C. Veinot. 2008. Whispering gallery modes in hollow cylindrical microcavities containing silicon nanocrystals. *Applied Physics Letters* 92:131119.
Rolver, R., B. Berghoff, D. Batzner et al. 2008. Si/SiO_2 multiple quantum wells for all silicon tandem cells: Conductivity and photocurrent measurements. *Thin Solid Films* 516 (20):6763–6766.
Rolver, R., M. Forst, O. Winkler, B. Spangenberg, and H. Kurz. 2006. Influence of excitonic singlet-triplet splitting on the photoluminescence of Si/SiO_2 multiple quantum wells fabricated by remote plasma-enhanced chemical-vapor deposition. *Journal of Vacuum Science & Technology A* 24 (1):141–145.
Rurali, R. 2010. Colloquium: Structural, electronic, and transport properties of silicon nanowires. *Reviews of Modern Physics* 82 (1):427–449.
Saeed, S., D. Timmerman, and T. Gregorkiewicz. 2011. Dynamics and microscopic origin of fast 1.5 μm emission in Er-doped SiO_2 sensitized with Si nanocrystals. *Physical Review B* 83:155323.
Saito, S., Y. Suwa, H. Arimoto et al. 2009. Stimulated emission of near-infrared radiation by current injection into silicon (100) quantum well. *Applied Physics Letters* 95 (24):241101–3.
Saito, S., T. Takahama, K. Tani et al. 2011. Stimulated emission of near-infrared radiation in silicon fin light-emitting diode. *Applied Physics Letters* 98 (26):261104–3.
Scalese, S., G. Franzò, S. Mirabella et al. 2000. Effect of O:Er concentration ratio on the structural, electrical, and optical properties of Si:Er:O layers grown by molecular beam epitaxy. *Journal of Applied Physics* 88:4091–4096.
Schittenhelm, P., C. Engel, F. Findeis et al. 1998. Self-assembled Ge dots: Growth, characterization, ordering, and applications. *Journal of Vacuum Science and Technology B* 16 (3):1575–1581.
Schittenhelm, P., M. Gail, J. Brunner, J. F. Nutzel, and G. Abstreiter. 1995. Photoluminescence study of the crossover from two-dimensional to three-dimensional growth for Ge on Si(100). *Applied Physics Letters* 67 (9):1292–1294.
Schlangenotto, H., H. Maeder, and W. Gerlach. 1974. Temperature dependence of the radiative recombination coefficient in silicon. *Physica Status Solidi A* 21 (1):357–367.
Schmidt, O. G., C. Lange, and K. Eberl. 1999. Photoluminescence study of the initial stages of island formation for Ge pyramids/domes and hut clusters on Si(001). *Applied Physics Letters* 75 (13):1905–1907.
Schmidt, P., R. Berndt, and M. I. Vexler. 2007. Ultraviolet light emission from Si in a scanning tunneling microscope. *Physical Review Letters* 99 (24):246103.
Schubert, E. F. 2006. *Light-emitting diodes*. 2nd ed. Cambridge; New York: Cambridge University Press.
Serna, R., M. Lohmeier, P. M. Zagwijn, E. Vlieg, and A. Polman. 1995. Segregation and trapping of erbium during silicon molecular beam epitaxy. *Applied Physics Letters* 66:1385–1387.
Shainline, J. M., and J. Xu. 2007. Silicon as an emissive optical medium. *Laser & Photonics Reviews* 1 (4):334–348.

Shaklee, K. L., and R. E. Nahory. 1970. Valley-orbit splitting of free excitons? The absorption edge of Si. *Physical Review Letters* 24 (17):942–945.
Shambat, G., S.-L. Cheng, J. Lu, Y. Nishi, and J. Vuckovic. 2010. Direct band Ge photoluminescence near 1.6 μm coupled to Ge-on-Si microdisk resonators. *Applied Physics Letter* 97:241102.
She, J. C., S. Z. Deng, N. S. Xu, R. H. Yao, and J. Chen. 2006. Fabrication of vertically aligned Si nanowires and their application in a gated field emission device. *Applied Physics Letters* 88 (1):013112–3.
Shimizu-Iwayama, T., T. Hama, D. E. Hole, and I. W. Boyd. 2001. Characteristic photoluminescence properties of Si nanocrystals in SiO_2 fabricated by ion implantation and annealing. *Solid-State Electronics* 45 (8):1487–1494.
Shiraki, Y., and A. Sakai. 2005. Fabrication technology of SiGe hetero-structures and their properties. *Surface Science Reports* 59 (7–8):153–207.
Song, D., E.-C. Cho, G. Conibeer, C. Flynn, Y. Huang, and M. A. Green. 2008. Structural, electrical and photovoltaic characterization of Si nanocrystals embedded SiC matrix and Si nanocrystals/c-Si heterojunction devices. *Solar Energy Materials and Solar Cells* 92 (4):474–481.
Soref, R. A. 1993. Silicon-based optoelectronics. *Proceedings of the IEEE* 81 (12):1687–1706.
Spitzer, W. G., F. A. Trumbore, and R. A. Logan. 1961. Properties of heavily doped n-type germanium. *Journal of Applied Physics* 32:1822–1830.
Spooner, M. G., T. M. Walsh, and R. G. Elliman. 2003. Effect of microcavity structures on the photoluminescence of silicon nanocrystals. *MRS Proceedings* 770:1.8.
Srinivasan, K., M. Borselli, O. Painter, A. Stintz, and S. Krishna. 2006. Cavity Q, mode volume, and lasing threshold in small diameter AlGaAs microdisks with embedded quantum dots. *Optics Express* 14:1094–1105.
Stoffel, M., U. Denker, and O. G. Schmidt. 2003. Electroluminescence of self-assembled Ge hut clusters. *Applied Physics Letters* 82 (19):3236–3238.
Sturm, J. C., H. Manoharan, L. C. Lenchyshyn et al. 1991. Well-resolved band-edge photoluminescence of excitons confined in strained $Si_{1-x}Ge_x$ quantum wells. *Physical Review Letters* 66 (10):1362–1365.
Sui, Z., H. H. Burke, and I. P. Herman. 1993. Raman scattering in germanium-silicon alloys under hydrostatic pressure. *Physical Review B* 48 (4):2162–2168.
Sun, J. M., S. Prucnal, W. Skorupa et al. 2006. Electroluminescence properties of the Gd^{3+} ultraviolet luminescent centers in SiO_2 gate oxide layers. *Journal of Applied Physics* 99:103102.
Sun, J. M., W. Skorupa, T. Dekorsy, M. Helm, L. Rebohle, and T. Gebel. 2004. Efficient ultraviolet electroluminescence from a Gd-implanted silicon metal–oxide–semiconductor device. *Applied Physics Letter* 85:3387–3389.
Sun, J. M., W. Skorupa, T. Dekorsy, M. Helm, L. Rebohle, and T. Gebel, 2005. Bright green electroluminescence from Tb^{3+} in silicon metal-oxide-semiconductor devices. *Journal of Applied Physics* 97:123513.
Sun, K., W. J. Xu, B. Zhang, L. P. You, G. Z. Ran, and G. G. Qin. 2008a. Strong enhancement of Er^{3+} 1.54 μm electroluminescence through amorphous Si nanoparticles. *Nanotechnology* 19:105708.
Sun, X., J. F. Liu, L. C. Kimerling, and J. Michel. 2008b. Optical bleaching of thin film Ge on Si. *ECS Transactions* 16:881–889.
Sun, X, J. F. Liu, L. C. Kimerling, and J. Michel. 2009a. Room-temperature direct bandgap electrolumineсence from Ge-on-Si light-emitting diodes. *Optics Letters* 34:1198–1200.
Sun, X., J. F. Liu, L. C. Kimerling, and J. Michel. 2009b. Direct gap photoluminescence of n-type tensile-strained Ge-on-Si. *Applied Physics Letters* 95:011911.
Sun, X., J. F. Liu, L. C. Kimerling, and J. Michel. 2010. Toward a germanium laser for integrated silicon photonics. *IEEE Journal of Selected Topics in Quantum Electronics* 16:124–131.
Sychugov, I., R. Juhasz, J. Valenta, and J. Linnros. 2005. Narrow luminescence linewidth of a silicon quantum dot. *Physical Review Letters* 94:087405.
Sykora, M., L. Mangolini, R. D. Schaller, U. Kortshagen, D. Jurbergs, and V. I. Klimov. 2008. Size-dependent intrinsic radiative decay rates of silicon nanocrystals at large confinement energies. *Physical Review Letters* 100 (6):067401–4.
Taguchi, A., K. Takahei, and J. Nakata. 1993. Electronic properties and their relations to optical properties in rare earth doped III–V semiconductors. *MRS Proceedings* 301:139–150.

Takahashi, Y., T. Furuta, Y. Ono, T. Ishiyama, and M. Tabe. 1995. Photoluminescence from a silicon quantum well formed on separation by implanted oxygen substrate. *Japanese Journal of Applied Physics* 34 (Part 1, No. 2B):950–954.

Takeoka, S., M. Fujii, and S. Hayashi. 2000. Size-dependent photoluminescence from surface-oxidized Si nanocrystals in a weak confinement regime. *Physical Review B* 62 (24):16820–16825.

Takeuchi, S., Y. Shimura, O. Nakatsuka, S. Zaima, M. Ogawa, and A. Sakai. 2008. Growth of highly strain-relaxed $Ge_{1-x}Sn_x$/virtual Ge by a Sn precipitation controlled compositionally step-graded method. *Applied Physics Letters* 93:231916.

Tewary, A., M. J. F. Digonnet, J.-Y. Sung, J. H. Shin, and M. L. Brongersma. 2007. Silicon-nanocrystal-coated silica microsphere thermooptical switch. *IEEE Journal of Selected Topics in Quantum Electronics* 12:1476–1479.

Thao, D. T. X., C. A. J. Ammerlaan, and T. Gregorkiewicz. 2000. Photoluminescence of erbium-doped silicon: Excitation power and temperature dependence. *Journal of Applied Physics* 88:1443–1455.

Thelander, C., P. Agarwal, S. Brongersma et al. 2006. Nanowire-based one-dimensional electronics. *Materials Today* 9 (10):28–35.

Thewalt, M. L. W., D. A. Harrison, C. F. Reinhart, J. A. Wolk, and H. Lafontaine. 1997. Type II band alignment in $Si_{1-x}Ge_x$/Si(001) quantum wells: The ubiquitous type I luminescence results from band bending. *Physical Review Letters* 79 (2):269–272.

Tilly, L. P., P. M. Mooney, J. O. Chu, and F. K. LeGoues. 1995. Near band-edge photoluminescence in relaxed $Si_{1-x}Ge_x$ layers. *Applied Physics Letters* 67 (17):2488–2490.

Timmerman, D., and T. Gregorkiewicz. 2009. Space-separated quantum cutting in differently prepared solid-state dispersions of Si nanocrystals and Er^{3+} ions in SiO_2. *Materials Science & Engineering, B: Solid-State Materials for Advanced Technology* 159–160:87–89.

Timmerman, D., I. Izeddin, P. Stallinga, I. N. Yassievich, and T. Gregorkiewicz. 2008. Space-separated quantum cutting with silicon nanocrystals for photovoltaic applications. *Nature Photonics* 2:105–108.

Tochikiyo, K. M. Fujii, S. Hayashi. 2003. Enhanced optical properties of Si nanocrystals in planar microcavity. *Physica E* 17:451–452.

Trupke, T., M. A. Green, and P. Wurfel. 2003a. Optical gain in materials with indirect transitions. *Journal of Applied Physics* 93 (11):9058–9061.

Trupke, T., J. H. Zhao, A. H. Wang, R. Corkish, and M. A. Green. 2003b. Very efficient light emission from bulk crystalline silicon. *Applied Physics Letters* 82 (18):2996–2998.

Tsimperidis, I., T. Gregorkiewicz, H. H. P. T. Bekman, and C. J. G. M. Langerak. 1998. Direct observation of the two-stage excitation mechanism of Er in Si. *Physical Review Letters* 81:4748–4751.

Tsu, R. 2011. *Superlattice to nanoelectronics*. 2nd ed. Amsterdam Boston: Elsevier.

Tsybeskov, L., and D. J. Lockwood. 2009. Silicon-germanium nanostructures for light emitters and on-chip optical interconnects. *Proceedings of the IEEE* 97 (7):1284–1303.

Tsybeskov, L., K. L. Moore, D. G. Hall, and P. M. Fauchet. 1996. Intrinsic band-edge photoluminescence from silicon clusters at room temperature. *Physical Review B* 54 (12):R8361–R8364.

Tu, H. 2007. High efficient infrared-light emission from silicon LEDs, University of Twente, Enschede, The Netherlands.

Valenta, J., R. Juhasz, and J. Linnros. 2002. Photoluminescence spectroscopy of single silicon quantum dots. *Applied Physics Letters* 80:1070–1072.

Van de Walle, C. G. 1989. Band lineups and deformation potentials in model-solid theory. *Physical Review B* 39:1871–1883.

Van de Walle, G. Chris, and R. M. Martin. 1986. Theoretical calculations of heterojunction discontinuities in the Si/Ge system. *Physical Review B* 34 (8):5621–5634.

van Driel, H. M., A. Elci, J. S. Bessey, and M. O. Scully. 1976. Photoluminescence spectra of germanium at high excitation intensities. *Solid State Communications* 20:837–840.

Vinh, N. Q. 2004a. Optical properties of isoelectronic centers in in crystalline silicon. PhD dissertation. University of Amsterdam, Amsterdam, The Netherlands.

Vinh, N. Q., N. N. Ha, and T. Gregorkiewicz. 2009. Photonic properties of Er-doped crystalline silicon. *Proceedings of the IEEE* 97 (7):1269–1283.

Vinh, N. Q., S. Minissale, H. Vrielinck, and T. Gregorkiewicz. 2007. Concentration of Er^{3+} ions contributing to 1.5 μm emission in Si/Si:Er nanolayers. *Physical Review B* 76: 085339.
Vinh, N. Q., H. Przybylicska, Z. F. Krasil'nik, and T. Gregorkiewicz. 2004b. Optical properties of a single type of optically active center in Si/Si:Er nanostructures. *Physical Review B* 70:115332.
Vrielinck, H., I. Izeddin, V. Y. Ivanov et al. 2005. Erbium doped silicon single and multilayer structures for LED and laser applications. *MRS Proceedings* 866:13.
Wagner, J., A. Compaan, and A. Axmann. 1983. Photoluminescence in heavily doped Si and Ge. *Journal de Physique* 44:61–64.
Wagner, J., G. Contreras, A. Compaan, M. Cardona, and A. Axmann et al. 1984. Germanium extremely heavily doped by ion-implantation and laser annealing: a photoluminescence study. *Materials Research Society Symposium Proceedings* 23:147–152.
Wagner, J., and L. Viña. 1984. Radiative recombination in heavily doped p-type germanium. *Physical Review B* 30:7030–7036.
Walavalkar, S. S., C. E. Hofmann, A. P. Homyk, M. D. Henry, H. A. Atwater, and A. Scherer. 2010. Tunable visible and near-IR emission from sub-10 nm etched single-crystal Si nanopillars. *Nano Letters* 10 (11):4423–4428.
Walters, R. J. 2007. *Silicon Nanocrystals for Silicon Photonics, Engineering and Applied Science*. California Institute of Technology, Pasadena, California, USA.
Walters, R. J., G. I. Bourianoff, and H. A. Atwater. 2005. Field-effect electroluminescence in silicon nanocrystals. *Nature Materials* 4 (2):143–146.
Wang, S., A. Eckau, E. Neufeld, R. Carius, and Ch. Buchalb. 1997. Hot electron impact excitation cross-section of Er^{3+} and electroluminescence from erbium-implanted silicon metal-oxidesemiconductor tunnel diodes. *Applied Physics Letters* 71:2824–2826.
Warga, J., R. Li, S. N. Basu, and L. Dal Negro. 2008. Electroluminescence from silicon-rich nitride/silicon superlattice structures. *Applied Physics Letters* 93 (15):151116–3.
Weber, J., and M. I. Alonso. 1989. Near-band-gap photoluminescence of Si-Ge alloys. *Physical Review B* 40 (8):5683–5693.
Weinstein, L. A. 1969. *Open Resonators and Open Waveguides*. Boulder, CO: The Golem Press.
White, I. M., H. Zhu, J. D. Suter, X. Fan, and M. Zourob. 2009. Label-free detection with the liquid core optical ring resonator sensing platform. *Methods in Molecular Biology* 503:139–165.
Williams, C. J., E. Corbin, M. Jaros, and D. C. Herbert. 1998. Auger recombination in strained Si_xGe_{1-x}/Si superlattices. *Physica B* 254 (3–4):240–248.
Wojdak, M., M. Klik, M. Forcales et al. 2004. Sensitization of Er luminescence by Si nanoclusters. *Physical Review B* 69:233315.
Wolkin, M. V., J. Jorne, P. M. Fauchet, G. Allan, and C. Delerue. 1999. Electronic states and luminescence in porous silicon *q*uantum dots: the role of oxygen. *Physical Review Letters* 82 (1):197–200.
Wu, Y., Y. Cui, L. Huynh, C. J. Barrelet, D. C. Bell, and C. M. Lieber. 2004. Controlled growth and structures of molecular-scale silicon nanowires. *Nano Letters* 4 (3):433–436.
Wu, X. Y., J. S. Kulkarni, G. Collins et al. 2008. Synthesis and electrical and mechanical properties of silicon and germanium nanowires. *Chemistry of Materials* 20 (19):5954–5967.
Würfel, P. 2009. Limits on light emission from silicon. *Chinese Optics Letters* 7 (4):268–270.
Würfel, P., S. Finkbeiner, and E. Daub. 1995. Generalized Planck's radiation law for luminescence via indirect transitions. *Applied Physics A: Materials Science and Processing* 60 (1):67–70.
Xia, Y. N., P. D. Yang, Y. G. Sun et al. 2003. One-dimensional nanostructures: Synthesis, characterization, and applications. *Advanced Materials* 15 (5):353–389.
Xiang, J., W. Lu, Y. Hu, Y. Wu, H. Yan, and C. M. Lieber. 2006. Ge/Si nanowire heterostructures as high-performance field-effect transistors. *Nature* 441 (7092):489–493.
Yassievich, I. N., and L. C. Kimerling. 1993. The mechanisms of electronic excitation of rare-earth impurities in semiconductors. *Semiconductor Science and Technology* 8 (5):718–727.
Yi, J. H. 2008. Silicon rich nitride for silicon based laser devices, Materials Science and Engineering, Massachusetts Institute of Technology.
Yoo, H. G., and P. M. Fauchet. 2008. Dielectric constant reduction in silicon nanostructures. *Physical Review B* 77 (11):115355–5.

Yu, P. Y., and M. Cardona. 2001. *Fundamentals of Semiconductors: Physics and Materials Properties*. 3rd, rev. and enlarged ed. Berlin, New York: Springer, Berlin-Heidelberg-New York.

Zacharias, M., J. Blasing, P. Veit, L. Tsybeskov, K. Hirschman, and P. M. Fauchet. 1999. Thermal crystallization of amorphous Si/SiO_2 superlattices. *Applied Physics Letters* 74 (18):2614–2616.

Zakharov, N. D., P. Werner, G. Gerth, L. Schubert, L. Sokolov, and U. Gosele. 2006. Growth phenomena of Si and Si/Ge nanowires on Si(111) by molecular beam epitaxy. *Journal of Crystal Growth* 290 (1):6–10.

Zhang, Qi, and S. C. Bayliss. 1996. The correlation of dimensionality with emitted wavelength and ordering of freshly produced porous silicon. *Journal of Applied Physics* 79 (3):1351–1356.

Zhang, R.-J., S.-Y. Seo, A. P. Milenin, M. Zacharias, and U. Gosele. 2006. Visible range whispering-gallery mode in microdisk array based on size-controlled Si nanocrystals. *Applied Physics Letters* 88:153120.

Zheng, B., J. Michel, F. Y. C. Ren, L. C. Kimerling, D. C. Jacobson, and J. M. Poate. 1994. Room-temperature sharp line electroluminescence at λ=1.54 μm from an erbium-doped, silicon light-emitting diode. *Applied Physics Letters* 64:2842–2844.

Žídek, K., I. Pelant, F. Trojánek et al. 2011. Ultrafast stimulated emission due to quasidirect transitions in silicon nanocrystals. *Physical Review B* 84 (8):085321–9.

Zhou, Z. Y., L. Brus, and R. Friesner. 2003. Electronic structure and luminescence of 1.1-and 1.4-nm silicon nanocrystals: Oxide shell versus hydrogen passivation. *Nano Letters* 3 (2):163–167.

9

Optical Modulation

Delphine Marris-Morini, Richard A. Soref, David J. Thomson, Graham T. Reed, Rebecca K. Schaevitz, and David A. B. Miller

CONTENTS

9.1 Introduction

Delphine Marris-Morini

The interest in silicon photonics continues to grow, as it is now considered to have potential applications in telecommunication and data communications [1]. One of the key challenges for the development of silicon photonics are high speed, low loss, and compact modulators. Research in this area has shown amazing growth just within the last decade. With the first silicon-based modulator showing 1 GHz bandwidth demonstrated fairly recently in 2004, most of the present results were unthinkable ten years ago [2]. In this chapter, we will discuss the physics and most recent progress of optoelectronic modulators based on silicon. However, before this discussion, we will review how a modulator works and key figures of merit that are important to determine the effectiveness of each type of modulator.

An optical modulator is an optoelectronic device that provides modulated optical signal at the output driven by an electrical command when a continuous input beam is provided at the input (Figure 9.1).

To evaluate the capabilities of silicon modulators to be used in future optical systems, figures of merit have to be defined and compared. The first two are the extinction ratio (ER), which is the ratio between on and off states and the insertion loss (IL). Both figures of merit are given by the following equations, where I_0 is the optical intensity at the modulator input, I_{max} is the maximum intensity of the modulated signal, and I_{min} is the minimum intensity of the modulated signal (Figure 9.1):

$$ER = 10 \times \text{Log}\left(\frac{I_{max}}{I_{min}}\right)$$

$$IL = 10 \times \text{Log}\left(\frac{I_{max}}{I_0}\right)$$

Optical modulation can be obtained either by absorption coefficient variation or by refractive index variation. In the first case, direct modulation of the optical intensity is obtained where ER and IL crucially depend on the absorption coefficient variation and the overlap of the active material with the guided mode. In the second case, an integrated interferometer is required to convert the phase modulation into intensity modulation (Mach–Zehnder, ring resonator, Fabry–Perot cavity, etc.). A Mach–Zehnder interferometry-based (MZI) modulator consists of an input waveguide, a splitter, two waveguide integrated phase shifters, and an output combiner. Maximum intensity (I_{max}) is achieved when there is no phase difference between the two signals from both arms of the MZI, and minimum intensity (I_{min}) is achieved for a phase shift of π radians. To decrease the modulator size, resonant structures can be used, such as a Fabry–Perot cavity or a ring resonator. In both cases, a compact and low-power modulator can be achieved thanks to resonance wavelength shifting by index variation inside the cavity, the drawback being a reduced

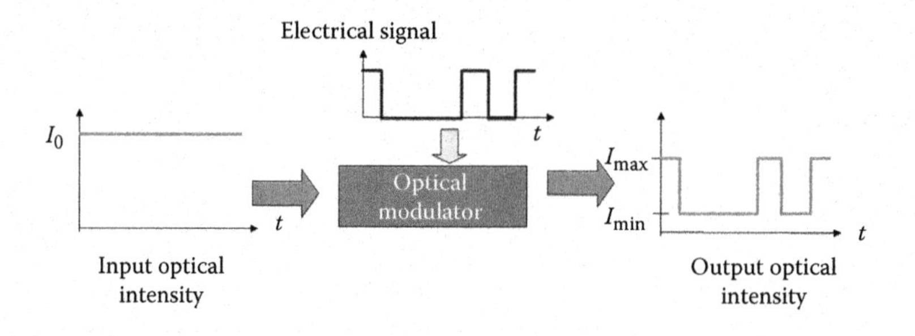

FIGURE 9.1
Optical modulator operating principle, where I_0 is the optical intensity at the modulator input, I_{max} is the maximum intensity of the modulated signal, and I_{min} is the minimum intensity of the modulated signal.

optical bandwidth and large temperature variation sensitivity. Since a Fabry–Perot cavity requires low loss mirrors, which can be difficult to obtain, ring-based modulators are generally preferred for waveguide integration.

For all the refractive index variation-based modulators, the $V_\pi L_\pi$ product can be used to evaluate the performance of the waveguide integrated phase shifters, without taking into account the interferometer performance. $V_\pi L_\pi$ is defined as the product between the length and the voltage required to achieve a π phase shift. As the $V_\pi L_\pi$ figure of merit reduces, the modulator becomes more efficient.

Finally, the latest figure of merit is the modulator speed or the maximum data rate of the output optical signal. The first limitation of the cutoff frequency (i.e., data rate) can come from the physical mechanism used to achieve optical modulation. Following this limitation is the RC time constant, as most optical modulators have an electrical structure that can be defined as a capacitance with a serial resistance. Reducing the RC time constant improves the modulator speed by increasing the bandwidth. Finally, the last speed limitation can come from the radiofrequency (RF) electrical signal propagation especially for long devices (greater than 1 mm typically).

This short analysis of modulator operation shows the breadth of optical modulator research, from physical mechanisms to applied physics and from electronics to integrated optics, making it a very exciting field of research. This chapter will meet expectations of readers who want to understand silicon modulators from the underlying physics to the state of the art results as well as the challenges that still remain for silicon based optical modulators (modulators based on III–V material integrated with silicon will be presented and analyzed by J. Bowers in Chapter 10). This chapter is organized as follows: the physical mechanisms to obtain optical modulation in silicon are presented in Section 9.2 by R. Soref. It will be shown that among the different possibilities, free carrier concentration variation is an efficient way to achieve fast refractive index variations in silicon. The different solutions that have been used to obtain optical modulators by carrier accumulation, injection, or depletion are detailed and compared in Section 9.3 by D. Thomson and G. Reed. Finally, Franz-Keldysh and quantum-confined Stark effects in Ge and SiGe materials recently paved the way for the demonstration of silicon-based electroabsorption modulators (EAMs). These results are presented in Section 9.4 by R. K. Schaevitz and D. A. B. Miller. The challenges that still have to be faced by silicon modulators will be presented in the conclusion.

9.2 Electrorefraction Mechanisms

Richard A. Soref

A variety of important modulation approaches have been developed since the inception of silicon photonics about 26 years ago. Among these techniques, the free carrier plasma dispersion effect (FCPD) and the thermo-optic (TO) effect were the first two to be exploited. The purpose of this section is to review the physical mechanisms that allow practical FCPD and TO modulation in silicon-based waveguided structures and to highlight improvements that have been made in the original FCPD theory. The Si-based Kerr and "Pockels-like" mechanisms were also discussed early on, but those techniques have not taken hold in the modulation industry because the effects are quite small.

In this section, we examine modulators that are controlled by fast, low-power electrical input signals (the information impressed upon the lightwave). There are mechanisms for optically controlled modulation such as the optical Kerr effect (the intensity-dependent index n_2), however we have elected not to discuss "all-optical" modulation here.

The waveguides of interest will be in most cases crystalline silicon "strips" and the wavelengths of interest will be near 1.31 and 1.55 μm, although the wavelength range of interest actually extends from the near infrared into the mid-infrared because group IV photonics is, or soon will be, practical in the mid-infrared as I have commented [3]. In addition to the well-known Si waveguide core, group IV photonic waveguides can be made of Ge [4] or of binary or ternary group IV alloys (SiGe, SiGeC, and SiGeSn) provided that the element or alloy is optically transparent at the desired wavelength. This means that the composition of the SiGe or SiGeC or SiGeSn must be chosen so that the fundamental bandgap of the alloy (whether indirect or direct) is greater than the photon energy. We will look at the physics of inducing changes in the real index Δn of the waveguide core, known as electrorefraction (or as thermorefraction in the TO case). Once the physics is known, the task is then to optimize the spatial overlap between the electromagnetic mode profile and the induced Δn spatial distribution. This is discussed elsewhere in this chapter. We are going to concentrate on bulk crystal effects rather than the quantum-confined responses discussed in later sections. There is experimental literature on FCPD optical modulation using amorphous Si and SiC. For those materials, we shall simply say that the carrier and thermal responses are similar to those of the crystalline case.

We shall look also at modulation due to nonlinear optical (NLO) effects, again using electrically controlled Δn. The second-order effect is known as the Pockels effect. Unfortunately, because the diamond cubic lattice of group IV crystalline alloys and elements has inversion symmetry, this second-order effect vanishes. However, in special cases, a second-order effect can be found. In Si(111), a small second-order response is found near the interface of Si with an SiO_2 layer because of built-in tensile strain and the "broken" Si bonds [5]. A literature report [6] indicates that a strain-inducing dielectric layer deposited upon a Si waveguide gives rise to a Pockels effect of $\chi(2) \cong 15$ pm/V, and quite recently (while this chapter was in press) two research groups in Europe have confirmed this Pockels-effect modulation result in strained silicon. The third situation is one in which a superlattice such as strained-layer superlattice of SiGe and Si is formed. Usually, SiGe is a random alloy. However, during "special kinds of epitaxy," the Si and Ge atoms within SiGe will become ordered in the unit cell. When that is achieved, a second-order NLO modulation capability is created [7, 8]. If a strong dc field is applied along Si(111), a small Pockels effect is induced along certain crystal axes. In summary, the second-order situations listed here—while fascinating—are not yet in the mainstream.

It is worth noting that the infrared phase shifting technique used in many modulators applies also to other kinds of on-chip waveguided components such as tunable infrared filters and optical spatial routing switches (both resonant and nonresonant) described as N × M switches. The TO effect can take on an added role in trimming "as-manufactured" group IV photonic components. In a feedback-control loop, such components can be "stabilized" using thermo-optics.

When assessing the various modulation techniques, we want to know whether it is necessary for the input light to have a linear polarization, and if so, whether the TE or TM mode is best. The answers depend upon (1) the isotropy of the physical mechanism and (2) the waveguide geometry such as rib or strip. Regarding (1), we can say that second-order NLO effects do require a definite polarization. However, the FCPD and TO effects are isotropic and polarization independent. The polarization dependence of the Franz–Keldysh effect

(FKE) is rather weak; thus, the TE and TM modes offer similar response. Regarding (2), the waveguide structure can sometimes favor one polarization.

There are several metrics for modulators. One is B/V, the useable bandwidth per unit voltage applied. Other metrics discussed below compare the depth of modulation to the insertion loss.

9.2.1 The Kerr and Franz–Keldysh Effects

These two modulation techniques were compared in the 1987 paper [9]. Both are actuated by an applied electric field rather than carrier transport. The RF Kerr effect is a third order NLO effect that exists in all group IV materials. The FKE is used typically in a reverse-based Si PIN waveguide diode where the electric field in the intrinsic region serves to tilt the valence and conduction band edge energy diagram along the field direction. This in turn promotes photon-assisted tunneling of electrons between the conduction band and the light hole (LH) and heavy hole (HH) valence bands. Our remarks on FKE in this section will supplement those of R. K. Schaevitz and D. A. B. Miller in Section 9.4.

The FKE can be thought of as a field-induced redshift of the intrinsic absorption spectrum (plus some ripples on the spectral curve) produced by tunneling. Thus FKE gives rise to a strong electro-absorption (EA) effect at photon energies $h\nu$ near the fundamental gap E_g. An FK electrorefraction effect is always present and is related to the EA by the Kramers–Kronig (K-K) relations. Let us examine the electrorefraction of FKE produced by a microwave or "radiofrequency" (RF) modulation signal $E_{rf} \geq 10^4$ V/cm. A physics analysis shows that the approximate field dependence of the Franz–Keldysh device is $\Delta n \propto E_{rf}^{2.5}$, whereas for the Kerr effect $\Delta n \propto E_{rf}^{2.5}$. That calculation suggests that the FKE in Si will always be much stronger than the Si Kerr effect at field strengths of several Volts per micrometer. The Kerr effect has a fairly flat wavelength dependence while FKE peaks at the band-gap wavelength λ_g, which is about 1.07 μm for Si since its indirect gap occurs at $E_{gi} = 1.16$ eV.

To make a quantitative comparison of FKE with Kerr, let us consider an RF field strength of 10^5 V/cm. Then, from the anharmonic oscillator model in [9], the predicted Δn for Si at $\lambda = 1.3$ μm is 10^{-6}, which is a factor of 14 smaller than the Δn of 1.4×10^{-5} predicted in [9] for the FKE in Si at $\lambda = 1.07$ μm. On an absolute basis, this Si-band edge FKE electrorefraction is still small and is even smaller at the 1.55-μm wavelength. (For the popular 1.31- and 1.55-μm telecom wavelengths, a smaller-gap group IV waveguide alloy, especially SiGe, will give the optimum FKE performance as discussed below). However, it is important to note that the EA part of FKE is rather large and thus practical for intensity modulation in a straight piece of waveguide. This beneficial property is seen clearly for the case of crystal germanium in the FKE modulation devices investigated by the MIT group [10]. Their work proves that the CMOS-compatible Ge-waveguided FKE EA 1.55-μm modulator is a fast, low-power, practical, integrable device. A 2011 paper [11] reports a 1600-nm FKE PIN EA modulator in Ge-on-Si achieving 6.5-dB extinction and <100 fJ/bit energy consumption at 3-V bias.

The superiority of Ge has to do with its band structure. In Ge, the 0.80-eV direct gap is only 0.14 eV above the indirect gap and the Ge experiments show that FKE EA modulation is really tied to the direct gap. The change in complex index peaks sharply near the "vertical" 0.80-eV portion of the absorption curve rather than reaching a maximum near the "slanted" 0.66-eV absorption tail.

Taking this "directness-dependent" FKE response a step further, we propose here that the binary group IV alloy GeSn has significant benefits for FKE EA waveguided modulators in the near- to mid-infrared range. (GeSn is really part of the Si-based photonics story). The anticipated strong EA is related to the fact that the band-gap of GeSn waveguide becomes truly direct as the concentration of tin is increased from zero to ~10%. In $Ge_{1-y}Sn_y$, the optimum wavelength for KFE modulation would vary from about 1.7 to 2.2 μm as y goes from zero to 0.1. The author of this section is planning to investigate GeSn FKE modulation on a 2012–2013 research grant. If we look more generally at the picture of FKE EA modulation over the wavelength range of 1.07 to 2.2 μm, we suggest that the ternary alloy of $Si_{1-x-y}Ge_xSn_y$ will—by suitable choice of x and y—yield an optimized figure-of-merit $\Delta\alpha(\lambda)/\alpha_o\lambda$ at a given operation wavelength λ (where $\Delta\alpha$ is the absorption increase with voltage and α_o is the zero-bias initial absorption). For $\lambda \leq 1.55$ μm, the EA modulator material would be essentially $Si_{1-x}Ge_x$. In the EA-dominant devices described here, the associated electrorefraction response is considered "incidental."

Elsewhere in this chapter, the V_π–L_π voltage–length product figure of merit is discussed. We focus here on merit figures for electrorefraction and EA that compare the depth of modulation to the background level of infrared absorption—the tradeoff between "signal and loss." Considering the device at zero bias, the waveguide core has a complex index $n + ik$ where k is the infrared extinction coefficient. The infrared loss coefficient α is related to k by the relation $\alpha = 4\pi k/\lambda$. The merit figure for electrorefraction is $\Delta n/k$, where k is the extinction in either the off or on state, whichever is larger.

We will close this subsection with a comment on the optical Kerr effect. Although the RF effect is decidedly weak as we have seen, the all-optical Kerr effect in group IV materials is rather strong and wavelength-dependent [12], quite useful in four-wave mixing applications.

9.2.2 Thermo-Optical Modulation

Physically, the TO effect is related to the expansion or contraction of the group IV material's atomic lattice when the material's temperature is raised or lowered. This volumetric effect changes the interband transition energies of the solid at various critical points in k-space, band modifications that may be thought of (in approximate way) as band-gap shrinkage, or dilation that modifies (mostly by shifting) the absorption spectrum of the material. Thus at photon energies near the fundamental band-gap, there will be a thermoabsorption effect as well as its important thermorefraction companion, the well-known TO effect—a change in index per unit temperature ($\Delta n/\Delta T$) that is evident and "available" at below-gap photon energies. The physical mechanism is discussed in more detail by G. Ghosh [13] who plots in his figure 3.45 the quantity $2n(\Delta n/\Delta T)$ versus wavelength for Si and Ge in the 1.4- to 14.0-μm wavelength range. The TO response is mostly flat over this λ range and rises slightly near the fundamental gap wavelength. Specifically, the thermo-optic coefficient $\Delta n/\Delta T$ for crystalline Si at 20°C and $\lambda = 1.55$ μm in the $h\nu < E_g$ transparency range is

$$\Delta n/\Delta T\ (\text{Si}) = 1.84 \times 10^{-4}\,^{\circ}\text{C}^{-1}$$

where ΔT is the temperature rise in degrees C. By comparison, the Ge TO coefficient at 20°C and $\lambda = 1.8$ μm is much larger [14]:

$$\Delta n/\Delta T\ (\text{Ge}) = 5.89 \times 10^{-4}\,^{\circ}\text{C}^{-1}$$

The coefficient for SiGe has a size between the Si and the Ge values. Coefficients of similar magnitude hold for amorphous group IV materials.

To make a TO modulator, one can deliberately heat the channel strip, or the TO might be present as an "incidental" side effect of FCPD modulation. By depositing a thin-film resistance-type heater like Cr atop the strip, the joule ΔT produces Δn from TO. Alternatively, within a FCPD injection device, temperature rise in the core will occur due to high levels of injection current. That gives some Δn due to the core's TO response. However, FCPD injection decreases the index while TO increases the index. Thus the two effects are opposed and "attempt" to cancel each other.

The attractiveness of the TO effect stems from its simplicity of modulator construction, ease of integration with other on-chip components, high stability, low cost, low insertion loss, and "moderate" drive power. In addition, the group IV waveguide core materials possess large thermal conductivity and a large TO coefficient. The disadvantages are twofold: the speed is comparatively low (rise and fall times of modulation measured in microseconds or milliseconds), and the drive power comparatively high (several milliwatts). However, the advantages often outweigh the disadvantages, and the literature contains many examples of useful TO modulators, both in Si and Ge; thus, it is fair to say that TO devices have practical applications.

The TO modulator-induced phase shift is not found simply by substituting the bulk Δn into a phase formula. An effective Δn must be put in the formula and Δn_{eff} is found by taking into account the temperature shift of each waveguide layer, the heat capacity, and thermal conductivity of the various layers in 3D, the mode overlap with the spatial Δn distribution, and the heater's optical loading.

The ultimate speed limits have been explored by Michael Geis and coworkers [15] who report: "Mach–Zehnder interferometer thermooptic switches were made using a thin-silicon-on-insulator material system. The switches use single-mode strip-Si waveguides, 0.26×0.4 μm, operating at 1.5 μm. The waveguides were heated directly by passing current through them, resulting in switching power of 6 mW, and a rise time of 0.6 μs. By heating both arms of the interferometer differentially, so one arm is cooling while the other heating, the switching signal input power is reduced to <100 μW. In this differential mode, the switching time can be decreased to 50 ns by pulsing the electrical input to 10 mW. Modeling indicates 10-ns-switching time would require ~23 mW of pulsed power."

Recent experimental results highlight TO modulation-and-switching capabilities. Qiu et al. [16] observed a resonance shift of 0.26 nm/mW of power applied to a silicon dual-microdisk modulator at 1.55 μm. An ultralow-crosstalk 2×2 TO switch was demonstrated in four interconnected Si-waveguided 1.56-μm Mach–Zehnder interferometers [17]. Their total switching power was less than 160 mW. In a $\lambda = 1563$ nm free-space reflective modulator, a combination of thermoabsorption and thermorefraction was employed in a Fabry–Perot resonator comprised of a Ge TO layer on SOI. For 1.2 V applied and 3.6 mW consumed, the IR contrast ratio was 8.7 dB [18].

9.2.3 Free-Carrier Plasma Dispersion Effects

Our early papers laid out for the first time the FCPD effect in Si and Ge as applied to electro-optical (EO) modulation. Optical absorption due to free carriers in the waveguide core arises from conduction- and valence-intraband transitions of electrons and/or holes. Always accompanying that $\Delta\alpha$ change in absorption is a change in index Δn, a linkage given by K–K. Within the material, both $\Delta\alpha$ and Δn are roughly proportional to the change in free electron concentration ΔN_e and/or to the change in free hole concentration ΔN_h. If we then use applied voltages and currents to modify electron and hole concentrations in spatial regions of the waveguide-core cross section, we shall produce electrorefraction and

EA in the core as desired for device operation. Core doping is a useful case for analysis. If the Si or Ge core is doped with *N*- or *P*-type impurity atoms, free electrons, or free holes come from ionized impurities. Therefore, as a convenience, we used in our early papers [9, 19] the literature reports of measured infrared absorption spectra in *N*-doped Si- and *P*-doped Si as spectral "surrogates" for the absorption that would be induced by injecting electrons and holes electrically into Si via a forward-biased PIN diode. The $\Delta\alpha$ *vs* λ for *N* and *P* material were determined by subtracting the α of undoped Si from tha t of doped Si. Having thereby determined the free carrier $\Delta\alpha$ spectra at various ΔN levels, we performed a K–K integration of those spectra to find the Δn spectra. Then, at two fixed wavelengths, we found the ΔN-dependence of Δn using a least squares fit to the K-K data points. That gave us the following semiempirical relationships for Si:

$$\text{At } \lambda = 1.3\ \mu\text{m},\ \Delta n = \Delta n_e + \Delta n_h = -[6.2 \times 10^{-22}\ \Delta N_e + 6.0 \times 10^{-18}\ (\Delta N_h)^{0.8}] \tag{9.1}$$

$$\text{At } \lambda = 1.55\ \mu\text{m},\ \Delta n = \Delta n_e + \Delta n_h = -[8.8 \times 10^{-22}\ \Delta N_e + 8.5 \times 10^{-18}\ (\Delta N_h)^{0.8}] \tag{9.2}$$

where the carrier concentrations are expressed in units of cm^{-3}. Over the years, these relations have been reasonably successful in predicting the performance of experimental silicon photonic devices. We placed these group IV findings within a "III–V context" in a 1986 paper [19], where we compared the above 1.3 μm Δn for Si to the Δn of GaAs and InP within their transparency range. We found the Si Δn to be slightly smaller but generally comparable.

Although Δn grows to several parts-per-thousand when ΔN is increased above 10^{18} cm^{-3}, at the same time $\Delta\alpha$ increases strongly with increasing ΔN. Happily, that $\Delta\alpha$ loss is not significant when ΔN is in the 10^{17}–10^{18}-cm^{-3} range—although the electrorefraction modulator designer must reach a compromise between Δn and $\Delta\alpha$.

The Drude model gives a good approximation of experimental behavior but has its limitations discussed below. To explore group IV beyond Si, we decided to apply the simple Drude model to crystal Si and Ge, and C (diamond) using the equations from our 1994 paper [20]:

$$\Delta n = \frac{-e^2\lambda^2}{8\pi^2 c^2 \varepsilon_0 n}\left(\frac{\Delta N_e}{m^*_{ce}} + \frac{\Delta N_h}{m^*_{ch}}\right) \tag{9.3}$$

$$\Delta k = \frac{e^3\lambda^3}{16\pi^3 c^3 \varepsilon_0 n}\left(\frac{\Delta N_e}{\mu_e (m^*_{ce})^2} + \frac{\Delta N_h}{\mu_h (m^*_{ch})^2}\right) \tag{9.4}$$

with $\Delta k = \lambda\Delta\alpha/4\pi$. In these MKS equations, e is the electron charge, c is the speed of light, $\varepsilon_0 = 8.85 \times 10^{-12}$ F/m, n is the real index of refraction at λ, m^*_{ce}, and m^*_{ch} are the conductivity effective masses and μ_e and μ_h are the electron and hole mobilities at the injected carrier concentrations ΔN_e and ΔN_h. The results for $\Delta N_e = \Delta N_h = 10^{18}$ cm^{-3} are plotted in figure 1 of [20]. Because the mid-infrared is a significant emerging area for group IV photonics, we decided to compare the predicted performance of FC modulators at a particular mid-infrared wavelength, 3.80 μm, with that at 1.55 μm. The results from Equations 9.3 and 9.4 are tabulated in Table 9.1. Some interesting conclusions are drawn from Table 9.1: (1) Diamond is an unexplored but excellent FC material at both near and mid-IR, (2) both

TABLE 9.1

Simple Drude Theory Results for Near- and Mid-Infrared $\Delta n + i\Delta k$ at Induced during Dual Injection of Electrons and Holes (at 10^{18} cm^{-3}) into an Elemental Group IV PIN Waveguide Modulator

Waveguide Core Material	Δn (1.55 μm)	Δn (3.8 μm)	Δk (1.55 μm)	Δk (3.80 μm)
Silicon	2.8×10^{-3}	1.6×10^{-2}	8.0×10^{-5}	1.2×10^{-3}
Germanium	Opaque	4.0×10^{-2}	Opaque	3.0×10^{-3}
Diamond	2.0×10^{-3}	1.1×10^{-2}	1.0×10^{-5}	1.4×10^{-5}

Note: This theory underestimates the experimental Δk.

Si and C are excellent electrorefraction modulator materials at telecom, (3) Ge and Si are excellent mid-infrared modulator materials for devices in which EA is dominant while electrorefraction is "neglected," and (4) diamond has good possibilities for an electrorefraction mid-infrared modulator.

As illustrated in figure 12–15 of [9], a good empirical fit to the experimental $\Delta\alpha$-*vs*-ΔN data points was achieved. However, the simple Drude theory does not adequately fit experiment because that straight line falls short of the points. This discrepancy was discussed recently in a paper by Aashish Singh [21] who pointed out that a scattering-dependent Drude model gives closer agreement between $\Delta\alpha$-theory and experiment over the 10^{17} to 10^{20} cm^{-3} ΔN range as shown in his figure 1. His model takes into account acoustic phonon scattering and ionized impurity scattering. The fit to $\Delta\alpha$ experiment is good except for N-Si at ΔN in the 10^{17} cm^{-3} range.

He argues that the scattering theory is a trustworthy predictor of Δn because of theory's success with $\Delta\alpha$. He used an analytical calculation to fit the various K-K Δn data points and unlike the findings of Equations 9.1 and 9.2 above, he finds that free electrons induce a larger Δn than free holes over the carrier concentration range. However, for free holes, there is a large discrepancy between his scattering theory Δn curve and the K–K Δn data at low ΔN. To determine whether Singh's Δn theory is more accurate than Equations 9.1 and 9.2 in predicting the behavior of actual electrorefraction devices, more experimental evidence is needed. At the moment, the evidence supports Equations 9.1 and 9.2. Regarding the FCPD wavelength dependence in the near- to mid-infrared, Equations 9.3 and 9.4 indicate that both Δn and $\Delta\alpha$ are proportional to λ^2. However, Singh's scattering theory suggests that $\Delta n \propto \lambda^{2.2}$ and $\Delta\alpha \propto \lambda^{3.2}$ when ΔN is 4×10^{17} cm^{-3}.

The versatile FCPD mechanism allows construction of modulators that use the injection, depletion, accumulation, or inversion of charge carriers. If we look into the historical roots of depletion, we find that a 1988 paper proposed a two-gate three-terminal Si waveguided FET modulator having two depletion regions [22]. The first recommendation for a PN junction carrier-depletion modulator was probably in the 1994 paper [20], which said that "Depletion, on the other hand, is faster than injection and requires less drive power. However, in a depletion-type waveguide modulator, we expect that the height of the waveguide (h) will be only 0.1 or 0.2 μm for a waveguide core doping of 2 to 4×10^{17} cm^{-3}. The height limit occurs because the induced modal phase shift is proportional to $1 - \exp(-d/h)$, where d is the height of the depletion layer; thus, we require that $d \sim h$". At that time, I did not know that the Si-strip single-mode channel-waveguide height would turn out to be ~0.25 μm at $\lambda = 1.55$ μm—which in the ensuring years really has allowed practical vertical PN-junction depletion modulators in Si waveguides. The lateral PN and PIPIN junction modulators have also proven to be quite practical. Reverse bias applied to the junction sweeps away the free carriers that were located near the junction at zero bias.

The MOS capacitor modulator is a significant new form of low-power two-terminal FCPD device. A dielectric gate is formed on top of the waveguide or this thin oxide gate layer is embedded in the middle of a group IV waveguide. In the latter geometry, the MOS modulation comprises accumulation of electrons and holes in Si (during bias) at the upper and lower oxide/Si interfaces—although there is a newer approach, MOS depletion of charge discussed below.

In electrorefraction modulators generally, a balance should be struck between electrorefraction and EA. For a depletion-mode device, the largest insertion loss occurs at zero bias where the α_o FC absorption spectra come into play. To limit that loss, the *P*- and *N*-type impurity concentrations are typically held to ~2×10^{17} cm^{-3} in depletion mode. For injection mode, α is highest at full bias. In an electrorefraction injector of length L, the infrared phase shift is $\Delta\phi = 2\pi\Delta nL/\lambda$, while the throughput loss is proportional to $\exp(\Delta\alpha L)$. At the high injection of $\Delta N_e = \Delta N_h = 10^{18}$ cm^{-3} at $\lambda = 1.55$ μm, we find from Equations 9.1 and 9.2 that $\Delta n = \Delta n_e + \Delta n_h = 3 \times 10^{-3}$, implying $\Delta\phi = \pi$ for $L = L\pi = 258$ μm, but there is a problem. The $\Delta\alpha$ obtained from inspecting figures 14 and 15 of [9] gives $\Delta\alpha_e + \Delta\alpha_h = 9$ cm^{-1} + 6 cm^{-1} = 15 cm^{-1}, which is 6.5×10^{-3} dB/μm. In this high injection example, the induced loss is 1.7 dB over the $L\pi$ length.

9.2.4 Recent Approaches to Free Carrier Modulators

To bring this mechanisms discussion up to date, let us review recent developments. Echoing the three-terminal theme mentioned above [22, 23], an SiGe/Si heterojunction bipolar "transistor waveguide" electrorefraction modulator can be designed to exploit free-carrier injection into a 5×10^{17}-cm^{-3} N-doped collector waveguide [24]. The past decade has witnessed tremendous progress in waveguided depletion-based silicon photonic modulators having either vertical or horizontal junction geometries. Innovative designs have been made, and experimental performance has been optimized. (There is voluminous technical literature that could be cited here.) An indication of how important depletion turned out to be in 2011 is given in the article from Intel Corporation that describes the sweep-out of free carriers from the junction region in only 7 ps using their traveling-wave depletion modulator [1].

The first MOS device came in 2004 from Intel. They used a three-layer composite waveguide of P-type poly-silicon upon a thin oxide upon N-type crystal silicon [2]. This buried-oxide approach is the "double MOS accumulator" described above. Their buried-oxide strategy was chosen because the alternative approach—placing the oxide on the top of the waveguide—appeared inferior. The top oxide was seen as producing unacceptably high levels of guided-mode attenuation. The idea was that the mode tail would tunnel through the oxide and touch the metal gate. Cary Gunn and Thomas Koch recently analyzed the buried-oxide device and compared its performance to that of a lateral-PN-junction depletion-type waveguide modulator [25]. They did not consider the insertion loss produced by the polysilicon waveguide portion (a significant issue) but they did examine the modulator length required to attain π-radians of phase shift in one arm of a Si waveguided Mach–Zehnder interferometer (a "non resonant" situation). The breakdown strength of the gate oxide is almost two orders larger than that of Si and this led them to conclude that the active length of the MOS device could be ~10× smaller than that of the junction device. However, the jury is still out on whether depletion devices are inferior to "accumulators." For example, Michael Watts has developed low-energy switching devices in silicon microdisk resonators that use a vertical PN junction to deplete the disk [26]. His 1.55-μm waveguided 4-μm-diameter disks are coupled to Si bus waveguides. He found experimentally a switching energy of only 10 fJ/bit.

We would now like to consider the new "single-sided" MOS depletion modulator, which has future promise. A lightly doped crystal-Si waveguide core has a top-surface oxide film plus a gate electrode, together with a bottom contact to the MOS waveguide capacitor. Sean Anderson and Philippe Fauchet investigated this "unburied" oxide MOS structure in a point-defect resonator formed within a silicon-membrane 2D photonic crystal [27]. To form an electro-optical modulator, the ultrasmall cavity was side-coupled to a line-defect waveguide. They simulated MOS depletion of the high-Q P-Si cavity, and they said: "We rely upon the depletion mode of operation because it allows a larger change in free carrier concentration and also lower capacitance than the accumulation mode. Further, the depletion mode has more favorable optical loss characteristics due to its removal of free carriers from the optically active region. We eliminate the inversion mode from consideration because it relies upon diffusion of carriers through the depletion region, which is inherently a slower process. Additionally, it requires operation at higher gate voltages (above the depletion-inversion threshold), making it less suitable for low-energy operation." For their 2-μm^3 resonators, they predict a switching energy of only 1 fJ/bit, which would be a world record. The MOS depleter is at the frontier of research, and I would suggest that this top-gate approach [25] can be applied to silicon-strip bus waveguides side-coupled to an MOS-depleted doped-silicon microdonut resonator in which the narrow-ring gate electrode and ring-shaped gate oxide film are positioned near the outer sidewall of the donut (which is where the fundamental mode is; the mode does not touch the smaller-diameter inner sidewall). This ultralow-energy modulator would be formed in an SOI wafer. The core doping density, gate dielectric thickness, and waveguide cross section could be engineered to yield high Q and acceptable values of modulation depth, loss, voltage, power, and speed. Until that has been done, the PN junction depletion modulator will remain the industry "workhorse."

9.3 Silicon Modulators Based on Free Carrier Concentration Variations

David J. Thomson and Graham T. Reed

Some of the most successful demonstrations of optical modulators formed in silicon in the previous decade have been based upon the plasma dispersion effect, which relates changes in free electron and hole concentrations to changes in refractive index and absorption. Generally, the change in refractive index is employed to produce phase modulation, which is later converted to intensity modulation in an interferometric or resonant structure. This approach, rather than direct intensity modulation via the change in absorption is preferred since the interaction length required to achieve a large modulation depth in silicon is much larger in the latter case. For example, for a change in electron and hole concentration of 5e18 cm^{-3}, the interaction length for a π radian phase shift is 64 μm, whereas, over the same interaction length, a modulation depth of only 2 dB would result via absorption. Controlling the density of free carriers interacting with the light propagating in the waveguide requires the implementation of an electrical structure. Over the past decade various electrical structures have been demonstrated, however, generally they can be categorized into one of three types: carrier injection, carrier depletion, or carrier accumulation. In the next sections, these three structures will be described together with a nonexhaustive history and description of the state of the art, concentrating mainly on carrier depletion structures where the majority of research effort has been focused in recent years. There are some general design considerations that are common to all three modulator variants and these will be described first. For example, the waveguide geometry and electrical contact positions have a strong influence on device performance. Assuming a fixed waveguide height, dictated by the SOI layer in which the device is fabricated, the waveguide width and slab height can be altered to optimize the optical performance of the waveguide in terms of the loss and modal properties. Varying the waveguide dimensions will also affect the performance of the phase shifter so both need to be considered in parallel when selecting appropriate values. A larger waveguide width generally results in a reduced optical loss due to decreased interaction of the optical mode with sidewall roughness. The mode will also be more confined within the waveguide lessening interaction with the electrical contact regions, which are, in most cases, are positioned in the slab regions on either side of the waveguide rib, thereby reducing loss. The increased confinement will also increase the interaction with the region of varying refractive index, increasing efficiency. The upper limitation of the waveguide width is the point at which higher-order modes are supported, which will degrade the performance of the device. Increasing the thickness of the slab will allow for a wider single mode waveguide, reduce waveguide sidewall interaction and reduce access resistance to the diode benefiting both the optical loss and the modulation bandwidth. On the other hand, the mode will be less confined within the waveguide, which can result in a decreased phase efficiency.

For the case of Mach–Zehnder (MZI)–based carrier depletion and accumulation modulators, the active length is required to be on the order of millimeters, and therefore, traveling waveguide coplanar waveguide electrodes are normally used to drive the device. The electrodes should allow the electrical signal to co-propagate along the device at a similar velocity to the light within the waveguide with minimal loss. The electrodes should also be designed to have an impedance of 50 Ω (when incorporated into a 50-Ω system), including the effects of the phase modulator, to reduce reflections of the RF signal and therefore maximize transfer of the RF signal from the source to the device.

9.3.1 Carrier Injection

Carrier injection structures are generally based upon a PIN diode where the waveguide is formed within the intrinsic region, as depicted in the example diagram of Figure 9.2. When the device is forward biased, electrons and holes are injected into the waveguide region of the device, causing a reduction in refractive index and a change in the phase of the light propagating through the device.

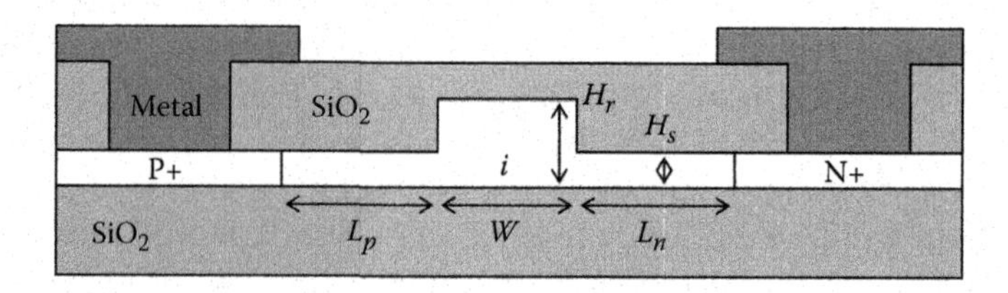

FIGURE 9.2
Example carrier injection PIN diode structure.

In the early years of silicon photonics, modulators were exclusively based upon carrier injection, which is well suited to the multi-micrometer size of waveguides under investigation at that time. Such sized waveguides eased fabrication requirements and ensured efficient coupling from fiber to waveguide. The first plasma dispersion-based device was reported by Soref and Bennett in the 1980s [19]. Their device, shown in Figure 9.3, was based upon carrier injection in a rib waveguide. Localized SiO_2 islands were placed under the waveguide rib to provide a lower boundary for optical confinement while allowing electrical contact to the back of the wafer.

As the years progressed, significant work went into optimizing device performances. It has been shown that the modulation bandwidth, phase efficiency, and optical loss are dictated by the positioning and doping concentrations of the P^+ and N^+ regions as well as the geometry of the waveguide.

For example, in 1994, Tang et al. reported that a 30% increase in the concentration of injected carriers could obtained by changing the angle of the waveguide sidewall from vertical to 54.7°, which also corresponds to the natural etch angle of silicon in KOH, providing convenience for fabrication [28]. The work of Tang et al. also concluded that three terminal devices, which have a contact of one doping type at the top of the waveguide and two of the opposite doping type in the slab on either side of the waveguide, are significantly more efficient that two terminal devices. The concentrations of the highly doped N- and P-type regions are also critical. Simulations performed by Hewitt et al., reported in 2000, concluded that for an example device with two terminals, the drive current could be decreased by a factor of 3 by increasing the N+ and P+ region doping concentrations by an order of magnitude [29]. A marginal reduction in switching time was also reported. The importance of having the highly doped regions close to the waveguide in terms of efficiency and speed was also reported. Any benefits of bringing the electrical contacts closer to the waveguide as well as the inclusion of a third contact within the waveguide must be considered together with the additional optical loss introduced due to the increased interaction of the propagating light with the doped regions. Hewitt et al. also proposed the

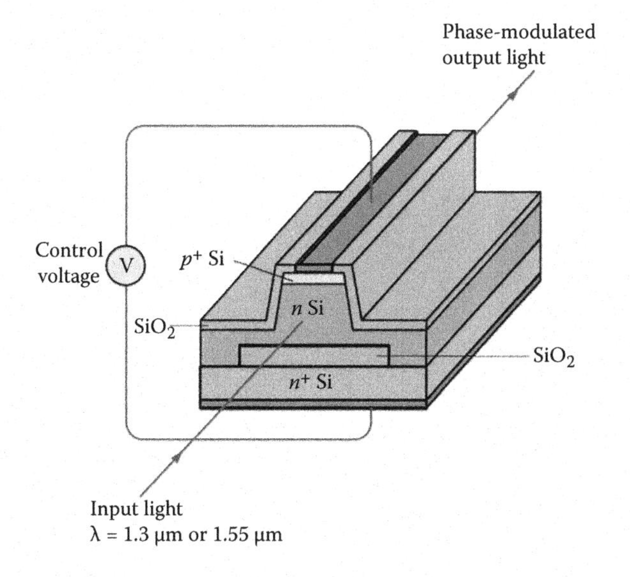

FIGURE 9.3
First plasma dispersion device reported by Soref et al.

use of an electrically insulating trench around the active region of the device to confine the injected free carriers to the waveguide region. The authors predicted that this confinement causes an improvement in the DC and transient performance by up to 74% and 18%, respectively [30]. Another advantage of this confinement is the reduction in crosstalk with adjacent devices. The same conclusion was drawn by Day et al. [31] who argued that to achieve highly efficient operation, it was necessary that all of the injected carriers in the silicon remain between the contacts. The device they proposed had the doped regions throughout the thickness of a thinned slab region therefore preventing the diffusion of free carriers away from the waveguide. The use of a thin insulating layer and defective silicon has also been proposed for the use of lateral free carrier confinement in carrier injection based devices [32, 33]. Scaling down the device dimensions also yields performance improvements. For example, Ang et al. [34] modeled a 500-MHz operation on devices similar yet smaller to those of Hewitt et al. and Tang et al.

Png et al. [35] later improved on this these results using devices of similar geometry but optimized in terms of the N+-doped regions. This device, as shown in Figure 9.4a, was the first device to propose gigahertz modulation in silicon.

Although the high-speed performance of carrier injection devices had progressively improved, ultimately, the modulation bandwidth of this type of device is limited by the minority carrier lifetime, which is relatively long in silicon. Modulation bandwidths generally in the MHz regime therefore result. One approach to overcome the speed limitation posed by the carrier lifetime is to use a pre-emphasis driving signal [35]. This involves shaping the drive signal such that the magnitude of the drive voltage is larger at the leading edge of the pulse as shown in Figure 9.4b. This allows the switching transitions to be driven harder while achieving the required injection concentration at steady state. Although, this can be used to dramatically increase the modulation speed, the complexity of the driving electronics is increased, which can be problematic especially as the modulation rate increases. Several demonstrations of the use of pre-emphasis driving have been published in the literature. For example, in 2007, Green et al. demonstrated 5- and 10-Gbit/s modulation in the PIN diode structure shown in Figure 9.5 along with the drive signal and resultant optical output [36]. Also in 2007, Xu et al. [37] demonstrated 12.5 Gb/s with an extinction ratio of around 9 dB from the ring resonator based carrier injection structure using pre-emphasis.

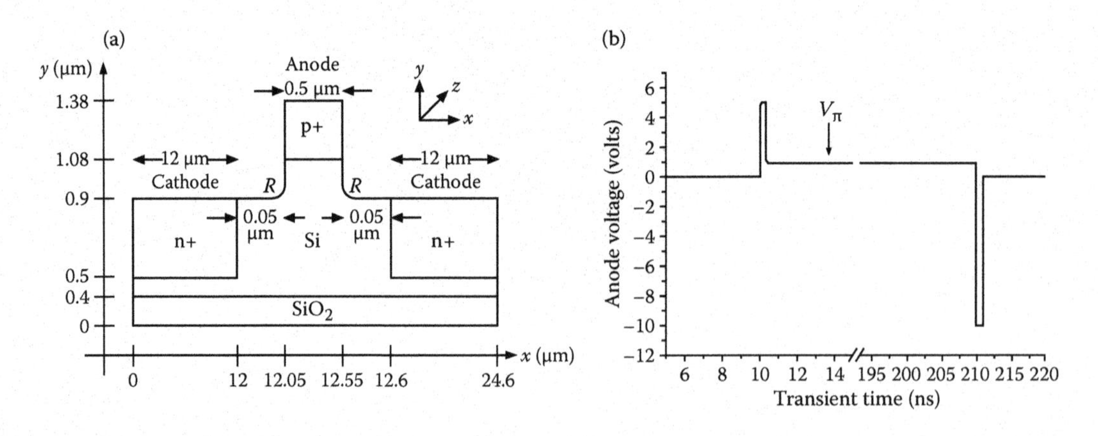

FIGURE 9.4
First proposed gigahertz modulator in silicon (a) and pre-emphasis driving signal (b).

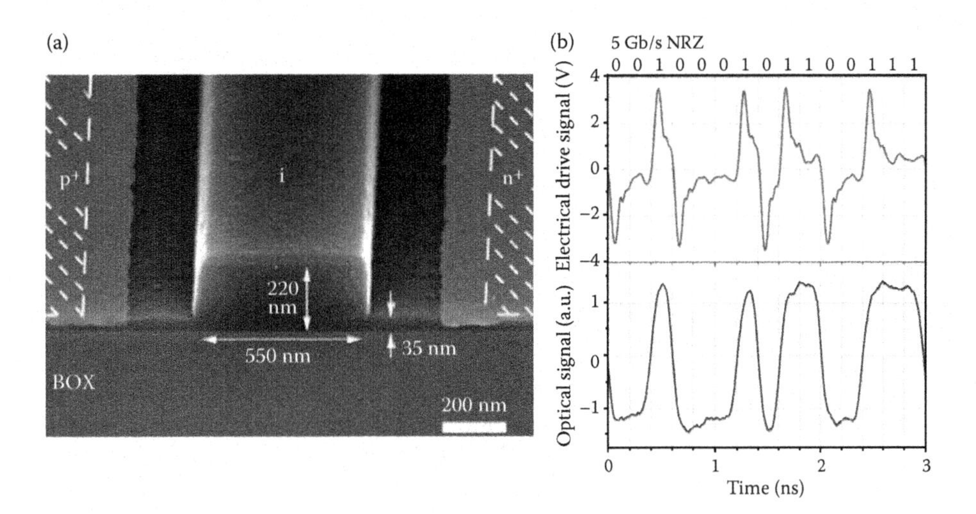

FIGURE 9.5
Annotated SEM image of the PIN diode structure (a) and electrical input/optical output signals (b).

Another approach to improve the modulation bandwidth of carrier injection based devices is to reduce the effective carrier lifetime. In 2009, Preston et al. [38] demonstrated modulation in a polycrystalline silicon based upon carrier injection. The defects (grain boundaries) in polycrystalline silicon effectively reduce the carrier lifetime of the material. Another advantage of implementing the device in polycrystalline silicon is that the concept of realizing photonics on a separate silicon layer to the electronics can be realized in a much more straightforward and low cost manner than if single crystal silicon were required. This saves real estate on the electronics layer providing a greater packing density. Modulation at 2.5 Gbit/s was demonstrated with a 10-dB extinction ratio.

9.3.2 Carrier Accumulation

Another approach to achieve a greater modulation bandwidth is to avoid techniques limited by the minority carrier lifetime. One method is, for example, carrier accumulation. Structures of this type use a thin insulating layer (or barrier) positioned in the waveguide such that when the device is biased, free carriers accumulate on either side of the barrier much like around the dielectric layer in a capacitor. An example structure is shown in Figure 9.6. The accumulation of free carriers again reduces the refractive index of the silicon and causes phase modulation. Since the change in free carrier density is more localized (around the barrier) than in carrier injection structures, they are much more efficient as device dimensions are scaled down. The thickness of the oxide barrier, which is on the order of nanometers, has a strong influence on the performance of the device. A thicker oxide barrier will reduce the capacitance of the device, which is good in terms of the high-speed performance, but reduces the phase efficiency as the carrier density at the barrier is decreased.

In 2004, the first experimental results from a silicon optical modulator exceeding the GHz barrier were reported by the Intel Corporation [2]. This device was based upon carrier accumulation in the structure shown in Figure 9.7a.

As can be seen in Figure 9.7a, the structure is more complex than typical carrier injection structures, which is a disadvantage from a fabrication point of view. One particular

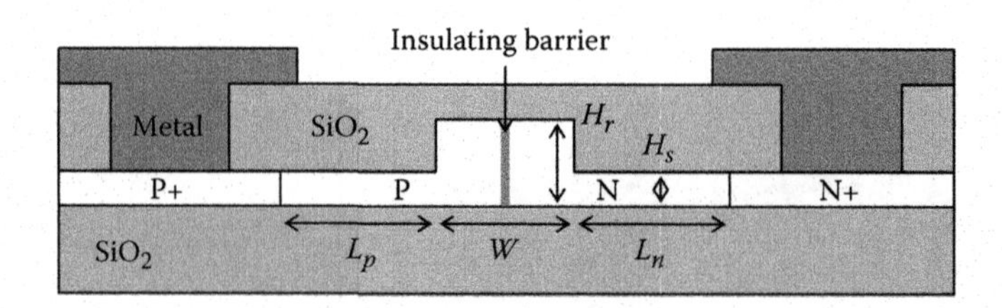

FIGURE 9.6
Cross-sectional diagram of example carrier accumulation.

challenge is forming the insulating layer in the waveguide while forming the entirety of the waveguide structure in single crystal silicon, which has low optical loss and good electrical properties. In the first iteration of Intel's device, polycrystalline silicon was used for the upper portion of the waveguide since the thin oxide barrier also meant that no seed was available to epitaxial growth of single crystal silicon. This was reported to account for most of the device loss. A phase efficiency of 8 V cm and 3-dB bandwidth of approximately 3 GHz was demonstrated. In an improved device, epitaxial lateral overgrowth was used to form the upper portion of the waveguide in single crystal silicon [39]. Device dimensions were also optimized resulting in an efficiency of 3.3 V cm along with data transmission at 10 Gbit/s with 3.8-dB modulation depth. In March 2009, Lightwire reported a different configuration of carrier accumulation device as shown in Figure 9.7b. The modulator operated at 10 Gbit/s with a 9-dB extinction ratio [40]; however, the key feature of this device was the high phase efficiency reported to be 0.2 V cm, permitting the use of a very low operating voltage or a small device footprint.

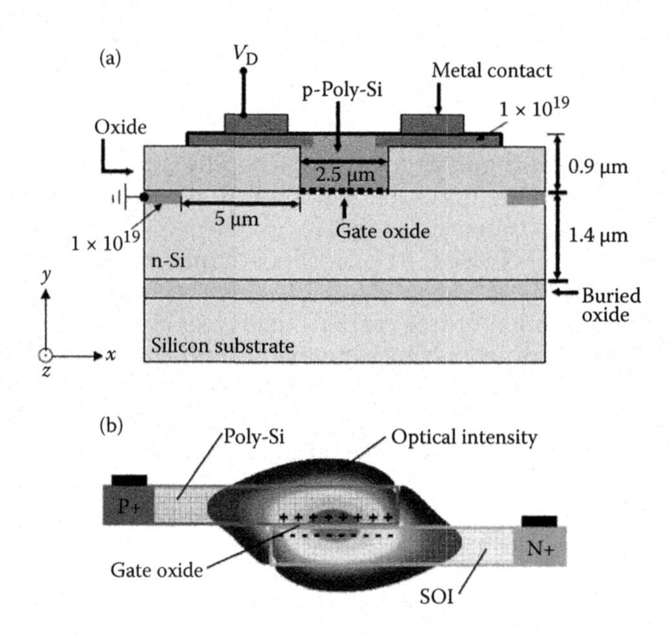

FIGURE 9.7
Cross-sectional diagram of first experimental GHz modulator in silicon (a) and carrier accumulation device by Lightwire Inc. (b).

9.3.3 Carrier Depletion

Carrier depletion is an alternative technique that is not limited by the minority carrier lifetime. Structures that employ this technique are normally based upon reversed biased PN junctions, an example of which is shown in Figure 9.8. The junction of the PN diode is positioned in interaction with the propagating light and as the device is reverse biased, the depletion region widens reducing the density of free carriers within the waveguide. As with accumulation-based devices, the region where the change in free carrier density occurs is localized (at the junction in this case) and, therefore, this technique also lends itself well to implementation in sub-micrometer waveguides. The concentration of active impurities and the positions of the doped regions will largely affect the optical loss, phase efficiency, and modulation bandwidth. For a large modulation bandwidth the access resistance and device capacitance should be low. A low access resistance is achieved with high doping concentrations and close proximity of the highly doped regions (small L_p and L_n). This conflicts with the requirement for low optical loss, and therefore, a tradeoff is required.

In terms of the phase efficiency, the device is highly dependant on the doping concentrations of the P- and N-type regions, as well as the junction position with respect to the waveguide. To obtain a large modulation efficiency, maximal overlap of the optical mode with the region of the device that becomes depleted during the application of a reverse bias is required. Increasing the doping concentrations of the p- and n-type regions will also increase the modulation bandwidth as well as the modulation efficiency at the expense of increased optical loss. The balance of the doping densities in these two regions will also dictate the modulation efficiency of the device. The work of Soref and Bennett [9] concluded that modulation by free holes provides a larger change in refractive index and lower optical absorption as compared with modulation by free electrons. For this reason, the doped regions within the waveguide are often positioned such that there is less interaction with the n type doping. The most successful demonstrations of modulation in the past few years have been based upon this technique. In 2006, carrier depletion was demonstrated experimentally in the all silicon planar waveguide structure shown in Figure 9.9a [41]. The different doped regions were epitaxially grown on an SOI wafer to an overall thickness of 500 nm. A modulation efficiency of 3.1 V cm was measured.

In 2005, Gardes et al. proposed the first rib waveguide based carrier depletion modulator. The layout, as shown in the cross-sectional diagram in Figure 9.9b, consists of a horizontal PN junction in a sub-micrometer size rib waveguide [42]. Modeling data predicted an unprecidented modulation bandwidth of 50 GHz with a phase efficiency of 2.5 V cm. In 2006, Gunn et al. of Luxtera [43] demonstrated waveguide-based modulators in both Mach–Zehnder interferometers (MZI) and ring resonators (RR). The former modulator operated at 10 Gbit/s and the efficiency was reported to be 3.6 V cm. The RR version of the

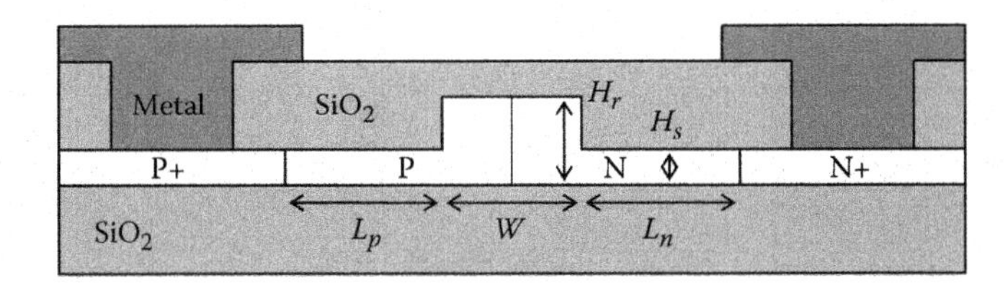

FIGURE 9.8
Cross-sectional diagram of an example PN diode–based carrier depletion modulator.

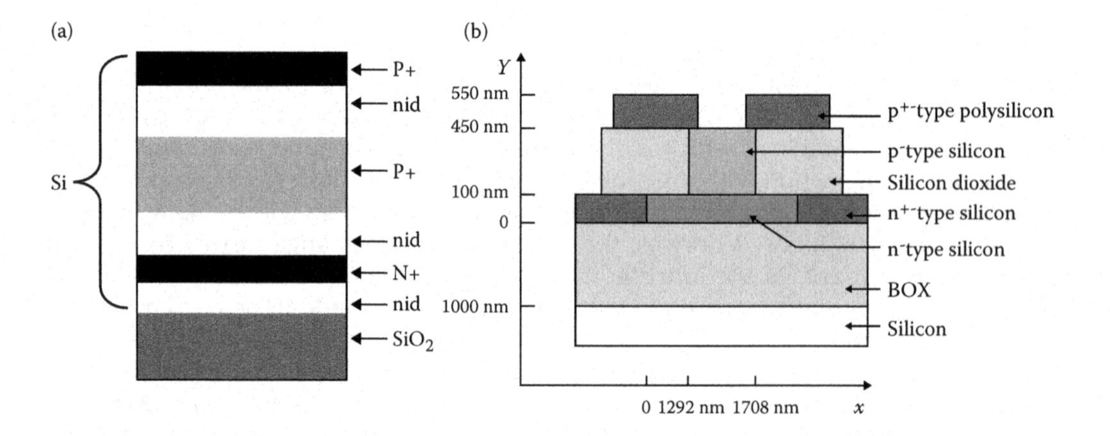

FIGURE 9.9
Experimental planar carrier depletion modulator (a) and first proposed rib waveguide carrier depletion modulator (b).

device also reported 10 Gbit/s, however with significantly reduced footprint compared with the MZI. The ring radius was 30 μm, whereas the MZI footprint was approximately 2 mm^2. Although no details were given in [43] regarding electrical structure, a later publication suggested that modulation was achieved via carrier depletion [44]. In 2007, the Intel Corporation realized a carrier depletion based device [45] similar to the one proposed by Gardes et al. [42]. The cross-sectional diagram of the device is shown in Figure 9.10a. The base and slab of the waveguide are doped p type (1.5e17 cm^{-3}) and the waveguide is capped with an n type region of varying doping of 3e18 cm^{-3} near the cap and 1.5e17 cm^{-3} near the junction. A modulation efficiency of 4 V cm was reported with a −3 dB roll off frequency of ~20 GHz and a data rate of 30 Gbit/s from a 1-mm-long device. Later that year, the authors improved the high-speed performance of the device by surface mounting a termination with reduced impedance (14 Ω) directly onto the modulator electrodes [46]. This yielded 40 Gbit/s modulation as shown in the right hand image of Figure 9.10b, although only 1 dB of modulation depth was achieved, making it impractical for most applications. However, 40 Gbit/s modulation speed was much faster than any other modulator reported at that time, and marked a significant achievement.

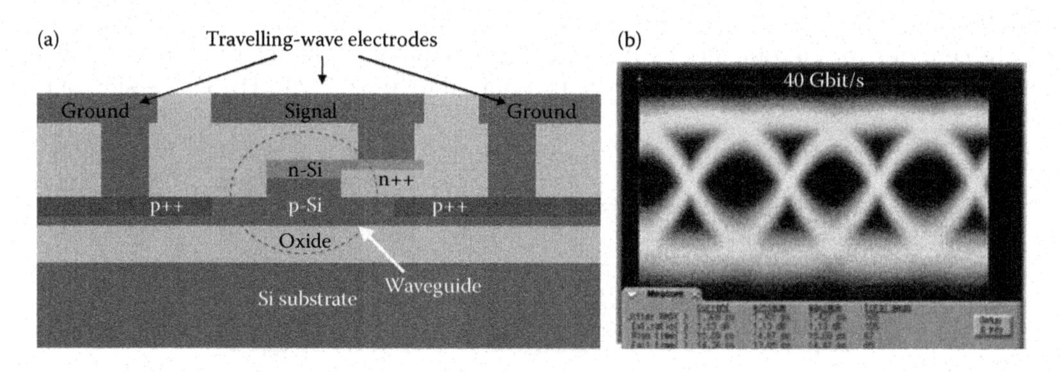

FIGURE 9.10
Cross-sectional diagram of phase shifter (a) and optical eye diagram showing data transmission at 40 Gbit/s (b).

In 2008, Marris-Morini et al. reported a carrier depletion modulator based upon a "PIPIN" diode as shown in Figure 9.11a [47]. The waveguide is undoped other than a p-type strip, fabricated vertically through the center of the waveguide and an n type strip at the waveguide edge. When the diode is reverse biased the p-type strip, which is positioned at the core of the optical mode, becomes depleted. This design allows the device to potentially achieve low optical loss and high phase efficiency simulataneously. A modulation bandwidth of 10 GHz was reported together with a modulation efficiency of 5 V cm and optical insertion loss of 5 dB from a 4-mm-long device. In 2011, Rasigade et al. reported further optimizations of the device, achieving data transmission at 10 Gbit/s (as shown in Figure 9.11b) with a modulation depth of 8.1 dB [48].

Also in 2008, Spector et al. reported the modulator structure shown in Figure 9.12a. Results for both forward bias (carrier injection) and reverse bias (carrier depletion) were reported, demonstrating the speed advantage of carrier depletion modulation, yet the superior modulation efficiency of carrier injection. A low modulation bandwidth of around 100 MHz was achieved in forward bias and around 18 GHz while in reverse bias [49].

From 2008 onwards, a variety of authors from different research groups around the world demonstrated a similar concept carrier depletion modulator where a vertical PN

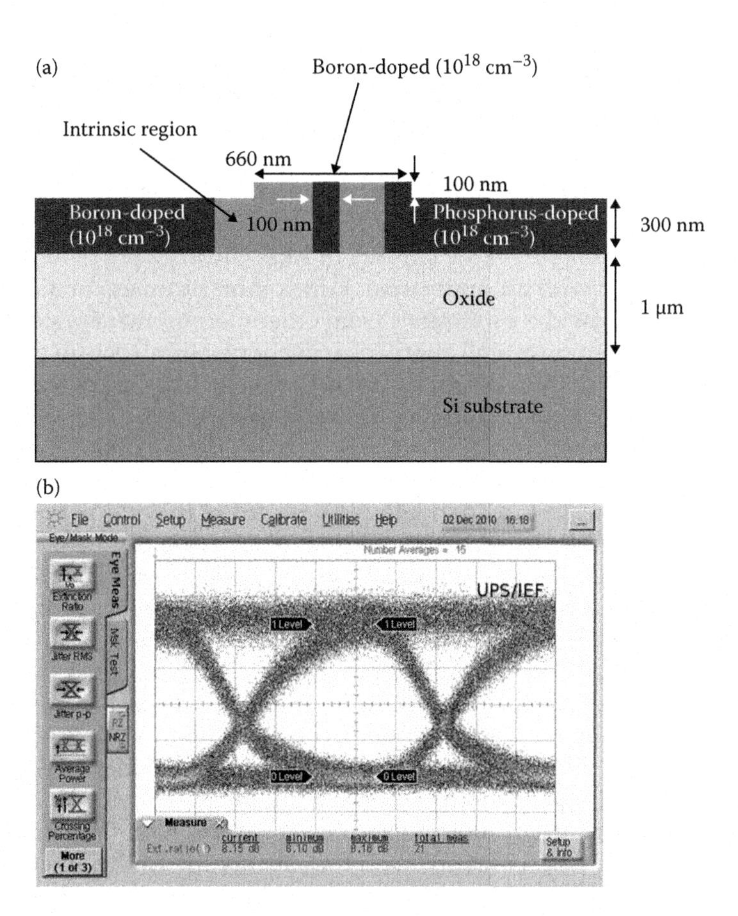

FIGURE 9.11
Cross-sectional diagram of PIPIN diode (a) and modulation optical eye diagram showing 10-Gbit/s modulation (b).

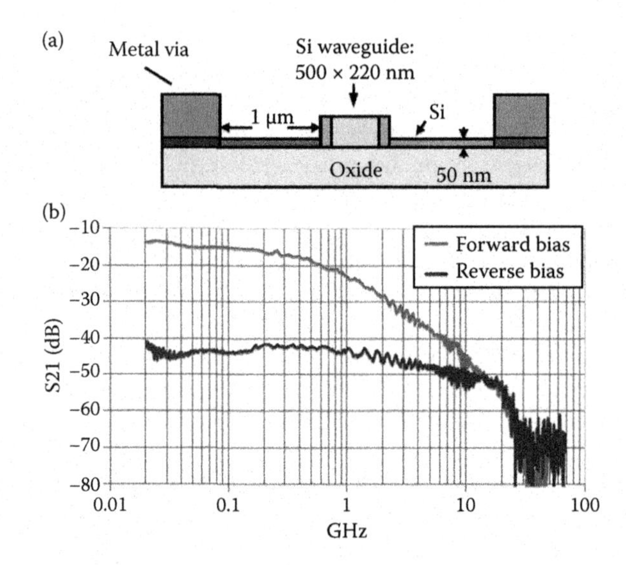

FIGURE 9.12
Cross-sectional diagram of modulator (a) and modulation bandwidth under forward and reverse bias operation (b).

junction is positioned inside the waveguide. Cross-sectional diagrams of seven example devices are shown in Figure 9.13.

In 2008, You et al. reported a PN depletion modulator in a ring resonator structure. A refractive data modulation at 12.5 Gbit/s and an index change of 5.41e−6/V was reported, corresponding to an efficiency of 14.3 V cm [50]. In 2009, Gill et al. [51] demonstrated modulation in a ring resonator with an impressive 3 dB roll off in excess of 35 GHz using the resonator to mitigate bandwidth limitations from other measurement system components. Later in 2009, Park et al. demonstrated another carrier depletion structure with the junction positioned in the middle of the waveguide. Modulation at 12.5 Gbit/s was demonstrated with an extinction ratio of 3 dB [52] and a phase efficiency of 1.8 V cm. In late 2009, Dong et al. reported a carrier depletion modulator placed in a ring resonator. A modulation efficiency of 1.5 V cm was reported with a 10-Gbit/s data rate (with 8 dB extinction ratio when driven with a 2-V RF drive) [53]. Also in late 2010, as part of the ePIXnet program, Gardes et al. demonstrated carrier depletion based modulation in a ring resonator structure with a 3-dB bandwidth of 19 GHz. The device was also based upon a carrier depletion modulator with the junction offset from the middle of the waveguide, to allow propagation mainly in the p-type material (which as previously mentioned is less lossy than n-type) [54]. In early 2010, Tsung-Yang et al. reported another carrier depletion MZI based modulator with vertical PN junction [55]. A modulation efficiency of 2.56 V cm was reported together with 10-Gbit/s data transmission with an extinction ratio of 6.1 dB (5 V RF drive). Finally, Feng et al. reported a carrier depletion modulator in 2010 with an efficiency of 1.4 V cm and data transmission at 12.5 Gbit/s, with an extinction ratio in excess of 7 dB (6 V RF drive) [56]. It can be seen that many optical modulators have been reported with the junction positioned within the waveguide. In terms of fabrication the positioning of the junction in this way can be problematic due to alignment tolerances, which to some extent are an unavoidable feature of lithography process. Performance variations and also a reduction in device yield can result. Within the HELIOS program, a project funded by the European Union,

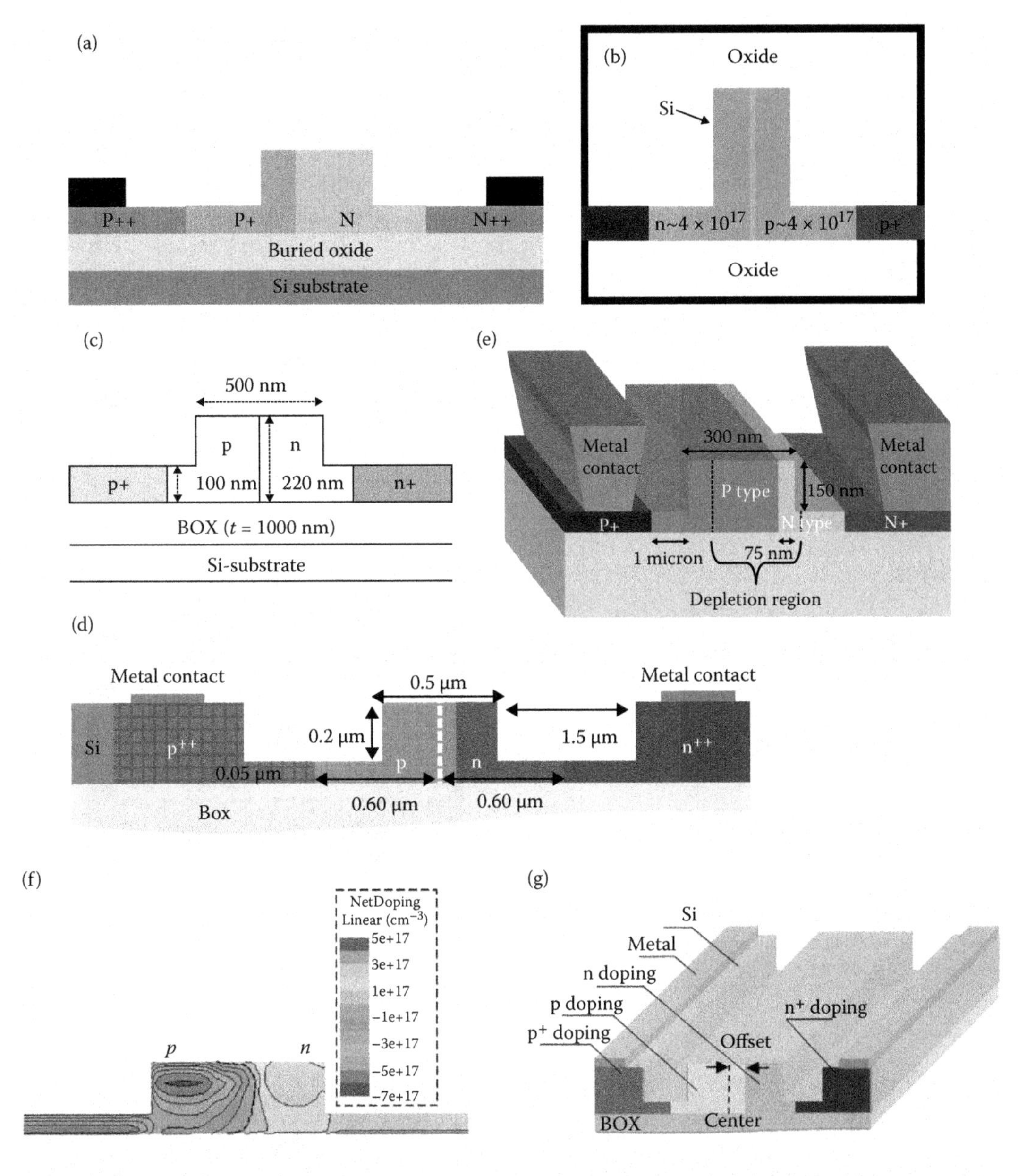

FIGURE 9.13
Carrier depletion modulators with vertical PN junction demonstrated by You et al. (a), Gill et al. (b), Park et al. (c), Dong et al. (d), Gardes et al. (e), Liao et al. (f), and Feng et al. (g). (From J.-B. You, M. Park, J.-W. Park, and G. Kim, *Opt. Express*, 16, 18340–18344, 2008; D. M. Gill, M. Rasras, K.-Y. Tu, Y.-K. Chen, A. E. White, S. S. Patel, D. Carothers, A. Pomerene, R. Kamocsai, C. Hill, and J. Beattie, *IEEE Photon. Technol. Lett.*, 21, 200–220, 2009; J.-B. You, M. Park, J.-W. Park, and G. Kim, *Opt. Express*, 16, 18340–18344, 2008; P. Dong, S. Liao, D. Feng, H. Liang, D. Zheng, R. Shafiiha, C.-C. Kung, W. Qian, G. Li, X. Zheng, A. V. Krishnamoorthy, and M. Asghari, *Opt. Express*, 17, 22484–22490, 2009; F. Y. Gardes, A. Brimont, P. Sanchis, G. Rasigade, D. Marris-Morini, L. O'Faolain, F. Dong, J.-M. Fedeli, P. Dumon, L. Vivien, T. F. Krauss, G. T. Reed, and J. Martí, *Opt. Express*, 17, 21986–21991, 2009; L. Liao, A. Liu, D. Rubin, J. Basak, Y. Chetrit, H. Nguyen, R. Cohen, N. Izhaky, and M. Paniccia, *Electron. Lett.*, 43, 1196–7, 2007; N.-N. Feng, S. Liao, D. Feng, P. Dong, D. Zheng, H. Liang, R. Shafiiha, G. Li, J. E. Cunningham, A. V. Krishnamoorthy, and M. Asghari *Opt. Express*, 18, 7994–7799, 2010. With permission.)

Thomson et al. have developed a carrier depletion modulator that can be formed using a self aligned fabrication process, thus avoiding this issue. In 2010, the authors reported 10-Gbit/s modulation with 6-dB extinction ratio and phase efficiency of 6 V cm [57]. In 2011, they reported improved results demonstrating 40-Gbit/s modulation with a 10-dB extinction ratio as well as an efficiency of 2.7 V cm [58]. A cross-sectional diagram of the device and 40-Gbit/s eye diagram are shown in Figure 9.14a and b, respectively.

Also in 2011, Gardes et al. demonstrated 40-Gbit/s modulation for both TE and TM polarizations from the device shown in Figure 9.15a [59]. The device can also be formed with a self-aligned process employing angled ion implantation steps to locate the doped regions on the sidewalls.

Clearly, tremendous progress has been made in silicon modulators based upon the plasma dispersion effect since the first demonstration over 25 years ago. In particular, over the previous decade device performance have improved dramatically with several modulation demonstrations at 10 Gbit/s and even 40 Gbit/s in silicon being reported recently. Consequently, the optical modulator is a significant success within the field of silicon

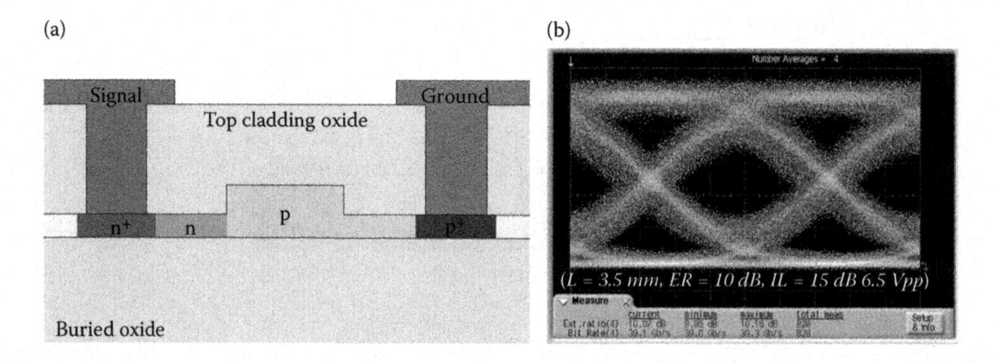

FIGURE 9.14
Cross-sectional diagram of carrier depletion optical modulator (a) and optical eye diagram demonstrating a data rate of 40 Gbit/s (b).

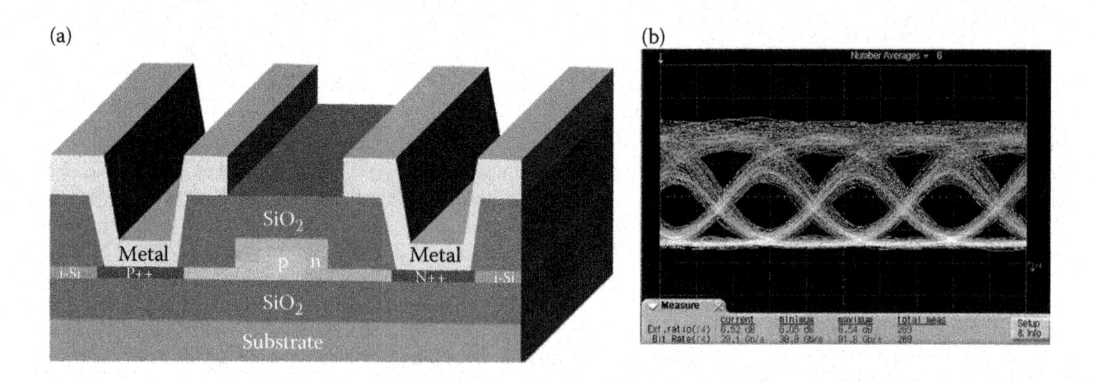

FIGURE 9.15
Cross-sectional diagram of carrier depletion optical modulator (a) and optical eye diagram demonstrating a data rate of 40 Gbit/s (b).

photonics, and numerous researchers will continue to attempt to improve these devices further.

9.4 Electroabsorption Modulators

Rebecca K. Schaevitz and David A. B. Miller

EAMs are a class of modulators that employ an electric field-dependent change in absorption to modulate the intensity of light (in contrast to the electrorefraction based devices discussed earlier in this chapter). There are two main mechanisms used in EAMs: the Franz–Keldysh effect (FKE) [60, 61] and the quantum-confined Stark effect (QCSE) [62]. FKE is a bulk material mechanism while QCSE is the quantum-confined version of FKE [63]. Although these effects are mainly seen in direct band gap semiconductors, such as GaAs or InP, recent work has shown their effectiveness in germanium [10, 11, 64–70]. Thus, Ge-based FKE and QCSE provide a platform for silicon-compatible high-speed, low-power, compact EAMs to alleviate the interconnect bottleneck [71, 72].

To understand FKE and QCSE in germanium, we will begin this section with an explanation of FKE and QCSE followed by a brief description of the Ge and SiGe material platform. We will conclude with proof-of-concept and best EAMs to-date using each of these mechanisms.

9.4.1 The Franz–Keldysh Effect and the Quantum-Confined Stark Effect

The Franz–Keldysh effect (FKE) and the quantum-confined Stark effect (QCSE) are very similar mechanisms seen in direct band gap semiconductor material. FKE is a bulk semiconductor material mechanism, and in its simplistic form (without excitonic contributions), it can be viewed as the tunneling of the electron and hole wavefunctions into the band gap with the application of electric field, allowing for absorption below the band edge. As electric field increases, the electron and hole wavefunction penetrate further into the energy band gap, allowing overlap between electron and hole wavefunctions even for photon energies below the band gap energy, hence shifting the absorption "tail" to lower energy (or longer wavelength) [73]. Figure 9.16 shows a cartoon depiction of the electron and hole wavefunctions tunneling into the band gap as well as the resulting electroabsorption spectrum one would expect without excitonic enhancement [74]. While FKE exhibits excitonic effects, they are negligible in the typical operating wavelengths (at energies far below the band gap) and electric field strengths [74].

Similar to FKE, QCSE is a shift toward lower energy absorption (longer wavelengths) with the application of electric field. When two direct band gap semiconductor materials (well and barrier) form a type I heterostructure, the electron and holes become confined within the well material. Type I confinement is where the "lowest" energy of both the electrons in the conduction band and holes (absences of electrons) in the valence bands lie in the same material in the heterostructure. Figure 9.17 shows a common quantum well material system, GaAs/AlGaAs, with type I confinement of the electrons and holes ("lowest" energy for holes corresponds to the "highest" energy in the valence band when viewed on such electron energy diagrams—holes can be viewed as analogous to bubbles that want to rise to the top of a liquid).

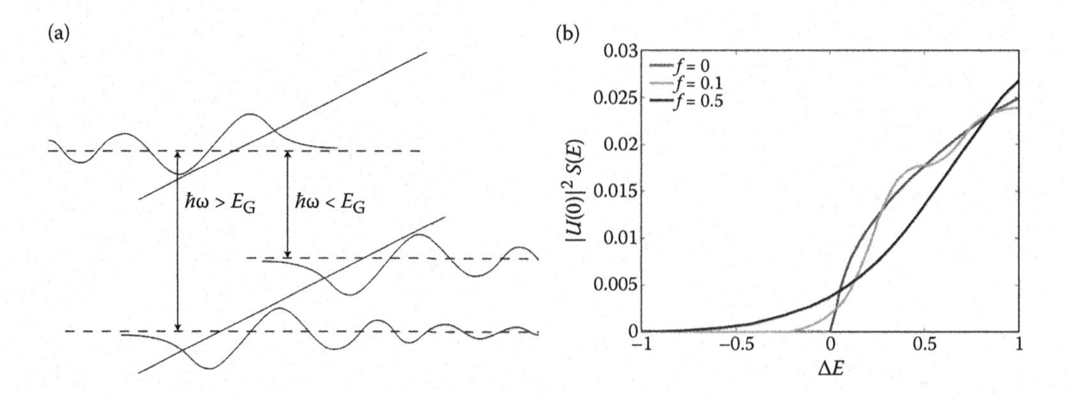

FIGURE 9.16
(a) Cartoon depiction of the characteristic airy functions used to describe the electron and hole wavefunctions. Two absorption regions are depicted: absorption below and above the band gap, where $\hbar\omega$ is the photon energy and E_G is the band gap energy. (b) Typical electroabsorption profile of the simplistic Franz–Keldysh model neglecting excitonic effects, where f is the electric field strength (in rydbergs per bohr radius) and E is the normalized band gap. (Based on calculations from H. I. Ralph, *J. Phys. C*, 1, 378–386, 1968. With permission.)

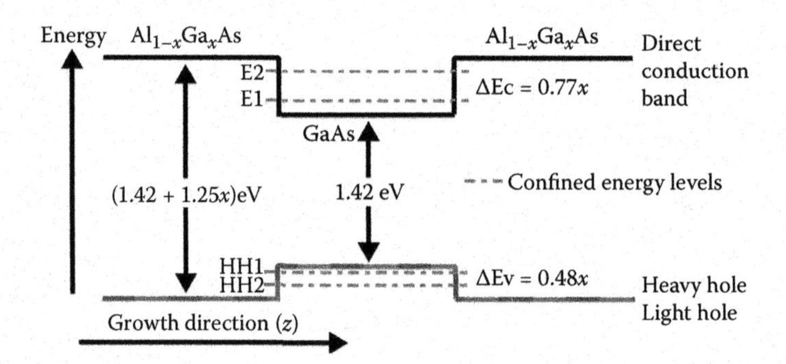

FIGURE 9.17
Typical quantum well material system of GaAs wells and AlGaAs barriers showing type I confinement, where the lowest energy of both the conduction and hole bands are in one material. (After W. R. Frensley, in *Heterostructures and Quantum Devices*, ed. N. G. Einspruch, and W. R. Frensley. San Diego, CA: Academic Press, 1994, pp. 1–24. With permission.)

The confinement of the electrons and holes results in a discretization of the density of states above the band edge (as shown in Figure 9.17 with the labels E1 and E2 for the first and second electron sub-band levels, respectively, and HH1 and HH2 for the first and second heavy hole sub-band levels, respectively). Additionally, the confinement enhances the Coulomb interaction between the electrons and holes by pushing them closer together, leading to stronger excitonic effects in the electroabsorption spectra. (Excitonic effects show up most obviously as absorption peaks near the separation energies between bands or sub-bands.) With the applied electric field, the electron and holes shift to lower energies with only a slight decrease in excitonic strength. Figure 9.18 shows the cartoon depiction of the band structure with the electron and hole wavefunctions under the influence of an electric field as well as sample electroabsorption spectra [76].

Polarization can strongly affect the FKE and QCSE electroabsorption profiles. For FKE, such polarization dependence can arise in strained bulk material; in contrast, polarization

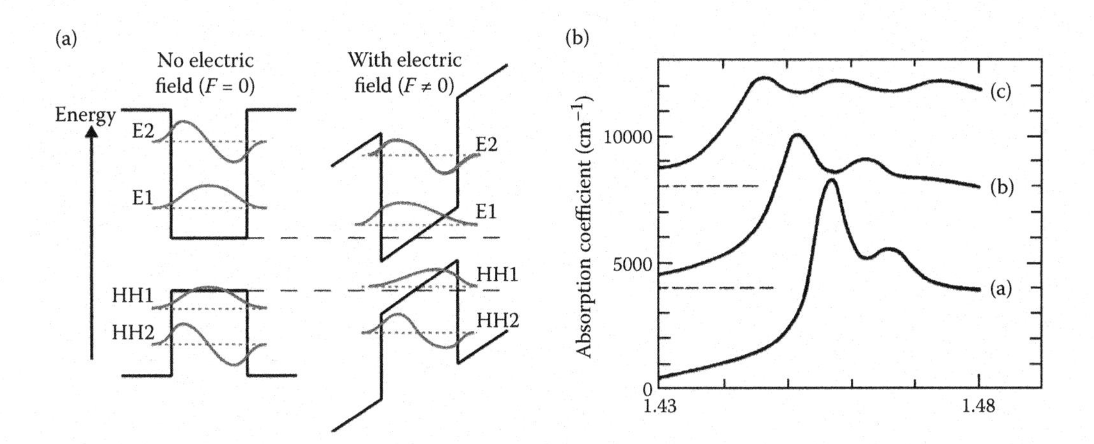

FIGURE 9.18
(a) Cartoon depiction of the influence of electric field on the confined electron (E1, E2) and heavy hole (HH1, HH2) wavefunctions in a heterostructure material system. (b) Electroabsorption profiles at various electric field strengths for TE polarized light. (From D. A. B. Miller, D. S. Chemla, T. C. Damen, A. C. Gossard, W. Wiegmann, T. H. Wood, and C. A. Burrus, *Phys. Rev. B*, 32, 2, 1043–1060, 1985. With permission.)

dependence is always present in QCSE spectra because the layered structure strongly breaks the symmetry of the approximately cubic crystal structure [77]. With polarization dependence, we see different absorption spectra for light polarized along the x- or y-directions (TE) versus the z-direction (TM) (if we consider the growth direction to be oriented along the z-axis). These differences are due to band selection rules from the conduction band to the light and heavy hole bands [77]. For TE polarized light, the electron to heavy hole (E-HH) transition is three times stronger than the electron to light hole (E-LH) transition. For TM polarized light, there is no E-HH contribution and the E-LH absorption is four times stronger than in the TE polarization. These polarization effects can play a significant role when integrating EAMs into waveguide optoelectronic devices.

With the change in absorption versus electric field, both FKE and QCSE are viable mechanisms to design EAMs for optical interconnect systems. However, to integrate these devices on silicon, it is ideal to use CMOS-compatible materials. Germanium is already incorporated in CMOS manufacture, so is already compatible. Being nearly a direct band gap material, it has demonstrated both FKE [78] and QCSE [65] behavior near its direct band gap. In the next section, we will describe the germanium material system used for FKE and QCSE EAMs.

9.4.2 Germanium Material System

To get FKE and QCSE in germanium, pure crystalline Ge or Ge-rich SiGe can be formed on a silicon substrate using a variety of growth techniques, although the most common presently is chemical vapor deposition (CVD). Germanium and silicon have approximately a 4% lattice mismatch, a difference so large that crystal defects readily form when trying to grow Ge on Si. As a result, growth techniques rely heavily on reducing defect density in the active device structure. To reduce defects, two techniques are typically used for growth: (a) a thin virtual substrate that is grown with cyclic anneals to prevent defects from propagating [10, 11, 64–68, 79] or (b) a thick-graded SiGe virtual substrate is grown in which the Ge content is increased by ~7%/μm to reach the desired composition [69, 70]. For FKE growths, the entire epitaxy of a single composition is grown prior to cyclic annealing to reduce the defect concentration [10, 11, 64, 79]. For QCSE, both growth techniques are used to form the virtual substrate prior to quantum well (QW) growth [65–70], although the strain-free thick graded growth is likely too thick to be very practical for forming active optoelectronic devices. For growth of the thin virtual substrate, pure Ge or Ge-rich SiGe is grown to thicknesses ~100 nm to several hundreds of nanometers, far beyond the critical thickness (which is usually just a few nanometers), causing defects to form. Then an anneal is performed to reduce the defect density. More than one such layer growth and annealing may be performed sequentially. Then quantum wells and/or other device layers may be grown, and the material is then cooled to room temperature for later device fabrication. This cooling process leads to thermally induced tensile strain (0.05% to 0.25%) of the epitaxial material since Ge has a larger thermal expansion coefficient than Si [64, 80]. This biaxial strain increases the in-plane lattice constant and distorts the electronic band structure (which we will discuss later).

While FKE devices use only bulk Ge or Ge-rich SiGe, and thus only have the residual thermally induced tensile strain, QCSE devices have a periodic structure of Ge/SiGe (well/barrier) grown on top of the tensile-strained "virtual substrate." Because the virtual substrate has a lattice constant much closer to the natural lattice constants of Ge and the SiGe barrier materials, the grown layers are pseudomorphic (i.e., having the same lattice constant as the virtual substrate) without forming additional defects, but will have some additional strain due to the remaining lattice mismatch between grown Ge and SiGe layers and the virtual substrate. The lattice mismatch induces compressive strain in the Ge well, while the SiGe barrier (which has more Si content then the "virtual substrate") gains tensile strain. For QCSE, doping the initial virtual substrate with boron and capping the growth with an n-type layer using dopants such as arsenic or phosphorus forms a PIN diode. A vertical PIN diode is necessary, as QCSE is only visible when the electric field is applied perpendicular to the plane of the quantum wells. For FKE, formation of the PIN diode can either occur in situ to create a vertical PIN diode or following growth using dopant implantation of the sidewalls. Figure 9.19 gives examples of a lateral PIN diode for FKE and a vertical PIN diode for QCSE.

For both FKE and QCSE structures, strain plays a significant role in the resulting electronic band structure and therefore the electroabsorption spectra. The strain present in

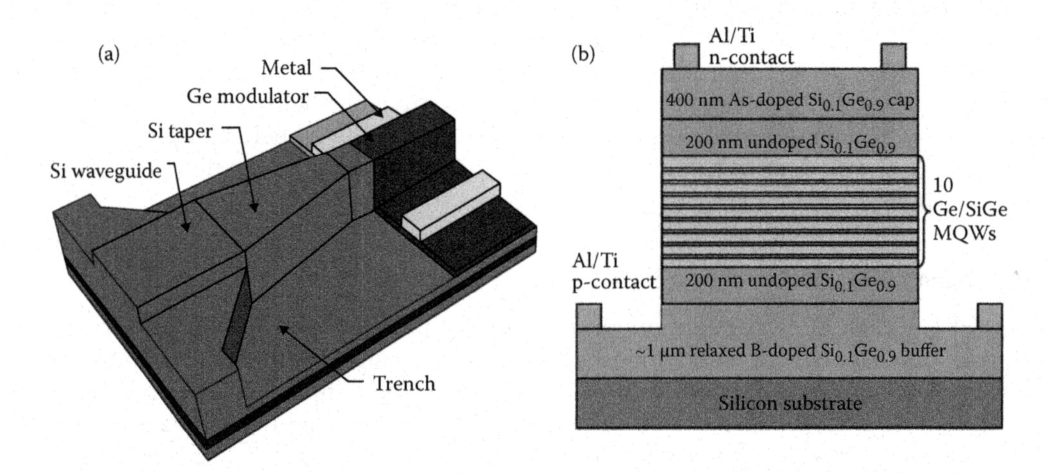

FIGURE 9.19
Material structures forming PIN diodes for (a) FKE [11] and (b) QCSE EAMs [67]. (From N.-N. Feng, D. Feng, S. Liao, X. Wang, P. Dong, H. Liang, C.-C. Kung, W. Qian, J. Fong, R. Shafiiha, Y. Luo, J. Cunningham, A. V. Krishnamoorthy, and M. Asghari, *Opt. Express*, 19, 8, 7062–7067, 2011; R. K. Schaevitz, J. E. Roth, S. Ren, O. Fidaner, and D. A. B. Miller, *IEEE J. Select. Top. Quant. Electron.*, 14, 4, 1082–1089, 2008. With permission.)

FKE devices is thermally induced by differential thermal contraction when cooling from the growth temperature, while for QCSE, the strain is a result of both thermal effects and lattice mismatch. Independent of the source of strain, the effects on the band structure are equivalent. For tensile strain, the overall direct band gap decreases and the degeneracy of the light and heavy hole bands lifts with the light hole shifting up and the heavy hole shifting down in energy. Compressive strain results in the opposite effect by increasing the direct band gap and shifting the light hole down and the heavy hole up. Not only are the overall energies affected by strain but also the general shape of the band structure. The effect is strongest for the valence bands, where the masses and density of states can change greatly. Figure 9.20 shows a cartoon depiction of the band structures of Ge and Ge-rich SiGe with and without compressive and tensile strain, and Figure 9.21 shows the band lineups of a typical Ge/SiGe heterostructure on a relaxed SiGe substrate [67].

While we now understand how strain affects the band structure and the overall material design for FKE and QCSE EAMs, we have not yet discussed all the absorption mechanisms present in Ge-based devices. To do so, we must first understand the complete band structures of both Si and Ge. While Ge-based FKE and QCSE devices use the direct band gaps as previously discussed, both Si and Ge materials are nominally indirect. The direct band gaps of Si and Ge, where the maxima of the valence bands and minimum of the conduction band are at the same (zero) momentum, are 4.0 and 0.8 eV, respectively [81, 82]. However, at lower energies are the "indirect" band gaps in both Si and Ge where the maxima of the valence bands and the minimum of the conduction band lie at different momenta. The minimum band gap in Ge is 0.66 eV [83] at the L-valley minimum (along the <111> direction of the Brillouin zone), which is ~140 meV lower than its direct band gap. Silicon has a much larger difference between its minimum band gap of 1.12 eV [84] at the Δ-valley (along the <100> direction) and its direct band gap of 4.0 eV. Additionally, Si

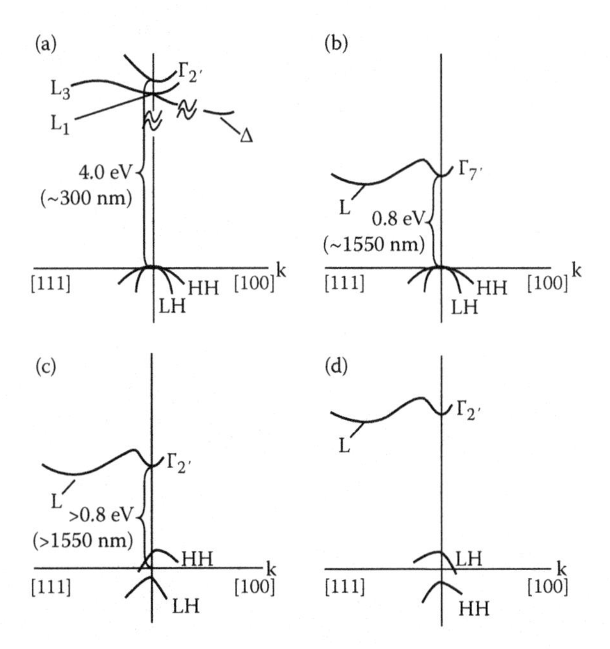

FIGURE 9.20
Cartoon depiction of the bulk unstrained band structure of (a) Si, (b) Ge as well as the shift with (c) compressive, and (d) tensile strain on the Ge and SiGe materials, respectively. (From R. K. Schaevitz, J. E. Roth, S. Ren, O. Fidaner, and D. A. B. Miller, *IEEE J. Select. Top. Quant. Electron.*, 14, 4, 1082–1089, 2008. With permission.)

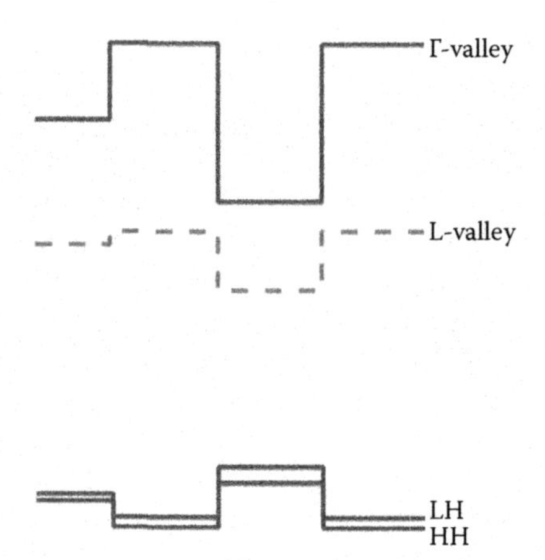

FIGURE 9.21
Typical band line up of a Ge/$Si_{1-x}Ge_x$ quantum well structure on a slightly tensile-strained $Si_{1-z}Ge_z$ substrate (with Si content between that of the well and the barrier) showing the direct, Γ, band, the indirect L band, and the splitting of the light hole (LH) and heavy hole (HH) bands.

has another L-valley minimum at 2.06 eV [85], which is also below its direct band gap. The equivalent Δ-valley in Ge is above its direct band gap at 0.82 eV, as determined theoretically [86]. The expected band alignment with the more dominant (lower energy) L-valley included is given in Figure 9.21 for a QCSE material structure.

Given the indirect band gaps in Si and Ge, the absorption at long wavelengths (low energies) begins with the indirect transitions (mainly the valence to the L-valley in Ge and Ge-rich SiGe materials), which are relatively weak compared with the direct transitions. Additionally, these transitions are substantially unaffected by electric field and thus do not contribute to the electroabsorption mechanism [81]. As the photon energy increases to ~0.8 eV, the absorption spectrum shows a sharp increase due to the direct transition in bulk Ge. The direct band gap absorption shifts with electric field and results in FKE and QCSE

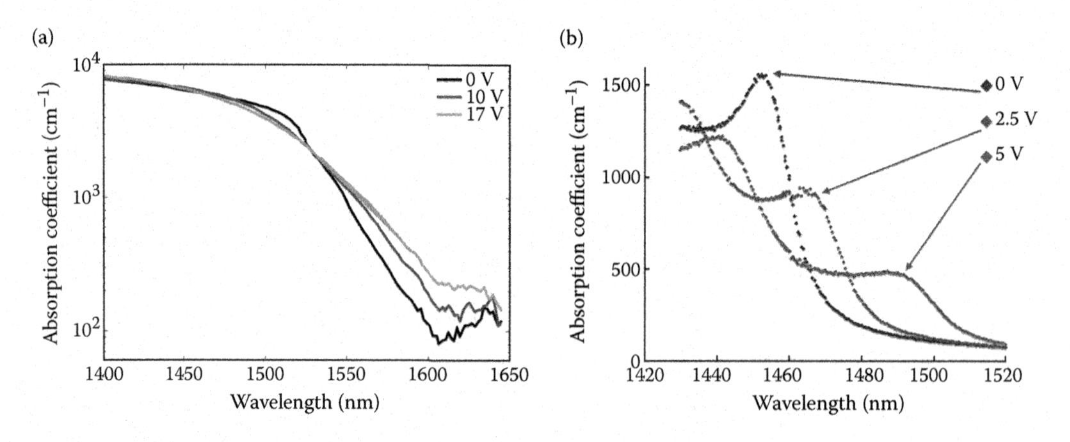

FIGURE 9.22
(See color insert.) (a) FKE spectra from 3-μm-thick $Si_{0.994}Ge_{0.006}$ grown on Si(100) wafers. (After Y. Luo et al., *Proc. SPIE*, 7944, 79440P, 2011. With permission.) (b) QCSE spectra for Ge/$Si_{0.16}Ge_{0.84}$ quantum wells on $Si_{0.1}Ge_{0.9}$ virtual substrate showing strong exciton enhancement and step-like absorption spectra over the underlying indirect absorption. (From R. K. Schaevitz, J. E. Roth, S. Ren, O. Fidaner, and D. A. B. Miller, *IEEE J. Select. Top. Quant. Electron.*, 14, 4, 1082–1089, 2008. With permission.)

behavior, where QCSE has more enhancement of the direct transition due to quantum confinement. Typical FKE and QCSE spectra in Ge-based devices are given in Figure 9.22. Given the underlying indirect band gap in the active Ge and Ge-rich SiGe, background absorption will always exist in FKE and QCSE EAMs at the operating energies (near the direct band gap), which contributes to the insertion loss and reduction in extinction ratios.

In the next section, we will discuss the design of Ge-based FKE and QCSE modulators and the best results to-date. We will also review the possibility of Ge EAMs for meeting the power, density, speed, and size requirements necessary for full integration of optical interconnects on chip.

9.4.3 Ge-Based Franz–Keldysh and Quantum-Confined Stark Effect EAMs

Given the relative simplicity of the bulk Ge and Ge-rich SiGe Franz–Keldysh effect modulators, great improvements have been achieved since the first demonstration in 2008 [10]. The quantum-confined Stark effect (QCSE) devices, while more difficult to grow and fabricate, have a greater potential for meeting aggressive speed and power requirements. In this section, we will review FKE and QCSE modulators demonstrated using the Ge material system followed by a discussion of their potential for meeting the power, speed, density, and size requirements necessary to alleviate the interconnect bottleneck.

9.4.3.1 Franz–Keldysh Effect EAMs

The first demonstration of an integrated germanium-based FKE modulator was in 2008 by Liu et al., using $Si_{0.008}Ge_{0.992}$ bulk material on an Si(100) substrate [10]. The material composition was chosen for C-band operation at 1550 nm since bulk Ge (with the thermal tensile strain) tends to operate usefully for wavelengths shorter than 1600 nm [11]. The addition of very little Si with its large 4.0-eV band gap efficiently shifts the wavelength of operation lower (higher energy), although with the addition of some more indirect absorption (insertion loss). The increase in indirect absorption is due to a larger separation of the indirect and direct band gaps with the addition of Si. With its very small active area of 30 μm^2 and energy consumption of 50 fJ/bit, this waveguide achieved an extinction ratio of greater than 7 dB, a 3-dB bandwidth of 1.2 GHz, an insertion loss of 3.7 dB, and an operating spectrum of 14 nm. Due to fabrication difficulties, the authors did not achieve the expected bandwidth of greater than 50 GHz. Figure 9.23 shows the extinction ratio (ER) and insertion loss (IL) of this C-band FKE optoelectronic modulator.

Since the initial demonstrate, Ge FKE modulators have significantly improved by showing high-speed modulation with 100 fJ/bit of energy [11]. The selective area growth technique was used to integrate pure Ge into a 3-μm-thick single mode waveguide on SOI. Following growth, the Ge ridge waveguide was defined and implant doping of the sidewalls formed the horizontal PIN diode (shown in Figure 9.19a). Figure 9.23 shows the typical transmission spectrum for various applied voltages as well as the associated extinction ratios (ER) and insertion loss (IL) for a 45-μm-long waveguide. The authors were able to achieve more than 4–7.5-dB ER with a 2.5–5-dB total IL for this waveguide. The loss associated with the facet between the Ge and Si waveguides was determined to be ~1.34 dB, giving an excess IL of 1.2–3.8 dB with 55 kV/cm of applied electric field (4 V_{pp}). With a −4-V bias, the device achieved a 3-dB bandwidth of 30 GHz, which is limited only by the RC time constant of the device (of which the capacitance is 25 fF). This bandwidth shows a possible device operation up to 40 Gb/s although the authors were limited in testing to 12.5 Gb/s. These results are comparable to the speeds achieved by Mach–Zehnder modulators in Si, which are much

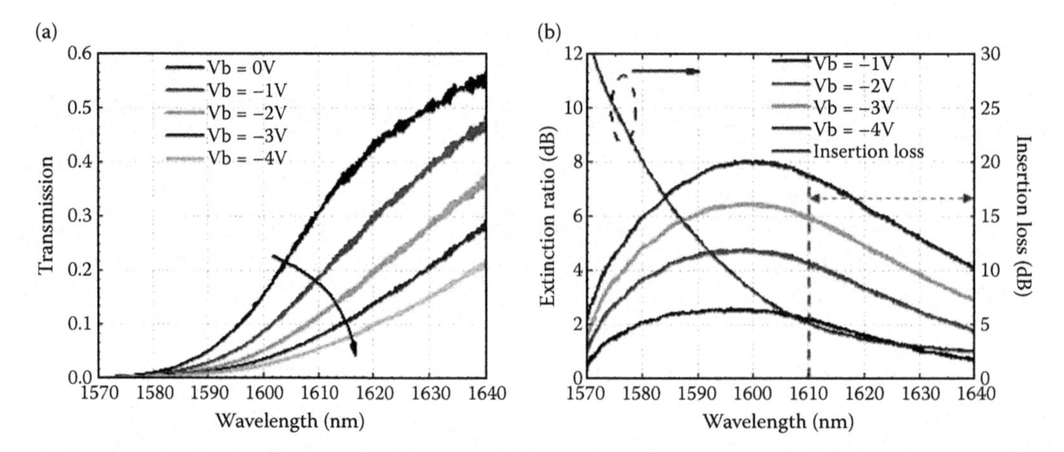

FIGURE 9.23
(See color insert.) (a) Transmission spectrum of the Ge FKE modulator versus voltage and (b) the associated extinction ratios and insertion loss for a 45-μm-long device. (From N.-N. Feng, D. Feng, S. Liao, X. Wang, P. Dong, H. Liang, C.-C. Kung, W. Qian, J. Fong, R. Shafiiha, Y. Luo, J. Cunningham, A. V. Krishnamoorthy, and M. Asghari, *Opt. Express*, 19, 8, 7062–7067, 2011. With permission.)

larger and consume at least 50 times the energy per bit [87]. At 12.5 Gb/s transmission rate, eye openings with >5 dB ER were achieved with 4 V reverse bias over >20 nm wavelength range.

9.4.3.2 QCSE EAMs

Although the demonstration of high-speed modulation has so far been limited in Ge-based QCSE modulators because of the more challenging device fabrication needed in this approach, demonstration of the possible extinction ratios (ER) and insertion loss (IL) continues to improve.

The first demonstration of a QCSE device with $Ge/Si_{0.16}Ge_{0.84}$ quantum wells in 2007 utilized an asymmetric Fabry–Perot (AFP) design for side-entry modulation [88]. The AFP design uses two mirrors of differing reflectivities and relies on the absorption of the material to create constructive or destructive interference upon reflection. The two mirrors for the initial device relied on oblique incidence and on the index difference between (1) air and SiGe (total internal reflection) as well as (2) SiGe and Si. A large change in absorption was necessary given the lower reflectivity of the second mirror, thus requiring 40 pairs of quantum wells (giving ~30% absorption) and a large reverse bias of 10 V_{pp} to achieve >3 dB ER over 20 nm of wavelength.

Subsequent growth on an SOI substrate with a 50 nm buried oxide (Figure 9.24) increases the reflectivity of the second mirror and thus reduces the number of quantum wells to 10 [68]. Fewer quantum wells allows for operation at CMOS-compatible voltages of just 1 V (compared with the previous 10-V swing), although with a narrower ~4-nm wavelength bandwidth for >3-dB ER (Figure 9.25a). To operate in the C-band of 1500–1550 nm, the device was heated to 100°C without degradation of the QCSE as shown in Figure 9.25b. Given the larger size of the device, there was a large tolerance for misalignment at the penalty of higher operation energy of ~270 pJ/bit due to its 33-pF capacitance. This side-entry device was the first demonstration of C-band optoelectronic modulation in the Ge/SiGe QW material system.

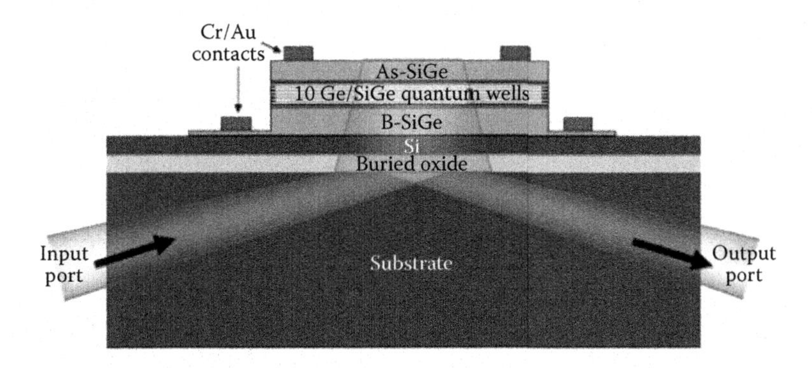

FIGURE 9.24
Device design showing 50 nm BOX layer in a side-entry Ge/SiGe modulator operating in the C-band. (From J. E. Roth, O. Fidaner, E. H. Edwards, R. K. Schaevitz, Y.-H. Kuo, N. C. Helman, T. I. Kamins, J. S. Harris, and D. A. B. Miller, *Electron. Lett.*, 44, 1, 49–50, 2008. With permission.)

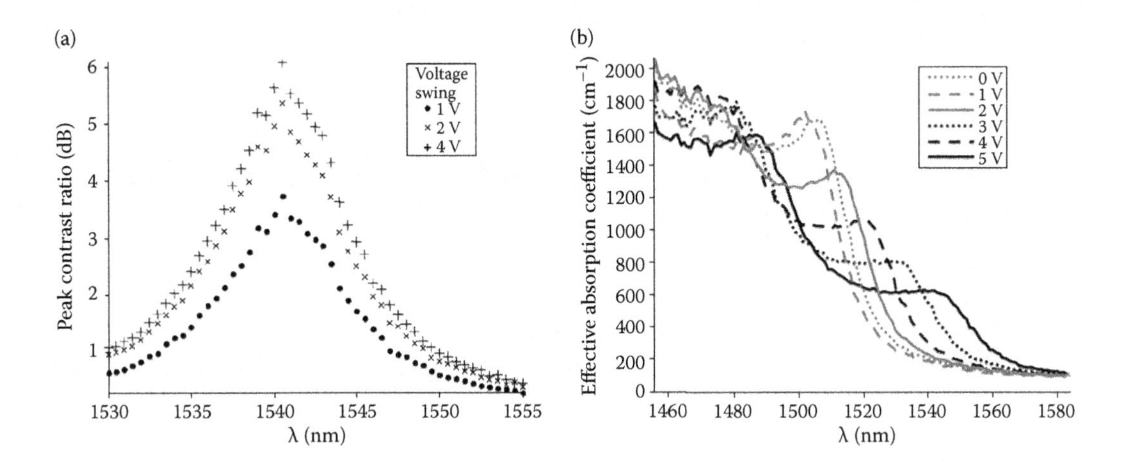

FIGURE 9.25
(a) Contrast ratios and (b) photocurrent of Ge/SiGe quantum wells at 100°C for a C-band modulator. (From J. E. Roth, O. Fidaner, E. H. Edwards, R. K. Schaevitz, Y.-H. Kuo, N. C. Helman, T. I. Kamins, J. S. Harris, and D. A. B. Miller, *Electron. Lett.*, 44, 1, 49–50, 2008. With permission.)

Other than a demonstration in principle of surface-normal modulation in a thick structure at high voltage [88], the first true surface-normal device was presented in 2010, which utilized an AFP architecture [89], although, this time, for light perpendicular to the layers. The authors presented two designs: a monolithic integration (Figure 9.26a) and a film transfer method (Figure 9.26b). The monolithic design consisted of the Ge/SiGe epitaxy forming a PIN diode grown on a double-buried oxide wafer (DSOI), which acts as the low reflectivity mirror in the AFP design. Following growth, the top surface is polished, the mesa is formed and a final high-reflectivity mirror is deposited using two to three pairs of Si/SiO_2 quarterwave stacks. The film transfer technique uses a Si handle wafer (not the DSOI wafer) for the Ge/SiGe epitaxial growth. The top surface is polished and a low-reflectivity mirror is deposited. The epitaxy is then bonded anodically to a quartz substrate and the Si handle wafer is removed through grinding and a final selective wet etch. The remaining surface is polished and defined into mesas followed by deposition of the high-reflectivity mirror, thus forming the AFP cavity. Independent of the initial monolithic

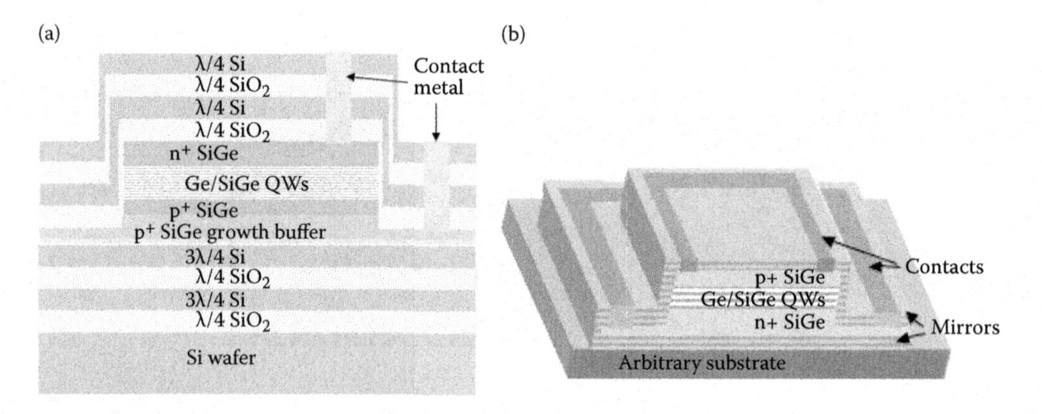

FIGURE 9.26
First surface modulator fabricated and designed to use QCSE in Ge/SiGe. (a) Cross-sectional view of the monolithic design and (b) shows a 3D view of the film transfer method. (From E. H. Edwards, R. M. Audet, Y. Rong, S. A. Claussen, R. K. Schaevitz, E. Tasyurek, S. Ren, T. I. Kamins, O. I. Dosunmu, M. S. Ünlü, J. S. Harris, and D. A. B. Miller, "Si-Ge surface-normal asymmetric Fabry–Perot quantum confined Stark effect electroabsorption modulator," IEEE Photonics Society Summer Topical Meeting, Playa del Carmen, Mexico, 2010, TuD2.4. With permission.)

or film transfer design, the final mesas formed then have vias etched through the mirrors to contact the p- and n-doped layers.

The reflectivity results of the film transfer technique are reprinted in Figure 9.27. At DC, the monolithic device achieved 3.8 dB ER over 2-V swing with 2-V pre-bias, whereas the film transfer method gave 4.2-dB ER over 2-V swing with 3-V pre-bias. A large component of the ER was from the refractive index change with applied bias. Although not optimized, these devices demonstrate the viability of surface-normal AFP optoelectronic modulation using the Ge/SiGe quantum well system and show a 500× improvement in energy/bit of ~500 fJ/bit compared with the side-entry devices. The potential for further reduction

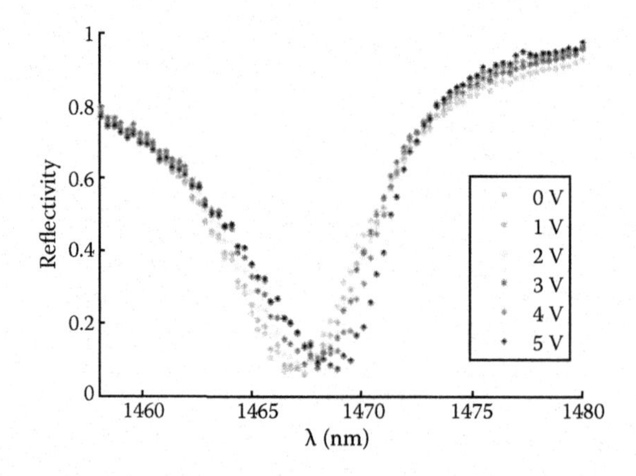

FIGURE 9.27
Reflectivity spectra of first surface modulator fabricated and designed using the film transfer technique. (From E. H. Edwards, R. M. Audet, Y. Rong, S. A. Claussen, R. K. Schaevitz, E. Tasyurek, S. Ren, T. I. Kamins, O. I. Dosunmu, M. S. Ünlü, J. S. Harris, and D. A. B. Miller, "Si-Ge surface-normal asymmetric Fabry–Perot quantum confined Stark effect electroabsorption modulator," IEEE Photonics Society Summer Topical Meeting, Playa del Carmen, Mexico, 2010, TuD2.4. With permission.)

in energy of such devices has been considered recently [90] through the possible use of smaller mesas and of vertical pillar resonator structures, with operating energies as low as 10 fJ predicted for optimized designs. Such QCSE modulators may be the only ones capable of reaching low operating energies while operating for beams perpendicular to the surface of the chips.

To test high-speed operation, simple square mesa devices ranging from 5 to 100 μm per side with 10 quantum wells of ~10-nm thickness were fabricated [91]. For the 100-μm mesa, a drop in response was observed ~3 GHz, whereas for a 30-μm device, drop-off occurred at 13 GHz. The 3-dB bandwidth was relatively narrow due poor contacts creating a large resistivity that is not inherent to the material system. These results are the first demonstration of high-speed measurements for optoelectronic modulation.

Polarization dependence on the ER and IL of the Ge quantum well modulators has also been tested in both TE and TM modes using 34- and 64-μm-long devices [92]. The 64-μm devices showed >5 dB ER over more than 20 nm from 1413 to 1439 nm with an <4-dB IL between 1428 and 1439 nm for a reverse bias of 6 V_{pp} in TE mode. The results for the TM polarized light were not as strong as compared with TE polarization; however, with a 0- to 8-V swing, the authors were able to achieve >5-dB ER with <4 dB IL from 1398 to 1409 nm for the 64-μm-long device.

The first selectively grown waveguide-integrated Ge/SiGe optoelectronic modulator was designed and fabricated in 2011 [93, 94]. Contrast ratios (CR) of the photocurrent from test structures showed a high CR of 2.84 at 1500 nm for a 1-V swing with a 5.5-V pre-bias as shown in Figure 9.28. The rib Si waveguide (310 × 800 nm) was fabricated on a 1-μm oxide SOI substrate and a selective area trench was opened into a section of the waveguide with lengths ranging from 5 to 400 μm and a depth through the oxide layer to the Si handle. The SiGe buffer material was grown slightly thicker than the oxide, and this material was followed by the QW region and an n-doped SiGe cap layer. The last layer is grown beyond the total thickness (of the oxide protected waveguide) to fill in the faceting regions. The final sample is polished flat and lithographically defined into a waveguide mesa structure (Figure 9.29a).

Given the device design, there exists a large mode mismatch between the rib waveguide and the very thick Ge/SiGe modulator. This mode mismatch results in a loss of ~12 dB in addition to the material loss, the input and output coupling losses, and any other

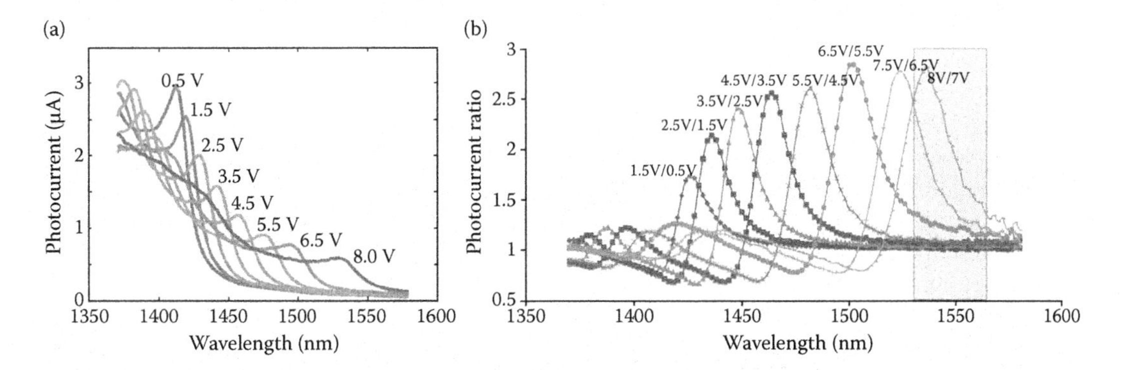

FIGURE 9.28
(a) Photocurrent [93] and (b) contrast ratios for epitaxially grown material with the C-band wavelength range highlighted [94]. (From S. Ren, "Ge/SiGe quantum well waveguide modulator for optical interconnect systems," PhD dissertation, Stanford University, 2011; S. Ren, Y. Rong, S. A. Claussen, R. K. Schaevitz, T. I. Kamins, J. S. Harris, and D. A. B. Miller, *IEEE Photon. Technol. Lett.* (in press), doi: 10.1109/LPT.2011.2181496. With permission.)

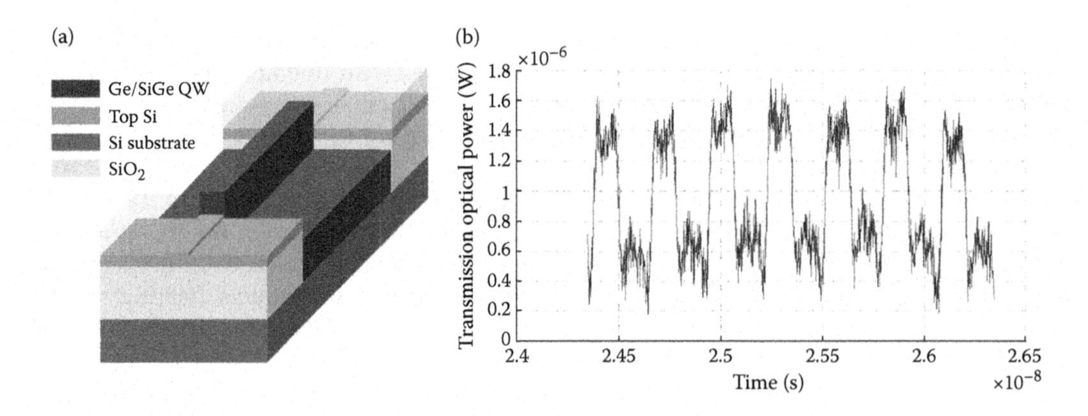

FIGURE 9.29
(a) Integrated optoelectronic Ge/SiGe modulator waveguide design and (b) high-speed oscilloscope trace of waveguide integrated Ge/SiGe modulator with 3 GHz (7 Gb/s if using non-return-to-zero modulation) operation. (From S. Ren, Y. Rong, S. A. Claussen, R. K. Schaevitz, T. I. Kamins, J. S. Harris, and D. A. B. Miller, *IEEE Photon. Technol. Lett.* (in press), doi: 10.1109/LPT.2011.2181496. With permission.)

waveguide losses. Due to the excessive insertion loss, significant power was lost in the device and testing was limited to the shortest waveguides of ~10 μm to increase the transmission of light. From these preliminary results, the rise and fall time was deduced to be no more than ~40 ps, which would allow operation up to at least ~8 GHz or 16 Gb/s with <20 fJ/bit of power dissipation as shown in Figure 9.29b. Future device concepts would reduce the insertion loss (especially due to the mode mismatch) and increase the n- and p-doping (reducing the RC time constant) to achieve significantly improved results.

9.4.4 Future of Ge-Based EAMs

EAMs based on germanium have great potential for solving the interconnect bottleneck for high-performance computing by meeting the speed, density, and energy constraints necessary for on-chip and short-distance off-chip interconnects [72]. Optical interconnects inherently have an advantage in meeting density requirements and use of electroabsorption devices typically achieves high-speed, with speed limits only from the RC time constant. Solving the interconnect bottleneck requires meeting the power dissipation targets at high speed, ideally with CMOS-compatible devices. Power available on chip is expected to saturate to 200 W due to constraints of heat removal, while operation speed continues to increase. Consequently, the total energy per bit available for interconnects decreases quickly and electrical solutions cannot meet the requirements. If optics is to supplant electrical interconnects by 2022, the total energy dissipation for the system should be ~100 fJ/bit for off-chip interconnects and significantly lower for on-chip connections [72]. These targets demand that optical output devices for off-chip and on-chip communication to meet energy per bit targets of 10–30 and 2–7 fJ/bit, respectively.

In CMOS-compatible material systems, we can use two mechanisms, electrorefraction and electroabsorption, to modulate light (if the laser source is assumed to be off chip due to power requirements and therefore not directly modulated). Electrorefraction is a relatively weak mechanism requiring either long path lengths (Mach–Zehnder interferometers, MZI) or highly resonant structures (ring resonators, RR). The best MZI device reported in silicon itself in terms of energy consumption still operates at 5 pJ/bit, two orders of magnitude greater than the tens of femtojoules per bit requirement [87]. While RR devices can initially meet the targets, with ~3 fJ/bit at 12.5 Gb/s reported thus far [95], their temperature sensitivity requires precise tuning and the resulting tuning may increase their energy consumption by an order of magnitude or more [56].

In contrast to electrorefraction, electroabsorption is a strong mechanism, so strong that it enables compact devices even without the use of resonators, and with the demonstration of both FKE and QCSE in Ge-based material structures, we have the capability of meeting the energy requirements in CMOS-compatible material. The simplicity of growth and fabrication of Ge FKE modulators has allowed for greater improvement of device performance since the first demonstration in 2008. The lowest reported energy per bit with high-speed testing at 12.5 Gb/s is 100 fJ/bit, which could be improved by optimizing the capacitance and operating voltages. However, these devices will be limited to operation greater than 1500 nm since shifting the band gap through addition of Si or compressive strain leads to poor performance due to indirect absorption dominating. Consequently, FKE in Ge or Ge-rich SiGe, while limited to C-band operation, provides a strong platform for meeting the power, speed, and size requirements for optical interconnects.

QCSE using the Ge material platform exceeds the performance of FKE modulators in meeting the energy requirements (with the potential of <10 fJ/bit) and has the potential to operate at high speed in both communication bands of 1310- and 1550-nm wavelengths [96]. Due to the quantum confinement and the exciton enhancement of the electroabsorption profile, the resulting strong electroabsorption should overcome the increase in indirect absorption at shorter wavelengths, allowing Ge quantum wells potentially to operate at 1310 nm in contrast to FKE devices. While high-speed modulation has not yet been fully demonstrated, Ge quantum well modulators have even greater potential for meeting speed, density, and power requirements for solving the interconnect bottleneck.

9.5 Conclusion

Delphine Marris-Morini

Throughout this chapter devoted to silicon-based optical modulators, the reader has been able to note the rapid and amazing evolution of this research topic in the recent years. This evolution has been possible thanks to fine analysis of physical mechanisms underlying electrorefraction and electroabsorption in silicon based materials. Among the different possibilities to get high-performance modulators, silicon depletion modulator, and Ge-based electroabsorption devices are today probably the most promising solutions. In the first case, large extinction ratio modulators have been recently demonstrated at 40-Gbit/s data rate [58], while for the second case, compact 45-μm-long waveguides with 30-GHz bandwidth have shown high-speed modulation requiring only 100 fJ/bit [11].

Despite these impressive results, there are still some challenges to be faced by the modulator to meet the requirements for telecommunications or data communications. Carrier depletion modulators suffer from high driving voltage, typically larger than 6 V, which increase the energy/bit consumption, while QCSE in Ge/SiGe MQW still requires an efficient waveguide integration scheme. Obtaining simultaneously low-loss, high-speed, compact, low-power consumption, and large optical bandwidth-silicon-based modulators is not straightforward, and innovative solutions have to be developed to revolutionize silicon photonics (for example, by the use of plasmonics to enhance optical and electrical interaction or by the integration of "new" materials on silicon). Finally, it seems that silicon-based optical modulation is still an open and exciting research area for the forthcoming years.

References

1. S. Koehl, A. Liu, and M. Paniccia, "Integrated silicon photonics: harnessing the data explosion," *OSA Opt. Photon. News*, 22, 24–39, 2011.
2. A. Liu, R. Jones, L. Liao, D. Samara-Rubio, D. Rubin, O. Cohen, R. Nicolaescu, and M. Paniccia, "A high speed silicon optical modulator based on a metal-oxide-semiconductor capacitor," *Nature*, 427, 615–618, 2004.
3. R. Soref, "Mid-infrared photonics in silicon and germanium" [invited commentary], *Nat. Photon.*, vol. 4, pp. 495–497, 2010.
4. R. A. Soref, "Towards silicon-based longwave integrated optoelectronics (LIO)," *SPIE Proc.*, vol. 6898, paper 6898-09, 2008.
5. J. H. Zhao, W. Su, Q. D. Chen, Y. Jiang, Z. G. Chen, G. Jia, and H. B. Sun, "Strain at native SiO_2/Si(111) interface characterized by strain scanning second-harmonic generation," *IEEE J. Quantum Electron.*, vol. 47, pp. 55–59, 2011.
6. R. S. Jacobsen, K. N. Andersen. P. L. Borel, J. Fage-Pedersen, L. H. Frandsen, O. Hansen, M. Kristensen, A. V. Lavrinenko, G. Moulin, H. Ou, C. Peucheret, B. Zsigri, and A. Bjarklev, "Strained silicon as a new electro-optic material," *Nature*, vol. 441, pp. 199–202, 11 May, 2006.
7. L. Friedman and R. A. Soref, "Second-order optical susceptibility of strained GeSi/Si superlattices," *J. Appl. Phys.*, vol. 61, p. 2324, 1987.
8. L. Friedman and R. A. Soref, "Second order susceptibilty and linear electrooptic effect in strained-layer GeSi/Si Superlattices," *SPIE Proc.*, vol. 792, pp. 222–227, 1987.
9. R. A. Soref and B. R. Bennett, "Electrooptical effects in Silicon," *IEEE J. Quantum Electron.*, vol. QE-23, pp. 123–129, 1987.
10. J. Liu, M. Beals, A. Pomerene, S. Bernardis, R. Sun, J. Cheng, L. C. Kimerling, and J. Michel, "Waveguide-integrated, ultralow-energy GeSi electroabsorption modulators," *Nat. Photon.*, vol. 2, pp. 433–437, 2008.
11. N.-N. Feng, D. Feng, S. Liao, X. Wang, P. Dong, H. Liang, C.-C. Kung, W. Qian, J. Fong, R. Shafiiha, Y. Luo, J. Cunningham, A. V. Krishnamoorthy, and M. Asghari, "30GHz Ge electro-absorption modulator integrated with 3 μm silicon-on-insulator waveguide," *Opt. Express*, vol. 19, no. 8, pp. 7062–7067, 2011.
12. N. K. Hon, R. A. Soref, and B. Jalali, "The third-order nonlinear optical coefficients of Si, Ge, and $Si_{1-x}Ge_x$ in the midwave and longwave infrared," *J. Appl. Phys.*, vol. 110, p. 011301, 2011.
13. G. Ghosh, *Handbook of Thermo-Optic Coefficients of Optical Materials with Applications*, Academic Press, Boston, 1998, ISBN-0-12-281855-5.
14. B. J. Frey, D. B. Leviton, and T. J. Madison, "Temperature dependent refractive index of Si and Ge," *NASA Goddard Memorandum*, 2007.
15. M. W. Geis, S. J. Spector, R. C. Williamson, and T. M. Lyszczarz, "Submicrosecond submilliwatt silicon-on-insulator thermooptic switch," *IEEE Photon. Technol. Lett.*, vol. 16, pp. 2514–2516, 2004.
16. C. Qiu, J. Shu, Z. Li, X. Zhang, and Q. Xu, "Wavelength tracking with thermally controlled silicon resonators," *Opt. Express*, vol. 19, pp. 5143–5148, 2011.
17. Y. Shoji, K. Kintaka, S. Suda, H. Kawashima, T. Hasama, and H. Ishikawa, "Low-crosstalk 2 × 2 thermo-optic switch with silicon wire waveguides," *Opt. Express*, vol. 18, pp. 9071–9075, 2010.

18. Y. H. Kuo, Y. A. Huang, and T. L. Chen, "A vertical Germanium thermooptic modulator for optical interconnects," *IEEE Photon. Technol. Lett.*, vol. 21, pp. 245–247, 2009.
19. R. A. Soref and B. R. Bennett, "Kramers–Kronig analysis of electrooptical switching in silicon," *SPIE Proc.*, vol. 704, pp. 32–37, September 16, 1986.
20. R. A. Soref and L. Friedman, "Electrooptical modulation in $Si_{1-x}Ge_x$/Si and related heterostructures," *Int. J. Optoelectron.*, vol. 9, no. 2, pp. 205–210, 1994.
21. A. Singh, "Free carrier induced refractive index modulation of crystalline silicon," paper P1.13, IEEE Photonics Society 7th International Conference on Group IV Photonics, Beijing, China, September 1–3, 2010.
22. L. Friedman, R. A. Soref, and J. P. Lorenzo, "Silicon double-injection electro-optic modulator with junction gate control," *J. Appl. Phys.*, vol. 63, pp. 1831–1839, 1988.
23. G. Breglio, A. Cutolo, A. Irace, P. Spirito, L. Zeni, M. Iodice, and P. M. Sarro, "Two silicon optical modulators realizable with a fully compatible bipolar process," *IEEE J. Select. Top. Quantum Electron.*, vol. 4, pp. 1003–1010, 1998.
24. S. Deng, Z. Rena Huang, and J. F. McDonald, "Design of high efficiency multi-GHz SiGe HBT electro-optic modulator," *Opt. Express*, vol. 17, pp. 13425–13428, 2009.
25. C. Gunn and T. L. Koch, "Silicon photonics." In *Optical Fiber Telecommunications V, Vol. A: Components and Subsystems*, Ed. by I. P. Kaminow, T. Li, and A. E. Willner, Elsevier, pp. 381–454, 2008.
26. W. A. Zortman, M. R. Watts, D. C. Trotter, R. W. Young, and A. W. Lentine, "Low-power high-speed silicon microdisk modulators," paper CThJ2014, Conference on Lasers and Electro-Optics (CLEO) and Quantum Electronics and Laser Science Conference (QELS), Technical Digest, Optical Society of America, 2010.
27. S. P. Anderson and P. M. Fauchet, "Ultra-low-power modulators using MOS depletion in a high-SiO_2-clad silicon 2-D photonic crystal resonator," *Opt. Express*, vol. 18, pp. 19129–19140, 2010.
28. C. K. Tang and G. T. Reed, "Highly efficient optical phase modulator in SOI waveguides," *Electron. Lett.*, vol. 31, pp. 451–452, 1995.
29. P. D. Hewitt and G. T. Reed, "Improving the response of optical phase modulators in SOI by computer simulation," *J. Lightwave Technol.*, vol. 18, pp. 443–450, 2000.
30. P. D. Hewitt and G. T. Reed, "Improved modulation performance of a silicon p-i-n device by trench isolation," *J. Lightwave Technol.*, vol. 19, pp. 387–390, 2001.
31. I. E. Day, S. W. Roberts, R. O'Carroll, A. Knights, P. Sharp, G. F. Hopper, B. J. Luff, and M. Asghari, "Single-chip variable optical attenuator and multiplexer subsystem integration," *Conf. Opt. Fiber Commun.*, vol. 70, pp. 72–73, 2002.
32. D. Thomson, F. Y. Gardes, G. Z. Mashanovich, A. P. Knights, and G. T. Reed, "Using SiO2 carrier confinement in total internal reflection optical switches to restrict carrier diffusion in the guiding layer," *J. Lightwave Technol.*, vol. 26, pp. 1288–1294, 2008.
33. D. J. Thomson, G. T. Reed, A. P. Knights, P. Y. Yang, F. Y. Gardes, A. J. Smith, and L. K. Litvinenko, "Total internal reflection optical switch in SOI with defect engineered barrier region," *J. Lightwave Technol.*, vol. 28, pp. 2483–2491, 2010.
34. T. W. Ang, P. D. Hewitt, A. Vonsovici, G. T. Reed, A. G. R. Evans, P. R. Routley, T. Blackburn, and M. R. Josey, "Integrated optics in Unibond for greater flexibility," *Electrochem. Soc. Proc.*, vol. 99-3, pp. 353–360, 1999.
35. C. E. Png, S. P. Chan, S. T. Lim, and G. T. Reed, "Optical phase modulators for MHz and GHz modulation in silicon-on-insulator (SOI)," *J. Lightwave Technol.*, vol. 22, pp. 1573–1582, 2004.
36. W. M. Green, M. J. Rooks, L. Sekaric, and Y. A. Vlasov, "Ultra-compact, low RF power, 10 Gb/s silicon Mach–Zehnder modulator," *Opt. Express*, vol. 15, pp. 17106–17113, 2007.
37. Q. Xu, S. Manipatruni, B. Schmidt, J. Shakya, M. Lipson, "12.5 Gbit/s carrier-injection-based silicon microring silicon modulators," *Opt. Express*, vol. 15, pp. 430–436, 2007.
38. K. Preston, S. Manipatruni, A. Gondarenko, C. B. Poitras, and M. Lipson, "Deposited silicon high-speed integrated electro-optic modulator," *Opt. Express*, vol. 17, pp. 5118–5124, 2009.

39. L. Liao, D. Samara-Rubio, M. Morse, A. Liu, D. Hodge, D. Rubin, U. Keil, and T. Franck, "High speed silicon Mach–Zehnder modulator," *Opt. Express*, vol. 13, pp. 3129–3135, 2005.
40. D. D'Andrea, "CMOS photonics today and tomorrow," http://www.ofcnfoec.org/osa.ofc/media/Default/PDF/2009/09-Dandrea.pdf.
41. D. Marris-Morini, X. Le Roux, D. Pascal, L. Vivien, E. Cassan, J.-M. Fédéli, J.-F. Damlencourt, D. Bouville, J. Palomo, and S. Laval, "High speed all-silicon optical modulator," *J. Lumin.*, vol. 121, pp. 387–390, 2006.
42. F. Y. Gardes, G. T. Reed, N. G. Emerson, and C. E. Png, "A sub-micron depletion-type photonic modulator in silicon on insulator," *Opt. Express*, vol. 13, pp. 8845–8854, 2005.
43. C. Gunn "CMOS photonics for high-speed interconnects," *Micro. IEEE*, vol. 26, pp. 58–66, 2006.
44. A. Huang, C. Gunn, G.-L. Li, Y. Liang, S. Mirsaidi, A. Narasimha, and T. Pinguet, "A 10 Gb/s photonic modulator and WDM MUX/DEMUX integrated with electronics in 0.13μm SOI CMOS," IEEE International Solid-State Circuits Conference (ISSCC), 2006.
45. A. Liu, L. Liao, D. Rubin, H. Nguyen, B. Ciftcioglu, Y. Chetrit, N. Izhaky, and M. Paniccia, "High-speed optical modulation based on carrier depletion in a silicon waveguide," *Opt. Express*, vol. 15, pp. 660–668, 2007.
46. L. Liao, A. Liu, D. Rubin, J. Basak, Y. Chetrit, H. Nguyen, R. Cohen, N. Izhaky, and M. Paniccia, "40 Gbit/s silicon optical modulator for high speed applications," *Electron. Lett.*, vol. 43, pp. 1196–7, 2007.
47. D. Marris-Morini, L. Vivien, J.-M. Fédéli, E. Cassan, P. Lyan, and S. Laval, "Low loss and high speed silicon optical modulator based on a lateral carrier depletion structure," *Opt. Express*, vol. 16, pp. 334–339, 2008.
48. G. Rasigade, M. Ziebell, D. Marris-Morini, J.-M. Fédéli, F. Milesi, P. Grosse, D. Bouville, E. Cassan, and L. Vivien, "High extinction ratio 10 Gbit/s silicon optical modulator," *Opt. Express*, vol. 19, pp. 5827–5832, 2011.
49. S. J. Spector, M. W. Geis, M. E. Grein, R. T. Schulein, J. U. Yoon, D. M. Lennon, F. Gan, G.-R. Zhou, F. X. Kaertner, and T. M. Lyszczarz, "High-speed silicon electro-optical modulator that can be operated in carrier depletion or carrier injection mode," Conference on Lasers and Electro-Optics (CLEO), 2008.
50. J.-B. You, M. Park, J.-W. Park, and G. Kim, "12.5 Gbps optical modulation of silicon racetrack resonator based on carrier-depletion in asymmetric p-n diode," *Opt. Express*, vol. 16, pp. 18340–18344, 2008.
51. D. M. Gill, M. Rasras, K.-Y. Tu, Y.-K. Chen, A. E. White, S. S. Patel, D. Carothers, A. Pomerene, R. Kamocsai, C. Hill, and J. Beattie, "Internal bandwidth equalization in a CMOS compatible Si ring modulator," *IEEE Photon. Technol. Lett.*, vol. 21, pp. 200–202, 2009.
52. J. W. Park, J.-B. You, I. G. Kim, and G. Kim, "High-modulation efficiency silicon Mach–Zehnder optical modulator based on carrier depletion in a PN diode," *Opt. Express*, vol. 17, pp. 15520–15524, 2009.
53. P. Dong, S. Liao, D. Feng, H. Liang, D. Zheng, R. Shafiiha, C.-C. Kung, W. Qian, G. Li, X. Zheng, A. V. Krishnamoorthy, and M. Asghari, "Low Vpp, ultralow-energy, compact, high-speed silicon electro-optic modulator" *Opt. Express*, vol. 17, pp. 22484–22490, 2009.
54. F. Y. Gardes, A. Brimont, P. Sanchis, G. Rasigade, D. Marris-Morini, L. O'Faolain, F. Dong, J.-M. Fedeli, P. Dumon, L. Vivien, T. F. Krauss, G. T. Reed, and J. Martí, "High-speed modulation of a compact silicon ring resonator based on a reverse-biased pn diode," *Opt. Express*, vol. 17, pp. 21986–21991, 2009.
55. L. Tsung-Yang, A. Kah-Wee, F. Qing, S. Jun-Feng, X. Yong-Zhong, Y. Ming-Bin, L. Guo-Qiang, and K. Dim-Lee, "Silicon modulators and germanium photodetectors on SOI: Monolithic integration, compatibility, and performance optimization," *IEEE J. Select. Top. Quantum Electron.*, vol. 16, pp. 307–315, 2010.
56. N.-N. Feng, S. Liao, D. Feng, P. Dong, D. Zheng, H. Liang, R. Shafiiha, G. Li, J. E. Cunningham, A. V. Krishnamoorthy, and M. Asghari "High speed carrier-depletion modulators with 1.4V-cm $V\pi L$ integrated on 0.25 μm silicon-on-insulator waveguides," *Opt. Express*, vol. 18, pp. 7994–7799, 2010.

57. D. J. Thomson, F. Y. Gardes, G. T. Reed, F. Milesi, and J.-M. Fedeli, "High speed silicon optical modulator with self aligned fabrication process," *Opt. Express*, vol. 18, pp. 19064–19069, 2010.
58. D. J. Thomson, F. Y. Gardes, Y. Hu, G. Mashanovich, M. Fournier, P. Grosse, J.-M. Fedeli, and G. T. Reed, "High contrast 40 Gbit/s optical modulation in silicon," *Opt. Express*, vol. 19, pp. 11507–11516, 2011.
59. F. Y. Gardes, D. J. Thomson, N. G. Emerson, and G. T. Reed, "40 Gb/s silicon photonics modulator for TE and TM polarizations," *Opt. Express*, vol. 19, pp. 11804–11814, 2011.
60. W. Franz, "Influence of an electric field on an optical absorption edge," *Z. Naturforsch.*, vol. 13a, p. 484, 1958.
61. L. V. Keldysh, "The effect of a strong electric field on the optical properties of insulating crystals," *Zh. Eksp. Teor. Fiz.*, vol. 34, p. 1138, 1958.
62. D. A. B. Miller, D. S. Chemla, T. C. Damen, A. C. Gossard, W. Wiegmann, T. H. Wood, and C. A. Burrus, "Band-edge electroabsorption in quantum well structures: The quantum-confined Stark effect," *Phys. Rev. Lett.*, vol. 53, no. 22, pp. 2173–2176, 1984.
63. D. A. B. Miller, D. S. Chemla, and S. Schmitt-Rink, "Relation between electroabsorption in bulk semiconductors and in quantum wells: The quantum-confined Franz–Kedlysh effect," *Phys. Rev. B*, vol. 33, no. 10, pp. 6976–6982, 1986.
64. D. D. Cannon, J. Liu, Y. Ishikawa, K. Wada, D. T. Danielson, S. Jongthammanurak, J. Michel, and L. C. Kimerling, "Tensile strained epitaxial Ge films on Si(100) substrates with the potential application in the L-band telecommunications," *Appl. Phys. Lett.*, vol. 84, no. 6, pp. 906–908, 2004.
65. Y.-H. Kuo, Y. K. Lee, S. R. Y. Ge, J. E. Roth, T. I. Kamins, D. A. B. Miller, and J. S. Harris, "Strong quantum-confined stark effect in germanium quantum-well structures on silicon," *Nature*, vol. 437, pp. 1334–1336, Oct., 2005.
66. Y.-H. Kuo, Y. K. Lee, Y. Ge, S. Ren, J. E. Roth, T. I. Kamins, D. A. B. Miller, and J. S. Harris, "Quantum-confined Stark effect in Ge/SiGe quantum wells on Si for optical modulators," *IEEE J. Select. Top. Quant. Electron.*, vol. 12, no. 6, pp. 1503–1513, Nov., 2006.
67. R. K. Schaevitz, J. E. Roth, S. Ren, O. Fidaner, and D. A. B. Miller, "Material properties of Si-Ge/Ge quantum wells," *IEEE J. Select. Top. Quant. Electron.*, vol. 14, no. 4, pp. 1082–1089, 2008.
68. J. E. Roth, O. Fidaner, E. H. Edwards, R. K. Schaevitz, Y.-H. Kuo, N. C. Helman, T. I. Kamins, J. S. Harris, and D. A. B. Miller, "C-band side-entry Ge quantum-well electroabsorption modulator on SOI operating at 1 V swing," *Electron. Lett.*, vol. 44, no. 1, pp. 49–50, 2008.
69. M. Bonfanti, E. Grilli, M. Guzzi, M. Virgilio, G. Grosso, D. Chrastina, G. Isella, H. von Känel, and A. Neels, "Optical transitions in Ge/SiGe multiple quantum wells with Ge-rich barriers," *Phys. Rev. B*, vol. 78, p. 041407, 2008.
70. M. Virgilio, M. Bonfanti, D. Chrastina, A. Neels, G. Isella, E. Grilli, M. Guzzi, G. Grosso, H. Sigg, and H. von Kanel, "Polarization-dependent absorption in Ge/SiGe multiple quantum wells: Theory and experiment," *Phys. Rev. B*, vol. 79, no. 7, p. 075323, 2009.
71. D. A. B. Miller, R. K. Schaevitz, J. E. Roth, S. Ren, and O. Fidaner, "Ge quantum well modulators on Si," *ECS Trans.*, vol. 16, no. 10, pp. 851–856, 2008.
72. D. A. B. Miller, "Device requirements for optical interconnects to silicon chips," *Proc. IEEE*, vol. 97, no. 7, pp. 1166–1185, 2009.
73. S. L. Chuang, *Physics of Optoelectronic Devices*. Wiley-Interscience, 1995.
74. H. I. Ralph, "On the theory of the Franz–Keldysh effect," *J. Phys. C*, vol. 1, pp. 378–386, 1968.
75. W. R. Frensley, "Heterostructure and quantum well physics." In *Heterostructures and Quantum Devices*, Ed. N. G. Einspruch and W. R. Frensley. San Diego, CA: Academic Press, 1994, pp. 1–24.
76. D. A. B. Miller, D. S. Chemla, T. C. Damen, A. C. Gossard, W. Wiegmann, T. H. Wood, and C. A. Burrus, "Electric field dependence of optical absorption near the band gap of quantum-well structures," *Phys. Rev. B*, vol. 32, no. 2, pp. 1043–1060, 1985.
77. D. A. B. Miller, J. S. Weiner, and D. S. Chemla, "Electric-field dependence of linear optical properties in quantum well structures: Waveguide electroabsorption and sum rules," *IEEE J. Quantum Electron.*, vol. QE-22, no. 9, pp. 1816–1830, 1986.

78. A. Frova and P. Handler, "Shift of optical absorption edge by an electric field: modulation of light in the space-charge region of a Ge p-n junction," *Appl. Phys. Lett.*, vol. 5, no. 1, pp. 11–13, 1964.
79. H.-C. Luan, D. R. Lim, K. K. Lee, K. M. Chen, J. G. Sandland, K. Wada, and L. C. Kimerling, "High-quality Ge epilayers on Si with low threading-dislocation densities," *Appl. Phys. Lett.*, vol. 75, no. 19, pp. 2909–2911, 1999.
80. Y. Luo et al., "Experimental studies of the Franz–Keldysh effect in CVD grown on GeSi epi on SOI," *Proc. SPIE*, vol. 7944, no. 79440P, 2011.
81. A. Frova and C. M. Penchina, "Energy gap determination in semiconductors by electric-field-modulated optical absorption," *Phys. Stat. Sol.*, vol. 9, no. 3, pp. 767–773, 1965.
82. J. S. Kline, F. Pollak, and M. Cardona, "Electroreflectance in Ge–Si alloys," *Helv. Phys. Acta*, vol. 41, no. 6–7, pp. 968–977, 1968.
83. G. G. MacFarlane, T. P. McLean, J. E. Quarrington, and V. Roberts, "Fine structure in the absorption-edge spectrum of Ge," *Phys. Rev.*, vol. 108, no. 6, pp. 1377–1383, 1957.
84. T. P. McLean, *Progress in Semiconductors*, vol. 5, A. F. Gibson, Ed. Heywood, London: John Wiley & Sons, 1960.
85. R. Hulthen and N. G. Nilsson, "Investigation of the second indirect transition of silicon by means of photoconductivity measurements," *Solid State Commun.*, vol. 18, pp. 1341–1342, 1976.
86. T. Low, Y. P. Feng, M. F. Li, G. Samudtra, Y. C. Yeo, P. Bai, L. Chan, and D. L. Kwong, "First principle study of Si and Ge band structure for UTB MOSFETs applications," Semicon. Device Research Symp., pp. 358–359, 2005.
87. W. M. J. Green, M. J. Rooks, L. Sekaric, and Y. A. Vlasov, "Ultra-compact, low RF power, 10 Gb/s silicon Mach–Zehnder modulator," *Opt. Express*, vol. 15, no. 25, pp. 17106–17113, 2007.
88. J. E. Roth, O. Fidaner, R. K. Schaevitz, Y.-H. Kuo, T. I. Kamins, J. S. Harris, and D. A. B. Miller, "Optical modulator on silicon employing germanium quantum wells," *Opt. Express*, vol. 15, no. 9, pp. 5851–5859, 2007.
89. E. H. Edwards, R. M. Audet, Y. Rong, S. A. Claussen, R. K. Schaevitz, E. Tasyurek, S. Ren, T. I. Kamins, O. I. Dosunmu, M. S. Ünlü, J. S. Harris, and D. A. B. Miller, "Si-Ge surface-normal asymmetric Fabry–Perot quantum confined Stark effect electroabsorption modulator," IEEE Photonics Society Summer Topical Meeting, Playa del Carmen, Mexico, 2010, TuD2.4.
90. R. M. Audet, E. H. Edwards, P. Wahl, and D. A. B. Miller, "Investigation of limits to the optical performance of asymmetric Fabry–Perot electroabsorption modulators," submitted to *IEEE J. Quant. Electron.*, vol. 48, no. 2, pp. 198–209, 2012.
91. Y. Rong, Y. Ge, Y. Huo, M. Fiorentino, M. R. T. Tan, T. I. Kamins, T. J. Ochalski, G. Huyet, and J. S. Harris, "Quantum-confined Stark effect in Ge/SiGe quantum wells on Si," *IEEE J. Select. Top. Quant. Electron.*, vol. 16, no. 1, pp. 85–92, 2010.
92. P. Chaisakul, D. Marris-Morini, G. Isella, D. Chrastina, X. L. Roux, S. Edmond, J.-R. Coudevylle, E. Cassan, and L. Vivien, "Polarization dependence of quantum-confined Stark effect in Ge/SiGe quantum well planar waveguides," *Opt. Lett.*, vol. 36, no. 10, pp. 1794–1796, 2011.
93. S. Ren, "Ge/SiGe quantum well waveguide modulator for optical interconnect systems," PhD dissertation, Stanford University, 2011.
94. S. Ren, Y. Rong, S. A. Claussen, R. K. Schaevitz, T. I. Kamins, J. S. Harris, and D. A. B. Miller, "Ge/SiGe quantum well waveguide modulator monolithically integrated with SOI waveguides, *IEEE Photon. Technol. Lett.* (in press), doi: 10.1109/LPT.2011.2181496.
95. W. A. Zortman, M. R. Watts, D. C. Trotter, R. W. Yong, and A. L. Lentine, "Low-power high-speed silicon microdisk modulators," CLEO, 2010, CThJ4.
96. R. K. Schaevitz, E. H. Edwards, J. E. Roth, E. T. Fei, Y. Rong, P. Wahl, T. I. Kamins, J. S. Harris, and D. A. B. Miller, "A simple quantum well electroabsorption calculator for designing 1310nm and 1550nm optoelectronic modulators with germanium quantum wells," *IEEE J. Quant. Electron.*, vol. 48, no. 2, pp. 187–197, 2012.

10

Photodetectors

Jurgen Michel, Steven J. Koester, Jifeng Liu, Xiaoxin Wang, Michael W. Geis, Steven J. Spector, Matthew E. Grein, Jung U. Yoon, Theodore M. Lyszczarz, and Ning-Ning Feng

CONTENTS

10.1 Introduction

Jurgen Michel

Photodetectors are an integral part of photonics systems. Since photodetectors are critical for the receivers by converting light into electrical signals, the performance requirements are usually very stringent. In fiber optical telecommunication systems, photodetectors are typically made from III–V semiconductor materials because of the superior noise and bandwidth performance of these discrete devices. Due to the indirect band-gap and transparency in the near infrared, silicon has played an insignificant role in telecom applications. When the silicon platform was being considered for high performance, low cost photonics circuits, monolithically integrated photodetectors were very important for the success of these systems. Since Si initially was only considered for applications in the visible, Ge became the semiconductor material of choice because of its excellent absorption properties in the near-infrared and the seamless monolithic integration into CMOS processes.

This chapter traces the development of Ge photodetectors, their performance, and eventual integration into Si photonics circuits, from the development of an adequate growth process of Ge on Si to the demonstration of high performance avalanche photodetectors (APDs). Today, monolithically integrated Ge photodetectors are ubiquitous and can be found in active optical cables or fully integrated with transimpedance amplifiers for telecom applications. Ge APDs are now outperforming III–V based APDs and will dominate the APD marked in the near future (Section 10.3). While Ge photodetectors have taken the lead in detector integration, Si-based photodetectors have also been developed for near-IR applications. Section 10.4 highlights the developments and points to markets where Si photodetectors can have an advantage over Ge photodetectors because of ease of integration and cost. Section 10.5 highlights some of the modeling developments that were necessary to design high efficiency waveguide integrated photodetectors.

10.2 Free Space Germanium Photodetectors

Steven J. Koester

Infrared photodetectors are essential components of fiber-optic communication and imaging systems. In both of these applications, the performance requirements of the photodetectors must be balanced with the overall cost. In most optical communications systems, the state-of-the-art photodetectors consist of III–V based materials. At a wavelength, λ, of 850 nm, GaAs components are typically utilized, while at $\lambda = 1300$ nm and 1550 nm, InGaAs detectors are the standard bearer. Similarly, infrared imagers operating in the eye-safe regime ($\lambda > 1300$ nm) utilize InGaAs components. It would be desirable to utilize silicon technology for these components, both to reduce cost as well as to allow monolithic integration with Si circuitry (this is particularly helpful in imaging systems). Unfortunately, Si is a rather poor optical absorber in the infrared, having an absorption coefficient of about 0.05–0.1 μm^{-1} at $\lambda = 850$ nm, and being transparent for $\lambda > 1100$ nm. To overcome this limitation, a significant effort has been undertaken in recent years to develop Ge photodetectors that can be integrated onto Si substrates. Ge is a nearly ideal material for infrared optical detection. It has an absorption edge of roughly $\lambda = 1800$ nm, and shows strong absorption for $\lambda \leq 1550$ nm, a wavelength that corresponds to the direct transition energy in Ge. Its absorption properties, combined with its high mobilities for electrons and holes, a unique property compared with nearly all other semiconductors, makes it an ideal candidate to replace III–V detectors for a variety of applications.

However, integration of Ge onto Si represents a significant materials challenge due to the large (4.2%) lattice mismatch between Si and Ge. The resulting dislocations degrade photodetector performance by increasing the dark current and reducing carrier mobility. Significant progress has been made in recent years on Ge-on-Si growth, and it is the purpose of this section to provide an overview of recent developments in Ge free space photodetectors for both communications and imaging applications. First, an overview of the trade-offs between free-space vs. waveguide-coupled photodetectors is provided. Next, the different types of Ge-on-Si high-speed photodetectors are reviewed. This discussion includes photodetectors grown on relaxed $Si_{1-x}Ge_x$ buffer layers, Ge grown directly on Si, Ge-on-insulator and Ge-on-silicon-on-insulator (SOI) detectors, resonant cavity enhanced (RCE) detectors and avalanche photodiodes (APDs). The performance capabilities and trade-offs of these different approaches are discussed. Next, a review of the infrared imaging applications of Ge-on-Si free-space detectors is described. Finally, an outlook for future research and applications of this technology is provided.

10.2.1 Comparison between Free-Space and Waveguide-Coupled Detectors

Ge free space photodetectors can be defined as detectors where incident light impinges on the sample directly from air or free space, as opposed to being coupled to the detector through a waveguide. Typically, the light is incident from the normal direction, either from a fiber in close proximity to the surface, or from ambient light in the case of imaging applications. Free-space photodetectors have several advantages over waveguide-coupled detectors. Firstly, they can typically operate over a wide range of wavelengths. They can also be made as large or as small as desired depending upon the application. Free-space detectors may be configured in either lateral or vertical junction geometries that can then be optimized for a particular application. The free-space configuration is particularly useful in fiber-optic communications since their use only requires an optical fiber to be placed in close proximity above the sample, and arrays of fibers can readily be positioned over the detectors for parallel communications applications. For infrared imagers, waveguide coupling is impractical due to the broad spectral range that needs to be collected, as well as the fact that the incident light impinges at a wide range of angles relative to the wafer surface.

Free-space detectors also have numerous drawbacks compared with their waveguide-coupled counterparts. The most important of these is that they typically suffer from a bandwidth responsivity trade-off. This is due to the fact that thick absorbing layers are needed to fully capture the light, but the increased thickness also increases the transit

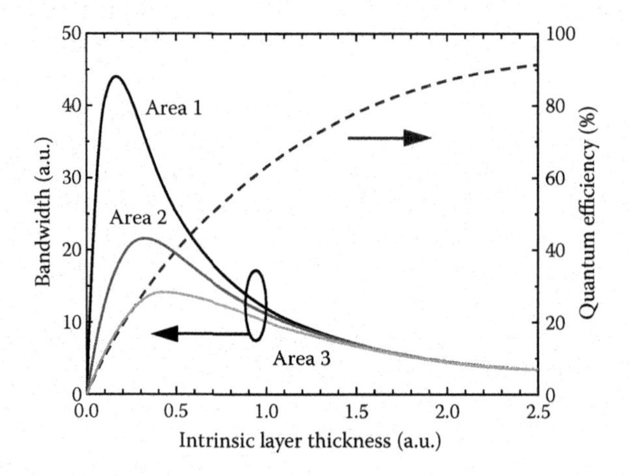

FIGURE 10.1
Illustration of fundamental bandwidth responsivity trade-off for normal-incidence photodetectors. Bandwidth decreases and quantum efficiency increases with increasing i-layer thickness. At very low intrinsic layer thicknesses, RC delays limit the bandwidth. In the above diagram, Area 1 < Area 2 < Area 3.

TABLE 10.1
Comparison of Free-Space vs. Waveguide Ge-on-Si Photodetectors

Parameter	Free Space	Waveguide
Responsivity	Requires thickness $\gg \alpha^{-1}$, limits bandwidth	Requires length $\gg \alpha^{-1}$, doesn't limit bandwidth
Bandwidth	High-speed if absorbing layer thin and area small	High-speed
BW × efficiency	Trade-off	No significant trade-off
Avalanche	Yes	Yes
Resonant cavity	Possible	Difficult (not necessary)
Imagers	Yes	No
Fiber coupling	Surface normal to detector	Edge or grating coupler
Wavelength	Wide band	Narrow band
Size	Variable	Constrained by waveguide

time and therefore reduces the bandwidth of the detectors. If high speeds are required, then thin absorbing layers, and thus lower responsivities must be accepted. This trade-off is shown graphically in Figure 10.1. Resonant cavity designs can overcome this problem to some degree, but with the trade-off of increased process complexity and the fact that high responsivity can only be achieved over a very narrow range of wavelengths. Free space photodetectors can also often be limited by RC delays, particularly for larger-area detectors, and minimizing the resistance often comes at the cost of thicker metal electrodes, which ultimately contribute to shadowing and loss of efficiency. A summary of the trade-offs between free-space and waveguide-coupled photodetectors is provided in Table 10.1. More details on waveguide-coupled photodetectors are presented in the Section 10.3.

10.2.2 Growth and Epitaxy

The potential of Ge for use in infrared photodetectors has been recognized for a long time. The combination of its high purity and long lifetimes, together with high mobilities for both carrier types—a unique property compared with nearly all other semiconductors—make it ideal for infrared optical detectors. Before about the late 1990s, research on Ge photodetectors was almost exclusively limited to stand-alone infrared detectors (Beiting, 1977). However, with the emergence of fiber-optic communications, efforts to integrate Ge detectors with Si receiver electronics were undertaken. In the early days of this research, it was assumed that defects associated with the lattice mismatch of Ge on Si had to be minimized in order to realize photodetectors with acceptable dark current, and therefore a large effort was expended to develop graded buffer layer technology for these devices. One of the first reports of Ge-on-Si detectors using a buffer-layer approach was reported by Luryi et al. (1984), and this work was improved upon considerably in subsequent years (Samavedam, 1998), helped by the coincident interest of using graded buffer layer technology for use in field-effect transistor (FET) applications (Fitzgerald et al., 1991). However, whereas, FET applications require buffer layers graded to Ge contents of 20% or so, Ge photodetectors require grading all the way to 100% Ge, a more substantial challenge. Samavedam et al. (1998) reported vertical p-i-n Ge photodetectors using a several-micron-thick graded buffer layer structure. This graded buffer layer was shown to have extremely low defect density of ~2×10^6 cm^{-2}, and photodetectors fabricated on these layers exhibited dark current density of 0.15 mA/cm^2, which was close to the theoretical pn diode reverse saturation current.

Despite the attractiveness of thick graded buffer layers to produce low defect densities and thus low-dark-current photodetectors, the graded buffer approach has generally fallen out of favor compared with the direct grown of Ge on Si, or the utilization of Ge layers with very thin buffer layers. There are several advantages for Ge grown directly on Si. First of all, graded buffer layers generally require a very thick set of layers to be grown (often several microns), creating significant nonplanarity between the detector top surface and the original substrate. This is not a problem for standalone detectors but is unwieldy for integration with CMOS. The graded buffer structure also does not allow integration with waveguides, or effective utilization of a buried oxide layer for carrier isolation and for use as a backside reflector (e.g., waveguide-coupled and resonant cavity photodetector designs are not possible using graded buffers). For these reasons, the trend in Ge integrated photodetector research has been to move the Ge absorbing layer as close to the Si substrate as possible. This is the case for both free-space and waveguide-coupled configurations.

The first reports of Ge-on-Si detectors fabricated using direct growth of Ge on Si were reported in the pioneering work of Colace et al. (1997). In this work, a two-step growth was utilized, where a thin (~50–100 nm) Ge seed layer was first grown at low temperatures

(~300°C–400°C), followed by a thicker layer grown at high temperatures. The nucleation layer is vital to ensure that Ge growth proceeds in a planar fashion, while the second layer, grown at high temperatures (~550°C–700°C), provides the increased growth rates needed to realize thick absorption layers. The original devices of Colace et al. (1997) utilized a metal-semiconductor-metal configuration and had very high defect densities (~1 × 10^9 cm^{-2}), leading to much higher dark currents than graded-buffer devices. However, advancements in the direct growth of Ge on Si (Masini et al., 1999) led to considerable reduction in defect densities. The densities achieved (~2 × 10^7 cm^{-2}) are only roughly 1 order of magnitude higher than graded buffer layers. In general, this optimization entails high-temperature (~900°C) thermal annealing, which helps to reduce the density of threading dislocations (Luan et al., 2001). The high temperature of the annealing, however, leads to substantial interdiffusion between Si and Ge, a particular problem for high-speed lateral pin geometries (Koester et al., 2006), and can cause integration difficulties with CMOS. Therefore, several alternatives to high-temperature annealing have been developed. Takada et al. (2010) used a hydrogen desorption technique to produce as-grown Ge p-i-n photodiodes on Si with dark current densities of 46 mA/cm^2. The suggested mechanism was a reduction in the surface roughness that effectively reduced the active junction area. Selective epitaxy of Ge films directly on silicon has also been shown (Yu et al., 2009) to produce layers with low dislocation density (~1 × 10^7 cm^{-2}) and p-i-n photodiodes with very low dark current density of 3.2 mA/cm^2 have been fabricated using this technique. While these defect densities are high by CMOS standards, the defects that result from Ge-on-Si growth do not appear to be a barrier to practical applications, with the possible exception of high-sensitivity imaging applications, which will be discussed later in this section. Numerous other low-temperature growth techniques for Ge-on-Si photodetectors have also been developed.

A critical finding in the development of Ge-on-Si photodetectors that is applicable to both free-space and waveguide-coupled geometries, was the realization that the growth of Ge directly on silicon can lead to the formation of tensile strain in the Ge layer (Ishikawa et al., 2005). This strain comes about as a result of the thermal mismatch between Si and Ge. As the wafer is cooled from the growth temperature, the silicon expands to a greater degree than the Ge, creating biaxial tensile strain in the final film. Tensile strain values of roughly 0.2% have been reported in (Liu et al., 2005a), which has the effect of reducing the Γ-point transition energy by 0.03 eV. This extends the energy edge for strong absorption from λ = 1550 nm to 1620 nm, allowing Ge-on-Si photodetectors to cover not only the C-band (1530–1560 nm), but also the L-band (1560–1620 nm), which is an important band for dense wavelength division multiplexing (DWDM) used in long-haul fiber-optic communications (Cannon et al., 2004). Further increases in the strain (0.25%) (Liu et al., 2005a) have been demonstrated by adding externally applied strain.

10.2.3 High-Speed Bulk Ge-on-Si Photodetectors

A great deal of progress on high-speed bulk Ge-on-Si photodetectors has been made since the original work of Colace et al. (1999). As shown in Figure 10.2, Ge-on-Si bulk photodetectors can be configured in a number of geometries, including vertical p-i-n, lateral p-i-n, and metal-semiconductor-metal (MSM). In addition, vertical p-i-n geometries have been reported using both an all-Ge structure as well as hybrid structures where contact is made to the Ge for one contact, and the Si for the other. A review of the recent progress on bulk Ge-on-Si photodetector results is provided in the subsections below.

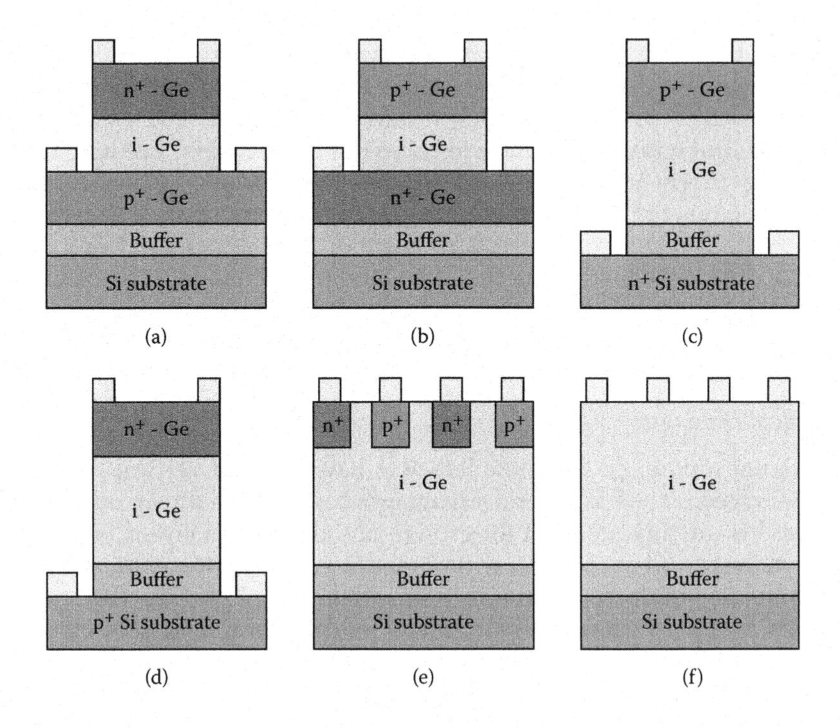

FIGURE 10.2
Diagram showing how basic Ge-on-Si photodetector geometrical variations can be configured in a number of geometries, including (a,b) all-Ge vertical p-i-n, (c,d) hybrid Ge/Si vertical p-i-n, (e) lateral p-i-n, (f) lateral metal-semiconductor-metal (MSM).

10.2.3.1 Lateral Detectors

The lateral MSM contact geometry is attractive due to its process simplicity in that only one lithographic step is needed to form the anode and cathode for a photodetector. Due to the fact that the contact only consists of metal, MSM detectors can have very low parasitic resistance compared with p-i-n devices. This is particularly true for large-area detectors, where metal fingers enable a low-resistance path from the contact pads to the interior of the detector. Finally, since they do not require ion implantation, MSM photodetectors can be fabricated using low temperature processing. However, MSM photodetectors lead to higher dark current than their p-i-n counterparts, and can only be practically fabricated using a lateral contact geometry. Therefore, while the MSM contact geometry is used routinely for wider-gap III–V semiconductors such as GaAs (Ito, 1986), early attempts to realized Ge MSM photodiodes have produced devices with excessively high dark current due to the low band gap of Ge (Colace et al., 1997). Therefore, most of the work on Ge-on-Si MSM photodiodes has involved various techniques at overcoming the dark current problem. For instance, Oh et al. (2004), utilized an amorphous Ge (α-Ge) barrier layer between the metal and semiconductor contact to reduce charge injection from the contacts, and this led to dark currents of 80 mA/cm^2 at −1 V, a value that was an > 100× improvement compared with devices without α-Ge. The device 3-dB bandwidth was observed to be 4.3 GHz at −4 V bias, but the speed of such an approach is likely to be limited by resistance associated with transport across the α-Ge boundary. An alternative approach has been to utilize NiGe Schottky contacts (Zang et al., 2008), where the n-type contact is subjected to a sulfur implantation. This leads to dopant segregation, and reduces the work function of the implanted contact, while the work function of the nonimplanted NiGe contact remains relatively high. Such an approach produced darks currents of ~100 mA/cm^2, similar to Colace et al. (1999), but with −3-dB bandwidths, as high as ~15 GHz. Additional approaches to reducing dark current using dual-work function metallization (Chui et al., 2003) and α-Si barrier (Ciftcioglu et al., 2010) layers have also been demonstrated. Due to the fact that nearly all of the techniques to reduce dark current involve increased process complexity by utilizing different contact materials for cathode and anode or by placing a barrier layer between the contact and the absorbing layer, MSM geometries have not been widely implemented for most Ge photodiodes reported in the literature.

10.2.3.2 Vertical Detectors

Most of the recent reports on Ge photodetectors have utilized vertical p-i-n geometries. The vertical p-i-n geometry has several advantages compared with lateral devices. First of all, since epitaxy is already required for growth of the Ge-on-Si layers, *in situ* doping can be used to form the p^+ and n^+ contacts, without the necessity for ion implantation (however several literature demonstrations have still utilized ion implantation, particularly for the top contact). Vertical p-i-n geometries also allow for a uniform field to be created through the absorbing region, and therefore carrier transport to the contacts is primarily dominated by drift, not diffusion. This feature improves the speed responsivity trade-off as it allows the absorption layer to be made thicker while maintaining good carrier collection properties. In addition, no fine-scale lithography or alignment to the implanted region is needed as in lateral p-i-n devices. However, vertical p-i-n geometries are subject to series resistance from the bottom contact associated with lateral transport through the heavily doped bottom contact. Therefore, high-speed operation is generally limited to relatively small-area devices. Vertical p-i-n detectors can also have high dark current since the junction area is equal to the detector area.

The variations in the Ge-on-Si photodiode vertical p-i-n photodiode geometries are shown in Figure 10.2a–d. The main geometric variations include all-Ge devices and hybrid Ge/Si devices. Of these geometries, the highest speed operation has been obtained from the all-Ge design. In the work of Jutzi et al. (2005), molecular beam epitaxy (MBE) was utilized to grow a p^+-Ge, i-Ge, n^+-Ge structure where the i-layer thickness was only 300 nm. At a wavelength of λ = 1550 nm, devices with 10-μm diameter had −3-dB bandwidth of 25.1 GHz (38.9 GHz) at a bias of 0 V (−2 V). As mentioned previously, series resistance is a limiting factor in the performance of vertical p-i-n detectors, and the devices in Jutzi et al. (2005) with larger diameters of 20 μm (30 μm) had zero-bias bandwidths of only 9.9 GHz (4.4 GHz). Further analysis in Oehme et al. (2006), confirmed the RC-limited performance of these devices, and indicated that optimized 10-μm-diameter devices with similar absorbing layer thicknesses could theoretically have −3-dB bandwidths as high as 62 GHz. The external quantum efficiency in these devices was measured to be 23%, 16%, and 2.8% for λ = 850, 1300, and 1550 nm, respectively. The reported dark current density at a bias voltage of −2 V was >200 mA/cm^2. More recently, design improvements to reduce series resistance have enabled maximum bandwidths to increase to 49 GHz (Klinger et al., 2009), and the device structure and bandwidth response from this work are shown in Figure 10.3. Despite these impressive performance achievements, the relatively thin absorbing layers utilized resulted in responsivities of only 0.2 A/W at λ = 1300 μm and 0.05 A/W at λ = 1550 nm. The low growth-temperature of 300°C in Jutzi et al. (2005) also likely resulted in a lack of significant tensile strain in the MBE-grown layers.

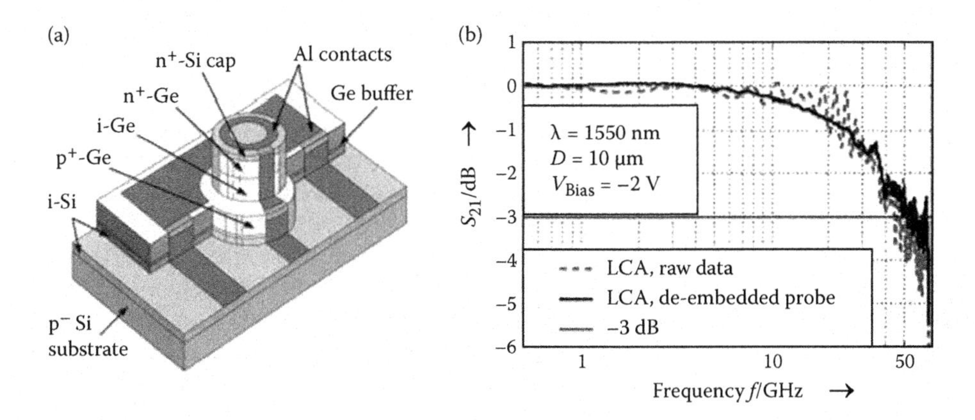

FIGURE 10.3
(a) Diagram of high-speed vertical Ge-on-Si p-i-n photodetector. The diameter is 10 μm (Klinger et al., 2009). (b) Frequency response of a device in (a) at −2 V. A −3-dB bandwidth of 49 GHz was observed (Klinger et al., 2009).

Despite the high bandwidths of the photodetectors described previously, the bandwidth-responsivity product at $\lambda = 1550$ nm is still relatively low. Photodetectors that incorporate tensile strain have been shown to have improved bandwidth responsivity trade-off, particularly at longer wavelengths. As mentioned in the previous section, Liu et al. (2005a), demonstrated Ge-on-Si detectors incorporating Ge layers that had 0.2% tensile strain. These devices demonstrated a −3-dB bandwidth of 8.5 GHz, but with very high responsivity of 0.87 A/W (0.56 A/W) at $\lambda = 1310$ nm ($\lambda = 1550$ nm), and these devices even demonstrated 0.11 A/W responsivity at $\lambda = 1605$ nm. Despite the lower speed, at $\lambda = 1550$ nm, the devices in Liu et al. (2005a) had a higher bandwidth responsivity product compared with the devices in Jutzi et al. (2005), a result that highlights the benefits of tensile strain for improving the long wavelength performance of Ge detectors. The devices in Liu et al. (2005a) also utilized the Si substrate for p-type contact, which may have contributed to the lower dark current density of 22 mA/cm^2 at −2 V bias.

More recently, very high-performance Ge p-i-n photodetectors have been realized that combine both high responsivity and very high-speed operation. In Suh et al. (2009), devices with 17-μm diameter were reported to have −3-dB bandwidth of 33 GHz at −1 V, 0.47 A/W responsivity at $\lambda = 1550$ nm (0.7 A/W at $\lambda = 1.3$ μm) and 75 mA/cm^2 dark current. The devices utilized 1.2-μm-thick Ge layers, and used the Si substrate as the n-type contact. The authors reported that the Ge layer was tensile strained at 0.16%, which could have contributed to the improved responsivity compared with the low-temperature–grown layers in Jutzi et al. (2005). These devices represent the highest bandwidth-responsivity product reported to date for normal-incidence Ge photodetectors at $\lambda = 1550$ nm.

While most applications for Ge photodiodes have focused on applications at $\lambda = 1300$ and 1550 nm, there remains interest in Ge detectors as a low-cost replacement for GaAs photodetectors operating at $\lambda = 850$ nm. Recently, Morse et al. (2006) evaluated the potential Ge-on-Si p-i-n photodetectors as a replacement for GaAs detectors in commercially available 4 Gbps receiver modules. The primary challenge at $\lambda = 850$ nm is to ensure that the Ge layers are grown sufficiently thick to avoid optical absorption in the underlying Si, which can lead to high-frequency "tails" that degrade the digital performance. Therefore, Ge photodiodes with thick (several micron) absorbing layers were grown to ensure full optical absorption by the Ge layer. These devices showed very low dark current of 6 mA/cm^2 at –1 V bias, and had responsivity of 0.59 A/W at λ = 850 nm, a value that is comparable to the responsivity of commercially available GaAs detectors. Measured bandwidths of 9 GHz at –2 V were demonstrated for 50-μm diameter detectors.

Finally, a novel approach to improve the bandwidth-efficiency trade-off in bulk, normal-incidence, Ge p-i-n photodiodes was reported by Chaisakul et al. (2011) who demonstrated high-speed Ge/SiGe multiple quantum wells (MQW) photodiodes using a surface-illuminated vertical p-i-n structure. The structure was developed to reduce the dark current density (0.2 A/cm^2) while still providing high speed operation (26 GHz at –4 V). However, the devices had relatively low responsivity of 0.05 A/W at λ = 1405 nm. This result highlights the difficultly of using the MQW design for normal-incidence devices. Not only is it difficult to build up sufficient thickness to fully absorb the incident light, but also large electric fields are needed to ensure efficient carrier escape from the quantum wells. This limitation was evident from the relatively strong dependence of the bandwidth on applied voltage in these devices.

10.2.4 High-Speed Ge-on-Insulator Photodetectors

Ge-on-SOI and Ge-on-insulator free-space detectors have several advantages compared with their bulk counterparts. First of all, a buried insulating layer can eliminate collection of carriers generated in the Si, therefore allowing the devices to use thinner Ge absorption layers at wavelengths where Si is absorbing, such as λ = 850 nm. The buried insulator can also act as a back-side reflector allowing improved efficiencies if the thicknesses of the buried oxide, Ge layer and anti-reflection coating are tuned correctly. When Ge is grown directly on SOI wafers, the thin Si can also reduce mixing between Si and Ge, since it limits the amount of Si available for interdiffusion. This is particularly important for lateral p-i-n geometries where it is desirable to make the absorbing layer extremely thin. Even a small amount of dilution of the Si into the Ge can dramatically reduce the absorption, since the Γ-point energy in Ge increases rapidly with increasing Si concentration. Finally, the Ge-on-SOI geometry is important in waveguide-coupled geometries, since the SOI-based waveguides require a buried insulator layer for optical confinement. In this section, the application of buried insulator layers for free-space Ge photodetectors is reviewed, while devices utilizing waveguide coupling will be described later in this chapter.

The initial work on Ge-on-SOI photodetectors was reported by Dehlinger et al. (2004). In this work, a 400-nm Ge layer was grown on a commercial SOI wafer using a two-step growth process, where the Si had been thinned to 15 nm. After the Ge growth, cyclic annealing at 900°C was performed to reduce the defect density. The p^+ and n^+ contacts were formed by ion implantation, and then contacted by interdigitated metal electrodes. It should be noted that this fabrication scheme is considerably more complex than simple vertical pin structures, and requires high-precision alignment of the metal fingers to the doped electrodes. Devices with area of 10 × 10 μm^2 and finger spacing of 0.4 μm, had –3-dB bandwidth of 29 GHz at –4 V and external quantum efficiency of 34% at λ = 850 nm. The devices had dark currents of, 20 mA/cm^2 at –2 V. Similar to the work of Jutzi et al. (2005), a very high zero-bias bandwidth of 25 GHz was reported. The buried oxide in these device acts as a back reflector that can be utilized to enhance the responsivity at a particular wavelength. However, the devices in Dehlinger et al. (2004) were not optimally tuned for their target wavelength of λ = 850 nm, and instead the devices had peak responsivity at λ = 895 nm. Subsequently, it was shown in Koester et al. (2006) how the corresponding thicknesses of the Ge, buried oxide, and anti-reflection coating could be tuned simultaneously to optimize the responsivity. For instance, simply by optimizing the thicknesses of the buried oxide and anti-reflection coating, external quantum efficiencies of >80% and 40% should be achievable for devices with the same absorbing layer thickness as in Dehlinger et al. (2004).

In Schow et al. (2006) and Koester et al. (2007), these photodetectors were integrated with transimpedance amplifiers (TIAs) fabricated using a commercial 0.18-μm CMOS process, and benchmarked for their bit-error-rate (BER) and power performance. The results showed that photodetectors integrated with high-gain CMOS transimpedance amplifiers (Figure 10.4) can achieve 15-Gbps operation with a sensitivity of −7.4 dB m at a BER = 10^{-12} while utilizing a single supply voltage of only 2.4 V. Error-free (BER < 10^{-12}) operation of receivers combining a Ge-on-SOI photodiode with a single-ended high-speed receiver front end was also demonstrated at 19 Gb/, and a low-power CMOS IC operating at 10 Gbps using a single 1.1-V supply while consuming only 11 mW of power. These results demonstrate the potential of Ge detectors for high-speed datacom applications.

As mentioned previously, the devices reported in Dehlinger et al. (2004) had external quantum efficiency of 25% at λ = 850 nm, but only 2.5% at λ = 1300 nm. Furthermore, the devices were actually transparent at λ = 1550 nm. This result is inconsistent with expectations for pure Ge, which should have moderate absorption at 1550 nm, and in apparent contradiction to Cannon et al. (2004), where enhanced absorption was observed at long wavelengths due to the tensile strain in the Ge layer. This seemingly contradictory result can be attributed to the very thin absorbing layers that were utilized in Dehlinger et al. (2004). Unlike the layers in Cannon et al. (2004), which were much thicker than the interdiffusion distance, in Koester et al. (2006), it was estimated that the thin Ge layer was diluted with Si by as much as 5% due to the high-temperature cyclic annealing process. Even a small increase in Si concentration in Ge increases the Γ-transition energy (Fischetti, 1996), which more than offsets the potential benefits of tensile strain.

Techniques to improve the long-wavelength response properties of Ge-on-SOI detectors were investigated by Rouviere et al. (2005). Lateral MSM devices were fabricated on Ge layers grown on thin SOI using a low-temperature growth process specifically designed to minimize interdiffusion. These devices, which had Si and Ge layer thicknesses of 50 and 320 nm, respectively, showed a measureable photoresponse at both 1300 and 1550 nm, and

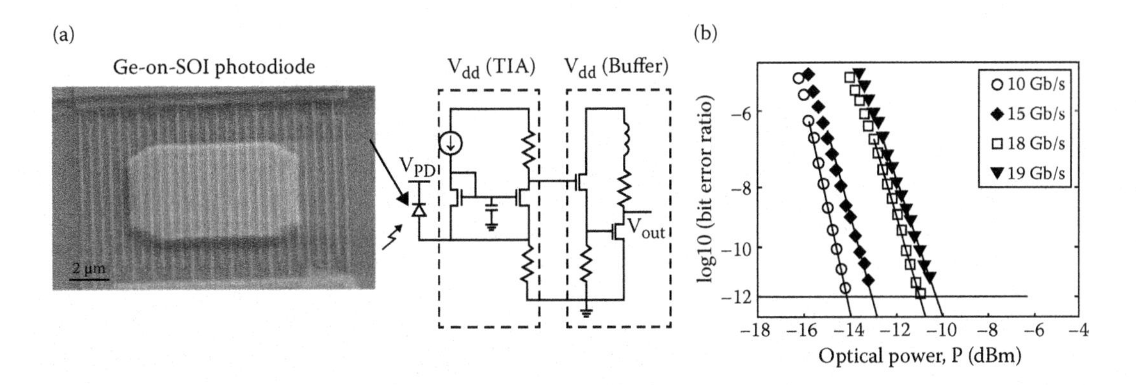

FIGURE 10.4
(a) Receiver circuit showing integration of a Ge-on-SOI with a high-speed CMOS transimpedance amplifier (Koester et al., 2006). (b) Bit-error rate measurements showing that receivers can achieve 19-Gbps operation with a sensitivity of −10.3 dB m at a BER = 10^{-12} (Koester et al., 2006).

at −2 V bias, devices with electrode spacing of 500 nm had −3-dB bandwidths as high as 35 GHz.

Numerous groups have also demonstrated vertical p-i-n devices on SOI. For vertical p-i-n detectors, to take advantage of the buried oxide as a back reflector, the thickness of the entire Ge layer stack above the buried insulator must be less than the absorption length. However, to date, a resonant enhancement of absorption has not been reported for simple single-layer buried oxide devices. In Loh et al. (2007a), a 17-GHz bandwidth vertical pin photodiode operating at λ = 1550 nm was demonstrated using selective-area growth and a thin $Si_{0.8}Ge_{0.2}$ buffer. The devices had very low dark current density of 2 mA/cm^2. Although the responsivity was not reported for these devices, the high-temperature annealing appeared to cause similar reduction to the responsivity at long wavelengths, similar to Koester et al. (2006). Vertical p-i-n devices on thick SOI were reported in Xue et al. (2010), and high optical power handling capability was observed. These devices had bandwidths of 13.4 GHz and responsivity of 0.65 A/W (0.31 A/W) at λ = 1300 nm (λ = 1550 nm).

10.2.5 Resonant Cavity and Avalanche Photodetectors

10.2.5.1 Resonant Cavity Enhanced Ge-on-Si Photodetectors

A logical extension of the resonant enhancement effect of the Ge-on-SOI device concept is to extend this to full quarter wavelength Bragg reflector. It is well-known that a resonant cavity enhanced (RCE) photodetector design can improve the responsivity of a photodetector over a narrow range of wavelengths (Unlu et al., 1992) and numerous examples of resonant-cavity photodetectors have been made using III–V (Heroux et al., 1999) and Si-based systems (Schaub et al., 1999). One of the challenges of group-IV resonant-cavity photodetectors compared with III–Vs is that single-crystal epitaxy cannot be used to grow the Bragg reflector. Therefore, insulating materials are generally needed, and this can lead to substantial integration difficulties. For instance, Schaub et al. (1999) reported Si RCE photodetectors where Si was laterally grown over a Si/SiO_2 Bragg reflecting mirror stack.

Before the initial work on Ge-resonant cavity photodetectors, a considerable amount of work was performed on multiquantum well structures. For instance, Li et al. (2000) fabricated backside illuminated $Si_{0.65}Ge_{0.35}$/Si multiple quantum-well resonant-cavity-enhanced photodetectors and demonstrated their operation at λ = 1305 nm. A top-side Si/SiO_2 layer Bragg reflector was utilized as the top mirror, while buried Si/SiO_2 single interface acted as the bottom mirror, and the devices displayed a responsivity of 0.03 A/W. However, it was not until the work of Dosunmu et al. (2004) that RCE photodetectors were realized using pure Ge layers. These devices utilized double SOI substrates designed for operation at λ = 1550 nm. Using a Ge layer thickness of 737 nm, the devices had a measured quantum efficiency of 14% at λ = 1550 nm at a bias voltage of −2 V. This value was over four times the value of 3.3% expected for a simple single-pass configuration, demonstrating the benefit of the resonant cavity design. Furthermore, simulations in Dosunmu et al. (2004) indicated that by using tensile-strained Ge and an optimized design, a quantum efficiency of 76% at λ = 1550 nm is possible.

An alternative RCE Ge photodetector device was reported in Dosunmu et al. (2005). As shown in Figure 10.5a, the detector was backside illuminated, and a metal top-side reflector as well as a buried insulator layer were used to enhance the optical absorption. These devices, which utilized a Ge layer thickness of 1.45 μm, exhibited quantum efficiencies of nearly 60% at λ = 1540 nm (Figure 10.5b), and had a spectral resonant peak full-width half maximum (FWHM) of 50 nm, which spanned the entire C-band used for long-haul

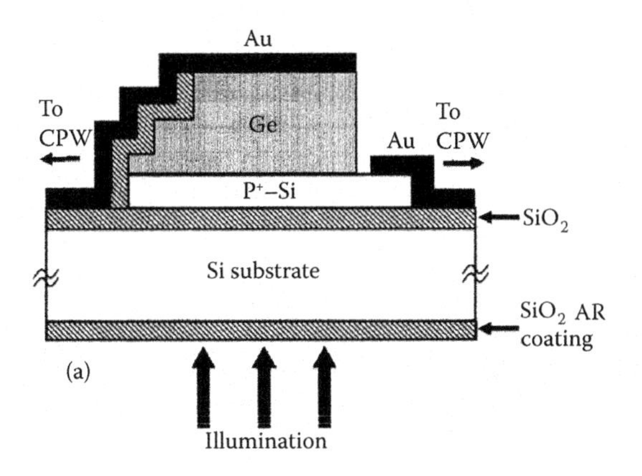

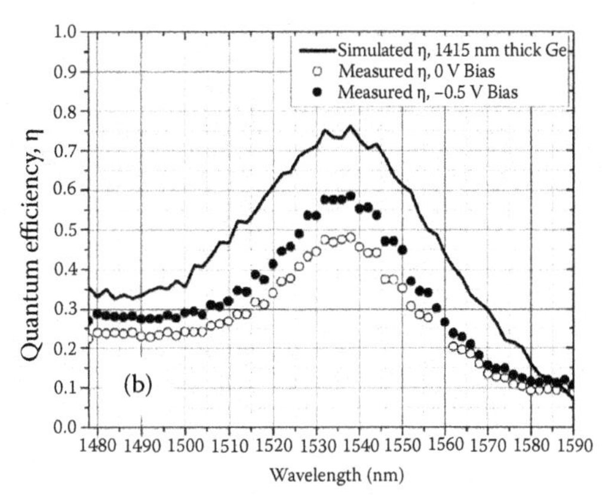

FIGURE 10.5
(a) Cross-sectional view of the back-illuminated Ge–SOI Schottky photodector design (Dosunmu et al., 2005). (b) Simulated and measured quantum efficiency for resonant cavity device shown in (a). Peak quantum efficiencies of nearly 60% were obtained at $\lambda = 1538$ nm (Dosunmu et al., 2005).

optical communications. The authors also reported −3-dB bandwidths of nearly 13 GHz. RCE photodetectors based upon Ge quantum dots have also been reported, but have generally demonstrated low responsivities relative to bulk detectors, despite using multilayer Bragg mirror stacks (Yu et al., 2006; Li et al., 2004). Although RCE photodetectors can help to address the bandwidth-efficiency trade-off, their narrow-band operation and process complexity remain as impediments to their use in practical applications, particularly for monolithic integration with CMOS circuits.

10.2.5.2 Ge-on-Si Avalanche Photodetectors

APDs have long provided a useful method to amplify an optical signal without the need for external electronic circuitry, and APDs have been realized in a wide range of material systems (Carrano et al., 2000; Campbell et al., 2004). However, bulk Ge APDs have proven impractical due to the large noise power that is created due to the avalanching of both electrons and holes. The Ge/Si heterostructure system, therefore, offers an excellent opportunity to create APDs with improved performance if separate absorption and multiplication (SAM) regions can be created. In Ge/Si SAM APDs, the Ge is used for absorption, while the avalanche multiplication takes place in the Si. One such device has been reported by Carroll et al. (2008). In this work, an APD with a nominal responsivity of 3.2×10^{-4} A/W at $\lambda = 1310$ nm and 4.5×10^{-5} A/W at $\lambda = 1550$ nm was realized using an APD structure with 200–400 nm of p^+ Ge grown directly on Si. The high doping was the likely cause of the low nominal responsivity. However, the device displayed a multiplication factor of 400× at a bias of −29 V. Improved results for Ge/Si SAM APDs were reported, where the configuration shown in Figure 10.6 was used (Kang, 2009a). The authors reported sensitivities of −28 dB m at 10 Gbps, for photodiodes operating at $\lambda = 1300$ nm. These devices also had bandwidth-efficiency product of 340 GHz, and low excess noise factor. Due to the interest of using the APD configuration for monolithic photonic integrated circuits, much of the work on Ge/Si APDs transitioned to waveguide coupled devices, and the progress on waveguide-coupled Ge/Si APDs will be reviewed later in this chapter.

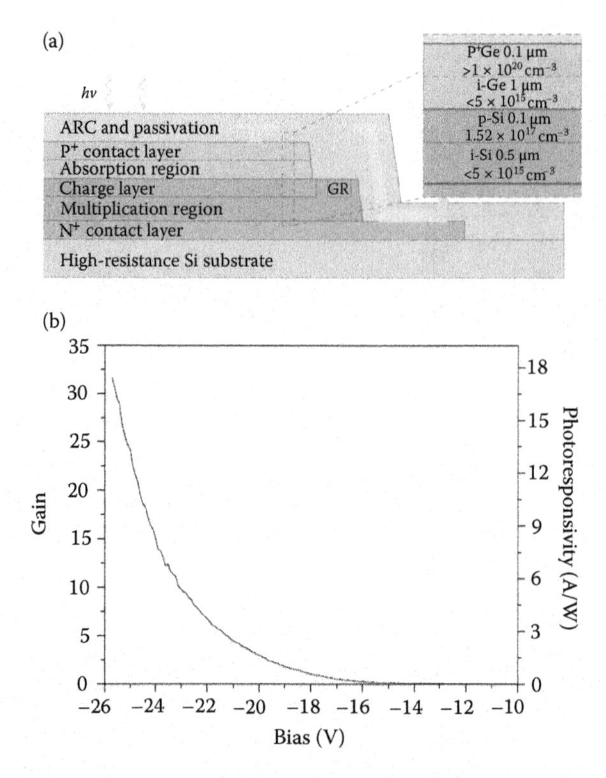

FIGURE 10.6
(a) Cross-sectional diagram of Ge-on-Si avalanche photodiode with separate absorption and multiplication layers. (b) Measured results from device in (a) showing multiplication factors of over 30 (Kang et al., 2009a).

10.2.6 Ge-on-Si Detectors for Near-Infrared Imaging

Perhaps one of the most compelling applications of Ge-on-Si free-space photodetectors is for use in near-infrared imaging systems. As mentioned previously, waveguide coupling is not practical for imaging systems, since a broad spectral range of light needs to be detected. The incident light can also have a wide range of incident angles. Si is a suitable absorber for imagers in the visible range, however, for near-infrared (NIR) imaging, especially in the eye-safer regime ($\lambda > 1300$ nm), Ge can provide a critical capability that is lacking with Si-only systems. Potential applications include NIR illuminated security systems and night-vision cameras. The potential of Ge for use in integrated infrared imagers has been recognized since the early days of Ge-on-Si research (Masini et al., 2002). However, the material quality required for the development of infrared imagers is much more stringent than for high-speed photodetectors. This is due to the fact that imagers are often required to operate in very low light levels, where the noise produced by the dark current can overwhelm the signal noise. Therefore, photodetectors for near-infrared camera applications must have very high quantum efficiency and low dark current. Unlike communication systems that utilize laser sources, infrared cameras also require response over a broad spectral range, and so it is desirable for the imaging pixels to provide uniform responsivity over a wide range of wavelengths. Some of the early work on Ge-on-Si infrared imagers utilized poly-Ge on Si (Masini et al., 2002; Colace et al., 2004, 2007). In Colace et al. (2007), the dark current problem

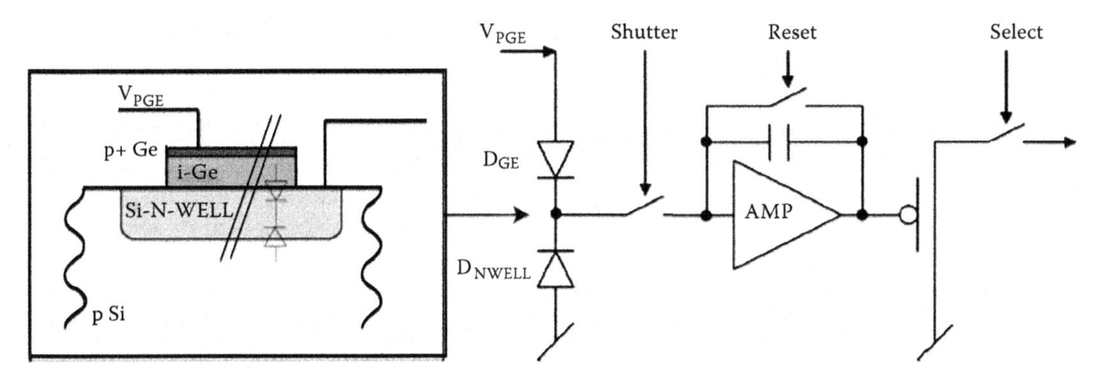

FIGURE 10.7
Schematic diagram of Ge photodiode and pixel circuit for infrared camera use (Kaufmann et al., 2011).

was addressed by utilizing a geometry where a poly-Ge absorbing layer was deposited on top of a standard Si pn junction. This configuration allowed relatively low dark currents of ~1 mA/cm^2 to be achieved, but at the expense of very low responsivity values of ~1 mA/W at λ = 1300 nm and 0.25 mA/W at λ = 1550 nm. These results are likely a result of the very fast recombination times in the defective poly-Ge. More recent work on Ge imaging arrays has focused on single-crystal material. In Kaufmann et al. (2011), a near-infrared image sensor using Ge grown by plasma enhanced chemical vapor deposition was reported. As shown in Figure 10.7, these devices utilized a 1.5-μm Ge layer that acted as the intrinsic layer of a p-i-n diode. The total thermal budget for the layer growth was 780°C, a value that is compatible with front-end CMOS technology. Here, devices with very flat spectral response, and high quantum efficiency of 30% were observed over a broad wavelength range (λ = 800 to 1500 nm). In addition, tensile strain of 0.17% in the Ge layer allowed an absorption edge of λ = 1580 nm to be achieved, and the devices had low dark current densities of 5–10 mA/cm^2. A final demonstration of the capability of Ge NIR imagers was reported in Ackland et al. (2009), where a novel selective growth method was used to reduce the defect density of Ge and allow integration with conventional CMOS circuitry. The devices utilized implanted n- and p-type electrodes, which occupied only a fraction of the total pixel area, thus allowing extremely low dark current densities in the nA/cm^2 range. A fully integrated 768 × 600-pixel camera was reported with capability for both visible and NIR imaging. A 180-nm foundry process enhanced with Ge photodiodes was developed with a unity signal-to-noise ratio of 1 at 10 nW/cm^2 input power. The devices had pixel yield is greater than 98%, and response good uniformity. The results for Ge imaging arrays are very encouraging from a technological perspective, although a compelling marketplace driver for Ge cameras needs to be developed before wide scale implementation can be achieved.

10.2.7 Future Directions

In this section, the future direction of Ge/Si free-space photodetector research is reviewed. A few key areas are highlighted, including the use of GeSn alloys to improve the long-wavelength response of Ge photodetectors, and the development of Ge photodetectors on a variety of different substrates, including flexible substrates.

Despite the success of tensile-strained Ge at extending the absorption window of Ge-on-Si photodetectors to longer wavelengths, the ability to reliably achieve long-wavelength operation using strain is very tenuous. It would therefore be desirable to have additional techniques to extend the absorption window even further to longer wavelengths. An interesting approach that has been realized in recent years is to grow Ge with a small concentration (1%–2%) of Sn, since the direct band gap in Ge is known to decrease even with very small concentrations of Sn (D'Costa, 2007). Extending the absorption edge would potentially allow GeSn to operate in the L-band (λ = 1560–1620 nm) and even the U-band (1620–1680 nm), which could be important for high-speed optical transport networks. To this end, a tremendous amount of work has been performed to develop techniques to grow GeSn films directly on Si substrates, as well as to develop methods to dope these materials for effective contact formation. D'Costa et al. (2009) measured the optical absorption properties of $Ge_{0.98}Sn_{0.02}$ films deposited using chemical vapor deposition. They measured values of α_{Ge} (λ = 1620 nm) = 206 cm^{-1} and α_{Ge} (λ = 1550 nm) = 625 cm^{-1}, while α_{GeSn} (λ = 1620 nm) = 4630 cm^{-1} and α_{GeSn} (λ = 1550 nm) = 6040 cm^{-1}, results that demonstrated the benefit of the GeSn for longer-wavelength absorption. Roucka et al. (2011) further improved this process and fabricated vertical p-i-n devices on $Ge_{0.98}Sn_{0.02}$ layers grown using in situ doping. As shown in Figure 10.8, devices with 900-nm-thick GeSn layers had external quantum efficiencies of 9% at λ = 1550 nm, and in the L-band (λ = 1565–1525 nm) had external quantum efficiencies over a factor 2 higher than the best previous reports in the literature for tensile-strained Ge detectors. Finally, Su et al. (2011), extended this work further and fabricated p-i-n photodetectors using 700-nm $Ge_{0.97}Sn_{0.03}$ layers and measured responsivity of 0.23 A/W (0.12 A/W) at λ = 1540 nm, (λ = 1640 nm), at −1-V bias. These results provided a strong indication that full coverage of the L-band can be achieved using GeSn alloys. The work on GeSn alloys is particularly compelling as these materials are also being investigated for group IV lasers (Chang, 2007), and if proven successful, could form the basis of an all-group-IV optical transceiver technology.

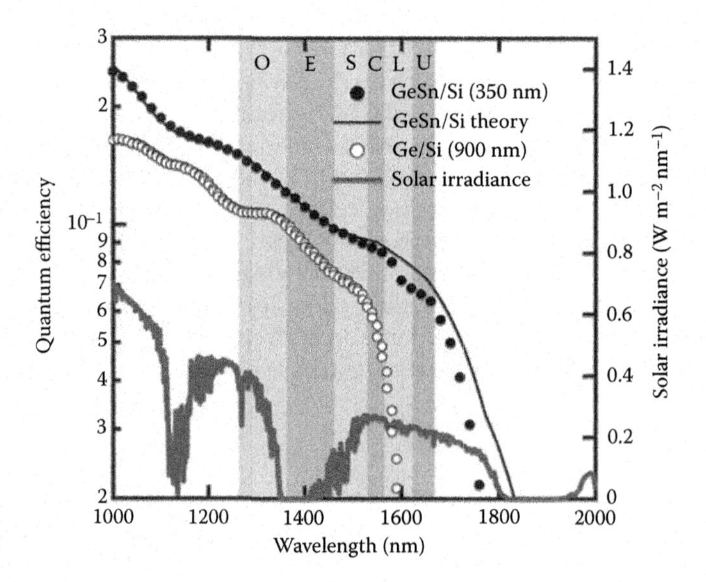

FIGURE 10.8
Comparison of external quantum efficiency for photodiodes on $Ge_{0.98}Sn_{0.02}$/Si (black circles) and Ge/Si (open circles) epi-layers. The black line is a theoretical fit for $Ge_{0.98}Sn_{0.02}$. The shaded vertical regions indicate the various telecom bands, while the dark gray line shows the solar irradiance spectrum. The results show the improvement long wavelength performance of GeSn layers compared with Ge (Roucka et al., 2011).

Another emerging area of research for Ge photodetectors involves the integration of these devices onto low-cost and flexible substrates. For instance, in Colace et al. (2010), using a layer transfer and wafer bonding approach, germanium-on-glass photodiodes were fabricated for both near-infrared (NIR) photodetector applications and for use in multijunction solar cells. The devices showed good responsivity (0.28 A/W at λ = 1550 nm) with very low dark current of ~0.1 mA/cm^2. The low dark current was made possible by the use of high-quality Ge grown on a Ge substrate, therefore eliminating defects associated with strain relaxation. The devices were also characterized for their potential use as the bottom cell in a multijunction solar cell stack. Here, external conversion efficiencies as high as 2.41% were obtained along with a fill factor of 53% under AM1.5 conditions, values similar to those obtained using bulk Ge. Another example of Ge photodiodes on alternative substrates was reported in Yuan et al. (2009), where flexible photodetectors on a plastic substrate were realized by transferring single-crystal germanium membranes grown on a Ge-on-insulator substrate onto a polyethylene terephthalate substrate. Although the infrared responsivity of the devices was not reported, an external quantum efficiency of 42% was reported for λ = 633 nm. The ability to grow/transfer germanium photodiodes onto flexible substrates could have numerous applications in realizing flexible infrared image sensors for military and biological applications, as well as for use in low-cost solar cell modules.

10.2.8 Summary

The recent and future trends in Ge-on-Si free-space photodetectors have been reviewed. The ability to grow high-crystal-quality Ge layers on Si with acceptable defect densities and low dark currents has led to rapid progress in the development of these devices for high-speed optical communications and infrared imaging applications. The basic trends in the material growth have been reviewed, and the wide range of normal-incidence photodetector designs has been described. Devices with bandwidths suitable for 40-Gbps operation, high responsivity and low dark current have been demonstrated using a variety of device configurations. Ge-on-Si use for infrared imaging applications has also been reviewed and recent results confirm the feasibility of Ge integration with CMOS electronics for infrared cameras. Future research looks to extend the application of Ge photodetectors to longer wavelengths, higher speeds, and integration onto a variety of different substrates. With bandwidth requirements for telecommunication and datacommunication applications continuing to increase, Ge free-space detectors are expected to continue to be relevant for both near- and long-term optoelectronic applications.

10.3 Waveguide-Integrated Ge-on-Si Photodetectors

Jifeng Liu

10.3.1 Advantages of Waveguide-Coupled Photodetectors

Waveguide coupling to photodetectors is not only a necessity to construct integrated photonic circuits but also an effective approach to improve photodetector performance in terms of higher bandwidth-efficiency product and lower dark current as has been pointed out in the previous section. In waveguide-integrated photodetectors, the optical absorption path is in the longitudinal direction along the waveguide, which can be as long as needed to fully absorb the input optical power. On the other hand, carrier collection can be implemented in a transverse direction perpendicular to the light propagation direction, with a dimension in the order of several hundred nanometers to minimize carrier transit time and increase bandwidth. Therefore, waveguide-integrated Ge-on-Si photodetectors have no trade-off between bandwidth and quantum efficiency (responsivity) as free-space detectors do. One can optimize the photodetector design for maximum bandwidth while maintaining a high quantum efficiency >90%.

In the most commonly used vertical p-i-n diode structures, there is an optimal intrinsic Ge layer thickness (d) at a given device area (A) for maximum bandwidth. This relation can be derived in the following way. The RC limited bandwidth increases linearly with d due to decrease in capacitance:

$$f_{RC} = \frac{1}{2\pi RC} = \frac{d}{2\pi R_L \varepsilon_{Ge} \varepsilon_0 A}, \tag{10.1}$$

where R_L is the load resistance, C is the capacitance, $\varepsilon_{Ge} = 16$ is the relative dielectric constant of Ge, and $\varepsilon_0 = 8.85 \times 10^{-14}$ F/cm is the dielectric constant of vacuum. Note that this equation describes the inherent RC limited bandwidth of the devices, without including series resistance and contact pad capacitance.

On the other hand, transit limited bandwidth is inversely proportional to d since increase in Ge layer thickness leads to a longer transit time (Sze, 1981):

$$f_{transit} = \frac{0.44 v_{sat}}{d}, \tag{10.2}$$

where $v_{sat} = 6.0 \times 10^6$ cm/s is the saturation drift velocity in Ge (Ryder, 1953; Jacoboni et al., 1981; Madelung, 1982). Since the overall 3-dB bandwidth is given by

$$f_{3\mathrm{dB}} = \sqrt{\frac{1}{1/f_{transit}^2 + 1/f_{RC}^2}}, \tag{10.3}$$

due to the trade-off between RC- and transit-limited bandwidths, we can find that the maximum bandwidth is

$$f_{\max} = \sqrt{\frac{0.11 v_{sat}}{\pi R \varepsilon_{Ge} \varepsilon_0 A}}, \tag{10.4}$$

Correspondingly, the optimal thickness d_{opt} for maximum bandwidth at a given device area A is given by:

$$d_{opt} = \sqrt{0.88 \pi R \varepsilon_{Ge} \varepsilon_0 A v_{sat}}, \tag{10.5}$$

The maximum bandwidth and corresponding Ge layer thickness vs. device area for vertical p-i-n diode structures are shown in Figure 10.9a. Figure 10.9b further shows theoretical and experimental bandwidth-efficiency product vs. device area for waveguide-integrated and free space vertical Ge p-i-n photodetectors. In the theoretical curves a load resistance of $R_L = 50\ \Omega$ is assumed and the Ge layer thicknesses were taken from Figure 10.9a for maximum bandwidth at given device area. The experimental data were obtained from

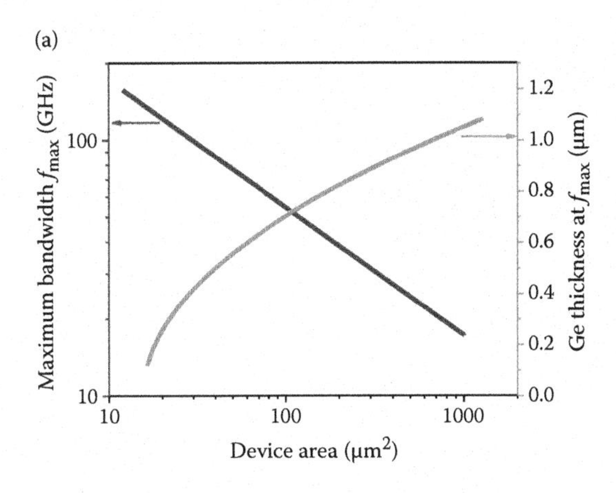

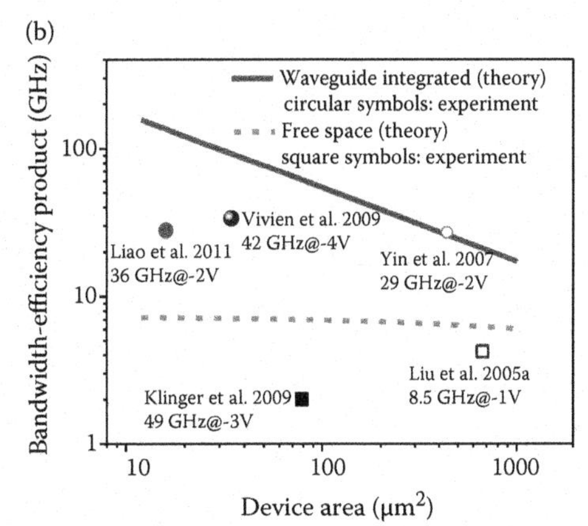

FIGURE 10.9
(a) Maximum bandwidth and corresponding Ge layer thickness vs. device area for vertical p-i-n diode structures. (b) Theoretical and experimental bandwidth-efficiency product vs. device area for waveguide-integrated and free-space vertical Ge p-i-n photodetectors. In the theoretical curves, a load resistance of $R_L = 50\ \Omega$ is assumed, and the Ge layer thicknesses were taken from (a) for maximum bandwidth at given device area.

Liu et al. (2005a), Yin et al. (2007), Klinger et al. (2009), Vivien et al. (2009), and Liao et al. (2011). While high-bandwidth waveguide-coupled devices have no trade-off in quantum efficiency, free space devices suffer from a significantly lower quantum efficiency for high-bandwidth operation due to insufficient absorption in very thin Ge layers. As a result, both theoretical and experimental data indicate that waveguide-coupled devices can achieve a bandwidth-efficiency product ~10× larger than free space devices.

Another advantage of waveguide-coupled photodetectors is that the active device area can be ~10× smaller than free space detectors, so the absolute dark current is also significantly lower considering a similar dark current density in the order of 10 mA/cm^2 (Yin et al., 2007; Vivien et al., 2009; Park et al., 2010; Liao et al., 2011). Since the photodetector noise is determined by the absolute dark current instead of dark current density, waveguide integration enhances the sensitivity of Ge photodetectors with higher responsivity (photocurrent) and lower dark current compared with free-space detectors. Furthermore, when waveguide-coupled photodetectors are selectively grown in narrow trenches (Liu et al., 2006a,b; Beals et al., 2008) the threading dislocation density can be significantly reduced since the dislocations can easily glide to the edge of the Ge mesa and annihilate (Luan et al., 1999). As a result, with adequate sidewall passivation the dark current density can be effective reduced to <1 mA/cm^2 in waveguide-coupled Ge photodetectors selectively grown within 600 nm wide trenches (Beals et al., 2008).

10.3.2 Waveguide Coupling Schemes

High-efficiency waveguide-to-detector coupling is critical to the performance of waveguide-integrated photodetectors. Evanescent coupling and butt coupling are the most commonly used waveguide coupling schemes, as schematically shown in Figure 10.10. Evanescent coupling uses the fact that light can be easily coupled evanescently from a lower index material (i.e., waveguide) to a higher index material (i.e., Ge detector) as long as the index difference is small enough. The electromagnetic mode and hence the optical power are typically transferred gradually to the high index material. On the other hand, butt coupling is more efficient since it directly couples input optical power into the Ge detector

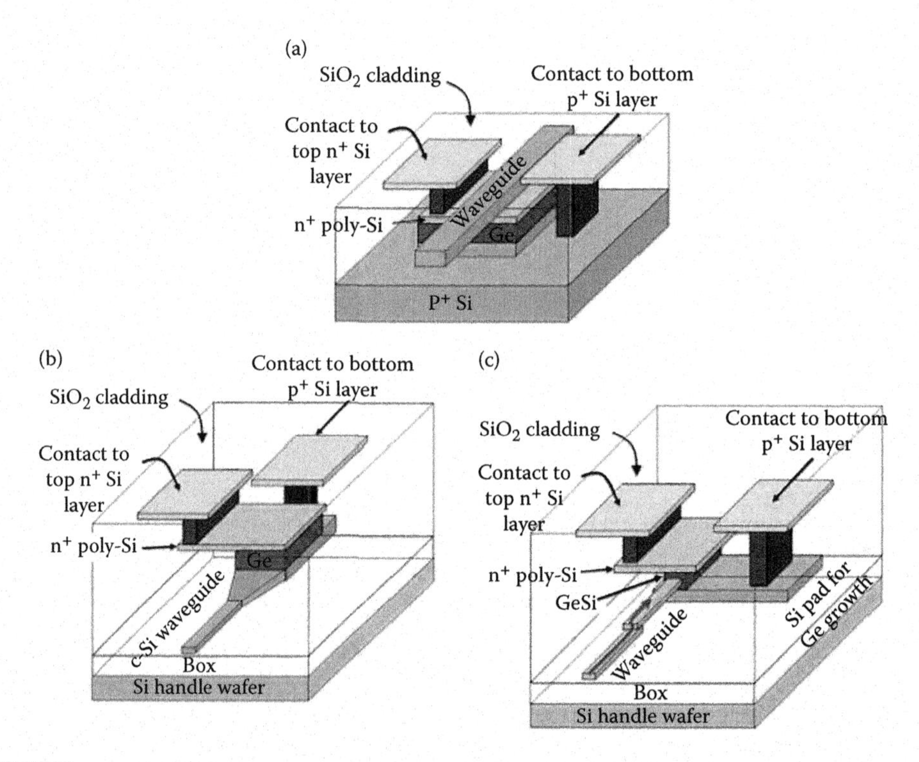

FIGURE 10.10
Schematics of waveguide coupling schemes to Ge photodetectors. Evanescent coupling can be achieved using (a) top or (b) bottom waveguides. The butt-coupling scheme is shown in (c). A vertical coupler can be applied to couple light from the crystalline Si waveguide to the upper amorphous Si waveguide for butt coupling with the Ge photodetector.

region, offering higher optical absorption per unit length. However, this coupling scheme typically requires a stringent design for mode matching conditions between waveguide and detector modes, so butt-coupled devices are generally less tolerant to fabrication errors than evanescently couples ones. For high-performance, single-crystalline Ge photodetectors, selective growth in pre-patterned trenches is required to fabricate butt-coupling structures (Liu et al., 2006a,b; Vivien et al., 2007, 2009; Liu et al., 2008; Beals et al., 2008; Ren, 2011), which is more complicated to implement compared with evanescently coupled detectors.

10.3.2.1 Evanescent Coupling

In evanescently coupled photodetectors, the waveguide is usually placed either on the bottom or the top of the Ge photodetector to simplify fabrication processes. We will call them "vertically coupled devices" in the later text. In bottom-waveguide–coupled structures, Ge photodetectors are grown epitaxially on crystalline Si waveguides fabricated on SOI substrates. By comparison, top-waveguide–coupled structures allow greater flexibility in choosing waveguide material since there is no epitaxy requirement. The waveguide material can be amorphous, such as amorphous Si (a-Si), SiON, or SiN. These amorphous waveguides are applicable to photonic circuits at upper interconnection level, which is a trend in future development of electronic-photonic integration.

Ahn et al. (2010) systematically studied coupling processes in vertically waveguide-coupled photodetectors, and identified three coupling mechanisms: (i) Direct coupling and scattering into the photodetector due to mode matching and abrupt effective index change when photons in the input waveguide reach the detector. (ii) Evanescent coupling at a constant rate and (iii) Backscattering of residual optical power in the waveguide into the detector at the end of waveguide/detector coupling region due to abrupt change in effective index and geometry. Most waveguide-coupled photodetectors are designed to absorb all input photons, so mechanism (iii) is negligible in most cases since there is hardly any residual optical power in the waveguide at the end of coupling region. Mechanism (i) is more efficient than mechanism (ii) in optical power coupling to the photodetector, so it is the preferred mechanism for maximal coupling efficiency.

Ahn et al. (2010) also investigated two major factors that affect coupling efficiency: (1) the refractive index difference between the photodetector material and the core of the waveguide materials, $n_{PD}-n_{core}$, and (2) the geometrical factor of the waveguide. Generally speaking, low index difference (i.e., small $n_{PD}-n_{core}$), small cross-sectional dimensions of waveguides, and low aspect ratio of waveguide height to width lead to mechanism (i)–dominated regime and greatly enhanced coupling efficiency. This phenomenon is due to the fact that lower index differences reduce impedance mismatch between waveguide and photodetector and facilitates coupling, while smaller waveguide cross-sectional dimensions or aspect ratio helps to increase optical confinement in the photodetector for more efficient optical absorption at the waveguide/detector coupling region.

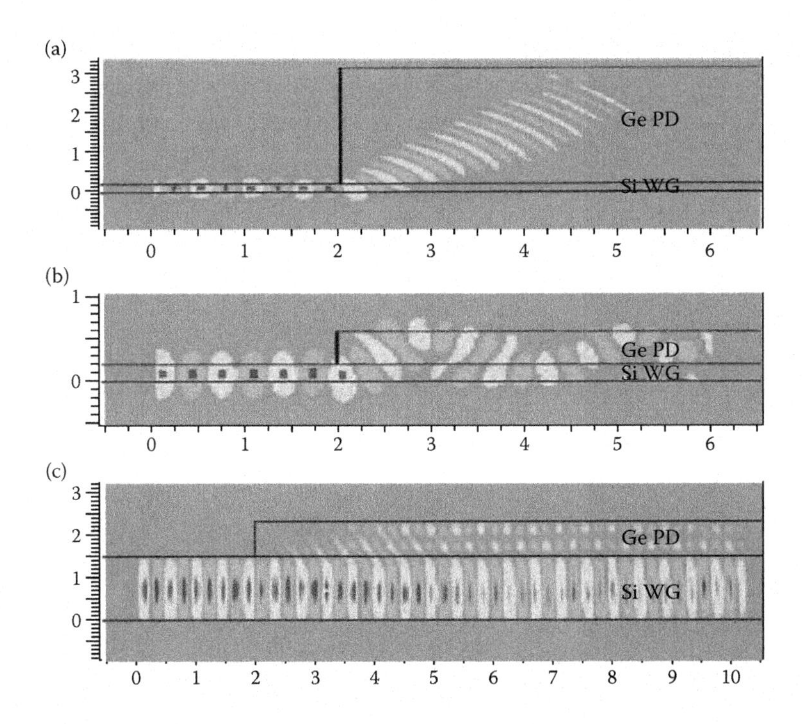

FIGURE 10.11
(See color insert.) 3D FDTD side-section view of photon propagation in bottom-waveguide–coupling structures, where (a) 3.0-μm- vs. (b) 0.4-μm-thick Ge photodetector is placed on a 0.2-μm-thick Si channel waveguide and (c) 0.8-μm-thick Ge photodetector is placed on a 1.5-μm-thick Si rib waveguide structure (Yin et al., 2007). Ge absorption coefficient of $\alpha_{Ge} = 4000\ cm^{-1}$ at $\lambda = 1550$ nm was assumed for simulation, i.e., considering 0.22% tensile strain in Ge (Liu et al., 2005b). As shown in the figures, when the thickness of Ge photodetector decreases relative to the Si waveguide, the coupling efficiency deceases due to a shift in coupling mechanism from direct coupling and scattering dominated regime in (a) to slow evanescent coupling-dominated regime in (c).

Figure 10.11 illustrates the effect of Si waveguide dimensions on coupling efficiency via 3D finite difference time domain (FDTD) simulation (Ahn et al., 2010). In Figure 10.11a, the Ge layer thickness is 3 μm~$1/\alpha_{Ge}$, where α_{Ge}= 4000 cm^{-1} is the absorption coefficient of 0.2% tensile strained Ge at 1550 nm (Liu et al., 2005b). This thickness is significantly larger than that of the Si channel waveguide (200 nm thick). Due to the small index difference ($n_{Ge} - n_{Si} = 0.5$) coupling occurs immediately as photons in the Si waveguide reach the Ge detector. Since the Ge layer is relatively thick, most photons are absorbed during the first pass and there is almost no optical power reflected back into the Si waveguide. This scenario represents the most efficient coupling dominated by mechanism (i) (direct coupling and scattering). In Figure 10.11b, the Ge layer thickness is reduced to 400 nm ≪ $1/\alpha_{Ge}$, so although the coupling still occurs immediately, there is some optical power reflected back to Si waveguide, creating an oscillating pattern similar to waveguide-to-waveguide coupling. In this case, mechanism (ii) contributes significantly to coupling in addition to mechanism (i). Correspondingly, the coupling length is increased to ~10 μm compared with ~5 μm in Figure 10.11a. This scenario has been confirmed by experimental results in Liu et al. (2006b). Figure 10.11c shows coupling from a 1.5-μm-thick Si ridge waveguide structure to an 800-nm-thick Ge photodetector. Since the Si waveguide is much thicker than the Ge detector, most of the optical mode is confined in the Si waveguide region, resulting in a low coupling efficiency dominated by mechanism (ii). Consequently, a relatively long photodetector of ~100 μm is required for complete optical absorption. This analysis is consistent with the experimental results reported by Yin et al. (2007). To enhance evanescent coupling efficiency between thick Si waveguides and Ge photodetectors, Wang and Liu (2011) have designed a step coupling structure to create mirror imaging modes at the Ge/Si interface using a vertical multimode interference (MMI) mechanism. The coupling length has been

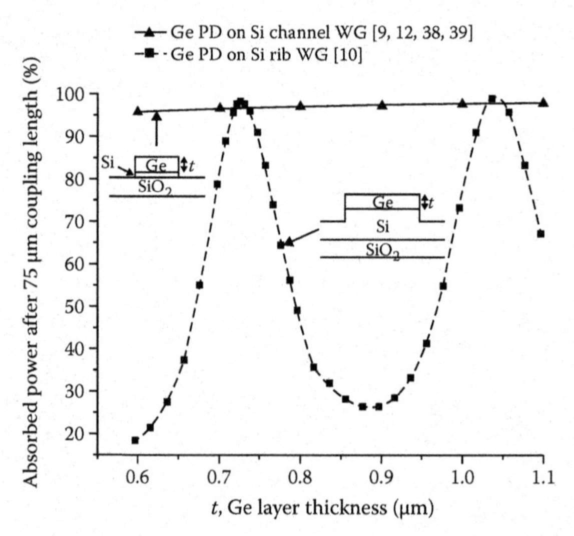

FIGURE 10.12
Dependence of absorption on Ge layer thickness in the two types of the Si-waveguide–coupled Ge photodetectors (Ahn, 2011). One employs the Si channel waveguide formed on SOI substrate before Ge photodetector is grown, as shown in Liu et al. (2006b), Wang (2008a), Michel et al. (2007), and Rouvière et al. (2005b). The other has Si rib waveguide structure, as shown in Yin et al. (2007). In the former, the Si waveguide thickness is 200 nm, whereas in the latter, it is >1 μm. While coupling from thin Si channel waveguides show little absorption dependence on Ge layer thickness, coupling from thick Si ridge waveguides show a strong absorption dependence on Ge thickness due to phase matching conditions. The film thickness dependence of Ge photodetectors evanescently coupled to thick Si ridge waveguides was also reported in Feng et al. (2009).

effectively reduced by 5× for 70% absorption and 3× for 90% absorption. More details about these couplers are presented in the next section on waveguide-coupled Ge-on-Si APDs.

For weak evanescent coupling dominated by mechanism (ii) due to thick Si waveguides or low index waveguide core materials such as SiON or SiN_x, the thickness of the Ge layer is a key design parameter since it determines phase matching conditions in vertical evanescent coupling. In these cases, waveguide-to-detector coupling can be treated as optical coupling between two waveguides (Ahn et al., 2011; Feng et al., 2010). In contrast, when the Ge photodetector thickness is far greater than the waveguide, phase matching conditions are easily satisfied so the coupling is relatively insensitive to photodetector layer thickness. For example, in optical coupling from thick Si ridge waveguides to Ge photodetectors, both Feng et al. (2010) and Ahn et al. (2011) showed that coupling efficiency varies periodically with Ge layer thicknesses, in which phase matching conditions are achieved at certain Ge thicknesses for maximum coupling efficiency (Figure 10.12). Similar periodically occurred Ge thicknesses for optimal coupling from SiON or Si_3N_4 waveguides to Ge detectors were also shown by Ahn et al. (2011). This phenomenon is in contrast to the case of coupling from thin Si channel waveguides to Ge photodetectors, where coupling efficiency becomes independent of Ge thickness as long as the Ge layer thickness is far greater than that of Si (see Figure 10.12) so that phase matching conditions can be satisfied easily.

10.3.2.2 Butt Coupling

As mentioned earlier, butt coupling launches optical power directly into the photodetector region, thereby offering higher coupling efficiency compared with evanescent coupling in most cases. Furthermore, these coupling structures can also be applied to efficient GeSi electro-absorption modulators (EAM) for monolithic photonic data links (Liu et al., 2007, 2008; Feng et al., 2011a,b). The fabrication process of butt-coupled devices is typically more elaborate than evanescently coupled ones, however, since Ge has to be selectively grown to fill predefined trenches to achieve butt coupling. Chemical mechanical polishing (CMP) is often required to flatten the top of the Ge mesas (Beals et al., 2008; Liu et al., 2008; Feng et al., 2009) since single crystalline selective growth usually results in faceting and a non-planar surface. An example of such a fabrication process is schematically shown in Figure 10.13, where cross-sectional scanning electron microscopy (SEM) images of Ge overgrown out of the trench and after planarization are also presented.

Similar to the case of evanescent coupling, a smaller index difference between waveguide core material and photodetector material reduces impedance mismatch for lower reflection losses at the waveguide/detector interface. To ensure a high coupling efficiency, three types of butt-coupling design have been reported:

1. Design Ge photodetector as a segment of a waveguide and optimize its dimensions for maximum modal overlap with that of the input waveguide (Liu et al., 2007, 2008). In this approach, optical modes are first solved for the input waveguide and the Ge photodetector, respectively. Then modal overlap is calculated based on the solved modes, and Ge photodetector dimensions are varied to optimize the modal overlap for maximum coupling efficiency. An example of such a design is shown in Figure 10.14, where modal overlap of a 600-nm-wide GeSi (0.8%Si) photodetector with a standard 500 nm (wide) × 200 nm (thick) single-mode Si channel waveguide is plotted as a function of GeSi photodetector thickness (H). As we can see from the plot, when the GeSi thickness decreases the modal overlap increases, leading to a smaller coupling loss. However, one should note that in this case the

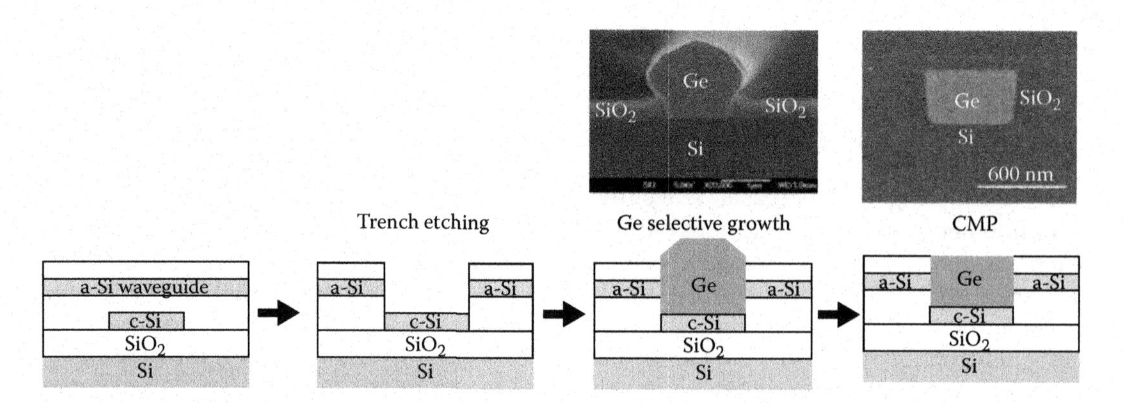

FIGURE 10.13
Schematic fabrication process of butt-coupled Ge photodetectors on SOI platform. "a-Si" and "c-Si" refer to amorphous Si and crystalline Si, respectively. Trench-filling selective growth is applied to achieve these structures. Cross-sectional SEM pictures of trench-filled Ge selective growth and Ge photodetector region after CMP are also shown.

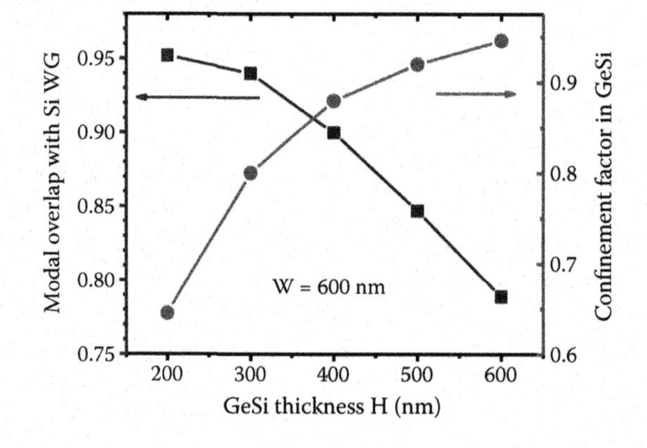

FIGURE 10.14
Modal overlap of a 600-nm-wide GeSi (0.8%Si) photodetector with a standard 500 nm (wide) × 200 nm (thick) single-mode Si channel waveguide plotted as a function of the GeSi photodetector thickness (H). The corresponding optical confinement factor in Ge region is also shown.

optical confinement factor in GeSi region also decreases, indicating that the length of the GeSi photodetector will need to be increased for complete optical absorption. Therefore, the overall device design should not only consider the coupling loss, but also trade-off in device footprint and possibly RC-limited bandwidth. This approach can achieve the most compact device area since the width of the Ge waveguide region can be as small as ~600 nm. Another advantage, as mentioned earlier, is that threading dislocation can be more easily annihilated in these very narrow selectively grown Ge mesas, leading to a better material quality. When the sidewalls of these narrow Ge mesas are well passivated by plasma-enhanced chemical vapor deposition (PECVD) SiO_2 with high hydrogen concentration, a very low dark current density of 0.7 mA/cm^2 has been achieved (Beals et al., 2008), compared with ~10 mA/cm^2 for common Ge-on-Si photodetectors. A disadvantage, however, is that this type of design requires a stringent control of fabrication process to implement the optimal device structure.

2. Use a taper as a mode converter to enhance waveguide-detector butt coupling. Feng et al. (2009) has reported a tapered mode converter to funnel optical power from a large-core Si rib waveguide efficiently into Ge photodetectors. The coupling efficiency is significantly higher than evanescent coupling shown in Figure 10.11c. A 10-μm-long butt-coupled device offers 90% quantum efficiency, compared with a ~100-μm photodetector length required for a similar quantum efficiency using evanescent coupling in Figure 10.11c. This mode-converter butt-coupling design is especially useful when there is a large mismatch between the cross-sectional dimensions of input waveguide and the Ge photodetector.
3. Use a wide patch of Ge photodetector to capture the optical power coupled and scattered into the Ge region (Vivien et al., 2007, 2009). This kind of optical design is somewhat similar to MMIs, except we do not care where the optical power couples out but rather choose a device length for complete optical absorption of input photons. Compared with the previous two approaches, this method is more tolerant to fabrication errors, especially misalignment in the lateral direction perpendicular to the waveguide. However, the Ge device area tends to be larger as a trade-off.

In all three butt-coupling approaches, a high quantum efficiency of >90% have been reported, indicating very efficient coupling.

10.3.3 Photodetector Structures

Similar to free-space coupled photodetectors, there are mainly two types of device structures for waveguide-coupled photodetectors: p-i-n diodes and metal–semiconductor–metal devices. Recently, a junction field effect transistor (JFET) structure has been applied to waveguide-coupled detectors, in which the large valence band offset between Ge and Si is applied to accumulate photogenerated holes in the Ge layer and gate the channel (Wang et al., 2011). As shown in Table 10.2, MSM structures tend to have a much larger dark current than p-i-n structures since it is hard to make a good Schottky contact to Ge due to Fermi-level pinning at the surface of Ge. Furthermore, p-i-n photodiodes can work at 0 bias under photovoltaic mode with practically no power consumption, while MSM devices have to be biased to work. Therefore, from noise and energy efficiency point of view p-i-n diodes are preferred. MSM structures do offer more flexibility in device fabrication though. For example, no implantation or high temperature dopant activation is needed for MSM structures, so they can potentially be fabricated using back-end-of-line CMOS process provided that high-quality Ge deposition can be achieved at low temperatures <450°C.

Another issue to be noted in waveguide-coupled photodetector design is that the position of metal contacts has to be well controlled to avoid optical absorption by metal layers. Metallic contacts are usually offset from the active Ge region to prevent direct interaction with the optical mode. The metal contact region can be completely shifted away from the Ge photodetector, as shown in Figure 10.10b,c, or it can be narrowed down and shifted to the edge of Ge photodetector region to further reduce contact resistance without interacting with the optical mode (Liao et al., 2011). While adequately designed metallic contacts adjacent to the Ge active region can provide surface plasma enhanced responsivity of waveguide-coupled MSM Ge photodetectors (Ren et al., 2010), such enhancement is highly polarization dependent due to the inherent nature of surface plasma. As a rule of thumb, for most applications metal contacts should be shifted away from the optical mode by at least 1 μm to avoid undesirable optical losses.

TABLE 10.2

Summary of Device Performance of Different Waveguide-Coupled Ge-on-Si Photodetectors

Responsivity (A/W) at 1550 nm		3-dB Bandwidth (GHz)		Dark Current Density (mA/cm²)	Dark Current (μA) at −1 V	Device Design	Reference
Max	0 Bias	Max	0 Bias				
1.0 at −3 V	–	4.5 at −3 V	–	0.7[a]	2×10^{-4}	Butt, p-i-n	Liu, 2006 Beals et al., 2008
1.08	1.08	7.2	6.6	1.3×10^{3} [a]	1	Top, p-i-n	Ahn et al., 2007
1	0	25 at −6 V	–	6.5×10^{5} [a]	130	Butt, MSM	Vivien et al., 2007
0.89	0.89	31 at −2 V	15.7	51 at 2 V	0.17 at −2 V	Bottom, p-i-n	Yin et al., 2007
0.85	0.85	26	–	–	3	Bottom, p-i-n	Masini 2008
0.65	–	18	–	125	0.06	Bottom p-i-n	Wang 2008b
1.1	–	37 at −3 V	17.5	1.6×10^{4}	1.3	Butt, p-i-n	Feng et al., 2009
1 at −4 V	0.2	42 at −4 V	12	60	0.018	Butt, p-i-n	Vivien et al., 2009
>1.1 at <1540 nm	–	50 at −5 V	–	8900 at −5 V[a]	4 at −5 V	Bottom, MSM	Chen and Lipson et al., 2009[b]
0.42	0	40 at −2.5 V	–	–	90	Bottom, MSM	Assefa[c] et al., 2010a, b
0.8	–	2.6 at −10 V	–	15	0.06	Bottom, p-i-n	Park et al., 2010 Tsuchizawa 2011
0.95	–	36	–	29	0.0046	Bottom, p-i-n	Liao et al., 2011
0.64	–	8	–	–	0.5	Bottom, JFET	Wang et al., 2011

Note: All Ge devices are fabricated by epitaxial growth on Si and performance data are for −1 V reverse bias, unless otherwise stated.

[a] Calculated and/or read from corresponding figures of the references.

[b] Ge layer fabricated by rapid melt growth.

[c] Ge layer fabricated by wafer bonding.

10.3.4 Summary of Device Performance

Since the first reports on waveguide-integrated Ge-on-Si photodetectors (Liu et al., 2006a,b; Ahn et al., 2007), the performance of these device have been rapidly improved in recent years. Table 10.2 summarizes the reported performance of waveguide-coupled Ge-on-Si photodetectors. The responsivity is typically ~1 A/W at λ = 1550 nm, or >80% quantum efficiency. When the Ge device is long enough, a wavelength-independent high responsivity has been extended to the L band (Ahn et al., 2007; Beals et al., 2008) due to the long absorption length and tensile strained enhanced absorption (Ishikawa et al., 2003; Liu et al., 2005b). The minimal dark current density reported is 0.7 mA/cm² with a very small absolute dark current of 0.2 nA (Beals et al., 2008), significantly lower than free-space photodetectors (~10 mA/cm²). The maximum bandwidth has exceeded 40 GHz (Vivien et al., 2009; Chen and Lipson, 2009a; Assefa et al., 2010a), and the maximum bandwidth-efficiency product has exceeded 30 GHz (Feng et al., 2009).

In recent years, bandwidths at 0 bias have also been improved significantly especially for waveguide-integrated p-i-n photodiodes since the thickness or width of the intrinsic Ge region can be reduced to obtain a stronger built-in electric field at 0 bias without affecting the optical absorption path length in the longitudinal direction. Full responsivity (Ahn et al., 2007; Yin et al., 2007) and >15 GHz bandwidth (Yin et al., 2007; Feng et al., 2009) have been demonstrated at 0 bias. Operations under photovoltaic mode at 0 bias are especially beneficial to achieve high-energy efficiency in large-scale electronic-photonic integrated circuits since the photodetectors practically consume no power at all.

As we have shown in Figure 10.9, theoretically waveguide-integrated Ge photodetectors can achieve a bandwidth efficiency produce exceeding 100 GHz. There are several reasons why experimental data have not reached this level yet: (1) Series resistance and additional capacitance associated with metal contact pads tend to decrease the bandwidth. This issue can be addressed by improving metal contact design and fabrication (2) Intrinsic Ge thickness/width for ultrahigh bandwidth operation (~100 GHz) is so small that efficient waveguide-detector coupling structures become more difficult to fabricate. As we can see in Figure 10.9a, for an ultrahigh bandwidth of ~100 GHz a very thin intrinsic Ge layer thickness <200 nm is required for vertical p-i-n didoes. As we have shown in Figure 10.11b, when the Ge layer thickness is comparable to that of standard single-mode Si channel waveguides (~200 nm) the evanescent coupling efficiency decreases such that longer devices are required to fully absorb the input light. The increase in device length can result in a larger device area that leads to higher capacitance and reduces RC-limited bandwidth. Butt coupling can be more efficient in this case, but due to fabrication complexity, butt coupling to such thin Ge layers have not been reported so far. If, on the other hand, one takes a lateral p-i-n configuration with an intrinsic Ge region width of ~200 nm, then tapered structures have to be adopted as mode converters to couple the light from 500 nm wide Si waveguide into 200 nm wide intrinsic Ge region. In any case waveguide coupling to these very thin or very narrow intrinsic Ge region requires more elaborate design and fabrication. (3) Measuring a bandwidth >60 GHz require advanced measurement equipment that are widely available. With progress in fabrication and measurement technologies all these barriers can be overcome in principle.

10.3.5 Waveguide-Coupled Ge-on-Si Photodetectors for System Integration

With the development of waveguide-coupled photodetectors, in recent years these devices have been coupled to other integrated photonic devices for system integration. Masini et al. (2008) has demonstrated a 10-Gbps optical receiver based on waveguide-coupled Ge photodetectors operating at low voltages, with a −0.3 V reverse bias for Ge photodetectors and 1.4 V driving voltage for the transimpedance amplifier (TIA) circuits. Pinguet et al. (2008) further demonstrated a 40-Gbps (4 × 10 Gbps) monolithic optical transceiver comprising Si Mach–Zehnder interferometer (MZI) modulators and waveguide-integrated Ge photodetectors. The transceiver demonstrated a good sensitivity of −20 dB m at a bit error rate (BER) of 10^{-12} with −1 V reverse bias for the Ge photodetector. Chen et al. (2009a) reported a 3-Gbps photonic link on Si comprising Si ring modulators coupled to Ge photodetectors. Park et al. (2010) reported integration and synchronous operation of Si variable optical attenuators (VOAs) with waveguide-coupled Ge photodetectors. Recently there has also been a surge of research on integrating Ge receivers with wavelength division multiplexing filters (WDM). Fang et al. (2010a) reported a 320-Gbps (32 × 10 Gbps) WDM multichannel receiver consisting of a Si arrayed waveguide grating (AWG) coupled to a Ge photodetector array. A sensitivity between 16 and 19 dB m was achieved at BER = 10^{-11}. In a later publication, Si ring resonator WDM filters coupled with Ge photodetectors were investigated (Fang et al., 2010b), demonstrating an 8-channel devices with 20 Gbps per channel. Assefa et al. (2010) reported 6-channel apodized-coupled ring resonator WDM filters integrated with Ge photodetectors and TIA circuits designed to operate at 7 Gbps per channel. Qian et al. (2011) reported a WDM receiver with a 12-channel echelle grating demultiplexer and an array of 12-channel Ge photodetectors. A bandwidth of 22 GHz was demonstrated at −1 V reverse bias. Doerr et al. (2011) demonstrated polarization-independent 8-channel SiO_2/Si_3N_4 AWG integrated with Si waveguides and Ge photodetectors, with a sensitivity of −20.3 dBm at BER = 10^{-12} and 2.5 Gbps data rate. Chen et al. (2011) further showed a 40-wavelength DWDM demultiplexer based on similar AWG devices, with single channel operation demonstrated up to 40 Gbps using Ge photodetectors and SiGe TIA. Clearly, waveguide-coupled Ge photodetectors on Si and SOI have become one of the most mature optical components in Si-based integrated photonic circuits.

10.4 APD Photodetectors

Xiaoxin Wang

APDs exploit avalanche multiplication of photogenerated carriers through impact ionization to amplify the detector current and improve the device sensitivity. Functionally, an APD photodetector can be viewed as an ordinary photodetector connected with a carrier multiplier. Physically, an APD photodetector consists of two separate regions: absorption region and multiplication region. A charge layer may be inserted between the two regions to control the electric field distribution inside the APD device. Thus, a typical APD photodetector features a separate-absorption-charge-multiplication (SCAM) configuration.

Recently, SCAM-configured Ge/Si APDs are developed to compete with group III–V counterparts for a market share in optical communications (Ang et al., 2010; Kang et al., 2007, 2009a, 2009b, 2011; Wang et al., 2009). For Ge/Si APDs, Ge exhibits excellent absorption in the telecommunications wavelength range of 1.3–1.6 μm (Liu et al., 2005a), while Si multiplication region offers low multiplication noise and high gain-bandwidth product (Hawkins et al., 1996). The achievement of high quality epitaxial Ge films on Si can leverage both advantages of Ge and Si materials for promising Ge/Si APDs (Kang et al., 2009a; Michel et al., 2010).

In this section, APD basic principles are provided, and then the sensitivity of APD receiver is introduced. The current research progress in free-space and waveguide-integrated Ge/Si APD photodetectors, finally the fabrication process is covered.

10.4.1 APD Basic Principles

In a semiconductor with the presence of a high electric field, the number of free carriers (electrons and holes) can be multiplied through a chain of impact ionization processes. Multiplication gain M is known as the number of extra electron–hole pairs generated by an initial carrier (electron or hole). In fact, carrier multiplication in semiconductors is a complex random process dependent on the electric field and involving different kinds of scattering mechanisms. The average number of electron–hole pairs created by a carrier per unit distance along the electric field direction is called impact ionization coefficient (α denoted as the impact ionization coefficient for electron, and β for hole). The ratio of α/β or β/α is defined as ionization coefficient ratio k, which is always equal to or smaller than 1. Ideally, k of the multiplication region should be minimized such that the multiplication process is dominated by either electron or hole to reduce the multiplication noise and improve the gain-bandwidth product. Among the multiplication materials used in the conventional group III–V APDs, InP has a k of 0.4–0.5 (Campbell et al., 1988), and InAlAs has a k of 0.1–0.2 (Kinsey et al., 2001). As a comparison, Si has a small k value ($k < 0.1$) with electron multiplication dominated (Chynoweth, 1958), while the k of Ge is close to 1.0 at room temperature (Mikawa et al., 1980). Hence, by combining a highly desirable Si multiplication region with the success of epitaxial Ge photodiodes, Ge/Si APDs promise to be a competitive candidate in high-speed optical communication systems with a higher gain-bandwidth product and better sensitivity over group III–V APDs. On the other hand, the design of Ge/Si APDs should prevent the multiplication process happening in Ge absorption region to degrade the device sensitivity.

Figure 10.15 shows a cross-sectional view of a Ge/Si APD with a SACM structure, together with its internal electric field distribution. In the Ge absorption region incident photons are absorbed and photocarriers are generated. Under the reverse bias the holes are swept toward the p^+ contact and electrons toward the n^+ contact. When traveling in the Si multiplication region with a high electric field, electrons primarily undergo a series of impact ionizations to create electron–hole pairs, which consequently amplifies the photocurrent. From the perspective of the high-sensitivity, high-speed APD design, the internal electric field distribution must meet four basic requirements. The electric fields of the absorption region and multiplication region are defined as E_a and E_m, respectively. Correspondingly, the requirements are summarized as the following:

- E_a is more than several kV/cm to ensure that carriers drift at their saturation velocities to have minimal carrier transit time in the Ge absorption region (Jacoboni et al., 1981).

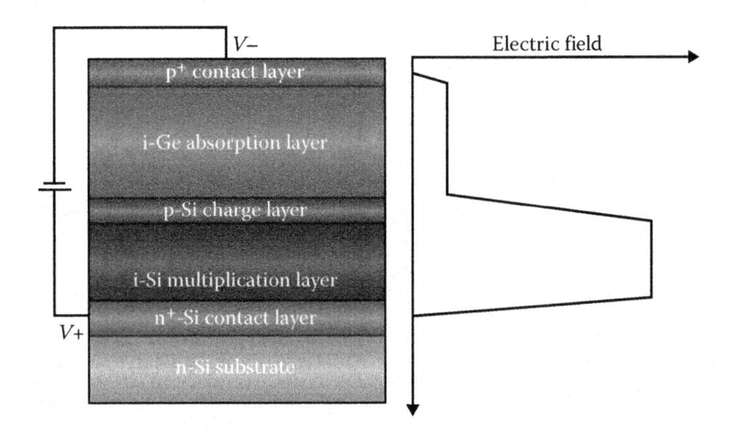

FIGURE 10.15
Schematic cross section of a Ge/Si APD with a SACM structure and its internal electric field distribution. (From Michel, J., Liu J. F., and Kimerling, L.C. 2010. *Nature Photonics* 4: 527–34. Reproduced with permission.)

- E_a is less than one hundred kV/cm to suppress the band-to-band tunneling current and the breakdown, as well as avoid the carrier multiplication in the Ge absorption region (Kyuregyan, 1989).
- E_m is more than several hundred kV/cm to ensure the occurrence of large electron multiplication in the Si multiplication region. The multiplication gain M is dependent on the electric field and the thickness of the multiplication region. Assuming the multiplication length is 1 μm, $E_m > 300$–500 kV/cm is sufficient to produce a substantially large gain (Lee et al., 1964).
- E_m is less than 700 kV/cm to keep the band-to-band tunneling current contribution in the Si avalanche region to a negligible level for high-performance APDs (Hurkx, 1982).

The requirements of the internal electric field distribution impose stringent constraints on the device design parameters, such as layer thickness and doping profile. Additionally, the design of an APD device should take the following specifications into account:

- *Primary responsivity R (A/W)* is defined as the ratio of the photocurrent to the input optical power at the multiplication gain $M = 1$, which is mainly determined by the thickness of the Ge absorption region. The primary responsivity can be experimentally obtained from a reference p-i-n photodiode with the same Ge absorption thickness.
- *Breakdown voltage V_{bd} (V)* is defined as the voltage at which the multiplication gain M goes to infinity. In practice, it is measured as the voltage at a dark current of 10 μA. The operating voltage is usually at 90% of the breakdown voltage ($0.9V_{bd}$) to ensure the device stability and reliability.
- *Breakdown voltage thermal coefficient* is defined as $\delta = (\Delta V_{bd}/V_{bd})/\Delta T$, where ΔV_{bd} and ΔT are the increments of V_{bd} and temperature, respectively. It reflects the material property of the multiplication region. For Ge/Si APDs, δ is 0.05% over a temperature range from 200 to 300 K, which is typical for other Si-based APDs and is about 30%–50% of the InP-based APDs (Kang et al., 2009a). The less breakdown voltage thermal coefficient offers Ge/Si APDs another advantage in facilitating the device temperature stabilization.
- *Punch-through voltage V_p (V)* corresponds to the voltage at which the depletion region penetrates into the Ge absorption region. If punch-through occurs before the field in the multiplication region is sufficient to produce gain, the photocurrent above punch-through will be relatively flat and can be used as a unity gain reference point. Accordingly, the punch-through voltage can be determined as the voltage where the current quickly jumps to a plateau. For APD practical applications, the operational bias ($0.9V_{bd}$) is always well above the punch-through voltage.
- *Multiplication gain M at $0.9V_{bd}$* is a practical gain defined at the operating voltage $0.9V_{bd}$. It can be experimentally determined by the ratio of the responsivity at $0.9V_{bd}$ to the primary responsivity.
- *3-dB Bandwidth (GHz)* is the frequency where the photodetector output electrical power has dropped 3 dB from a low-frequency reference. The limiting factors for Ge/Si APD bandwidth are carrier transit time, resistive-capacitive delay (RC delay), and avalanche build-up effect at a high multiplication gain M.

To harness the performance enhancement, it is crucial for a careful optimization of the doping profile and thickness of the respective layer of the Ge/Si APD structure in Figure 10.15. Particularly, it is found that the doping charge density σ_c of the p-Si charge layer is critically important. Examples of the simulated electric field in the SACM-configured Ge/Si APDs with different parameters of the p-Si charge layer are shown in Figure 10.16. For simulations, the structure in Figure 10.15 is specified as: p$^+$-Si (200 nm, $1 \times 10^{20}/\text{cm}^3$)/i-Ge (1 μm, $1 \times 10^{15}/\text{cm}^3$)/p-Si ($t_c$, N_c)/i-Si (500 nm, $1 \times 10^{15}/\text{cm}^3$)/n$^+$-Si (1 μm, $2.5 \times 10^{19}/\text{cm}^3$). The simulation voltage is set at 25 V, close to the breakdown voltage. When the thickness of the charge layer t_c is fixed at 100 nm, E_a is decreased and E_m is enhanced with the increase of the charge layer doping concentration N_c, as shown in Figure 10.16a. For the doping concentration $N_c = 3 \times 10^{17}/\text{cm}^3$, i-Ge absorption region cannot be depleted completely at 25 V. On the other hand, for $N_c = 1 \times 10^{17}/\text{cm}^3$, the electric field E_m is not sufficient for the avalanche multiplication in Si while the high electric field E_a may lead to the unfavorable band-to-band tunneling and breakdown in Ge. Hence, the optimal N_c window is 1.5–2.0 × $10^{17}/\text{cm}^3$ for a fixed charge layer thickness of 100 nm. Since the total charge density of the p-Si charge layer, σ_c, equals the product of the doping concentration N_c and the thickness t_c, another set of simulations is performed where both N_c and t_c are changed while keeping their product, σ_c, fixed at $1.5 \times 10^{12}/\text{cm}^2$. As shown in Figure 10.16b, it is clear that the electric field profile remains almost the same as long as σ_c stays the same except for a few insignificant details. This verifies that the total charge density σ_c is the determining parameter in controlling the electric field profile of SCAM-configured APDs.

A typical simulated current–voltage curve of a Ge/Si APD with a charge density of $2 \times 10^{12}/\text{cm}^2$ is shown in Figure 10.17a. The punch-through voltage V_p and breakdown voltage V_{bd} are indicated by arrows. At $V = V_p$, the current quickly jumps to a plateau, while at $V = V_{bd}$, the current increases significantly. Figure 10.17b summarizes V_{bd} and V_p as a function of the total charge density σ_c of the charge layer. It is divided into three charge density regions based on the electric field profile and the voltage difference ΔV between V_{bd} and V_p. Region A ($\sigma_c < 1 \times 10^{12}/\text{cm}^2$) is the Ge breakdown tunneling region where ΔV is very large and thus a high multiplication electric field exists in Ge under the operating condition. In region B ($1 \times 10^{12}/\text{cm}^2 < \sigma_c < 3 \times 10^{12}/\text{cm}^2$), $V_p < V_{bd}$, i-Si breakdown happens

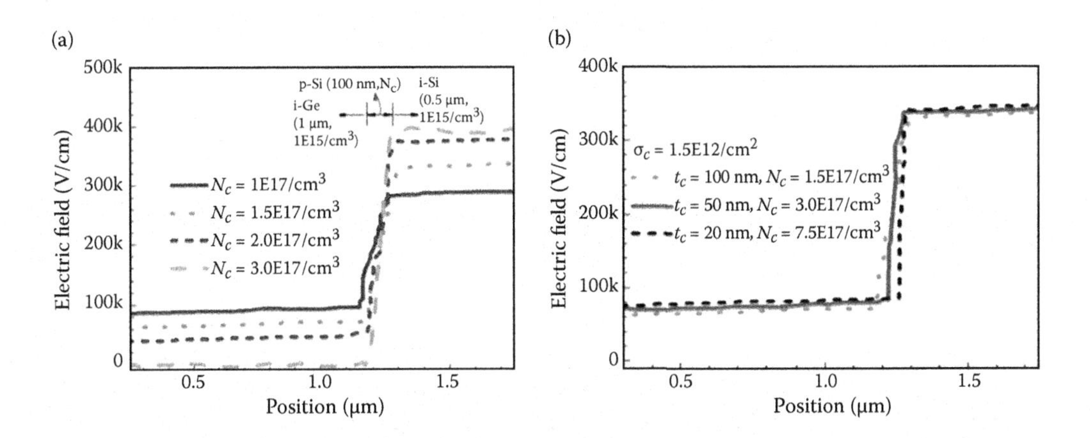

FIGURE 10.16
Simulated electric field distribution at 25 V in SACM-configured Ge/Si APDs with different p-Si charge layer parameters (a): thickness t_c is fixed at 100 nm and doping concentration N_c = 1, 1.5, 2.0, 3.0 × $10^{17}/\text{cm}^3$, (b) total charge density σ_c is fixed at $1.5 \times 10^{12}/\text{cm}^2$ with both t_c and N_c changed.

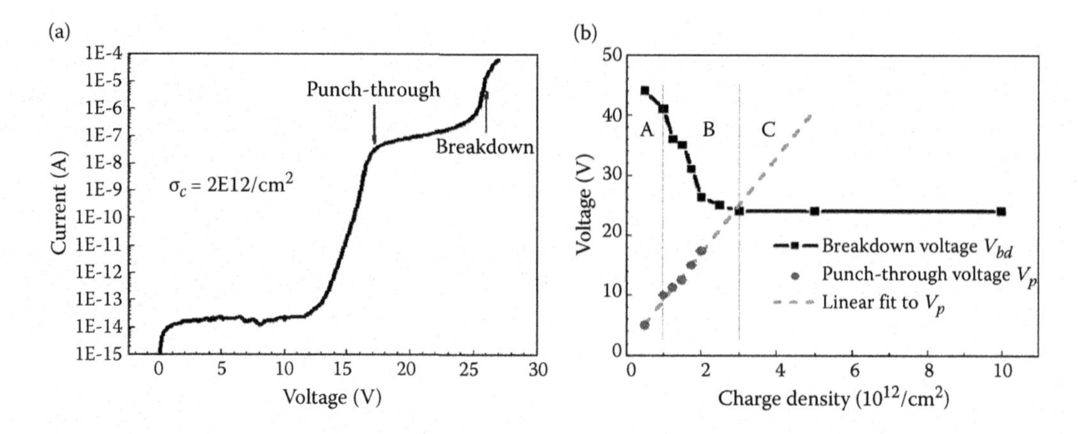

FIGURE 10.17
(a) A typical simulated current–voltage curve of a Ge/Si APD with a charge density of $2 \times 10^{12}/\text{cm}^2$. The punch-through and breakdown voltages are indicated by arrows. (b) Breakdown voltage and punch-through voltage as a function of the total charge density of the charge layer.

after the i-Ge depletion and the electric field E_m is very large. In region C ($\sigma_c > 3 \times 10^{12}/\text{cm}^2$), $V_p > V_{bd}$, i-Si breakdown happens before i-Ge punch-through, which means that Ge absorption layer is not completely depleted at V_{bd} thus limiting the photocarrier transit. Therefore, region B is favorable to SACM-Ge/Si APDs from the perspective of the working mechanism. It should be noted that ΔV decreases with increasing the charge density in region B. Practically, the optimal σ_c is chosen in the range 1.5–2.0 × $10^{12}/\text{cm}^2$ for the case of p^+-Si (200 nm, 1 × $10^{20}/\text{cm}^3$)/i-Ge (1 μm, 1 × $10^{15}/\text{cm}^3$)/p-Si (t_c, N_c)/i-Si (500 nm, 1 × $10^{15}/\text{cm}^3$)/n^+-Si (1 μm, 2.5 × $10^{19}/\text{cm}^3$).

10.4.2 Analysis of APD-Receiver Sensitivity

In terms of APD performance and application, APD receiver sensitivity is an overall figure of merit, which reveals the relationship among gain-bandwidth product, responsivity, excess multiplication noise and dark current of APD devices. The sensitivity is the minimal input optical power that APD device can detect.

Under zero (ideal) extinction ratio assumption, APD receiver sensitivity can be expressed as (Kressel, 1982)

$$Sen_{APD} = 10 \times \log \left\{ \frac{Q}{R} \left[\left(\frac{\langle i_0^2 \rangle_T}{M^2} + 2qF(M)I_{DM}I_2B \right)^{\frac{1}{2}} + qQF(M)I_1B \right] \times 1000 \right\} \text{dBm} \qquad (10.6)$$

where q is electron charge (=1.6 × 10^{-19} C); I_1 is a Personic integral, $I_1 = 0.5$ for rectangular input pulses; I_2 is an integral determined by circuit transfer function, $I_2 = 0.55$ for cosine output pulses; M is the multiplication gain; B is the bandwidth (Hz); R is the primary APD responsivity (A/W) at $M = 1$; $\langle i_0^2 \rangle_T$ is equivalent mean-square amplifier noise, mainly coming from the front-end amplifier in a receiver, and is a function of the system's bit rate; I_{DM} is APD dark current that undergoes multiplication, also called primary dark current; Q is photocurrent signal-to-average-noise ratio for a given bit error rate (BER), $Q = 6.4$ for $BER = 10^{-10}$; $F(M)$ is McIntyre excess noise factor, described in Equation 10.7:

$$F(M) = \frac{\langle M^2 \rangle}{\langle M \rangle^2} = M\left[1-(1-k)\frac{(M-1)^2}{M^2}\right] \tag{10.7}$$

k is electron and hole ionization coefficient ratio, $k \approx 0.1$ for Si.

Note that total dark current I_D of APD are divided into two parts, $I_D = I_{DU} + I_{DM}M$, I_{DU} is unmultiplied dark current, I_{DU} short noise contribution can be neglected in the case of good surface passivation and insignificant sidewall/surface leakage.

According to Equation 10.6, a low M reduces excess noise factor $F(M)$ and multiplication noise but increases the receiver amplifier noise effect. A high multiplication gain M achieves lower receiver amplifier noise but gives more multiplication noise. This trade-off determines an optimal operational gain M for the best APD receiver sensitivity.

As a comparison, the sensitivity of p-i-n receiver can be simplified as

$$Sen_{PIN} = 10 \times \log\left(\frac{Q}{R} \times \sqrt{\langle i_0^2 \rangle}_T \times 1000\right) \text{dBm} \tag{10.8}$$

Under the conditions of 10 Gbps, BER at 10^{-10}, root-mean-square noise of 1 μA and R of 0.5 A/W, the theoretical p-i-n receiver sensitivity is estimated –18.9 dBm.

Figure 10.18 shows theoretical 10 Gbps Ge/Si APD receiver sensitivity at $BER = 10^{-10}$ as a function of M for different primary dark currents I_{DM}. This figure assumes that the primary responsivity of Ge/Si APD at 1310 nm is 0.5 A/W, which corresponds to 1.0-μm-thick Ge absorption layer. It also assumes that Ge/Si APD is used with a very sensitive transimpedance amplifier (TIA) at present, which input-referred root-mean-square (RMS) noise is typically ~1 μA. Clearly at $I_{DM} < 1$ μA and $M > 10$, Ge/Si APD receiver sensitivity is theoretically better than commercial III–V APD receivers (–28~–26 dBm) and more sensitive than p-i-n-receivers by >10 dBm. The experimental data of Ge/Si APD receiver is

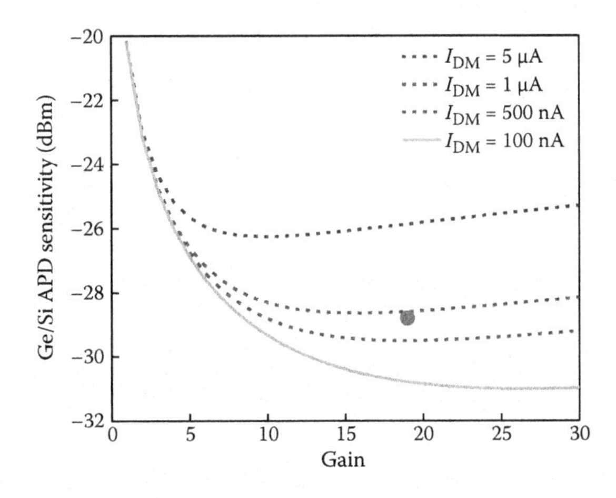

FIGURE 10.18
Theoretical 10 Gbps Ge/Si APD receiver sensitivity at a bit-error rate of 10^{-10}, wavelength of 1310 nm, and an ionization ratio of $k = 0.1$, as a function of gain for different primary dark currents, I_{DM}. The dot shows the Ge/Si APD receiver measurement from Kang et al. (2009b). (From Michel, J., Liu J. F., and Kimerling, L.C. 2010. *Nature Photonics* 4: 527–34. Reproduced with permission.)

also shown in Figure 10.18 as a dot. At the data rate of 10 Gbps, the sensitivity at 1310 nm is measured to be −28.7 dBm at a gain of 19 and $BER = 10^{-10}$, which is in good agreement with the theoretical model. It is anticipated that the APD receiver sensitivity can be further improved by 3 dBm by means of enhancing the primary responsivity to typical 0.85 A/W and reducing the primary dark current to 100 nA.

10.4.3 Free-Space Ge/Si APD Photodetectors

The potential advantages over group III–V APDs have attracted great interest in Ge/Si APDs. Recently considerable efforts are made in the research and development of free-space (normal incidence) and waveguide-integrated Ge/Si APDs.

For a typical free space Ge/Si APD, the thicknesses of Ge absorption region and Si multiplication region are 1 and 0.5 μm, respectively (Kang et al., 2009a,b). Their concentrations are required less than $5 \times 10^{15}/\text{cm}^3$ for better device performance, otherwise their electric fields are quite nonuniform. The key charge density of the p-Si charge layer is $1.5–2 \times 10^{12}/\text{cm}^2$, formed by either in situ doping or implantation (Kang et al., 2007; Wang et al., 2009). It has been demonstrated a monolithically grown Ge/Si APD with a gain–bandwidth product of 340 GHz, a k_{eff} of 0.09 and a sensitivity of −28 dBm at 10 Gbps operating at the wavelength of 1300 nm (Kang et al., 2009a).

The room temperature dark current and photocurrent of a typical 30-μm-diameter APD are shown in Figure 10.19a. The breakdown voltage V_{bd} is −25 V, defined at a dark current of 10 μA. The punch-through is not clear from current curves in Figure 10.19a, indicating that multiplication may occur prior to punch-through. The multiplication gain M is 8 at $0.9V_{bd}$, obtained from a reference *p-i-i-n* structure with a primary responsivity of 0.55 A/W (Figure 10.19b). Liu et al. (2009) proposed an alternative method to determine the gain through excess noise analysis for the case of the punch-through gain greater than unity instead of using a reference structure (Liu et al., 2009).

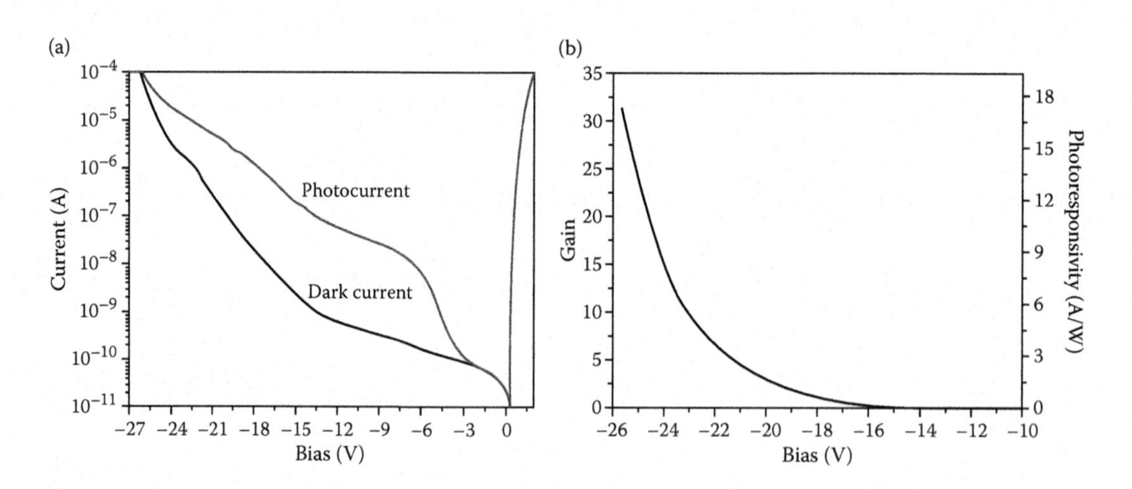

FIGURE 10.19
Direct current characteristics of a 30-μm-diameter Ge/Si APD: (a) the measured dark current (black curve) and total photocurrent (gray curve) at room temperature; (b) the measured multiplication gain and photoresponsivity as a function of bias at a wavelength of 1.3 μm. (From Kang, Y.M., Liu, H.D., Morse, M. et al. 2009a. *Nature Photonics* 3: 59–63. Reproduced with permission.)

The effective ionization coefficient ratio is estimated to be 0.09 based on McIntyre's theory, verifying the low noise advantage of using Si as a multiplication region (Kang, 2009a). From Figure 10.19b, it is clear that the maximum measured bandwidth of 30-μm-diameter device is 11.5 GHz for gains up to 20, limited by the RC delay and transit time. As the gain is increased beyond 20, the bandwidth drops owing to the avalanche build-up time effect. The highest gain-bandwidth product in Figure 10.20 is 340 GHz, exceeding the previously reported III–V compound APDs (Kang et al., 2009a). The same structure exhibits dramatically higher gain-bandwidth products at increased bias voltage and under lower input optical power, and a value of 868 GHz is shown in Figure 10.21 (Sfar Zaoui et al., 2009). The effect of the input power is very important on the gain and the speed of the APD, since the number of charge carriers determines the gain and the buildup of the multiplication process in the avalanche region. Lower input power produces higher gain, but on the other hand lower speed performance due to avalanche buildup time and the gain-bandwidth product will saturate at one value.

More efforts are devoted to understanding the dark current generation mechanism in Ge/Si APDs since they undergo a high electric field at a much higher voltage (>20 V) compared with Ge photodetectors operating at <5 V (Michel et al., 2010). For free space Ge/Si APDs, the measured dark current increases linearly with the device active area, suggesting that it is dominated by the bulk leakage current. Temperature-dependent dark current is investigated to identify the origin of the bulk leakage current (Dai et al., 2010). Figure 10.22 shows the activation energy extracted from dark current versus temperature, using the relation, $I_D \propto T^2 \exp(-E_a/kT)$. At higher bias, the activation energy E_a is around half of the Ge band-gap (0.66 eV). Since the depletion region extend from the Si multiplication layer to the Ge absorption layer typically at and above the punch-through voltage for Ge/Si APDs, it appears that the generation–recombination (G-R) in the depletion region makes primary contribution to the dark current at higher bias (Ando et al., 1978). Similar to epitaxial Ge-on-Si *pin* photodetectors, the G-R current is tightly associated with threading dislocations at the Ge/Si interface and inside the Ge epi-layer. The decrease in E_a with

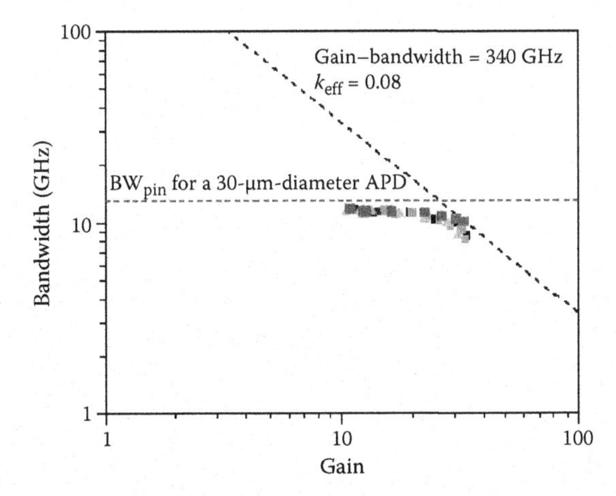

FIGURE 10.20
Gain dependence of the excess noise and 3-dB bandwidth of the Ge/Si APDs. Measured 3-dB bandwidth vs. gain of 30-μm-diameter Ge/Si APDs at a wavelength of 1300 nm. The colored symbols are measured bandwidths. (From Kang, Y.M., Liu, H.D., Morse, M. et al. 2009a. *Nature Photonics* 3: 59–63. Reproduced with permission.)

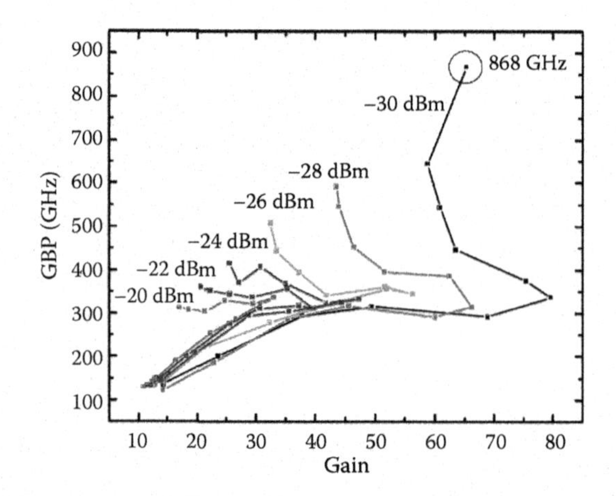

FIGURE 10.21
Gain-bandwidth-product (GBP) vs. the gain of the APD under different input power values between −20 and −30 dBm. (From Sfar Zaoui, W., Chen, H., Bowers, J. E. et al. 2009. Origin of the gain-bandwidth-product enhancement in separate-absorption-charge-multiplication Ge/Si avalanche photodiodes. *Optical Fiber Communication Conference,* OSA Technical Digest (CD) (Optical Society of America, 2009), paper OMR6. Reproduced with permission.)

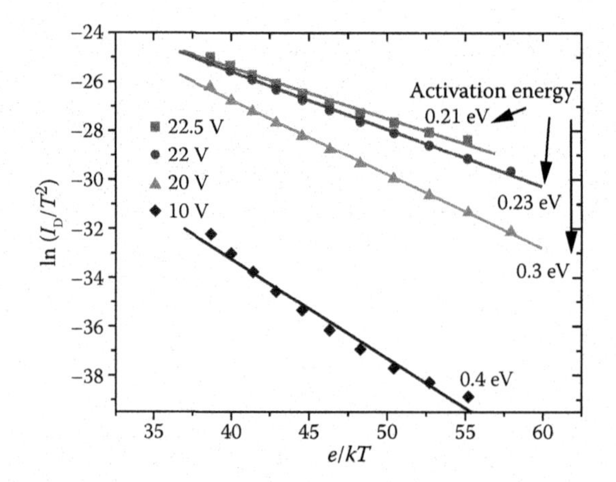

FIGURE 10.22
Activation energy for Ge/Si APDs at different biases. (From Dai, D. X., Bowers, J. E., Lu, Z. W. et al. 2010. Temperature dependence of Ge/Si avalanche photodiodes. *2010 Device Research Conference* (DRC): 231–2. Reproduced with permission.)

increasing bias voltage is consistent with an energy distribution of those defects around the mid-gap and their voltage-dependent activation/deactivation (Colace et al., 2008). It should be possible to reduce the dark current of Ge/Si APDs by optimizing epitaxial growth and the high-temperature treatment during the device fabrication to improve the Ge crystal quality (Michel et al., 2010).

Since the maximum operating temperature is typically 85°C for APD applications, temperature-dependent sensitivity of APD receivers is evaluated here. Considering the dominant dark current of Ge/Si APDs is the G-R current, it gives at least one order of magnitude rise in dark current from 25°C to 85°C (Dai et al., 2010; Colace et al., 2008). From Kang et al.

(2009a), the primary dark current I_{DM} at room temperature is about ~440 nA; thus, I_{DM} at 85°C is about 5 μA. If it assumes TIA input-referred RMS noise does not increase and the gain does not change, the Ge/Si APD receiver sensitivity degrades by ~3 dBm at 85°C, referring to Figure 10.18. Therefore, it is necessary to reduce the dark current and enhance the primary responsivity for higher sensitive Ge/Si APDs to compete with commercially available III–V group APDs at high operating temperatures.

10.4.4 Waveguide-Integrated Ge/Si APD Photodetectors

Waveguide integration is a very attractive approach to achieve high performance Ge/Si APDs with low dark current, high primary responsivity, high bandwidth, and high sensitivity. It overcomes the tradeoff between efficiency and bandwidth using a long horizontal optical path to absorb most of the photons and a short vertical carrier collection path to minimize carrier transit time, similar to waveguide-coupled Ge *p-i-n* photodiodes. Besides, the small footprint of the active detection part can reduce the device dark current. Recently, it has been successfully demonstrated a waveguide-integrated 10 Gbps Ge/Si APD with −31 dBm sensitivity at $BER = 10^{-10}$ (Kang et al., 2009b). The highest reported 3-dB bandwidth for a waveguide-coupled Ge/Si APD is 29.5 GHz at the unit gain bias (Kang et al., 2011).

The waveguide-type Ge/Si APDs adopt the same SACM structure except that thinner Ge layer is used due to the horizontal absorption. Typically, the SOI waveguide thickness is equal to the thickness of the Si avalanche stack, which includes n^+-Si contact layer, i-Si multiplication layer, and p-Si charge layer. Two waveguide-detector integration schemes are possible. The first is the evanescent-coupling configuration, in which the light will be gradually coupled up into the Ge absorption layer as it propagates into the lower-index Si waveguide underneath the higher-index Ge layer (Figure 10.23a). The second configuration is butt coupling, in which Ge absorber is in direct contact with the Si rib output facet and becomes part of the rib waveguide (Figure 10.23b). Part of the light will horizontally enter the Ge layer, and the rest of the light can be evanescently coupled into Ge. Efficient butt-coupling requires mode-matching conditions for waveguide and detector modes; otherwise, the return loss caused by the mode-mismatch cannot be neglected. In addition, the butt-coupling configuration may add more complexity into the Ge/Si device processing,

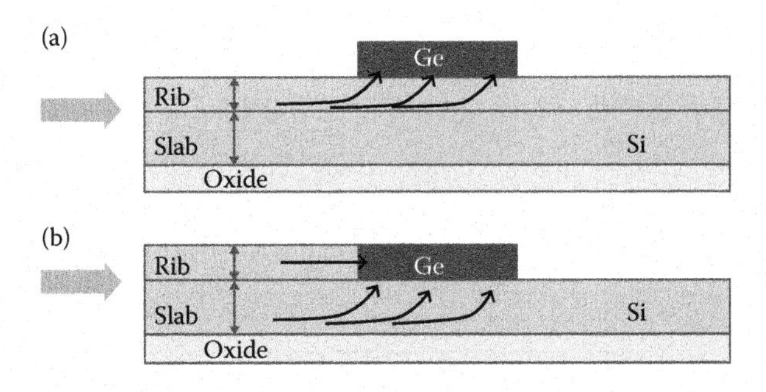

FIGURE 10.23
Schematic diagrams of (a) evanescent- and (b) butt-coupled Ge/Si waveguide APDs along the light propagation direction. (From Kang, Y., Huang, Z., Saado, Y. et al. 2011. High performance Ge/Si avalanche photodiodes development in Intel. *Optical Fiber Communication Conference*, OSA, Technical (CD): OWZ1. Reproduced with permission.)

such as Ge selective growth followed by chemical–mechanical polishing (CMP) planarization. Therefore, evanescent coupling is more often used because of the ease of integration into a CMOS process, although butt coupling is more efficient coupling mechanism that allows for a very short Ge detector.

On the other hand, it is very important to get a small detector footprint with a high coupling efficiency for more sensitive Ge/Si APDs. A novel step-coupler approach for more efficient evanescent coupling is proposed, which can achieve >70% absorption with a 10-μm-long Ge absorption region at a wavelength of 1550 nm, 5× reduction in device length compared with the conventional evanescent coupling scheme (Wang et al., 2011). The conventional reference structure and the new step-coupler structure of waveguide-integrated Ge/Si APDs are shown schematically in Figure 10.24a,b, respectively. The coupling efficiency is investigated by finite difference time-domain (FDTD) simulations as a function of three important parameters Ge absorption length (L_{Ge}), Si waveguide II height (H), and step WG coupler length (L_{Step}), respectively (Figure 10.24).

As shown in Figure 10.25a, H = 0.2 μm is the optimal height of Si waveguide II for efficient absorption at L_{Ge} = 10 μm, which also matches the typical Si layer thickness on SOI substrates for single-mode Si channel waveguides, suggesting that the step waveguide coupler structure can be easily integrated into Si photonic circuits on a 0.2-μm-thick SOI platform. In Figure 10.25b, the absorption oscillates between 55% and 74% with the increase of L_{Step}. This phenomenon is similar to the vertical MMI coupler based on the general interference self-imaging mechanism. The vertical MMI can enhance light scattering toward the Ge active region and creates mirror images of optical modes close to the Ge layer, thus increasing the coupling efficiency between SOI waveguide and Ge/Si APD with a step coupler (Wang et al., 2011). It is also found that the optimal step coupler is temperature-insensitive and can achieve >70% coupling efficiency in the interesting

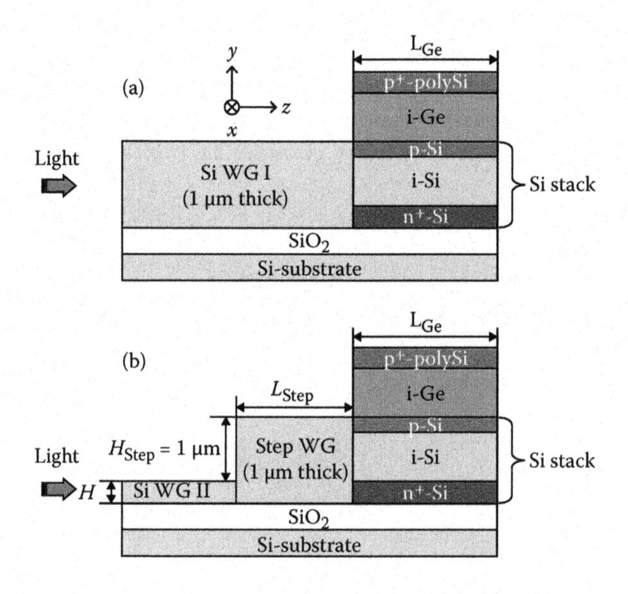

FIGURE 10.24
Schematic cross section (in the y–z plane) of waveguide-integrated Ge/Si APD structures: (a) reference structure with Si WG I, (b) new structure with Si WG II and step WG coupler. (From Wang, J., Yu, M., Lo, G., Kwong, D.-L., and Lee, S. 2011. *IEEE. Photon. Technol. Lett.* 23: 765. Reproduced with permission.)

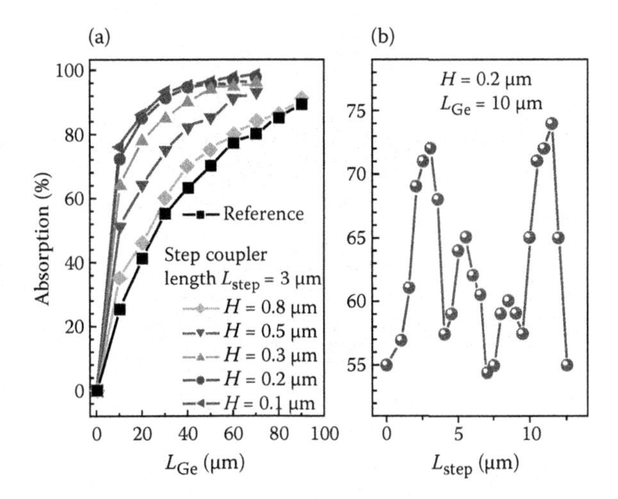

FIGURE 10.25
(a) Simulated absorption as a function of the length of Ge absorption region (L_{Ge}) for the reference and step WG coupler structures (H). (b) Simulated absorption for the new step WG coupler structures as a function of L_{Step} at $H = 0.2$ μm and $L_{Ge} = 10$ μm. (From Wang, J., Yu, M., Lo, G., Kwong, D.-L., and Lee, S. 2011. *IEEE. Photon. Technol. Lett.* 23: 765. Reproduced with permission.)

wavelength range 1300~1550 nm. The return loss of the optimal step coupler structure is estimated less than −30 dB (Wang et al., 2011). With a similar step coupler method, Ang et al. demonstrated an internal primary responsivity of 0.8 A/W (quantum efficiency QE = 64%) at 1550 nm with 15 μm-long evanescent-coupled APD (Ang et al., 2010). As a comparison, the measured primary responsitivies are 0.81 A/W (QE = 64.8%) and 0.6 A/W (QE = 48%) at 1550 nm for 50-μm-long butt-coupled and conventional evanescent-coupled APDs (Kang et al., 2009b). Hence, this novel step coupler approach offers a more efficient coupling mechanism, which can be applied to other waveguide evanescently coupled devices.

When it comes to the dark current of waveguide APDs, the primary source is device perimeter and surface leakage instead of the bulk leakage originated inside Ge, which is different from the normal incident Ge/Si APDs (Kang et al., 2011). More surface passivation strategies are required to prevent device perimeter leakage. The Ge/Si APD structure should be optimized to reduce the electric field strength along the device perimeter, such as floating guard ring (Kang et al., 2009a) and beveled mesa (Ang et al., 2010).

10.4.5 Fabrication of Ge/Si APD Photodetectors

Ge/Si APDs can be fabricated either by Ge wafers bonded to Si wafers or epitaxial growth of Ge on Si. The drawback of wafer bonding can be the formation of Ge native oxide at the Ge/Si interface, which may introduce interface traps created by the dangling bonds of the Ge native oxide (Byun et al., 2011). With epitaxy, dislocations can be formed at the Ge/Si heterojunction interface due to 4% mismatch in the lattice constants of Ge and Si. However, the progress in the high-quality epitaxial Ge growth and the development in CMOS-compatible Ge processing make it desirable and feasible to achieve high-performance Ge photodetectors and APDs.

Epitaxial Ge/Si APD fabrication processing is described as the following: the processing is done on bulk Si or SOI wafers. The epitaxial growth of intrinsic Si is followed by the formation

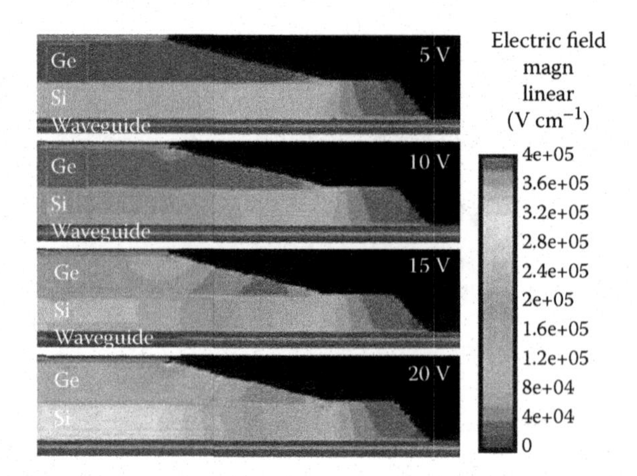

FIGURE 10.26
(See color insert.) Electric field distribution of a waveguide-integrated beveled Ge/Si APD. (From Ang, K. W., Ng, J. W., Lim, A. E. et al., 2010. Waveguide-integrated Ge/Si avalanche photodetector with 105 GHz gain-bandwidth product. *Opt. Fiber Commun. Conf.*, OSA, Technical (CD) JWA36. Reproduced with permission.)

of the charge layer either by in situ doping or implantation. Then Ge is epitaxially grown on Si. Circular mesas are then formed by etching through the Si epitaxial film. Finally, top and bottom contacts are made. Anti-reflection coating may be deposited on top of the mesa. Ge epitaxial film can be deposited by blanket or selective growth. With the selective growth, there are steps of the oxide deposition and patterning for Ge growth window. The selective growth can reduce the threading dislocation density if the selective growth region is small enough (<40 μm), as dislocations can glide to the edge of the mesas and annihilated (Luan et al., 1999). The selectively grown Ge mesas have ~30° sidewalls due to {311}{111} facet growth (Loh et al., 2007b). The formation of beveled Ge sidewall was observed to contribute favorably in reducing the electric field intensity at the mesa edge to prevent early edge breakdown. Figure 10.26 shows the electric field distribution of Ge-slanted mesa at different bias voltages. The less strong electric field along the device perimeter may also reduce the surface leakage.

To take advantage of low excess noise and high gain-bandwidth product of Si multiplication region, some Ge material-related issues should be addressed to improve overall performance of Ge/Si APDs. To name a few, new techniques or methods are in great need to reduce dislocation density inside Ge film and at Ge/Si interface; carrier concentration in Ge film should be much less than $5 \times 10^{15}/\text{cm}^3$, otherwise, some high nonuniform electric fields in Ge film may initiate avalanche multiplication inside Ge itself, which may raise the excess noise due to the largest k of Ge ($k \sim 1$). Additionally, further device structure optimization and reliability issues should be put into consideration for high performance Ge/Si APDs.

10.5 Near-Infrared Silicon Photodetectors

Michael W. Geis, Steven J. Spector, Matthew E. Grein, Jung U. Yoon, and Theodore M. Lyszczarz

The unprecedented investments in both silicon complementary-metal-oxide-semiconductor, Si CMOS, technology, and near-infrared optical communication, wavelengths 1260 to 1675 nm, have motivated research to combine both these technologies into one integrated process providing low-cost high-performance photonic systems. Since the photon energy, at these wavelengths, is less than Si's band-gap of 1.12 eV, Si waveguides can be used to transport light on the integrated circuit and have been used to make optical components, modulators, and optical filters. However, because Si is transparent to this radiation, transforming the light intensity into an electrical signal can be problematic. Several technologies have been used to enhance Si response to this radiation, which include using surface states between Si and other materials, modifying the Si surface by forming porous or black Si, generating mid-band-gap states in the bulk of the semiconductor, and exploiting two-photon absorption (TPA) phenomena. This chapter reviews sub-band-gap optical technologies emphasizing those techniques that are compatible with CMOS fabrication. Specifically, it describes the use of mid-band-gap states, and TPA to enhance the response of silicon in the near-IR.

10.5.1 Introduction

The unprecedented investment in silicon complementary-metal-oxide-semiconductor, Si CMOS, technology has resulted in mass production capability with high yields, $>10^9$ transistors in a single microprocessor, while using structures <30 nm in width. A similar investment in near-infrared optical communication, wavelengths 1260 to 1675 nm, has resulted in high communication rates >50 Gbs, timing with femtosecond resolution and degrees of freedom with data processing not easily realized in Si electronics alone. Research has been directed to combine these two technologies into one integrated process having the advantages of both. Since the photon energy, at these wavelengths, is less than Si's band-gap, <1.12 eV, Si waveguides can be used to transport light on the integrated circuit.

Optical filters, modulators, and optical routers have been fabricated using low-loss Si waveguides, but on-chip detectors are the missing device for many applications. Unfortunately, Si is transparent to this radiation and transforming light intensity into an electrical signal can be problematic. Other smaller band-gap semiconductors (Ge and InGaAs), with varying compatibility with CMOS technology, are often used to detect these Si sub-band-gap wavelengths. This chapter reviews recent developments in efficient sub-band-gap Si optical detectors. We discuss two physical phenomena that enable optical detection of sub-band-gap radiation: the creation of mid-band-gap states with tailored crystal defects in the Si that absorb this radiation and generate a photocurrent and the simultaneous absorption of two photons that generate energy in excess of the band-gap, creating a hole–electron pair. Both these properties of Si have been known for more than 50 years, but recent developments in submicron-integrated optics enable practical applications for these effects. Because of this, much of the experimental data given here will be limited to Si optical waveguides with dimensions from 0.1 to 5 μm. Other approaches to sub-band-gap radiation detection use the interface of Si with other materials or Si–Ge alloys. These include: porous Si (Raissi, 2002), laser-modified black Si (Carey et al., 2005; Huang et al., 2006), Schottky diodes (Casalino et al., 2011; Zhu et al., 2009), Si/SiO_2 interfaces states (Baehr-Jones et al., 2008; Chen et al., 2009b), and Si–Ge photodiodes (Michel et al., 2010). Some of these near-infrared detector approaches have been summarized elsewhere (Casalino et al., 2010; Knights, 2008; Michel et al., 2010).

10.5.2 Generating Tailored Crystalline Defects in Si

Before discussing Si defect sub-band-gap detectors a brief description of the defects caused by ion implantation and annealing is required. Self-ion implantation, implantation of Si^+ into single crystal Si, or energetic neutrons are primarily considered here, since implantation with other ions that remain in the crystal have the complication of doping the semiconductor. Early experiments to characterize these defects in damaged Si were performed by Fan and Ramdas in 1959 using ~1-MeV fast neutrons. They reported that radiation-damaged Si contains mid-band-gap states that produce optical absorption and photocurrents for wavelengths as large as 3.9 μm, 0.31 eV, as shown in Figure 10.27. The photo absorption and photocurrents are dependent upon the Fermi energy of the semiconductor as shown by the comparison of Figures 10.27 and 10.28.

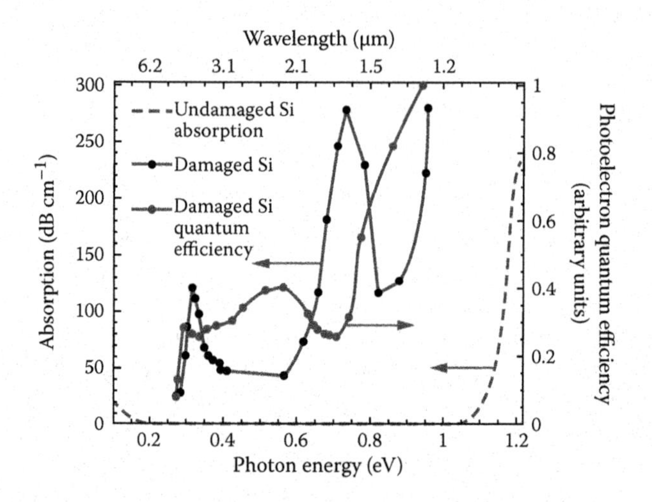

FIGURE 10.27
Undamaged Si absorption and absorption and photoelectron quantum efficiency spectrum at 90 K of neutron-irradiated, flux 3.5×10^{18} cm^{-2}, p-type Si with the Fermi energy less than 0.25 eV above the top of the valence band. The 3.9- and the 1.8-μm absorption bands are evident. The 1.8-μm band generates no photocurrent and causes a reduction in the photon quantum efficiency around 1.8 μm (Fan and Ramdas, 1959).

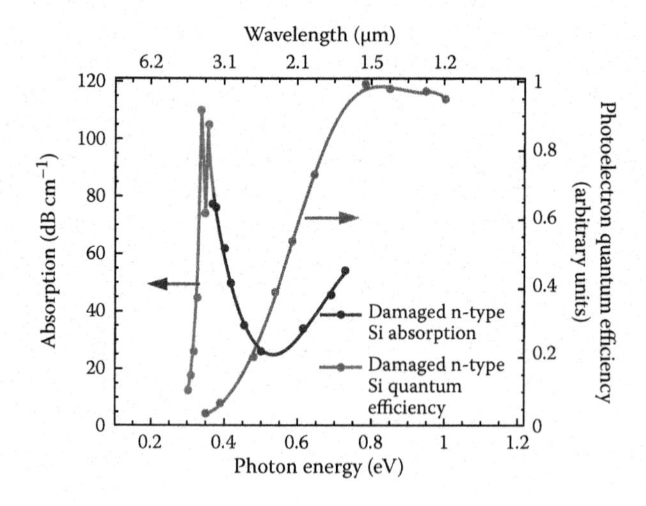

FIGURE 10.28
Absorption and photoelectron quantum efficiency spectrum of neutron-irradiated n-type Si with the Fermi energy 0.25 eV above the top of the valence band. Radiation at 3.9 μm produces no measurable photocurrent. A 3.3-μm absorption band is present but generates no photocurrent. Other traps are responsible for the photocurrent at shorter wavelengths, higher energies (Fan and Ramdas, 1959).

The Fermi energy does not substantially influence defect formation, but it does determine its charge state and its ability to absorb light and to generate a photocurrent. Fan speculated (substantiated by Cheng et al. in 1966) that the photocurrent in Figure 10.27 at 3.9 μm is caused by a divacancy trap, V_2, which is known to have an electronic transition from a positive state to a neutral state, $V_2^+ + e^- \rightarrow V_2^0, V_2(0/+)$. This transition is depicted in the left edge of Figure 10.30. This transition requires ~0.25 eV to move an electron from the top of the valence band to the V_2^+. If the Fermi energy is less than 0.25 eV above the top of the valence band as in p-type Si, then photocurrent is generated when a photon promotes an electron from the valence band, thus neutralizing the positive divacancy. The electron then tunnels to the conduction band leaving behind V_2^+. In n-type Si the Fermi energy is above the 0.25-eV level from the top of the valence band, the defect is neutral, and no photocurrent is generated. This model is consistent with the photo absorption and photoelectron efficiency shown in Figures 10.27 and 10.28.

Since the work of Fan et al. in 1959 there are very few reports of infrared absorption combined with photocurrent. Most of the recent characterizations of defects in Si since 1990 have relied on photoluminescence, transmission electron microscopy (TEM), and deep-level transient spectroscopy (DLTS). This section will review the DLTS measured energy levels of ion-implantation–generated defects in Si. Several groups have characterized defects formed in Si by implantation and annealing using different ion species, energies, and implantation temperatures. However, the initial defect formation is critically dependent upon the substrate temperature and ion implantation flux. An increase in ion flux by a factor of 4 can increase the defect density by a factor of ten for the same ion dose. Similarly an increase of the ion implantation temperature by 25°C to 40°C can decrease the defect density by a factor of ten (Pelaz et al., 2004; Chelyadinskii, 2003). Unfortunately, most implantation reports do not give the ion flux and some neglect to report the implantation temperature.

Using selected reports of the authors (Benton et al., 1998; Coffa et al., 2000; Libertino et al., 1997, 1999, 2000, 2001; Pan et al., 1997; Robertson et al., 2000; Schmidt et al., 2000) that use similar experimental procedures to generate and characterize defects, a map of defect energy levels as a function of implant dose and annealing temperature can be created as shown in Figure 10.29. The map is divided into regions depending upon the character of the defect for a given ion dose and annealing temperature. Boundaries between the different regions will vary in their position depending on the details of defect formation. During room temperature, 1.2-MeV Si^+ implantation, ~2400 vacancy interstitials are generated per ion, but most recombine during implantation leaving ~60 defects/ion (Benton, 1998). The vacancies form pairs, V_2, or complex with oxygen, VO and V_2O, and are the dominant optical and electrical defect for annealing temperatures <300°C. Between 300°C and 400°C, these vacancies recombine with interstitials or form larger vacancies clusters (Fleming et al., 2007), which are no longer electrically active. Above 400°C, interstitials at a density of 0.1 to 3 interstitials/ion are the dominant defect (Benton et al., 1998). As the annealing temperature increases they either diffuse away leaving no DLTS detectable signal or, if the density is high enough, they aggregate and form clusters, {311} rods and at still higher temperatures <111> twin loops.

Shown in Figure 10.30 are the DLTS-determined transition energies between the charged states of some defects (Benton et al., 1998; Coffa et al., 2000; Libertino et al., 1997, 1999, 2000, 2001). The values vary by a few tens of millivolts between references. Transition energies are defined in electron volts either from the top of the valence band, E_V, or from the bottom of the conduction band, E_C. Since DLTS usually measures the charge-changing thermal activation energy of the ground states of the trap, DLTS and optical transition energies

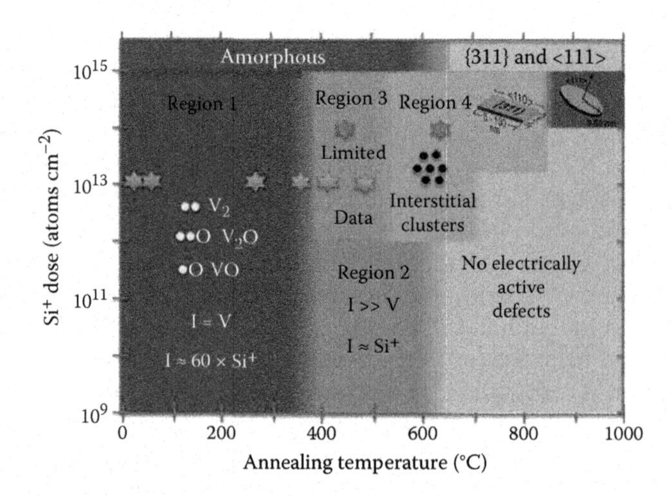

FIGURE 10.29
A roadmap of crystal defects formed with room temperature Si^+ implantation as a function of implantation dose and annealing temperature. Data for regions 1, 2, 3, 4, and {311} are from Benton et al. (1998), Coffa et al. (2000), and Libertino et al. (1997, 1999, 2000, 2001) using 1.2 MeV Si^+ and 30-minute annealing times. Additional data for {311} and <111> are from Schmidt et al. (2000) using 5.6 MeV Si^{3+} and 30-minute annealing times and Stolk et al. (1997) using various ion energies and times. Data for annealing the amorphous region to {311} and <111> are from Robertson et al. (2000) and Pan (1997). Below 350°C, the defects consist of vacancies, V, and interstitials, I. Above 400°C, the electrically active defects are interstitials, interstitial clusters, and I. At still higher temperatures {311} rods and eventually <111> twin loops are formed. The stars indicate the dose-annealing schedules used to make the Si sub-band-gap photodetectors, which are summarized in this review.

may not coincide. For example, the DLTS measured energy for the transition $V_2^+ + e^- \rightarrow V_2^0, V_2(0/+)$, is 0.25 eV, while the minimum photon energy for the same transition is 0.31 eV, 3.9 μm, as shown in the left edge of Figure 10.30. The traps may have excited energy levels, which are not characterized by DLTS.

10.5.3 Ion Implanted and Annealed Silicon Detectors

Knights et al., in 2003, first demonstrated that the photo response reported by Fan et al. in 1959 was of practical value. Using Si-integrated optics, a pin diode was fabricated in an ion-implanted waveguide that allowed optically generated hole-electron pairs to produce a current. The low photo quantum yield of ~2% restricted this photodiode's use to optical power monitoring (Bradley et al., 2005).

A cross section of the diode waveguide reported by Knights et al. (2006) is shown in the insert of Figure 10.31. Divacancy defects, region 1 in Figure 10.29, were formed by 170-keV H^+ implantation. Knight and Bradley observed, as shown in Figure 10.31, that as the width of the intrinsic region of the pin diode was reduced the efficiency of converting photons into photocurrent increased. After Knight's publications, several groups reported sub-band-gap optical detectors based on pin photodiodes using defects or traps formed by several techniques (Doylend et al., 2010; Geis et al., 2007a,b, 2009; Haret et al., 2010; Knights et al., 2006; Liu et al., 2006c; Logan, 2011; Logan et al., 2010, 2011, 2012; Mao et al., 2011; Park et al., 2011; Preston et al., 2011; Shafiiha et al., 2010; Tanabe et al., 2010). An electron-to-photon ratio, quantum efficiency, ~40% of the theoretical maximum at 1.55 μm (Logan, 2011) has been reported by reducing the waveguide structure to submicron geometries. A further

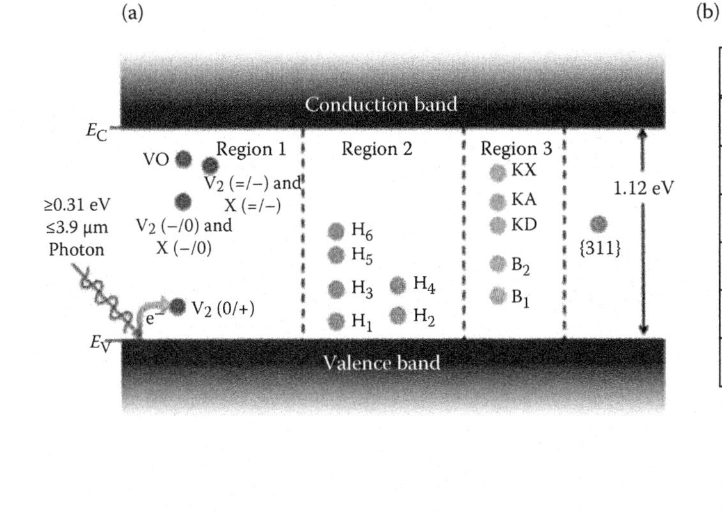

(b)

Region 1		Region 2	
V_2 (0/+)	E_V+0.25	H_1	E_V+0.08
V_2 (−/0)	E_C−0.43	H_2	E_V+0.13
X (−/0)	E_C−0.43	H_3	E_V+0.23
V_2 (=/−)	E_C−0.23	H_4	E_V+0.25
X (=/−)	E_C−0.23	H_5	E_V+0.39
VO	E_C−0.18	H_6	E_V+0.53

Region 3		(311)	
B_1	E_V+0.33	{311}	E_V+0.5
B_2	E_V+0.52		
KD	E_C−0.58		
KA	E_C−0.29		
KX	E_C−0.05		

FIGURE 10.30
(a) Map of trap energy distributions with reference to the bottom of the conduction band bottom, E_C, or the top of the valence band top, E_V, for the regions show in Figure 10.29. Shown on the left edge of the figure is an example of a photon-induced charge transition from a positively charged V_2^+ to a neutral V_2^0 by a photon exciting an electron from the valence band. (b) Table of energy levels determined by DLTS. The energy levels were obtained from Benton et al. (1998), Coffa et al. (2000), Libertino et al. (1997, 1999, 2000, 2001).

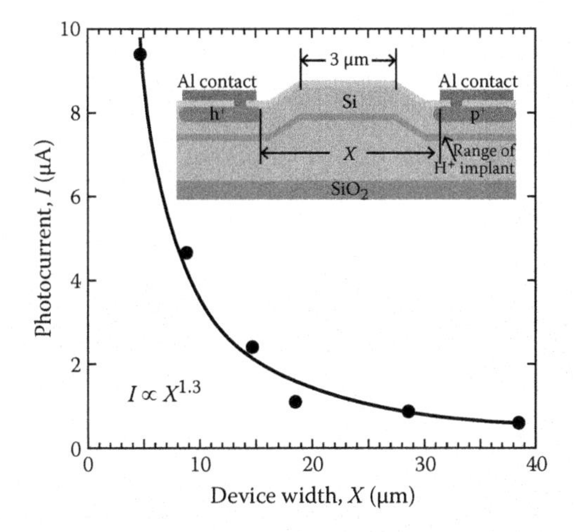

FIGURE 10.31
Photocurrent at 0 V bias as a function of junction separation, X. The waveguide pin photodiode is 6 mm long with an on-chip power of 3 mW. The curve is a power law fit to the data. The insert shows a cross section of the waveguide (Bradley et al., 2005; Knights et al., 2006). Photo response generating defects are vacancy dominated in region 1 of Figure 10.29. The quantum efficiency for $X = 5$ μm is ~2%.

increase in quantum efficiency was discovered by using interstitial defects (region 3 of Figure 10.29) instead of divacancies (Geis et al., 2007b).

10.5.4 Photodiode Characterization

Near-infrared defect photodiodes are still in development, so a complete recipe of device geometry, implantation, and annealing that results in a device with specific properties does not yet exist. In fact, different groups, which use superficially similar recipes, do not generate photodiodes with the same properties. The remainder of this section discusses in some detail the results out of MIT Lincoln Laboratory, with which the authors are most familiar. Impressive research results out of McMaster University in Hamilton, Ontario, Canada, are also discussed, much of which is covered in more detail by Logan (2011).

The defect roadmap, Figure 10.29, can be used to help analyze the near-IR photodiode results. The experimental parameters reported here are indicated by the stars in Figure 10.29 showing the implant doses and annealing temperatures, ≤600°C, for which waveguide photodiodes were made and characterized. The data shown in Figure 10.29 for annealing temperatures ≤600°C were obtained from references using 1.2 MeV Si^+. How the defect map changes when the ion energy is changed from 1.2 MeV Si^+ to 190 keV is unknown, but defect density is relatively insensitive to ion energy. This is probably because most of the defects are formed near the end of the ion's range (Pelaz et al., 2004). While the map in regions 1, 2, 3, and 4 represent 30-minute annealing times, the annealing times for the stars are ~15 minute for temperatures <450°C and 2 minute for diodes annealed to 475°C and 1 minute for those annealed to 600°C. It is unknown how to compare these shorter annealing times, 1 to 2 minute, with the previous longer annealing times, 30 minute, reported in the references.

10.5.4.1 Optical Absorption and Quantum Efficiency

A waveguide diode structure, shown in Figure 10.32, which was designed for use as an optical modulator (Spector et al., 2008), was found to be an efficient platform for photodiode

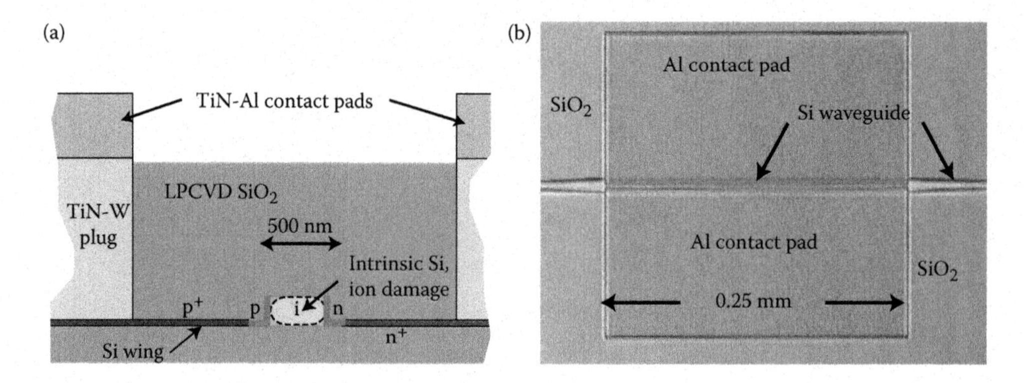

FIGURE 10.32
(a) Schematic cross-section drawing of waveguide pin photodiode. The junction separation is 0.5 μm (Spector et al., 2008). (b) Top view optical micrograph of a 0.25-mm-long Si waveguide photodiode. Quantum efficiencies (e $photon^{-1}$) of ~40% of the theoretical maximum were obtained with small junction separation and defects dominated by interstitials, region 3 of Figure 10.29 (Geis et al., 2007a).

characterization as a function of implantation dose and annealing. Figures 10.33 and 10.34 show the effect of annealing on absorption and photo response for two room-temperature ion-implantation doses of 10^{13} and 10^{14} Si^+ cm^{-2}, respectively. The implantation was performed using a ~1-mm-diameter ion beam scanned over a 6-inch wafer with ion beam

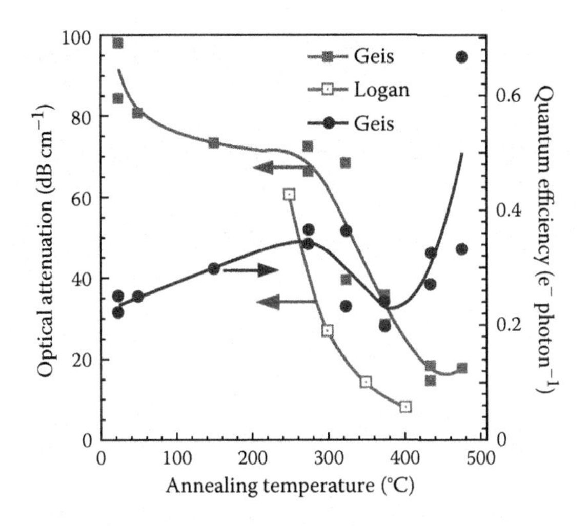

FIGURE 10.33
Photodiode quantum efficiencies of 1550 nm light at 5 V bias as a function of annealing temperature after defect formation by 10^{13}-cm^{-2}, 190-keV Si^+ implantation. For temperatures 25°C to 435°C, annealing was in vacuum, $<5 \times 10^{-4}$ Pa, for ~15 minutes. The 475°C anneal was in flowing nitrogen for 2 minutes (Geis et al., 2007a,b, 2009). Logan's optical attenuation is from "Insertion loss (dB)" column of table I in Logan (2011) divided by the photodiode's length, 0.25 mm. The curves are drawn to follow the data points.

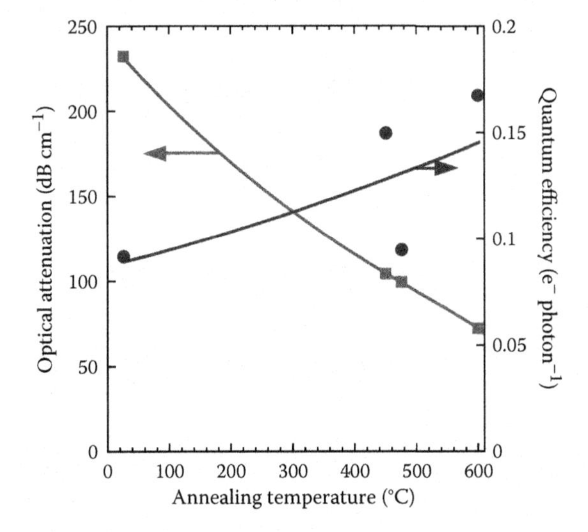

FIGURE 10.34
Photodiode quantum efficiencies of 1550 nm light at 5 V bias as a function of annealing temperature after defect formation by 10^{14}-cm^{-1}, 190-keV Si^+ implantation. For temperatures 25°C to 435°C, annealing was in vacuum, $<5 \times 10^{-4}$ Pa, for ~15 minutes. The 475°C anneal was in flowing nitrogen for 2 minutes and the 600°C anneal in flowing nitrogen for 1 minute. The curves are drawn to follow the data points (Geis, 2007a,b, 2009). The 600°C annealed diodes were used in an optical analog-to-digital converter (Grein, 2011; Khilo, 2011).

currents of 2 to 7 μA for the 10^{13}-cm^{-2} dose and 32 to 84 μA for 10^{14} cm^{-2}. After ion implantation, the crystal contains a variety of defects, but even a 15-minute anneal a few degrees above room temperature, 50°C, removes many of these defects, and the absorption decreases. It is speculated that the photo response is due to the V_2 (0/+) transition in region 1 of Figure 10.29 for diodes annealed to <300°C and interstitials and interstitial clusters for those annealed to >400°C.

As the annealing temperature is increased, the divacancies either form electrically inactive large clusters (Fleming et al., 2007) or recombine with interstitials and the absorption decreases. Other more temperature-stable and photoactive interstitial defects remain and exhibit a photo response. Figure 10.33 shows data from Logan (2011) and Geis et al. (2007a,b, 2009). Although Logan and Geis measure similar absorption vs. temperature characteristics between 250°C and 400°C, they measure different photo response. Logan reports 0.05 e $photon^{-1}$ at 5 V, while Geis reports a photo response of between 0.2 and 0.3 e $photon^{-1}$ for diodes annealed to 350°C. This variation in photo response could be due to details in ion implantation parameters or the presence of impurities, C, O, B, or P, in the silicon.

10.5.4.2 Photocurrent and Bias Voltage

In addition to the sub-band-gap quantum efficiency as discussed above, the photodiodes have other characteristics that are a function of implantation and annealing. Diodes implanted with minimal annealing <300°C exhibit a significant increase in quantum efficiency with bias voltage. As the annealing temperature is increased the absorption decreases, Figures 10.33 and 10.34, and the quantum efficiency increases and in general becomes less dependent upon the bias voltage as shown in Figures 10.35 and 10.36 (Geis et al., 2007a,b, 2009). Similar results have been reported by Logan in 2011. The increase in quantum efficiency with bias voltage may be the result of field-enhanced ionization (Yong et al., 2011). Once a photon interacts with a defect to change its charge state, it has to return to its original state to continue generating a photocurrent. The transition mechanism back

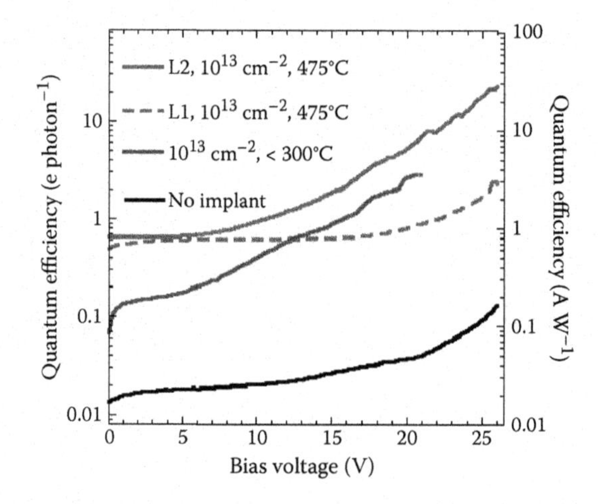

FIGURE 10.35
Photodiode quantum efficiency, the ratio of photocurrent to light absorbed for devices implanted with Si^+ to 10^{13} cm^{-2} as a function of bias voltage, for several annealing temperatures (Geis et al., 2007a). The annealing temperatures and diode states, L_1 L_2, are in the legend.

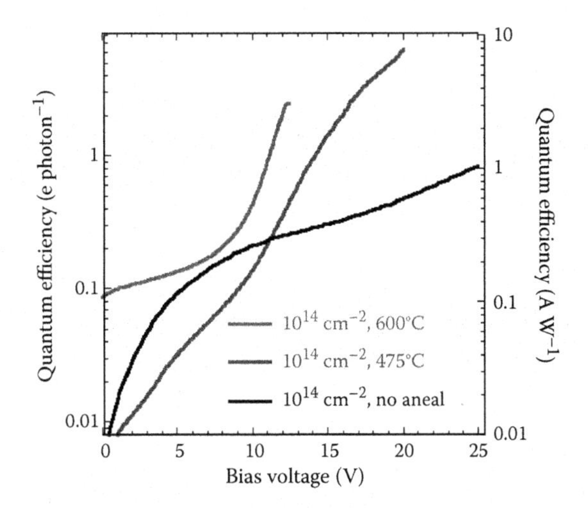

FIGURE 10.36
(See color insert.) Photodiode quantum efficiency, the ratio of photocurrent to light absorbed, for devices implanted with Si^+ to 10^{14} cm^{-2} as a function of bias voltage, for several annealing temperatures (Geis et al., 2007a,b, 2009). The annealing temperatures are in the legend.

to the equilibrium charge state is poorly understood but can be enhanced by an electric field.

Photodiodes implanted to 10^{13} cm^{-2} and annealed to 475°C exhibit two stable states, L_1 and L_2. After fabrication the diode is in the L_1 state and exhibits a photocurrent bias voltage curve, as shown Figure 10.35. If the diode is forward biased for a few minutes with a current per diode length >100 mA cm^{-1}, then the diode is transformed to the L_2 state. The process is reversed by heating the diode to 250°C in air for a few seconds. The remarkable increase in photo response, >20, at 20-V bias for the same photodiode as it transitions between the L_1 and L_2 states is shown in Figure 10.37. The difference in photocurrent

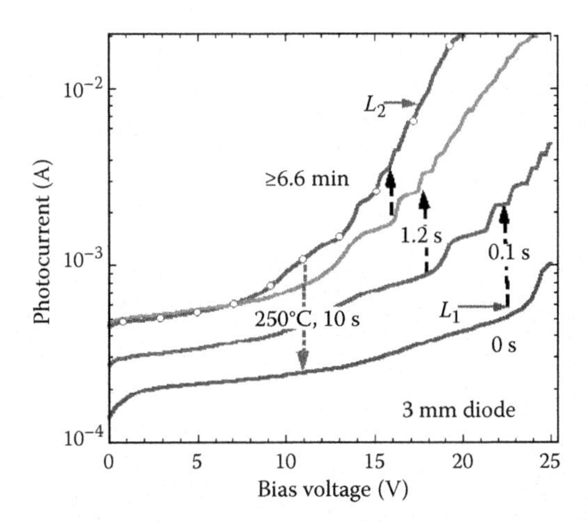

FIGURE 10.37
(See color insert.) Sequence of measurements of photocurrent vs. bias voltage for a single photodiode at fixed optical power. At time 0 s, the diode is in state L_1. Operation in forward bias of 100 mA/cm continuously transforms the diode to the L_2 state, saturating in 6 minutes. Heating the diode to 250°C for 10 s returns it to the L_1 state (Geis et al., 2007a).

between L_1 and L_2 for bias voltages <10 V is the difference in their optical absorption, both having about the same quantum efficiency. For a 3-mm-long diode in the L_1 state with an optical absorption of ~8 dB cm^{-1}, only 30% of the incoming light is absorbed in the diode. L_2 having an optical absorption of ~18 dB cm^{-1} absorbs 80% of the incoming light. Above 10 V, the quantum efficiency of the L_2 state increases faster with bias voltage than the L_1 state, possibly due to enhanced avalanche gain. Similar bistability has been observed in Si waveguides by Shafiiha et al. in 2010 and in the bulk by others (Fleming et al., 2007; Giri, 2001; Junkes et al., 2010).

Increasing the ion doses to 10^{14} cm^{-2} changes the character of the diode, as shown in Figures 10.34 and 10.36. The optical absorption increases with dose and remains high despite annealing to 600°C. There is no evidence of hysteresis or bistability. The photodiode's response for voltage <2 V is only of practical value for diodes annealed to 600°C.

10.5.4.3 Bandwidth and Linearity

Photodiodes implanted to 10^{13} cm^{-2} and annealed to 475°C exhibit a frequency response >35 GHz and a transient response <13 ps, as shown in Figure 10.38. However, these high frequency measurements were made on a 0.25-mm-long photodiode, which only absorbs ~10% of the incoming light even in the L_2 state. If it is required to absorb 90% of the incoming light then the diode must be 6 mm long and the transit time of light through the diode is ~90 ps. However, since the light exponentially decreases in intensity as it moves through the diodes the effective pulse width is ~30 ps, resulting in a frequency response ~5 GHz. Pulse photocurrents >100 mA with a bias voltage of 20 V have been obtained with 0.5-mm-long photodiodes using higher light intensity pulses than used for the data in Figure 10.38. Photocurrent as a function of optical input power at a constant voltage for a diode implanted with 10^{13} ions cm^{-2} and annealed to 475°C is shown in Figure 10.39. The response is slightly super linear at low voltages <10 V, which becomes sublinear at higher voltages, with no indication of two-photon photo response.

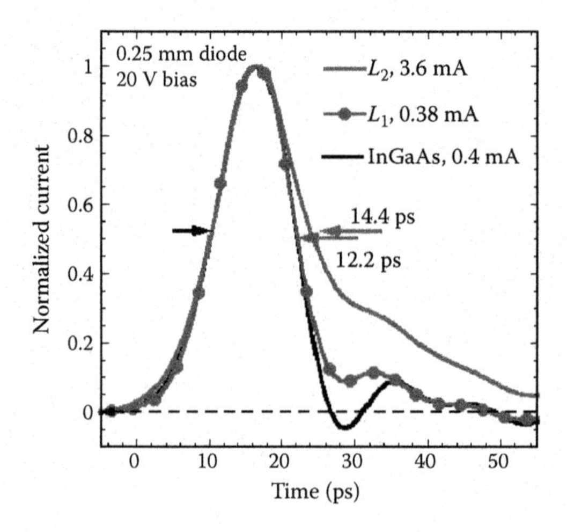

FIGURE 10.38
Transient response to a 1550-nm 0.25-ps light pulse for the same diode in the L_1 and L_2 states and for a 50-GHz bandwidth InGaAs photodiode. The peak current for the diodes is shown in the legend. The same light intensity was used for the diode in the L_1 and L_2 states, but it was attenuated for the InGaAs photodiode (Geis et al., 2007a).

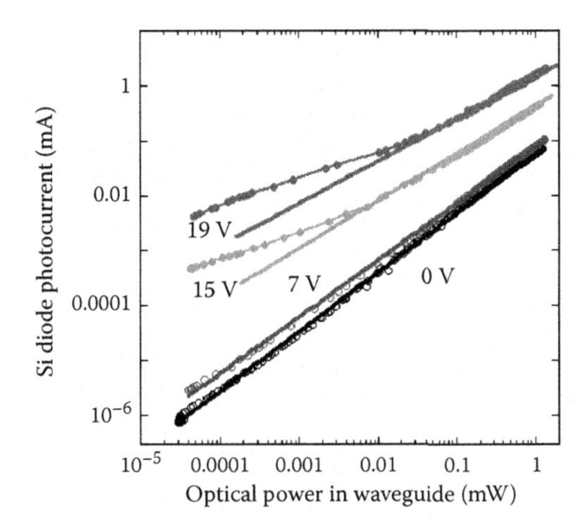

FIGURE 10.39
Plot of photocurrent as a function of optical power entering a 0.25-mm-long photodiode in the L_2 state for several bias voltages. A commercial InGaAs photodiode was use as linear standard. The curves were fit to a power function, photocurrent = (light input)^X. The exponents for the fitted curves between 10 μW and 1 mW of 1550-nm light are x = 1.07, 1.04, 0.87, and 0.76 for bias voltages of 0, 7, 15, and 19 respectively (Geis et al., 2007a, Holzwarth et al., 2009).

10.5.4.4 Phototransistors

Figure 10.40 shows the quantum efficiency of Si nin and pip phototransistors formed with interstitial defects in Region 9.3 of Figure 10.30. The nin transistor exhibits gain with quantum efficiencies approaching 100, while the efficiency of the pip transistors never exceeds 1 e photon^{-1}. The difference can be explained by a model that has the photons exciting electrons from interstitial defects to the conduction band, generating a positive charge in

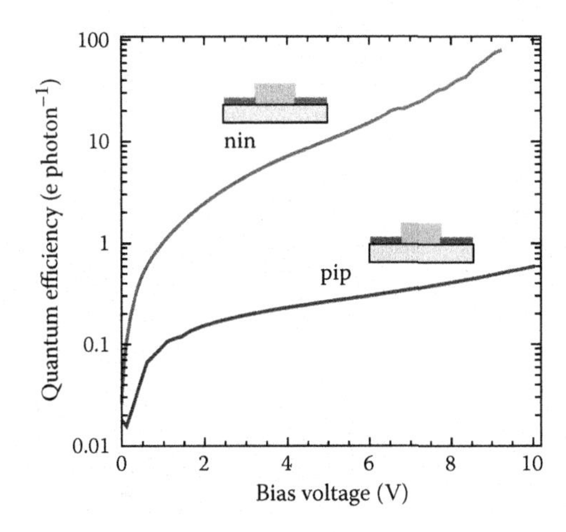

FIGURE 10.40
Photocurrent through nin and pip phototransistors implanted with 190 keV Si$^+$ at 10^{14} cm^{-2} and annealed to 475°C (Geis et al., 2009).

the intrinsic region, I. This positive charge temporally dopes the I region n-type, allowing electrons to flow between the n-doped waveguide contacts. When the contacts are p-type, the positive charge inhibits the current and the quantum efficiency is limited to hole–electron pairs generated by the absorbed photon. Nin phototransistors have orders of magnitude higher quantum efficiency than the pip transistors for both implant doses of 10^{13} and 10^{14} cm^{-2} and annealing temperatures of 475°C and 600°C, regions 2, 3, and 4 of Figure 10.29. No hysteresis or bistability has been observed in any of the pip or nin phototransistors (Geis et al., 2009).

10.5.5 Two-Photon Absorption

If the photons have energy greater than half the band-gap, >0.56 eV or <2.2 μm wavelength, then two photons can combine with sufficient energy to create a hole–electron pair and generate a photocurrent. TPA is dependent upon the photon wavelength and the square of the light intensity and is characterized by the coefficient, β. Three-photon absorption, with even lower energy photons, can also generate a photocurrent, but the cross section for this process is too small to be of general interest (Dinu et al., 2003). Figure 10.41 shows several of the published values for β as a function of wavelength. The coefficient, β, is defined by

$$dI/dz = -\alpha I - \beta I^2, \tag{10.9}$$

where I is the light intensity propagating in the z direction and α is the linear optical absorption coefficient. For most conditions, β is too small to be important. However, the small dimensions of Si waveguides allow for optical power concentration into a small region. At 1550 nm in a Si waveguide with a cross section of 0.5 × 0.2 μm, it takes ~1 GW cm^{-2}, ~400 mW, for TPA to equal the linear optical absorption of 3 dB cm^{-1}, a value common

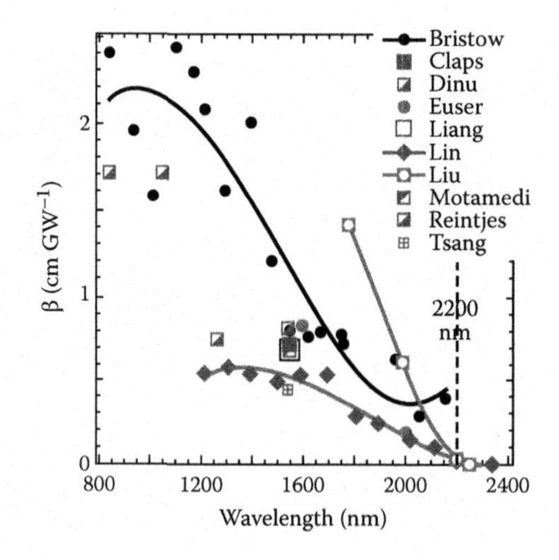

FIGURE 10.41
Measured values of the two-photon absorption, TPA, coefficient, β, as a function of wavelength from Bristow et al. (2007), Claps et al. (2004), Dinu et al. (2003), Euser and Vos (2005), Liang and Tsang (2004), Lin et al. (2007), Liu et al. (2011), Motamedi et al. (2011), Reintjes and McGroddy (1973), and Tsang et al. (2002). The lines are third polynomial fit to the data. The 2200-nm dotted line is the maximum wavelength allowed for TPA.

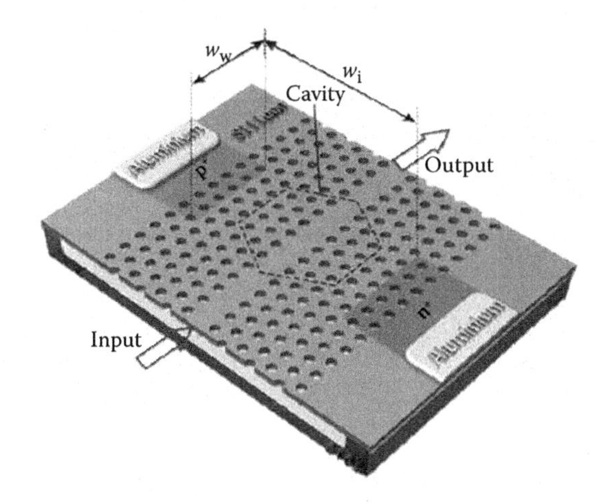

FIGURE 10.42
Schematic illustration of a pin photonic crystal nanocavity. The lattice constant a, hole radius r, slab thickness t, w_i, and w_w are 420 nm, 108 nm, 204 nm, 8.72 μm, and 8.4 μm, respectively. The input and output waveguides with widths of 1.05 a are in-line connected with the cavity through barrier line defects with widths of 0.98 a. The length of the barrier line defect is d = 11 a.

to such waveguides. However, light absorbing free carriers generated by TPA and inherent nonlinear effects start to become evident at power levels >50 mW (Osgood et al., 2009).

Tanabe et al. (2010) and Notomi et al. (2011) demonstrated a TPA detector using a high-Q photonic crystal resonator, as shown in Figure 10.42. The high Q enhances the circulating power in the resonant structure to levels where TPA becomes important, even with low power inputs, 10 nW. Since the photodiode contained no traps its leakage current was very small, 15 pA, while still having a reasonable photo response of ~0.1 e photon^{-1} (Notomi et al., 2011).

Liang et al. (2002) used TPA to generate an autocorrelation current to map out the time profile of a light pulse with sub-picosecond resolution. Figures 10.43 and 10.44 show the autocorrelation setup and some of their experimental results. The light pulse to be characterized was split into two beams with a polarizing cube. A TPA signal was obtained by injecting the two light pulses into the same Si photodiode and varying the time between the two pulses with a movable mirror.

For many applications, TPA is a source of parasitic optical loss both by its optical absorption and by the free carrier absorption of the TPA-generated hole-electron pairs. The free carrier absorption is especially problematic for Si waveguide Raman lasers (Boyraz, 2004; Chen et al., 2006; Rong et al., 2005) where the carriers must be limited in concentration as they inhibit laser operation. One mitigation technique is to fabricate a pin diode in the waveguide, similar to that in Figure 10.32 and remove the carriers with an electric field generated by reverse biasing the diode (Rong et al., 2005).

10.5.6 Discussion

10.5.6.1 Optical Absorption and Bandwidth

Defect sub-band-gap Si detectors are CMOS compatible, can be linear in photo response, and have high frequency bandwidth, >35 GHz. The optical absorption at 1550 nm varies

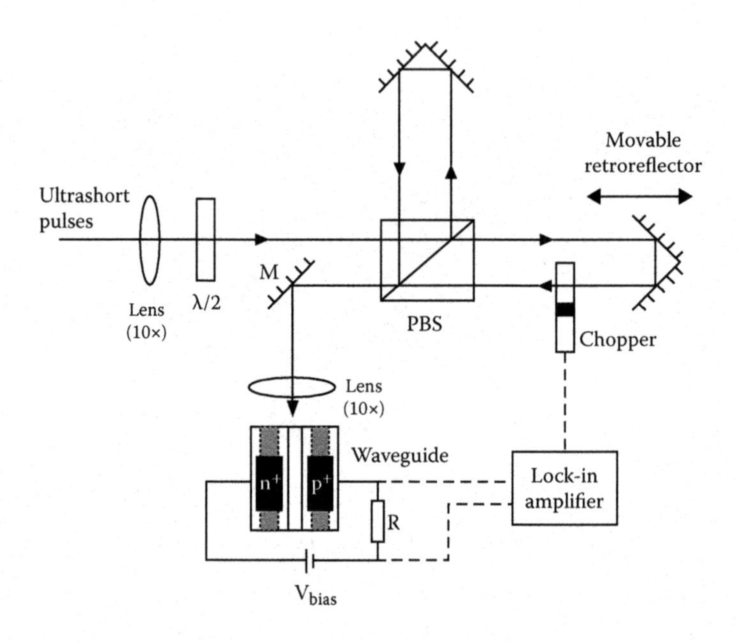

FIGURE 10.43
Experimental setup for autocorrelation measurements. PBS is the polarization beam splitter, R is a biasing resistor, and M is a mirror. The waveguide photodiode is 17 mm long, giving an autocorrelation time resolution of ~100 fs (Liang et al., 2002).

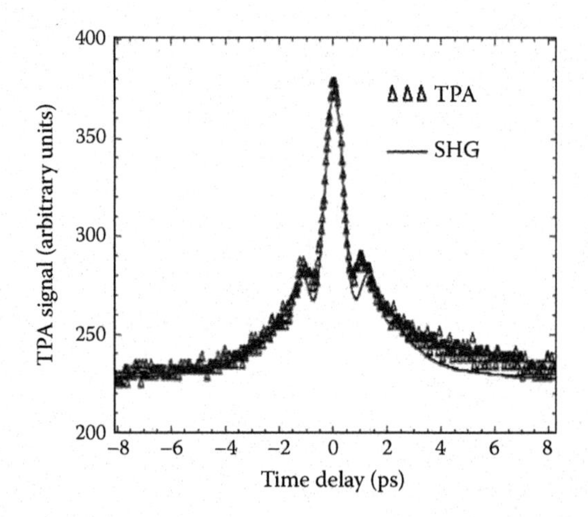

FIGURE 10.44
Autocorrelation trace obtained by two-photon absorption, TPA, in a silicon waveguide using optical soliton pulses (~530 fs FWHM). Second harmonic generation, SHG, obtained separately is shown for comparison. Figure 10.43 is a schematic drawing of the experimental setup for TPA (Liang et al., 2002).

from 8 to 100 dB cm^{-1}, requiring diode lengths of 12 to 1 mm to absorb 90% of the incoming light. For the waveguide shown in Figure 10.32, the speed of 1550-nm light is ~6.7×10^9 cm s^{-1} and the transmission line formed by the diode and its metal contacts has a propagation velocity of 1.3×10^{10} cm s^{-1} (Geis et al., 2009). A photodiode with sufficient length to absorb most of the incoming light will have its bandwidth limited by the effective transit time of the light in the diode. Considering only the light's transit time in the waveguide the bandwidth frequency, f, in GHz is given by

$$f = 0.25 \text{ (optical attenuation)} \quad (10.10)$$

where the optical attenuation is the absorption of light in the waveguide (dB cm^{-1}). For a waveguide photo detector, the optical absorption must be >80 dB cm^{-1} to have a bandwidth >20 GHz.

10.5.6.2 Comparison of Photodetectors

Table 10.3 is a summary of several reports of sub-band-gap Si waveguide photodetectors. Although only defects formed by ion implantation and TPA have been characterized as

TABLE 10.3

List of the Properties of pin Photodiodes and nin and Phototransistors

Device	Defect for Absorption	Absorption (cm^{-1})	Quantum Efficiency (A/W)	Leakage Current for Device Length to Absorb Half of Incoming Light	Bandwidth (GHz)	Reference
pin	Si–SiO_2 surface states	5	0.034 at 11 V		0.06	Baehr-Jones et al., 2008
pin	Si–SiO_2 polysilicon ring resonator	6	0.15 at 13 V	40 nA at 13 V[a]	2.6	Preston et al., 2011
pin	Surface and TPA		~0.1 at 3 V	15 pA at 3 V[a]	2.7	Notomi et al., 2011
pin	B^+ 10^{13} cm^{-2} implantation, ring resonator	45	0.14 at 10 V	0.2 nA at 10 V[a]	–	Doylend et al., 2010
pin	Implant Si^+ 10^{13} cm^{-2}, annealed 475°C, L_1 and L_2 states	8–10 L_1 18–20 L_2	0.5–0.8 at 5 V 1–2 at 20 V L_1 6–10 at 20 V L_2	0.5 nA at 5 V L_1 10 nA at 20 V L_1 3 mm 4 nA at 5 V L_2 0.4 μA at 20 V L_2 1.5 mm	>35[b]	Geis et al., 2007a, 2009
pin	Implant Si^+ 10^{14} cm^{-2}, annealed 600°C	73	0.2 at 6 V	10 nA, at 6 V 0.4 mm	>10	
nin	Implant Si^+ 10^{14} cm^{-2}, annealed 475°C	>100	50 at 8 V	0.1 mA at 8 V <0.3 mm	0.2	Geis et al., 2009
pin	Ge on Si waveguide	~5 × 10^3	0.97 at 1 V	4.6 nA at 1 V[a]	36	Liao et al., 2011

Note: These measurements were made with ~1550-nm radiation. For 1550-nm radiation, the quantum efficiency in e/photon = 0.8 (A/W).

[a] Leakage current of actual device.

[b] Bandwidth for a short pin device where light transit time does not affect bandwidth.

a source of photo response, several other techniques are included; Si-SiO_2 interface states (Baehr-Jones et al., 2008; Chen et al., 2009b), polysilicon photodiodes (Preston et al., 2011), selenium-doped Si (Mao et al., 2011), and a Ge diode on a Si waveguide (Michel et al., 2010; Liao et al., 2011). Some of the best waveguide results were obtained from Ge diodes, which are included for comparison.

The last entry in Table 10.3 is for a Ge 1550-nm photodiode grown on a Si waveguide. The Ge device exhibits simultaneously excellent photo response and bandwidth. It responds to wavelengths up to 1620 nm (Liao et al., 2011). Si defect detectors have response to 1750 nm (Geis et al., 2009), and from published data (Fan, 1959), a response extending as far as 3300 nm could be expected.

TPA is a source of parasitic optical loss for Si waveguide Raman lasers. However, the same properties have been used by Liang in 2002 for optical autocorrelation and by Notomi et al. in 2011 with a photodetector having exceptionally low leakage current, 15 pA, as shown in Table 10.3.

10.5.6.3 Stability of Defect Photodiodes in Applications

Operational stability is a requirement for application of these devices. Vacancies used for photo response may not be stable for extended times at high temperatures, >100°C. Over a period of years, the vacancies may diffuse, recombine with interstices, or form optically inactive defects. Detectors that exhibit two states, like L_1–L_2, may be of scientific interest, but may have limited applications. Of the Si defect pin detectors, those implanted with 10^{14} Si^+ cm^{-2} and annealed between 400°C and 600°C show consistent properties, sufficient optical absorption to have frequency bandwidths >20 GHz, Equation 10.10, and have been used successfully in an all-Si photo analog-to-digital converter (Grein et al., 2011; Khilo et al., 2012). Although all the Si detectors reported here have a linear response to light at constant bias voltage, in an application, the diode is usually in series with a load resistance and its bias voltage varies with optical input. Detectors with a large increase in photo response with bias voltage, like the diode implanted 10^{14} cm^{-2} Si^+ and annealed to 475°C, Figure 10.36, will have inferior linearity to the same diode annealed to 600°C that has a photo response less dependent upon bias voltage.

10.5.6.4 Present Status and Future Developments

Si defect photodiodes and transistors are still in development and are approaching the theoretical maximum photo response. They have demonstrated >35-GHz bandwidth, optical response up to 1750 nm. Practical stable Si photodiodes with response of 0.2 A/W at 1500 nm and a bandwidth >10 GHz and Si phototransistors with optical gain of ~50 have been demonstrated. These devices require 100 to 200 keV Si^+ implanted to 10^{14} cm^{-2} and an annealing temperature of 600°C, processes common to most CMOS foundries and compatible with standard integrated circuit fabrication. With an optical absorption of ~100 dB cm^{-1}, >1 mm of waveguide length is required to absorb 90% of the incoming light.

Future developments may address the major limitations of low optical absorption, and the existence of defects that absorb light without generating a photocurrent. The implantation-annealing map of Figure 10.29 shows that most of the experimental space has not been characterized and a more efficient detector implantation-annealing region is likely to be discovered. Implantation temperature is a variable not even explored. Hot implantation where the vacancies are annealed during implantation is expected to generate a different defect profile. Other means of generating mid-band-gap states, at the grain boundaries of polycrystalline Si or with dopants like Se, may result in improved devices performance. Reasonable goals would be optical absorption >100 dB cm^{-1} and a quantum efficiency >0.8 e $photon^{-1}$ or a photo response to wavelengths >1650 nm with reasonable quantum efficiency.

10.6 Modeling and Simulation of Photodetectors

Ning-Ning Feng

One of the key issues when designing waveguide-integrated photodetectors is the light coupling efficiency. Both evanescent and butt coupling schemes have been used to achieve high efficient light coupling from waveguides (silicon or SiON) to silicon (Si) or germanium (Ge) absorption layers. When integrated with small core Si waveguides, light can be easily coupled into the Ge films due to the much stronger evanescent field. However, small-waveguide–based photodetectors suffer from high fiber-coupling loss and tight fabrication tolerance, which make further integration of wavelength-division-multiplexing (WDM) components rather challenging, hence, prevent them from being widely employed in optical links and interconnects. As for large core waveguide-integrated photodetectors, due to the weak-evanescent field outside the Si waveguide core, it is preferred to use a butt coupling scheme (Feng et al., 2009). A recent study shows that the optical mode center shift between Si and Ge waveguide is the major source causing significant coupling loss in butt-coupled waveguide photodetectors (Feng et al., 2011b).

Besides the waveguide coupling issue, the light–matter interaction is another important aspect that requires more attention when designing photodetectors. Such interaction has direct impact on the device's active performance, including responsivity and high-speed performance. Given the nature of the different optical structures, photodetectors integrated on small and large core waveguides behave very differently in terms of both passive and active performances.

10.6.1 Evanescent Coupled Waveguide-Integrated Photodetectors

Thin-film Ge photodetectors integrated on SOI substrates have become a potential candidate building block to construct next generation high performance optical links. Waveguide-integrated photodetectors can overcome the performance limitation in conventional surface-normal photodetectors by decoupling the photon-absorption path and carrier-collection path. Details can be found in Section 10.2. Different choices of the waveguide core material (Si or SiON) lead to device configurations with either top- or bottom-evanescent–coupled structures depending on process integration.

The waveguide-to-photodetector coupling efficiency strongly depends on the device structure. A parameter, named waveguide geometry factor (WGF), has been defined to characterize the device structural parameters (Ahn et al., 2010). A design map has been plotted in Ahn et al. (2010) to demonstrate that there exists two distinguished regimes, namely, the slow coupling and fast coupling regime, as shown in Figure 10.45. To intuitively understand the design map, a series of simulations, including 3D FDTD simulation, have been carried out and results for devices with WFG located in the three regimes: fast, intermediate, and slow coupling regime, have been demonstrated in Ahn et al. (2010). The

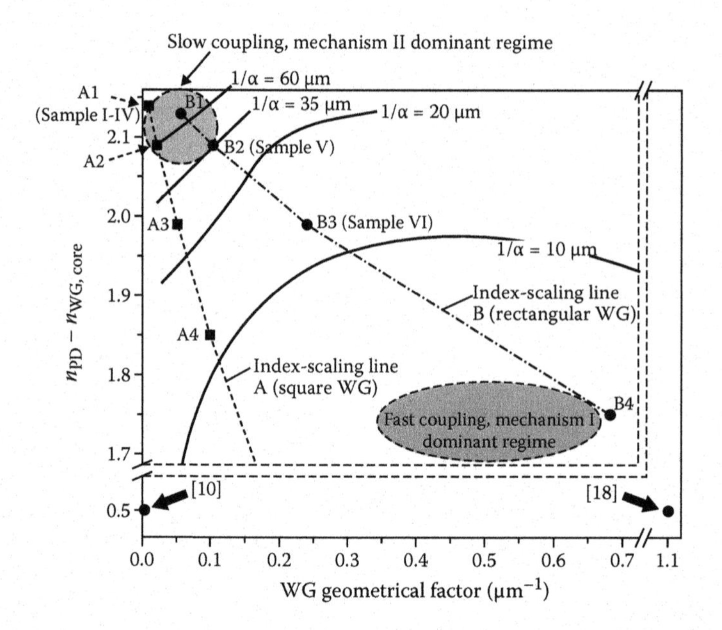

FIGURE 10.45
The evanescent coupling design map showing coupling behavior dependence on the variations of waveguide structure. Figure 10.11 shows the corresponding 3D FDTD simulation results of light propagation in coupling structures located in (a) fast, (b) intermediate, and (c) slow coupling regimes. (Ahn, D. H., Kimerling, L. C., and Michel, J. *J. Lightwave Technol.* 28: 3387–94. ©2010 IEEE.)

results reveal that when impinging into the device with larger WGF), the light is immediately coupled into photodetector absorption region. This process is so fast that light is completely coupled into the absorption region in merely a few micrometers. On the other hand, for the devices with smaller WGF, much longer coupling length is required to achieve high coupling efficiency. When dealing with devices with intermediate WGF, although the coupling occurs immediately at the beginning of the coupling region, the mode oscillates between absorption and waveguide layer and therefore needs a longer interaction length before completely absorbed. The latter is more common in small core waveguide-integrated Ge photodetectors.

10.6.2 Small Core Waveguide-Integrated Ge Photodetectors

Photodetectors integrated on small core waveguides usually employ an evanescent coupling scheme. A three-dimensional schematic view of the described Ge photodetector is shown in Figure 10.46a. As illustrated in the figure, the intrinsic Ge absorption layer is integrated on top of the silicon waveguide to form a vertically oriented p-i-n structure. Figure 10.46b illustrates the schematic cross-sectional view of the device (Feng, 2010a).

Optical mode simulation is the most important and fundamental modeling step when designing a waveguide-integrated photodetector. It helps to understand the optical mode evolution and interaction with the optical structures. From the schematic views shown in Figure 10.46, it can be assumed that the light is incident from the bottom silicon waveguide then evanescently couples back and forth between the top Ge layer and the bottom silicon

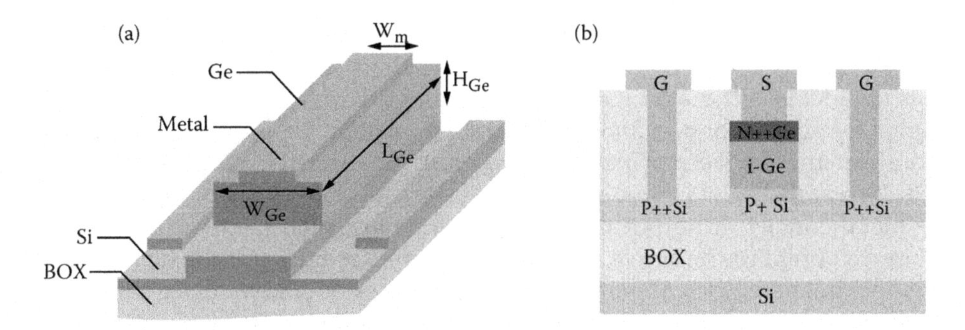

FIGURE 10.46
(a) 3D schematic view of an evanescently coupled waveguide integrated photodetector, and (b) corresponding cross-sectional view. (Reproduced from Feng, N.-N., Liao, S., Dong, P. et al., 2010, Proc. SPIE 7607, 760704-6, by permission of SPIE.)

waveguide. Figure 10.47a,b shows the optical modes of the bottom silicon waveguide and top Ge waveguide, respectively. Due to the small size of the waveguide core, a significant portion of the optical field (evanescent tail) penetrates the Si and Ge interface. Such a structure has an intermediate WGF, therefore, the mode beating length of such a structure is expected to be very short due to the strong coupling.

To further illustrate the coupling interplay, a beam propagation simulation was carried out to simulate the light coupling and absorption in this coupled waveguide system with a typical Ge width of 1.6 μm and thickness of 0.5 μm (Feng, 2010a). The simulated field distribution along the longitudinal direction of the device is shown in Figure 10.47c. The monitor value plot shown on the right side of Figure 10.47c reveals mode coupling interplay among total power (1), power in the Si layer (2), and power in the Ge

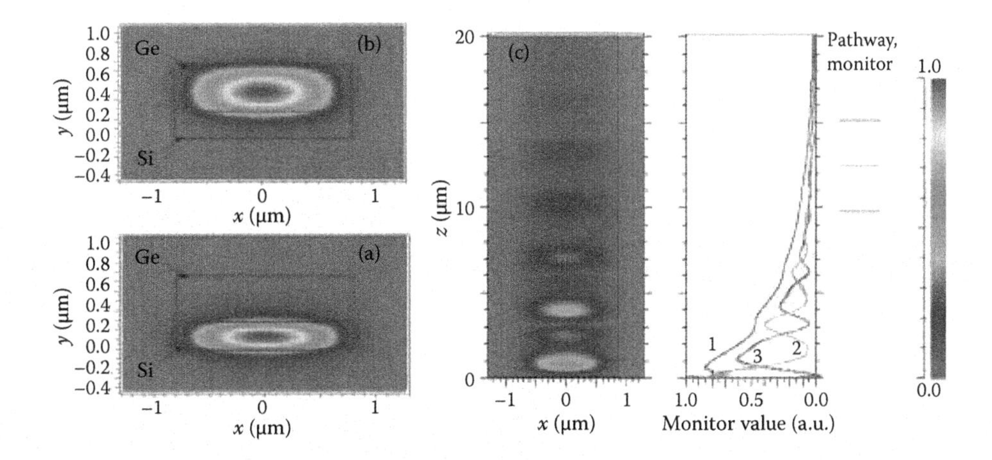

FIGURE 10.47
(a) Optical mode in the input end silicon waveguide, (b) optical mode in Ge waveguide, (c) evanescent power coupling from Si to Ge waveguides. (Reproduced from Feng, N.-N., Liao, S., Dong, P. et al., 2010, Proc. SPIE 7607, 760704-6, by permission of SPIE.)

layer (3) versus the propagation distance. It is found that about 92% and 97% power is absorbed by the Ge layer with device lengths less than 10 and 15 μm, respectively.

Despite its ideality, the above simulation offers a simple picture of the mode coupling behavior in a strong evanescent coupling system. In practice, many other factors can affect the device performance. When evaluating the cross-sectional view in Figure 10.47b carefully, it is apparent that the p-type doping in the bottom Si waveguide and the metal contact on top of the Ge waveguide are such factors that can have significant impact on both the active and the passive performance of the device. A balance between optical and electrical performances is another important design task in photodetector design.

Figure 10.48a shows the simulation results of the additional optical loss caused by p-type doping. With higher doping levels, this additional loss can be as high as 0.5 dB, which translates into a 10% responsivity reduction. Top contact metal absorption is a second factor causing responsivity reduction. From the mode simulation shown in Figure 10.47b, it can be observed that the optical field indeed penetrates into the top Ge interface and enters into the metal contact layer. This portion of optical power will be absorbed by the top contact metal. The power absorption caused by the metal is determined by the complex refractive index of the metal stack used for the contact. The real part determines how much power is in the metal and the imaginary part determines how much power is absorbed. A simulation with the metal absorption present shows that the metal absorption can reduce the responsivity by another 10%–15% depending on the contact design. The metal absorption becomes more severe in the case of a narrow device. In previous work (Feng, 2010a), a photodetector integrated on a 0.25-μm-thick Si waveguide has been demonstrated to have a responsivity of 0.56 A/W at 1.55 μm wavelength. In that particular design, a very compact 1.6 × 10 μm Ge active region was used with a Ge thickness 0.5 μm. The metal absorption was believed to be responsible for the relatively low responsivity.

There are two possible ways to effectively reduce the metal absorption–induced performance degradation: (a) increasing the Ge layer thickness and (b) deliberately shifting metal contact away from the waveguide center. Both are aiming at reducing the mode overlap with the metal contact. The simulation suggests that with the combined effect of

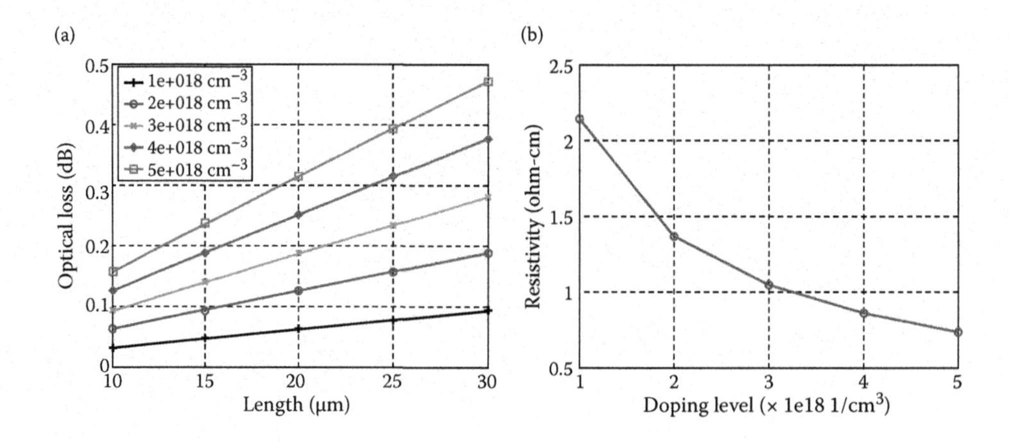

FIGURE 10.48
(a) Additional optical loss vs. device length and doping level and (b) resistivity due to bottom silicon p-type doping.

these two approaches, the impact of metal absorption on responsivity can be reduced to less than 3%.

To deal with the additional doping loss in the bottom silicon waveguide, the most effective way is to reduce the doping level. However, this can jeopardize the photodetector's performance, specifically, its high speed performance. Since the Ge layer thickness H_{Ge} usually is small in waveguide-integrated photodetectors, the transit frequency $f_{transit}$ (inversely proportional to H_{Ge}) is much larger than the RC frequency f_{RC} (inversely proportional to RC constant). Therefore, careful design of the device RC constant is an important task to realize a high-speed device. A balance between optical and electrical performances is another important design task in photodetector design.

Figure 10.48b suggests that the resistivity of the doping area can be reduced by 1/3rd from a low doping level (1×10^{18} cm^{-3}) to a medium doping level (5×10^{18} cm^{-3}). This reduction indicates a trade-off between optical and electrical performances of the device. It is worthwhile to mention that increasing the Ge thickness can increase device speed as a result of reduced capacitance. However, a thick Ge layer increases the carrier drift time and can slow the device when the transit limited frequency $f_{transit}$ becomes dominant. A balance between optical and electrical performances is another important design task in photodetector design. An evanescently coupled Ge detector with 36-GHz 3-dB bandwidth and 0.95-A/W responsivity has recently been reported by taking most of the factors mentioned above into account (Liao et al., 2011).

10.6.3 Large Core Waveguide-Integrated Ge Photodetectors

Evanescent coupled structures are simple and easy to fabricate and are therefore widely used in waveguide-integrated photodetector structures. Advantages have been demonstrated above when using small core waveguide devices, where the field penetrates more outside the waveguide core and leads to a very short coupling length. Despite the smaller WGF in the large core waveguide case, where the evanescent field is much weaker, it does not mean the evanescent coupling scheme cannot be considered. Actually, some previous work has demonstrated that with careful design and fabrication, it is possible to realize high-speed photodetectors on large core waveguides with careful design and fabrication. Figure 10.49a,b illustrates the three-dimensional schematic views of the larger core waveguide-integrated Ge photodetectors with an evanescent and a partial butt-coupled structure (Feng, 2010b).

Figure 10.49c,d shows the simulation results of transmitted power at the output end of the device versus Ge layer thickness. In both cases, the transmitted power shows a strong relationship with the Ge thickness, which is the signature behavior of a weakly coupled system with smaller WGF. When the phase matching condition is satisfied, for instance $H_{Ge} = 0.88$ μm, almost one hundred percent of the optical power is absorbed by the Ge layer. Compared with the small core case, the metal absorption is negligible due to the good optical mode confinement in the waveguide core. However, to realize a high responsivity, precise control of the device parameters, such as Ge layer thickness, is crucial. It is noticed that the partial butt-coupled structure does not improve the parameter sensitivity of the power absorption except for reducing the device length. In addition, the partial butt-coupled structure is much more challenging to fabrication than the evanescent coupled case. A fabricated device with a 0.7-A/W external responsivity has been demonstrated in Feng et al. (2010b) using the evanescent coupled structure. A 3-dB bandwidth of 8.3 GHz has also been demonstrated by the same device with a length of 200 μm.

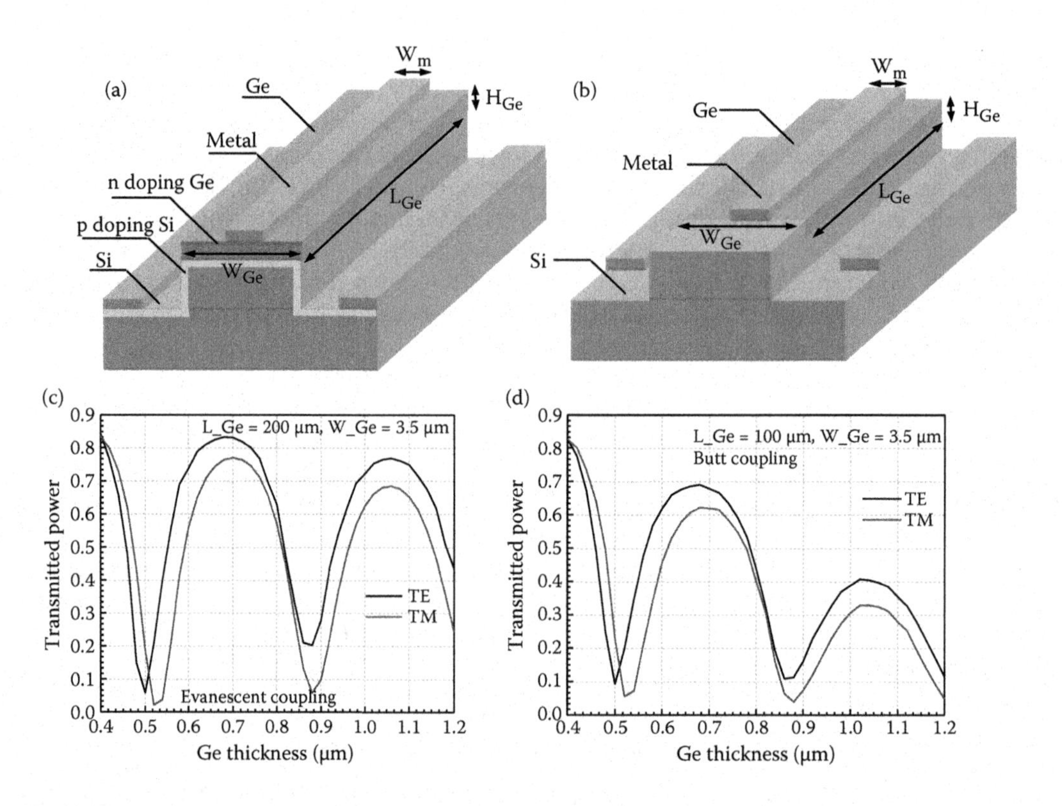

FIGURE 10.49
Schematics of a vertical pin Ge waveguide photodetector integrated on a large core SOI waveguide for (a) an evanescent and (b) a partial butt-coupled structures. Simulation results of the transmitted power at the output end of the devices with (c) an evanescent and (d) a partial butt-coupled structures. (Reproduced from Feng, N.-N., Dong, P., Zheng, D. et al., 2010, *Optics Express* 18: 96–101, by permission of OSA.)

10.6.4 Butt-Coupled Waveguide-Integrated Photodetectors

A fully butt-coupled structure can further improve the device performance and make the device much more compact. Several successful approaches have been demonstrated (Liu et al., 2006a,b; Vivien et al., 2009; Ren et al., 2011). Generally, great care has to be taken of the design of the contacts to avoid significant losses due to the interaction of the optical mode with the metal contacts. Especially, butt-coupling into rib waveguide structures reduces the device bandwidth due to small transit-time–limited frequency response. For this case, a horizontally oriented p-i-n structure has been proposed in Feng et al. (2009) to resolve this issue.

A schematic view of the waveguide-integrated Ge photodetector/EA modulator is illustrated in Figure 10.50a. The device was fabricated on a six inch SOI wafer with 0.375 μm thick buried oxide (BOX) and 3-μm-thick epitaxial-Si layer. The 2.7-μm-high Ge waveguide was grown in a deep-etched recess trench with 0.3-μm-thick Si remaining. The Ge waveguide width is designed to be 1.0 μm to have better fabrication tolerance to a low-loss transition between the Ge and Si waveguides. A 2.4-μm-deep Ge ridge waveguide was fabricated with a 0.3-μm remaining Ge slab sitting on top of the Si slab as shown in Figure 10.50a. The cross-sectional views of the deep-etched Si and Ge waveguides are sketched in Figure 10.50b–d. The figures also illustrate the optical modes of the Si and the Ge waveguides (with two Ge

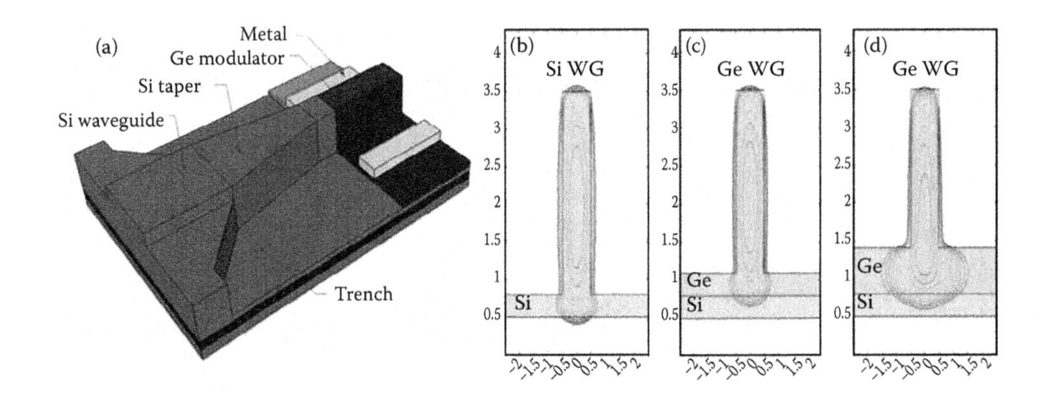

FIGURE 10.50
(a) Schematic view of the Ge photodetector/EA modulator integrated with large core single-mode SOI waveguide. (b–d) Cross-sectional views and optical mode profiles of the Si and the Ge waveguides with (c) 0.3 and (d) 0.6 μm Ge slab thicknesses. The amplitude difference between two adjacent contours is 3 dB. (Reproduced from Feng, N.-N., Liao, S., Feng, D. et al., 2011b, *Optics Express* 19: 8715–20, by permission of OSA.)

slab thicknesses, 0.3 and 0.6 μm). From the optical mode profiles shown in Figure 10.50b,c, it is observed that although the Ge waveguide has a 0.3-μm additional slab on top of the Si slab, the optical mode centers of the Si and Ge waveguides have very little offset. This indicates that the transition between these waveguides will be smooth because the higher order mode excitation is highly suppressed. However, for thicker Ge slab cases (>0.3 μm), the Ge waveguide mode center shows a significant downward shifting and the mode is pushed into the Ge slab as shown in Figure 10.50d, where a 0.6-μm Ge slab case is shown as an example. Significant transition loss is expected from devices with thicker Ge slabs.

Simulations have been carried out to study the impact of the Ge slab thickness on the transition loss of the device with various Ge waveguide widths and slab thicknesses (Feng, 2011b). The results are shown in Figure 10.51a. The observation is that there is a cutoff Ge width for each different Ge slab thickness. Devices can have significant transition loss when their widths are narrower than the cutoff widths. In practice, narrower devices are usually preferable to achieve driving efficiency. The simulation shows that devices with thinner Ge slab remaining have smaller cutoff width. For example, the device with 0.3-μm Ge slab has a cutoff width of 0.7 μm. Hence, devices with width >0.7 μm have negligible transition loss. However, with a 0.6-μm-thick Ge slab, a device with 1.0 μm width can have about 1.3 dB transition loss according to the simulation. A beam propagation simulation can reveal the physical insight into the transition mechanism and the results are shown in Figure 10.51b,c. The optical mode evolution was simulated for devices with 0.3- and 0.6-μm thick Ge slab, respectively. For the thin Ge slab case, due to good mode matching, negligible higher order modes are excited and lead to a very smooth transition with less than 0.15-dB transition loss. For devices with thicker Ge slabs, significant amount power is coupled to higher order modes, which eventually will contribute to device loss. As a result, the transition loss (two interfaces) increases to 1.3 dB for a device with 0.6-μm-thick Ge slab.

Recently, a step coupler structure was proposed to directly couple light from a small core Si waveguide to a Ge/Si APD (Wang et al., 2011). Such a structure was developed to improve the low coupling efficiency of a waveguide-photodetector system that has large height difference between waveguide and photodetector. Efficient light coupling from a small core Si waveguide to a stacked layer APD is obviously a very challenging task due to

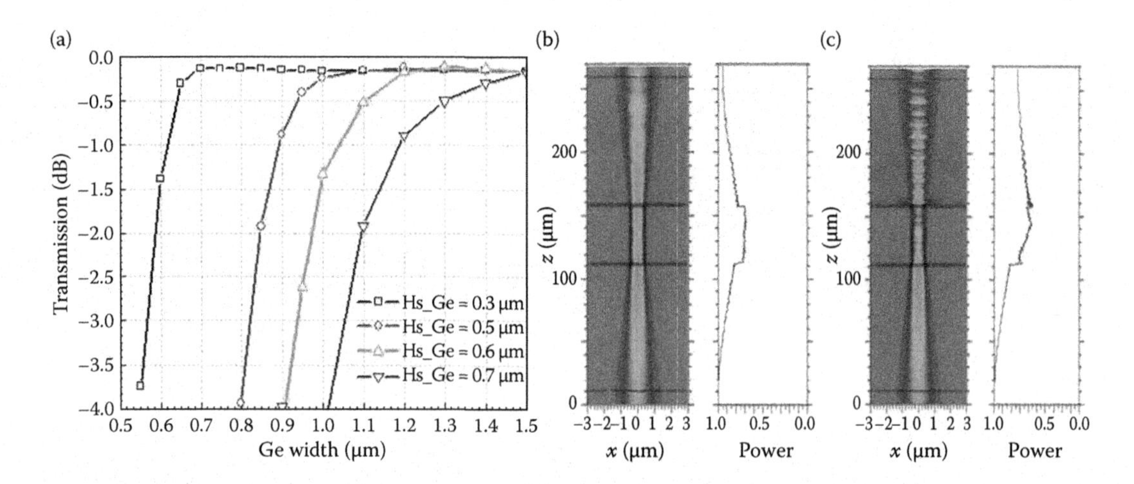

FIGURE 10.51
(a) Simulated transition loss between Ge and Si waveguides excluding Ge absorption. (b) Beam propagation simulation of the devices with 0.3 μm and (c) 0.6 μm thick Ge slab, respectively. (Reproduced from Feng, N.-N., Liao, S., Feng, D. et al., 2011b, *Optics Express* 19: 8715–20, by permission of OSA.)

the height difference. A step waveguide is introduced to improve the coupling efficiency through the so-called self-imaging of a vertical MMI coupler. For the typical parameters listed in Wang et al. (2011), the simulation results of the new structure demonstrate a quantum efficiency improvement compared with the reference structure without a step coupler structure.

10.6.5 Conclusions

Waveguide integrated photodetectors are crucial components to realize the next-generation high-speed optical links. Using several successful examples, it was shown that coupling efficiency is one of the most important design parameters and comprehensive modeling and design of the coupling structure is crucially important to realize high performance waveguide integrated photodetector devices. It is equally important to study the impact of the variations of structural parameters to the active performances of the device while managing to improve the passive waveguide performances. With the increase of device complexity, simulation and modeling of optoelectronic devices will play an increasing role in new device development. Eventually, photonic device simulation has to be integrated with current electronic device simulation to enable fully integrated electronic photonic integrated circuits.

References

Ackland, B., Rafferty, C., King, C. et al. 2009. A monolithic Ge-on-Si CMOS imager for short wave infrared. Presented at the 2009 *International Image Sensor Workshop* (IISW), Bergen, Norway, 2009.

Ahn, D.H., Hong, C.Y., Liu, J.F. et al. 2007. High performance, waveguide integrated Ge photodetectors. *Optics Express* 15: 3916–21.
Ahn, D.H., Kimerling, L.C. and Michel, J. 2010. Evanescent coupling device design for waveguide-integrated group IV photodetectors. *Journal of Lightwave Technology* 28: 3387–94.
Ahn, D.H., Kimerling, L.C. and Michel, J. 2011. Efficient evanescent wave coupling conditions for waveguide-integrated thin-film Si/Ge photodetectors on silicon-on-insulator/germanium-on-insulator substrates. *Journal of Applied Physics* 110, 083115.
Ando, H., Unbe, H., Imura, T. et al. 1978. Characteristics of germanium avalanche photodiodes in the wavelength region of 1–1.6μm. *IEEE Journal of Quantum Electronics* 14: 804–9.
Ang, K.W., Ng, J.W., Lim, A.E. et al. 2010. Waveguide-integrated Ge/Si avalanche photodetector with 105GHz gain-bandwidth product. *Optical Fiber Communication Conference*, OSA, Technical (CD) JWA36.
Assefa, S., Xia, F., Green, W.M.J. et al. 2010a. CMOS-integrated high-speed MSM germanium waveguide photodetector. *Optics Express* 18: 4986–99.
Assefa, S., Xia, F., Green, W.M.J., Schow, C.L., Rylyakov, A.V. and Vlasov, Y.A. 2010b. CMOS-integrated optical receivers for on-chip interconnects. *IEEE Journal of Selected Topics in Quantum Electronics* 16: 1376–85.
Baehr-Jones, T., Hochberg, M. and Scherer, A. 2008. Photodetection in silicon beyond the band edge with surface states. *Optics Express* 16: 1659–68.
Beals, M., Michel, J., Liu, J.F. et al. 2008. Process flow innovations for photonic device integration in CMOS. *2008 Proceedings of the International Society for Optical Engineering* (SPIE) 6898: 689804.
Beiting, E.J. and Feldman, P.D. 1977. Fabrication and performance of intrinsic germanium photodiodes. *Applied Optics* 16: 800–2.
Benton, J.L., Halliburton, K., Libertino, S., Eaglesham, D.J. and Coffa, S. 1998. Electrical signatures and thermal stability of interstitial clusters in ion implanted Si. *Journal of Applied Physics* 84: 4749–56.
Boyraz, O. and Jalali, B. 2004. Demonstration of a silicon Raman laser. *Optics Express* 12: 5269.
Bradley, J.D.B., Jessop, P.E. and Knights, A.P. 2005. Silicon waveguide-integrated optical power monitor with enhanced sensitivity at 1550 nm. *Applied Physics Letters* 86: 241103.
Bristow, A.D., Rotenberg, N., van Driel, H.M. et al. 2007. Two-photon absorption and Kerr coefficients of silicon for 850–2,200 nm. *Applied Physics Letters* 10: 1063.
Byun, K.Y., Ferain, I., Hayes, J. et al. 2011. Surface activation using oxygen and nitrogen radical for Ge-Si Avalanche photodiode integration. *Microelectronic Engineering* 88: 522–5.
Campbell, J.C., Demiguel, S., Ma, F. et al. 2004. Recent advances in avalanche photodiodes. *IEEE Journal of Selected Topics in Quantum Electronics* 10: 777–87.
Campbell, J.C., Tsang, W.T., Qua, G.J. et al. 1988. High-speed InP/InGaAsP/ InGaAs avalanche photodiodes grown by chemical beam epitaxy. *IEEE Journal of Quantum Electronics* 24: 496–500.
Cannon, D.D., Liu, J.F., Ishikawa, Y. et al. 2004. Tensile strained epitaxial Ge films on Si(100) substrates with potential application in L-band telecommunications. *Applied Physics Letters* 84: 906–8.
Carey, J.E., Crouch, C.H., Shen, M. and Mazur, E. 2005. Visible and near-infrared responsivity of femtosecond-laser microstructured silicon photodiodes. *Optics Letters* 30: 1773–5.
Carrano, J.C., Lambert, D.J.H., Eiting, C.J. et al. 2000. GaN avalanche photodiodes. *Applied Physics Letters* 76: 924–6.
Carroll, M.S., Childs, K., Jarecki, R., Bauer, T. and Saiz, K. 2008. Ge-Si separate absorption and multiplication avalanche photodiode for Geiger mode single photon detection. *Applied Physics Letters* 93: 183511–3.
Casalino, M., Coppola, G., Gioffrè, M. et al. 2011. Microcavity silicon photodetectors at 1.55 μm. Hindawi Publishing Corporation, *Advances in OptoElectronics*, Article ID 965967, doi:10.1155/2011/965967.
Casalino, M., Coppola, G., Iodice, M., Rendina, I. and Sirleto, L. 2010. Near-infrared sub-bandgap all-silicon photodetectors: state of the art and perspectives. *Sensors* 10: 10571–600.
Chaisakul, P., Marris-Morini, D., Isella, G. et al. 2011. Ge/SiGe multiple quantum well photodiode with 30 GHz bandwidth. *Applied Physics Letters* 98: 131112–3.

Chang, S.W. and Chuang, S.L. 2007. Theory of optical gain of Ge-SixGeySn1-x-y quantum-well lasers. *IEEE Journal of Quantum Electronics* 43: 249–56.

Chelyadinskii, A.R. and Komarov, F.F. 2003. Defect-impurity engineering in implanted silicon. *Physics - Uspekhi* 46: 789–820.

Chen, H., Luo, X. and Poon, A.W. 2009b. Cavity-enhanced photocurrent generation by 1.55 μm wavelengths linear absorption in a p-i-n diode embedded silicon microring resonator. *Applied Physics Letters* 95: 171111.

Chen, L. and Lipson, M. 2009. Ultra-low capacitance and high speed germanium photodetectors on silicon. *Optics Express* 17: 7901–6.

Chen, L., Preston, K., Manipatruni, S., and Lipson, M. 2009a. Integrated GHz silicon photonic interconnect with micrometer-scale modulators and detectors. *Optics Express* 18: 15248–56.

Chen, L., Doerr, C.R., Buhl, L., Baeyens, Y., and Aroca, R.A. 2011. Monolithically integrated 40-wavelength demultiplexer and photodetector array on silicon. *IEEE Photonics Technology Letters* 23: 1041–3.

Chen, X., Panoiu, N.C. and Osgood, Jr., R.M. 2006. Theory of Raman-mediated pulsed amplification in silicon-wire waveguides. *IEEE Journal of Quantum Electronics* 42 (2): 160–70.

Cheng, L.J., Corelli, J.C., Corbett, J.W. and Watkins, G.D. 1966. 1.8-, 3.3-, 3.9-μm bands in irradiated silicon: correlation with the divacancy. *Physics Review* 152 (2): 761–74.

Chui, C.O., Okyay, A.K. and Saraswat, K.C. 2003. Effective dark current suppression with asymmetric MSM photodetectors in Group IV semiconductors. *IEEE Photonics Technology Letters* 15: 1585–7.

Chynoweth, A.G. 1958. Ionization rates for electrons and holes in Silicon. *Physical Review* 109: 1537–40.

Ciftcioglu, B., Zhang, J., Sobolewski, R. and Wu, H. 2010. An 850-nm normal-incidence germanium metal-semiconductor-metal photodetector with 13-GHz bandwidth and 8-mu a dark current. *IEEE Photonics Technology Letters* 22: 1850–2.

Claps, R., Raghunathan, V., Dimitropoulos, D. and Jalali, B. 2004. Influence of nonlinear absorption on Raman amplification in silicon waveguides. *Optics Express* 12: 2774–80.

Coffa, S., Libertino, S. and Spinella, C. 2000. Transition from small interstitial clusters to extended {311} defects in ion-implanted Si. *Applied Physics Letters* 76: 321–313.

Colace, L., Balbi, M., Sorianello, V. et al. 2008. Temperature-dependence of Ge on Si p-i-n photodetectors. *Journal of Lightwave Technology* 26: 2211–4.

Colace, L., Masini, G. and Assanto, G. 1999. Ge-on-Si approaches to the detection of near-infrared light. *IEEE Journal of Quantum Electronics* 35: 1843–52.

Colace, L., Masini, G., Cencelli, V.O., de Notaristefani, F. and Assanto, G. 2004. Polycrystalline germanium enables near-IR photodetectors integrated with silicon CMOS electronics—a near-IR imaging chip based on polycrystalline germanium on silicon opens new frontiers for integrated optoelectronics. *Photonics Spectra* 38: 88.

Colace, L., Masini, G., Cozza, S., Assanto, G., DeNotaristefani, F. and Cencelli, V. 2007. Near-infrared camera in polycrystalline germanium integrated on complementary-metal-oxide semiconductor electronics. *Applied Physics Letters* 90: 011103–3.

Colace, L., Masini, G., Galluzzi, F. et al. 1997. Ge/Si (001) photodetector for near infrared light. *Solid State Phenomena* 54: 55–8.

Colace, L., Sorianello, V., Assanto, G., Fulgoni, D., Nash, L. and Palmer, M. 2010. Germanium on glass: a novel platform for light-sensing devices. *IEEE Photonics Journal* 2: 686–95.

Dai, D.X., Bowers, J.E., Lu, Z.W. et al. 2010. Temperature dependence of Ge/Si avalanche photodiodes. *2010 Device Research Conference* (DRC): 231–2.

D'Costa, V.R., Cook, C.S., Birdwell, A.G. et al. 2006. Optical critical points of thin-film Ge1-ySny alloys: A comparative $Ge_{1-y}Sn_y/Ge_{1-x}Si_x$ study. *Physical Review B* 73: 125207.

D'Costa, V.R., Fang, Y.Y., Mathews, J. et al. 2009. Sn-alloying as a means of increasing the optical absorption of Ge at the C- and L-telecommunication bands. *Semiconductor Science and Technology* 24: 115006.

Dehlinger, G., Koester, S.J., Schaub, J.D., Chu, J.O., Ouyang, Q.C. and Grill, A. 2004. High-speed germanium-on-SOI lateral PIN photodiodes. *IEEE Photonics Technology Letters* 16: 2547–9.

Dinu, M., Quochi, F. and Garcia, H. 2003. Third-order nonlinearities in silicon at telecom wavelengths. *Applied Physics Letters* 82: 2954–6.

Doerr, C.R., Chen, L., Buhl, L.L. and Chen, Y.-K. 2011. Eight-channel SiO_3 /Si_3N_4 /Si/Ge CWDM receiver. *IEEE Photonics Technology Letters* 23: 1201–3.

Dosunmu, O.I., Cannon, D.D., Emsley, M.K. et al. 2004. Resonant cavity enhanced Ge photodetectors for 1550 nm operation on reflecting Si substrates. *IEEE Journal of Selected Topics in Quantum Electronics* 10: 694–701.

Dosunmu, O.I., Cannon, D.D., Emsley, M.K., Kimerling, L.C. and Unlu, M.S. 2005. High-speed resonant cavity enhanced Ge photodetectors on reflecting Si substrates for 1550-nm operation. *IEEE Photonics Technology Letters* 17: 175–7.

Doylend, J.K., Jessop, P.E. and Knights, A.P. 2010. Silicon photonic resonator-enhanced defect mediated photodiode for sub-bandgap detection. *Optics Express* 18: 14671–8.

Euser, T.G. and Vos, W.L. 2005. Spatial homogeneity of optically switched semiconductor photonic crystals and of bulk semiconductors. *Journal of Applied Physics* 97: 043102.

Fan, H.Y. and Ramdas, A.K. 1959. Infrared absorption and photoconductivity in irradiated silicon. *Journal of Applied Physics* 30: 1127–34.

Fang, Q., Liow, T.Y., Song, J.F. et al. 2010a. WDM multi-channel silicon photonic receiver with 320 Gbps data transmission capability. *Optics Express* 18: 5106–13.

Fang, Q., Phang, Y.T., Tan, C.W. et al. 2010b. Multi-channel silicon photonic receiver based on ring-resonators. *Optics Express* 18: 13510–5.

Feng, D., Liao, S., Dong, P. et al. 2009. High-speed Ge photodetector monolithically integrated with large cross-section silicon-on-insulator waveguide. *Applied Physics Letters* 95: 261105.

Feng, N.-N., Liao, S., Dong, P. et al. 2010a. A compact high-performance Germanium photodetector integrated on 0.25 μm thick silicon-on-insulator waveguide. *Proceedings of the SPIE* 7607: 760704–6.

Feng, N.-N., Dong, P., Zheng, D. et al. 2010b. Vertical p-i-n germanium photodetector with high external responsivity integrated with large core Si waveguides. *Optics Express* 18: 96–101.

Feng, N.-N., Feng, D., Liao, S. et al. 2011a. 30GHz Ge electro-absorption modulator integrated with 3μm silicon-on-insulator waveguide. *Optics Express* 19: 7062–7.

Feng, N.-N., Liao, S., Feng, D. et al. 2011b. Design and fabrication of 3μm silicon-on-insulator waveguide integrated Ge electro-absorption modulator. *Optics Express* 19: 8715–20.

Fischetti, M.V. and Laux, S.E. 1996. Band structure, deformation potentals, and carrier mobility in strained Si, Ge, and SiGe alloys. *Journal of Applied Physics* 80: 2234–52.

Fitzgerald, E.A., Xie, Y.H., Green, M.L. et al. 1991. Totally relaxed Ge_xSi_{1-x} layers with low threading dislocation densities grown on Si substrates. *Applied Physics Letters* 59: 811–3.

Fleming, R.M., Seager, C.H., Lang, D.V., Cooper, P.J., Bielejec, E. and Campbell, J.M. 2007. Effects of clustering on the properties of defects in neutron irradiated silicon. *Journal of Applied Physics* 102: 043711.

Geis, M.W., Spector, S.J., Grein, M.E. et al. 2007b. CMOS-compatible all-Si high-speed waveguide photodiodes with high responsivity in near-infrared communication band. *IEEE Photon Technology Letters* 19: 152–4.

Geis, M.W., Spector, S.J., Grein, M.E. et al. 2007a. All silicon infrared photodiodes: photo response and effects of processing temperature. *Optics Express* 15: 16886–95.

Geis, M.W., Spector, S.J., Grein, M.E., Yoon, J.U., Lennon, D.M. and Lyszczarz, T.M. 2009. Silicon waveguide infrared photodiodes with >35 GHz bandwidth and phototransistors with 50 AW–1 response. *Optics Express* 17: 5193–204.

Grein, M.E., Spector, S.J., Khilo, A. et al. 2011. Demonstration of a 10 GHz CMOS-Compatible Integrated Photonic Analog-to-Digital Converter. *Conference of Lasers and Electro Optics (CLEO)* Baltimore, Maryland. Optical Signal Processing (CThI).

Giri, P.K. and Mohapatra, Y.N. 2001. Evidence of metastability with athermal ionization from defect clusters in ion-damaged silicon. *Physical Review B* 62: 16561–5.

Haret, L.-D., Checoury, X., Han, Z., Boucaud, P., Combrié, S. and De Rossi, A. 2010. All-silicon photonic crystal photoconductor on silicon-on-insulator at telecom wavelength. *Optics Express* 18: 23965–72.

Hawkins, A.R., Wu, W., Abrham, P. et al. 1996. High gain-bandwidth-product silicon heterointerface photo detector. *Applied Physics Letters* 70: 303–5.
Heroux, J.B., Yang, X. and Wang, W.I. 1999. GaInNAs resonant-cavity-enhanced photodetector operating at 1.3 μm. *Applied Physics Letters* 75: 2716–8.
Holzwarth, C.W., Amatya, R., Araghchini, M. et al. 2009. High speed analog-to-digital conversion with silicon photonics. *Procedings of the SPIE* 7220: 72200B–1.
Huang, Z., Carey, J.E., Liu, M., Guo, X., Mazur, E. and Campbell, J.C. 2006. Microstructured silicon photodetector. *Applied Physics Letters* 89: 033506.
Hurkx, G.A.M., Klaassen, D.B.M. and Knuvers, M.P.G. 1992. A new recombination model for device simulation including tunneling. *IEEE Transactions on electron devices* 39: 331–8.
Ishikawa, Y., Wada, K., Cannon, D.D., Liu, J.F., Luan, H.C. and Kimerling, L.C. 2003. Strain-induced direct bandgap shrinkage in Ge grown on Si substrate. *Applied Physics Letters* 82: 2044–6.
Ishikawa, Y., Wada, K., Liu, J.F. et al. 2005. Strain-induced enhancement of near-infrared absorption in Ge epitaxial layers grown on Si substrate. *Journal of Applied Physics* 98: 9.
Ito, M. and Wada, O. 1986. Low dark current GaAs metal-semiconductor metal (msm) photodiodes using WSi_x contacts. *IEEE Journal of Quantum Electronics* 22: 1073–7.
Jacoboni, C., Nava, F., Canali, C. and Ottaviani, G. 1981. Electron drift velocity and diffusivity in germanium. *Physics Review B* 24: 1014–26.
Junkes, A., Eckstein, D., Pintilie, I., Makarenko, L.F. and Fretwurst, E. 2010. Annealing study of a bistable cluster defect. *Nuclear Instruments and Methods in Physics Research A* 612: 525–9.
Jutzi, M., Berroth, M., Wohl, G., Oehme, M. and Kasper, E. 2005. Zero biased Ge-on-Si photodetector on a thin buffer with a bandwidth of 3.2 GHz at 1300 nm. *Materials Science in Semiconductor Processing* 8: 423–7.
Kang, Y., Huang, Z., Saado, Y. et al. 2011. High performance Ge/Si avalanche photodiodes development in Intel. *Optical Fiber Communication Conference*, OSA, Technical (CD): OWZ1.
Kang, Y., Litski, S., Sarid, G. et al. 2007. Ge/Si avalanche photodiodes for 1.3 μm optical fiber links. *2007 4th IEEE International Conference On Group IV Photonics* (IEEE, Piscataway, NJ, USA, 2007) 294–6.
Kang, Y.M., Liu, H.D., Morse, M. et al. 2009a. Monolithic germanium/silicon avalanche photodiodes with 340 GHz gain-bandwidth product. *Nature Photonics* 3: 59–63.
Kang, Y.M., Morse, M., Paniccia, M.J. et al. 2009b. Monolithic Ge/Si avalanche photodiodes. in *Proceedings of the 2009 6th IEEE International Conference on Group IV Photonics* (GFP) 25–7.
Kaufmann, R., Isella, G., Sanchez-Amores, A. et al. 2011. Near infrared image sensor with integrated germanium photodiodes. *Journal of Applied Physics* 110: 023107–6.
Khilo, A., Spector, S.J., Grein, M.E. et al. 2012. Photonic ADCs come of age: overcoming the bottleneck of electronic jitter. *Optics Express* 20: 4454.
Kinsey, G.S., Campbell, J.C., Dentai, A.G. et al. 2001. Waveguide avalanche photodiode operating at 1.55 μm with a gain–bandwidth product of 320 GHz. *IEEE Photonics Technology Letters* 13: 842–4.
Klinger, S., Berroth, M., Kaschel, M., Oehme, M. and Kasper, E. 2009. Ge-on-Si p-i-n photodiodes with a 3-dB bandwidth of 49 GHz. *IEEE Photonics Technology Letters* 21: 920–5.
Knights, A.P. and Bradley, J.D.B. et al. 2008. Optical detection technologies for silicon. In *Silicon photonics: the state of the art.* ed. G.T. Reed, (John Wiley & Sons, Ltd), chapter 6, 191–217.
Knights, A.P., Bradley, J.D.B., Gou, S.H. and Jessop, P.E. 2006. Silicon-on-insulator waveguide photodiode with self-ion-implantation-engineered-enhanced infrared response. *Journal of Vacuum Science and Technology A* 24: 783–6.
Knights, A.P., House, A., MacNaughton, R. and F. Hopper, F. 2003. Optical power monitoring function compatible with single chip integration on silicon-on-insulator. *Conference on Optical Fiber Communication, Technical Digest Series* 86: 705–6.
Koester, S.J., Schaub, J.D., Dehlinger, G. and Chu, J.O. 2006. Germanium-on-SOI infrared detectors for integrated photonic applications. *IEEE Journal of Selected Topics in Quantum Electronics* 12: 1489–502.
Koester, S.J., Schow, C.L., Schares, L. et al. 2007. Ge-on-SOI-detector/Si-CMOS-amplifier receivers for high-performance optical-communication applications. *Journal of Lightwave Technology* 25: 46–57.

Kressel, H. 1982. *Semiconductor Devices for optical communication*. Springer-Verlag: Berlin.
Kyuregyan, A.S. and Yurkov, S.N. 1989. *Soviet Physics: Semiconductors* 23: 1126–32.
Lee, C.A., Logan, R.A., Batdorf, R.J. et al. 1964. *Physical Review* 134: A761.
Li, C., Yang, Q.Q., Wang, H.J. et al. 2000. Back-incident SiGe-Si multiple quantum-well resonant-cavity-enhanced photodetectors for 1.3 mm operation. *IEEE Photonics Technology Letters* 12: 1373–5.
Li, C.B., Mao, R.W., Zuo, Y.H. et al. 2004. 1.55 mm Ge islands resonant-cavity-enhanced detector with high-reflectivity bottom mirror. *Applied Physics Letters* 85: 2697–9.
Liang, T.K. and Tsang, H.K. 2004. Nonlinear absorption and raman scattering in silicon-on-insulator optical waveguides. *IEEE Journal of Selected Topics Quantum Electronics* 10: 1149–53.
Liang, T.K., Tsang, T.K., Day, I.E., Drake, J., Knights, A.P. and Asghari, M. 2002. Silicon waveguide two-photon absorption detector at 1.5 μm wavelength for autocorrelation measurements. *Applied Physics Letters* 81: 1323–5.
Liao, S., Feng, N.-N., Feng, D. et al. 2011. 36 GHz submicron silicon waveguide germanium photodetector. *Optics Express* 19: 10967–72.
Libertino, S., Benton, J.L., Jacobson, D.C. et al. 1997. Evolution of interstitial- and vacancy-type defects upon thermal annealing in ion-implanted Si. *Applied Physics Letters* 71: 389–91.
Libertino, S., Coffa, S. and Benton, J.L. 2001. Formation, evolution and annihilation of interstitial clusters in ion-implanted Si. *Physical Review B* 63: 195206.
Libertino, S., Coffa, S., Benton, J.L., Halliburton, K. and Eaglesham, D.J. 1999. Formation, evolution and annihilation of interstitial clusters in ion implanted Si. *Nuclear Instruments and Methods in Physics Research* B 148: 247–51.
Libertino, S., Coffa, S., Spinella, C., Benton, J.L. and Arcifa, D. 2000. Cluster formation and growth in Si ion implanted c-Si. *Material Science and Engineering* B 7: 137–42.
Lin, Q., Zhang, J., Piredda, G., Boyd, R.W., Fauchet, P.M. and Agrawal, G.P. 2007. Dispersion of silicon nonlinearities in the near infrared region. *Applied Physics Letters* 91: 021111.
Liu, H.D., Pan, H.P., Hu, C. et al. 2009. Avalanche photodiode punch-through gain determination through excess noise analysis. *Journal of Applied Physics* 106: 064507–4.
Liu, J.F., Ahn, D., Hong, C.-Y. et al. 2006b. Waveguide integrated Ge p-i-n photodetectors on a silicon-non-insulator platform, in *Proceedings of the Optical Valley of China International Symposium on Optoelectronics*, 2006, pp. 1–4.
Liu, J.F., Beals, M., Pomerene A. et al. 2008. Waveguide integrated, ultra-low energy GeSi electroabsorption modulators. *Nature Photonics* 2: 433–7.
Liu, J.F., Camacho-Aguilera, R., Sun, X. et al. 2011. Ge-on-Si optoelectronics. *Thin Solid Films* (in press).
Liu, J.F., Cannon, D.D. and Wada, K. 2005b. Tensile strained Ge p-i-n photodetectors on Si platform for C and L band telecommunications. *Applied Physics Letters* 87: 011110–3.
Liu, J.F., Dong, P., Jongthammanurak, S., Wada, K., Kimerling, L.C. and Michel, J. 2007. Design of monolithically integrated GeSi electroabsorption modulators and photodetectors on an SOI platform. *Optics Express* 15: 623–8.
Liu, J.F., Michel, J., Giziewicz, W. et al. 2005a. High-performance, tensile-strained Ge p-i-n photodetectors on a Si platform. *Applied Physics Letters* 87: 103501–3.
Liu, J.F., Pan, D., Jongthammanurak, S. et al. 2006a. Waveguide-Integrated Ge p-i-n Photodetectors on SOI Platform. *3rd IEEE International Conference on Group IV Photonics* (Ottawa, Canada, Sep. 2006).
Liu, Y., Chow, C.W., Cheung, W.Y. and Tsang, H.K. 2006c. In-line channel power monitor based on helium ion implantation in silicon-on-insulator waveguides. *IEEE Photon Technology Letters* 18: 1882–4.
Logan, D.F. 2011. Defect-enhanced silicon photodiodes for photonic circuits. Doctor of Philosophy Thesis, McMaster University 2011.
Logan, D.F., Knights, A.P., Jessop, P.E. and Tarr, N.G. 2011. Defect-enhanced photo-detection at 1550 nm in a silicon waveguide formed via LOCOS. *Semiconductor Science and Technology* 26: 045009.
Logan, D.F., Velha, P., Sorel, M., De La Rue, R.M., Jessop, P.E. and Knights, A.P. 2012. Monitoring and tuning micro-ring properties using defect enhanced silicon photodiodes at 1550 nm. *IEEE Photonics Technology Letters 24: 261.*

Logan, D.F., Velha, P., Sorel, M., De La Rue, R.M., Knights, A.P. and Jessop, P.E. 2010. Defect-enhanced silicon-on-insulator waveguide resonant photodetector with high sensitivity at 1.55 μm. *IEEE Photon Technology Letters* 22: 1530–2.

Loh, T.-H., Wang, J., Nguyen, H.-S. et al. 2007a. High speed selective-area-epitaxial ge-on-SOI PIN photo-detector using thin low temperature $Si_{0.8}Ge_{0.2}$ buffer by ultra-high-vacuum chemical vapor deposition. *Proceedings of the 4th International Conference on Group IV Photonics,* Tokyo, Japan.

Loh, T.H., Xiong, Y.Z., Lee, S.J. et al. 2007b. Impact of local strain from selective epitaxial germanium with thin Si/SiGe buffer on high-performance p-i-n photodetectors with a low thermal budget. *IEEE Electron Device Letters* 28: 984–6.

Luan, H.-C., Lim, D.R., Lee, K.K. et al. 1999. High-quality Ge epilayers on Si with low threading-dislocation densities. *Applied Physics Letters* 75: 2909–11.

Luan, H.C., Wada, K., Kimerling, L.C., Masini, G., Colace, L. and Assanto, G. 2001. High efficiency photodetectors based on high quality epitaxial germanium grown on silicon substrates. *Optical Materials* 17: 71–3.

Luryi, S., Kastalsky, A. and Bean, J.C. 1984. New infrared detector on a silicon chip. *IEEE Transactions on Electron Devices* 31: 1135–9.

Madelung, O. 1982. *Physics of Group IV Elements and III–V Compounds, Landolt-Börnstein: Numerical Data and Functional Relationships in Science and Technology*. Springer. vol. 17a.

Mao, X., Han, P., Gao, L. et al. 2011. Selenium doped silicon-on-insulator waveguide photodetector with enhanced sensitivity at 1550 nm. *IEEE Photonics Technology Letters* 23: 1517.

Masini, G., Cencelli, V., Colace, L., De Notaristefani, F. and Assanto, G. 2002. Monolithic integration of near-infrared Ge photodetectors with Si complementary metal-oxide-semiconductor readout electronics. *Applied Physics Letters* 80: 3268–70.

Masini, G., Colace, L., Assanto, G., Luan, H.C., Wada, K. and Kimerling, L.C. 1999. High responsivity near infrared Ge photodetectors integrated on Si. *Electronics Letters* 35: 1467–8.

Masini, G., Sahni, S., Capellini, G., Witzens, J. and Gunn, C. 2008. High-speed near infrared optical receivers based on Ge waveguide photodetectors integrated in a CMOS process. *Advances in Optical Technology* 2008: 196572.

Michel, J., Liu, J.F., Ahn, D. et al. 2007. Advances in fully CMOS integrated photonic devices. *Proceedings of the SPIE* 6477: 64770P.

Michel, J., Liu, J.F. and Kimerling, L.C. 2010. High-performance Ge-on-Si photodetectors. *Nature Photonics* 4: 527–34.

Mikawa, T., Kagawa, S., Kaneda, T. et al. 1980. Crystal orientation dependence of ionization rates in germanium. *Applied Physics Letters* 37: 387–9.

Morse, M., Dosunmu, O., Sarid, G., and Chetrit, Y. 2006. Performance of Ge-on-Si p-i-n Photodetectors for Standard Receiver Modules. *IEEE Photon Technology Letters* 18: 2442–4.

Motamedi, A.R., Nejadmalayeri, A.H., Khilo, A., Kärtner, F.X. and Ippen, E.P. 2011. Ultrafast nonlinear optical processes and free-carrier lifetime in silicon nanowaveguides. *Conference of Lasers and Electro Optics (CLEO)* Baltimore, Maryland. (CFO2).

Notomi, M., Shinya, A., Nozaki, K. et al. 2011. Low-power nanophotonic devices based on photonic crystals towards dense photonic network on chip. *Institution of Engineering and Technology* 2: 84–93.

Oehme, M., Werner, J., Kasper, E., Jutzi, M. and Berroth, M. 2006. High bandwidth Ge p-i-n photodetector integrated on Si. *Applied Physics Letters* 89: 071117–3.

Oh, J., Banerjee, S.K. and Campbell, J.C. 2004. Metal-germanium-metal photodetectors on hetero-epitaxial Ge-On-Si with amorphous Ge Schottky barrier enhancement layers. *IEEE Photonics Technology Letters* 16: 581–3.

Osgood Jr., R.M., Panoiu, N.C., Dadap, J.J. et al. 2009. Engineering nonlinearities in nanoscale optical systems: physics and applications in dispersion engineered silicon nanophotonic wires. *Advances in Optics and Photonics* 1: 162–235.

Pan, G.Z. and Tu, K.N. 1997. Transmission electron microscopy on {113} rod-like defects and {111} dislocation loops in silicon-implanted silicon. *Journal of Applied Physics* 82: 601–8.

Park, S., Tsuchizawa, T., Watanabe, T. et al. 2010. Monolithic integration and synchronous operation of germanium photodetectors and silicon variable optical attenuators. *Optics Express* 18: 8412–21.

Park, S., Yamada, K., Tsuchizawa, T. et al. 2011. All-silicon and in-line integration of variable optical attenuators and photodetectors based on submicrometer rib waveguides. *Optics Express* 19: 11969–76.

Pelaz, L., Marqués, L.A. and Barbolla, J. 2004. Ion-beam-induced amorphization and recrystallization in silicon. *Journal of Applied Physics* 96: 5947–76.

Pinguet, T., Analui, B., Masini, G., Sadagopan, V., Gloeckner, S. 2008. 40 Gbps monolithically integrated transceivers in CMOS photonics. *Proceedings of the SPIE* 6898: 689805.

Preston, A.K., Lee, Y.H.D., Zhang, Mi. and Lipson, M. 2011. Waveguide-integrated telecom-wavelength photodiode in deposited silicon. *Optics Letters* 36: 52–4.

Qian, W., Feng, D.Z., Liang, H. et al. 2011. Integration of high speed Ge PIN photodetectors with WDM filter on SOI platform. *2011 IEEE Photonics Society Summer Topical Meeting* paper Thc3 (Jul. 2011, Montreal, Canada).

Raissi, F. and Far, M.M. 2002. Highly sensitive PtSi/porous Si Schottky detectors. *IEEE Sensors Journal* 2: 476–81.

Reintjes, J.F. and McGroddy, J.C. 1973. Indirect two-photon transitions in Si at 1.06 μm. *Physical Review Letters* 30: 901–3.

Ren, F.-F., Ang, K.-W., Song, J. et al. 2010. Surface plasmon enhanced responsivity in a waveguided germanium metal-semiconductor-metal photodetector. 97: 091102–3.

Ren, S., Kamins, T.I., and Miller, D.A.B. 2011. Thin dielectric spacer for the monolithic integration of bulk germanium or germanium quantum wells with silicon-on-insulator waveguides. *IEEE Photonics Journal* 3: 739–47.

Robertson, L.S., Jones, K.S., Rubin, L.M. and Jackson, J. 2000. Annealing kinetics of {311} defects and dislocation loops in the end-of-range damage region of ion implanted silicon. *Journal of Applied Physics* 87: 2910–3.

Rong, H., Liu, A., Jones, R. et al. 2005. An all-silicon Raman laser. *Nature* 433: 292–4.

Roucka, R., Mathews, J., Weng, C. et al. 2011. High-performance near-IR photodiodes: a novel chemistry-based approach to Ge and Ge-Sn devices integrated on silicon. *IEEE Journal of Quantum Electronics* 47: 213–22.

Rouviere, M., Vivien, L., Le Roux, X. et al. 2005. Ultrahigh speed germanium-on-silicon-on-insulator photodetectors for 1.31 and 1.55 mm operation. *Applied Physics Letters* 87: 3.

Rouvière, M., Halbwax, M., Cercus, J.-L. et al. 2005b. Integration of germanium waveguide photodetectors for intrachip optical interconnects. *Optical Engineering* 44: 75402–6.

Ryder, E.J. 1953. Mobilities of holes and electrons in high electric fields. *Physics Review* 90: 766–9.

Samavedam, S.B., Currie, M.T., Langdo, T.A. and Fitzgerald, E.A. 1998. High-quality germanium photodiodes integrated on silicon substrates using optimized relaxed graded buffers. *Applied Physics Letters* 73: 2125–7.

Schaub, J.D., Li, R., Schow, C.L., Campbell, J.C., Neudeck, G.W. and Denton, J. 1999. Resonant-cavity-enhanced high-speed Si photodiode grown by epitaxial lateral overgrowth. *IEEE Photonics Technology Letters* 11: 1647–9.

Schmidt, D.C., Svensson, B.G., Seibt, M., Jagadish, C. and Davies, G. 2000. Photoluminescence, deep level transient spectroscopy and transmission electron microscopy measurements on MeV self-ion implanted and annealed *n*-type silicon. *Journal of Applied Physics* 88: 2309–17.

Schow, C.L., Schares, L., Koester, S.J., Dehlinger, G., John, R. and Doany, F.E. 2006. A 15-Gb/s 2.4-V optical receiver using a Ge-on-SOI photodiode and a CMOS IC. *IEEE Photonics Technology Letters* 18: 1981–3.

Sfar Zaoui, W., Chen, H., Bowers, J.E. et al. 2009. Origin of the gain-bandwidth-product enhancement in separate-absorption-charge-multiplication Ge/Si avalanche photodiodes. *Optical Fiber Communication Conference*, OSA Technical Digest (CD) (Optical Society of America, 2009), paper OMR6.

Shafiiha, R., Zheng, D., Liao, S. et al. 2010. Silicon waveguide coupled resonator infrared detector. Optical Society/Optical Fiber Communication Conference (OSA/OFC/NFOEC) San Diego, California, March 21, 2010, in Silicon Photonic Devices (OMI8.pdf).

Spector, S.J., Geis, M.W., Zhou, G.-R. et al. 2008. CMOS-compatible dual-output silicon modulator for analog signal processing. *Optics Express* 16: 11027–31.

Stolk, P.A., Gossmann, H.-J., Eaglesham, D.J. et al. 1997. Physical mechanisms of transient enhanced dopant diffusion in ion-implanted silicon. *Journal of Applied Physics* 81: 6031–50.

Su, S.J., Cheng, B.W., Xue, C.L. et al. 2011. GeSn p-i-n photodetector for all telecommunication bands detection. *Optics Express* 19: 6408–13.

Suh, D., Kim, S., Joo, J. and Kim, G. 2009. 36-GHz high-responsivity Ge photodetectors grown by RPCVD. *IEEE Photonics Technology Letters* 21: 672–4.

Sze, S.M. *Physics of Semiconductor Devices*, 2nd ed. (Wiley, New York, 1981).

Takada, Y. et al. 2010. As-grown Ge pin photodiodes on Si with low dark current achieved by hydrogen desorption technique. *7th International Conference on Group IV Photonics*, Beijing, China: 266–8.

Tanabe, T., Sumikura, H., Taniyama, H., Shinya, A. and Notomi, M. 2010. All-silicon sub-Gb/s telecom detector with low dark current and high quantum efficiency on chip. *Applied Physics Letters* 96: 101103.

Tsang, H.K., Wong, C.S. and Liang, T.K. 2002. Optical dispersion, two-photon absorption and self-phase modulation in silicon waveguides at 1.5 μm wavelength. *Applied Physics Letters* 80: 416 –8.

Tsuchizawa, T., Yamada, K., Watanabe, T. et al. 2011. Monolithic integration of silicon-, germanium-, and silica-based optical devices for telecommunications applications. *IEEE Journal of Selected Topics in Quantum Electronics* 17: 516–25.

Unlu, M.S., Kishino, K., Liaw, H.J. and Morkoc, H. 1992. A theoretical-study of resonant cavity-enhanced photodetectors with Ge and Si active regions. *Journal of Applied Physics* 71: 4049–58.

Vivien, L., Osmond, J., Fédéli, J.M. et al. 2009. 42 GHz p.i.n germanium photodetector integrated in a silicon-on-insulator waveguide. *Optics Express* 17: 6252–7.

Vivien, L., Rouvicre, Fédéli, J.M. et al. 2007. High speed and high responsivity germanium photodetector integrated in a silicon-on-insulator microwaveguide. *Optics Express* 15: 9843–8.

Wang, J., Loh, W.Y., Chua, K.T. et al. 2008a. Evanescent- coupled Ge p-i-n photodetectors on Si-waveguide with SEG-Ge and comparative study of lateral and vertical p-i-n configurations. *IEEE Electron Device Lettets* 29: 445–8.

Wang, J., Loh, W.Y., Chua, K.T. et al. 2008b. Low-voltage high-speed (18 GHz/1 V) evanescent-coupled thin-film-Ge lateral PIN photodetectors integrated on Si waveguide. *IEEE Photonics Technology Letters* 20: 1485–7.

Wang, J., Yu, M., Lo, G., Kwong, D.-L. and Lee, S. 2011. Waveguide integrated germanium JFET photodetector with improved speed performance. *IEEE Photonics Technology Letters* 23: 765.

Wang, X.X., Chen, L., Chen, W. et al. 2009. 80 GHz bandwidth-gain-product Ge/Si avalanche photodetector by selective Ge growth. *Optical Fiber Communication Conference*, OSA Technical (CD), OMR3.

Wang, X.X. and Liu, J.F. 2011. Step-coupler for efficient waveguide coupling to Ge/Si avalanche photodetectors. *IEEE Photonics Technology Letters* 23: 146–8.

Xue, H.Y., Xue, C.L., Cheng, B.W., Yu, Y.D. and Wang, Q.M. 2010. High-saturation-power and high-speed Ge-on-SOI p-i-n photodetectors. *IEEE Electron Device Letters* 31: 701–3.

Yin, T., Cohen, R., Morse, M.M., Sarid, G., Chetrit, Y., Rubin, D., and Paniccia, M.J. 2007. 31 GHz Ge n-i-p waveguide photodetectors on silicon-on-insulator substrate. *Optics Express* 15: 13965–71.

Yong, Z., Chao, X., Wan-Jun, W. et al. 2011. Photocurrent effect in reverse-biased p-n silicon waveguides in communication bands. *Chinese Physics Letters* 28: 074216–3.

Yu, H.Y., Ren, S., Jung, W.S., Okyay, A.K., Miller, D.A.B. and Saraswat, K.C. 2009. High-efficiency p-i-n photodetectors on selective-area-grown Ge for monolithic integration. *IEEE Electron Device Letters* 30: 1161–3.

Yu, J., Kasper, E. and Oehme, M. 2006. 1.55-mm resonant cavity enhanced photodiode based on MBE grown Ge quantum dots. *Thin Solid Films* 508: 396–8.

Yuan, H.C., Shin, J.H., Qin, G.X. et al. 2009. Flexible photodetectors on plastic substrates by use of printing transferred single-crystal germanium membranes. *Applied Physics Letters* 94: 013102–3.

Zang, H., Lee, S.J., Loh, W.Y. et al. 2008. Dark-current suppression in metal-germanium-metal photodetectors through dopant-segregation in NiGe–Schottky barrier. *IEEE Electron Device Letters* 29: 161–4.

Zhu, S., Lo, G.Q., Yu, M.B. and Kwong, D.L. 2009. Silicide Schottky-barrier phototransistor integrated in silicon channel waveguide for in-line power monitoring. *IEEE Photonics Technology Letters* 21: 185–7.

11

Hybrid and Heterogeneous Photonic Integration

Martijn J. R. Heck and John E. Bowers

CONTENTS

11.1 Introduction

The research on silicon photonics in the silicon-on-insulator (SOI) platform is driven by the motivation to realize large-scale and low-cost photonic integrated circuits (PICs), owing to a mature and highly accurate fabrication infrastructure compatible with complementary metal oxide semiconductor (CMOS) technology [1]. This technology allows for compact and highly integrated PICs operating in the telecommunication windows at wavelengths of approximately 1.3 and 1.55 μm. However, SOI-based technology by itself is rather limited in providing high-quality optical sources, detectors, and modulators.

Silicon modulators have been reported using free carrier plasma dispersion in Mach–Zehnder interferometric [2,3] and ring configurations [4]. Under forward bias, speeds of 10 Gb/s using pre-emphasis have been obtained [5]. Under reverse bias, the bandwidth goes up to 30 GHz [6,7]. The integration of strained germanium or silicon germanium, compatible with CMOS fabrication technology, greatly increases the possibilities because the bandgap can be pushed into the 1.55-μm regime. Electroabsorption modulators (EAMs) based on the quantum confined Stark effect in strained silicon germanium have been shown [8]. However, especially in the field of optical detection, the use of germanium has proven very useful in making CMOS-compatible photodetectors (PDs) at wavelengths of 1.55 μm [9].

The fabrication of efficient optical sources on silicon remains far more of a challenge, however. By using band engineering, some promising first results have been obtained with germanium-on-silicon lasers [10]. This approach overcomes the bottleneck that silicon has using only an indirect bandgap, preventing efficient gain and light generation. An overview of the numerous efforts is presented in Liang and Bowers [11]. Electrically pumped efficient sources on silicon remain an unsolved challenge, however.

Using III–V-based integration technologies, sources, detectors, and modulators operating efficiently and at high speeds are common. InP-based electrorefractive modulators can operate in the 40- to 80-Gb/s range at efficiencies of approximately 1 V·mm [12–14]. InP-based EAMs are able to achieve even higher speeds well above 100 Gb/s and up to 500 Gb/s [15,16]. PDs are commercially available at up to 100 GHz bandwidths [17], and published work has shown operational bandwidths into the hundreds of gigahertz regime [18]. Laser diodes and semiconductor optical amplifiers (SOAs) are finally widely available in III–V materials and have been the first choice for efficient and low-cost systems in which single-mode or tunable sources are required.

It seems obvious that for future applications, like photonic interconnects, silicon PICs are very promising, but they will require low power, high-speed performance of sources, and modulators similar to what has been achieved with III–V devices [19]. There are several ways in which III–V-based sources can be combined with silicon photonics. Prefabricated lasers can be coupled to silicon PICs, for example, by using grating couplers [20]. However, because each laser needs to be aligned individually, this approach does not scale well beyond attaching a few laser diodes only. More sophisticated hybrid integration schemes are also possible [21], but again these do not scale well for large-scale integration.

The most promising approach seems to be the recently demonstrated approaches in which III–V wafers are bonded to a silicon wafer either by molecular bonding, oxide bonding, or benzocylobutene (BCB) bonding [22]. With this approach, active III–V sections can be fabricated with lithographic precision and alignment accuracy, thereby avoiding pick-and-place active device bonding issues and enabling large-scale integration. Moreover, this approach allows for a back-end III–V fabrication process at limited temperatures of 350°C maximum. This means that contamination of a CMOS foundry is avoided and the III–V fabrication process is, in principle, fully compatible with the CMOS fabrication process.

In this chapter, we will review the status of hybrid (also called heterogeneous) integration of III–V devices on an SOI-based PIC. In Section 11.2, we will review the concepts in these integration schemes and their fabrication technologies. In Section 11.3, we will present an overview of the spectrum of devices that have been demonstrated, emphasizing the complete photonic toolbox that this technology offers. In Section 11.4, the specific integration issues will be addressed. The trade-offs for large-scale integration and complex PIC fabrication will be discussed. In Section 11.5, we will conclude and show an outlook for the future of this technology as a mature platform for highly complex and integrated PICs.

11.2 Technology

Integration technologies of GaAs and InP, the materials-of-choice for photonic applications, and CMOS-compatible SOI-based photonics have been well studied. However, they have only rarely been successfully integrated. The large mismatch in lattice constant and thermal expansion coefficient make monolithic integration very difficult. Wafer bonding–based hybrid integration is not limited by lattice mismatch [23], but still needs to tackle the issue of thermal expansion coefficient mismatch. In the following sections, we first review three low-temperature wafer bonding techniques. Thereafter, we discuss how these techniques can be used to integrate III–V-based photonic components on an SOI-based PIC, effectively creating hybrid integration platforms. This approach has been studied most intensely with InP/Si integration, but applies as well to GaAs/Si, GaN/Si, and InAs/Si integration for a variety of other applications.

11.2.1 Bonding Technology

In this section, we give an overview of three low-temperature wafer bonding techniques [22]. The first two bonding methods are O_2 plasma–assisted and SiO_2 covalent direct bonding, both sharing similar bonding mechanisms and falling into the molecular (or hydrophilic) bonding category. Because only inorganic materials are involved in the integration process, we discuss them in the inorganic-to-inorganic bonding section. The third method uses polymers as an adhesive to "glue" silicon and III–V wafers together, and will be discussed in the organic-to-inorganic bonding section.

11.2.1.1 O_2 Plasma–Assisted/SiO_2 Covalent Direct Bonding

For conventional direct bonding, high temperature is typically required to strengthen the bonding. It is therefore often referred to as "fusion bonding." In other bonding applications, this has proven highly effective. However, a special development process is required when a high-temperature anneal is strictly prohibited in III–V-to-silicon bonding. O_2 plasma surface treatment emerged as an attractive approach to obtain high bonding strength under low-temperature (<400°C) annealing [24,25]. This bonding mechanism is discussed in a later section. SiO_2 covalent bonding, the dominant process in fabricating microelectronics-grade SOI wafers of up to 300 mm [26], is a relatively old approach, but with careful surface treatment [27–30], can also be modified to meet the same low-temperature, high-strength criteria.

Figure 11.1 shows the schematic process flow of the O_2 plasma–assisted and SiO_2 covalent wafer bonding. After sample cleaning, the native oxide on SOI and InP are removed in buffered hydrogen fluoride (HF) solution and NH_4OH, respectively, resulting in clean and hydrophobic surfaces. In an O_2 plasma–assisted process, the samples then undergo an O_2 plasma surface treatment to grow a thin layer of plasma oxide (~15 nm) [31], which leads to very smooth hydrophilic surfaces (root mean square roughness <0.5 nm) [32]. For SiO_2 covalent direct bonding, a clean hydrophilic surface comes by deposited or thermally grown SiO_2 on both surfaces. Both bonding methods require a final activation step to passivate the two surfaces with a

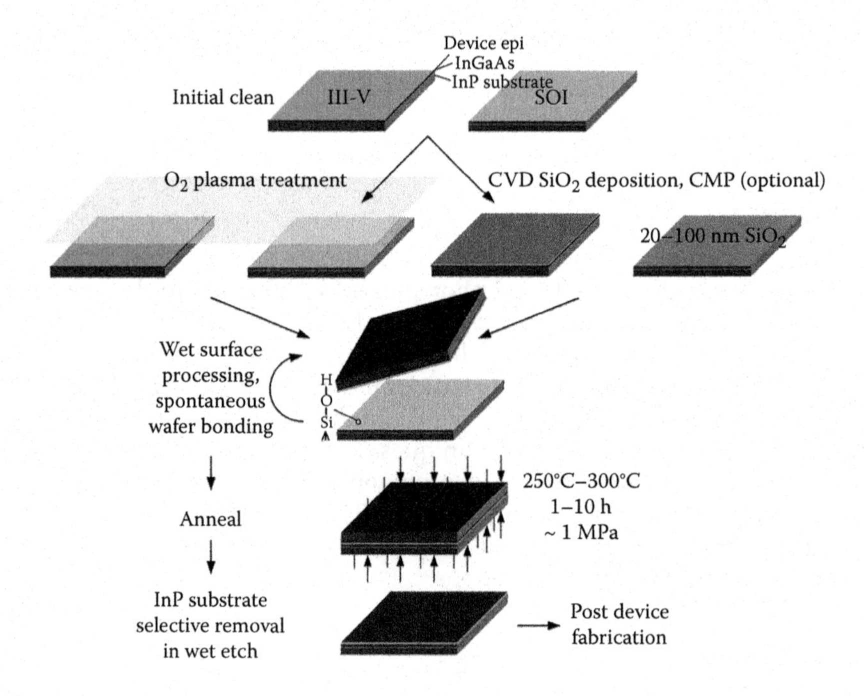

FIGURE 11.1
Schematic process flow for O_2 plasma–assisted and SiO_2 covalent wafer bonding. (Picture taken from Liang, D. et al., *Materials* 3:1782–1802, 2010. doi:10.3390/ma3031782.)

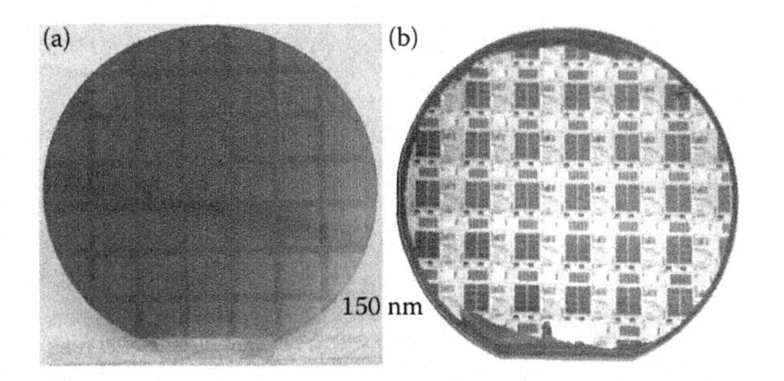

FIGURE 11.2
Photograph of processed 150 mm diameter O_2 plasma–assisted bonded wafer showing individual device dies (a). Infrared image of 150 mm SiO_2 covalent bonded wafer with CMOS devices successfully transferred (b). (Picture taken from Liang, D. et al., *Materials* 3:1782–1802, 2010. doi:10.3390/ma3031782.)

high density of polar hydroxyl groups (–OH), bridging bonds between the mating surfaces, enabling spontaneous bonding at room temperature. After immediate physical mating, typically in air at room temperature, the bonded sample is annealed at 300°C with external coaxial pressure (1–2 MPa) for an hour or more to form strong covalent bonds through polymerization reactions. After annealing and cooling, the InP substrate is selectively removed to leave thin (<2 μm) InP-based epitaxial layers on Si that are ready for further processing.

Prohibition of elevated temperature anneals results in H_2O and H_2 outgassing as a major issue in all low-temperature hydrophilic bonding [33]. Embedding a thick layer of porous material such as thermal SiO_2 or plasma-enhanced chemical vapor deposition (PECVD) dielectrics is an efficient outgassing medium for diffusion and absorption [28,33], which is the motivation for using SiO_2 covalent bonding here. However, it is not applicable for situations in which integration with a high proximity of two mating materials is needed, or in which optical, electrical, or thermal interactions between mating materials are desired and, therefore, a vertical outgassing channel design was developed to tackle this outgassing issue [22].

Up to 150 mm in diameter III–V-to-Si bonding, presently the largest available III–V epitaxial wafer, was demonstrated using O_2 plasma–assisted and SiO_2 covalent bonding methods. Figure 11.2a shows 150 mm diameter thin III–V epilayers transferred onto the SOI by an O_2 plasma–assisted bonding process. More than 98% area transfer and mirror-like III–V surface with a typical root mean square roughness of 0.6 to 0.7 nm was demonstrated [32,34]. A 150-mm diameter SiO_2 covalent bonded wafer is also exhibited in Figure 11.2b, in which thin Si devices were transferred onto InP substrates for CMOS-on-III–V mixed material integration [35]. The infrared image shows void-free interface and transferred SOI CMOS circuits.

11.2.1.2 Adhesive Wafer Bonding Technology

Adhesive bonding is an alternative that can be used to transfer III–V epitaxial layers onto a SOI waveguide circuit. There are various types of thermosetting adhesives, but the most promising results have been obtained using divinylsiloxane-disbenzocyclobutene (DVS-BCB) as an adhesive bonding agent [36–41]. DVS-BCB is superior in terms of high bonding strength and bonding quality (because no by-products are created during curing), its high degree of planarization, and its resistance to all sorts of chemicals used in III–V processing. In the subsequent section, the DVS-BCB adhesive wafer bonding process is reviewed for III–V epitaxial layer transfer to a SOI waveguide circuit.

The DVS-BCB adhesive die-to-wafer bonding process is schematically outlined in Figure 11.3. The III–V epitaxial layer structures are cleaved into individual dies (typical dimensions are in the range of 25–100 mm^2) and are temporarily mounted on a glass carrier using either a thermoplastic adhesive or thermal release tape. The mounting on a glass carrier serves two purposes: first of all, it allows easy handling of the III–V dies during cleaning of the die surface. Second, the glass carrier allows mounting of multiple III–V dies at the same time, thereby allowing multiple die-to-wafer bonding. The most important part of the die-to-wafer bonding procedure consists of the cleaning of the SOI waveguide wafer and III–V dies. Because ultrathin bonding layers (typically <100 nm) are required for a good optical and thermal coupling between the III–V and SOI in the hybrid device platforms, commercially provided DVS-BCB oligomer solutions (Dow Chemicals, Midland, MI) need to be diluted using mesitylene to achieve the required bonding layer thickness. Although the topography of the SOI waveguide wafer (typically 220 nm for nanophotonic silicon rib waveguides) is larger than the required spacing between the top of the SOI waveguide and the III–V epitaxial layer structure, good planarization can still be achieved.

After spin-coating, the SOI is heated to 150°C for 1 min to evaporate the remaining mesitylene solvent in the DVS-BCB film. This is required to avoid the generation of voids at the bonding interface. After the evaporation of the mesitylene, the III–V die is attached, epitaxial layers facing down, to the SOI waveguide circuit. This can be done either in clean room air (manually) or in the vacuum chamber of a commercial automatic wafer bonding tool. Hereafter, the stack is cured at 250°C and the result is a void-free bond. After bonding, the original InP growth substrate is removed, leaving the epitaxially grown III–V layer stack attached to the SOI waveguide circuit.

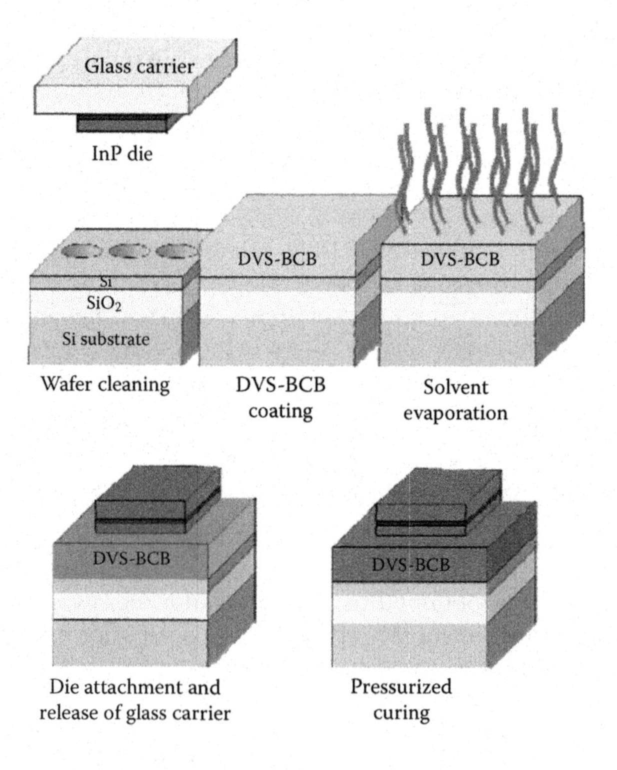

FIGURE 11.3
Overview of the DVS-BCB die-to-wafer bonding process. (Picture taken from Liang, D. et al., *Materials* 3:1782–1802, 2010. doi:10.3390/ma3031782.)

11.2.2 Integration Platforms

11.2.2.1 Hybrid Silicon Platform

The hybrid silicon platform, developed at the University of California Santa Barbara, uses covalent wafer bonding. The cross-section of the hybrid silicon waveguide device is shown in Figure 11.4a [42,43]. It consists of a III–V multiple quantum well (QW) epitaxial layer structure bonded to an SOI rib waveguide. The device fabrication process can be divided into three major parts. First, the Si waveguides and any other desired (passive) Si devices, such as arrayed waveguide gratings (AWGs), couplers, and splitters, are fabricated in a CMOS fabrication process. Next, the III–V epitaxial layer structure is transferred to the Si waveguides through the O_2 plasma–assisted, low-temperature bonding process described previously. Finally, postprocessing of the III–V layers is done after bonding to control the flow of current through the structure and confine the optical mode to ensure efficient optical gain to the waveguide mode.

By tuning the bandgap of the III–V QW lasers, EAMs and phase modulators can be made. AlGaInAs-based QWs are advantageous for uncooled laser operation at elevated temperatures [44]. An alternative approach is to use InGaAsP-based QWs. Sun et al. [45] have reported good performance of hybrid silicon lasers using this approach. Threshold current density and threshold voltage in this work are 30% to 40% lower than those reported by Fang et al. [42]. More details of the transferred epitaxial structure and cross-section design are mentioned in later sections.

As stated previously, the optical mode characteristics are determined by the Si rib waveguide dimensions. Figure 11.5 shows the calculated optical mode with a fixed waveguide height for various waveguide widths. It can be seen that as the Si waveguide becomes wider, the mode is pulled more into the Si region, with the same trend being seen for variation of

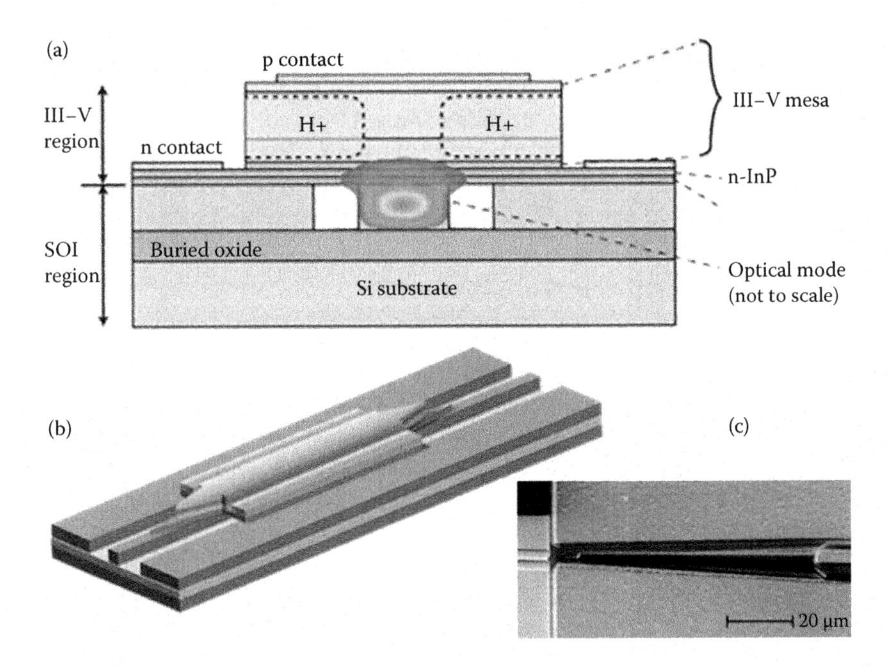

FIGURE 11.4
Cross-section of the hybrid Si device (a). Schematic of the transition taper of a passive silicon waveguide to an active hybrid section and vice versa (b). SEM picture of a taper (c) [94]. (Picture taken from Heck, M.J.R. et al., *IEEE J. Sel. Top. Quant. Electron.* 17(2):333, 2011.)

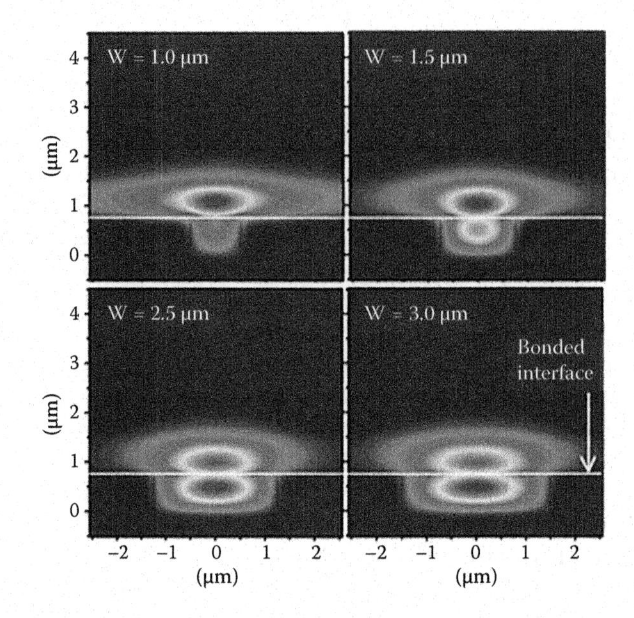

FIGURE 11.5
Calculated optical mode for waveguide widths of 1.0 μm up to 3.0 μm using the RSoft BeamPROP simulation software. (Picture taken from Heck, M.J.R. et al., *IEEE J. Sel. Top. Quant. Electron.* 17(2):333, 2011.)

the waveguide height. This feature can be used to tailor each device's QW confinement and hence, its optical gain characteristics. For example, lasers and modulators could be designed with narrower waveguides, which increase modal gains to achieve lower thresholds, whereas amplifiers in an adjacent section of the wafer can be designed to have wide waveguide widths to increase the saturation power of the amplifier. Tapers can be used to move the optical mode from the silicon waveguide to the hybrid waveguide and vice versa, as shown in Figure 11.4b and c, and as discussed in detail in the next section. Alternatively, such hybrid silicon devices can, in principle, be fabricated using DVS-BCB bonding, as presented in Stankovic et al. [46,47].

11.2.2.2 Heterogeneous III–V/SOI Platform

Another hybrid III–V-on-Si platform developed by Ghent University possesses similar device structure, although the III–V material and the Si waveguide perform relatively independent functions [48]. A three-dimensional schematic is shown in Figure 11.6 to depict the photon generation, optical feedback, and coupling to the Si waveguide. The III–V epitaxial layer transfer is achieved by DVS-BCB adhesive bonding, as discussed previously. A typical DVS-BCB layer is in the order of several hundreds of nanometers with a refractive index of approximately 1.5 at λ = 1.55 μm. A waveguide in the InP layers and a relatively thick low-index medium between III–V and Si prevents photons generated in the III–V active region from coupling into the Si waveguide instantly. Lasing is achieved through the gain provided by a III–V active region and reflection at the etched laser facets. As the stimulated emission leaves the edge of the laser diode, an additional coupling structure is required for efficient coupling to the SOI waveguide. An optimal adiabatic inverted taper structure is employed to achieve good coupling efficiency and fabrication tolerance. The concept of the inverted adiabatic structure is to butt-couple the bonded laser diode to a

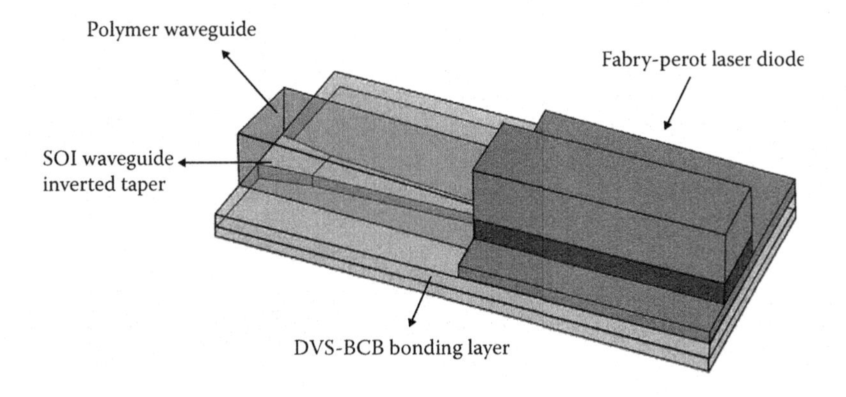

FIGURE 11.6
Schematic of the layout of the optical coupling scheme for efficient and fabrication-tolerant coupling between a bonded Fabry–Perot laser diode and an underlying SOI waveguide circuit using an inverted adiabatic taper approach [49]. (Picture taken from Liang, D. et al., *Materials* 3:1782–1802, 2010. doi:10.3390/ma3031782.)

polymer waveguide, after which the optical mode is gradually transformed into that of the SOI waveguide by increasing the cross-sectional area of the Si waveguide. The polymer waveguide is self-aligned to the laser ridge, eliminating any possible source of coupling efficiency reduction arising from the misalignment between the waveguides. The Si inverted taper structure is buried underneath the polymer waveguide. The inverted taper tip width has to be sufficiently small for the fundamental optical waveguide mode at the tip to resemble the waveguide mode of the polymer waveguide closely [49]. The formation of III–V mesa and electrodes are similar to the platform.

Another design in which the optical confinement is primarily in the III–V layer is the microdisk laser (MDL) [50]. In this case, the disk couples evanescently to the SOI waveguide, as shown in Figure 11.7. In principle, this approach allows for micron-sized cavities and enables the large-scale integration of low-power sources on an SOI chip.

The microdisk is etched in a thin InP-based layer bonded on top of a SOI waveguide wafer. The fundamental optical resonances in such a structure are whispering-gallery modes, which are confined to the edges of the microdisk. Therefore, a top metal contact can be placed in the center of the microdisk, without adding extra optical losses. The bottom contact is positioned on a thin lateral contact layer. A tunnel junction with low optical

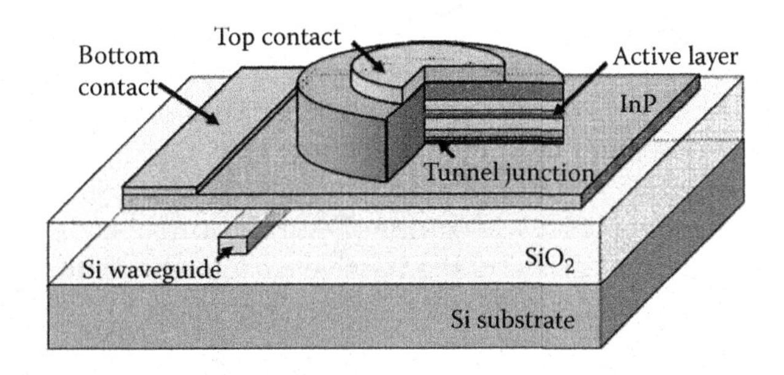

FIGURE 11.7
Schematic drawing of the heterogeneous MDL structure, showing the disk cavity, SOI wire waveguide, bottom contact layer, tunnel junction, and metal contacts. (Picture taken from Liang, D. et al., *Materials* 3:1782–1802, 2010. doi:10.3390/ma3031782.)

loss, in combination with another n-type contact instead of low-bandgap p-type contact layer, was implemented [51]. The laser resonance is evanescently coupled to the underlying SOI waveguide, which is vertically aligned with the edge of the microdisk. Further fabrication details can be found in Van Campenhout et al. [50].

Although the work presented here reflects the main body of work on hybrid and heterogeneous integration techniques and approaches, other important works are presented in the references [52–55].

11.3 Devices

In this section, we will review work on the basic active building blocks for a hybrid photonic platform, that is, optical sources, modulators, and PDs. We will review recently published work on the various integration platforms as presented in the previous section. This overview shows that with a hybrid integration approach, components that can compete with purely III–V-based devices in terms of modulation speed, detection bandwidth, and energy efficiency of sources can be realized. Hence, we show that full III–V functionality can be integrated into the SOI platform.

Due to the abundance and importance of the work on optical sources, we have split the review into two sections, that is, the more conventional distributed feedback (DFB), distributed Bragg reflector (DBR) and ring lasers on the one hand, and a class of lasers that is especially suitable for SOI-compatible and high-integration density type applications on the other hand: microring and MDLs.

11.3.1 Optical Sources

Although Fabry–Pérot type lasers having cleaved or polished facets are the most commonly used laser diodes, they cannot be integrated into a chip with other components. As a result, these lasers are not useful for further integration, for example, on-chip interconnects. In the following section, we review the work done on sources that can be connected to and integrated with other components on the same chip.

11.3.1.1 Hybrid Silicon DFB Lasers

The DFB laser presented in Fang et al. [56] consists of a 360-μm-long quarter wavelength shifted hybrid silicon grating with a grating κ of approximately 247 cm^{-1}, and reflectivity peak at approximately 1600 nm. Figure 11.8a shows the device layout. The laser has a 200-μm-long gain region, with a cross-section as shown in Figure 11.4. Tapers from the gain region to the passive silicon waveguide are 80 μm long. They are formed by linearly narrowing the III–V mesa region above the silicon waveguide (Figure 11.4). This adiabatically transforms the mode from the hybrid waveguide to the passive silicon waveguide, allowing for losses on the order of 1.2 dB per taper and reflections on the order of 6×10^{-4}. Hybrid silicon PDs are placed on both sides of the laser to enable on-chip testing of the DFB laser performance. The PDs are 240 μm long, including the two 80-μm-long tapers. The detector to the right is placed 400 μm away to allow room for dicing and polishing for off-chip spectral tests [56].

The light-current (LI) characteristics of the DFB laser are measured on-chip by collecting light out of both sides of the laser with the integrated PDs. To determine the laser power

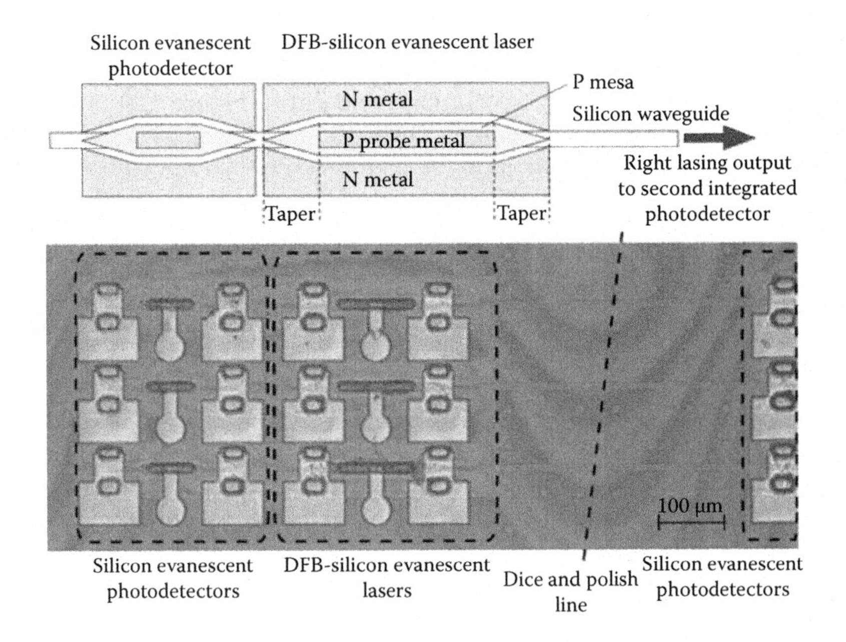

FIGURE 11.8
Hybrid silicon DFB device layout (top). Microscope image of the DFB laser with integrated PDs (bottom). (Picture from Fang, A.W. et al., *IEEE J. Sel. Top. Quant. Electron.* 15(3):535–544, 2009.)

output, 100% internal quantum efficiency of the PDs is assumed to conservatively assess the laser performance. It can be seen from Figure 11.9 that at 10°C, the lasing threshold is 25 mA, with a maximum output power of 4.3 mW. This corresponds to a threshold current density of 1.4 kA/cm^2. The maximum lasing temperature is 50°C. The laser has a 13-Ω device series resistance. This value scales appropriately with the 4.5-Ω resistance measured on 800-μm-long Fabry–Pérot lasers with similar III–V mesa dimensions [42].

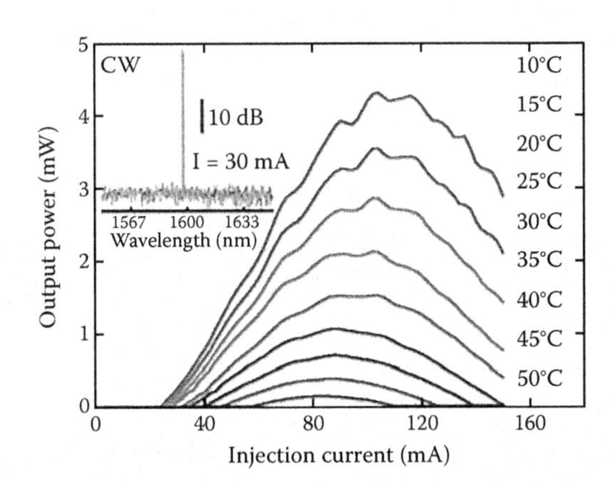

FIGURE 11.9
LI curve for stage temperatures of 10°C to 50°C. Inset: the lasing spectrum at 30 mA injection current, showing a single-mode operation span over a 100-nm wavelength range. (Picture from Fang, A.W. et al., *IEEE J. Sel. Top. Quant. Electron.* 15(3):535–544, 2009.)

The lasing spectrum is taken by dicing off the right PD, polishing, and antireflection coating the silicon waveguide output facet. Light is collected with a lensed fiber into a spectrum analyzer with a 0.08-nm resolution bandwidth. Figure 11.9 (inset) shows the optical spectrum at 30 mA injection current. The laser has a lasing peak of 1599.3 nm and a side mode suppression ratio of 50 dB. It can be seen that the laser operates at a single mode over a 100-nm span. The laser linewidth is measured by using the delayed self-heterodyne method [57]. A minimum linewidth is measured at a laser output power of 1.8 mW, with a convoluted Lorentzian linewidth of 7.2 MHz corresponding to a 3.6-MHz linewidth, a typical value for commercial DFB lasers.

11.3.1.2 DBR Lasers

The DBR laser presented in Fang et al. [56] consists of two passive Bragg reflector mirrors placed 600 μm apart to form an optical cavity, as shown schematically in Figure 11.10. The gratings have an etch depth and duty cycle of 25 nm and 75%, respectively, leading to a grating strength, κ of 80 cm^{-1}. The back and front mirror lengths are 300 and 100 μm, resulting in power reflectivity of 97% and 44%, respectively. A 440-μm-long silicon evanescent gain region and two 80-μm-long tapers are placed inside the cavity. The tapers are electrically driven in parallel with the gain region to minimize absorption. More details can be found in Fang et al. [56].

The continuous wave laser output power is measured with an integrating sphere at the front mirror of the laser. The front mirror output LI characteristic is shown in Figure 11.11. The device has a lasing threshold of 65 mA and a maximum front mirror output power of 11 mW, leading to a differential quantum efficiency of 15%. The taper transmission loss can have a significant effect on the threshold current and therefore it affects many important laser characteristics such as wall plug efficiency and resonance frequency. If we use our estimations of the material and laser properties, calculations show that the taper loss of 1.2 dB increases the threshold current by a factor of 2 due to the accumulated loss through

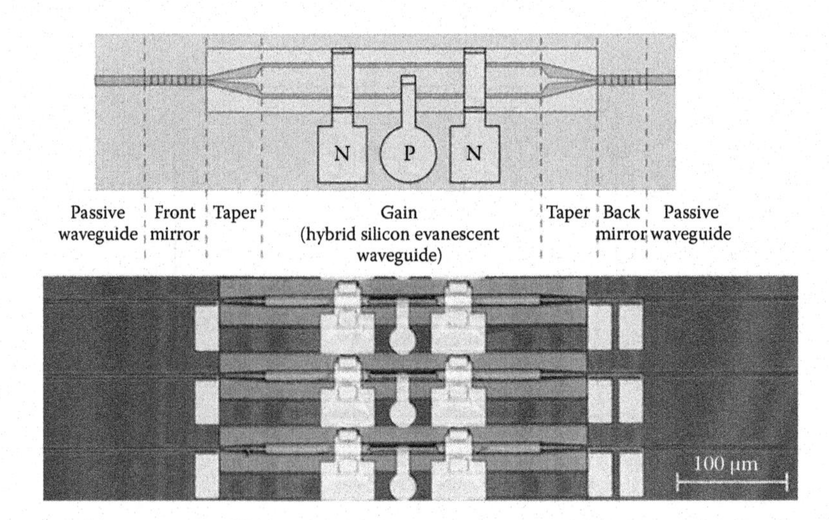

FIGURE 11.10
DBR laser schematic (top) and microscope image (bottom). (Picture from Fang, A.W. et al., *IEEE J. Sel. Top. Quant. Electron.* 15(3):535–544, 2009.)

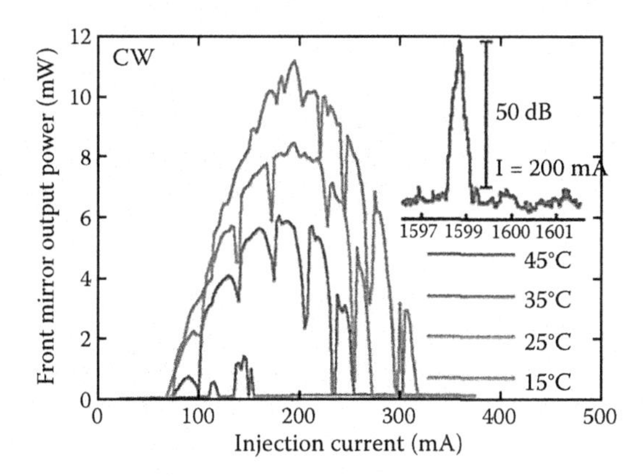

FIGURE 11.11
DBR laser LI curve for various temperatures measured at the front mirror. Inset: the lasing spectrum at 200 mA injection current, showing a single-mode operation with 50 dB side mode suppression ratio. (Picture from Fang, A.W. et al., *IEEE J. Sel. Top. Quant. Electron.* 15(3):535–544, 2009.)

four taper transitions in one round-trip through the cavity. We estimate that a reduction in the single-pass taper losses to 0.5 dB would reduce this factor to 1.2. The laser operates up to a stage temperature of 45°C. The kinks in the LI are from mode hopping [56]. The device has a lasing turn-on voltage of 2.6 V and a series resistance of 11.5 Ω. The lasing spectrum is shown in Figure 11.11, with a lasing peak at 1597.5 nm when driven at 200 mA.

Figure 11.12 shows the photodetected electro-optic response of the laser combined with all connected components under small signal modulation. A 2-pF device capacitance was extracted from the S11 measurement, resulting in an RC-limited bandwidth (RC is the product of the circuit resistance and circuit capacitance) of 7 GHz. Figure 11.12 (inset) shows the resonance frequency versus the square root of direct current (DC) drive current above threshold, which has a roughly linear dependence as expected. Under higher modulation powers, the resonance peak becomes significantly dampened. The 3-dB electrical

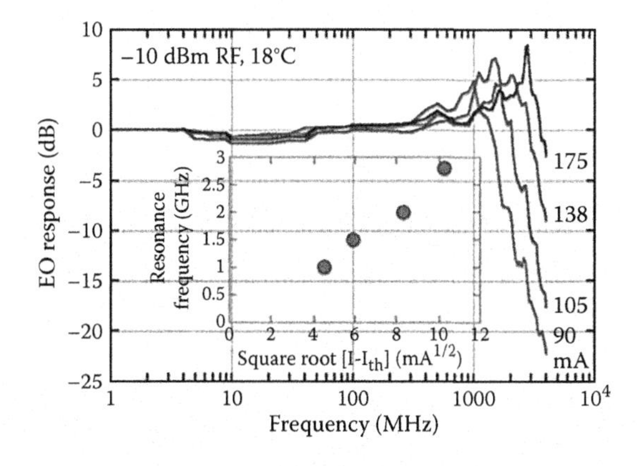

FIGURE 11.12
Frequency response of the DFB laser for three different bias currents with a stage temperature of 18°C and plot of resonance frequency versus the square root of current above threshold (inset). (Picture from Fang, A.W. et al., *IEEE J. Sel. Top. Quant. Electron.* 15(3):535–544, 2009.)

bandwidth at 105 mA is approximately 2.5 GHz. This laser has been successfully modulated at 2.5 and 4.0 Gb/s with extinction ratios of 8.7 and 5.5 dB, respectively [56]. Improving the laser design to decrease the threshold current, decrease the cavity length, and increase the differential gain will significantly improve the modulation bandwidth.

11.3.1.3 Mode-Locked Laser Sources

Mode-locked lasers are able to generate short optical pulses. The optical spectrum of these pulses can be broadband and the modes have a fixed spacing, defined by the cavity length. This makes these lasers ideal sources for a multiwavelength mode-comb that can be used for wavelength-division multiplexed (WDM) applications [58,59]. Such a frequency comb laser eliminates the need for multiple laser drivers and stabilizers and wavelength lockers, as are typically used nowadays in WDM systems in which multiple single-frequency lasers are employed, and hence is far more energy efficient.

Mode-locked lasers have been realized in the hybrid silicon platform both in FP-type configurations at 10 and 40 GHz [60], and in a racetrack configuration at 30 GHz [61]. They have been operated under both passive mode-locking and hybrid mode-locking, that is, applying an radio-frequency (RF)-signal to the saturable absorber. The racetrack mode-locked laser is shown in Figure 11.13. This laser shows a stable operation under passive mode-locking, with RF linewidths in the order of 2 to 3 MHz at −20 dB as shown in Figure 11.14a. Hybrid mode-locking narrows down the RF linewidth, indicating a more stable operation due to the synchronization with the electrical clock. The timing jitter (1 kHz–100 MHz integration range) is 0.36 ps at 14 dBm of RF power. Typical pulse durations for these lasers are in the order of 6 to 10 ps.

For application as a multiwavelength source for Dense WDM (DWDM) systems, it is important that the output of the mode-locked laser has a large optical bandwidth, that is, the width of the mode-comb. For the 30-GHz racetrack laser, this 3 dB bandwidth is however limited to less than 1 nm (Figure 11.14c). This is not sufficient because this means that only two to three modes can be used in the WDM system without having the need for excessive equalization. Better results were obtained with the longer cavity 10 GHz FP laser, which shows a 9-nm optical bandwidth at a relatively large injection current (Figure 11.14b). The combination of increased self-phase modulation in the longer gain cavity and the larger number of modes that can lase with the higher injection current are assumed to be responsible for the broad optical spectrum. This spectrum can support 100 modes spaced at 10 GHz for WDM applications. Such lasers can be operated at 10 mW output

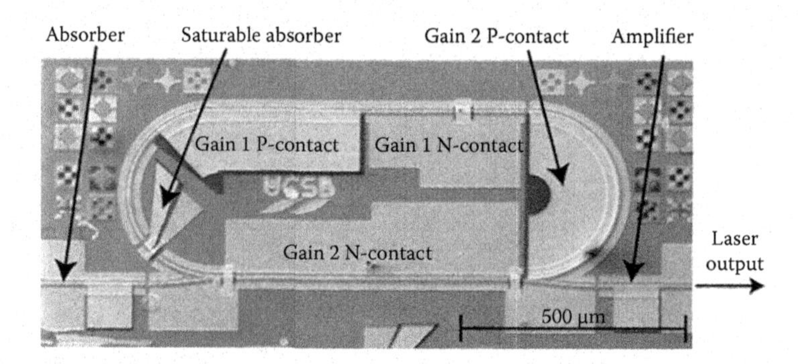

FIGURE 11.13
Scanning electron micrograph of a racetrack mode-locked silicon evanescent laser.

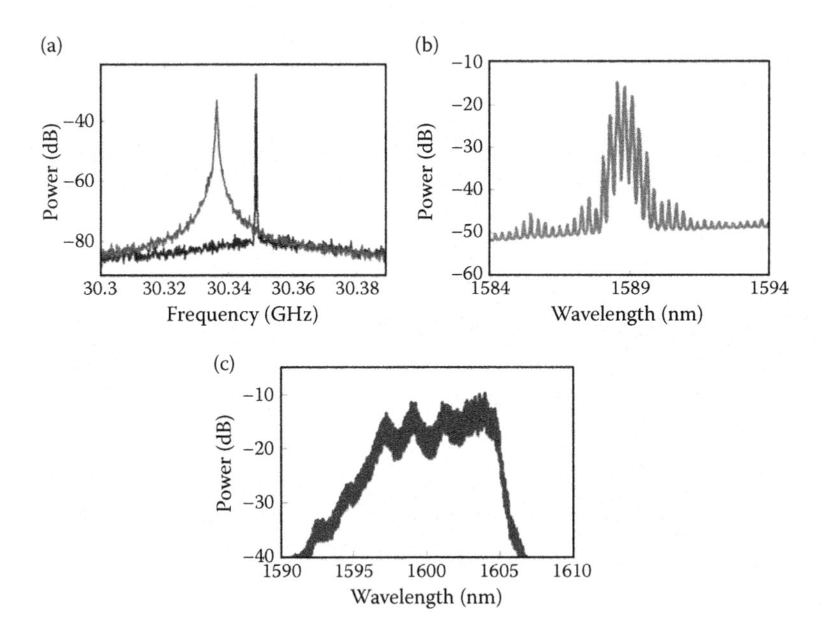

FIGURE 11.14
The 100-MHz RF frequency span around the mode-locking frequency for passive and hybrid mode locking of the 30-GHz racetrack laser (a). Logarithmic scale optical spectrum of the 30-GHz racetrack laser (b). Logarithmic scale optical spectrum of the 10-GHz FP laser (c).

power and approximately 250 mW electrical power consumption [42,56]. If more than 100 modes are available, which can be modulated at 10 Gb/s, for example, by an EAM, this means that the energy consumption per bit is in the order of 0.1 to 0.3 pJ/bit.

Measurements of the optical linewidths show that these are broad at approximately 225 MHz. However, injection with a narrow-linewidth continuous-wave (CW) laser can decrease the linewidth for all the modes. Preliminary experiments have shown that the optical linewidth could be decreased down to 100 kHz, that is, the linewidth of the injected CW mode. Therefore, it can be stated that the hybrid silicon mode-locked lasers have promising characteristics for use as a mode-comb source.

11.3.2 Micron-Sized Sources

Lasers with ring or disk resonator geometries having micron-sized diameters are attractive single-frequency on-chip light sources because they require no gratings or facets for optical feedback. Practical use of such a device in an interconnect requires power-efficient, CW operation, high-speed direct modulation, and operation at elevated temperatures. In the following section, we review the work on such micron-sized sources realized in different technology platforms.

11.3.2.1 Hybrid Silicon Microring Lasers

Figure 11.15 shows the schematic of the hybrid silicon microring laser [62]. The laser consists of a III–V ring resonator on top of a silicon disk with the same diameter. The fundamental whispering-gallery mode shifts toward the resonator edge as shown by a beam

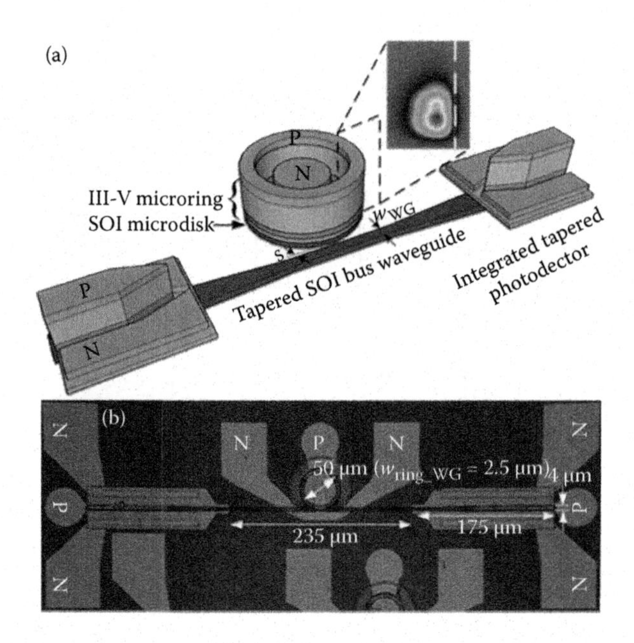

FIGURE 11.15
Schematic of compact hybrid silicon ring resonator laser with beam propagation method mode profile and integrated, tapered PDs (a). Variables of coupling gaps and bus waveguide width w_{WG} are labeled. The microscopic image of a finished device with critical dimension labeled (b). (Picture from Liang, D. et al., *Opt. Express* 17(22):20355–20364, 2009.)

propagation method simulated mode profile in the inset of Figure 11.15a. This mode has confinement factors of 15.2% and 51.7% in the active region and silicon, respectively. A silicon bus waveguide connects the microring laser with two integrated PDs with a length of 180 μm. More details about the device design and fabrication can be found by Liang et al. [62].

Figure 11.16 shows the LI characteristics of devices with (a) 150 nm and (b) 250 nm coupling gaps and a 0.6-μm bus waveguide width. For these coupling gaps, the minimum threshold currents are 8.37 and 5.97 mA at 10°C, respectively. These correspond to current densities of 2.02 and 1.43 kA/cm^2, assuming uniform carrier distribution in the active region. Devices with coupling gaps of 150 and 250 nm lase up to stage temperatures of 40°C and 65°C, respectively.

Figure 11.17 shows the calculated 3 dB bandwidth for injection current 10, 20, and 30 mA as a function of device diameter (cavity length), using parameters and assumptions as mentioned by Liang et al. [62]. These estimates result in approximately 5 mA threshold current for a 50-μm microring laser. No thermal effect is taken into account in this calculation. To achieve 3 dB bandwidth of 10 GHz, a 50-μm diameter device needs to be driven with a 20-mA bias current, that is, $4 \times I_{th}$. Devices with smaller dimension have smaller threshold and require less injection current to reach the 10-GHz, 3-dB bandwidth. Clearly, employing short cavity devices (e.g., microring lasers) is an efficient approach to further increase the direct modulation bandwidth without sacrificing low-power dissipation. Increased thermal impedance poses a major obstacle for all compact devices.

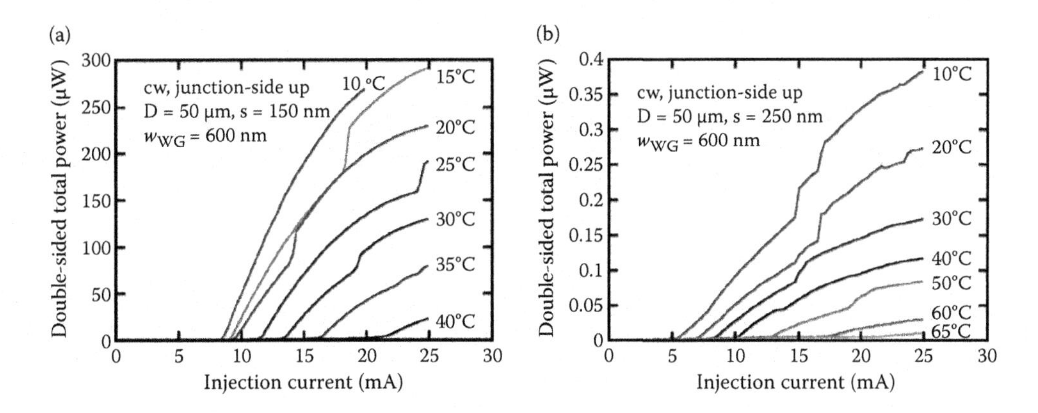

FIGURE 11.16
LI characteristic of microring lasers with coupling gap of (a) s = 150 nm and (b) s = 250 nm at various stage temperatures. Output power is the sum of the response of both integrated photodiodes assuming 1 A/W responsivity. (Picture from Liang, D. et al., *Opt. Express* 17(22):20355–20364, 2009.)

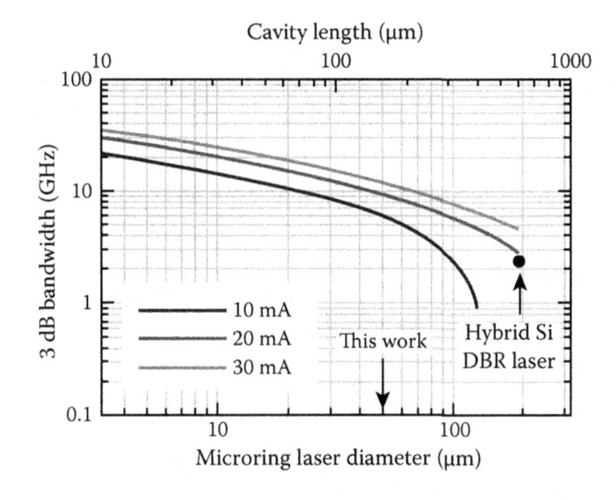

FIGURE 11.17
Calculated 3 dB bandwidth as a function of microring laser diameter (cavity length) for 10, 20, and 30 mA injection current. Black dot represents experimental 3 dB bandwidth of 2.5 GHz measured on a hybrid Si DBR laser without approximately 1.6 × I_{th} bias current [69]. (Picture from Liang, D. et al., *Opt. Express* 17(22):20355–20364, 2009.)

11.3.2.2 Heterogeneously Integrated MDLs

The DVS-BCB bonding approach has been shown to be very successful in the fabrication of MDLs [64]. Using a configuration as shown in Figure 11.7, CW room temperature operation was achieved for a disk laser with a 7.5-μm radius [50]. Threshold currents were 0.5 mA, as shown in Figure 11.18. Self-heating limited the output power to approximately 5 μW at a current of 1 mA. As can be seen in Figure 11.18b, the laser operates at single-mode with a large side-mode suppression due to its large free spectral range.

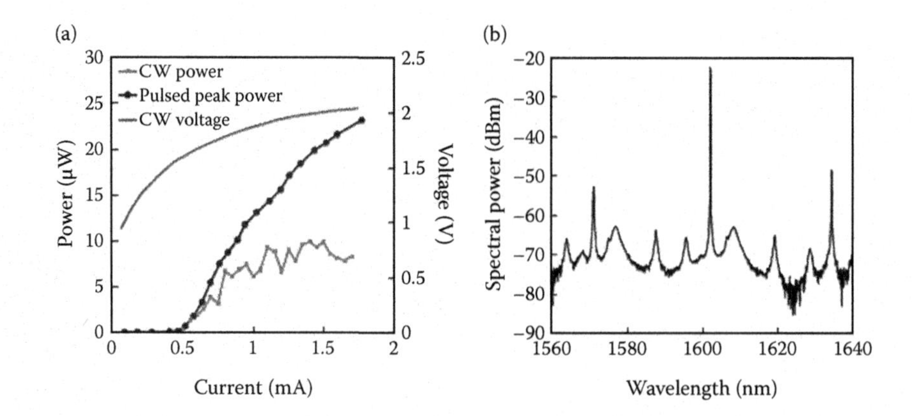

FIGURE 11.18
Lasing characteristics at 20°C for a 7.5-μm disk (a). CW lasing spectrum for 1.4 mA, normalized for the fiber-coupler efficiency and on-chip propagation loss (b). (Taken from Van Campenhout, J. et al., *Opt. Express* 15(11):6744, 2007.)

Improving the design by using the top contact as a heat sink, that is, by using a 600-nm-thick Ti/Pt/Au metal layer, the maximum output power increased to 120 μW and the threshold current went down to 350 μA [65]. Heating in these disk lasers using DVS-BCB bonding is mainly caused by the low thermal conductivity of the thick buried oxide layer and the BCB layer [66].

By cascading an array of MDLs and coupling them to a single SOI waveguide, a multi-wavelength source for WDM can be achieved, as shown in Figure 11.19a [70]. The spectra from the two output sides are shown in Figure 11.19b and c, for instances in which the four lasers are operated simultaneously at the same current. The designed diameter of adjacent disk lasers increases by 44 nm and hence their outputs are equally spaced at 8 nm. Coupling to other (downstream) lasers causes nonuniform output of the four channels. Thermal crosstalk was shown to be negligible, even though the distance between adjacent microdisks was only 33 μm.

Furthermore, these disk lasers have successfully been used as optical flip-flops [67] and for high-speed wavelength conversion [68].

For microdisk and microring lasers, there are several fundamental limits for decreasing the threshold. These include the volume of the active region, which is related to the device dimension, the internal cavity loss and modal gain, which are related to the confinement factor, and the injection efficiency and mirror loss, which are related to the power outcoupling of the rings. If these factors are kept constant, the active region volume will be the dominant factor that determines threshold.

Compared with InP disks, BCB bonded on SOI [70], the major difference is vertical confinement of the light. In BCB-bonded disks, most of the light is confined in the III–V disk, and hence, their modal gain is higher than the hybrid silicon microring. Limitations on BCB-bonded disks are the high thermal impedance and limited output power.

11.3.3 Modulators

In this section, we review two types of modulator, namely, an EAM and a Mach–Zehnder type modulator (MZM) based on phase modulators. Both approaches were demonstrated on the hybrid silicon platform.

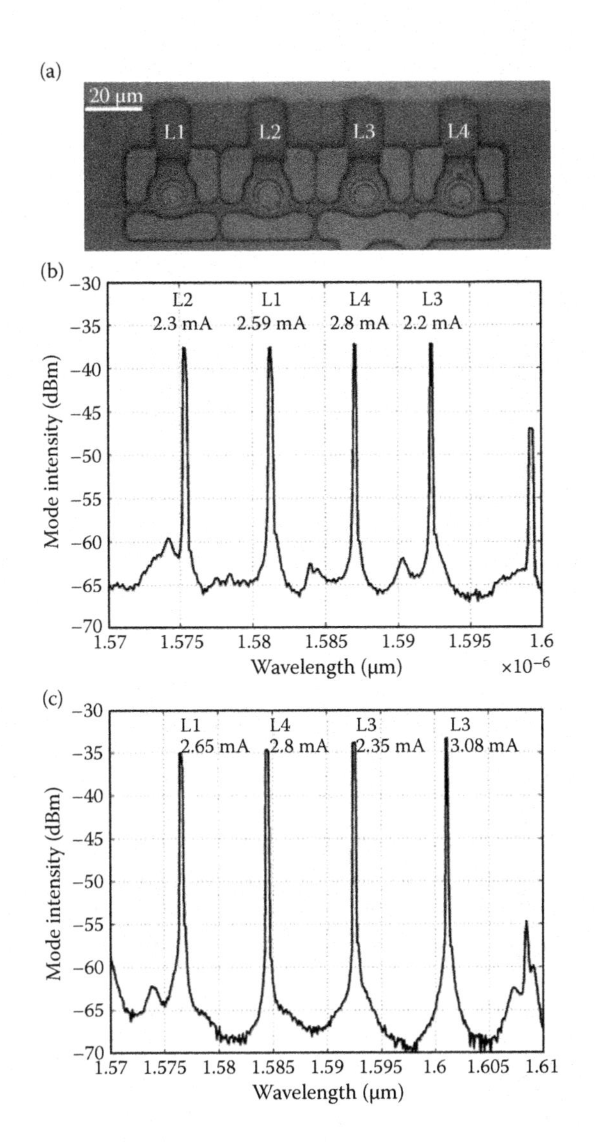

FIGURE 11.19
Fabricated MWL before metallization (a), composed of four MDLs on a single silicon bus waveguide. Spectra of MWLs with 6 and 8 nm channel spacing (b and c). The bias current of each MDL and the laser number is indicated on the corresponding lasing peak [69]. (Picture from Van Thourhout, D. et al., *IEEE J. Sel. Top. Quant. Electron.* 16(5):1363, 2010.)

11.3.3.1 Electroabsorption Modulator

A cross-section of a hybrid silicon EAM modulator is shown in Figure 11.20a. Two 60-μm hybrid tapers, laterally tapered in both silicon and III–V layers, are used to minimize reflection and mode mismatch loss through adiabatically transforming the optical mode. The *p*-InP cladding mesa is 4 μm wide, whereas the QW and separate confinement heterostructure (SCH) layers are undercut to reduce the total device capacitance [71]. In general, the EAM has a very small footprint of approximately 100 μm, such that the device can be easily operated at more than 10 GHz with careful design of the active region to control the

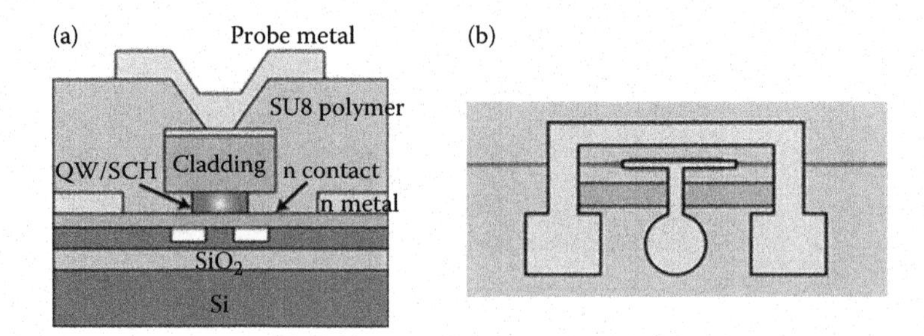

FIGURE 11.20
Cross-section of both EAM and MZM at the hybrid section (a). Schematic top view of an EAM (b). (Pictures from Van Thourhout, D. et al., *IEEE J. Sel. Top. Quant. Electron.* 16(5):1363, 2010.)

overall RC cutoff frequency. Figure 11.20b shows the top view of a single EAM with simple DC probe pads for both *p*- and *n*-contact. Further details can be found in Kuo et al. [72].

Figure 11.21 shows the relative extinction ratio at a wavelength of 1550 nm under various reverse biases. An extinction ratio of more than 10 dB can be achieved with less than 4 V bias for a 100-μm-long device. For a longer device with 250 μm absorber, it only takes 2.5 V to achieve a 10-dB extinction ratio. The insertion loss is approximately 3 dB mainly due to the excess loss from both tapers.

To investigate the high-speed performance of the EAM, the frequency response was measured. As shown in Figure 11.22, two 100-μm-long EAMs with different QW undercuts were measured. The device with a 3-μm-wide QW section has a series resistance of approximately 30 Ω and a capacitance of 0.2 pF at 2 V bias, which corresponds to a cutoff frequency of approximately 10 GHz. The modulators are also driven with a 2^{31}–1 pseudorandom bit sequence (PRBS) to explore the performance of large signal modulation. Peak-to-peak driving voltage of 0.82 V is used to produce the clear eye diagram with 5 dB extinction ratio. The speed can be further improved with a more aggressive QW undercut. The device with a 2-μm-wide QW section dropped the capacitance to approximately 0.1 pF and produces a 3-dB bandwidth over 16 GHz. The downside is the reduction of the QW volume, which leads to increased driving voltage due to a decrease of the overlap of the optical mode with the QWs.

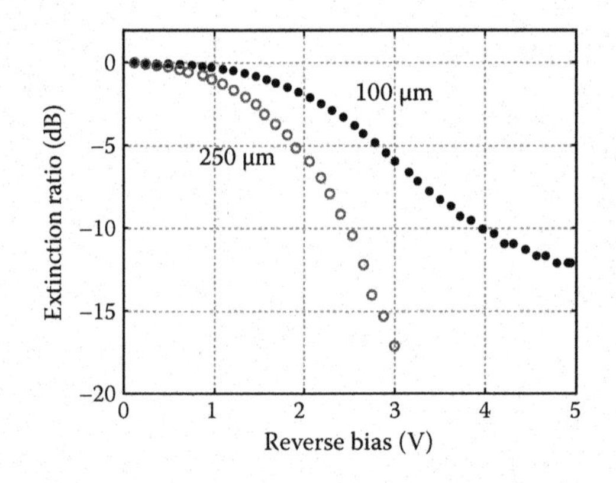

FIGURE 11.21
Extinction ratio at 1550 nm for 100-μm-long and 250-μm-long EAMs. (Pictures from Kuo, Y.H. et al., *Opt. Express* 16(13):9936–9941, 2008.)

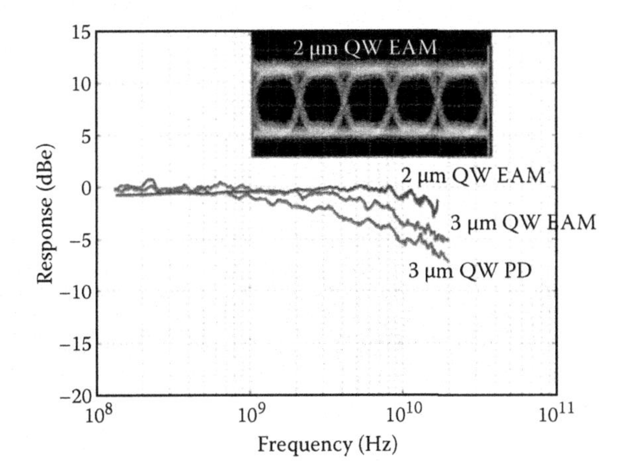

FIGURE 11.22
Response curves of two 100-μm-long EAMs with different QW section widths of 2 and 3 μm, The response of a PD is also given. Inset shows the eye diagrams. (Pictures from Kuo, Y.H. et al., *Opt. Express* 16(13):9936–9941, 2008.)

The voltage swing of this EAM can be less than 1 V at 5 dB extinction ratio, which is essential for compatibility with CMOS technology. The energy consumption of such EAMs is estimated to be approximately 20 fJ/bit.

The 16-GHz modulation bandwidth of this EAM was limited by the lumped electrode design [72]. By using a travelling wave (TW) electrode design, faster operation can be achieved. In Tang et al. [73], a TW-EAM is presented that reaches a high figure of merit of 23 GHz/V and has a clear open eye up to 50 Gb/s. The TW electrode design overcomes the RC-limit due to a distributed circuit configuration, as shown in Figure 11.23a. Details of the material are similar to those presented above for the work by Kuo et al. [72] and the fabrication is similar to the work by Chen et al. [75]. A cross-section of the hybrid section is shown in Figure 11.23b. Ground–signal–ground contact pads with 100 μm pitch were used for the input and output RF ports, as depicted in Figure 11.23c. The device has a compact footprint of 240 × 430 μm.

The TW-EAM insertion loss is 5 dB. For wavelengths of approximately 1550 nm, a higher than 11 dB extinction ratio is achieved for a voltage change from −2 to −4 V. Figure 11.24 shows the small-signal electro-optical modulation response and a 3-dB bandwidth of 42 GHz is obtained, using a bias voltage of −3 V. Large-signal performance was evaluated using a nonreturn-to-zero (NRZ) PRBS signal with a 2-V swing on the TW-EAM under a −3 V bias. As shown in Figure 11.25, a clear open eye was observed at 50 Gb/s with a dynamic extinction ratio of 9.8 dB, which is sufficient for practical applications.

11.3.3.2 Mach–Zehnder–Based Modulator

MZMs are commonly designed using a coplanar waveguide electrode design. However, for a coplanar waveguide incorporating a PIN diode, the electrical field penetrates into the diode and propagation losses of the electrical signal tend to be large. To reduce the propagation loss, it is important to ensure that the propagating electrical fields have minimal overlap with the doped semiconductor. In this design, a capacitively loaded TW electrode is implemented. As illustrated in Figure 11.26a, the small pads extending from the transmission line can provide the necessary electrical signal to drive the device while the TWE

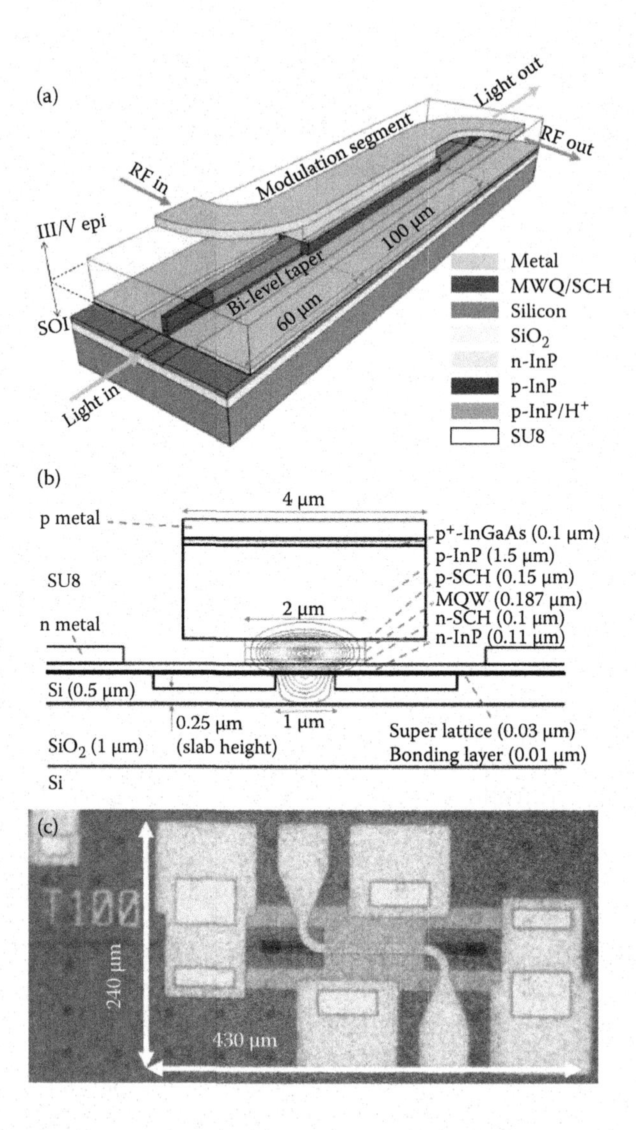

FIGURE 11.23
(**See color insert.**) Schematic structure of a hybrid silicon TW-EAM (a). Cross-section of the modulation segment with a superimposed fundamental optical mode (b). Top-view photograph of a fabricated TW-EAM (c). (Tang, Y. et al., *Opt. Express* 19(7):5811, 2011.)

is kept away from the semiconductor. Furthermore, the phase velocity of the electrical signal can be adjusted by changing the distributed capacitance of the transmission line. This can help reduce velocity mismatch between the electrical and optical signals. The cross-section of the loaded region is depicted in Figure 11.26b. More details about the fabrication can be found in Chen et al. [74], who also subsequently reported on an improved version [75].

The modulation bandwidth at −3 V reverse bias voltage and with 25 Ω termination was measured to be 27 GHz, as shown in Figure 11.27b. With this device, large signal modulation at 25 and 40 Gb/s with 16 and 11 dB extinction ratios were shown, respectively (Figure 11.28).

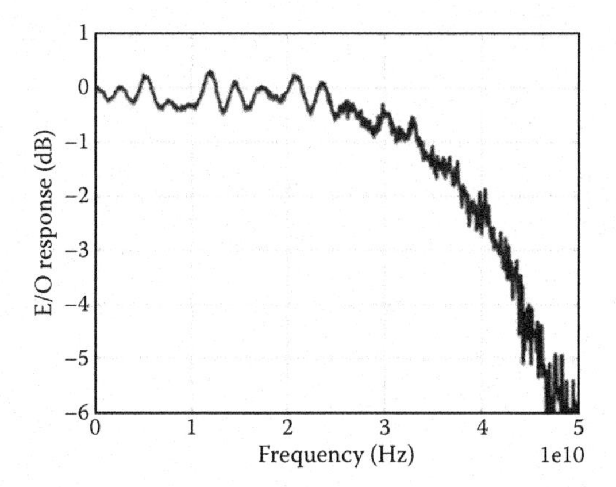

FIGURE 11.24
Small signal E/O response for a hybrid silicon TW-EAM with a 100-μm active segment. (Picture from Tang, Y. et al., *Opt. Express* 19(7)5811, 2011.)

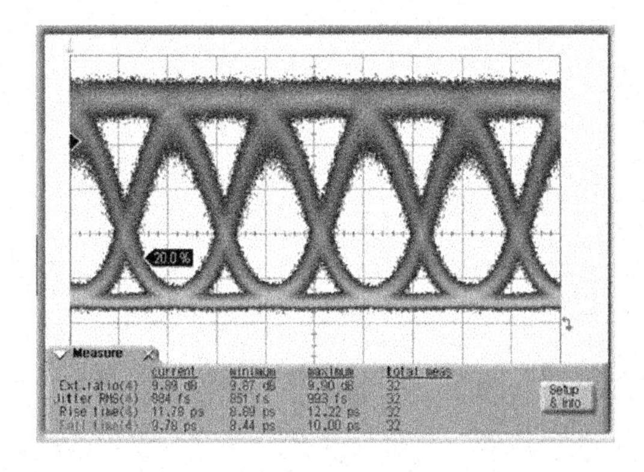

FIGURE 11.25
NRZ eye diagram (50 Gb/s) at λ = 1550 nm with 2^{31}–1 PRBS pattern for a 100-μm-long hybrid silicon TW-EAM. The modulator was biased at –3 V with a driving voltage swing of only 2 V. (Picture from Tang, Y. et al., *Opt. Express* 19(7):5811, 2011.)

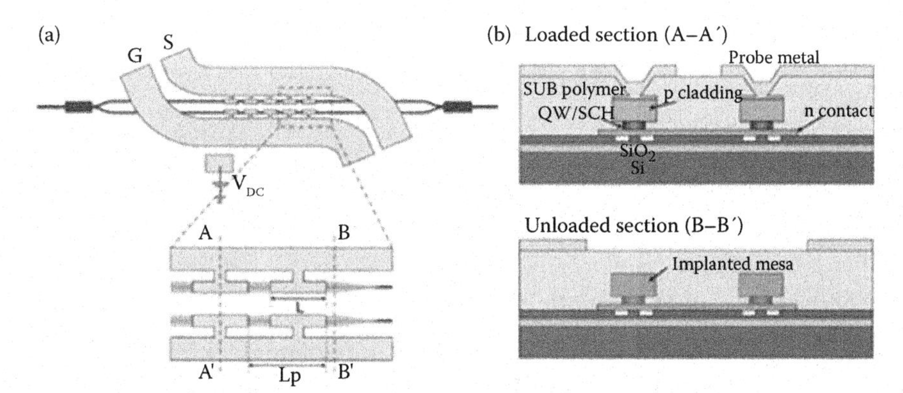

FIGURE 11.26
Top view of a device with a capacitively loaded slotline electrode (a). Cross-section of loaded (along A–A′) and unloaded sections (along B–B′) of the hybrid waveguide (b). (Pictures from Chen, H.-W. et al., *Opt. Express* 18(2):1070–1075, 2010.)

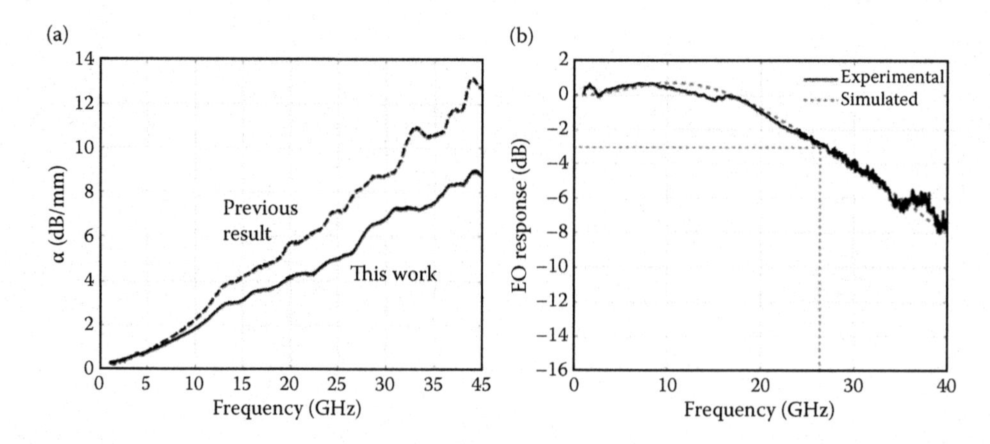

FIGURE 11.27
Propagation loss improvement of a MZM with L_a = 500 μm (a). Modulation bandwidth measured at −3 V with 25 Ω termination (b). (Pictures from Chen, H.-W. et al., *Opt. Express* 19(2):1455, 2011.)

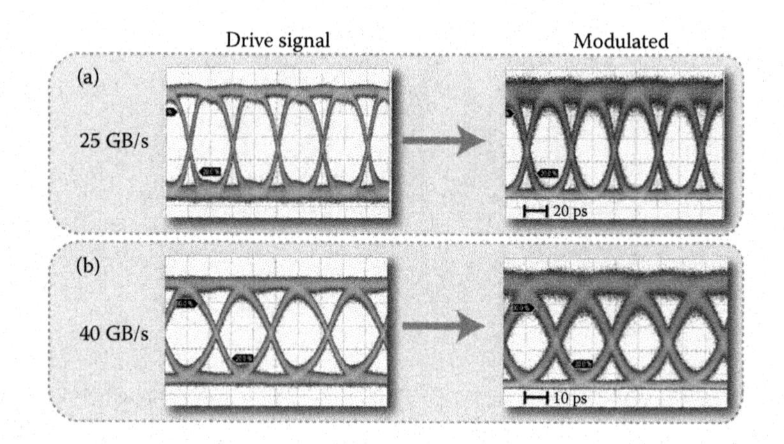

FIGURE 11.28
Left: 25 Gb/s electrical driven signal with 231-1 NRZ PRBS. Right: signal after modulator with 15.5 dB extinction ratio (a). Left: 40 Gb/s electrical driven signal with 231-1 NRZ PRBS. Right: signal after modulator with 11.4 dB extinction ratio (b). (Pictures from Chen, H.-W. et al., *Opt. Express* 19(2):1455, 2011.)

In terms of energy efficiency, these hybrid silicon modulators will be important for high-speed links because the absolute power consumption of the interconnect is relatively independent of modulation speed, whereas the efficiency (energy per bit) increases with speed [76].

11.3.3.3 Microdisk-Based Electrooptic Modulator

In Liu et al. [77], a carrier injection–based microdisk modulator fabricated on the III–V/SOI heterogeneous platform was presented. Such a resonant modulator has a dimension of less than 10 μm and the transmission is modulated by the current-dependent gain of the active material in the microdisk cavity (MDC).

The presented MDC has a diameter of 7.5 μm, a thickness of 1 μm, and has three strained InAsP QWs. The MDC couples evanescently to an underlying SOI wire waveguide

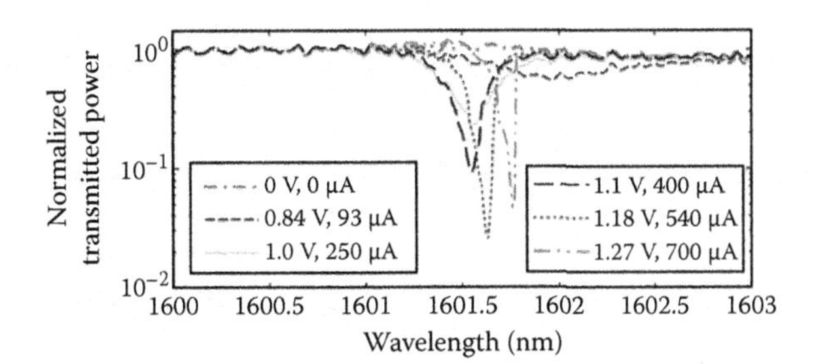

FIGURE 11.29
Static transmission spectra through the MDC with different bias voltages and currents. The lasing threshold current of the MDC is 800 μA. (Picture from Liu, L. et al., *Opt. Lett.* 33:2518–2520, 2008.)

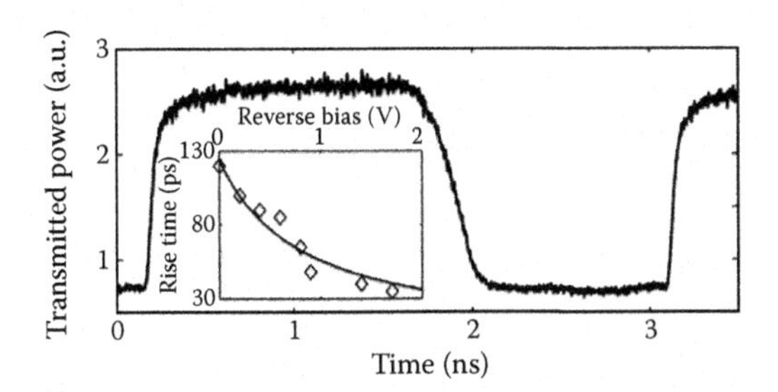

FIGURE 11.30
Optical response of the modulator under a square wave signal with a voltage level of 0 to 1.1 V and a frequency of 340 MHz. The inset shows the rise time as a function of the reverse bias voltage. (Picture from Liu, L. et al., *Opt. Lett.* 33:2518–2520, 2008.)

with dimensions of 500 × 220 nm. The design is similar to the MDL presented by Van Campenhout et al. [66]. Light input at the SOI waveguide interacts with the whispering-gallery mode in the MDC and experiences loss or gain, depending on the injected current into the MDC.

Figure 11.29 shows the transmission spectra for different injection currents. No dip can be observed for zero bias. With increasing injection current, the loss decreases and a transmission dip becomes more prominent, with the highest extinction ratio of 16 dB obtained at 540 μA. Furthermore, the dip blueshifts first as the carrier density increases, followed by a redshift as the device heats up. Increasing the input power causes a decrease of the extinction ratio due to carrier depletion and it saturates at approximately 10 dB.

The dynamic operation is shown in Figure 11.30, in which the MDC is driven with a 340-MHz square wave from 0 to 1.1 V, and with a rise and fall time of 30 ps. The achieved modulation depth is 6 dB, due to noise limitations of the setup used [77]. The rise and fall times of the signal in Figure 11.30 are 120 and 350 ps, respectively, mainly limited by the carrier injection speed. Operation at 2.73 Gb/s was achieved with an NRZ signal without special driving techniques. This speed can be improved by using a pre-emphasis technique.

11.3.4 PDs and Receivers

11.3.4.1 Metal–Semiconductor–Metal PD on SOI Waveguides

In Brouckaert et al. [78], an InAlAs–InGaAs metal–semiconductor–metal PD, coupled to a silicon waveguide was reported. Such PDs offer the advantage of lower capacitance per

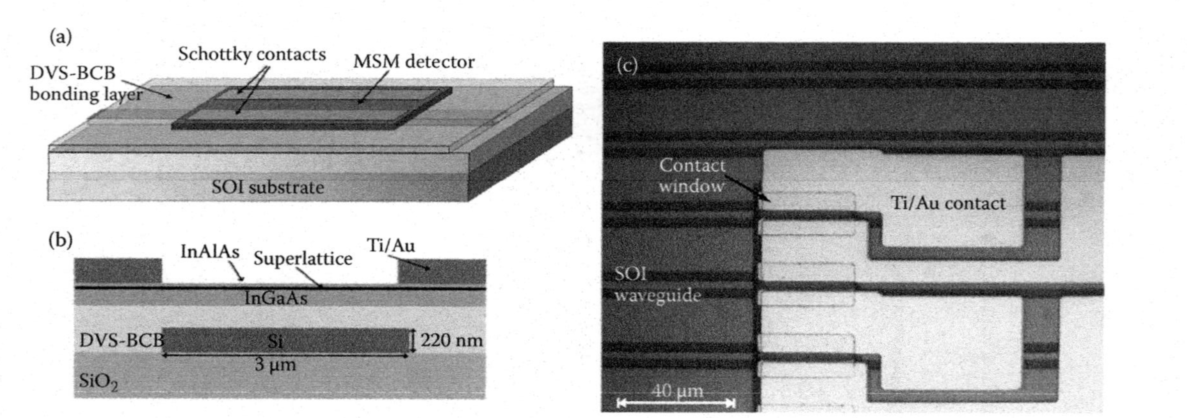

FIGURE 11.31
Three-dimensional (a) and cross-sectional (b) schematic views of the waveguide-integrated metal–semiconductor–metal detector. Top view on three 40-μm-long thin-film metal–semiconductor–metal PDs on top of 3-μm-wide SOI waveguides (c). The waveguide pitch is 30 μm. (Pictures from Brouckaert, J. et al., *IEEE Photon. Technol. Lett.* 19(19):1484, 2007.)

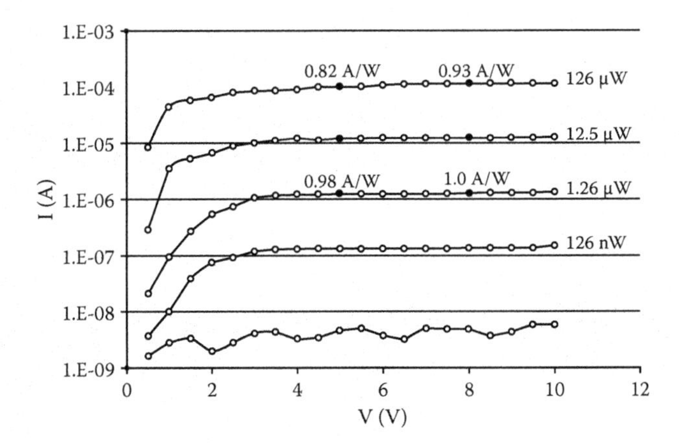

FIGURE 11.32
I–V characteristics of a 30-μm-long detector at a wavelength of 1.55 μm. The measured dark current and photocurrents for different SOI waveguide powers are plotted. Optical power is increased in steps of 10 dB. (Picture from Brouckaert, J. et al., *IEEE Photon. Technol. Lett.* 19(19):1484, 2007.)

unit area as compared with p-i-n PDs. These devices were realized using DVS-BCB bonding of a III–V die to a processed SOI wafer. The device layout is shown in Figure 11.31a and b. Coupling from the 220-nm-thick and 3-μm-wide silicon waveguide to the PD is achieved by directional coupling. The PD stack consists of a 40-nm InAlAs Schottky barrier enhancement layer, a 20-nm InAlAs–InGaAs graded superlattice to decrease carrier trapping, and a 145-nm InGaAs absorption layer. Two Ti–Au Schottky electrodes are deposited on top. Lateral confinement of light in the PD waveguide is obtained by the two coplanar Ti–Au Schottky contacts and loss in these contacts is calculated to be less than 10% [78,79]. The metal–semiconductor–metal geometry allows for independent optimization of optical path length and carrier transit time, that is, by changing the Schottky contact spacing.

PDs with lengths of 25 to 40 μm were measured at wavelengths of approximately 1.55 μm for TE polarized input light. Dark currents are 4.5 nA for 5 V bias voltage. In Figure 11.32, it is shown that an internal responsivity of 1.0 A/W is measured, corresponding to a quantum efficiency of 80%, in agreement with the simulations [79]. As can also be seen in Figure 11.32, a small saturation takes place at the higher input powers due to carrier screening effects. At a waveguide power of 126 μW, the saturation is 0.8 dB at 5 V bias, and 0.3 dB at 8 V bias. Efficient detection was shown to be possible for wavelengths up to 1.65 μm, that is, the bandgap wavelength of InGaAs.

11.3.4.2 p-i-n PD on SOI Photonic Circuitry

Compact PDs are required for low dynamic power consumption, that is, through low diode capacitance, and for high integration density. As a trade-off, the PD efficiency needs to be sufficiently high. To achieve this, a PD is reported which is defined on top of a 300-nm SiO_2 layer that is deposited on a silicon photonic wire as shown in Figure 11.33 [80]. The silicon waveguide is 500 nm wide and 220 nm thick. An InP membrane input waveguide is used to couple the light from the silicon waveguide to the PD, which consists of a p-i-n junction in which the optical power is absorbed. The device footprint is 5 × 10 μm^2 and the thickness is 1 μm, which is a trade-off between efficiency and speed. In this structure, the

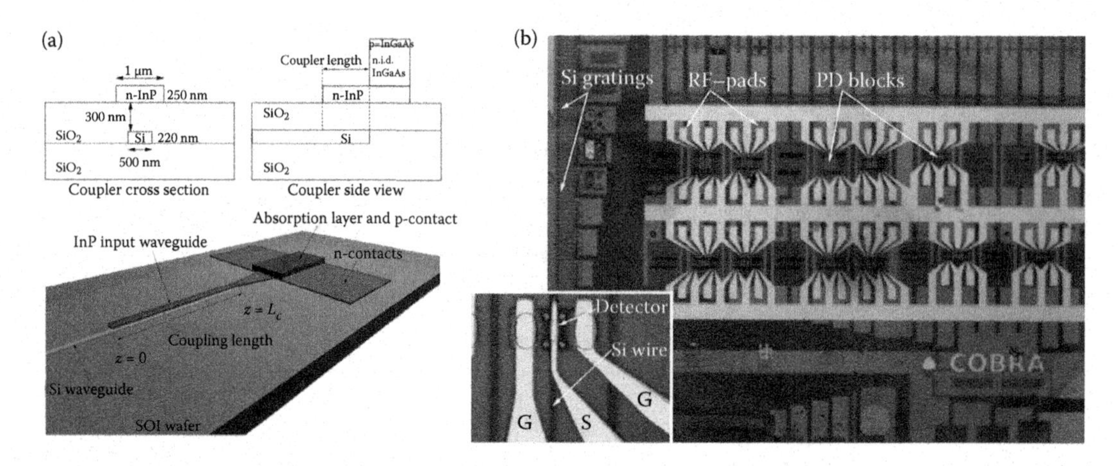

FIGURE 11.33
PD structure (a). The coupling from the Si photonic waveguide layer to the PD is realized through the InP membrane input waveguide, on top of which the detector is stacked. Picture of the fabricated detectors with improved metallization pattern (b). A close-up view of a detector is shown in the inset. (Pictures from Binetti, P.R.A. et al., *IEEE Photon. J.* 2(3):299, 2010.)

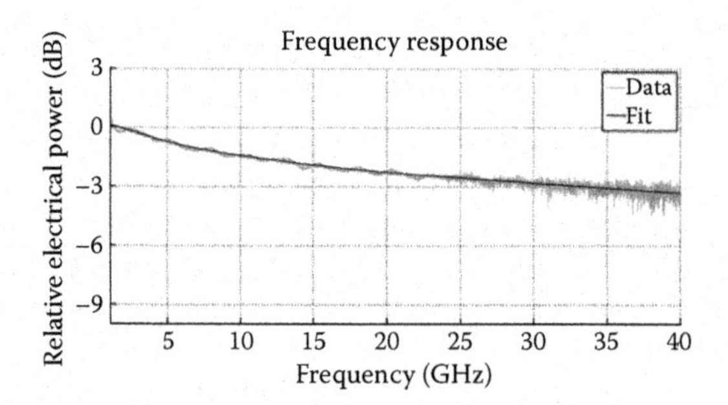

FIGURE 11.34
Detector RF frequency response. (Pictures from Binetti, P.R.A. et al., *IEEE Photon. J.* 2(3):299, 2010.)

speed is limited by photocarrier transit time in the depletion region. The PD layer stack was grown on a III–V wafer and bonded to the processed SOI wafer using molecular bonding with a SiO_2 interface [81].

The device was first characterized at a wavelength of approximately 1550 nm under DC operation and a dark current of 1.6 nA at −4 V bias voltage was found. The responsivity was 0.45 A/W, corresponding to a total quantum efficiency of 35%. This efficiency includes the internal quantum efficiency and the InP membrane coupler. The junction capacitance is less than 10 fF. By optimizing the RF pad layout by placing them close to the PDs using 100-μm-long tapers, a 33-GHz electronic bandwidth was obtained for these PDs, in agreement with simulations (Figure 11.34) [82].

11.3.4.3 Integration of InP/InGaAsP PDs onto SOI Waveguide Circuits

Thick bonding layers give a higher yield because these are more tolerant against imperfections of the mated surfaces. By using grating couplers, silicon waveguides can be coupled to PDs on top of a thick bonding layer, as depicted conceptually in Figure 11.35 [83]. The silicon layer has a thickness of 220 nm and the 10-μm-long grating has a period of 610 nm, a duty cycle of 50%, and an etch depth of 50 nm. The device efficiency depends on the bonding layer and BOX thicknesses. In Figure 11.36, the simulated absorbed power fraction is shown. This fraction depends on the InGaAs absorption spectrum and on the grating coupler bandwidth. Also, the spurious reflection back into the silicon waveguide is shown, that is, where the grating coupler acts as a pure second-order grating. Efficient detection up to 1600 nm wavelength can be achieved.

PDs of 10 × 10 μm were fabricated. Dark currents are 0.3 nA at a bias voltage of 1 V with a responsivity of 0.02 A/W at 1550 nm. This value is limited by a nonoptimized absorbing layer and simulations show that 0.4 A/W can be achieved by optimization. These PDs were also fabricated on SOI circuitry to be used as wavelength-selective filters. In Figure 11.37a,

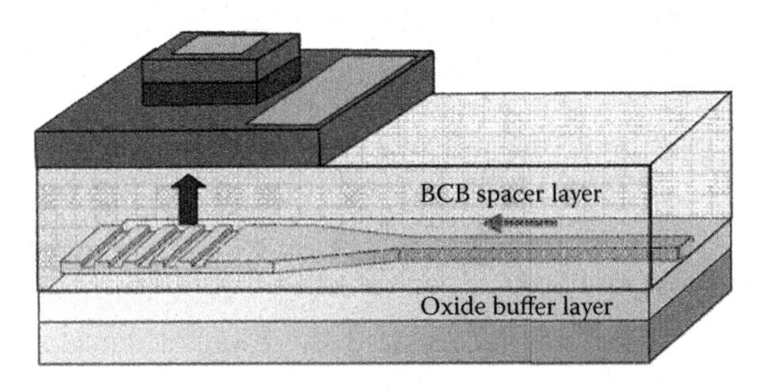

FIGURE 11.35
Coupling scheme for III–V PDs bonded to SOI waveguide circuitry. (Picture from Roelkens, G. et al., *Optics Exp.* 13:10102–10108, 2005.)

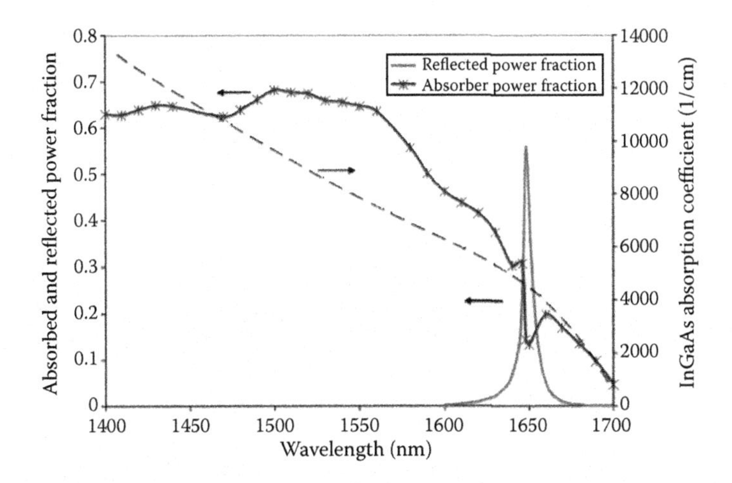

FIGURE 11.36
Wavelength dependence of the absorbed power fraction in the detector and reflected power back into the SOI waveguide. A 50-μm-long device and optimum SiO_2 buffer layer thickness and BCB bonding layer thickness are assumed. The wavelength dependence of the InGaAs absorption coefficient is also shown. (Picture from Roelkens, G. et al., *Optics Exp.* 13:10102–10108, 2005.)

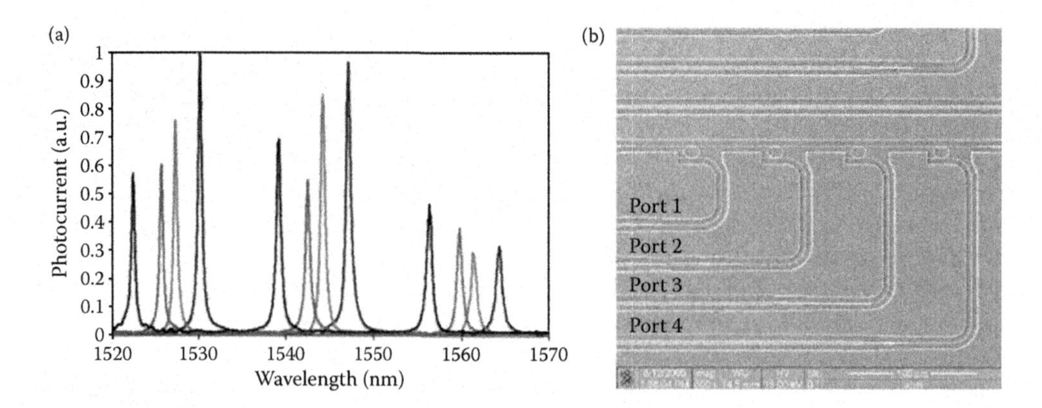

FIGURE 11.37
Response of four PDs integrated onto a four-racetrack resonator filter (a) and an SEM view of the fabricated SOI waveguide structure (b). (Picture from Roelkens, G. et al., *Optics Exp.* 13:10102–10108, 2005.)

the responses of four PDs bonded to a wavelength-selective filter (Figure 11.37b) circuit are shown.

11.3.4.4 Wavelength-Selective PDs

Wavelength-selective detectors can be realized by resonant detectors. In Liu et al. [84], a compact resonant metal–semiconductor–metal PD was shown by the integration of an absorbing III–V mesa onto an SOI ring resonator. Peak responsivities of approximately 1 A/W were achieved. To tune the resonant wavelength, a heater can also be added, as shown in Figure 11.38a [85]. In these devices, an InGaAs absorbing layer was bonded to the silicon ring using DVS-BCB bonding, with a layer stack similar to the detectors published in Brouckaert et al. [79]. The wavelength-dependent absorption and photocurrent are shown in Figure 11.39, having peak responsivities of approximately 0.47 A/W for this device. Tuning of this peak to more than 3 nm was achieved by varying the current through the heater, without degradation of responsivity or dark current.

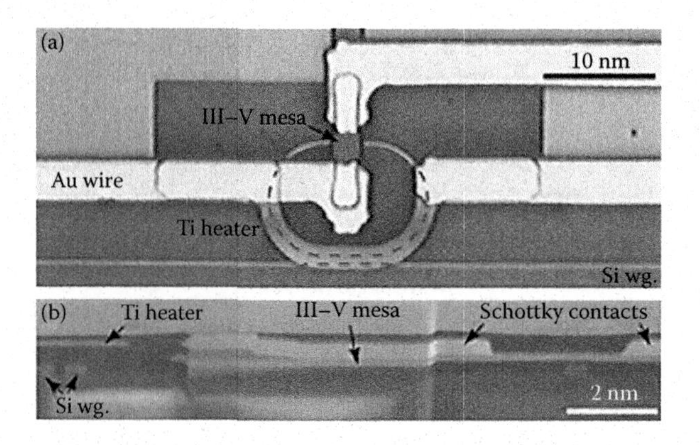

FIGURE 11.38
Top view (a) and cross-sectional (b) view of a fully fabricated resonant detector.

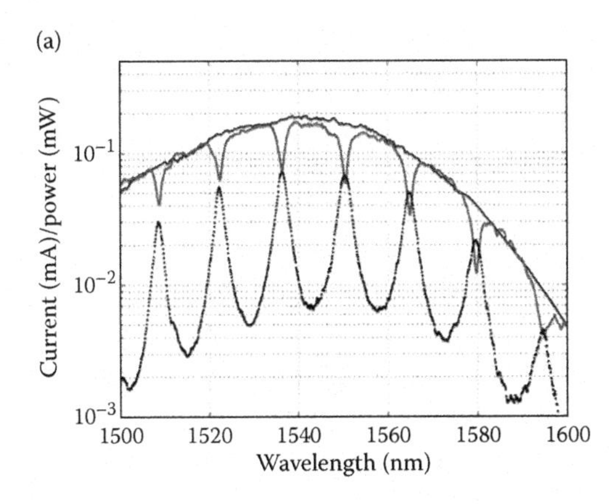

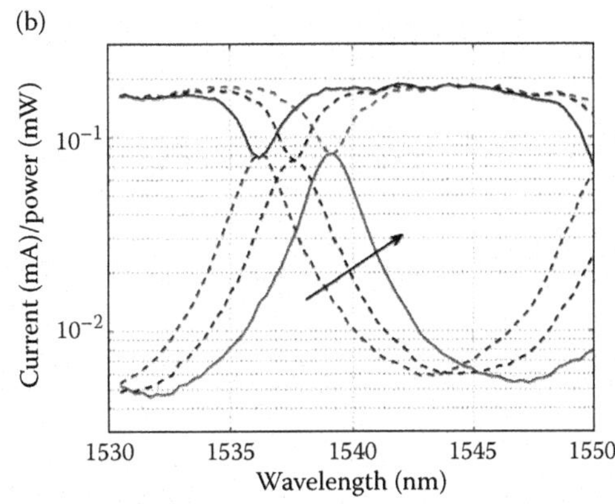

FIGURE 11.39
Measured transmission spectrum and detected current (a) of a fabricated device for a bias voltage of 7 V. Solid line indicates the incident power spectrum, dashed lines indicate the transmission spectrum, and dotted lines indicate the detected current. Because grating couplers were used to interface to the access fibers, the input light follows a Gaussian spectrum (b). Transmission and current at different electrical powers applied to the heater. Along the arrow direction, the applied powers are 0, 4.2, and 8.2 mW, respectively. (Picture from Van Thourhout, D. et al., *IEEE J. Sel. Top. Quant. Electron.* 16(5):1363, 2010.)

11.3.4.5 A Hybrid Silicon Preamplifier and PD

Combining hybrid silicon amplifiers and PDs on a single platform can lead to more practical, better-performing optoelectronic photoreceivers. Hybrid silicon PDs are interesting because their absorption edge can easily be extended beyond the 1600-nm regime by engineering III–V QWs.

In Figure 11.40, a hybrid silicon preamplified receiver is shown [86]. It is comprises an optical amplifier and a waveguide PD. The transition between the passive silicon waveguide and the hybrid waveguide of the amplifier is formed by tapers, as shown in Figure 11.4. The same III–V epitaxial structure is used for the amplifier and the PD. The total length of the amplifier and the detector is 1240 and 100 μm, respectively. Figure 11.40b shows an array of these receivers. Further device details can be found in Park et al. [86].

For a 1.2-mm-long structure, the maximum gain is 9.5 dB at 300 mA. The quantum efficiency of the 100-μm PD is 50%. By putting the two together, the responsivity of the receiver increases to 5.7 A/W with preamplification. The device shows 0.5 dB saturation at a photocurrent of 25 mA.

The device bandwidth is measured as 3 GHz by time domain impulse response (Figure 11.41). Although the resistance–capacitance–limited bandwidth is estimated to be 7.5 GHz. This indicates that the device speed is primarily constrained by the current III–V layer design. A higher bandwidth can be achieved by using QWs with a smaller valence band offset and a thinner separated confinement heterostructure layer to reduce the hole transit time. Bit error rate measurements for a NRZ 2.5 Gb/s PRBS shows a receiver sensitivity of −17.5 dBm [86].

An improved bandwidth using an InGaAs/InP-based hybrid silicon PD was reported by Chang et al. [87]. The bandwidth was measured to be 6 GHz and open eye diagrams were obtained up to 12.5 GHz. Higher speeds are achievable by careful design of the photodiode. For example, wafer-bonded InP/InGaAs-based PDs having a bandwidth of 20 to 25 GHz were reported by Binetti et al. [88,89], showing the potential of the hybrid silicon approach.

(a)

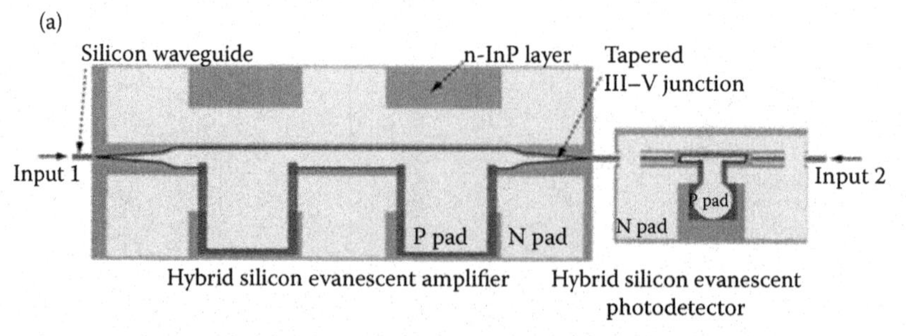

(b)

FIGURE 11.40
Top view of a hybrid silicon evanescent preamplified receiver (a). SEM picture of eight integrated devices with amplifiers (b). (Pictures from Park, H. et al., *Optics Exp.* 15(21):13539–13546, 2007.)

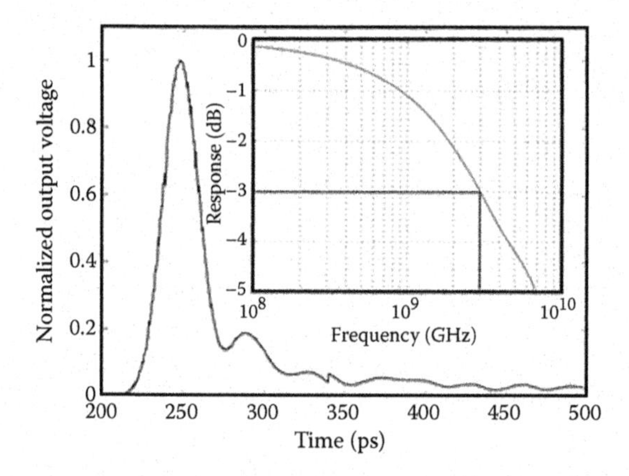

FIGURE 11.41
Impulse response (0.6 ps input pulse) of the detector. The inset shows the Fourier transform of the impulse response. (Pictures from Park, H. et al., *Optics Exp.* 15(21):13539–13546, 2007.)

11.4 Integration Platforms

In the previous section, we reviewed the development of basic fabrication and component technology. Here, we discuss three integration platforms that enable the integration of active and passive components, as well as active components with different epitaxial structures or bandgaps (or both). First, the integration of hybrid silicon components with passive SOI circuitry is reviewed. Second, the integration of different-bandgap hybrid silicon components using QW intermixing is discussed. And finally, we review some work on selective-die bonding, a technology that enables the integration of components with different epitaxial layer stacks.

11.4.1 Integration of Hybrid Silicon III–V-Based Devices with Passive SOI Circuitry

For the integration of hybrid silicon devices (such as lasers, amplifiers, modulators, and detectors) with passive SOI-based silicon devices (e.g., waveguides or multiplexers), low-loss mode conversion from the silicon waveguide mode to the hybrid mode and vice versa is essential. To achieve this, tapered structures can be used. In this section, we first present an experimental study on two taper designs. Then, we show some PICs that make use of these tapers to achieve a larger density integration of active and passive components.

11.4.1.1 Low-Loss Hybrid Silicon Tapers

Hybrid silicon structures are realized by first defining silicon waveguides on an SOI wafer. Waveguides are etched 0.35 μm into a 0.70-μm silicon layer. The buried oxide layer is 1.0 μm. The passive silicon waveguide width was 0.8 μm to ensure single-transverse mode operation and taper excitation. Hereafter, a III–V die with an epitaxial layer design, corresponding to that of Park et al. [90], is bonded to the SOI. Finally, in a three-step etch process, the mesas and tapers are etched into the III–V die. In this work, we have shifted the bandgap of the QWs to 1.40 μm, that is, well below the wavelength of 1.55 μm we use for characterization. By doing this, we eliminate QW absorption and the need for electrical pumping.

We consider two taper designs, as shown in Figure 11.42. In type 1 tapers, the 0.8-μm silicon waveguide is tapered out to 2 μm in which the III–V taper starts. Tapering of the *p*-InP layer starts almost immediately after the start of the other two taper levels, allowing a 1.2-μm margin of the *n*-InP and SCH levels for fabrication tolerance purposes. This taper has been widely used, for example, by Park et al. [86]. In more recent work [91], a type 2 taper was used, in which the mode is more gradually converted from the silicon mode to the hybrid mode. In this case, the three taper levels start at different positions, requiring a total taper length of 160 μm. In both taper types, the *p*-InP taper flares out to 4 μm, after which the wide III–V mesa starts. Mesa width is 24 μm for type 1 and 14 μm for type 2.

To vary the overlap of the hybrid mode with the QWs, we use two different widths of silicon waveguide in the hybrid section, that is, 1.0 and 1.5 μm. This is realized by tapering the silicon width from 2.0 μm down over the length of the III–V taper.

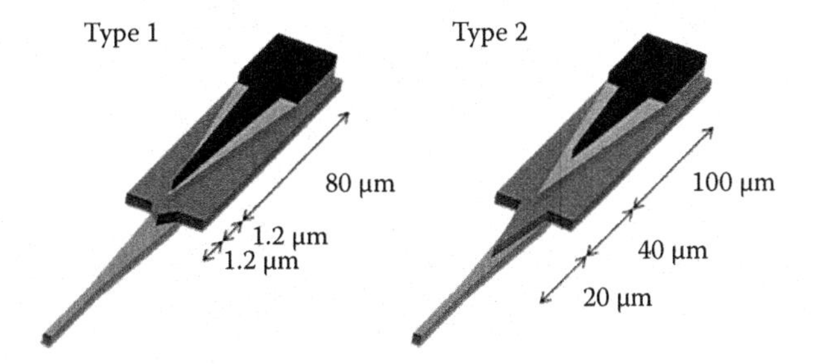

FIGURE 11.42
Type 1 and type 2 tapers showing the silicon waveguide (bottom layer), *n*-InP (bottom center), SCH layer containing QWs (top center), and *p*-InP top (top layer). Taper start offsets are indicated.

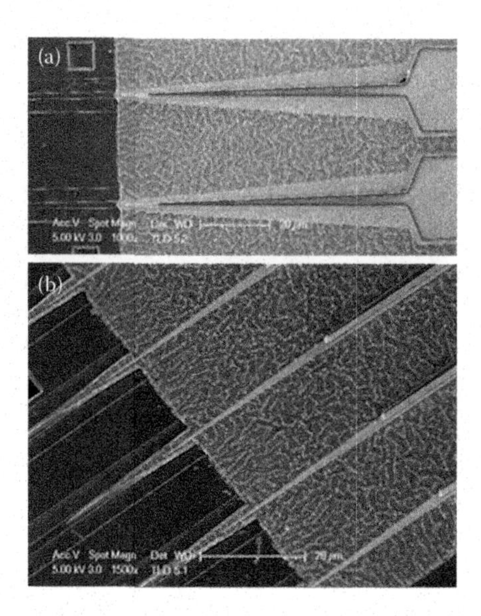

FIGURE 11.43
SEM pictures of type 1 (a) and type 2 (b) tapers.

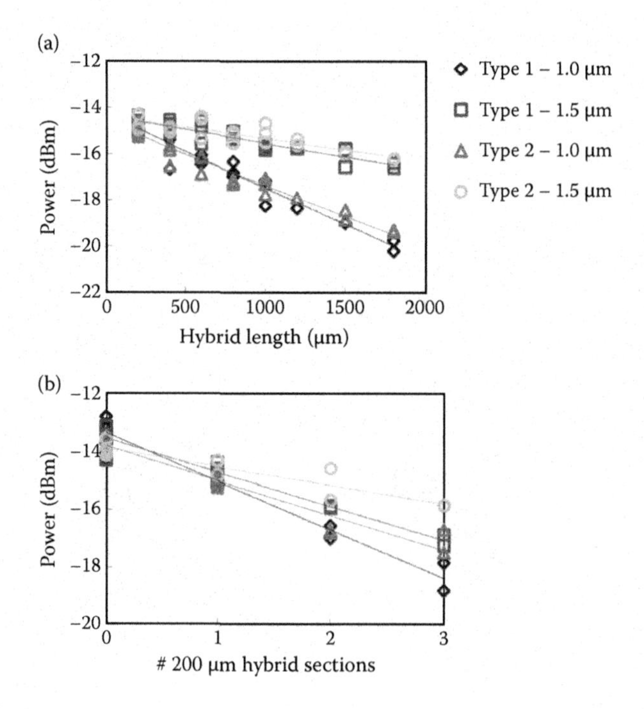

FIGURE 11.44
Measured in-fiber output power after propagation through structures with various hybrid waveguide lengths (a) and various numbers of 200 μm hybrid sections (b).

TABLE 11.1

Experimental Loss Values

	Hybrid Loss (dB/mm)	Taper Loss (dB)[a]	Taper Loss (dB)[b]
Type 1: 1.0 μm	3.2 ± 0.2	0.52 ± 0.05	0.47 ± 0.11
Type 1: 1.5 μm	1.2 ± 0.1	0.47 ± 0.04	0.42 ± 0.07
Type 2: 1.0 μm	2.7 ± 0.1	0.33 ± 0.05	0.43 ± 0.07
Type 2: 1.5 μm	1.0 ± 0.1	0.22 ± 0.05	0.25 ± 0.08

[a] As determined from the slope of Figure 11.44b.
[b] As determined from the offset of Figure 11.44a.

The devices have been realized with a fabrication process, as mentioned by Kurczveil et al. [91]. Due to finite lithography resolution, the tapers have a tip radius of approximately 0.5 μm, as determined from a scanning electron microscope (SEM) picture. In Figure 11.43 (top view), SEM pictures of the realized tapers are shown. As can be seen in Figure 11.43a, the taper start position varies somewhat due to lithography limitations on the narrow stripe (0.5 μm starting width on mask).

Taper losses and mesa propagation losses are then extracted by measuring propagation losses through devices with varying lengths and repeats [92]. In Figure 11.44a, the transmitted power as a function of hybrid device length is plotted. A linear fit through the data gives the hybrid waveguide loss. The offset in Figure 11.44a and the slope in Figure 11.44b give the taper losses. The results are summarized in Table 11.1. The tapers show losses of 0.5 and 0.3 dB per taper transition for the two taper designs presented, that is, they add 1.0 to 0.6 dB to the insertion loss of hybrid silicon devices integrated on passive SOI integrated circuits. These losses are significantly lower than the 1- to 2-dB coupling loss values that are generally obtained in other hybrid integration approaches [21], showing the clear advantage of this integration approach.

11.4.1.2 An Integrated Recirculating Optical Buffer

Silicon waveguide can, in principle, be low-loss, enabling PICs like integrated optical buffers. In Park et al. [93], such a device is presented, integrating long passive silicon waveguides with hybrid silicon amplifiers. A schematic of the device is shown in Figure 11.45. The device has a 9-cm-long delay line and a hybrid silicon gate matrix switch because of low crosstalk and high extinction ratio [94]. The presented design is a recirculating buffer, which means that optical packets are stored in the delay line until the gate matrix reroutes them to the output. Booster amplifiers in the delay line add gain to overcome the passive losses of the long silicon waveguides. The PIC makes use of type 1 tapers as presented previously.

On a 0.6 × 1-cm^2 chip, four devices are integrated by interleaving the four delay lines. The fabricated device is mounted on an aluminum nitride carrier and the electrical pads of the device are wire-bonded to the carrier for device probing. An image of the mounted device is shown in Figure 11.46.

The gain of the switch and booster amplifiers is high enough to overcome the 15 dB of passive silicon delay line loss. The crosstalk and DC extinction ratio of the gate matrix are measured to be −34 dB and 30 dB, respectively. The performance as a buffer was evaluated with a return-to-zero PRBS at 40 Gb/s and 1560 nm. The minimum power penalty with a 1.1-ns delay is 2.5 dB with an input dynamic range of 9 dB. Further improvements on the waveguide loss and the amplifier gain should improve the device performance, leading to a longer buffering time, and will provide enough functionality as integrated optical buffers for all optical packet-switched networks.

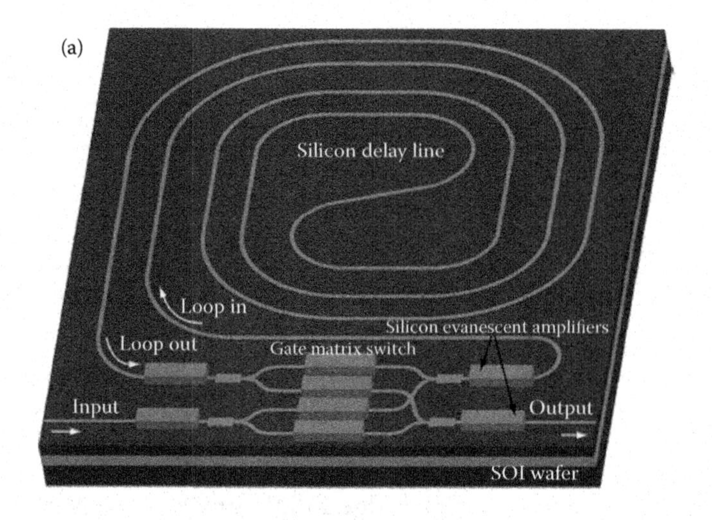

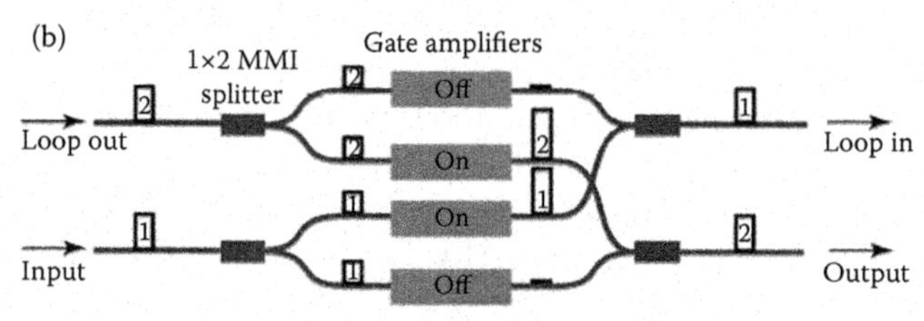

FIGURE 11.45
Device layout integrated buffer (a) and gate matrix switch (b). The cross-over operation is illustrated as an example. (Picture from Park, H. et al., *Optics Exp.* 16(15):11124, 2008.)

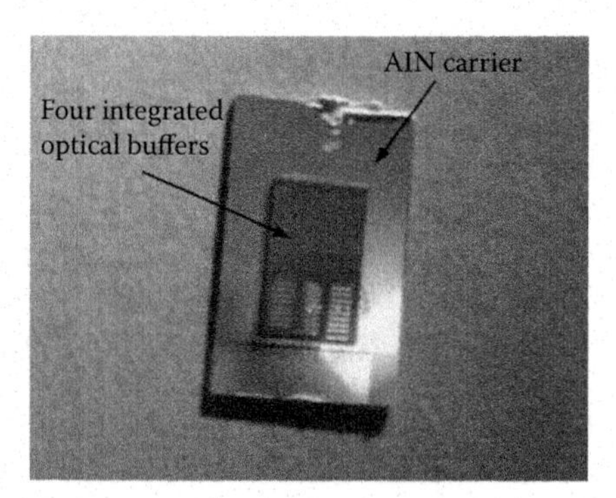

FIGURE 11.46
Microscope image of the four integrated buffers mounted on an aluminum nitride carrier. (Picture from Park, H. et al., *Optics Exp.* 16(15):11124, 2008.)

11.4.1.3 An Integrated Hybrid Silicon Multiwavelength AWG Laser

Another example that makes use of the integration of hybrid silicon actives with passive SOI-based circuitry is the AWG laser presented by Kurczveil et al. [91]. This is the first realization of a device in which an AWG is integrated on the hybrid silicon platform. The mature CMOS technology allows, in principle, for high-quality and high-resolution AWGs, making this approach interesting. The layout of this laser is shown in Figure 11.47a and b. The AWG

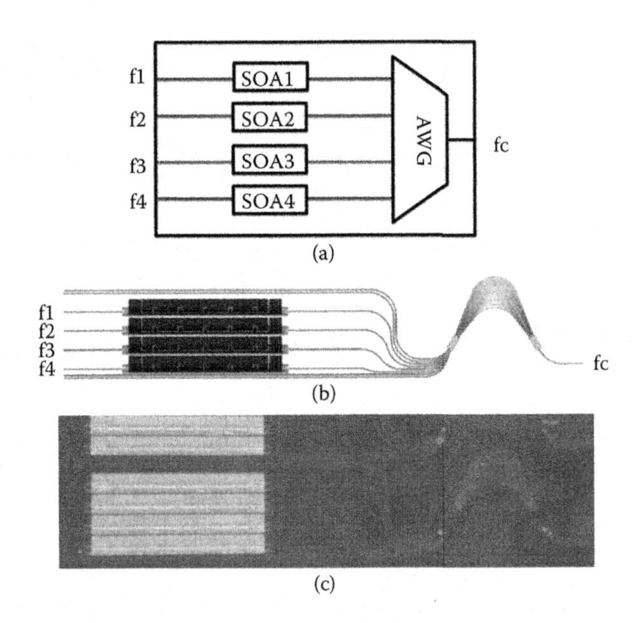

FIGURE 11.47
Schematic diagram of an AWG laser with four channels (a). Schematic diagram of the fabricated chip (b). The AWG has eight channels. Four of the eight channels have hybrid silicon SOAs. Optical photograph of the fabricated chip (c). The sample is covered with thick SU8 polymer that decreases the contrast of the image. (Picture taken from Kurczveil, G. et al., *IEEE J. Sel. Top. Quant. Electron.* 17(6):1521–1527, 2011.)

has eight channels with a channel spacing of 360 GHz. The 1-mm hybrid silicon SOAs are coupled to the silicon waveguides using a type 2 taper, as introduced in the previous section. By biasing SOAs 1 to 4 above threshold, four lasing cavities are formed between facets *f*1 to *f*4 and the common output facet *fc* in Figure 11.47b. An optical photograph of the fabricated device is shown in Figure 11.47c. Further details can be found in Kurczveil et al. [91].

The AWG transmission spectrum of the device is shown in Figure 11.48. These data were collected by forward biasing one SOA at a time below the threshold and measuring the output of the common output waveguide *fc* (Figure 11.47b). The AWG passbands degraded slightly due to the exposure of the silicon waveguides to III–V etches, which

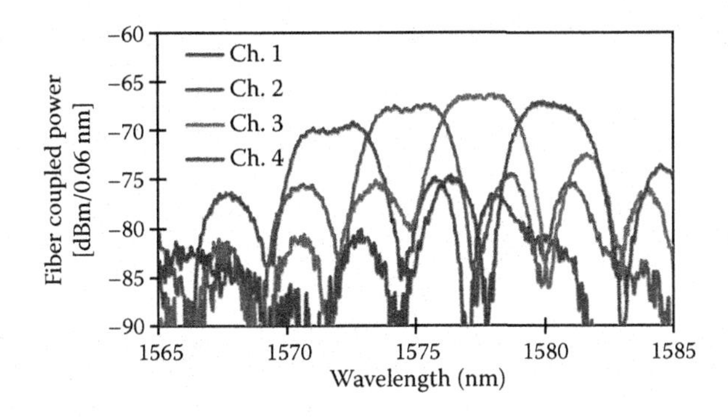

FIGURE 11.48
(See color insert.) ASE output of the common output waveguide for an SOA bias of 75 mA. (Picture taken from Kurczveil, G. et al., *IEEE J. Sel. Top. Quant. Electron.* 17(6):1521–1527, 2011.)

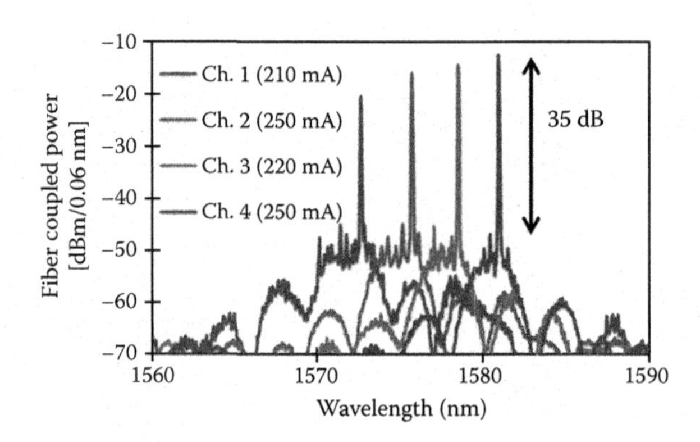

FIGURE 11.49
(**See color insert.**) Optical spectrum of the four channels optimized for maximum side mode suppression ratio. (Picture taken from Kurczveil, G. et al., *IEEE J. Sel. Top. Quant. Electron.* 17(6):1521–1527, 2011.)

slightly attacked the silicon, introducing phase errors. This can be avoided by using an approach such as that shown by Park et al. [93]. All four channels show lasing, as depicted in Figure 11.49. Lasing thresholds are between 113 and 123 mA for channels 2 to 4, with channel 1 slightly higher at 150 mA due to poorer AWG transmission (Figure 11.48). Taking chip-to-fiber coupling losses into account, the output power from the common facet is approximately 0.3 mW. The output power can be almost doubled by applying a high-reflection coating on one side of the chip. The AWG has an insertion loss of 3.5 dB in this realization. Side-mode suppression ratio is optimum in channel 4 at 35 dB, as can be seen in Figure 11.49. Direct modulation at 1 and 2.5 GHz shows open eyes in both cases with extinction ratios of 7.8 and 3 dB, respectively, making the AWG laser, in principle, useful as a WDM transmitter. Although this device is limited to four channels, it can be easily scaled up to several tens of channels as mature CMOS manufacturing technology allows the fabrication of high-quality AWGs on SOI.

11.4.2 Selective Die Bonding for Multiple Bandgap Device Integration

11.4.2.1 Integrated Hybrid Silicon Triplexer

A triplexer for fiber-to-home networks that provides a triple play service (data, voice, and video), requires three key components: a demultiplexer which can separate 1310, 1490, and 1550 nm channels, a 1310-nm laser, and two PDs operating at the wavelengths of 1490 and 1550 nm. By selective area wafer bonding, these different-bandgap components can, in principle, be integrated on the same chip. In Chang et al. [95], a successful realization of a fully integrated triplexer making use of the hybrid silicon platform was presented, with footprints less than 4 mm^2.

A selective area wafer bonding technique was used to integrate the 1310-nm laser and the longer wavelength PDs onto a patterned SOI chip. The process is similar to the one described previously, with the addition of a thin and compressible aluminum plate in the bonding fixture to compensate for the thickness differences (typically tens of microns) between these different III–V dies. Figure 11.50 shows an overview of the bonding process.

Figure 11.51a shows the successful bonding of two lasers and one PD III–V die onto a 1-cm^2 SOI chip. Figure 11.51b and c show the schematic and top view microscope images of the realized triplexer. An multi-mode interference (MMI) multiplexer cascaded with a

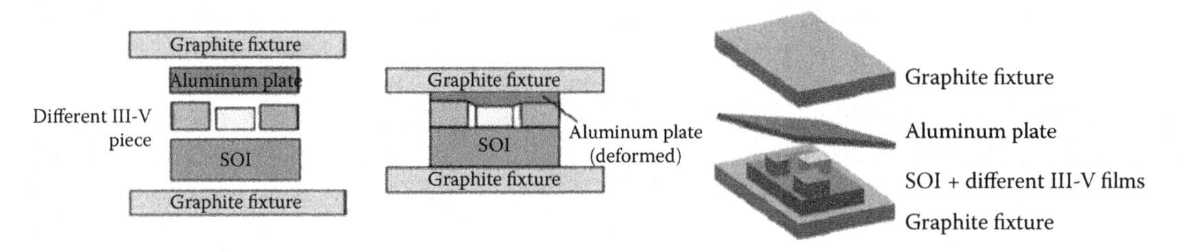

FIGURE 11.50
Selective area wafer bonding diagram. (Picture taken from Chang, H.-H. et al., Integrated hybrid silicon triplexer. *Optics Exp.* 18(23):23891–23899.

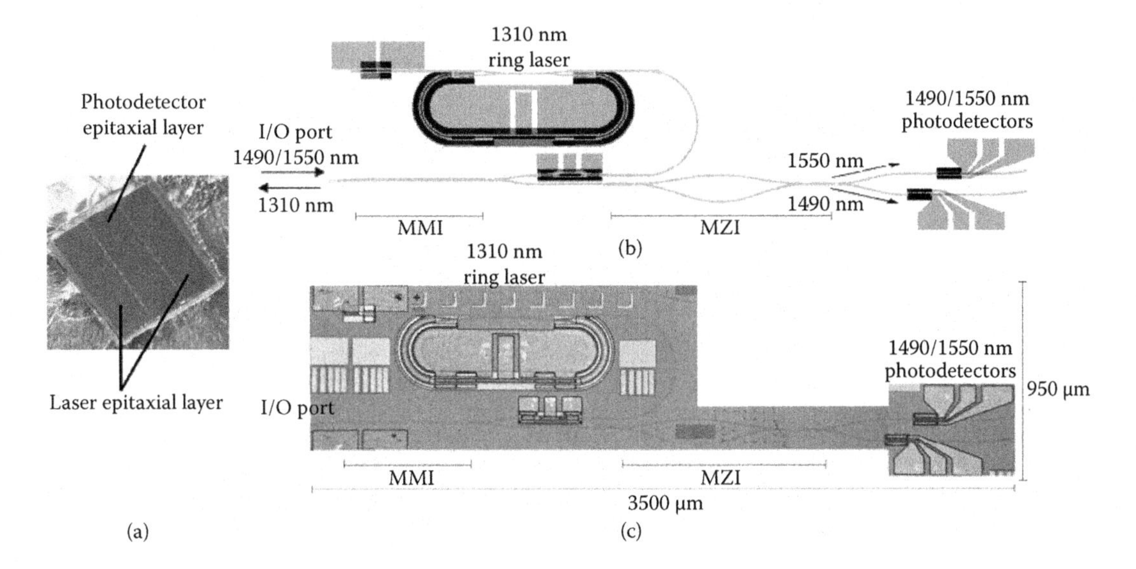

FIGURE 11.51
Selective area wafer bonding results: two laser and one PD III–V films (a) integrated on an SOI triplexer chip after substrate removal; schematic view (b) and top view (c) of an integrated triplexer. (Picture taken from Chang, H.-H. et al., Integrated hybrid silicon triplexer. *Optics Exp.* 18(23):23891–23899.)

Mach–Zehnder interferometric demultiplexer is used to separate or combine the different channels. The nominally 1310 nm laser is designed in a similar way as the one presented by Chang et al. [96]. The lasing wavelength is centered at 1348 nm and the threshold current is approximately 330 mA with a maximum on-chip measured power of 6.3 mW at 540 mA current injection. Its 3 dB modulation bandwidth is 2 GHz, which is sufficient for the required 1.25 GHz standard. The PD bandwidth is approximately 16 GHz and open eyes are measured for 12.5 Gb/s PRBS input.

This result shows that an International Telecommunication Union (ITU)-compliant fully integrated triplexer can be realized in the hybrid silicon platform using selective die bonding.

11.4.2.2 Low-Footprint Optical Interconnect on an SOI Chip through Heterogeneous Integration of InP-Based MDLs and Microdetectors

MDL and PD epitaxial designs can be optimized separately for on optical interconnect on a chip when selective die bonding is used. In Van Campenhout et al. [97], a proof-of-principle is presented in which light from an electrically driven MDL is coupled evanescently to a silicon waveguide in the interconnection layer. Downstream, the light is coupled to a PD using an InP membrane coupler, as shown schematically in Figure 11.52.

The MDL and PD are designed and fabricated similar to the work by Van Campenhout et al. [50] and Binetti et al. [80,82], respectively. Two III–V wafers with the MDL and PD layer stacks were epitaxially grown and diced to 9 × 4.5 mm² dies, which were then bonded to a patterned SOI wafer, as shown in Figure 11.53a. Fiber grating couplers are used to monitor

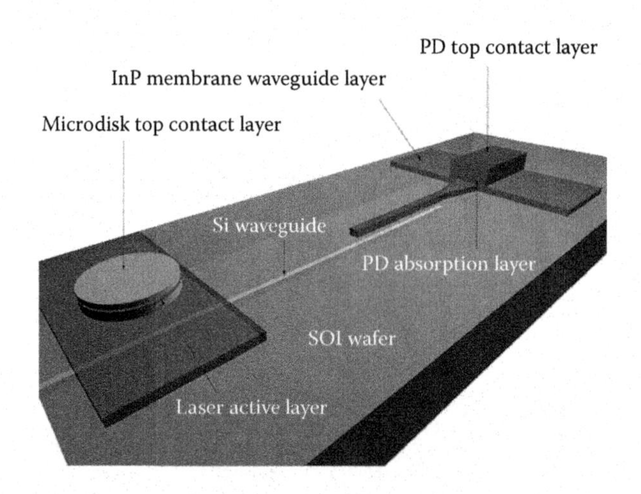

FIGURE 11.52
Schematic diagram of a photonic link on an SOI substrate: the electrically pumped MDL (left) launches an optical signal into the Si waveguide. Light is collected by the detector structure (right). (Picture taken from Van Campenhout, J. et al., *IEEE Photonics Technol. Lett.* 21(8):522, 2009.)

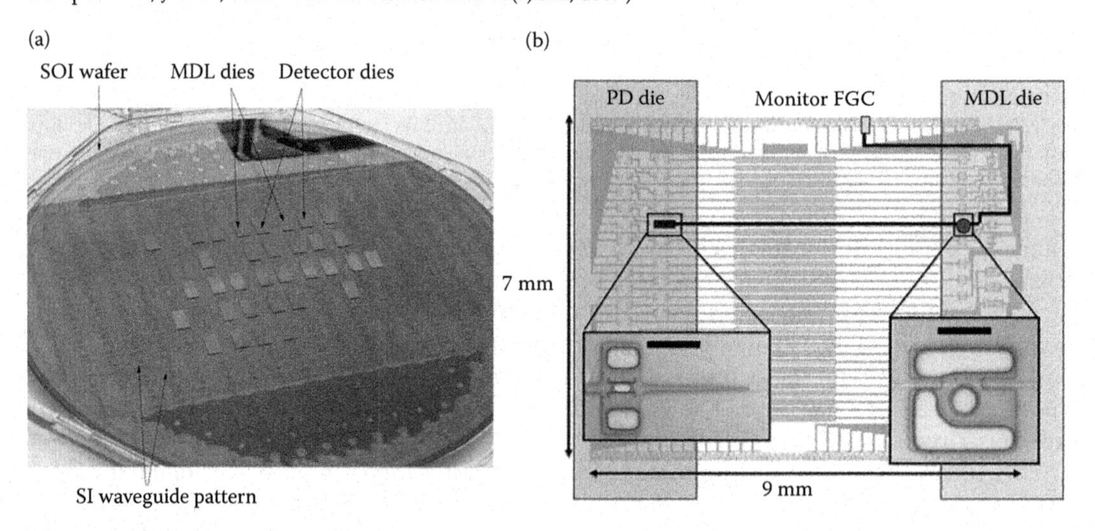

FIGURE 11.53
Laser and PD InP-based dies bonded on a 200-mm SOI wafer (a). Layout of the test chip, showing the position of the bonded laser and PD die, the Si photonic waveguide pattern (background) and the fiber grating couplers at the edges of the chip (b). Pictures of an MDL and a PD taken with an optical microscope before the chip metallization are shown in the inset (scale bar, 10 μm). (Picture taken from Van Campenhout, J. et al., *IEEE Photonics Technol. Lett.* 21(8):522, 2009.)

the output. A SiO_2 layer was deposited on the SOI and III–V wafers to obtain molecular bonding. PDs and MDLs on the same chip were fabricated separately by covering one die with a protective layer of photoresist while processing the other and vice versa. BCB planarization and metallization occurred for MDLs and PDs simultaneously. The mask layout and pictures of fabricated devices are shown in Figure 11.53b.

A link consisting of a 7.5-μm diameter MDL, a 7-mm silicon waveguide, and a 30 × 5-μm PD was tested. To avoid heating, the link is first tested under pulsed electrical operation and the results are shown in Figure 11.54a. Laser threshold is at 55 μA, that is, at peak currents of 700 μA. The PD responsivity was estimated to be 0.33 ± 0.04 A/W.

Under CW electrical current, the laser threshold is approximately 600 μA and the slope efficiency 1.6 μW/mA. Due to self-heating at currents of more than 1.5 mA, the lasing mode shifts to a lower order whispering-gallery mode at 1625 nm and the detected power fluctuates significantly, as shown in Figure 11.54b. Moreover, mode competition causes fluctuation between clockwise and counterclockwise operation.

This proof-of-principle demonstration of an optical interconnect on an SOI chip shows a link efficiency of approximately 0.027%, which can be optimized to efficiencies of approximately 10% at a data rate of 10 Gb/s and higher [97].

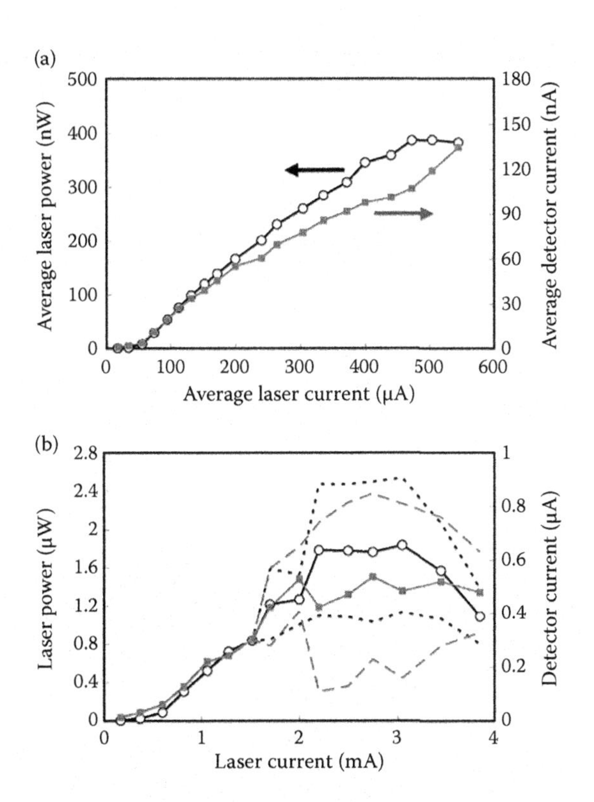

FIGURE 11.54
Optical power arriving at the fiber grating coupler (black) and detector current (gray), measured as a function of laser drive current for the pulsed experiment (a) and the CW experiment (b). The measured optical power (detector current) fluctuated between the values indicated by the dotted (dashed) lines. (Picture taken from Van Campenhout, J. et al., *IEEE Photonics Technol. Lett.* 21(8):522, 2009.)

11.4.3 QW Intermixing

In this section, an approach based on QW intermixing is presented. This technology can be used, for example, to integrate a source with a modulator. A sampled grating (SG) DBR laser integrated with an InGaAsP/InP EAM is shown. Details of this work can be found in Sysak et al. [98].

11.4.3.1 QW Intermixing

The QW intermixing process in this work is based on implant enhanced intermixing in combination with selective removal of an InP buffer [99]. Details of the QW intermixing process and the as-grown hybrid laser III–V base structure are shown in Figure 11.55a through 11.55e. By selectively masking and implanting the QWs, three different bandgaps are created in the wafer.

Figure 11.56 shows photoluminescence spectra from the three bandgaps in the SG-DBR-EAM. Good uniformity of the photoluminescence peak at full-width half-maximum can be seen for all three bandgaps, indicating consistent material quality for the modulator, gain, and mirror regions [100]. The EAM has a 50-nm blue-shifted bandgap and the low-loss mirrors of the SG-DBR have an 80-nm blue-shifted bandgap. It has to be noted that the QW intermixing takes place before wafer bonding and hence it is fully compatible with standard CMOS processing.

In Jain et al. [101], a similar approach was chosen for InGaAsP-based epitaxial material operating at wavelengths of approximately 1.3 μm. Four bandgaps spread over 60 nm are generated across a single chip.

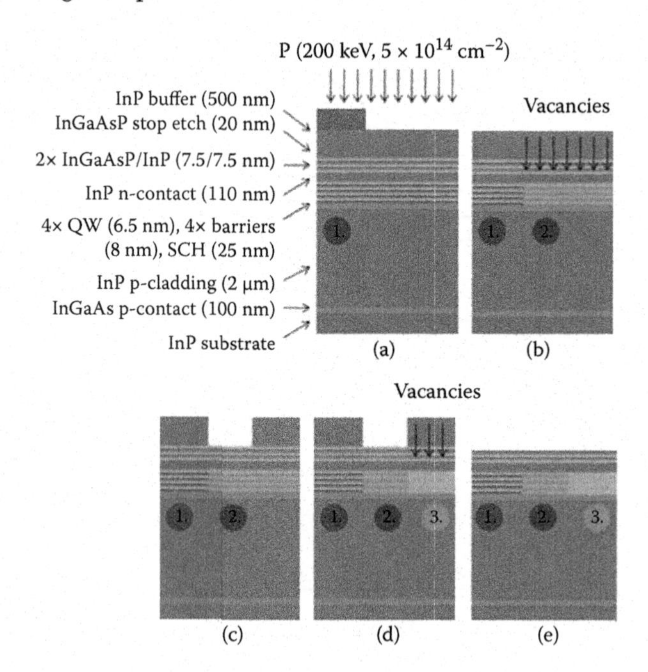

FIGURE 11.55
Overview of the QW intermixing process used for the hybrid laser. The three bandgaps realized are numbered 1, 2, and 3. Implantation of P into InP buffer with SiNx mask to preserve the as-grown bandgap (a). Diffusion of vacancies through QWs and barriers via RTA for bandgap 2 (b). Removal of InP buffer layer to halt intermixing (c). Diffusion of vacancies via RTA for bandgap 3 (d). Removal of InP buffer layer and InGaAsP stop etch layer (e).

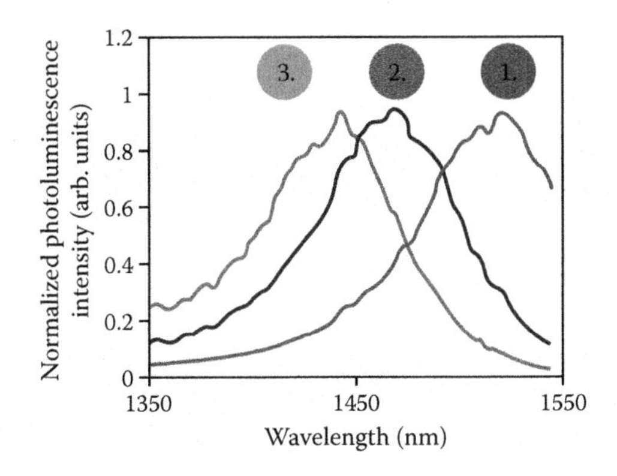

FIGURE 11.56
Normalized photoluminescence spectra from the three bandgaps utilized in the SG-DBR-EAM devices.

11.4.3.2 SG-DBR-EAM Laser

A cross-section of the completed SG-DBR-EAM is shown in Figure 11.57. The grating was etched into the III–V wafer before bonding. The laser portion of the integrated SG-DBR-EAM consists of five electrically isolated sections. A taper is used to transition the optical mode from the hybrid waveguide with both III–V and Si to a purely silicon waveguide. The design of the integrated SG-DBR-EAM uses a 650-μm backside absorber, 760-μm-long rear mirror, an 80-μm-long phase section, a 550-μm-long gain region, a 780-μm-long front mirror, a 200-μm-long EAM, and a 100-μm-long taper.

The laser uses a 20-μm-wide III–V mesa in the gain, mirror, phase, and backside absorber regions and a 4-μm-wide mesa in the modulator. The measured loss and κ of the III–V gratings was 165 cm^{-1} and 3 dB/100 μm, respectively, for a 2-μm-wide Si waveguide and 100 nm etch depth (into the III–V).

The integrated SG-DBR-EAM continuous wave LI characteristics are shown in Figure 11.58. For a device with a 2.5-μm-wide Si waveguide, CW operation is achieved up to 45°C with output power up to 0.5 mW at 10°C and 170 mA of gain current. Dips in the LI characteristics are due to temperature-induced cavity mode hops. With this laser tuning, more than four supermodes can be achieved at wavelengths of 1524, 1518, 1512, and 1554 nm with side mode suppression of more than 35 dB [98].

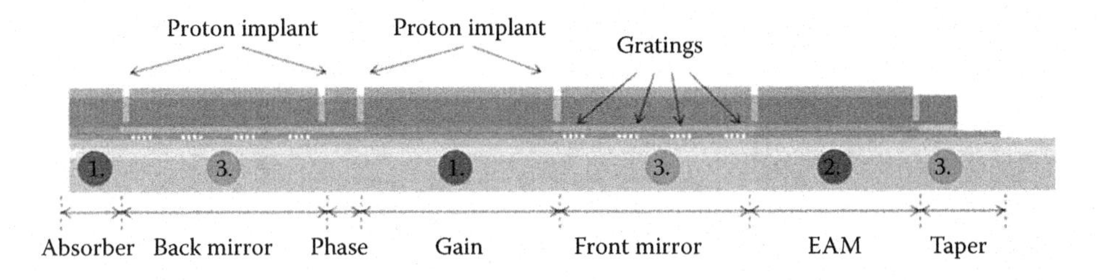

FIGURE 11.57
Cutaway of a hybrid silicon SG-DBR-EAM shown with four front mirror and back mirror grating bursts. Proton implantation is used for electrical isolation between various laser sections. The active, modulator, and passive bandgaps are labeled 1, 2 and 3, respectively.

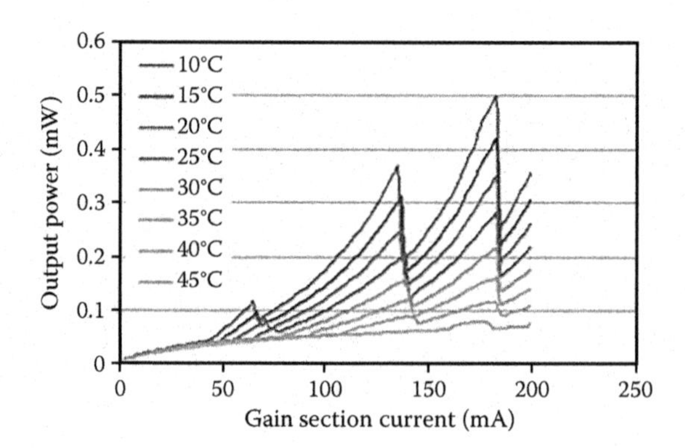

FIGURE 11.58
(See color insert.) Output power-gain current measurement results for SG-DBR at stage temperatures from 10°C to 45°C.

The integrated EAMs show the extinction of more than 5 dB at −6 V reverse bias depending on the wavelength of operation. Shorter wavelengths show more efficient operation than longer wavelengths due to the proximity between the modulator band edge and the operating wavelength. The bandwidth of the integrated modulators depends largely on the applied DC reverse bias achieving greater than 2 GHz with a 6-V bias, as can be seen in Figure 11.59. The series resistance of the modulator is 40 Ω. The large variation in frequency response is due to carrier diffusion effects, as photocurrent is generated outside the 4-μm ridge and diffuses toward the center of the structure. It is possible to improve both the bandwidth and extinction performance of the modulator by increasing the number of QWs in the base structure and using an undercut III–V QW design as in Kuo et al. [72].

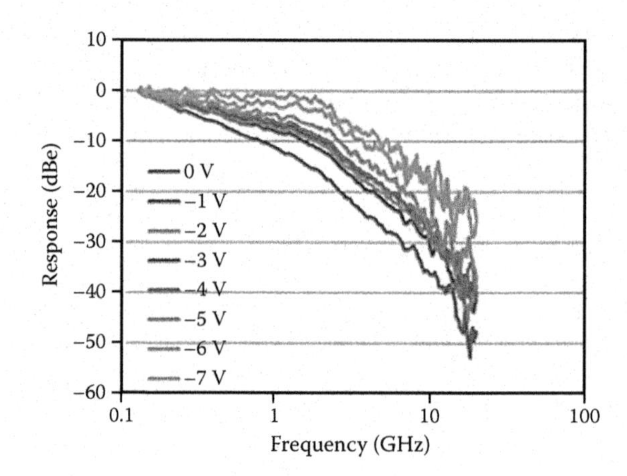

FIGURE 11.59
(See color insert.) Electrical to optical small signal response of integrated EAMs in the SG-DBR-EAM with a 2.5-μm-wide silicon waveguide.

11.4.3.3 Broadband Integrated Hybrid Silicon DFB Laser Array

In Jain et al. [101] a broadband DFB laser array that uses two bandgaps with a 17-nm shift to compensate for gain roll-off is presented. As a result, the range of operating wavelengths is extended with a usable gain bandwidth of more than 90 nm from 1255 to 1345 nm. A picture of such a chip is shown in Figure 11.60.

CW operation is achieved up to 60°C. Single-side output power is 4 mW at room temperature and side-mode suppression is more than 40 dB [101]. Figure 11.61a plots the variation

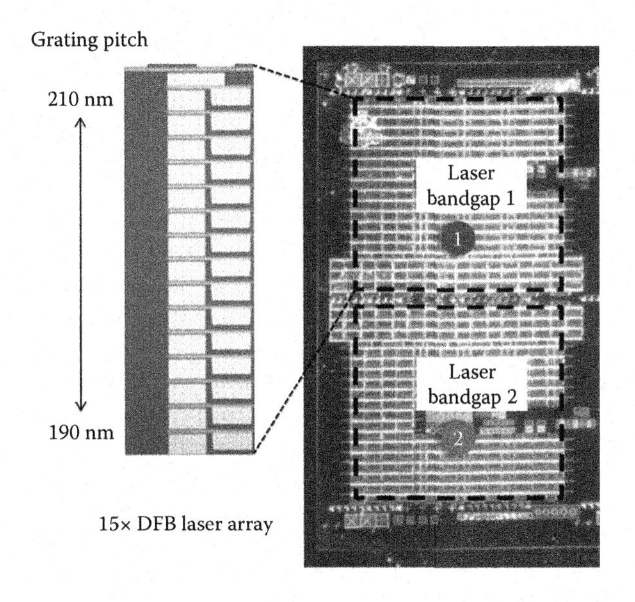

FIGURE 11.60
Top view of the fabricated chip. To the left is a zoomed-in image of 15 DFB laser array with varying grating pitch. Similar laser arrays are defined using laser bandgaps (nos. 1 and 2). (Picture from Jain, S.R. et al., *Optics Exp.* 19(14):13692, 2011.)

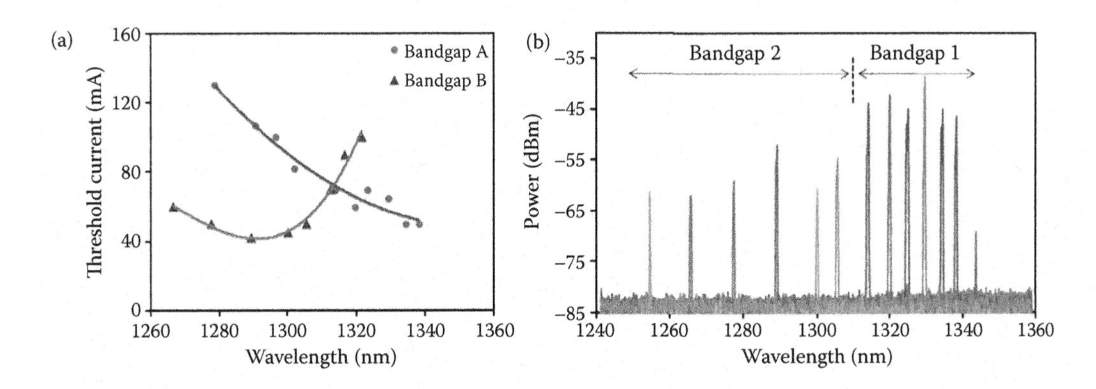

FIGURE 11.61
Variation of threshold current with lasing wavelength of 300-μm-long DFBs (a). Collective spectra of 13 DFB lasers selected from lasers over two bandgaps with 40 to 70 mA threshold current (b). (Picture from Jain, S.R. et al., *Optics Exp.* 19(14):13692, 2011.)

in threshold for all operational 300 μm lasers, with the lasing wavelength using either bandgap 1 or 2 as the gain medium. Lasers over one bandgap (no. 1) could operate at more than 60 nm (1280–1340 nm), with threshold variation from 50 to 125 mA, but with an additional bandgap (no. 2) to compensate for gain roll-off, operating wavelength range can be increased to 75 nm with smaller variation in threshold current (40–70 mA). Figure 11.61b plots the collective lasing spectra of all functioning DFBs over bandgaps 1 and 2 with threshold current between 40 and 70 mA.

11.5 Conclusions and Future Perspectives

11.5.1 Summary of the Current State of Hybrid Silicon Technology

In this chapter, we have reviewed the hybrid (or heterogeneous) integration of III–V materials on silicon photonic circuitry. This effectively adds functionality to the toolbox of silicon photonics that could otherwise only be obtained by III–V-based components and PICs. First, we have reviewed the technologies used for hybrid integration by wafer bonding. Especially SiO_2 to SiO_2 wafer bonding that is a very mature technology used, for example, in SOI wafer fabrication and CMOS image sensor manufacturing. Molecular bonding using ultrathin layers of oxide and DVS-BCB bonding approaches are ubiquitously used in research environments, as highlighted by this chapter; however, the numbers on yield, uniformity, and reproducibility have not yet been quantified. Hybrid silicon technology is currently being commercialized however, so there seems to be a clear development trajectory [102].

Hereafter, we have reviewed the suite of components that have been demonstrated making use of hybrid integration technology. DFB and DBR sources show power output levels of approximately 10 mW, that is, within the typical range commonly obtained with lasers integrated on InP-based PICs. Optical linewidths of a few megahertz were shown, comparable with commercial DFB lasers. The abundance of work on microdisk and microring lasers, having output powers approaching 1 mW and threshold currents of less than 1 mA, shows that hybrid integration is a real enabler for such tightly confined and small-sized optical resonators. EAM and MZM modulator bandwidths of 42 and 27 GHz were obtained, enabling transmission speeds of 50 and 40 Gb/s, respectively. Finally, we reviewed various designs of PDs, which showed bandwidths up to 33 GHz.

A full integration platform requires that all these different components can be integrated on the same PIC. We have discussed three integration schemes. Tapers enable the integration of III–V-based components with SOI circuitry and enable complex PICs, such as fully integrated optical packet buffers and AWG-based multiwavelength lasers. The integration of components that require different epitaxial layer stacks or bandgaps can be achieved by techniques such as QW intermixing and selective die bonding. Successful application of these technologies has enabled the realization of tunable lasers integrated with EAMs, integrated triplexers, and full photonic interconnects on a silicon chip.

Therefore, in summary, it can be stated that III–V hybrid integration on SOI circuitry offers a complete toolbox for high-speed optical links with sufficient optical power levels. The hybrid technology allows for the integration of any combination of these components with silicon photonics, enabling highly integrated and multifunctional PICs.

11.5.2 Comparison with Other Technologies

Although in principle the hybrid silicon integration technology allows for the realization of (complex) PICs, it should also compare favorably to alternative technologies or solutions. In the following section, we give a qualitative analysis of the merits of different available and mature PIC technologies and we compare those to the platform presented in this chapter. We will limit the discussion to the InP-based, SOI and silica-on-silicon platforms because these are the most commonly used.

Hybrid silicon technology is an addition to the silicon photonics toolkit, but the silicon photonics family, or more specifically, silicon-germanium, is already quite complete with high-speed and efficient modulators, germanium and SiGe-based PDs, and recently, also Ge-based sources. It can be argued that for most applications, these PDs and modulators are fast enough. For example, the next generation of 40 and 100 Gb/s optical data communications will probably not require more than 25 Gb/s per channel, speeds that can be obtained with silicon modulators and Ge-based PDs. III–V hybrid integration technology, however, offers far more efficient and far higher-quality electrically pumped sources than are available in SOI technology. This gap in quality seems to be too large to be bridged in the future—if ever. By using grating couplers or other means of spot size conversion, separate laser diodes can be relatively easily bonded to SOI PICs, and this technology is at a commercial stage at the moment.

Silica-on-silicon technology is basically fully complementary to III–V hybrid integration technology, providing high-quality and low-loss passive components. It can be foreseen that this platform is far from a competitor but rather a driver for III–V hybrid technology because both are, in principle, compatible with SOI photonics and with CMOS fabrication technologies [103].

InP-based photonic integration technology has been the technology-of-choice for PIC requiring integration of actives such as lasers, LEDs, and amplifiers. However, because the hybrid silicon platform makes use of a bonded III–V epitaxial layer, there is basically no fundamental reason why there would be a difference in performance between these platforms. The suite of components that have been reviewed shows a performance that comes close to state-of-the-art InP-based PIC performance, in terms of laser output power and linewidths, modulator efficiency and speed, and PD bandwidth. One might even argue that integration approaches such as selective die bonding offer a better alternative to wafer regrowth techniques that are necessary to achieve maximum performance InP PICs. Moreover, silicon substrates allow for efficient fabrication because they have typical diameters of 200 or 300 mm, whereas InP wafers are limited to 150 mm maximum. Another point to mention here is that although the III–V processing obviously cannot be done with CMOS processes, the fact that silicon substrates are used, instead of more fragile InP-substrates, allows for wafer handling technologies that are commonly used in CMOS environments.

There is a clear driver for silicon photonics (and its CMOS compatible family including silica-on-silicon and germanium) in terms of low cost and high quality due to the mature fabrication infrastructure, supported by an annual $45 billion Research and Development (R&D) investment in electronic Integrated circuit (IC) [104]. This technology has all but replaced passive InP-based PICs for telecommunication and datacom applications. So, just like for electronic ICs, the adage for PICs seems to be: if you can make it on silicon in a CMOS fabrication, why do anything else?

Our quick overview has shown that basically, only the gain elements for amplifiers and lasers cannot be made on silicon, and so here lies the great opportunity for III–V hybrid integration. Whether this technology can compete with the established InP-based integration platforms, despite its potential advantages (as outlined in the previous sections), will strongly depend on the applications (as outlined in the next section).

11.5.3 In Which Applications Will Hybrid Silicon Integration Have an Impact?

Although PICs have seen only limited applications in communications, 40 Gb/s (4 × 10 Gb/s) and 100 Gb/s (4 × 25 Gb/s) ethernet standards will change this. Long-term requirements for significant reductions in the cost, size, and power of such transceivers will lead to a broad demand for photonic integration [105]. State-of-the-art PIC technologies that use silica-on-silicon and SOI technologies with bonded laser diodes [20,106] or fully integrated InP-based PICs [106] can meet these requirements. However, hybrid silicon technology is catching up quickly. In Koch et al. [107], a fully integrated hybrid silicon transmitter was presented as part of a complete silicon photonics link. This PIC has four hybrid silicon lasers integrated with a TW optical modulator array operating at 4 × 12.5 Gb/s, as shown in Figure 11.62.

The case for fully integrated transmitters becomes more and more favorable when the channel count is scaled up, and it can be expected that for future terabit ethernet applications, fully integrated solutions will become the technology of choice.

As compared with InP-based PICs, the advantage of hybrid silicon PICs is a matter of economies of scale. Fabrication technologies for both approaches have an approximately identical complexity and such PICs have, in principle, similar functionality and performance. Infinera (Sunnyvale, CA) has set the trend with large-scale integrated InP-based PICs for long-haul telecommunication applications [108]. It is unclear, however, whether this market justifies economies-of-scale with volumes of 1 ks to 10 ks per year [109]. When datacom with volumes of 100 ks to 1 Ms per year moves into the hundreds of gigabits or even terabit range, this volume increase of two orders of magnitude might well topple the balance toward hybrid silicon approaches, with their 300 mm silicon substrates and CMOS compatible fabrication technologies.

Hybrid silicon technology is also an enabler for novel applications. More energy efficient interconnect technologies can have a large effect on global energy consumption. This can be achieved by replacing electronic interconnects with optical ones. IBM has predicted that this might even reach the chip level, that is, having on-chip photonic transmitters and receivers [110]. Obviously, this requires CMOS compatible technologies, and hybrid silicon technology is the prime candidate for this, especially where sources are concerned.

In Figure 11.63, the recently obtained hybrid silicon results are mapped on the roadmap presented by Miller [111]. Components like EAMs are well able to meet energy footprint requirements. The real challenge is to make sources with low enough energy consumption. Hybrid-integrated microdisk and microring lasers are promising candidates to meet these targets, as discussed in this chapter. But they have to compete with technologies like off-chip multiwavelength sources [112]. This battle is twofold though, and even if hybrid silicon technologies prove to be energy efficient, the fabrication yield should be sufficiently high as not to be detrimental to the total electronic IC yield.

The picture becomes more fuzzy in other fields, such as sensor technology, microwave photonics, photonic and RF beam steering, and biomedical applications. PICs for these applications are currently being investigated in academia. Killer applications will, however, be required for PICs to be cost-effective, with one such example being silicon ring-based sensors [113]. Foundry-like efforts such as ePIXfab in Europe [114] and OPSIS in the United States [115] might shift the paradigm however significantly, opening up silicon photonics and, in its wake, hybrid integration technologies to a larger community at reduced costs.

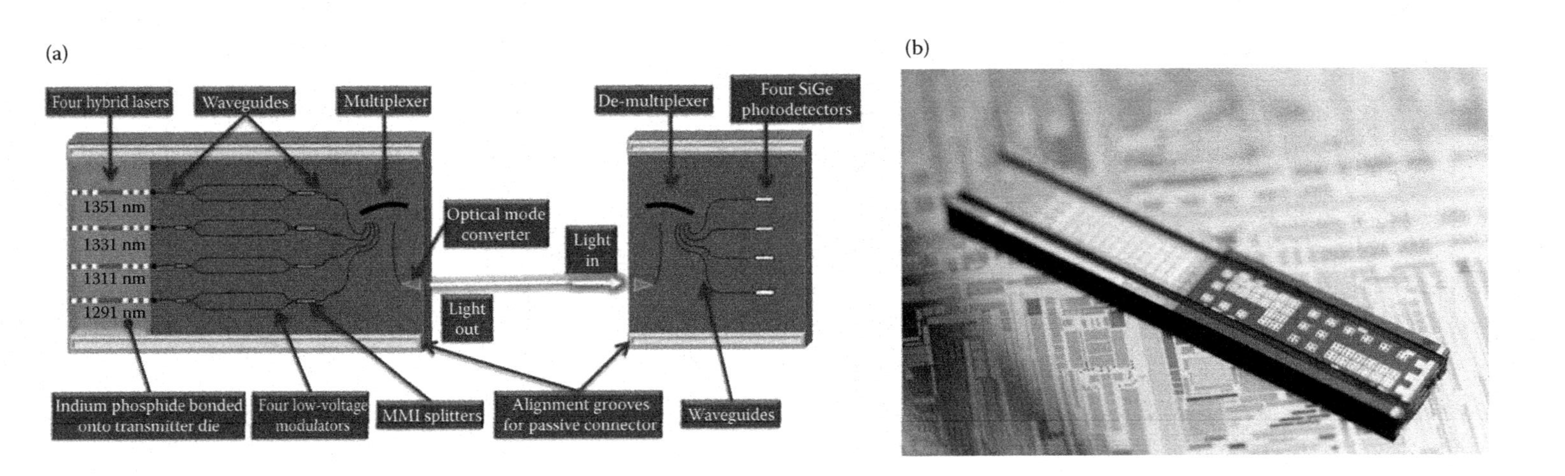

FIGURE 11.62
(a) Schematics of the integrated four-channel CWDM silicon photonics transmitter die (left) and receiver die (right). (b) Picture of the realized transmitter chip. (a, Picture from Koch, B. et al., A 4 × 12.5-Gb/s CWDM Si photonics link using integrated hybrid silicon lasers. In *Proceedings of the CLEO 2011*, CThP5; b, Picture courtesy of Intel.)

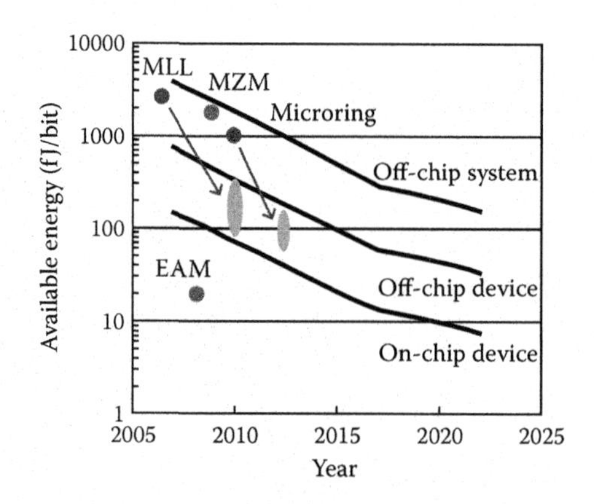

FIGURE 11.63
Estimated roadmap for the available energy for off-chip interconnects (top line) and for off-chip and on-chip optical output devices (middle and bottom lines), as presented by Miller [111]. The hybrid silicon sources presented in this work (circles), their expected performance increase (arrows), and the hybrid silicon modulators (circle) are mapped on this roadmap. (Picture from Heck, M.J.R. et al., *IEEE J. Sel. Top. Quant. Electron.* 17(2):333, 2011.)

11.5.4 Conclusion and Future Trends

In this chapter, we have reviewed the work on hybrid integration in silicon for photonic applications. We have limited the discussion to III–V epitaxial layer integration using wafer bonding. Other materials, however, can also be bonded to silicon for photonic applications. For example, magneto-optic materials can be bonded to make optical isolators in both Mach–Zehnder interferometric [116] and ring configurations [117]. On the other hand, III–V hybrid integration does not only allow for photonic applications but can also basically extend III–V technologies to silicon and silica technology compatible fields such as microfluidics and nanofluidics.

The question remains whether epitaxial growth of III–V layers on silicon substrates becomes a viable alternative for wafer bonding. Although wafer bonding is, in principle, a mature technique, III–V substrates are limited in size and are not compatible with the silicon substrates used commercially. Moreover, the bulk of the III–V wafer is removed and the material is lost. Demonstrations include InGaAs quantum dot lasers [118], InGaSb QW lasers [119], and recently, InAs/GaAs quantum dot lasers monolithically grown on silicon [120]. Techniques like epitaxial layer overgrowth are quite promising in this respect [121]. However, efficient coupling schemes remain a challenge because most of the lasing mode is in the III–V layers.

All in all, we have shown that hybrid silicon integration can expand to silicon photonics toolbox with III–V capabilities, most notably, lasers and amplifiers. This technology might well replace current integration platforms like monolithic InP-based integration technologies for large-volume applications such as datacom. Moreover, micron-sized laser devices in this platform open up possibilities for novel nanophotonic applications, heralding the potential for on-chip photonic interconnects to solve future energy efficiency bottlenecks in computing and communication.

References

1. Park, H., A.W. Fang, D. Liang, Y.-H. Kuo, H-H. Chang, B.R. Koch, H.-W. Chen, M.N. Sysak, R. Jones, and J.E. Bowers. 2008. Photonic integration on hybrid silicon evanescent device platform. *Advances in Optical Technologies* 2008, id. 682978.
2. Liu, A., R. Jones, L. Liao et al. 2004. A high-speed silicon optical modulator based on a metal-oxide-semiconductor capacitor. *Nature* 427(6975):615–618.
3. Marris-Morini, D., X.L. Roux, L. Vivien et al. 2006. Optical modulation by carrier depletion in a silicon PIN diode. *Optics Express* 14(22):10838–10843.
4. Xu, Q., B. Schmidt, S. Pradhan, and M. Lipson. 2005. Micrometre scale silicon electro-optic modulator. *Nature* 435(7040):325–327.
5. Green, W.M.J., M.J. Rooks, L. Sekaric, and Y.A. Vlasov. 2007. Ultra-compact, low RF power, 10 Gb/s silicon Mach–Zehnder modulator. *Optics Express* 15(25):17106–17113.
6. Liao, L., A. Liu, D. Rubin, J. Basak, Y. Chetrit, H. Nguyen, R. Cohen, N. Izhaky, and M. Paniccia. 2007. 40 Gbit/s silicon optical modulator for high-speed applications. *Electronics Letters* 43(22):1196–1197.
7. Liu, A., L. Liao, D. Rubin, J. Basak, Y. Chetrit, H. Nguyen, R. Cohen, N. Izhaky, and M. Paniccia. 2008. Recent development in a high-speed silicon optical modulator based on reverse-biased pn diode in a silicon waveguide. *Semiconductor Science and Technology* 23(6):064001.
8. Roth, J.E., O. Fidaner, R.K. Schaevitz et al. 2007. Optical modulator on silicon employing germanium quantum wells. *Optics Express* 15(9):5851–5859.
9. Ahn, D., C.-Y. Hong, J. Liu et al. 2007. High performance, waveguide integrated Ge photodetectors. *Optics Express* 15(7):3916–3921.
10. Liu, J., X. Sun, R. Camacho-Aguilera, L.C. Kimerling, and J. Michel. 2010. Ge-on-Si laser operating at room temperature. *Optics Letters* 35:679–681.
11. Liang, D., and J.E. Bowers. 2010. Recent progress in lasers on silicon. *Nature Photonics* 4:511.
12. Zhang, L., J. Sinsky, D. Van Thourhout, N. Sauer, L. Stulz, A. Adamiecki, and S. Chandrasekhar. 2004. Low-voltage high-speed travelling wave InGaAsP–InP phase modulator. *IEEE Photonics Technology Letters* 16(8):1831.
13. Klein, H.N., H. Chen, D. Hoffmann, S. Staroske, A.G. Steffan, K.-O. Velthaus. 2006. 1.55 µm Mach–Zehnder modulators on InP for optical 40/80 Gbit/s transmission networks. *Proc. Indiun Phosphide Relat. Mater. Conf.* 171–173.
14. Ohe, H., H. Shimizu, and Y. Nakano. 2007. InGaAlAs multiple-quantum-well optical phase modulators based on carrier depletion. *IEEE Photonics Technology Letters* 19(22):1816.
15. Chaciński, M., U. Westergren, B. Stoltz, L. Thylén, R. Schatz, and S. Hammerfeldt. 2009. Monolithically integrated 100 GHz DFB-TWEAM. *Journal of Lightwave Technology* 27:3410–3415.
16. Kodama, S., T. Yoshimatsu, and H. Ito. 2004. 500 Gbit/s optical gate monolithically integrating photodiode and electroabsorption modulator. *Electronics Letters* 40:555–556.
17. www.u2t.com.
18. Beling, A., and J.C. Campbell. 2009. InP-based high-speed photodetectors. *Journal of Lightwave Technology* 27(3):343.
19. Heck, M.J.R., H.-W. Chen, A.W. Fang, B.R. Koch, D. Liang, H. Park, M.N. Sysak, and J.E. Bowers. 2011. Hybrid silicon photonics for optical interconnects. *IEEE Journal of Selected Topics in Quantum Electronics* 17(2):333.
20. Gunn. C. 2006. CMOS photonics for high-speed interconnects. *Micro, IEEE* 26(2):58–66.
21. Poustie. A. 2008. Hybrid integration for advanced photonic devices. In *Proceedings of SPIE* 7135:713502. doi: 10.1117/12.803089.
22. Liang, D., G. Roelkens, R. Baets, and J.E. Bowers. 2010. Hybrid integrated platforms for silicon photonics. *Materials* 3, 1782–1802. doi:10.3390/ma3031782.

23. Black, A., A.R. Hawkins, N.M. Margalit, D.I. Babic, A.L. Holmes, Y.-L. Chang, P. Abraham, J.E. Bowers, and E.L. Hu. 1997. Wafer fusion: Materials issues and device results. *IEEE Journal of Selected Topics in Quantum Electronics* 3:943–951.
24. Pasquariello, D., C. Hedlund, and K. Hjort. 2000. Oxidation and induced damages in oxygen plasma in situ wafer bonding. *Journal of the Electrochemical Society* 147:2699–2702.
25. Pasquariello, D., M. Lindeberg, C. Hedlund, and K. Hjort. 2000. Surface energy as a function of self-bias voltage in oxygen plasma wafer bonding. *Sensors and Actuators A* 82:239–244.
26. Soitec Home Page. http://www.soitec.com (accessed January 12, 2010).
27. Singh, R., I. Radu, R. Scholz, C. Himcinschi, U. Gosele, and S.H. Christiansen. 2006. Low temperature InP layer transfer onto Si by helium implantation and direct wafer bonding. *Semiconductor Science and Technology* 21:1311–1314.
28. Tan, C.S., A. Fan, K.N. Chen, and R. Reif. 2003. Low-temperature thermal oxide to plasma-enhanced chemical vapor deposition oxide wafer bonding for thin-film transfer application. *Applied Physics Letters* 82:2649–2651.
29. Tong, Q.Y., Q. Gan, G. Fountain, G. Hudson, and P. Enquist. 2004. Low-temperature bonding of silicon-oxide-covered wafers using diluted HF etching. *Applied Physics Letters* 85:2762–2764.
30. Tong, Q.Y., Q. Gan, G. Hudson, G. Fountain, and P. Enquist. 2004. Low temperature InP/Si wafer bonding. *Applied Physics Letters* 84:732–734.
31. Liang, D., D. Chapman, Y. Li, D. Oakley, T. Napoleone, P. Juodawlkis, C. Brubaker, C. Mann, H. Bar, O. Raday, and J.E. Bowers. 2010. Uniformity study of wafer-scale InP-to-silicon hybrid integration. *Applied Physics A* 103(1):213–218.
32. Liang, D., J.E. Bowers, D.C. Oakley, T. Napoleone, D.C. Chapman, C.-L. Chen, P.W. Juodawlkis, and O. Raday. 2009. High-quality 150 mm InP-to-silicon epitaxial transfer for silicon photonic integrated circuits. *Electrochemical and Solid-State Letters* 12:H101–H104.
33. Zhang, X., and J.P. Raskin. 2005. Low-temperature wafer bonding: A study of void formation and influence on bonding strength. *IEEE Journal of Microelectromechanical Systems* 14:368–382.
34. Liang, D., and J.E. Bowers. 2008. Highly efficient vertical outgassing channels for low-temperature InP-to-silicon direct wafer bonding on the silicon-on-insulator (SOI) substrate. *Journal of Vacuum Science & Technology B* 26:1560–1568.
35. Warner, K., D. Oakley, J. Donnelly, C. Keast, and D. Shaver. Layer transfer of FDSOI CMOS to 150 mm InP substrate for mixed-material integration. In *IEEE Conference on Indium Phosphide and Related Materials (IPRM), Princeton, NJ, USA, May 2006*, 226–228.
36. Niklaus, F., G. Stemme, J.Q. Lu, and R.J. Gutmann. 2006. Adhesive wafer bonding. *Journal of Applied Physics* 99:031101.
37. Niklaus, F., H. Andersson, P. Enoksson, and G. Stemme. 2001. Low temperature full wafer adhesive bonding of structured wafers. *Sensors and Actuators A* 92:235–241.
38. Garrou, P., R. Heistand, M. Dibbs, T. Mainal, C. Mohler, T. Stokich, P. Townsend, G. Adema, M. Berry, and I. Turlik. 1993. Rapid thermal curing of BCB dielectric. *IEEE Transactions on Components, Hybrids, and Manufacturing Technology* 16:46–52.
39. Kwon, Y., J. Seok, J.Q. Lu, T.S. Cale, and R.J. Gutmann. 2005. Thermal cycling effects on critical adhesion energy and residual stress in benzocyclobutene bonded wafers. *Journal of the Electrochemical Society* 152:G286–G294.
40. Kwon, Y., J. Seok, J.Q. Lu, T.S. Cale, and R.J. Gutmann. 2006. Critical adhesion energy of benzocyclobutene-bonded wafers. *Journal of the Electrochemical Society* 153:G347–G352.
41. Niklaus, F., R. Kumar, J.J. McMahon, J. Yu, J.Q. Lu, T.S. Cale, and R.J. Gutmann. 2006. Adhesive wafer bonding using partially cured benzocyclobutene for three-dimensional integration. *Journal of the Electrochemical Society* 153:G291–G295.
42. Fang, A.W., H. Park, O. Cohen, R. Jones, M.J. Paniccia, and J.E. Bowers. 2006. Electrically pumped hybrid AlGaInAs-silicon evanescent laser. *Optics Express* 14:9203–9210.

43. Fang, A.W., H. Park, Y.-H. Kuo, R. Jones, O. Cohen, D. Liang, O. Raday, M.N. Sysak, M.J. Paniccia, and J.E. Bowers. 2007. Hybrid silicon evanescent devices. *Materials Today* 10(7–8):28.
44. Zah, C.-E., R. Bhat, B.N. Pathak, F. Favire, W. Lin, M.C. Wang, N.C. Andreadakis, D.M. Huang, M.A. Koza, T.-P. Lee, Z. Wang, D. Darby, D. Flanders, and J.J. Hsieh. 1994. High-performance uncooled 1.3-μm $Al_xGa_yIn_{1-x-y}As$/InP strained-layer quantum-well lasers for subscriber loop applications. *Journal of Quantum Electronics* 30(2):511–523.
45. Sun, X.K., A. Zadok, M.J. Shearn, K.A. Diest, A. Ghaffari, H.A. Atwater, A. Scherer, and A. Yariv. 2009. Electrically pumped hybrid evanescent Si/InGaAsP lasers. *Optics Letters* 34:1345–1347.
46. Stankovic, S., R. Jones, J. Heck, M. Sysak, D. Van Thourhout, and G. Roelkens. 2011. Die-to-die adhesive bonding procedure for evanescently-coupled photonic devices. *Electrochemical and Solid-State Letters* 14(8):H326–H329.
47. Stankovic, S., G. Roelkens, D. Van Thourhout, R. Jones, M. Sysak, and J. Heck. 2010. Evanescently-coupled hybrid III-V/silicon laser based on DVS-BCB bonding. In *15th Annual Symposium of IEEE Photonics Society Benelux Chapter, Netherlands*, 77–80.
48. Roelkens, G., J. Brouckaert, D. van Thourhout, R. Baets, R. Notzel, and M. Smit. 2006. Adhesive bonding of InP/InGaAsP dies to processed silicon-on-insulator wafers using DVS-bis-benzocyclobutene. *Journal of the Electrochemical Society* 153:G1015–G1019.
49. Roelkens, G., D. van Thourhout, R. Baets, R. Notzel, and M. Smit. 2006. Laser emission and photodetection in an InP/InGaAsP layer integrated on and coupled to a silicon-on-insulator waveguide circuit. *Optics Express* 14:8154–8159.
50. Van Campenhout, J., P. Rojo-Romeo, P. Regreny, C. Seassal, D. Van Thourhout, S. Verstuyft, L. Di Cioccio, J.-M. Fedeli, C. Lagahe, and R. Baets. 2007. Electrically pumped InP-based microdisk lasers integrated with a nanophotonic SOI waveguide circuit. *Optics Express* 15(11):6744.
51. Boucart, J., C. Starck, F. Gaborit, A. Plais, N. Bouche, E. Derouin, J.C. Remy, J. Bonnet-Gamard, L. Goldstein, C. Fortin, D. Carpentier, P. Salet, F. Brillouet, and J. Jacquet. 1999. Metamorphic DBR and tunnel-junction injection: A CW RT monolithic long-wavelength VCSEL. *IEEE Journal of Selected Topics in Quantum Electronics* 5:520–529.
52. Maruyama, T., T. Okumura, S. Sakamoto, K. Miura, Y. Nishimoto, and S. Arai. 2006. GaInAsP/InP membrane BH-DFB lasers directly bonded on SOI substrate. *Optics Express* 14:8184–8188.
53. Yariv, A., and X. Sun. 2007. Supermode Si/III-V hybrid lasers, optical amplifiers and modulators: a proposal and analysis. *Optics Express* 15:9147–9151.
54. Yuya, S., M. Tetsuya, Y. Hideki, I.W. Hsieh, and R.M. Osgood, Jr. 2008. Magneto-optical isolator with silicon waveguides fabricated by direct bonding. *Applied Physics Letters* 92:071117.
55. Chen, L., P. Dong, and M. Lipson. 2008. High performance germanium photodetectors integrated on submicron silicon waveguides by low temperature wafer bonding. *Optics Express* 16:11513–11518.
56. Fang, A.W., M.N. Sysak, B.R. Koch, R. Jones, E. Lively, Y.-H. Kuo, D. Liang, O. Raday, and J.E. Bowers. 2009. Single-wavelength silicon evanescent lasers. *IEEE Journal of Selected Topics in Quantum Electronics* 15(3):535–544.
57. Derrickson, D. 1998. *Fiber Optic Test and Measurement*, 185. Upper Saddle River, NJ: Prentice Hall.
58. Akrout et al. Error-free transmission of 8 WDM channels at 10 Gbit/s using comb generation in a quantum dash based mode-locked laser. In *Proceedings of the 34th European Conference on Optical Communication, 2008 (ECOC 2008).*
59. Kovsh, A.R., G.L. Wojcik, D. Yin, A.E. Gubenko, I. Krestnikov, S.S. Mikhrin, and D.A. Livshits. 2010. Cost-effective DWDM optical interconnects enabled by quantum dot comb laser. In *Proceedings of the SPIE Photonics West* [7607-31].
60. Koch, B.R., A.W. Fang, O. Cohen, and J.E. Bowers. 2007. Mode-locked silicon evanescent lasers. *Optics Express* 15(18):11225–11233.

61. Fang, A.W., B.R. Koch, K-G. Gan, H. Park, R. Jones, O. Cohen, M.J. Paniccia, D.J. Blumenthal, and J.E. Bowers. 2008. A racetrack mode-locked silicon evanescent laser. *Optics Express* 16(2):1393–1398.
62. Liang, D., M. Fiorentino, T. Okumura, H.-H. Chang, D. T. Spencer, Y.-H. Kuo, A.W. Fang, D. Dai, R.G. Beausoleil, and J.E. Bowers. 2009. Electrically-pumped compact hybrid silicon microring lasers for optical interconnects. *Optics Express* 17(22):20355–20364.
63. Fang, A.W., B.R. Koch, R. Jones, E. Lively, D. Liang, Y.-H. Kuo, and J.E. Bowers. 2008. A distributed Bragg reflector silicon evanescent laser. *IEEE Photonics Technology Letters* 20(20):1667–1669.
64. Morthier, G., R. Kumar, P. Méchet, T. Spuessens, L. Liu, K. Huybrechts, G. Roelkens, J. Van Campenhout, D. Van Thourhout, and R. Baets. 2010. Microdisk lasers heterogeneously integrated on silicon for low-power, high-speed optical switching. *IEEE Photonics Newsletter,* 5.
65. Spuesens, T., L. Liu, T. De Vries, P. Rojo-Romeo, P. Regreny, D. Van Thourhout. 2009. Improved design of an InP-based microdisk laser heterogeneously integrated with SOI. In *6th IEEE International Conference on Group IV Photonics, United States*, p. FA3.
66. Van Campenhout, J., P. Rojo-Romeo, D. Van Thourhout, Ch. Seassal, P. Regreny, L. Di Cioccio, J.-M. Fedeli, and R. Baets. 2007. Thermal characterization of electrically injected thin-film InGaAsP microdisk lasers on Si. *Journal of Lightwave Technology* 25(6):1543.
67. Liu, L., R. Kumar, K. Huybrechts, T. Spuesens, G. Roelkens, E.-J. Geluk, T. de Vries, P. Regreny, D. Van Thourhout, R. Baets, and G. Morthier. 2010. An ultra-small, low-power, all-optical flip-flop memory on a silicon chip. *Nature Photonics* 4:182–187.
68. Raz, O., L. Liu, D. Van Thourhout, P. Rojo-Romeo, J.M. Fédéli, and H.J.S. Dorren. 2009. High speed wavelength conversion in a heterogeneously integrated disc laser over silicon on insulator for network on a chip applications. In *Proceedings of the 35th European Conference on Optical Communication (ECOC 2009) 20–24 September 2009, Vienna.*
69. Van Campenhout, J., L. Liu, P.R. Romeo, D. Van Thourhout, C. Seassal, P. Regreny, L. Di Cioccio, J.M. Fedeli, and R. Baets. 2008. A compact SOI-integrated multiwavelength laser source based on cascaded InP microdisks. *IEEE Photonics Technology Letters* 20(16):1345–1347.
70. Roelkens, G., L. Liu, D. Liang, R. Jones, A. Fang, B. Koch, and J.E. Bowers. 2010. III–V/silicon photonics for on-chip and intra-chip optical interconnects. *Laser & Photonics Reviews* 1–29.
71. Fukano, H., T. Yamanaka, M. Tamura, and Y. Kondo. 2006. Very-low-driving-voltage electroabsorption modulators operating at 40 Gb/s. *Journal of Lightwave Technology* 24:2219–2224.
72. Kuo, Y.-H., H.-W. Chen, and J.E. Bowers. 2008. High speed hybrid silicon evanescent electroabsorption modulator. *Optics Express* 16(13):9936–9941.
73. Tang, Y., H.-W. Chen, S. Jain, J.D. Peters, U. Westergren, and J.E. Bowers. 2011. 50 Gb/s hybrid silicon traveling-wave electroabsorption modulator. *Optics Express* 19(7):5811.
74. Chen, H.-W., Y.-H. Kuo, and J.E. Bowers. 2010. 25Gb/s hybrid silicon switch using a capacitively loaded traveling wave electrode. *Optics Express* 18(2):1070–1075.
75. Chen, H.-W., J.D. Peters, and J.E. Bowers. 2011. Forty Gbs hybrid silicon MZm with low chirp. *Optics Express* 19(2):1455.
76. Miller, D.A.B. 2009. Device requirements for optical interconnects to silicon chips. *Proceedings of the IEEE* 97(7):1166–1185.
77. Liu, L., J. Van Campenhout, G. Roelkens, R.A. Soref, D. Van Thourhout, P. Rojo-Romeo, P. Regreny, C. Seassal, J.-M. Fédéli, and R. Baets. 2008. Carrier-injection-based electro-optic modulator on silicon-on-insulator with a heterogeneously integrated III-V microdisk cavity. *Optics Letters* 33:2518–2520.
78. Brouckaert, J., G. Roelkens, D. Van Thourhout, and R. Baets. 2007. Compact InAlAs–InGaAs metal–semiconductor–metal photodetectors integrated on silicon-on-insulator waveguides. *IEEE Photonics Technology Letters* 19(19):1484.
79. Brouckaert, J., G. Roelkens, D. Van Thourhout, and R. Baets. 2007. Thin-film III–V photodetectors integrated on silicon-on-insulator photonic ICs. *Journal of Lightwave Technology* 25(4):1053–1060.

80. Binetti, P.R.A., X.J.M. Leijtens, T. de Vries, Y.S. Oei, L. Di Cioccio, J.-M. Fedeli, C. Lagahe, J. Van Campenhout, D. Van Thourhout, P.J. van Veldhoven, R. Nötzel, and M.K. Smit. 2010. InP/InGaAs photodetector on SOI photonic circuitry. *IEEE Photonics Journal* 2(3):299.
81. Di Cioccio, L. 2006. New result obtained at LETI on 3D heterostructures. In *210th Meeting of The Electrochemical Society, Cancun, Mexico, Oct. 29–Nov. 3, 2006*, 602, Abstract No. 1649.
82. Binetti, P., R. Orobtchouk, X. Leijtens, B. Han, T. de Vries, Y. Oei, L. Di Cioccio, J.-M. Fedeli, C. Lagahe, P. van Veldhoven, R. Nötzel, and M. Smit. 2009. InP-based membrane couplers for optical interconnects on Si. *IEEE Photonics Technology Letters* 21(5):337–339.
83. Roelkens, G., J. Brouckaert, D. Taillaert, P. Dumon, W. Bogaerts, D. Van Thourhout, R. Baets, R. Nötzel, and M. Smit. 2005. Integration of InP/InGaAsP photodetectors onto silicon-on-insulator waveguide circuits. *Optics Express* 13:10102–10108.
84. Liu, L., J. Brouckaert, G. Roelkens, D. Van Thourhout, and R. Baets. 2009. Compact, wavelength-selective resonant photodetector based on III–V/silicon-on-insulator heterogeneous integration. Presented at the *2009 Conference on Lasers and Electro-Optics/Quantum Electronics and Laser Science Conference (CLEO/QELS 2009)*, Baltimore, MD. Paper CTuV3.
85. Van Thourhout, D., T. Spuesens, S. Kumar Selvaraja, L. Liu, G. Roelkens, R. Kumar, G. Morthier, P. Rojo-Romeo, F. Mandorlo, P. Regreny, O. Raz, C. Kopp, and L. Grenouillet. 2010. Nanophotonic devices for optical interconnect. *IEEE Journal of Selected Topics in Quantum Electronics* 16(5):1363.
86. Park, H., Y.-H. Kuo, A.W. Fang, R. Jones, O. Cohen, M.J. Paniccia, and J.E. Bowers. 2007. A hybrid AlGaInAs-silicon evanescent preamplifier and photodetector. *Optics Express* 15(21): 13539–13546.
87. Chang, H.-H., Y.-H. Kuo, H.-W. Chen, R. Jones, A. Barkai, M.J. Paniccia, and J.E. Bowers. 2010. Integrated triplexer on hybrid silicon platform. *Proc. Optical Fiber Communication Conference*, San Diego, California, March 21, 2010, paper OThC (accepted for publication).
88. Binetti, P., X. Leijtens, T. de Vries, Y. Oei, O. Raz, L. Di Cioccio, J.-M. Fedeli, C. Lagahe, R. Orobtchouk, J. Van Campenhout, D. Van Thourhout, P. van Veldhoven, R. Nötzel, and M. Smit. 2008. Indium phosphide based membrane photodetector for optical interconnects on silicon. In *Proceedings of the LEOS Annual, Newport Beach, CA, Nov. 2008*, 302–303, Paper TuT4.
89. Binetti, P.R.A., X.J.M. Leijtens, A.M. Ripoll, T. de Vries, E. Smalbrugge, Y.S. Oei, L. Di Cioccio, J.-M. Fedeli, C. Lagahe, R. Orobtchouk, D. Van Thourhout, P.J. van Veldhoven, R. Nötzel, and M.K. Smit. 2009. InP-based photodetector bonded on CMOS with Si3N4 interconnect waveguides. In *Proceedings of the 2009 IEEE LEOS Annual Meeting Conference, (LEOS '09) October 4–8, 2009, Belek-Antalya.*
90. Park, H., A.W. Fang, O. Cohen, R. Jones, M.J. Paniccia, and J.E. Bowers. 2007. An electrically pumped AlGaInAs-silicon evanescent amplifier. *IEEE Photonics Technology Letters* 19:230–232.
91. Kurczveil, G., M.J.R. Heck, J.D. Peters, J.M. Garcia, D. Spencer, and J.E. Bowers. 2011. An integrated hybrid silicon multiwavelength AWG laser. *IEEE Journal of Selected Topics in Quantum Electronics* 17(6):1521–1527.
92. Pintus, P., M.J.R. heck, G. Kurczveil, and J.E. Bowers. 2011. Low-loss hybrid silicon tapers. In *Proceedings of the Group IV Photonics, London.*
93. Park, H., J.P. Mack, D.J. Blumenthal, and J.E. Bowers. 2008. An integrated recirculating optical buffer. *Optics Express* 16(15):11124.
94. Burmeister, E.F., and J.E. Bowers. 2006. Integrated gate matrix switch for optical packet buffering. *IEEE Photonics Technology Letters* 18:103–105.
95. Chang, H.-H., Y.-H. Kuo, R. Jones, A. Barkai, and J.E. Bowers. Integrated hybrid silicon triplexer. *Optics Express.*
96. Chang, H.-H., A.W. Fang, M.N. Sysak, H. Park, R. Jones, O. Cohen, O. Raday, M.J. Paniccia, and J.E. Bowers. 2007. 1310 nm silicon evanescent laser. *Optics Express* 15(18):11466.

97. Van Campenhout, J., P.R.A. Binetti, P. Rojo Romeo, P. Regreny, C. Seassal, X.J.M. Leijtens, T. de Vries, Y-S. Oei, P.J. van Veldhoven, R. Nötzel, L. Di Cioccio, J.-M. Fedeli, M.K. Smit, D. Van Thourhout, and R. Baets. 2009. Low-footprint optical interconnect on an SOI chip through heterogeneous integration of InP-based microdisk lasers and microdetectors. *IEEE Photonics Technology Letters* 21(8):522.
98. Sysak, M.N., J.O. Anthes, J.E. Bowers, O. Raday, and R. Jones. 2008. Integration of hybrid silicon lasers and electroabsorption modulators. *Optics Express* 16(17):12478–12486.
99. Skogen, E.J., J.S. Barton, S.P. Denbaars, and L.A. Coldren. 2002. A quantum-well-intermixing process for wavelength-agile photonic integrated circuits. *IEEE Journal of Selected Topics in Quantum Electronics* 8:863–869.
100. Nie, D., T. Mei, H.S. Djie, M.K. Chin, X.H. Tang, and Y.X. Wang. 2005. Implementing multiple bandgaps using inductively coupled argon plasma enhanced quantum well intermixing. *Journal of Vacuum Science & Technology B* 23:1050–1053.
101. Jain, S.R., M.N. Sysak, G. Kurczveil, and J.E. Bowers. 2011. Integrated hybrid silicon DFB laser-EAM array using quantum well intermixing. *Optics Express* 19(14):13692.
102. Alduino, A. et al. 2010. Demonstration of a high speed 4-channel integrated silicon photonics WDM link with hybrid silicon lasers. In *Integrated Photonics Research, Silicon and Nanophotonics*, OSA Technical Digest (CD) (Optical Society of America), paper PDIWI5.
103. Yamada, K., T. Tsuchizawa, T. Watanabe, H. Fukuda, H. Shinojima, H. Nishi, S. Park, Y. Ishikawa, K. Wada, and S. Itabashi. 2011. Silicon photonic devices and their integration technology. In *Optical Fiber Communication Conference*, OSA Technical Digest (CD) (Optical Society of America), paper OWQ6.
104. Liang, D., and J.E. Bowers. 2009. Photonic integration: Si or InP substrates? *Electronics Letters* 45:12.
105. Cole, C., B. Huebner, and J.E. Johnson. 2009. Photonic integration for high-volume, low-cost applications. *IEEE Communications Magazine* S16.
106. Kato, K., and Y. Tohmori. 2000. PLC hybrid integration technology and its application to photonic components. *IEEE Journal of Selected Topics in Quantum Electronics* 6(1):4.
107. Koch, B. et al. A 4 × 12.5 Gb/s CWDM Si photonics link using integrated hybrid silicon lasers. In *Proceedings of the CLEO 2011*, CThP5.
108. Nagarajan, R. et al. 2005. Large-scale photonic integrated circuits. *IEEE Journal of Selected Topics in Quantum Electronics* 11(1):50–65.
109. Finisar – C. Cole - http://www.parallaxgroup.com/media/cole_ucsb.pdf (accessed December 28, 2012).
110. Benner, A., M. Ignatowski, J.A. Kash, D.M. Kuchta, and M.B. Ritter. 2005. Exploitation of optical interconnects in future server architectures. *IBM Journal of Research and Development* 49(4/5):755–775 (http://www.zurich.ibm.com/st/photonics/interconnects.html).
111. Miller, D.A.B. 2009. Device requirements for optical interconnects to silicon chips. *Proceedings of the IEEE* 97(7):1166–1185.
112. Ahn, J., M. Fiorentino, R.G. Beausoleil, N. Binkert, A. Davis, D. Fattal, N.P. Jouppi, M. McLaren, C.M. Santori, R.S. Schreiber, S.M. Spillane, D. Vantrease, and Q. Xu. 2009. Devices and architectures for photonic chip-scale integration. *Applied Physics A* 95:989–997.
113. Iqbal, M., M.A. Gleeson, B. Spaugh, F. Tybor, W.G. Gunn, M. Hochberg, T. Baehr-Jones, R.C. Bailey, and L.C. Gunn. 2010. Label-free biosensor arrays based on silicon ring resonators and high-speed optical scanning instrumentation. *IEEE Journal of Selected Topics in Quantum Electronics* 16(3):654.
114. http://www.epixfab.eu/ (accessed December 28, 2012).
115. http://depts.washington.edu/uwopsis/ (accessed December 28, 2012).
116. Shoji, Y., T. Mizumoto, H. Yokoi, I.-W. Hsieh, and R.M. Osgood, Jr. 2008. Magneto-optical isolator with silicon waveguides fabricated by direct bonding. *Applied Physics Letters* 92:071117.
117. Tien, M.-C., T. Mizumoto, P. Pintus, H. Kromer, and J.E. Bowers. 2011. Silicon ring isolators with bonded nonreciprocal magneto-optic garnets. *Optics Express* 19(12):11740.

118. Mi, Z., P. Bhattacharya, J. Yang, and K.P. Pipe. 2005. Room-temperature self-organised In0.5Ga0.5As quantum dot laser on silicon. *Electronics Letters* 41(13):742–744.
119. Balakrishnan, G., S.H. Huang, A. Khoshakhlagh, P. Hill, A. Amtout, S. Krishna, G.P. Donati, L.R. Dawson, and D.L. Huffaker. 2005. Room-temperature optically-pumped InGaSb quantum well lasers monolithically grown on Si(100) substrate. *Electronics Letters* 41(9):531–532.
120. Wang, T., H. Liu, A. Lee, F. Pozzi, and A. Seeds. 2011. 1.3-μm InAs/GaAs quantum-dot lasers monolithically grown on Si substrates. *Optics Express* 19(12):11381–11386.
121. Olsson, F., M. Xie, S. Lourdudoss, I. Prieto, and P.A. Postigo. 2008. Epitaxial lateral overgrowth of InP on Si from nano-openings—Theoretical and experimental indication for defect filtering throughout the grown layer. *Journal of Applied Physics* 104, 093112.

12

Fabrication of Silicon Photonics Devices

Francisco López Royo

CONTENTS

12.1 Introduction

12.1.1 Chapter Contents

This chapter has been written on the assumption that the reader is approaching the subject of silicon photonics with a basic background in science or engineering, perhaps even some specific knowledge in the field of photonics, but not necessarily any previous knowledge of microfabrication. It attempts to complement the rest of the book by presenting a basic foundation on micro-nanofabrication technology, materials, and processes, and a discussion of the most challenging aspects involved. Most other chapters focus instead on specific photonic devices, although they may also briefly discuss some aspects of their fabrication.

To this end, it starts by providing a broad overview of clean rooms (Figure 12.1) and the basic microfabrication technology employed in the fabrication of the devices discussed in other chapters (Section 12.2). With this initial section, it is thus hoped that, despite the limited space available, the microfabrication concepts, processes, and equipment tools most likely to be encountered elsewhere in this book, in the technical literature of silicon photonics and certainly in the other sections of this chapter, are thereby briefly introduced.

Next, in Section 12.3, the processing and properties of the various silicon based material platforms are reviewed. Section 12.4 describes a generic silicon-on-insulator (SOI) process flow and a few specific examples focusing on novel techniques. Lastly, Section 12.5 discusses some of the main issues encountered in the fabrication of silicon photonic devices.

That target audience described above is thought to be numerous and therefore the inclusion of Sections 12.2 and 12.4.1 should make this chapter of benefit to a larger number of

FIGURE 12.1
A technologist or process engineer wearing clean room apparel is visually inspecting a 6-in. wafer in an ISO Class 5 clean room. (All clean room photographs and SEM micrographs in this chapter, unless specified otherwise, courtesy of the Nanophotonics Technology Center, Universitat Politècnica de València, Spain [1].)

people. On the other hand, readers who already have a solid background in microfabrication, whether for photonics, microelectronics, or microsystems, may want to skip Section 12.2 entirely and perhaps also the first part of Section 12.4, although the subsection on innovative process flows might also be of interest to them. Sections 12.3 and 12.5 provide balance by being more specialized, reviewing the latest technical literature in this field.

It should be noted that formulas providing formal definitions can be found in most books on microfabrication but have been deliberately avoided here, with an aim to focus on a more practical discussion of the general concepts and on fabrication procedures. Also, as in other technical fields, the use of acronyms in microfabrication is widespread, but an effort has been made to ensure the full names are provided at least the first time they are introduced.

12.1.2 CMOS-Compatible Fabrication

Students first approaching the field of silicon photonics from any background other than electronic engineering or semiconductor physics might be intrigued by the ubiquity of a term in the technical literature that may be new to them: CMOS compatibility. They might get the impression that this seems to be a highly desirable tag that most silicon photonics research groups are eager to attach to their devices and fabrication processes. CMOS stands for Complementary Metal Oxide Semiconductor, and is currently the main technology being employed for producing electronic integrated circuits.

The reason is as simple as it is appealing: economics. The very *raison d'être* of silicon photonics is the promise it holds of offering the desirable features of photonics technology, such as high performance and potentially low heat dissipation, at the same low cost (for reasonable volumes) of silicon microelectronics by using the existing technology and vast infrastructure of CMOS manufacturing. Therefore, it should come as no surprise that people take great care to ensure that any fabrication processes involved in the fabrication of their proposed photonic devices are compatible with CMOS processing.

Additionally, CMOS compatibility is required for monolithic integration, in which both electronic and photonic devices are manufactured on the same wafer. Indeed, for most applications, the coexistence of photonics with electronics on the same chip is required (there are exceptions, e.g., disposable photonic chips for some sensing applications), and therefore some degree of integration is needed. In general, for very high volumes, the higher the level of integration, the lower the cost, and hence monolithic integration, is attractive. However, it has also been argued that photonic–electronic monolithic integration is not always necessarily the optimum solution, as it may be difficult or impossible to optimize the fabrication parameters for both the electronic integrated circuit and the photonic integrated circuit simultaneously [2]. Moreover, it might actually make better economic sense to always employ the latest CMOS technology for the electronic integrated circuit (IC) for lowest cost and energy consumption whereas utilizing older generation processes and (potentially amortized) *fabs* for photonics, which does not necessarily require better than 90 nm lithography.

Either way, even in the case in which photonics and electronics are not monolithically integrated, some level of CMOS compatibility will be required to ensure that those fabs (a common name for semiconductor manufacturing facilities) can indeed be used to produce the photonic ICs with minimal modifications to the available processes and equipment to achieve low cost.

So what exactly does "CMOS-compatible" mean? Depending on the context, it can refer to several concepts and to varying levels of stringency. Ideally, no change at all should be brought to the CMOS manufacturing line so that photonics devices would be fabricated seamlessly within a standard microelectronics process flow at no extra cost, either financial or technological. As it turns out, this is hardly realistic, hence the more reasonable target is to try and change as few aspects as possible (or as little as possible), while analyzing the likely effects of these changes on the yield, throughput, reliability, and cost of the fabrication processes or any constraints on the specifications of the final product. Hence, the dependency on the particular context: in each case, a state of minimal modifications is sought before declaring "compatibility." The reader should bear in mind that even when no modifications to a standard CMOS process are reported, the very SOI substrate with 2 μm buried oxide typically used in silicon photonics is not fully compatible with the latest CMOS technology on SOI [3]. Furthermore, 92% of CMOS logic production on 300 mm wafers uses conventional silicon wafers rather than SOI [4].

From the point of view of concepts, CMOS-compatible may be referring to substrates, fabrication materials, process steps, or process flows. Some materials are simply not allowed in a CMOS manufacturing line (e.g., gold or III–V compound substrates) as they diffuse into silicon, creating energy levels that will ruin device performance whereas others (e.g., polymers) would not withstand even the lowest (back-end) processing temperatures. Yet, the list of CMOS-compatible materials has grown over time from just a handful a few decades ago, thus providing an example of differing levels of stringency toward what is considered compatible. Another example of this is provided by the fact that some academic laboratories implement color codes to differentiate equipment and items such as wafer cassettes and tweezers used for III–V wafers otherwise used in the same premises as their silicon counterparts, whereas a small production facility may actually assign a separate zone for these materials, and larger silicon fabs may not even consider keeping these materials under the same roof.

In terms of process steps, some are not generally considered CMOS compatible merely because they are not currently employed in a standard CMOS process (e.g., the lift-off process) whereas some others are not compatible for a specific reason, such as being incompatible with the expected throughput (e.g., electron beam lithography; EBL) or because its introduction may alter, in unexpected ways, other processes and therefore would need thorough feasibility studies.

Regarding process flows, some processes, which are part of a CMOS flow, cannot be used at certain other steps along that process flow, such as high-temperature processes after metallization. A fabrication process flow for photonic devices that contravenes any of these rules would not be deemed CMOS compatible.

In general, most of the materials, processes, and process flows mentioned in this chapter are either CMOS compatible or can be easily modified to be CMOS compatible. More detailed information on this subject, including the systems' perspective, may be found in the chapter dedicated to photonic–electronic integration.

12.1.3 Access to Silicon Photonics Fabrication

When people speak or write about the low-cost production possibilities enabled by CMOS compatibility, it is not always made sufficiently clear to novice readers that they are referring to the price per chip for very high volumes. Semiconductor equipment machinery is very expensive to purchase and maintain, and so are clean rooms. For small volumes (in semiconductor manufacturing, a few tens of thousands of devices, which may be contained in a few hundred wafers, is considered a very small volume), the price per device will certainly not be as low, if a suitable foundry willing to take your request can be found at all, something which should not be given for granted, as the author can attest.

When just a few devices are required to test a new design idea, the cost can actually be prohibitive for a small company or research group. To tackle this issue, (at least) a couple of multiproject wafer initiatives for sharing the cost of production runs using common design rules have been set up in the last few years. They are perhaps modeled in a similar arrangement in microelectronics, which emerged in the late 1970s and became today's MOSIS organization [5]. The longest running service is ePIXfab [6], organized by the european research institutes Interuniversity Microelectronics Centre (IMEC) and Laboratoire d'Électronique et de Technologies de l'Information (LETI) and open to users worldwide, established as a structural multiproject wafer service since 2006 [7]. More recently, a similar service has been organized in the United States [8].

These services typically operate several shuttle runs at different times each year and are an excellent option open to researchers and fabless companies for fabricating a small but significant number of devices with critical dimensions attainable by deep ultraviolet (DUV) lithography. For higher resolution or quicker turnaround, another option, which is particularly attractive when only a few samples are needed, is that of EBL fabrication services. These are offered by a few research institutes such as Cornell Nanofabrication Facility (CNF) in the United States [9], Valencia Nanophotonics Technology Center (NTC) in Europe [1], and a few commercial organizations.

12.1.4 Word of Warning on Safety Issues

It should be stressed that some of the gases and chemicals involved in silicon processing are extremely hazardous. This chapter is intended to be an introduction to the subject of microfabrication of silicon photonic devices from a descriptive point of view. Proper training should be sought before attempting any of the laboratory work described in this chapter.

12.2 Silicon Microfabrication/Nanofabrication Technology

The process of creating microstructures on a planar substrate is called *patterning*. Patterning is often a subtractive process and, in its most basic form, involves deposition of the material for making the microstructures, lithography to cover and protect selected areas (the pattern) of that material with a second sacrificial material (often resist) while leaving other parts unprotected, selective etching to remove the intended material from the unprotected areas, and finally, stripping to remove the sacrificial material. These three main steps to patterning, deposition, lithography, and etching and other microfabrication processes will be briefly reviewed in this section after a short introduction to clean rooms, where all these operations take place (Figure 12.1).

In silicon photonics, the starting material is often the top silicon layer in SOI wafers and, in this case, the initial deposition step does not apply. For waveguide-based structures, the basic steps described above are often followed by the deposition of an upper cladding material, usually silicon dioxide (SiO_2), mainly to protect the waveguides and, following planarization, to allow for further layers of patterning to be realized on top.

Patterning can also be an additive process (such as electroplating or the lift-off process), in which no etching is involved and lithography comes before deposition, leaving open regions where the desired material will be deposited whereas the remaining area stays covered with resist. See Figure 12.2 for a simplified schematic of a fabrication

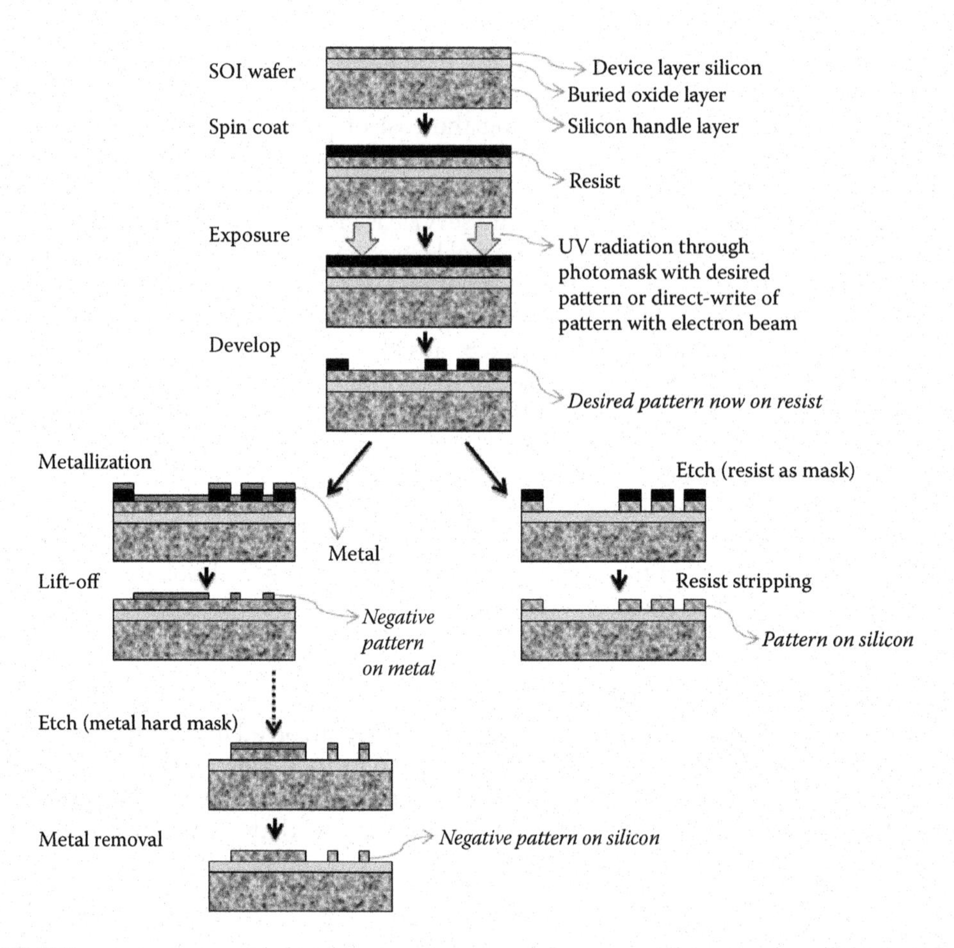

FIGURE 12.2
A simplified schematic of a microfabrication process route illustrating both a lift-off process (route to the left) and the standard etching process (route to the right). The lift-off process may also be employed as part of an etching process (bottom). Baking steps are not shown.

process flow. Whereas this section focuses on describing each of the individual process *steps*, a detailed description of a generic process *flow* is presented in Section 12.4.1.

12.2.1 Clean Rooms

In the world of microstructures and nanostructures, a foreign particle of just a few microns is actually comparatively large and can have destructive consequences. Also, even the lowest concentrations of unwanted materials may be regarded as contamination as they can easily have an effect on the optical or electrical behavior of the fabricated structures, such as propagation loss in optical waveguides due to metallic impurities.

For these reasons, the utmost care must be taken during the fabrication stages to keep wafers clean and to avoid contamination. Ideally, the microfabrication equipment is housed in a clean room where all the fabrication operations are performed. This is invariably the case for a production environment in which high-standard clean rooms are the norm (ISO 3–ISO 5, defined in the next paragraph), just as in the microelectronics or microelectromechanical systems (MEMS) industry. For small research laboratories, the high cost of building and maintaining a clean room may impose the cheaper option of a lower standard clean room or even minienvironments, in which each piece of fabrication equipment is enclosed in plastic curtains and a large flow of clean air is forced inside. In this case, wafers are only taken out of their protective boxes or cassettes within these enclosed areas when being transported from one process to the next.

A clean room, as the name implies, is a room in which all surfaces are specially treated to avoid the generation of particles, are maintained clean, and the amount of airborne particles is kept to a minimum by constantly recirculating clean air (Figures 12.1 and 12.3). To this end, a proper clean room must be purposely built to a given design and must be equipped with powerful air circulation systems capable of pushing large quantities of air through several stages of filtering, the final stage being high efficiency filters at the clean room ceiling or wall. Clean rooms are classified according to standards which set the maximum particulate concentration limits for each clean room class. For instance, in the ISO 14644 standard, ISO Class 3 has a limit of 1000 suspended particles larger than 0.1 μm per cubic meter and ISO Class 4 has a limit of 10,000. These are roughly equivalent to Class 1 and Class 10 respectively in the US standard, which is based on the number of particles larger than 0.5 μm per cubic foot. The clean room's air-handling system must also control parameters such as temperature, humidity, and pressure within a tight range. These parameters are also important in microfabrication because, for example, temperature may affect certain chemical processes whereas humidity has an effect on photoresists, as will be discussed later. A certain percentage of air being circulated is taken from outside the clean room to refresh it and to maintain a slightly higher pressure in the clean room with respect to surrounding areas to prevent contaminated air from entering the room. All items that may be introduced to the clean room (including furniture and clothing) must be especially designed to minimize particle generation. As an illustrative example, standard paper is not allowed inside a clean room: notebooks must be made of a special clean room-certified paper.

There are two main types of clean rooms, conventionally ventilated and unidirectional airflow clean rooms. In a conventionally ventilated clean room, the air change rate is usually between 20 and 60 air changes per hour (compared with 2 to 10 in a standard office or shop), although the important parameter which defines its performance is the volume of air supplied per unit of time. This must be sufficiently high to offset the particles being

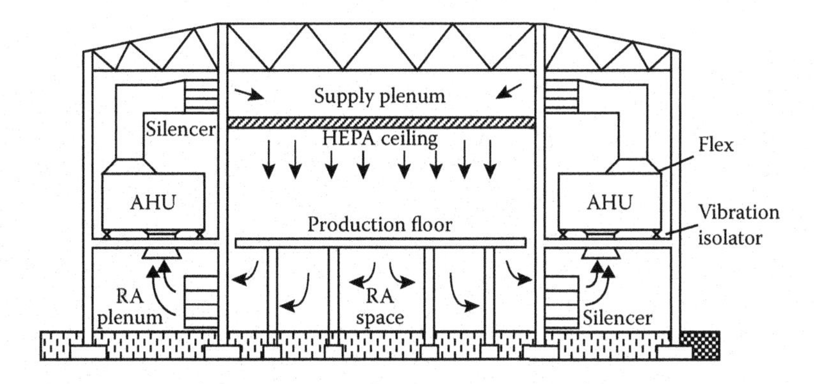

FIGURE 12.3
A typical clean room configuration. RA, return air; AHU, air-handling unit. (Whyte, W.: *Cleanroom Technology. Fundamentals of Design, Testing and Operation*. 2010. Copyright Wiley-VCH Verlag GmbH & Co. KGaA. Reproduced with permission.)

generated by people and machinery, which in the case of particles larger than 0.5 μm, this can be millions per minute even when proper apparel is worn. For higher standard clean rooms (around ISO 5 and better), regions where turbulence is created must be avoided for more effective removal of particles. This is achieved in unidirectional (or laminar) airflow clean rooms in which air filters make up the entire ceiling (or sometimes an entire wall) and air is aspirated from the floor (or opposite wall), thus the air is flowing in one direction only. In this type of clean room, the air volume supplied is still 1 or 2 orders of magnitude higher; however, the most important parameter in this case is air velocity, which must be kept between 0.3 and 0.45 m/s to ensure that particles do not deposit on surfaces at air obstructions [10]. The volume of air that needs to be circulated and the number of filters required are the main reasons why unidirectional flow clean rooms are more expensive to build and maintain than conventional ones.

Clean rooms are desirable for a wide range of purposes, from operating theaters in hospitals to paint shops in car factories. Higher-standard clean rooms are needed by the biotechnology and medical device industry, whereas the absolute highest standards (ISO 3 and better) are necessary in the semiconductor industry. Apart from the general clean room concept described above, each industry may have specific requirements such as bacterial contamination control in the former or ultrapure water supply and chemical waste handling systems in the latter. Silicon photonics boasts of its ability to be compatible with standard CMOS processing and, as such, does not have any additional requirements to those of the microelectronics industry in terms of clean room specifications.

12.2.2 Wafer Cleaning Procedures

A lot of effort has been devoted in the semiconductor industry to developing effective wafer cleaning procedures. Clean surfaces free of contaminants are paramount in microfabrication for many reasons, including film adhesion properties, topography integrity, device performance, and keeping fabrication tools clean and contaminant-free to avoid subsequent contamination. In very basic terms, it is easy to understand that when the device being fabricated is comparable in size to, if not smaller than, a foreign body in its vicinity, it may be damaged or otherwise affected even if that foreign body may be a tiny particulate from our (human sized) point of view. As seen in the previous subsection, this is in fact one of the reasons for the need of clean rooms for microfabrication, aside from other considerations such as controlled temperature, humidity, and clean air free of chemical contaminants.

Standard cleaning procedures involve immersing the wafers in chemicals such as sulfuric acid, ammonium hydroxide, hydrogen chloride, and hydrogen fluoride sometimes diluted in hydrogen peroxide or deionized water in specific mixtures, sequences, and lengths of time. The *RCA* or *SC1/SC2* cleaning procedure (standard cleaning 1 and 2, developed at the laboratories of Radio Corporation of America) is one of the better known cleaning procedures and has been proven to remove organic residues, alkali ions, particulates, and metals from the wafer surface. It is of particular importance prior to furnace diffusion processes. Another well-known cleaning procedure is the so-called *piranha clean* (also known as sulfuric-peroxide mixture), which is a strong oxidizer and is particularly useful for removing organic contamination such as stubborn photoresist stains. These are all fairly standard cleaning procedures in the semiconductor industry and so the recipes for the chemical mixtures and the bath times will not be reproduced here as they may be easily found on the Internet or in any microfabrication reference book.

12.2.3 Silicon Oxide Growth

Among the materials mentioned in the next subsection dealing with deposition, silicon dioxide deserves particular attention as a material not only for its significant role in silicon photonics but also for its uniqueness in that it may not only be deposited but also grown by simply oxidizing silicon in a high-temperature furnace. The oxidation process yields a high-quality oxide (in terms of absence of impurities, density, hydrogen content, etc.) which is often called *thermal oxide* and grows very slowly (~48 h for a 1-μm-thick oxide at 1100°C) as the oxygen molecules must diffuse through the oxide layer being grown to continue oxidizing the underlying silicon. This growth process is called *dry oxidation* in contrast to an alternative process in which water molecules are introduced (e.g., with an oxyhydrogen flame inside the furnace) which, being smaller than the oxygen molecule, can diffuse more easily thus resulting in a faster *wet oxidation* (albeit still slow at ~130 min for the same conditions as above).

Wet oxides are of slightly inferior quality than dry oxides, but are very similar for most purposes and still higher quality (and slower to obtain) than deposited silicon dioxide. Nevertheless, the rationale to use growth or deposition will usually be given not by speed and material choice alone but by many other considerations: deposition can be performed at much lower temperatures, it can be potentially applied to specific regions on the wafer, and it does not affect the underlying topography. Conversely, oxide growth must be performed at high temperatures, it will occur unselectively at any silicon surface reached by the oxidizing molecules (and hence may affect structures and surfaces other than the intended ones including, of course, the back of the wafer) and because the thickness of the newly produced oxide is approximately 2.2 times that of the consumed silicon, any affected topography will deform, perhaps in an undesirable fashion.

It is interesting to note that silicon wafers kept in standard conditions always have a thin silicon oxide layer on their surface known as *native oxide*. This layer may be removed by a dip in diluted hydrogen fluoride, which is seen to "slide off" the surface of the wafer in a peculiar way as it is taken out of the liquid because bare silicon is hydrophobic. If the wafer is then kept in air, it will again quickly grow a native oxide of approximately 1.5 to 2 nm.

12.2.4 Deposition

Regardless of the intended application of the microfabricated structures—whether microelectronics, MEMS, photonics, or others—high-quality uniform thin-film materials with minimized imperfections and very low contaminant levels are required. The semiconductor industry has optimized deposition technologies to achieve this goal using high-purity source materials, which are generally gases but sometimes vaporized liquids and solids, often in a low pressure or vacuum environment. These technologies can be classified according to the reaction mechanism as either chemical vapor deposition (CVD) or physical vapor deposition (PVD). There are other deposition methods, such as spin-on-glass coating and electroplating, which have specific applications in microfabrication.

In CVD, vapor-phase reactants transported near a heated substrate are adsorbed to its surface where they react forming a solid film. There are various types of CVD processes depending on whether this reaction occurs at atmospheric pressure (APCVD), in low pressure chambers (LPCVD), or in a setup in which plasma provides some of the energy required for the chemical reactions as in plasma-enhanced CVD (PECVD; Figure 12.4a).

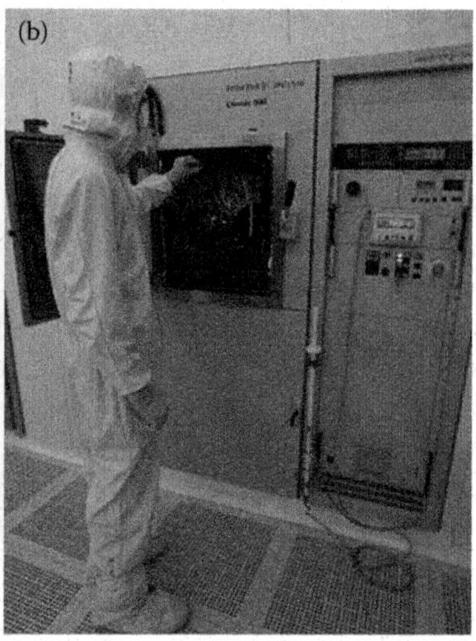

FIGURE 12.4
Two multi-chamber PECVD cluster tools in the foreground, an ion implanter in the background in a ISO Class 7 clean room area (a). Note the reduced number of air filters on the ceiling and perforated floor tiles as compared with the ISO Class 5 area shown in Figures 12.1 and 12.4b. A process engineer operating a PVD system (evaporator) (b).

APCVD has the advantage of being lower temperature than LPCVD, hence it can be used with substrates sensitive to high temperature, and is comparatively simple (Figure 12.14 in Section 12.4.1). LPCVD offers higher purity, uniformity, throughput and better step coverage (also referred to as conformality), which is the ability to deposit the same thickness in vertical and horizontal surfaces (illustrated also in Figure 12.14). The more complex PECVD affords most of the advantages of LPCVD except for film quality at even lower temperatures than APCVD, by providing some of the energy required for the chemical reactions through a radiofrequency-induced plasma. Some common materials deposited by CVD include amorphous or polycrystalline silicon; dielectrics such as silicon dioxide, which may be deposited pure or "doped" with minor concentrations of boron, phosphorus, or germanium; and also silicon nitride or oxynitride (SiON), the refractive index of which may be tuned from that of SiO_2 to that of Si_3N_4, which makes it especially useful for designing and fabricating taper couplers (Figure 12.5).

In PVD, the material to be deposited is introduced in solid form in a vacuum chamber in which it is evaporated by heating or ejected (sputtered) by high-energy ion impacts, creating a plume of vaporized material that condenses on the substrate. PVD, whether through evaporation or sputtering, is often selected for materials that are difficult or impractical to deposit through CVD such as metals and metal oxides. Another advantage of PVD is the possibility of depositing on a low-temperature substrate, such as at room temperature (Figure 12.4b).

Deposition by evaporation yields very high purity films because it is performed in high vacuum ($<10^{-3}$ Pa). The source material may be heated by a tungsten resistor or through electron beam bombardment resulting in vaporization. The latter method is usually preferred because higher temperatures can be achieved and also because the highest temperature point is always localized in the source material, thus avoiding contamination from the crucible or from the tungsten filament.

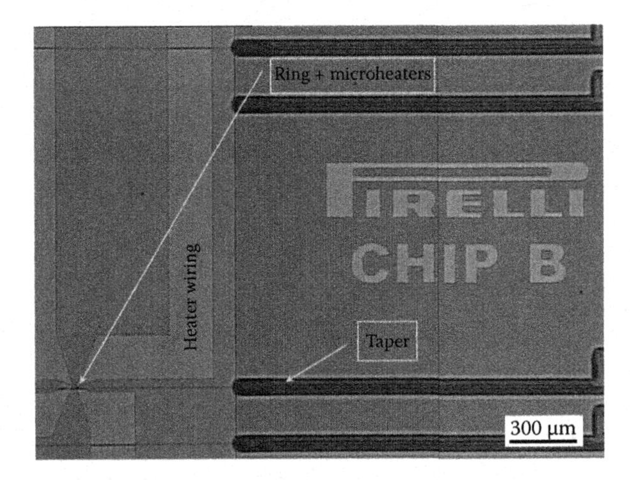

FIGURE 12.5
A chip fabricated in 2007 by silicon photonics pioneer Pirelli Labs (Milan, Italy) involving seven mask levels, more than 60 process steps, and a critical dimension of 130 nm. (Courtesy of Pirelli Labs.)

If the source material were to be bombarded by a downpour of heavy ions instead of a beam of electrons, rather than melting and evaporating, target atoms become dislodged (or sputtered) from the source material as a result of the collisions. An entirely different equipment setup is needed for this purpose: a *sputtering* system. The sputtering process maintains a plasma of ionized argon atoms between the target and the substrate. The target is kept at sufficiently low voltage with respect to the rest of the chamber and receives the impacts of the Ar^+ ions from the plasma, which sputters the target atoms, forming the vapor that will then travel and condense on the substrate.

Standard direct current (DC) sputtering requires that the target be a conductor, such as metals and silicides. Radio frequency (RF) sputtering uses a high-frequency alternating voltage (usually 13.56 MHz) to overcome this limitation, allowing sputtering of virtually any material. Also, in RF sputtering a lower pressure can be used, as the plasma is maintained more efficiently.

Except in cases wherein either the ion impacts or minor argon contamination can be a problem for the substrate, sputtering has advantages over evaporation, and is more commonly used in modern microfabrication. Some materials cannot be evaporated whereas others are more easily sputtered (e.g., metals with a high melting point). Sputtering affords better uniformity and better process control, especially when depositing alloys (because of the different vapor pressures of the constituent metals) or for reactive PVD (e.g., nitrides and oxides of metals). It yields better step coverage because the target atoms land at wider angles due to collisions with argon atoms. It also allows *sputter etching* of the substrate prior to deposition, which may be useful for certain processes, for example, to condition the substrate surface or to remove native oxides without breaking the vacuum.

Manufacturers of deposition systems include Applied Materials, Novellus (recently acquired by Lam Research) both based in Silicon Valley (California), and SPTS (Newport, UK). The latter is a company formed by mergers and acquisitions over the years of some well-known brand names in microfabrication such as Watkins-Johnson, STS, Trikon Technologies, Thermco, SVG, ASML Thermal, and Aviza.

12.2.5 Optical Lithography and Resists

We may think of optical lithography (or photolithography) as a process akin to photography, wherein an image is transferred from the photomask to a photoresist-coated wafer.

The image, in this case, is of course the pattern that we wish to eventually reproduce on the wafer (or the negative thereof). This pattern is delineated with metal lines, usually chromium, on a transparent substrate, usually fused-quartz (or soda lime in lower-end applications) called the *photomask* or simply the mask, which is generally ordered from specialized suppliers who fabricate it by laser beam or electron beam writing. EBL is discussed in the next subsection.

The photoresist is basically a photosensitive polymer, the solubility of which changes when exposed to ultraviolet (UV) radiation. By coating the wafer with a thin layer of photoresist and then shining UV light through the mask onto the coated wafer, only the regions on the wafer which are not shadowed by the chromium on the mask are exposed to UV radiation, and only those sections undergo the chemical process that changes their solubility. A heat treatment is then performed on the exposed wafer to complement the chemical process. Finally, continuing with the analogy to photography, the exposed resist is *developed* in a special solvent that will dissolve regions in the resist leaving the desired pattern on the wafer.

Wafers are generally coated in a spin-coater (although other techniques such as spray coating can also be employed) by dispensing the resist on the wafer placed on a rotating chuck. Spinning starts initially at a low speed, such as 200 rpm for the resist to spread all over the wafer, and then speeds up at a precisely controlled acceleration to a few thousand revolutions per minute, thus forcing excess resist out of the rotating wafer, leaving a thin uniform layer, the thickness of which is dictated by the speed of rotation for a given resist viscosity. Part of the solvent evaporates during spinning and part is driven off the wafers in a subsequent bake in an oven or hot plate at temperatures of approximately 80°C to 150°C. Wafers are then exposed to UV radiation, subjected to a further bake (postexposure bake), and then developed.

Photoresists fall into two main categories: positive and negative tone resists. In the former, the exposed regions dissolve during development as the UV radiation breaks down the sensitizer, which holds the molecules together, whereas in the latter, the chemistry of the resist is such that the unexposed sections dissolve while the exposed regions harden during exposure and baking (through the creation of molecular cross-links) becoming much less soluble, remaining in place after development. The most important properties that characterize a resist include sensitivity, contrast, etch resistance, line edge roughness (LER), viscosity, and shelf-life. When a photoresist is exposed to UV light, there is a threshold irradiation dose (energy per unit area) before which no change in resist solubility occurs. Then, the resist starts becoming soluble and at a slightly higher dose, a complete change in solubility is achieved that results in all of the exposed resist to dissolve (positive resists) or to stay (negative resists) during development. That dose is defined as the *sensitivity* of the resist. High sensitivity or fast resists (i.e., lower required dose) are generally desirable for a faster lithographic process, although it is important to note, in particular for silicon photonics applications, that higher sensitivities are associated with higher LER and lower dimensional control. As discussed in Section 12.5.2, the ability to produce smooth photoresist walls is of particular importance in silicon photonics, as any roughness will be transferred to our photonic structures (e.g., waveguides); hence, having an effect on propagation losses due to radiation scattering.

The contrast of a resist is related to the difference between the two dose levels mentioned previously, such that the smaller that difference, the higher the contrast. Thus, higher contrasts are indicative of a more binary relationship between dose and solubility. This is generally beneficial as it can be correlated to higher resolution and more vertical sidewalls, except in certain applications, such as gray scale lithography, in which a more linear relationship is necessary to create three-dimensional (3-D) structures. As discussed later in the subsection dedicated to etching, photoresists are often used as etch masks, and in that case, high etch resistance is also a desirable property for resists because it affords better etching performance or allows thinner resist layers to be used, which can improve resolution. A resist's viscosity, which is inversely proportional to the amount of solvent it contains, will determine the spin speed required to obtain a certain thickness. Finally, because photoresists tend to age quickly, a longer shelf-life is important for the resist to be useful for a longer span of time, especially for sporadic use in research laboratories in which several different resists may be in use in a given period. As the resist stability will depend on the storage conditions (mainly temperature and exposure to light), it is not uncommon for the properties of a resist to vary significantly even before its nominal expiry date. Therefore, after a long period since a resist was last used or in demanding applications, dose tests may be necessary before the definitive lithography.

Traditionally, mercury lamps have been used as sources of UV radiation for optical lithography. As the microelectronics industry has been shrinking the size of transistors to increase performance and reduce cost over the past few decades, optical lithography has gradually migrated to lower wavelengths to avoid the limits to the available resolution set by the laws of diffraction. Thus, the g-line in the mercury emission spectrum at 436 nm was soon replaced first by the i-line at 365 nm, and then by 248 nm, thus entering the so-called DUV region, reaching 193 nm technology, which is still in use today in state-of-the-art fabs, albeit with certain enhancements such as immersion lithography. This transition to lower wavelengths has brought about the introduction of excimer laser sources (KrF for 248 nm and ArF for 193 nm), which have replaced the mercury lamp for higher irradiation power. Before the widespread use of laser sources, there was a need to develop photoresists with higher sensitivities at DUV wavelengths because the radiation at those wavelengths from mercury lamps was too weak. Hence, *chemically amplified* resists were developed for that reason by IBM in the 1980s. These resists only require a relatively low exposure dose to activate a chemical that provokes a catalytic chain reaction that takes place during the postexposure bake, thus enhancing the resist's sensitivity. With the move from 248 nm to 193 nm lithography, resists based on aromatic polymers, which had a good performance at 248 nm, were found to have too much absorption at 193 nm; hence, other photoresists based on acrylate polymers such as poly(methyl methacrylate) or PMMA were developed their main drawback being their low etch resistance. Finally, with the advent of immersion lithography, new resists have been developed that maintain good stability in contact with the immersion liquid (often water).

Our interest being to create micrometer- or even nanometer-sized structures, resolution is of course a critical factor by which the performance of any lithographic process must be judged. The highest resolution that can be achieved by a particular process is often expressed in terms of the minimum linewidth in a pattern of lines and spaces of equal width (rather than in an isolated line in which overexposure or underexposure might be affecting the result). In fact, resolution is, along with overlay alignment accuracy for successive layers, one of the two main performance metrics in general for lithographic processes.

An important issue in submicron optical lithography is that of *proximity effects*, which manifest themselves when the features are of similar size to the illuminating wavelength

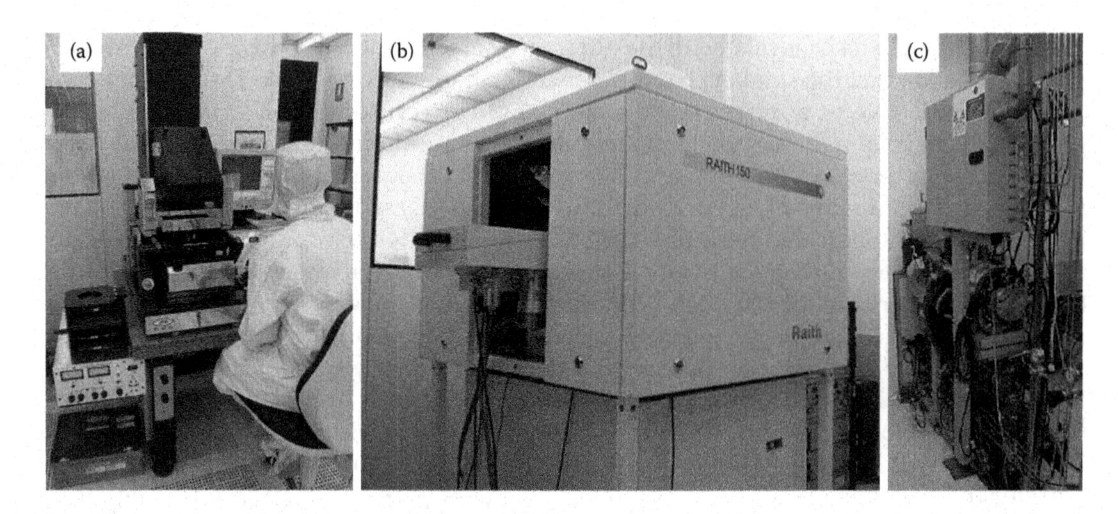

FIGURE 12.6
A technologist operating a mask aligner (a). Partial view of an e-beam system, showing the loadlock chamber on the left for introducing wafers (b). Backside of an etch tool showing various gas lines coming into the gas box for precise delivery into the chamber by mass flow controllers (c).

and are due to intensity variations because of constructive or destructive interference. These fidelity errors become apparent when features are in proximity of each other, including parts of the same feature such as two lines forming a corner. What is meant by fidelity errors is that the exact position of the edges of those features in the resist, once developed, will differ from that intended. They can be compensated to a certain extent by predistorting feature dimensions on the mask through so-called optical proximity correction (OPC) algorithms.

Depending on the application requirements, three main operating modes are possible in optical lithography: contact, proximity, and projection. In contact mode, as the name implies, the mask and the coated substrate are brought into intimate contact for best resolution; however, this leaves a residue on the mask and shortens its life. In proximity mode, a microscopic gap (from a few microns to 50 μm) separates the coated wafer from the mask, hence avoiding the drawbacks of contact mode and being more suited for production, although the maximum resolution that can be achieved is not as high. In projection lithography, which requires a different equipment setup, an image of the mask is projected onto the wafer through free space optics. Projection lithography is thus superior as it is noncontact and achieves high resolution by reducing the mask pattern through the projection optics at the expense of a more complex and more expensive system. These systems are called *steppers* or *scanners*. In the former, the image is projected on separate regions on the wafer sequentially or in "steps," whereas in the latter, the image is projected continuously on a moving wafer. The prices of these systems have experienced a tremendous increase in the last couple of decades as the technology has become more complex and is fast approaching the $100 million mark for a state-of-the-art system. Manufacturers include ASML (the Netherlands), and Nikon and Canon (Japan). For contact and proximity lithography, the system used is called a *mask aligner* (Figure 12.6a). The typical price for this type of system might be approximately half a million dollars, although total cost can vary greatly with tool capabilities and options. Mask aligner manufacturers include EVG of Austria and SUSS MicroTec (previously Karl Suss) of Germany.

12.2.6 Electron Beam Lithography

There is a more elementary way to expose the resist than that described above and that is to use a precisely controlled beam of either laser light or, as discussed here, energetic electrons. As opposed to exposure through a mask, which is parallel in nature, this lower level of series writing allows the patterning of any arbitrary shape using apposite software for accurate control of the position, scan speed, size, and intensity of the beam. This is in fact the technology used to fabricate the photomasks used in standard optical lithography.

Although the main use of EBL in the semiconductor industry is precisely that of mask writing, its use for direct patterning of substrates is widespread in research laboratories due to its flexibility and its superior resolution capabilities. Direct-write EBL, as it is commonly referred to when used for this purpose, will now be described, whereas a direct comparison with optical lithography and more details on its capabilities are presented in Section 12.5.3.

Electron beam (or e-beam) lithography (EBL) originates from scanning electron microscopy (SEM), where an energetic e-beam impinges upon an object and bouncing electrons are used to provide an image of nanoscopic scale features, thus circumventing the limits that the laws of diffraction pose on electromagnetic-radiation-based microscopy because the wavelength of electrons at typical EBL energies are on the atomic scale. EBL uses basically the same principles used in SEM systems to steer a beam of energetic electrons to expose special e-beam resist, which changes its solubility when exposed to their impacts (akin to the way photoresist behaves when exposed to incident UV radiation). In fact, many e-beam systems commercialized today are based on modified SEM columns. Figure 12.6b shows a partial view of such an EBL system.

Although electron diffraction is not a problem in EBL, other issues are present such as electrons backscattered from the substrate and from the resist itself, which limits resolution and produce proximity effects. This is because backscattered electrons from neighboring structures at short distances will add to the nominal dose received by any particular point in the resist. In EBL, they can be minimized via process characterization by applying a bias to specific features or structures and also by adjusting the dose. Some e-beam systems make use of automatic algorithms, which adjust exposure dose to compensate for proximity effects. Other strategies involve employing lower electron energies overall or positioning dummy structures to avoid proximity effects in critical areas.

As an example, Figure 12.7a shows a photonic crystal structure in which all holes are exposed with identical nominal dose. Proximity effects cause the holes at the top and bottom rows of each block to be smaller than those in the bulk because they only receive backscattered electrons from one side. Likewise, the holes in the leftmost and rightmost columns are larger than those in the bulk because they are receiving extra backscattered electrons from the trenches. Figure 12.7b shows a similar structure in which proximity effects have been successfully corrected.

Common e-beam resists used for silicon photonics applications include PMMA (positive tone) and hydrogen silsesquioxane or HSQ (negative tone). They are both capable of very high resolution patterns. PMMA has the drawback of having low etch resistance but, being a positive resist, it is often used for the lift-off process (described subsequently), assisted also by the fact that the resist profile resulting from EBL is generally undercut due to electron backscattering, that is, the open regions are wider at the bottom (contrary to the general case in optical lithography) which is advantageous for lift-off. HSQ is discussed in some detail in Section 12.5.3.

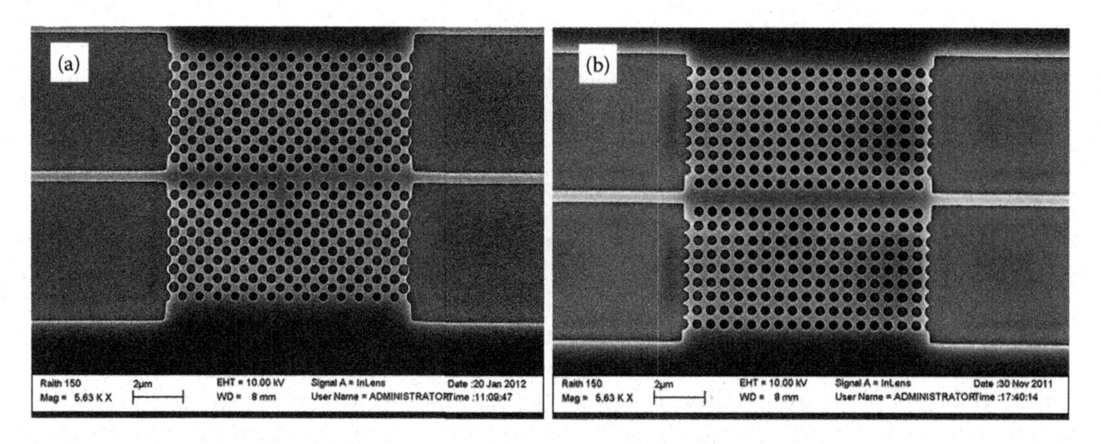

FIGURE 12.7
Proximity effects: all holes are exposed to the same nominal dose (a), resulting in different sizes depending on the presence or absence of neighboring structures. Proximity effects have been successfully compensated (b). (Courtesy of Dr. Amadeu Griol; Nanophotonics Technology Center, Universitat Politècnica de València, Valencia, Spain.)

E-beam systems can be categorized as either Gaussian or variable-shape beams as well as raster or vector scan. In variable-shape e-beam systems, the beam passes through an aperture in the electron optics generally of square shape. This shape allows these systems to expose corners accurately with relatively large beams, which in turn makes them faster and are the usual choice for mask making. Gaussian beams on the other hand have a smaller beam, which allows them to reach a higher resolution. In raster scan systems, the beam is continuously scanned along the entire write-field and a blank is used to block the beam except in regions where it is meant to expose the resist, whereas newer vector scan systems generally save time by positioning the beam directly over the area to be exposed.

The wafer or sample to be exposed is placed in intimate contact with the stage to allow discharging the electrical build-up from electron impacts. Depending on the electrical conductivity of the material stack being exposed, this electrical charge can become an issue because it may deflect the incoming beam. The stage movement is precisely controlled by an interferometric system typically in steps, although in some special EBL system configurations, it moves continuously while writing takes place. In general, at each step, the substrate stays still while the beam is deflected by magnetic lenses to reach every position on the substrate within a certain area called the *write-field*. The size of the write-field is given by the distance between stage movement steps and may be selected from several values depending on the system, typically between 0.01 and 1 mm^2. Despite the high-precision interferometric mechanism for controlling the movements of the stage, a common issue in EBL arises as a result of the finite accuracy in the juxtaposition of bordering write-fields and is referred to as *stitching* error. It causes a slight discontinuity at the edge of structures stretching from one write-field into another, which can be less than a nanometer to several nanometers depending on the system's specifications and calibration. The consequences of these defects depend of course on their severity, but also on the particular structure being affected. It may be irrelevant in the case of, for example, an electrical contact pad, detrimental but tolerable in an input waveguide, or critical if it falls on a resonant structure such as a ring resonator. More details on this issue and on techniques to minimize it are given in Section 12.5, in which the main downside of EBL is also discussed, that is, its very low processing speed (or throughput) compared with photolithography which, despite its merits, renders it inadequate for high volume manufacturing.

Companies that manufacture dedicated e-beam systems include Crestec, Elionix, JEOL, and Vistec. Raith GmbH sells more affordable SEM-based systems well-suited for smaller research laboratories.

12.2.7 Etching

As already mentioned, etching is one of the basic steps of microfabrication. It consists of the removal of a thin-film material via chemical or physical means, and often through a combination of both.

Etching may be performed on a *blanket* film to reduce the overall thickness of the layer, or to remove it completely as when removing native oxides, which is explained later. However, it really becomes an essential part of the patterning process when done selectively after lithography on areas not covered by photoresist. This allows the transfer of the pattern from the exposed resist, a sacrificial layer, to the actual material of choice.

We speak of *wet* etching when the substrate is immersed in liquid etchants and the unprotected areas are subjected to chemical attack. Most modern processes, however, involve *dry* etching in vacuum chambers in which gas phase etchants in plasma are involved (O, F, Cl, or Br based). Positively charged species in the plasma are accelerated by controlled electric fields toward the substrate, where their impacts may exert physical erosion and, in the case of reactive species, be adsorbed to the surface where they react with substrate material resulting in desorption of volatile byproducts. We refer to the former mechanism as sputter etching, whereas the latter is chemical etching. Often, a synergistic combination of both is employed, such as in reactive ion etching, which dramatically enhances the etch rate to several times the individual etch rates of each process. A common etcher configuration for R&D is the inductively coupled plasma-reactive ion etching (ICP-RIE) (Figure 12.6c), which allows ion energy to be controlled independently of ion density.

Important parameters that determine the characteristics and performance of a particular etching process include selectivity, directionality, byproduct mobility, and etch rate. These parameters will have an effect on such properties as wall verticality and smoothness, profile control uniformity across the wafer, and process efficiency.

For most purposes, a directional etching is desired as it will faithfully reproduce the lateral dimensions of the features in the photoresist pattern in the underlying material. Therefore, the etch rate must be fast in one direction (usually perpendicular to the surface) and small or negligible in any other directions. This is called *anisotropic* etching as opposed to *isotropic* etching in which the etch rates are similar in all directions. Wet etching and chemical plasma etching are usually isotropic (except when crystallographic orientations of different etch rates are involved), whereas when the dominant mechanism of etching is physical rather than chemical, it becomes anisotropic.

An important concept in etching is selectivity, defined as the ratio of the etch rates of the various materials being etched under the same conditions. It is of particular importance that the selectivity between the mask material (usually photoresist but it may also be other sacrificial materials in what is called a *hard mask*) and the material to be patterned is high; otherwise, we risk consuming the photoresist before the etch has reached the required depth. Even before we arrive at this extreme case, low selectivity will also mean that the walls of the photoresist are being etched, thus jeopardizing the dimensions of the lateral dimensions of the etched pattern.

Quite often, the etch process is meant to finish only when the underlying material is reached, that is, the full depth of the layer being etched. In this case, the etching process is generally allowed to continue a while longer than strictly necessary (this is called an *overetch*) to ensure that the desired etch depth has been achieved all over the substrate. In these circumstances, the selectivity between the material being etched and the underlying material must also be high to avoid potentially undesired etching of the latter.

Efficient byproduct transport during etching is important to avoid or reduce unwelcome effects such as *microloading*, which refers to observed lower etch rates in areas of higher feature density, and *aspect ratio-dependent etching*, that is, lower etch rates in narrow trenches (Figure 12.8a). Aspect ratio is a common term in microfabrication and refers to height/width ratios for both features and trenches. In plasma etching, efficient transport is influenced by byproduct volatility and parameters affecting gas flow dynamics (such as pressure), whereas in wet etching, it will depend on byproduct solubility and, if applicable, stirring or ultrasonic agitation. Figure 12.8b illustrates another phenomenon referred to as *notching,* produced by the accumulation of positive charge from incident ions at the bottom of trenches, when a different material layer such as the buried oxide is reached. This charge causes incoming ions to be electrostatically deflected toward the sides of the trenches, thus increasing the etch rate at the sides where the notches appear. However, notching only becomes an issue for higher aspect ratios in time-multiplexed etch systems and is therefore not a problem in standard silicon photonics structures.

In addition to the main considerations discussed thus far, other important issues regarding etch processes include substrate damage, particulate contamination, and external considerations such as cost-efficiency (directly related to etch rate), safety, and environmental impacts.

Some of the main manufacturers of etch systems include LAM Research (in the United States), Oxford Instruments (in the United Kingdom), and TEL and Hitachi (in Japan).

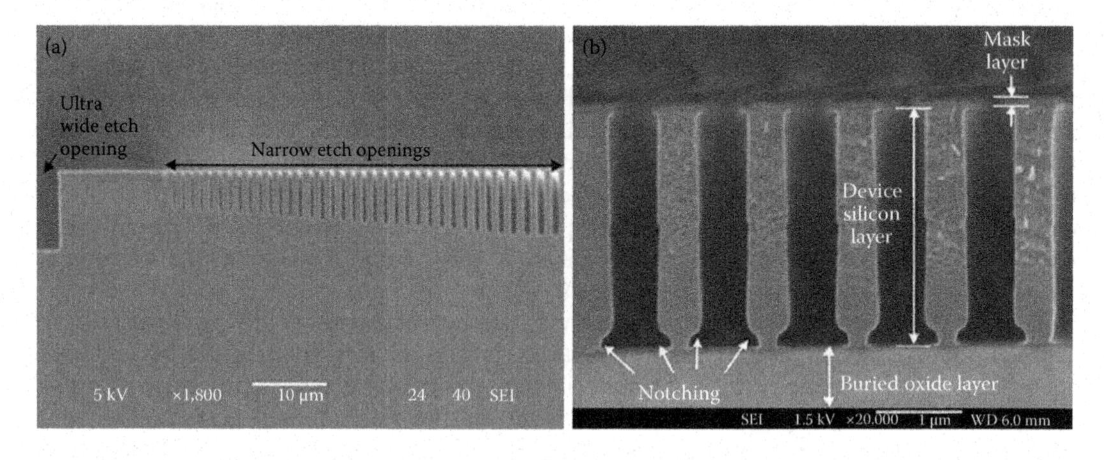

FIGURE 12.8
Etch processes' unintended effects: aspect ratio–dependent etching (a) and notching effect in thick device silicon SOI (b). (Reprinted from *Sens Actuators*, 133, Teo, S. et al. Hole-type two-dimensional photonic crystal fabricated in silicon on insulator wafers, 388–394. Copyright 2007, with permission from Elsevier.)

12.2.8 The Lift-Off Process

This process, which is not common in CMOS processing, is an additive patterning method whereby a material (often a metal) is deposited by PVD over a pattern that has been previously created on the resist. The wafer is then immersed in a solvent, sometimes in an ultrasonic/megasonic bath or sprayed at high pressure to dissolve the resist (Figure 12.13a; right). This causes the film of metal that was lying on top of it to peel off the substrate (or *lift-off*), whereas in the open regions, the metal lies in direct contact with the surface of the substrate and remains attached as intended.

When lift-off is used for patterning, sputtering is usually avoided as the deposition method because it is more conformal than evaporation, that is, the sputtered metal will cover the resist walls thus hindering physical contact between the solvent and the resist (Figure 12.13a; right). If sputtering must be used, for example, for materials that cannot be evaporated or if an evaporator is not available, then ideally a multilayer of resists should be used where the bottom layer dissolves in the developer irrespective of exposure, thus creating an overhang that allows the solvent to make direct contact with the resist.

12.2.9 Other Process Steps

Some other fabrication processes that the reader may encounter in this book are briefly described subsequently such as implantation, annealing, chemical-mechanical polishing/planarization (CMP), and nanoimprint lithography. Metrology techniques such as SEM, profilometry, ellipsometry, etc., are also important concepts in microfabrication but fall beyond the scope of this text.

Ion implantation, as the name implies, is the process of generating, accelerating, and directing energetic ions toward the substrate to implant them into a solid, in our case, generally silicon. By adjusting dose and energy, ion implanters allow doping silicon with impurities at precise concentrations and depth, and with the required profiles—a flexibility not available with other methods such as doping by diffusion. Most commonly boron, phosphorous, or arsenic are implanted for creating PN junctions, although other elements are useful for certain applications such as hydrogen (e.g., in fabricating SOI wafers) and oxygen (as exemplified in a fabrication process in Section 12.4.2). As shown in Figure 12.9, resist is usually employed to expose the regions where implantation is desired while masking the rest of the substrate. A standard process flow will be described in more detail in Section 12.4.1.

Thermal annealing is a basic process involving thermal treatment at process-dependent temperatures up to 1200°C performed on thin films (usually after deposition) to achieve one or several of the following objectives on the thin film material: densification, dehydration, outgassing, reflow (perhaps including planarization), releasing film stress, repairing the crystal structure, interdiffusion, enhancing contact between layers (or substrates in wafer bonding), or simply homogenizing a certain layer in wafers to a common known state by a combination of these effects. It is also performed after ion implantation to drive in and activate the dopants. Conventional thermal annealing is performed in semiconductor tube furnaces with resistive heaters capable of maintaining a precisely controlled temperature along the tube which follows user-specified ramps during both heating and cooling. Conversely, *rapid thermal annealing* is carried out in specific tools featuring high-power lamps used to heat wafers in just a few seconds. Annealing is also done to form ohmic contacts between silicon and metals by creating a silicide layer. In this case, rapid thermal annealing is sometimes preferred to keep interdiffusion to short ranges.

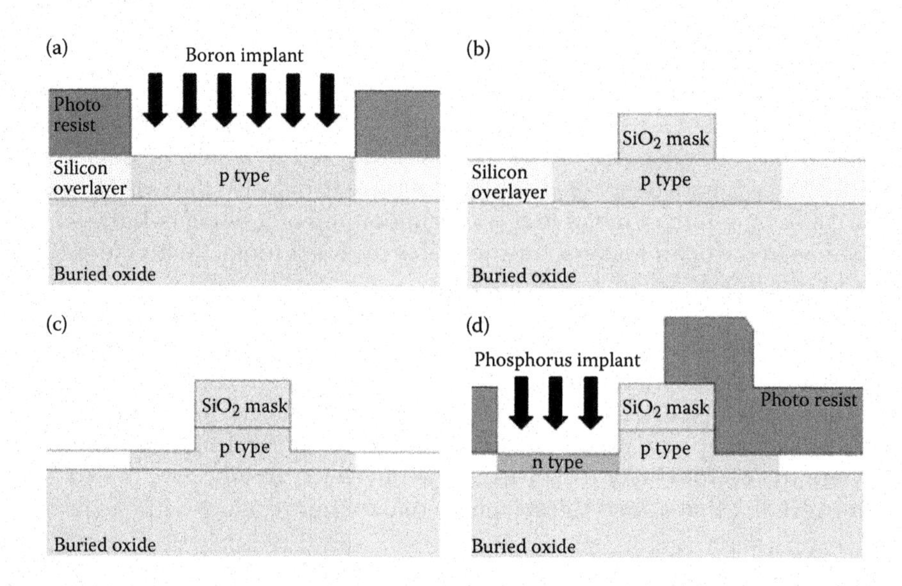

FIGURE 12.9
A self-aligned process for creating a PN junction by ion implantation. Photoresist is used to mask the silicon layer, leaving open regions where boron is first implanted (a). A SiO_2 mask is patterned (b) and used as a hard mask for etching (c). Another photoresist layer is used to mask the phosphorus implant (d). (Reprinted from Thomson, D. J. et al., *Opt. Express* 19(12):11507–11516, 2011. With permission from the Optical Society of America.)

CMP is used to planarize or reduce the topography on a wafer using a process in which the rotating wafer is pressed against a large-diameter pad (that is also rotating) in the presence of a chemical slurry. Thus, the wafer surface is subjected to both physical abrasion and chemical corrosion, which effectively planarizes its topography.

Nanoimprint lithography is a nonconventional technique that uses stamps to reproduce a pattern on substrates coated with a special resist by a combination of pressure and either heat treatment, or curing with UV light, which is irradiated through the stamp. The stamps can be made of quartz or silicon carbide and are generally patterned by EBL. This technique is capable of a resolution well below 50 nm, can be high-throughput, and it has the potential of being a much lower cost alternative to projection DUV lithography [14]. The reference provided is a good review article with further details, although there are also specific reports of photonic components fabricated using this promising nanofabrication technique [15].

12.3 Waveguiding Materials: Processing and Properties

The main silicon-based materials that are commonly used for waveguiding will be reviewed in this section. First, brief introductions will be provided for SOI and silicon nitride (SiN). The core material in the case of SOI, crystalline silicon, is an integral part of the substrate that is generally purchased. Therefore, it is not part of the in-house fabrication process. SOI wafers are briefly described in Section 12.3.1, but not in as much detail in this section as the other platforms because they will be the main protagonists of the rest of the chapter. Silicon nitride (Si_3N_4 for stoichiometric, Si_xN_y to specify nonstoichiometric or, in general, simply SiN), reviewed in Section 12.3.2, is a well-established material platform but is not electrically active. More emphasis will be placed on deposited silicon materials which have lately generated a great deal of interest: amorphous and polycrystalline silicon, which will be covered in Sections 12.3.3 and 12.3.4, respectively.

There are other silicon-based materials, however these will not be reviewed because they are either out of the scope of this chapter or for space reasons. An example of the former is doped glasses, although certainly silicon-based materials, they have been fully developed since the 1980s and then commercialized, and are not what one would generally refer to as silicon photonics. Moreover, although the material itself may be CMOS compatible, the unfeasibility of dense integration due to large bend radii and the thick layers required for optical isolation are not. Other materials that the interested reader might want to investigate further elsewhere include silicon-rich SiO_2, silicon nanocrystals, porous silicon, silicon oxynitride, Ge and SiGe, and functional polymers that may be used to fabricate, for example, athermal devices or high-speed modulators.

In this section, and in the remainder of the chapter, process recipes are sometimes provided but the reader and in particular students should note that all process recipes are equipment-specific and are given for illustrative purposes only. Standard process recipes should be provided by the equipment vendor during the course of the so-called *acceptance test* after the installation of a new (or refurbished) tool. It is the job (and often the enjoyment) of process or development engineers to realize *design of experiment* studies adjusting the process parameters to obtain the desired film properties or deposition parameters such as temperature or deposition rate. This is generally carried out based on general process knowledge that can be acquired through other researchers' experience as reported in the literature, but ultimately, the right recipe that will fulfill a particular requirement in a specific tool can only be found empirically. This way, a *baseline process* is achieved for that tool which can later be further optimized or fine tuned to future requirements. This section on waveguiding materials deals primarily with deposition, but the above mentioned arguments regarding recipes apply to almost all other processes.

12.3.1 SOI Wafers

Apart from the advantages of silicon itself as a material for photonics (Section 12.5.1), the fabrication of photonic devices on silicon has also benefited enormously from the wide availability of SOI wafers. These substrates have quickly become the main material platform for high-index contrast silicon photonics.

Originally developed for the microelectronics industry, SOI wafers are readily available commercially. A detailed review of the different ways to manufacture them and their properties from a photonics perspective can be found in one of the early books on silicon photonics by Graham Reed [16]. Nowadays, Unibond SOI wafers seem to be most favored by the research community.

SOI wafers consist of a stack of three layers: the *handle layer*, which acts as the mechanical substrate at the bottom, the *device layer* usually a few hundred nanometers to 2 to 3 μm thick (for photonics applications) on top, and a silicon oxide layer buried between the two. The *buried oxide* (BOX) layer of these wafers can be selected of an adequate thickness (generally 2 μm or more for standard strip waveguides) to suppress radiation leakage into the substrate, thus acting as a lower cladding to light propagating in waveguides patterned on the device layer, which should be high-quality, high-resistivity single crystal silicon (c-Si). Crystalline silicon is indeed, in most cases, the preferred phase of silicon for photonics applications. Amorphous and polycrystalline silicon have also received a great deal of attention lately and will be discussed later, along with silicon nitride. However, in the remainder of this chapter the main focus is placed on the SOI platform, as this is by far the most common platform in silicon photonics nowadays. The reader should assume it is the waveguiding material used in any of the work mentioned in this chapter, unless otherwise noted.

12.3.2 Silicon Nitride

As discussed in the previous subsection, SOI wafers have contributed in the last 15 years or so to a wealth of scientific research and to great technological advances in the field of silicon photonics and will likely continue to be its main technological platform in this respect.

However, there are two important limitations to the SOI platform. One of the most often cited benefits of silicon-based materials in general and SOI wafers in particular for their use in photonics is their compatibility with mainstream silicon microelectronics. Yet, in the case of SOI wafers commonly required in Very Large Scale Integration (VLSI) processes, the buried oxide layer thickness is much thinner than the 2 μm that is necessary to avoid radiation loss from a standard strip waveguide [3]. Furthermore, as mentioned in Section 12.1, 92% of CMOS logic production on 300 mm wafers uses conventional silicon wafers and does not provide a patternable single-crystal Si layer for photonics [4]. Another important issue is that the SOI material platform is, needless to say, a planar material platform limited to two dimensions. As silicon photonics technology continues to progress and device integration becomes denser and more complex, there arises the need for a way to extend into 3-D integration via multilayer stacking. This is particularly the case for applications in which integration involves electronics and the need to minimize the use of silicon real estate.

To create multiple waveguiding layers, high index silicon-based materials that can be easily deposited are required. Silicon nitride is a very good candidate. It is fully CMOS-compatible and constitutes a high index contrast technology with SiO_2 as a lower index cladding. It has the great advantage of showing very low propagation loss, a very desirable feature for any application. It is also transparent to visible radiation and is a popular material for biosensing applications.

Apart from not being electrically active, one of the main hurdles with SiN has been its high level of film stress, which makes it difficult to deposit thick layers of this material with the lower-impurity LPCVD technique [18]. PECVD SiN can be lower stress but generally experiences higher losses, primarily due to absorption by N–H bonds. The hydrogen content can be reduced by high-temperature annealing, and hence the propagation loss due to absorption. However, CMOS back-end compatibility is then lost because of the high temperature, whereas this is an important advantage of PECVD SiN (deposited at temperatures lower than 400°C) compared with LPCVD SiN (deposited at much higher temperatures). Therefore, work is being concentrated on reducing the hydrogen content of PECVD films as deposited, that is, without the need to anneal the films. Recent work at IME (A*STAR) in Singapore [19] has demonstrated that hydrogen content in the film can indeed be reduced by optimizing the gas flow ratios and other PECVD parameters. Several recipes were experimented with and the best results for their specific equipment was achieved with a SiH_4/N_2 flow ratio of 80 sccm/4000 sccm, a temperature of 350°C, and RF power of 400 W. In this recipe, the more traditional NH_3 was substituted with N_2 as the

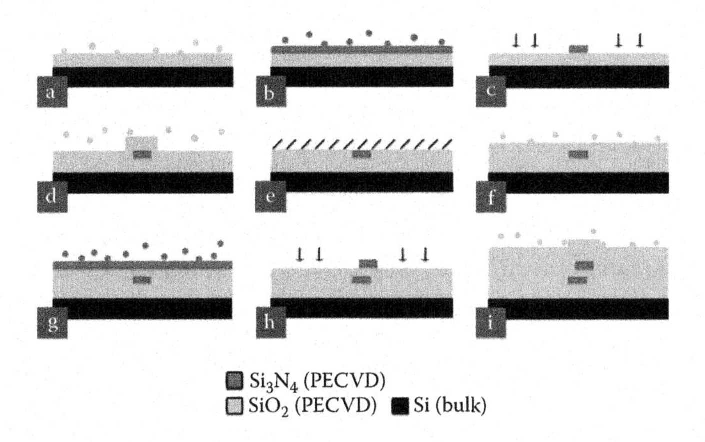

FIGURE 12.10
SiO_2 deposited as BOX (a), Si_3N_4 deposited for layer L1 (b), lithography and etch L1 (c), SiO_2 deposited as buffer (d), chemical mechanical planarization (e), SiO_2 deposited as a spacer (f), Si_3N_4 deposited for layer L2 (g), lithography and etch L2 (h), and SiO_2 deposited as top cladding (i). (Reprinted from Sherwood-Droz, N., and M. Lipson. *Opt. Express* 19(18):17758–17765, 2011. With permission from the Optical Society of America.)

source of nitrogen for the chemical reactions to form SiN. The fabricated 700 nm × 400 nm waveguides resulted in a propagation loss of approximately 2.1 ± 0.2 dB/cm at 1550 nm.

SiN has provided the technology platform for a great deal of outstanding work in silicon photonics. A recent example is the double photonic layer featuring interlayer optical coupling to microring resonators [17] fabricated by Prof. Lipson's group. Figure 12.10 illustrates the various process steps in a simplified way. By following the steps listed in the caption, the schematic becomes fairly self-explanatory. Nevertheless, more details on those process steps and any missing steps, such as spin-coating, stripping, etc., will be given in Section 12.4.1, albeit for a generic fabrication process flow for SOI rather than SiN waveguides. The deposition recipe they used is reported in their publication and another sample recipe has been provided in the previous paragraph. It is worth noting that i-line lithography was utilized, something not uncommon when working with SiN owing to usually slightly larger designs. However, little information was provided on the level of overlay precision achieved.

The reader may find additional publications on fabrication with SiN in the work of, among others, the following groups: propagation loss and waveguide geometries studies at Prof. Pavesi's group at the University of Trento, Italy [18,20]; Prof. Poon's group at Hong Kong University of Science and Technology [21] the work of Prof. Smith's group at the Massachusetts Institute of Technology (MIT), some of which is reviewed in Sections 12.5.1 and 12.5.2; the Triplex technology waveguides based on alternating Si_3N_4 and SiO_2 layers presented jointly by the group of Prof. Melloni in Milan, Italy, and a Dutch company [22], which is now being commercialized by the latter; and the work on SiN-based biosensors by several European groups within the SABIO project [23].

12.3.3 Polysilicon

Due to the emerging need for 3-D integration discussed previously, and the desirability of employing silicon to achieve this goal, considerable attention has been given in the last decade to amorphous and polycrystalline silicon. These materials can be deposited using simple standard microfabrication techniques. Thanks to these research efforts, these two embodiments of silicon, which were once considered unsuitable for waveguiding due to very high losses near 1550 nm, can now be deposited in such conditions that will render the propagation loss comparable (or nearly comparable, in the case of polysilicon) to that of c-Si for both strip and rib single mode waveguides.

Ideally, of course, crystalline silicon would be used throughout, also for multilayer stacking. It affords even lower optical losses and has superior effective electrical carrier mobility, which facilitates fabricating active devices such as electro-optic modulators. However, to achieve 3-D integration with c-Si, nonstandard fabrication techniques, such as wafer bonding, epitaxial overgrowth, or oxygen implantation, would be required [24]. Moreover, both the wafer bonding [25] and the oxygen implantation [26] fabrication techniques are still at the proof-of-concept stage in their application for 3-D integration of silicon waveguides and have important limitations in terms of complexity and implementation of active devices, respectively. Hence, the interest in deposited silicon, whether polycrystalline or amorphous.

Polycrystalline silicon (usually referred to as *poly*) can be deposited by LPCVD and is a standard and relatively simple process, commonly used for some of the most basic microelectronic structures. In microelectronics, it has also been considered as an alternative to single crystal silicon, for example, in thin-film transistors [27].

Early efforts to reduce propagation losses include the work of Liao at MIT on minimizing bulk absorption and scattering as well as surface-roughness-induced scattering [28]. The former is strongly influenced by grain size and therefore by deposition conditions and also by the presence of dangling bonds and defects.

Figures 12.11a and b show transmission electron microscopy images in which the polycrystalline grains and boundaries can be easily recognized. In the waveguides shown in Figure 12.11c, the crystal grains have been made clearly visible with standard SEM by first oxidizing the polysilicon, which advances at a faster rate in the grain boundaries where silicon is locally amorphous, and then etching with hydrofluoric acid.

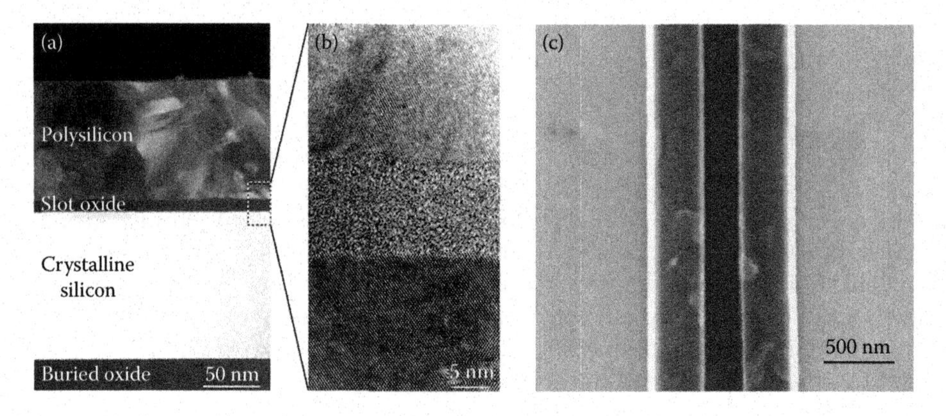

FIGURE 12.11
TEM images of polysilicon (a and b). SEM image of polysilicon waveguides (c). (a and b reprinted from Preston, K., and M. Lipson. *Opt. Express* 17(3):1527–1534, 2009. With permission from the Optical Society of America; c reprinted from Preston, K. et al., *Opt. Express* 15(25):17283–17290, 2007. With permission from the Optical Society of America.)

Deposition temperature and layer thickness both have an effect on grain size. It has been found that lower growth temperatures lead to smaller grain size and therefore lower surface roughness but also more grain boundaries and therefore increased bulk scattering [30]. Through hydrogenation, monoatomic hydrogen is introduced into the poly-Si film, which diffuses in the positive charge state, thus passivating dangling bonds and defects and relieving network strain.

As for surface roughness, CMP is an effective method to reduce roughness as demonstrated by atomic force microscopy (AFM) measurements. In Liao's work [28], both hydrogenation and CMP had a positive effect in reducing losses. However, under some conditions likely related to the degree of roughness, CMP applied on hydrogenated amorphous silicon has been demonstrated to have a deleterious effect on losses [31], perhaps by breaking Si–H or creating Si–OH bonds.

Lower surface roughness polysilicon can also be achieved by depositing amorphous silicon at 560°C (the amorphous to polycrystalline transition is approximately 590°C in typical processes) and then recrystallizing it into polysilicon through a 600°C anneal by solid phase crystallization. An additional high-temperature anneal of 1100°C was shown to further reduce losses by improving the degree of crystallinity, inducing larger grain size and thus yielding fewer grain boundaries and reducing light-absorbing dangling bonds [32].

Recent studies have demonstrated the possibility of somewhat reducing the required solid phase crystallization temperature (while still high at 1000°C), minimizing wall roughness by performing the crystallization after patterning, and showing that the effects of phosphorus doping on losses are lower than in corresponding SOI waveguides [33]. With the need to achieve higher degrees of device integration in mind, recent work has also focused on establishing the reduction in propagation loss that can be achieved in submicrometer waveguides. TE mode propagation loss of 6.45 ± 0.3 dB/cm at 1550 nm within the entire C-band has been demonstrated for a 700 × 250-nm polysilicon waveguide [34].

Beyond low loss waveguides, some important passive building blocks fabricated in polysilicon have already been realized, and have demonstrated good performance. These include optical filters [35], out-of-chip couplers [34], and high-Q resonators with vertical coupling to underlying c-Si [29].

Despite the formidable progress made by these groups in reducing optical losses in polysilicon, thus greatly enhancing its usability as an optical material, even better results have been obtained in the case of amorphous silicon (a-Si), achieving losses comparable with c-Si as will be discussed in the next subsection.

Before discussing a-Si, however, it is important to note that polysilicon retains a major advantage with respect to a-Si: its superior properties as an electrically active material. These include its moderate effective carrier mobility, which at approximately 100 $cm^2/V{\bullet}s$ is only an order of magnitude lower than c-Si but is two orders of magnitude higher than a-Si [36,37] and its shorter carrier lifetime enabling fast electro-optic devices [38], such as a 2.5-Gbps modulator demonstrated by Preston et al. [39]. Recently, a photodetector was fabricated in polysilicon exploiting subband absorption caused by intrinsic defects [40], operating at 2.5 Gbps with 0.15 A/W responsivity, and 40 nA dark current.

These advances open the path for deposited silicon to be used as a single material platform (as an alternative to SOI and crystalline Ge-on-Si) to fabricate both modulators and detectors. The former by suppressing absorption using hydrogen passivation and the latter by enhancing responsivity through the amount and type of doping. Following a scheme proposed in Preston et al. [40], further elaborated in an article by researchers from both Cornell and Columbia [41], SiN and polysilicon could be used in a combined platform such that SiN would be used for passive devices for its low loss, thus compensating at the system level for polysilicon's lower performance in active devices (compared with c-Si and Ge counterparts).

12.3.4 Amorphous Silicon

Amorphous silicon, apart from achieving lower material loss than polysilicon, as mentioned previously, has the major advantage of being CMOS back-end compatible as it can be deposited by PECVD or sputtering at temperatures lower than 300°C.

As a result of the research efforts since the mid-1990s, when we speak of amorphous silicon as an optical material today, we are generally speaking of *hydrogenated* amorphous silicon (a-Si:H). When deposited by PECVD at the appropriate process conditions (substrate temperature, RF power, chamber pressure, and gas flows), hydrogen atoms are introduced in the film, saturating silicon dangling bonds and removing undesirable energy states that have a crucial contribution to optical loss.

Using this technique, which also allows adjusting the refractive index by the addition of carbon or germanium to the alloy, a-Si:H film with an absorption coefficient of the order of 0.1 cm^{-1} was deposited and employed to fabricate multimode rib waveguides with losses of approximately 0.7 dB/cm and an interferometric thermoptic modulator [43]. A few years later, similar results were obtained by a different group for multimode waveguides (<0.5 dB/cm) while also fabricating single mode rib waveguides with 2 dB/cm loss [44]. Other groups have recently demonstrated submicron strip waveguides with losses as low as 4.5 dB/cm [45], 3.46 dB/cm [46], and 3.2 dB/cm for TE polarization [31], that is, basically limited by sidewall roughness scattering and comparable to SOI waveguides.

At IMEC, Selvaraja and colleagues [46] fabricated racetrack resonators and Mach–Zehnder interferometers (MZI) and provide, in their communication, some interesting details regarding the optimum PECVD process conditions to obtain a low-loss, high-quality film. They found that the propagation loss remained constant or within the measurement error (0.2 dB/cm) for samples kept in standard conditions as long as tested (up to 55 days). It also remained constant after they had been subjected to 400°C during the deposition of SiO_2 upper cladding. However, in their work, Zhu et al. [31] discusses some of the reasons for the differences in propagation loss among the results reported by different groups and brings attention to the fact that even mild temperature treatment of a-Si:H waveguides will increase propagation loss, an effect attributed to hydrogen out-diffusion. This may be a limitation for some applications unless a way to prevent the out-diffusion can be found. They also demonstrated that attenuation has a significant dependence on wavelength even within the c-band.

A sample PECVD recipe used by Della Corte et al. [42] for depositing intrinsic a-Si:H (and doped amorphous silicon carbide, a-SiC:H) is provided in Table 12.1. The main optical and electrical parameters of the deposited films are also presented. As always, it should be noted that recipes are equipment-specific.

From a technologist's point of view, the interest of this compliable material goes far beyond the low loss waveguides and standard passive devices that have been demonstrated. Hydrogenated amorphous silicon lends itself to exploring new innovative fabrication routes because it can be deposited relatively easily (compared with epitaxy) and at low temperature. At the risk of overlooking many others, some examples of recent reports in which this versatility is being exploited in passive structures include:

- Vertical couplers for transmitting optical signals from one level to another in a multi-stack configuration [47]

- High-efficiency CMOS-compatible grating couplers (–1.6 dB) thanks to the extra degree of freedom for the design provided by a deposited silicon overlay [48]
- a-Si 3-D tapers fabricated with shadow masks by both PECVD [49] and RF sputtering [50] (Figure 12.12)

The latter group had previously demonstrated [51] a graded index semicylindrical spot-size converter fabricated with sputtered hydrogenated a-SiO_x in which the oxygen content in the film was gradually adjusted by controlling the oxygen flow rate during deposition. As a result, the final structure featured a refractive index which varied continuously between 1.7 (silicon-rich silicon dioxide) and 3.4 (amorphous silicon) and was made to follow a square-law profile. Its propagation loss was measured to be 1.75 dB/mm, scaling to a negligible loss for a spot-size converter of a few tens of microns. Figure 12.12a shows a schematic of the vertical taper spot-size converter demonstrated in 2007, whereas Figure 12.12b shows an SEM micrograph of this structure.

The growing interest in photonic–electronic integration schemes and the ability to deposit a-Si:H at low temperatures with relative ease will surely spur the creativity of researchers and technologists to continue to develop many novel structures, process routes, and alternative ways to preserve the hydrogen passivation [52].

TABLE 12.1

Typical PECVD Process Recipes for Depositing a-Si:H and Doped a-SiC:H

Material	f (MHz)	Power (W)	Pressure (hPa)	T (°C)	Deposition Rate (Å/s)	Layer Thickness (μm)	Process Gas (sccm)	n	k	σ (S/cm)
a-SiC:H (p)	13.56	4	0.8	120	2.74	2	SiH_4 6, CH_4 6, B_2H_6* 10, H_2 20	2.99 ± 0.02	5.3 × 10^{-3}	2.3 × 10^{-6}
a-Si:H (i)	100	50	0.3	120	23.5	2	SiH_4 20, H_2 20	3.20 ± 0.02	2.4 × 10^{-6}	10^{-10}
a-SiC:H (n)	13.56	4	0.9	170	2.85	0.1	SiH_4 25, CH_4 35, PH_3 3	2.89 ± 0.02	2.8 × 10^{-3}	1.9 × 10^{-8}

Source: Della Corte et al., *Opt. Express* 19(4):2941–2951, 2011. With permission from the Optical Society of America.
Notes: Refractive index (n) and extinction coefficient (k) measured at 1550 nm. *, 0.5% diluted in H_2.

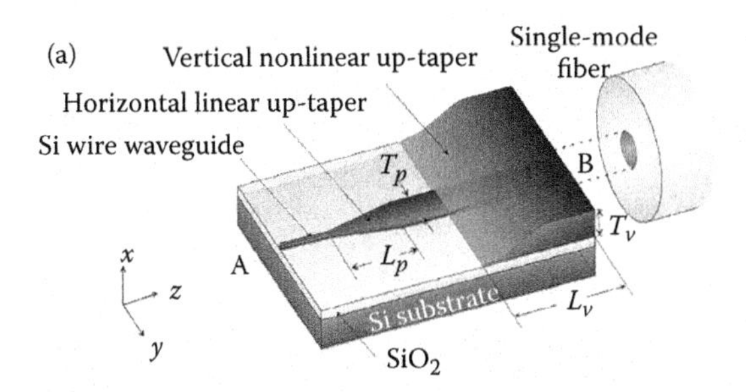

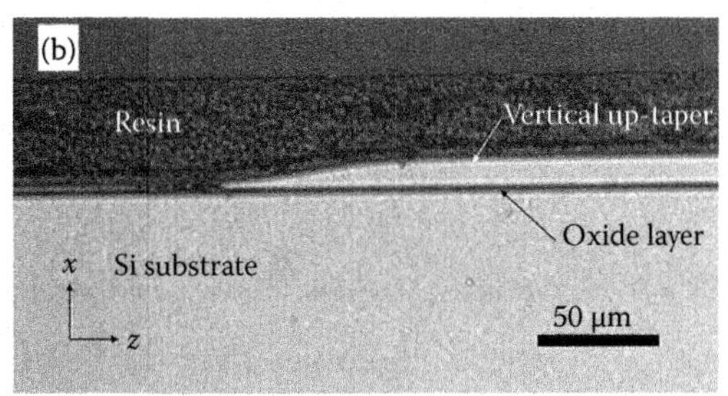

FIGURE 12.12
Schematic of the vertical taper SSC fabricated by RF sputtering of a-Si:H using a shadow mask technique (a) and SEM micrograph of this structure (b). (Reprinted from Shiraishi, K. et al., *Appl. Phys. Lett.* 91(14):141120, 2007. With permission from the American Institute of Physics.)

Regarding active devices, as mentioned before, besides the intrinsic limitations of silicon such as the lack of strong electro-optic effects comes a much lower carrier mobility in the case of a-Si, putting it at a disadvantage with respect to poly and c-Si. Despite this fact, light-emitting diodes [53], photodetectors [54], and digital optical switches [55] have been demonstrated. For modulators, the thermo-optic (up to a few Mb/s) or all-optical effects are generally employed with limited success. To fully exploit its nonlinear effects, its stability issues need to be overcome [56]. Nevertheless, recent experimental work [42, 57–59] continues to make progress toward useful a-Si:H based active devices, at the very least for applications such as switching and probably higher functionality devices in the future, as these efforts point toward future leveraging of its nonlinear effects and short carrier lifetime.

In summary, deposited silicon is poised to have a bright future as a photonic material. It has certainly been able to attract the attention of an impressive list of research groups (AIST, Cornell, IMEC, LETI, RIT, A*STAR, I-Shou, TUHH, UCSD, VTT, etc.) who in the last few years have joined those who started reporting their work in the 1990s (Italy's CNR and MIT). All of these groups are now successfully employing PECVD and other deposition techniques to fabricate poly and a-Si:H photonic structures. Excellent results have been obtained for passive photonic devices fabricated with a-Si:H and polysilicon. The former can be deposited at temperatures lower than 400°C, thus being CMOS back-end compatible and researchers are finding new possibilities for it to be used as an active material. As for polysilicon, despite not being back-end compatible due to high crystallization temperatures, it is a common front-end process and has already proven to be a good electrical material with several active optical devices already demonstrated. These advances are paving the way to monolithic large-scale integration of photonic networks on microprocessor chips.

12.4 Silicon Processing of Photonics Devices

In Section 12.2, the main microfabrication techniques available in the field of semiconductor processing were briefly reviewed. Now we will focus on a more detailed discussion of fabrication procedures in silicon processing. First, the fabrication of generic passive photonic devices on SOI will be described step-by-step by following a standard process flow such as that employed in an academic laboratory or in small batch manufacturing, including the following processing steps: wafer cleaning, dehydration and priming, covering with resist, prebake, exposure, postexposure bake, development, etch, resist strip, and cladding deposition. A schematic showing the standard process sequence is shown in Figure 12.2 and may be helpful for readers new to this subject. Then, in Section 12.4.2, a few examples of fabricated devices published in the literature are given.

12.4.1 A Generic Process Flow for Fabricating Structures on SOI

Although silicon photonics thus far lacks the well-defined design standards that benefit other technologies such as microelectronics, the two most common designs for single mode waveguides on SOI are (1) strip or channel waveguides (sometimes referred to as photonic wires) usually 220 nm thick and approximately 500 nm wide, and (2) rib or ridge waveguides, which are generally thicker. Rib waveguides are usually necessary for some active devices such as modulators based on the plasma dispersion effect. They have lower propagation losses for straight waveguides, whereas strip waveguides provide better radiation confinement and therefore permit higher integration (shorter radii bends). The latter are also more robust to fabrication variability–induced changes of the effective refractive index as a result of etch depth variations, because the Si layer is fully etched. Details on the design considerations of these well-established structures are discussed in other chapters in this book.

In either case, and indeed also in the case of photonic crystal structures, the SOI platform provides the lower cladding (the BOX oxide layer) and the core (device silicon layer) for waveguiding. Therefore the first processing step toward fabricating photonic structures is not deposition or oxide growth as in other material platforms (e.g. SiN, deposited silicon or doped silica) but the lithography needed to pattern the structure in the device layer.

First, the SOI wafers need to be thoroughly cleaned. In the case of brand new wafers in a laboratory setting, cleaning with acetone, methanol, and a subsequent IPA (isopropanol) rinse may be sufficient. An oxygen plasma is another possibility, which also serves to dehydrate the surface of the substrate as discussed in the next paragraph. More meticulous cleaning procedures involving acids (Figure 12.13b) were discussed in Section 12.2.2.

Because silicon dioxide (including native oxide discussed earlier) is hydrophilic, a silicon wafer kept in standard room conditions always has a thin film of moisture on its surface, which should be removed to improve the adhesion of the photoresist. This dehydration can be accomplished by applying an oxygen plasma or baking the wafer in an oven at approximately 200°C or on a hot plate for a few minutes. For most resists, it is beneficial to apply a chemical called hexamethyldisilazane, much better known by its acronym HMDS, before dispensing the resist. This chemical reacts with the silicon surface to further enhance resist adhesion. Ideally, a purpose-built oven can be used for the prebake, which incorporates a bubbler with HMDS, or alternatively, it can be applied in liquid form after dehydration. The wafer is then left to cool for a few minutes before placing it on a spinner to dispense the resist. In the case of optical lithography, sometimes a bottom antireflective coating is applied before the resist, or a top antireflective coating may be applied on top after prebaking. The purpose of these coatings is to minimize a standing wave effect caused by a reflection at the resist–wafer interface, which results in corrugated resist walls.

FIGURE 12.13
A spin coater (left) and a lift-off (right) system (a); a single-wafer spin processor for wet etching and cleaning processes (b).

The choice of the resist as well as the optimum process parameters is a crucial part of the development of a new microfabrication process. In a research environment, it is generally accomplished through a combination of previous experience, advice from fellow technologists, reviewing both technical and commercial (from resist manufacturers) literature, and most important of all, trial and error experimentation. Considerations for the initial choice of resist include whether it will be used for e-beam or UV lithography, UV wavelength, positive or negative resist, for a lift-off process or as an etch mask, wet or dry etch, required thickness, required resolution, etc. Table 12.2 presents the resists, hard mask materials, lithography tools, and etch gases used by a sample of the institutions with groups working on silicon photonics. Although 365 nm radiation (i-line photolithography) is sufficient for fabricating sophisticated silicon photonics devices with critical dimensions of approximately 300 nm, as demonstrated by groups such as Prof. Andrew Poon's [60], in general, either e-beam or DUV lithography are employed to achieve the necessary resolution due to the submicrometer lateral dimension of waveguides and even narrower features likely to be included in the design. These may include inverted taper tips, which usually must be smaller than 100 nm by design and sometimes even smaller than 60 nm [61], and coupling regions generally 100 to 300 nm between waveguides and ring resonators or in directional couplers. Later in this chapter, it will become apparent that apart from high resolution and overlay accuracy, for silicon photonics, excellent linewidth uniformity and low LER are also required from the lithography process.

Regarding the choice of resist thickness, the first consideration to be made is that in the case of substrates with initial topography (i.e., 3-D structures from previous processing as opposed to a flat surface) the thickness of the resist to be applied must be higher than the original topography as a general rule in standard processing. However, this is not necessarily always the case and depends among other things on the type of lithography and depth of focus available. In fact, with spray coating, the extreme opposite can be achieved both with photolithography [72] and EBL [73].

Thicker resist layers are also less susceptible to particle contamination. When used as an etch mask, the required thickness will partly be imposed by the thickness of the underlying material to be etched and by the selectivity of the resist to that material. The resist that is protecting the underlying layer must be thick enough to withstand the etching process with a reasonable safety margin. For lift-off processes, the resist needs to be much thicker than the material to be deposited over it to ensure that the solvent will be able to lift the resist. As a general rule of thumb, it must be at least three times thicker for evaporated metals.

These considerations point toward a thicker resist, yet if the resist is unnecessarily thick, it will mean lower resolution, more slanted sidewalls, or in some cases it may risk mechanical collapse above a certain aspect ratio. Hence, a compromise must be found for the right resist thickness. This in turn narrows down the choice as some resists can only be spun to a thickness within a specific range whereas in other cases a resist with a specific formulation must be chosen for the desired thickness. If a resist that fits the required compromise cannot be found, a different strategy will need to be implemented such as using a sacrificial layer (e.g., silicon nitride or silicon dioxide) as a hard mask for etching, or a multiple resist layer, which can be a useful strategy for both etch masks and lift-off processes.

Examples of specific resists commonly used for fabricating photonic structures on SOI include HSQ and PMMA for e-beam patterning and Shipley UV210, a chemically amplified resist for DUV lithography.

After pouring the resist, the wafer is spun at a controlled acceleration and speed determined by the desired thickness, as long as it is within a certain range as explained

TABLE 12.2

Lithography, Etch Masks, and Etch Gases Used by Some of the Institutions with Research Groups Working in Silicon Photonics

Group	Lithography	Etch Mask	Etch Gases	Loss[a]	Additional Information
IME (A*STAR)	Nikon S203B KrF 248 nm stepper	Shipley UV210(CAR) + SiO_2 hard mask	SF_6+C_4F_8	–	Time multiplexed etch and passivation cycling
CEA-LETI	ASML 1100 193 nm stepper	SiO_2 hard mask etched with C_4F_8	HBr-based	~2	Thermal oxidation of 10 nm to smooth edges
Cornell U.	E-beam	Ma-N 2403 + SiO_2 hard mask	Cl_2-based	–	Surpass 3000 as adhesion promoter
IBM	Leica/Vistec VB6-HR	SiO_2 hard mask etched with CF_4/CHF_3/Ar	HBr-based	1.7	Double oxidation (first 10 nm Si, removed, again 10 nm, left)
IMEC	ASML 5500/1100 193 nm stepper	Photoresist etch mask	Cl_2+HBr+CF_4+02// HBr+O_2	2.7	Multiple-step etch // 0.013 dB/90° bend $R = 5\ \mu m$
NTC- UPVLC	Raith 150 e-beam	HSQ	SF_6+C_4F_8	~4	HSQ is dried in vacuum (no prebake) for increased contrast
NTT	EB-X3 (X-ray mask writer e-beam)	ZEP-520 + SiO_2 hard mask	SF_6+CF_4 (ECR)	2.8	Roughness < 2 nm
Pirelli Labs[b]	Vistec VB6-HR	NEB-35 + SiO_2 hard mask	SF_6+O_2 (cryo)	3.5	DuPont EKC chemicals used to reduce roughness to 1.5 nm
U. of Glasgow	Vistec VB6 UHR EWF at 1700 uC/cm^2	HSQ (Fox16 50% with MIBK)	SF_6+C_4F_8	0.9	SF_6 30 sccm, C_4F_8 90 sccm, 10 mT, Source 600 W, Bias 12 W
U. of St. Andrews	E-beam	ZEP-520A	SF_6+CHF_3	–	–
UCDAVIS	ASML 5500/90 248 nm stepper	Shipley UV210	HBr+Cl_2	–	Std SOI 420 nm resist thickness

Sources: IME (Teo et al. Hole-type two-dimensional photonic crystal fabricated in silicon on insulator wafers. *Sens. Actuators* 133:388–394, 2007), LETI (Fedeli, J. M. Private Communication. CEA-LETI, 2012), Cornell (Cardenas, J. et al. Wide-bandwidth continuously tunable optical delay line using silicon microring resonators. *Opt. Express* 18(25):26525–26534, 2010), IBM (Green et al. Silicon photonic wire circuits for on-chip optical interconnects. In *Proceedings of SPIE* 6883:688306–688310, 2008), IMEC (Selvaraja et al. Fabrication of photonic wire and crystal circuits in silicon-on-insulator using 193-nm optical lithography. *J. Lightwave Technol.* 27(18):4076–4083, 2009.), NTT (Tsuchizawa et al. Microphotonics devices based on silicon microfabrication technology. *IEEE J. Sel. Top. Quantum Electron.* 11(1):232–240, 2005), Pirelli (Sardo et al. Line edge roughness (LER) reduction strategy for SOI waveguides fabrication. *Microelectron. Eng.* 85:1210–1213, 2008), Glasgow (Samarelli et al. Optical characterization of a hydrogen silsesquioxane lithography process. *J. Vacuum Sci. & Technol. B* 26(6):2290–2294, 2008 and Gnan et al. Fabrication of low-loss photonic wires in silicon-on-insulator using hydrogen silsesquioxane electron-beam resist. *Electron. Lett.* 44(2):44–45, 2008), St. Andrews (Debnath et al. Slowlight enhanced photonic crystal modulators. In *IEEE International Conference on Group IV Photonics GFP* 14, art. no. 6053728:98–100, 2011), and UCDAVIS (Zhou et al. Design and evaluation of an arbitration-free passive optical crossbar for on-chip interconnection networks. *Appl. Phys. A* 95:1111–1118, 2009) [62–71].

Note: Not a comprehensive list of research groups working in this field. Some of the groups may use additional processes to those reported in the articles cited

[a] Loss in dB/cm for strip waveguides where provided.

[b] Pirelli's Silicon Photonics operations have been suspended.

previously (Figure 12.13a; left). The resist datasheet will provide a graph or table in which the required rotational speed for a particular thickness may be found. Any sign of non-uniformity on the resist after spinning, such as bubbles or comets, is indicative of a non-optimized spinning process or an issue with the resist, or the substrate. Resists are usually kept refrigerated at a specified temperature, their shelf life being dependent on proper storage conditions. The temperature at which they are used is also important because this will have an effect on the rate of evaporation of the solvent and therefore on the final thickness [74]. Even a simple fact such as keeping the lid of the spinner open or closed has an effect (because of the vapor pressure of the solvent being vaporized) and it is one of the process parameters sometimes cited in technical literature when describing the spinning process. In microfabrication, especially during process development, it is paramount to be methodical and to pay close attention to every detail of the process, using run sheets to take notes of any relevant information and equipment logs for tool status and events. The information collected will prove crucial when it comes to attempting to troubleshoot a process issue.

In the case of e-beam resists, a conductive polymer by the name of Aquasave can be applied on top of the resist to mitigate the effects of charging [75], that is, electrical charge build-up due to the arrival of electrons and to poor electrical conductivity of the resist or the underlying layer which affects the trajectory of incoming electrons.

After spinning the resist to the desired thickness, the wafer is normally baked in an oven or preferably a hot plate to vaporize some of the solvent. By decreasing the solvent content, this *prebake* reduces the chances of bubble formation, the risk of mask contamination in the case of mask aligners (resist sticking to the mask), and it generally improves resist adhesion to the substrate. Accurate control of baking time and temperature is particularly important for critical processes such as subwavelength photonic structures.

The resist-coated wafer is then exposed to either electron bombardment in an e-beam system or to UV radiation in a mask aligner, stepper or scanner tool. The chosen imaging method will transfer the two-dimensional pattern that we wish to fabricate from the GDS file (e-beam) or mask (UV) to the resist on the wafer. GDSII is a common computer file format for lithography layouts to be transferred to a mask that can also be read by e-beam systems. The details for this critical step are provided in Sections 12.2 and 12.5 for both e-beam and optical lithography.

The next step is the postexposure bake to enhance or, in the case of chemically amplified resists, to finalize the chemical process that will make some of the resist soluble in the following step. It may reduce the effect of standing waves in the resist and enhance adhesion, but depending on the resist, this step is not always necessary. In the case of chemically amplified resists, it is not only necessary but also very sensitive to any variations to its conditions.

The resist is then developed by either manually immersing the wafer in a suitable developer (usually supplied by the same manufacturer as the resist) for a precisely controlled time or by an automatic tool called a *developer track*, which also includes hot plates for the bake steps. The developer dissolves the exposed regions in the resist in the case of positive tone resists or the unexposed regions in the case of negative tone resists, thus leaving the intended pattern on the wafer. Finally, a last postdevelopment bake or *hardbake* is sometimes also performed to increase the stability of some resists, although in the case of HSQ, exposure to oxygen plasma is a more efficient way of hardening the resist [74].

It is very important to note that the amount of time elapsed between each of the processes described above can have an effect on the resist performance and therefore on both the final image and the wafer-to-wafer (WTW) and run-to-run uniformity of the process.

Whether the delay between prebake and exposure or between exposure and postexposure bake or any other is the most critical one depends on the specific resist being used.

The pattern on the resist is now open and ready to be transferred to the underlying silicon. Before the etch step, a descum step may be necessary to remove resist residues from the open regions. This must be sufficiently gentle to avoid damage to the resist in covered areas and can be achieved with either an oxygen-reactive ion etching in just a few seconds or in about a minute in a plasma barrel etcher.

The wafer is then introduced into an etch chamber, typically an inductively coupled plasma reactive ion etching briefly described in Section 12.2. Typical etch gases commonly used in silicon photonics fabrication are listed in Table 12.2. An advanced etch recipe optimized for the fabrication of nanophotonic circuits can include the descum step and several other etch steps, each with a different purpose. In a publication reporting on the fabrication of photonic wires and crystals using state-of-the-art 193 nm lithography, Selvaraja and colleagues [65] report on a six-step etch process. The first step uses a Br/F based chemistry to etch the antireflective coating and is followed by a Br plasma to increase resist selectivity during the Si etch and to reduce roughness. The third step is a breakthrough etch with CF_2 and CH_2F_2 to remove the native oxide. In the fourth step, Si etch begins using a Cl_2/HBr/CF_4/O_2 recipe, which aims to reduce lateral critical dimension bias between isolated and dense regions on the die. The fifth step is also a silicon etch and uses HBr/O_2 chemistry, which is highly selective for SiO_2 (thus stopping at the BOX layer) whereas also highly passivating, reducing sidewall roughness. Finally, a common overetch step is applied to ensure all silicon has been etched. Further details on the etch recipes can be found in the references cited. It is our experience at NTC with our SF_6/C_4F_8 etches that lateral bias is not generally an issue [76].

During the etch process, the resist thickness will diminish (as discussed when the concept of selectivity was introduced) but some should remain on the wafer at the end of the etch process and therefore must be removed. This step is called *resist stripping* and can be done by either immersing the wafer in a solvent aptly named a stripper (or remover), subjecting the wafer to an oxygen plasma in a tool called an *asher*, or sometimes by both steps in sequence to remove stubborn resist. In the case of HSQ, which essentially becomes silicon dioxide after a hard bake, an etch recipe that will remove SiO_2 or a dip in buffered hydrofluoric acid may be necessary if it needs to be removed.

In a research environment, in which an upper cladding might not be available or desirable (such as for testing prototypes in sensing applications), the waveguide dimensions must be designed accordingly. In general, however, photonic structures are designed to have an upper cladding, which is often silicon oxide or sometimes a polymer of similar refractive index (to keep index symmetry with the lower cladding). An upper cladding helps to protect the nanophotonic circuits and it is certainly necessary if there are more layers to be deposited over the fabricated circuits, such as for patterning microheaters over frequency-selective devices for thermal tuning. In that case, approximately 1 μm of silicon oxide is usually deposited, although an optimum compromise in terms of thickness must be found between thermal contact and optical isolation to the microheater material (usually metal or doped polysilicon).

If the wafer topography needs to be planarized, the CMP process discussed in Section 12.2.9 is a possible solution. Alternatively, a doped glass such as borophosphosilicate glass (BPSG) can be deposited over the structures followed by a thermal anneal which will cause it to soften and reflow, thus yielding a flat surface (Figure 12.14a). Both phosphorous and boron decrease the glass transition temperature of silicon oxide, whereas phosphorous increases the refractive index and boron lowers it slightly; therefore, a dopant combination

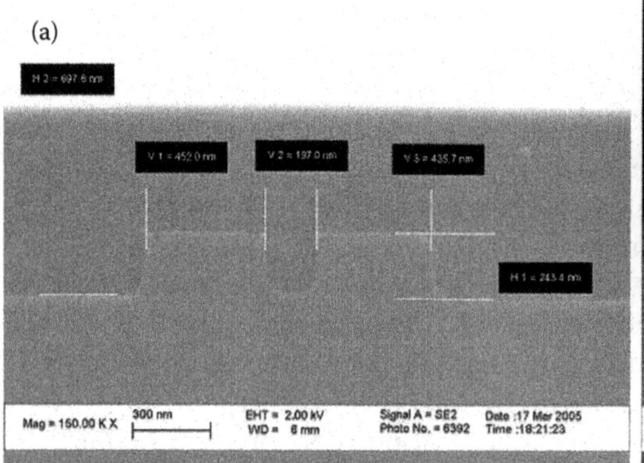

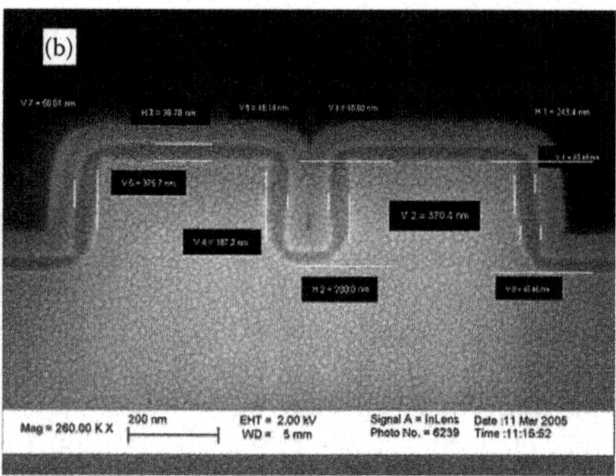

FIGURE 12.14
Planarized surface by BPSG reflow (a). Step coverage of two conformal layers of APCVD TEOS-based oxides achieving good gap-filling (this illustrative example shows a small void still present between the two waveguides) (b). (Courtesy of Pirelli Labs.)

that preserves the refractive index of undoped silicon oxide while reflowing at a much lower temperature can be achieved. As annealing borophosphosilicate glass over silicon is itself a traditional way to diffuse dopants into silicon, a thin layer of undoped SiO_2 should first be deposited to prevent such interdiffusion, which may otherwise lead to an increase in propagation loss in the silicon core (Figure 12.14b shows two conformal layers of deposited oxides over the coupling region between two waveguides). Another possibility for depositing an upper cladding which directly yields a planar surface, if the underlying structures can tolerate high temperatures, is by applying HSQ followed by annealing. HSQ with an optimized annealing has been demonstrated to exhibit equivalent optical properties to TEOS-based glass, while maintaining superior gap-filling and planarization characteristics and also being a more affordable, simpler process [77]. Once a flat surface has been achieved, it can be lowered to the desired thickness by blanket etching, a technique known as *etch-back*, which can also be applied selectively to a certain region on the wafer by protecting with photoresist [78].

In principle, the wafer is ready after planarization to repeat the fabrication process from the start for depositing and processing further layers, and this is, in general, the case in microelectronics in which many layers are patterned on the same substrate. This is also partly the reason for the huge interest recently aroused by deposited silicon, which was discussed in the previous section. Conversely, in SOI photonics, we are limited to fabricating photonics devices on the device layer. Nevertheless, there are three common examples in which further lithography levels are required: (1) when more than one etch depth is necessary, (2) when a top layer must be processed for fabricating heaters on top of the underlying devices, and (3) when other active devices are being fabricated. Here, we have briefly described a basic process for the photonic layer only. Today, complex state-of-the-art silicon photonics technology features both passive and active devices as well as integrated electronic functions. These chips involve many mask levels and process steps. One such example is the chip shown in Figure 12.15, fabricated within the framework of European FP7 project HELIOS [79].

Microheaters are typically fabricated with metals such as Ti and NiCr, although polysilicon is also a possibility. For CMOS-compatible electrical contacts, a metal stack of Ti/TiN/AlCu is generally deposited by PVD. The bottom titanium layer acts as a contact layer for

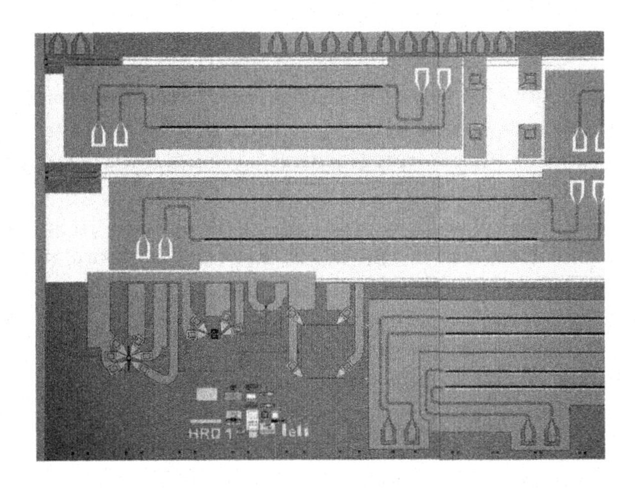

FIGURE 12.15
View of a QAM modulator including ring resonators with heaters, Si PN modulators, and Ge photodiodes. (Courtesy of DAS Photonics, Valencia, Spain (www.dasphotonics.com) and CEA-LETI, Grenoble, France (www.cea-leti.fr).)

its good adhesion properties to the substrate, the TiN interlayer is used as a barrier layer against interdiffusion, and a small percentage of Cu in Al helps in reducing electromigration [80]. As a rule, after the top metal is deposited and the contacts are patterned, an anneal in a N_2 atmosphere is performed to improve ohmic contact and release stress in the films. Finally, a layer of SiO_2 is deposited over the metal for protection against oxidation and physical damage. It is sometimes said that metallization is as much an art as it is a science, as a good metal contact is very sensitive to cleaning and other process variations, and it must fulfill a long list of requirements that are interdependent [81]. Further details fall outside the scope of this chapter but can be found in the references.

When a process involves more than one lithographic step, *alignment marks* must be used to align the structures from one level to the next. These are geometric figures such as squares and crosses strategically placed on each mask such that the figures on the wafer from mask "n" will allow the operator (or the software algorithm in case of automatic alignment) to be able to properly align the wafer to mask "n + x" by ensuring the figures on a subsequent mask overlap those on the wafer. In the case of EBL, there are two types of marks: those relative to the wafer or sample and those for the individual chips. The latter should be as close to critical structures as possible for best registration accuracy. The operator must find the location coordinates of the alignment marks and input this information into the system's software. In some e-beam systems, the machine can automatically return to the marks to correct for beam drift.

There is a second type of structure that is usually added to the fabrication process. These are the *process control monitors,* which are structures that are fabricated along with the intended devices and serve the purpose of witnessing the quality of each monitored process step at several positions on the wafer. They may be used to measure resolution, alignment accuracy, etch depth, sidewall verticality, etc. When precise measurement of a certain process parameter cannot be obtained from the process wafer, for instance, in the case of film thickness or refractive index in a wafer with multiple layers, a blank wafer is processed as a process monitor just before or after the device wafer. Even in the case of certain measurements that are possible on the device wafer, the use of a monitor wafer is sometimes preferred to spare the device wafer from unnecessary handling.

12.4.2 Some Examples of Innovative Process Flows

After the generic process flow for standard waveguide-based devices described in the previous section, a few illustrative case studies will be briefly reviewed in this subsection, focusing on innovative fabrication strategies that have been demonstrated in the last few years for fabricating photonic structures.

A remarkable and yet conceptually simple structure, which will serve to introduce two processes not mentioned thus far, is the silica microcavity fabricated by Prof. Vahala's group in California Institute of Technology [84]. These toroid-shaped microcavities have achieved Q factors in excess of 100 million and have been used as micromechanical oscillators that can be cooled using radiation pressure (Figure 12.16a) [85].

The silicon underneath the SiO_2 disks was etched with XeF_2 at a pressure of 3 Torr. This gas etches silicon isotropically without the need for a plasma at a faster rate than SF_6-based inductively coupled plasma (Section 12.2.7) and with better selectivity to SiO_2 masks [86].

To create the toroid-shaped SiO_2 disks, an innovative process was employed consisting of localized heating of the silica disks with a CO_2 laser of 10.6 μm wavelength to the point when the SiO_2 reflows. This allows its surface tension to both smoothen the surface and collapse the disk into a toroid shape in a self-quenching manner in such a way that the final diameter of the disk is determined by the prior lithography and etch processes. Localized heating has also been proposed for planarization [87] and sidewall roughness reduction [88], a subject that will be discussed in detail in Section 12.5.2.

Another innovative process flow in which XeF_2 was used was devised by researchers at MIT to break the dependence on SOI substrates [89,90]. They fabricated waveguides using the polysilicon layer employed in commercial bulk CMOS processing for transistor gates and polysilicon resistors, which are deposited above the shallow trench isolation layers (Figure 12.16b). The necessary high index contrast was achieved by creating air tunnels with an XeF_2 etch beneath the waveguides. Waveguides were first defined in the polysilicon layer deposited on a thermally grown oxide, as shown in the inset of Figure 12.16b. Then, a top oxide layer was deposited to protect them and 3-μm diameter holes, 8 μm apart, were wet-etched along the waveguide in the oxide to expose the silicon substrate. Finally, several pump–etch–pump cycles were carried out in an XeF_2 etcher to remove the silicon underneath the waveguides, thus creating the air holes.

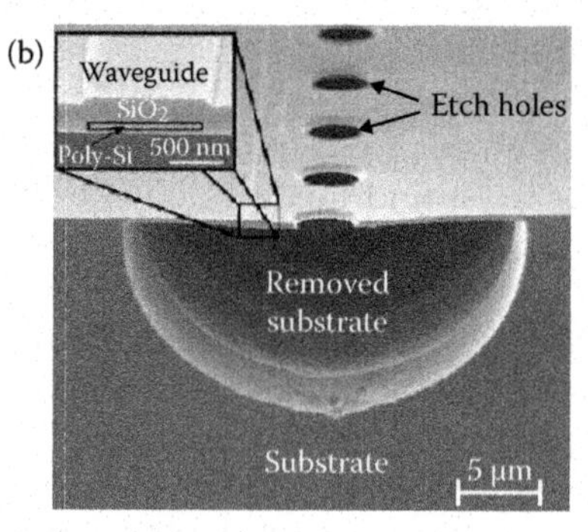

FIGURE 12.16
SEM of a toroid microcavity (a) and SEM of air tunnels (b), inset shows the polysilicon waveguide. (a, adapted from Schliesser, A. et al., *Phys. Rev. Lett.* 97(24):1–4, 2006. Available under Creative Commons Attribution 3.0 licence, doi: 10.1103/PhysRevLett.97.243905, 2006. The American Physical Society; b, reprinted from Holzwarth, C.W. et al. In *2008 Conference on Lasers and Electro-Optics*, 2008, p. CThKK5. With permission from the Optical Society of America.)

Silicon etching with XeF_2 is particularly popular for MEMS microfabrication [91]. When silicon sacrificial layers are wet-etched to release freestanding structures such as cantilevers, a common issue known as *stiction* is often encountered. Stiction is a neologism derived from "static friction" and it refers to the static friction that needs to be overcome to move or separate surfaces that have come into contact. For example, a cantilever may bend down and stick to the underlying surface due to the surface tension of the acid or water as it dries after the sacrificial layer has been etched. Sometimes, drying with supercritical CO_2 is used as a solution but the problem can more easily be circumvented by etching with XeF_2 instead.

Inspired by the need to find new ways of integrating monolithically optical devices on silicon substrates without consuming precious real-estate used for electronics, Jalali's group [92] at the University of California, Los Angeles used the so-called SIMOX (separation by implanted oxygen) 3-D sculpting technique to fabricate photonic structures in a buried silicon layer. This leaves the top surface unoccupied and available for the realization of electronic devices. The technique is described in more detail in a previous publication [93] and is schematically illustrated in Figure 12.17. Basically, it involves oxygen implantation through an implant thermal oxide mask, which is followed by a high-temperature anneal at 1300°C to cure the implantation damage and produce a continuous SiO_2 layer above the BOX layer. This SiO_2 layer splits the silicon device layer into two: a bottom layer with buried photonic structures and a top layer available for patterning of electronic devices. The SiO_2 mask decelerates the oxygen ions and therefore, by selecting a specific thickness, the bottom silicon layer thickness can be controlled. Using this technique, they demonstrated laterally coupled microdisk resonators with quality factors of 2000 and extinction ratios higher than 20 dB. In a subsequent publication [94], they showed a further development of the technique and demonstrated a number of 3-D integrated photonic devices, including vertically coupled microdisk resonators, in which they noted the possibility of very precise

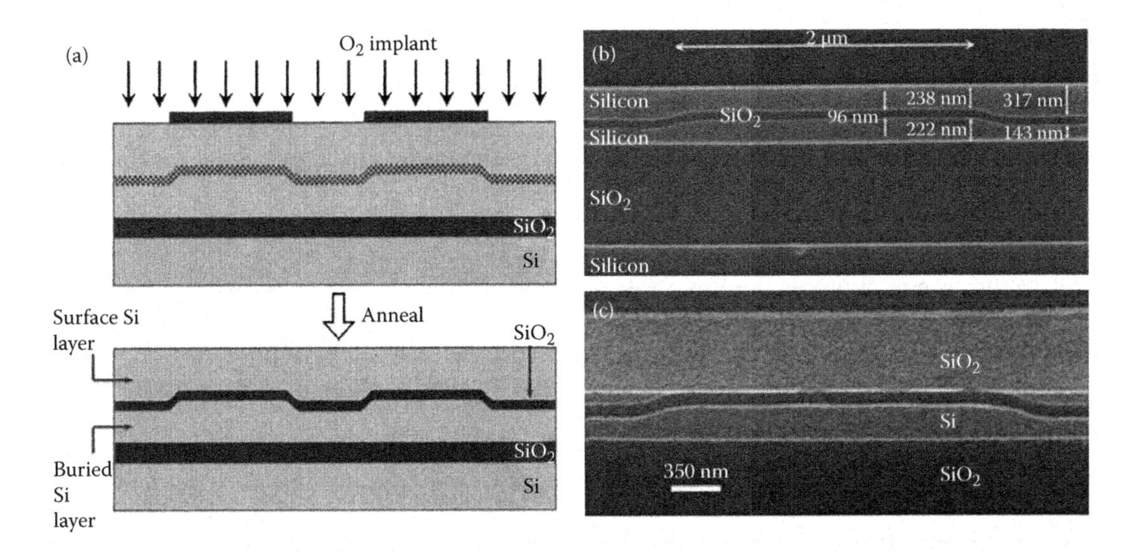

FIGURE 12.17
Schematic of SIMOX 3-D sculpting technique (a) (see text for details), SEM view of fabricated structure (b), and same after oxidation of top silicon layer (c). (Reprinted from Indukuri, T. et al., *Appl. Phys. Lett.* 158–160, 2005. With permission from the American Institute of Physics.)

control of the coupling distance. This is because the critical dimension can be controlled with better precision in the vertical direction. The main drawback of this technique is the difficulty of producing active devices in the buried layer, apart from limitations of pattern control in the horizontal direction.

The reader may refer to the following publications for a few more examples of innovative fabrication process flows:

- Advantageous use of the normally undesirable aspect ratio-dependent etching effect mentioned in Section 12.2 [95]
- Horizontal slot waveguide to avoid fabrication difficulties in the standard vertical slot configuration [96]
- Waveguides on silicon-on-oxidized-porous-silicon via proton beam irradiation [97]
- Ultrasmooth etchless waveguides fabricated by selective oxidation [98]
- Slots of 100 nm fabricated by DUV using a double patterning technique and an ultrasharp tip of a few nanometers with a slanted lithography window [99]

Although all the fabrication reports mentioned above are meritorious in one way or another, they have been chosen based on the diversity of the proposed fabrication routes. This should give the reader an idea of the endless possibilities available to designers and technologists alike. There are certainly many other instructive examples deserving mention that could have been selected from the literature.

12.5 Main Fabrication Issues

12.5.1 Advantages and Disadvantages of Silicon

When we consider the advantages of silicon over other materials for fabricating microelectronic devices, we end up with an impressive list: many of its electrical, optical, and mechanical properties; the existence of a stable oxide with excellent properties that can be easily grown from it; its abundance in the Earth's crust and the ability to purify it to a high degree; the huge knowledge base and economies of scale created around it… to name a few. In the case of photonic devices, most of these advantages remain on the list, whereas new items, this time advantages specific to photonics, join it: optical transparency in the telecommunications wavelengths; high refractive index enabling submicrometer waveguides, and short bend radii, hence, the ability to miniaturize; and a high thermo-optic effect, which facilitates useful functionalities.

There are also certainly downsides associated with using silicon for photonics, which are well publicized and easy to encounter when reviewing the technical literature. Some are what one might call "genuine" disadvantages: its indirect band gap structure, which precludes the fabrication of efficient optical sources and the lack of a strong electro-optic effect, which complicates the development of fast modulators. Then there are other problems that are in fact a direct consequence of the features we considered "advantages" above: it is not easy to make a detector for those same wavelengths for which silicon is transparent; high index contrast and miniaturization implies challenging input/output

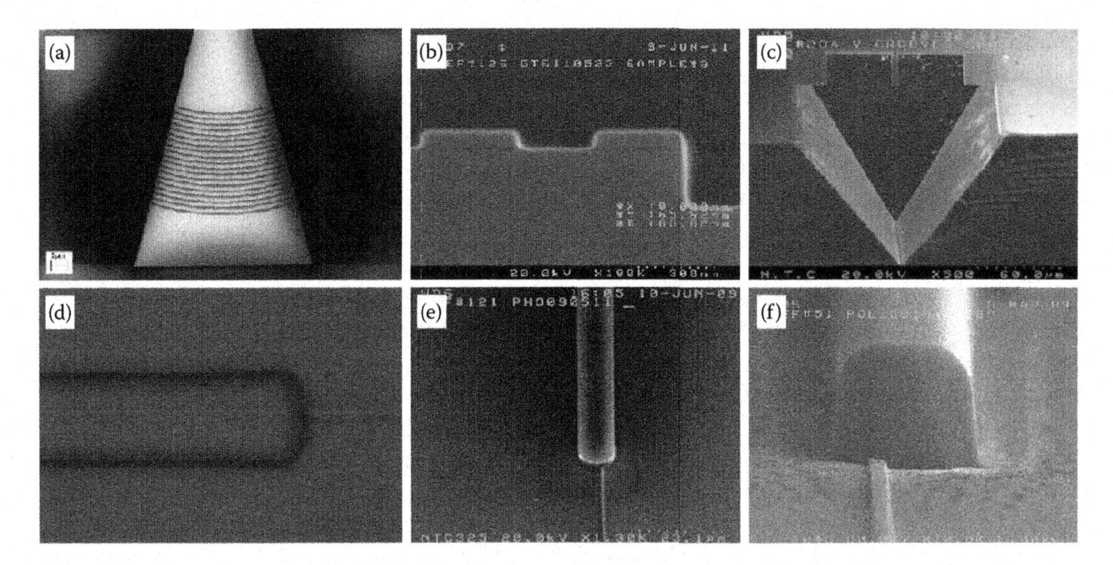

FIGURE 12.18
Fiber input/output structures fabricated at NTC-UPVLC [1]. Top view (a) (this image was the 1st prize winner of Raith's Award 2011; www.raith.com) and cross-section of grating couplers (b); V-groove coupler fabricated by wet etch with KOH (c) (note freestanding waveguide fabricated in the buried oxide, which couples to an inverted taper silicon wire, see Galán et al. [102]); optical micrograph (d), and SEMs of polymer waveguides over a silicon strip waveguide (e and f). Waveguides aligned with an EVG-620 mask aligner.

coupling (Figure 12.18), higher roughness-related losses, polarization dependence, and extreme sensitivity to fabrication variability; and the large thermo-optic effect can also work against us when anomalous device performance is caused by a change in ambient conditions.

Out of all the challenges mentioned above, efficient silicon-based sources, high-speed modulation, and high responsivity detection are very wide-ranging issues and the state-of-the-art solutions are covered in dedicated chapters elsewhere in this book. As for thermal and polarization dependence issues as well as input/output coupling, the most effective way to deal with them should begin with advanced designs, and only then can clever fabrication techniques and accurate processing also play their role. However, the issues of propagation loss and fabrication nonuniformity, which ultimately affect both passive and active devices, are inherently fabrication related. They will be reviewed in detail in the remainder of this section.

We will first discuss the causes of propagation loss, such as sidewall roughness, and ways to minimize it in the next subsection. Then, the fabrication tolerances in waveguide dimensions will be discussed in the next two subsections, both from the point of view of R&D work, which is commonly based on EBL, and also for the case of processes analogous to those currently in use in CMOS production lines. In some of the research work mentioned subsequently, silicon nitride rather than silicon is the waveguiding material, but the results being highlighted are relevant to both high-index contrast platforms.

12.5.2 Waveguide Propagation Loss

One of the most important performance parameters of a photonic device is insertion loss because optical power budgets in photonic integrated circuits are, needless to say, limited.

The waveguide is the main constituent of more complex devices, and its propagation loss will add to any other loss mechanisms intrinsic to the device. Hence, it is paramount to keep waveguide propagation losses as low as possible. In silicon photonics, in which the aim is quite often to employ high confinement waveguides for maximum device integration, this is not always easy for reasons that will become apparent shortly and becomes an important fabrication challenge. In fact, propagation losses measured in decibels per centimeter of a straight waveguide of a particular design is often used as a benchmark for the quality of fabrication processes.

From a broader perspective, undoped single crystal bulk silicon has very low intrinsic loss for photon energies below its band gap (1.1 eV) and the remaining potential contributors to propagation loss in waveguides are radiation leakage loss, absorption/scattering by impurities and defects, and scattering by roughness in the waveguide's surfaces.

Strip waveguides provide high radiation confinement, which allows for small radii designs. The smallest radius before there is noticeable loss depends on the waveguide dimensions and cross-sectional profile. For standard strip waveguides of approximately 500 × 220 μm, this threshold value has been demonstrated for fabricated waveguides at approximately 3 to 5 μm [66]. In bends, polarization cross talk is also a source of propagation loss. Researchers at IMEC have recently shown that by optimizing the waveguide etch to obtain high verticality sidewalls, the bend loss can be reduced by 25% for radii less than 5 μm. Full details of their findings including the etch recipes employed can be found in their publication [100]. For the case of ridge waveguides, much larger radii or corner mirrors are necessary [101].

Therefore, for a given waveguide design, the appropriate selection of the BOX thickness [103] and bend geometry and radius will set an upper bound to the radiation leakage to the substrate. Radiation leakage to other surrounding structures will likewise be avoided through proper design by leaving enough distance between structures and, if there are other structures on top (e.g., microheaters), a sufficiently thick upper cladding. Bulk absorption and scattering by impurities and defects in the core layer is not usually considered to be an issue if high-quality SOI wafers with high resistivity are used.

On the other hand, absorption and scattering at the interfaces of the core and the cladding material (generally silicon dioxide or air) are much more problematic, both in global terms and from the point of view of fabrication, and therefore they will be discussed in more detail. Although fabrication can also potentially have an effect on radiation loss and bulk absorption in SOI waveguides, this is far less common because it would generally be a consequence of substandard fabrication practices (or simply oversights) resulting, for instance, in lack of fidelity to design specifications or unintentional doping because of contamination in annealing furnaces.

Surface absorption can originate from surface states to the extent that photocurrent attributed from this effect has been used to fabricate an all-silicon photodetector at California Institute of Technology [104], albeit with a relatively low responsivity of 36 mA/W. But surface absorption can also originate from metal contamination at or near the interfaces of waveguides. Indeed, surface absorption by metal contaminants is likely to be more commonplace in SOI photonics than generally acknowledged, according to a recent investigation by Barwicz et al. [105] at MIT. As they noted, there seems to be a consistent discrepancy between the values for waveguide propagation loss that one would expect from sidewall roughness alone and the values measured by most groups working in silicon photonics. Their study suggests that metal contamination from etch chambers and, more notoriously, from metal masks is to be blamed for this discrepancy. They showed that silicides can form at temperatures as low as 90°C, perhaps due to additional energy received from ion bombardment in the etch chamber, and found a direct relationship between the metal mask used and the measured propagation loss. The presence of metals on the waveguides was confirmed with a scanning transmission electron microscope equipped with an energy dispersive X-ray spectrometer.

Regarding surface scattering, it may be caused once again by impurities or particles at any waveguide surface but most conspicuously by sidewall roughness as a result of the fabrication process, whether the waveguiding principle is total internal reflection or the photonic band gap effect. Roughness at the top of SOI wafers is in the order of 0.1 nm, but due to limitations in line edge resolution during the patterning process (whether through etch or lift-off) it is usually more than 1 nm at the sidewalls. In fact, sidewall roughness has been demonstrated to be the major source of scattering loss for (sub)micron-sized waveguides in the Si/SiO_2 platform because scattering loss is directly proportional to the square of the root-mean-square roughness [106].

Sidewall roughness usually measured as LER is therefore the main issue to be confronted by the fabrication team with respect to reducing waveguide losses in silicon photonics. It is also the main cause of backscattering, which is typically more than 3 orders of magnitude higher than in low index–contrast waveguides and can be a showstopper for some applications [107]. To reduce LER, process engineers should pay a great deal of attention to any process parameter that may have an effect on sidewall roughness while at the same time envision new fabrication techniques that may minimize it.

A great deal of research has been carried out in the last few years to develop such fabrication techniques. One way to tackle the issue is of course to come up with process enhancements or new process steps that will result in minimum LER, whereas another route is to devise methods to reduce it after waveguide fabrication. Progress can be quantified by either direct AFM measurements of the root-mean-square roughness in fractions of a nanometer, focused-ion-beam transmission electron microscopy, or by statistical analysis of waveguide profiles from SEM images whether on-line or off-line, which provides the relevant LER parameters of root-mean-square roughness and correlation length [110]. Indirect measurements of relative LER by measuring attenuation losses in fabricated waveguides is another possibility.

The relationship between sidewall roughness and propagation loss has been well characterized, for example, by researchers at the NTT [66], where successive improvements in fabrication lowered the resulting roughness and hence the measured attenuation. Initial fabrication with no optimization resulted in waveguides with approximately 10 nm roughness and 60 dB/cm loss. Waveguides fabricated with improved processes, which reduced surface roughness down to approximately 5 nm roughness resulted in 13 dB/cm loss. The process was optimized further to less than 2 nm roughness as estimated from the focused-ion-beam transmission-electron microscopy images, thus achieving 2.8 dB/cm loss. The authors pointed out that the final value could be further reduced by oxidation smoothing, which was not carried out in this work.

The right choice of process and resist are the first steps to ensuring low LER. The lift-off process discussed below is not the best choice for minimizing LER, although it is has other advantages in certain circumstances. Conversely, for e-beam fabrication of SOI waveguides, HSQ is a good resist choice if it fulfills all the other specific process requirements, and in fact the waveguides with the lowest-reported propagation loss were fabricated with this resist [69]. For other resists, there are techniques to keep LER as low as possible. One such technique successfully applied to ZEP resists consists of using a lower dose to expose the central part of the waveguide, leaving a narrow region unexposed and then using a higher dose at the edges of the waveguides to break up aggregates of resist polymer at the edges, thus achieving a lower sidewall roughness [111,112]. A related strategy followed at NTC [108] is to adapt the GDS files by including open paths as a frame for all areas. With this technique and an optimized etch recipe, approximately 2 nm sidewall roughness is routinely achieved (Figure 12.19).

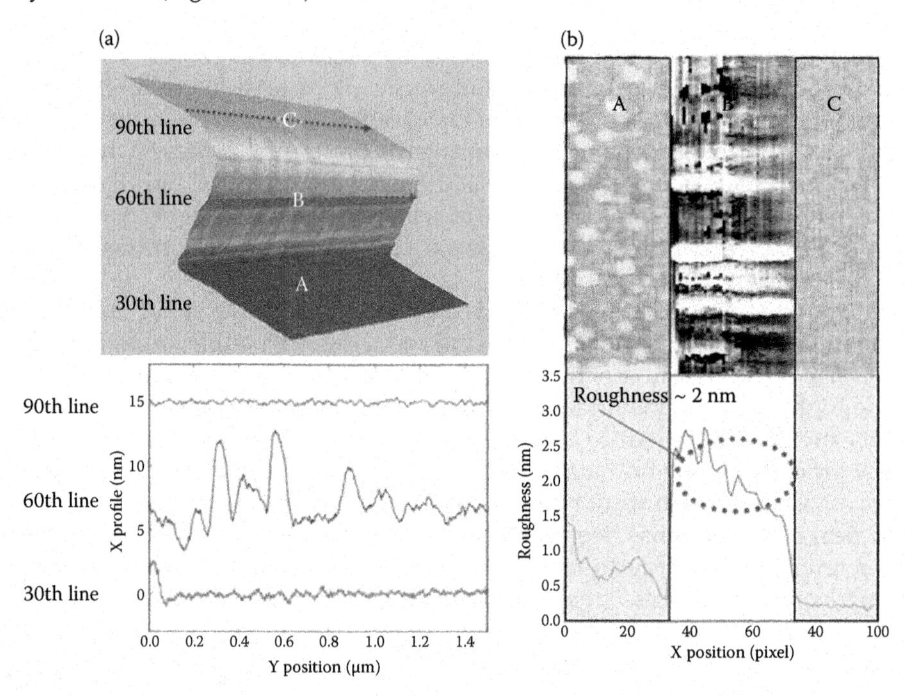

FIGURE 12.19
(See color insert.) AFM images of low sidewall roughness of approximately 2 nm for waveguides fabricated at NTC-UPVLC [108] obtained by an optimized process. Region A (a) is the top surface of the BOX oxide, region B (b) is the sidewall, region C is the top surface of the silicon waveguide. (Images courtesy of Park Systems, KANC 4F, Iui-Dong 906-10 Suwon, South Korea 443-270 (www.parkafm.com).)

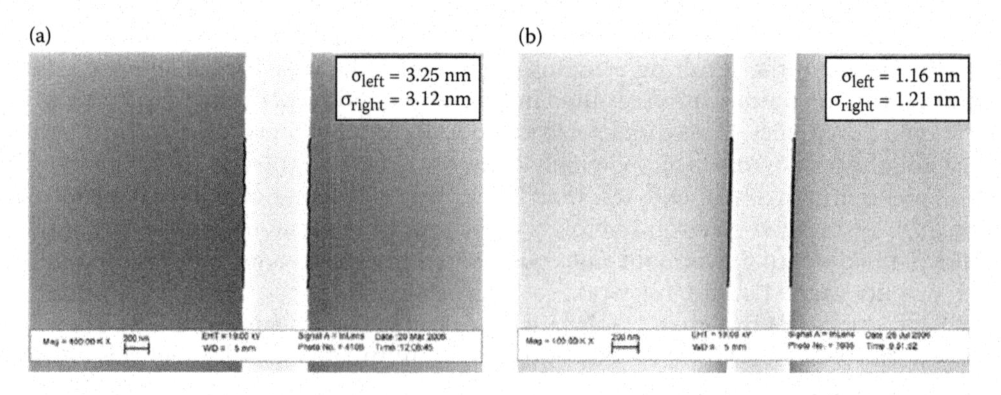

FIGURE 12.20
LER analysis using top-down SEM inspection of waveguides fabricated using chemically amplified negative resist HN-432 developed with TMAH (a) and KOH (b). In these conditions, LER was reduced by a factor of 3. (Reprinted from Ubaldi, M. C. et al. Roughness reduction in submicron waveguides by low-molecular weight development. *Photonics and Nanostructures* 5:145–148, 2007. With permission from Elsevier.)

Other reported process optimizations to minimize LER include reflowing the photoresist [113] to obtain smoother edges and resist development optimization using low molecular size chemicals (such as KOH instead of tetramethylammonium hydroxide; TMAH) [114]. In the latter study, LER analysis using top-down SEM inspection was directly compared with AFM for silicon waveguides and was shown to be a viable non-intrusive alternative as mentioned earlier (Figure 12.20). The etch process must also be optimized, such as the right combination of etching and sidewall passivation and using low forward RF power [115].

As for methods to reduce roughness after silicon patterning, the most prominent include thermal oxidation smoothing [116], wet chemical smoothing [117], and lowering the silicon surface energy by hydrogen annealing [118]. A very interesting novel technique is that of material-selective ultrafast local heating of silicon waveguides by an XeCl excimer laser, which was demonstrated at Princeton [88]. Figure 12.21 shows the results obtained using this technique for 250-nm-wide Si gratings on SiO_2. Although the shape of the smooth structures shown in Figure 12.21b was modified after local heating, the authors reported that a flat top can be maintained by a modified technique [119].

In the case of thermal oxidation, researchers from the NTT found that 1100°C is the optimum temperature to ensure that the relaxation time of the viscous flow of SiO_2 is shorter than the oxidation time, thus achieving very smooth and flat surfaces avoiding the distortion of the waveguide cross-section occurring at temperatures of approximately 900°C to 1000°C [120].

At Pirelli Labs, for our etching processes, the best results in terms of reducing LER were obtained using a SiO_2 hard mask, which was etched with CHF_3 and O_2 in M0RI mode (helicon source), with low ion bombardment and less passivation, removing etch residues after mask etching with DuPont EKC chemicals. For silicon etching, an optimized $SF_6 + O_2$ cryogenic process proved to give LER values comparable to the best HBr-based processes reported in the literature at the time. Surface roughness with standard deviation of 1.5 nm and a correlation length of 13 nm were obtained, translating into 3.5 dB/cm propagation loss waveguides. We also tried reducing roughness through the oxidizing and etching effect of RCA cycles, but found no beneficial effect [67].

When the lift-off process is used for patterning (which is quite common in the case of silicon nitride waveguides due to low selectivity between SiN and SiO_2 hard masks or

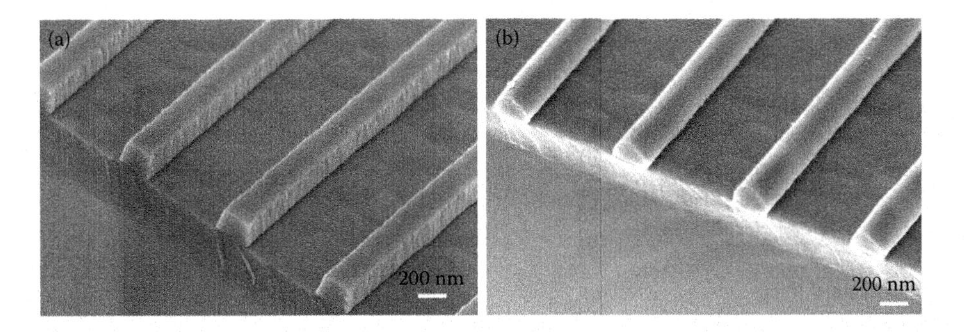

FIGURE 12.21
Si gratings (250 nm) with rough sidewalls (a). Smoothed structures on a single laser pulse exposure at a laser fluence of 490 mJ cm^{-2} (b). (Reprinted from Xia, Q. et al., *Nanotechnology*, 20, 345302, 2009. With permission from IOP Publishing.)

HSQ), the LER of the metal hard mask will replicate that of the lift-off resist but it will also be affected by the hard mask material's microstructure and coverage of the resist sidewall. To minimize these effects, the evaporation parameters and resist sidewall angle, respectively, must be optimized [121]. In this work supported by Pirelli, researchers from MIT found that creating a small undercut in the resist by increasing the dose was beneficial and yet, a larger undercut, as can be obtained with a double-layer resist, became detrimental. It was found that if the undercut was sufficiently large, a second thin metal line was formed along the side of the hard mask strip, which is then unintentionally sputtered during etch and contributes to sidewall roughness. Another interesting result of this study was that a small amount of (controlled) mask erosion was found to be beneficial for LER reduction, presumably due to a polishing effect of the nickel mask edge. Yet another clever strategy used in this work was to optimize the e-beam dose for minimum sidewall roughness, and then the patterns were dimensionally biased in the layout to obtain the desired dimensions.

A recent comprehensive study by Michael Hochberg's group and JEOL USA, Inc., well validated by statistical data has recently been reported and is worth being discussed here in detail [122]. Their research included data from a total of 4840 waveguides in 484 dies, which were measured with an automated wafer-scale system through input/output grating couplers, allowing them to appreciate even small performance variations with good statistical confidence. Their study was focused on the effect on propagation losses of various e-beam writing strategies through their effect on (mostly) sidewall roughness, while also noting any increase in process time. The results were observed and analyzed from direct measurement of optical attenuation rather than physical characterization such as AFM or SEM.

They used an optimized standard etch process and modified the e-beam process parameters under study with the intention of reducing LER, but they did not add any additional process steps that may further reduce the LER (such as those described earlier) to avoid introducing added process variability to the results. The baseline process parameters are summarized below.

- Substrates were 25 × 25 mm samples cleaved from 150 mm SOI wafers with 205- or 220-nm-thick device silicon on 3 μm BOX
- Electron beam exposure with a JEOL JBX-6300FS system at 100 kV, 2 nA beam current, and 500 μm field size (machine grid, 1 nm). Beam stepping grid 4 nm. Exposure dose 2800 $\mu C/cm^2$
- Development in 25% TMAH for 4 min, then DI water rinse for 60 s and IPA rinse for 10 s and N_2 blow dry
- Plasma etching in an Oxford Plasmalab System 100, using 20 sccm Cl_2, 20 mT, inductively coupled plasma power of 800 W, bias power of 40 W, and platen temperature of 20°C resulting in bias voltage of 185 V. Chips were mounted on a 100-mm Si carrier wafer using perfluorinated polyether (PFPE) vacuum oil
- Samples were clad with 2 μm of 950 K average molecular weight PMMA for measurement

Each chip contained between 50 and 120 test cells, as shown in their report. This baseline process had been employed for more than 5 months and yielded a reproducible waveguide propagation loss of 4 dB/cm, with a standard deviation of 0.7 dB/cm.

For straight waveguides, no significant difference was found when the beam current was increased from 2 to 8 nA, whereas in bends, an improvement from 12.6 to 10.6 dB/cm was found. The authors suggest that smoother sidewalls are achieved due to the increased beam diameter and process blur from the higher current. The exposure time per die was also improved by a factor of 3. However, this writing mode can be expected to cause lower ultimate resolution which, although not noticeable in these particular structures, could be pernicious for smaller features in other photonic structures such as slot waveguides or taper tips.

Waveguide losses were consistently lower using a field shift averaging technique available in their e-beam system, in which partial exposure at a certain sample location was done at different times and different stage locations, thus achieving temporal and spatial averaging, respectively, for noise and aberrations of the deflection system including field stitch boundaries. This technique increases writing time by only a small percentage in conditions wherein process duration is dominated by shape writing rather than stage moves. As shown in detail in this article, the results were better than temporal averaging alone. Intrafield distortion is discussed further in the next subsection.

As for lens mode, the baseline process uses the standard, high-speed lens, which has a field size of 500 μm and a 1-nm machine grid. Better performance was obtained with lens 5, which has a shorter working distance and a field size of 62.5 μm and 0.125 nm machine grid, albeit at the cost of doubling the exposure time.

Some of the other parameters evaluated were the generation of the pattern design file, the writing grid or shot pitch, and the effect of simulated stitching errors. Because the GDS-II file does not support curved shapes natively, these had to be approximated by polygons. As one might expect, a higher points-per-circle count gave better results, as also reported by other groups [123]. In particular, a maximum error of 0.3 nm for a 2400-point approximation, which translated to 11.5 dB/cm for a curved waveguide, as opposed to 12.7 dB/cm in the case of a 600-point approximation, which produced a deviation from the designed curve of 3.5 nm. Also, they found that the exposure time was the same for either approximation. It should be noted that other e-beam systems' software use a modified GDS version that accepts concentric circles.

Another unsurprising result was the effect of the writing grid variation, at several steps, from 2 to 32 nm, a study which confirmed that a finer writing grid produced smoother sidewalls (as measured in terms of attenuation loss) as one might expect for both linear and curved waveguides. There was no increase in exposure time for either increasing the polygon digitization or using a finer writing grid, so it was concluded that both strategies are advisable.

Finally, the simulation of stitching errors by introducing intentional offsets in the pattern resulted in an estimated 0.1 dB/cm additional loss per nanometer. Considering the difference in write-fields, this is in fair agreement with the calculations of Gnan et al. [124]. These researchers used a numerical model to simulate the loss introduced by stitching errors as a function of their displacement and fabricated waveguides with intentional stitching errors to experimentally measure the loss introduced. The length of the wires was kept to less than 1.2 mm, that is, within a single write-field, to avoid additional unintentional stitching. The experimental results were in good agreement with those predicted by the model and highlighted the important role of stitching errors in propagation losses for waveguides fabricated by EBL. For example, for a worst case displacement error of 60 nm, which is the specification at 3σ of a Raith 150 (a less sophisticated but very reliable e-beam system, commonplace in many R&D laboratories) the attenuation value estimated from these results is approximately 0.15 to 0.20 dB. Although actual average stitching displacements in a well-calibrated Raith 150 at 20 kV are much lower than that [126], a write-field of 100 × 100 μm will result in the addition of many stitching errors in a waveguide of several milimeters and may represent a significant attenuation factor unless an effort is made to minimize them.

At the time of writing this text, the lowest propagation loss reported for a submicron strip silicon waveguide (0.92 ± 0.14 dB/cm) had been achieved by the group of researchers last mentioned (Gnan et al.) with HSQ, the use of which as an EBL resist has gained enormous popularity for silicon photonics applications [69]. The authors used a 250-nm-thick HSQ layer, a base dose of 1700 μC/cm^2 and 100 keV electron energy in a Vistec VB6 UHR EWF e-beam system. They report taking extreme care to reduce the effect of stitching errors by ensuring substrate flatness and using a tilt compensation technique. This technique and the ways to optimize process reproducibility with HSQ are discussed in detail in the next section.

Despite the research efforts summarised in this section, reducing the propagation loss of standard strip waveguides to less than 1 to 3 dB/cm remains a difficult issue [127]. These waveguides, also referred to as photonic wires, are generally preferred as they allow the highest level of dense integration, which is ultimately one of the main advantages of silicon photonics. Alternatively, single-mode rib waveguides lead to much lower straight-line losses, down to an order of magnitude lower [101], and are advantageous for the realization of low loss crossings [128] and particularly for the fabrication of active devices such as switches, modulators, and detectors [134]. Therefore, in addition to making efforts to reduce sidewall roughness during fabrication or developing postfabrication smoothing techniques, a third way to contend with the issue would be using rib waveguides for their lower loss in straight sections and strip waveguides for their higher confinement in bends. Such hybrid integration has been demonstrated, for instance, in the European HELIOS project by the group of Université Paris Sud (UPS), using a 50-μm-long rib-to-strip transition adding only 0.1 dB loss [79].

12.5.3 Dimensional Control and Process Variability in R&D Laboratories

Direct-write EBL, described in Section 12.2.6, has always played a central role in silicon photonics R&D and deserves to be discussed in more detail from a process point of view. There are two main reasons for EBL's ubiquity in silicon photonics research.

First, it is much more flexible than photolithography because no mask is needed and therefore one can start patterning a newly designed device as soon as the layout on the GDS file is finalized. If the next day a new idea or modification to the design is desired, perhaps as a result of the outcome of the optical characterization of the previous design, one can pattern a new sample immediately after modifying the GDS file. This extremely fast turnaround from design to fabrication cannot be matched by optical lithography. For this type of prototyping, in which small samples with new designs are patterned one at a time, it is both much faster than photolithography and also cheaper in terms of operational costs, as ordering a mask can typically take at least a week and cost more than $1000 (much more in the case of DUV masks).

Second, EBL can achieve higher resolution than optical lithography systems. Despite technological advances such as immersion lithography and multiple patterning, the resolution of photolithography systems is ultimately limited by the wavelength of the radiation they use. On the other hand, lines as narrow as 10 nm have been created by EBL, the resolution of the best e-beam systems being limited by factors such as resist chemistry and beam stability rather than diffraction, as the wavelength of electrons at typical e-beam energies is in the picometer range. Although many practical and experimental designs in silicon photonics can be created within the resolution limits of DUV steppers, the superior resolution afforded by EBL allows for a wider range of possibilities, lower LER in waveguides and other structures, and potentially better repeatability at the resolution limits of DUV.

The main shortcoming of EBL is throughput, that is, the rate at which it can pattern structures, which is orders of magnitude slower than photolithography. This issue makes EBL unsuitable for high-volume production, although it is seldom a crucial problem for an R&D lab, usually being outweighed by the benefits. EBL is also used for low-volume production of III–V photonic devices on 2-in. and 3-in. wafers. It has been used for the production of silicon photonics structures on 6-in. silicon wafers at low volumes, for example, in the fabrication of tunable lasers.

Another important issue with EBL is that of stitching, briefly described in Section 12.2.6. It is certainly of secondary importance compared with throughput because, unlike that issue, it would not prevent the use of EBL for mass production by its own, as there are strategies that may be applied to minimize its effects. Nevertheless, it is an important problem that can affect both propagation losses and reproducibility and must be dealt with both at the design level by trying to keep critical structures within the dimension of write-fields, and during fabrication when the e-beam session is set up, by implementing appropriate writing strategies.

One such strategy is the tilt compensation technique reported by researchers from the University of Glasgow in a publication mentioned in the previous subsection [124]. The technique is particularly effective in reducing stitching errors with substrates that are not perfectly flat (for example, because of film stress). It involves first experimentally obtaining the relationship between the tilt, which can be measured using the system's height meter, and the geometrical distortion error for various tilt values. When a substrate was subsequently loaded into the system, the tilt was measured and the known linear relationship was used to calculate the necessary correction to be applied.

SEM images were used to measure the stitching errors (with an estimated error of ±5 nm) and a graph was plotted with the measured stitching values as a function of distance to the center of the write-field boundary, showing a linear relationship.

This tilt correction technique has the advantage of only increasing pattern writing times by a small amount, but deals only with stitching errors caused by non-flat substrates. There are other approaches that can be explored to minimize stitching errors, such as superimposing multiple exposures with different field sizes [125,129], methods to reduce the field size [130] and MEBES-like writing strategies [131]. An example of the latter is Raith's own fixed beam moving stage mode in which the stage moves continuously as opposed to staying still while each write-field is written, thus avoiding stitching altogether [132]. However, they all have their own complexities and some require special hardware and software modifications.

Additional sources of process variability in EBL will now be discussed, including geometrical drift, e-beam current drift, intrafield distortion, and proximity effects. Resist-related issues, such as storage conditions and ageing are also important: the precision of dose tests at different points through the resist's shelf-life has a large effect on run-to-run variability (Figure 12.22). Although not directly connected with EBL processing, the varying thickness of the waveguiding material, usually the device layer in SOI, is also an important source of variability, particularly over large distances across the wafer and for run-to-run uniformity.

In long exposures, geometrical precision and process uniformity can be improved by the use of scripts in the e-beam software to force the system to perform several automatic adjustments such as refocusing, realigning the Cartesian grid to the alignment marks,

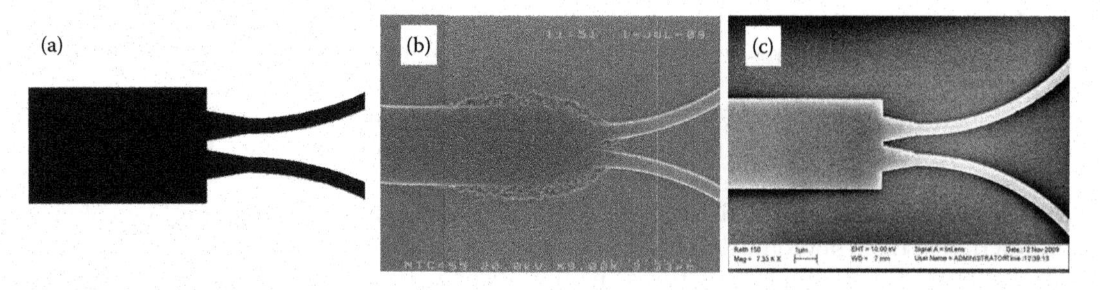

FIGURE 12.22
Manual adjustments of dose in e-beam lithography. Desired pattern of MMI device (a), dose optimized for waveguides only (b), and dose optimized separately for each element (c). (Courtesy of Fabrication team Nanophotonics Technology Center, Univeristat Politècnica de València, Valencia, Spain, 2009.)

and recalibrating the dose by measuring the current in long exposures. Also, the choice of layout orientation to expose the most critical structures first to minimize the effect of drift can have a significant effect, particularly in systems with wider drift tolerance specifications [133].

Intrafield distortion, which was also mentioned in the previous subsection, is due to systematic beam-deflection error (Figure 12.23) within a single field, and can be caused by both electron–optics aberrations and digital–analog converter error [135]. It was shown to be the main cause of frequency mismatch between high-order SiN ring-resonator–based filters exposed with a Raith 150 e-beam system. The authors presented a method to reduce its effect by mapping the distortion introduced by the system and correcting for it by predistorting the beam position.

For correcting proximity errors, Barwicz et al. [121] suggested the use of dummy structures noticing that the use of standard proximity error correction algorithms for ring resonators in their e-beam system could stress the system into higher inaccuracy. They also emphasized the need to choose all device dimensions to be multiples of the step size to avoid discretization errors. In this work, accurate etch depth control of the SiN waveguides was achieved by stopping the process a few times to measure the step height with a profilometer. In terms of linewidth, by adjusting the exposure dose of the middle ring in series-coupled third-order filters, a dimensional control of 5 nm was demonstrated and the average ring-waveguide width achieved was within 26 pm from the target relative width offset.

Researchers in the same group later demonstrated a 20-channel second-order dual filter bank with average channel spacing of 83 GHz (80 GHz target) and a standard deviation of 8 GHz in SiN. This corresponds to a dimensional precision of the relative change in average waveguide width in the ring of 75 pm and a random variation of 210 pm. Again, this was achieved by fine-tuning the waveguides' width by adjusting the exposure dose. They first calibrated the frequency shift obtained by changing the radii of the rings by one pixel (6 nm) and by modifying the dose by up to 20% with respect to optimum dose. Then, they used the calibrated data to fabricate the filter-bank with calculated values for the ring radii for coarse adjustment and modifying the dose (a maximum of 5.7% change was sufficient) for fine-tuning. Calibration experiments were also performed to characterize the exact intrafield distortion and proximity effects on the rings so that they were also corrected for by dose adjustment [136].

At IBM, researchers [137] fabricated an optical delay line of 56 nominally identical rings of 5 μm diameter using a Leica VB6-HR at 100 keV and an optimized process including

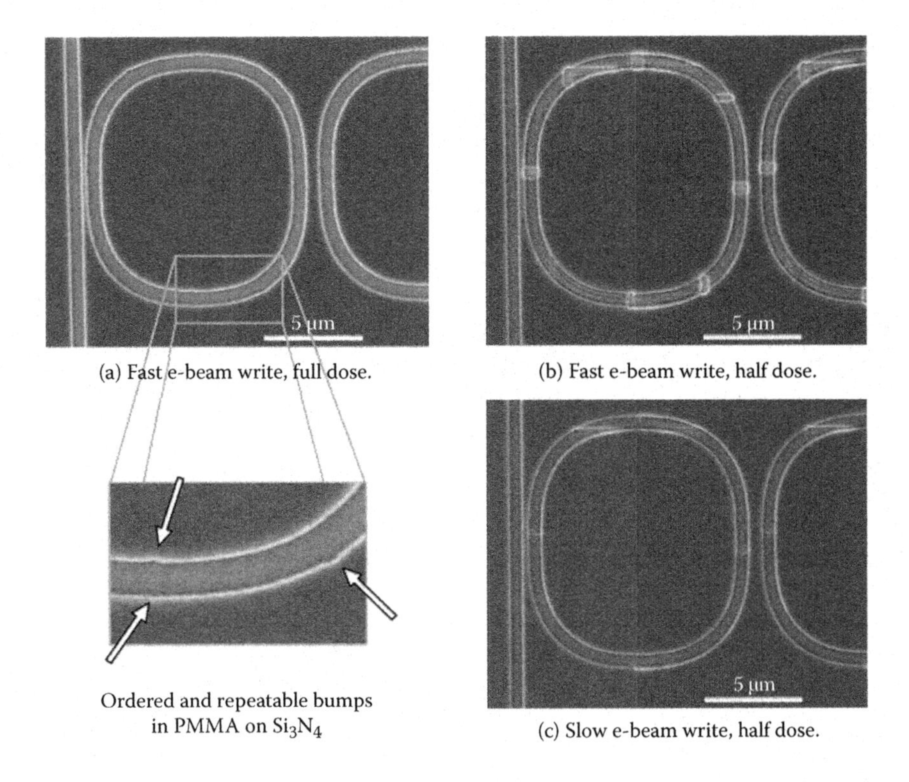

(a) Fast e-beam write, full dose.
(b) Fast e-beam write, half dose.
(c) Slow e-beam write, half dose.

FIGURE 12.23
Electron micrographs demonstrating e-beam deflection errors introduced by a Raith 150 e-beam system in racetrack resonators written in PMMA on Si_3N_4. A fast write speed generates 50 to 100 nm bumps on the racetracks (a). The low-dose exposure shows that the bumps are mainly due to double-exposed resist (b). A slow write speed reduces the deflection errors (c). (Courtesy of Dr. Tymon Barwicz. Reprinted from T. Barwicz's MIT PhD Thesis, Copyright 2005, available at MIT's website, with permission from MIT Technology Licensing Office.)

double oxidation steps to reduce sidewall roughness [64]. The waveguides showed a very low loss of 1.7 ± 0.1 dB/cm and a sidewall roughness as measured by AFM of only $\sigma = 1.1$ nm. By data-fitting, they showed that the variation of resonant wavelengths had a standard deviation of 0.4 nm.

In addition to the complex and sometimes monotonous exercises discussed in some of the reports previously mentioned, other steps necessary to achieve high dimensional control and maintain high process uniformity in EBL involve more simple but methodical actions. These include a careful setup of the exposure session with special attention to substrate cleanness, a thoughtful design of the layout to keep the most sensitive structures within single write-fields whenever possible, the use of as many alignment marks as necessary in the most appropriate locations, performing precise exposure dose tests, and other manual adjustments. The steps involving resist processing before and after exposure are also of great importance as exemplified later.

HSQ has been used as an interlayer dielectric material due to its low dielectric constant, and good performance in terms of gap fill and planarization capabilities. It is also known by its commercial name Fox or Flowable Oxide, although diluted solutions intended for lithography are now sold under a different name (currently XR-1541 by Dow Corning Midland, Michigan). It is not sensitive to UV radiation at more than

157 nm [138] but researchers at NTT demonstrated that it had good characteristics as a negative tone e-beam resist [139]. It has since become very popular due to its high resolution with moderate sensitivity, low LER, good etch resistivity, and the possibility of being imaged on silicon by SEM without gold coating. It can be thought of as either a siloxane-based polymer that, when cured, undergoes the scission of Si–OH and Si–H bonds and forms a Si-O-Si network, or as a caged oligomer that, during curing, opens and forms a network structure, thus gradually reducing the caged/network ratio [140,141]. We will not discuss its chemistry in detail, but it is important to know that what this basically means is that HSQ, when exposed and cured, essentialy becomes silica (as the name Flowable Oxide suggests).

Although some reproducibility issues have been reported with this resist [142], these can be minimized through an awareness of the main factors that affect its performance and through good laboratory practices including systematic fabrication procedures, such as precise timing between processes.

For instance, in the article discussed in the previous subsection, in which Gnan et al. [69] demonstrated 0.92 dB/cm propagation loss, the repeatability of the process was also studied by fabricating four standard photonic wire Bragg gratings over a period of 37 days. From the measurement variations, they concluded that there was a dimensional change, which was 1.5 nm on average. Later that year, they published a report, in which they showed that by preparing fresh dilutions of HSQ immediately before its use, reproducibility increased dramatically, resulting in waveguide width variations of less than 0.5 nm (which was also the 3σ precision of the measurement method) for 500 nm waveguides over a period of 70 days [68]. They also highlighted the remarkable thickness uniformity of their SOI wafers, although they did mention in that article that they were using samples from the same region of a 200-mm wafer. Good thickness uniformity over short distances, unlike that across the whole wafer, was also reported in some of the research reviewed in the next subsection. Incidentally, they also found a consistent width deviation between specific positions in the write-field, which is also compatible with the findings of Prof. Smith's group at MIT, and the results relative to write-field averaging obtained by Hochberg's group, which were both previously reviewed.

Apart from the handling of the resist and of course the crucial exposure step, other lithographic process steps are also important. In an extensive study of the interplay of the various process parameters that affect processing with HSQ, researchers at AMO [143] found that by increasing the TMAH developer concentration from 2.38%, which at the time was reportedly a commonly used concentration, to 25%, both the contrast and the reproducibility were improved. They also found that a 14-day delay between baking and exposure led in general to lower sensitivity and higher contrast but its effect was only significant at temperatures higher than 150°C and particularly at lower TMAH concentrations. For the various baking temperatures that they considered before exposure (90°C, 120°C, 150°C, 180°C, and 220°C all for 40 min), the lowest temperature, that is, 90°C produced the best results in terms of reproducibility and contrast. They noted that baking can be considered as a kind of pre-exposure of the whole HSQ resist. This trend of higher contrast at lower temperatures is in agreement with our own experience at the NTC, and in fact, we found that drying the resist in a vacuum chamber with no prebake at all further increases the contrast with no noticeable deleterious effects [74]. This has also been reported by other researchers [144], who also demonstrated that contrast was increased to approximately 50% when development with TMAH at 25% concentration was realized at 60°C rather than at room temperature. A further advantage of drying the resist in vacuum is the reduction of surface roughness, an effect most noticeable for low resist thickness [145].

The studies discussed above are, in some cases, specific to particular resists or e-beam systems. A broader treatment of this subject is not possible due to space constraints and perhaps it would be impossible to cover all cases of interest to silicon photonics. The references provided are a good starting point to learn more on these issues. Nevertheless, the most important message the neophyte reader or student should identify is the fact that variations in even the most basic process steps in nanofabrication may have significant consequences on the results and should therefore be explored to continuously search for improved recipes and procedures. For the same reasons, they should be kept strictly unchanged when process reproducibility is sought, unless the process window (the range within which process parameters may be changed with no noticeable effect on the results) for each process is well known. To this end, the accurate fabrication necessary for high performance silicon photonics devices requires a methodical mindset while in the clean room, and the utmost attention to every detail in following fabrication procedures.

12.5.4 Dimensional Variability in CMOS Lines

As discussed earlier, the high index contrast between the silicon core and the silica cladding in the SOI platform allows for high levels of integration but it also causes extreme sensitivity to nanometer-sized variations in feature dimensions, including unintended deviations due to fabrication tolerances. For wavelength-selective devices in SOI waveguides, 1 nm tolerance in waveguide dimensions leads to a spectral shift of the same order of magnitude in resonant wavelength, which gives an idea of the high degree of sensitivity involved (the exact value depends on the waveguide geometry, see Figure 12.24 for the case of ring resonators in standard strip waveguides and [147] for a specific rib waveguide design, including etch depth sensitivity). This low tolerance makes it virtually impossible to achieve accurate and reproducible devices in an ad hoc laboratory experiment, let alone in a large-volume mass production environment. In fact, in SOI photonics, this prerequisite would not be satisfied from the onset because typical commercial SOI device layer thickness variations can be almost 10 nm for 220 nm SOI (Figure 12.25). Furthermore, except for applications involving a precise thermal control, ambient thermal variations will also induce a displacement of the resonant frequency of SOI devices. Therefore, so-called *trimming* and *tuning* strategies will need to be implemented for most applications, that is, postprocessing techniques to modify the geometry or behavior of these devices whether permanently for fabrication-related deviations (trimming), dynamically for thermal variations (tuning), or both.

The implementation of these strategies is at best very complex, time-consuming and, in the case of active tuning, a drain to power consumption budgets. The fact that techniques such as trimming each device using an AFM tip (via electric field–induced nano-oxidation of silicon) are being proposed [148] as a possibility should illustrate the extent of this power consumption problem. Hence, to minimize the effort required from tuning and trimming techniques, it will be necessary to perform thorough investigations of the effects of actual fabrication tolerances on the desired device performance and functionality, and to use this information to devise novel ways for targeted improvements of the fabrication processes.

Enticed by the benefits that CMOS compatibility will eventually bring to the fabrication of silicon photonic devices, a great deal of research [149] and technology development [150] efforts have been expended on new concepts, designs, and processing techniques for new photonic structures and devices as well as on the improvement of the performance of existing ones. Surprisingly, relatively few attempts at studying and addressing the issues of fabrication tolerances, uniformity, reproducibility, yield, or mass manufacturability have

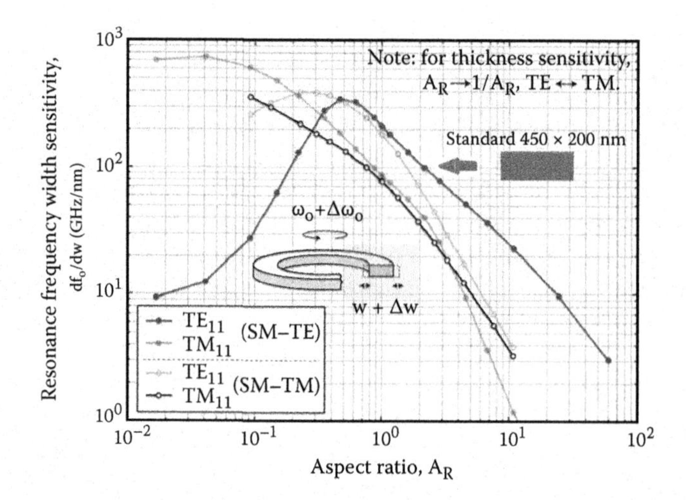

FIGURE 12.24
Resonant-frequency dimensional sensitivity in strip waveguide ring resonators. (Adapted from Popović et al. In *Conference on Lasers and Electro-Optics (CLEO)*, 2006. With permission from the Optical Society of America.)

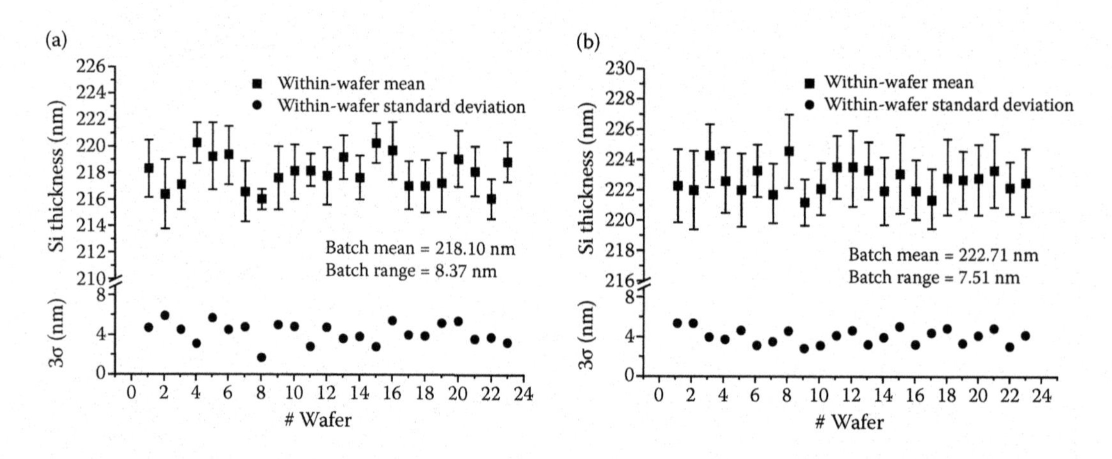

FIGURE 12.25
Within-batch and batch-to-batch device layer Si thickness nonuniformity for 200 mm SOI wafers. The data shown on (a) and (b) correspond to two different wafer batches. (Reprinted from S. Selvaraja's PhD thesis, available online, with permission from Intec, University of Ghent, Belgium.)

been made until recently. Yet, a deep understanding of these issues and of the possibilities of matching silicon photonics manufacturing requirements to current CMOS processes will be required to acquire a realistic view of what is achievable today and to determine what viable adaptations or new strategies may be necessary in the future. In the remainder of this section, the main results of some of those recent reports will be reviewed. We can expect a large amount of further work to be carried out in this area in the next few years.

Early work on this subject includes that of Pieter Dumon [151] for his 2007 PhD thesis at the University of Ghent (Belgium). Working at IMEC's 200 mm CMOS pilot line and employing 248 nm DUV lithography, he investigated *within-die* (WID) uniformity at shorter and longer scales by placing ring resonators and MZIs a few tens of microns and a few milimeters apart, respectively. He also measured different dies to study *die-to-die* (or *within-wafer*, WIW) uniformity. The resonant frequency deviations that were found were demonstrated to be larger over longer distances, and the difference was shown to be greater than that attributable to SOI thickness variations alone. Thus, the main cause was fluctuations in waveguide linewidth over various scales and in sidewall slopes. These fluctuations were in turn suggested to be due to a combination of mask inaccuracies, local variations in both resist thickness and illumination dose, and the effect of etch nonuniformity. Hence, he predicted that a more accurate process would improve the die-to-die uniformity, which was confirmed in subsequent publications. It was determined that the frequency shifts measured within a specific wafer region were consistent for both filter types, ring resonators and MZIs. However, no correlation was found amongst specific wafer regions across wafers (such as top to bottom or as a function of distance from the center). Many of these early findings have been confirmed in subsequent studies (including those of other groups) as discussed later.

More recently, also at IMEC, Selvaraja et al. [152] have developed a state-of-the-art fabrication process in a CMOS line with 193 nm lithography and etch recipes tailored for photonic devices and then used it to investigate uniformity at various levels. They had shown the remarkable benefits that switching from 248 nm lithography to 193 nm lithography and improved etching had on linearity, process window, and proximity effects in a previous publication [65] whereas they now demonstrated how the increase in process uniformity is absolutely critical for achieving a less than 1% nonuniformity (1σ). They reported the results of thickness measurements by ellipsometry for 200 mm SOI wafers with a 220-nm device layer (Figure 12.25) as well as a bottom antireflective coating and photoresist films. The device layer thickness was mapped as a function of position on the wafer and showed radial symmetry as could be expected from the CMP step in the production of SOI wafers [153].

They also reported the standard deviations of an automated CD-SEM characterization of linewidths both after lithography (σ = 2.0 nm) and after etch (σ = 2.6 nm) across a 200-mm wafer, at 175 locations from a previous work [154]. These measurements are reproduced in a different wafer-mapping format in Figure 12.26. The photoresist linewidth map (Figure 12.26a) shows no clear symmetry on the wafer, which is symptomatic of good resist

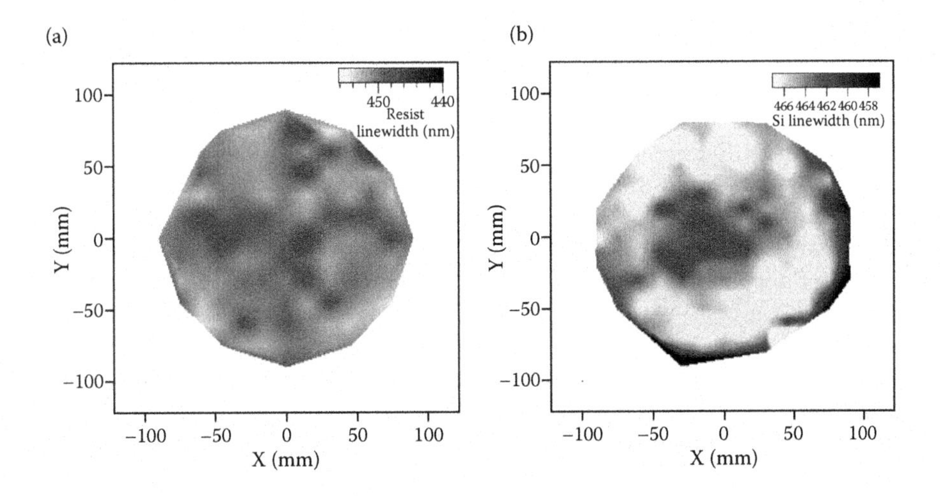

FIGURE 12.26
Photoresist linewidth uniformity after projection photolithography (a) and silicon waveguide linewidth after etch (b). (Courtesy of Dr. Shankar Selvaraja, Intec, University of Ghent, Belgium.)

and bottom antireflective coating thickness uniformity, whereas Figure 12.26b shows two distinct regions (center and periphery), a type of symmetry that is typical of plasma etch.

It was noted that the average linewidth along the device is actually what defines its performance rather than that measured at specific points, and hence, optical characterization is the only way to measure device uniformity. To this end, they investigated the intra-die (WID) and inter-die (WIW) uniformity of the spectral response of three wavelength-selective devices, namely, racetrack ring-resonators, MZI, and arrayed waveguide gratings, placed at various locations on a die, the same die repeated at several locations on a wafer. A significant (double-digit) number of devices were measured. The WID resonance wavelength nonuniformity had a mean standard deviation of $\sigma = 0.15$ nm for racetrack resonators separated by 25 μm and $\sigma = 0.55$ nm for devices 1.7 mm apart, whereas for MZIs, the mean σ was slightly higher. WIW nonuniformity at 10 and 20 mm separation was $\sigma = 1.3$ nm and $\sigma = 1.8$ nm, respectively, for ring resonators and, in this case, slightly lower for MZI devices. For arrayed waveguide gratings, the WIW mean standard deviation was similar to that obtained for WID measurements at near 0.5 nm.

The analysis of their measurements, including plots of the resonant frequencies versus device location within a die for various dies, allowed them to conclude that WID nonuniformity is mainly caused by mask errors whereas WIW nonuniformity is mainly caused by etch plasma and wafer thickness variations. Additionally, it led them to suggest a process flow model to improve device uniformity in production environments featuring a send-ahead wafer and exposure dose compensation for dry etch nonuniformity. In addition to the articles referenced here, a detailed account of their results and ideas can be found in Dr. Selvaraja's PhD thesis [155].

In 2010 researchers at HP Laboratories presented a report on a process uniformity study realized in an external foundry with a 248-nm DUV lithography process [156]. They investigated the effects of mask accuracy by employing two different masks with ± 15 nm and ± 30 nm CD tolerance, both with a 20-nm grid on the mask (4x stepper reduction). They studied ring resonators with various nominal diameters and coupling gaps in a rib-waveguide design 450 μm wide and 200 nm deep, leaving a 50-nm silicon layer (250 nm device layer SOI).

One type of device had 8 rings with 10 μm nominal diameter which were designed with an equal channel spacing of 80 GHz (~0.44 nm at 1310 nm). The measured spacing varied between 10 and 108 GHz. They also studied devices with a single ring of 10 μm nominal diameter and found a WID standard deviation for the resonant frequencies of ± 0.4 nm for 500 × 500 μm dies. The WIW standard deviation was 2 nm [157].

The effect of mask CD tolerance was illustrated by a plot of quality factor versus extinction ratio of 32 rings of 5 μm diameter on a single device, which showed a clearly tighter spread for the more accurate mask.

Later that year, Zortman et al. [158] published their work on silicon photonics manufacturing using Sandia National Laboratories' 0.35 μm process with 248 nm lithography. They investigated wafer-to-wafer (WTW), WIW, and WID resonant frequency uniformity of 40 nominally identical 6-μm diameter microdisk resonators on two SOI wafers (identified as W1 and W2 in Figure 12.27) with 240 nm device silicon on 3 μm buried oxide. There were eight dies distributed along one direction on the wafer from the bottom of the wafer (where the flat or notch which indicates crystal orientation lies) to the opposite side, which we shall call the top of the wafer. Each die had five microdisks coupled to waveguides by a 320-nm separation, the microdisks were placed left to right with a 75-μm separation to study short range fabrication variations (the farthest distance between resonators in a die is therefore 300 μm).

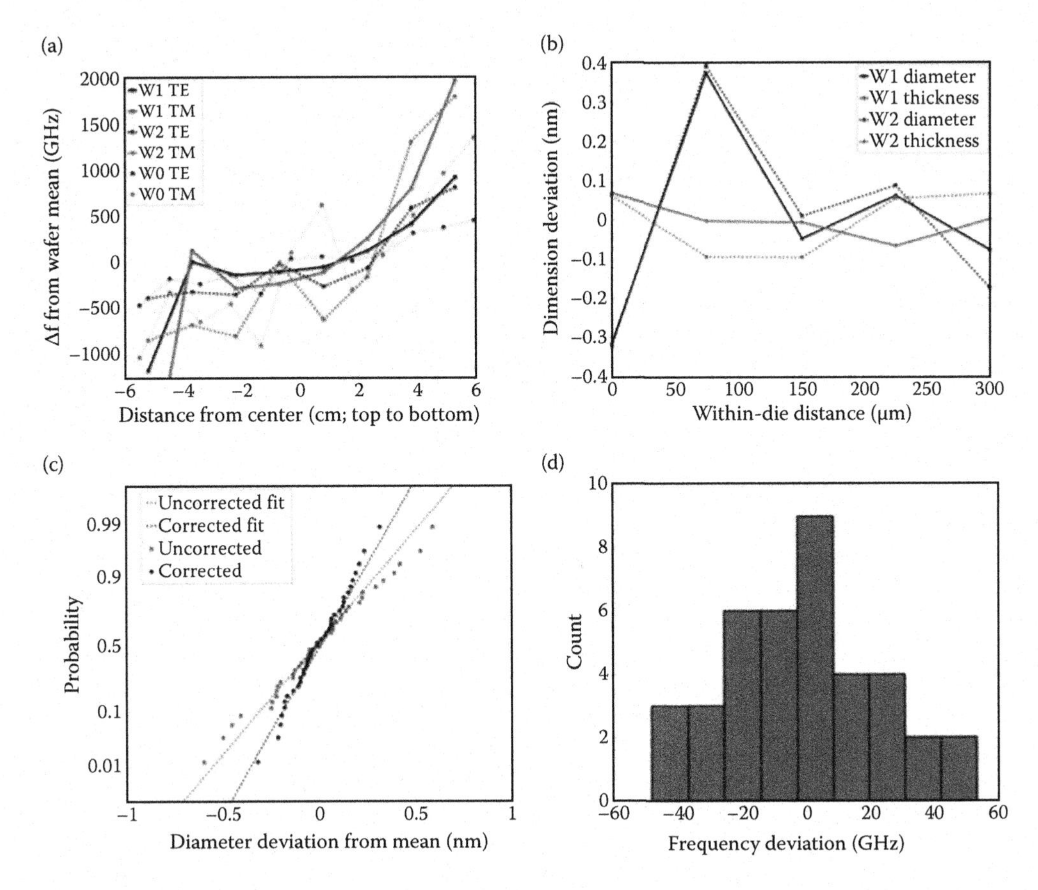

FIGURE 12.27
Results from the resonant frequency variations study of Zortman et al. [158], see text for details. Frequency deviation by wafer location (a). Within-chip dimension deviation (b). Probability plot of within-die width variation (c). Diameter-corrected frequency deviation histogram (d). (Reprinted from Zortman, W. et al., *Opt. Express* 18(23):23598–23607, 2010. With permission from the Optical Society of America.)

Microdisks were chosen for this study because their resonant frequencies depend only on their thickness and diameter, that is, two unknowns that can be solved in a linear system of two equations derived using mode solver predictions and measured resonant frequency information from both TE and TM polarizations.

All 80 microdisks (5 × 8 dies × 2 wafers) were measured and data from a third wafer (W0) from a previous work was also provided. They reported the variations from the mean and not from the design resonant frequency, pointing out that any known bias of a well-characterized, albeit variable process can be compensated in the design.

Figure 12.27a shows, for all three wafers, the TE and TM resonant frequency variations from the wafer mean. Each data point for W1 and W2 represents a die and gives the mean of the five resonators. W0 and W1 had gone through the full process, whereas W2 did not have metal contacts or doped regions. No significant differences in the measurements were found that could be attributed to the different processing of the wafers. However, a clear trend can be seen in all three wafers—that of increasing frequency from top to bottom, which was attributed to thickness variation. The difference between wafer means was more than 1 THz, whereas the standard deviations were both near 500 GHz.

Given the number of dies and wafers used in this study, the most relevant results were the WID variations. This is in fact what the authors were most interested in because they argued that if small differences in resonant frequency among the devices in a single die were corrected by thermal tuning or otherwise resolved, the solution to larger differences between dies, wafers, and runs could be device *binning*. This is the process of testing individual dies and then sorting them according to performance, an automated process that is already employed in high-volume production of microelectronics devices.

For each of the 40 microdisks in a wafer, the WID variation was calculated by comparing its resonant frequency to the die average. Then, through the linear system equation they had derived, for each die, the average WID dimensional die variation is calculated (these are illustrated in Figure 12.27b). The data in this graph shows a small ±0.1 nm thickness variation which should be no surprise over such small distances; however, the ±0.35 nm diameter variation was unexpected and was thought to be the result of a mask error, which is consistent with the fact that it seems to affect the two leftmost disks only and both wafers in a similar fashion. In fact, on inspection of the CAD file, a drawn error of ±0.7 nm due to the grid snap error was found.

An analysis of variance test and residual analysis were performed with this data, yielding a new set of diameter-corrected data in which the effect of the mask error had been removed. This is shown as a probability plot with respect to frequency variation in Figure 12.27c, and for the case of diameter-corrected data also in histogram format in Figure 12.27d.

From the corrected data, the capability of the process was found to be 105 GHz for dies in which maximum separation between devices was 300 μm. Hence, in a high-volume manufacturing environment in which dies were sorted by device binning, the resonant frequency of the devices would still need to be tuned 50 GHz on average.

Considering a heater efficiency of 4.4 μW/GHz as previously demonstrated [159], they estimated a power consumption of 231 μW to overcome fabrication-induced variations and a further 1.1 mW required for temperature-induced resonant frequency displacement for the case of a 25°C temperature range. Both of these values are much larger than the switching energy required for a modulator and therefore the need to increase thermal tuner efficiency is highlighted. To this end, new clever designs will need to be conceived and innovative fabrication processes will need to be developed.

In one of the most comprehensive studies to date, researchers from Oracle Labs and Luxtera [147] analyzed characterization results from more than 500 four-channel microring arrays from several wafers in four different lots, hence, obtaining results that were also relevant for WTW and run-to-run uniformity. They used 1 × 4 cascaded microring arrays fabricated in an 8-in, 0.130-μm CMOS process line using rib waveguides of 300 nm thickness, 360 nm width and 150 nm slab height with various ring diameters and coupling gaps. In a first wafer lot, nominally identical rings were spaced 500 μm apart to study chip-level nonuniformity and they found that the variation of the ring resonances was very large as predicted, but the chip-level spread was less than 1.6 nm (200 GHz).

In subsequent lots, the four cascaded rings were designed with incremental radii for a channel spacing of 1.6 nm at 1550 nm and were positioned 45 μm away from each other. In one of the lots, the wafers underwent the full CMOS flow including back-end metals and interlayer dielectrics for comparison, but no difference in the results was reported. They studied the free spectral range (FSR), channel spacing, and central wavelength. Once again, it was confirmed that the designed resonant wavelength could be off by more than 10 nm due to substrate thickness variations and fabrication tolerances even within the same wafer. Moreover, these variations were shown to be essentially random because no systematic pattern of resonance shift was found across the wafers when the variations were mapped as a function of position on the various wafers. This is consistent with the findings of Pieter Dumon's early work described above. However, the WIW and even WTW variation in FSR was very small at 0.48 and 0.66 nm (6σ), respectively, for 7.5 μm radius rings and 0.27 and 0.60 nm for 10 μm radius rings. Just as importantly, it was observed that it was also possible to control the channel spacing (6σ variations stayed at less than 0.80 nm in all cases for this set of data) for these closely spaced rings, an outcome which is consistent with the low resonance-wavelength spread found in the first lot of wafers. Although all the results shown corresponded to a specific coupling gap, it was reported that similar results were obtained from the other data groups.

The controllability of both FSR and channel spacing led them to propose a method to reduce the required tuning range of ring resonators in a properly designed system in which FSR ≈ (number of channels * channel spacing). Instead of tuning each ring to its designed wavelength λ_1, one could tune another ring the resonance of which is closest to λ_1 and continue assigning consecutive channels to the neighboring rings such that a WDM link is established just once at system initialization. This way, the tuning range is minimized to a fraction of the FSR. The idea was experimentally demonstrated with an eight-channel synthetic resonant comb with an FSR of 12.8 nm and 1.6 nm channel spacing using built-in silicon resistors [160], resulting in an average tuning requirement of less than 1.2 nm/channel.

Using these experimental results and Monte Carlo simulations, they calculated that, for an eight-channel microring array in a transmission link at 20 Gb/s, the thermal tuning energy per bit can be as low as 15 fJ/bit in the case of undercut free-standing strip-waveguide rings which had been recently demonstrated [161] to consume as little as 2.4 mW per FSR (Figure 12.28). This would be an important milestone as it is in line with the power budget targeted for optical interconnects [162].

Table 12.3 summarizes the main results of the investigations reviewed above. Although information such as typical separation between devices has been provided for completeness, too many other crucial variables are missing to even attempt to make a comparison between these results. Each group pursued a particular aim with their study, therefore different experimental methodologies have been followed, from the way uniformity statistics

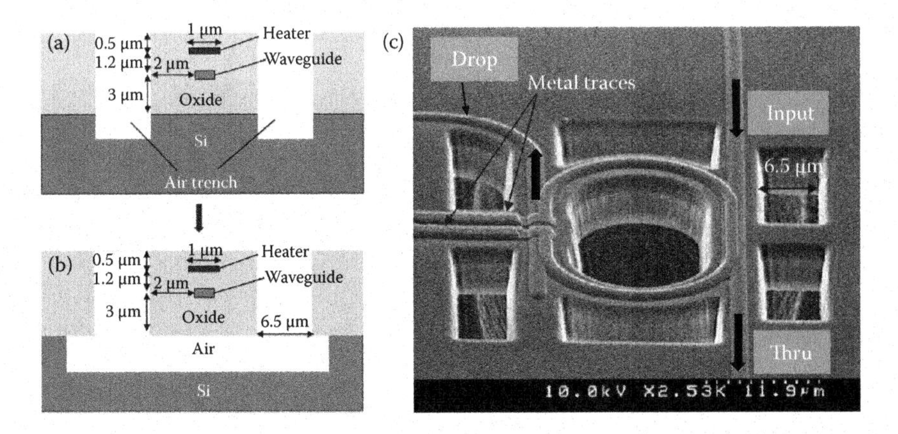

FIGURE 12.28
Cross-section of the ring-resonator waveguide and integrated heater showing trenches (a), which are used to etch the silicon beneath the ring (b) thus providing high thermal isolation. SEM micrography of the fabricated device (c). (From Dong, P. et al., *Opt. Express* 18(19):20298–20304, 2010. With permission.)

TABLE 12.3

Summary of the Main Numeric Results of the Resonant Frequency Uniformity for Silicon Photonic Wavelength-Selective Devices Fabricated in CMOS Lines

Publication	W-I-D Res. Freq.	W-I-W Ch. Spacing	W-I-W Res. Freq.	W-T-W Ch. Spacing	Range/6σ	Separation btw Devices	Additional Information
Selvaraja et al.	0.9 (3.3)	–	7.8–10.8 (10–20 mm sep.)	–	6a	25 μm (1700 μm)	Strip waveguides 450 × 220 nm, racetrack resonators r = 4 μm, CL = 4 μm
Peng et al.	2.4	–	~12	–	6σ	<500 μm	250-nm rib waveguides with 50-nm slab, 450 nm wide, r = 5 μm rings, 1310-nm radiation
Zortman et al.	0.84	–	~8	–	range	75–225 μm	Residual-analysis corrected data, r = 3 μm Microdisks
Krishnamoorthy et al. (r = 7.5 μm)	–	0.54–078 (~0.5–1)	(~12)	0.66 (~1.2)	6σ (range)	45–135 μm	300-nm rib waveguides with 150-nm slab, 360 nm wide, r = 7.5 μm
Krishnamoorthy et al. (r = 10 μm)	–	0.42–0.72 (~0.4–0.6)	(~10)	0.57 (~0.8)	6σ (range)	45–135 μm	300-nm rib waveguides with 150-nm slab, 360 nm wide, r = 10 μm

Notes: Separation between devices refers to distances (in microns) within a die. All resonant frequency and channel spacing values in nanometers; results not directly comparable (see text for details).

are calculated to the differing mask accuracy levels, device designs, waveguide types, die size, number of devices per die, etc. With all these considerations in mind, the table does provide a fairly congruous general picture of what these groups have found to be achievable in CMOS lines.

Sections 12.5.3 and 12.5.4 have dealt primarily with the causes and consequences of fabrication tolerances for silicon photonics devices, whether in an R&D environment or in CMOS fabrication lines. As a final note, the need for photonic devices designs to be as fabrication-tolerant as possible should be highlighted. To this end, it is important for designers to be aware of the degree to which some fabrication processes and specific device features are prone to stochastic variations or particularly challenging to fabricate in a reproducible manner. It is also important for designers to keep an open mind and continuously challenge one's assumptions as to what is viable for fabrication and what is not whenever new circumstances arise or specific applications are targeted. This argument is also valid for technologists regarding the fabrication techniques to be employed. As an example that illustrates these suggestions, we could consider the case of a new grating coupler design in which the silicon is fully etched and, for this reason, easier (in principle) to fabricate reliably but the design of which is abandoned for insufficient bandwidth or low efficiency. If and when new applications are encountered in the future in which bandwidth is not a requirement or a situation arises in which grating couplers are needed only as a wafer-level test structure, then, at that point, the more fabrication-tolerant design should be reconsidered.

A few published examples of fabricated devices featuring designs tolerant to fabrication variations include:

- The optical modulator discussed in Section 12.4.2, which was fabricated with a self-aligned process tolerant to overlay inaccuracies [163]
- Beam splitters [164] and polarization splitters [165] tolerant to waveguide width variations
- Diffraction gratings tolerant to sidewall verticality [166]
- and the fully-etched grating couplers mentioned in the previous paragraph [167]

The latter have better reproducibility than grating couplers in which the etch is timed to a precise depth [167], assuming the required lithography precision is available because, in this particular case, some of the difficulty is transferred from the etch to a more demanding lithographic step.

12.6 Summary

A general overview has been provided of four fundamental aspects of silicon photonics fabrication: the most common tools of microfabrication technology, the main material technology platforms, a generic and a few illustrative microfabrication process flows, and the main technological issues that silicon photonics confronts from the point of view of fabrication. A few subjects have been omitted due to space constraints, including processes such as epitaxy or electroplating and the relevant fabrication equipment, and such materials as polymers (for functional claddings), germanium, porous silicon, and silicon nanocrystals. More emphasis has been placed on mainstream CMOS-compatible processes and materials rather than on those which are simply "silicon-based." The criteria for selecting the number of topics and the level of detail could, of course, have been different, but the intention has been to make this chapter of benefit to as large an audience as possible, while also attempting to strike a balance between breadth and the level of detail necessary for some topics.

In many occasions, in which publications have been cited, including the few process flow examples provided in Section 12.4, many alternative options are available in the literature. The reader or student is encouraged to expand the knowledge just acquired by identifying those articles in which new fabrication issues or techniques are presented. The proceedings of the *Conferences on Group IV Photonics* is a good place to start. For readers most interested in specific fabrication aspects, the references therein will eventually lead them to journals such as the *Journal of Vacuum Science and Technology A and B, Applied Physics Letters, Microelectronic Engineering*, etc.

Acknowledgments

The author thanks all his present and former colleagues who shared their technical knowledge over the years at NEWI (Wrexham, UK), Pirelli Cables (Bishopstoke, UK), Pirelli Labs (Milan, Italy) in particular Sergio Doneda and Massimo Gentili, MIT (Cambridge, Massachusetts), and Nanophotonics Technology Center (Valencia, Spain) in particular the fabrication team. Special gratitude goes to Prof. Hoyt (MIT) for her permission to attend her lectures on *Physics of Microfabrication* as a visiting scientist in 2003. Her handouts (course 6.774) have served as a catalyst for ideas for Section 12.2 of this chapter. Any inaccuracies are my sole responsibility. I also thank Francesco Della Corte (Università Mediterranea Reggio Calabria, Italy) for his helpful comments on Sections 12.3.3 and 12.3.4 and those who helped through private communications (see references).

References

1. NTC-UPVLC. Nanophotonics Technology Center, Universitat Politècnica de València, Valencia, Spain (www.ntc.upv.es).
2. Paniccia. M. 2010. Integrating silicon photonics. *Nature Photonics* 4(8):498–499.
3. Fedeli, J. M., L. D. Cioccio, L. Vivien, R. Orobtchouk, C. Seassal, and F. Mandorlo. 2008. Development of silicon photonics devices using microelectronic tools for the integration on top of a CMOS wafer. *Hindawi Publishing Corporation—Advances in Optical Technologies* 2008(412518).
4. Orcutt, J. S. et al. 2011. Nanophotonic integration in state-of-the-art CMOS foundries. *Optics Express* 19(3):2335–2346.
5. Hochberg, M., and T. Baehr-Jones. 2010. Towards fabless silicon photonics. *Nature Photonics* 4(8):492–494.
6. ePIXfab. [Online]. Available: www.epixfab.eu.
7. Dumon, P., W. Bogaerts, R. Baets, J.-M. Fedeli, and L. Fulbert. 2009. Towards foundry approach for silicon photonics: Silicon photonics platform ePIXfab. *Electronics Letters* 45(12):581.
8. Opsis. University of Washington, Seattle, WA [Online]. Available: depts.washington.edu/uwopsis/.

9. CNF Cornell University. Ithaca, NY (www.cnf.cornell.edu/cnf_remotework.html).
10. Whyte, W. 1999. *Clean Room Design*, 2nd ed. John Wiley & Sons.
11. Whyte, W. 2010. *Clean Room Technology. Fundamentals of Design, Testing and Operation*. John Wiley & Sons.
12. Teo, S. H. G., A. Q. Liu, J. Singh, and M. B. Yu. 2007. Hole-type two-dimensional photonic crystal fabricated in silicon on insulator wafers. *Sensors and Actuators* 133:388–394.
13. Thomson, D. J. et al. 2011. High contrast 40Gbit/s optical modulation in silicon. *Optics Express* 19(12):11507–11516.
14. Guo, L. J. 2007. Nanoimprint lithography: Methods and material requirements. *Advanced Materials* 19(4):495–513.
15. Plachetka, U. et al. 2008. Fabrication of photonic ring resonator device in silicon waveguide technology using soft UV-nanoimprint lithography. *IEEE Photonics Technology Letters* 20(7):490–492.
16. Reed, G. T. 2004. *Silicon Photonics: An Introduction*. John Wiley & Sons.
17. Sherwood-Droz, N., and M. Lipson. 2011. Scalable 3D dense integration of photonics on bulk silicon. *Optics Express* 19(18):17758–17765.
18. Melchiorri, M. et al. 2005. Propagation losses of silicon nitride waveguides in the near-infrared range. *Applied Physics Letters* 86(12):121111.
19. Mao, S. C., S. H. Tao, Y. L. Xu, X. W. Sun, M. B. Yu, and G. Q. Lo. 2008. Low propagation loss SiN optical waveguide prepared by optimal low-hydrogen module. *Optics Express* 16(25):20809–20816.
20. Daldosso, N. et al. 2004. Comparison among various Si 3 N 4 waveguide geometries grown within a CMOS fabrication pilot line. *Journal of Lightwave Technology* 22(7):1734–1740.
21. Zheng, S., H. Chen, S. Member, A. W. Poon, and A. Member. 2006. Microring-resonator cross-connect filters in silicon nitride: Rib waveguide dimensions dependence. *IEEE Journal of Selected Topics in Quantum Electronics* 12(6):1380–1387.
22. Morichetti, F. et al. 2007. Box-shaped dielectric waveguides: A new concept in integrated optics? *Lightwave* 25(9):2579–2589.
23. Maire, G. et al. 2008. High efficiency silicon nitride surface grating couplers. *Optics Express* 16(1):0–5.
24. Preston, K., and M. Lipson. 2009. Slot waveguides with polycrystalline silicon for electrical injection. *Optics Express* 17(3):1527–1534.
25. Brooks, C. J., J. K. Doylend, A. P. Knights, and P. E. Jessop. 2006. Vertically-stacked SOI waveguides for 3-D photonic circuits. *Proceedings of IEEE Conference on Group IV Photonics (Institute of Electrical and Electronics Engineers, New York)*:72–74.
26. Koonath, P., and B. Jalali. 2007. Multilayer 3-D photonics in silicon. *Optics Express* 15(20):12686–12691.
27. Agarwal, A. M., L. Liao, J. S. Foresi, M. R. Black, X. Duan, and L. C. Kimerling. 1996. Low-loss polycrystalline silicon waveguides for silicon photonics. *Journal of Applied Physics* 80:6120–6123.
28. Liao, L. 1997. *Low Loss Polysilicon Waveguides for Silicon Photonics*. MIT.
29. Preston, K., B. Schmidt, and M. Lipson. 2007. Polysilicon photonic resonators for large-scale 3D integration of optical networks. *Optics Express* 15(25):17283–17290.
30. Saynatjoki, J., H. Riikonen, A. Lipsanen, and J. Ahopelto. 2003. Optical waveguides on polysilicon-on-insulator. *Journal of Materials Science Materials in Electronics* 14(5/7):417–420.
31. Zhu, S., G. Q. Lo, and D. L. Kwong. 2010. Low-loss amorphous silicon wire waveguide for integrated photonics: Effect of fabrication process and the thermal stability. *Optics Express* 18(24):25283–25291.
32. Liao, L., D. R. Lim, A. M. Agarwal, X. Duan, K. K. Lee, and L. C. Kimerling. 2000. Optical transmission losses in polycrystalline silicon strip waveguides: Effects of waveguide dimensions, thermal treatment, hydrogen passivation, and wavelength. *Journal of Electronic Materials* 29(12):1380–1386.
33. Zhu, S., Q. Fang, M. B. Yu, G. Q. Lo, and D. L. Kwong. 2009. Propagation losses in undoped and n-doped polycrystalline silicon wire waveguides. *Optics Express* 17(23):20891–20899.

34. Fang, Q., J. F. Song, S. H. Tao, M. B. Yu, G. Q. Lo, and D. L. Kwong. 2008. Polycrystalline silicon waveguide integrated with efficient SiON waveguide coupler. *Optics Express* 16(9):6425–6432.
35. Orcutt, J. S., A. Khilo, M. A. Popović, C. W. Holzwarth, and B. Moss. 2008. Demonstration of an electronic photonic integrated circuit in a commercial scaled bulk CMOS process. In *Conference on Quantum Electronics and Laser Science Conference on Lasers and Electro-Optics, CLEO/QELS* 3, CTuBB3.
36. Street, R. A. 1991. *Hydrogenated Amorphous Silicon*. Cambridge University Press.
37. Kamins, T. 1998. *Polycrystalline Silicon for Integrated Circuits and Displays*, 2nd ed. Kluwer, Norwell, Massachussetts.
38. Preston, K., P. Dong, B. Schmidt, and M. Lipson. 2008. High-speed all-optical modulation using polycrystalline silicon microring resonators. *Applied Physics Letters* 92(15):151104.
39. Preston, K., S. Manipatruni, A. Gondarenko, C. B. Poitras, and M. Lipson. 2009. Deposited silicon high-speed integrated electro-optic modulator. *Optics Express* 17(7):5118–5124.
40. Preston, K., Y. Ho, D. Lee, M. Zhang, and M. Lipson. 2011. Waveguide-integrated telecom-wavelength photodiode in deposited silicon. *Optics Letters* 36(1):52–54.
41. Biberman, A., K. Preston, J. S. Levy, and M. Lipson. 2011. Photonic network-on-chip architectures using multilayer deposited silicon materials for high-performance chip multiprocessors. *Journal on Emerging Technologies in Computing Systems* 7(2).
42. Corte, F. G. D., S. Rao, G. Coppola, and C. Summonte. 2011. Electro-optical modulation at 1550 nm in an as-deposited hydrogenated amorphous silicon p-i-n waveguiding device. *Optics Express* 19(4):2941–2951.
43. Cocorullo, G., F. G. D. Corte, R. D. Rosa, I. Rendina, A. Rubino, and E. Terzini. 1998. Amorphous silicon-based guided-wave passive and active devices for silicon integrated optoelectronics. *IEEE Journal of Selected Topics in Quantum Electronics* 4(6):997–1002.
44. Harke, A., M. Krause, and J. Mueller. 2005. Low-loss singlemode amorphous silicon waveguides. *Electronics Letters* 41(25):41–43.
45. Han, B. et al. 2007. Comparison of optical passive integrated devices based on three materials for optical clock distribution. In *European Conference on Integrated Optics*. ThF3.
46. Selvaraja, S. K., E. Sleeckx, M. Schaekers, W. Bogaerts, D. V. Thourhout, and P. Dumon. 2009. Low-loss amorphous silicon-on-insulator technology for photonic integrated circuitry. *Optics Communications* 282(9):1767–1770.
47. Sun, R., M. Beals, A. Pomerene, J. Cheng, C.-Yin Hong, and J. Michel. 2008. Impedance matching vertical optical waveguide couplers for dense high index contrast circuits. *Optics Express* 16(16):11682–11690.
48. Vermeulen, D. et al. 2010. High-efficiency fiber-to-chip grating couplers realized using an advanced CMOS-compatible silicon-on-insulator platform. *Optics Express* 18(17):18278–18283.
49. Harke, A., T. Lipka, J. Amthor, O. Horn, M. Krause, and Jö. Muller. 2008. Amorphous silicon 3-D tapers for Si photonic wires fabricated with shadow masks. *IEEE Photonics Technology Letters* 20(17):1452–1454.
50. Shiraishi, K., H. Yoda, A. Ohshima, H. Ikedo, and C. S. Tsai. 2007. A silicon-based spot-size converter between single-mode fibers and Si-wire waveguides using cascaded tapers. *Applied Physics Letters* 91(14):141120.
51. Shiraishi, K., and C. S. Tsai. 2005. A micro light-beam spot-size converter using a hemicylindrical GRIN-slab tip with high-index contrast. *Journal of Lightwave Technology* 23(11):3821–3826.
52. Sun, R. et al. 2009. Transparent amorphous silicon channel waveguides with silicon nitride intercladding layer. *Applied Physics Letters* 94(14):141108-1–141108-3.
53. Gusev, O. B. et al. 1997. Room-temperature electroluminescence of erbium-doped amorphous hydrogenated silicon. *Applied Physics Letters* 70(2):240.
54. Okamura, M., and S. Suzuki. 1994. Infrared photodetection using a-Si:H photodiode. *IEEE Photonics Technology Letters* 6(3):412–414.
55. Sirleto, L., M. Iodice, F. G. D. Corte, and I. Rendina. 2007. Digital optical switch based on amorphous silicon waveguide. *Optics and Lasers in Engineering* 45:458–462.
56. Kuyken, B. et al. 2011. Nonlinear properties of and nonlinear processing in hydrogenated amorphous silicon waveguides. *Optics Express* 19(26):146–153.

57. Corte, F. G. D., S. Rao, M. A. Nigro, F. Suriano, and C. Summonte. 2008. Electro-optically induced absorption in α-Si:H/α-SiCN waveguiding multistacks. *Optics Express* 16(10):7540–7550.
58. Narayanan, K., and S. F. Preble. 2010. Optical nonlinearities in hydrogenated-amorphous silicon waveguides. *Optics Express* 18(9):8998–9005.
59. Shoji, Y. et al. 2010. Ultrafast nonlinear effects in hydrogenated amorphous silicon wire waveguide. *Optics Express* 18(6):5668–5673.
60. Zhou, L., and A. W. Poon. 2006. Silicon electro-optic modulators using p-i-n diodes embedded 10-micron-diameter microdisk resonators. *Optics Express* 14(15):593–595.
61. Shoji, T., T. Tsuchizawa, T. Watanabe, K. Yamada, and H. Morita. 2002. Low loss mode size converter from 0.3um square Si wire waveguides to singlemode fibres. *Electronics Letters* 38(25):1669–1670.
62. Fedeli, J. M. 2012. Private Communication. CEA-LETI.
63. Cardenas, J. et al. 2010. Wide-bandwidth continuously tunable optical delay line using silicon microring resonators. *Optics Express* 18(25):26525–26534.
64. Green, W. M. J., F. Xia, S. Assefa, M. J. Rooks, L. Sekaric, and Y. A. Vlasov. 2008. Silicon photonic wire circuits for on-chip optical interconnects. In *Proceedings of SPIE* 6883:688306–688310.
65. Selvaraja, S. K., P. Jaenen, W. Bogaerts, D. Van Thourhout, P. Dumon, and R. Baets. 2009. Fabrication of photonic wire and crystal circuits in silicon-on-insulator using 193-nm optical lithography. *Journal of Lightwave Technology* 27(18):4076–4083.
66. Tsuchizawa, T., K. Yamada, H. Fukuda, T. Watanabe, J-Ichi Takahashi, and M. Takahashi. 2005. Microphotonics devices based on silicon microfabrication technology. *IEEE Journal of Selected Topics in Quantum Electronics* 11(1):232–240.
67. Sardo, S. et al. 2008. Line edge roughness (LER) reduction strategy for SOI waveguides fabrication. *Microelectronic Engineering* 85:1210–1213.
68. Samarelli, A., D. S. Macintyre, M. J. Strain, R. M. D. L. Rue, M. Sorel, and S. Thoms. 2008. Optical characterization of a hydrogen silsesquioxane lithography process. *Journal of Vacuum Science & Technology B* 26(6):2290–2294.
69. Gnan, M., S. Thoms, D. S. Macintyre, R. M. D. L. Rue, and M. Sorel. 2008. Fabrication of low-loss photonic wires in silicon-on-insulator using hydrogen silsesquioxane electron-beam resist. *Electronics Letters* 44(2):44–45.
70. Debnath, K., L. O. Faolain, and T. F. Krauss. 2011. Slowlight enhanced photonic crystal modulators. In *IEEE International Conference on Group IV Photonics GFP* 14, art. no. 6053728:98–100.
71. Zhou, L., S. S. Djordjevic, and R. Proietti. 2009. Design and evaluation of an arbitration-free passive optical crossbar for on-chip interconnection networks. *Applied Physics A* 95:1111–1118.
72. Pham, N. P., J. N. Burghartz, and P. M. Sarro. 2005. Spray coating of photoresist for pattern transfer on high topography surfaces. *Journal of Micromechanics and Microengineering* 15:691–697.
73. Linden, J., C. Thanner, B. Schaaf, S. Wolff, B. Lägel, and E. Oesterschulze. 2011. Spray coating of PMMA for pattern transfer via electron beam lithography on surfaces with high topography. *Microelectronic Engineering* 88(8):2030–2032.
74. Hurtado, J. 2007. Internal Report, Nanophotonics Technology Center, Univeristat Politècnica de València, Valencia (Spain).
75. Barwicz, T. et al. Microring-resonator-based add-drop filters in SiN: Fabrication and analysis. *Optics Express* 12(7):1437–1442, 2004.
76. Bellieres, L., and J. Ayucar. 2011. Internal Report, Nanophotonics Technology Center, Univeristat Politècnica de València, Valencia (Spain).
77. Holzwarth, C. W., T. Barwicz, and H. I. Smith. 2007. Optimization of hydrogen silsesquioxane for photonic applications. *Journal of Vacuum Science & Technology B* 25(6):2658–2661.
78. Sardo, S. et al. 2008. Fabrication of ultra high aspect ratio Bragg gratings for optical filter. *Microelectronic Engineering* 85(7):1511–1513.
79. HELIOS. *European Union's Framework Programme 7*. [Online]. Available: www.helios-project.eu.
80. Mayer, J. W., and S. S. Lau. 1990. *Electronic Materials Science: For Integrated Circuits in Si and GaAs*. New York: MacMillan.

81. Chen, L. J. 2004. *Silicide Technology for Integrated Circuits*. The Institution of Electrical Engineers.
82. DAS Photonics. DAS Photonics, Valencia, Spain (www.dasphotonics.com).
83. CEA-LETI. CEA-LETI, Grenoble, France (www.cea-leti.fr).
84. Armani, K. J., D. K., Kippenberg, T. J., Spillane, and S. M. Vahala. 2003. Ultra-high-Q toroid microcavity on a chip. *Nature* 421:925–928.
85. Schliesser, A., P. Del'Haye, N. Nooshi, K. Vahala, and T. Kippenberg. 2006. Radiation pressure cooling of a micromechanical oscillator using dynamical backaction. *Physical Review Letters* 97(24):1–4.
86. Zhu, T., P. Argyrakis, E. Mastropaolo, K. K. Lee, and R. Cheung. 2007. Dry etch release processes for micromachining applications. *Journal of Vacuum Science & Technology B* 25(6):2553–2557.
87. Delfino, M., and T. A. Reifsteck. 1982. Laser activated flow of phosphosilicate glass in integrated circuit devices. *IEEE Electron Device Letters* 3(5):116–118.
88. Xia, Q., P. F. Murphy, H. Gao, and S. Y. Chou. 2009. Ultrafast and selective reduction of sidewall roughness in silicon waveguides using self-perfection by liquefaction. *Nanotechnology, IOP Publishing* 20, 345302. doi:10.1088/0957-4484/20/34/345302: 1–5.
89. Holzwarth, C. W. et al. 2008. Localized substrate removal technique enabling strong-confinement microphotonics in bulk Si CMOS processes. In *2008 Conference on Lasers and Electro-Optics*, p. CThKK5.
90. Batten, C. et al. 2008. Building manycore processor-to-DRAM networks with monolithic silicon photonics. *2008 16th IEEE Symposium on High Performance Interconnects*, 21–30.
91. Nielson, G. N. et al. 2005. Integrated wavelength-selective optical MEMS switching using ring resonator filters. *IEEE Photonics Technology Letters* 17(6):1190–1192.
92. Indukuri, T., P. Koonath, and B. Jalali. 2005. Subterranean silicon photonics: Demonstration of buried waveguide-coupled microresonators. *Applied Physics Letters* 158–160.
93. Koonath, P., T. Indukuri, and B. Jalali. 2004. Vertically-coupled micro-resonators realized using three-dimensional sculpting in silicon. *Applied Physics Letters* 85(6):1018.
94. Koonath, P., T. Indukuri, and B. Jalali. 2006. Monolithic 3-D silicon photonics. *Journal of Lightwave Technology* 24(4):1796–1804.
95. Tang, Y., Z. Wang, L. Wosinski, U. Westergren, and S. He. 2010. Highly efficient nonuniform grating coupler for silicon-on-insulator nanophotonic circuits. *Optics Letters* 35(8):1290–1292.
96. Jordana, E. et al. 2007. Deep-UV lithography fabrication of slot waveguides and sandwiched waveguides for nonlinear applications. In *IEEE International Conference on Group IV Photonics GFP* 1, 222.
97. Teo, E. J. et al. 2009. Fabrication of low-loss silicon-on-oxidized-porous-silicon strip waveguide using focused proton-beam irradiation. *Optics Letters* 34(5):659–661.
98. Cardenas, J., C. B. Poitras, J. T. Robinson, K. Preston, L. Chen, and M. Lipson. 2009. Low loss etchless silicon photonic waveguides. *Optics Express* 17(6):4752–4757.
99. Zhang, H., J. Zhang, S. Chen, and J. Song. 2012. CMOS-compatible fabrication of silicon-based sub-100-nm slot waveguide with efficient channel-slot coupler. *IEEE Photonics Technology Letters* 24(1):10–12.
100. Selvaraja, S. K., W. Bogaerts, and D. Van Thourhout. 2011. Loss reduction in silicon nanophotonic waveguide micro-bends through etch profile improvement. *Optics Communications* 284(8):2141–2144.
101. Lardenois, S. et al. 2003. Low-loss submicrometer silicon-on-insulator rib waveguides and corner mirrors. *Optics Letters* 28(13):1150–1152.
102. Galán, J. V., P. Sanchis, G. Sánchez, and J. Martí. 2007. Polarization insensitive low-loss coupling technique between SOI waveguides and high mode field diameter single-mode fibers. *Optics Express* 15(11):7058–7065.
103. Grillot, F., L. Viv, S. Laval, and E. Cassan. 2006. Propagation loss in single-mode ultrasmall square silicon-on-insulator optical waveguides. *Journal of Lightwave Technology* 24(2):891–896.
104. Baehr-Jones, T., M. Hochberg, and A. Scherer. 2008. Photodetection in silicon beyond the band edge with surface states. *Optics Express* 16(3):1659–1668.
105. Barwicz, T., C. W. Holzwarth, P. T. Rakich, M. A. Popović, E. P. Ippen, and H. I. Smith. 2008. Optical loss in silicon microphotonic waveguides induced by metallic contamination. *Applied Physics Letters* 92(13):131108.

106. Lee, K. K., D. R. Lim, H.-C. Luan, A. Agarwal, J. Foresi, and L. C. Kimerling. 2000. Effect of size and roughness on light transmission in a Si/SiO[sub 2] waveguide: Experiments and model. *Applied Physics Letters* 77(11):1617.
107. Morichetti, F., A. Canciamilla, C. Ferrari, M. Torregiani, A. Melloni, and M. Martinelli. 2010. Roughness induced backscattering in optical silicon waveguides. *Physical Review Letters* 104: 033902.
108. Griol, A., N. Sánchez-Losilla, and L. Bellieres. 2010. Internal Report, Nanophotonics Technology Center, Univeristat Politècnica de València, Valencia (Spain).
109. Park Systems. KANC 4F, Iui-Dong 906-10 Suwon, South Korea 443-270 (www.parkafm.com).
110. Patsis, G. P., V. Constantoudis, A. Tserepi, E. Gogolides, and G. Grozev. 2003. Quantification of line-edge roughness of photoresists. I. A comparison between off-line and on-line analysis of top-down scanning electron microscopy images. *Journal of Vacuum Science & Technology B: Microelectronics and Nanometer Structures* 21(3):1008.
111. Yamazaki, K., T. Yamaguchi, and H. Namatsu. 2003. Edge-enhancement writing for electron beam nanolithography. *Japanese Journal of Applied Physics* 42(6):3833–3837.
112. Inoue, K., D. Plumwongrot, N. Nishiyamay, S. Sakamotoz, and H. Enomoto. 2009. Loss reduction of Si wire waveguide fabricated by edge-enhancement writing for electron beam lithography and reactive ion etching using double layered resist mask with C_{60}. *Japanese Journal of Applied Physics* 48(3):030208.
113. Borselli, M., T. Johnson, and O. Painter. 2005. Beyond the Rayleigh scattering limit in high-Q silicon microdisks: Theory and experiment. *Optics Express* 13(5):1515–1530.
114. Ubaldi, M. C., V. Stasi, U. Colombo, D. Piccinin, and M. Martinelli. 2007. Roughness reduction in submicron waveguides by low-molecular weight development. *Photonics and Nanostructures* 5:145–148.
115. Shearn, M. et al. 2009. Advanced silicon processing for active planar photonic devices. *Journal of Vacuum Science & Technology B* 27(6):3180–3182.
116. Lee, K. K., D. R. Lim, and L. C. Kimerling. 2001. Fabrication of ultralow-loss Si-SiO2 waveguides by roughness reduction. *Optics Letters* 26(23):1888–1890.
117. Sparacin, D. K., S. J. Spector, and L. C. Kimerling. 2005. Silicon waveguide sidewall smoothing by wet chemical oxidation. *Journal of Lightwave Technology* 23(8):2455–2461.
118. Lee, M.-C. M., and M. C. Wu. 2006. Thermal annealing in hydrogen for 3-D profile transformation on silicon-on-insulator and sidewall roughness reduction. *Journal of Microelectromechanical Systems* 15(2):338–343.
119. Chou, S. Y., and Q. Xia. 2008. Improved nanofabrication through guided transient liquefaction. *Nature Nanotechnology* 3(5):295–300.
120. Takahashi, J.-I., T. Tsuchizawa, T. Watanabe, and S.-I. Itabashi. 2004. Oxidation-induced improvement in the sidewall morphology and cross-sectional profile of silicon wire waveguides. *Journal of Vacuum Science & Technology B* 22(5):2522–2525.
121. Barwicz, T., M. A. Popovi, M. R. Watts, P. T. Rakich, E. P. Ippen, and H. I. Smith. 2006. Fabrication of add–drop filters based on frequency-matched microring resonators. *Journal of Lightwave Technology* 24(5):2207–2218.
122. Bojko, R. J., J. Li, L. He, T. Baehr-jones, M. Hochberg, and Y. Aida. 2011. Electron beam lithography writing strategies for low loss, high confinement silicon optical waveguides. *Journal of Vacuum Science & Technology B* 29(6):1–6.
123. Niehusmann, J., A. Vörckel, P. H. Bolivar, T. Wahlbrink, and W. Henschel. 2004. Ultrahigh-quality-factor silicon-on-insulator microring resonator. *Optics Letters* 29(24):2861–2863.
124. Gnan, M., D. S. Macintyre, M. Sorel, R. M. De La Rue, and S. Thoms. 2007. Enhanced stitching for the fabrication of photonic structures by electron beam lithography. *Journal of Vacuum Science & Technology B* 25(6):2034–2037.
125. Goodberlet, J. G., J. T. Hastings, and H. I. Smith. 2001. Performance of the Raith 150 electron-beam lithography system. *Journal of Vacuum Science & Technology B* 19(June):2499–2503.
126. Vlasov, Y. A., and S. J. Mcnab. 2004. Losses in single-mode silicon-on-insulator strip waveguides and bends. *Optics Express* 12(8):1622–1631.

127. Bogaerts, W., P. Dumon, D. V. Thourhout, and R. Baets. 2007. Low-loss, low-cross-talk crossings for silicon-on-insulator nanophotonic waveguides. *Optics Letters* 32(19):2801–2803.
128. Fedeli, J. M., L. Fulbert, D. Van Thourhout, and P. Viktorovitch. 2010. HELIOS: Photonics electronics functional integration on CMOS. In *Proceedings of the SPIE* Art 771907.
129. Ohki, S., T. Matsuda, and H. Yoshihara. 1993. X-ray mask pattern accuracy improvement by superimposing multiple exposures using different field sizes. *Japanese Journal of Applied Physics* 32(12):5933–5940.
130. Dougherty, D. J., R. E. Muller, P. D. Maker, and S. Forouhar. 2001. Stitching-error reduction in gratings by shot-shifted electron-beam lithography. *Journal of Lightwave Technology* 19(10):1527–1531.
131. Albert, J. et al. 1996. Minimization of phase errors in long fiber Bragg grating phase masks made using electron beam lithography. *IEEE Photonics Technology Letters* 8(10):1334–1336.
132. Komukai, T., and M. Nakazawa. 1998. Long-phase error-free fiber Bragg gratings. *IEEE Photonics Technology Letters* 10(5):687–689.
133. Kahl, M. Zero stitching error using fixed beam moving stage (FBMS) mode. [Online]. Available: www.raith.de (accessed January 15, 2011).
134. Griol, A. 2009. Internal Report, Nanophotonics Technology Center, Univeristat Politècnica de València, Valencia (Spain).
135. Sun, J., C. W. Holzwarth, M. Dahlem, J. T. Hastings, and H. I. Smith. 2008. Accurate frequency alignment in fabrication of high-order microring-resonator filters. *Optics Express* 16(20):15958–15963.
136. Holzwarth, C. W. et al. 2008. Fabrication strategies for filter banks based on microring resonators. *Journal of Vacuum Science & Technology B* 26(6):2164–2167.
137. Xia, F., L. Sekaric, and Y. Vlasov. 2007. Ultracompact optical buffers on a silicon chip. *Nature Photonics* 1:65–71.
138. Peuker, M., M. H. Lim, H. I. Smith, R. Morton, and A. K. V. Langen-suurling. 2002. Hydrogen silsesquioxane, a high-resolution negative tone e-beam resist, investigated for its applicability in photon-based lithographies. *Microelectronic Engineering* 61:803–809.
139. Namatsu, H., Y. Takahashi, K. Yamazaki, T. Yamaguchi, and M. Nagase. 1998. Three-dimensional siloxane resist for the formation of nanopatterns with minimum linewidth fluctuations. *Journal of Vacuum Science & Technology B* 16(1):69–76.
140. Hacker, N. P. 1997. Organic and inorganic spin-on polymers for low-dielectric-constant applications. *MRS Bulletin* 22(10):33–38.
141. Loboda, M. J., C. M. Grove, and R. F. Schneider. 1998. Properties of a-SiOx:H thin films deposited from hydrogen silsesquioxane resins. *Journal of the Electrochemical Society* 145(8):2861–2866.
142. Falco, C. M., and J. M. Van Delft. 2002. Delay-time and aging effects on contrast and sensitivity of hydrogen silsesquioxane. *Journal of Vacuum Science & Technology B* 20(6):2932–2936.
143. Henschel, W., Y. M. Georgiev, and H. Kurz. 2003. Study of a high contrast process for hydrogen silsesquioxane as a negative tone electron beam resist. *Journal of Vacuum Science & Technology B* 21(5):2018–2025.
144. Häffner, M., A. Haug, A. Heeren, M. Fleischer, H. Peisert, and D. P. Kern. 2007. Influence of temperature on HSQ electron-beam lithography. *Journal of Vacuum Science & Technology B* 25(6):2045–2048.
145. Sidorkin, V., A. Grigorescu, H. Salemink, and E. V. D. Drift. 2009. Microelectronic engineering resist thickness effects on ultra thin HSQ patterning capabilities. *Microelectronic Engineering* 86(4–6):749–751.
146. Popović, M. A., T. Barwicz, E. P. Ippen, and F. X. Kärtner. 2006. Global design rules for silicon microphotonic waveguides: Sensitivity, polarization and resonance tunability. In *Conference on Lasers and Electro-Optics (CLEO)*, p. CTuCC1.
147. Krishnamoorthy, A. V. et al. 2011. Exploiting CMOS manufacturing to reduce tuning requirements for resonant optical devices. *IEEE Photonics Journal* 3(3):567–579.
148. Shen, Y., I. B. Divliansky, D. N. Basov, and S. Mookherjea. 2011. Electric-field-driven nano-oxidation trimming of silicon microrings and interferometers. *Optics Letters* 36(14):2668–2670.

149. Soref, R. 2010. Silicon photonics: A review of recent literature. *Electronics Letters* 10(3):1–6.
150. Romagnoli, M., L. Socci, and L. Bolla. 2008. Silicon photonics in Pirelli. In *Proceedings of SPIE 6996, Silicon Photonics and Photonic Integrated Circuits, 699611* (May 1, 2008); doi:101117/12.786539.
151. Dumon. P. 2007. Ultra-compact Integrated Optical Filters in Silicon-on-Insulator by Means of Wafer-Scale Technology. PhD Thesis, available at their website. University of Ghent (Belgium).
152. Selvaraja, S. K., W. Bogaerts, P. Dumon, D. Van Thourhout, and R. Baets. 2010. Subnanometer linewidth uniformity in silicon nanophotonic waveguide devices using CMOS fabrication technology. *IEEE Journal of Selected Topics in Quantum Electronics* 16(1):316–324.
153. Aspar, B., M. Bruel, H. Moriceau, C. Maleville, T. Poumeyrol, and A. M. Papon. 1997. Basic mechanisms involved in the Smart-Cut* process. *Microelectronic Engineering* 36(1–2):233–240.
154. Selvaraja, S. K., W. Bogaerts, D. V. Thourhout, and R. Baets. 2008. Fabrication of uniform photonic devices using 193nm optical lithography in silicon-on-insulator. In *Proceedings of the 14th European Conference on Integrated Optics and Technical Exhibition* 2, 359.
155. Selvaraja, S. K. 2011. Wafer-Scale Fabrication Technology for Silicon Photonic Integrated Circuits. PhD Thesis, available at their website. University of Ghent (Belgium).
156. Peng, Z., D. Fattal, M. Fiorentino, and R. G. Beausoleil. 2010. Fabrication variations in SOI microrings for DWDM networks. In *IEEE International Conference on Group IV Photonics* (c):120–122.
157. Fiorentino. M. 2012. Private communication. HP Laboratories.
158. Zortman, W. A., D. C. Trotter, and M. R. Watts. 2010. Silicon photonics manufacturing. *Optics Express* 18(23):23598–23607.
159. Watts, M. R., W. A. Zortman, D. C. Trotter, G. N. Nielson, D. L. Luck, and R. W. Young. 2009. Adiabatic resonant microrings (ARMs) with directly integrated thermal microphotonics. In *Conference on Lasers and Electro-Optics* c, p. Art. CPDB10.
160. Zheng, X. et al. 2010. A tunable 1 × 4 silicon CMOS photonic wavelength multiplexer/demultiplexer for dense optical interconnects. *Optics Express* 18(5):5151–5160.
161. Dong, P. et al. 2010. Thermally tunable silicon racetrack resonators with ultralow tuning power. *Optics Express* 18(19):20298–20304.
162. Miller, D. A. B. 2009. Device requirements for optical interconnects to silicon chips. *Proceedings of the IEEE* 97(7):1166–1185.
163. Thomson, D. J., F. Y. Gardes, G. T. Reed, F. Milesi, and J.-M. Fedeli. 2010. High speed silicon optical modulator with self aligned fabrication process. *Optics Express* 18(18):19064–19069.
164. Rasigade, G., X. Le Roux, D. Marris-Morini, E. Cassan, and L. Vivien. Compact wavelength-insensitive fabrication-tolerant silicon-on-insulator beam splitter. *Optics Letters* 35(21):3700–3702, Nov. 2010.
165. Hosseini, A., S. Rahimi, X. Xu, D. Kwong, J. Covey, and R. T. Chen. 2011. Ultracompact and fabrication-tolerant integrated polarization splitter. *Optics Letters* 36(20):4047–4049.
166. Feng, D. et al. 2008. Novel fabrication tolerant flat-top demultiplexers based on etched diffraction gratings in SOI. In *IEEE International Conference on Group IV Photonics* (1):386–388.
167. Schmid, B., A. Petrov, and M. Eich. 2009. Optimized grating coupler with fully etched slots. *Optics Express* 17(13):11066–11076.

13

Convergence between Photonics and CMOS

Thierry Pinguet and Jean-Marc Fedeli

CONTENTS

13.1 Introduction

Silicon-based photonics has generated an increased amount of interest in recent years, mainly for optical telecommunications or for optical interconnects in high-performance microelectronic circuits. The development of elementary passive and active components (input/output couplers, hybrid sources, modulators, passive functions, and photodetectors) has reached such a performance level that the combination of these building blocks can lead to the development and commercialization of high-performance transceivers such as a new generation of active optic cables [1]. Photonics chips with active devices are connected to electronic drivers or amplifiers and thus the integration challenge of silicon photonics with microelectronic circuits has been studied for a long time [2]. The rationale of silicon photonics is the reduction of the cost of photonic systems through the integration of photonic components and an electronic integrated circuit (EIC) on a common chip. In the longer term, the introduction of an optical network between cores of a high-performance circuit will require this cointegration for the enhancement of the data rate with lower power consumption. Therefore, many integration schemes have been studied and developed. Each one has its figures of merits and can fulfill a special application with a particular packaging. For example, biosensing and high-performance computing systems will not necessarily share the same integration schemes because the specifications and the system packaging differ strongly.

If we consider chip-to-chip connection, mature technologies such as wire bonding, stud bumping, and flip-chip technologies can be used. The same technologies are also available in the case of die-to-wafer connections. The silicon photonics circuit can be considered as a board where the different circuits are attached (laser diodes, drivers, photodetectors, transimpedance amplifiers, etc.).

Integration at the wafer level is considered for higher-performance solutions (reduction of the parasitics) or for higher miniaturization. The challenge now is the combination of photonic technologies with microelectronics technologies. Considering the cross-section of an electronic wafer shown in Figure 13.1, three main options can be sketched:

- The first option is often called three-dimensional (3D) or above integrated circuit (AIC) integration, in which the photonic layer is realized at the metallization levels with back-end of the line technology.
- The second option combines the fabrication of transistors and the photonic devices at the front-end fabrication level.
- The third option takes advantage of the flat surfaces at the backside of electronic wafers, but requires double-sided processing.

Each option leads to various technological solutions and so this chapter will describe these integration schemes without any guarantee of being exhaustive.

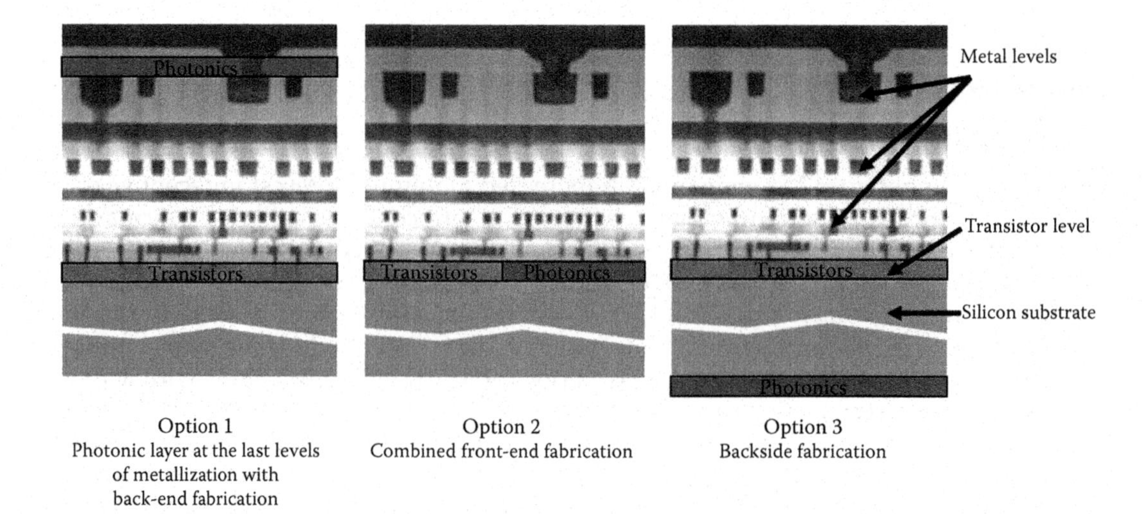

FIGURE 13.1
Cross-section of a photonic EIC showing different integration options.

13.2 Integration with Metallic Bonding

Photonic integrated chips and electronic integrated chips can be fabricated on separate wafers and on different manufacturing lines. Each circuit can be tested independently and only the "good" circuits will be selected for common connections. Therefore, different maturity technologies can be used and mixed.

The wire-bonding technology is the most flexible way to connect two chips in a package. As an example, Figure 13.2 describes a 10-G receiver with a Ge photodiode connected by wire bonding to a transimpedance amplifier. This EIC is then also connected to the pads of the quad flat no-leads package. With an air cavity package, operations of up to 20 GHz can be developed. The two common techniques for wire bonding are ball bonding and wedge bonding. The latter, with a rectangular cross-section of wire, provides better radio-frequency (RF) operation at more than 30 GHz.

For RF applications with a large number of wires, wire bonding can lead to some difficulty in the reproducibility of the wiring, which gives some randomness in the achieved performance. Therefore, solutions with a reduced length of connections are sought. Flip-chip assembly answers this need with different solutions. Stud-bumping is derived from the ball bonding techniques in which gold stud bumps are realized on the pads of a chip (Figure 13.3) by cutting the wire right after ball formation. Then the other chip is flip-chipped and bonded by pressing it with heating. The size of the bonding pads can be defined down to 80 μm, but the limitation comes from the number of pads to connect.

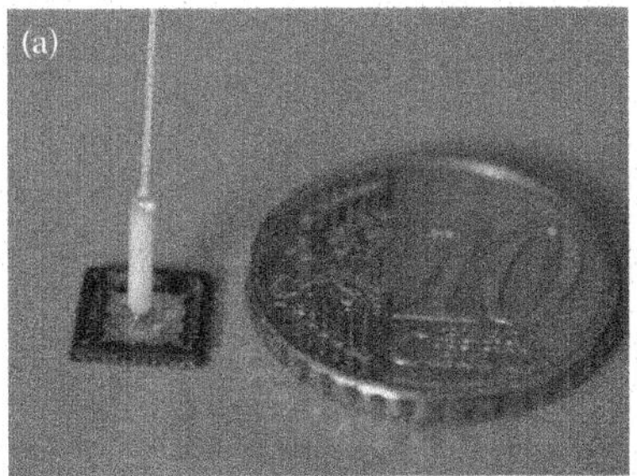

FIGURE 13.2
(a) Pigtailed optical high-speed receiver, embedding a Ge-on-Si photodiode and (b) related electronics in a quad flat no leads package multichip module.

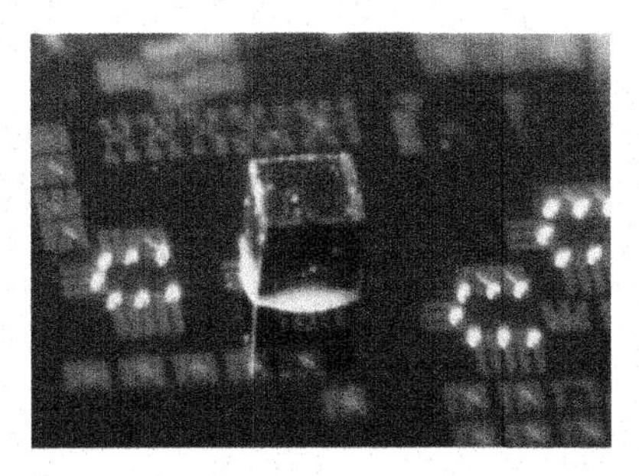

FIGURE 13.3
Waveguided Ge-on-Si photodiode with transimpedance amplifier connected by stud bumping.

The flip-chip technique is also a face-down assembly technique originally developed by IBM and is now widely used especially for imagers. Millions of simultaneous connections can be performed with a fine pitch (<15 μm) and a high accuracy (<1 μm) self-alignment. The indium bump flip-chip assembly process, for instance, consists of several steps. First, metallic structures are deposited and etched at the future location of the bumps. This structure, called under bump metallization (UBM), uses gold as the top layer. Then, indium is deposited on the full wafer using vapor deposition combined with a photolithographic process, as described in Figure 13.4.

After the lift-off step, controlled indium cylinders remain on the UBMs. They are then reflowed by heating the wafer up to the indium liquidus temperature (156°C) to obtain truncated balls, the basis of which is limited to the UBM pattern. The next step is the flip-chip assembly of the top and bottom chips. It requires a high-accuracy flip-chip equipment to get an initial prealignment accuracy of ±5 μm, in accordance with the typical bump pitch (10–50 μm). This accuracy is obtained using alignment patterns such as crosses that are processed at the UBM level. When a mechanical contact is detected between the two chips, the chips are heated, leading again to indium reflow and connection achievement. Self-alignment occurs during this step, due to centering forces, as illustrated by Figure 13.5.

Using a similar approach, ORACLE developed a silicon-photonic bridge chips comprising a 40-nm bulk CMOS VLSI chip hybrid integrated with silicon nanophotonic devices. To enable this integration, microsolder bumps were process-integrated onto 40 nm technology

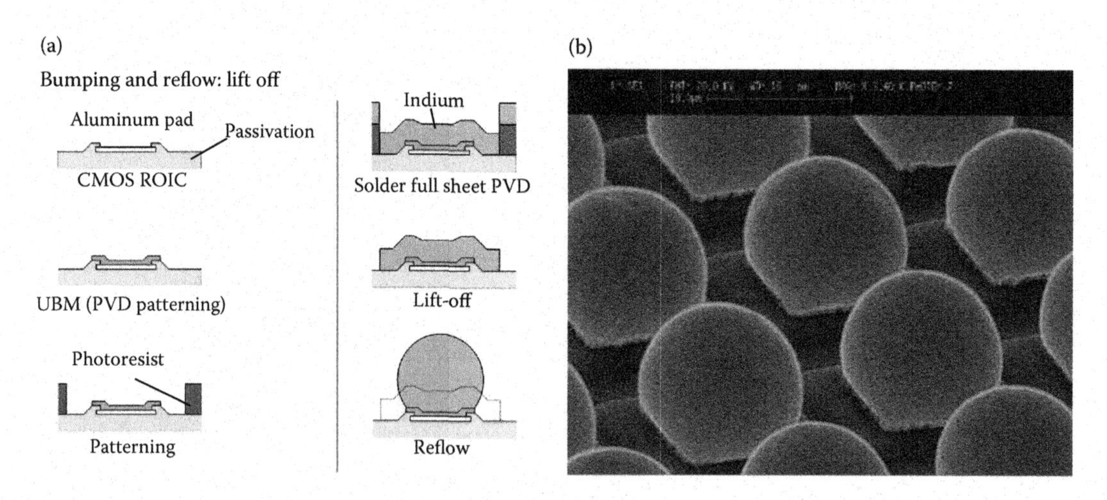

FIGURE 13.4
(a) Indium deposition and reflow process, (b) sea of balls.

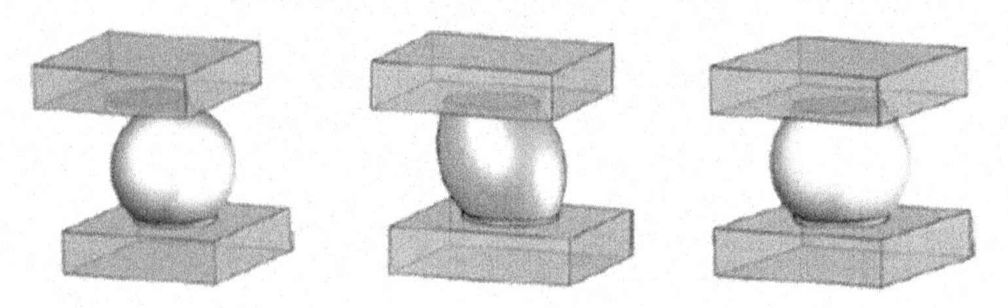

FIGURE 13.5
Self-alignment phenomenon.

CMOS chips by implementing an e-less Ni/Au UBM process and a novel batch-processing method [3].

With this self-alignment, assembly accuracy of less than 1 μm can be obtained. Therefore, during the assembly, optical coupling of the two chips can also be used either by mirrors [4] or by gratings [5].

Two photonic wafers including silicon waveguides and grating couplers were processed with the UBM structures and the indium bumps. After dicing, the "bridge" chip is flip-chipped onto the "carrier" chip, as seen in Figure 13.6. Flip-chip assembly enables accurate relative positioning of these two grating couplers that are facing each other.

A huge development in 3D packaging for pure electronic application is ongoing and the goal is to pile different EICs using through-silicon vias (TSV) and bonding via Cu pillars as sketched in Figure 13.7. This pillar technology is solder-free and allows a finer pitch and height. Therefore, for high-frequency operation, there is a reduction in the RC constant for the connection between the photonics and electronics parts.

To go one step further in the reduction of the length of connection and on the pitch, one has to reduce the height of the copper pillar. This can be successfully done using direct bonding of Cu [6].

Typical thickness of the Cu layer is approximately 1 μm. A chemical–mechanical polishing (CMP) step is used to lower the roughness to less than 0.5 nm root mean square (RMS) on the two wafers or dies that will be bonded together. After surface preparation, wafers are put into contact (face to face) at room temperature, atmospheric pressure, and ambient air to achieve copper wafer direct bonding as seen in Figure 13.8.

Therefore, as soon as the fabrication yield of electronic wafers and photonic wafers is high enough, this technique can be used as sketched in Figure 13.9, corresponding to

FIGURE 13.6
(See color insert.) Photonic dies coupled by grating couplers using the flip-chip technique.

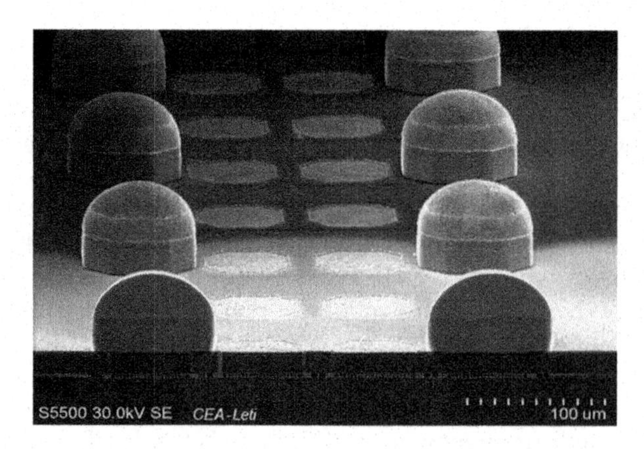

FIGURE 13.7
Copper pillars for die-to-wafer metallic bonding.

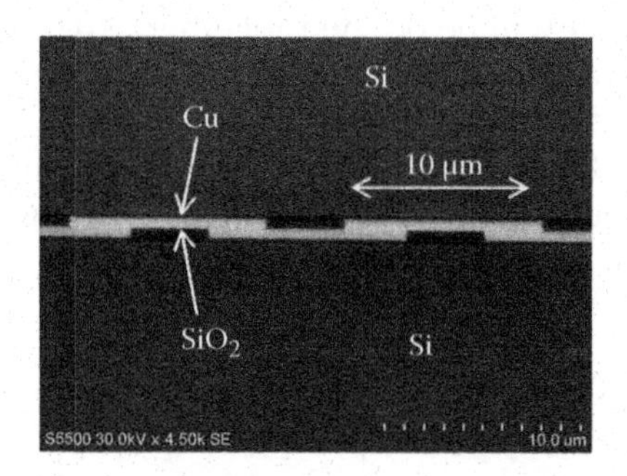

FIGURE 13.8
Cross-cut of a trial Cu bonding for wafer-to-wafer integration.

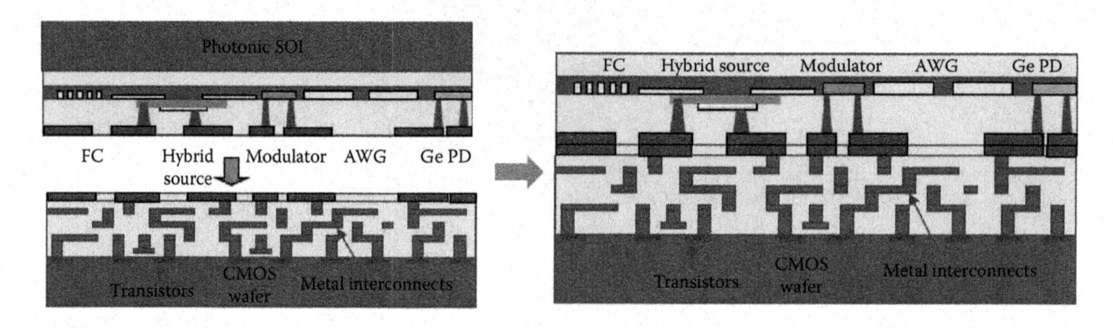

FIGURE 13.9
Cu bonding of photonic wafer and electronic wafer.

option 1 in Figure 13.1. The two wafers are prepared separately from the copper layers. The photonic wafer is then aligned to the electronic one and is directly bonded with alignment accuracy on the order of 2 μm. The silicon photonic substrate is removed down to the buried oxide (BOX), which acts as a stopping layer for the final chemical etching of Si.

13.3 Integration above the Metallization Layer of the EIC

In this chapter, we will consider integration using option of Figure 13.1 with only integration at the wafer level to end up with the sketch of Figure 13.10. Compared with the direct Cu bonding, we will use regular metal connections between one of the last metal interconnecting layers of the EIC to the electrodes of the active building blocks of the photonic circuit. This will shorten the length of the connections, thus reducing the RC figure.

The main advantage of this integration approach, compared with option 2, is the independence of electronics and photonics layers that avoid any change in the electronics library design. Moreover, any EIC technology (e.g., CMOS, SiGe, and analog) can be used: it is open to any standard front-end electronic technologies and full integration of III–V on Si is available. Depending on the devices to be processed, two suboptions are considered. The first consists of building the photonic layer with only low-temperature processes (<400°C). The second consists of fabricating the photonic functions on a separate wafer and then bonding it to the electronic wafer.

13.3.1 Back-End Technology

Above IC fabrication means that the photonic part of the circuit will be treated as complementary process steps after the first metallization of the CMOS process flow using a back-end of the line process. The thermal budget is limited to 400°C to avoid any degradation of the EIC, mostly of the interconnections underneath. For the passive circuitry, one has to rely on a deposited waveguide process such as SiON, SiN, or preferably, hydrogenated amorphous Si layer, which has the same contrast index as monocrystalline silicon of the surrounding material [7]. The most important feature for Si circuitry is the possibility of easily piling up layers and therefore opening up new design concepts and allowing designs such as crossings or couplings. However, the main disadvantage is the limited silicon processing that can be used (lack of ion implantation for doping, silicide, etc.). For

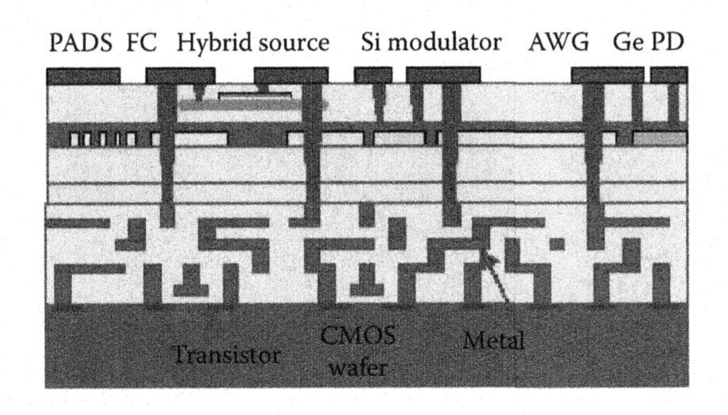

FIGURE 13.10
Illustration of a photonic layer integration at the last levels of metallization above the IC layer.

the active parts, efficient III–V compounds materials can be bonded and processed to form lasers, modulators, and photodetectors coupled to the passive circuitry [8]. Similar to the introduction of copper in a microelectronics circuit, contamination has to be controlled and some parts of the process have to take place in a dedicated portion of the microelectronics clean room. After the amorphous silicon waveguide circuitry is defined, silica deposition followed by a CMP planarization and a surface functionalization are performed. The formation of active devices relies on the heterogeneous approach, in which III–V material is bonded in the proximity of the amorphous silicon waveguide. Therefore, the process consists of dicing a III–V wafer (usually InP-based) with all the heteroepitaxial layers grown upside down with respect to standard epitaxy. Then, the dice are bonded to the required places (Figure 13.11), and the InP substrate is selectively removed to leave only the active thin films (0.5 to 3 μm typically thick) attached to the EIC wafer, thus enabling the processing of InP components on a dedicated 200 or 300 mm fabrication line.

Using this approach (Figure 13.12) in which the III–V dice are bonded unprocessed, no precise positioning (±10 μm) is required during the bonding step, which can be performed at high speed.

Further processing steps, such as the etching of the InP heterostructure, are performed resulting in sources and detectors connected to the metallic interconnections of the integrated circuit.

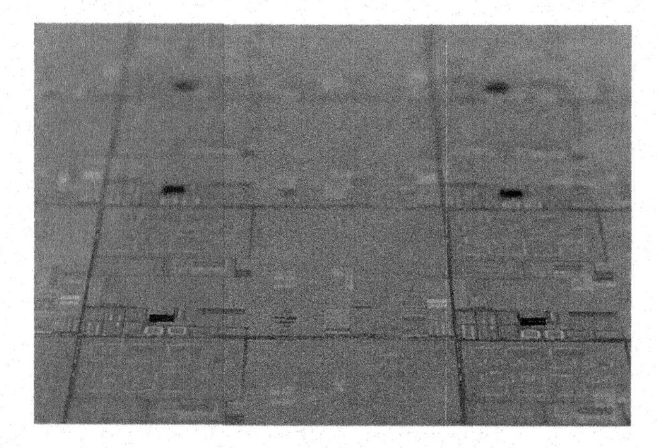

FIGURE 13.11
InP dies bonded onto a CMOS circuit wafer at the metal interconnect layer before the next step of processing.

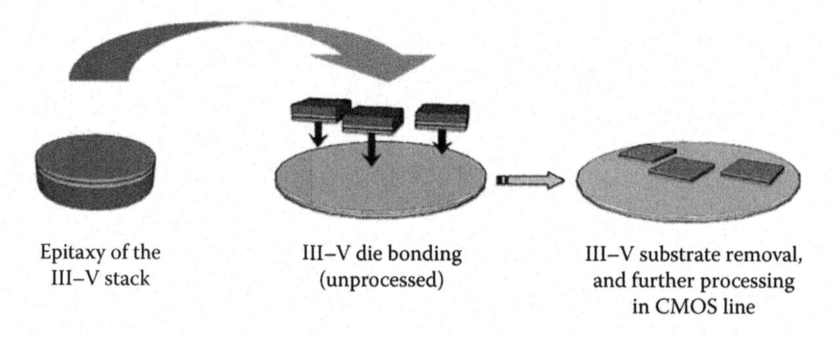

FIGURE 13.12
Schematics of the die-to-wafer process.

For compatibility with microelectronics processes, and for better thermal behavior, the die-to-wafer bonding technology is preferably direct bonding with oxide interfaces. This means, for example, that a thin layer of SiO_2 is deposited on the III–V dice before bonding on a planarized SiO_2 surface of the EIC. SiO_2/SiO_2 molecular bonding enables good bonding quality without any additional adhesive materials [9], which could inhibit efficient optical coupling. Furthermore, molecular bonding satisfies better the requirements in terms of thermal conductivity and dissipation, transparency at the device's working wavelengths, and mechanical resistance. SiO_2/SiO_2 molecular bonding was chosen because it is fully compatible with both microelectronics (SiO_2 is a standard material) and the above IC route in terms of thermal budget (the bonding step occurs at room temperature and the bonding strength can support a thermal budget of 400°C maximum). However, it requires quasi-perfect interfaces; meaning that surface roughness, defect density, particular contamination, and wafer bowing are critical issues. The required flatness and uniformity can be obtained using CMP. The additional role of CMP is to adjust the thickness of the silicon dioxide cladding layer to satisfy the optical coupling conditions. The III–V part needs much more "care" because dicing/cleaving of the III–V material can generate a lot of particles, and because III–V epitaxy generally results in emerging defects. The adhesion of the thin SiO_2 layer deposited on the III–V compound, as well as its roughness, also has to be well controlled.

After bonding, the InP substrate of the dies are mechanically and chemically removed down to a sacrificial layer needed as an etch stop. Therefore, only the heterostructure remains on the CMOS wafer, as seen in Figure 13.13.

After bonding, continuous wave operation microdisk lasers, photodetectors, and wavelength-selective circuits have been demonstrated using a CMOS compatible III–V/silicon technology at 200 mm wafer scale [10]. A specific process adapting and modifying the standard III–V material process steps to comply with a CMOS environment was developed using the first option described previously and as sketched in Figure 13.14. Using the same epitaxy stack, InP lasers and InGaAs photodetectors can share a common process. The use of deep ultraviolet lithography allows a very precise alignment between the silicon waveguides and the III–V active devices. The subsequently III–V processing, including wet and CH_4–H_2-based reactive ion etching, oxide isolation layer deposition, and metallization is optimized at 200 mm wafer scale.

FIGURE 13.13
Top view of a processed InP heterostructure bonded on silicon photonic devices.

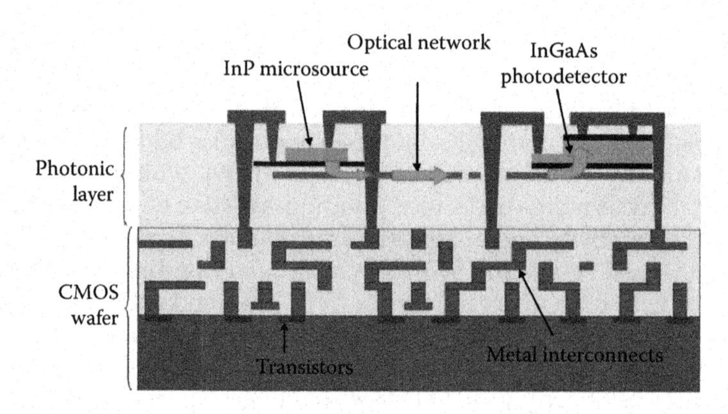

FIGURE 13.14
PEIC for intrachip optical transmission.

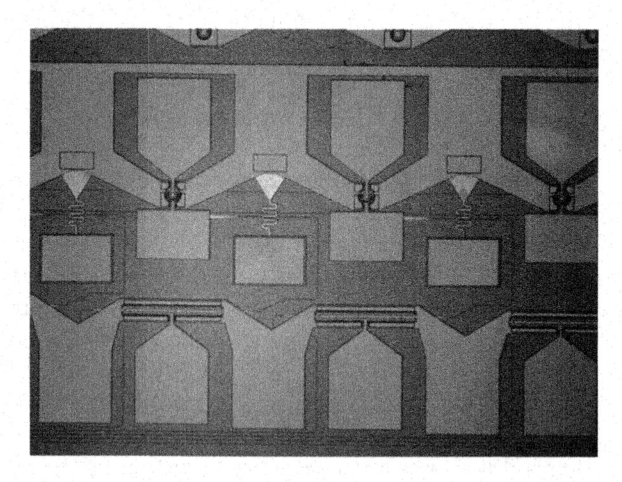

FIGURE 13.15
InP laser, RR with heaters, and InGaAs photodetectors fabricated on 200 mm microelectronics line.

The main challenge for the processing of III–V materials with microelectronics tools can be summarized as contamination, which is overcome with careful cleaning procedures. Another issue is the development of III–V dry etching with large-sized wafer equipment. Regarding electrical contacts, the standard gold-based metallization and the usual lift-off technique to define the contact area are discarded. Instead, new metallization free of gold, such as a Ti/TiN/AlCu metal stack, are deposited at less than 400°C. After a lithography step, the metal stack is dry-etched with a chlorine-based chemistry down to the oxide isolation layer, which also acts as an etching stop layer.

Microdisk lasers with diameters ranging from 7.5 to 40 μm have been fabricated with evanescently coupled waveguide detectors that are 20 μm in length on a silicon network with heaters for wavelength division multiplexing, as reported in Figure 13.15.

13.3.2 SOI Photonics and Electronic Wafer Bonding

An alternative way to AIC fabrication and Cu bonding is based on wafer-to-wafer bonding at the last levels of metallization with back-end fabrication and in the final process of metallization to connect photonic devices and electronic devices. High-temperature processing and monocrystalline Si layer can be used for the fabrication of photonic functions (e.g., Si-based modulators and Ge-based photodiodes) on the silicon-on-insulator (SOI) photonic wafer. After the high-temperature processing steps, III–V dies can be mounted on top of the waveguides by die-to-wafer bonding as discussed in a previous section. Further processing steps are performed for the fabrication of hybrid source parts coupled to the silicon waveguide circuitry.

In parallel, after fabrication of one of the last metal layers of the EIC, a planarized surface can be formed by first coating the surface with a deposited oxide, and then by performing CMP. After cladding the optical wafer with oxide and planarization with CMP, perfect cleaning of both wafers facilitates their molecular bonding at room temperature. The substrate of the SOI wafer is then removed by mechanical grinding, followed by a chemical etching of Si to stop at the BOX interface. Then, the electro-optical and electrical components are connected using silica etching followed by metal deposition and etching (Figure 13.16). The alignment between the electrical and the photonic wafers can be performed as precisely as ±2 μm. Therefore, the design rules between the metal layers have to take this alignment margin into account. This technique is often called 3D heterogeneous integration because the electric part is separated from the photonic part without any silicon surface waste at the transistor level.

To illustrate this process, under the HELIOS project [11], a photonic wafer with a wavelength division multiplexing receiver was bonded by two means on a CMOS wafer. After the fabrication of metal 4 in the CMOS process, the planarized surface was coated with a deposited oxide. The planarity and the microroughness have to match the specifications for bonding. In parallel, on a SOITEC optical SOI wafer with 220 nm Si on a 2-μm BOX, a photonic layer with surface grating couplers, arrayed waveguide grating (AWG), and germanium photodetectors was processed. Two different eight-channel AWGs were selected with 200 or 400 GHz spacing. Germanium lateral PIN photodiodes were coupled to the eight outputs of the AWG [12]. Cladding with oxide and planarization of the optical wafer with CMP was then performed to prepare the wafers for bonding. Perfect cleaning of both wafers facilitated their direct bonding at room temperature. After bonding, grinding, and chemical etching of the backside of the Si optical wafer, a flat surface of thermal oxide remained on the top of this PEIC circuit embedded in the photonic layer. Figure 13.17 shows the photonic-electronic integrated circuit (PEIC) composed of the photonic layer (AWG and Ge photodiodes) bonded on a trial CMOS wafer.

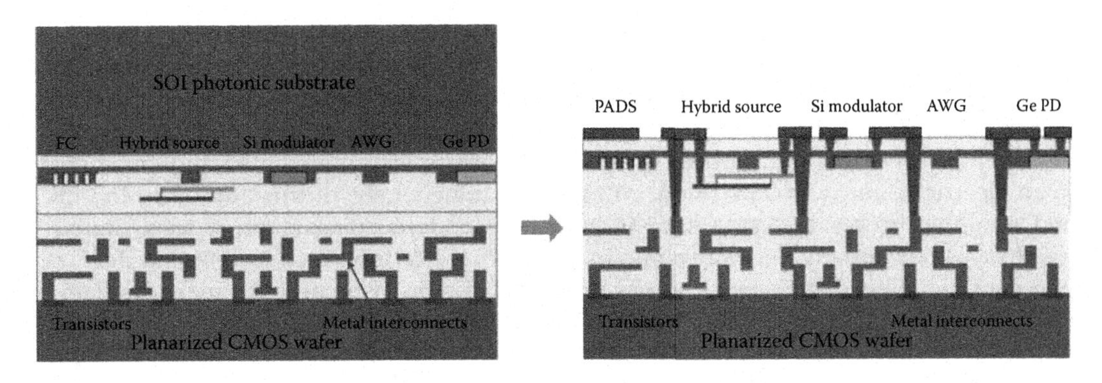

FIGURE 13.16
Sketch of photonics–electronics integration with wafer bonding.

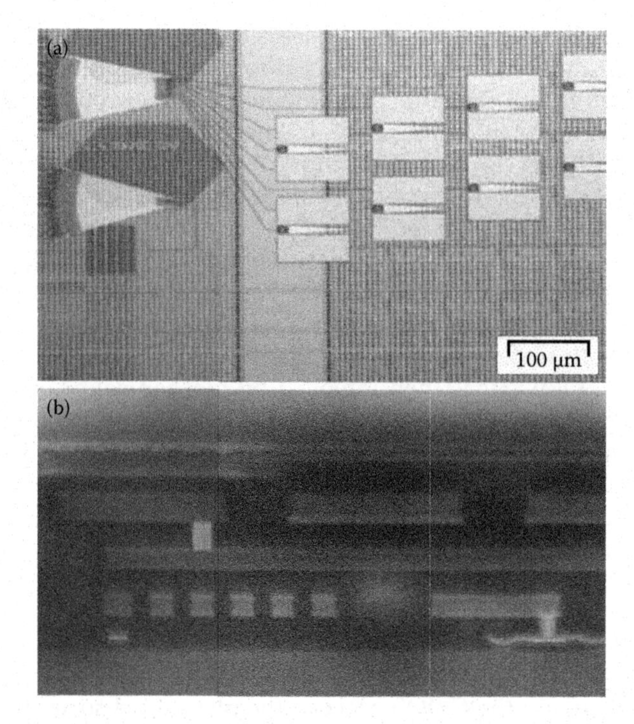

FIGURE 13.17
(See color insert.) Photonic–electronic wafer with AWG and Ge photodiodes. Cross-cut with flip fiber coupler.

13.4 Backside Fabrication

The fabrication of a photonic layer on the backside of the EIC (option 3 in Figure 13.1) can be considered and developed thanks to TSV. First, the CMOS wafer is processed up to the last metal layer and the backside is thinned and fine-polished to prime wafer surface quality. Then, the solutions described for option 2 with either wafer bonding (Figure 13.18) or low-temperature processing are applicable. Afterward, electrical interconnects between the CMOS and photonic layers are obtained using TSV.

Deep silicon etching is performed down to the metal layer of the photonic layers where subsequent TSV isolation and metallization are deposited. Finally, the top metal, which connects the TSV with the IC, and the passivation are deposited and structured. A demonstration of this kind of integration has been achieved under the HELIOS project, in which ring resonators and optical wafers with heaters were assembled and connected to an electric wafer with TSV (Figure 13.19). Advantageously, in this approach, the EIC and photonic processing are rather independent and the packaging of such double-sided chips has already been developed for other applications such as imaging devices. However, double-sided thermal management may be an issue and the main challenge lies in the ability to form TSV with a high aspect ratio (depth versus width). A low ratio limits the frequency operation to the megahertz range, which is incompatible with applications of silicon photonics for communications or optical interconnects that operate in the 10-GHz range. However, for other applications, such as sensing, the circuit can be

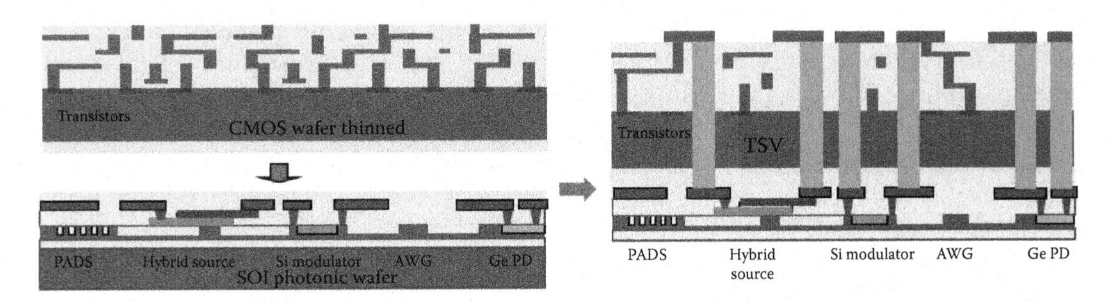

FIGURE 13.18
Backside integration with photonic wafer direct bonding.

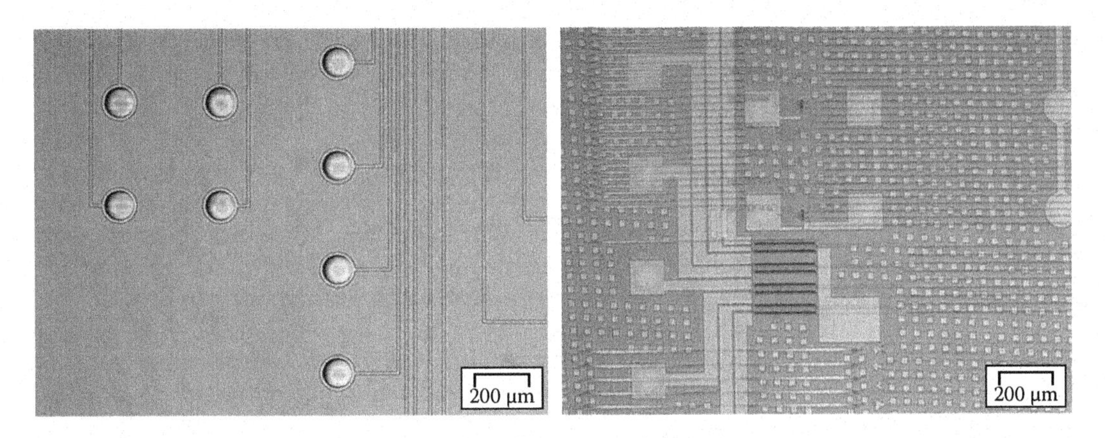

FIGURE 13.19
The two sides of a wafer with backside integration: front side of the wafer with electrical connections and upper TSVs (left). Backside of the wafer with the photonic layer with ring resonators and heaters connected to the TSVs (right).

connected by flip-chip on a board, leaving the backside accessible for the elements to be detected (biomolecules, gases, etc.) or for assembly with microtubes (microfluidics, lab-on-a chip, etc.).

13.5 Front-End of Photonics–Electronics CMOS Processes

The last approach we will evaluate in this chapter is that of monolithic, front-end integration of electronics and photonics. This can be compared with traditional System-on-Chip (SoC) integration in electronics, in which performance/cost benefits are gained from the integration of previously disparate technologies—for instance, logic (microprocessor), memory, and RF circuits for wireless applications.

Similar to the case of established monolithic SoC integrations, however, there are constraints and trade-offs to be taken into account before deciding whether this is an appropriate approach for a specific application. In the case of electronic SoC, very high volumes end up providing enough of a driver to perform such a technically complex integration. In the case of photonics–electronics integration, it is unclear if performance benefits from small parasitics and being in proximity with the electronics overcomes the technical complexity and costs associated with the approach. In the following section, we will discuss the constraints and trade-offs and explore examples of such monolithic integration.

13.5.1 Basic Approach

The main examples of such an integration that are available today are those reported by Luxtera [13], MIT [14], and IBM [15]. An example of the Luxtera process is shown in Figure 13.20, but it is generally applicable to all monolithic integrations. In its most simple embodiment, the technology is a modification of an existing CMOS SOI process, with added steps to enable photonic devices. Typical additions include different silicon trenches, additional implants, and the introduction of a pure Ge module for photodetection.

Photonic devices are usually built out of the same silicon used to form the body of the SOI transistors, with waveguiding provided by the index contrast between Si as the core of the waveguides and the surrounding low index materials: silicon dioxide of the BOX layer, or the trench fill (oxide deposited between the various etched silicon features), and other low index dielectric present in the back-end. In CMOS SOI processes, a full etch of silicon is used to create complete isolation of transistors; this etch can be used to create very tightly confined waveguides in a single, combined step with transistor bodies. Another approach can consist of adding one or more etches to create partially etched silicon areas that are optimized for the performance of different optical devices, as shown in Figure 13.20.

Although the SOI layer is often the primary device layer for photonic devices, typical CMOS processes also include other layers that can be used for photonic devices, such as the gate poly-silicon layer [14] or high-index dielectrics such as silicon nitride layers [16]. The use of noncrystalline silicon layers for photonics creates severe limitations in its usefulness: poly-silicon waveguides typically experience very high propagation losses (for instance, MIT reported 55 dB/cm) [14], whereas silicon nitride–based waveguides can only be used in passive devices (no modulation or detection due to lack of implantation).

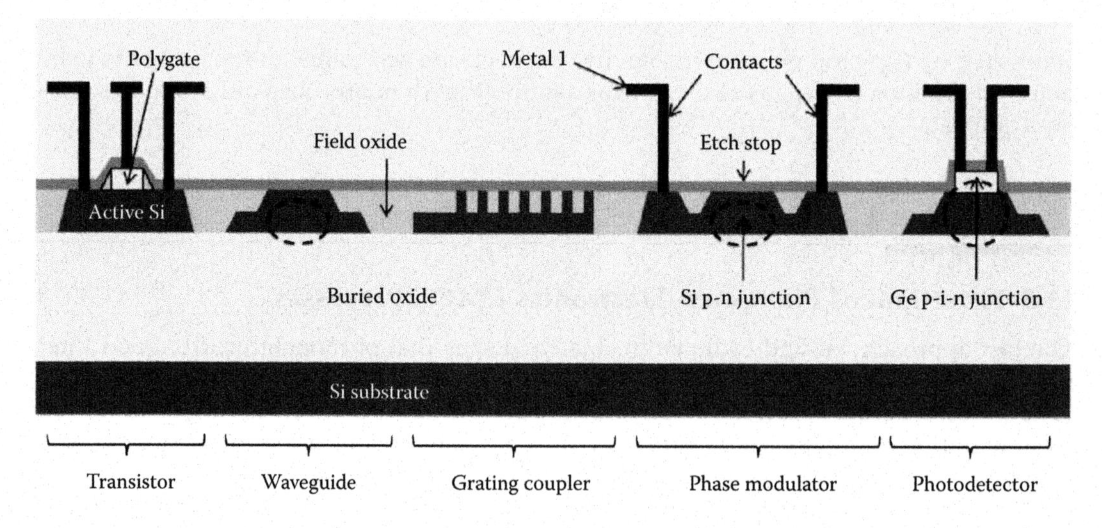

FIGURE 13.20
Cross-section of a typical CMOS/photonic front-end monolithic integration [13].

Although most passive functionality (waveguiding, splitting, combining, etc.) can be achieved with a single etch or a combination of etches, active functionality (modulation and detection) requires additional steps. Luxtera [13] and IBM [15] reported the use of pn or pIn junctions as phase/amplitude shifters for modulation, and required the use of implantation; implants can be similar to those used in the CMOS flow or can be customized for the optimization of performance of the devices, but essentially constitute the simplest process change because they typically have no effect on the rest of the flow and on the CMOS performance, and can use already existing thermal activation steps.

Photodetection is accomplished by the integration of pure, 100% Ge deposition and implantation on silicon because Ge absorbs light in the traditional windows of interest (1310–1550 nm). Ge photodiodes typically consist of either vertical pIn SiGe heterojunction [15] or lateral Ge-only pIn junctions [13].

13.5.2 Typical Constraints in Monolithic Front-End Integration

The concept of monolithic integration stems primarily from the benefits of having high-performance electronics located closely to the photonic devices, allowing optoelectronic circuits that were not previously feasible. One of the basic assumptions made is that the CMOS flows on which the integrated processes are based on must result in minimal disruption to the CMOS device performance and, ideally, to the models that have been created for these devices, that is, that the introduction of changes to the flow enabling photonics should not result in a shift in performance of CMOS benefiting from the infrastructure already established for said CMOS flow.

Therefore, in this paradigm, photonic device performance has to be a secondary concern after ensuring that CMOS devices are not perturbed in a significant manner. In this section, we explore examples of ways in which photonic devices are affected by this requirement, or vice versa, trade-offs that have been made that end up changing or degrading CMOS device performance, and potential ways to improve on these situations.

13.5.2.1 Substrate Selection and Effect on CMOS

SOI substrates are typically the standard of choice for integrated photonics in silicon, due to the presence of the BOX, which provides an obvious cladding for waveguiding: for the purpose of monolithic integration, that would seem to be an ideal situation due to the existence of SOI-based CMOS processes and the emergence over the last decade of SOI as a technology for high-performance logic applications. However, there are fundamental incompatibilities in the requirements for SOI specifications between photonics and electronics.

If we assume that the two main specifications for a substrate that matter for photonics and electronics performance are the BOX thickness and the top oxide thickness, we can make the following observations:

- BOX and top silicon thickness are trending increasingly toward smaller and smaller values as CMOS technologies scale into deep submicron territories. Typical thicknesses for top Si at 130 nm are in the 100-nm range, whereas 50 nm or smaller will become common at 45 nm and smaller technologies [17].
- Unfortunately, for best photonics performance, top silicon thickness must be in the sweet spot in the range of 200 to 300 nm to provide optimal confinement for parameters such as bending radius, optical propagation loss, and efficiency of optical

modulators. Because the wavelength of the light used in Si photonics systems does not change with time, this requirement is not likely to change with time either.
- The BOX thickness must be thick enough (typically in the 1-μm range) to ensure that optical modes confined in the top silicon layer do not experience significant leakage into the substrate and thus optical propagation loss.

Because of this, an SOI substrate considered appropriate for photonics performance would likely have layer thicknesses ranging from 2 to 10 times as thick as in existing SOI CMOS processes, leading to a compromise having to be made either in degrading optical performance or changing the performance of CMOS devices.

For instance, Luxtera [13] reports that in their 130 nm SOI CMOS-photonics technology, fully isolated SOI transistors behave much more like bulk devices due to the thick silicon layer compared with the original process the technology was based on. Needless to say, this entails a significant investment in terms of efforts and cost to recharacterize electrical devices and extract new models for them.

It should be noted that these are constraints in the case of a "simple" SOI substrate, one in which there is only one silicon layer and one BOX layer, and they are shared between device types. There are ways to get around these limitations, by increasing the complexity of the substrate fabrication process or that of the CMOS process itself.

Figures 13.21 and 13.22 show examples of approaches that could provide essentially different silicon and BOX thicknesses to photonic and electronic devices on the same substrate, even providing a way to use a bulk silicon substrate to create photonic devices. Using a bulk substrate in particular would remove one of the critical issues—that of having to recharacterize electrical models of CMOS devices due to a different silicon thickness. Bulk devices would presumably behave exactly the same, whether a nearby part of the substrate is removed locally or not, assuming there are no stress effects or at least the range of such effects is small enough.

It should be noted that SOI substrates are already priced at a premium over standard bulk silicon substrates, and any additional complexity that takes a "photonic" substrate further into nonmainstream fabrication process flows is undesirable at best.

13.5.2.2 Effect of Front-End Layers and Processes on Photonic Performance

Typical CMOS processes present a number of processes and layers that can affect photonic device performance, for instance:

- Silicon sidewall etching and passivation
- Front-end planarization (CMP)
- Sacrificial oxidation and clean cycles
- Gate oxidations
- Gate (poly-silicon) deposition and removal
- Gate liners
- Salicidation and salicidation blocking layers
- Contact layers (e.g., etch stop)

Figure 13.23 shows examples of the local environment of a waveguide in two different processes. All of the process steps described previously play a role in the ultimate performance of the waveguide and other photonic devices. Due to the requirements of

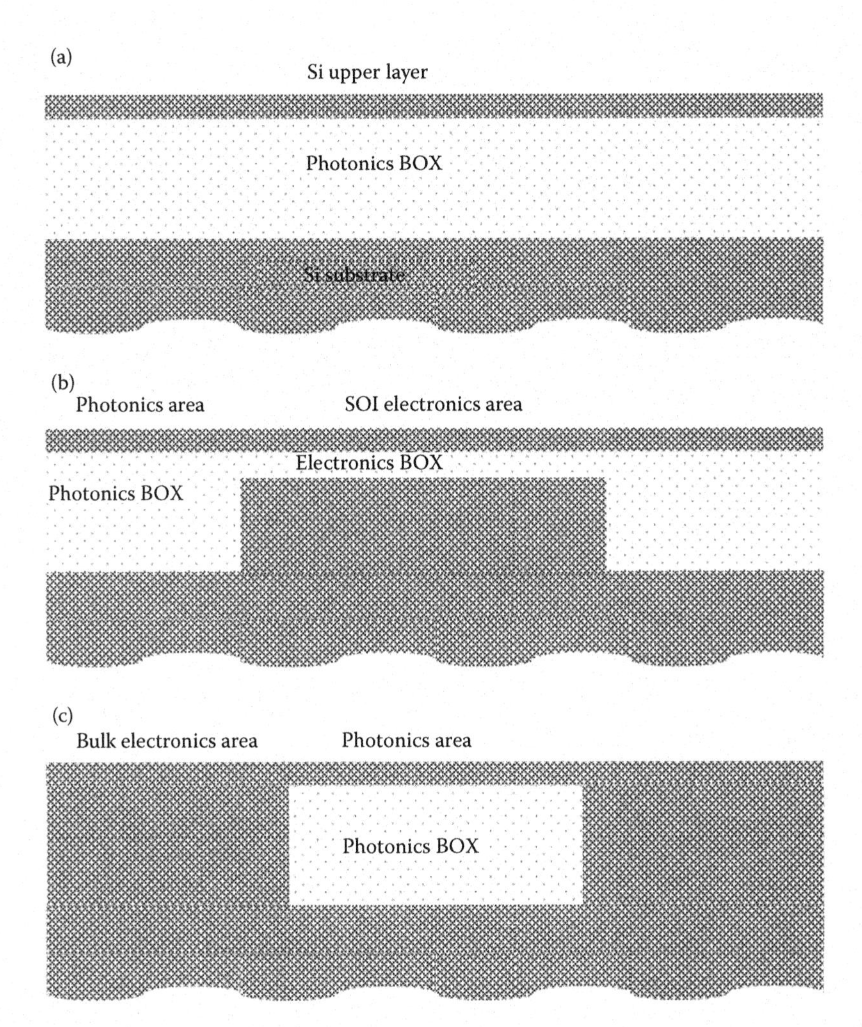

FIGURE 13.21
Substrates for combined fabrication. The same layers for both electronics and photonics (a), locally different BOX thicknesses (b), and local BOX in a bulk silicon substrate (c).

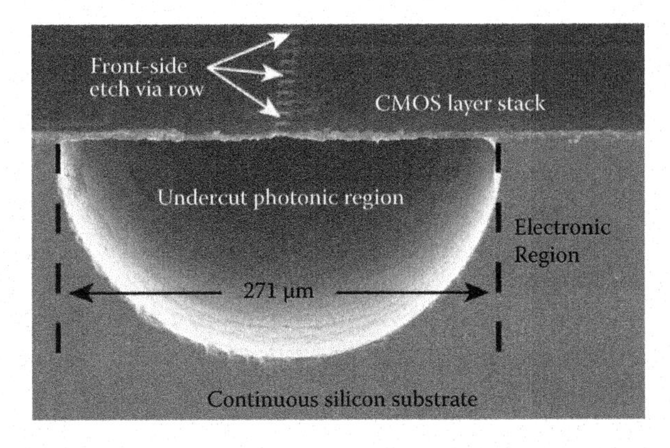

FIGURE 13.22
Local removal of silicon in a bulk substrate to create air cladding [14].

(a)

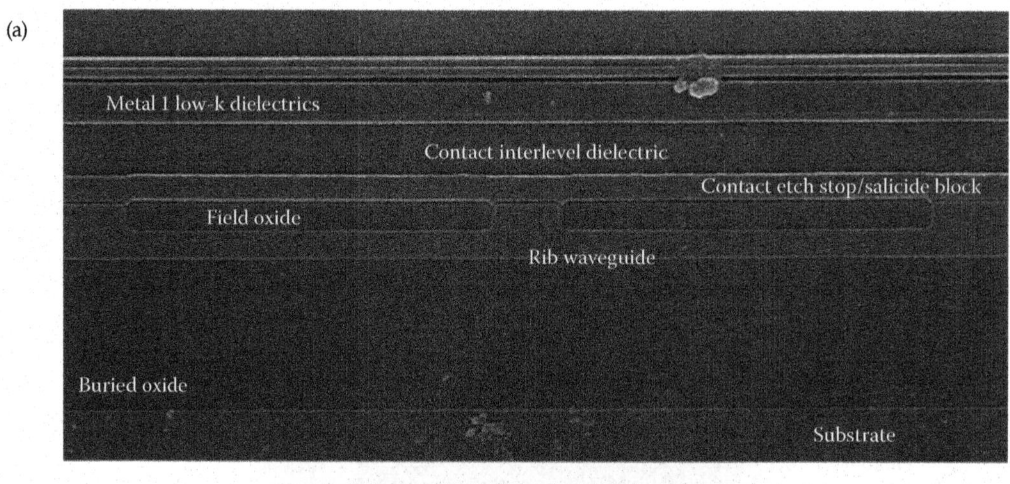

(b)

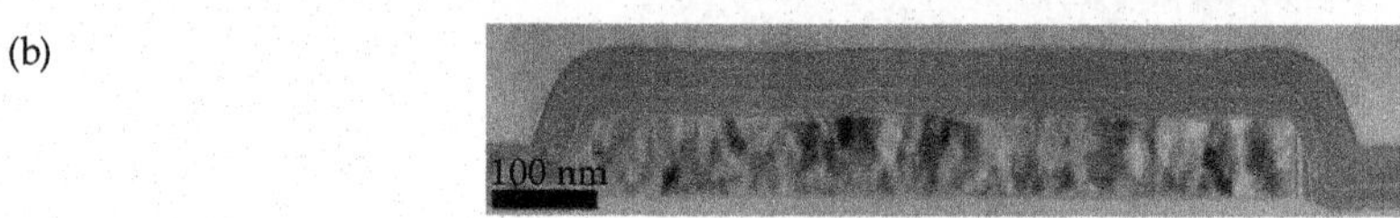

FIGURE 13.23
Cross-section of waveguide in the Luxtera process (a) and in the MIT work (b) (poly-silicon waveguide) showing the surrounding environment.

minimal changes to the CMOS flow, the creators of these processes have not had the ability to include steps that could benefit optical performance. For instance, very low waveguide losses have been demonstrated in silicon with processes targeted specifically at improving surface roughness [18], but these processes may not be compatible with standard CMOS sidewall passivation, and thus are simply not available. Instead, photonic devices must rely on simple liner oxidation and strip cycles.

Other steps present in the flow, such as cycles of gate oxidation/strip cycles (in the case of two or more gate thicknesses, as present in typical CMOS flows), deposition and removal of poly-silicon (or lately, metal) gates, and the presence of dielectric layers very close to the optical mode (salidication blocking layers, contact etch stop layers, gate liner layers, etc.) have a definite effect on performance, creating variability in thickness, width, or roughness on the various exposed layers. This results in increased waveguide loss or reduction in coherence length. For instance, Luxtera's typical waveguide loss for a single-mode waveguide is 2 dB/cm, and studies have shown how the various process steps affect waveguide loss [19].

CMOS processes also include planarization steps. In particular, there is typically a planarization step performed immediately after trench filling in which the oxide deposited to fill the etched regions between silicon "islands" forming the bodies of transistors is polished back down to essentially the silicon surface. During the design phase, a layout must be carefully planned to match requirements in global and local density of features to allow for this planarization step to succeed. Examples of failures could be insufficient removal of oxide over silicon or, conversely, overpolishing into silicon. The addition of photonic features with widely different geometries, feature densities, etc., compared with standard CMOS, along with the introduction of different trench depths, for instance, makes matching density rules extremely difficult or, in the worst case, may require the development and validation of updated density rules or a change in the polishing process itself (or both).

13.5.2.3 Germanium Module Integration

The availability of epitaxy tools used for SiGe deposition in CMOS processes (for instance, as stressors) makes it easy to envision integrating a 100% Ge deposition module in a CMOS process. However, there are both thermal and geometrical constraints to introducing such a module.

Achieving high-quality Ge growth on silicon is obviously a difficult procedure, but there are extensive examples of good device results in the literature [20]. Doing so while in the constraints of a CMOS process thermal budget is challenging indeed. Figure 13.24 shows examples of the Luxtera and IBM integration approaches.

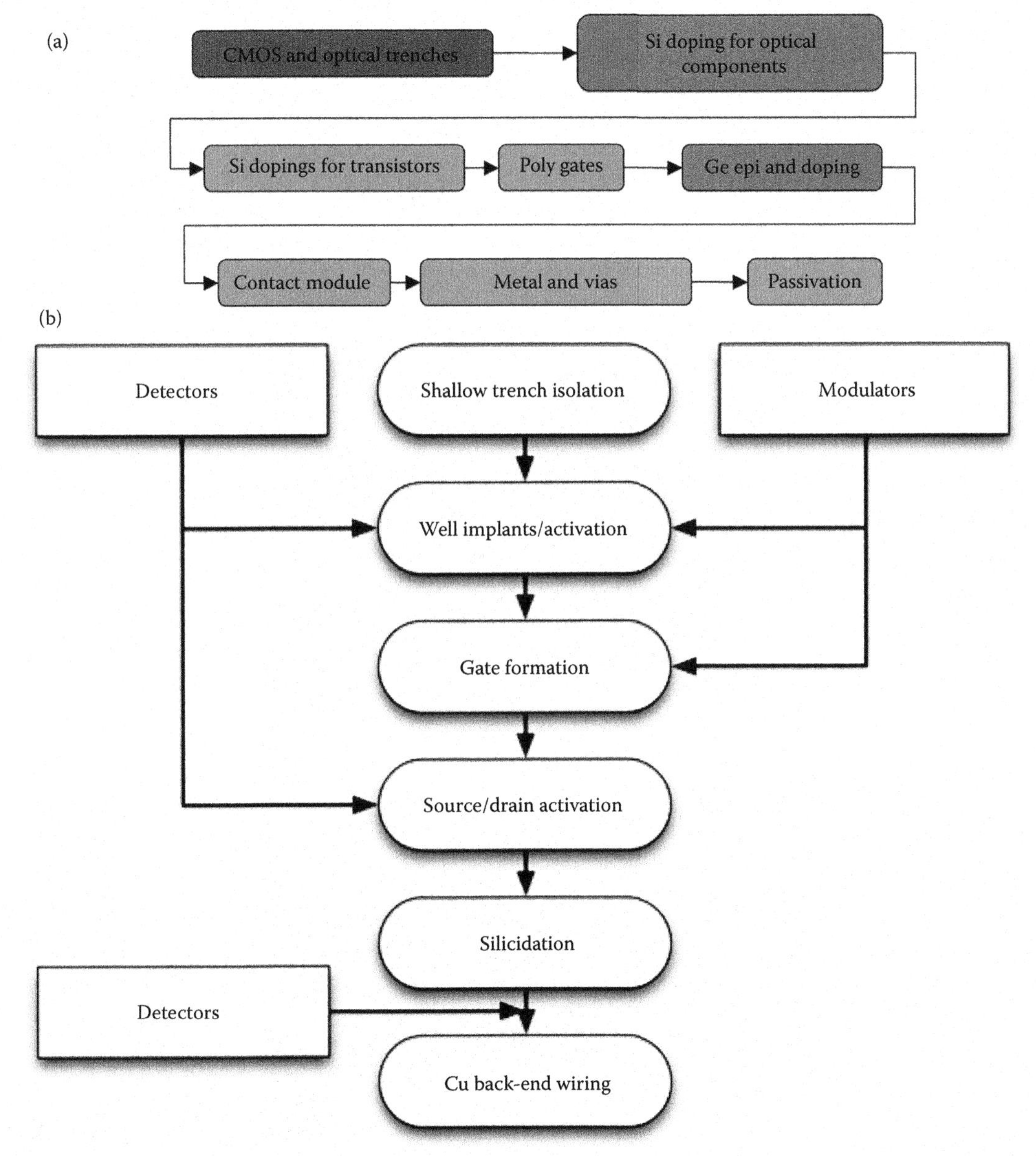

FIGURE 13.24
Luxtera (a) and IBM (b) approaches for integration of Ge and other photonic enabling steps into the flow.

In the Luxtera process, the thermal budget constrains the placement of the Ge epitaxy step after the annealing steps for silicon implants and for the gate module because Ge could not withstand the gate formation temperature. On the other hand, low-k dielectric back-ends are typically not compatible with the temperatures required during Ge epitaxy, so that Ge growth must be introduced before the contact module, and after the poly gate formation. Specifically, in the Luxtera process, Ge growth is performed after salicidation to minimize interactions with the critical transistor spacers.

In the IBM work, an alternative method was developed to get around the thermal constraints. The Ge module is integrated before the activation of source/drain implants, and uses a rapid melt growth technique using the same step as the implant anneal to produce single-crystal Ge.

In both approaches, the geometrical headroom provided by the contact module forces the use of waveguide-photodetector geometries because the Ge thickness cannot exceed the thickness of the gate of the process by a wide margin. Indeed, the contact module is used to contact the Ge, gate, and silicon, and thus the Ge surface cannot be many microns thick. Typical thicknesses are in the 300- to 500-nm range. Figure 13.25 shows a 3D view of the Luxtera photodetector and a cross-section of the IBM process for Ge.

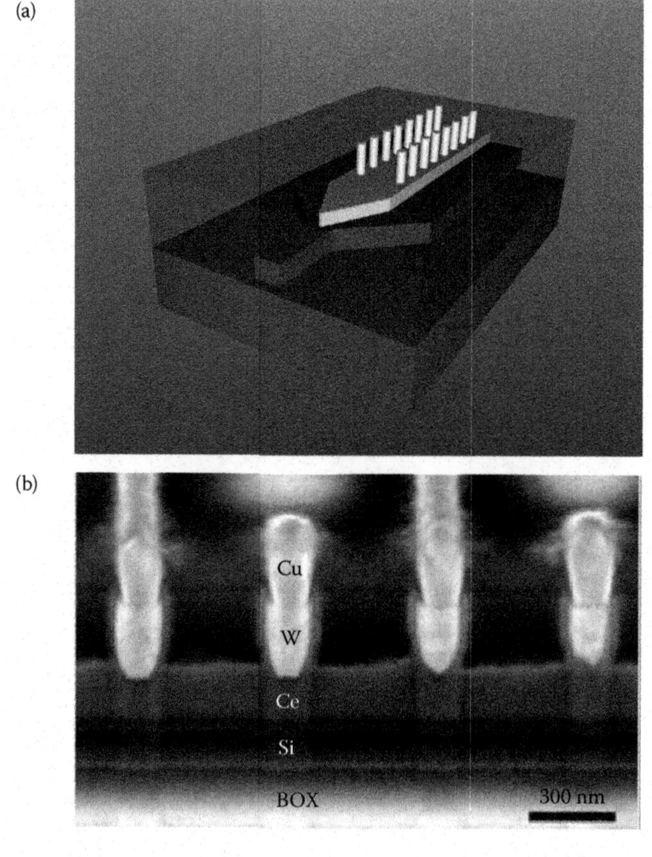

FIGURE 13.25
3D rendering of the Luxtera photodetector (a) [13], and cross-section SEM of the IBM Ge process (b) [15].

13.6 Effect of Back-End on Optical Performance

The back-end of the CMOS process is carefully crafted to optimize the density of features and minimize parasitics. As such, dielectric layer compositions and thicknesses cannot be changed easily without significant effects on the performance of the back-end. In the assumption that no changes are allowed, the layer thicknesses can affect performance in different ways.

The presence of low-level metal interconnects over waveguide or other optical devices can negatively affect performance by creating additional loss due to optical mode interaction with metals. This may require the exclusion of low-level metals over photonic areas, resulting in issues with meeting local densities of features in metal layers, or vice versa, having to space photonic devices further apart because of metal density requirements.

Some silicon photonics technologies have also made the choice of using vertical, or close-to-vertical, optical input/output interfaces, primarily using grating couplers. In that situation, light has to travel through the back-end layers between the silicon devices and the surface of the chip. A complex combination of layers of varying indices in the back-end (and particularly their interfaces) can result in additional optical loss during propagation through this stack. Removing the back-end locally above the grating couplers could be an alternative solution (Figure 13.26).

13.6.1 Lack of Scalability

As CMOS nodes evolve to ever more deeply scaled submicron nodes, their physical characteristics can change dramatically, with smaller and smaller features, thinner films, shallower junctions, etc.... Unfortunately, photonics do not tend to scale down in size, and benefits, in terms of performance of the fabrication improvements, are much more difficult to quantify. It is likely that finer feature sizes, for instance, can be leveraged into either better performing devices, or can enable new photonic devices altogether, but it is very difficult to predict. Additionally, CMOS improvements are optimized for the features of relevance to electronics and may result in fabrication processes that do not provide benefits or, in the worst case, are incompatible with photonics.

All this suggests that front-end monolithic integration is essentially unscalable due to incompatible physical properties (dimensions primarily) and the difficulty of understanding yet unknown issues that will come up with every new node. Additional complicating factors include the exponentially increasing cost of development with advanced toolsets and the lack of understanding of fundamental photonic performance by the staff of traditional CMOS fabrication facilities. The costs and efforts required to perform this type of integration on a rolling basis for each new node can be staggering, unless photonics integration is a requirement at the beginning of every new node's development.

13.6.2 Benefits

As we have seen, there is enormous complexity and cost associated with monolithic front-end integration, but the potential benefits are still worth considering and can be categorized in terms of performance, and ultimately, fabrication yields and costs.

When it comes to performance, we have already established that the constraints of CMOS compatibility can only result in, at best, matching of, and at worst, degradation of performance compared with a hypothetical "photonics-only" platform of similar complexity

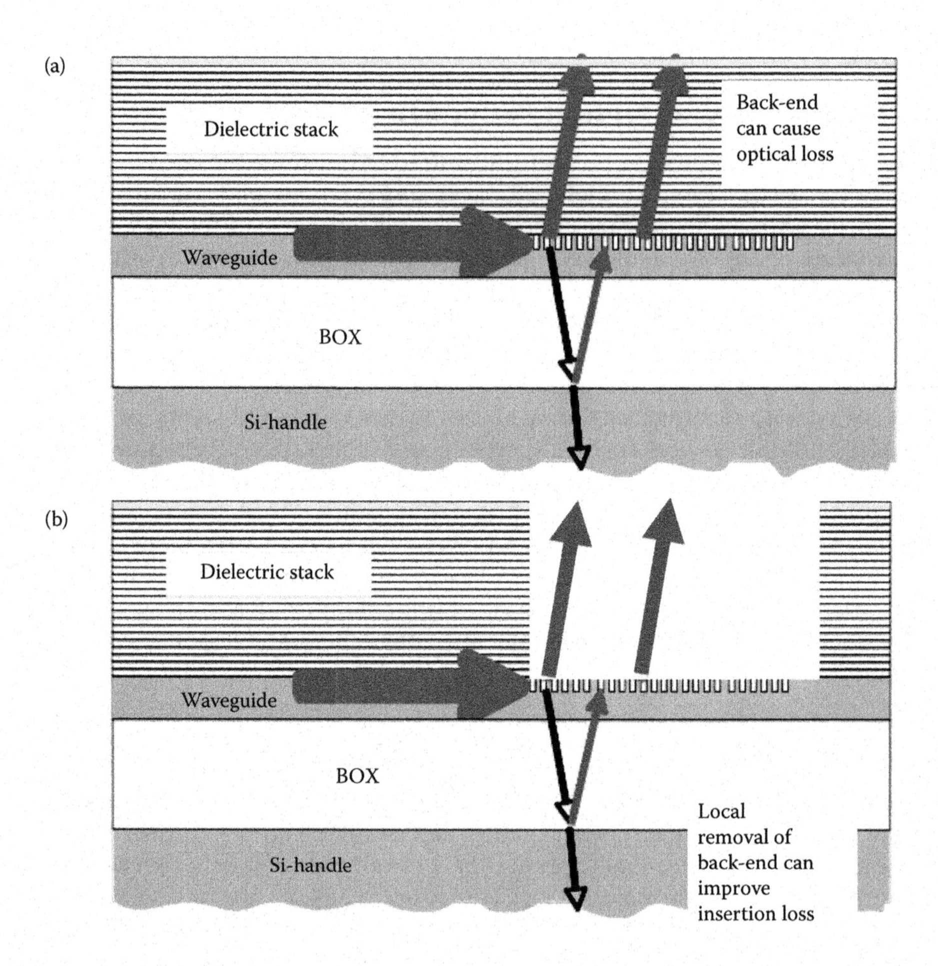

FIGURE 13.26
Effect of the back-end on optical performance. Grating coupler performance can be degraded when light travels through the back-end (a), and removing the back-end locally can improve optical coupling (b).

with similar fabrication capabilities. The tight integration with CMOS electronics must then be the source of performance benefits.

Examples of such benefits can be found in Luxtera's work:

- The extremely small capacitance of the on-chip photodetectors (one order of magnitude smaller than standard, stand-alone III–V photodiodes), along with the extremely small parasitics of the on-chip interconnects, can be leveraged to achieve an extremely low sensitivity receiver; in that work, Luxtera demonstrated a better than –21 dBm sensitivity at the Ge photodiode for BER of 10^{-12} (Figure 13.27) [21].
- The combination of on-chip photodetectors, low-speed phase modulators, and mixed-signal circuitry allows for automated biasing of optical elements using on-chip feedback loops containing A/D, D/A, and logic to run the control algorithms [21]. One could envision that keeping control local, on-chip, is probably an advantage due to the use of highly sensitive analog circuits with low-level signals, for instance, that may be difficult to handle when crossing interconnect barriers into a different chip.

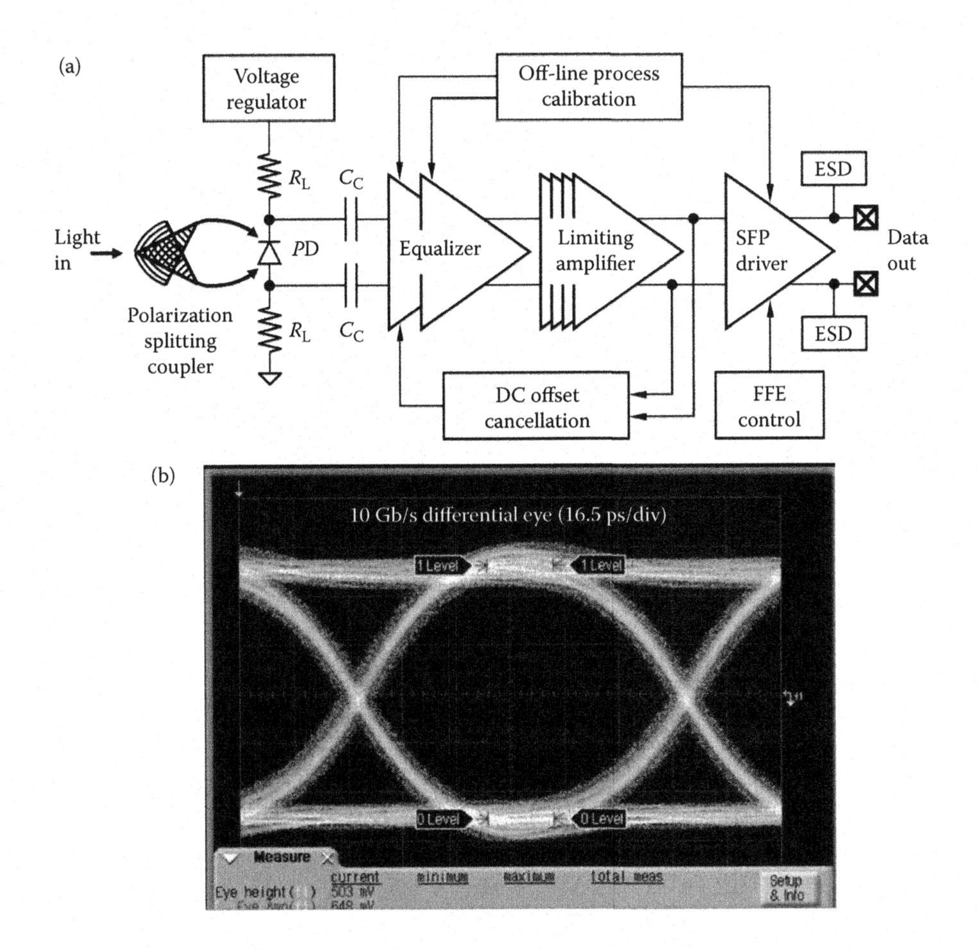

FIGURE 13.27
(a) Example of 10 Gb/s integrated receiver in the Luxtera process with (b) representative output electrical eye diagram, achieving –21 dBm sensitivity at 10^{-12} BER at the receiver.

Ultimately, just as in pure electronics SoCs, the benefits of integration can only be seen in one of two ways:

- Is a product or application enabled by the integration that would not be possible without the existence of that technology? This is a hard question to answer and must be evaluated for any specific application field.
- Is the integration resulting in higher yields and lower costs than a more hybrid solution? Again, this may be the case, depending on the specifics of the application. In an application in which many interfaces to the chip have to be taken into account, such as fibers, lasers, electrical connections, and thermal interfaces for heat sinking, then tighter integration in the front-end may provide relief in terms of design of the overall system and thus provide cost and yield advantages. Examples of different approaches for overall system integration are shown in Figure 13.28.

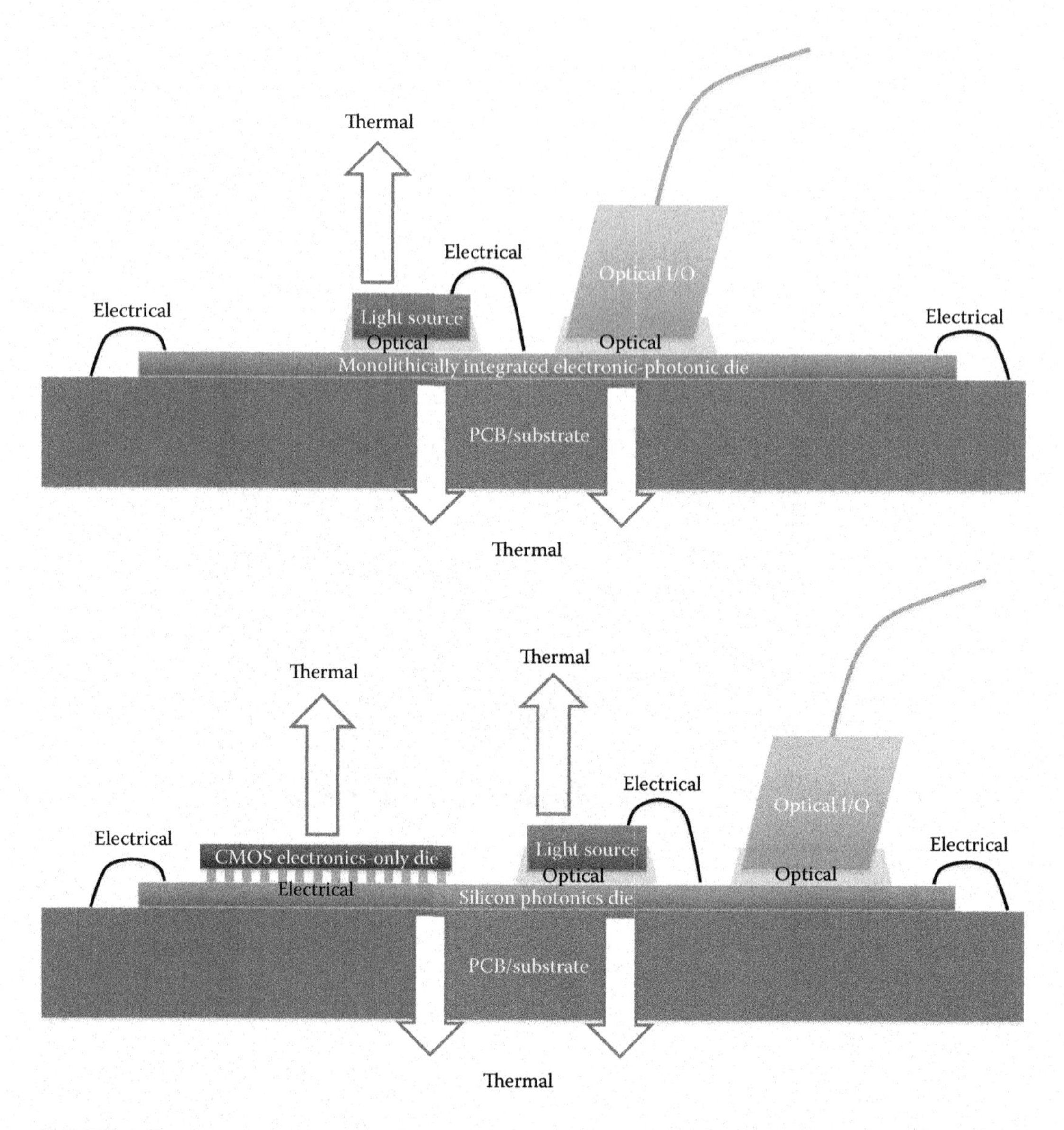

FIGURE 13.28
Examples of overall system integration with optical, electrical, and thermal interfaces in the case of front-end fabrication level integration versus hybrid integration.

13.7 Conclusion

Several different approaches to the integration of a photonic circuit with an electronic circuit have been detailed. In Table 13.1, a comparative analysis has been attempted. Different solutions for photonics–electronics lead to different technologies with their own merits and drawbacks. Depending on the applications and the associated fabrication volumes, the system designers will choose the best way to make their desired system with the necessary building blocks.

TABLE 13.1

Comparative Analysis of Different Connecting Technologies

	D2D	D2D	W2W or D2W	Wafer	W2W	W2W	Wafer
Integration	Wire bonding	Flip-chip Cu bumps	Cu Bonding	AIC low temperature	AIC high temperature	Backside fabrication	Combined fabrication
Pros	Known process	Footprint	Footprint	Footprint	Footprint	Footprint	Common electrode process steps
	Heat separation	Different die technologies and sizes	Different wafer technologies	High density	Different wafer technologies	Different wafer technologies	Very low RC
	Separate dies	High density	High density	Low RC		Separate sides	
	Multiple EIC nodes	Multiple EIC nodes	Low RC	Multiple EIC nodes	Multiple EIC nodes	Multiple EIC nodes	
			Multiple EIC nodes				
Cons	Large	Thermal cross talk	Similar die sizes	Hybrid technology	Similar die size	RF limitation	Footprint
	Parasitics	RF limitation	Thermal management	Yield of added process steps	Thermal management	TSV fabrication	Single EIC node integration
	Reproducibility			Similar die sizes			Cointegration of photonic and electronics

Note: D2D, die-to-die connection; W2W, wafer-to-wafer bonding; wafer, single wafer fabrication.

References

1. Pinguet, T., B. Analui, E. Balmater, D. Guckenberger, M. Harrison, R. Koumans, D. Kucharski, Y. Liang, G. Masini, A. Mekis, S. Mirsaidi, A. Narasimha, M. Peterson, D. Rines, V. Sadagopan, S. Sahni, T.J. Sleboda, D. Song, Y. Wang, B. Welch, J. Witzens, J. Yao, S. Abdalla, S. Gloeckner, and P. De Dobbelaere. 2008. Monolithically integrated high-speed CMOS photonic transceivers. In *The 5th International Conference on Group IV Photonics, September 17–19, 2008*.
2. Fedeli, J.M., R. Orobtchouk, C. Seassal, L. Vivien. Integration Issues of a Photonic Layer on Top of a CMOS Circuit (http://lib.semi.ac.cn:8080/tsh/dzzy/wsqk/SPIE/vol6125/61250H.pdf).
3. Thacker, H.D., I. Shubin, Y. Luo, J. Costa, J. Lexau, X. Zheng, G. Li, J. Yao, J. Li, D. Patil, F. Liu, R. Ho, D. Feng, M. Asghari, T. Pinguet, K. Raj, J.G. Mitchell, A.V. Krishnamoorthy, and J.E. Cunningham. 2011. Hybrid integration of silicon nanophotonics with 40 nm-CMOS VLSI drivers and receivers. In *2011 ECTC Conference*. IEEE 61st pp. 829–835.
4. Lee, D.C., X. Zheng, D. Feng, C.-C. Kung, J. Fong, W. Qian, J. Yao, G. Li, K. Raj, J.E. Cunningham, A.V. Krishnamoorthy, and M. Asghari. 2011. Dual-layer WDM routing for wafer-scale packaging of photonically-interconnected computing systems. *ECTC Conference*. 2011 IEEE 61st pp. 812–818, May 31–June 3, 2011.
5. Bernabé, S., C. Kopp, M. Volpert, J. Harduin, J.-M. Fédéli, and H. Ribot. 2012. Chip-to-chip optical interconnections between stacked self-aligned SOI photonic chips. *Optics Express* 20(7):7886–7894.
6. Gueguen, P., C. Ventosa, L. Di Cioccio, H. Moriceau, F. Grossi, M. Rivoire, P. Leduc, and L. Clavelier. 2010. Physics of direct bonding: Applications to 3D heterogeneous or monolithic integration. *Microelectronic Engineering* 87:477–484.
7. Orobtchouk, R., N. Schnell, T. Benyattou, and J.-M. Fedeli. 2005. Compact building block for optical link on SOI technology. In *Proceedings of 12th European Conference on Integrated Optics ECIO, 221–224.*
8. Romeo, P.R., J. Van Campenhout, P. Regreny, A. Kazmierczak, C. Seassal, X. Letartre, G. Hollinger, D. Van Thourhout, R. Baets, J.-M. Fedeli, and L. Di Cioccio. 2006. Heterogeneous integration of electrically driven microdisk based laser sources for optical interconnects and photonic ICs. *Optics Express* 14(9):3864–3871.
9. Kostrzewa, M., P. Regreny, M.-P. Besland, J.-L. Leclercq, G. Genet, P. Rojo-Romeo, E. Jalaguier, P. Perreau, H. Moriceau, O. Marty, G. Hollinger. 2003. High quality epitaxial growth on new InP/Si substrate. In *Conference on Indium Phosphide and Related Materials*, May 12–16, 2003, 12–16.
10. Fedeli, J.-M., L. Liu, L. Grenouillet, D. Bordel, F. Mandorlo, N. Olivier, T. Spuesens, P. Régreny, P. Grosse, P. Rojo-Romeo, R. Orobtchouk, and D. Van Thourhout. 2011. Towards Optical Networks-on-Chip with 200 mm hybrid technology. OFC Paper.
11. Results from the European HELIOS project: http://www.helios-project.eu/.
12. Fedeli, J.M., E. Augendre, J.M. Hartmann, L. Vivien, P. Grosse, V. Mazzocchi, W. Bogaerts, D. Van Thourhout, and F. Schrank. 2010. Photonics and electronics integration in the HELIOS project. *IEEE Proceedings of the GFP Conference 2010.*
13. Mekis, A., S. Gloeckner, G. Masini, A. Narasimha, T. Pinguet, S. Sahni, and P. De Dobbelaere. 2011. A grating-coupler-enabled CMOS photonics platform. *Journal of Selected Topics in Quantum Electronics* 17(3):597–608.
14. Orcutt, J.S., A. Khilo, C.W. Holzwarth, M.A. Popovic, H. Li, J. Sun, T. Bonifield, R. Hollingsworth, F.X. Kartner, H.I. Smith, V. Stojanovic, and R.J. Ram. 2011. Nanophotonic integration in state-of-the-art CMOS foundries. *Optics Express* 19(3):2335–2346.
15. Green, V., S. Assefa, A. Rylyako, C. Schow, F. Horst, and Y. Vlasov. 2010. CMOS integrated silicon nanophotonics: Enabling technology for exascale computation. *Semicon Tokyo* available at www.research.ibm.com/photonics.

16. Melchiori, M., N. Daldosso, F. Sbrana, L. Pavesi, G. Pucker, C. Kompocholis, P. Belluti, A. Lui. 2005. Propagation losses of silicon nitride waveguides in the near-infrared range. *Applied Physics Letters* 86:121111.
17. Cheng, K., A. Khakifirooz, P. Kulkarni, S. Ponoth, J. Kuss, L.F. Edge, A. Kimball, S. Kanakasabapthy, S. Schmitz, A. Reznicek, T. Adam, H. He, S. Mehta, A. Upham, S.-C. Seo, J.L. Herman, R. Johnson, Y. Zhu, P. Jamison, B.S. Haran, Z. Zhu, S. Fan, H. Bu, D.K. Sadana, P. Kozlowski, J. O'Neill, B. Doriss, and G. Shahidi. 2010. Extremely thin SOI (ETSOI) technology: Past, present, and future. *IEEE 2010 SOI Conference.*
18. Lee, K.K., D.R. Lim, and L.C. Kimerling. 2001. Fabrication of ultralow-loss Si/SiO2 waveguides by roughness reduction. *Optics Letters* 26(23):1888–1890.
19. Pinguet, T., S. Gloeckner, G. Masini, and A. Mekis. 2010. CMOS photonics: A platform for optoelectronics integration. In *Silicon Photonics II: Components and Integration (Topics in Applied Physics)*, edited by Lockwood, D.J., and L. Pavesi. Heidelberg: Springer.
20. Vivien, L., A. Polzer, D. Marris-Morini, J. Osmond, J.M. Hartmann, P. Crozat, E. Cassan, C. Kopp, H. Zimmermann, and J.M. Fédéli. 2012. Zero-bias 40Gbit/s germanium waveguide photodetector on silicon. *Optics Express* 20:1096–1101.
21. Kucharski, D., D. Guckenberger, G. Masini, S. Abdalla, J. Witzens, and S. Sahni. 2010. 10 Gb/s 15 mw optical receiver with integrated germanium photodetector and hybrid inductor peaking in 0.13mm SOI CMOS technology. In *2010 IEEE Internal Solid-State Circuits Conference Digest of Technical Papers* 360–361.

14

Silicon Photonics for Biology

Dan-Xia Xu, Siegfried Janz, Adam Densmore, André Delâge, Pavel Cheben, Jens H. Schmid, Ryan C. Bailey, Adam T. Heiniger, Qiang Lin, Philippe M. Fauchet, Qi Wang, and Yimin Chao

CONTENTS

14.1 What Silicon Photonics Have to Offer to Biology: A Brief History

Dan-Xia Xu

Silicon, in the same element group as carbon, has long been speculated as an alternative basis for life. In the opening address to the British Association for the Advancement of Science in 1893, James Emerson Reynolds, president of the Chemistry Section, pointed out that the heat stability of silicon compounds might allow life to exist at very high temperatures (Reynolds 1893). Although not a basis of life on this planet, elemental silicon has become indispensable in our information age. From playing music on a radio or iPod that awakes us in the morning, to the computers we depend on at work, to the cars and planes that take us home, silicon microelectronic chips are ubiquitous in every aspect of our daily life. After more than two decades of research, today silicon is also accepted as a versatile material for integrated optics. Primarily driven by demands in communication applications, fundamental photonic functionalities, such as light guiding, switching, modulation and detection, have all been demonstrated. Breakthroughs in performance, integration density, and functionalities continue. The term "silicon photonics," a phrase not often uttered only 15 years ago, has come into common use. Today, silicon photonics is poised to overcome the data flow bottleneck by bridging the gap between electronics and photonics. Does silicon photonics also have an advantageous place in biological applications? From the research advances reviewed in the following four sections, we come to a positive outlook.

Biological cells are the basic functional units of all known living organisms. Science has attempted to understand and manipulate their structure, composition, and ultimately their function. Deoxyribonucleic acid (DNA) and ribonucleic acid (RNA) molecules in the interior of a cell encode the genetic information and provide unique fingerprints of each life-form, whereas proteins carry out the biological functions. Proteins embedded in the cell membrane act as channels and pumps that move different molecules into and out of a cell. Surface membranes also contain receptor proteins that allow cells to detect external signaling molecules such as hormones. How these proteins bind and dissociate with drug molecules is the basis of discovering new drugs. Bacteria often consist of many structurally similar strains that have very different bioactivities. For example, *Escherichia coli*, a common food contaminant, consists of hundreds of varieties of which only a few strains are harmful to the human body. These strains can be differentiated through the identification of particular sections of the cell's DNA structure or particular proteins on the cell walls. Cells express particular proteins when under stress. These signaling products, or biomarkers, are useful signs for the early detection of illness. For example, stroke and cancer patients have a low level of particular proteins in the bloodstream or saliva before the onset of visible symptoms. If these biomarkers can be detected at low concentrations, early intervention becomes possible.

Many technologies for monitoring the presence and interaction of biomolecules have been developed over the last several decades. The state of biomolecules and cells can be revealed by attaching probes with fluorescent tags to particular target sites, a technique used in enzyme-linked immunosorbent assay and cell immunofluorescence. Fluorescent tags that are bright, stable, and harmless to cells are some of the desired properties. Optical label-free sensors are another class of powerful tools for monitoring biomolecular interactions (Vo-Dinh and Allain 2003; Passaro et al. 2007; Baird and Myszka 2001; Lazcka et al. 2007). In common with several other surface-sensing techniques, a surface is first coated with linker molecules that anchor receptor (or probe) molecules to the sensor surface. Receptor molecules are chosen for their specific binding preference to the molecules of interest (target or analyte). Specific binding between the receptor and target occurs when target molecules are brought in contact with the sensor surface in an aqueous solution. These binding events modify the optical properties of the sensor, which transduces an easily measurable signal. Because radioactive or fluorescent tags are not required to induce measurable signals, they are therefore referred to as label-free sensors (Fan et al. 2008). Label-free sensors based on surface plasmon resonance are one of the well-established commercial tools, providing information on the concentration and binding kinetics of a range of biomolecules (Homola 2006, 2008; Karlsson 2004). Dielectric waveguides, particularly optical fibers, have also been used for molecular sensing applications. Silicon planar waveguide sensors are a rather new entry, but they have seen rapid development in recent years, rivaling surface plasmon resonance in sensitivity and multiplexing capabilities.

Silicon was first investigated as a material for biological applications before it became well-established for integrated optics. In 1990, Canham reported that silicon quantum wire arrays can be prepared by electrochemical etching of boron-doped Si wafers (Canham 1990). These mesoporous Si layers of high porosity exhibited visible light emission under unfocused green or blue laser excitation. This phenomenon is commonly attributed to quantum size effects, which could produce emission well above the band gap of bulk crystalline silicon. It was later shown that these films display well-resolved Fabry–Perot fringes in its reflectometric interference spectrum (Doan and Sailor 1992). Several years later, the same group reported an optical interferometric biosensor using a single layer of porous silicon (Lin et al. 1997). Detection of DNA and protein binding was achieved by monitoring the wavelength shift in its reflection fringe pattern. Since then, biological applications have been the main area for porous silicon research. Relatively simple fabrication, tunable pore size, and large internal area for hosting analyte molecules are some of the advantages. Both optical and electrochemical transducers have been investigated. Porous silicon multilayers of different pore sizes have also been used to size-select certain analyte molecules or reaction products (Jane et al. 2009; Lawrie et al. 2010; Orosco et al. 2009). The interrogation method is still generally in a reflection configuration. The detection sensitivity continues to improve, but the difficulty in producing large arrays of sensors remains. Combining molecule sorting and preconcentration using multilayers of porous silicon with real-time interrogation methods may be the strength of this type of sensor. The status of porous silicon optical transducers is already well-reviewed (Fauchet 2008; Miller 2010) and therefore is not covered in this chapter. The interested reader may consult these references.

A porous silicon layer can be further processed into nanoparticles, or quantum dots, by sonication. These nanoparticles retain the photoluminescent properties, albeit sometimes with modified emission wavelengths. They have lower luminescence quantum efficiencies compared with other semiconductor quantum dots based on heavy metals, such as CdSe. However, they are chemically more stable, having a high photobleaching threshold, and pose no known cytotoxicity to cells (O'Farrell et al. 2006; Wang et al. 2011). They are seen as promising candidates for fluorescent labeling in cell immunofluorescence applications.

In the form of silicon-on-insulator (SOI), silicon was explored as a material platform for integrated optics in the 1990s. Optical communication applications were the main driving force. Initially active Si layers of 2 to 5 μm were used by most groups due to the ease of fabrication and a high efficiency for fiber to waveguide light-coupling (Pavesi and Lockwood 2004). As material qualities and fabrication technologies improved, more and more photonic functionalities were realized with high performance. The research has also moved to the use of silicon layers at a submicron thickness (200–400 nm), in which channel waveguides (also called wires) can be made truly single mode. Due to the high-refractive index contrast in the SOI platform, device sizes are drastically reduced as wire waveguides can guide light around bends with a radius of a few microns. The highly concentrated optical field led to efficient optical modulation and new solutions in nonlinear optical processes (Lockwood and Paveisi 2010). Optical communications at a range of distances continues to be the central focus of silicon photonics research, establishing high-quality fabrication infrastructures. It was soon realized that these submicron silicon nanophotonic devices were also promising candidates for sensing applications due to its strong evanescent field (Densmore et al. 2006).

The advantage of using high index contrast waveguides for evanescent field sensing was understood long ago. Tiefenthaler and Lukosz reported an integrated-optical chemical sensor using a thin film of TiO_2 ($n \approx 1.8$) on a SiO_2 ($n \approx 1.47$) substrate (Tiefenthaler and Lukosz 1989). The sensitivity was nearly two orders of magnitude higher than those of $Ti:LiNbO_3$ waveguides (core $n = 2.30$ and substrate $n = 2.29$). They had pointed out "Not a high film index n_F itself but a high value of the difference n_F–n_S of the indices of waveguiding film F and substrate S leads to maximum sensitivities." Even though this principle was clearly stated, it was only when the material quality and fabrication processes for SOI advanced that silicon-based biosensors become truly attractive. Compact sensors with high sensitivity to the surface adsorption of, for example, DNA and protein have been developed by many research groups using a variety of device configurations (Densmore et al. 2008; Xu et al. 2008; Washburn et al. 2009; De Vos et al. 2009b). Real-time and multiplexed sensing is readily achieved because large sensor arrays can be easily implemented using integrated optics methods (Densmore et al. 2009, 2011; De Vos et al. 2009a; Xu et al. 2010a; Washburn et al. 2010). These fully integrated sensors exhibited a similar level of sensitivity to the more exotic sensing methods. High-quality SOI wafers are commercially available, and the feasibility of volume manufacturing has been validated by several groups using a CMOS photonics foundry such as epixFab (http://www.epixfab.eu/).

Sensor elements are only a small fraction of the final sensing system cost and technology requirements. For silicon nanophotonic sensors to truly penetrate the biological sensing and point-of-care diagnostics market, the innovation in silicon photonics must play a significant role in systems instrumentation. Recent work has shown that temperature drift–induced noise can be accurately compensated by using reference sensors (Xu et al. 2010b), thereby removing the need for temperature stabilization, which is a stringent requirement for commercial label-free sensing instruments using, for example, surface plasmon resonance. The coupling of light in and out of the submicron sensing waveguides is enabled by efficient grating couplers, giving large alignment tolerances (De Vos et al. 2009a; Iqbal et al. 2010; Densmore et al. 2011). There are continuing efforts to integrate biological sample preparation and handling functions on-chip using microfluidics. Whereas the deployment of silicon-based label-free sensing chips is on the horizon, more sophisticated functionality built from this platform is probably still to come.

The high index contrast of SOI waveguides, which benefits evanescent field sensing applications, also makes this platform advantageous for optical manipulation and transport of biomolecules and nanoparticles. Nanoscopic matter can be located or moved to particular locations by optical gradient force and radiation pressure. High optical field strength and fast decay near the waveguide surface give rise to a large gradient force, allowing molecules of less than 100 nm in dimension to be manipulated effectively. These effects are further enhanced by using slot waveguides and resonant structures (Yang et al. 2009; Lin et al. 2010).

The existence of a high-quality passivation material, that is, silicon dioxide (SiO_2), to prevent surface charge states is one of the key attributes that has made silicon the dominant material in microelectronics. This property is also important when it comes to biosensing applications. In air or aqueous solutions, a native silicon surface is easily oxidized to form an oxide layer. Unlike in many other materials, however, this oxidation process is self-limiting and stops after the formation of a thin layer of oxide on the order of a nanometer. A thicker (≤3 nm) and better controlled oxide layer can also be formed by treating the silicon surface with oxygen plasma. These oxide surfaces are chemically similar to the glass slides used in conventional microarray techniques, allowing the use of well established functionalization chemistries to anchor desired receptor molecules to the surface.

This chapter is organized as follows. Section 14.2 reviews the progress of planar silicon waveguide molecular affinity sensors, whereas the strategies of surface functionalization and bioconjugation are discussed in Section 14.3. The manipulation and transport of biomolecules using silicon nanostructures is reviewed in Section 14.4, and the status of bioimaging using silicon nanoparticles is reviewed in Section 14.5.

"Silicon is a material where the extraordinary is made ordinary" (Leamy and Wernick 1997). What silicon brought to electronics is astounding capabilities by executing millions upon millions of simple operations. Sophisticated manufacturing technologies are made affordable by producing myriads of functionalities in large volume using similar process flows. Planar waveguide–based sensors using various other materials have been investigated for several decades, yet the commercial penetration is limited. When technologies for biological applications take full advantage of the infrastructure and knowledge of microelectronics and silicon photonics, silicon can yet again do extraordinary things.

A silicon-based life-form may stay in the world of science fiction, but silicon, the most abundant solid element in Earth's crust, is predicted to play a vital role in understanding and maintaining carbon-based life.

14.2 Silicon Planar Waveguide Biomolecular Sensors

Siegfried Janz, Dan-Xia Xu, Adam Densmore, André Delâge, Pavel Cheben, and Jens H. Schmid

14.2.1 Introduction

Silicon planar waveguide sensors are a new entry in the field of label-free molecular sensing. Even a simple millimeter-long photonic wire evanescent field (PWEF) sensor can have a response that is more than one order of magnitude larger (Densmore et al. 2006; Janz et al. 2009) than established instruments based on surface plasmon resonance (SPR; Homola 2006; Karlsson 2004). The silicon platform also leverages the enormous silicon electronic device infrastructure. Silicon manufacturing processes can be adopted almost "as is" to fabricate complex optical circuits that combine many sensors in dense arrays and allow them to be interrogated both simultaneously and in real time. No other label-free sensing platform offers the combination of high sensitivity and ability to integrate complex functionality on a single platform while lending itself naturally to mass production.

Waveguide sensors in general have been in use for more than two decades (Tiefenthaler and Lukosz 1989; Lukosz 1991; Luff et al. 1996, 1998; Weisser et al. 1999; Prieto et al. 2003; Boyd and Heebner 2001; Horvath 2005; Densmore et al. 2008; Xu et al. 2008), but the silicon platform has been demonstrated and developed only within the last 5 years (Densmore et al.

2006, 2008; De Vos et al. 2007; Iqbal 2010; Xu et al. 2008, 2010b). The advantages of waveguides with subwavelength dimensions using high-refractive index contrast materials for sensing were clear from earlier studies (Tiefenthaler and Lukosz 1989; Parriaux and Veldhuis 1998), but silicon waveguide sensors have only recently become feasible with the availability of very high-quality silicon-on-insulator wafers for making low loss waveguides, and the advances in silicon manufacturing methods that make the fabrication of devices with 100 nm feature sizes routine. The following sections will provide a brief overview of why silicon waveguides are well suited for molecular sensing, the methods for converting individual sensor responses into a measurable signal, and finally the integration of sensors arrays, input and output couplers, and microfluidics necessary to create a viable sensor technology.

14.2.2 Waveguide Molecular Sensors

Optical label-free molecular sensors operate by detecting and measuring the local refractive index change caused when target molecules (the analyte) bind to probe molecules attached to the sensor surface. The probe molecule is chosen for its ability to selectively and exclusively bind to the target molecule of interest. Some well-known examples of probe–target pairs in biology are an antibody and its antigen, and a single strand of DNA and its complementary strand. The probe molecules are bound to the sensor surface before the measurement. If the corresponding target molecules are present in the liquid flowing over the sensor, they will bind to the probe molecules until either all the probe molecule binding sites are occupied or equilibrium is reached between bound target molecules and free target molecules in solution. The measurement problem is simply that of determining the change in refractive index in the surface molecular layer as the target molecules bind to the surface. Because these molecular layers are only a few nanometers thick in general, the resulting changes in ordinary reflection or transmission at the surface are of the order 10^{-4} or less, a range at the limits of conventional thin-film optical metrology. Instead, waveguide and SPR sensors use bound optical modes that propagate parallel to the surface rather than through it. As molecules bind to the surface, the phase velocity of the surface mode changes because of the higher index of refraction of the bound molecules compared with the water they displace. In the case of an SPR sensor, this manifests itself as a shift in the surface plasmon mode resonance wavelength. For waveguides, the change in mode phase velocity can be determined by several different interferometric or resonance techniques. Figure 14.1 illustrates the schematic configuration for a simple evanescent field waveguide molecular sensor.

The molecule-induced perturbation of the optical mode phase velocity depends on how tightly the mode is localized to the surface layer. In most waveguides, the optical mode is largely confined to the waveguide core, with a small evanescent tail extending into the cladding material surrounding the waveguide. It is only this evanescent field that can interact with the molecules at the waveguide surface to change the effective index N_{eff} of the waveguide mode. The effective index in turn determines the phase velocity of the guided mode $v = c/N_{\text{eff}}$, and hence, the optical phase $\varphi = N_{\text{eff}}(2\pi/\lambda)L$ for light passing through a sensor of length L. Perturbation theory (Kogelnik 1990) can predict the effective index change δN_{eff} induced by the small local refractive index perturbation Δn caused by molecules binding to the surface.

$$\delta N_{\text{eff}} = c\int \Delta\varepsilon E\, E^{*}\, dy\, dx = 2c\varepsilon_0 \int n\Delta n E E^{*}\, dy\, dx \tag{14.1}$$

Here, E denotes the normalized waveguide electric field profile for a mode power of unity. From Equation 14.1, the strategy for optimizing waveguide designs for sensing is

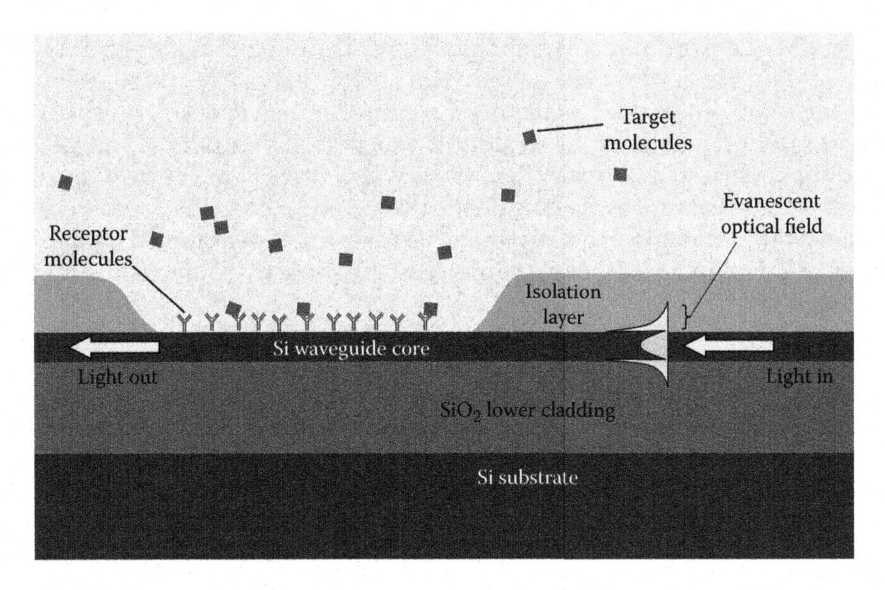

FIGURE 14.1
A cross-section of a silicon waveguide molecular sensor. As the target molecules bind to the receptor or probe molecules, the change in refractive index at the surface causes the phase velocity of the guided light to change.

to concentrate the electric field mode at the layer in which the induced molecular index change Δn is non-zero. This layer is comparable to the nanometer dimensions of the molecules involved—much smaller than both the waveguide mode width and the waveguide core. If the mode is largely contained within the waveguide core, the coupling of light to the molecules at the surface is very weak. On the other hand, if the guided light extends too far into the cladding, the waveguide properties will be dominated by the bulk properties of the liquid or gas above the waveguide.

The mode profile of the 260 × 450-nm silicon photonic wire in Figure 14.2 makes it clear why silicon photonics is so attractive for molecular sensing (Densmore et al. 2006). Here, most of the mode field skims along immediately above the surface of the waveguide, extending not more than a few hundred nanometers from the surface with a maximum amplitude at the

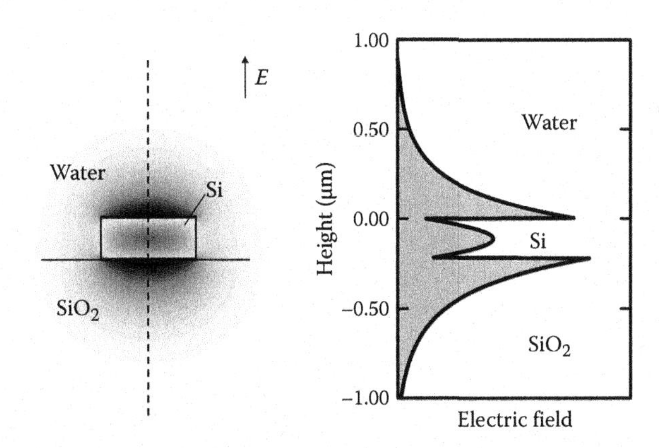

FIGURE 14.2
The TM polarized waveguide mode profile of a 260 × 450-nm silicon photonic wire sensor shown in cross-section, and as a field amplitude plot along the dashed line.

surface precisely where the molecular binding occurs. This unique mode profile gives silicon photonic wire waveguides of thicknesses near 220 nm the highest δN_{eff} response to molecular binding of any commonly available channel waveguide system (Densmore et al. 2006).

A number of effects conspire to create this mode shape (Janz et al. 2009). When the waveguide thickness is much smaller than the wavelength of light, much of the waveguide mode power is forced to propagate outside the waveguide core in the evanescent tails. The evanescent field decay length of approximately $\delta = 160$ nm is determined by the difference between the effective index of the waveguide and the refractive index of the cladding.

$$\frac{1}{\delta} = \left(\frac{2\pi}{\lambda}\right)\sqrt{N_{eff}^2 - n_{clad}^2} \tag{14.2}$$

For silicon photonic wires, $N_{eff} \sim 2$ for transverse magnetic (TM) polarized light at $\lambda =$ 1550 nm, and the refractive index of water is approximately $n = 1.32$. The high concentration of electric field at the surface molecular binding layer is thus a direct result of the high core-cladding index contrast of the silicon waveguide system. Finally, by choosing the light polarization to be perpendicular to the sensing surface (i.e., the TM-like mode), the field $E(y^+)$ immediately above the surface is amplified relative to the mode field just below the surface $E(y^-)$ by the ratio of core to cladding dielectric function

$$\frac{E(y^+)}{E(y^-)} = \frac{\varepsilon_{core}}{\varepsilon_{clad}}\left(\frac{n_{core}}{n_{clad}}\right)^2 \tag{14.3}$$

This is just the result of the electromagnetic boundary conditions at the interface, and gives a sevenfold amplification of the surface field in a silicon waveguide with water cladding.

Although the effective index concept is not commonly used in the surface plasmon literature, it is useful here in comparing the two sensor platforms. The phase velocity v of a surface plasmon or a dielectric waveguide mode can both be described in terms of an effective index ($v = c/N_{eff}$). Waveguide and SPR mode calculations reveal that change in effective index (δN_{eff}) or mode phase velocity in response to a given bound molecule density for a photonic wire waveguide is approximately the same as for SPR (Janz et al. 2009). The advantage of the Si waveguide platform arises instead from the fact that guided light can travel unlimited distances in dielectric waveguides (neglecting inevitable scattering losses due to surface imperfections). In contrast, propagating surface plasmons are rapidly attenuated by the metal after at most only a few tens of micrometers. As will be discussed in more detail later in the section, the minimum level of detection for any optical sensor is inversely proportional to the light propagation length within the sensor, or equivalently, the lifetime of light in the sensor (Schmid et al. 2009). Thus, a silicon sensor with only a millimeter-long waveguide has a response at least an order of magnitude larger than a typical SPR system. This very general rule arises directly from the wave vector–position (or time–frequency) uncertainty relation for measurements on any wavelike excitation.

14.2.3 Silicon Slot Waveguides

The response of a silicon waveguide sensor to surface molecular binding can be further enhanced by using a slot waveguide configuration. A slot waveguide sensor element is formed by two silicon channel waveguides brought very close together, so that they are separated by a slot of a few hundred nanometers or less (Almeida et al. 2004; Barrios et al.

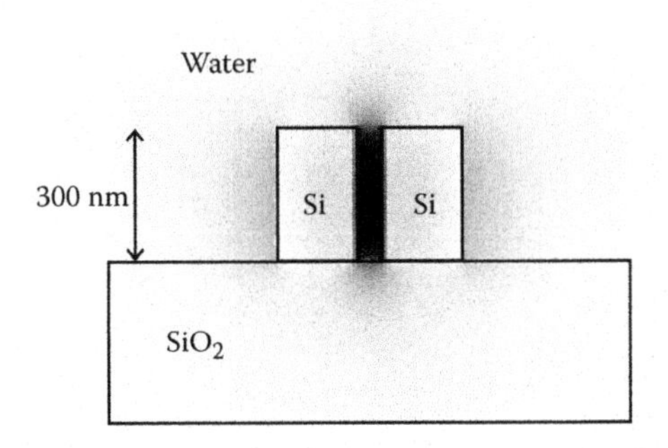

FIGURE 14.3
A cross-section of a silicon slot waveguide consisting of two Si channel waveguides separated by a 60-nm slot. The horizontally polarized component of the electric field for the TE mode profile is shown as a gray scale contour plot (dark = high field amplitude).

2008; Yang et al. 2009). For appropriately chosen silicon channel and slot dimensions, the optical mode in a slot waveguide is localized almost entirely within the slot gap, as shown in Figure 14.3.

This unique field structure is again a result of the high index of refraction of silicon relative to the gap material, combined with the small size of the silicon waveguides. Waveguide mode calculations show that the effective index response of such a waveguide to a molecular layer can be several times stronger than the response for a simple silicon channel waveguide as described previously. In the simple channel waveguide of Figure 14.2, the mode field overlap with the molecular surface layer is limited by the evanescent field decay length δ as determined by Equation 14.2, and therefore, cannot easily be changed from its typical value of $d \sim 160$ nm. On the other hand, in a slot waveguide, the extent of the field outside the waveguide surface is simply equal to the slot width, which can be as small as a few tens of nanometers. Slot waveguide sensors have been demonstrated in the silicon planar waveguide platform as well as in other waveguide systems. Although the response of a slot waveguide to molecular binding is much improved over the simple channel waveguide of Figure 14.2, the fabrication of narrow slots remains more challenging. The best reported waveguide losses for slot waveguides are typically 10 dB/cm or more (Baehr-Jones et al. 2005; Claes et al. 2009), which are much higher than the loss of 1 to 2 dB/cm typical of simple silicon channel waveguides (Lardenois et al. 2003; Dumon et al. 2004). Fluid handling can also be more challenging. For example, if the walls of the slot are hydrophobic, the fluid–surface interface energy can make it very difficult for fluid to penetrate the slot. Such problems can be alleviated by ensuring that surface functionalization produces a hydrophilic surface on the slot sidewalls.

14.2.4 Interrogation of Waveguide Sensors

The fundamental response of a simple waveguide sensor is the change in effective index δN_{eff}, and the corresponding phase delay $\delta\varphi = \delta N_{eff} L k_0$ induced in a waveguide sensor of length L, where k_0 is the wave number. The Mach–Zehnder interferometer (MZI) of Figure 14.4a is the simplest configuration to convert phase delay into a measurable intensity change.

Light in the input waveguide is split between two separate waveguide paths, which later rejoin, and the optical power coupled in the output waveguide will vary with the phase of

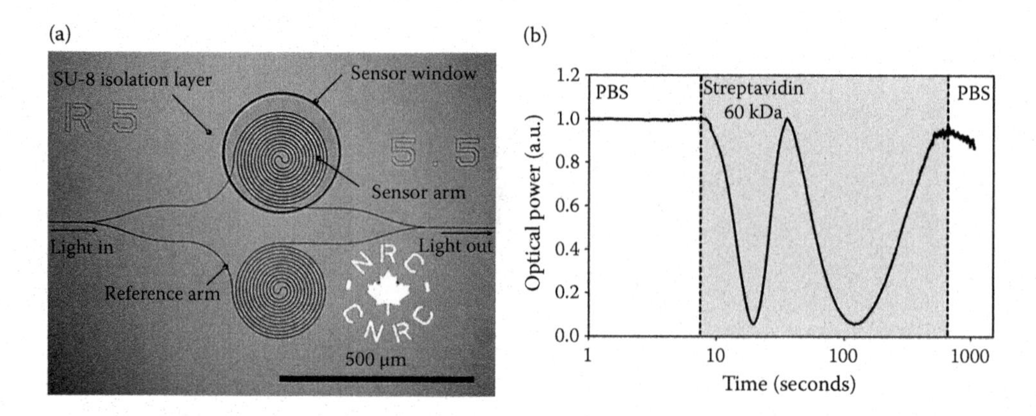

FIGURE 14.4
(a) A top-view of a MZI with folded spiral arms, where the arm in the sensor window is exposed to the overlying fluid for optical sensing. (b) The output intensity of the MZI as a monolayer of streptavidin protein binds to the surface. The shaded region indicates the time interval over which the sensor is exposed to streptavidin.

the light arriving at the terminal MZI combiner from the two arms. When one arm of the MZI includes a sensor waveguide section of length L_s, changes δN_{eff} in the sensor section will cause the output intensity to vary as

$$I = \cos^2\left(\delta N_{eff}\left(2\pi/\lambda\right)L_s + \varphi_0\right), \tag{14.4}$$

where φ_0 is the baseline phase difference shift between light traveling through the two arms of the interferometer, and is determined by the relative path lengths and waveguide effective indices in each arm of the MZI. Thus, molecular binding (i.e., the resulting change δN_{eff}) produces an oscillatory change in transmitted intensity for any fixed light wavelength, as in the example streptavidin binding data in Figure 14.4b. One can also monitor the output spectrum of the MZI because the periodic interference pattern shifts by a wavelength increment $\Delta\lambda$ approximately proportional to δN_{eff}, and hence bound molecular density. For a MZI with the sensor length L_s much longer than the reference arm length, this wavelength shift is given by $\delta\lambda = \lambda\delta N_{eff}/N_g$, where N_g is the sensing waveguide mode group index. On the other hand, MZI can also be made with precisely balanced arms, to eliminate wavelength variation in the output signal and suppress response temperature drifts. In balanced geometries the sensor is read by monitoring output intensity alone because the output is almost wavelength independent. The Mach–Zehnder configuration has been used many times for evanescent field sensors (Luff et al. 1998; Densmore et al. 2006, 2008, 2009), in silicon and other waveguide systems.

The first cosine argument in Equation 14.4, $\delta\varphi = \delta N_{eff}(2\pi/\lambda)L_s$, is the phase response of a simple waveguide sensor and scales with sensor length. To obtain the largest possible response from a silicon photonic wire sensor in an MZI configuration, the sensing waveguides can be folded into tightly wound spirals as shown in Figure 14.4 (Densmore et al. 2008). A single spiral can contain a sensor up to several millimeters long, but the entire sensor element can occupy a circular area on the chip only a couple hundred micrometers or less across. This sensor design is only practical in the silicon photonic wire platform. The high-refractive index contrast enables silicon photonic wires to follow curved paths with radii as small as 5 μm with no measurable bend loss. In other waveguide platforms, the high bend loss precludes the use of such compact sensors.

Instead of direct interferometric methods, resonators are often used to convert an effective index change δN_{eff} into a measurable intensity variation. In classical optics, the transmission resonance wavelengths of a two-mirror Fabry–Perot resonator vary with the refractive index of the medium inside the optical cavity. A ring resonator is the waveguide equivalent of the classic Fabry–Perot cavity, and its resonances vary similarly with the effective index of the ring waveguide.

The ring resonator consists of a waveguide in a closed ring path, shown in Figure 14.5, coupled to one or two straight waveguides through a directional coupler. When the round-trip optical phase delay

$$\varphi = (2\pi/\lambda)N_{eff}L \quad (14.5)$$

around a ring of length L is a multiple of 2π, light coupled to the ring from the input waveguide constructively interferes to build up a large intensity within the ring. At the same time, the transmitted light in the output or through waveguide decreases. At such a resonance, this power is transferred from the ring either to the second straight waveguide, if present, or simply dissipated by waveguide loss mechanisms in the ring cavity. The output intensity spectrum of a ring resonator in Figure 14.5 therefore shows the comb of intensity minima, each minima corresponding to the successive resonances separated in wavelength by the free spectral range

$$\Delta\lambda_{FSR} = \frac{\lambda^2}{N_g L}. \quad (14.6)$$

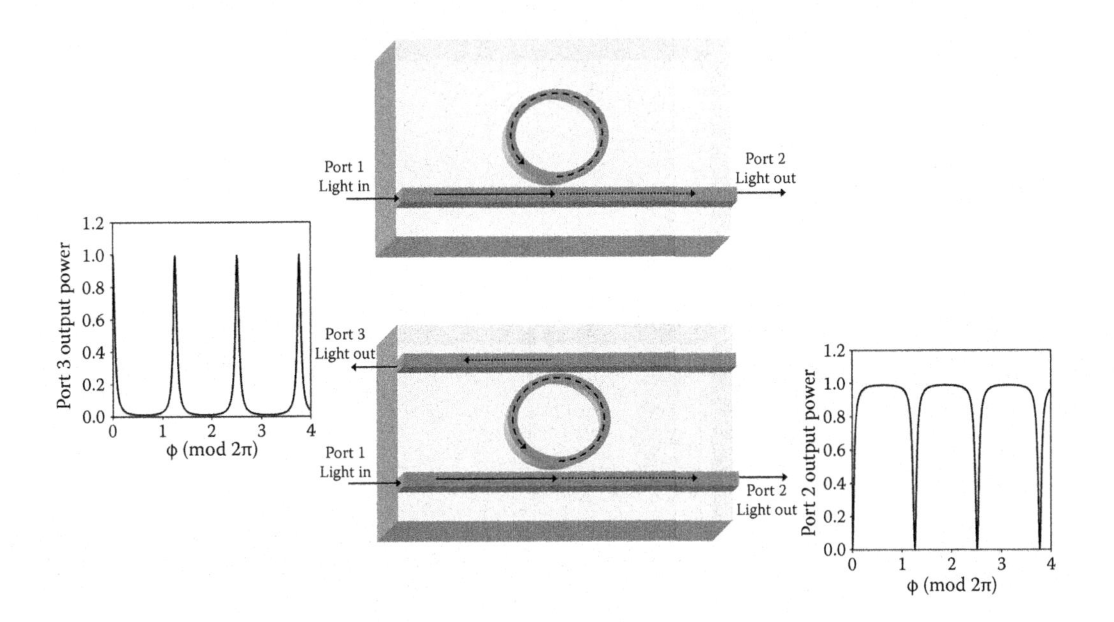

FIGURE 14.5
A schematic top view of ring resonators with one and two coupled waveguides. The plots show typical output spectra for the drop port (port 2) and the add port (port 3).

The round-trip phase accumulation from Equation 14.5 depends on the effective index, so that the resonant wavelengths shift by $\delta\lambda$ in response to a perturbation δN_{eff} induced by molecular binding to the surface. Just as for the unbalanced MZI, the shift is given by $\delta\lambda = \lambda\delta N_{eff}/N_g$ and is directly proportional to the molecular density on the waveguide surface. The group index N_g can be calculated from Equation 14.6. Waveguide ring resonators have also been used extensively for sensor transduction in many different waveguide platforms and fiber sensors. There are many reports describing the operating principles behind applications of waveguide ring resonators (Little et al. 1997; Yariv 2000; Rabiei et al. 2002; Ksendzov et al. 2004; Xu et al. 2005, 2008; Yalcin et al. 2006; De Vos et al. 2007), and design issues of particular relevance to ring sensors have been described by Delâge et al. (2009). Figure 14.6a shows a silicon ring resonator in a spiral configuration, and Figure 14.6b shows the shift in resonance as a streptavidin protein binds to the surface of the ring. The maximum shift shown here corresponds to approximately 10% surface coverage by a monolayer of protein (Xu et al. 2008).

Because sensor response scales with waveguide length, the ring cavity in Figure 14.6a is extended using a folded spiral geometry, just as is done for the MZI sensors of Figure 14.4. But in a ring resonator, light may also circulate many times around the ring before being dissipated or coupled out of the ring. Therefore, the light interacts many times with molecules at the sensor surface, so that a ring sensor with comparable response to a simple sensor in a MZI configuration can be much smaller. A more detailed analysis shows that the effective sensor length of a ring is $L_{eff} = (F/\pi)L$ (Schmid et al. 2009), where L is the physical length of the ring waveguide and F/π is the number of times light circulates around the ring before being dissipated or coupled out. The finesse F is determined by resonator cavity loss, and can be used to characterize both optical ring resonators (Delâge et al. 2009) and Fabry–Perot cavities (Fowles 1975). In waveguide ring resonators and photonic crystal resonators, the resonance performance is more commonly characterized by the quality factor Q and resonance linewidth $\Delta\lambda_{3dB}$, which are both measures of the amount light energy stored within the resonator relative to the rate of energy loss or dissipation. For a ring resonator the three quantities are connected through the relations

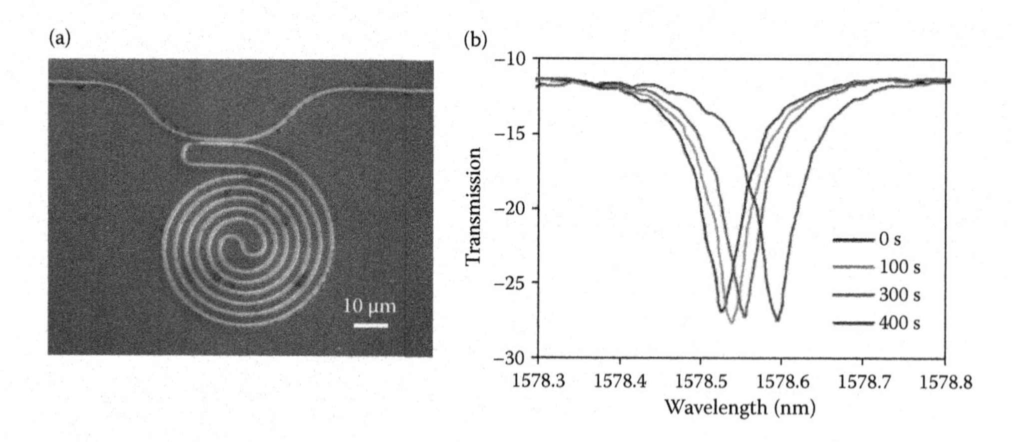

FIGURE 14.6
(See color insert.) (a) A spiral ring resonator sensor element formed by silicon photonic wire waveguides. (b) The wavelength shift of a resonance as streptavidin protein molecules bind to the surface. The maximum shift at 400 s corresponds to approximately 10% of a protein monolayer. (From Xu, D.-X. et al., *Opt. Express* 16:15137–15148, 2008. With permission.)

$$Q = \frac{\lambda}{\Delta\lambda_{3dB}} = \left(\frac{N_g L}{\lambda}\right) F \tag{14.7}$$

where N_g is again the group index of the resonator waveguide.

Another approach to silicon planar waveguide sensors is to use the unique spectral properties of photonic crystals for signal transduction (Chow et al. 2004; Schmidt et al. 2004; Lee and Fauchet 2007a,b; Buswell et al. 2008). One- and two-dimensional photonic crystals can be formed in thin silicon waveguides by etching a row or lattice of holes through the waveguide layer, similar to the structures shown in Figure 14.7. When the size and arrangement of the holes is chosen correctly, the photonic crystal will have an optical band gap—a range of wavelengths (or frequencies) over which light has no propagating modes within the photonic crystal waveguide. The transmission of light in this wavelength (frequency) range along any direction through a finite length of such a structure is therefore very small.

The missing row of holes in the photonic crystal shown in Figure 14.7a can form a waveguide (Buswell et al. 2008) that provides a conduit for light to pass through the photonic crystal section. This waveguide will, however, have a cutoff wavelength above which light is not transmitted. The exact wavelength of this transmission cutoff depends on the refractive index of the materials immediately adjacent to the missing row waveguide. Hence, the transmission cutoff shifts measurably as molecules bind to the silicon surface near the waveguide. For example, the adsorption of a monolayer of streptavidin protein to the photonic crystal in Figure 14.7a results in a cutoff wavelength shift of 0.86 nm (Buswell et al. 2008).

Another approach is to create a crystal defect formed by a hole of different size or shape in the photonic crystal (Chow et al. 2004; Schmidt et al. 2004; Lee and Fauchet 2007a,b), as in Figure 14.7b. At a certain resonance wavelength within the optical band gap, the defect can trap light and build up a high optical intensity localized around the defect, much as light intensity builds up in a ring resonator or Fabry–Perot cavity at resonance. By introducing such a localized optical state in the photonic band gap frequency

FIGURE 14.7
(a) Two photonic crystal sensor configurations. A missing row of holes forms a waveguide with a transmission cutoff wavelength that shifts with molecular binding. (b) A defect hole acts as a resonator that couples light from the input to the output waveguide. The resonance wavelength shifts with molecular attachment in and around the defect hole. (Panel a from Buswell, S. C. et al., *Opt. Express* 16:15949–15957, 2008; panel b from Lee, M.R. and Fauchet, P.M., *Opt. Letters* 32:3284–3286, 2007. With permission.)

range region, light incident from one side of a finite photonic crystal, as in the case of Figure 14.7b, can couple weakly to the defect state and build up power. The light in the resonant state can in turn couple weakly to the output side of the photonic crystal. The defect state is observed as an enhancement of the transmission through the finite photonic crystal at the resonance wavelength.

The resonance wavelength of a photonic crystal defect state, and therefore, the transmission peak, is again determined by the optical constants of the materials near the defect structure. Therefore, the resonance wavelength shifts as molecules bind to the surface of the waveguide near the defect, or on the walls of a hole forming the defect in Figure 14.7b. Figure 14.8 shows an example of such a resonance shift. Photonic crystal structures are also sensitive to the refractive index of the material filling the defect hole, and can be used to detect a single nanoparticle or even a virus (Lee and Fauchet 2007a; Schmidt et al. 2004), provided such a single nanoscale object can be directed into the hole.

The operating mode and interrogation of defect state-based photonic crystal sensors and a ring resonator sensor are very similar. Both are resonators and transduction relies on the shift in resonant wavelength caused by molecular binding (Lee and Fauchet 2007a,b; Xu et al. 2008, 2010a). The resonator formed by a photonic crystal defect state is, however, much smaller than the ring resonator cavity. A defect created by a single hole is a few hundred nanometers across, and the electric field distribution associated with the defect state extends only in the order of 1 μm around the defect (Schmidt et al. 2004; Lee and Fauchet 2007a). As a result, the absolute number of molecules required to induce a given resonant wavelength shift can, in principle, be much smaller in a photonic crystal sensor than for any extended waveguide sensor. On the other hand, the same electric field concentration in photonic crystal resonators also causes high scattering losses, and these resonators tend to have relatively low-quality factors. Nonetheless, resonator sensors with a measurable response to as little as 1 fg or less of molecular mass have been reported in both ring and photonic crystal configurations (Lee and Fauchet 2007a; Xu et al. 2010a).

The last approach to interrogating silicon waveguide sensors discussed here is the use of surface coupling gratings. Waveguide gratings are in fact one of the first methods used to interrogate waveguide sensors (Tiefenthaler and Lukosz 1989). A grating is fabricated on the surface of a waveguide with a line pitch Λ_g chosen so that light incident on the grating

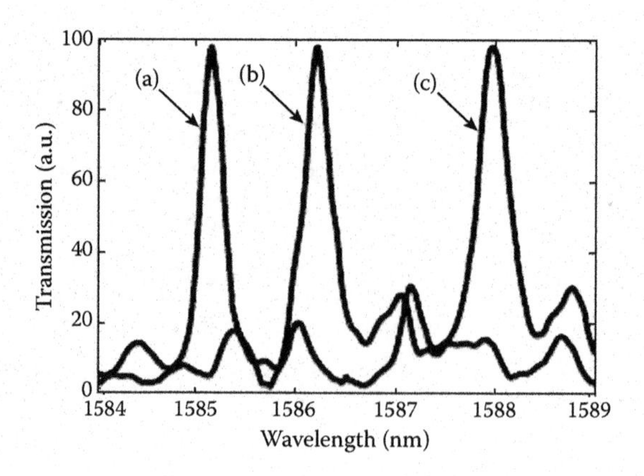

FIGURE 14.8
Photonic crystal defect resonance spectrum for (a) the initial clean sensor, (b) the sensor after exposure to glutaraldehyde, and (c) and after BSA protein molecules are bound to the sensor. (From Lee, M.R. and Fauchet, P.M., *Opt. Express* 15:4530–4534, 2007. With permission.)

from above is coupled into the waveguide as shown in Figure 14.9. The coupling condition is determined by the grating equation

$$N_{\text{eff}} = \sin\theta + m\left(\frac{\lambda}{\Lambda_g}\right) \tag{14.8}$$

which relates the coupling angle θ, wavelength of incident light λ, and the effective index of the waveguide. If the waveguide effective index is changed by molecular binding to the grating waveguide surface, the optimal coupling wavelength at a given incident angle will shift, or alternatively, the optimal angle at a fixed wavelength will change. Either quantity can be used as the transduction parameter for a waveguide sensor. Alternatively, both can be held fixed and the variation in optical coupling efficiency can be measured, as was the case for the first demonstrated waveguide sensors (Tiefenthaler and Lukosz 1989).

In the silicon platform, a grating configuration sensor was demonstrated using the wavelength resonance of the coupling grating (Schmid et al. 2009). The device is shown in Figure 14.9a. The waveguide has the same silicon waveguide thickness of approximately

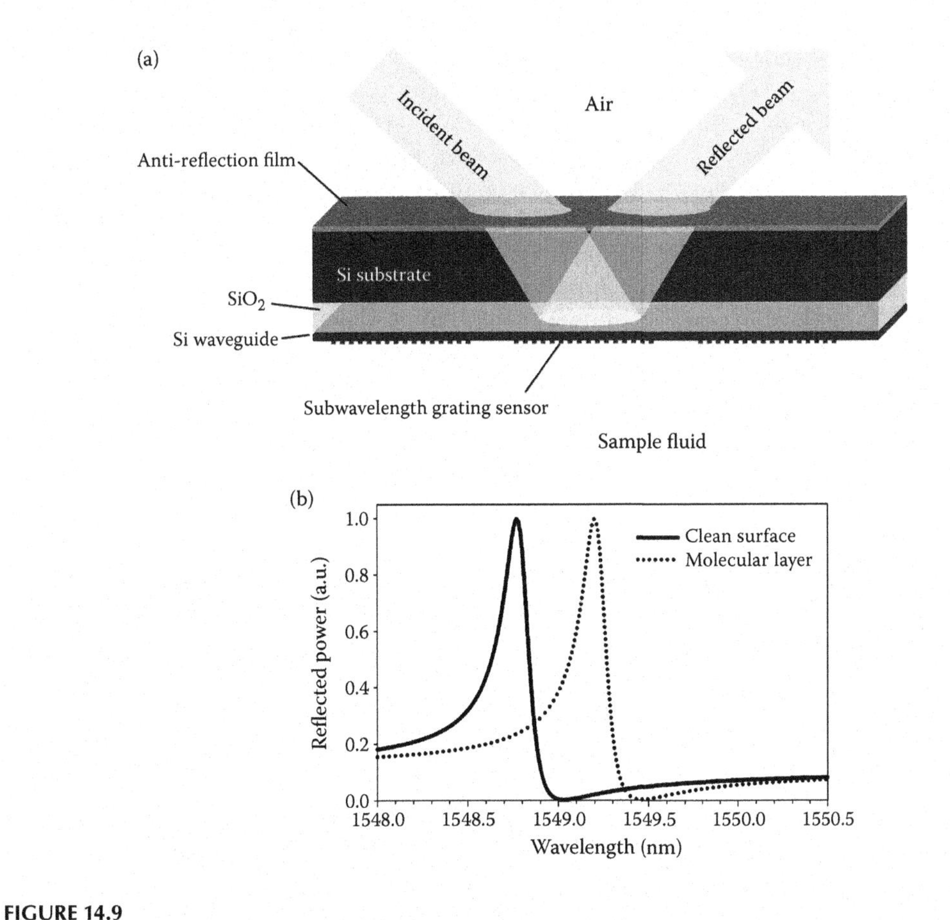

FIGURE 14.9
(a) A schematic diagram of a grating based silicon evanescent field sensor. (b) The calculated grating reflectivity spectra for a clean surface and with a 2-nm-thick layer with index n = 1.5 deposited on the sensor. This layer emulates the optical properties of a typical (e.g., streptavidin) protein monolayer.

250 nm to be near the maximum sensitivity point (Densmore et al. 2006). The grating sensor described by Schmid et al. (2009) is formed by stripes of silicon dioxide patterned into 0.6 μm lines and spaces on the Si waveguide surface. Light is incident through the back of the silicon-on-insulator wafer that has been covered with an antireflection coating. The light passing through the Si wafer, buried oxide, and Si waveguide layer reflects from the grating on the waveguide surface. When the grating coupling condition is satisfied, some light is also coupled into the Si waveguide and then diffracted back out into the reflected beam direction. This results in a sharp resonance in the reflected beam power shown in Figure 14.9b when the wavelength and angle of incidence satisfy Equation 14.8. As molecules bind to the waveguide grating surface, the reflection resonance shifts with the induced effective index perturbation δN_{eff} according to the relation

$$\delta\lambda = \lambda\left(\frac{\delta N_{\text{eff}}}{N_{\text{g}} - \sin\theta}\right) \tag{14.9}$$

Note that because sin(θ) is much smaller than the group index N_g, the wavelength shift is comparable to that for unbalanced MZI and a ring resonator sensor. The waveguide grating can, in principle, retain the high sensitivity of the silicon photonic wire platform, but is much simpler to fabricate and can be interrogated using a very simple reflection arrangement. In fact, the measurement geometry is essentially the same as a Kretschmann SPR sensor geometry. However, the grating sensor area must be relatively large (i.e., >1 mm^2) to achieve comparable sensitivity to a ring resonator or MZI sensor, limiting the number of sensors and other optical elements that can be combined on a single chip.

Given the number of different sensor configurations that can all be implemented on the same silicon planar waveguide platform, it is natural to consider what the relative advantages are for each. There is no all-encompassing answer, and the best path depends on the ultimate sensor application and technology constraints. Usually, the first comparison is the relative sensitivity or minimum level of detection, for the different sensor configurations. Setting photonic crystals aside for now, the wavelength shift for a given molecule induced effective index perturbation δN_{eff} is the same for an unbalanced MZI, ring resonator, and waveguide grating sensor. In each case, this shift is approximately $\delta\lambda \sim \lambda(\delta N_{\text{eff}}/N_g)$. The minimum level of detection is determined by the precision in the measurement of wavelength shift or intensity change, which in turn depends on the ratio of induced wavelength shift to the effective linewidth $\Delta\lambda 3_{\text{dB}}$ of the spectral feature monitored during sensor operation. The response of all three structures has already been analyzed and compared by Schmid et al. (2009). The response of a sensor interrogation geometry is characterized by a sensitivity parameter

$$S = \frac{1}{\Delta\lambda_{3\text{dB}}}\left(\frac{\delta\lambda}{\delta N_{\text{eff}}}\right) \tag{14.10}$$

The MZI spectrum has the sinusoidal variation of Equation 14.4, for which effective 3 dB width of each peak depends only on the physical length of the sensor arm. For ring resonators and waveguide grating sensors, the linewidth $\Delta\lambda_{3\text{dB}}$ depends on the total photon path length L_{eff} (or equivalently, the photon lifetime $\tau = N_{\text{eff}}L_{\text{eff}}/c$ lifetime) within the sensor structure. This in turn depends on the waveguide loss and coupling strength of the ring or grating cavity. In all three cases, the sensitivity scales as the ratio of effective sensor length to wavelength: $S \sim L_{eff}/\lambda$. The maximum intensity change induced by a given δN_{eff} in a ring sensor with finesse F is the same as a Mach–Zehnder sensor with sensor waveguide length approximately $F/4$ times that the physical ring circumference (Janz et al. 2009). This is not surprising given that a photon effectively circulates F/π times around the ring before being attenuated or coupled out. However, the sensitivity and size advantages of resonator configurations are achieved only if scattering loss and out coupling rate of a resonator are sufficiently low. Fortunately, this is generally the case in the current silicon photonics technology. The sensitivity of a Mach–Zehnder sensor is relatively independent of waveguide loss.

The achievable sensitivity of any given waveguide interrogation method depends on the photon lifetime or path length within the structure, so the advantage of resonators is clear. Because light circulates around a ring resonator many times, the same sensitivity can be achieved in a physically smaller sensor device. In recent work, ring resonator sensors occupying a chip area of only 50 μm in diameter have been used to monitor molecular binding of IgG molecules at concentrations as low as 20 pM (Xu et al. 2010a) and binding of streptavidin molecules has been measured as low as 60 fM (Iqbal et al. 2010). Small sensor size also allows more sensors to be integrated into a smaller area and, at least in principle, a smaller absolute number of molecules are required to induce a measurable sensor response.

Photonic crystal sensors also rely on resonance, and the sensitivity is again determined by the ratio of molecule-induced wavelength shift $\delta\lambda$ to resonance linewidth $\Delta\lambda_{3dB} = \lambda/Q$. The sensitivity of a photonic crystal is somewhat more difficult to characterize in general terms because $\delta\lambda$ for each sensor example case depends in a complex way on the exact electric field distribution in and around the crystal defect, rather than simply the change in effective index of the silicon waveguide layer. As discussed previously in the section on photonic crystals, where data is available, the ratio of wavelength shift to resonance linewidth for photonic crystal resonators and ring sensors is comparable. Improving the sensitivity for either device depends on the ability to reduce resonance linewidth (i.e., increase Q) by improved fabrication and coupling design. Photonic crystal sensors take the size advantage of resonators to the extreme, with sensor dimensions approaching 1 μm if one does not include the extent of the surrounding photonic crystal lattice. In addition, the electric field in certain photonic crystal sensors can be concentrated in the space inside a defect hole, so that the optical coupling with molecules or particles in the hole and the resulting resonance shift can be large. On the other hand, photonic crystal structures with high Q resonances are much less forgiving of fabrication than simple channel waveguide sensors. The capture probability of target molecules is also lower for smaller devices.

14.2.5 Silicon Waveguide Sensor Integration

For most practical applications, sensor integration and instrumentation are of greater importance than the individual sensor design. In many applications, it is desirable to run many simultaneous molecular binding measurements in a single operation. In diagnostic and public safety applications in which false negatives and false positives can cause direct harm, test reliability is crucial. This can be addressed by using numerous sensors individually functionalized to provide redundancy checks, cross-reactivity control tests, and reference tests for calibration. In molecular research applications, it is often desirable to screen molecules of interest for binding with many different target compounds at the same time. Sensor arrays for such applications require many sensors to be integrated onto a single chip in such a way that they can be simultaneously addressed and interrogated.

Furthermore, practical sensor instrumentation requires automated delivery of sample fluids to all the sensors on the chip. Light must be coupled onto the chip and directed through the sensor elements, and the output light must be coupled off the chip to an appropriate set of detectors. Finally, if a silicon sensor technology is to be adopted by the general biomedical community, an instrument that can read the sensor chips automatically is required. An operator must be able to insert a chip into the instrument so that the optics and microfluidic connections are made and correctly aligned to the chip in a few minutes.

Silicon waveguide sensors lend themselves naturally to the integration of many sensors on a single chip. By using the folded spiral waveguide sensor layout, individual waveguide sensors several millimeters long can be contained in sensor area of 200 μm or less in diameter. Appropriately designed ring resonators can be less than 50 μm in diameter, and single photonic crystal sensors approach the 1-μm length scale. Fitting tens or hundreds of sensors on a silicon chip of approximately 1 cm in size is therefore already well within the realm of existing technology. The crux of the problem is how to address multiple sensors on a chip simultaneously.

The simplest approach is to use a broadcasting scheme of Figure 14.10, whereby light is coupled into a single input waveguide, which then splits into N waveguides, each leading to one of the N sensors in the array. In this example, the sensor array consists of 16 PWEF sensors in a spiral ring resonator configuration. After passing through the sensor waveguide sections, the output light from each PWEF sensor element is carried by a dedicated waveguide to one of N output couplers that direct light off the chip. The optical power in all the output waveguides can then be simultaneously monitored using a photodetector array, several of which are available commercially. Densmore et al. (2009) have shown that the broadcast approach works surprisingly well. In the first multisensor PWEF array demonstration, the binding curves shown in Figure 14.11 for IgG target molecules binding to six monolithically integrated sensors were acquired simultaneously and in real time. Each sensor was independently functionalized with appropriate antigen IgG molecules using a commercial spotting tool. The anticipated shortcoming of $1 \times N$ broadcasting is that the power to each sensor is reduced by the factor N. However, in practice, it was found that when a sensor array was monitored using a commercially available InGaAs detector array, an input power of only 100 nW per sensor was sufficient to provide 35 dB dynamic range in

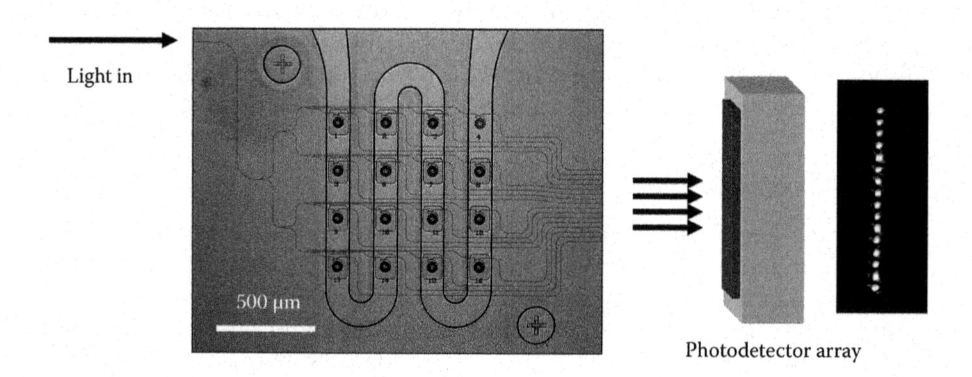

FIGURE 14.10
A section of a 16-sensor PWEF chip viewed from above, illustrating the broadcast approach to sensor array interrogation. Light is coupled into a single input waveguide at left and distributed to 16 sensors through a series of cascaded 1 × 2 Y-junctions. The camera image at right shows 16 output waveguide array at the facet of a 16-sensor chip.

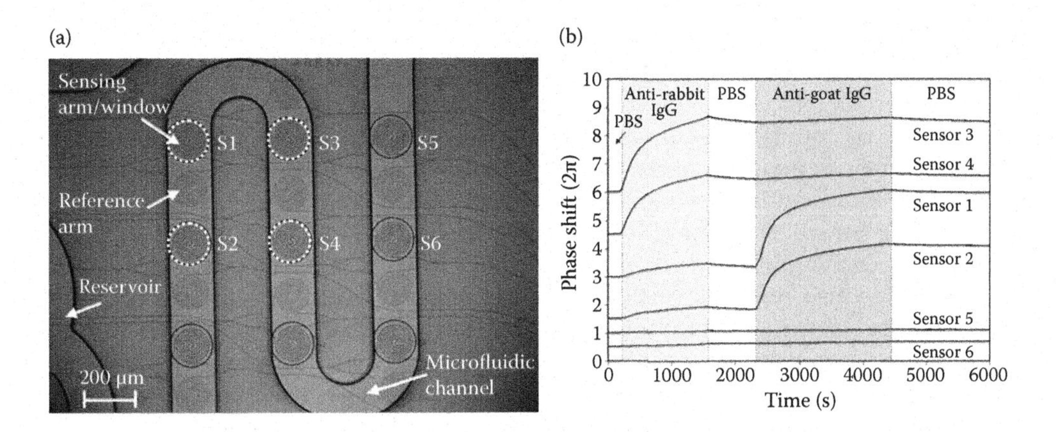

FIGURE 14.11
(a) A six-sensor MZI sensor array in which sensor elements were spotted with goat IgG antibody (S1, S2), rabbit IgG antibody (S3, S4), and BSA (S5, S6) to act as control sensors. (b) The six binding curves collected simultaneously from the chip as solutions containing the corresponding antigen IgG molecules were flowed through the fluid channel. (From Densmore, A. et al., *Opt. Letters* 34:3598–3600, 2009. With permission.)

the measured intensity variation (Densmore et al. 2009). The 1-mW power output of a typical telecommunication laser is therefore sufficient to interrogate more than 1000 sensors in parallel. Since these first experiments, this group has successfully used the $1 \times N$ broadcast approach to address 16 and 64 sensor arrays (see Figures 14.10 and 14.12).

An alternative approach to sensor integration is to provide one input and one output waveguide for each sensor element. Each input/output waveguide then has a dedicated coupler to bring light on and off the chip. To address multiple sensors, the input light beam is then scanned across the chip using movable mirrors to address each sensor in sequence (Iqbal et al. 2010). A movable mirror is also used to direct the output of each sensor to a single photodetector. This approach avoids the need for complex on-chip optical distribution

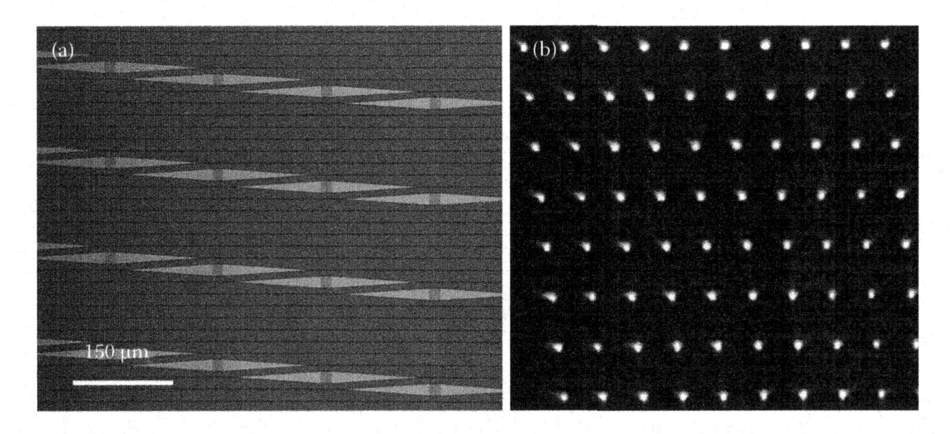

FIGURE 14.12
A 64-coupler grating array. (a) A microscope image of a subset of 16 grating couplers showing the channel to coupler waveguide tapers and the gratings (dark 15 × 15 μm squares). (b) An infrared camera image of λ = 1550 nm light is coupled off the chip by the full 8 × 8 coupler array.

and a multipixel photodetector array, but does require a continuously running mechanical scanning mirror assembly to correctly and rapidly address each sensor.

Once sensors are placed on the chip and connected to input and output waveguides, light must still be coupled on-chip and optical signals coupled off the chip to a detector for readout. Light is usually coupled into silicon photonic wires by aligning an optical fiber with the end of an inverse tapered waveguide (Shoji et al. 2002; Almeida et al. 2003; Lee et al. 2005). For silicon photonic wires, however, the required alignment tolerances for inverse taper coupling between chip and fiber are a few micrometers at best. Coupling to the chip edge also requires vertical, optically smooth facets, which can only be produced by careful cleaving or facet polishing. Direct facet coupling is clearly not practical for applications requiring rapid replacement and alignment of the sensor chip, or where chip manufacturing cost must be low enough that chips can be considered disposable (i.e., most medical and laboratory test applications).

Several groups have turned their attention to surface coupling using diffraction gratings formed on the chip surface, which couple light between the waveguide and beams incident onto the chip surface (Taillaert et al. 2003; Roelkens et al. 2006, 2007; Van Laere et al. 2007) as shown schematically in Figure 14.13. The gratings are typically 10 μm or more across so optical alignment tolerances to an external optical fiber or free space beam are much larger than for facet coupling. The photonic wire waveguide can be adiabatically expanded from its nominal width of a few hundred nanometers to match these large gratings. Gratings can be formed by etching shallow grating grooves into the chip surface (Van Laere et al. 2007; Taillaert et al. 2003). However, sensor chip manufacture can be considerably simplified using the subwavelength grating approach illustrated in Figure 14.14, which was first described by Halir et al. (2009, 2010).

These gratings are formed using rows of subwavelength dimension holes etched completely through the silicon waveguide layer. Each row of holes acts as a homogeneous grating groove, but the contrast between grating line (silicon) and groove (row of holes) can be

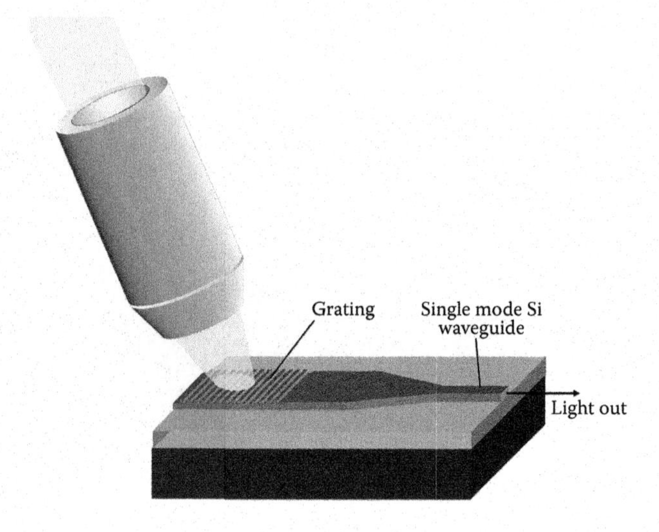

FIGURE 14.13
A schematic diagram (not to scale) showing the configuration for coupling light into a silicon waveguide using a grating patterned on the waveguide surface. The grating initially couples light into a wide slab waveguide section, which then tapers down adiabatically to direct the light into the single mode photonic wire at right.

FIGURE 14.14
Electron microscope views of a subwavelength waveguide grating coupler formed by rows of holes etched completely through the silicon waveguide layer.

freely adjusted by tuning the silicon-to-hole ratio along the row. As a result, it is possible to fabricate gratings of arbitrary length, and apodize the grating (Halir et al. 2010) to eliminate back-reflections and control the emitted beam profile. Unlike conventional gratings, subwavelength gratings can be manufactured on-chip in the same etch step as the sensor waveguides, using wafer-scale deep ultraviolet lithography, and no chip facet preparation is needed apart from ordinary singulation using a dicing saw. Arrays of coupling gratings have now been demonstrated on integrated multisensor photonic wire chips, with reported coupling efficiencies between −4 and −5 dB. Figure 14.12 shows microscopic images of such an array along with an infrared image of the array coupling light off a 64-sensor chip.

Finally, sample fluids carrying the biological materials of interest must be delivered to the sensor elements. Given the individual sensor size, this requires microfluidic channels of the order of a few hundred micrometers or less in height and width. A network of such channels can easily be molded into a gasket of material such as polydimethylsiloxane (PDMS) or thermoplastic. The gasket can then be applied to the surface of the sensor chip with the microfluidic channels aligned to the sensor elements (Kanda et al. 2004; Iqbal et al. 2010). The silicon platform also lends itself to the fabrication of microfluidic channels directly on-chip as part of the semiconductor fabrication process. For example, fluid channels with a height of 50 μm and a few hundred micrometers' width have been built on-chip using the photosensitive polymer SU-8 (Densmore et al. 2009). Examples of such monolithic fluid channels are visible on the PWEF sensor arrays in Figures 14.10 and 14.11. Channel patterning is then done through optical lithography, so that alignment of the channels to the sensor elements with better than ±1 μm positioning tolerance is easily achieved. The lid of the fluidic channel is part of the sensor instrument itself and the lid, with its external fluid connections, can be easily aligned to the 1-mm-wide reservoirs that serve as the inlet and outlet ports on the sensor chip.

A problem that all sensor technologies eventually encounter is the cross-reactivity to both molecular species other than the target molecule of interest (i.e., nonspecific binding), as well as changes in environmental factors such as temperature and vibration. Nonspecific binding is determined by the sample composition and details of the surface chemistry. A more detailed discussion can be found in Section 14.4. The cross-reactive response of a silicon wire sensor to a temperature drift of only 1°C is comparable to the signal arising from the formation of a significant fraction of a monolayer of some typical biomolecules (e.g., streptavidin and IgG; Xu et al. 2010b). To mitigate temperature effects, some commercial molecular sensors are housed in an environmental chamber that controls temperature to within 0.01°C of a given set point. A less costly solution may be to use a dedicated on-chip sensor to monitor temperature, and then normalize the signals from other sensors with an appropriate correction factor (Xu et al. 2010b; Iqbal et al. 2010). The small size and ability to manufacture many sensors on a single silicon chip at essentially no extra cost suggests that the latter solution can indeed be the most cost-effective and reliable.

The earliest example of on-chip temperature correction uses the silicon MZI sensor of Figure 14.4 (Densmore et al. 2008), in which the sensor arm and reference arm incorporate identical sensor structures. The reference arm is isolated from direct contact with the sample fluid by a polymer layer. The temperature-induced phase change in both arms is approximately equal so the common mode temperature response is cancelled out. More recently, ring resonator sensor arrays have been demonstrated with temperature correction (Xu et al. 2010b; Iqbal et al. 2010). In both examples, ring resonator reference sensors were buried in polymer to isolate them from chemical interaction with the sample fluid. The reference sensor resonance wavelength shift was therefore only dependent on temperature and could be used to correct the response of all the other sensors on-chip. Using this approach, Iqbal et al. (2010) reported measuring streptavidin protein binding at concentrations as low as 60 fM. Xu et al. (2010a) reported the detection of IgG binding for concentrations better than 20 pM, and the ability to monitor reactions over periods as long as 3 h with effective sensor drift comparable to a system with 0.01°C temperature control.

14.2.6 Summary

Silicon waveguides can provide a much higher sensitivity to the binding of target molecules to the waveguide sensor surface than other well-established optical label-free sensors. The sensitivity arises from the enhanced electric field at the surface in thin (<300 nm) silicon waveguides, in combination with the ability to guide light over long distances so that a large phase response can accumulate as molecules bind to the surface. The fundamental phase response must be converted to a measurable intensity. This section has reviewed a number of transduction configurations using common integrated optic circuit elements: MZIs, ring resonators, photonics crystals, and waveguide gratings. Ultimately, all these configurations are simply means of measuring the small refractive index change that occurs when molecules attach themselves to a silicon surface.

The ability to integrate complex optical circuits on a silicon sensor chip is the key to implementing a practical silicon sensor technology. Sensor arrays of tens or hundreds of individual sensor elements can be fabricated on a single 1 cm^2 silicon chip, and structures can be built to couple light on and off the chip and direct sample fluids to the sensors. All this is possible because the silicon waveguide sensor platform combines decades of development in integrated optics with silicon electronics manufacturing infrastructure. It is worth noting at this point that many of the sensor devices and structures described in this section have been fabricated at the wafer scale using deep ultraviolet lithography (e.g., the 16-sensor array of Figure 14.10 and the grating couplers of Figure 14.14), so that the cost of individual sensor chips is negligible and they can be considered disposable—a prerequisite for deploying label-free sensing technology out of the research laboratory and into volume applications in medical diagnostics and public safety.

14.3 Surface Functionalization and Molecular Biology Aspects of Silicon-Based Biosensors

Ryan C. Bailey

14.3.1 Chemistry of Oxide-Passivated Silicon

In addition to the advantages of silicon waveguide structures in terms of optical properties and ease of fabrication, facile surface functionalization is a further benefit of this material system for biochemical analysis. Exposed to an oxygen- and moisture-containing environment, silicon reacts to form a stable, native oxide layer that is approximately 10 Å (1 nm) in thickness (Morita et al. 1990). This oxide surface reacts in similar manner to bulk silica, the material that makes up glass slides commonly used for many conventional microarray applications, which is important in that silicon photonic structures can be readily functionalized with biomolecular receptors using well-established functionalization methods. Oxide coatings can also be generated by treatment of the silicon surface with O_2 plasma (Xu et al. 2010b) and, in some cases, a thicker oxide coating may be intentionally deposited onto the sensor surface to serve as an etch mask during the silicon photonic microstructure fabrication process.

The oxide surface on a silicon photonic structure can be chemically heterogeneous, as schematically shown in Figure 14.15, and due to different bonding schemes present within the "bulk" oxide, it is often referred to as SiO_x rather than as the stoichiometric SiO_2. At the exposed surface, silicon atoms are bonded to oxygen atoms in several forms: free silanols, germinal silanols, and siloxanes. Furthermore, adjacent free silanols on the surface can interact weakly via hydrogen bonding to form associated silanols. The silanol moieties are in equilibrium between their protonated and deprotonated states, and thus SiO_x surfaces typically have a net negative surface charge in the physiologically relevant pH range. To maximize the number of silanol groups on the oxide for subsequent chemical derivatization, surfaces are often treated with acidic (low pH) aqueous solutions, such as hydrochloric acid or piranha solution (typically a 3:1 volumetric mixture of concentrated sulfuric acid and 30% aqueous hydrogen peroxide; piranha solutions should be used with care as they can react explosively with trace amounts of organic compounds).

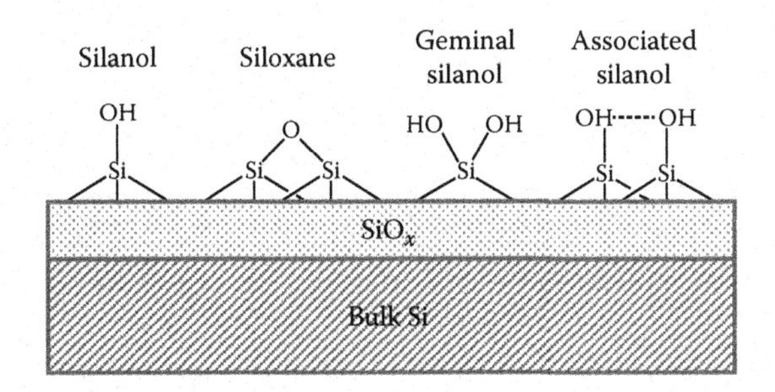

FIGURE 14.15
The silicon dioxide surface coating that clads silicon photonic waveguide structures is composed of multiple species, including silanols, siloxanes, germinal silanes, and associated silanols. Silanols, which are commonly the species that reacts with organofunctional silane reagents, can be increased in number on the surface by treatment with acid.

Although chemical treatments of oxide surfaces vary dramatically, the most common method to introduce additional chemical functionality is through the reaction with organofunctional silane coupling reagents. Organofunctional silane coupling reagents typically consist of three parts: a silicon atom bonded to a hydrolyzable group(s) (halogens and alkoxy groups are the most common), a linker moiety, and an organofunctional group that serves as a handle through which additional chemical or biomolecular ligation can be achieved. There exists a wide variety of commercially available silane coupling reagents having anywhere from one to three hydrolyzable groups as well as a large number of unique organic functionalities. Silane coupling reagents can be deposited from solution or the vapor phase and react through surface silanol groups. The most common hydrolyzable groups on commercial silane reagents are alkoxy (e.g., methoxy-, ethoxy-) and chloro groups. Upon reaction with surface silanols, alkoxysilane coupling reagents liberate the corresponding alcohol whereas chlorosilanes liberate small amounts of hydrochloric acid. Figure 14.16 shows the result of a reaction between a very common silane coupling reagent, aminopropyltriethoxysilane and a free silanol on a native oxide passivated silicon surface. The amino functionality, and other commercially available organofunctional groups, can then be subsequently reacted using a range of chemistries, some of which are described in Section 14.16.

Another interesting recent report by Fukuyama and coworkers (2010) described the use of a recombinantly expressed protein capture agent, which presented a "Si-tag" moiety that facilitated direct oriented attachment of the capture agent to the silicon dioxide surface. The potential to avoid multiple chemical reactions is attractive and thus this technology, although requiring recombinant protein expression for every needed capture agent, may be attractive for some applications.

Reagents with two or three hydrolyzable groups can react not only with the surface silanols but also with adjacent silane reagents to form a silane network that can be multiple layers thick, which may not be desirable. By contrast, monohydrolyzable silane reagents are only capable of making a single cross-link to the surface and thus are more likely to form monolayer coverages. Despite this control, trihydrolyzable reagents are the most commonly used as they typically give the highest density of organofunctional groups. Surface functionalization methods can often be optimized by simply measuring the

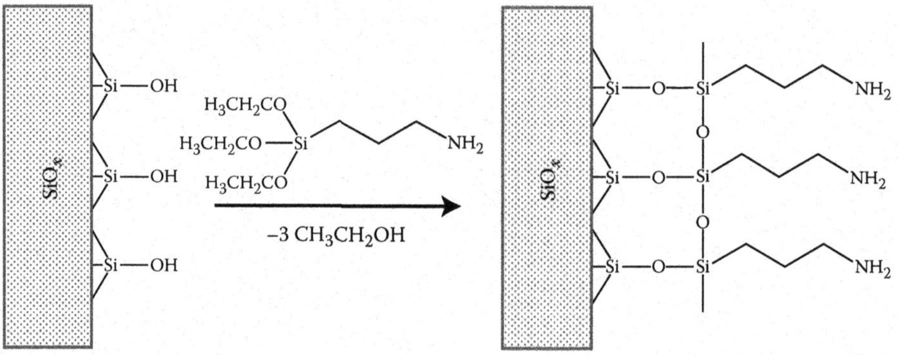

FIGURE 14.16

Free silanols are readily reacted with organofunctional silane-linking reagents to introduce bioconjugate handles to silicon photonic devices. Shown here is the immobilization of the common silane, 3-aminopropyltriethoxysilane, which results in the liberation of 3 mol equivalents of ethanol. Silanes can often be cast from either the solution or the vapor phase.

contact angle of a drop of water on a native oxide interface, with contact angles increasing dramatically upon the addition of more hydrophobic organofunctional groups. Although this characterization method cannot typically be applied to the submicron silicon photonic waveguides used for biosensing, native oxide–coated bulk silicon wafers can be used to optimize silanization reactions and subsequent derivatization chemistries.

14.3.2 Bioconjugate Strategies for Attaching Protein and Nucleic Acids to Sensor Surfaces

Silicon photonic waveguide structures, by themselves, are incapable of serving as selective chemical or biosensing agents. Rather, specificity for particular antigens of interest is conferred by immobilizing a protein or nucleic acid capture agent onto the waveguide surface. The simplest method for introducing these capture agents is through nonspecific physisorption, which is facilitated by van der Waals or electrostatic interactions between the capture agent and the surface. This approach is commonly used in the fabrication of DNA microarrays on glass microscope slides. In this case, slides are modified to present positively charged surface (often via derivatization with poly-L-lysine) and electrostatic interactions with the negatively charged backbone of single-stranded capture DNA strands result in immobilization. On a silicon photonic biosensing platform, surfaces modified with amino groups such as 3-aminopropylmethyl-diethoxysilane were used to noncovalently capture multiple antibodies for multiplexed sensing applications (Densmore et al. 2009; Xu et al. 2010a). Although there are many successful examples of this type of immobilization, covalent bonds between capture agents and the sensor surfaces are an attractive alternative to ensure irreversible attachment.

Through functionalization with an organofunctional silane, many different types of covalently reactive moieties can be introduced to silicon photonic sensors for subsequent bioconjugation to analyte-specific receptors, and each has a different pallet of chemistries that are useful for covalent immobilization of protein or nucleic acid capture agents. For brevity, only a brief selection of several chemistries that have been previously applied to silicon photonic biosensors will be covered here. Alternative bioconjugate chemistries can be accessed through other more exhaustive sources (Hermanson 1996).

An extremely common covalent reactive group used on both the sensor surface and capture agent is the amine group. Amines can be introduced to the surface via reactions with readily available silanes, such as aminopropyltriethoxysilane. Amines are also present on protein capture agents in the form of the side chains of lysine amino acid residues, and amines are also easily introduced into solid phase synthesized DNA sequences. For bioconjugate purposes, amine groups are often reacted with aldehydes, activated carboxylic acid esters (commonly N-hydroxysuccinimide [NHS] esters), epoxides, and isothiocyanates. Aldehydes react with amine groups to form reversible imine linkages (Schiff bases) that should be chemically reduced to a stable secondary amine using reagents such as $NaCNBH_3$. Activated carboxylic acid esters, which can be generated *in situ* through the reaction of a carboxylic acid with 1-ethyl-3-[3-dimethylaminopropyl] carbodiimide (EDC) and NHS, or purchased commercially as the NHS ester of a particular linker reagent, readily react with amines to form stable amide linkages; however, one must carefully optimize reaction conditions as NHS esters hydrolyze back to the unreactive carboxylic acid moieties, sometimes leading to variable surface derivatization densities unless excesses of reagents are used. Epoxides, which can be directly introduced to surfaces through the use of organofunctional silanes such as 3-glycidyloxypropyltrimethoxysilane, can directly attach to biomolecules by forming secondary amine linkages (Claes et al. 2009;

Ramachandran et al. 2008). Isothiocyanates, which can be directly introduced onto silicon photonic surfaces using reagents such as 3-isocyanatepropyltriethoxysilane, react with amines present on biomolecules to form isothiourea bonds (García-Rupérez et al. 2010). A number of alternative amine reactive chemistries are also available, although they are not generally as widely employed as reactions with aldehydes, NHS esters, and epoxides.

To link an amine-presenting biomolecule to an amine-presenting surface, homobifunctional linkers, often based on aldehydes (glutaraldehyde) or bis-NHS esters can be used. Given the broad applicability of coupling through side chain lysines on proteins or amine-modified synthetic capture DNA strands, and the ease of introducing stable amino silanes onto silicon photonic waveguide surfaces, these symmetric bioconjugation strategies are quite common. Barrios and co-workers (2008) have even used glutaraldehyde to link antibodies to native oxide–stripped Si_3N_4 slot waveguide structures.

Analogous to homobifunctional linkers, heterobifunctional linkers can also be used to form asymmetric linkages in between groups such as amines and thiols (Kirk et al. 2011). Maleimide groups, which are commercially available alongside NHS esters in several heterobifunctional linkers, preferentially reacts with thiol groups over amines, and is also more stable to hydrolysis compared with the NHS moiety. A range of other heterobifunctional linkers are available, some of which can also be used to engender other preferential chemical properties, such as the resistance to protein biofouling by implementation of poly(ethylene glycol) units (De Vos et al. 2009a,b).

Because the instability of NHS reagents, in particular, during the actual conjugation reaction between the biomolecule and the surface can be of concern, an alternative two-step linker scheme has been employed on silicon photonic devices (Washburn et al. 2009, 2010; Luchansky and Bailey 2010; Iqbal et al. 2010). This strategy, which involves forming a hydrazone linkage, relies on the reaction between an alpha-effect nitrogen group, such as a hydrazide or amino-oxy moiety, and an aryl aldehyde moiety, both of which can be attached to either the silicon waveguide surface or biomolecule of interest via one of several commercial linkers such as HyNic, S-4FB, and HyNic silane, among others. The hydrazone bond formation is a quantitative linkage and thus the net yield of surface functionalization can be well-defined because the less controllable reactions with NHS esters or maleimides are not directly involved in the surface-to-biomolecule linkage itself. Moreover, the efficiency of hydrazone ligations can be further improved through the use of aniline as a biomolecularly benign catalyst. Byeon et al. (2010) directly showed improved ligation between a representative antibody capture agent and a silicon photonic biosensor using hydrazone bond formation as compared with a similar NHS ester reaction scheme.

As an alternative to covalent immobilization, another common biofunctionalization strategy relies on the high-affinity interaction between the small molecule biotin (vitamin H) and streptavidin (and its uncharged equivalent neutravidin). Commercially available NHS esters of biotin have been immobilized onto silicon photonic biosensor surfaces and subsequently used to capture streptavidin (Xu et al. 2008; Densmore et al. 2008; Luchansky et al. 2010; De Vos et al. 2007).

14.3.3 Array-Based Functionalization Strategies

Major advantages of silicon photonic devices for biosensor applications include their inherent mass producibility, scalability, and multiplexing capability, all afforded by their genesis in silicon fabrication technology. However, there have been, thus far, only a limited number of demonstrations of multiplexed detection. Given a multielement sensor array, a major challenge is attaching different analyte-specific capture agents to different spatially distinct sensors elements. The simplest method for creating multiplexed arrays involves simple spotting by hand of 10 microliter drops of capture agents onto different locations across a sensor substrate (Qavi et al. 2011a,b). However, spotting methods that involve allowing the drop of capture agents to dry on the sensor substrate have potential complications. Although drying is often adequate for nucleic acid–based capture agents, proteins are notoriously unstable in these conditions. An alternate method for creating biomolecular sensing arrays with modest multiplexing capability on silicon photonic microstructures involves the use of micron-scale microfluidic channels that allow the deposition of capture agents in a fully hydrated state, thus avoiding the potential complications associated with drying (Washburn et al. 2010).

Despite the potential complications associated with drying out protein capture agents, microspotting technologies, previously developed for printing both DNA and protein microarrays, are the most promising tools available for the creation of highly multiplexed silicon photonic biosensing architectures. Specifically, noncontact, piezoelectric microspotters are extremely attractive as they are capable of depositing extremely small volumes of reagents with good spatial resolution. Furthermore, noncontact microspotters, as opposed to commonly available pin-based contact spotters, completely avoid any potential damage to the underlying waveguide structures. There have been a handful of nice examples to date in which small arrays of protein-based capture agents have been deposited onto silicon photonic biosensors using noncontact microspotters (Xu et al. 2010a,b; Densmore et al. 2009; De Vos et al. 2009a). Notably, Kirk and co-workers (2011) have further expanded the multiplexing of silicon photonic biosensors by creating a six-element array with proteins, glycoproteins, and neoglycoconjugates as target-specific reagents.

14.3.4 Applications of Silicon Photonics for Biomolecule Detection and Interaction Screening

Due to the potential for highly multiplexed, ultrasensitive, and cost-effective biosensors, the majority of reports of silicon photonic biosensors to date have focused on the development of *in vitro* diagnostic tools. The label-free detection capabilities of these technologies make them highly amenable to the detection of both protein and nucleic acid biomarkers. As these applications have been previously addressed in this chapter, they will not be considered further here.

In addition to sensitive, quantitative detection, silicon photonic devices are also promising for a number of label-free bioanalysis applications, many of which have been previously explored with the more well-established label-free technique of surface plasmon resonance. In particular, multiplexed silicon photonic devices are well-suited to applications in (bio)molecular screening, with several recent examples demonstrating the ability to quantitatively evaluate the relative binding affinities of multiple capture agents in parallel (Washburn et al. 2011; Byeon and Bailey 2011). Another promising, but yet to be fully realized opportunity for silicon photonics is in the analysis of different modes of biomolecular interactions. Notably, Kirk and co-workers (2011) showed that protein–carbohydrate interactions can be monitored in an array-based format, paving the way toward silicon photonic-based glycomics.

Although in their infancy compared with more well-established surface analysis methodologies, such as surface plasmon resonance, the intrinsic scalability, multiplexing capabilities, cost-effective fabrication, and often real-time analytical capabilities afforded by silicon photonic sensor arrays make these structures extremely promising for a range of known, emerging, and yet-to-be-discovered applications. In parallel with advances in device geometry and read-out instrumentation development, researchers should freely investigate new applications of this powerful, and potentially revolutionary, suite of biosensor technologies.

14.4 Manipulation and Transportation of Biomolecules Using Optical Forces in Silicon Photonic Structures

Adam T. Heiniger, Qiang Lin, and Philippe M. Fauchet

14.4.1 Introduction

In 1970, Arthur Ashkin showed that submicrometer dielectric spheres suspended in solution could be attracted to and propelled by a focused Gaussian beam, and he demonstrated the use of a counterpropagating beam to form a stable trap (Ashkin 1970). These were significant advances for the field of biology. Since 1970, particle manipulation using Ashkin's optical tweezers has allowed for a better understanding of biological processes by measuring the force on or between various biological molecules, and optical tweezers have been employed for applications including imaging and spectroscopy of single cells and cell patterning, growth, and sorting (Stevenson et al. 2010).

Unfortunately, in some situations, free-space optical beams cannot be utilized for particle manipulation. A beam cannot be focused to a spot smaller than approximately half the optical wavelength, and so manipulation of particles with a radius smaller than 100 nm would require the use of ultraviolet wavelengths. Viruses, DNA, and individual proteins can thus be too small for effective trapping and transport by free-space visible or infrared beams. Additionally, there is a drive under way to transfer biological techniques from research laboratories to cheap and portable labs-on-chip (Lazcka et al. 2007). If particle manipulation could be performed without free-space optics, then these devices could be made less bulky and expensive.

Fortunately, planar photonic devices can be used to overcome both of these shortcomings. They can be fabricated on-chip, and the manipulation of particles smaller than 100 nm has been demonstrated using optical waveguides and resonators. For example, consider a ridge waveguide in a microfluidic environment. Total internal reflection confines light in the high-refractive index core region, whereas evanescent tails extend into the fluid cladding. As we will see in Section 14.4.2, one component of the optical force attracts particles to regions of high optical intensity. Thus, particles suspended in the fluid will be attracted to the optical waveguide. As light propagates along the waveguide, the travelling photons may be scattered or absorbed by the particle. This induces a scattering force, which propels particles along the waveguide. These scattering and gradient forces are the tools for particle manipulation in photonic devices. In the following sections, we will expand on the theory of optical forces on submicrometer particles and we will review progress in the field of photonic particle manipulation.

14.4.2 Theory of Optical Forces

14.4.2.1 The Maxwell Stress Tensor

The optical force on a dielectric body is the superposition of the forces acting on the charged particles that make up the body. This is the Lorentz force, given by

$$\mathbf{F}(\mathbf{r},t)=\frac{d\mathbf{P}_{\text{mech}}}{dt}=\int_V (\rho(\mathbf{r},t)\mathbf{E}(\mathbf{r},t)+\mathbf{J}(\mathbf{r},t)\times\mathbf{B}(\mathbf{r},t))\,d^3r. \tag{14.11}$$

Here, ρ and **J** are the charge and current densities, **E** and **B** are the electric and magnetic fields, and $\mathbf{P}_{\text{mech}}$ is the mechanical momentum of the particle.

Using Maxwell's equations, we eliminate ρ and **J** from Equation 14.11. Furthermore, we restrict consideration to harmonic fields of the form $\mathbf{E}(r, t) = [\mathbf{E}(r)\exp(-i\omega t) + \mathbf{E}^*(r)\exp(i\omega t)]/2$, which are typically encountered in studies of photonic devices. We can then find the average force $\langle \mathbf{F} \rangle$ by integrating **F** over the optical period $T = 2\pi/\omega$. Putting this all together, we find

$$\langle \mathbf{F} \rangle = \oint_S \langle \overleftrightarrow{\mathbf{T}} \rangle \cdot \hat{\mathbf{n}}\, da \tag{14.12}$$

where

$$\langle T_{\alpha\beta} \rangle = \frac{1}{2}\varepsilon_m\varepsilon_0 E_\alpha(\mathbf{r})E_\beta^*(\mathbf{r}) + \mu_m\mu_0 H_\alpha H_\beta^* - \frac{1}{4}\left(\varepsilon_m\varepsilon_0 \left|\mathbf{E}(\mathbf{r})\right|^2 + \mu_m\mu_0\left|\mathbf{H}(\mathbf{r})\right|^2\right)\delta_{\alpha\beta} \tag{14.13}$$

is the αβ element of the time-averaged Maxwell stress tensor. Here, $\hat{\mathbf{n}}$ is the outward pointing normal of the surface S enclosing volume V, ε_0, and μ_0 are the permittivity and permeability of free space, and ε_m and μ_m are the relative permittivity and permeability of the background medium. E_γ and H_γ are the γ components of the macroscopic electric and magnetic fields, and $\delta_{\alpha\beta}$ is the Kronecker δ. The Maxwell stress tensor represents the momentum flux into the region V through the closed surface S (Jackson 1998).

14.4.2.2 Force on a Small Particle

The Maxwell stress tensor gives accurate results for force, but the equations require knowledge of the total field in the presence of the particle and they do not illuminate the underlying physics. Researchers have developed methods to calculate the optical force from the incident field alone. For spherical particles, Mie theory can be used to derive the scattered field from the incident field. The total field then can be substituted into Equation 14.13, and the force can be determined analytically in terms of Mie coefficients (Almaas and Brevik 1995). For particles of arbitrary shape, the discrete-dipole approximation can be used. It treats an extended body as a grid of point dipoles and calculates the self-consistent field everywhere due to the incident field and the field scattered from each dipole (Draine and Flatau 1994). The force on the total particle is then the sum of the force of the total field acting on each dipole.

Researchers have also developed instructive expressions for the force on particles much smaller than the optical wavelength (Albaladejo et al. 2009). The approximate force is given by

$$\langle \mathbf{F} \rangle = \frac{1}{4}\varepsilon_p\varepsilon_0\Re[\alpha]\nabla\left|\mathbf{E}_0(\mathbf{r})\right|^2 + \varepsilon_p\varepsilon_0\frac{\sigma}{c}\langle \mathbf{S}(\mathbf{r})\rangle + i\varepsilon_p\varepsilon_0\sigma c\nabla \times \langle \mathbf{L}_s(\mathbf{r})\rangle. \tag{14.14}$$

ε_p is the relative permittivity of the particle, c is the speed of light in vacuum, $\mathbf{E}_0$ is the incident electric field, $\langle \mathbf{S} \rangle = \frac{1}{2}\Re\left[\mathbf{E}_0 \times \mathbf{H}_0^*\right]$ is the time-averaged Poynting vector, and $\langle \mathbf{L}_s \rangle = \frac{\varepsilon_0}{4i\omega}\mathbf{E}_0 \times \mathbf{E}_0^*$ is the time-averaged spin density of a transverse electromagnetic field. The particle polarizability α is calculated using (Draine and Flatau 1994)

$$\alpha_0 = 3V\frac{\varepsilon_p - \varepsilon_m}{\varepsilon_p + 2\varepsilon_m} \tag{14.15a}$$

$$\alpha = \frac{\alpha_0}{1 + in_p^3 k_0^3 \alpha_0/(6\pi)}. \tag{14.15b}$$

n_p is the particle refractive index and k_0 is the free-space wave vector. The scattering cross-section is $\sigma = k_0\Im[\alpha]/\varepsilon_0$. $\Re$ and $\Im$ denote the real and imaginary parts of the enclosed quantities. A comparison of all the force calculation methods mentioned previously can be found in Dienerowitz et al. (2008).

The first term of Equation 14.14 is called the gradient force, and it attracts particles to regions of high optical intensity (if $\varepsilon_p > \varepsilon_m$). For photonic devices in a microfluidic environment, the gradient force plucks particles from the flow and attracts them to the device. The gradient force can be increased by increasing the input optical power, or by using a structure that more tightly confines light (thus increasing the gradient). For example, as discussed in Section 14.4.3.1, larger optical forces are present in silicon slot waveguides, which confine light in an approximately 100 nm slot compared with 500-nm-wide solid-core waveguides.

The second term of Equation 14.14 is the radiation pressure, and is due to the transfer of linear momentum from photons that are scattered or absorbed by the particle. This force is responsible for optical propulsion of particles along waveguides or travelling wave resonators. Radiation pressure is proportional to peak intensity, so again, it can be maximized with maximum optical power, or by confining light as tightly as possible into the traveling wave mode of the waveguide.

The third term of Equation 14.14 is the spin force, and is relevant when the particle is in a field distribution with nonuniform helicity. Particles can follow nearly closed trajectories when trapped in optical vortices (Albaladejo et al. 2011). Wong and Ratner (2006) only recently discovered the significance of the spin force.

14.4.3 Progress to Date

Progress in optofluidic particle manipulation can generally be sorted into three categories of devices: waveguides, resonators, and plasmonic devices. Optical waveguides can also serve as "particle guides," driving particles from place to place on a chip. Optical resonators can trap and store particles in stationary positions or in bound orbits. Waveguides and resonators can also be realized from plasmonic devices; these use metals instead of or in addition to low-loss dielectric materials. Plasmonic devices tend to have even smaller optical mode volumes than dielectric photonic devices, often leading to a larger optical gradient force. However, here we restrict our attention to dielectric photonic devices, and direct the interested reader to other sources on plasmonic particle manipulation (Erickson et al. 2011).

14.4.3.1 Trapping and Transport Using Waveguides

Optical waveguides tightly confine light in the transverse direction while light travels in the longitudinal direction. As seen in Section 14.4.2.2, particles are attracted to high-intensity regions by the gradient force and are pushed by radiation pressure. This makes waveguides ideal roadways for particle manipulations on-chip; particles are attracted from a microfluidic environment to the waveguide by the gradient force, and radiation pressure pushes particles along the waveguide. To our knowledge, Kawata and Tani (1996) first demonstrated trapping and propulsion of dielectric and metal particles on optical waveguides. This work was followed by various other approaches to move artificial particles (Ng et al. 2000, 2002; Tanaka and Yamamoto 2000; Grujic et al. 2005) and biological cells (Gaugiran et al. 2005) on waveguides. All these devices employed channel waveguides defined using ion exchange in glass, and in all these cases, the systems operated in a stationary fluid. In 2007, the Erickson and Lipson research groups at Cornell demonstrated capture and transport of flowing particles from a microfluidic channel onto a solid-core SU-8 waveguide (Schmidt et al. 2007).

These relatively low-index, solid-core waveguides are capable of particle manipulation on-chip, overcoming one shortcoming of particle manipulation using free-space beams. Unfortunately, like free-space beams, they are unable to trap and stably transport small (r_p < ~100 nm) particles. Yang and Erickson (2008) showed that silicon waveguides have better trapping stability for smaller particles because light is confined more tightly in the silicon waveguide core due to the higher refractive index of silicon. Again, in accordance with our discussion of the gradient force in Section 14.4.2.2, this smaller mode area creates a steeper (although faster decaying) gradient, and thus a larger force. However, even solid-core silicon waveguides are incapable of manipulating r_p < ~100 nm particles. This requires further tailoring the shape and location of the optical mode.

In solid-core waveguides, the optical intensity is concentrated in the core, and particles can interact only with the mode's evanescent tail extending above the waveguide. Yang et al. instead fabricated slotted silicon waveguides. In the transverse direction, the optical intensity was concentrated in the approximately 100 nm slot, allowing for even tighter confinement and steeper gradients than the smallest solid-core waveguides. This also allowed particles to directly access the intensity maximum, leading to faster transport velocities compared with solid-core waveguides. They demonstrated capture and stable transport of 75 nm diameter particles (Yang et al. 2009b). In Figure 14.17, we show images from their experiments demonstrating the trapping of these small particles in slot waveguides. Additionally, numerical studies suggest that these devices are capable of transporting even smaller particles of radius 5 to 10 nm (Yang et al. 2009a).

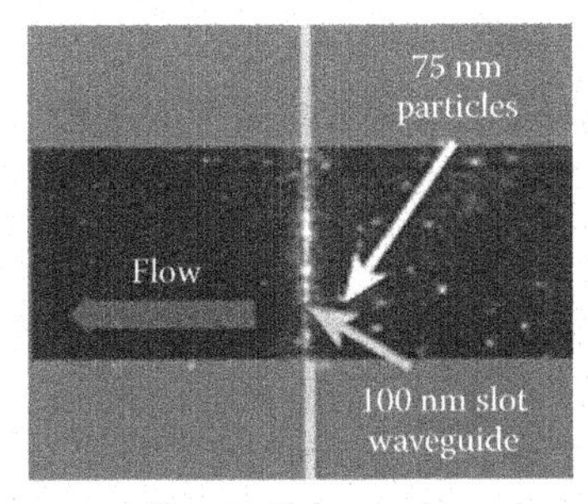

Nanoparticles passing over the waveguide collect in the slot

Laser turned off: particles released

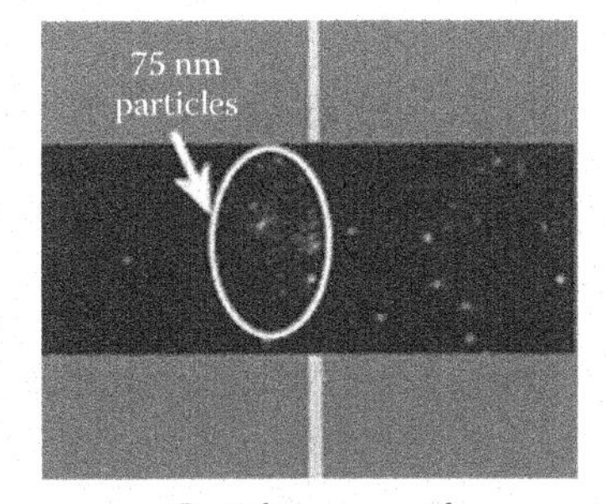

Particles converted downstream with flow

FIGURE 14.17
Microfluidic flow brings 75 nm diameter particles near a slot waveguide in which they are captured by optical forces. Particles can be released by switching off the laser. (Reprinted from Yang, A.H.J. et al., *Nature* 457(7225):71–75, 2009. With permission from Macmillan Publishers Ltd.)

FIGURE 14.18
Resonator geometries for trapping. (a) Microsphere, (b) ring resonator, and (c) photonic crystal resonators. (Panel a from Arnold, S. et al., *Opt. Express* 17(8):6230–6238, 2009. With permission. Panel b adapted from Lin, S. et al., *Nano Lett.* 10:2408–2411, 2010. With permission. Panel c reprinted from Lee, M.R. and P.M. Fauchet, *Opt. Letters* 32(22):3284–3286, 2007. With permission.)

14.4.3.2 Trapping and Storage Using Resonators

Many of the applications listed in the Introduction require traps that hold particles in stable positions. Optical resonators are more suited for this than waveguides, as they can trap particles in orbits or with no movement at all. Furthermore, from Equation 14.14, one way to increase the optical force is to increase the electric field amplitude. In optical resonators, resonant light makes multiple round-trips before exiting, creating much larger intensities and thus stronger forces than are found in waveguides for the same input power.

Particle trapping has been demonstrated in on-chip optical resonators in three configurations—microspheres, rings, and photonic crystal resonators (Figure 14.18). Microspheres are formed by melting optical fibers. This is an off-chip approach that employs glass fibers instead of silicon, and so we direct the interested reader to the work by Arnold et al. (2009). Ring resonators are waveguides that form a closed cavity, into which light can couple evanescently from an adjacent waveguide. Particles with r_p = 550 nm have been successfully transported by waveguides and then added to the ring resonator cavity, which has a high optical intensity, keeping the particles circulating the cavity (Lin et al. 2010).

Photonic crystal resonators can have much smaller mode volumes than ring resonators and correspondingly larger optical forces. Photonic crystals are systems with periodically varying refractive index. In such systems, bands of light may scatter and interfere in such a way that propagation through the crystal is forbidden. Line and point defects may be formed in the crystal to guide or trap light with frequency in the band gap (Joannopoulos et al. 2008). The trapping of particles with $r_p < 100$ nm in photonic crystal resonators has been theoretically suggested (Rahmani and Chaumet 2006; Lin et al. 2009) and experimentally verified (Mandal et al. 2010).

14.4.4 Conclusion

Silicon photonic devices have emerged as tools to transport and trap biological molecules in situations inaccessible to free-space optical beams. Particles with radii smaller than 100 nm have been successfully manipulated by silicon waveguides and resonators, all within an on-chip platform that leverages the cost and efficiency of silicon fabrication lines perfected by the microelectronics industry.

14.5 Silicon Nanoparticles for Bioimaging

Qi Wang and Yimin Chao

14.5.1 Introduction

Bioimaging techniques using fluorescent labeling allow the investigation of both cellular components and processes (Sharma et al. 2006; Wang et al. 2006). Traditionally, organic dyes and fluorescent proteins are used in most biological studies as fluorescent labels because they are highly water-soluble and easily bioconjugated (Wang et al. 2006). However, the rapid development of biomedical sciences demands new advanced techniques and instruments. The development of synthesis strategies for the fabrication of nanoparticles is anticipated to lead to advancements in understanding biological processes at the molecular level in addition to progress in the development of diagnostic tools and innovative therapies (Arya et al. 2005). Silicon nanoparticles (SiNPs) have overcome many of the limitations of organic dyes, such as poor photostability, high cytotoxicity, and insufficient *in vitro* and *in vivo* stability. The main focus of this section is to provide an overview of synthesis and properties of SiNPs in bioimaging research (Selvan et al. 2010). This technology promises future breakthroughs in cancer diagnosis, intracellular tracking, drug delivery, and gene delivery. However, successful and wide application of this technology relies greatly on robust nanoparticle synthesis methodologies, which involve design, synthesis, surface modification, and bioconjugation. It is desirable to obtain SiNPs that are highly sensitive, stable, and biocompatible. Here, we focus on SiNPs that have been developed for biosensing and bioimaging applications.

14.5.2 Synthesis Routes

SiNPs can be synthesized by a variety of "chemical" or "physical" routes, including solution routes using a variety of reducing agents (Neiner et al. 2006; Zhang et al. 2007), the microemulsion technique (Tilley and Yamamoto 2006; Yu et al. 2006), ultrasonic dispersion of electrochemically etched silicon (Chao et al. 2005, 2007; Lie et al. 2002), laser-induced pyrolysis of silane (Hua et al. 2005, 2006), synthesis in inverse micelles (Warner et al. 2005), and so on. Chemical routes tend to produce rather large amounts of material, whereas physical routes generally produce small quantities mainly for physical or electronic applications. Nanoparticles produced by the chemical routes may also be compatible with the conjugation of biological molecules at the particle surface, which is therefore emphasized in our study.

A common starting point for nanoparticle preparations is porous Si, which was first discovered by Leigh Canham (1990). He showed that porous Si materials produced by electrochemical etching of Si wafer in fluoride solution could have large photoluminescence efficiency at room temperature in the visible region. Porous Si layers exhibited red photoluminescence, which is observable under green or blue laser excitation. Heinrich et al. (1992) then demonstrated that porous Si can be broken up into individual nanoparticles by several methods including ultrasound, which is one of the simplest routes to Si nanoparticles. Therefore, individual Si nanoparticles with orange–red emission, but broad size distribution, could be easily produced from porous Si by sonication. Wolkin et al. (1999) found that the photoluminescence of SiNPs could be tuned from the near-infrared to the ultraviolet depending on the size when the surface was passivated with Si–H bonds.

However, after exposure to oxygen, the photoluminescence intensity declined and the fluorescence shifted to the red, which suggested that both quantum confinement and surface passivation determined the electronic states of SiNPs. The surface of hydrogen-terminated SiNPs can be further stabilized by a variety of compounds, including alkenes (Chao et al. 2005, 2007; Lie et al. 2002), amine (Shiohara 2010; Warner et al. 2005), etc. Our group (Wang et al. 2011) produced polyacrylic acid–grafted water-dispersible SiNPs, which were stable and exhibited strong red photoluminescence at room temperature.

14.5.3 Confocal Imaging

Laser scanning confocal microscope was developed in the 1980s, and is based on a fluorescence microscope fitted with an imaging laser scanning device. The use of computer image processing greatly enhances the resolution of optical imaging. The subcellular structures of the cell or tissue can be observed by ultraviolet or visible light excitation using the fluorescent probe. Figure 14.19 is the basic setup of a confocal microscope, in which a laser is used to provide the excitation light. The laser light reflects off a dichroic mirror, and then hits two mirrors, which scan the laser across the sample. The emitted light from the fluorescent probe in the sample gets descanned by the same mirrors that are used to scan the excitation light. The emitted light passes through the dichroic and is focused onto the pinhole, and is then measured by a detector (Prasad et al. 2007). The detector is attached to a computer, which builds up the image one pixel at a time. By having a confocal pinhole, the microscope is efficient at rejecting fluorescent light that is out of focus. As a result, a sharp fluorescent image of the cell or tissue can be obtained.

Ideally, luminescent probes for biological imaging should emit in a region of the spectrum in which cells do not, stable, water-soluble, and have a high luminescence quantum yield. Poly-acrylic acid (PAAc) grafted SiNPs, which emit orange–red light, have all the

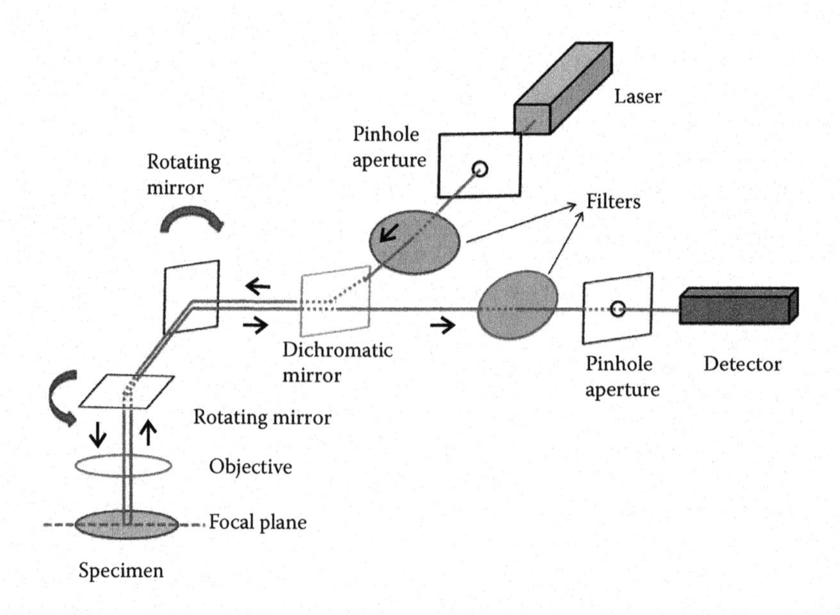

FIGURE 14.19
Basic setup of a confocal microscope. Light from the laser is scanned across the specimen by the vertical and horizontal rotating scanning mirrors. Arrows indicate the light directions.

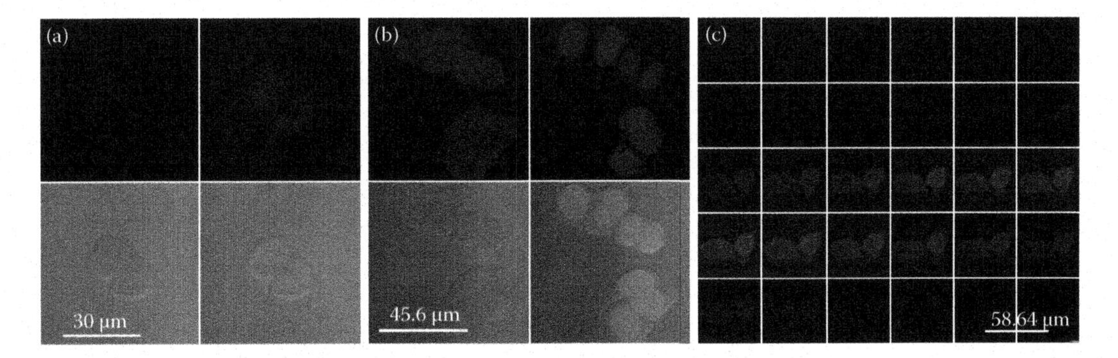

FIGURE 14.20
(See color insert.) Confocal images of MCF-7 cells incubated for 2 h with (a) medium only, (b) 0.5% v/v PAAc-SiNPs in distilled water, (c) *z*-stack image. Images were collected at 0.37-μm intervals with the 488-nm laser to create a stack in the Z axis. In (a) and (b), the top left corner shows images obtained from the red channel, which is the fluorescence from SiNPs; the top right corner presents images from the blue channel, indicating nuclei staining with DAPI; at the bottom left is the bright field; and at the bottom right is a combination of the three. (From Wang, Q. et al., *J. Nanoparticle Res.* 13:405–413, 2011. With permission.)

required characteristics as well as a much lower inherent toxicity over fluorescent dyes. Red emitters are more suitable if autofluorescence is the major problem. In addition, after PAAc capping, the SiNP is also stable against photobleaching, which is another advantage over molecular fluorophores. Figure 14.20 shows the confocal images of MCF-7 cells (a breast cancer cell) stained with PAAc-grafted SiNPs. Compared with the control, which was cultured in no-particle medium, red fluorescence was observed brightly in the MCF-7 cells cultured with the SiNPs. From these results, the red fluorescence has arisen from the emission of SiNPs. Importantly, there were no signs of morphological damage to the cells upon treatment with the PAAc grafted SiNPs (Wang et al. 2011). These findings indicated the possibility of the SiNPs as chromophores in bioimaging.

14.5.4 Flow Cytometry

Fluorescent SiNP is also useful in flow cytometry detection of live cell signals (Prina-Mello et al. 2010). Flow cytometry was developed in the 1970s to rapidly quantify the physical, chemical, and biological properties of cells and other biological particles (cell size, DNA/RNA content, cell surface antigen expression, etc.). It draws on fluorescence microscopy and blood cell counting principles, whereas the use of fluorescent dyes, laser technology, monoclonal antibody technology, and computer technology greatly improve the detection speed and accuracy of statistics. The sample is injected into the center of a sheath flow. Similar to confocal microscope, lasers are often used as a light source in flow cytometry. Cells are focused on the laser by the principles of hydrodynamics. One unique feature of flow cytometry is that it measures fluorescence per cell or particle. In our study, both $HepG_2$ and 3T3-L1 cells can take up SiNPs when exposed to high concentrations. However, the uptake of PAAc-SiNPs in both 3T3-L1 and $HepG_2$ cells was found to show a time-dependent behavior, which implies an activated mechanism involved in the uptake of PAAc-SiNPs (Figure 14.21). Endocytosis was suggested to be the major mechanism in the cellular uptake of PAAc-SiNPs, as the inhibitions of such processes were found to demonstrate a very significant suppression in the uptake rate and on the extent of intracellular accumulation in various cell types (Alsharif et al. 2009).

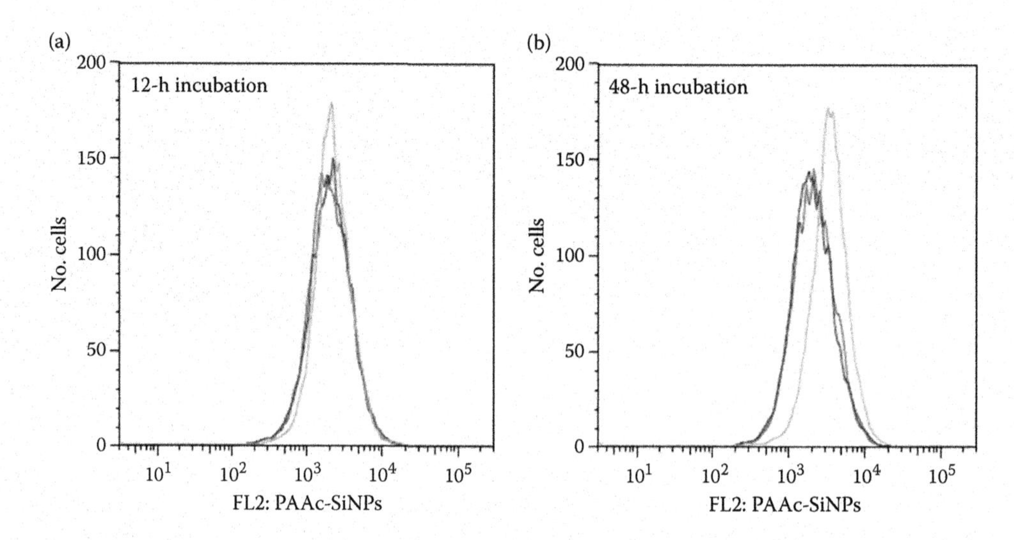

FIGURE 14.21
(See color insert.) Uptake efficiency of SiNPs in $HepG_2$ cells with various incubation times. $HepG_2$ cells were exposed to PAAc-SiNPs for 12, 24, 36, 48, 60, and 72 h at two concentrations of 10 or 100 μg/mL. Red, control; blue, 10 μg/mL PAAc-SiNP; green, 100 μg/mL PAAc-SiNP: (a) 12 h incubation; (b) 48 h incubation.

14.5.5 Summary

SiNP is a promising research tool in many biological assays and fluorescence imaging techniques because of several factors: (i) the robust and flexible surface chemistry of Si facilitates the conjugation of DNA or protein probes, (ii) red light emission at small particle size, (iii) preliminary indications are that they do not show the same toxicity as heavy metal–based nanoparticles, and (iv) they are photostable and have tunable luminescence. However, to increase their utility, the preparation techniques necessary to reliably produce controlled particle sizes with appropriate biochemical functionality need to be improved. Accurate information on the cytotoxicity study is also necessary.

References

Albaladejo, S. et al. 2009. Scattering forces from the curl of the spin angular momentum of a light field. *Physical Review Letters* 102(11):113602.

Albaladejo, S., M.I. Marqués, and J.J. Sáenz. 2011. Light control of silver nanoparticle's diffusion. *Optics Express* 19(12):11471–11478.

Almaas, E., and I. Brevik. 1995. Radiation forces on a micrometer-sized sphere in an evanescent field. *Journal of the Optical Society of America B* 12(12):2429–2438.

Almeida, V.R., R.R. Panepucci, and M. Lipson. 2003. Nanotaper for compact mode conversion. *Optics Letters* 28:1302–1304.

Almeida, V.R., Q. Xu, C.A. Barrios, and M. Lipson. 2004. Guiding and confining light in void nanostructure. *Optics Letters* 29:1209–1211.

Alsharif, N.H., C.E.M. Berger, S.S. Varanasi et al. 2009. Alkyl-capped silicon nanocrystals lack cytotoxicity and have enhanced intracellular accumulation in malignant cells via cholesterol-dependent endocytosis. *Small* 5:221–228.

Arnold, S. et al. 2009. Whispering gallery mode carousel—a photonic mechanism for enhanced nanoparticle detection in biosensing. *Optics Express* 17(8):6230–6238.
Arya, H., Z. Kaul, R. Wadhwa et al. 2005. Quantum dots in bio-imaging: Revolution by the small. *Biochemical and Biophysical Research Communications* 329:1173–1177.
Ashkin, A. 1970. Acceleration and trapping of particles by radiation pressure. *Physical Review Letters* 24(4):156–159.
Baehr-Jones, T., M. Hochberg, C. Walker, and A. Scherer. 2005. High-Q optical resonators in silicon-on-insulator-based slot waveguides. *Applied Physics Letters* 86:081101–0801103.
Baird, C.L., and D.G. Myszka. 2001. Current and emerging commercial optical biosensors. *Journal of Molecular Recognition* 14:261–268.
Barrios, C.A., M.J. Banuls, V. Gonzalez-Pedro et al. 2008. Label-free optical biosensing with slot-waveguides. *Optics Letters* 33:708–710.
Boyd, R.W., and J.E. Heebner. 2001. Sensitive disk resonator photonic biosensor. *Applied Optics* 40:5742–5747.
Buswell, S.C., V.A. Wright, J.M. Buriak, V. Van, and S. Evoy. 2008. Specific detection of proteins using photonic crystal waveguides. *Optics Express* 16:15949–15957.
Byeon, J.-Y., and R.C. Bailey. 2011. Multiplexed evaluation of capture agent binding kinetics using arrays of silicon photonic microring resonators. *Analyst* 136:3430–3433.
Byeon, J.-Y., F.T. Limpoco, and R.C. Bailey. 2010. Efficient bioconjugation of protein capture agents to biosensor surfaces using aniline-catalyzed hydrazone ligation. *Langmuir* 26:15430–15435.
Canham, L. 1990. Silicon quantum wire array fabrication by electrochemical and chemical dissolution of wafers. *Applied Physics Letters* 57:1046–1048.
Chao, C.Y., and L.J. Guo. 2003. Biochemical sensors based on polymer microrings with sharp asymmetrical resonance. *Applied Physics Letters* 83:1527–1529.
Chao, Y., S. Krishnamurthy, M. Montalti et al. 2005. Reactions and luminescence in passivated Si nanocrystallites induced by vacuum ultraviolet and soft-x-ray photons. *Journal of Applied Physics* 98:044316.
Chao, Y., L. Šiller, S. Krishnamurthy et al. 2007. Evaporation and deposition of alkyl-capped silicon nanocrystals in ultrahigh vacuum. *Nature Nanotechnology* 2:486–489.
Chow, E., A. Grot, L.W. Mirkarimi, M. Sigalas, and G. Girolami. 2004. Ultracompact biochemical sensor built with two-dimensional photonic crystal microcavity. *Optics Letters* 29:1093–1095.
Claes, T., J.G. Molera, K. De Vos, E. Schacht, R. Baets, and P. Bienstmann. 2009. Label free biosensing with slot-waveguide-based ring resonator in silicon-on-insulator. *IEEE Photonics Journal* 1:197–204.
De Vos, K., I. Bartolozzi, E. Schacht, P. Bienstman, and R. Baets. 2007. Silicon-on-insulator microring resonator for sensitive and label-free biosensing. *Optics Express* 15:7610–7615.
De Vos, K., J. Girones, T. Claes, Y. De Koninck, S. Popelka, E. Schacht, R. Baets, and P. Bienstman. 2009a. Multiplexed antibody detection with an array of silicon-on-insulator microring resonators. *IEEE Photonics Journal* 1:225–235.
De Vos, K., J. Girones, S. Popelka, E. Schacht, R. Baets, and P. Bienstman. 2009b. SOI optical microring resonator with poly(ethylene glycol) polymer brush for label-free biosensor applications. *Biosensors and Bioelectronics* 24:2528–2533.
Delâge, A., D.-X. Xu, R.W. McKinnon et al. 2009. Wavelength-dependent model of a ring resonator sensor excited by a directional coupler. *Journal of Lightwave Technology* 27:1172–1180.
Densmore, A., M. Vachon, D.-X. Xu, S. Janz, R. Ma, Y.-H. Li, G. Lopinski, A. Delâge, J. Lapointe, C.C. Luebbert, Q.Y. Liu, P. Cheben, and J.H. Schmid. 2009. Silicon photonic wire biosensor array for multiplexed real-time and label-free molecular detection. *Optics Letters* 34:3598–3600.
Densmore, A., D. Xu, S. Janz, P. Waldron, T. Mischki, G. Lopinski, A. Delâge, J. Lapointe, P. Cheben, and B. Lamontagne. 2008. Spiral-path high-sensitivity silicon photonic wire molecular sensor with temperature-independent response. *Optics Letters* 33:596–598.
Densmore, A., D.X. Xu, N. Sabourin, H. McIntosh, P. Cheben, J. Schmid, R. Ma, M. Vachon, A. Delage, and W. Sinclair. 2011. A fully integrated silicon photonic wire sensor array chip and reader instrument. 8th International Conference on Group IV Photonics. London, IEEE.

Densmore, A., D.-X. Xu, P. Waldron, S. Janz, P. Cheben, J. Lapointe, A. Delâge, B. Lamontagne, J. Schmid, and E. Post. 2006. A silicon-on-insulator photonic wire based evanescent field sensor. *IEEE Photonics Technology Letters* 18:2520–2522.
Dienerowitz, M., M. Mazilu, and K. Dholakia. 2008. Optical manipulation of nanoparticles: A review. *Journal of Nanophotonics* 2(1):021875.
Doan, V.V., and M.J. Sailor. 1992. Luminescent color image generation on porous silicon. *Science* 256:1791.
Draine, B.T., and P.J. Flatau. 1994. Discrete-dipole approximation for scattering calculations. *Journal of the Optical Society of America A* 11(4):1491–1499.
Dumon, P., W. Bogaerts, V. Wiaux et al. 2004. Low-loss SOI photonic wires and ring resonators fabricated with deep UV lithography, *IEEE Photonics Technology Letters* 16:1328–1330.
ePIXfab: The Silicon Photonics Platform. Available at: http://www.epixfab.eu/.
Erickson, D. et al. 2011. Nanomanipulation using near-field photonics. *Lab on a Chip* 11:995.
Fan, X., I.M. White, S.I. Shopova, H. Zhu, J.D. Suter, and Y. Sun. 2008. Sensitive optical biosensors for unlabeled targets: A review. *Analytica Chimica Acta* 620:8–26.
Fauchet, P. 2008. *Porous Silicon Optical Label-Free Biosensors*. New York: Springer Verlag.
Fowles, G.R. 1975. *Introduction to Modern Optics*. New York: Holt, Rinehart and Winston.
Fukuyama, M., S. Yamatogi, H. Ding et al. 2010. Selective detection of antigen-antibody reaction using Si ring optical resonators. *Japanese Journal of Applied Physics* 49:04DL09 1–4.
García-Rupérez, J., V. Toccafondo, M.J. Bañuls et al. 2010. Label-free antibody detection using band edge fringes in SOI planar photonic crystal waveguides in the slow-light regime. *Optics Express* 18:24276–24286.
Gaugiran, S., S. Gétin, and J.M. Fedeli. 2005. Optical manipulation of microparticles and cells on silicon nitride waveguides. *Optics Express* 13(18):6956–6963.
Grujic, K. et al. 2005. Sorting of polystyrene microspheres using a Y-branched optical waveguide. *Optics Express* 13(1):1–7.
Halir, R., P. Cheben, S. Janz et al. 2009. Waveguide grating coupler with subwavelength microstructures. *Optics Letters* 34:1408–1410.
Halir, R., P. Cheben, J.H. Schmid et al. 2010. Continuously apodized fiber-to-chip surface grating coupler with refractive index engineered subwavelength structure. *Optics Letters* 35:3243–3245.
Hermanson, G.T. 1996. *Bioconjugate Techniques*. San Diego, CA: Academic Press.
Heinrich, J.L., C.L. Curtis, G.M. Credo et al. 1992. Luminescent colloidal silicon suspensions from porous silicon. *Science* 255:66–68.
Homola, J. 2006. *Surface Plasmon Based Resonance Sensors*. Berlin: Springer Verlag.
Homola, J. 2008. Surface plasmon resonance sensors for detection of chemical and biological species. *Chemical Reviews* 108:462–493.
Hua, F.J., F. Erogbogbo, M.T. Swihart, and E. Ruckenstein. 2006. Organically capped silicon nanoparticles with blue photoluminescence prepared by hydrosilylation followed by oxidation. *Langmuir* 22:4363–4370.
Hua, F.J., M.T. Swihart, and E. Ruckenstein. 2005. Efficient surface grafting of luminescent silicon quantum dots by photoinitiated hydrosilylation. *Langmuir* 21:6054–6062.
Hunsperger, R.G. 1991. *Integrated Optics: Theory and Technology*, 3rd ed. Berlin: Springer-Verlag.
Iqbal, M., M. Gleeson, B. Spaugh, F. Tybor, W. Gunn, M. Hochberg, T. Baehr-Jones, R. Bailey, and L. Gunn. 2010. Label-free biosensor arrays based on silicon ring resonators and high-speed optical scanning instrumentation. *IEEE Journal of Selected Topics in Quantum Electronics* 16:654–661.
Jackson, J.D. 1998. *Classical Electrodynamics*, 3rd ed., Wiley.
Jane, A., R. Dronov, A. Hodges, and N.H. Voelcker. 2009. Porous silicon biosensors on the advance. *Trends in Biotechnology* 27:230–239.
Janz, S., A. Densmore, D.-X. Xu et al. 2009. Silicon photonic wire waveguide sensors. In *Advanced Photonic Structures for Chemical and Biological Sensing*. Berlin: Springer-Verlag.
Joannopoulos, J.D. et al. 2008. *Photonic Crystals: Molding the Flow of Light*, 2nd ed., Princeton, NJ: Princeton University Press.

Kanda, V., J.K. Karluki, D.J. Harrison, and M.T. McDermott. 2004. Label-free reading of microarray-based immunoassays with surface plasmon resonance imaging. *Analytical Chemistry* 76:7257–7262.
Karlsson, R. 2004. SPR for molecular interaction analysis: A review of emerging application areas. *Journal of Molecular Recognition* 17:151–161.
Kawata, S., and T. Tani. 1996. Optically driven Mie particles in an evanescent field along a channeled waveguide. *Optics Letters* 21(21):1768–1770.
Kirk, J.T., G.E. Fridley, J.W. Chamberlain et al. 2011. Multiplexed inkjet functionalization of silicon photonic biosensors. *Lab on a Chip* 11:1372–1377.
Kogelnik, H. 1990. Theory of optical waveguides. In *Guided Wave Optoelectronics*, edited by Tamir, T., 2nd ed., 7–87. Berlin: Springer-Verlag.
Ksendzov, A., M.L. Homer, and A.M. Manfreda. 2004. Integrated optics ring-resonator chemical sensor with polymer transduction layer. *Electronics Letters* 40:63–65.
Lardenois, S., D. Pascal, L. Vivien et al. 2003. Low-loss submicrometer silicon-on-insulator rib waveguides and corner mirrors. *Optics Letters* 28:1150–1152.
Lawrie, J.L., Y. Jiao, and S.M. Weiss. 2010. Size-dependent infiltration and optical detection of nucleic acids in nanoscale pores. *IEEE Transactions on Nanotechnology* 9:596–602.
Lazcka, O., F.J.D. Campo, and F.X. Muñoz. 2007. Pathogen detection: A perspective of traditional methods and biosensors. *Biosensors and Bioelectronics* 22(7):1205–1217.
Leamy, H.J., and J.H. Wernick. 1997. Semiconductor silicon: The extraordinary made ordinary. *MRS Bulletin* 22:47–55.
Lee, K.K., D.R. Lim, D. Pan et al. 2005. Mode transformer for miniaturized optical circuits. *Optics Letters* 30:498–500.
Lee, M.R., and P.M. Fauchet. 2007a. Nanoscale microcavity sensor for single particle detection. *Optics Letters* 32:3284–3286.
Lee, M.R., and P.M. Fauchet. 2007b. Two-dimensional silicon photonic crystal based biosensing platform for protein detection. *Optics Express* 15:4530–4534.
Lie, L.H., M. Duerdin, E.M. Tuite, A. Houlton, and B.R. Horrocks. 2002. Preparation and characterisation of luminescent alkylated-silicon quantum dots. *Journal of Electroanalytical Chemistry* 538:183–190.
Lin, S. et al. 2009. Design of nanoslotted photonic crystal waveguide cavities for single nanoparticle trapping and detection. *Optics Letters* 34(21):3451–3453.
Lin, S., E. Schonbrun, and K. Crozier. 2010. Optical manipulation with planar silicon microring resonators. *Nano Letters* 10:2408–2411.
Lin, V.S.Y., K. Motesharei, K.P.S. Dancil, M.J. Sailor, and M.R. Ghadiri. 1997. A porous silicon-based optical interferometric biosensor. *Science* 278:840.
Little, B.E., S.T. Chu, H.A. Haus, J. Foresi, and J.-P. Laine. 1997. Microring resonator channel dropping filters. *Journal of Lightwave Technology* 15:998–1005.
Lockwood, D.J., and L. Paveisi. 2010. *Silicon Photonics II: Components and Integration.* London, New York: Springer Verlag.
Luchansky, M.S., and R.C. Bailey. 2010. Silicon photonic microring resonators for quantitative cytokine detection and T-cell secretion analysis. *Analytical Chemistry* 82:1975–1981.
Luchansky, M.S., A.L. Washburn, T.A. Martin et al. 2010. Characterization of the evanescent field profile and bound mass sensitivity of a label-free silicon photonic microring resonator biosensing platform. *Biosensors and Bioelectronics* 26:1283–1291.
Luff, B.J., R.D. Harris, J.S. Wilkinson, R. Wilson, and D.J. Schiffrin. 1996. Integrated-optical directional coupler biosensor. *Optics Letters* 21:618–620.
Luff, B.J., J.S. Wilkinson, J. Piehler, U. Hollenback, J. Ingenhoff, and N. Fabricius. 1998. Integrated optical Mach–Zehnder biosensor. *Journal of Lightwave Technology* 16:583–591.
Lukosz, W. 1991. Principles and sensitivities of integrated optical and surface plasmon sensors for direct affinity sensing and immunosensing. *Biosensors and Bioelectronics* 6:215–225.
Mandal, S., X. Serey, and D. Erickson. 2010. Nanomanipulation using silicon photonic crystal resonators. *Nano Letters* 10(1):99–104.
Miller, B.L. 2010. *Nanostructured Silicon Optical Sensors*. London, New York: Springer Verlag.

Morita, M., T. Ohmi, E. Hasegawa, M. Kawakami, and M. Ohwada. 1990. Growth of native oxide on a silicon surface. *Journal of Applied Physics* 68:1272.

Neiner, D., H.W. Chiu, and S.M. Kauzlarich. 2006. Low-temperature solution route to macroscopic amounts of hydrogen terminated silicon nanoparticles. *Journal of the American Chemical Society* 128:11016–11017.

Ng, L.N. et al. 2000. Manipulation of colloidal gold nanoparticles in the evanescent field of a channel waveguide. *Applied Physics Letters* 76(15):1993.

Ng, L.N. et al. 2002. Propulsion of gold nanoparticles on optical waveguides. *Optics Communications* 208(1–3):117–124.

O'Farrell, N., A. Houlton, and B.R. Horrocks. 2006. Silicon nanoparticles: Applications in cell biology and medicine. *International Journal of Nanomedicine* 1:451.

Orosco, M.M., C. Pacholski, and M.J. Sailor. 2009. Real-time monitoring of enzyme activity in a mesoporous silicon double layer. *Nature Nanotechnology* 4:255–258.

Parriaux, O., and G.J. Veldhuis. 1998. Normalized analysis for the sensitivity optimization of integrated optical evanescent-wave sensors. *Journal of Lightwave Technology* 16:573–582.

Passaro, V., F. Dell'Olio, and B. Casamassima. 2007. Guided-wave optical biosensors. *Sensors* 7:508–536.

Pavesi, L., and D.J. Lockwood. 2004. *Silicon Photonics*. Berlin: Springer Verlag.

Prasad, V., D. Semwogerere, and E.R. Weeks. 2007. Confocal microscopy of colloids. *Journal of Physics: Condensed Matter* 19:113102.

Prieto, F., B. Sepulveda, A. Calle et al. 2003. An integrated optical interferometric nanodevice based on silicon technology for biosensor applications. *Nanotechnology* 14:907–912.

Prina-Mello, A., A.M. Whelan, A. Atzberger et al. 2010. Comparative flow cytometric analysis of immunofunctionalized nanowire and nanoparticle signatures. *Small* 6:247–255.

Qavi, A.J., J.T. Kindt, M.A. Gleeson, and R.C. Bailey. 2011a. Anti-DNA:RNA antibodies and silicon photonic mircoring resonators: Increased sensitivity for multiplexed microRNA detection. *Analytical Chemistry* 83:5949–5956.

Qavi, A.J., T.M. Mysz, and R.C. Bailey. 2011b. Isothermal discrimination of single-nucleotide polymorphisms via real-time kinetic desorption and label-free detection of DNA using silicon photonic microring resonator arrays. *Analytical Chemistry* 83:6827–6833.

Rabiei, P., W. Steier, C. Zhang, and L.R. Dalton. 2002. Polymer micro-ring filters and modulators. *Journal of Lightwave Technology* 20:1968–1975.

Rahmani, A., and Chaumet, P.C. 2006. Optical trapping near a photonic crystal. *Optics Express* 14(13):6353–6358.

Ramachandran, A., S. Wang, J. Clarke et al. 2008. A universal biosensing platform based on optical micro-ring resonators. *Biosensors and Bioelectronics* 23:939.

Reynolds, J.E. 1893. Opening address to the British Association for the Advancement of Science. *Nature* 48:477–481.

Roelkens, G., D. Van Thourhout, and R. Baets. 2006. High efficiency silicon-on-insulator grating coupler based on a poly-Silicon overlay. *Optics Express* 14:11622–11630.

Roelkens, G., D. Thourhout, and R. Baets. 2007. High efficiency grating coupler between silicon-on-insulator waveguides and perfectly vertical optical fibers. *Optics Letters* 32:1495–1497.

Schmid, J.H., W. Sinclair, J. Garcia et al. 2009. Silicon-on-insulator guided mode resonant grating for evanescent field molecular sensing. *Optics Express* 17:18371–18380.

Schmidt, B., V. Almeida, C. Manolatou, S. Preble, and M. Lipson. 2004. Nanocavity in silicon waveguide for ultrasensitive nanoparticle detection. *Applied Physics Letters* 85:4854–4856.

Schmidt, B.S. et al. 2007. Optofluidic trapping and transport on solid core waveguides within a microfluidic device. *Optics Express* 15(22):14322–14334.

Selvan, S.T., T.T.Y. Tan, D.K. Yi, and N.R. Jana. 2010. Functional and multifunctional nanoparticles for bioimaging and biosensing. *Langmuir* 26:11631–11641.

Sharma, P., S. Brown, G. Walter, S. Santra, and B. Moudgil. 2006. Nanoparticles for bioimaging. *Advances in Colloid and Interface Science* 123–126:471–485.

Shiohara, A., S. Hanada, S. Prabakar, K. Fujioka, T.H. Lim, K. Yamamoto, P.T. Northcote, and R.D. Tilley. 2010. Chemical reactions on surface molecules attached to silicon quantum dots. *Journal of the American Chemical Society* 132:249–253.

Shoji, T., T. Tsuchizawa, T. Watanabe, K. Yamada, and H. Morita. 2002. Low loss mode size converter from 0.3 mm square Si wire waveguides to single mode fibers. *Electronics Letters* 38:1669–1670.

Stevenson, D.J., F. Gunn-Moore, and K. Dholakia. 2010. Light forces the pace: Optical manipulation for biophotonics. *Journal of Biomedical Optics* 15(4):041503.

Taillaert, D., P. Bienstman, and R. Baets. 2003. Compact efficient broadband grating coupler for silicon-on-insulator waveguides. *Optics Letters* 29:2749–2751.

Tanaka, T., and S. Yamamoto. 2000. Optically induced propulsion of small particles in an evanescent field of higher propagation mode in a multimode, channeled waveguide. *Applied Physics Letters* 77(20):3131.

Tiefenthaler, K., and W. Lukosz. 1989. Sensitivity of grating couplers as integrated-optical chemical sensors. *Journal of the Optical Society of America B* 6:209–220.

Tilley, R.D., and K. Yamamoto. 2006. The microemulsion synthesis of hydrophobic and hydrophilic silicon nanocrystals. *Advanced Materials* 18:2053–2056.

Van Laere, F., G. Roelkens, M. Ayre et al. 2007. Compact and highly efficient grating couplers between optical fiber and nanophotonic waveguides. *Journal of Lightwave Technology* 25:151–156.

Vo-Dinh, T., and L. Allain. 2003. Biosensors for medical applications. *Biomedical Photonics Handbook.*

Wang, F., W.B. Tan, Y. Zhang, X. Fan, and M. Wang. 2006. Luminescent nanomaterials for biological labelling. *Nanotechnology* 17:R1.

Wang, Q., H. Ni, A. Pietzsch et al. 2011. Synthesis of water-dispersible photoluminescent silicon nanoparticles and their use in biological fluorescent imaging. *Journal of Nanoparticle Research* 13:405–413.

Warner, J.H., A. Hoshino, K. Yamamoto, and R.D. Tilley. 2005. Water-soluble photoluminescent silicon quantum dots. *Angewandte Chemie (International ed. in English)* 44:4550–4554.

Washburn, A.L., J. Gomez, and R.C. Bailey. 2011. DNA-encoding to improve performance and allow parallel evaluation of the binding characteristics of multiple antibodies in a surface-bound immunoassay format. *Analytical Chemistry* 83:3572–3580.

Washburn, A.L., L.C. Gunn, and R.C. Bailey. 2009. Label-free quantitation of a cancer biomarker in complex media using silicon photonic microring resonators. *Analytical Chemistry* 81:9499–9506.

Washburn, A.L., M.S. Luchansky, A.L. Bowman, and R.C. Bailey. 2010. Quantitative, label-free detection of five protein biomarkers using multiplexed arrays of silicon photonic microring resonators. *Analytical Chemistry* 82:69–72.

Weisser, M., G. Tovar, S. Mittler-Neher et al. 1999. Specific biorecognition reactions observed with and integrated Mach–Zehnder interferometer. *Biosensors and Bioelectronics* 14:405–411.

Wolkin, M.V., J. Jorne, P.M. Fauchet et al. 1999. Electronic states and luminescence in porous silicon quantum dots: The role of oxygen. *Physical Review Letters* 82:197–200.

Wong, V., and M.A. Ratner. 2006. Gradient and nongradient contributions to plasmon-enhanced optical forces on silver nanoparticles. *Physical Review B* 73(7):075416.

Xu, D.-X., A. Densmore, A. Delâge, P. Waldron, R. McKinnon, S. Janz, J. Lapointe, G. Lopinski, T. Mischki, E. Post, P. Cheben, and J.H. Schmid. 2008. Folded cavity SOI microring sensors for high sensitivity and real time measurement of biomolecular binding. *Optics Express* 16:15137–15148.

Xu, D.-X., M. Vachon, A. Densmore, R. Ma, A. Delâge, S. Janz, J. Lapointe, Y. Li, G. Lopinski, D. Zhang, Q.Y. Liu, P. Cheben, and J.H. Schmid. 2010a. Label-free biosensor array based on SOI ring resonators addressed using the WDM approach. *Optics Letters* 35:2771–2773.

Xu, D.X., M. Vachon, A. Densmore, R. Ma, S. Janz, A. Delâge, J. Lapointe, P. Cheben, J.H. Schmid, E. Post, S. Messaoudène, and J-M. Fédéli. 2010b. Real-time cancellation of temperature induced resonance shifts in SOI wire waveguide ring resonator label-free biosensor arrays. *Optics Express* 18:22867–22879.

Xu, Q., B. Schmidt, S. Pradhan, and M. Lipson. 2005. Micrometre-scale silicon electro-optic modulator. *Nature* 435:325–327.

Yalcin, A., K.C. Popat, J.C. Aldridge et al. 2006. Optical sensing of biomolecules using microring resonators. *IEEE Journal of Selected Topics in Quantum Electronics* 12:148–155.

Yang, A.H.J., and D. Erickson. 2008. Stability analysis of optofluidic transport on solid-core waveguiding structures. *Nanotechnology* 19:045704.

Yang, A.H.J., T. Lerdsuchatawanich, and D. Erickson. 2009a. Forces and transport velocities for a particle in a slot waveguide. *Nano Letters* 9(3):1182–1188.

Yang, A.H.J., S.D. Moore, B.S. Schmidt, M. Klug, M. Lipson, and D. Erickson. 2009b. Optical manipulation of nanoparticles and biomolecules in sub-wavelength slot waveguides. *Nature* 457(7225):71–75.

Yariv, A. 2000. Universal relations for coupling of optical power between microresonators and dielectric waveguides. *Electronics Letters* 36:321–323.

Yu, Z.R., M. Aceves-Mijares, and M.A.I. Cabrera. 2006. Single electron charging and transport in silicon rich oxide. *Nanotechnology* 17:3962–3967.

Zhang, X.M., D. Neiner, S.Z. Wang, A.Y. Louie, and S.M. Kauzlarich. 2007. A new solution route to hydrogen-terminated silicon nanoparticles: Synthesis, functionalization and water stability. *Nanotechnology* 18:095601.

15

Silicon-Based Photovoltaics

Mario Tucci, Massimo Izzi, Radovan Kopecek, Michelle McCann, Alessia Le Donne, Simona Binetti, Shujuan Huang, and Gavin Conibeer

CONTENTS

15.1 Introduction

Simona Binetti

Today, after more than 60 years of continued discovery and development of silicon, approximately 90% of cumulative installed photovoltaic (PV) modules are based on crystalline silicon technology. PV devices based on silicon are the most commonly used solar cells currently being produced and, thanks to silicon technology, the worldwide market for solar electric energy has grown by 40% a year over the last decade.

The use of silicon solar cells resulted in the achievement of a 25% record efficiency, very close to the theoretical maximum efficiency of 31% for a single-junction PV device. This record, as its holder Prof. M. A. Green wrote, represents the latest step in the more than 60-year history of silicon cell development, which shows three major phases (Green 2009). The first major improvements occurred in the 1950s with the development of bulk crystal growth, whereas the second occurred in the 1970s with the development of shallow junctions, metallization, texturing, and antireflection coating (ARC) processes. The third phase started in the 1980s and has lasted up until now: an increase of bulk diffusion length, the improvement of surface passivation, and enhancement of device technologies being among its main achievements.

An additional step in the development of silicon solar cells is ongoing and is related to a further efficiency improvement through defect control, device optimization, surface modification, and nanotechnology approaches. These researches are being aimed at increasing or overcoming the conventional efficiency limit.

The importance of crystalline silicon for solar energy explains the need for a chapter in this handbook devoted to silicon-based solar cells. For this same reason, the following sections emphasize monocrystalline and multicrystalline silicon solar cells, currently the most diffused technologies available. This chapter attempts to review the recent advances and current technologies used to produce crystalline silicon solar devices (Sections 15.2 and 15.3), and at the same time, the most challenging and promising strategies increasing the ratio between the efficiency and the cost of silicon solar cells (Section 15.4). Finally, the effect and the potential of using a nanotechnological approach in a silicon-based solar cell is described in the last section.

Despite the arrival of new materials and the opening of new outlooks into the future of solar energy, silicon should continue to be the leading material, especially if we consider its peculiar properties and unique advantages, such as availability, nontoxicity, long lifetime, and sustainability. Development, innovation, and new device concepts in silicon solar cells are taking place, driving down the costs of solar technologies and making them even more cost-competitive with conventional sources.

15.2 High-Efficiency Monocrystalline Silicon Solar Cells: Reaching the Theoretical Limit

Mario Tucci and Massimo Izzi

15.2.1 Introduction

By the end of 2011, approximately 60 GW_p of PVs are expected to be installed all over the world, mainly driven by the feeding tariff fixed in several countries to stimulate the PV market. In particular, crystalline silicon (c-Si)–based PVs have reached 85% of the total market. This is mainly due to the well-established and reliable technology of silicon-based modules. From a general point of view, silicon is the most favorable PV material, being the most abundant element on the Earth's surface and having a bandgap with a nearly ideal match with the solar spectrum.

15.2.2 Silicon Solar Cell Limits

According to Tiedje et al. (1984), the ultimate efficiency of silicon solar cell as a function of silicon substrate thickness can be evaluated, taking into account the c-Si absorption as a function of the wavelength. The absorbance within the semiconductor can be considerably enhanced by texturing both wafer surfaces thus deflecting the incident light and trapping it inside the material until it is either absorbed or scattered back into the escape cone. If the surface texturing is able to produce a Lambertian surface (Sheng 1984), the light-trapping effect increases the mean path length up to $4n^2w$, with n being the silicon refractive index (Yablonovitch 1982) and w the wafer thickness, which practically means 50 times the wafer thickness in the near-infrared wavelength range of interest for PV applications of c-Si. The theoretical short circuit current density J_{sc} under AM1.5 G sunlight as a function of wafer thickness is shown in Figure 15.1a for textured (solid black line) and flat silicon surface (dashed black line). In this case, only absorption and radiative recombination are considered.

In real semiconductors, other loss mechanisms are present. Indeed, photons can be absorbed by free carriers, which can dissipate energy into heat interacting with the crystalline lattice. Furthermore, electron–hole pairs in equilibrium at a given temperature can recombine radiatively or nonradiatively, mainly depending on the band alignment in terms of quantum momentum. The radiative recombination in c-Si was evaluated as being 2.1×10^{-15} cm^3/s (Pankove 1971). This recombination process promotes a reabsorption mechanism, which competes with luminescent emission from the surface of the

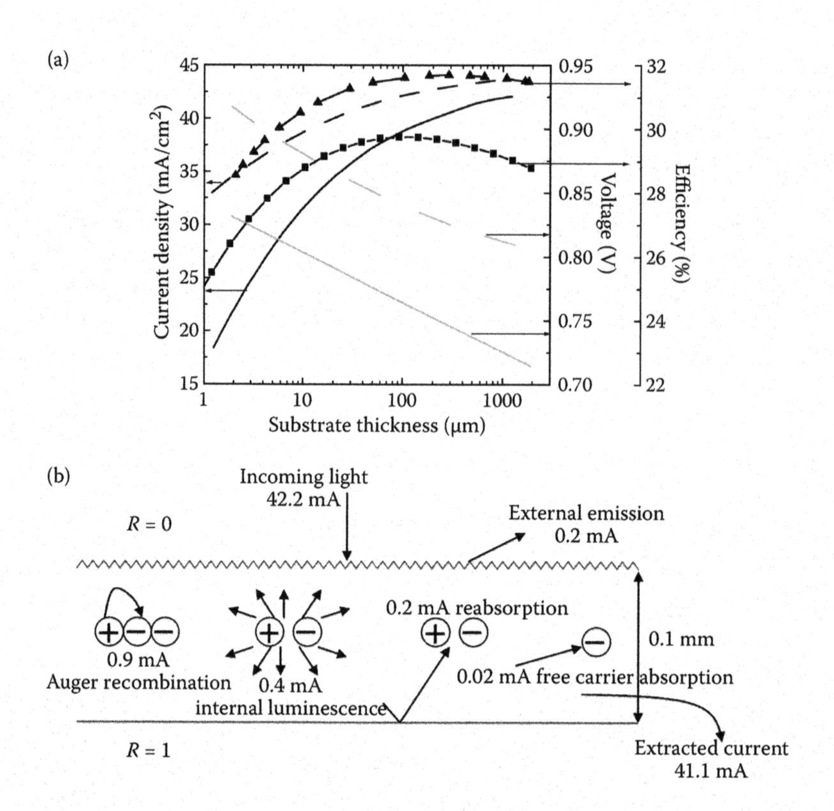

FIGURE 15.1
Short circuit current density (black lines) for textured (dashed line) and flat surface (solid line) based on AM1.5G solar spectrum (a). Open circuit voltage (gray lines) and cell efficiency (symbols and black line) as a function of cell thickness for radiative recombination only (dashed lines) and for radiative, Auger, and free carrier absorption (solid lines). Relative magnitude of loss processes in the 0.1-mm-thick silicon solar cell at the maximum power point (b). (From Tiedje, T. et al. *IEEE Trans. Electron. Devices* 31(5):711, 1984. With permission.)

semiconductor. On the other hand, nonradiative recombination can take place even in the absence of defect-assisted Shockley–Read–Hall type recombination, as in the Auger recombination mechanism. The Auger recombination coefficient was estimated as being $C = C_n + C_p = 3.88 \times 10^{-31}$ cm^6 s^{-1} for low doped silicon in which both electrons and holes are comparable under illumination, as reported in (Svantesson and Nilsson 1979). Open circuit voltage (V_{oc}) as a function of silicon thickness is shown in Figure 15.1a for radiative cases only (dashed gray line), with the addition of free carrier absorption and Auger recombination (solid gray line), the latter being the most relevant loss mechanism. Solar cell efficiency as a function of silicon thickness at room temperature is also shown in Figure 15.1a. Auger mechanism and free carrier absorption strongly reduce the peak efficiency from 32.9% down to 29.8% for a 100-μm-thick silicon cell. This last value is produced by V_{oc} = 769 mV, J_{sc} = 42.2 mA/cm^2, and FF = 89%. The loss mechanisms in the 100-μm-thick silicon solar cell are resumed in Figure 15.1b. Neither bandgap narrowing nor carrier concentration gradients have been included in the abovementioned theoretical limits because the first produces a voltage reduction of only 3.4 mV (Lanyon and Tuft 1979) and the latter has a very negligible effect on efficiency. The ultimate efficiency limit of single-bandgap p–n junction silicon solar cells under AM1.5G can be moved forward taking into account the AM1.5G spectrum normalized to 100 mW/cm^2 and the ideal light trapping properties of the cell. Then, it can generate a short-circuit current density of approximately 46.0 mA/cm^2 in the one-phonon limit (Green 1987). This effect assumes the assistance of only a single phonon in the absorption process and thus shifts the absorption edge for generation from 1110 to approximately 1200 nm at room temperature. This assumption excludes both the generation of more than one electron–hole pair by high-energy photons (Green 1987; Kolodinski et al. 1993) and the generation of electron–hole pairs by sub-bandgap photons involving more than one phonon or deep defect level in the bandgap (Guttler and Queisser 1970; Keevers and Green 1994). Therefore, in the one-phonon limit, the ultimate AM1.5G efficiency of a single-bandgap p–n junction silicon solar cell with ideal light trapping properties is approximately 33% at room temperature.

However, such efficiency calculations do not concern specific cell issues related to the manufacturing processes or to the absorber material, and therefore, can be considered as an ideal case. In a real silicon-based solar cell, a reasonable estimation for the technologically achievable efficiency limit is approximately 28% (J_{sc} = 42.5 mA/cm^2, V_{oc} = 760 mV, FF = 87%; Aberle et al. 1992). In fact, several issues, such as bulk lifetime, emitter formation, optical confinement, metal contacts, and surface recombination of both front and rear wafer surfaces, limit cell efficiency. Each of them can be optimized to reduce the loss mechanisms, even if such optimization affects the manufacturing cost of the cell. In the following section, each of these issues will be evaluated through several examples proposed in the literature to achieve the best efficiency.

15.2.3 Bulk Lifetime

It is apparent that the high quality of the absorber material is mandatory to achieve high efficiency; therefore, great efforts are continuously devoted to reducing impurities as well as defect densities in the silicon matrix. Usually, metallic impurities are present in either dissolved or precipitated form. They mainly arise from the crystal growth environment (both the growth and rotation rate), the crystal diameter, and the cooling rate (Ravi 1981). Neutron activation analysis (Davis et al. 1980; Harada et al. 1995) revealed considerable quantities of metals in silicon wafer. These contaminants become severely detrimental when occurring at interstitial or substitutional lattice sites because they can be distributed throughout the wafer and then strongly enhance the recombination through deep levels within the material bandgap. Generally, interstitial impurities occur in much higher concentrations than the substitutional ones. Interstitial Fe is one of the most relevant metallic contaminants (Istratov et al. 1999), which shows donor behavior with an activation energy of 0.38 eV from the valence band. Such a contaminant has a much larger capture cross-section for electrons than for holes, its effect being therefore more detrimental in case of p-type doped silicon ($\sigma_n = 5 \times 10^{-14}$ cm^2, $\sigma_p = 7 \times 10^{-17}$ cm^2; MacDonald and Geerligs 2004a). A comparison between the minority carrier lifetime of n-type and p-type Si is shown in Figure 15.2a and b for Fe and Ni, respectively (MacDonald and Geerligs 2004a,b). The lifetime of the n-type material is quite independent from the injection level due to the relatively low capture cross-section of minority carrier. In turn, in p-type material, the low injection regime is dominated by the relatively high capture cross-section of the minority carriers, whereas the high injection regime is limited both by electron and hole capture rates.

As well, it has generally been found that the interstitial metals are more detrimental in p-type Si, whereas the substitutional metals strongly affect n-type Si (Graff 2000; Itsumi 1993) due to the higher electron capture cross-section. A comparison of electron–hole capture cross-section ratios for different interstitial and substitutional metallic impurities in

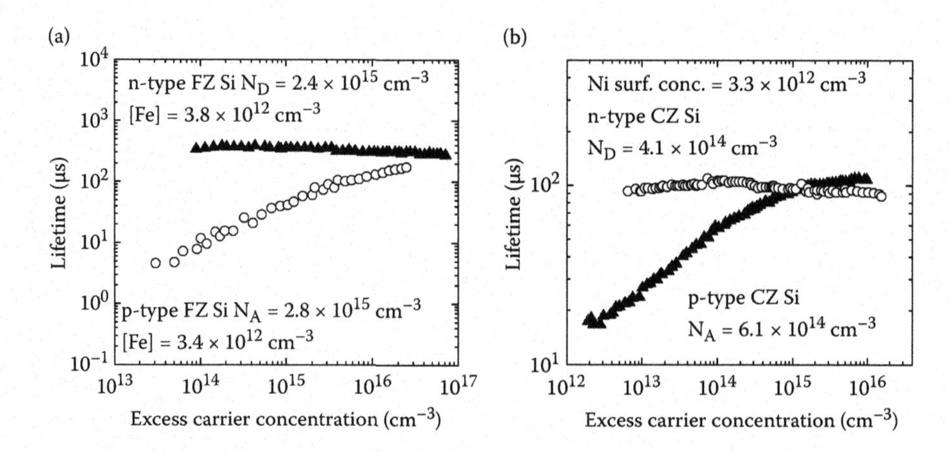

FIGURE 15.2
Minority carrier lifetime as a function of excess carrier concentration in two FZ Si wafers doped with Fe (a) and Ni (b), respectively. (Reprinted from MacDonald, D. and Geerligs, L.J. *Proc. 19th EUPVSEC, Paris* 492–495, 2004. With permission.)

Si as a function of their position within the Si bandgap is shown in Figure 15.3 (MacDonald and Geerligs 2004a).

Metallic impurities can diffuse along the crystal during the high-temperature process needed to form the solar cell device (Davis et al. 1980). However, phosphorus diffusion at high temperature is a well-known gettering process due to the enhanced solubility of gettered species in the phosphorus diffused layer. Such a gettering process is particularly effective in reducing the Fe concentration. In turn, further thermal annealing processes at temperatures higher than the gettering temperature tends to drive Fe back into the wafer if the phosphorus diffused layer is not etched. However, such poisoning does not occur at lower temperatures. Figure 15.4 shows the phosphorus gettering effect on Fe concentration in FZ 65 Ωcm p-type silicon wafer, and the effect of subsequent thermal annealing steps.

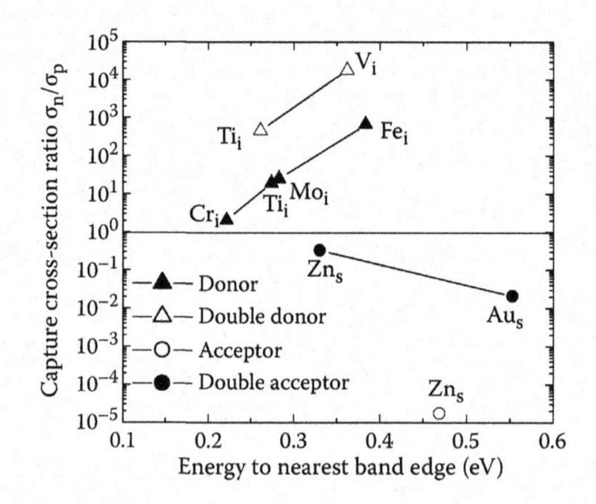

FIGURE 15.3
Capture cross-section ratio as a function of energy level depth for different impurities (i = interstitial; s = substitutional). (Reprinted from MacDonald, D. and Geerligs, L.J. *Appl. Phys. Lett.* 85:4061, 2004. With permission.)

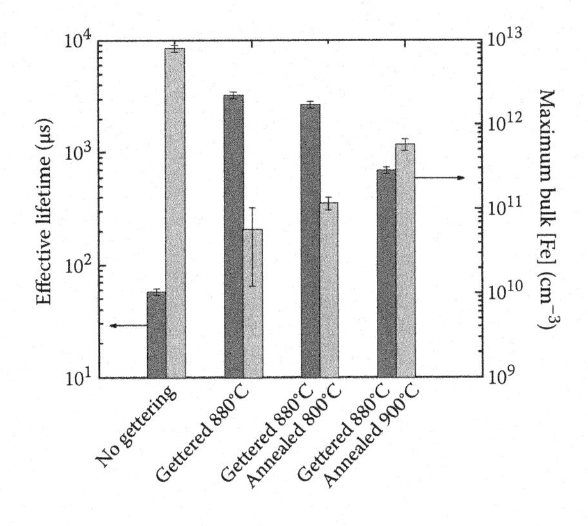

FIGURE 15.4
Effective lifetime and iron concentration after different thermal treatments. (Data from MacDonald, D. et al. *Proc. 3rd WCPVSC*, 1336, 2003. With permission.)

More than 99% of iron can be gettered thus resulting in higher minority carrier lifetime after the thermal treatment (MacDonald et al. 2003). Also, aluminum, commonly used as back contact in p-type silicon-based cells, is known for being effective in the gettering of metallic impurities even at relatively low temperatures (Thompson and Tu 1978). The very expensive ultrapure technology is useful in reducing the surface iron contamination down to 10^{11} cm^{-2}, mainly by chemical gettering with HCl (Davis et al. 1980).

Although metallic impurities can be kept at concentrations lower than 10^{12} cm^{-3} in high-quality silicon wafers, carbon and oxygen content can in turn be in the order of 10^{16} cm^{-3} and 10^{18} cm^{-3}, respectively. They mainly originate from the silicon growth system (graphite-based heaters and quartz crucible) and they assume a radial distribution within the wafer. Fortunately, the oxygen and carbon active concentrations in Czochralski Si crystals are two orders of magnitude lower and usually reveal their presence after thermal treatment at temperatures higher than 300°C. In particular, thermal treatments in the range between 300°C and 500°C give rise to the formation of oxygen-based aggregates acting as donors. Annealing at higher temperatures, such as 1000°C, gives rise instead to the formation of SiO_x precipitates within the silicon matrix and, in the meantime, to oxygen out-diffusion from the wafer surface with the formation of a depletion region (Davis et al. 1980) This procedure is known as intrinsic gettering (Sardana 1985) and, as expected, results in high-quality surface but poor quality bulk wafer, which is not really useful for solar cell devices.

When oxygen meets boron within the Si crystal, metastable B–O defects originate (Schmidt et al. 1997, 2001a,b), inducing a severe degradation of the minority carrier lifetime under illumination or carrier injection (Knobloch et al. 1995; Fischer and Pschunder 1973). To overcome this problem, several alternatives are presently available at the R&D level, not yet scaled up for industrial applications: (1) the use of low doped p-type wafer (Glunz et al. 2001; Rein et al. 2000); (2) the use of a suitable regeneration process (Herguth et al. 2006); (3) the use of magnetic Czochralski (Hoshi et al. 1985) silicon, which demonstrated very high performance (Glunz et al. 1999, 2000b) due to very low oxygen and carbon contents;

(4) the use of FZ-PV, which showed a very low oxygen content resulting in very high and stable lifetimes for both p-type and n-type wafers (Vedde et al. 2003); (5) the use of gallium-doped silicon, which showed no degradation (even if a homogeneous doping concentration over the entire silicon ingot is, however, quite difficult to obtain due to the poor Ga segregation coefficient in Si). To better evaluate the degradation induced by oxygen, lifetime as a function of doping and oxygen content ratio on different kinds of silicon

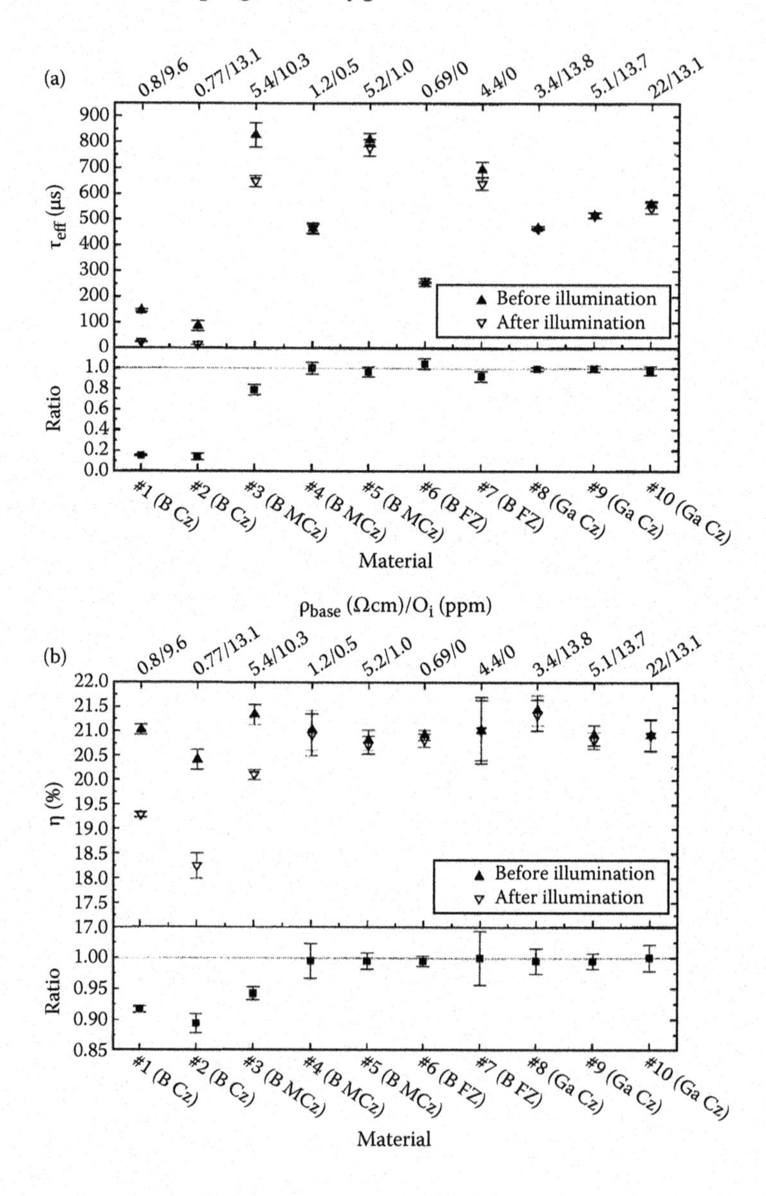

FIGURE 15.5
Effective lifetime decay before and after illumination as a function of doping and oxygen content ratio within different silicon substrates (a). Passivating emitter and rear contact solar cell efficiencies before and after illumination as a function of doping and oxygen content ratio within different silicon substrates (b). The bottom graphs give the ratio between the values before and after illumination. (Glunz, S.W. et al.: Comparison of boron- and gallium-doped p-type Czochralski silicon for photovoltaic application. *Progress in Photovoltaics: Research and Applications*. 1999. 7. 463–469. Copyright Wiley-VCH Verlag GmbH & Co. KGaA. Reproduced with permission.)

substrates doped with B or Ga are shown in Figure 15.5a before and after illumination. In Figure 15.5b, the efficiencies of passivating emitter and rear contact solar cells fabricated on the same substrates as in Figure 15.5a are reported as a function of doping and oxygen content ratio. In the same figures, the lifetimes and the efficiencies ratio before and after illumination are also depicted (Glunz et al. 1999).

As mentioned previously, a possible way to overcome lifetime degradation is the use of n-type silicon wafers. Despite the theoretical lower diffusion length with respect to the p-type Si due to higher effective mass for holes, n-type Czochralski Si shows no minority carrier lifetime degradation even in the presence of oxygen (Schmidt et al. 1997) and, as already mentioned, the majority of metallic impurities have higher capture cross-sections for electrons than for holes, resulting in lower detrimental effect on n-type silicon. Because it is not compatible with conventional manufacturing processes, the n-type doped Si is adopted only for the fabrication of high-efficiency solar cells, such as Sanyo (Sakata et al. 2010) or SunPower (Cousins et al. 2010) heterojunction with intrinsic thin layer (HIT) cells. Moreover, less complicated structures with Al diffused rear emitters have demonstrated relevant performances (V_{oc} = 621 mV, J_{sc} = 37 mA/cm^2, FF = 79.5%, Eff = 18.3% on 4 Ωcm n-type doped CZ 280 μm thick; Schmiga et al. 2005).

On the other hand, the material quality nowadays is becoming less critical because the PV silicon wafers are getting thinner, thus increasing the diffusion length-to-thickness ratio, which ensures a high collection probability (Munzer et al. 1998).

15.2.4 Surface Passivation

Surface recombination is the most relevant loss mechanism in a PV device. As the wafer thickness decreases, the diffusion length becomes less relevant with respect to the surface passivation, thus this point must be carefully accounted to achieve high efficiency. Basically, thermally grown SiO_2 layers have been demonstrated to be effective in passivating the c-Si surface; therefore, reducing the effective surface recombination velocity S_{eff} down to 1 cm/s. High-quality SiO_2 has been successfully used in both n-type and p-type c-Si surfaces as in the cell schemes reported by Zhao et al. (1998, 2006). Because SiO_2 is formed at a temperature of approximately 1050°C, usually very thin SiO_2 layers are deposited to reduce the thermal stress of the Si wafer and to avoid diffusion of the emitter dopant. To form the ARC ZnS/MgF_2 are sequentially deposited on thin SiO_2 layers. However, high-quality SiO_2 is far from industrial scale. Therefore, the laboratory scale J_{sc} value of 42.2 mA/cm^2, which almost reaches the silicon limit as shown in Figure 15.1a, strongly decreases to 37 mA/cm^2 in industrial manufacturing process.

The best c-Si cells from Zhao et al. (1998, 2006), indeed, use a double ARC of ZnS/MgF_2. Non-stoichiometric silicon nitride (SiN_x) grown by plasma-enhanced chemical vapor deposition (PECVD) at temperatures below 400°C has been proven to provide an outstanding c-Si surface passivation, which achieves the best performances if SiN_x is silicon-rich (Schmidt et al. 1996; Lauinger et al. 1997). It also can form an effective ARC. Two mechanisms are responsible for the surface passivation by SiN_x films: (i) an interface states (D_{it}) reduction, obtained by the chemical compatibility between SiN_x and Si; (ii) a field effect induced by the positive charge, related to the hydrogen ions' presence. The latter effect induces the accumulation of electrons in the underlying n-type c-Si. Best performances indicate that typical values for SiN_x, values are $D_{it} = 10^{11}$ cm^{-2} eV^{-1} and $Q_f = 10^{11}$ cm^{-2}, which can be compared with typical values obtained using SiO_2: $D_{it} = 10^{10}$ cm^{-2} eV^{-1} and $Q_f = 10^{10}$ cm^{-2} (Aberle 1999). SiN_x has also been demonstrated to be effective in passivating p-type c-Si surface, inducing an inversion layer. Conversely, when SiN_x is used to passivate the rear side of p-type c-Si cell, its beneficial effect vanishes due to the hole injection, which can compensate for the inversion layer (Dauwe et al. 2002, 2003; Glunz et al. 2000a). This problem can be overcome developing SiN_x, the passivation properties of which are more related to interface states reduction than the field effect. An alternative approach to D_{it} reduction is represented by low temperature (<850°C) SiO_2 layers grown via rapid thermal oxidation of the c-Si surface before SiN_x deposition (Schultz et al. 2005). This double layer has been proved to be crucial in achieving a cell efficiency of 20.5% (Schultz et al. 2005). Stacked SiO_2/SiN_x layers have been reported to be able to passivate c-Si surface also forming ARC (Schmidt et al. 2001b). A comparison of effective lifetimes used to evaluate the passivation degree of 1 Ωcm FZ p-type doped c-Si is shown in Figure 15.6 for several SiO_2 thicknesses grown at two different temperatures with and without SiN_x coating.

Very effective c-Si surface passivation can be achieved with an amorphous silicon (a-Si:H) layer (De Wolf and Kondo 2007) deposited at temperatures lower than 200°C. To avoid hydrogen effusion from the film, a low-temperature step is needed to deposit the ARC or, in the case of heterojunction cells, to produce the emitter layer as widely used by Sanyo HIT cells (Schaper et al. 2005; Hofmann et al. 2006). Moderate thermal annealing treatments can also improve the c-Si surface passivation degree (Schulze et al. 2009). On the other hand, a-Si:H layer absorbs high-energy photons thus filtering the cell photocurrent. Nevertheless, silicon carbide (SiC), which has a higher gap with respect to a-Si:H, has also been investigated (Martin et al. 2001).

Low temperature (<300°C) Al_2O_3, obtained not only by plasma-assisted Atomic layer deposition (ALD) but also via metallorganic PECVD, has recently gained great attention since it was demonstrated to be effective in passivating p-type c-Si emitter as well as c-Si wafer due to field effects induced by negative charges related to oxygen content within the nonstoichiometric layer (Hoex et al. 2008; Li et al. 2011). A comparison of different surface passivation degrees is shown in Figure 15.7 in terms of effective lifetime for both p-type (Figure 15.7a) and n-type (Figure 15.7b) c-Si wafers (Dingemans et al. 2011).

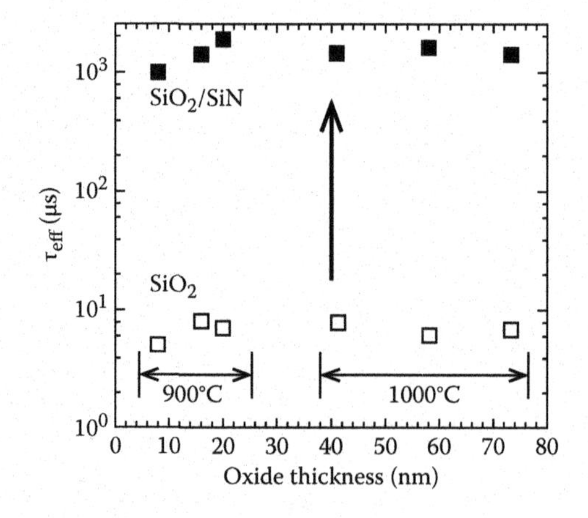

FIGURE 15.6
A comparison of effective lifetimes used to evaluate the passivation degree of 1 Ωcm FZ p-type doped c-Si for several SiO_2 thicknesses grown at two different temperature with and without SiN_x coating. (Reprinted from Schmidt, J. et al. *Semicond. Sci. Technol.* 16:164, 2001. Copyright 2001, IOP Publishing Ltd.)

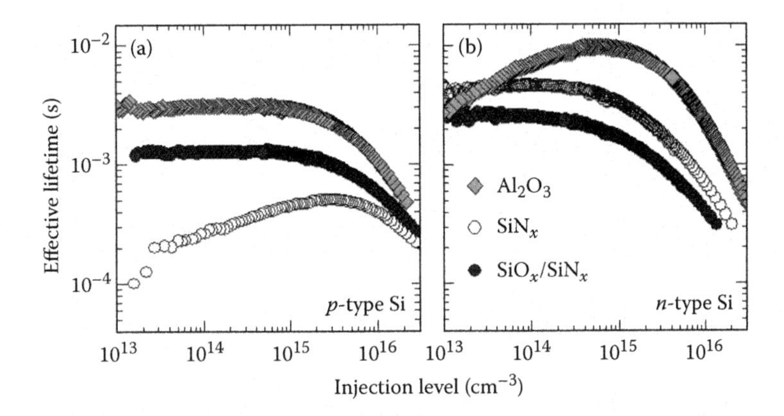

FIGURE 15.7
Comparison of effective lifetime measured on 50/70 nm SiO_x/SiN_x stack (filled circles), 70 nm single layer SiN_x (open circles) and 30 nm Al_2O_3 passivating the surfaces of 2 Ωcm p-type and 3.5 Ωcm n-type c-Si surface in (a) and (b), respectively. (Reprinted from Dingemans, G. et al. *Appl. Phys. Lett.* 98: 222102, 2011. Copyright 2011, American Institute of Physics.)

Except for Sanyo HIT cells, manufacturing industries nowadays mostly use screen-printed and thermally fired Al contacts for p-type c-Si–based cells. Al forms a back surface field (BSF) but induces an unwanted wafer bowing, which can be detrimental due to wafer shrinking. These cells suffer from high rear surface recombination velocity and parasitic absorption of the full area aluminum metallization. In turn, all the high-efficiency cells use dielectric layers as back surface passivation and a rear contact electrode with reduced metallization fractions (point contacts). Two main technologies based on such structures have been presented in the literature: passivated emitter rear locally diffused (PERL) and laser-fired contact (LFC) cells. In the first scheme, starting from p-type c-Si, photolithographic apertures are performed on the oxide layer used to passivate the rear surface of the c-Si wafer and then a boron p^+ diffusion is performed in small regions followed by a metallic contact realization (Zhao et al. 2002). In the second scheme, the passivation layer is covered by Al and then a local Al diffusion is promoted by Nd:YAG pulsed laser (1064 nm; Schneiderlochner et al. 1999). In this approach, different kinds of c-Si surface passivation can also be adopted, such as a-Si:H/SiN_x (Schaper et al. 2005; Tucci et al. 2008a). A comparison among internal quantum efficiency (IQE) of c-Si–based cells fabricated with different solutions for the back electrode is shown in Figure 15.8. Data related to LFC, PERL, and HIT cells were obtained from Zhao et al. (2002, 2006) and Sakata et al. (2010), respectively.

When dielectric layers are used as surface passivation, the right process sequence in cell manufacturing is still under investigation due to thermal instability of surface passivation degree. Indeed, when a dielectric layer is deposited at low temperatures on the cell's rear side, the emitter diffusion should be processed before this dielectric passivation to avoid thermal stress. Otherwise, heterojunction structures would be preferred. In this case, indeed, the BSF can also be obtained at low temperatures using the n-c-Si/i-a-Si:H/n-a-Si:H structures, in which the valence bandgap mismatch between c-Si and a-Si:H allows a good collection of electrons and a mirroring effect for holes.

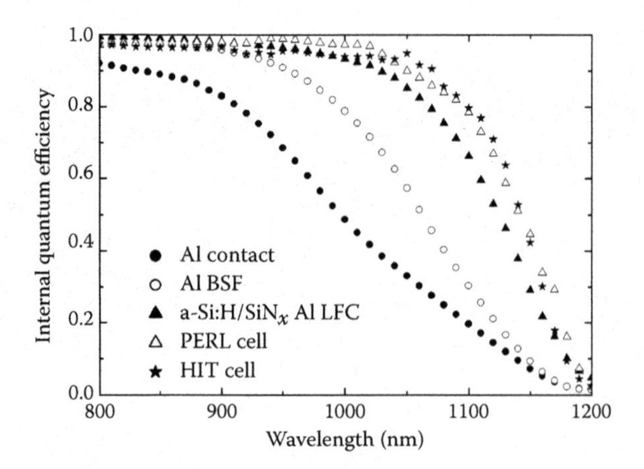

FIGURE 15.8
IQE comparison in the near-infrared spectrum of different back contact solutions. Data related to LFC, PERL, and HIT cell were collected from Zhao et al. (2002, 2006) and Sakata et al. (2010), respectively.

15.2.5 Emitter

Two strategies can be adopted to form an emitter layer on Si wafers: the first, known as homojunction, can be obtained by thermal diffusion of a dopant element that is able to realize an opposite conductivity with respect to that in the bulk; the second, known as heterojunction, is based on the growth of a material compatible with the bulk (i.e., a-Si) but showing opposite doping. Nowadays both strategies, commonly showing the emitter located at the front side of the cell, have been exploited to their limits.

15.2.5.1 Homojunctions

In this case, a very shallow region of the wafer is converted into opposite doping. The higher the doping is, the lower the series resistance is but the filtering effect of the impinging sunlight is stronger, then sharp doping and thin emitter is the main rule. Therefore, a grid-shaped metal contact is needed as the front cell electrode. The contact coverage has a detrimental effect on the open circuit voltage. If the contact coverage is high, a heavy or deep diffusion is needed to shield the bulk from the high surface recombination velocity at the silicon/metal interface under the contact openings. Unfortunately, heavy diffusions have a high emitter saturation density, which also negatively affects the open circuit voltage (V_{oc}). To overcome this as well as the undesired filtering effect, a selective emitter has been proposed in the literature and also investigated for large-area manufacturing processes (Horzel et al. 2000). It consists of high doping under the front metal grid to reduce the related contact resistance and low doping for the rest of the emitter. Then, the uncovered emitter becomes easy to passivate. This concept was demonstrated by the PERL cell (Zhao et al. 1998), shown in Figure 15.9. This cell still shows the highest efficiency for laboratory scale silicon-based solar cells: V_{oc} = 706 mV, J_{sc} = 42.7 mA/cm^2, FF = 82.8%, Eff = 25% (the J_{sc} value is taken on the cell active area; Horzel et al. 2000). It could be interesting to evaluate the losses of such top cell efficiency with respect to the ideal limit. Indeed, 3.6 mA/cm^2 is lost due to long wavelength radiation escape. The series resistance due to front metal fingers, the emitter resistivity and the current-crowding effect near the rear

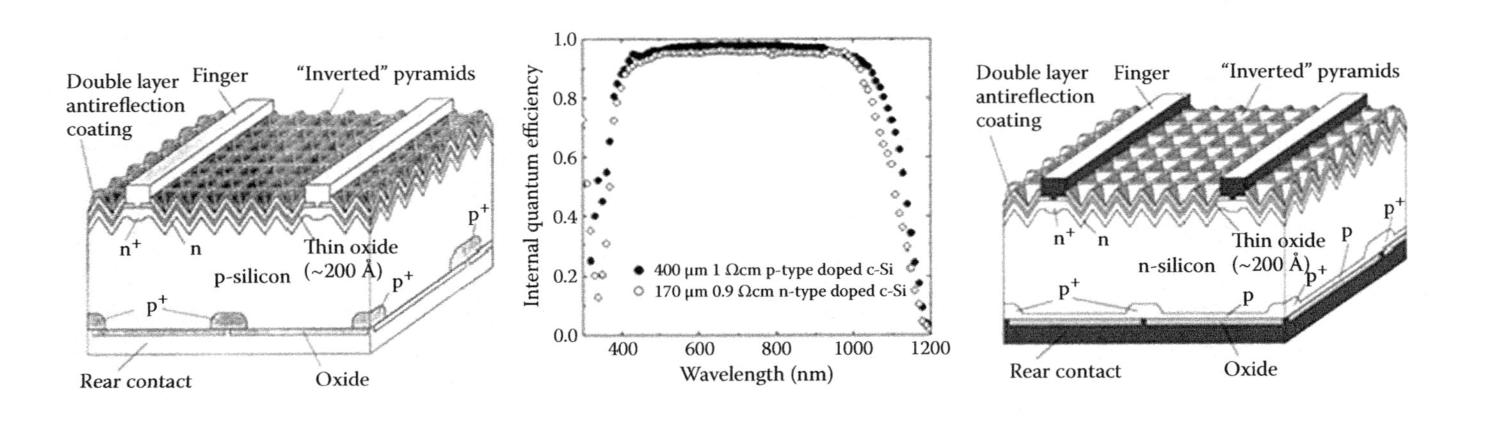

FIGURE 15.9
Schemes of PERL (left) and re-PERT cells (right). In the middle, a comparison between the IQEs of the two cells. (Reprinted from Zhao, J. et al. *Appl. Phys. Lett.* 73:1991, 1998; and Zhao, J. and Wang, A. *Appl. Phys. Lett.* 88:242102, 2006. Copyright 1998 and 2006, American Institute of Physics).

point contacts affect the FF. A Shockley–Read–Hall recombination through defects occurs at the rear and front oxidized surfaces (for lower cell photovoltage) and in the base (for higher cell photovoltage), producing junction ideality factor higher than 1 thus reducing the V_{oc} (Aberle et al. 1992).

To appeal to manufacturing industries, the complexity of the emitter design should be compensated by higher quantum efficiency for lower wavelength photons and then by higher J_{sc}. A bright example of the selective emitter technology in large-area production is currently proposed by Suntech with Pluto cell (V_{oc} = 632 mV, J_{sc} = 38.2 mA/cm^2, FF = 78.8%, Eff = 19%; Shi et al. 2009) and by Innovalight, which is able to enhance the c-Si–based cell efficiency up to 19% by printing a grid-shaped pattern of silicon-based ink over the c-Si wafer. Such ink is able to promote a selective emitter during the emitter diffusion (Innovalight).

As addressed previously, n-type materials do not experience light-induced degradation (LID; Glunz et al. 2001). The main challenges in processing n-type material for solar cells are the formation of p-type emitters by boron diffusion and the passivation of these highly doped p-type emitter regions. Boron diffusion requires much higher temperatures than the conventional phosphorus diffusion, making the simultaneous formation of the emitter and BSF a challenge. Furthermore, the passivation of highly doped layers is not straightforward. One of the best laboratory scale n-type–based solar cells is the rear emitter-passivated emitter rear totally diffused (re-PERT) silicon solar cell. Phosphorus has been slightly diffused into the front surface of this cell, as shown on the right side of Figure 15.9, to generate a front surface field improving the surface passivation and cell stability. The BBr_3 boron diffusion with very low surface damage has been used to form the cell emitter over the entire rear surface. Both front and rear surfaces have been further passivated by high-quality thermally grown SiO_2. The re-PERT cell (Zhao et al. 2006) fabricated on FZ-Si (1.5 Ωcm, 170 μm, 22 cm^2) provided an efficiency value of approximately 22.7% (V_{oc} = 702 mV, J_{sc} = 40.1 mA/cm^2).

A simple technology in large-area (148.9 cm^2), fully screen-printed n-type rear junction, shallow front surface field, SiN_x passivated solar cell with an efficiency of 17.4% (V_{oc} = 636 mV, J_{sc} = 35.1 mA/cm^2) is reported in Froitzheim et al. (2005). Among n-type Si-based solar cells using an aluminum alloyed rear side emitter, Glunz et al. (2004) reported an efficiency of 19.4% (V_{oc} = 646.5 mV, J_{sc} = 39.8 mA/cm^2) for a laser-fired local Al emitter on 100 Ωcm FZ-Si (Figure 15.10). The rear surface is covered with a thermal oxide and a 2-μm-thick aluminum layer. The back-junction is created by local laser-firing of the aluminum through the oxide layer resulting in a local p^+ emitter.

The performances of these cells are still limited by the damage introduced by the laser process. In the case of n-type cells, in fact, the local Al profile acts as an emitter and the

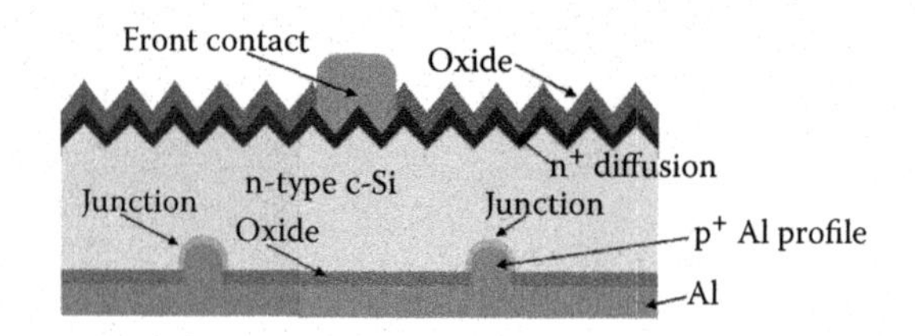

FIGURE 15.10
Structure of the n^+np^+ laser-fired local Al emitter cell.

quality of this junction and that of the nearby area has a stronger effect on the device's performance than in the case of p-type cells, in which the local aluminum firing acts as a local BSF point.

The simplest way to fabricate n-type back-junction solar cells on an n-type c-Si substrate is the use of full-back aluminum, commonly used as back contact for p-type Si-based cells, as the p^+ emitter in Figure 15.11, in which the structure developed at Fraunhofer ISE (Glunz et al. 2004; Schmiga et al. 2009) is depicted. The front surface field is slightly and uniformly doped. A special contact front paste, able to contact such weakly doped emitters, is necessary. A special aerosol silver paste inkjet system has been developed for this purpose (Mette et al. 2007). With a proper passivation of the cell's rear side, the efficiency was increased, reaching values of approximately 19.5% with a-Si/SiO_x and 20.1% with Al_2O_3/SiO_x rear passivation stacks (Schmiga et al. 2008). A further approach based on the formation of a front-side boron diffusion emitter could also lead to high performances associated with a low-cost process. The potentialities of such an architecture were recently highlighted by Benick et al. (2010), with an efficiency of 23.9% (V_{oc} = 703.6 mV, J_{sc} = 41.2 mA/cm^2) on FZ-Si (1 Ωcm, 4 cm^2). A negative-charged dielectric Al_2O_3 was applied as the surface passivation layer on p^+ front emitter and a V_{oc} value of 703.6 mV was achieved. Figure 15.12 shows the structure of such a cell, similar to that of a PERL device with no selective emitter. For PERT solar cells with a front-diffused boron emitter, Zhao et al. (2006) reported an efficiency of 21.9% on 0.9 Ωcm FZ-Si.

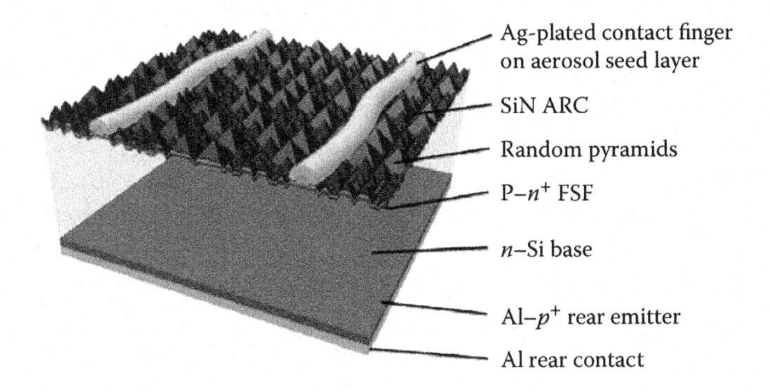

FIGURE 15.11
Structure of near-industrial n^+np^+ on n-type doped c-Si solar cell with screen-printed Al alloyed as rear p^+ emitter. (Reprinted from Schmiga, C. et al. *Proc. 24th EUPVSEC, Hamburg, Germany*, 1167, 2009. With permission.)

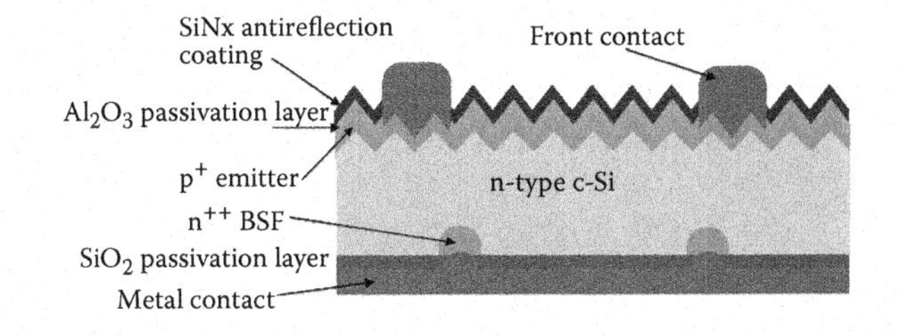

FIGURE 15.12
Cell on n-type c-Si with front boron diffusion and rear local diffusion.

The application of large-area (239 cm^2) n-type CZ-Si in industrial scale solar cells has been developed by Burgers et al. (2010), who reached an efficiency of 18.65% (V_{oc} = 635 mV, J_{sc} = 37.5 mA/cm^2). The front side has a boron diffusion of 60 Ω/square, covered with suitable SiN_x ARC. On the back side an n^+ phosphorus diffused layer acting as a BSF, coupled with a SiN_x as passivation layer and an ultrathin 1.5 nm silicon dioxide layer between the emitter and the silicon nitride ARC film. Such ultrathin oxides are grown at room temperature by soaking the silicon wafers in a solution of nitric acid before the deposition of the silicon ARC film. The open rear structure can be used for bifacial modules, as a consequence, further enhancing the absorption of the cell.

15.2.5.2 Heterojunctions

The choice of heterojunction structures instead of high-temperature emitter diffusion to form a homojunction shows several advantages: (i) a high open circuit voltage due to the potentiality of heterojunction and due to the good silicon surface passivation promoted by the a-Si:H layer, which results in high cell efficiency, as demonstrated not only by Sanyo but also by other competitors (Sakata et al. 2010); (ii) an outstanding stability with respect to a-Si:H thin film cells because the a-Si:H layer in the heterojunction technology are extremely thin, thus avoiding the Staebler–Wronski degradation effect, which still affects a-Si:H devices (Staebler and Wronski 1977); (iii) a better response as a function of the operating temperature with respect to conventional silicon-based cells (−0.3%/°C vs. −0.5%/°C; Taguchi et al. 2000); (iv) a lower cost related to the lower thermal budget and the reduced time needed for cell manufacturing; and (v) the use of very thin c-Si wafers (Henley et al. 2008) that can strongly reduce the PV cost.

Nevertheless, several drawbacks still affect the market breakthrough of this technology: (a) problems in the optimization of the cleaning procedure for the c-Si wafer as well as the a-Si:H passivation layer, these issues strongly affect the device's performance; (b) n-type c-Si wafers presently have higher prices than p-type c-Si. A heterojunction on p-type c-Si does not reach efficiency comparable to that of n-type c-Si–based heterojunctions, thus making possible application in a production line disadvantageous. In 1994, Sanyo overcame the 20% efficiency boundary with a heterojunction solar cell based on n-type CZ-Si textured absorber (Sawada et al. 1994; Taguchi et al. 2000), and in 2009, achieved a 23% efficiency (Sakata et al. 2010; Taguchi et al. 2009). A sketch of the Sanyo HIT cell is shown in Figure 15.13.

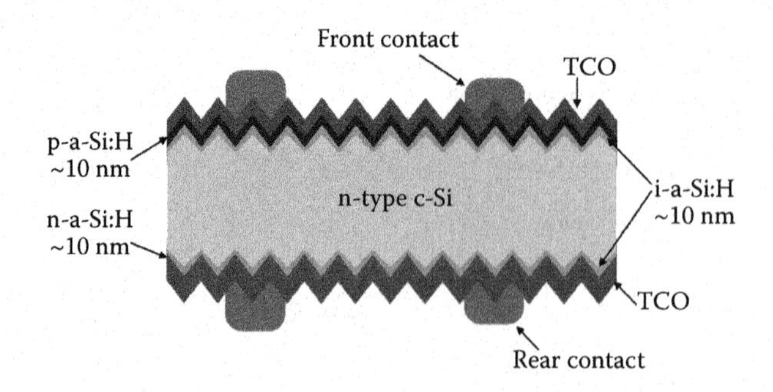

FIGURE 15.13
Schematic view of the Sanyo HIT cell (Sanyo).

15.2.6 Internal Reflection and Texturing

Besides the rear surface passivation, a further important concern that has to be considered is the internal reflectivity of the cell, with the aim of enhancing photon absorption. The increase in the measured reflectance to more than 1040 nm is due to silicon increasing transparency at these wavelengths. In fact, after having performed at least one double-pass through the PV device, a considerable fraction of the light entering the cell escapes through the front surface. The common c-Si/Al contact shows very low reflectivity, thus reducing the cell J_{sc}. Conversely, dielectric passivating layers covered by metal contacts and patterned similar to the LFC or PERL structures show better performances that can lead up to 95% of internal reflection. It should be remarked that the reflection-related losses increase when thin wafers are used (Green 1998). As an example, the PV parameters of a simulated heterojunction cell as a function of the silicon wafer thickness, normalized to the initial value, are shown in Figure 15.14. The V_{oc} enhancement is related to the recombination reduction along the wafer, which induces higher built-in voltage when the wafer thickness decreases. In turn, the thinner the wafer, the lower the photocurrent is. It is apparent in Figure 15.14 that the photocurrent reduction is stronger than the photovoltage enhancement, thus affecting cell efficiency, which decreased as the wafer became thinner. These results are in good agreement with those reported by Sanyo (Taira et al. 2007). It should therefore be noted that the use of thinner wafers ensures a cost reduction due to the reduced material quantity required for device production, but also induces a cell efficiency reduction. A possible way to overcome this undesired condition is the exploitation of the mirroring effect, consisting of dielectric Bragg reflectors located on the rear side of the cells, which allow an internal reflection as high as 97% depending on the number of coupled dielectric films adopted in the dielectric Bragg reflector structure (Tucci et al. 2009). The evolution of PV parameters as a function of the wafer thickness in the presence of dielectric Bragg reflectors is also shown in Figure 15.14. In the case of such high reflection, the cell efficiency is almost constant for a wafer thickness in the range between 200 and 60 μm. Several structures have been proposed and compared with the aim of enhancing the internal reflection of high-efficiency c-Si–based cells (Hofmann et al. 2006).

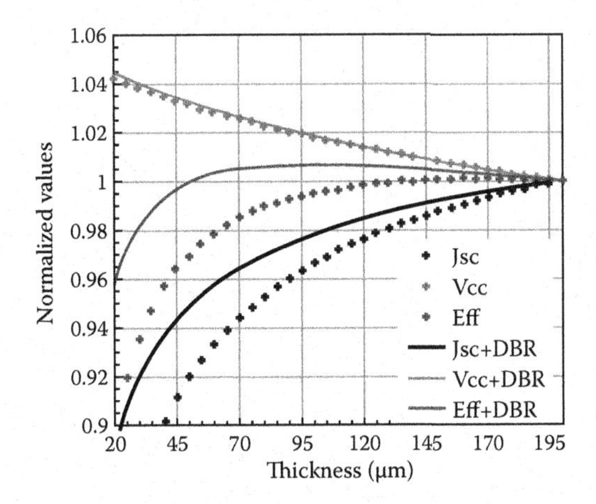

FIGURE 15.14
Normalized photovoltaic parameters of the heterojunction cell as a function of the wafer base thickness with and without mirroring effect. (Reprinted from Sakata, H. et al. *25th EUPVSEC*, 1102–1106, 2010. With permission.)

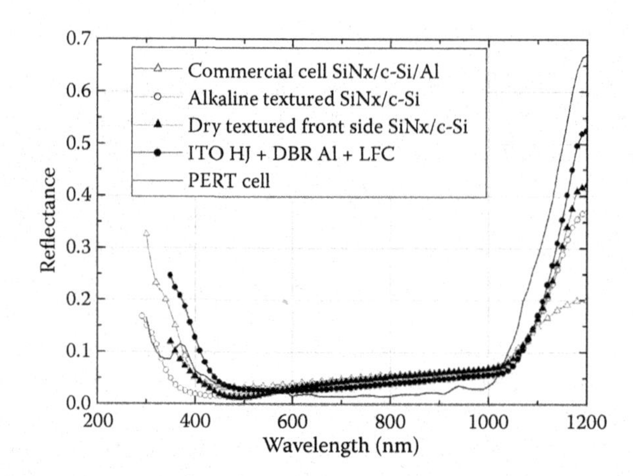

FIGURE 15.15
Comparison among cell reflectances as measured on different texturing and rear side contacts.

An effective strategy for a suitable light trapping is the realization of a randomized textured front surface and a flat rear surface (Campbell et al. 2001). To this aim, different treatments of the two cell surfaces are needed, as can be deduced from the PERL cell scheme depicted in Figure 15.9, in which the rear side of the cell is almost flat whereas the front side of the cell is textured by well-ordered inverted pyramids. In the case of laboratory scale PV devices, specific and expensive techniques can be used to obtain the best texturing, whereas in the case of large-area production, pyramidal textures based on intersecting <111> equivalent crystallographic planes etched from a <100> starting surface are commonly used for c-Si. These pyramids show average sizes of approximately 10 μm on both sides of the cell. It should be, however, remarked that a double texturing can provide a suitable light trapping only in the case of well-ordered and spaced pyramids. Otherwise, the light reflected from the rear surface of the cell undergoes multiple reflections before being transmitted back across the device, thus reducing the reflectance of the rear surface (Campbell et al. 2001). As a result of the common c-Si surface texturing and ARC treatments, the front surface reflectance remains approximately constant at long wavelengths but increases considerably in the blue region of the visible spectrum. The comparison between the front surface reflectance and the AM1.5G spectrum indicates that an extra 5% of current would be available for the use of an ideal ARC such as $MgF_2/ZnS/SiO_2$. At present, dry etching techniques that are able to perform single-side randomized textured surfaces are under development. The main challenge concerns the passivation degree of the surface after the texturization treatment (Moreno et al. 2010; Agarwal et al. 2011). A comparison of cell reflectances, as measured on different texturing and rear-side contacts, is shown in Figure 15.15.

15.2.7 Metal Contacts

The screen printing process is the most common technique used in Si-based industrial cell production. Silver and aluminum are used as front and rear cell electrodes, respectively. Despite the poor gridline aspect ratio and line resistance, the run-up in the price of Ag is presently focusing attention on the development of different approaches for metal contacts. An interesting solution for large-area production based on a laser-grooved buried contact cell scheme has been proposed by the National Renewable Energy Centre. The high-doped, laser-grooved trenches are filled with a Ni–Cu–Ag multilayer instead of being screen-printed (Bruton et al. 2003). The Ni film ensures a very low resistance contact, the Cu layer ensures high line conductivity, and the Ag layer caps the Cu layer and enhances line conductivity, allowing an easy soldering process. For the thickening of the Ni seed layer, light-induced plating was developed to exploit the PV effect of the solar cell, obtaining the bias needed in the solution for electroplating (Mette et al. 2006). In the laser-grooved buried contact scheme, no metal adhesion problems arise because the metal contacts are buried. In turn, the adhesion problems of plated contacts still need a valid solution, even if some new ideas involving porous silicon are under development. Aside from the screen-printing technique, metal aerosol inkjet is a further solution suitable for the deposition of metal layers with reduced aspect ratios and the front grid shadowing effect (Mette et al. 2007).

15.2.8 Back Contact Cells

Once the recombination processes have been limited, the main limitation affecting the short-circuit current is the front grid shadowing of the cell area. Although the PERL cell shows reduced shadowing as low as 3.5%, a front contact grid pattern with bus bars on commercial solar cells leads to surface shadowing values of 7% to 8%. Such losses have been strongly reduced by SunPower, which developed the interdigitated back contact (IBC) solar cells initially proposed by Lammert and Schwartz (1977), in which the sunward side is completely exposed, thus maximizing absorption. Both the emitter and base contact, produced with alternate diffusion processes at high temperature, are located at the back side following an interdigitated pattern.

The SunPower IBC key features consist of an n-type doped region to form the front surface field, a SiO_2 passivation layer to reduce surface recombination losses, a backside mirror to reduce back light absorption and to induce light trapping, backside gridlines to reduce shadowing, a high coverage metal to reduce resistance losses, and localized n^+ and p^+ contacts to reduce contact recombination losses. The most important dimensional issues are the contact coverage fraction, the size of the base diffusion, and the pitch of the diffusions.

The only solution is a size limitation (by as much as possible) of the contact openings and a careful optimization of the backside diffusions. The effect of the diffusion pitch, defined as the distance between two base diffusions, has a large effect on the internal series resistance. When the pitch is larger than the wafer thickness, the majority carriers above the emitter drift laterally and are collected by the base diffusion. When the pitch decreases, the internal resistive losses are reduced, whereas recombination losses above the base diffusion increase. When the width of the base diffusion is larger than the wafer thickness, the minority carriers generated above a base diffusion will diffuse laterally to be collected by the nearest emitter diffusion. This diffusion current is possible only if a gradient of minority carriers is present, this phenomenon raises the minority carrier concentration above the base diffusions. To limit these losses, the width of the base diffusion must be made as narrow as the printing technology allows and the area coverage of the base diffusion must be kept at a minimum value. This requirement unfortunately clashes with minimizing the series resistance losses generated by the lateral transport of the majority carriers. A compromise must be found to minimize the overall losses. The optimum pitch is a trade-off between the lateral resistive losses and the additional recombination over the base diffusion. The SunPower record cell, based on an n-type FZ wafer, shows 24.2% efficiency (V_{oc} = 721 mV; J_{sc} = 40.5 mA/cm^2; FF = 82.8%; area = 155.1 cm^2; Cousins et al. 2010). With this kind of cell, SunPower fabricates modules with 21.5% record efficiency. The schematic structure of SunPower record solar cell is illustrated in Figure 15.16.

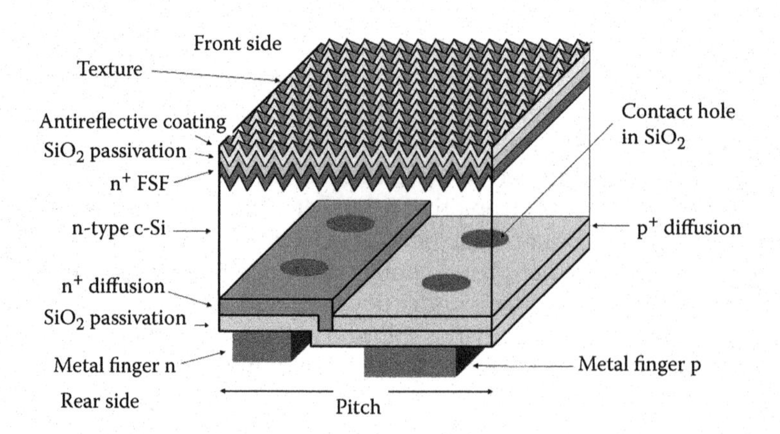

FIGURE 15.16
Schematic structure of the SunPower record solar cell.

Nowadays, the laser-grooved solar cell technology is largely used for the implementation of the rear junction backside contact solar cells. Laser-grooved interdigitated backside-buried contact solar cells with an efficiency of 19.2% (V_{oc} = 664 mV; J_{sc} = 37.9 mA/cm²), whose structure is illustrated in Figure 15.17, has been realized by Guo et al. (2005).

Recently, a very interesting solar cell was presented by Woehl et al. (2011). The device is an all-screen–printed back-contact, back-junction solar cell with aluminum-alloyed emitter made on 190-μm-thick 10 Ωcm FZ-Si. Using an emitter coverage of 58% with respect to the total 25 cm² area, a relevant efficiency of 19.7% has been achieved (V_{oc} = 641 mV; J_{sc} = 38.8mA/cm²; FF = 79.3%). A schematic cross-section of this cell is shown in Figure 15.18.

Unfortunately, all these technologies can be applied only to high-quality c-Si; therefore, the cell cost still remains higher than that of standard cells. To overcome this problem, a back-contacted cell technology called metal wrap-through (MWT) has been proposed (Guillevin et al. 2010). Here, a collecting emitter on the cell front side makes the device less sensitive to low diffusion length (Kray et al. 2001). To reduce the shadowing of the front metal grid, the emitter has been collected to the rear side of the cell. The cell structure includes a boron emitter and an open rear side metallization suitable for thin wafers. The passivation process of the highly doped boron emitter is obtained by industrial equipments and provides an outstanding passivation quality of industrial emitters. The solar cells have been prepared from 200-μm-thick CZ-Si wafers (238 cm²). The process gives an efficiency of 18.7% (V_{oc} = 638 mV, J_{sc} = 38.3 mA/cm²). Presently, the fill factor and the series resistance are the main limiting factors to the overall performances of n-type MWT solar cells. A sketch of the MWT cell is shown in Figure 15.19. A similar process has been developed on p-type doped large-area FZ-Si leading to conversion efficiencies of up to 20.2% (Thaidigsmann et al. 2011). The MWT

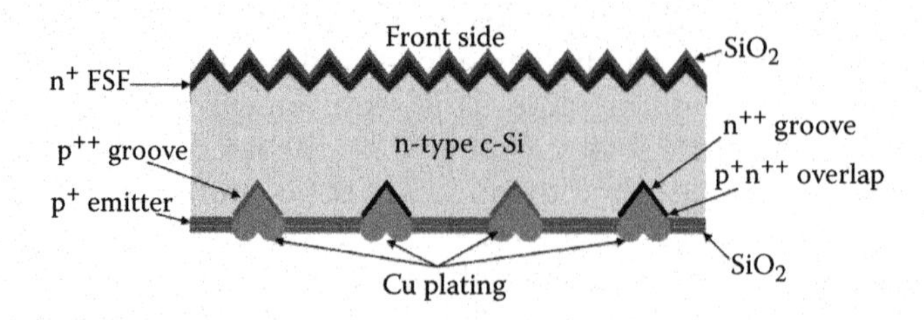

FIGURE 15.17
Cross-section of the n-type interdigitated backside buried contact solar cell.

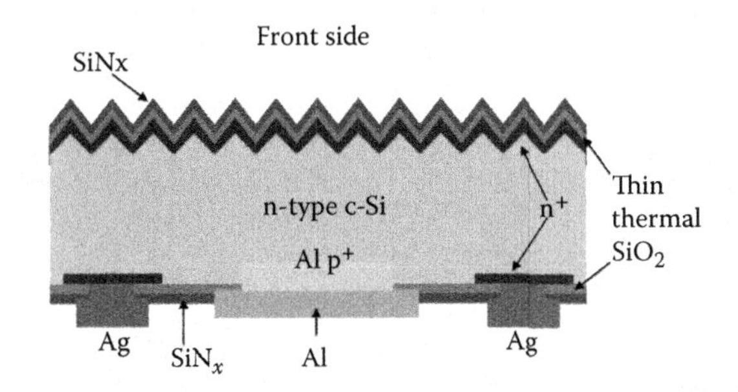

FIGURE 15.18
All-screen–printed back-contact, back-junction silicon solar cell with aluminum-alloyed emitter.

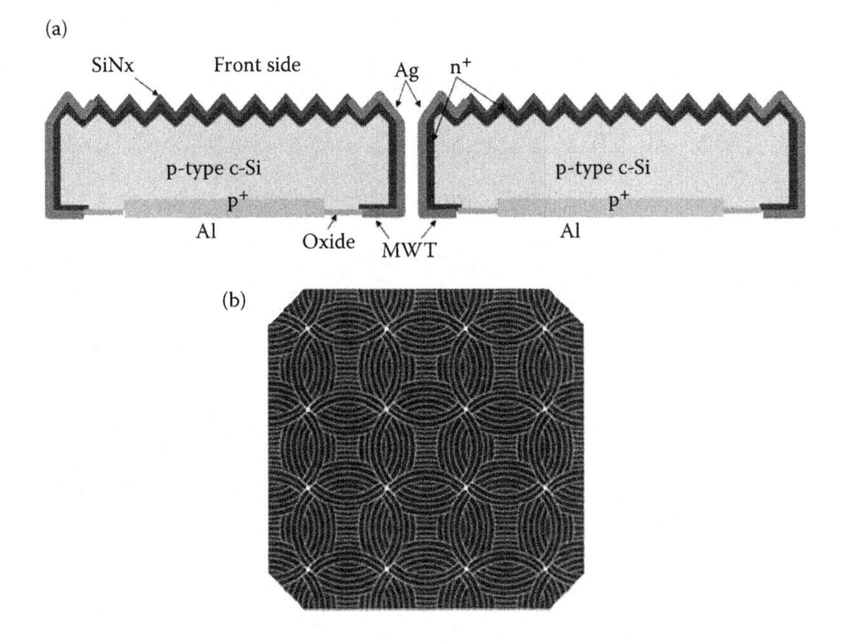

FIGURE 15.19
MWT cell structure: (a) cell cross section; (b) cell front side. (Reprinted from Guillevin, N. et al. *25th EUPVSEC, Valencia, Spain*, 2010. With permission.)

is relatively easily adapted to existing manufacturing cell production lines. These cells have been used to compose the Pin Up Module recently developed by the Energy Research Centre of the Netherlands (Bultman et al. 2001). In EWT cells, the front side metal grid has been completely transferred to the rear side using VIA holes (Gee et al. 1993). Due to the limited lateral conductivity of the front emitter, more VIAS are needed with respect to the MWT structure. The EWT cell sketch is shown in Figure 15.20. The best efficiency is 18.8% (V_{oc} = 617 mV; J_{sc} = 40 mA/cm^2; FF = 76.4%; area 16.67 cm^2; Mingirulli et al. 2011a).

To further improve the cell efficiency, a challenging idea is to apply the heterojunction concept to the IBC cell, to profit from the low-temperature process, which allows the use of thin c-Si wafer as well as high V_{oc} as demonstrated by Sanyo's record. Taking into account the high FF that can be obtained by IBC cells, it is possible to consider an IBC heterojunction cell as one of the most attractive and promising technologies for c-Si–based solar cell production aimed at reaching up to 26% efficiency (Diouf et al. 2008, 2009). The a-Si:H doped layer conductivity,

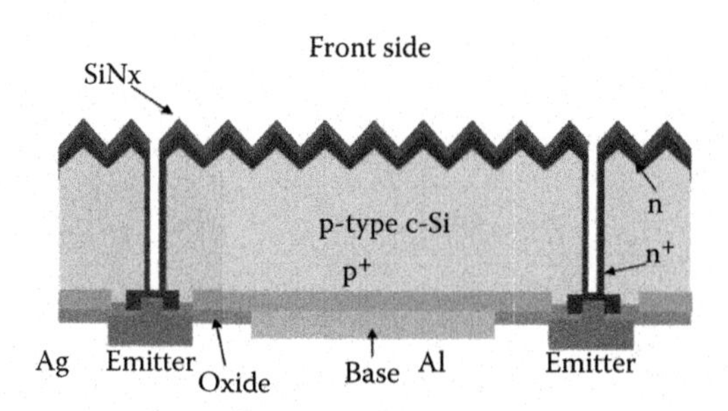

FIGURE 15.20
EWT cell structure.

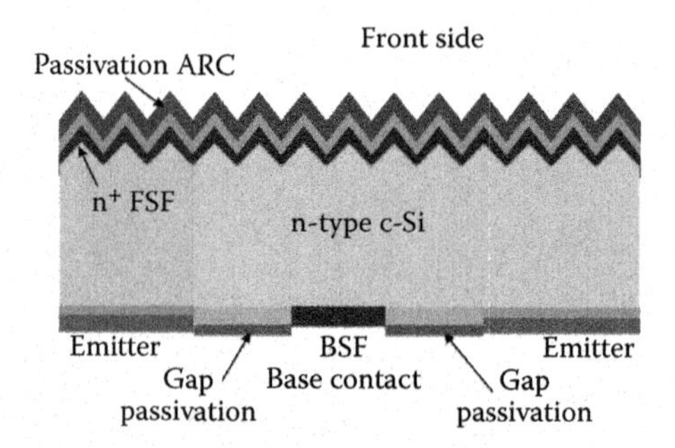

FIGURE 15.21
Heterojunction-IBC record cell structure.

the gap passivation between the two interdigitated contacts, and the patterning and pitch between the contacts have been carefully investigated to achieve high efficiency (Lu et al. 2011). Moreover, the possibility of avoiding high temperature and photolithographic steps has been exploited (Tucci et al. 2008a,b). At present, 20.2% (V_{oc} = 673 mV, J_{sc} = 39.7 mA/cm^2, FF = 75.7%) is the highest efficiency result for IBC heterojunction (Mingirulli et al. 2011b). It has been achieved on a 1-cm^2 device area, fabricated with photolithography as well as with some high-temperature steps. Starting on an n-type 3 Ωcm FZ substrate, the front side is textured and phosphorus diffused to form a front surface field, it is then passivated with silicon oxide and silicon nitride. The back side has 60% area coverage by emitter strips (p-type), 28% BSF strips (n-type), and 12% of gap space between that is passivated by a SiO_2/SiN_x stack, with pitch in the range of millimeters. The metal contact is aluminum for both emitter and BSF regions. A schematic section of the IBC heterojunction cell is depicted in Figure 15.21.

15.3 Multicrystalline Silicon Solar Cells: Standard Processes and Trends

Radovan Kopecek and Michelle McCann

15.3.1 Introduction: Dominance of Multicrystalline Silicon Technology

As discussed in the previous section, the PV market is dominated by crystalline silicon (c-Si)—85% of all solar cells produced in the last 10 years were made with c-Si technology. Among c-Si, multicrystalline (mc)-Si technology is the more commonly used material.

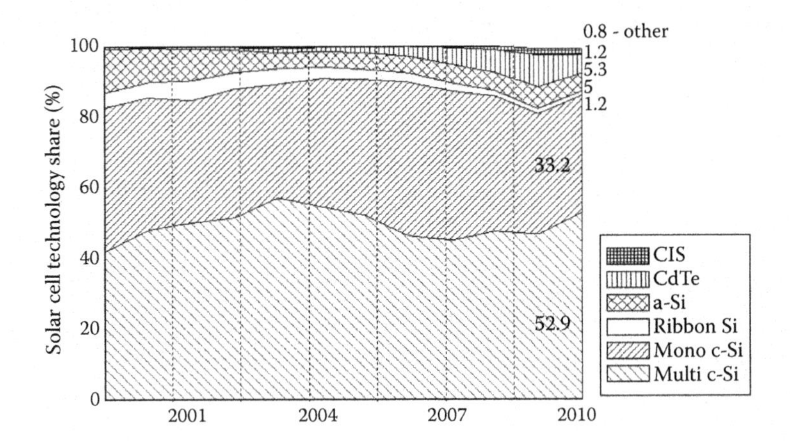

FIGURE 15.22
Shares in solar cell technologies from the years 2000 to 2010. The numbers shown on the right-hand side are for 2010. (Reprinted from *PHOTON International*, 2011. With permission.)

Figure 15.22 shows the market share held by the different solar cell materials from 2000 until 2010 (Photon 2011). We can see that every second module on the market is based on mc-Si technology. This trend is ongoing and mc-Si modules will continue to make up a large portion of the market in the future.

This dominance is due to a very cost-effective and simple solar cell processing technique that has been developed over the years to yield low costs per Watt peak (W_p). More details on the processing techniques as well as in the most important milestones and breakthroughs for mc-Si solar cells are reported in Budhraja et al. (2010). In the following sections, we will summarize the current technological status and the properties of mc-Si wafers and solar cells. In addition, we will reveal some trends in this technology that may gain importance in the future, further lowering the costs and increasing module efficiency.

15.3.2 Crystallization

The production of mc-Si wafers is characterized by a fast crystallization process and the ability to produce large ingots. Although 50% of the material is subsequently lost during wafering, this is still the best technology for producing large-area, fully square wafers with a relatively high yield.

15.3.2.1 Crystallization Techniques for mc-Si Ingots

mc-Si wafers are sawn from 300 kg ingots grown using directional solidification techniques. The wafers typically have an area of 156 × 156 mm^2 and a thickness of approximately 200 μm. Figure 15.23 shows an ingot and some of the important properties.

Once an ingot has been formed, it is cut into 25 bricks and the bricks are sawn into approximately 500 wafers. The properties of the Si ingots are dependent on the crystallization conditions, including the purity of the starting material and the growing speed. On the right of Figure 15.23, we see the typical properties of a directionally solidified, boron-doped mc-Si ingot as a function of height. Due to a segregation coefficient of boron that is close to 1, the resistivity is relatively homogeneous within the entire ingot and is typically approximately 1.5 Ωcm. The metal impurity concentration is lowest in the center of the

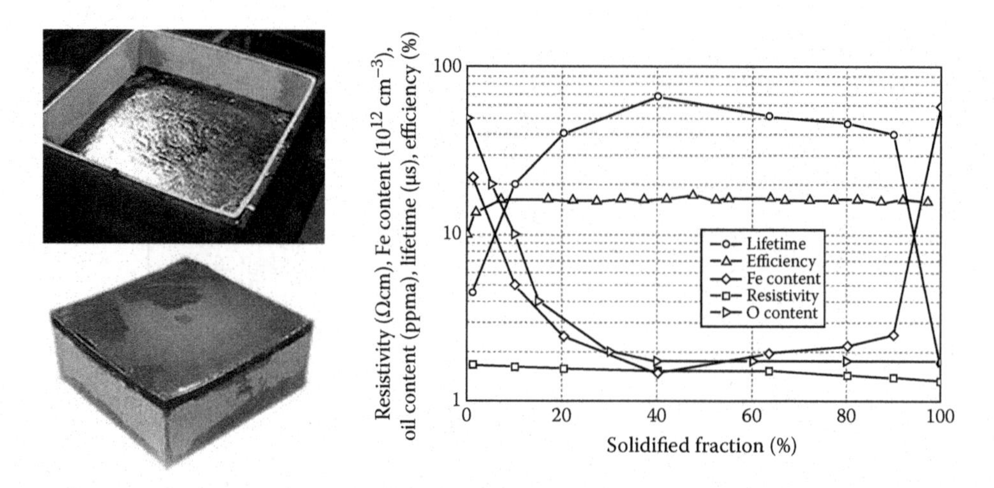

FIGURE 15.23
Mc-Si ingot (left) and its properties as a function of solidified fraction (right).

ingot and highest at the bottom (0%) and top (100%). The high concentration on the bottom is due to an in-diffusion of metals from the crucible, and the high concentration on the top is due to a high segregation of metallic impurities (MacDonald et al. 2005). Conversely, the lifetime of minority charge carriers is lowest at the top and bottom of the ingot because metal impurities (e.g., Fe_i) serve as recombination centers and reduce the material lifetime. A similar trend can be observed for the solar cell efficiency but it is much less marked. The "smoothing" of the curve occurs because on the one hand, some metal impurities can be removed (gettered) by the solar cell process, and on the other hand, the solar cell process itself limits the efficiency.

15.3.2.2 Cutting of and Defects in mc-Si Material

As already mentioned, the ingot is cut into 25 bricks. Figure 15.24 depicts a box with such bricks missing at the edge, middle, and corner.

FIGURE 15.24
Bricks from one ingot (left) and lifetime measurement on a brick from the edge (right).

A few centimeters from the bottom and top of the ingot have been already removed. These parts will be recycled and will enter a new crystallization process. On the right-hand side, a lifetime scan of an edge brick is shown. Light (and dark) regions show areas where the lifetime of minority charge carriers is high (and low). The low lifetimes of the dark regions originate from the in-diffusion of the impurities from the crucible during the crystallization (cooling) process. As the bottom of the ingot solidifies first, the impurities have longer to in-diffuse from both the bottom and the edges of the crucible. Such lifetime pictures are a good measure for cutoff criteria of low lifetime, and therefore low cell efficiency regions.

The cut bricks are then glued to a bar and mounted into a wire saw in which 200-μm-thin wafers are cut as shown on the left in Figure 15.25. Approximately 500 wafers are cut per brick and they are then washed and separated. Some saw damage results from this process, but this is later removed in the solar cell process.

Figure 15.26 depicts two spatially resolved pictures of the same wafer. On the left, the lifetime is shown, in which the bright regions represent regions with high lifetimes. On

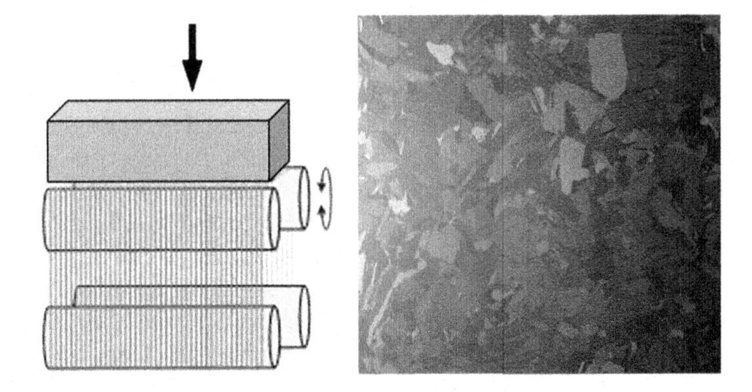

FIGURE 15.25
Schematic drawing of a wire saw (left) and top view on a mc-Si wafer (right).

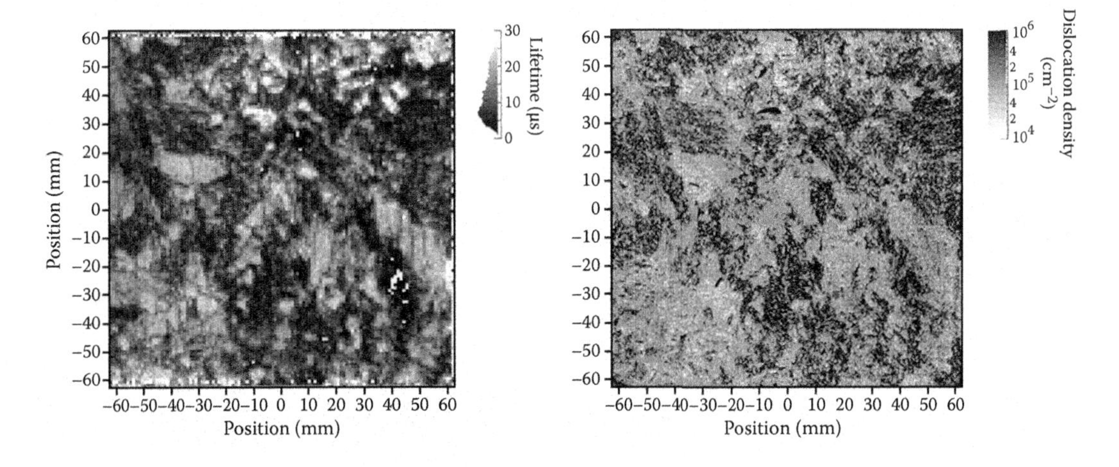

FIGURE 15.26
Laterally resolved lifetime (left) and dislocation density (right) measurements of a mc-Si wafer. (Reprinted from Bentzen, A. et al. *J. Appl. Phys.* 99:093509, 2006. Copyright 2006, American Institute of Physics.)

the right, the dislocation density map reveals that such regions have the lowest dislocation density.

We can generally state that the magic of a good mc-Si wafer is to have a crystal with low dislocation density. If this is the case, any metallic impurities can be removed later by the solar cell process. More about mc-Si material properties can be read in Cuevas and MacDonald (2005).

15.3.3 Standard Solar Cell Process

In this section, we will describe the current standard mc-Si solar cell process. Individual processing steps will be discussed in more detail later.

15.3.3.1 Process Flow and Results

Figure 15.27 shows the most important steps in the mc-Si solar cell process. The sequence starts with an acidic surface texture from both sides of the mc-Si substrate. This can be done using an inline apparatus or in a batch process. The texture reduces the reflectance and removes any saw damage from the surface of the wafer, so that it can be better passivated by the dielectrics that are applied later. Before entering the P-diffusion furnace, the wafer is cleaned in solutions of HCl and HF. HCl removes metals from the surface and HF strips the thin native oxide, which may also contain some metals. The wafers are dried, loaded in a quartz boat, and in most cases diffused on both sides. During this high-temperature step (typically done at ~860°C), a phosphorus glass is formed. This is removed by an HF dip, after which the wafers are immediately loaded into the graphite boat of a PECVD SiN_x furnace. A SiN_x dielectric is then deposited at a temperature of approximately 450°C. Because this is a one-sided deposition, after this step, the "sunny side" of the wafer is defined. Metal pastes are then screen-printed on the front and rear side. On the front side, silver paste is printed as can be seen in Figure 15.27 on the right. Depending on the design of the solar cell, two or three bus bars are used. These are then soldered to the neighboring cell during the process of making the module. The rear side is almost fully covered by an aluminum paste with just some openings opposite the front bus bars where silver/aluminum paste is printed. These enable the rear side to be contacted using a solder. The pastes

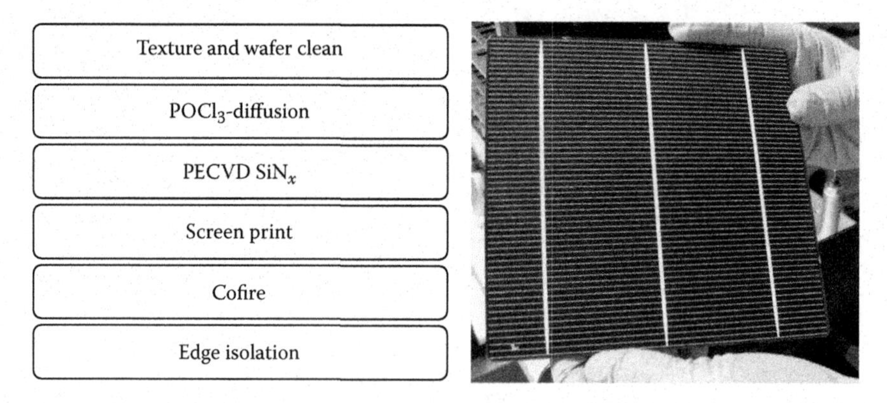

FIGURE 15.27
Process flow chart (left) and textured mc-Si solar cell with three busbars (right).

on the cell are dried and then fired in a belt furnace to obtain a good electrical and physical contact on both sides. As the last step, a laser is used to cut a groove approximately 0.5 mm from the edge of the cell. This process is called "edge isolation" and serves to separate the emitter from the rear contact. Alternatively, this edge isolation can be done directly after the $POCl_3$ diffusion in an inline wet chemical apparatus. In this case, the wafers float on an acidic solution and the rear side is etched.

15.3.3.2 Single Processing Steps

In the following paragraphs, the single steps from the flow chart in Figure 15.27 are described in more detail.

15.3.3.2.1 Surface Texture

The most widely used method for texturing mc-Si substrates is to immerse them in an acidic wet chemical bath consisting of HNO_3 and HF that is cooled to 6°C. To achieve a homogeneous, and therefore effective texture, the substrate has to be used as-cut because the etching process is triggered at regions with large saw damage. The etching is faster in regions with a high dislocation density. Therefore, for material with a high dislocation density, a less aggressive etching solution has to be used (Figure 15.28).

15.3.3.2.2 $POCl_3$ Diffusion

The diffusion is carried out in a tube furnace that is flooded by N_2, O_2, and $POCl_3$ gas at atmospheric or low pressure (Figure 15.29). In the first stage, a phosphorus silicate glass is formed on the wafer surface. This is then in-diffused at temperatures of approximately 860°C resulting in a sheet resistance of approximately 70 to 80 Ω/sq.

Depending on diffusion time and temperature, the profile can be tailored for specific needs. There are many tricks, for example, to reduce the P-atom concentration of the surface and drive in the active P-atoms more deeply. After diffusion and before the PECVD SiN_x deposition, the phosphorus silicate glass is removed by a short (~1 min) HF dip in a dilute (2%) HF solution.

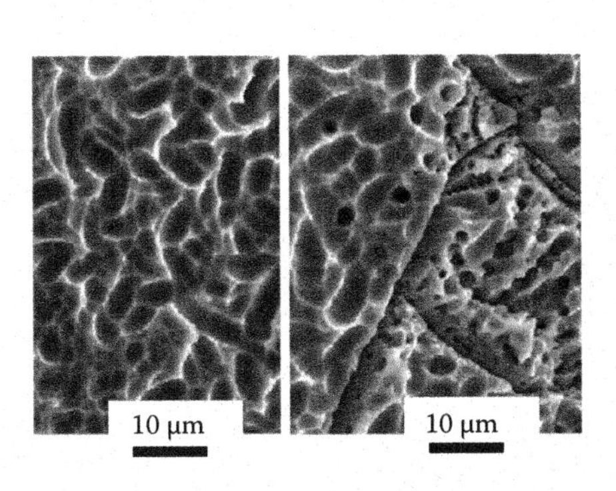

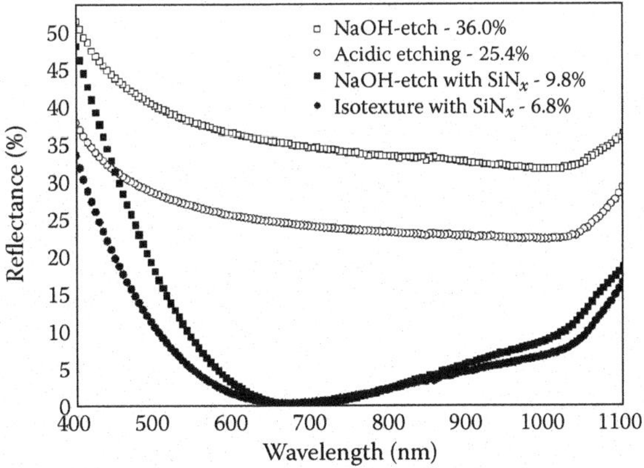

FIGURE 15.28
Isotexture of mc-Si surface on a single grain and on grain boundaries (left). Reduction of reflectance of isotextured surface compared with a NaOH etched surface before and after SiN_x deposition (right).

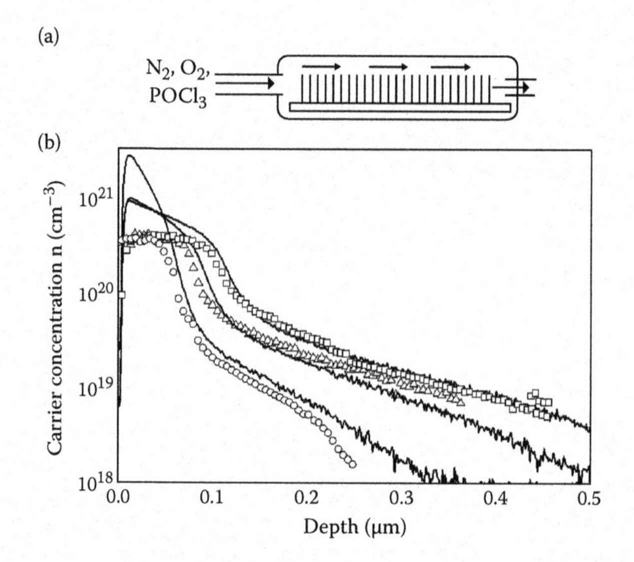

FIGURE 15.29
Schematic drawing of a diffusion tube furnace (a) and different emitter profiles (b).

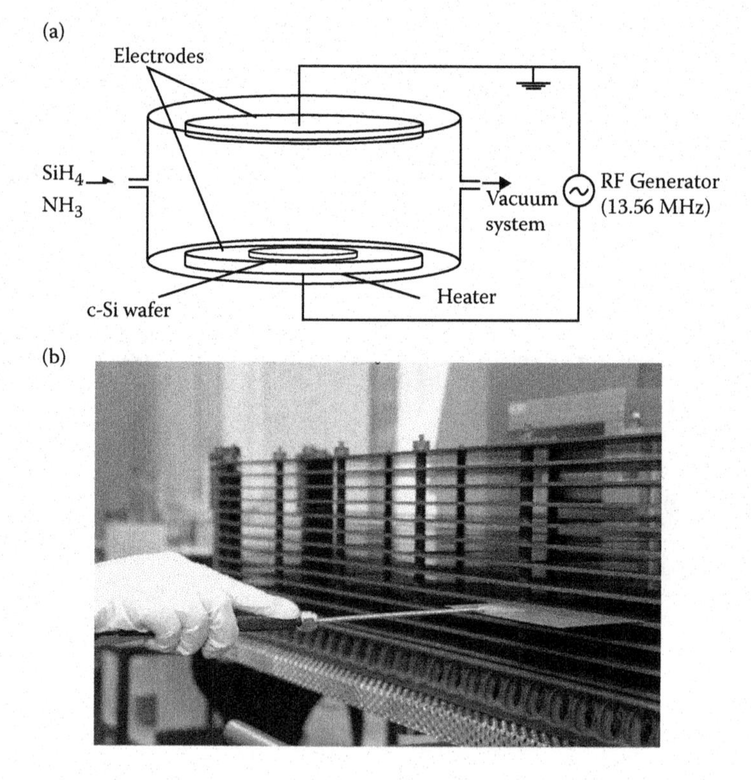

FIGURE 15.30
Schematic drawing of a PECVD reactor (a) and picture of centrotherm's horizontal carbon boat (b).

15.3.3.2.3 PECVD SiN_x Deposition

To reduce the reflectance of the front surface, an ARC is deposited on the sunny side of the solar cell. The state of the art ARC is a deep blue SiN_x layer that is formed in a PECVD reactor as shown in Figure 15.30a. Silane (SiH_4) and ammonia (NH_3) gases are introduced into a vacuum system and SiN_x is formed on the wafer surface. To reach the deep blue color and thereby minimize the reflectance, a layer thickness of approximately 70 nm is deposited. The dielectric layer has additional benefits for the solar cell. Because the layer is hydrogen (H)–rich, the surface and the bulk is passivated by H as it attaches to Si dangling bonds.

In Figure 15.31b, a vertical boat for the PECVD reactor is visible. This is loaded into a quartz boat that is heated to temperatures of approximately 450°C. The coated surface has

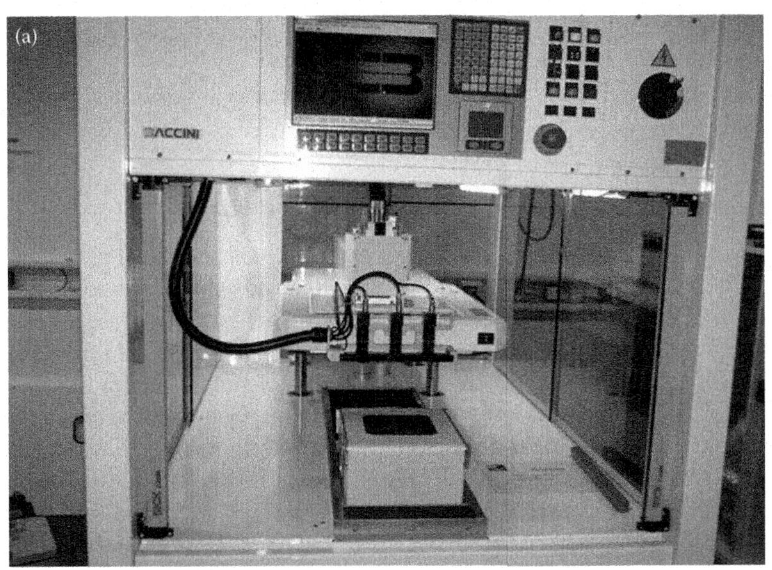

FIGURE 15.31
Picture of a Baccini screen printer (a) and topography of a typical finger contact taken by a laser scanning microscope (b).

an overall reflectance of approximately 7% compared with a bare surface of 36% (Figure 15.28). For comparison, a textured CZ-Si surface coated with SiN_x has a reflectance of approximately 4%.

15.3.3.2.4 Screen-Printed Metallization

As the second to the last step, metal contacts have to be "attached" to the front and rear. The most widely used and cost-effective metallization technique for both front and rear is a screen print method, similar to that used for other applications, such as printing on T-shirts.

Figure 15.31 depicts a modern screen-printing apparatus with a visual alignment system. After each printing step (front Ag finger, rear AgAl pads, and Al contact), the metal paste is dried and then cofired in a belt firing furnace with several different temperature zones. The last zone is the hottest and is set to approximately 860°C. The wafers remain there for less than 1 s but the most important processes happen in this time. The finger contact penetrates through the dielectric and the hydrogen is released and passivates both the silicon-dielectric interface and the silicon bulk.

In the very last step of the cell process, the p–n junction is separated by a laser process on the edge of the cell.

15.3.3.3 Average and Highest Solar Cell Efficiencies and LID

Average efficiencies achieved in industrial production of standard (156 × 156 mm^2, two or three bus bars, homogeneous emitter, and totally covered Al-BSF) mc-Si cells are approximately 16.7%. On monocrystalline CZ-Si substrates, a similar process results in efficiencies of approximately 18%. The difference is mostly due to a lower short circuit current (typically ~2 mA/cm^2) and a lower open circuit voltage (typically ~15 mV) for the mc-Si material. These are caused by a more effective (alkaline) surface texture and a base material with a higher lifetime and lower dislocation density in the mono c-Si material. Advanced processes, such as those described in the last chapter for the mono c-Si substrates, do not tend to have a positive effect on the performance as the mc-Si material is very inhomogeneous within the ingot (compare the differences between bottom and top positions and the corner and middle bricks shown in Figure 15.24). In Section 15.3.4, we will describe possible advanced processes for mc-Si materials.

In Table 15.1, we list the record efficiencies achieved on small and large-area mc-Si wafers. One advantage of mc-Si over mono-Si solar cells is the lower light induced degradation (LID) during operation, which is due to lower oxygen levels in the mc-Si material. Figure 15.32 shows as an example of a study of Q-cells (Hoffmann et al. 2008).

Whereas CZ-Si solar cells show a large LID due to BO_{2i}, mc-Si cells degrade only 0.2% to 0.3% in absolute efficiency. The highest degradation is observed in the bottom of the ingot as the O-content is the highest here due to in-diffusion from the crucible.

TABLE 15.1

Record Efficiencies for mc-Si Solar Cells and Modules on Small and Large Areas

Institution	Cell/Module (mc-Si Substrate)	Area (cm^2)	J_{sc} (mA/cm^2)	V_{oc} (mV)	*FF* (%)	η (%)
Fraunhofer ISE	Small cell	1.002	38.0	664	80.9	**20.4 ± 0.5** (Schulz et al. 2004, Green et al. 2011)
Q-cells	Large cell	242.7	39.0	652	76.7	**19.5 ± 0.4** (Green et al. 2011)
Q-cells	60 cell module	14,920	9.04 A	38.86 V	75.7	**17.8 ± 0.4** (Green et al. 2011)

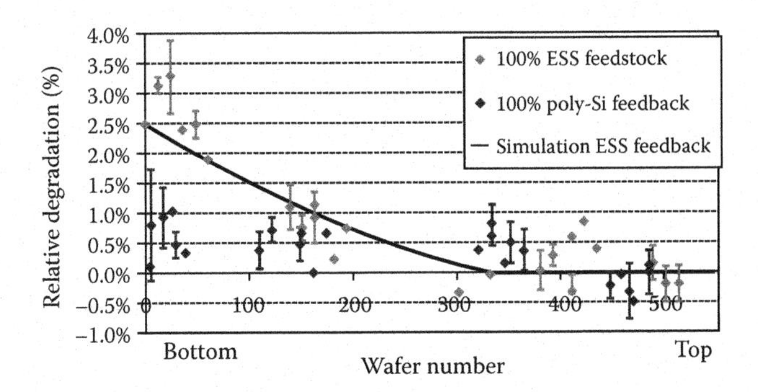

FIGURE 15.32
LID in mc-Si solar cells due to BO_{2i} complexes as a function of wafer position in the ingot (Reprinted with permission from Hoffman, V. et al. First results on industrialization of Elkem Solar Silicon at Pillar and Q-Cells. 23rd European Photovoltaic Solar Energy Conference and Exhibition, Valencia, Spain, 1117–1120, 2008). ESS feedstock is SoG-material from Elkem Solar.

15.3.4 Trends in Material, mc-Wafer, and mc-Cell Processing

In addition to the standard processing sequence, there are several trends for processing of mc-Si solar cells, but these take a different route to trends for mono c-Si solar cells. The main drawback of mc-Si material is the inhomogeneous wafer quality from different suppliers and within one ingot and even on a single wafer, which means that the high-efficiency processes (selective emitter and passivated rear side) discussed in the previous chapter do not lead to high efficiencies on every wafer.

15.3.4.1 Feedstock-Compensated Material

During the Si feedstock crisis that started in 2005, the metallurgical purification routes taken by different companies (Elkem, 6N Silicon, Dow Corning, Fesil, Timminco, and others) attracted much attention due to the scalability and fast setup times for factories. This Si feedstock typically includes more metallic impurities, but the main difference between pure poly-Si and SoG-Si produced by this metallurgical route is the coexistence of boron and phosphorus impurities. The material is called compensated material and has different properties compared with, for example, pure boron-doped Si.

Depending on the quantity of the dopants and on the compensation ratio, Rc, the resistivity profile of the mc-Si ingot looks very different from that of pure poly-Si, even having large regions of n-type material due to the different segregation coefficients of B and P. Figure 15.33 shows a resistivity profile of a SoG-Si ingot.

The art of making high-quality compensated ingots is to have insight into the compensation behavior of the Si material and to understand the resulting characteristics of the solar cells. Within the last 5 years, detailed studies have been performed on compensated materials from different suppliers resulting in the following findings:

1. Moderate compensation can lead to high-efficiency solar cells with even better performance than reference cells with the same resistivity (Peter et al. 2008)
2. Minority carrier lifetime increases with compensation ratio (Libal et al. 2008)
3. Majority as well as minority mobility decreases with compensation ratio (Libal et al. 2008; Rougieux et al. 2010)

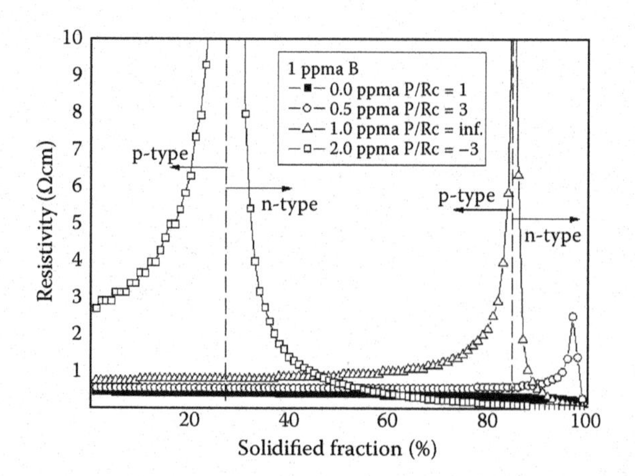

FIGURE 15.33
Resistivity as a function of compensation ratio for boron and phosphorus doping.

4. LID based on BO_{2i} complexes depends on net B doping and not on the total B doping (Kopecek et al. 2008; MacDonald 2011), and
5. n-Type compensated material shows LID as well (Schutz-Kuchly et al. 2011)

15.3.4.2 Mono Cast-Crystallization

In addition to the compensated material that has been on the market for the last 5 years, since 2010, a new method for crystallization has entered the market—mono casting. The technology is based on the mc-Si crystallization technique using a square crucible, but mono c-Si seeding wafer material is distributed on the bottom of the ingot, which initiates an oriented crystallization. The crystallization has to be performed more slowly than for a typical mc-Si crystallization to assure a good quality crystal without structural loss. The entire ingot is not monocrystalline, the edges and top of the ingot have a mc-Si structure, but approximately 70% of the ingot is monocrystalline. The remainder of the wafers have a mixed form as shown in Figure 15.34.

FIGURE 15.34
Picture of a typical monocast wafer (left) and resulting solar cells depending on the solar cell process (middle and right).

Figure 15.34 shows a quasi-mono wafer with mono- and multi-parts on the left. Depending on the solar cell process (mostly on the texture) the solar cells seem homogeneous (right) or stay inhomogeneous (middle). The large companies nowadays are optimizing the crystal growth conditions to minimize the mc-Si regions and reduce the oxygen content. Such wafers could soon become very important because this could be a very inexpensive method to produce large-area, fully square mono c-Si wafers with low oxygen content.

15.3.4.3 Advanced Solar Cell Processes

Advanced solar cell processes for mono c-Si solar cells were discussed in the previous paragraph. When processing mc-Si solar cells, the most important and diverging issue from mono c-Si is that the Si bulk has to be purified (gettered) and passivated during the solar cell process. In addition, surface improvements such as selective emitters and passivated rear sides are very difficult to apply as the material quality differs from wafer to wafer.

15.3.4.3.1 Processes Adapted to SoG-Si Material

As already mentioned, the homogeneity of a mc-Si wafer very much depends on the Si feedstock quality, crystallization conditions, and position in the ingot as shown in Figure 15.23. Therefore, it is essential to include processing steps in the solar cell process that improve the quality of the wafer. One of the most important "tuning steps" is the emitter diffusion. In particular, the end of the process can be prolonged. This results in an enhanced gettering by the diffusion as more metal impurities are captured from the bulk into the emitter region. Nevertheless, the recombination velocity in the emitter is high, which means that the metal impurities do less damage here than in the bulk. The emitter diffusion process can be adapted in such a way so that the emitter sheet resistance is almost unaffected and only the tail of the profile is diffused deeper into the Si bulk as shown in Figure 15.35.

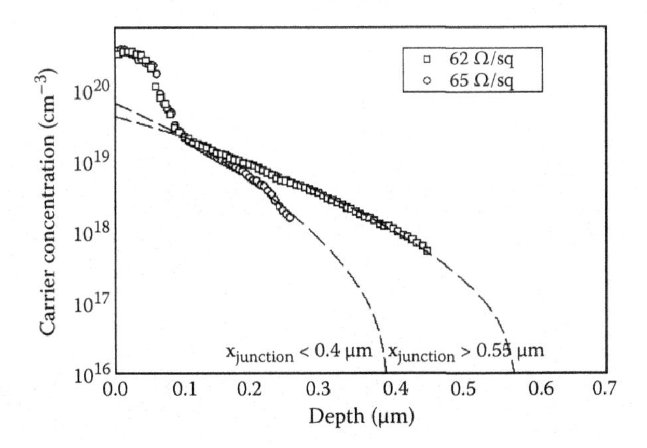

FIGURE 15.35
Standard emitter profile compared with prolonged gettering emitter for processing of high-efficiency SoG-Si solar cells.

For a solar cell with an emitter profile such as that shown in Figure 15.35, the blue response is not significantly reduced but the gettering is much more effective as the mobile metal impurities are captured in the prolonged tail region and the red response of the cell is enhanced. Using the prolonged gettering process, solar cell efficiencies are increased by almost 0.5% absolute.

In addition, for SoG-Si material, a good hydrogen passivation is important and therefore the PECVD SiN_x has to be H-rich. H is released during the belt furnace firing and passivates ungettered metal impurities that are stuck at grain boundaries and dislocations. Also, the Al BSF that is formed during belt furnace firing getters impurities from the silicon bulk (Meemongkolkiat et al. 2006).

15.3.4.3.2 Front Side and Rear Side Improvements

The varying crystal orientations on the surface, grain boundaries, dislocations, and different species of impurities inherent in mc-Si material mean that it is very difficult to improve the solar cell efficiency of mc-Si material using the same advanced solar cell processes used for mono c-Si. High efficiencies can be achieved on very good mc-Si wafers, but the averages for all wafers from an entire ingot are low—sometimes even lower than for cells processed with "standard process" as the gettering and H-passivation processes are sometimes excluded or have a reduced effect with the advanced processes (e.g., lower P-gettering due to shallow emitter or lower Al gettering due to local BSF).

15.3.4.3.3 Rear Contacted Solar Cells: MWT Technology

One important technology, which is an exception to the above and that is already on the market is metallisation wrap through (MWT) solar cells (Figure 15.36), embedded in special modules with conductive back sheets, which were discussed in the mono c-Si section. Several solar cell manufacturers have had this technology for some time including Photovoltek, Solland, and Sunways. Until recently, no module manufacturer was ready to implement the technology as the conductive back sheet was very expensive and conductive adhesives were proven not to be stable over time. This situation is now changing as there are solutions for both problems. There are now two different techniques on the market for attaching the MWT cells to the back sheet. One is laser soldering by Solland and the other is using conductive adhesives stencil printed on the conductive back sheet (e.g., Eurotron, Canadian Solar). Both are done using mc-Si wafers preferably.

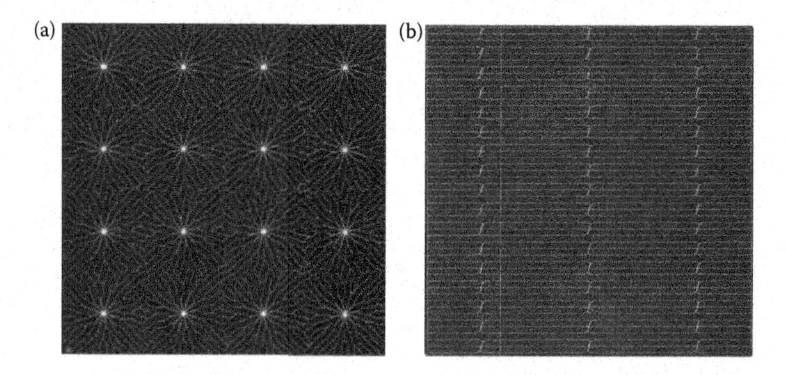

FIGURE 15.36
Different MWT solar cell front grid geometries. (a) type used at Soll and Solar Tech GmbH from; (b) type used at Photovoltech nv-sa. The busbars are located on the rear.

15.3.5 Summary

In contrast to mono c-Si solar cell technology, the most important processing steps for mc-Si solar cell production are sequences dedicated to the improvement of the material such as P, Al gettering or H passivation. When these steps are left out or when the effectiveness is reduced (selective emitter, local BSF, dielectric passivation layers without H-content), the average efficiencies on wafers across an entire ingot are reduced.

Therefore, the trends in mc-Si solar cell processing are focused on material issues, such as the use of compensated material or improvements in the mc-Si crystallization process. One very promising technique is the crystallization of mono c-Si material in large ingots previously used for mc-Si crystallization. It is likely that in the future, we will find only mono c-Si solar cells on the market and the wafers will be made using both the CZ method and cast in large ingots.

15.4 Solar Spectrum Modification to Enhance Silicon Solar Cell Efficiency

Alessia Le Donne and Simona Binetti

15.4.1 Introduction

Despite the many advantages that have made crystalline silicon–based solar cells the dominant PV devices used through most of the latter half of the last century, first-generation solar cells suffer from high cost of manufacturing and installation (Kamat 2007; Miles et al. 2007). The present research activity on first-generation devices is therefore devoted to the development of several ways to increase their efficiency-to-cost ratio to make PV competitive with the conventional energy resources.

Among the approaches to increase the efficiency of first-generation PV devices is light harvesting strategy. From a general point of view, this method is based on a better exploitation of those solar spectrum regions which show no good match with the spectral response of Si PV devices (Figure 15.37). According to the detailed balance model by Shockley et al. (1961) the theoretical efficiency limit for a single-junction solar cell with E_g equal to 1.1 eV is 31%. The largest portion of such a 69% energy loss can be ascribed to thermalization-related losses (in the spectral region A, Figure 15.37) and to sub-bandgap photon transmission–related losses (in the spectral region B, Figure 15.37). A modification of the incident solar spectrum, which allows a better exploitation of these spectral regions, is therefore expected to provide a theoretical efficiency of more than 40% (Trupke et al. 2002a,b).

15.4.2 Modification of the High-Energy Side of the Solar Spectrum

Due to thermalization-related losses, the high-energy side of the solar spectrum is not efficiently converted from Si. In fact, thermalization occurs when the solar converter absorbs photons with much higher energies than the energy gap: in this case, the generated charge carriers relax in a short time to the edge of the conduction and valence bands, respectively, and the energy difference ($E_{phot} - E_{gap}$) is lost to heat in the device. As a consequence, Si solar cells show a low spectral response in the region below 500 nm, as clearly pointed out in Figure 15.37.

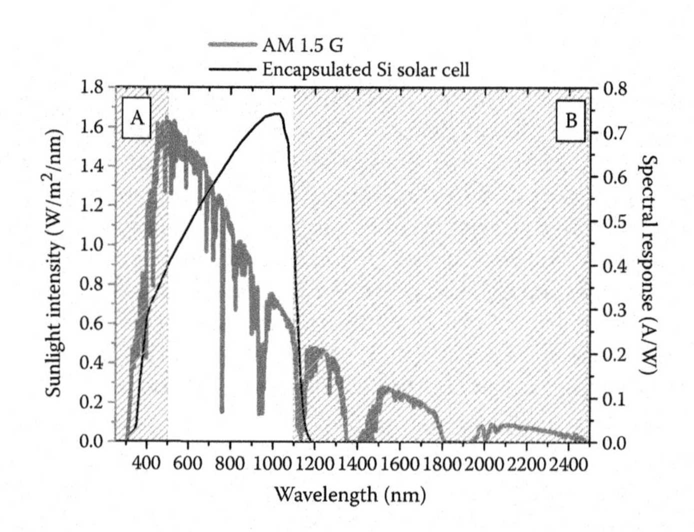

FIGURE 15.37
Comparison between the terrestrial solar spectrum in the case of solar zenith angle equal to 48.2° (AM 1.5 G-ASTM G173-03) and the spectral response of an encapsulated crystalline Si solar cell.

Different options have been proposed in the literature for a better exploitation of the solar spectrum region below 500 nm. They differ from each other in the number of electron–hole (e–h) pairs or photons created by a single-incident high-energy photon.

Multiple exciton generation (MEG) and space-separated quantum cutting (SSQC) result in the generation of multiple e–h pairs in the cell. In the MEG process, one photon with energy $E_{phot} \geq nE_{gap}$ yields up to n e–h pairs with energy equal to E_g. MEG has been estimated to enhance the efficiency of single-junction solar cells up to 44%, but this limit can be only reached if an e–h pair is created for every increase by E_g in the photon energy (Hanna and Nozik 2006; Nozik 2002). Even though e–h pair generation efficiencies of 700% for photon energies of $8E_g$ have been reported (Meijerink 2008), MEG has been shown to be hindered by several long-investigated but still unresolved issues, namely, competing de-excitation processes, recombination of excitons without carrier separation, and high-impact ionization threshold (Shpaisman et al. 2008). In the SSQC process, an absorbed higher-energy photon divides into two or more e–h pairs through the interaction of two spatially separated neighboring nanocrystals. Even though SSQC could reduce the loss of energy due to thermalization in solar cells by increasing the number of charge carriers created per absorbed photon, a deeper knowledge on this topic is required to establish the possible efficiency of charge carrier generation.

Down-conversion (DC; Badescu and De Vos 2007; Abrams et al. 2011), or quantum cutting (QC), results instead in the generation of two low-energy photons from the cutting of one higher-energy photon with energy exceeding 2 E_g. When the solar cell absorbs these two low-energy photons, current doubling can be obtained in the high-energy region of the solar spectrum, therefore reducing the energy loss due to thermalization of hot charge carriers (Trupke et al. 2002a). DC is therefore similar to MEG and SSQC, but multiple excitons are created by photon doubling before the absorption into the PV device rather than being created by one high-energy photon absorbed into the solar cell. The theoretical gain which could be obtained by DC/QC for non-concentrated solar irradiation as a function of the absorber bandgap has been widely investigated (Trupke et al. 2002a; Badescu and De Vos 2007; Abrams et al. 2011). According to Trupke et al. (2002a) and Badescu and De Vos

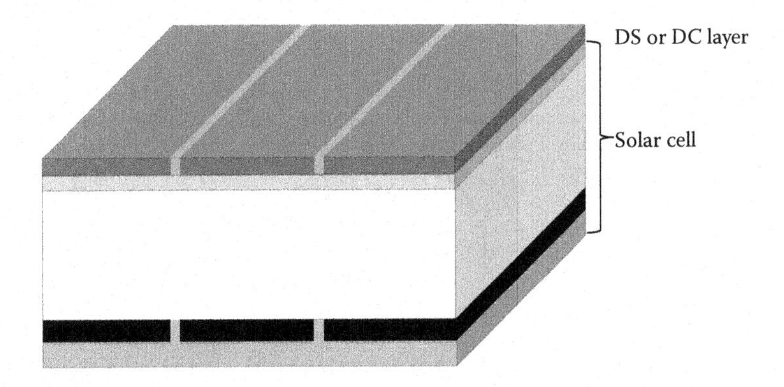

FIGURE 15.38
Schematic structure of a Si solar cell coupled with a system for DS or DC.

(2007), the Shockley–Queisser limit of 31% for Si single-junction PV devices increases by up to 37% when the down-converter is located at the front surface of the solar cell and by up to 40%, when the converter is built onto the rear surface. The latter option is, however, unsuitable for Si PV devices because a better exploitation of high-energy photons requires that such photons reach the down-converter before being absorbed by the solar cell. Instead, Abrams et al. (2011) investigated the more realistic solution of placing a down-converter onto the front cell surface, showing that the total absolute efficiency enhancement related to the DC/QC process is up to 7% in such an ideal system.

A further approach to modify the high-energy side of the solar spectrum is down-shifting (DS), but in this case, the enhanced efficiency of the PV device could never exceed the Shockley–Queisser limit (Shockley and Queisser 1961). In the DS process, in fact, one low-energy photon is created by one high-energy photon absorbed into a proper molecular system emitting around the maximum quantum efficiency value of the c-Si cell. Down-shifters, as well as down-converters, are usually located at the front cell surface, as schematically illustrated in Figure 15.38.

15.4.3 Modification of the Low-Energy Side of the Solar Spectrum

The solar spectrum region above 1100 nm could never be exploited from Si-based solar cells because photons with energy lower than the E_g are not absorbed. Such transmission-related loss can be reduced by the up-conversion (UC) process (Strümpel et al. 2007), whereby one or two low-energy photons are converted into one high-energy photon, which can be absorbed from Si (Shpaisman et al. 2008; Auzel 2004; Shalav et al. 2007; Badescu 2008). Two main types of UC mechanisms are known, which differ in the number of photons involved (Strümpel et al. 2007; Auzel 2004). The anti-Stokes shift involves only one photon, the additional energy being provided by phonons. Conversely, the main group of UC mechanisms, described in detail by Auzel (2004), is based on the absorption of two photons. They all are schematically illustrated in Figure 15.39 and summarized in the following:

- Energy transfer up-conversion (ETU) or addition de photon par transferts d'energie (APTE): efficiency approximately 10^{-3}
- Ground state absorption followed by an excited state absorption (GSA+ESA): efficiency approximately 10^{-5}
- Cooperative sensitization: efficiency approximately 10^{-6}

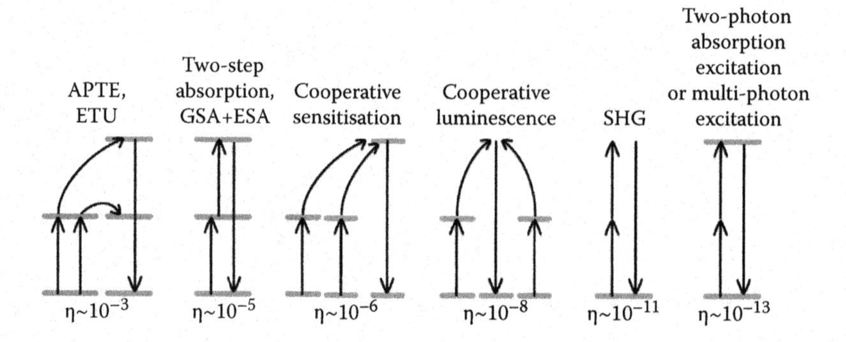

FIGURE 15.39
UC mechanisms based on the absorption of two photons. The horizontal lines represent real energy levels, the vertical arrows indicate excitation (or de-excitation) of energy levels and the arrows connecting different ions indicate energy transfer. (Reprinted from Strümpel, C. et al. *Solar Energy Materials and Solar Cells* 91:238–249, 2007. Copyright 2007. With permission from Elsevier.)

- Cooperative luminescence: efficiency approximately 10^{-8}
- Second harmonic generation (SHG): efficiency approximately 10^{-11}
- Two-photon absorption or multiphoton excitation: efficiency approximately 10^{-13}

The energy transfer up-conversion, ground state absorption + excited state absorption, and cooperative sensitization mechanisms show higher efficiencies because they involve real, existing energetic levels. Conversely, the cooperative luminescence, second harmonic generation, and two-photon absorption/multiphoton excitation processes involve virtual levels, high excitation energies being therefore required. In an actual up-converter many of the above mentioned mechanisms take place simultaneously.

UC systems are usually located at the rear side of bifacial devices, as schematically shown in Figure 15.40, since they have to convert of the low-energy radiation transmitted by the Si solar cell.

The gain which could be obtained by UC has been theoretically investigated by Trupke et al. (2002b) and Badescu (2008). The latter work showed that adding an up-converter onto the rear side of the PV device is beneficial just in the case of top quality solar cells operating under spectral concentration of the solar radiation (Badescu 2008).

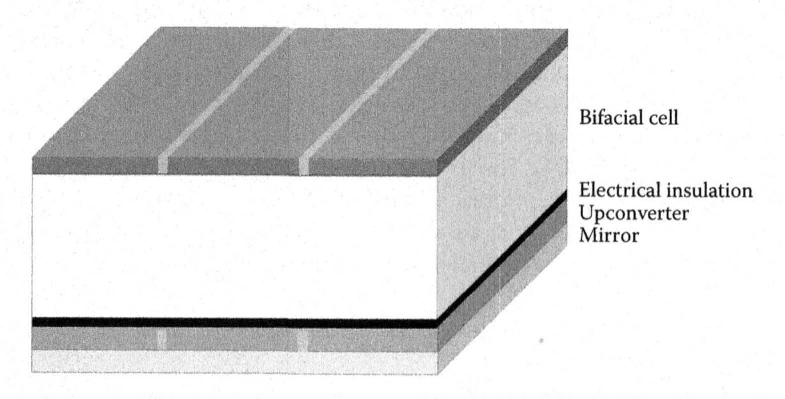

FIGURE 15.40
Schematic structure of a bifacial Si solar cell coupled with an UC. The presence of an electrical insulation layer between the cell and the UC is required to avoid additional recombinations at the rear cell surface.

15.4.4 Main Classes of Materials for the Solar Spectrum Modification

15.4.4.1 Lanthanide-Based Systems

The possible use of rare earths in PV was proposed for the first time in 1979 by Dexter (1979), who suggested the combination of a PV device with an organic dye that would harvest high-energy photons and transfer the absorbed energy, allowing an enhancement of the solar cell efficiency. In the next few years, Munz and Bucher (1982) and Reisfeld (1983) further elaborated on this idea. Afterward, early attempts to increase the efficiency of Si PV devices both by organic (Jin et al. 1997) and inorganic (Kawano et al. 1997) lanthanide-based compounds dealt with DS. Among them, as an example, surface coating of c-Si cells with 1-mm-thick organically modified silicate containing $[Eu(phen)_2]Cl_3$ or $[Tb(bpy)_2]Cl_3$ resulted in a relative increase in the energy conversion by up to 10% (Jin 1997). The implementation of such systems in a real PV module manufacturing process was, however, hampered by their typical realization time, which is not compatible with industrial requirements. More recently, relative enhancements of the total delivered power up to 2.9% have been reported for Si solar cells coated with 250-nm-thick $[Eu(phen)_2](NO_3)_3$–doped PVA (Marchionna et al. 2006), 2-μm-thick $Eu(dbm)_3phen/Eu(tfc)_3$–doped PVA (Le Donne et al. 2009), and 1-μm-thick $Eu(tfc)_3$/4,4′bis(diethylamino)benzophenone–doped EVA (Le Donne et al. 2011). Such organo-lanthanide–based down-shifters being included in polymeric layers suitable as PV modules encapsulating matrix, they could be integrated into the PV module production without introducing strong modifications of the industrial manufacturing process.

A more efficient increase of Si solar cell efficiency can be obtained by Ln^{3+} ion–based compounds suitable for QC (Trupke et al. 2002a; Wegh et al. 1999). Early attempts concerning $LiGdF_4$ doped with Eu^{3+} (Wegh et al. 1999) provided quantum yields between 160% and 190%, whereas more recently, values close to 200% have been obtained (Zhang and Huang 2010). Three different QC mechanisms, described in detail by Richards (2006), are known in rare earth doped materials:

- QC on host lattice states: this mechanism is based on the impact ionization process, whereby one incident high-energy photon originates multiple lower energy e–h pairs. The generation of a second e–h pair in Si has been, however, demonstrated to require an electron energy of at least 2.6 eV (Boer 1990). Therefore, a large increase of the IQE requires that short wavelengths are used (IQE $< 5\%$ for $\lambda = 350$ nm, IQE $\cong$ 50% at $\lambda = 250$ nm; Richards 2006), hampering possible terrestrial applications.
- QC on single rare earth ions: this mechanism involves well-separated energy levels, which can be used to create more than one visible photon out of one high-energy photon. Early investigations on this mechanism dealt with Pr^{3+}-based materials. In this case, the light absorption at 185 nm involves the $^1S_0$ level, resulting in two different emissions at 405 and 620 nm with an efficiency of approximately 140% (Piper et al. 1974).
- QC on rare earth ions pairs: this mechanism involves two interacting rare earth ions, one of them being able to show QC. By interaction with the other ion in the pair, part of the energy is transferred to the QC ion, avoiding possible IR and UV losses (Wegh et al. 1999). A typical example is the absorption of one UV photon by Gd^{3+}, followed by energy transfer onto Eu^{3+}, which gives rise to the emission of two 612 nm photons with an IQE close to 200% (Wegh et al. 1999). However, due to very weak absorption of Gd^{3+}, the overall efficiency is limited to 32%, this value being improved up to 110% by Er^{3+} codoping and using Tb^{3+} instead of Eu^{3+}.

In the last few years, the optical properties of many lanthanide-based materials were thoroughly investigated in view of their potential application as DC/QC or DS systems in Si solar cells (Chung et al. 2007; Zhang et al. 2007; van der Kolk 2008; Deng et al. 2011; van Wijngaarden et al. 2010; Meijer et al. 2010; van der Ende et al. 2009; Lin et al. 2010). Recently, very promising works reported on the DC/QC (Smedskjaer et al. 2011; Miritello et al. 2010; Lakshminarayana et al. 2008), or DS (He et al. 2010) capability of different kinds of rare earth doped glasses. Besides acting as a support or a protective sheet, such glasses could also become an active component of the PV device by incorporating the proper system for solar spectrum modification.

Strong efforts were also devoted to the investigation of lanthanide-based materials, usually involving Er^{3+}, as UCs suitable for applications in Si PV devices (Strümpel et al. 2007; Shalav et al. 2007; Sewell et al. 2010; Fischer et al. 2010; Richards and Shalav 2007; Suyver et al. 2005; Ivanova and Pelle 2009).

Considering the above mentioned limitations of UCs working under non-concentrated solar irradiation (Badescu 2008), fluorescent sensitizers have been recently proposed (Strümpel et al. 2007) to increase the performances of lanthanide-based UC systems. Otherwise, due to the weak and narrow absorption of rare earth–based UCs, most of the low-energy solar radiation could remain unused, therefore limiting the ability of lanthanide-based systems to reduce sub-bandgap losses.

Since both DC/QC and UC processes are available for many rare earth ions (Auzel 2004), lanthanide-based materials show great potential for a strong improvement of Si solar cell performances by simultaneous implementation of both light-harvesting strategies, as recently proposed by Eilers et al. (2010).

15.4.4.2 Quantum Dots–Based Systems

Quantum dots (QD) materials are a promising solution as systems for the modification of the high-energy side of the solar spectrum. Due to reduced dimensionality, QDs exhibits quantization of their electronic energy levels, with a consequent blue-shift of the optical absorption edge. The QDs electronic energy levels and optical absorption being dependent on size, the effective bandgap can be easily tuned for solar spectrum modification purposes (Norris et al. 1996).

In the last few years, MEG has been reported for different QD systems (e.g., CdSe, PbSe, and PbS; Nozik 2002; Schaller et al. 2005a,b, 2006; Klimov et al. 2007; Luque et al. 2007; Schaller and Klimov 2004), including the remarkable seven-exciton generation per single incident photon shown by Schaller et al. (2006). Timmerman et al. (2008) have recently reported the occurrence of the SSQC process in Si nanocrystals, which also showed that the SSQC efficiency can be further optimized by changing the separation between individual nanocrystals.

The application of QDs as systems for light-harvesting strategies, however, requires new inexpensive and scalable techniques for producing QDs with the desired optical properties. To this purpose, colloidal synthesis could be a possible attractive solution, this approach being easily scalable for relatively inexpensive large-scale production (Su et al. 2003). Huang et al. (2010) recently showed promising results on the action of ZnS QDs produced by colloidal synthesis, both as down-shifters and ARC for light trapping of the incident radiation.

15.4.4.3 Organic Dye–Based Systems

Even though organic dyes have been used as conventional DS systems both in PMMA (McIntosh et al. 2009) and EVA (Klampaftis and Richards 2011) layers built onto the front

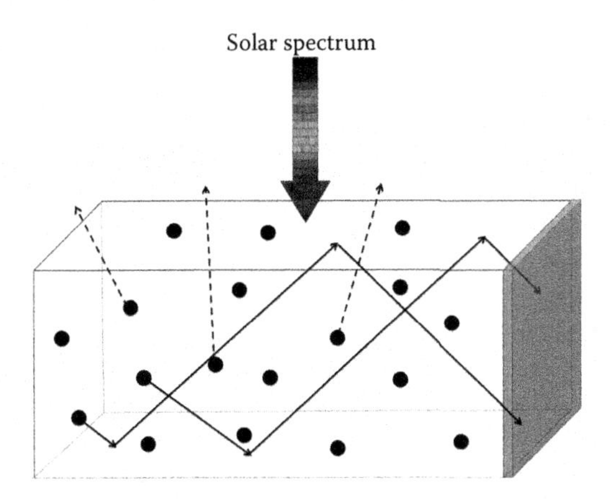

FIGURE 15.41
Schematic structure of a LSC system. A portion of the light emitted by the luminescent particles (black dots) reaches the solar cell (gray slice on the right side of the transparent matrix) by total internal reflection (solid black arrows). Another portion of the emitted light escapes from the LSC at angles larger than the critical angle for total internal reflection (dashed black arrows).

surface of Si PV devices, their main application was in luminescent solar concentrators (LSCs), which are also based on the DS action of proper luminescent systems.

A LSC system consists of a transparent host material, usually a glass or plastic plate, coupled at one or more sides with solar cells (Figure 15.41). Luminescent particles, such as organic dyes, able to partially absorb the incident solar spectrum, are dispersed in the transparent matrix. A portion of the light emitted by the luminescent particles is guided toward the solar cells by total internal reflection. Because the edge surface area is smaller than the front surface area, sunlight concentration effects can be achieved.

The LSC concept was proposed in the late 1970s (Weber and Lambe 1976; Goetzberger and Greubel 1977) and widely investigated in the last decades (Zastrow 1994; van Sark et al. 2008; Goldschmidt et al. 2009; Rowan et al. 2008; Slooff et al. 2008). Nevertheless, up to now, this approach has not delivered on its promise. Even though strong concentration effects have been theoretically estimated (Smestad et al. 1990), the best LSC-based PV system obtained thus far showed a conversion efficiency of 7.1% and reduced the PV cell area by a factor of 2.5 (Slooff et al. 2008). Two main reasons are behind the difficulty in obtaining the theoretically predicted performances. First of all, organic dyes with quantum yield close to 100% and long-time durability under solar irradiation are required for LSCs with higher performances. Secondly, relevant losses from escaping luminescent light from the concentrator at angles larger than the critical angle for total internal reflection have been observed (de Boer 2010).

15.5 Silicon QDs for the Next Generation of Silicon-Based Solar Cells

Shujuan Huang and Gavin Conibeer

15.5.1 Introduction

Third-generation approaches aim to achieve high efficiency for PV devices by circumventing the Shockley–Queisser limit for single-bandgap devices. The concept is to do this with only a small increase in areal costs and hence reduce the cost per watt peak (Green 2003). Also, in common with the silicon-based second-generation thin film technologies, these will use abundant and nontoxic materials.

Tandem cells use the strategy of increasing the number of energy levels to more efficiently convert the energy of a wider range of photons absorbed to electricity. Solar cells consisting of p–n junctions in different semiconductor materials of increasing bandgap are placed on top of each other, such that the highest bandgap intercepts the sunlight first (Figure 15.42). This approach was first suggested by Jackson (1955) using both spectrum splitting and photon selectivity. The particle balance limiting efficiency depends on the number of subcells in the device. For 1, 2, 3, 4, and ∞ subcells, the limiting efficiency η is 31.0%, 42.5%, 48.6%, 52.5%, and 68.2% for unconcentrated sunlight (Green 2003; Brown and Green 2002).

Tandem cells have already been used in III–V PV and some amorphous thin film PVs. The former is the highest quality, and hence, highest efficiency tandem devices using very expensive epitaxial processes, up to 43.6% at 400 suns (Green et al. 2011). The latter produces lower efficiencies, for instance, in the Micromorph cell at approximately 12% and hence, are less able to leverage the basic material cost and balance of the system.

One tandem solar cell approach based entirely on silicon (Si), making use of the quantum confinement effect in a silicon nanostructure, has been proposed (Green et al. 2005). By constraining silicon in one or more dimensions at less than the Bohr exciton radius of bulk crystalline Si (~5 nm), quantum confinement causes its effective bandgap to increase (Brus et al. 1984). If Si is constrained in all three dimensions such as in QDs, a stronger confinement effect can be achieved. Hence, bandgap tuning by adjusting the size of Si QDs allows the fabrication of higher bandgap solar cells that can be used as tandem cell elements. Figure 15.42 illustrates a schematic of the QD-based "all-silicon" tandem solar cell concept. As can be seen, two Si QD cells with higher effective bandgaps are stacked onto a standard single-junction Si solar cell. The effective bandgaps of the two upper cells could be achieved as a result of quantum confinement by adjusting the size of Si QDs embedded in a higher bandgap dielectric material, such as SiO_2, Si_3N_4, or SiC. For a three-cell "all-silicon" tandem with Si as the bottom cell, the optimal effective bandgap for the middle

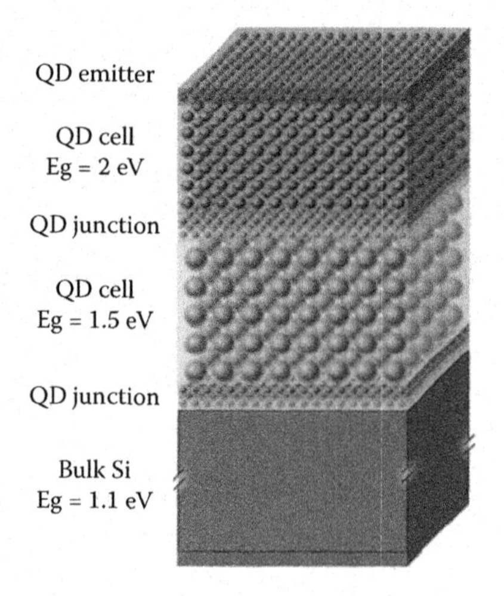

FIGURE 15.42
Schematic of the silicon QD-based all-silicon tandem solar cells as proposed by Green. (Adapted from Green, M.A. et al. *Proceedings of 20th European Photovoltaic Solar Energy Conference, Barcelona, Spain*, 3–6, 2005. With permission.)

and upper cells would be 1.5 and 2.0 eV, respectively, to push the theoretical efficiency limit to 47.5%. In the case of a two-cell "all-silicon" tandem, a top cell with an effective bandgap of approximately 1.7 eV would be optimal to reach an efficiency of 42.5% (Meillaud et al. 2006; Green et al. 2006).

To realize the "all-silicon" tandem solar cell concept, there are several aspects that need to be established. First of all, certain features of Si QD materials need to be demonstrated including dot size, bandgap control, enhanced optical absorption, and adequate electronic transport (Brus 1984; Green et al. 2006). This is then followed by the formation of QD solar cells, which involves doping to make materials of two carrier types or some other rectification mechanism, such as the formation of a heterojunction. After achieving a working QD cell, the final step would be the interconnection of different cells in the tandem stack, which is most likely achieved using defect or tunnel junction interlayers.

15.5.2 Si QD Formation

Engineering wider bandgaps for Si-based materials using quantum confinement in nanostructures is the key issue in developing "all-silicon" tandem cells. This bandgap engineering can be done using either quantum wells (QWs) or QDs of Si sandwiched between layers of a dielectric based on Si compounds such as SiO_2, Si_3N_4, or SiC as shown in Figure 15.43. For sufficiently close spacing of QWs or QDs, a true miniband is formed, creating an effectively larger bandgap as shown in Figure 15.44. For QDs of 2 nm (QWs of 1 nm), an effective bandgap of 1.7 eV results, ideal for a tandem cell element on top of Si. These layers are grown by thin film sputtering or chemical vapor deposition processes followed by a high-temperature anneal to crystallize the Si QWs/QDs (Zacharias et al. 2002; Conibeer et al. 2008; Kurokawa et al. 2010). The matrix remains amorphous, thus avoiding some of the problems of lattice mismatch.

Si QDs constrained between dielectric thin layers seems to be much more efficient in terms of bandgap engineering. This approach provides the best control of QD growth, which involves alternating deposition of Si-rich and stoichiometric dielectric thin layers. Si QDs precipitate from the supersaturated phase within the Si-rich layers upon thermal annealing, as shown schematically in Figure 15.43. Ideally, the growth of these QDs is expected to be constrained by the adjacent dielectric layers (also referred to as barrier layers), and hence, is spherical. In addition, interface energy minimization will tend to favor the formation spherical nanocrystals because of their minimum surface area to volume ratio. Hence, spherical Si QDs of uniform size will tend to form embedded in the dielectric matrix.

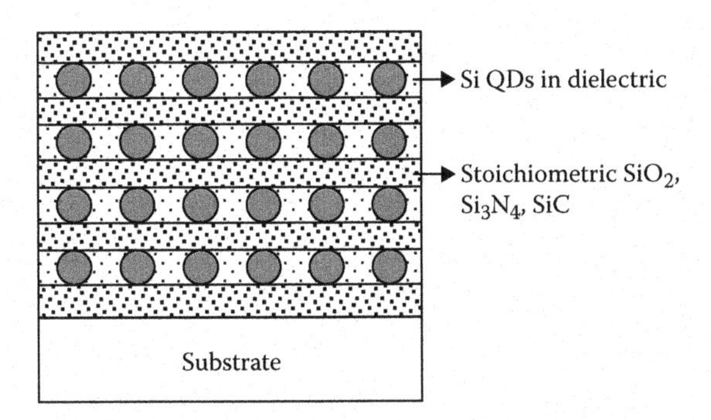

FIGURE 15.43
Multilayer deposition of alternating Si-rich dielectric and stoichiometric dielectric in layers of a few nanometers.

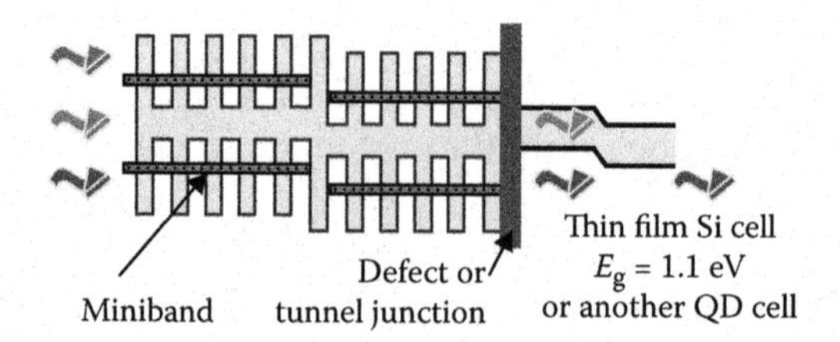

FIGURE 15.44
Band structure of "all-silicon" tandem cell: the nanostructure cell consists of Si QWs or QDs in an amorphous dielectric matrix connected by a defect tunnel junction to a thin film Si cell.

15.5.2.1 Crystallization Temperature

In practice, the shape and size control is more complicated. In the solid phase, if the temperature is high enough, a complete phase separation can occur (Zacharias et al. 2002):

$$SiO_x \rightarrow \frac{x}{2} SiO_2 + \left(1 - \frac{x}{2}\right) Si$$

where $0 < x < 2$.

The silicon nanocrystalline phase will only precipitate if both the temperature is high enough to promote nucleation and if the Si excess is high enough to provide a driving force. In terms of temperature, it varies depending on the preparation process. For PECVD growth, Si nanocrystals or QDs can form at temperatures between 800°C and 1050°C, depending on the excess Si content (Nesbit 1985), whereas the sputtered structures require 1000°C to 1100°C (Conibeer et al. 2008; So et al. 2011a,b,c). So and coworkers have reported that the average size of Si QDs increases from 1000°C to 1100°C in a multilayer structure in which Si QD size is expected to be defined within the thickness of Si-rich layers (So et al. 2011c). When annealing temperature is 900°C or lower, only amorphous nanoclusters form from annealing.

15.5.2.2 Effect of Excess Si Content

The Si excess or richness in the Si-rich layers also plays significant role in shape and size control. As suggested by Nesbit (1985), the classic theory of diffusion-controlled growth is adaptable for the formation of Si QDs in the dielectric matrix. Therefore, in a thick monolayer, the Si QD size increases with increasing silicon content (decreasing O/Si, N/Si, or C/Si ratio) for a given annealing time and temperature. As mentioned in a previous section, a multilayer structure is believed to provide the best control of QD growth, which is expected to be constrained within the thin Si-rich layers by the adjacent stoichiometric dielectric layers; hence, spherical shaped Si QDs form embedded in the matrix. However, an investigation of the effects of the O/Si ratio on the growth of Si QDs in a SiO_2/SRO/SiO_2 multilayer structure (SRO refers to silicon-rich SiO_2 layer) demonstrates that the average Si QD diameter increases as the O/Si ratio decreases (Hao et al. 2009c), that is, the higher the Si excess, the larger the QD size. In addition, the QD shape is elongated rather than spherical when the O/Si ratio is 1.0 or smaller, as shown by the TEM image in Figure 15.45b. This indicates that even though the Si QD growth is constrained vertically by the adjacent stoichiometric layers, it is not constrained laterally in-plane. When the Si excess is high, the interface energy minimization is no longer sufficient to ensure the formation of spherical nanocrystals and the precipitated Si nanocrystals tend to coalesce to form extended shapes (either discs or ellipsoids), suggesting a critical O/Si ratio of 1.0, above which spherical QDs

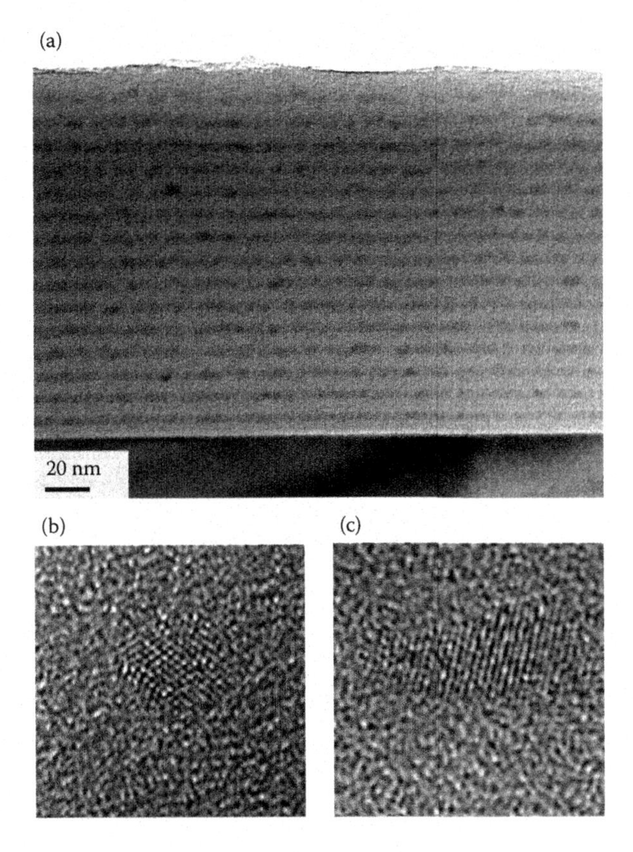

FIGURE 15.45
TEM images showing SiO_2/SRO/SiO_2 multilayer structure deposited on Si substrate (a), a spherical QD when Si excess is low (O/Si = 1.3) (b), and an elongated QD when Si excess is high (O/Si = 0.86) (c).

form. On the other hand, when the O/Si ratio is too large, for instance, 1.3 or higher, the Si QDs cannot fully grow to the size defined by the thickness of the Si-rich layer due to the small Si excess. Therefore, an optimized range of O/Si ratio between 1.0 and 1.3 seems to be required to produce uniform spherical Si QDs.

15.5.2.3 Effect of Dielectric Matrices

Because the growth of Si QDs is a diffusion-controlled growth process, the matrix materials have a significant effect on the precipitation and crystallization of embedded Si QDs. Si_3N_4 has a dense and rigid structure that is quite different from SiO_2. As a consequence, Si atoms have a lower diffusivity in this matrix when annealed at the same temperature. This results in a more uniform and spherical shape of the Si QDs embedded in Si_3N_4. The cross-sectional TEM images in Figure 15.46 clearly shows the difference of Si QDs formed in SiO_2 and Si_3N_4 multilayer structures, with a similar Si excess content (35 vol%) and the same annealing conditions. Because SiO_2 is porous and flexible, it provides a larger diffusion coefficient for Si to move and coalesce to form an elongated shape within the thin Si-rich layer.

The SiC matrix has been considered the best choice in terms of carrier transport because it offers lower energy barrier height compared with SiO_2 and Si_3N_4 (see next section). However, when annealing a multilayer structure consisting of Si-rich SiC and stoichiometric

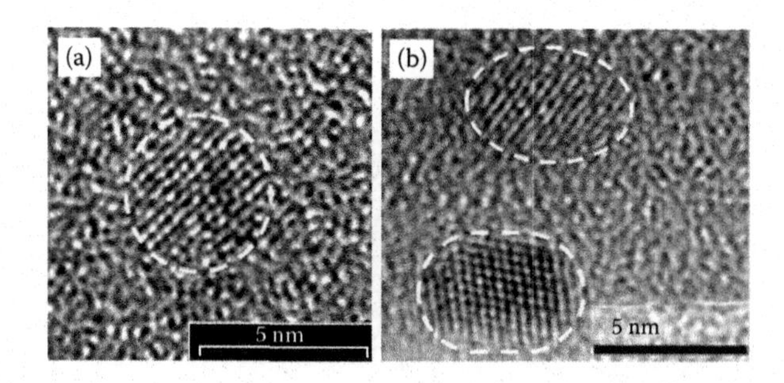

FIGURE 15.46
TEM images comparing Si QD shape in Si_3N_4 matrix (a) and in SiO_2 matrix (b) when Si excess is approximately 35 vol%.

SiC layers, nanocrystalline β-SiC also forms in the SiC barrier layers at 1100°C. Hence, the SiC barrier layers fail in the function of constraining the Si QDs growth within the Si-rich layers. As a result, the superlattice structure is compromised by substantial Si interdiffusion across the SiC layers (Song et al. 2008; Wan et al. 2011a). A pure SiC film requires much higher temperatures to crystallize—approximately 1400°C. The formation of β-SiC nanocrystals, therefore, is believed to be via a heterogeneous nucleation and crystallization mechanism induced by the initial formation of Si nanocrystals. This is, in itself, an interesting and potentially useful effect for the formation of crystalline SiC. However, its advantage in this context is not clear, so reduction of SiC crystallization is preferable, at least for now. The short ramping-up time in rapid thermal annealing has been reportedly effective in reducing β-SiC nanocrystal formation, hence, improving the structural properties (Wan et al. 2011a).

15.5.2.4 Effect of Dielectric Barrier Layers

As described previously, the multilayer structure provides the best control of QD growth, which consists of alternating Si-rich and stoichiometric dielectric thin layers. Besides the function of enclosing Si QDs, the dielectric layers also work as carrier tunneling barriers between Si QDs in the vertical direction. Different barrier materials provide different tunneling probabilities that heavily depend on the height of this barrier, as shown in Figure 15.47. Si_3N_4 and SiC give lower barrier heights than SiO_2, allowing better carrier transport for a given spacing between the QDs.

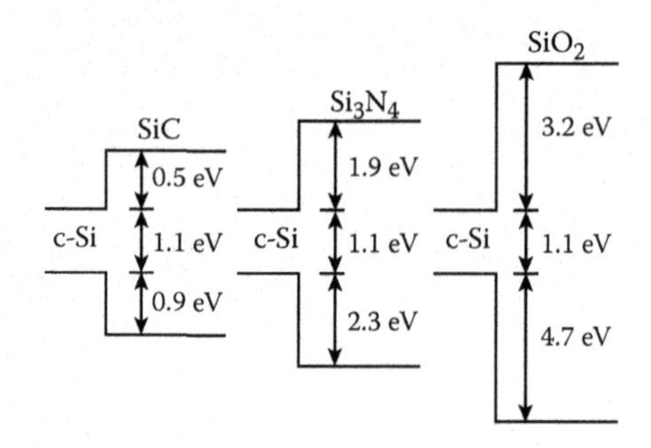

FIGURE 15.47
Conduction and valence band offsets between crystalline silicon and SiC, Si_3N_4, and SiO_2.

However, as stated previously, the Si-rich/SiC multilayer structures face the potential problem of nonuniformity of Si QDs due to the formation of β-SiC NCs.

SiO_2 layers are robust and can constrain Si QD growth between them when the thickness is larger than 2 nm. When a thinner thickness is used, which is ideal for carrier transport, the Si QDs tend to coalesce across the SiO_2 layers and form larger sizes. This is attributed to the fact that sputtered SiO_2 thin layers are quite porous, providing diffusion channels for Si to diffuse across.

Si_3N_4 layers are the best choice in terms of effective antidiffusion of Si during annealing. So and coworkers (2011a) have investigated the effect of ultrathin silicon nitride (Si_3N_4) barrier layers on the formation of Si QDs in Si-rich nitride (SRN)/Si_3N_4 multilayer structures. The layered structures, composed of alternating layers of SRN and Si_3N_4, were prepared using magnetron sputtering followed by furnace annealing. The formation of uniformly sized Si QDs was confirmed by TEM and x-ray diffraction measurements. In particular, the 1-nm-thick Si_3N_4 barrier layers were found to be sufficient in restraining the growth of Si QDs within the SRN layers upon high-temperature annealing, as indicated by x-ray reflection and TEM measurements (So et al. 2011a).

With the rationale of taking the advantages of low barrier height of SiC and high stability of Si_3N_4 thin layers, Wan and coworkers (2011b) have proposed a hetero-multilayer structure by incorporating Si_3N_4 barrier layers to sandwich Si-rich carbide layers. This work reveals that the ultrathin Si_3N_4 (0.8–2.0 nm) barrier layers have significantly suppressed the formation of β-SiC NCs hence sufficiently inhibiting Si interlayer diffusion.

To maximize all quantum confinements, barriers to diffusion, and electrical transport in the vertical direction, Di et al. (2010a) have used a hetero-interlayer structure consisting of SRO layers, to maximize quantum confinement, interspersed with stoichiometric Si_3N_4 interlayers, to minimize diffusion and maximize vertical transport. The results certainly show greater control of size uniformity for Si nanocrystals, including both P- and B-doped material. They also indicate significantly higher conductivities for the nitride interlayer materials, certainly for B-doped material (Di et al. 2010b).

15.5.3 Bandgap Tunability and Photoluminescence

For all-Si tandem solar cell applications, quantum confinement of Si QD materials allows the fabrication of higher bandgap solar cells that can be used as upper cells on top of normal silicon cells. The effective mass approximation (EMA) is often used in determining the absolute confined energy levels for isolated nanocrystals (Brus 1984):

$$E_g(r) = E_g + \Delta E = E_g + \frac{\hbar^2\pi^2}{2r^2}\left(\frac{1}{m_e} + \frac{1}{m_h}\right)$$

where E_g is the bandgap of the bulk material, ΔE is the bandgap increase, r is the radius of the nanocrystals, and m_e and m_h are the effective mass of electrons and holes. Although this method relies on the semiconductor nature of the bulk materials and is arguably valid in the ultrasmall size regime, it gives a relatively correct prediction of the bandgap widening as QD size decreases. Figure 15.48 compares the EMA calculation of Si QDs with some reported photoluminescence (PL) energies of Si QDs embedded in SiO_2 and Si_3N_4 (Zacharias et al. 2002; Nesbit 1985; Wan et al. 2011b; Di et al. 2010a,b; Kim et al. 2004, 2006; Cho 2003, Cho et al. 2005; Takagi et al. 1990). This shows a decreasingly inaccurate

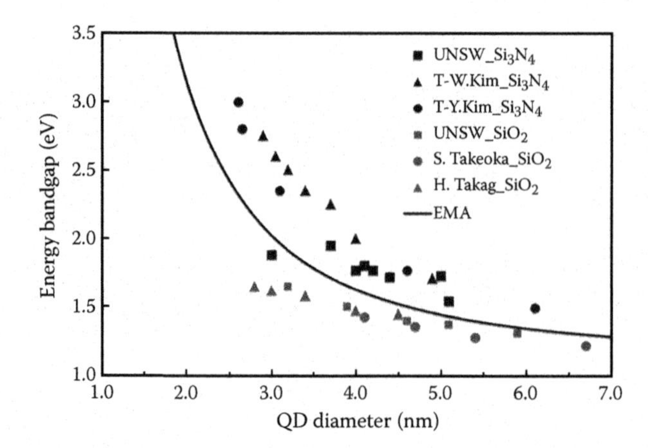

FIGURE 15.48
Measured photoluminescence energies of Si QDs in SiO_2 and Si_3N_4 (300°K) as a function of QD size, from several authors. Also shown is the EMA calculation of quantum confined energy levels for spherical QDs. (From Brus, L.E. *J. Chem. Phys.* 80:4403–4409, 1984. With permission.)

prediction of the confined energy level by the EMA as the QD size decreases (<3.5 nm in diameter). It also indicates that the trend of the data is better predicted for QDs in nitride than in oxide in the smaller size regime.

The PL results shown in Figure 15.48 for Si QDs from different authors are in good agreement where the matrix is the same, but are quite different for QDs in oxide (Cho 2003; Takagi et al. 1990; Takeoka et al. 2000) as compared with nitride (Nesbit 1985; Kim et al. 2004; Cho 2003, Cho et al. 2005) particularly for small QDs. They are also qualitatively consistent with the results from *ab initio* modeling (König et al. 2008a,b), which have been carried out for the confined energy levels in Si QDs consisting of a few hundred atoms. This uses Gaussian03, a density functional, Hartree–Fock-based *ab initio* program. Calculations have been carried out on the gaps between the highest occupied molecular orbital and the lowest unoccupied molecular orbital. These gaps are the ground state confined energy levels for Si nanocrystals of various sizes terminated with either –H (the closest to vacuum for a terminated surface), –OH groups, or $–NH_2$ groups (using a nearest-neighbor assumption these two are assumed analogous to a SiO_2 and Si_3N_4 matrix). The modeling shows the expected increasing confinement energy with decreasing QD size but also shows the reduction in energy on going from a QD effectively in a vacuum to one embedded in a dielectric (König et al. 2008a,b). It is also seen that the amino-terminated QDs (simulating nitride) have energies approximately 0.5eV greater than the hydroxyl terminated ones (simulating oxide). This is qualitatively consistent with the PL results for QDs in oxide and nitride shown in Figure 15.48. This can be explained by the more significant effect of the strong polar interface of the Si–O bond on the electronic structure in the case of SiO_2 matrix as compared with the much lower polarity of the Si–N bond for the Si_3N_4 matrix, especially when the QD size is smaller than approximately 3.7 nm.

There are, however, other factors which should be taken into account with PL data. The larger discrepancy of PL energies for small Si QDs embedded in SiO_2 may also be attributed to the wide size distribution of amorphous Si clusters coexisting with Si QDs, which may dominate the PL property of the samples (So et al. 2011c). This also implies that PL energy may not be a good measure for bandgap energy in some cases, particularly where an amorphous phase is involved. Other problems associated with spurious PL emission from defects and the convolution of exciton binding energy with confined energy levels in PL (exciton binding energy also increasing as QD size decreases) mean that PL data must be considered very carefully and can probably only be taken as being indicative of trends.

15.5.4 Doping and Carrier Transport in Si QD Materials

A requirement for a tandem cell element is the presence of some form of junction for carrier separation. The impurities in bulk crystalline silicon play an important role in a semiconductor devices. Dopants such as phosphorous and boron alter the conductivity of bulk Si by several orders of magnitude. There are several questions about impurity doping in a low-dimensional structure (Ossicini et al. 2006). It is not certain if dopants will continue to play a role similar to that in bulk semiconductors, or whether alternative methods of work function control will be required. It is not clear whether or not the doping of Si nanocrystals currently provides the generation of free charge carriers (Polisski et al. 1999). Also, because the interface area to volume ratio increases significantly with the decrease of nanocrystal size, dopants are expected to easily diffuse out of nanocrystals, thus making nanocrystal doping kinetically unfavorable (Erwin et al. 2005). Moreover, Cantele et al. (2005) have reported that the formation energies of neutral impurities within Si QDs is higher than that of bulk Si, thereby making impurity doping of nano-dimensional structures energetically unfavorable. Despite all these difficulties, several researchers have reported successful doping of nanocrystals (Hao et al. 2009a,b; Park et al. 2009; Perego et al. 2010; So et al. 2011b; Song et al. 2007).

15.5.4.1 n-Type Doping

Phosphorus (P) is an excellent dopant in bulk Si as it has a high solid solubility at the annealing temperature. Hence, it is a good initial choice to study the doping in Si QDs, although as discussed previously, there are reasons to suppose it will not dope them in the same way as bulk Si. P incorporation into the Si QDs/SiO_2 superlattices has been achieved using P_2O_5 cosputtering during the deposition of SRO layers followed by a 1100°C anneal. Transfer length measurements were used to calculate both the conductivity at 300 K and its temperature dependence, with plots of log R versus $1000/T$. The values of the activation energy E_A can be calculated from the relation $R \approx \exp(E_A/kT)$. As the P doping level increases from 0 to 0.1 at.%, the activation energy decreases from 0.527 to 0.101 eV, together with a very significant increase in conductivity of seven orders of magnitude (Hao et al. 2009b). This implies an effective doping of presumably n-type.

Antimony (Sb) is another widely used n-type dopant for the bulk Si. To avoid bringing oxygen into other dielectric matrices, such as Si_3N_4, Sb doping has been attempted using cosputtering (So et al. 2011b). The effect of Sb doping on the Si nanocrystals films was investigated in terms of structural, optical, and electrical properties. The results of TEM and XRD suggest that low Sb concentration (0.54 at.%) induced negligible Si QD size variation in the Sb-doped SRN films (So et al. 2011b). XPS analysis revealed the presence of Sb–Si bonds, which implies Sb atoms were either incorporated within the Si QDs or located at the interface of QDs and the Si_3N_4 matrix. n-type electrical behavior in the Sb-doped Si QDs film was observed by Hall measurements. Conductivities measured by the transfer length measurement show that for a 0.54 at.% Sb-doped sample (~2.8×10^{-2} S/cm), the conductivity is six orders of magnitude higher than for an undoped one (~7.3×10^{-8} S/cm), which could be attributed to the increase in carrier concentration. The result from Hall measurements, together with the observed enhancement in conductivity, indicates that

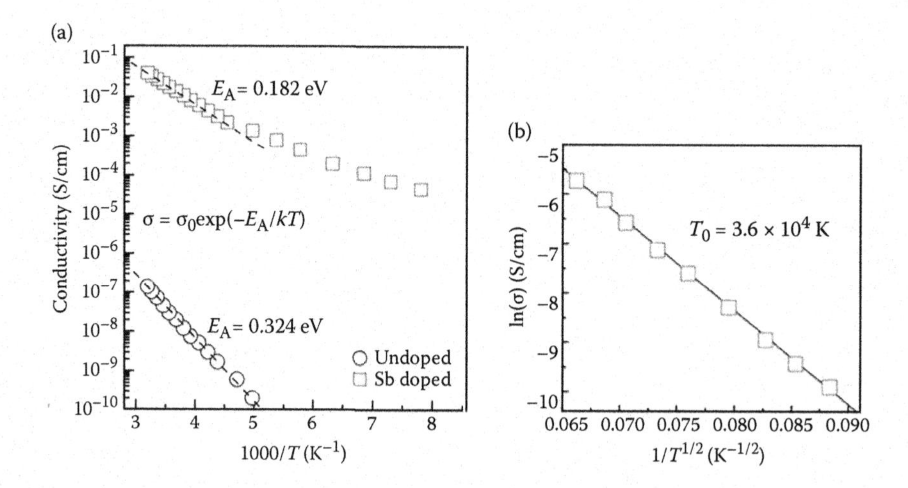

FIGURE 15.49
Temperature-dependent conductivities for undoped and 0.54 at.% Sb-doped SRN films (a). ln(σ) versus $1/T^{1/2}$ plot from 120 to 220 K, for the 0.54 at.% Sb-doped SRN film (b). The solid line is a least square fit to the data.

the improvement of electrical properties could be due to effective Sb doping in Si QDs films. The temperature-dependent conductivities of both undoped and Sb-doped SRN films are shown in Figure 15.49a. As can be seen, doping of the Si QDs strongly influences the electronic transport properties of the films. Arrhenius-like temperature dependence was observed in the temperature range between 220 and 320 K, attributable to thermally activated conduction. From the slope of the Arrhenius plot, the activation energy E_A was found to decrease to 0.182 eV for the Sb-doped Si QDs film, suggesting effective doping. Nevertheless, the extracted E_A is much larger than that of bulk Si with the same dopant. One possible explanation for the large E_A observed would be the deeper donor level expected in Si QDs. Also, it is possible that the number of free carriers available for conduction may be limited due to a trap density distributed within the bandgap that is comparable to the doping density. In addition, it is noted that the E_A of the undoped film (~0.324 eV) was shifted and lower than that expected for intrinsic Si of approximately 0.5 eV, which is around half the bulk Si energy bandgap. This implies the existence of traps and defects in the Si_3N_4 matrix as well as those arising from the interface or in the QDs. Interestingly, the conductivity for the Sb-doped SRN film at $T < 220$ K deviates from the Arrhenius behavior and is best described by the expression, $\sigma = \sigma_0 \exp[-(T_0/T)^{1/2}]$ as illustrated in Figure 15.49b. Here, σ_0 is a pre-exponential factor and the material constant T_0 is 3.6×10^4 K from the data fitting. This behavior of $\ln(\sigma) \propto T^{-1/2}$ suggests that the percolation-hopping model explains the low temperature (<220 K) conduction in Sb-doped Si QD films (Simánek 1981).

15.5.4.2 p-Type Doping

Boron (B) doping has been achieved by cosputtering from B, Si, and SiO_2 targets with subsequent annealing to form the Si QDs in SiO_2 matrix [35]. XPS results indicate that the chemical environment of B in both as-deposited and annealed B-doped SRO films is highly dependent on the O/Si ratio of the SRO layer. A tendency for greater B–O bonding upon high-temperature annealing, indicating B out-diffusion from B–B and B–Si to B–O, was found to be more pronounced in the high-oxygen content SRO film. The results suggest a higher probability of effective B doping in SRO films with high Si (low oxygen) content. Using transfer length measurements, it was found that increasing the B sputtering plasma power, and hence, the B concentration, causes a dramatic increase in conductivity (~10^{-2} Scm^{-1}), to a similar extent as that with P doping. The activation energy also decreases from 0.527 to 0.099 eV (Hao et al. 2009a). This implies an effective doping of presumably p-type material.

B-doping has also been attempted for Si QDs in Si_3N_4 matrix using similar cosputtering and annealing processes (So 2011). The introduction of B (~1.03 at.%) leads to an increase in the conductivity of the film to 10^{-6} Scm^{-1} at room temperature. This is nearly 100 times higher than that of the undoped film (~10^{-8} Scm^{-1}), which could be a result of an increase in carrier concentration due to doping. However, it has been noted that the improvement of conductivity with the B doping is less significant than the Sb doping discussed above or of that in B-doped Si QDs in oxide. The observed reduction in activation energy (~0.2 eV) of the B-doped film could be explained by the shift in the Fermi level as a cumulative effect of defects, traps, and activated dopants in the film. The charge transport mechanism in the films could be attributed to the thermally activated nearest hopping conduction as revealed by the $\ln(\sigma) \propto T^{-1}$ relationship by temperature-dependent measurement. From XPS measurements, it was found that a large fraction of B atoms are bonded with N. Thus, the less pronounced improvement in conductivity could be due to the inefficient formation of B–Si bonding.

Further independent evidence that true n- and p-type material can be made comes from the fabrication of MOS Si QD multilayer structures in SiO_2 matrix, with doping by either P or B from spin-on sources followed by an appropriate diffusion anneal (Ma et al. 2011; Lin et al. 2011). C-V on these structures clearly shows a complete reversal of sign from P to B doped material, with negative bias giving inversion in P-doped and accumulation in B-doped material. This clearly indicates n- and p-type behavior, respectively.

15.5.4.3 Doping Mechanisms

In the fabrication of p- and n-type QD materials, the doping mechanisms are not clear. Direct doping of the Si QDs is very difficult due to the exclusion of impurities from the nanocrystals discussed in Section 15.4.4. Hence, free carriers are possibly introduced by either doping of the matrix or of the interface between matrix and QDs. The interface is the most likely place to which these impurity atoms will migrate because of the relatively larger interstitial and defect sites at these locations. The defects at the interface are also associated with multiple charge states. Hence, formation of appropriately charged states such that free carriers are given up and captured by the QDs seems plausible. But these locations are also associated with many defects in the bandgap of the QDs and hence would be expected to enhance recombination dramatically. However, the passivation of these regions, which seems important for the device performance (So 2011; Perez-Wurfl et al. 2011; Di et al. 2010c), presumably reduces the effect of these defects. Furthermore, the tendency of the strong local fields, associated with the interface regions, to sweep free carriers into the QDs, will mitigate the recombination somewhat.

15.5.5 Si QDs Solar Cell

Both homojunction and heterojunction PV devices have been fabricated based on Si QDs in Si dielectric matrices. Figure 15.50 is a schematic diagram of a p-i-n device fabricated on a quartz slide. The device consists of sequentially grown multilayers of P-doped SRO layers, followed by multilayers of undoped (i layer) and then B-doped SRO layers, all interspersed

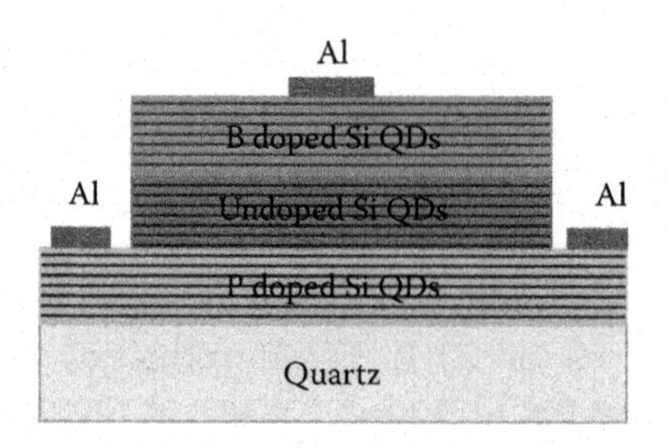

FIGURE 15.50
Schematic diagram of the fabricated p-i-n Si QDs solar cell.

with stoichiometric SiO_2 layers (Perez-Wurfl et al. 2011; Di et al. 2010c; Hao et al. 2009d). The top B-doped bilayers and i layers were selectively etched to create isolated p-type mesas of approximately 0.1 cm^2, thus allowing access to the buried P-doped bilayers. Aluminum contacts were deposited by evaporation, and then patterned and sintered to create ohmic contacts on both p- and n-type layers, as shown in Figure 15.50.

IV measurements in the dark and under one-sun illumination indicate a good rectifying junction and generation of an open circuit voltage, V_{oc}, up to 492 mV (Perez-Wurfl et al. 2011; see Figure 15.51). The high sheet resistance of the deposited layers, in conjunction with the insulating quartz substrates, causes an unavoidable extremely high series resistance in the devices. The high resistance severely limits both the short circuit current and the fill factor of the cells, particularly under illumination. This also makes it necessary to include effects due to in-plane current flow in the analysis of the measured electrical characteristics (Annual Report of the Photovoltaics Centre of Excellence 2008). It should be noted that such high lateral series resistances are not an inherent problem for a final vertically integrated tandem cell device (although high resistivity is still an issue). High lateral resistance is however an issue for these interim test devices on transparent quartz substrate. Methods to get around this, by using transparent conducting substrates, are under investigation.

Further evidence that this PV effect occurs in a material with an increased bandgap is given by temperature-dependent dark I-V measurements, from which an electronic

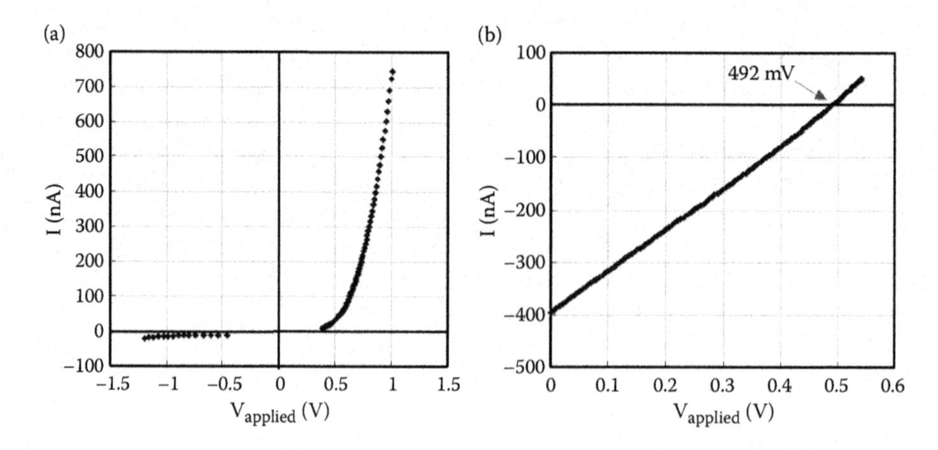

FIGURE 15.51
Electrical characteristics of a p-i-n device with 4 nm SRO/2 nm SiO_2 bilayers with nominal O/Si = 0.8: dark (a) and illuminated (b) I-V measurements showing V_{oc} = 492 mV.

bandgap for the Si QD nanostructure materials can be extracted. A bandgap of 1.8 eV was extracted for a structure containing Si QDs with a diameter of 4 nm. The extracted bandgap is larger than that of bulk silicon, highlighting the ability to alter the bandgap of a semiconductor using these multilayer nanostructures (Perez-Wurfl et al. 2009).

Current in devices is very small, due principally to the very high resistance, but also because of the small amount of absorption in the thin material used—but both these problems are being addressed. Hence, together with a further increase in the V_{oc}, this represents a promising approach to the fabrication of a solar cell with an engineered bandgap. A full tandem cell would then require devices of different engineered bandgap (different QD sizes) to be grown on top of each other with a suitable connection between them. This connection would need to allow excited carriers resulting from photogeneration at one wavelength, to cross into the adjacent cell and be available to absorb another photon at another wavelength. This can be achieved using very thin layers with large defect densities, which allow a recombination of electrons and holes but not a relaxation in their energy (Annual Report of the Photovoltaics Centre of Excellence 2008).

15.5.6 Conclusion

The approach of third-generation PV is to provide high conversion efficiency of photon energy with low manufacturing cost. The Si QD–based "all-Si" tandem solar cell combines the methodology of using multiple energy thresholds and of low cost deposition processes with abundant nontoxic materials to offer significant leverage in achieving this goal.

The principal property exploited in the use of Si QDs for solar cells is the confined energy levels. Control of the QD size and shape allows a material with a larger bandgap than the bulk material to be engineered through the overlap of levels to create a miniband. This in turn allows the fabrication of doped p–n junction solar cells in which the bandgap can be optimized for the best extraction of energy from a particular part of the solar spectrum. The optimized superlattice structures of Si QDs in Si dielectric matrices and barrier layers can be achieved by choosing the appropriate annealing temperature, excess Si content, and barrier materials. Although the impurity doping mechanism is not clearly understood, P, Sb, and B doping have successfully shown a great enhancement of the conductivities of the materials and evidence for appropriate carrier types. Homojunction devices have been fabricated based on Si QD multilayer structures, which demonstrate both increased effective bandgap and promising open circuit voltages. Improvement in carrier transport in these cells is needed. For a full tandem cell device, Si QD structures with different bandgaps (QD sizes) need to be grown on top of each other with appropriate connections. These two requirements should be met with further developments.

References

Aberle, A.G. 1999. Crystalline Silicon Solar Cells: Advanced Surface Passivation and Analysis. The University of New South Wales, Sydney, Australia.

Aberle, A.G., S., Glunz, and W. Warta. 1992. Limiting loss mechanisms in 23% efficient silicon solar cells. *Journal of Applied Physics* 77:3491.

Abrams, Z.R., A. Niv, and X. Zhang. 2011. Solar energy enhancement using down-converting particles: A rigorous approach. *Journal of Applied Physics* 109:114905.

Agarwal, G., S. De Iuliis, L. Serenelli et al. 2011. Dry texturing of mc-Si wafers. *Physica Status Solidi C* 8(3):903–906.

Annual Report of the Photovoltaics Centre of Excellence. 2008. University of New South Wales, Sydney. http://www2.pv.unsw.edu.au/nsite-files/anreports/unsw_Photovoltaics_an_report_2008.pdf.

Auzel, F. 2004. Upconversion and anti-Stokes processes with f and d ions in solids. *Chemical Reviews* 104:139–173.

Badescu, V. 2008. An extended model for up-conversion in solar cells. *Journal of Applied Physics* 104:113120.

Badescu, V., and A. De Vos. 2007. Influence of some design parameters on the efficiency of solar cells with down-conversion and down shifting of high-energy photons. *Journal of Applied Physics* 102:073102.

Benick, J. et al. 2010. Effect of a post-deposition anneal on Al_2O_3/Si interface properties. *35th PVSC, Honolulu, Hawaii.*

Bentzen, A., A. Holt, R. Kopecek et al. 2006. Gettering of transition metal impurities during phosphorus emitter diffusion in multicrystalline silicon solar cell processing. *Journal of Applied Physics* 99:093509.

Bentzen, A., E.S. Marstein, R. Kopecek et al. 2004. Phosphorus diffusion and gettering in multi-crystalline silicon solar cell processing. In *The 19th European Photovoltaic Solar Energy Conference and Exhibition, Paris, France*, 935–938.

Boer, K.W. 1990. *Survey of Semiconductor Physics: Electrons and Other Particles in Bulk Semiconductors.* New York: Van Nostrand Reinhold.

Brown, A.S., and M.A. Green. 2002. Limiting efficiency for current-constrained two-terminal tandem cell stack. *Progress in Photovoltaics* 10:299–307.

Brus, L.E. 1984. Electron–electron and electron–hole interactions in small semiconductor crystallites: The size dependence of the lowest excited electronic state. *Journal of Chemical Physics* 80:4403–4409.

Bruton, T.M., N.B. Mason, S. Roberts et al. 2003. Towards 20% efficient silicon solar cells manufactured at 60 MWp per annum. *Proceedings of the 3rd WCPEC, Osaka, Japan*, 1899.

Budhraja, V. et al. 2010. Advancements in PV multicrystalline silicon solar cells from 1980 to 2010—An overview. *Proceedings of the 37th IEEE PV Conference, Seattle, USA.*

Bultman, J.H., M.W. Brieko, A.R. Burgers et al. 2001. Interconnection through VIAS for improved efficiency and easy module manufacturing of crystalline silicon solar cells. *Solar Energy Materials and Solar Cells* 65(1–4):339–345.

Burgers, A.R. et al. 2010. 19% efficient n-type Si solar cell made in pilot line production. *Proceedings of the 25th EUPVSEC, Valencia, Spain.*

Campbell, P., and M.A. Green. 2001. High performance light trapping textures for monocrystalline silicon solar cells. *Solar Energy Materials and Solar Cells* 65:369.

Cantele, G., E. Degoli, E. Luppi et al. 2005. First-principles study of n- and p-doped silicon nanoclusters. *Physical Review* B 72:113303.

Cho, E.-C. 2003. Optical Transitions in SiO_2/Crystalline Si/SiO_2 Quantum Wells and Nanocrystalline Silicon (nc-Si)/SiO_2 Superlattice Fabrication. PhD Thesis, University of New South Wales, Sydney.

Cho, Y.-H., E.-C. Cho, Y. Huang et al. 2005. *Proceedings of the 20th European Photovoltaic Solar Energy Conference, Barcelona, Spain.*

Chung, P., H. Chung, and P.H. Holloway. 2007. Phosphor coatings to enhance Si photovoltaic cell performance. *Journal of Vacuum Science & Technology A* 25(1):61–66.

Conibeer, G., M.A. Green, E-C. Cho et al. 2008. Silicon quantum dot nanostructures for tandem photovoltaic cells. *Thin Solid Films* 516:6748–6756.

Cousins, P.J., D.D. Smith, H.C. Luan et al. 2010. Generation III: Improved performance at lower cost. *Proceedings of the 35th IEEE PVSC, Hawaii*, 275–278.

Cuevas, A., and D. MacDonald. 2005. Multicrystalline silicon: A review of its electronic properties. *International Photovoltaic Science and Engineering Conference (PVSEC 2005), Shanghai*, 521–524.

Dauwe, S., L. Mittelstadt, A. Metz et al. 2002. Experimental evidence of parasitic shunting in silicon nitride rear surface passivated solar cells. *Progress in Photovoltaics: Research and Applications* 10(4):271–278.

Dauwe, S., L. Mittelstadt, A. Metz et al. 2003. Low-temperature rear surface passivation schemes for >20% efficient silicon solar cells. *Proceedings of the 3rd WCPVSEC, Osaka, Japan*, 1395–1398.
Davis, J.R., A. Rohatgi, R.H. Hopkins et al. 1980. Impurities in silicon solar cells. IEEE Transactions on Electronic Devices 27:677–687.
de Boer, D.K.G. 2010. Luminescent solar concentrators: The road to low-cost energy from the sun. *SPIE Newsroom* 21. doi: 10.1117/2.120100.
De Wolf, S., and M. Kondo. 2007. Abruptness of a-Si:H/c-Si interface revealed by carrier lifetime measurements. *Applied Physics Letters* 90:042111.
Deng, K., T. Gong, L. Hu et al. 2011. Efficient near-infrared quantum cutting in NaYF4: Ho3+, Yb3+ for solar photovoltaics. *Optics Express* 19(3):1749–1754.
Dexter, D.L. 1979. Two ideas on energy transfer phenomena: Ion-pair effects involving the OH stretching mode, and sensitization of photovoltaic cells. *Journal of Luminescence* 18:779–784.
Di, D., I. Perez-Wurfl, G. Conibeer et al. 2010a. Formation and photoluminescence of Si quantum dots in SiO_2/Si_3N_4 hybrid matrix for all-Si tandem solar cells. *Solar Energy Materials and Solar Cells* 94:2238–2243.
Di, D., I. Perez-Wurfl, G. Conibeer et al. 2010b. Fabrication and characterisation of silicon quantum dots in SiO2/Si3N4 hybrid matrix. In *Proceedings of SPIE—Optics and Photonics—Next Generation (Nano) Photonic & Cell Technologies for Solar Energy Conversion, San Diego.*
Di, D., I. Perez-Wurfl, A. Gentle et al. 2010c. Impacts of post-metallisation processes on the electrical and photovoltaic properties of Si quantum dot solar cells. *Nanoscale Research Letters* 5:1762–1767.
Dingemans, G., M.M. Mandoc, S. Bordihn et al. 2011. Effective passivation of Si surfaces by plasma deposited SiOx/a-SiNx:H stacks. *Applied Physics Letters* 98:222102.
Diouf, D., J.P. Kleider, T. Desrues et al. 2008. Interdigitated back contact a-Si:H/c-Si heterojunction solar cells modelling: Limiting parameters influence on device efficiency. *Proceedings of the 23rd EUPVSEC, Valencia, Spain, 1949.*
Diouf, D., J.P. Kleider, T. Desrues et al. 2009. Study of interdigitated back contact silicon heterojunction solar cells by two-dimensional numerical simulations. *Materials Science and Engineering B* 159–160:291–294.
ECN, Energy Research Centre of the Netherlands, www.ecn.nl.
Eilers, J.J., J.T. van Wijngaarden, K. Krämer et al. 2010. Efficient visible to infrared quantum cutting through down-conversion with the Er3+–Yb3+ couple in Cs3Y2Br9. *Applied Physics Letters* 96:151106.
Erwin, S.C., L. Zu, M.I. Haftel et al. 2005. Doping semiconductor nanocrystals. *Nature* 436:91–94.
Fischer, H., and W. Pschunder. 1973. Investigation of photon and thermal induced changes in silicon solar cells. *Proceedings of the of the 10th IEEE PVSC, Palo Alto, CA, USA*, 404–411.
Fischer, S., J.C. Goldschmidt, P. Löper et al. 2010. Enhancement of silicon solar cell efficiency by up-conversion: Optical and electrical characterization. *Journal of Applied Physics* 108:044912.
Fraunhofer ISE. Institute for Solar Energy. http://www.ise.fraunhofer.de/.
Froitzheim, A., K.A. Münzer, K.H. Eisenrith et al. 2005. n-type silicon solar cell based on industrial technology. *Proceedings of the 20th EUPVSEC, Barcelona, Spain*, 594.
Gee, J.M., K.W. Schubert, and P.A. Basore. 1993. Emitter wrap through solar cell. *Proceedings of the 23rd IEEE PVSC, Louisville, USA*, 265.
Glunz, S.W., R. Preu, S. Schaefer et al. 2000a. New simplified methods for patterning the rear contact of RP-PERC high efficiency solar cells. *Proceedings of the 28th IEEE PVSC, Anchorage, Alaska, USA*, 168–171.
Glunz, S.W., S. Rein, J. Knobloch et al. 1999. Comparison of boron- and gallium-doped p-type Czochralski silicon for photovoltaic application. Progress in Photovoltaics: Research and Applications 7:463–469.
Glunz, S.W., S. Rein, and J. Knobloch. 2000b. Stable Czochralski silicon solar cells using gallium-doped base material. *Proceedings of the 16th EUPVSEC, Glasgow, UK*, 1070–1075.
Glunz, S.W., S. Rein, J.Y. Lee et al. 2001. Minority carrier lifetime degradation in boron-doped Czochralski silicon. *Journal of Applied Physics* 90(5):2397–2404.

Glunz, S.W., E. Schneiderlöchner, D. Kray et al. 2004. Silicon solar cells on p-type and n-type substrates. *Proceedings of the 19th EUPVSEC, Paris, France*, 408–411.

Goetzberger, A., and W. Greubel. 1977. Solar energy conversion with fluorescent collectors. *Applied Physics A* 14:123–139.

Goldschmidt, J.C., M. Peters, A. Bosch et al. 2009. Increasing the efficiency of fluorescent concentrator systems. *Solar Energy Materials and Solar Cells* 93:176–182.

Good, E.A., R. Kopecek, and J. Arumughan. 2008. Characterising device efficiency potential from industrial multi-crystalline cell structures composed of solar grade silicon. *23rd European Photovoltaic Solar Energy Conference and Exhibition, Valencia, Spain*, 1219–1224.

Graff, K. 2000. *Metal Impurities in Silicon Device Fabrication (Springer Series in Materials Science)*, 2nd ed., Berlin: Springer Verlag.

Green, M.A. 1987. *High Efficiency Silicon Solar Cells*. Aedermannsdorf, Switzerland: Trans. Tech Publications.

Green, M.A. 1998. Solar Cells: Operating Principles, Technology and System Applications. The University of New South Wales, Sydney, Australia.

Green, M.A. 2003. *Third Generation Photovoltaics—Advanced Solar energy Conversion*. Berlin: Springer-Verlag.

Green, M.A. 2009. The path to 25% silicon solar cell efficiency: History of silicon cell evolution. *Progress in Photovoltaics: Research and Applications* 17:183–189.

Green, M.A., E.-C. Cho, Y. Cho et al. 2005. All-silicon tandem cells based on "artificial" semiconductor synthesised using silicon quantum dots in a dielectric matrix. *Proceedings of 20th European Photovoltaic Solar Energy Conference, Barcelona, Spain*, 3–6.

Green, M.A., G. Conibeer, D. Konig et al. 2006. Progress with all-silicon tandem cells based on silicon quantum dots in a dielectric matrix. *Proceedings of 21st European Photovoltaic Solar Energy Conference, Dresden, Germany*.

Green, M.A., K. Emery, Y. Hishikawa et al. 2011. Solar cell efficiency tables (version 38). *Progress in Photovoltaics* 19:565–572.

Guillevin, N. et al. 2010. High efficiency n-type metal wrap trough Si solar cells for low cost industrial production. *25th EUPVSEC, Valencia, Spain*.

Guo, J.-H., B.S. Tjahjono, and J.E. Cotter. 2005. 19.2% efficiency n-type laser-grooved silicon solar cells. *Proceedings of the 31st IEEE PVSC, Orlando, USA*, 983.

Guttler, G., and H.J. Queisser. 1970. Impurity photovoltaic effect in silicon. *Energy Conversion* 10:51.

Hanna, M.C., and A.J. Nozik. 2006. Solar conversion efficiency of photovoltaic and photoelectrolysis cells with carrier multiplication absorbers. *Journal of Applied Physics* 100:074510-1-074510-8.

Hao, X.J., E.-C. Cho, C. Flynn et al. 2009a. Synthesis and characterization of boron-doped Si quantum dots for all-Si quantum dot tandem solar cells. *Solar Energy Materials and Solar Cells* 93:273–279.

Hao, X.J., E-C. Cho, G. Scardera et al. 2009b. Phosphorus doped silicon quantum dots for all-silicon quantum dot tandem solar cells. *Solar Energy Materials and Solar Cells* 93:1524–1530.

Hao, X.J., I. Perez-Wurfl, G. Conibeer et al. 2009d. Study on properties of Si QDs junction in oxide matrix for "all-silicon" tandem solar cells. *Proceedings of the PVSEC 19, Korea*.

Hao, X.J., A. Podhorodecki, Y.S. Shen et al. 2009c. Effects of Si-rich oxide layer stoichiometry on the structural and optical properties of Si QDs/SiO_2 multilayer film. Nanotechnology 20:485703.

Harada, K., H. Tanaka, J. Matsubara et al. 1995. Origins of metal impurities in single-crystal Czochralski silicon. *Journal of Crystal Growth* 154:47–53.

He, D., C. Yu, J. Cheng et al. 2010. Intense near-infrared emission of Yb3+ related with charge transfer in phosphate glass. *Solid State Communications* 150:2354–2356.

Henley, F., A. Lamm, S. Kang et al. 2008. Direct film transfer (DFT) technology for ker-free silicon wafering. *Proceedings of the 23rd EUPVSEC, Valencia, Spain*, 2BO.2.3.

Herguth, A., G. Schubert, M. Kaes et al. 2006. Avoiding boron–oxygen related degradation in highly boron-doped CZ silicon. *Proceedings of the 21st EUPVSEC, Dresden, Germany*, 530–537.

Hering, G. 2011. The year of the tiger. *Photon International* 3:186–218.

Hoex, B., J.J.H. Gielis, M.C.M. van de Sanden et al. 2008. On the c-Si surface passivation mechanism by the negative-charge dielectric Al2O3. *Journal of Applied Physics* 104:113703.

Hoffmann, V., K. Petter, J. Djordjevic-Reiss et al. 2008. First results on industrialization of Elkem Solar Silicon at Pillar and Q-Cells. *23rd European Photovoltaic Solar Energy Conference and Exhibition, Valencia, Spain*, 1117–1120.

Hofmann, M., S.W. Glunz, R. Preu et al. 2006. 21%-efficient silicon solar cells using amorphous silicon rear side passivation. *Proceedings of the 21st EUPVSEC Dresden, Germany*, 609.

Horzel, J., J. Szlufcik, and J. Nijs, 2000. High efficiency industrial screen printed selective emitter solar cells. *Proceedings of the 16th EUPVSEC, Glasgow, UK*, OB1.2.

Hoshi, K., N. Isawa, T. Suzuki et al. 1985. Czochralski silicon crystals grown in a transverse magnetic field. *Journal of the Electrochemical Society* 132:693–700.

Huang, C.Y., D.Y. Wang, C.H. Wang et al. 2010. Efficient light harvesting by photon down-conversion and light trapping in hybrid ZnS nanoparticles/Si nanotips solar cells. *ACS Nano* 4(10):5849–5854.

Innovalight, http://www.innovalight.com.

Istratov, A.A., H. Hieslmair, and E.R. Weber. 1999. Iron and its complexes in silicon. *Applied Physics. A* 69:13–44.

Itsumi, M. 1993. Method of determining metal contamination by combining p-type Si and n-type Si recombination lifetime measurements. *Applied Physics Letters* 63:1095.

Ivanova, S., and F. Pelle. 2009. Strong 1.53 μm to NIR-VIS-UV up-conversion in Er-doped fluoride glass for high-efficiency solar cells. *Journal of the Optical Society of America. B, Optical Physics* 26:1930–1938.

Jackson, E.D. 1955. Areas for improvement of the semiconductor solar energy converter. *Transactions of the Conference on the Use of Solar Energy* 5:122–126.

Jin, T. 1997. Studies on Luminescence Properties of the Lanthanide Complexes Incorporated into Silica-Based Glass Matrices. PhD Thesis, Department of Applied Chemistry, Faculty of Engineering, Osaka University, Japan.

Jin, T., S. Inoue, K. Machida et al. 1997. Photovoltaic cell characteristics of hybrid silicon devices with lanthanide complex phosphor-coating film. *Journal of the Electrochemical Society* 144:4054–4058.

Kamat, P. 2007. Meeting the clean energy demand: Nanostructure architectures for solar energy conversion. *Journal of Physical Chemistry C* 111:2834–2860.

Kawano, K., K. Arai, H. Yamada et al. 1997. Application of rare-earth complexes for photovoltaic precursors. *Solar Energy Materials and Solar Cells* 48:35–41.

Keevers, M., and M.A. Green. 1994. Efficiency improvements of silicon solar cells by the impurity photovoltaic effect. *Journal of Applied Physics* 75:4022.

Kim, T.-W., C.-H. Cho, B.-H. Kim et al. 2006. Quantum confinement effect in crystalline silicon quantum dots in silicon nitride grown using SiH4 and NH3. *Applied Physics Letters* 88:123102.

Kim, T.Y., N.M. Park, K.H. Kim et al. 2004. Quantum confinement effect of silicon nanocrystals in situ grown in silicon nitride films. *Applied Physics Letters* 85:5355.

Klampaftis, E., and B.S. Richards. 2011. Improvement in multi-crystalline silicon solar cell efficiency via addition of luminescent material to EVA encapsulation layer. *Progress in Photovoltaics: Research and Applications* 19:345–351.

Klimov, V.I., S.A. Ivanov, J. Nanda et al. 2007. Single-exciton optical gain in semiconductor nanocrystals. *Nature* 447:441–446.

Knobloch, J., S.W. Glunz, V. Henninger et al. 1995. 21% efficient solar cells processed from Czochralski grown silicon. *Proceedings of the 13th EUPVSEC, Nice, France*, 9–12.

Kolodinski, S., J.H. Werner, T. Wittchen et al. 1993. Quantum efficiencies exceeding unity due to impact ionization in silicon solar cells. *Applied Physics Letters* 63:2405.

König, D., J. Rudd, M.A. Green et al. 2008a. Impact of interface on effective band gap of Si quantum dots. *Solar Energy Materials and Solar Cells* 93:753–758.

König, D., J. Rudd, M.A. Green et al. 2008b. Role of the interface for the electronic structure of silicon quantum dots. *Physical Review B* 78:035339.

Kopecek, R., J. Arumughan, K. Peter et al. 2008. Crystalline Si solar cells from compensated material: Behaviour of light induced degradation. *23rd European Photovoltaic Solar Energy Conference and Exhibition, Valencia, Spain*, 1855–1858.

Kray, D., S. Rein, D. Oswald et al. 2001. High-efficiency emitter wrap-through cells. *Proceedings of the 17th EUPVSEC, Munich, Germany*, 1299.

Kurokawa, Y., A. Yamada, S. Miyajima et al. 2010. Effects of oxygen addition on electrical properties of silicon quantum dots/amorphous silicon carbide superlattice. Current Applied Physics 10:S435–S438.

Lakshminarayana, G., H. Yang, S. Ye et al. 2008. Co-operative down-conversion luminescence in Tm3+/Yb3+: SiO2–Al2O3–LiF–GdF3 glasses. Journal of Physics D: Applied Physics 41:175111.

Lammert, M.D., and R.J. Schwartz. 1977. The interdigitated back contact solar cell—a silicon solar cell for use in concentrated sunlight. *IEEE Transactions on Electronic Devices* 24:337.

Lanyon, H.P.D., and R.A. Tuft. 1979. Bandgap narrowing in moderately to heavily doped silicon. *IEEE Transactions on Electronic Devices* 26:1014.

Lauinger, T., A.G. Aberle, and R. Hezel. 1997. Comparison of direct and remote PECVD silicon nitride films for low-temperature surface passivation of p-type crystalline silicon. *Proceedings of the 14th EUPVSEC, Barcelona, Spain*, 853.

Le Donne, A., M. Acciarri, D. Narducci et al. 2009. Encapsulating Eu3+ complex doped layers to improve Si-based solar cell efficiency. Progress in Photovoltaics: Research and Applications 17(8):519–525.

Le Donne, A., M. Dilda, M. Crippa et al. 2011. Rare earth organic complexes as down-shifters to improve Si-based solar cell efficiency. *Optical Materials* 33:1012–1014.

Li, T.-T., and A. Cuevas. 2011. Role of hydrogen in the surface passivation of crystalline silicon by sputtered aluminum oxide. *Progress in Photovoltaics: Research and Applications* 19:320–325.

Libal, J., S. Novaglia, M. Acciarri et al. 2008. Effect of compensation and of metallic impurities on the electrical properties of CZ-grown solar grade silicon. *Journal of Applied Physics* 104:104507.

Lin, D., L. Ma, and G. Conibeer. 2011. Study on electrical properties of Si quantum dots based materials. *Physica Status Solidi B* 248:472–476.

Lin, H., S. Zhou, H. Teng et al. 2010. Near infrared quantum cutting in heavy Yb doped Ce0.03Yb3xY(2.97–3x)Al5O12 transparent ceramics for crystalline silicon solar cells. *Journal of Applied Physics* 107:043107.

Lu, M., U. Das, S. Bowden et al. 2011. Optimization of interdigitated back contact silicon heterojunction solar cells: Tailoring hetero-interface band structure while maintaining surface passivation. Progress in Photovoltaics: Research and Applications 19:326.

Luque, A., A. Marti, and A.J. Nozik. 2007. Solar cells based on quantum dots: Multiple exciton generation and intermediate bands. *MRS Bulletin* 32:236–241.

Ma, L., D. Lin, G. Conibeer et al. 2011. Introducing dopants by diffusion to improve the conductivity of silicon quantum dot materials in 3rd generation photovoltaic devices. *Physica Status Solidi C* 8:205–208.

MacDonald, D. 2011. The impact of dopant compensation on the boron-oxygen defect in p- and n-type crystalline silicon. *Physica Status Solidi A* 208(3):559–563.

MacDonald, D., A. Cheung, and A. Cuevas. 2003. Gettering and poisoning of silicon wafers by phosphorus diffused layers. *Proceedings of the 3rd WCPVSC*, 1336.

MacDonald, D., A. Cuevas, A. Kinomura et al. 2005. Transition-metal profiles in a multicrystalline silicon ingot. *Journal of Applied Physics* 97(3):1–7.

MacDonald, D., and L.J. Geerligs. 2004a. Recombination activity of interstitial iron and other transition metal point defects in p- and n-type crystalline silicon. *Applied Physics Letters* 85:4061.

MacDonald, D., and L.J. Geerligs. 2004b. Recombination activity of iron and other transition metals in p and n-type silicon. *Proceedings of the 19th EUPVSEC, Paris* 492–495.

Marchionna, S., F. Meinardi, M. Acciarri et al. 2006. Photovoltaic quantum efficiency enhancement by light harvesting of organo-lanthanide complexes. *Journal of Luminescence* 118:325–329.

Martin, I., M. Vetter, A. Orpella et al. 2001. Surface passivation of p-type crystalline Si by plasma enhanced chemical vapor deposited amorphous SiCx:H films. *Applied Physics Letters* 79(14):2199–2201.

McCann, M., B. Raabe, W. Jooss et al. 2006. 18.1% efficiency for a large area, multi-crystalline silicon solar cell. *IEEE 4th World Conference on Photovoltaic Energy Conversion, Hawaii, USA*.

McIntosh, K.R., G. Lau, J.N. Cotsell et al. 2009. Increase in external quantum efficiency of encapsulated silicon solar cells from a luminescent down-shifting layer. Progress in Photovoltaics: Research and Applications 17:191–197.

Meemongkolkiat, V., K. Nakayashiki, D.S. Kim et al. 2006. Factors limiting the formation of uniform and thick Al-back surface field and its potential. *Journal of the Electrochemical Society* 153(1):G53–G58.

Meijer, J.M., L. Aarts, B.M. van der Ende et al. 2010. Down-conversion for solar cells in YF3:Nd3+, Yb3+. *Physical Review B* 81:035107.

Meijerink, A. 2008. Exciton dynamics and energy transfer processes in semiconductor nanocrystals. In *Semiconductor Nanocrystal Quantum Dots: Synthesis, Assembly, Spectroscopy and Applications*, edited by Rogach, A.L. Wien–New York: Springer.

Meillaud, F., A. Shah, C. Droz et al. 2006. Efficiency limits for single-junction and tandem solar cells. *Solar Energy Materials and Solar Cells* 90:2952–2959.

Mette, A., C., Schetter, D. Wissen et al. 2006. Increasing the efficiency of screen-printed silicon solar cells by light-induced silver plating. *Proceedings of the 4th WCPVSC, Hawaii USA*, 1056.

Mette, A., P.L. Richter, M. Hörteis et al. 2007. Metal aerosol jet printing for solar cell metallisation. Progress in Photovoltaics: Research and Applications 15(7):621–627.

Miles, R.W., G. Zoppi, and I. Forbes. 2007. Inorganic photovoltaic cells. *Materials Today* 11(10):20–27.

Mingirulli, N., D. Stuwe, J. Specht et al. 2011a. Screen printed emitter wrap through solar cell with single step side selective emitter with 18.8% efficiency. Progress in Photovoltaics: Research and Applications 19:336.

Mingirulli, N., J. Haschke, R. Gogolin et al. 2011b. Efficient interdigitated back-contacted silicon heterojunction solar cells. Physica Status Solidi. *Rapid Research Letters* 5(4):159.

Miritello, M. R. Lo Savio, P. Cardile et al. 2010. Enhanced down conversion of photons emitted by photoexcited ErxY2–xSi2O7 films grown on silicon. *Physical Review B* 81:041411

Moreno, M., D. Daineka, I. Roca, and P. Cabarrocas. 2010. Plasma texturing for silicon solar cells: From pyramids to inverted pyramids-like structures. *Solar Energy Materials and Solar Cells* 94:733.

Munz, P. and E. Bucher. 1982. The use of rare earths in photovoltaics. *The Rare Earths in Modern Science and Technology* 3:547–553.

Munzer, K.A., K.T. Holdermann, R.E. Schlosser et al. 1998. Improvements and benefits of thin crystalline silicon solar cells. *Proceedings of the 2nd WCPVSEC, Vienna, Austria*, 1214–1219.

National Renewable Energy Centre (NaREC). http://www.narec.co.uk/.

Nesbit, L.A. 1985. Annealing characteristics of Si-rich SiO films. *Applied Physics Letters* 46:38.

Norris, D.J., and M.G. Bawendi. 1996. Measurement and assignment of the size-dependent optical spectrum in CdSe quantum dots. *Physical Review B* 53:16338–16346.

Nozik, A.J. 2002. Quantum dot solar cells. *Physica E* 14:115–120.

Ossicini, S., F. Iori, E. Degoli et al. 2006. Understanding doping in silicon nanostructures. *IEEE Journal of Selected Topics in Quantum Electronics* 12:1585.

Pankove, J.I. 1971. *Optical Processes in Semiconductors*. Englewood Cliffs, NJ: Prentice-Hall.

Park, S., E.-C. Cho, D. Song et al. 2009. n-Type silicon quantum dots and p-type crystalline silicon heteroface solar cells. *Solar Energy Materials and Solar Cells* 93:684–690.

Perego, M., C. Bonafos, and M. Fanciulli. 2010. Phosphorus doping of ultra-small silicon nanocrystals. *Nanotechnology* 21:025602.

Perez-Wurfl, I., X.J. Hao, A. Gentle et al. 2009. Si nanocrystal p-i-n diodes fabricated on quartz substrates for third generation solar cell application. *Applied Physics Letters* 95:153506.

Perez-Wurfl, I., L. Ma, D. Lin et al. 2012. Silicon nanocrystals in an oxide matrix for thin film solar cells with 492 mV open circuit voltage. *Solar Energy Materials and Solar Cells* 100:65–68.

Peter, K., R. Kopecek, A. Soiland et al. 2008. Future potential for SoG-Si feedstock from metallurgical process route. *23rd European Photovoltaic Solar Energy Conference and Exhibition, Valencia, Spain*, 1109–1115.

Piper, W.W., J.A. De Luca, and F.S. Ham. 1974. Cascade fluorescent decay in Pr3+-doped fluorides: Achievement of a quantum yield greater than unity for emission of visible light. *Journal of Luminescence* 8(4):344–348.

Polisski, G., D. Kovalev, G. Dollinger et al. 1999. Boron in mesoporous Si—Where have all the carriers gone? *Physica B* 273:951.
Ravi, K.V. 1981. *Imperfection and Impurities in Semiconductor Silicon*. New York: Wiley.
Rein, S., W. Warta, and S.W. Glunz. 2000. Investigation of carrier lifetime in p-type CZ-Si: Specific limitations and realistic prediction of cell performance. *Proceedings of the 28th IEEE PVSC, Anchorage, Alaska, USA*, 57–60.
Reisfeld, R. 1983. Future technological applications of rare earth doped materials. *Journal of the Less-Common Metals* 93:243–251.
Richards, B.S. 2006. Luminescent layers for enhanced silicon solar cell performance: Down-conversion. *Solar Energy Materials and Solar Cells* 90:1189–1207.
Richards, B.S., and A. Shalav. 2007. Enhancing the near-infrared spectral response of silicon optoelectronic devices via up-conversion. *IEEE Transactions on Electronic Devices* 54:2679–2684.
Rougieux, F., D. MacDonald, A. Cuevas et al. 2010. Electron and hole mobility reduction and Hall factor in phosphorus-compensated p-type silicon. *Journal of Applied Physics* 108:013706.
Rowan, B.C., L.R. Wilson, and B.S. Richards. 2008. Advanced material concepts for luminescent solar concentrators. *IEEE Journal of Selected Topics in Quantum Electronics* 14:1312–1322.
Sakata, H., Y. Tsunomura, H. Inoue et al. 2010. R&D progress of next-generation very thin HIT solar cells. *25th EUPVSEC*, 1102–1106.
Sanyo, http://panasonic.net/sanyo/.
Sardana, D.K. 1985. Gettering in processed silicon. *Semiconductor International* 362–368.
Sawada, T., N. Terada, S. Tsuge et al. 1994. *Proceedings of the 1st WCPEC*, 1219.
Schaller, R.D., V.M. Agranovich, and V.I. Klimov. 2005a. High-efficiency carrier multiplication through direct photogeneration of multi-excitons via virtual single-exciton states. *Nature Physics* 1:189–194.
Schaller, R.D., and V.I. Klimov. 2004. Non-poissonian exciton populations in semiconductor nanocrystals via carrier multiplication. *Physical Review Letters* 92:186601.
Schaller, R.D., M.A. Petruska, and V.I. Klimov. 2005b. Effect of electronic structure on carrier multiplication efficiency: Comparative study of PbSe and CdSe nanocrystals. *Applied Physics Letters* 87:253102.
Schaller, R.D., M. Sykora, and J.M. Pietryga et al. 2006 Seven excitons at a cost of one: Redefining the limits for conversion efficiency of photons into charge carriers. *Nano Letters* 6:424–429.
Schaper, M., J. Schmidt, H. Plagwitz et al. 2005. 20.1%-efficient crystalline silicon solar cell with amorphous silicon rear-surface passivation. Progress in Photovoltaics: Research and Applications 13(5):381–386.
Schmidt, J., A.G. Aberle, and R. Hezel. 1997. Investigation of carrier lifetime instabilities in CZ-grown silicon. *Proceedings of the IEEE 26th PVSC, Anaheim, CA, USA*, 13–18.
Schmidt, J., A. Cuevas, S. Rein et al. 2001a. Impact of light induced recombination centres on the current voltage characteristic of CZ silicon solar cells. Progress in Photovoltaics: Research and Applications 9:249–255.
Schmidt, J., M. Kerr, A. Cuevas. 2001b. Surface passivation of silicon solar cells using plasma-enhanced chemical-vapour-deposited SiN films and thin thermal SiO_2/plasma SiN stacks. *Semiconductor Science and Technology* 16:164.
Schmidt, J., T. Lauinger, A.G. Aberle et al. 1996. Record low surface recombination velocities on low-resistivity silicon solar cell substrate. *Proceedings of the 25th IEEE PVSC Washington, DC*, 413.
Schmiga, C., A. Froitzheim, M. Ghosh et al. 2005. Solar cells on n-type Si materials with screen-printed rear aluminium-p+ emitter. *Proceedings of the 20th EUPVSEC, Barcelona, Spain*, 918–921.
Schmiga, C., M. Hermle, and S.W. Glunz. 2008. Towards 20% efficiency n-type silicon solar cells with screen printed aluminum alloyed rear emitter. *Proceedings of the 23rd EUPVSEC, Valencia, Spain*, 982.
Schmiga, C. et al. 2009. Large area n-type silicon solar cells with printed contacts aluminum alloyed rear emitter. *Proceedings of the 24th EUPVSEC, Hamburg, Germany*, 1167.
Schneiderlochner, E., R. Preu, R. Ludemann et al. 1999. Laser-fired rear contacts for crystalline silicon solar cells. *Progress in Photovoltaics: Research and Applications* 7:471–474.
Schultz, O., S.W. Glunz, and G.P. Willeke. 2004. Multicrystalline silicon solar cells exceeding 20% efficiency. *Progress in Photovoltaics: Research and Applications* 12(7):553–558.

Schultz, O., M. Hofmann, S.W. Glunz et al. 2005. Silicon oxide/silicon nitride stack system for 20% efficient silicon solar cells. *Proceedings of the 31st IEEE PVSC Orlando, Florida*, 872–876.
Schulze, T.F., H.N. Beushausen, T. Hansmann et al. 2009. Accelerated interface defect removal in amorphous/crystalline silicon heterostructures using pulsed annealing and microwave heating. *Applied Physics Letters* 95:182108.
Schutz-Kuchly, T., S. Dubois, J. Veirman et al. 2011. Light-induced degradation in compensated n-type Czochralski silicon solar cells. *Physica Status Solidi A* 208(3):572–575.
Sewell, R.H., A. Clark, R. Smith et al. 2010. Silicon solar cells with monolithic rare-earth oxide up-conversion layer. *34th IEEE Photovoltaic Specialists Conference (PVSC), 2009*, 002448–002453.
Shalav, A., B.S. Richards, and M.A. Green. 2007. Luminescent layers for enhanced silicon solar cell performance: Up-conversion. *Solar Energy Materials and Solar Cells* 91:829–842.
Sheng, P. 1984. Optical absorption of thin film on a Lambertian reflector substrate. *IEEE Transactions on Electronic Devices* 31:634–636.
Shi, Z., S.R. Wenham, J. Ji. 2009. Mass production of the innovative PLUTO solar cell technology. *34th IEEE PVSC, Philadelphia, Pennsylvania*, 1922.
Shockley, W., H.J. Queisser. 1961. Detailed balance limit of efficiency of p–n junction solar cells. *Journal of Applied Physics* 32:510–519.
Shpaisman, H., O. Niitsoo, I. Lubomirsky et al. 2008. Can up- and down-conversion and multi-exciton generation improve photovoltaics? *Solar Energy Materials and Solar Cells* 92:1541–1546.
Simánek, E. 1981. The temperature dependence of the electrical resistivity of granular metals. *Solid State Communications* 40:1021–1023.
Slooff, L.H., E.E. Bende, A.R. Burgers et al. 2008. A luminescent solar concentrator with 7.1% power conversion efficiency. Physica Status Solidi. *Rapid Research Letters* 2(6):257–259.
Smedskjaer, M.M., J. Qiu, J. Wang et al. 2011. Near-infrared emission from Eu–Yb doped silicate glasses subjected to thermal reduction. *Applied Physics Letters* 98:071911.
Smestad, G., H. Ries, R. Winston et al. 1990. The thermodynamic limits of light concentrators. *Solar Energy Materials* 21:99–111.
So, Y.H. 2011. Silicon Nitride as Alternative Matrix for Silicon Quantum Dot Based All-Silicon Tandem Solar Cells. PhD Thesis, University of New South Wales.
So, Y., S. Huang, G. Conibeer et al. 2011a. Formation and photoluminescence of Si nanocrystals in controlled multilayer structure comprising of Si-rich nitride and ultrathin silicon nitride barrier layers. *Thin Solid Films* 519:5408–5412.
So, Y., S. Huang, G. Conibeer et al. 2011b. N-type conductivity of nanostructured thin film composed of antimony-doped Si nanocrystals in silicon nitride matrix. *Europhysics Letters* 96:17011.
So, Y.-H., A. Gentle, G. Conibeer et al. 2011c. Size dependent optical properties of Si quantum dots in Si-rich nitride/Si3N4 superlattice synthesized by magnetron sputtering. *Journal of Applied Physics* 109:064302.
Song, D., E.-C. Cho, Y.-H. Cho et al. 2008. Evolution of Si (and SiC) nanocrystal precipitation in SiC matrix. *Thin Solid Films* 516:3824–3830.
Song, D., E.-C. Cho, G. Conibeer et al. 2007. Fabrication and electrical characteristics of Si nanocrystal/c-Si heterojunctions. *Applied Physics Letters* 91:123510.
Staebler, D.L., and C.R. Wronski. 1977. Reversible conductivity changes in discharge produced amorphous Si. *Applied Physics Letters* 31:292.
Strümpel, C., M. McCann, G. Beaucarne et al. 2007. Modifying the solar spectrum to enhance silicon solar cell efficiency—An overview of available materials. *Solar Energy Materials and Solar Cells* 91:238–249.
Su, K.H., Q.H. Wei, X. Zhang et al. 2003. Interparticle coupling effects on plasmon resonances of nanogold particles. *Nano Letters* 3(8):1087–1090.
SunPower, http://us.sunpowercorp.com/.
Suyver, J.F., A. Aebischer, D. Biner et al. 2005. Novel materials doped with trivalent lanthanides and transition metal ions showing near-infrared to visible photon up-conversion. *Optical Materials* 27:1111–1130.
Svantesson, K.G., and N.G. Nilsson. 1979. The temperature dependence of the Auger recombination coefficient of undoped silicon. *The Journal of Physical Chemistry* 12:5111.

Taguchi, M., K. Kawamoto, S. Tsuge et al. 2000. HIT cells—High-efficiency crystalline Si cells with novel structure. *Progress in Photovoltaics: Research and Applications* 8:503.

Taguchi, M., Y. Tsunomura, H. Inoue et al. 2009. High-efficiency HIT solar cell on thin (<100 μm) silicon wafer. *Proceedings of the 24th EUPVSEC, Hamburg, Germany*, 1690.

Taira, S., Y. Yoshimine, T. Baba et al. 2007. Our approach for achieving HIT solar cells with more than 23% efficiency. *Proceedings of the 22nd EUPVSEC, Milan, Italy*, 932–935.

Takagi, H., H. Ogawa, Y. Yamazaki et al. 1990. Quantum size effects on photoluminescence in ultrafine Si particles. *Applied Physics Letters* 56:2379.

Takeoka, S., M. Fujii, and S. Hayashi. 2000. Size-dependent photoluminescence from surface-oxidized Si nanocrystals in a weak confinement regime. *Physical Review B* 62:16820.

Thaidigsmann, B., E. Lohmüller, U. Jäger et al. 2011. Large-area p-type HIP-MWT silicon solar cells with screen printed contacts exceeding 20% efficiency. Physica Status Solidi. *Rapid Research Letters* 5(8):286–288.

Thompson, R.D., and K.N. Tu. 1978. Low temperature gettering of Cu, Ag and Au across a wafer of Si by Al. *Applied Physics Letters* 33:238–240.

Tiedje, T., E. Yablonoitch, G.D. Cody et al. 1984. Limiting efficiency of silicon solar cells. *IEEE Transactions on Electronic Devices* 31(5):711.

Timmerman, D., I. Izeddin, P. Stallinga et al. 2008. Space-separated quantum cutting with silicon nanocrystals for photovoltaic applications. *Nature Photonics* 2:105–108.

Trupke, T., M. Green, and P. Würfel. 2002a. Improving solar cell efficiencies by down-conversion of high-energy photons. *Journal of Applied Physics* 92:1668–1674.

Trupke, T., M. Green, and P. Würfel. 2002b. Improving solar cell efficiencies by up-conversion of sub-band-gap light. *Journal of Applied Physics* 92:4117–4122.

Tucci, M., E. Talgorn, L. Serenelli et al. 2008a. Laser fired back contact for silicon solar cells. *Thin Solid Films* 516:6767–6770.

Tucci, M., L. Serenelli, E. Salza et al. 2008b. Back enhanced heterostructure with interdigitated contact-BEHIND-solar cell. *Proceedings of the IEEE Conference on Optoelectronic and Microelectronic Materials and Devices, Sidney, Australia, 2008*, 242.

Tucci, M., L. Serenelli, E. Salza et al. 2009. Bragg reflector and laser fired back contact in a-Si:H/c-Si heterostructure. *Materials Science and Engineering B* 159–160:48–52.

van der Ende, B.M., L. Aarts, and A. Meijerink. 2009. Near-infrared quantum cutting for photovoltaics. *Advanced Materials* 21:3073–3077.

van der Kolk, E., P. Dorenbos, K. Kramer et al. 2008. High-resolution luminescence spectroscopy study of down-conversion routes in NaGdF4:Nd3+ and NaGdF4:Tm3+ using synchrotron radiation. *Physical Review B* 77:125110

van Sark, W.G., K.W.J. Barnham, L.H. Slooff et al. 2008. Luminescent solar concentrators: a review of recent results. *Optics Express* 16:21773–21792.

van Wijngaarden, J.T., S. Scheidelaar, T.J.H. Vlugt et al. 2010. Energy transfer mechanism for down-conversion in the (Pr3+, Yb3+) couple. *Physical Review B* 81:155112.

Vedde, J., T. Clausen, and L. Jensen. 2003. Float-zone silicon for high volume production of solar cells. *Proceedings of the 3rd WCPVSEC, Osaka, Japan*, 943–946.

Wan, Z., S. Huang, M.A. Green et al. 2011a. Rapid thermal annealing and crystallization mechanisms study of silicon nanocrystal in silicon carbide matrix. *Nanoscale Research Letters* 6:129.

Wan, Z., R. Patterson, S. Huang et al. 2011b. Ultra-thin silicon nitride barrier implementation for Si nano-crystals embedded in amorphous silicon carbide matrix with hybrid super-lattice structure. *Europhysics Letters* 95-6:67006.

Weber, W.H., and J. Lambe. 1976. Luminescent greenhouse collector for solar radiation. *Journal of Applied Optics* 15:2299–2300.

Wegh, R.T., H. Donker, K.D. Oskam et al. 1999. Visible quantum cutting in LiGdF4: Eu3+ through down conversion. *Science* 283:663–666.

Woehl, R., J. Krause, F. Granek et al. 2011. 19.7% efficient all-screen-printed back-contact back-junction silicon solar cell with aluminum-alloyed emitter. *IEEE Electron Device Letters* 32(3):345–347.

Yablonovitch, E. 1982. Statistical ray optics. *The Journal of the Optical Society of America* 12:899.

Zacharias, M., J. Heitmann, R. Scholz et al. 2002. Size-controlled highly luminescent silicon nanocrystals: a SiO/SiO2 superlattice approach. *Applied Physics Letters* 80:661–663.

Zastrow, A. 1994. Physics and applications of fluorescent concentrators: a review. *Proceedings of the SPIE* 2255:534–547.

Zhang, Q.Y., and X.Y. Huang. 2010. Recent progress in quantum cutting phosphors. *Progress in Materials Science* 55:353–356.

Zhang, Q.Y., G.F. Yang, and Z.H. Jiang. 2007. Cooperative down-conversion in GdAl3(BO3)4: RE3+,Yb3+ (RE = Pr, Tb and Tm). *Applied Physics Letters* 91:051903.

Zhao, J., and A. Wang. 2006. Rear emitter n-type passivated emitter, rear totally diffused silicon solar cell structure. *Applied Physics Letters* 88:242102.

Zhao, J., A. Wang, M.A. Green et al. 1998. Novel 19.8% efficient "honeycomb" textured multicrystalline and 24.4% monocrystalline silicon solar cells. *Applied Physics Letters* 73:1991.

Zhao, J., A. Wang, M.A. Green. 2002. 24.5% efficiency silicon PERT cell on MCZ substrates and 24.7% efficiency PERL cell on FZ substrates. *Progress in Photovoltaics: Research and Applications* 10(1):29–34.

Index

Page numbers followed by f and t indicate figures and tables, respectively.

D

E

H

M

N

O

Q

R

S